*Proceedings of Symposium on Qinghai-Xizang (Tibet) Plateau (Beijing, China)*

# Geological and Ecological Studies of Qinghai-Xizang Plateau

VOLUME ONE OF TWO VOLUMES

## Geology, Geological History and Origin of Qinghai-Xizang Plateau

SCIENCE PRESS, BEIJING 1981
GORDON AND BREACH, SCIENCE PUBLISHERS, INC.
NEW YORK

Published by Science Press, Beijing

Distribution rights throughout the world, excluding the People's Republic of China, granted to Gordon and Breach, Science Publishers, Inc., New York

Printed by C & C Joint Printing Co., (H.K.) Ltd.



First published 1981

ISBN 0-677-60210-3
ISBN of the 2-volumes set 0-677-60230-8

Science Press Book No. 2474-13

The reception of the foreign guests and the participants of the Symposium by Vice Premiers Deng Xiao-ping and Fang Yi

# PREFACE

The Qinghai-Xizang (Tibet) Plateau is the largest plateau in the world rich in natural resources. Studies on the natural conditions and evolution of the plateau are of great scientific significance. Thus, it has long been the common aspirations of scientists and scholars to reveal the mysteries of nature in the plateau; however, in the past, only preliminary studies were made by some pioneers. In recent thirty years, seven large-scaled multi-disciplinary scientific expeditions to Xizang were organized by Academia Sinica and a wealth of results were obtained. Meanwhile, the foreign scientists have made considerable studies on the Himalayan and Karakorum regions outside China. The Symposium on Qinghai-Xizang (Tibet) Plateau sponsored by Academia Sinica was held at Beijing from May 25 to June 1, 1980, which offered a rare opportunity for both Chinese and foreign scientists to exchange scientific results and research ideas. It is quite natural that some of the scholars spoke of the Symposium as a milestone of the research on Qinghai-Xizang Plateau.

More than two hundred papers submitted and read at the Symposium, which involve all the important results of the research on the plateau by Chinese and foreign scholars in the last thirty years, are compiled in this Proceedings in two volumes. The central theme of the Proceedings is "The rise of Qinghai-Xizang (Tibet) Plateau and its effects on the environments and man". The main topics discussed include the geophysical conditions, the geological history and the origin of formation of the plateau; the characteristics and evolution of the fauna and flora of the plateau and the effects of its rise on them; and the formation, evolution and differentiation of the geographic environments of the plateau.

In the course of the editorial work, earnest help was given by a number of Chinese and foreign scholars. On behalf of Academia Sinica, I wish to avail myself of this opportunity to extend to them my heartfelt thanks.

钱三强

Qian San-qiang

Vice-President of Academia Sinica

# CONTENTS



## Volume I
## Geology, geological history and origin of Qinghai-Xizang Plateau

## Volume II
## Environment and Ecology of Qinghai-Xizang Plateau

# OPENING SPEECH FOR SYMPOSIUM ON QINGHAI-XIZANG (TIBET) PLATEAU


Qian San-qiang*

*(Vice-President of Academia Sinica)*


Ladies and gentlemen,
Friends and comrades,

The Symposium on Qinghai-Xizang (Tibet) Plateau opens today.

This symposium is sponsored and organized by the Academia Sinica, which has met warm support from the scholars of various countries since we sent out our notice. More than 100 scholars expressed their wishes to participate the symposium in their letters to us. But due to our limited capacities, we have not been able to invite all of them. For this we feel very sorry. Present at today's opening ceremony are about 80 scholars coming from the countries as Australia, Bangladesh, Canada, France, Federal Republic of Germany, India, Italy, Nepal, Netherlands, New Zealand, pakistan, Switzerland, Turkey, the United Kingdom, the United States of America and Yugoslavia, while the Chinese scientists involved in this symposium amount to more than 180.

We are especially happy over the fact that we also have quite a number of world-famous scholars with us today, who have already made many years of research work in the Himalaya region and will give lectures during the symposium.

So please allow me, on behalf of the Academia Sinica and the Organizing Committee of Symposium, to express our warm welcome to all the guests and scientists both at home and abroad.

Qinghai-Xizang (Tibet) Plateau, known as the "Roof of the World", is the highest, yet youngest and most gigantic plateau. Its existence is greatly influential not only to the plateau proper but also to the natural environment and human activities in the adjacent regions. The formation and evolution of the plateau and its effects have long been the important subjects which attract great attention of the scientific circle. The scientists regard the plateau as one of the key areas in studying such theoretical problems as the crust movement, the origin of fauna and flora, etc.

For a long period, a large number of survey and research work on the plateau and its perifery has been carried out by the scientists of various countries who have achieved significant results and made a lot of theoretical contributions related to the formation, development and evolution of the plateau.

Since 1950s, China has organized 7 multi-disciplinary integrated surveys on Qinghai-Xizang (Tibet) Plateau, with a view to accumulate basic informations, to explore the basic theories concerning the formation and evolution of the plateau and to find out about the scientific basis for the exploration and utilization of the natural resources as well as the control of natural calamities.

---

* Chien San-chiang

The relative results of the research work for the first few surveys have been published in some special works and Scientia Sinica, and the data about the integrated survey since 1973 have been published in succession in our newspapers and scientific journals in recent years.

We understand that many foreign scientists concerned all have strong desires to know about Qinghai-Xizang (Tibet) Plateau and they hope some day they can go there by themselves. So we will arrange a scientific excursion to Xizang after the symposium.

We hope that the broad exchanges and discussions will be beneficial to increasing our mutual understanding and friendly contacts as well as to promoting the further development of the scientific research work on Qinghai-Xizang (Tibet) Plateau.

Finally, I wish the symposium a great success, and I wish all of you good health!

Thank you.

# A REVIEW OF THE MULTI-DISCIPLINED SCIENTIFIC RESEARCH ON THE QINGHAI-XIZANG PLATEAU BY CHINESE SCIENTISTS IN RECENT YEARS

Liu Dong-sheng*
(Institute of Geology, Academia Sinica)

Sun Hong-lie
(Integrated Survey Commission of Natural Resources, Academia Sinica)

Mr. Chairman, Comrades and Friends,

It is a great pleasure to meet here with the learned colleagues from 17 countries to explore the important problem on the "Formation and evolution of the Qinghai-Xizang Plateau and its influences on the environments and human activities."

The upheaval of the Qinghai-Xizang Plateau is an event of the greatest significance in Asia during the last several million years. The Plateau has fundamentally changed the natural landscape from what it had been, bringing about profound influences on the neighboring areas, and drawing the attention of scientists from various fields. Most of the Plateau lies within the boundaries of China, starting from the Kunlun Mts. in the north to the Himalayas in the south and from the Karakorum in the west to the Hengduan Mts.(Meridional Range) with great gorges in the east, covering an area of 2,400,000 km$^2$, almost ¼ of China's territory. It is the home of the Xizang nationality of the Chinese people who have well adapted to the unique environment, and have greatly improved their living conditions since liberation. It is clear then that a thorough investigation of the Plateau will not only add to the treasury of scientific information, but also lead the way in making contributions to humanity.

Academia Sinica, aiming at a summarization of all available data concerning the formation and evolution of the Qinghai-Xizang Plateau as well as the formulation of a proposal concerning the management of natural resources and the protection against natural catastrophes, has been sponsoring since the 50s a number of expeditions to the Plateau, i.e. (1) the 1951-1952 expedition in geology, geography and agriculture of central and eastern parts of the Xizang Autonomous Region; (2) the 1959 expedition of Mt. Qomolangma (the first time); (3) the 1960-1961 expedition in geology, geography, biology and agriculture of central and southern parts of Xizang; (4) the 1958-1961 expedition in geography, agriculture and river conservancy of the eastern part of the Plateau, including the gorge district; (5) the 1964 expedition of Mt. Xishabangma; (6) the 1966-1968 expedition of Mt. Qomolangma (the second time); (7) the 1973-1979 expedition in geophysics, geology, geography, biology and agriculture of the whole Xizang; (8) the 1975 expedition of Mt. Qomolangma (the third time). One after the other, these multi-disciplinary expeditions amplify and intensify the understanding of the Plateau. The 1973-1979 expedition lasted longer with more disciplines than the other expeditions, when, for 7 years, over 50 disciplines were involved and over 400 scientists and technicians participated in the

* = Liu Tung-sheng

field work and a personnel of over 1,200 in the laboratory work. Based on the results of these expeditions, a series of "Monographs of the Qinghai-Xizang Plateau" including 32 volumes and a pictorial "Exploring the Secrets of the Roof of World" have been, or are being published in succession later this year.

In addition to the works by Academia Sinica, various institutes of geology, seismology, meteorology and agriculture have shared in the researches and have published a number of monographs; the journals of scientific societies have also published many papers on different topics of Qinghai-Xizang research.

Comrades and friends, allow me, now, to make a summary of the major achievements by the Chinese scientists, particularly since 1973.

## A WORKING HYPOTHESIS ON THE EVOLUTION OF THE PLATEAU, THE MECHANISM OF THE UPHEAVAL, AND ON THE CHARACTERISTICS OF THE DIFFERENT GEOPHYSICAL FIELDS

The Qinghai-Xizang plateau may be considered one of the best places to study the earth interior and the mechanism of crustal movement. The Chinese have made a commendable start in this respect through the researches in stratigraphy, paleontology, petrology, geothermics, geochemistry, geotectonics and geophysics. The fossils are extremely rich in the Plateau. Already 30 or more families totally 3,800 species have been discovered in recent years, comparing with only 300 species mentioned in literature prior to the liberation. According to paleontological and sedimentary informations, Xizang may be divided into 4 stratigraphic regions and 11 subregions, for each of which a complete stratigraphic profile has been established. The 4 regions are: 1) the Himalayas, 2) the Gandise-Nyainqengtanglha, 3) the Tanggula-Hengduan gorge district and 4) the Hohxil-Kunlun. In the Himalayan region, fossils of <u>Glossopteris</u> and cold-water biota have been found in the Late Paleozoic beds, while to the north of the Yarlung Zangbo River, fossils of <u>Gigantonoclea</u> and warm-water biota have been found in the Permian formation. That means, the Yarlung Zangbo valley lies on an important boundary of paleontological significance.

Petrographic analysis shows that the granitoids of Xizang may chronologically be grouped into 3 belts: the Gandise belt (120-70 m.y., belonging to I-type), the Lhagoi Kangri belt (50-30 m.y.), and the Himalayan belt (20-10 m.y. belonging to the S-type). A series of progressive changes exists in the 3 belts, with the result that from N to S, the alkali-feldspar coefficient increases, the An value in the plagioclase decreases, the differentiation index and K, Si contents increase and the Ti, Fe, Mg and Ca contents decrease, the distribution pattern of rare earth elements differs, and finally, the ratio of $^{87}Sr/^{86}Sr$ increases from 0.704 to 0.747. Time and space distribution of granitoids indicates that the early magmatic activities originated partly from the remelting of the sediments atop the oceanic crust. Later, the collision and compression resulted in the remelting of the sial material of the continental crust and in the differentiation of magmatic pockets. The ophiolite belt which extends W-E for about 1,000 km along the southern bank of the Yarlung Zangbo River has also been examined in petrographical details. This belt connects in the west, the ophiolite belt of Indus River and in the east, beyond the great river bend, that at the Burmese border. The ophiolite might have emplaced sometime between Eocene and Miocene, accompanied by intense folds of flysch and by development of melange and exotic blocks. The occurrence of ophiolites and a series of linear features associated seems to confirm that the composite structural line along the Yarlung Zangbo River is the suture zone between two plates.

The strong Himalaya movement is also echoed by geothermal phenomena. The greatest thermal belt of China's mainland occurs here starting from W. Ngari, Xizang, down the Yarlung Zangbo valley, crossing the Hengduan Mts., before joining the geothermal region of W. Yunnan. Altogether, there are about 600 hydrothermal areas, including 11 areas with explosions, 3 geysers and tens of boiling springs. According to the data of shallow drilling to a few hundred meters in the thermal field of Yangbajain, about 80 km NW of Lhasa, the geothermal gradient in the upper part of the crust averages to $2.9^{o}$-$3.5^{o}C/m$, which is about 100 times the average value of the whole crust. Chances are that the heat source of this thermal belt comes from some sort of shallow magmatic activities beneath the ground surface.

For the purpose of exploring the deeper structure in the crust and upper mantle, geophysical researches of different kinds have been initiated, such as seismology, gravimetry, paleomagnetism and magneto-telluric probing. Deep seismic sounding confirmed the layered structure and disclosed the existence of a low velocity layer in the crust. Paleogeomagnetic study revealed that the areas to the north and south of the Yarlung Zangbo valley belong to 2 different plates: to the north, the crust is about 70 km thick and to the south, about 50 km. The isostatic anomaly in the Mt. Qomolangma area is +120 mg, indicating that the crust is overcompensated. But the uplift is still continuing. This is due to the compression effect caused by the pushing northward of the Indian plate.

All the above-mentioned phenomena suffice to show that the Qinghai-Xizang Plateau and the Himalayan system are the products of the collision of the Indian plate with the Eurasian. The Himalayan system might have been completed between Miocene and Pliocene; but the <u>Hipparion</u> fauna found in Xizang implies that even in Late Tertiary, the Plateau might not have exceeded 1,000 m in elevation. Strong upheaval at the end of Pliocene, with 3 successive lifts of large amplitude accelerated the rise. For the last 100,000 years since Pleistocene, a rise of 10 mm/yr has been estimated.

## AN EXPLANATARY DESCRIPTION OF THE INFLUENCES OF THE PLATEAU'S UPHEAVAL ON THE ENVIRONMENT, AND AN UNDERSTANDING OF THE CHARACTERISTICS, EVOLUTION AND DIFFERENTIATION OF THE ENVIRONMENT

In the course of the Plateau's upheaval, there occurred at least 4 glacial periods separated by interglacial. In Late Pleistocene, when the Himalayas have reached a great height, the warm wet monsoon was blocked to the south of the Plateau, resulting in cold aridity over Xizang. Permafrost appears, glaciers retreat, lakes dwindle and the uniqueness of the environment strengthens.

For many years, the Chinese scientists have systematically analysed the natural features, particularly their types, characteristics, distribution and modification due to the upheaval of the Plateau. Regarding meteorology and climatology, the existence of the atmospheric boundary of the Plateau makes its dynamic effect far exceed that of the actual area of the Plateau. The activities of the low vortices and the shear lines within the boundary, as well as the important influences of the South Asiatic High (at 100 mb) on weather and climate have been studied. The intensity of heat source and sink of the Plateau have also been observed and calculated. The monsoonal phenomena over the Plateau, being a separated system from the ordinary monsoon over the lowlands, is motivated by the seasonal variation of heat source and the consequent change in the upper air circulation. In addition, the formation, development and maintainence of the circulation over the Plateau are simulated by both models and numerical experiments. These phenomena may even affect the central Pacific and intrude into the Southern Hemisphere. The physico-geographer is of the opinion that the surface relief and its impact on the atmospheric circulation are decisive in differentiating the physical environment. The gradual worsening of hydrothermal conditions from SE toward NW is echoed by a vegetal change as spaced in the following zones: forest-meadow-steppe-desert. This zonation may be taken as a vertical expression of the horizontal zonality. By integrating the vertical with the horizontal zonation, 9 physical belts have been suggested within the Plateau each with its distinctive characteristics.

In line with local conditions, a tentative scheme has been made regarding the agricultural development. Thus, north of the Gandise-Nyainqentanglha Range is best suited for yak and sheep raising and the Yarlung Zangbo valley for cultivation of winter wheat and Qingke (a kind of huskless barley). The upper limit of these 2 crops reaches respectively 4,200 m and 4,800 m, being the highest in the world. So far as cultivation is concerned, the Plateau tops all other croplands, thanks to its high intensity of radiation, long duration of growing temperature, favorable combination of light and warmth and adequate period for photophilous and cool-adapted crops to tiller and milk; hence the yield of winter wheat reaches as high as 13,000 kg/ha, being one of the highest in China today. In the forest belt from the south slopes of the Himalayas to the gorge district in East Xizang, the dark coniferous forests grow as far up as 4,100 m - 4,600 m, yielding 2,000 $m^3$/ha, a rare record in the world.

## AN ANALYSIS OF THE FLORA AND FAUNA, AND THE ADAPTATION OF PLANTS AND ANIMALS TO THE PLATEAU ENVIRONMENT

The upheaval of the Plateau has retained, on the one hand, some ancient forms of the biota, while on the other hand, has induced the formation of new species. The Plateau has long been called the "Treasure House" of biological resources. Already 530 species of birds and 190 of mammals have been collected, 7 being new species and 11 new subspecies. This exceeds the total number of species recorded prior to the founding of New China. As to the insects, 100,000 specimens have been collected and 2,300 species so far identified. Among them there are 20 new genera, 400 new species and subspecies, about ¼ as many as the new species identified since the founding of New China. Furthermore, in SE Xizang, specimens of the order *Zoraptera* have been first recorded for China.

On the botanical side, the newly finished monograph *The Flora of Xizang* describes more than 5,000 species of higher plants, including several hundred new species and 5 new genera. Among plants, fungi, mosses and lichens, there are several new species discovered in the Plateau.

Based on the analyses of a great quantity of data, the zoogeographers, agreeing with the phytogeographers, reject the simple idea of taking Xizang as merely an ancient or a recent fauna. True, the scarcity of endemic genera of vertebrates indicates the relatively short history of evolution, but the 23 endemic species of birds and 44 of mammals verify their survival of the glaciation. The botanists observe that the Xizang flora has originated from the Hengduan Mts. and have evolved during the uplift of the Himalayas. As in East Asia and North America, the Plateau is a center of differentiation and evolution of some major species or species-groups in the north temperate zone.

Observations have also been made by the Chinese scientists regarding the adaptations of men and animals to the environment of the Plateau. The long dwelt inhabitants of the Plateau have been taken as an example in comparing with the lowland dwellers concerning their physiological changes and their adaptability to the high altitudes. An electrocardiogram of Ms. Pando, the well known Tibetan woman mountaineer, was taken by telemetry, when she was up Mt. Qomolangma in 1975.

Comrades and friends, looking back to our previous work, I feel that we have only taken the first step in the long journey ahead of us. A good number of important problems, such as the nature of the major tectono-sedimentary belts, the stratigraphy in relation to the composition and evolution of the paleo-biota, the characteristics and dynamic system of the geophysical fields, the influences of the Plateau on the weather and climate in minor localities, the physiological and ecological adaptations as well as the genetic variations of the various plants and animals to the Plateau, the rational management of the natural resources and the measures taken in overcoming the unfavorable factors, all of these require further investigations.

We are, indeed, glad to see that more and more attention has been paid to the research work of the Plateau by the scientists all over the world. We know that numerous magnificent results have been achieved through insistent work by our science colleagues on the Himalayas, the Korakorum and vicinities. We shall listen to your lectures and reports with greatest interests. It is our sincere hope that this symposium will pave the way for constant exchanges of views and that, further cooperation will facilitate our researches in a common effort of making greater contributions to humanity.

Thank you for attention.

# THE CLOSING SPEECH

Zhao Bei-ke
(Deputy Secretary - General of Academia Sinica)

Ladies and gentlemen,
Friends and comrades,

The Symposium on Qinghai-Xizang (Tibet) Plateau is now scheduled to a close. Thanks to the common efforts by the foreign and Chinese scientists, we have achieved fruitful results and a great success.

Nearly 300 foreign and Chinese scientists attended the symposium, with 18 countries involved. During the seven-day sessions, 257 papers were presented by 238 foreign and Chinese scholars and experts. They have made excellent lectures and held warm academic discussions about the series of subjects of the uplift of the Himalayas and its impact to the physical environment and human activites there. This symposium, from beginning to end, was permeated with an atmosphere of friendship and proceeded in a spirit of daring to explore new scientific fields and in a style of free academic debating. They have greatly facilitated the academic exchanges and the expanding of scientific cooperation. We all think that the symposium is a successful one with its special style and high level.

It is true that our Chinese scientists have made some work and achieved certain results in their scientific researches in Qinghai-Xizang (Tibet) Plateau. And we are very glad that our foreign friends highly appreciate our work. Nevertheless, we have learned a lot from foreign scientists and we have benefited a great deal from the excellent academic lectures and many pertinent comments presented by foreign scholars at the symposium. Moreover, some foreign friends have also given us many valuable science books. So, on behalf of the Academia Sinica and the Organizing Committee of the Symposium, I would like to express our heartfelt thanks to our foreign scholars and friends.

Friends and comrades,

This is the first time for our Academia Sinica to organize a symposium on Qinghai-Xizang (Tibet) Plateau, we hope it is a good start. We also hope that through this symposium our international academic exchanges and scientific cooperation will be promoted. Let us go ahead hand in hand to make further contributions to our joint efforts in unveiling the mysteries of Qinghai-Xizang (Tibet) Plateau and in scaling the scientific heights of the Himalayas. We hope too that the seeds of friendship sown among our foreign and Chinese scientists this time will blossom and bear fruits and will be handed down from generation to generation.

After the symposium, the foreign and Chinese scientists will go to Xizang together for the scientific excursion. This will be a very interesting and significant trip. I wish you a pleasant and successful journey!

Now concludes the Symposium on Qinghai-Xizang (Tibet) Plateau!
Thank you.

## SPEECHES OF THE REPRESENTATIVES OF FOREIGN SCIENTISTS AT THE CLOSING CEREMONY

Ardito Desio
(Istituto di Geologia, Universita Deglu Studi di Milano, Italy)

Mr. Chairman, Ladies and Gentlemen,

As a dean of the participants to the Symposium on the Qinghai-Xizang Plateau I am sure of interpreting the thought of my colleagues of foreign countries, thanking heartily the Academia Sinica which sponsored the Symposium, the Organizing Committee under the leadership of Prof. Qian San-qiang and Secretary General Prof. Liu Tung-sheng and his staff for the perfect organization of the Symposium in Beijing.

We are really in admiration at the extension and the deepness of the studies made by the Chinese colleagues in the Qinghai-Xizang Plateau in many scientific fields. The reports of the Symposium, when printed, will be an important base for the further studies and the practical development of that remote but very interesting area.

Our best wishes to the progress of the new China in every fields.

Xiexie nimen!

Hiroshi Hara
(Department of Botany, University of Tokyo)

Mr. Chairman, Ladies and Gentlemen

I am a botanist from Japan, one of your neighbouring countries in Asia. First of all I wish to express my deep gratitude to Academia Sinica for kindly inviting us to this grand Symposium. I was greatly impressed and surprised not only by the fact that the Symposium was so well prepared and organized but also that the Symposium was unique in covering nearly all branches of natural sciences. Most of scientific informations presented by the Chinese scientists during the Symposium were quite new to us. It is certain that this Symposium made a great contribution to progress and development in various fields of sciences. I believe that this is the first chance to promote intimate collaboration between the Chinese scientists and ours, and I am sincerely hoping that this friendly relationship will last for ever. Finally I should like to express my hearty thanks again to all staffs of Academia Sinica, all Chinese scientists and all other persons who have helped this Symposium for giving us facilities and hospitality during the Symposium. Thank you.

Ulrich Schweinfurth
(Department of Geography, South Asia Institute,
University of Heidelberg)

Mr. President,
colleagues, friends,

After a week of concentrated work, you are, all of a sudden, confronted with the privilege — or shall I say: the burden - it depends, how you feel about it — of trying to express in a few words, what may be the reaction of some 70-80 people of various backgrounds to this unique occasion: the first Symposium on the Tibetan Plateau — quite an impossible request to meet satisfactorily, I may add.

I am fully aware, that it is quite impossible to do justice, to all the preparatory work invested in this very week, which went on so smoothly — at least for us, the guests.

I remember well: exactly a year ago, to the day, before the symposium started, I received the invitation and we can only ponder about the processes of discussion and decision making, which went on before.

In our case: we, the outsiders, lived a year in anticipation; for many of us, if not most of us, it became the first opportunity to visit China - in many cases: a chance longed for over many years (in my own, if I may add, for over a quarter of a century!).

Once here, we have been confronted with the results of 30 years of research on the plateau by our Chinese colleagues, and we have been, to say the least, deeply impressed by what has been done and what has been achieved.

We from the outside, contributed to the best of our knowledge, and, I am sure, I am speaking for all of us: we learned a lot.

We sincerely hope, that Academia Sinica will come to the conclusion, that the contributions from inside and from outside made for a good blend - worth the effort.

We shall return to our various institutions, seats of learning, chairs of teaching enriched, not only with books and maps — think only of the 1375 pp. volume of the Vegetation of China! — but also enriched with an unforgettable experience.

And when we, once back home, begin dissipating, multiplying, what we have learned here during this week, I am sure, it will not only be facts and figures, species and strata, which we will relate, it will also be some of this wonderful spirit of openness and enterprise, which we have met here right from the beginning, which signified the atmosphere and, I am sure, I am speaking for all the foreign guests, has been greatly appreciated.

Not to forget: our basic needs have been looked after most efficiently: everyone will agree that Jing Xi Guest House proved to be a most congenial habitat in every respect; and everyone familiar with conferences of this sort knows, how important it is, not to waste time in covering distances between one meeting place and the next one - this all has been splendidly organised and the arrangement worked to full satisfaction.

After a period of, to us — the outsiders, unspecified duration of contemplation and decision making, whether to have the symposium at all or not, — the time of active preparation to follow, — the very week of the symposium we are about to close now, - and the excursion to the plateau ahead.

Academia Sinica decided to shoulder the additional burden of publishing the proceedings of the symposium; everyone vaguely familiar with an undertaking of this kind feels full sympathy with the editorial board; all of us wish to see this volume published, as a source of reference, - as a source of enlightenment, and as a lasting memento to this unique occasion: the first symposium of the Qinghai-Xizang Plateau.

We hope - we have given the editorial board the necessary freedom, it needs, for this onerous task; we wish the editorial board a full success; and we pledge full cooperation to get this task done within reasonable time.

I am sure, there are many more things, the one or the other would have liked to stress specifically; I am most humbly aware not to be up to the situation to know what everyone of us would have liked to say in particular: I offer my apologies: for feelings, sentiments, wishes, I have not expressed, but an address of this kind, out of necessity, is a compromise.

We would like, Academia Sinica, our Chinese hosts, to know, once again, how greatly we appreciate to have been invited to attend this symposium, to learn, first-hand, about your work in a part of the globe, we are interested in ourselves deeply.

With our thanks, we combine our most sincere wishes to carry on the good work begun so efficiently and splendidly; as far as we — from the outside — are concerned: we offer our full cooperation - and, perhaps, one day, there might even be a chance to meet again.

And now, in the spirit of scientific cooperation, let me sum up in the time-honoured academic well-wishing: "Academia Sinica – vivat, crescat, floreat! "

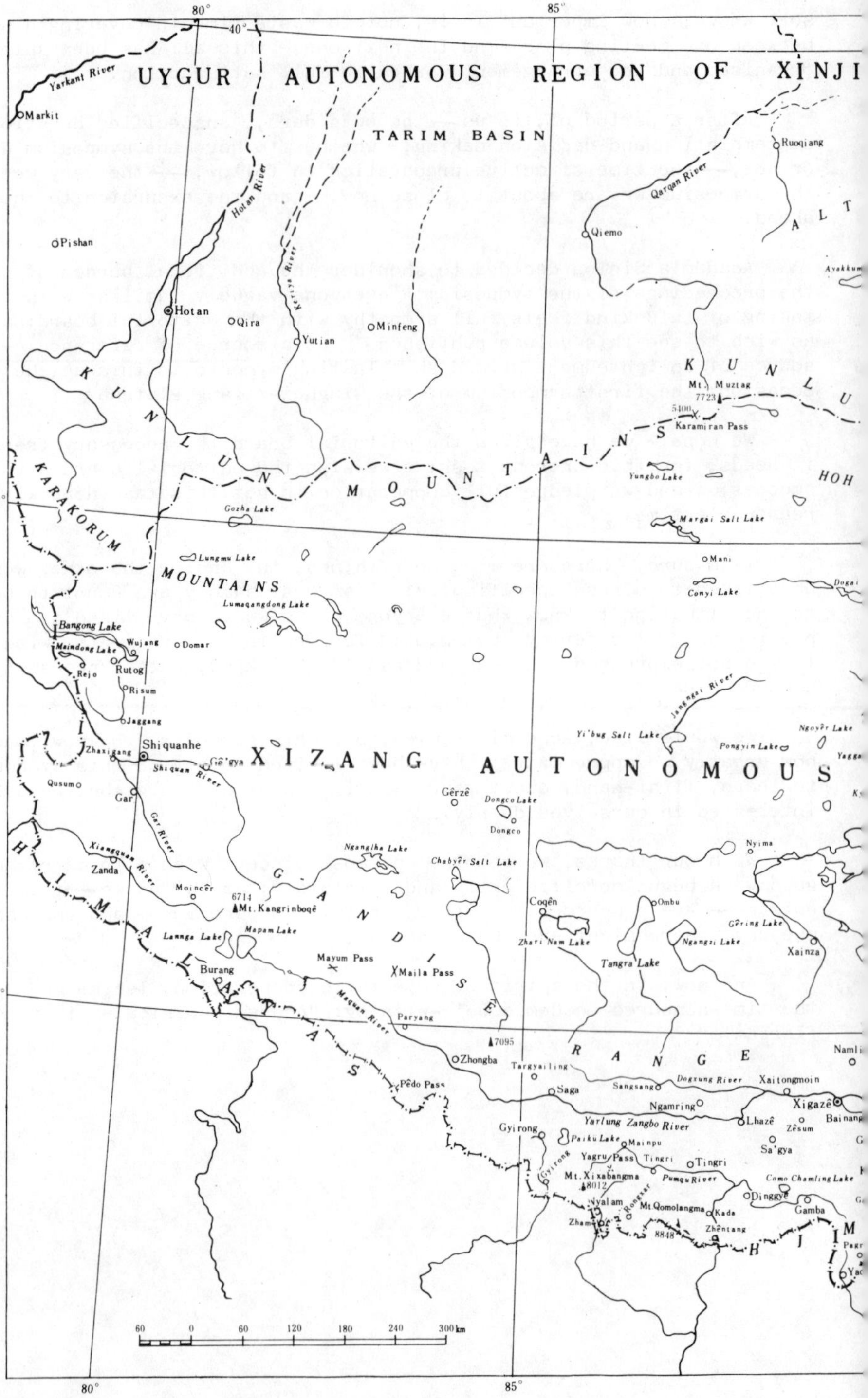


**The boundaries of this map according to "Map of the People's Republic of China" Published by the Atlas Press, 1972 (the same below).**

The geographic map

ghai-Xizang Plateau

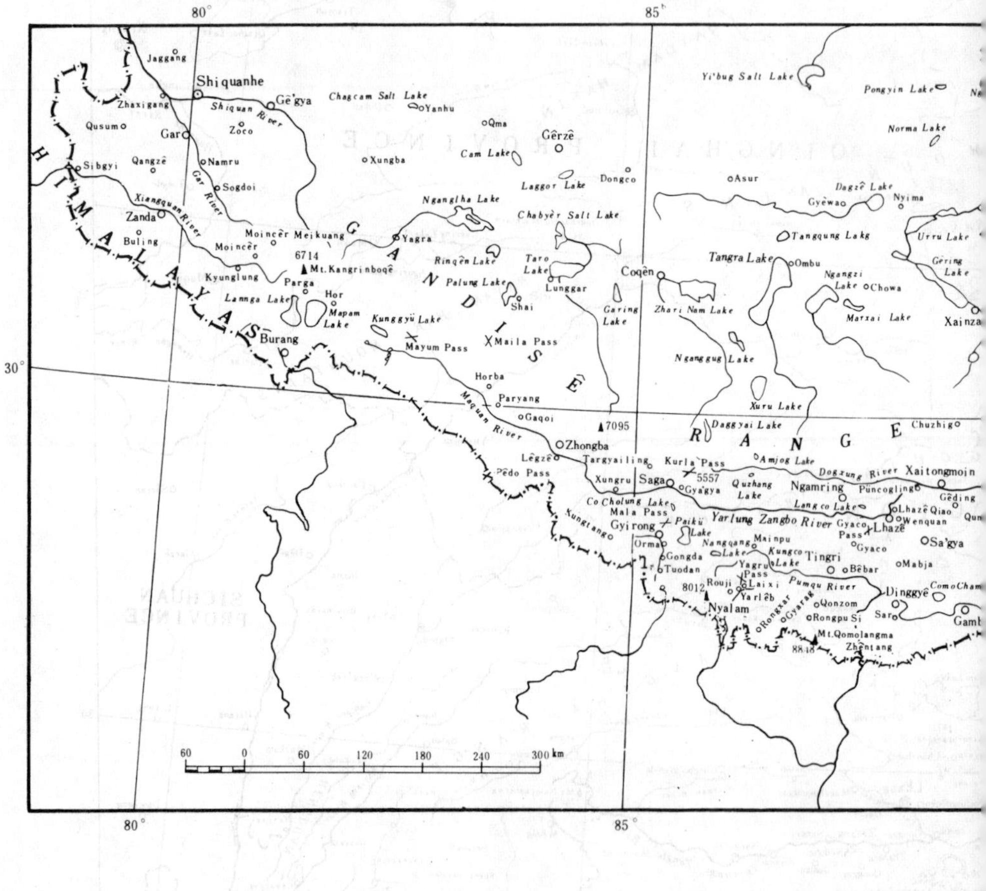


The geographic

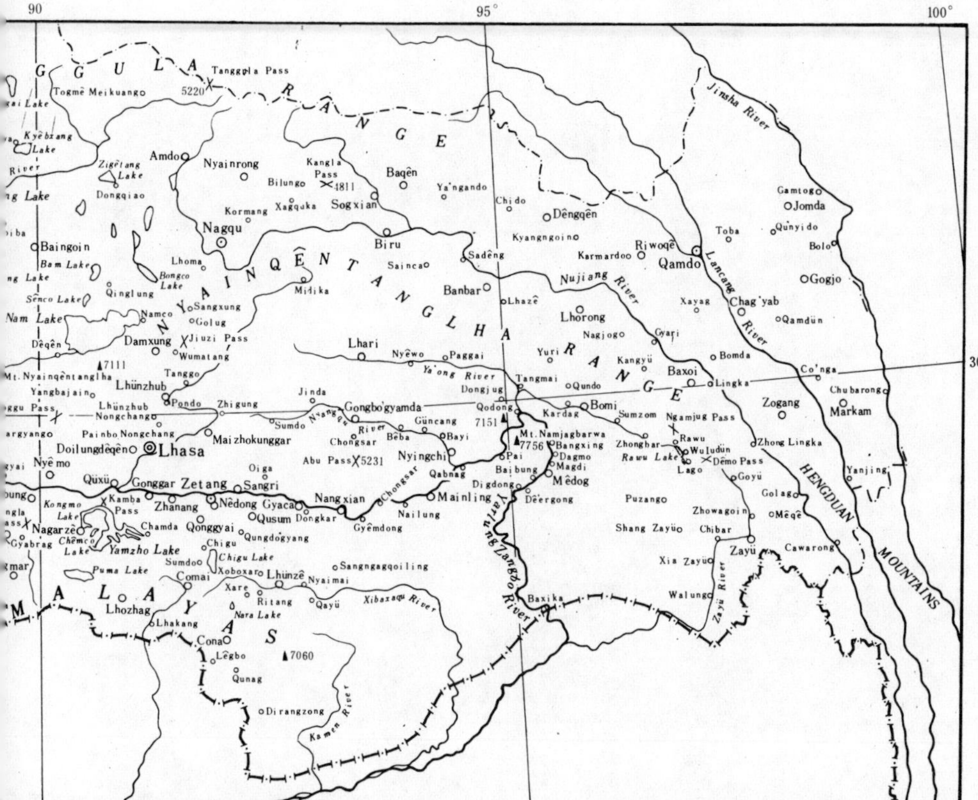


southern Xizang

# A BRIEF DISCUSSION ON THE TECTONIC EVOLUTION OF QINGHAI—XIZANG PLATEAU

Chang Cheng-fa, Pan Yu-shen
(Institute of Geology, Academia Sinica)

Since 1960 we have made several geological investigations in the Qinghai-Xizang Plateau, and some regional works and articles have been published. Here in this paper an attempt is made to analyse the tectonic evolution of Qinghai-Xizang Plateau based on the new data collected in the last decade. The Himalayan belt and the Yarlung Zangbo suture zone are better known through recent investigations by the Scientific Expedition organized by Academia Sinica. Although the accuracy is not high, the area north of the Tanggula Range has been mapped. We have also made a number of reconnaiasance expeditions across the Qinghai-Xizang Plateau, including both the geotraverse from the Himalayas to the Kunlun Mts. in the western part of the Xizang Plateau and the geological structural section from Yadong to Kalamuloin Pass, as well as the geological section from Lhakang to Golmud (Fig. 1).

## MAJOR STRUCTURAL UNITS AND OUTLINE OF TECTONIC EVOLUTION

Fundamental differences in crustal character ( continental versus oceanic ), in the age of basement consolidation, in sedimentary characteristics of platform cover and biogeographic considerations of the sediments in the age of main orogenic episodes, and in the age and nature of magmatism are used to subdivide the region into five tectonic units presumably separated by sutures. They are the Kunlun folded system, the Hoh Xil-Bayan Har folded system, the Qangtang-southern Qinghai structural region, the northern Xizang structural region, and the Himalayan structural region. The ages of the main orogenic episodes of these structural regions successively decrease from north to south (Fig. 2). The sutures which separate above-mentioned are southern Kunlun, Litian Lake-Margai Caka-Jinsha River, Bangong Lake-Dong Qiao-Nujiang River and the Yarlung Zangbo suture zones, respectively. The Qinghai-Xizang Plateau seems to have been formed by the successive accretion of continental and/or island arc fragments between Late Paleozoic and Late Cretaceous. Thus, apparently it was not a stable block for a long period of time.

### The Himalayan Structural Region

The Himalayan structural region, of continental character, has a crystalline basement that was consolidated in Precambrian time but has been recycled by Tertiary metamorphism. The basement has yielded radiometric ages of 640-660 m.y. in the Qomolangma area in China. Elsewhere in the Himalayas, metamorphic basement events have been recorded at 1,800-1,400 m.y.(19, 11). The Tethyan sequence of Late Precambrian to Eocene age constitutes its sedimentary cover. It is a sequence mainly of marine, essentially unmetamorphosed sediments in its upper parts and with many horizons rich in fossils in successions from Ordovician to

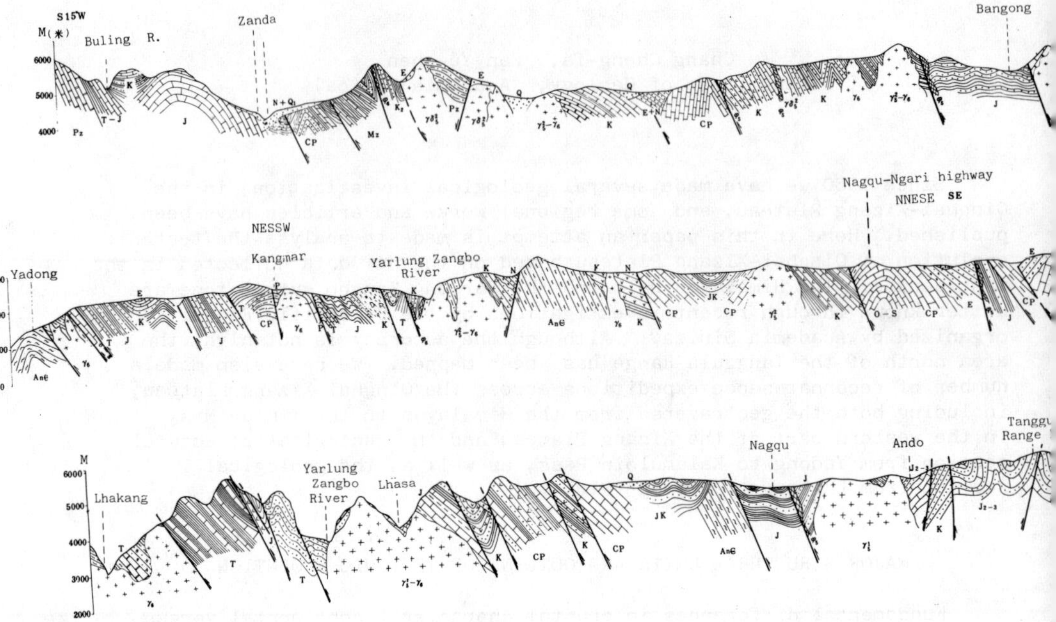


Fig. 1 Geological - structu

a. The geological-structural section from Buling R. to border pass

b. The geological structural section from Yadong to Kalamuloin Pas

1. sandstone 2. conglomerate 3. shale 4. argillaceo
schist 9. granite 10. granodiorite 11. gneissoid
14. ultramafic rocks 15. migmatite 16. schist 17. sl
22. unconformity

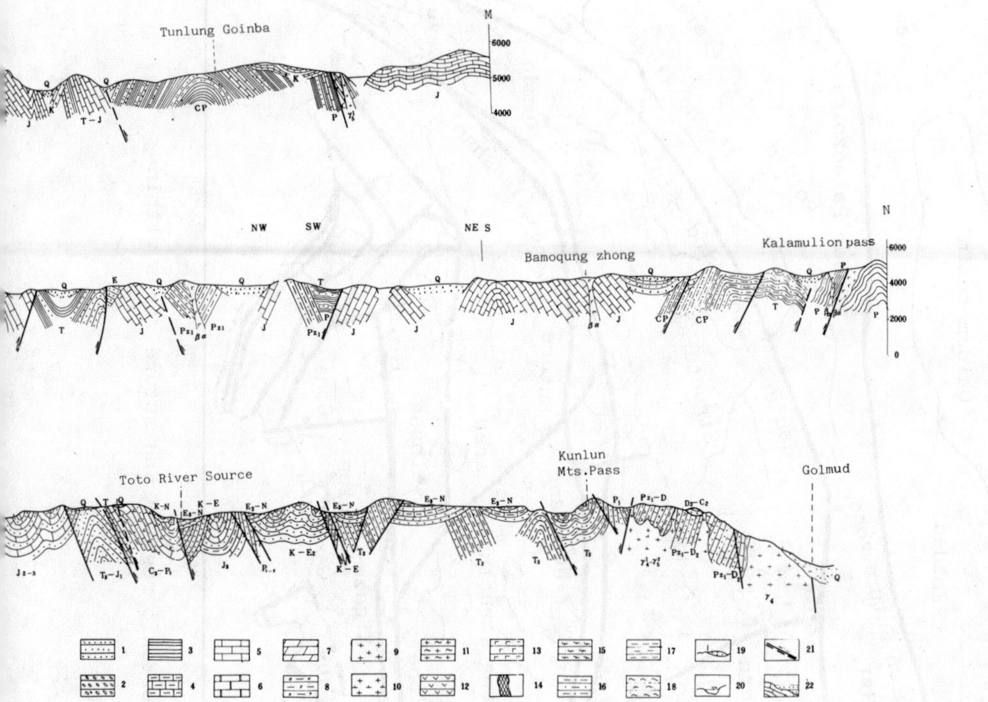


ctions in Qinghai-Xizang Plateau

ween Xizang and Xinjiang in the western part of Qinghai-Xizang Plateau
The geological section from Lhakang to Golmud
estone 5. limestone 6. marble 7. marl 8. hornblende
nite 12. intermediate volcanics 13. basic volcanics
18. gneiss 19. Quaternary gravels 20. lake 21. fault

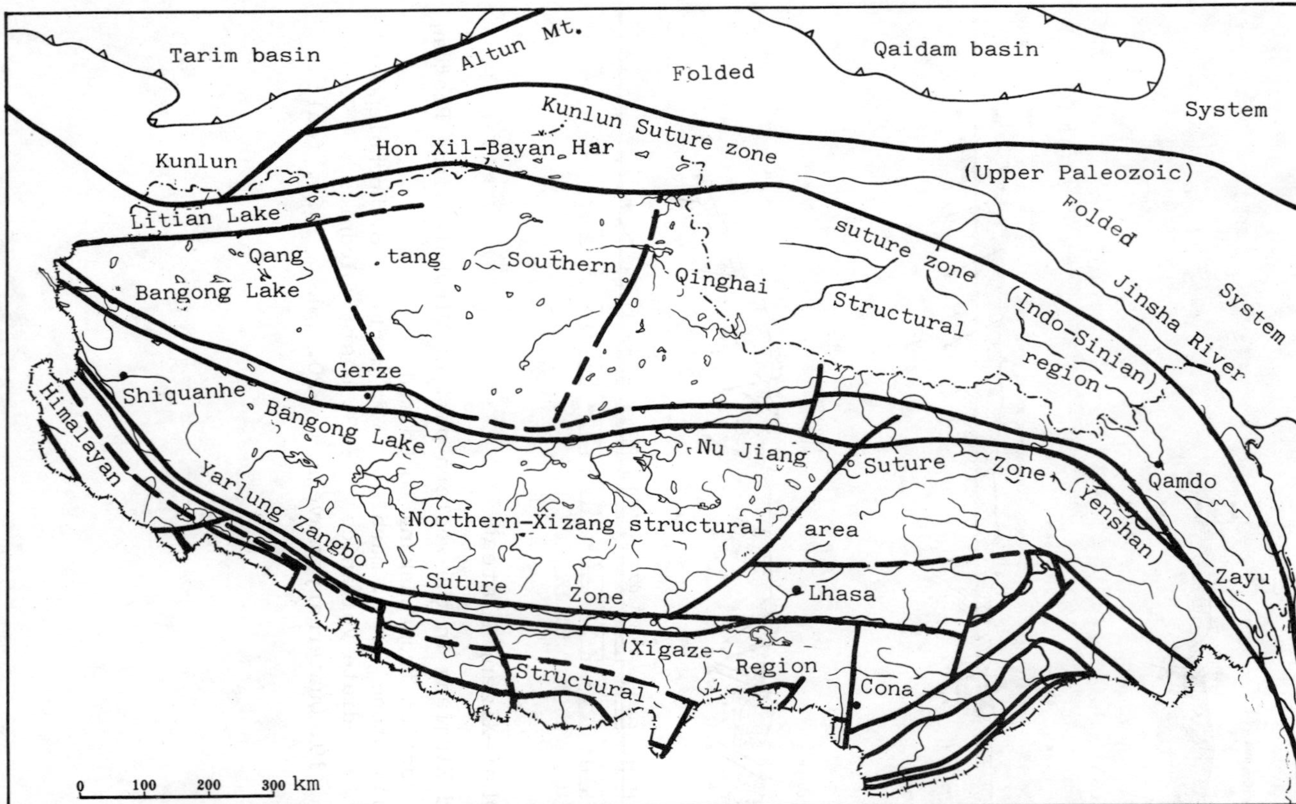


Fig. 2 Map showing the tectonic regions and suture zones within the Qinghai-Xizang Plateau

Eocene. It has been shown that the fossilifereous Ordovician to Eocene sediments are not geosynclinal in nature but represent sedimentation on outer margins of the Indian platform(9,2). Along the Main Central Thrust ( MCT ) the crystalline basement and its sedimentary cover overthrust southwards the sedimentary and the low-grade metamorphic rocks of the Lower Himalayan zone. Between the basement and the fossiliferous Ordovician-Eocene sediments in the Qomolangma area of China, there are low-grade meta-sedimentary rocks, which constitute the lower part of the Tethyan sequence and tentatively correspond to the Vindhyan Series on the Indian platform and the Midland meta-sedimentary Group in Nepal. In the Mt. Qomolangma area this is called the North Col Formation(22). They are Late Precambrian to Early Paleozoic in age. The overlying Ordovician-Early Eocene sedimentary cover has been stripped off and thrown into nappes. The thrust which separates the low-grade meta-sedimentary rocks ( North Col Formation ) and the unmetamorphosed sedimentary rocks is called Main Northern Thrust ( MNT ) by Wang and Zheng(21)- Mylonites and drag folds are well-developed within the thrust zone. In this area, the existence of an Early Paleozoic orogeny in Himalayas cannot be excluded, although no pronounced unconformity is apparent between the crystalline basement and the overlying sediments.

In the region north of Qomolangma, there are deposits with Glossopteris flora(13). Stepenoviella fauna, reflecting a cold environment, as well as marine glacial deposits yielding striated pebbles of Upper Carboniferous - Lower Permain were discovered. Vestiges of these glacial deposits are quite pronounced and extend far northward to the vicinity of Kangmar County about 50 km south of the Yarlung Zangbo River, where widespread pebbly slates can be seen. The pebbles set in the slates are quite angular with a lack of sorting. Some pebbles are oriented with their long axes perpendicular to the bedding plane. The lithofacies and fauna of these sediments are similar to the coeval Gondwanian sediments occurring farther south. Thus, the interpretation of Himalayas as an integral part of Gondwana is now widely accepted. However, the site of the northern boundary of the Gondwana is disputed(5,3). Similarities of the besement, the stratigraphic succession of Paleozoic rocks, the lithofacies and the fauna (as the Stepenoviella has been found) suggest similar Carboniferous and Lower Permian series in the Himalayas and Northern Xizang. All of these observations reveal similar paleogeographic conditions in the two regions. We thus suggest that the northern Indian boundary of the Gondwana continent was possibly situated along the Bangong Lake-Dongqiao-Nujiang River line until the Permian (see Fig. 2).

Tectonism in the Himalayas occurred in Early Tertiary but reached its climax in Middle Miocene. Here, particularly in the High Himalayas, we can see the classical features of an orogeny, such as polyphase deformations, metamorphism and granitization. The orogenic deformation in Himalayas is the result of continental-continental collision. After the continental collision, deformation shifted to the south and the Main Central Thrust (MCT) and then the Main Boundary Thrust (MBT) were successively formed in Miocene-Oligocene and in Pleistocene, respectively. Meanwhile, the sedimentary cover was stripped off from its basement and thrown into nappes.

## The Yarlung Zangbo Ophiolite Belt

The Yarlung Zangbo ophiolite belt, of oceanic crustal character, extends from the Gyemdong area near the bend of the Yarlung Zangbo River and then continues to the east of the Yarlung Zangbo valley in scattered metabasic outcrops along a major fault zone to the Burma-Andaman arc. The ophiolite belts are uninterrupted but extend westwards as a discontinuous zone along the Yarlung Zangbo valley to Upper Indus valley, and then westwards along the southern front of Hindu Kush- Karakorum and around the Nanga Parbat spur to join the ophiolite of the Middle Indus valley. Based on recent observations (7,8,18), two sub-belts are recognized in Pakistan, Afghanistan and Iran. The southern sub-belt is characterized by extensive coherent sheets mainly of peridotite resting on Mesozoic oceanic sediments. They are associated with huge exotic limestone blocks among which Permo-Triassic limestones predominate. The northern sub-belt is characterized by Late Cretaceous ophiolite melanges accompanied by thick Upper Cretaceous-Tertiary flysch successions.

In recent years some work has been done in the Yarlong Zangbo belt. Here the ophiolite belt borders the Tethyan Himalayas on the south and the Upper Cretaceous flysch on the north. Both are in tectonic contact. The Yarlung Zangbo ophiolite belt is not a single fracture,but is represented by a narrow, largely vertical, intensely tectonized zone. The vertical zone contains flysch with thrusted sheets of ultramafic rock associated with deep sea sediments, mafic volcanics, pillow lavas and radiolarian cherts and with complicated blocks, which lithologically and faunistically are similar to the Permian rocks discovered both in Tethyan Himalayas and Lhasa area of Xizang. They all are not geosynclinal in nature but represent sedimentation on the outer margin of the Indian platform.

As far as our knowledge is concerned the ophiolites within the Yarlung Zangbo belt may be one of the well preserved ophiolites in the world and appears to be composed of an orderly sequence. From bottom to top we have ultramafic rocks, in which harzburgite predominates, gabbros and/or gabbro swarm dikes, massive mafic volcanics, pillow lavas and radiolarian cherts with some spilites (Fig. 3, 4). The described orderly sequence is similar to the Troodos massif and the Papua massif as well as the sequence of ideal oceanic crust.

In the northern Himalayas there is a zone with the succession in time from Late Permian to Late Cretaceous of neritic limestone, pelagic and turbid sedimentary deposits, intermediate-basic volcanics, an ophiolite suite and flysch series. This succession reflects the characteristics of an evolutionary process of platform disintegration and the installation of a divergent plate boundary. Thus, a process of opening and the development of an oceanic Tethys with new oceanic crust in the inner part of Gondwana during Triassic is suggested (Fig. 5). In this case, the Yarlung Zangbo belt seem to be of intragondwana origin.

Northern Xizang was united with the mother Gondwanaland and separated from the northern continent (Eurasia) in Paleozoic. It became detached from Gondwana in Early Mesozoic, then became attached to Eurasia in Late Mesozoic and finally rejoined Gondwanian India in Early Cenozoic(18, 3). It is necessary to point out that the time of closure of northern Xizang along the Bangong Lake-Dongqiao-Nujiang River line was approximately simultaneous with the time of the major spreading of the southern ( new ) Tethys, which later on became the Yarlung Zangbo suture zone.

In spite of the intragondwana origin, the Yarlung Zangbo belt appears to be a real suture with all the typical features. Well-developed

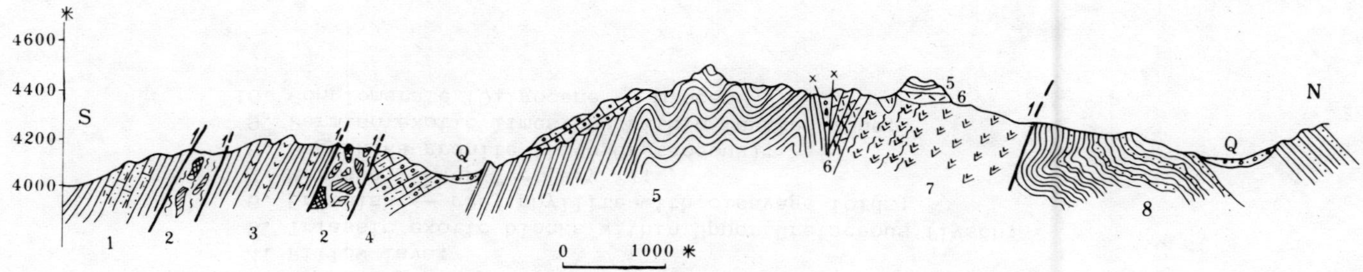


Fig. 3 Generalized ophiolite section from Tangga to Yinzen

1. Triassic flysch;
2. Ophiolitic melange;
3. Purple and greenish — grey siliceous shale with spilite;
4. Purplish — red conglomerate and sandy conglomerate (Oligocene — Miocene ? );
5. Purplish — red cherts;
6. Dark greenish — grey mafic volcanic rocks;
7. Intensely tectonized ultramafic rocks;
8. Flysch (Xigaze Group — Upper Cretaceous);
9. Quaternary gravel

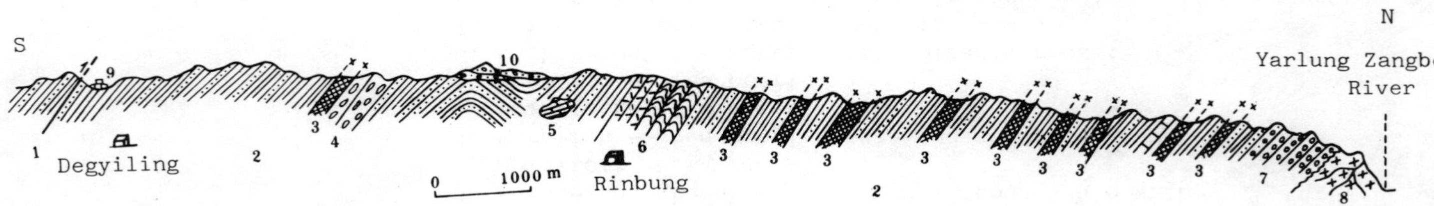


Fig. 4 The geological section from Yarlung Zangbo to Degyiling

1. Triassic flysch;
2. Upper Cretaceous flysch (Xigaze Group);
3. Intensely tectonized and serpentinized peridotite;
4. Pillow lava;
5. Triassic exotic blocks within Upper Cretaceous flysch;
6. Greenish — grey phyllite with cleavage folds;
7. Molásse (Lower Tertiary ?);
8. Dag Zhuka granite (Late Cretaceous);
9. Permian exotic limestone within faulted zone;
10. Conglomerate (Oligocene — Miocene ?)

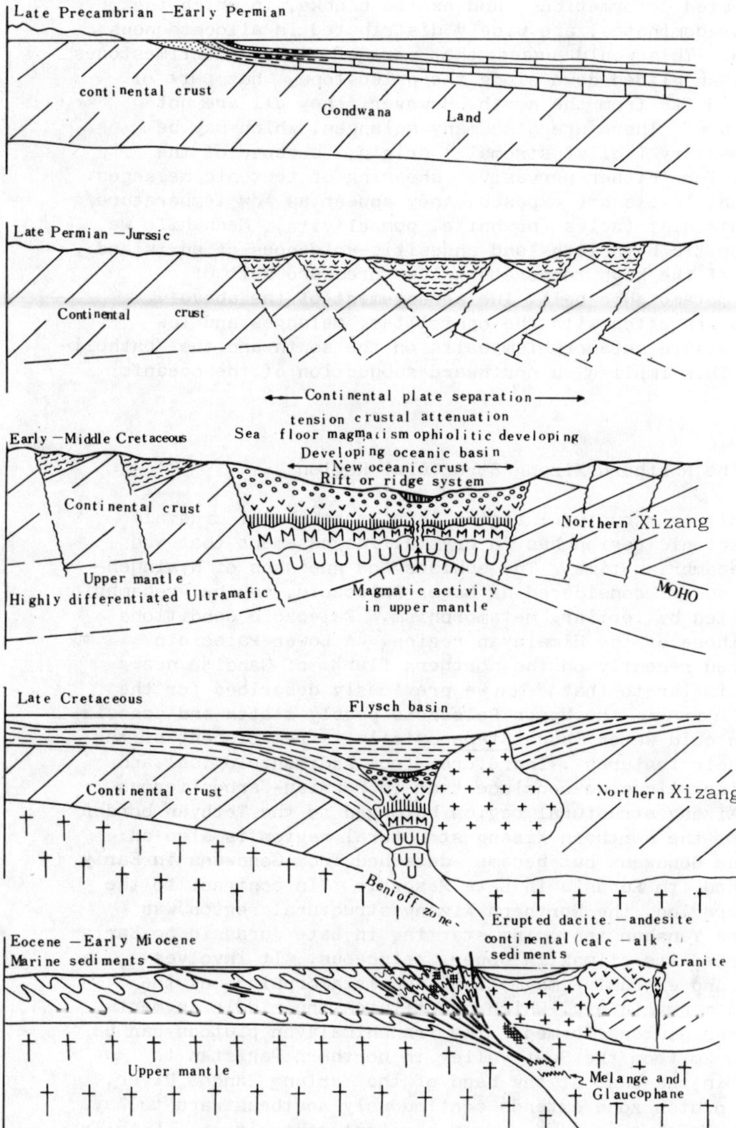


Fig. 5 Diagrammatic illustration of the principal stages of the evolution of the plaeogeographic and geodynamics in the Himalayas and Yarlung Zangbo River

U: Ultra mafic intrusions M: Mafic intrusions — gabbro V V Mafic volcanics and diabase complex ⁰ ⁰ Pillow lavas |||| Siliceous rocks (radiolarian cherts) Sedimentary rocks ophiolites, complicated deformation, and exotic blocks, in which the Permian limestone predominates, are widely distributed in allochthonous masses to the south. This would suggest that part of the exotic limestones were uplifted and than glided down along the paleoslope, but part of them were transported far from the north. However, they all are not geosynclinal in nature. There are also many melanges, which may be attributed to sedimentary ( olistostromal ) origin, because of the absence of evidence for either pervasive shearing or tectonic melanges. When deeper erosional levels are exposed, they appear as low temperature/ high pressure green schist facies (prehnite, pumpellyite). Meanwhile we have large granodioritic batholiths and andesitic volcanoes of an island arc situated north of the high temperature/low pressure belt of Miyashiro's paired metamorphic belt. The arrangement of the above-mentioned belts appears often with the ophiolites, melanges and low temperature/high pressure metamorphism belts on the south and the batholiths on the north. This implies a northward subduction of the oceanic lithosphere.

## The Northern Xizang Structural Region

The northern Xizang structural region is of continental crustal character. This tectonic region has a crystalline basement that was consolidated in Precambrian time. The schists and gneisses of Nyainqentanglha belt, previously considered as Lower Paleozoic, may be Precambrian that has been recycled by Tertiary metamorphism. Paleozoic conditions appear similar to those of the Himalayan region. A Lower Paleozoic succession discovered recently on the northern flank of Gandise near Xainza is broadly similar to that which we previously described for the Himalayan region. Moreover, the Upper Paleozoic pebbly slates and associated sediments with cold water fauna Stepanoviella is reminiscent of the Gondwana type in their features . Therefore, from sedimentological and biogeographic considerations we conclude that in Permian-Carboniferous time the northern Xizang structural region belonged to the Tethyan border of Gondwana. Thus, the northern Xizang structural region is also an integral part of the Gondwana but became detached from Gondwana in Early Mesozoic and attached to Eurasia in Late Mesozoic. In contrast to the Himalayan tectonic region, the Northern Xizang structural region was subjected to intense Yenshan tectonism starting in Late Jurassic to Early Cretaceous and reached its climax in Upper Cretaceous. It involves folding, intrusive and extrusive magmatism, and metamorphism, as the Gandise- Bomi grand magmatic arc, which constitutes the middle segments of the Transhimalayan plutons formed. The Transhimalayan plutons can be followed over 2,500 km from the Swat valley in northern Pakistan to Namche Barwa Group in the east at the bend of the Yarlung Zangbo River, and from there the pluton zone extends continuously southeastward to Zayu area. Regionally these plutons have a very constant composition. Locally rock types vary from peridotite (rare) to norites, and from gabbros to the widespread tonalites and hornblende granites. The geometric relationships indicate an intrusive succession from basic to acidic(14, 11). Within the widespread tonalites and granites, regional hornblendisation of the original pyroxene is common. The subsequent dacite-andesites ( calc-alc. ) follow in a belt to the north of it. Absolute ages of the plutonic and volcanic rocks fall into the span of time from 110 m.y. to 45 m.y. with the bulk between 45-80 m.y. That span is essentially the entire period of pre-collisional subduction of oceanic lithosphere. Geochemically, the volcanics have very high initial value of $^{87}Sr/^{86}Sr$ ( 0.708 ). This fact suggests that these parts of the volcanics were generated by melting of a continental crust. Before the continental

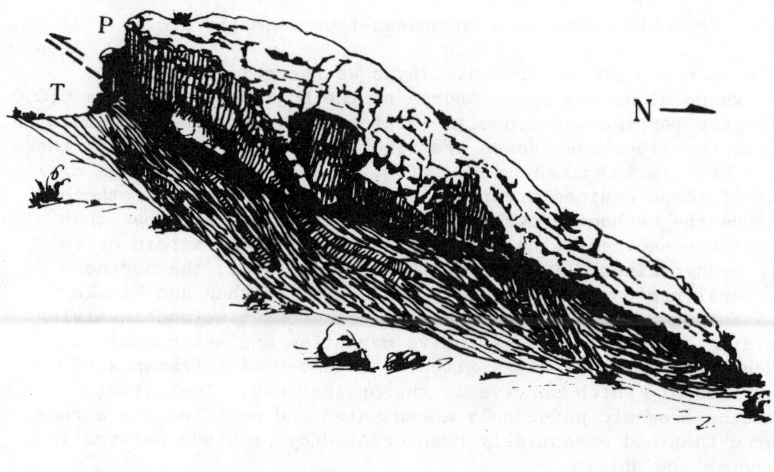


Fig. 6 The Permian exotic blocks near Quxar between Xigaze and Lhaze

T. Triassic sandstone and shale with complicated drag folds. There is a crushed zone between P and T

P. Grey and pink Permian limestone with *Stenoscisma giganten* (Diener), *Marginifer* sp., *Spiriferella raiah* (Salter), *Neospirifer* sp., and *Amplexocarinia* sp.

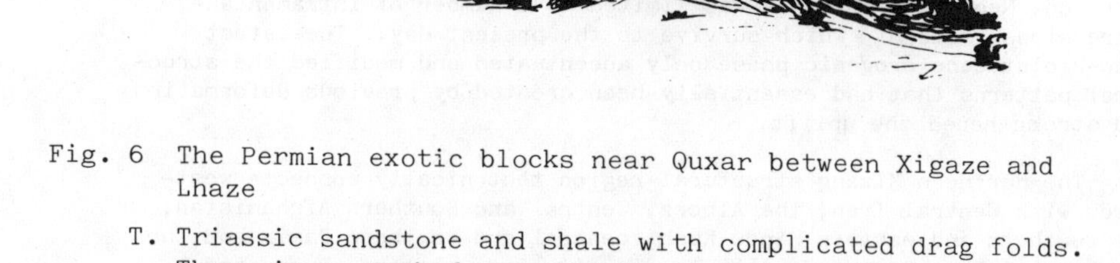


Fig. 7 The exotic blocks near Laser La of western Himalayas, Xizang

1. Permian grey and red limestone with *Neoschwagerina marigartae* Deprat, *Verbeekina* sp., *Lantschichites minima* (Chen) and *Iranophyllum* sp., *Wenzellella* sp., *Tachylasma* sp.,
2. Jurassic sandstone, marl and shales with Belemonoids and Ammonites

collision the Transhimalayas were an andean-type magmatic arc.

In the area north of the Transhimalayas we recognized a belt of subsidence, where along the upper course of the Nujiang valley some 5,000 meters of largely terrigenous deposits (slates and quartzites) accumulated. In some places few limestone lenses are included and in many places there are plenty of basic-acid volcanics and tuffs. Based on the findings of faunas, most of these sediments vary in age from Jurassic to Cretaceous, but may include the Carboniferous in the lower part. We presume that they represent the sedimentation on an active continental margin or in pericratonic geosynclines. In contrast to the Himalayas, the northern Xizang structural region was subjected to intense Yenshan and Himalayan tectonism, starting in Late Jurassic-Early Cretaceous time and involving repeated folding, intrusive and extrusive magmatism and metamorphism. Later on, Neogene subsidence was limited to a number of intramontane depressions, some of which survive to the present day. The latest Plio-Pleistocene orogenic phase only accentuated and modified the structural patterns that had essentially been created by previous deformation and strengthened the uplift.

The northern Xizang structural region tectonically connects westwards with Central Iran, the Alborz, Central and Southern Afghanistan, the southern and eastern Hindu Kush, central and southern Pamir and the Karakorum. Southeastwards it goes directly into eastern Burma and Thailand.

## The Bangong Lake-Dongqiao-Nujiang River Belt

The Bangong Lake-Dongqiao-Nujiang River Belt is characterized by a series of scattered ultramafic rocks, which may be older and poorly preserved relics of oceanic crust. The Bangong Lake-Dongquio-Nujiang River belt extends southwards through medial Malaya and joins with a suture in central Thailand. It is characterized by intermittent peridotite outcrops in Lower to Middle Jurassic sediments and is associated with cherts and melanges in some places. On the northern side, a belt of flysch, probably Triassic-Jurassic, always can be followed, and there are some granodiorite batholiths north of this belt. The ultramafic bodies are overlain by Upper Jurassic-Lower Cretaceous sediments with plant remains and faunas in the Dongqiao area ( Fig. 8 ). Thus, the process led to the emplacement of the ultramafic rocks which took place in Middle to Late Jurassic. The Bangong Lake-Dongqiao-Nujiang River ultramafic belt may reflect a Tethys, which in the Paleozoic-Early Mesozoic separated a northern continental fragment from the Indian Gondwana in the south.

## The Qangtang - Southern Qinghai Structural Region

The Qangtang-Southern Qinghai structural region, of continental crustal character, is sharply separated from the northern Xizang by the Bangong Lake-Dongqiao-Nujiang River line in the south and from Hoh Xil-Bayan Har fold system by the presumed Litian-Margai Caka-Jinsha River suture line. The Basement of this region is buried under the Paleozoic and Mesozoic sediments and is elevated only in a few faulted blocks. It is represented by schists. They are meta-sedimentary rocks probably of Lower Paleozoic age or older. From the Lower Carboniferous onward this region does not differ much in geological development from the Northern

Xizang structural region, and is characterized by platform type Paleozoic development. Jurassic sedimentary rocks occupy a large area in the south and are intruded by granodioritic plutons. Much of the northern region is occupied by a thick sedimentary succession of Triassic age, which may be divided into epicontinental terrigenous deposits in the south and volcano-sedimentary rocks in the north. They were deformed into gentle to tight folds. This structural region was subjected to intense Indosinian tectonism, involving folding starting in Late Triassic time . In Qamdo area Upper Triassic continental deposits are found. They overlie the Lower-Middle Triassic formations with a pronounced unconformity. This region was broadly affected by the Yenshan and Himalayan tectonism, which appears as folding and faulting and as uplift, respectively. The Qangtang-southern Qinghai structural region, particularly in its northern and western parts, are the least known parts under review. Through reconnaissance work performed by Sven Hedin previously and through our recent expeditions, a greater understanding has become possible. This region is noted for its widespread Tertiary, Quaternary and recent, largely alkali and peralkali volcanism(6). It has manifested itself as a zone of continuing mobility in spite of its stabilization during the Early Yenshan phase.

## Litian Lake-Margai Caka-Jinsha River Suture Line

The Litian Lake-Margai Caka-Jinsha River suture line extends from Litian Lake eastwards through Margai Caka, then along the upper course of Jinsha River gorges in Sichuan and Yunnan into the region of Red River zone in Yunnan and Vietnam. Ophiolitic rocks and various melanges extending discontinuously throughout the belt may be interpreted as manifestations of a suture between Qangtang-Southern Qinghai block and the Hoh Xil-Bayan Har block, indicating the site of closure of the Tethyan ocean in the Triassic time. The ophiolitic rocks are widespread and known in some detail from the Jinsha River gorges(23) and in Yushu area(24), where they include tectonized flysch, melanges, ultramafic rocks, pillow lavas and radiolarian cherts. Much of the sediments along the suture is of Upper Triassic in age. The Tethys, which separated the Qangtang-Southern Qinghai in the south and the Hoh Xil-Bayan Har in the north with oceanic crust may have existed from Paleozoic until it closed in the Upper Triassic.

## The Hoh Xil-Bayan Har Fold System

The Hoh Xil-Banyan Har fold belt, of continental crustal character, is separated from the Qangtang-Southern Qinghai structural region by the Litian Lake-Margai Caka-Jinsha River suture in the south and the Southern Kunlun suture in the north. This latter suture may have existed during the Paleozoic until the closure in Late Paleozoic of another ocean. The basement is buried deep under the Paleozoic and Mesozoic sediments. Based on the few Upper Paleozoic outcrops, the Upper Palaeozoic is predominantly shallow marine carbonate rocks. Much of the region is occupied by a thick monotonous flysch succession of Upper Triassic. This structural region was subjected to intense Indosinian tectonism, starting in Late Triassic, involving folding and intrusive magmatism. This thick series is often thoroughly metamorphosed.

## The Kunlun Folded Belt

Along the southern margin of the Kunlun, oceanic crustal relics could intervene between Kunlun and Hoh Xil-Bayan Har foldbelt in a zone which we call the Southern Kunlun suture zone. Based on available data, the Paleozoic (including the Late Precambrian) deposits are fully developed in the crestal part of Kunlun Mts. and include crystalline limestones, marbles, schists and submarine volcanics, which are metamorphosed into green schists. Based on many fossil findings it is known to be largely of Lower Paleozoic age. These rocks are strongly folded and are intruded by granites. In Bulghan Buda Mt. there is an important unconformity between Lower and Middle Devonian deposits (Compilation Group of the Geological Map of Asia, 1976, 1978). The Carboniferous series locally amounts to about 10 km in thickness. The Kunlun fold system was deformed and largely consolidated by strong Hercynian (middle segments) and Early Indosinian folding. Yenshan and Himalayan folding were superimposed on Hercynian and Indosinian structures. The western continuation of the Kunlun seems to connect with the north Pamir-West Hindu Kush-Parapamisus and the South Caspian depression foldbelt(18) and seems to continue eastwards in the Qinling system.

Thus, a migration of the main orogenic processes from north to south can be clearly discerned, as mentioned above. Hercynian folding and magmatism dominate in the Kunlun foldbelt, Indosinian folding and magmatism play the decisive role in Hoh Xil-Bayan Har fold system, and Early Yenshan tectonism is the strongest in Qangtang-Southern Qinghai structural region. Late Yenshan folding and magmatism predominate in Northern Xizang structural region, where the unconformity between Lower and Upper Cretaceous is very pronounced, but Himalayan tectonism including thrusting and strong intrusive and extrusive magmatism also severely affected this region. Tertiary tectonism, which affected all of the regions discussed, reached its climax in the Himalayas with thrusting the formation of nappes, metamorphism and granitization.

The migration can be tentatively explained by successive northward drift of different continental fragments and their collision with the Eurasia. In other words, Asia appears to have grown by successive addition to the Angara nucleus(2, 18).

These accretionary mountain ranges also may be formed by a combination of accretion and alternating opening and closing of associated Mediterranean seas.

# THE FORMATION AND UPLIFT OF THE QINGHAI-XIZANG PLATEAU

Much attention has been devoted to the formation and uplift of the Qinghai-Xizang Plateau, because of their effect on the natural environment and their significance for understanding of geodynamics. In the last decade much work has been done in the Xizang Plateau by geologists and geophysicists. Nevertheless, an understanding regarding this problem is still rather limited.

As mentioned above, the Qinghai-Xizang Plateau seems to have been formed by the successive accretion of continental and/or island arc fragments between Late Paleozoic and Late Cretaceous. Following the collision of India with Xizang along the Yarlung Zangbo in Eocene-Oligocene, Xizang experienced intense horizontal compression, resulting in very large crustal shortening and thickening. After the India-Eurasian collision, with

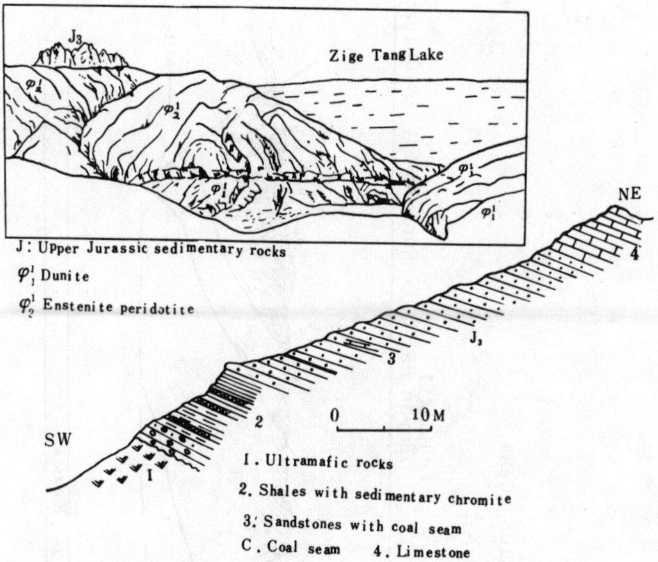


Fig. 8 The transgressive unconformity between Upper Jurassic — Lower Cretaceous and ultramafic rocks near Dongqiao

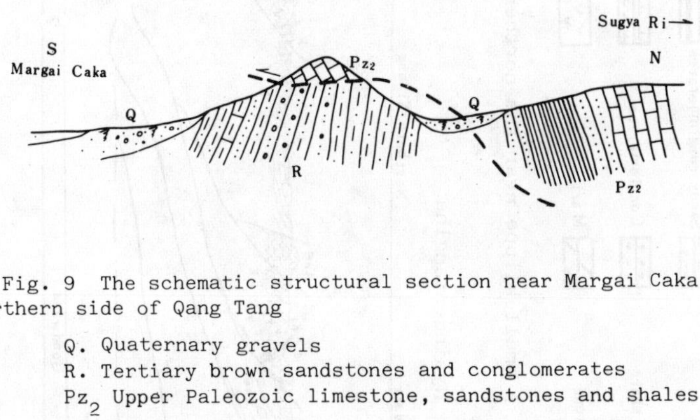


Fig. 9 The schematic structural section near Margai Caka on northern side of Qang Tang

Q. Quaternary gravels
R. Tertiary brown sandstones and conglomerates
$Pz_2$ Upper Paleozoic limestone, sandstones and shales

continued northward movement of India a new thrust the Main Central Thrust, developed within the Indian Plate in Oligocene. Movement of at least 100 km along this thrust is generally presumed. Metamorphism and emplacement of Cenozoic granites have been related to movement of Higher Himalayan rocks along the thrust(1).

Middle Tertiary uplift resulting from the thrusting led to the deposition of the Late Cenozoic Siwalik Molasse succession to the south. Following this and possibly accompanying it thrusting developed on the Main Boundary Fault(15).

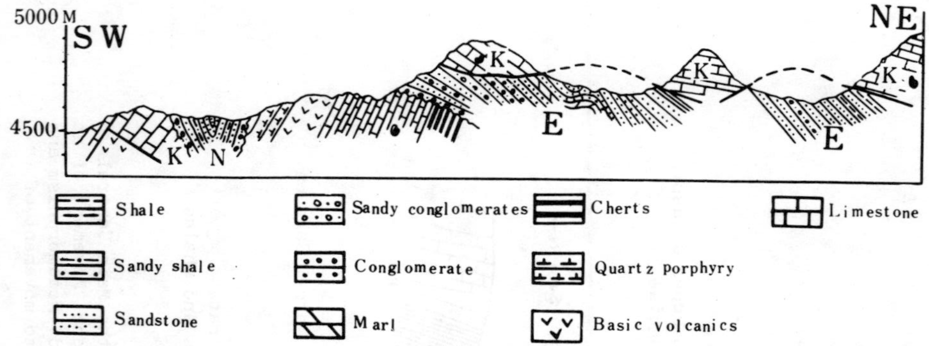


Fig. 10 A small klippe near the northern side of Zhilung Caka

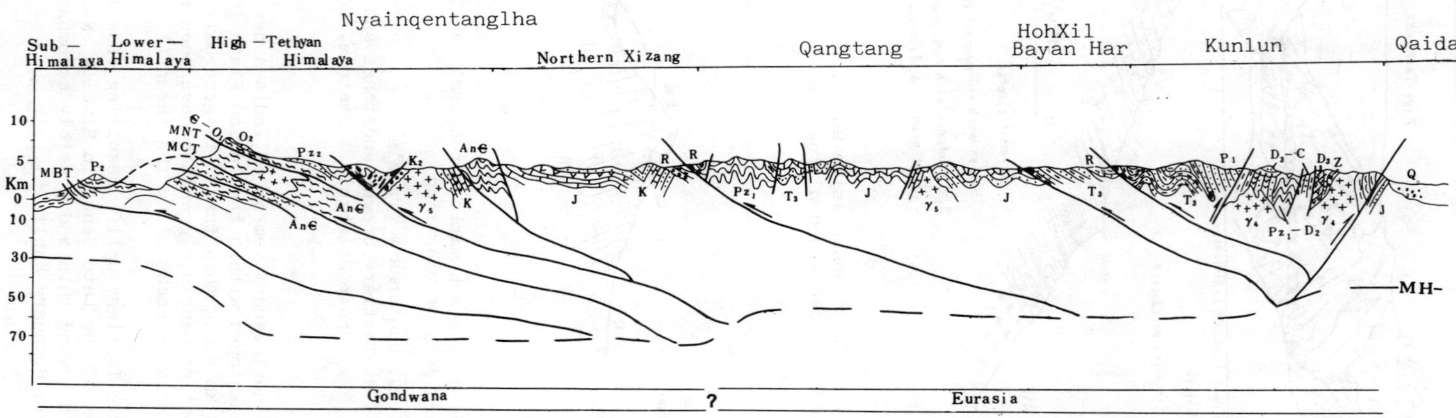


Fig. 11 Sketch map of the crustal structure of the Qinghai-Xizang Plateau

In northern Xizang, thrusting is also the most conspicuous tectonic feature (Fig. 9, 10 ). Observations from a regional geological expedition disclosed that the tectonic units constituting the Qinghai-Xizang Plateau were successively overthrust one over another from north to south. Thus, the idea of piling up crustal wedges might be applicable to the creation of the thick crust of the Qinghai-Xizang Plateau ( Fig. 11 ).

Some authors advocate that normal faulting on north-south faults is prevalent in Xizang(16, 17). Indeed, there are many transverse faults in Xizang Plateau, but they are particularly prominent in Himalayas(20). Based on our observations the movement along these north-south faults seems to be strike-slip in the Chinese Himalayas. This means that the Indian subcontinent was probably divided into a number of blocks which moved northwards with different speeds. In northern Xizang the existence of normal faulting along approximately north-south faults and east-west extension cannot be excluded, but more work needs to be done to substantiate these speculations.

Based on the results of explosive seismic sounding in Xizang by Institute of Geophysics, Academia Sinica , the crust of the Xizang Plateau is composed of alternate high-and low velocity layers. The crustal thickness reaches 10-15 km in the region north of the Yarlung Zangbo River, but it is only 68-45 km south of the Yarlung Zangbo River, and gradually tilts to south. Here, we have discovered low velocity channel about 30 km (south) to 50 km (north) below the surface.

Recent volcanism and the geothermal activity are other very conspicuous tectonic features in Xizang Plateau. So far, we know that there are about 600 sites with hot springs and geysers within the Xizang Plateau. The crustal media of the plateau and its southern margins show strong abosorptive property to the energy of seismic wave and the Lg wave is relatively weak and even gradually vanishes. All these suggest a very low shear wave velocity in Xizang and might require melting or partial melting as well as the presence of magma bodies in the lower crust and upper mantle. In this case, the formation of the low velocity channel under the Qinghai-Xizang Plateau seems to be a result of a splitting within the continental crust and its shearing along thrust faults. The tremendous uplift of the Qinghai-Xizang Plateau since Late Cenozoic is the consequence of the crustal thickening and isostatic adjustment. Of course, heating and thermal expansion of the lithosphere could also add some effect.

References

(1) Andrieux, J., Brunel, M., Hamet., Metamorphism and relation with the Main Central Thrust in Central Nepal, $^{87}Rb/^{86}Sr$ age determinations and discussions, in Jest, C. (ed.), Himalaya : Science de la Terre, Colloques Internationaux du Centre National de la Recherche Scientifique, 268, Paris, 31-40 (1977).

(2) Chang Chengfa and Zheng Hsilan, Some tectonic features of the Mt. Jolmo Lungma area, southern Tibet, China. Scientia Sinica, 16 (2), 257-260 (1973).

(3) Chang Chengfa et al., The geological history, tectonic zonation and origin of uplifting of the Himalayas. (Peking,separate) (1977).

(4) Compilation Group of Geological Map of Asia, The geologic development and tectonic frame of Asia. Bull. Geol. Soc. China (3) and Chinese Acad. Geol. Sci., 1-36 (1976, 1978).

(5) Crawford, A. R., The Indus suture line, the Himalaya, Tibet and Gondwanaland, Geol. Mag. 3, 369-383 (1974).

(6) Den Wanming, A preliminary study on the petrology and petrochemistry of the Quaternary volcanic rocks of northern Tibet Autonomous region, Acta Geologica Sinica, 2, 148-162 (1977) (in Chinese ).

(7) Desio, A., The occurrence of blueschists between the middle Indus and the Swat valleys as an evidence of subduction (north Pakistan). Rend. Accad. Naz. Lincei, 62 (149) Roma (1977).

(8) Frank, W., et al, Geological observations in the Ladakh area (Himalayas). A preliminary report, Schw. Mineral. Petro. Mitt., 57 (1), 89-113 (1977).

(9) Gansser, A., Geology of the Himalayas, Interscience Pub., Wiley & Sons, London, 269 p. (1964).

(10) Gansser, A., The ophiolitic suture zones of the Ladakh and the Kailas region - A comparison , Jour. Geol. Soc. of India, 20 (June) 277-281 (1979).

(11) Gansser, A., The Himalayas - a fascinating geological challenge. Episodes, 17-20 (1979).

(12) Gansser, A., The significance of the Himalayan suture zone. Tectonophysics. 62, 37-52 (1980).

(13) Hsu Jen, On the discovery of a Glossopteris flora in southern Xizang and its significance in geology and palaeogeography. Scientia Geol. Sinica, 4, 323-331 (1976) (in Chinese).

(14) Jin Chengwei , Zhou Yunsheng, Igneous rocks belts in the Himalayas and the Gandise arc and their genetic model. Scientia Geol. Sinica. 4, 297-312 (1978) (in Chinese).

(15) Mitchell, A. H. G., Guides to metal provinces in the Central Himalaya collision belt; the value of regional stratigraphic correlations and tectonic analogies (1979) (manuscript).

(16) Molnar, P., Tapponnier, P., Cenozoic tectonics of Asia : effects of a continental collision, Science, 189, 419-426 (1975).

(17) Molnar, P., Tapponnier, P., Active tectonics of Tibet. Jour. Geophys. Res., 83, 5361-5375 (1978).

(18) Stocklin, J., Structural correlation of the Alpine ranges between Iran and central Asia. Mem. H. Ser. Soc. Geol. Sinica, 8, 333-353 (1977).

(19) Thakur, V. C., Divergent isogrades of metamorphism in some part of higher Himalaya zone, in Jest, C. (ed.), Himalayas. Sciences de la Terro: Colloques Internationeux du Centre National de la Recherche Scientifique, 268, Paris, 433-442 (1977).

(20) Valdiya, K. S., Himalayan transverse faults and folds and their parallelism with subsurface structures of north Indian plains. Tectonophysics, 32 (3/4), 353-386 (1976).

(21) Wang Yipeng, Zheng Shilan., The imbricate structure of the northern slope of Jolmo Lungma and the uplift of Jolmo Lungma: In reports of Scientific Expedition to Mt. Qomo Lungma, (1975), Geology. Science Press. 199-221 (1979).

(22) Yin Jixiang and Guo Shizeng, The stratigraphy in area of Jolmo Lungma and its northern slope. Scientia Sinica, 90-102 (1978).

(23) Zhang Zhimeng, Jin Meng, Two kinds of melange and their tectonic significance in Xiangcheng-Dorong area, southeastern. Sichuan Scientia Geol. Sinica. 3, 205-214 (1979) (in Chinese).

(24) Zheng Yenzong, A preliminary research on the geologic-structures of the Tanggula Range and Yushu area (1980) (reprints in Chinese).

# THE GEOLOGY OF THE UPPER SHAKSGAM VALLEY, KARAKORUM, XINJAN (SINKIANG)

ARDITO DESIO

Institute of Geology, University of Milan (Italy)

## Abstract

My report is a summary of a book that I have the honour to present as first copy at the Simposium to Qinghai-Xizang (Tibet) Plateau. It is the $8^{th}$ volume of the Scientific Reports of my expeditions in the Karakorum. Four of them are devoted to the Geology, two to the Paleontology, Zoology and Botany, one to the Geophysics and another to the Paleoethnology and Ethnography. Other four volumes are in preparation.

The Shaksgam River is one of the left tributaries of the Yarkand River which, together with the Kashgar, flows into the Tarim.

As to the geographical position, the Shaksgam Valley is placed between the Karakorum (south) and the Aghil (north) ranges and between lat. 30°31' and 36°50'N, long. 76°08' and 77°30' E from Greenwich, that is at the extreme NW corner of the Xizang (Tibet) Plateau. The southern watershed dividing the Shaksgam Valley from the Baltoro and Siachen valleys runs along the high crest of the Karakorum range reaching on the top of $K^2$ 8611 m.

The Shaksgam Valley can be divided into two main sections. The Upper Shaksgam from the head to the Sarpo Laggo confluence; the Lower Shaksgam from this confluence to the junction with the Yarkand River near Toquz Bulaq. The whole length is about 245 km, 129 of them belong to the upper section, which is desert and uninhabited.

Under the point of view of the stratigraphy the Shaksgam area is to be divided into two section; the western section is mostly composed of igneous and metamorphic rocks, the eastern section of sedimentary rocks.

The tributary valleys of Skamri, Sarpo Laggo, $K^2$, and Sughet belong to the western section, the main Shaksgam valley and the other tributary valleys belong to the eastern section.

Western section. The more spread rock complexes in this section are the followings:

1. Baltoro granite of the Karakorum axial batholith;
2. Peribatholithic plagioclase gneiss (locally migmatitic gneiss);
3. $K^2$ Gneiss: biotite porphyroblastic gneiss, augen gneiss, granite gneiss with lenses of paraschists and marbles. Over 2000 m;
4. Muztagh Tower Gneiss: Biotite-amphibole and tonalitic gneiss (= Sarpo Laggo gneiss). About 2000 m;
5. Sarpo Laggo Slates with siltstone and sandstone interbeds. Some 1000 m.

These rock complexes have been described in the volume on the "Geology of the Baltoro basin" (Desio & Zanettin, 1970). The age is known only for some of them, that is for the Baltoro granite which was dated as Miocene-Pliocene; the $K^2$ Gneiss which seem to represent an advanced stage of granitization of the Singhié and Shaksgam formations (Permo-Carboniferous); the Sarpo Laggo Slates which are the metamorphic facies of the Singhié Shales (Carboniferous).

Eastern section. The lithostratigraphic units outcropping in this section are the following from top to bottom:

7. Bdongo-la sequence, thickness unknown. Late Jurassic;
6. Aghil Limestone, about 3000 m. Late Triassic and probably Liassic;
5. Urdok Conglomerate, some 100 m } Late Triassic
4. Chikchi-ri Shales, some 100 m
3. Kyagar cherty limestone, 300 m. Middle and possibly Early Triassic;
2. Shaksgam Formation, about 1000 m. Permian;
1. Singhié Shales, some 1000 m. Carboniferous.

Only the Shaksgam and the Aghil formations show well defined characters. The Urdok Conglomerate might represent the basal bed of the Aghil Limestone, a homotaxial deposit of the Chichi-ri unit. Probably in the Shaksgam Valley there are more than one conglomerate horizons similar to each other; they seem to represent different rock units. The same could be true for the Kyagar cherty limestone, which could represent a real formation, at least in the area between the Staghar and Kyagar glaciers.

Under the chronostratigraphic point of view the Shaksgam formation is to be divided into two chronozones, representing the Early and the Late Permian. To the first one belong numerous fossil brachiopoda and some foraminifera (among them Parafusulina) which are referred to the Karachatyrian and Darvasian stages. To the second one the fossil content is characterized by other species of brachiopoda, but also by pelecypods, bryozoa and, among the foraminifera, some Paraschwagerina. This zone represent the Murgabian and the Pamirian stages.

The more widespread units in our area are the Singhié, the Shaksgam and the Aghil formations. The former two formations show flysch facies, the last one reefal facies. Between them a conglomeratic and/or a lagoonal bed is inserted. Up to now in the Shaksgam Valley no Cretaceous and Paleogene marine deposits were found, but they are present along the borders, at the North Rimu Glacier snout and on the Falchan Kangri (Baltoro basin). The above mentioned formations stretch toward south-east, in the Linghzi-thang (Western Tibet).

Up to now strata older than the Singhié Shales are unknown in the Shaksgam Valley and probably also in the Eastern Karakorum. The early authors considered the metamorphic rocks, which occur in hundred-kilometre-long belts between the granitic batholiths of the Eastern Karakorum (but very reduced in the Shaksgam), as belonging to the Archeozoic and Silurian (Lydekker, 1883). More recently a metamorphic complex characterized by thick layers of marble and quartzite associated with different types of gneiss and crystalline schists, were assigned to the Early Paleozoic, mostly Silurian and Devonian (Dainelli, 1935). Not having supplied fossils the age of these formations was deduced by the presumed equivalence with Spiti sedimentary formations and the supposed extension toward west of the Ordovician fossiliferous beds at Qizil Pass, in Aq-sai-chin. Nevertheless the same metamorphic formations were more recently correlated by Desio and Zanettin (1970) with the Cretaceous and Eocenic formations of Nanga Parbat, studied by P. Misch (1949). Moreover in one occurrence labelled in the previous geological maps as " crystalline schists of indefinite age, prevailingly pre-Silurian or older", I collected a small fossil fauna belonging to the Late Cretaceous. Also the metamorphic rocks (gneiss and slates) of the western Shaksgam area, which were referred to the Precambrian and Paleozoic, represent the metamorphic facies of the Shaksgam and Singhié formations (Permo-Carboniferous).

In spite of the previous statements, I cannot exclude that rocks older than Carboniferous are present in the Eastern Karakorum.

The lithofacies of the early part of the Late Paleozoic is represented by black shales containig inferiorly arenaceous and locally also conglomeratic intercalations and superiorly also calcareous. These euxinic sediments lead us to suppose that during the Carboniferous that area took part of the oceanic basin stretched south of the Eurasiatic continent and limited towards the ocean by a ridge. The quartz and feldspar clasts of the sandstone interlayers indicate an approachment to the source area of clasts located, according to Norin, to the south.

In conclusion, taking into account also the relatively great thickness of these deposits, the Singhié Shales, together with the sediments of the Shaksgam formation, could represent a distal geosynclinal deposit of flysch type, in a subsiding basin with active sedimentation. In time the rate of subsidence gradually decreased so much as to cause the progressive infilling of the basin; but it is also possible that the movement was reversed, and an uplifit of the sea-bottom replaced the subsidence. In the Late Permian a sedimentation, first prevailingly calcareous, then marly-arenaceous with brachiopods, pelecypods and fusulinids took the place of the black shales. The calcareous facies indicates a warm neritic marine environment near the continent, but with reefal episodes and characteristic sedimentary features varying both horizontally and vertically.

In the neighbouring regions, proceeding from north to south, the black shales facies becomes more and more prevalent until it represents most of the Permian, like in the Upper Siachen basin.

From above, it can be deduced that during the Late Permian the geosynclinal phase (more developed to the south) ended in the Shaksgam area and the subsiding movement was reversed becomming the uplift of the Hercynian orogeny most active at the north (North Pamirs and Tien Shan) and north-east (Kun Lun).

The stratigraphic data of the Triassic series are scarse. The Kyagar cherty limestone of Kyagar type may be interpreted as pelagic sediment deposited at a deep over 200 m. It seems that at the beginning of the Triassic occurred a phase of rapid subsidence which was followed by a new uplift of the sea-bottom with the spreading of reefs represented by massive, reddish dolomites of Middle Triassic and perhaps with partial emergences.

In Late Triassic the small basins between the reefs and the continent were gradually filled with the clastic sediments of the Chikchi-ri shales untill they became lagoons. East of the Shaksgam Valley these beds include gypsiferous interlayers which confirm our interpretation. This event occurred in Carnian as it indicated by the Heterastridium found just east of the Shaksgam-la.

In the middle Shaksgam Valley and especially in the Urdok one the lagoonal formation is replaced by a locally very thick calcareous conglomerate which represents a deltaic deposit, perhaps partly continental. It marks the end of the Carnian age.

At the beginning of Norian a marin regime with clear waters and abundant calcareous-dolomitic sedimentation, very analogous both in the lithofacies and in the biofacies to the Hauptdolomit of the Southern Alps, was established in the examined area.

The sea-bottom instability of Carnian time swiftly reduced at the beginning of the Norian. The relatively great thickness of the Aghil limestone is explained by slow, steady uniform subsidence of the basin. This situation continued throughout Rhetian and Lias, while in the Middle and Late Jurassic the presence of a nearby landmass subjected to active degradation is indicated by the clastic horizons and by the red colour in the rocks. The Cretaceous clastic sediments of the Falchan and Baltoro Kangri are the sign of a phase of active erosion of the nearby landmass and the beginning of the Alpine orogeny.

As to the tectonics, the Shaksgam Valley is located near the contact of the Muztagh and Aghil tectonic zones, but mostly enters in the second one. It is characterized by a bundle of folds partly faulted, locally overthrusted, and generally trending NW-SE, with small deviations. This bundle become progressively wider toward SE and NE, that is both along the strike of the tectonic axes (Upper Shayok and Depsang Plateau) and towards the north (Aghil Plateau). In the opposite direction the folds become progressively squeezed and faulted, evidencing that in that direction the compressional tectonic stresses were greater.

In the eastern part of the Baltoro Valley the formations of the Shaksgam area, such as the Singhié Shales and the limestones and shales of the Shaksgam formation, become slates and even gneiss (Falchan Gneiss and $K^2$ Gneiss), while the limestones become marble, sometime with recognizable fossil remains inside (Early Permian fusulinids).

In this connection it should be added that in the same direction the Karakorum axial batholith outcrops more extensively, affecting not only the texture of the rocks, but also the tectonic setting. The vergence of the beds, in fact, generally shows opposite directions on the two sides of the batholith.

Apart from local tectonic disruptions depending on the different competence of the lithotypes, there are clues of overthrusts, whose magnitude and importance is difficult to evaluate. Three examples can be given here: one mentioned by J.B. Auden (1938) near the Aghil Pass, another on the Skyang Kangri and a third in the Falchan Kangri group. The latter two were recognized only in the Baltoro side (Desio and Zanettin, 1970). Possibly these dislocations may represent a collateral effect of the geotectonic events which lead to the present setting of the Central Asiatic area (Desio 1976). Prior to the Eocene, the Indian plate was moving toward the north and later toward north-east. The Tethys sediments in front of the advancing plate were pushed toward the north-northeast and obliged of setting inside the great "gulf" between the old platforms of Tarim (east) and Qarakum (west).

As lateral effect of this shifting, strike-slip-faults were produced west and east, between the mobile belt and the stable platforms, that is the Middle Badakhshan fault to the west and the main Pamir-Karakorum fault to the east (Desio 1979). Our area was involved in the last one. The Aghil tectonic zone, which shows close stratigraphic affinities with South-east Pamir tectonic zone, is divided from it along the Pamir-Karakorum fault of more than 100 km and this represent one effect of the shifting of the Pamir block toward the north.

## References


Auden, J.B., Resumé of geological results, Shaksgam expedition 1937. Himal. Journal. X, 40-48 (1938).

Dainelli, G., Resultati Geol. e Geogr. Spedizione Ital. De Filippi 1913-14. II La serie dei terreni (II) (1934).

Desio, A., Geological works of the Italian Expedition to the Karakorum. Geogr. Journal 75, 402-411 (1930).

Desio, A., Some geotectonic problems of the Kashmir Himalaya-Karakorum-Hindu Kush and Pamir area. Int. Colloquium on the Geotectonics of the Kashmir Himalaya etc., Accad.Naz.Lincei, 115-129 (1977).

Desio, A., Geologic evolution of the Karakorum. Geodynamics of Pakistan. 111-124 (1979).

Desio, A., Geology of the Upper Shaksgam Valley, North-East Karakorum, Xinjiang (Sinkiang). Scient. Reports of Desio's Italian Expeditions to the Karakorum ($K^2$) and Hindu Kush, III, 4th, 1-196 (1980).

Desio, A., Zanettin, B., Geology of the Baltoro basin. Ibiden, III, 2nd, 1-308 (1970).

Lydekker, R., The Geology of the Kashmir and Chamba territories and the British district of Chagan. Mem.Geol.Survey of India, 22, 1-344 (1883).

Misch, P., Metasomatic granitization of batholitic dimensions. Amer.Journal of Science, 274, 4 & 10 (1949).

Norin, E., Geological exploration in Western Tibet. Reports from the Scient. Expedition in NW prov. of China u. leadersh. of Sven Hedin, III, 7, 1-214 (1946).

# THE TIMING AND SIGNIFICANCE OF OROGENIC EVENTS IN THE HIMALAYA

Augusto Gansser
(Geological Institute of the ETH, Zürich)

## INTRODUCTION

The spectacular mountain range of the Himalaya has evolved through a long and complicated history. This history, in some respects similar to other important mountain ranges on our globe, shows however aspects typical for the Himalaya only. The high mountains which we admire today represent the very last phase of this history, while during the earlier phases the mountains have been morphologically rather insignificant.

Very active orogenic episodes and rather passive epirogenic events began over 1800 my ago and culminated in the last major phases about 25 my ago. The present earthquakes indicate that the last event is still going on and has changed from orogeny to epirogeny. Unique for the Himalaya is the fact that orogenic activities were practically absent over a span of over 1000 my, from the first multiphase events in the late Precambrian to a renewed major activity in the Neogene.

The main and better known Himalayan orogenies were caused by the collision and post collision phases of Gondwanic India with Eurasia. The Himalaya as wė see it today is an intercontinental orogen which went however through the stages of a marginal range during the time of a relatively narrow Tethyan sea, later consumed during collision. The history prior to this collision is still uncertain. The fragmentation of a northern Gondwanic continental margin has certainly played an important role. This period involved a Paleotethys, a mysterious ocean of unknown dimensions.

The timing and the type of Himalayan structural events is compiled in the enclosed tectogenetic diagram of the Himalaya (Fig.1). It will form the base for our discussions. For the location of these events within the wider tectonic frame we refer to the well known regional subdivision of the range (Fig.2). From south to north we distinguish:

1. <u>The northward plunging Indian shield and the Himalayan foreland basin</u>. Late Precambrian Vindhyan clastic sediments transgress the northeastwards trending Aravalli crystallines and increase towards the Himalaya (Vindhyan basin).

2. <u>Subhimalaya</u>. Siwalik molasse belt. Limited towards the Vindhyan basin by the Main Frontal Thrust (MFT) with its recent tectonic activity. Thrusting of the frontal molasse on Quaternary terraces and warping of the latter.

3. <u>The Lesser Himalaya</u>. Thrust along the Main Boundary Thrust (MBT) on the Subhimalaya. Lower to Middle Pleistocene orogeny. Incorporated material from the Indian shield and the Vindhyan basin. Subdivided by thrusts with progressively older rocks from south to north and covered by Precambrian crystalline nappes with a reversed Himalayan metamorphism. No real basement is known.

4. The High Himalaya. Forms thick crystalline Precambrian base of Tethyan sediments, thrust along the Main Central Thrust (MCT) on Lesser Himalaya. Metamorphic and orogenic events are Precambrian and Himalayan only, with intermittent epirogenic phases. Does not always coincide with the topographically highest range.

5. The Tethys Himalaya. Conformable, mostly platform type sedimentary pile from Cambrian to Eocene. A northern flysch belt from Trias to Cretaceous reflects the incipient Himalayan orogeny.

6. The Indus-Tsangpo Suture Zone. Complex, steeply thrusted ophiolitic belt with consistent ophiolitic melanges and southwards thrusted ophiolitic nappes with ultramafic mantle rocks as highest structural elements. Represents mantle, ocean floor as well as island arcs, compressed and "obducted" during collision of Greater India with fragmented southern border of Eurasia, which includes slices of Gondwana rocks. The ophiolitic belt is thrust northwards on conspicuous molasse predominantly of Paleogene age which transgresses on Transhimalayan plutons of oceanic origin.

7. The Tibetan backland. To the north the Transhimalayan batholite intrudes or is transgressed by mostly platform type sediments of the highly elevated and structurally complicated Tibetan block. The southern belt (Nyenchen Tangla) is rich in acid volcanics ranging from Upper Cretaceous to subrecent, and is separated from the main block by a steep, partly ophiolitic thrust zone, the "Nyenchen Tangla Line".

## THE TECTOGENETIC EVENTS IN THE HIMALAYA

The oldest events are represented by the crystalline base of the High Himalaya above the Main Central Thrust. These crystalline rocks, 15 to 20 km thick, with ages between 1800 to 1400 my (Bhanot et al.1977) are overprinted by strong Himalayan orogenies. They belong to a northern slice of the Indian shield which originally formed the floor of the southern Tethyan sea. The lithofacies compares to the old marbles (Rialo and Kushalgarh) and the thick quartzites of the Aravallis and Delhis (Gansser,1964) which have been deformed and metamorphosed by multiple orogenic events dominated by the Aravalli phase. It produced the NNE directed structures of once dominant mountain ranges. This important alignement, truncated by the WNW-ESE trending Himalaya, has locally been rejuvenated during the youngest Himalayan phases. The transgressive Vindhyan sediments (Latest Precambrian) make up large sections of the Lesser Himalaya (Valdiya, 1969) here with evaporitic end stages and trap like volcanics (western Lesser Himalaya), while a more northern distal facies in a deeper environment represents the 4-6000 m of argillaceous sediments at the base of the highly fossiliferous Tethyan sedimentary pile.

Epirogenic events are following the late Precambrian orogenies. They are pronounced in the Cambrian, here with 500 my old granite intrusions, in the Devonian and during the Permo-Carboniferous. Important upheavals and a pronounced unrest during this latter period are responsible for a widespread glaciation followed by basic to intermediate volcanics such as the Panjal traps in the west and the Abor volcanics in the east. Their composition is tholeiitic, the environment shallow marine in contrast to the late Precambrian basic volcanics of the Lesser Himalaya extruded in deeper waters.

Orogenic events, which during the Paleozoic where practically nonexistent, begin with the Trias. Rifting of the northern Gondwanaland has probably begun in the Upper Permian and part of the Trias shows already a flysch facies, locally with olistolitic blocks of platform limestones of Trias and Permian age. This flysch facies is particularly well developped in the northern part of the western and eastern Tethys Himalaya. A part of the Trias was developed in a pecurliar pelagic limestone facies of the Hallstätter type, rich in ammonites (Heim and Gansser 1939). A characteristic volcanism is related to the Trias in the form of highly vesicular phonolitic alkali basalts. Both, sediments and volcanics are so far known only in exotic blocks related to the ophiolitic melange of the Indus-Tsangpo Suture Zone (Gansser 1974). It is a surprising fact that all along the Suture Zone the exotic limestone facies and geochemical composition of the related volcanics seems the same. This pecurliar Triassic volcanism is unlike the underlying Permian Panjal traps or the younger ophiolitic volcanics.

During Jurassic and particularly during the Cretaceous orogenic activities are localized along the present Himalayan Suture Zone, todays scar of the collision of Greater India with a complex Eurasian continent or plate. This Suture Zone represents one of the outstanding structural features on our globe and stretches from Karachi in the SW for over 5000 km to the Andaman Islands in the SE. The composition of this Suture Zone, locally highly complicated, is regionally surprisingly constant. Actually it is the only continuous structural element from one end to the other of the Alpine-Himalayan belt. Beautifully exposed, but difficult of access, it has stimulated a multitude of often contractictory theoretical opinions.

In spite of the numerous available facts, timing and type of the successive events which lead to the present Suture Zone are still difficult to assess. Following the break up in the late Permian and Trias the ophiolitic rocks with its widespread ophiolitic melanges and allochthonous mantle slabs suggest a rather rapid opening of steep, deep but probably narrow oceanic basins (1000 km ?) including island arcs. In spite of widespread occurrence of radiolarites it seems unlikely that these basins were stable for any length of time but that a certain unrest continued intermittently during the Jurassic and particularly during the early Cretaceous when subduction began. The timing of this subduction coincides with the formation of the very consistent and widespread (over 2500 km length) Transhimalayan plutons which follow the present Suture Zone to the N. Their ages range from 110 to 40 my beginning with older basic and ending with younger acid rocks. The average composition is granodioritic to tonalitic and the characteristic low $Sr^{87}/Sr^{86}$ initials as well as the structural control suggest an oceanic origin. The substantial volume of the plutonic rocks may be related to the size of the consumed oceanic crust. The plutonic phase was followed to the N by a widespread acid volcanism.

We assume that the collision occurred during the end of the Cretaceous and at the beginning of the Paleocene. During a subsequent orogenic calm in the Eocene the orogeny was substituted by epirogenic uplifts which raised the Transhimalayan plutons and the volcanic hinderland producing the Kailas molasse. This was deposited to the S of the plutonic range and has its type locality in the holy Mount Kailas where 4000 m of conglomerates and sandstones transgress the granodioritic rocks. The acid volcanic pebbles of this molasse gave a good Rb/Sr isochron of 40 my (Dietrich et al. in prep.) a possible age for the northern volcanics. Epirogenic movements continued during the Oligocene when the northwards drift of the Indian plate was minor. In the Miocene a sudden renewal of activities produced the largest orogenic event of the Himalaya. An intense compression along the Suture Zone obducted the ophiolitic rocks along steep thrusts with a strong tectonisation of the ophiolitic melange zones. A substancial amount of oceanic crust and mantle was squeezed into south vergent nappes where large ultramafic slabs of mantle rocks from the highest structural elements (Spongtang and Jungbwa nappes, Fuchs 1979, Gansser 1979).

After a consolidation of the near vertical Suture Zone the continuing north drift of the Indian plate broke the Himalaya up into intracrustal thrusts. The dominant feature is the Miocene Main Central Thrust, the crystalline base of the Tethyan sediments, with a visible minimal shortening of over 100 km. At the same time high PT conditions produced the intense Himalayan metamorphism which overprinted the older Precambrian phases. This metamorphism was syn to post kynematic, and, curiously enough, was mostly concentrated in the already metamorphosed basement rocks of 1800 and 1400 my, and affected only the lower part of the Tethyan sediments. An exception are the Trias infolded in the crystalline of the Nun Kun high (Ladakh region Dietrich et al.) as well as in the easternmost Tethyan flysch belts (Chang et al. 1977). The Himalayan metamorphic age is surprisingly constant with an average of 25 my. At the end of this main metamorphic event intruded in the northern Himalaya, south of the Suture Zone, the syn to mostly post genetic leucogranites. They cut all known structures but locally they can be slightly foliated. Ages range from 30 to 15 my. Larger bodies appear concentrated in certain areas within and at the northern edge of the crystalline units. They are frequently related to high and highest elevations such as Nanga Parbat in the W, Badrinath and Kedernath in the Kumaon Himalaya, Mustang in west Nepal, Manaslu and its continuation into Sisha Pangma in northern Nepal and southern Tibet, Chomolungma (Everest) and Makalu region and Chomolhari with Kankar Pünzum region in northern Bhutan. In 5 areas the granites reach 8000 m and in the rest they surpass the 7000 m elevation (Gansser 1964, Le Fort 1973). Most of the discordant granite bodies expose a halo of smaller stocks, lenses, apophyses, dykes and sills. The composition is surprisingly constant with muscovite dominating over biotite. Black tourmalines are frequent. Geochemically they are distinct from older Himalayan granites. The initial $Sr^{87}$ $Sr^{86}$ ratio is extremely high (Hamet a.Allègre 1976).

Following the main orogenic Miocene phase the south Himalayan molasse, the Siwaliks, were deposited during the calmer Pliocene and the active Lower Pleistocene. Here we note a renewal of the Himalayan orogenic phase producing the intracrustal thrusts of the Lesser Himalaya and the final thrusting of this unit over the Lower Pleistocene Siwaliks along the complicated Main Boundary Thrust. Some of the Siwalik sections have been locally eroded prior to the thrusting and this fact produced some interesting relief-thrusting.

Towards the end of the Pleistocene and particularly during the Quaternary and up to recent the still active orogeny was overtaken by a strong epirogeny resulting in a substancial and still active uplift of the Himalaya, the well known morphogenic phase. At this time the Siwaliks were thrust over the Quaternary terraces along the, only locally visible, Main Frontal Thrust. At the same time the Quaternary in front of the foothills was warped and tilted. Step like terraces along the main rivers within and along the south margin of the foothills reflect the youngest uplifts of this morphogenic phase. The famous Karewa lake beds in Kashmir have been tilted and uplifted to nearly 3000 m. The present erosion is not as yet balanced. Even the total volume of the Siwalik molasse represents only a small part of the visible volume of the present Himalaya in contrast to the Alps, where the volume of the molasse corresponds to that of the Alps. The detritus of recent river terraces is considerably coarser and badly sorted to fanglomeratic if compared to the youngest deposits of the Siwaliks. Much of the recent detritus is still trapped in large basins (Hundes) and terraces along the northern Himalaya. The largest rivers such as Indus, Sutlej, Arun and Tsangpo which originated north of the range were not able to transport much of this detritus to the S. They have cut the rising Himalaya in deep gorges at precisely such areas with the highest culminations (Nanga Parbat, Everest/Kangchendzönga and Namche Barwa). From all available data a minimum of 5 mm uplift per current year can be accepted for the Himalaya, five times the rate measured in the Alps.

Considering the present seismicity of the Himalaya we note the difficulty in comparing it with the known structural features. The Suture Zone as well as the Main Central Thrust are practically inactive in spite of the numerous hot springs which follow these important trends. Most activity is now observed along the Main Frontal Thrust, but here the epicenters hardly follow the visible alignement. Recent microseismic investigations in the western Himalaya foothills suggest strikingly aberrant trends of unrest (Seeber a. Armbruster 1979). These and a recent fracturing seem to indicate a reactivation of an old pre-Himalayan structural pattern, such as the Aravalli trend of the northern Indian Shield. It will be one of the main future tasks in Himalayan and Tibetan research to reconcile the differences in the geological and the geophysical pictures.

## References

(1) Bhanot, V.B., Singh, V.P., Kansal, A.K., and Thakur, V.C., Early Proterozoic Rb-Sr whole rock age for Central Crystalline gneiss of Higher Himalaya, Kumaun. Jour.Geol. Soc. India, 18 (2), 90-91 (1977)

(2) Dietrich, V., Frank, W., Gansser, A., Honegger, K., Thöni, M., Trommstorff, V., Petrological and geochemical investigations in the Ladkh region (in preparation)

(3) Chang, C., Cheng, X., Pan, Y., The Geological History, Tectonic Zonation and Origin of Uplifting of the Himalayas. Institute of Geology, Academia Sinica Peking, 17 p. (1977)

(4) Fuchs, G., On the Geology of Western Ladakh. Jahrb.Geol. B.-A., 122 (2), 513-540 (1979)

(5) Gansser, A., Geology of the Himalayas. Wiley-Interscience, London, 289 p. (1964)

(6) Gansser, A., The Ophiolitic Mélange, a World-wide Problem on Tethyan Examples. Eclogae geol. Helv. 67/3, 479-507 (1974)

(7) Gansser, A., The Ophiolitic Suture Zones of the Ladakh and the Kailas region - a comparison. Jour.Geol. Soc.India, 20, 277-281 (1979)

(8) Gansser, A., The Peri-Indian Suture Zone. In:Geology of the Alpine chains born of the Tethys. Mém. B.R.G.M. 115, 140-148 (1980)

(9) Hamet, J. and Allègre, C., Rb-Sr systematics in granite from central Nepal (Manaslu): Significance of the Oligocene age and high $^{87}Sr/^{86}Sr$ ratio in Himalayan orogeny. Geology, 4, 470-472 (1976)

(10) Heim, Arn. and Gansser, A., Central Himalaya, Geological Observations of the Swiss Expedition 1936. Denkschr. schweiz. natf. Ges., 73(1), 245 p (1939)

(11) Le Fort, P., Les leucogranites de l'Himalaya, sur l'exemple du granite du Manaslu (Népal central). B.S.G.F. 7, (15), 555-561 (1973)

(12) Seeber, L. and Armbruster, J., Seismicity of the Hazara Arc in Northern Pakistan: Decollement vs. Basement Faulting. Geodynamics of Pakistan (Farah and De Jong Ed.) Geological Survey of Pakistan, Quetta, 131-142 (1979)

(13) Valdiya, K.S., Stromatolites of the Lesser Himalayan carbonate formations and the Vindhyan. Jour. Geol. Soc. India, 10, 1-25 (1969)

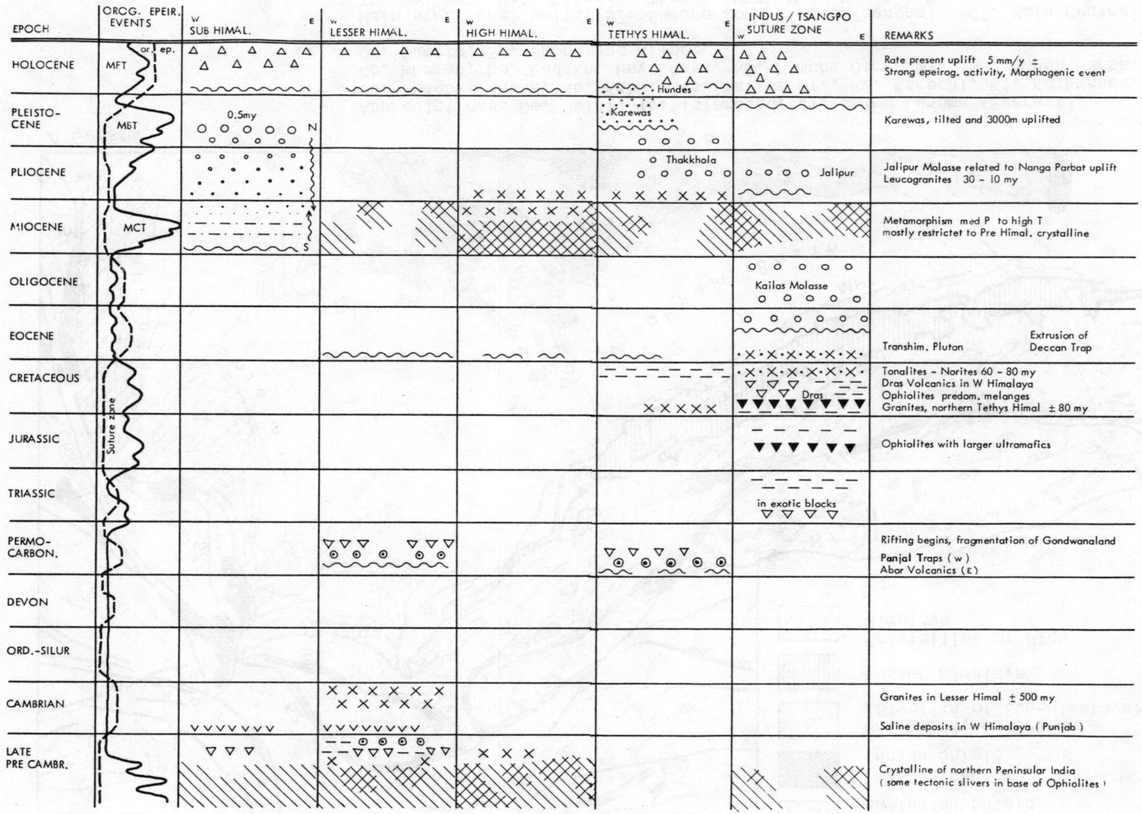


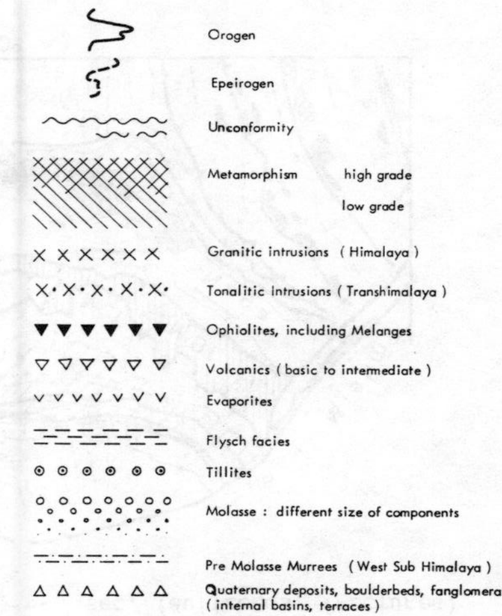


**Tectogenetic Diagram of the Himalaya**

by A. Gansser

Fig. 1

Geotectonics of the Himalaya

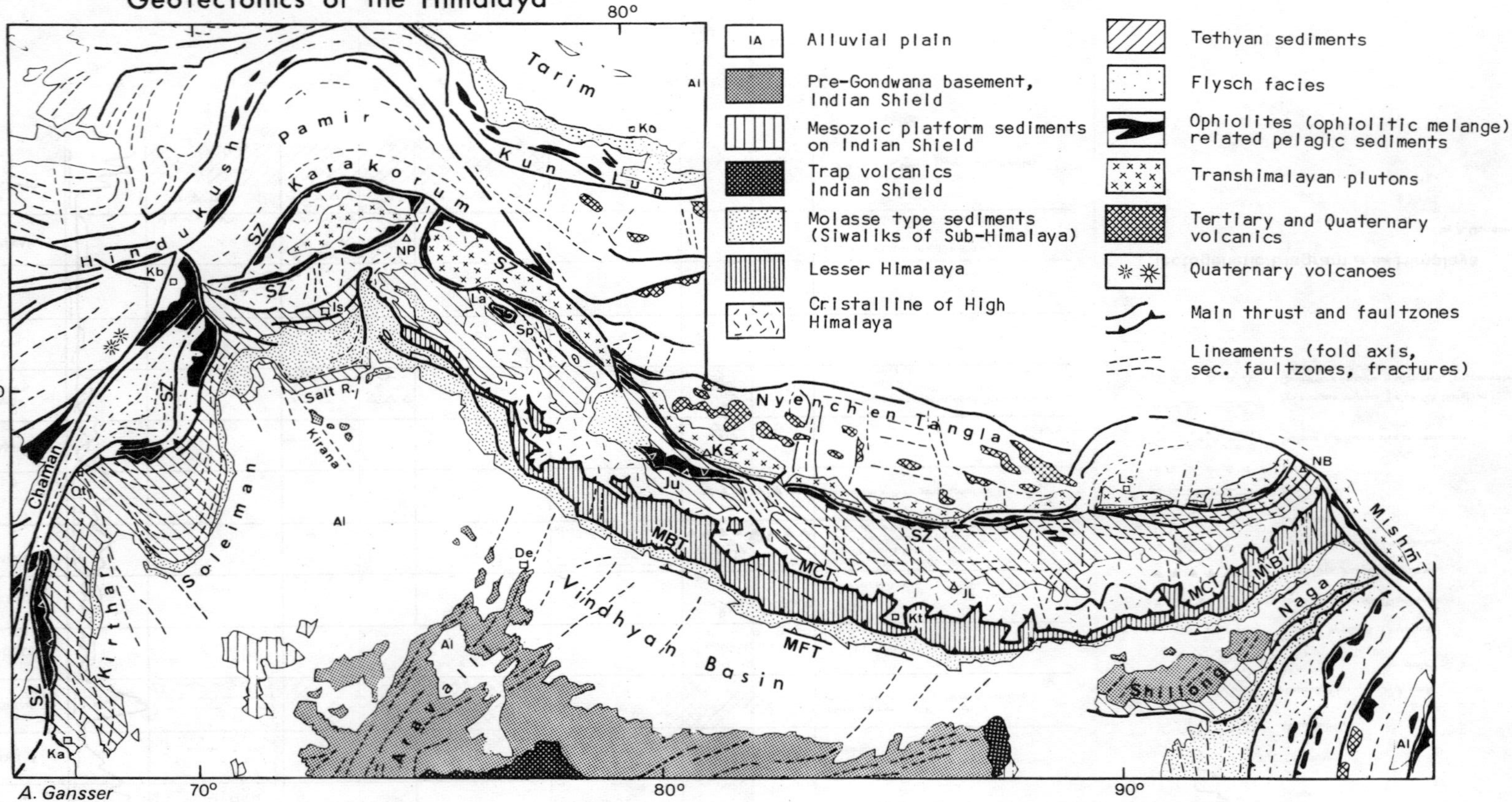


Abbreviations: De. Delhi, Is. Islamabad, Jl. Jolmo Lungma (Everest), Ju. Jungbwa nappe, Kb. Kabul, Ks. Kailas Mt., Ka. Karachi, Kt. Kathmandu, Ko. Khotan, La. Ladakh, Ls. Lhasa, NB. Namche Barwa Mt., NP. Nanga Parbat Mt., Qt. Quetta, Sp. Spongtang klippe.

Main structural units: SZ: Suture Zone (Indus/Tsangpo), MCT: Main Central Thrust, MBT: Main Boundary Thrust, MFT: Main Frontal Thrust.

Fig. 2

# A PRELIMINARY NOTE ON THE FRACTURE ZONES AND FAULT-BLOCK TECTONICS OF CHINA AND THEIR BEARING UPON THE ORIGIN OF QINGHAI-XIZANG PLATEAU

Zhang Wen-you, Li Yin-huai
(Institute of Geology, Academia Sinica)

The fracture zones of China are of planetary in nature. When the vector of earth's rotational force coincides with that of internal forces, essentially of gravitational contraction and thermal expansion, the lithospheric fracture zones due to rotational effects might be happened to control the crustal structures through geological times. There are two main fracture systems in Chinese continent with one running in nearly NS direction adapting to NNE-NNW X-shaped conjugate shear planes of global nature and another almost in EW direction adapting to NEE-NWW X-shaped conjugate shear planes also of global nature. According to our experiments, the NE and NW trending fractures are usually adapting the NNE-NEE planetary shear planes and the NNW-NWW ones respectively. We think that the NNE-NNW and NEE-NWW shear planes might be due to earth rotation resulted from NS and EW compressions and fundamental structures of lithosphere controlling all tectonic features through geological ages. The nearly NS and nearly EW as well as the NE and NW fractures might be derived from those fundamental fractures in the NNE-NNW and NEE-NWW directions(1). Chinese continent developed as platform was cut by these fracture zones into fault-blocks with nearly rhombic form having long axis in nearly EW direction in West China and in nearly NS direction in East China. These fracture zones once occupied by oceanic crust are developed as geosynclines and folded into mountains of different geological ages.

The Qinghai-Xizang fault-block is an eye-shaped plateau which is controlled by a rhombic basement of continental crust.To the north it is bordered by the Kunlun Mts. convex to north along the southern border of Tarim and Qaidam rhombic blocks. To the south it is bordered by the Himalayas convex to south along the northern border of Indian block. From stratigraphical developments, the Indian, the Qinghai-Xizang, the Tarim blocks are connected in Lower Paleozoic time but splited during Upper Paleozoic. According to Chang Cheng-fa and Pan Yu-sheng(2), there are three suture lines with small ultrabasic intrusions in the Qinghai-Xizang Plateau: the northern one of Late Paleozoic occurring along the Kunlun Mts., the middle one of Mesozoic along the Tanggula Range and the southern one of Cenozoic along the Yarlung Zangbo River. These suture lines are doubtlessly traces of lithospheric fractures, however, whether they are true ophiolites or not. Thus, the Qinghai-Xizang Plateau consists of two blocks: the northern one, namely the Qiangtang and the southern one, the Nyainqentanglha, which have combined together as one block since the Yanshan movement of Late Mesozoic. The Himalayan arc added to the southern border of the Qinghai-Xizang fault-block coming from India due to Cenozoic movements known as the Himalayan. From the synthesis of the double back arc structures of the Himalayan arc at Pamir, we can readily see on the geological map of Asia(3) two arcs of Pre-Cambrian complex with Paleozoic and Mesozoic sedimentary covers in between. To the east, the northern back arc is connected with Pre-Cambrian belt in the western Kunlun, while the southern one, connected with that of the Himalayas. The eastern double back arcs of the Himalayan arc might be superimposed each other into one probably located to the south of the great bent course of the Yarlung Zangbo River. We think that the age of basement rock of the Qinghai-Xizang Plateau might be same as that of the Shan Plateau of Burma.

Our model experiment(4) has shown that the Himalayan arc structure push toward the south overriding the Indian block with latitudinal tension fracture of Yarlung Zangbo River suture instead of longitudinal compressive back-bone of J.S.Lee's $\theta$-type structure.

According to Huan Wen-lin(5), most of the focal zones of earthquakes in the Qinghai-Xizang Plateau are usually dipped to south instead of that to north imaged by plate tectonicians. On the southern slope of the Himalayas, a series of overthrusts are overriding the Indian block, while on the northern slope, the ovethrusts overriding the foot hills along the southern side of the Yarlung Zangbo valley. Thus, the Himalayas look like a fan-shaped orogen. The Gandise Range occurs as the southern edge of the Qinghai-Xizang Plateau with overthrusts towards south, while the Kunlun Mts. as the northern edge with overthrusts towards north. This leads us to consider the Qinghai-Xizang Plateau consisting of the Nyainqengtanglha and the Qiangtang blocks to be another fan-shaped orogen. Between the northward push of the Indian block and the southward push of Tarim block occur two fan-shaped orogens, the Himalayas and the Qinghai-Xizang Plateau, which might be due to longitudinal compression induced by earth's rotation and the regional high heat flow in or under the earth's crust incorporated with interlayer gliding between high and low velocity layers in the lithosphere as well.

## References


(1) Chang Wen-you, Zhong Jia-you, On the development of fracture system in China, Scientia Geologica Sinica, (3), (1977) (in Chinese).

(2) Chang Cheng-fa, Pan Yu-sheng, A brief discussion on the tectonic evolution of Qinghai-Xizang Plateau, In this Proccedings of Symposium on Qinghai-Xizang (Tibet) Plateau, Academia Sinica, (1981).

(3) Geological Map of Asia, scale 1 : 5,000,000, compiled by the Chinese Academy of Geological Sciences, (1975) (in Chinese).

(4) Huan Wen-lin et al., The distribution of earthquakes foci and plate motion in the Qinghai-Xizang Plateau, Acta Geophysica Sinica, 23 (3), (1980) (in Chinese).

(5) Chang Wen-you, Li Yin-huai, Zhong Jia-you; The origin of arcuate structures and their relation to the smaller-focus earthquakes, Kexue Tongbao 18 (2), (1973) (in Chinese).

# TECTONICS OF THE CHINHAI-TIBET HIGH PLATEAU

T. Boldizsàr
(Technical University, Miskolc, Hungary)

The most magnificent and striking feature of the Earth's surface is the Chinhai-Tibet high plateau surrounded by the greatest Tertiary mountain systems of the Himalaya, Transhimalaya and Karakorum from the south and Altyn Tagh, Kunlun and Nansan from the north. Geologic data seem to suggest, that folding took place in late Paleozoic and Mesozoic in the northern part and in the Cenozoic in the south. Scarce geophysical data indicate crust thicknesses of 65-70 km and recent seismic data show normal faults in the NS direction. The large territory is not well mapped and the Landsat imagery is an important contribution to geography, but few geological information can be derived from it.

The explanation of the recent alleged east-west extension would be the supposed colliasion of India and Eurasia. The result of this colliasion would be the building up of the Himalayas and the elevation of the Tibet high plateau. The stress and displacement is considered transmitted across Asia, through Tarim basin, Tien Shan and Siberia /e.g. Molnár and Tapponnier/.

Present author do not believe that the continental crust or lithospheric blocks are able to push each other and to elevate mountains and thicken the crust or litosphere. The elevation of Chinhai-Tibet plateau and other Alpine mountain features all over the world is the consequence of isostatic adjustment of the thickening crust. The increase of crust or lithospheric thickness is taking place in seismically active zones. Where mountains are increasing in height, there are always neighbouring basins which are subsiding in the same time. The slow plastic mantle flow in the asthenosphere from the oceans and intermontane basin toward the emerging continents and mountain ranges gives support to the elevation, and is working against the erosion. The continents are elevating while the ocean and intercontinental basins are sinking and being filled up by sediments from the eroded mountain rocks. The main feature of mountain building is vertical uplift with normal and reverse faults. The slow asthenospheric plastic flow results sometimes in strike-slip faults, but high mountains and plateaus cannot be produced by them.

The energy causing the change of the face of the earth is the internal heat, that is why geothermal investigations are of extraordinary importance to establish the heat energy changes in the mantle.

# THE APPLICATION OF REMOTE SENSING IMAGES TO THE COMPILATION OF THE GEOLOGICAL MAP OF QINGHAI-XIZANG PLATEAU

Liu Xin-ji, Yu Xue-zheng, Chen Yin-xiang
(The Geological Remote Sensing Center, The Ministry of Geology)

In 1979, co-operating with the Geological Institute of Qinghai-Xizang Plateau and other institutions, we compiled the Geological Map of Qinghai-Xizang Plateau on a scale of 1:1,500,000. The area mapped is about 2.4 million square kilometers.

During this mapping, based on geological data available at present, we used extensively the techniques of remote sensing including landsat images and aerophotographs.

The new geological map compiled with remote sensing images, in contrast with the old ones, has many new features, for instance the distribution and extent of some large scale abyssal faults in the whole area, and some regional faults in restricted areas are more clearly seen. Many concealed faults were mapped from large areas in Qinghai Province, as were the folds in Qaidam Basin.Many of the Palaeozoic blocks are separated from the Mesozoic over a large area, and Quaternary basaltic lavas associated with volcano-tectonic structures were discovered in northern Xizang. A series of brachy-axis anticlines in the low-water-shed of southern Xizang, and the shapes and distribution of ultrabasic rocks were determined, etc.

## SOME FACTORS AFFECTING THE INTERPRETATION

The method of interpretation depends mainly on unaided eyes.

The remote sensing images that we used include the black and white landsat images (MSS 4,5,6,7) and their color compositions on a scale of 1 : 1,000,000, aerophotographs on 1 : 60,000 and a mosaic of landsat images on 1 : 500,000.

The contrast among different bands of black and white landsat images has enhânced the interpretation. Methods of optical enhancement, false color composition and density slicing, have further clarified the geologic information. The combination of the images from landsat and airplanes as well as the mosaic, combining "huge-view" with "micro-view", supplement each other and have also raised the level of interpretation. These favourable steps help us to find and extract more geological information from landsat images in solving the problems of discriminating and locating various geological bodies on the map. New light is shed on some regional geological problems. By interpreting and mapping, some knowledge is gained about image characteristics of various geological bodies such as strata, igneous rocks and tectonics in the mapped area, and primary interpreting keys are set up.

It should be pointed out that, it is difficult to build more systematic interpreting keys because of the extensive area mapped, the complicated geological conditions and a disparity in research level. In North Xizang Plateau, there are many lakes, the differentiation is not evident and the exposures of geological bodies are insufficient. The high altitude and cold climate together make weathered rocks and shelter the essential features of geological bodies, let alone the ice-covered area where no interpretation can be actually made. So it is very difficult to find their direct interpreting keys. Mostly we have to outline their shapes roughly according to indirect keys such as features of the landscape and analyses of the geological background, so that the mapping quality is more or less affected.

## FEATURES OF LANDSAT IMAGES AND THEIR EFFECTS ON MAPPING

### Strata

It is difficult to interpret rocks and strata from remote sensing images. However, based on the available stratigraphic data, we have still gained some new knowledge about the strata in the area mapped, which provides some clues for further stratigraphical classification of the plateau.

1. Clarifying the distrubution of the Mesozoic beds. From landsat images we find that the Jurassic beds are distributed extensively in northern Xizang. These marine limestones intercalated with gypsum layers are characterized by pale tone, fine bedding and continuous extent. They can be discriminated so easily that we can trace them from east to west in the western part of northern Xizang according to their features on the images.

The Cretaceous beds,however, are quite different. They consist of brown conglomerates and muddy sandstones forming a contrast with the linear Jurassic that extends continuously by their typical monadnock-shaped landscape. In some places we can find the Cretaceous unconformably overlying the Jurassic.

2. Many Palaeozoic blocks are separated from the Mesozoic beds over a large area. The Palaeozoic strata mainly consist of schists, phyllites, metamorphic sandstones, quartzites and marbles, but it is difficult to distinguish them in the images. Their landscape features are mostly presented as block mountains with darker tones, higher altitudes and no stratification,and high snow covered mountains, controlled by some large faults, are often formed.

According to above image keys, many Palaeozoic strata are separated from the vast area of the Mesozoic beds, such as the Carboniferous and the Permian separated from the Cretaceous in the north of Tangra Lake.

3. A distinct distribution of the Permo-carboniferous slightly metamorphosed rocks develops in the western Kunlun Mountains. These Permo-carboniferous strata consist of sandstones and slates bearing special image features. For example, near Yanghu Lake, due to the interbedding of soft and hard rocks, the stratification is very clear. The dark belt of the Permian is clearly distinguished from Permo-carboniferous with its obscure stratification and darker tone. The schistosity and flexure of these strata are very well developed as a result of slight metamorphism. Especially, a group of joints with a NE direction roughly parallel each

other, show an opposite torsion direction of the two large faults with EW directions, which rotate the strata clockwise (Fig. 1).

4. The ancient metamorphic rocks of the Precambrian occur at the margins of the plateau and the deeply cut regions of high mountains. The Precambrian medium to deeply metamorphosed rocks constitute the crystalline basement of the plateau. Since their occurrence is mainly on the tops of high mountains or in the deep-cut canyons where people can hardly reach, severe difficulties are met when surveys are made. However, we have taken an over-view of the Precambrian distribution, looking for their image features and then discovering new areas of their distribution. For example, the varieties of schists and gneisses of Nyalam group (i.e. formerly Jolmo Lungma group) often form steep mountains covered with snow, developing dendritic gullys and bearing no stratification. Some individual areas show plumose gully patterns which may be developed from the schistosity and gneissosity. In some area of gneisses, lens-shaped forms show that strong lateral compression has taken place.

## Igneous Rocks

1. Granites. Generally, the granites in this district have undergone multistage activities, and most are complex,so it is very difficult to summarize the unified interpreting keys. In addition, the magmatic rock zones roughly distribute along the main ridges of high mountains with heavy snow-cover so that the image features of the granites are sheltered. Also some small granite bodies have lost their granitic characteristics, since either they have not yet been exposed owing to insuffiicent erosion or they have become negative topography by weathering. So the interpretations are mostly made by landscape indirect keys as follows:

(1) The topography of high mountains covered with snow (such as Gandise Complex Belt). These are distinguished from sedimentary rocks by characteristics of large dendritic gullys, forceps-shaped heads of gully, massive structure and no stratification (Fig. 2). The Nianbao granite in Qinghai Province shows also the same landscape.

(2) The topography of hills and lowland. Suffering severe weathering, medium to small granite bodies have lost their original structure, and their tones have become paler but their boundaries clearer.

(3) Sometimes two groups of joints of the granites or the chequer-shaped landscape in aerophotographs may also become an indirect key for deducing granites.

2. Ultrabasic rocks. A series of ultrabasic rock bodies distributed along the southern side of Yarlung Zangbo River show as circular, elliptical or narrow bands of massive bodies with positive topography and darker tone. Sometimes they are controlled by two faults (Fig. 3).

On the basis of these image characteristics, some basic-ultrabasic rocks can be circled out from two sides of the deep-great fault lying at the northwest of Shiquan River, which shows that the Yarlung Zangbo River ultrabasic belt extends westward continuously.

3. Basalts. Intense Quaternary volcanisms occurred in the Kunlun Mountains region, so we may circle out many basalts or andesite basalts. Field surveys prove that these volcanic rocks are mostly lava sheets or flows, which show darker or paler tone according to their different

components. They present a large area of lava sheets due to their low viscosity and good fluidity, forming so called plateau basalts. They often show large areas of blocks with uneven boundaries. Undulational lava terraces are formed, and their margins often constitute cliffs or steep slopes. Since the basalts are hard, they play a protective role for the underlying rocks from erosion (Fig. 4 Right).

Also, we have discovered a lot of volcanic vents from landsat images. Images of new and old volcanic vents present circular, colored points or rings of darker tones and positive circular figures. At the top, due to the shadow or weathered fragments in the vents, usually the circular volcanic vent depressions show as black points. In this figure we may see a group of volcanic vents distributed along the Altun Fault ( Fig. 4 Left ).

## Tectonic

1. Faults. The most effective way to interpret fault structures probably is from landsat images. Especially, owing to the "huge-view" character of landsat images they are particularly favourable for observing the extent of large regional faults. In sections of faults sheltered by Quaternary deposits, they may also be shown in MSS 7 images, so that we can still trace the faults.

The results for interpreting the faults are summarized as follows:

(1) Some large scale abyssal faults and large regional faults are shown as clear images and good continuity, especially in mosaics of landsat images. The faults may be traced for long distances from some hundreds to a thousand and more kilometers, and their accurate position and extent may be determined.

(2) The north-eastern tectonic lines are added to the western Sichuan region. In southern Xizang, the combination of the two groups of faults in NE or NEE and NW or NWW directions cut this region into many rhombic blocks. And in northern Xizang, arcuate or circular faults develop well, forming many rotational structures.

(3) Many concealed faults that have never been seen before are interpreted from landsat images. Fig. 5 shows some concealed faults near the Taijinaier Lake in Qinghai Province. For example, a group of buried faults with NW directions controls the distribution of lakes and diluvial fans. One of the faults in the south even cuts through an alluvial fan, showing that it is an active fault of very young age.

2. Folds. Generally, the folds show parallel color belts with different tones, and form circular, elliptical or irregular shapes in landsat images.

Looking for a folded arc is a key to interpreting folds. In aerophotographs many folds occur in Mesozoic strata, showing distinct folded arcs. Some interrupted folds are often found in landsat images. Owing to the destruction of faults, we can only find one limb or one folded arc.

According to the above keys we have discovered a series of brachy-axis anticlines in the Lhagoi Kangri Range (the lower watershed) in southern Xizang, such as the Lagulalong Mountain and Kangmar anticlines. The fold consisted of Cretaceous strata near Yamzhog Lake is even more evident there.

3. Neo-Tectonic Movements. The effect of Himalayan Movement on the neo-tectonic movements in Qinghai-Xizang Plateau is very intensive revealing itself as follows:

(1) The common occurrence of the fault basins of Tertiary and Quaternary.

(2) The shrinkage of lakes and the strong activity and evolution of water bodies.

(3) The occurrence of many inherited concealed faults and active faults.

(4) A large area of extrusion of Quaternary basalts, modern volcanism, earthquakes and geothermal activities.

(5) The phenomena of the Siwalic Series (Neogene) sliding southward (Fig. 6).

Researches in neo-tectonic movements help us inquire further into the geological structure and uplift mechanism of Qinghai-Xizang Plateau.

Fig. 1 Permo-carboniferous slightly metamorphosed rocks near Yanghu Lake

Fig. 2 High mountains covered with snow of Gandise complex belt

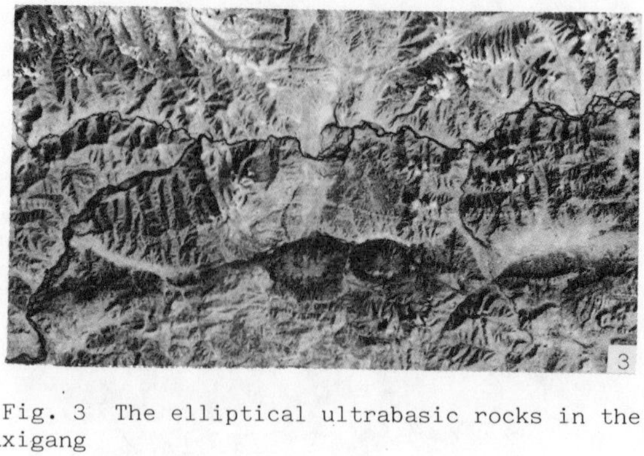

Fig. 3 The elliptical ultrabasic rocks in the south of Zhaxigang

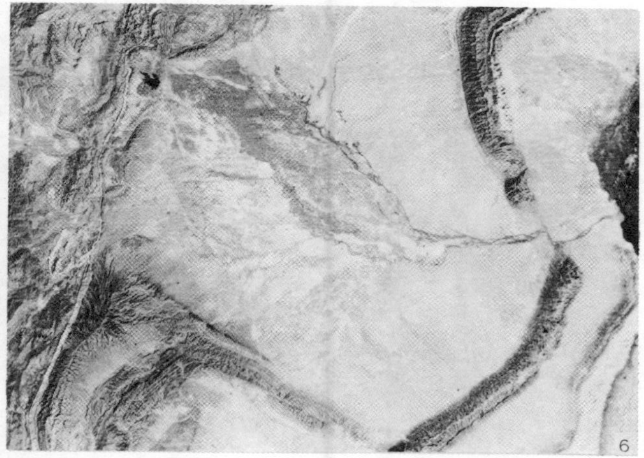

Fig. 4 The Quaternary basaltic lava sheet near Huangyang gully

Fig. 5 Some concealed faults nearby the Taijinaier Lake in Qinghai

Fig. 6 Phenomenon of the Siwalic Series(Neogene) sliding southward

# ACTIVE NORMAL FAULTING IN TIBET

Paul Tapponnier

Institut de Physique du Globe, 4 place Jussieu, 75230 Paris Cédex 05, France

Peter Molnar and Wang-Ping Chen

Massachusetts Institute of Technology, Department of Earth and Planetary Sciences, Cambridge, Massachusetts, U.S.A.

Abstract

From an interpretation of Landsat imagery of Tibet the most recent structures appear to be normal faults that trend approximately north-south. Brief visits to a few localities support this inference. Moreover, fault plane solutions of 15 earthquakes in the central part of the Xizang plateau indicate large components of normal faulting. Most solutions are not well constrained, but for the most reliable ones the T axes are oriented approximately east-west. One of these events occurred at 90 km depth. Only on the margins of the high plateau, where elevations are lower, do fault plane solutions show active thrust faulting. These observations imply an east-west extension of most of Tibet at the present time. We relate this pattern to the collision of India and Eurasia and to deformation of a relatively weak Tibetan crust and upper mantle. India applies a pressure to Eurasia that maintains Tibet at a high uniform altitude, and the hydrostatic head caused by this elevation transmits the pressure northwards Accordingly, Tibet is the pressure gauge of Asia. At the same time a small east-west strain (or flow) of the lower crust and upper mantle of Tibet would stretch the surface layer. Tibet's position relative to India at the present time is analogous to that of a "dead" triangle in a plastic material indented by a rigid indenter (India).

Collision between India and Asia appears to be responsible for the active tectonics throughout much of Asia. Within the Tibetan plateau sismicity is particularly intense and diffuse. Major earthquakes which have occurred in Tibet and in adjacent regions are plotted on figure 1, together with fault plane solutions for earthquakes large enough since 1961 and major active faults observed on the Landsat photographs. We here summarize briefly data and conclusions that have been discussed in more detail elsewhere (Tapponnier and Molnar, 1977, Molnar and Tapponnier, 1978, Chen et al, 1981).

The most recent tectonic regime in Tibet appears to be dominated by East West extension. Although extension in Tibet is a surprise because it appears to contradict prevailing hypotheses on how the plateau was formed (Argand, 1922, Dewey and Burke, 1973), two independent sets of observations support it strongly : fault plane solutions of earthquakes, and the study of active faults on the Landsat photos.

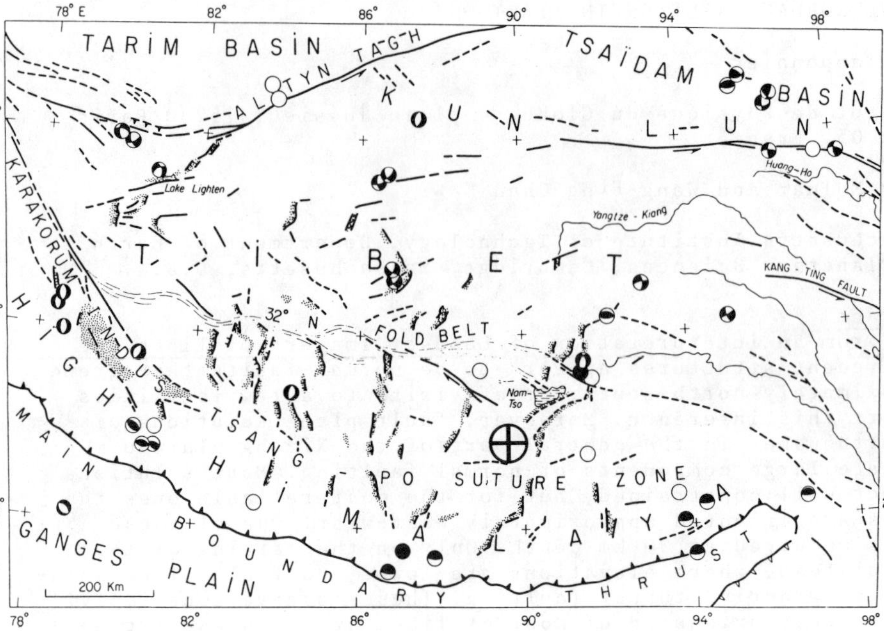


Figure 1 : Simplified map of active tectonics in Tibet and surrounding regions. Heavy lines are major active faults (dashed when less certain) recognized on Landsat imagery (from Tapponnier and Molnar, 1977, and Molnar and Tapponnier, 1978). For fault plane solutions (lower hemisphere) solid areas are compressional quadrants. Open circles represent historical earthquakes with magnitude above 7, solid circles historical earthquakes with magnitude above 8. Circle with cross is for the intermediate earthquake of September 14, 1976 (Wang Ping Chen et al, 1981). Dotted areas are sediment filled basins due to normal faulting. Thin dashed lines are a schematic outline of the 32°N fold belt. Cenozoic volcanism has not been represented. North-South normal faulting appears to be the dominant mode of present crustal deformation within the Tibetan plateau, in contrast with over-thrusting in the Himalayas and large-scale left lateral strike slip faulting to the northeast.

Landsat photographs. North South trending normal faults are particularly spectacular on the Satellite photos, as shown in Figures 2, 3 and 4.

Figure 2 : Landsat image of prominent normal fault west of Baiku Co (white arrow). The fault cuts east west folds in the Spiti Shales.

Figure 3 : Lansat image of the Kung Co fault system. The normal faults extend from Cho Oyu to the south to the Kangdese range to the north. They cut all the folds in the diverse sedimentary series of the suture zone (Spiti Shales, Tsang Po flysch Xigaze group...) and a younger granite (immediately south of white arrow).

Figure 4 : The Yadong-Karila-Angang fault system. Structurally similar to those in Figure 2 and 3, this fault system is part of long rift zone that extends more than 500 hundred kilometers to the north into central Tibet.

The faults cut more ancient structures, such as east-west trending folds in the "Spiti" series, the Triassic flysch, and the Xigaze group, or granites intrusive in them (Figure 2 and 3). The three Landsat photos shown here are located immediately north of the high Himalayan range. It is clear that the faults cut across the Yarlung-Tsangbo suture zone, but stop approximately at the high range to the south.

To the north, the normal fault systems extend several hundred kilometers (Figure 1), well into central Tibet. On the Landsat photos, normal faulting seems to be limited to an area that corresponds to that of maximum average elevation (above 4000 meters) and no such faulting can be observed in the Tarim and Tsaïdam basins or the Eastern Kun Lun to the north and east. There is evidence that normal faulting is more important in southern Tibet (south of the Bang Gong-Nu Chiang suture, or 32° north fold-belt) (Figure 1), where it appears to be connected with right lateral strike-slip faults trending about NW (Karakorum fault, Kyaring Tso fault, Peng Tso fault) (Molnar and Tapponnier, 1978). In northeastern Tibet, normal faults give way to giant strike-slip faults (Altyn Tagh, Kun Lun, Kang Ting faults) which "guide" most of the intracontinental deformation of central China (Tapponnier and Molnar, 1977).

Fault plane solutions. This large scale kinematic scheme is corroborated by fault plane solutions of 15 earthquakes within the Tibetan plateau. 14 of the earthquakes are at shallow depth in the crust, but one of them (shown with a large crossed circle on Figure 1) appears to be about 90 km deep in the upper mantle below the 75 km thick crust (Chen et Al, 1981).

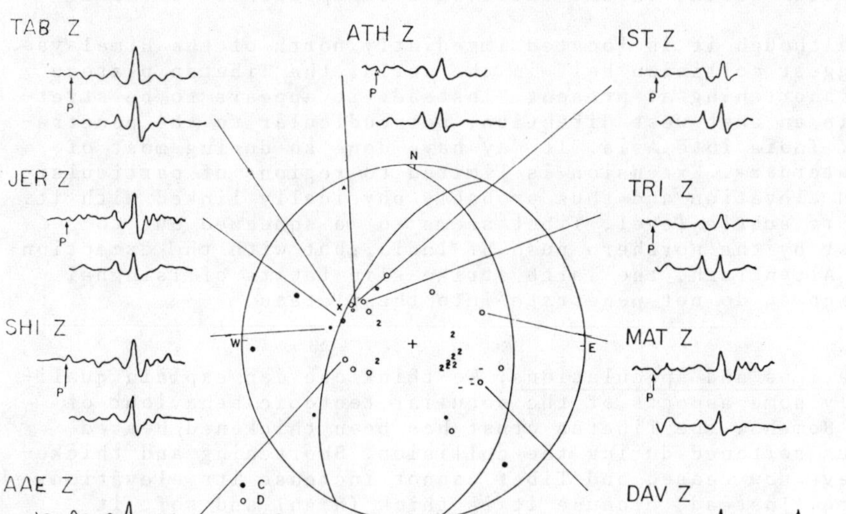


September 14, 1976 Tibet 29.81° N, 89.57°E 90 km

Figure 5 : Equal area projection of the lower focal sphere showing the nodal planes of the double couple component of the seismic moment tensor obtained from the linear inversion of the amplitudes of the P phases. Also plotted are the P wave first motion polarities (solid circles for compressional, open circles for dilatational, and X for nodal) pP polarities (+ for compressional and -for dilatational), and positions of the sP reflections (with the symbol '2'). Digitized vertical component long-period WWSSN seismograms (top trace) and synthetic seismograms (bottom trace) are shown for each of the 9 stations used in the inversion. For the synthetic seismograms, we used double couple solution 3 (Chen Wang Ping et Al, 1981) at 90 km depth with a seismic moment of $7 \times 10^{24}$ dyne-cm. The beginning of the synthetic seismograms marks the onsets of the direct P phases. The lack of clarity of the initial P phases is due to the digitizing, which does not allow the change in intensity of the original trace at the P onset to be seen, but the originals are clear (Chen Wang Ping et Al, 1981).

Although most of the nodal planes are not well constrained and some variable amount of strike-slip component may exist, the T axes are oriented roughly east-west on the average. The pattern of east-west extension drastically changes south and northeast of the Tibetan plateau. South of the Himalayan range, fault plane solutions reflect the underthrusting of India under Tibet. Northeast of the Kun Lun mountains in the Tsaïdam basin, thrust faulting also appears to predominate.

Thus, although it is located immediatly north of the Himalayas, the biggest collision belt in the world, the Tibetan plateau is not shortening at present. Instead, it appears to be stretching in an east-west direction, perpendicular to the penetration of India into Asia. It may have done so during most of the Quaternary. Extension is limited to regions of particularly high elevation and thus probably physically linked with it. At a more subtle level, Tibet seems to be squeezed out to the east by the northern push of India, but with the exception of the Altyn Tagh, the large strike slip faults of Tsinghaï and Szechwan do not penetrate into the plateau.

Implications and speculations. We think one can explain qualitatively some aspects of the peculiar tectonic behaviour of Tibet. Somehow the Tibetan crust has been thickened, heated and thus softened during the collision. Shortening and thickening have now ceased and Tibet cannot increase its elevation any more. Instead, because it is thick (high) and soft it now expands and thins, much as ripe cheese on a table. It cannot do so in the north-south direction because it is strongly buttressed by boundary forces applied by the rigid precambrian shields of India and Tarim, but it is freeer to stretch in an east west direction; hence, north south normal faults have developped through the more brittle parts of the crust and mantle.

In a way, Tibet behaves as the pressure gauge of the India-Eurasia collision: the forces applied by India onto Asia maintain the plateau at a high <u>uniform</u> altitude. In other words, the lithostatic pressure, increased by crustal thickening, balances the horizontal push of India. This "hydrostatic head" also helps transmit the convergence forces far north into the heart of Asia (Tien Shan, Altaï mountains).
Finally, an analogy with two dimensional experiments where a "plastic" material is indented by a rigid die (Figure 6) (Tapponnier and Molnar, 1976) may help explain why large scale strike slip faulting does not extend into Tibet.

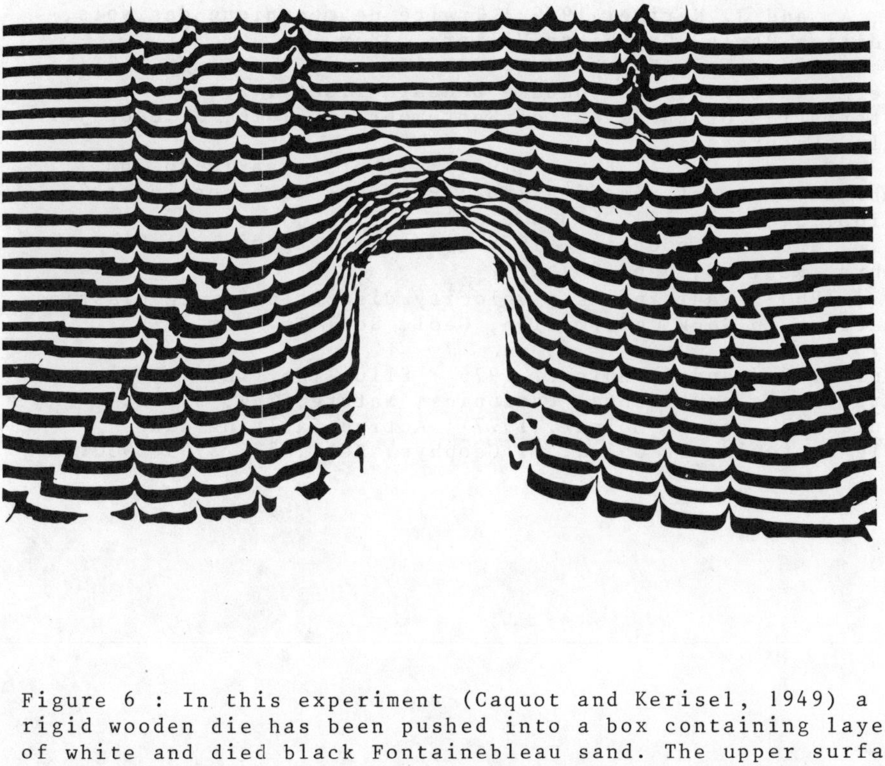

Figure 6 : In this experiment (Caquot and Kerisel, 1949) a rigid wooden die has been pushed into a box containing layers of white and died black Fontainebleau sand. The upper surface of the box was left free and vertical markers were made before the experiment by pulling a wire through the sand from top to bottom (the view presented here is upside down). This is a symmetrical, two-dimensional (plane strain) experiment. The die penetrated a distance equivalent to its width into the sand. The "dead" triangle and the slip lines or faults which offset both vertical and horizontal markers are clearly visible. The experiment was made popular by Ode's paper on "Faulting as a velocity discontinuity in plastic deformation" (1960).

A common character of experiments of the type shown in Figure 6 is the existence, in front of the die, of a sheltered triangular region where, although at yield stress, the deformable material experiences minor deformation because it "does not know whether to flow to the left or right. We believe that, in a sense, the Tibetan plateau is comparable to part of this "dead" triangle. Zones of intense shear occur along its boundaries and strike slip faults "emanate" from it but do not penetrate into it.

References

Argand, E, 1924 : La tectonique de l'Asie, 13th International Geological Congress, Brussels, 1922, Comptes rendus, I, Pt.5, 171-372.

Caquot, A. and J. Kérisel, 1949 : Traité de mécanique des sols, 2ème édition, Paris, Gauthiers Villars, 385p.

Chen Wang Ping, T.J. Fitch, J.L. Nabelek, and P. Molnar, 1981 : An intermediate depth earthquake beneath Tibet : Source characteristics of the event of September 14, 1976, J. Geophys. Res., in the press.

Dewey, J.F., and K.C.A. Burke, 1973 : Tibetan, variscan and precambrian basement reactivation : Products of a continental collision, J. Geol., 81, 683-692.

Molnar, P. and P. Tapponnier, 1978 : Active tectonics of Tibet, J. Geophys. Res, 83, 5361-5375.

Odé, H., 1960 : Faulting as a velocity discontinuity in plastic deformation, in Rock Deformation, Geol. Soc. Am. Mem., 79, 293-321

Tapponnier, P., and P. Molnar, 1976 : Slip line field theory and large scale continental tectonics, Nature, 264, 319-324.

Tapponnier, P. and P. Molnar, 1977 : Active faulting and cenozoic tectonics of China, J. Geophys. Res., 82, 2905-2930.

# THE EVOLUTION OF PALAEO-TETHYS IN THE TIBETAN SEGMENT OF THE ALPIDES

A.M.C. Şengör
(Department of Geological Sciences
State University of New York at Albany
Albany, New York 12222 U.S.A.)

## INTRODUCTION

Since the last quarter of the nineteenth century it has been known that the Alpine-Himalayan system of Eurasia formed as a result of the late Mesozoic -Cenozoic demise of an oceanic realm - Tethys (1). Within the framework of plate tectonics this oceanic realm had originally been equated with the large, triangular embayment of Pangaea of Permo-Triassic age, and the Alpine-Himalayan mountain-building during the late Mesozoic-Cenozoic had been ascribed to the closure of this oceanic area as a result of the dispersion of Pangaea (2)(3). It has been shown recently, however, that the evolution of the "Tethyan domain" comprises the life cycles of two distinct oceanic areas: a ?Permian to Liassic Palaeo-Tethys, and a Triassic to Cenozoic Neo-Tethys (4)(5). Palaeo-Tethys closed by the anti-clockwise rotation of a strip of continental fragment(s), called the "Cimmerian continent(s)" (4), away from the Tethyan margin of Gondwana-Land. This rotation simultaneously opend Neo-Tethys between the Cimmerian continent and the rest of Gondwana-Land during the Triassic to the end of Lias. The Palaeo-Tethyan suture extends from northern Greece / southern Bulgaria and north Dobrudja through northern Turkey, Greater Caucasus, northern Iran, Hindu Kush, central and ?southern Pamirs, central and northern Tibet to the Red River and to the Dien Bien Phu sutures (4)(5). This paper briefly reviews the Palaeo-Tethyan geology of Tibet and its immediately neighbouring regions and attempts to improve a previously presented (4)(5) interpretation.

## TETHYAN SUTURES AND CONTINENTAL FRAGMENTS IN TIBET AND SURROUNDING REGIONS

Figure 1 is a simplified palaeotectonic map of Tibet and its immediately neighbouring regions, which illustrates the present geometry of the Tethyan sutures and continental fragments in the area. In the extreme north, the horizontal ruling represents the Laurasian platform. Different parts of this heterogeneous continental mass were consolidated at different times ranging from the Precambrian (e.g. the Tarim Basin) to the latest Palaeozoic (e.g. the Turan Platform). What is common to all such regions is that they formed a united "foreland" for the predominantly Indosinian and some Yenshanian deformations along their southern border, i.e. north of the Wanch-Akbaytal Lineament (6) and the northern border of the Sungpan-Kantze

(Songpan-Ganzi) foldbelt (7)(8). The Sungpan-Kantze system is predominantly made up of Permo-Triassic flysch and magmatic rocks (Fig.1, column 7)(9) and possibly has no pre-Permian continental basement except for small "microcontinents". It represents an "oceanic hole", imperfectly closed between Laurasia, North China Platform, South China Platform and the North Tibetan Block, filled with oceanic lithologies plus ?ensimatic island arc remnants. Although severed by the Karakorum Fault (10), it continues westward into the Wanch-Akbaytal lineament and separates the northernmost "Gondwana-Land" elements from the southernmost "Laurasia" elements.

On all three sides the Sungpan-Kantze system is surrounded by thin zones of ophiolitic lithologies. In the Kokohsili mountains (Hoh Xil Shan) the ophiolitic zone includes a Triassic ophiolitic mélange with blocks ranging from the Devonian to the Permian in age (9). This Kokohsili-Yangtse suture is bounded on the south-west by a Permo-Triassic arc constructed apparently directly atop the Gondwana basement of the North Tibetan Block (Fig.1, col. 6). It continues to the south-east and near the Shan Plateau joins the Dien Bien Phu and the Ailao-Shan/ Red River sutures through the east Tibetan virgation (V in Fig. 1).

The largest portion of the Tibetan Plateau is underlain by the North and the South Tibetan Blocks, separated by the Tanglha (Tanggula Shan) suture. Columns 3-5 in Fig. 1 summarise the basic stratigraphy of these regions. Both of these blocks are here interpreted to have been originally pieces of Gondwana-Land but rifted from the latter during the Triassic to form a part of the Cimmerian continent(s). Sometime also during the Triassic the Tanglha ocean began opening and separated the northern block from the southern one (Fig.1, col. 3-5). The recent suggestion, that the Tanglha ocean was directly connected with the Farah-Rud zone in Afghanistan (11)(Fig.1, col. 1) is an attractive possibility, but has yet to be demonstrated through the Pamirs.

## PALAEO-TETHYAN EVOLUTION OF TIBET

Figure 2 summarizes, in a schematic fashion, the Permian to present tectonic evolution of Tibet roughly along the Lanzhou-Delhi cross-section. During the Permian Palaeo-Tethys separated Laurasia from Gondwana-Land. The latter included both the south and the north Tibetan blocks. That the subduction of the floor of the Palaeo-Tethys, today represented by the ophiolitic zones that rim the Sungpan-Kantze system, was going on both northwards beneath Laurasia and southwards beneath Gondwana-Land is indicated by granitic and granodioritic intrusives, andesitic volcanics and mélanges and flysch of generally Permo-Triassic age along the respective margins (9) (12). The eruption of the Panjal Traps (13)(Fig.2, P) during the Permian may be taken as the initiation of the opening of Neo-Tethys. During the Triassic both the Neo-Tethys, along the present Indus-Tsangpo (Indus-Yarlung Zangbo) suture (13) and the Tanglha oceans (Fig.1, TB) along the Tanglha suture opened. The "Lamayuru Flysch" in Ladakh (14) and the probably equivalent Triassic clastics south of the Tsangpo (15)(Fig.2, L) are here interpreted as continental rise aprons. At the end of the Triassic the opposing accretionary prisms flanking the active margins of Laurasia and Gondwana-Land collided and began to be uplifted. During the early Jurassic, the Tanglha and the Neo-Tethyan oceans continued to expand while the Sungpan-Kantze system became terminally consolidated. During the late Jurassic the same picture persisted with the exception that sometime between the mid-Jurassic and the mid-Cretaceous the Tanglha basin also closed. This closure was accompanied by the southward (onto the South Tibetan Block) obduction of a part of the floor of the Tanglha ocean, while the rest of it was subducted northward (beneath the North Tibetan Block). For the later, the Neo-Tethyan, evolution of the region, I refer the reader particularly to Le Fort (16) and to the numerous contributions in this symposium proceedings.

## CONCLUSIONS

The Palaeo-Tethyan geology of Tibet shows the same general organizational pattern recognised all along the Asiatic Alpides from Turkey to China (4)(5). This involves a Laurasian landmass in the north (Permian to Cretaceous) and a Gondwana-Land in the south (Precambrian to Cretaceous). They were separated by a "Tethyan domain" consisting of a ?Permian to Liassic Palaeo-Tethys in the north separated from a Triassic to Cenozoic Neo-Tethys by a Cimmerian Continent. The Cimmerian continent rifted from the northern margin of Gondwana-Land to open Neo-Tethys in its wake while its counter-clockwise rotation closed Palaeo-Tethys in front of it. During its rotation numerous mini-oceans opened within the Cimmerian continent (e.g. the Tanglha basin), formed perhaps as marginal basins. Neo-Tethys itself may have begun opening, locally, as a Palaeo-Tethyan marginal basin.

## References


(1) Suess, E., Das Antlitz der Erde, Tempsky, Vienna III/1 (1901)

(2) Smith, A.G., Alpine deformation and the oceanic areas of the Tethys, Mediterranean, and Atlantic. Geol. Soc. America Bull. 82, 2039-2070 (1971)

(3) Dewey, J.F., Pitman, W.C., III., Ryan, W.B.F., Bonnin, J., Plate tectonics and the evolution of the Alpine System. Geol. Soc. America Bull. 84, 3137-3180 (1973)

(4) Şengör, A.M.C., Mid-Mesozoic closure of Permo-Triassic Tethys and its implications, Nature 279, 590-593 (1979)

(5) Şengör, A.M.C., Yilmaz, Y., Ketin, I., Remnants of a pre-late Jurassic ocean in northern Turkey: fragments of Permo-Triassic Palaeo-Tethys ? Geol. Soc. America Bull. (in press)

(6) Stöcklin, J., Structural correlation of the Alpine ranges between Iran and Central Asia. Soc. Geol. France, Mém. h. sér. 8, 333-353 (1977)

(7) Stöcklin, J., Geology of Nepal and its regional frame. J.

Geol. Soc. London 137, 1-34 (1980)
(8) Huang, T.K., Die geotektonischen Elemente im Aufbau Chinas, I. Teil: Die geotektonischen Einheiten Chinas und ihre Merkmale. Geologie 9, 715-733 (1960)
(9) Bally, A. et al., Tectonics of Tibet and surrounding areas. USGS Open File Report, 50380 (1980)
(10) Molnar, P., Tapponier, P., Cenozoic tectonics of Asia: effects of a continental collision. Science 189, 419-426 (1975)
(11) Bassuolet, J.-P. et al., L'évolution des domaines téthysiens au pourtour du Bouclier indien du Carbonifére au Crétacé. Mém. BRGM, 115, 180-198 (1980)
(12) Tectonic Map of China, 1/10,000,000 Beijing (1976)
(13) Gansser, A., Geology of the Himalayas. Wiley, Interscience, New York (1964)
(14) Frank, W., Gansser, A., Trommsdorff, V., Geological observations in the Ladakh area (Himalayas). Schweiz. Mineral. Petrog. Mitt. 57, 89-113 (1977)
(15) A Scientific Guidebook to South Xizang (Tibet). Org. Comm. Sym. Qinghai-Xizang (Tibet) Plateau. Acad. Sinica (1980).
(16) Le Fort, P., Himalayas, the collided range: present knowledge of the continental arc. Am. Jour. Sci. 275-A, 1-44 (1975)

Note: In this paper, the term "Alpide" is used to designate the orogenic belts that descended from Tethys. "Tethys", in general, or "Tethyan domain" refer to the large, triangular, westward-narrowing embayment of Pangaea that separated Laurasia from Gondwana-Land and that include both Palaeo- and Neo-Tethys and their margins. As the Greek suffix "-ide" implies descendence, it would have been perhaps more appropriate to speak of "Tethysides" instead of "Alpides", but I continue to use the latter term for obvious historical reasons. My usage, however, deviates from the original employment of Stille and Kober in that it is geographical as opposed to theirs, which was temporal.

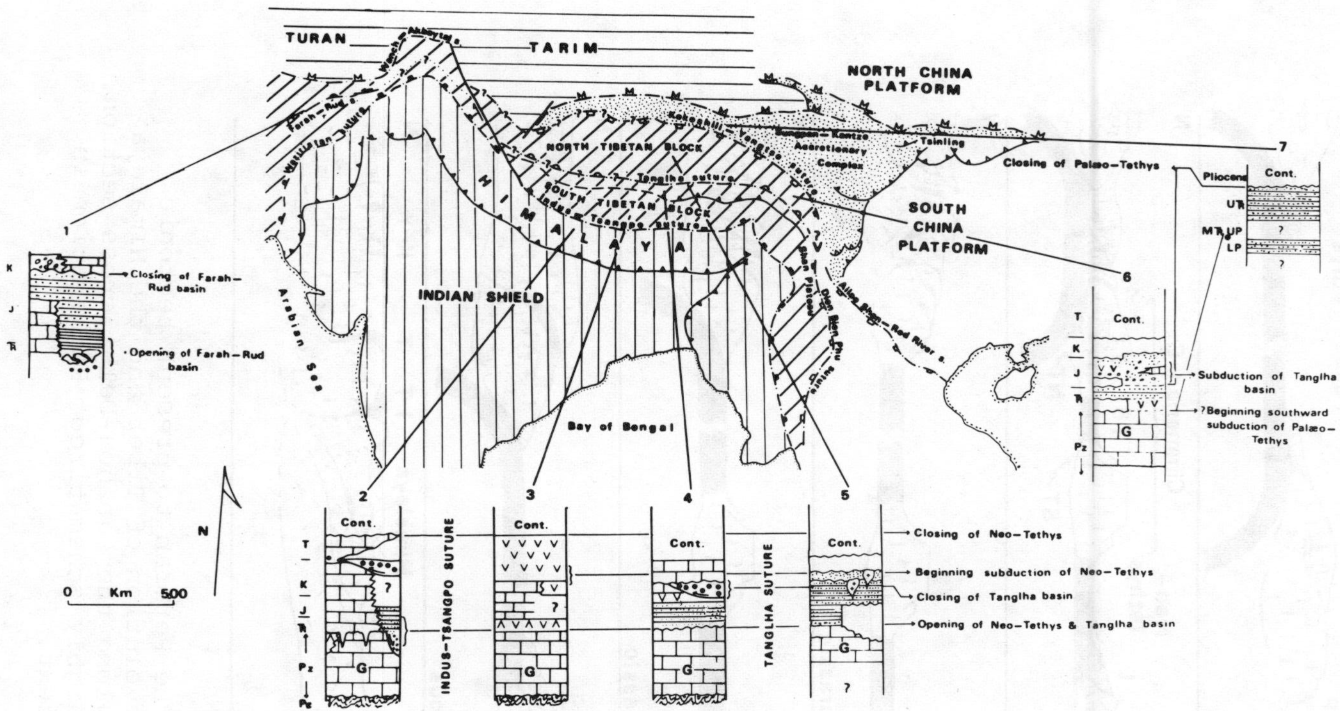


Fig. 1 A simplified palaeotectonic map of Tibet and surrounding regions. Horizontally ruled areas are parts of Laurasia, obliquely ruled ones those of the Cimmerian continent(s), and vertically ruled ones those of Gondwana-Land. Stippled region is the accretionary complex of the Sungpan-Kantze system. The columns 1-7 show the greatly simplified stratigraphies of the tectonic zones in the area.

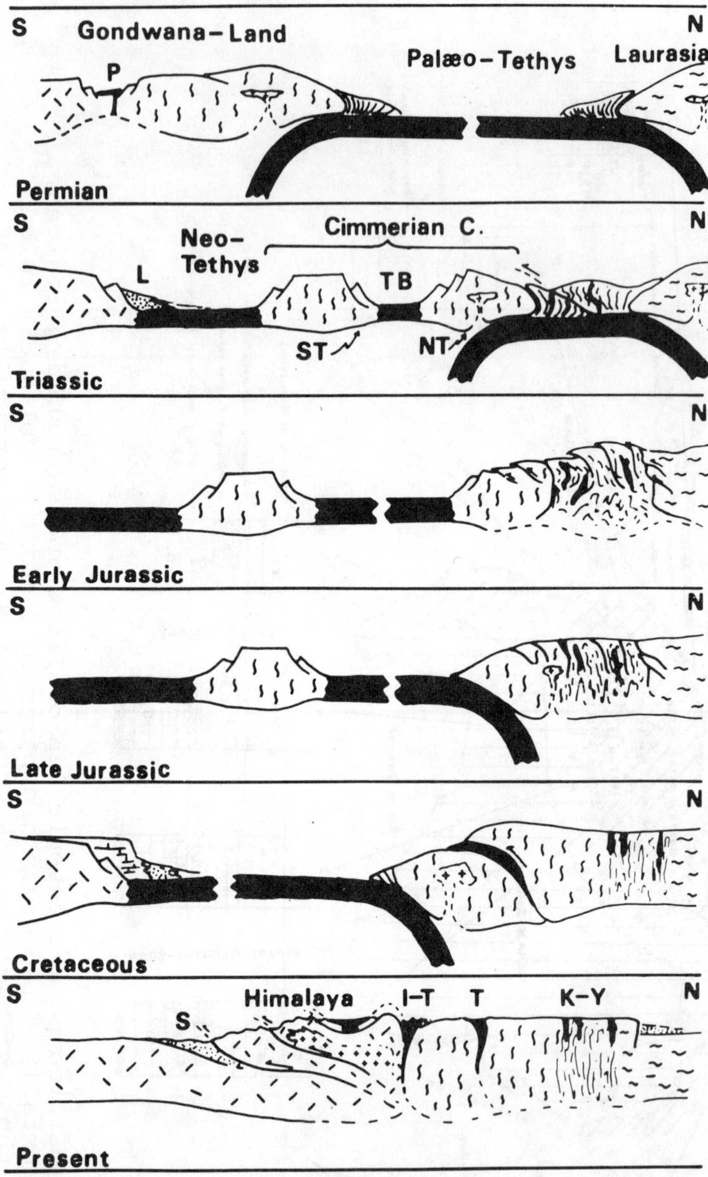


Fig. 2 The Permian to present tectonic evolution of Tibet and the Himalaya along the Lanzhou-Delhi cross section. Highly schematized. Explanation in text.

# A PRELIMINARY INVESTIGATION ON TECTONIC FEATURES IN TANGGULA-YUSHU REGION, QINGHAI

Zheng Yan-zhong
(Comprehensive Research Group of Fifteenth Geological Team,
Geological Bureau of Qinghai)

The so-called Tanggula-Yushu region is situated in Qinghai Province, south of Hoh Xil-Tongtian River and north of the Tanggula Range. This region approximately coincides with the northern segment of the "San Jiang (Three Rivers) Fold System" classified by the Chinese Academy of Geological Sciences in 1976. As one of the important parts of Qinghai-Xizang Plateau, this region reflects the geological structure and tectonic evolution of the plateau.

## ANALYSIS OF THE CRUSTAL ACCRETION PROCESS OF THIS REGION FROM THE FEATURES OF THE LATE TRIASSIC ASYMMETRICAL FAULT TROUGH

The stratigraphic sequences here range from Lower Palaeozoic to Quaternary, and therein the Carboniferous, Permian and Upper Triassic are the important ones, in which two different facies — the platform type Jieza Group and the geosyncline type Batang Grou — coexist. The linear NW trending strata, varying from older to younger northward, can be divided into six depositional cycles that are similar to one another. From bottom up, their components are: purplish-red clastic rocks to thick bedded limestones to coal-bearing clastic rocks to limestone (Fig. 1). All of these indicate that in this region, deposited on the Early Palaeozoic continental block from Carboniferous through Permian and up to Triassic, an intermittently developing epicontinental geosyncline shifted from south to north. This implies an expansion of the continental block. This expansion of continent through the shift of the so-called epicontinental geosyncline can be exemplified by the outline of the fault trough of the Late Triassic (Figs. 2, 3).

The Late Triassic asymmetric fault trough formed by intense activity on the Xijinwulan Lake-Jiegu fault. At the northern margin, a mobile-belt, i.e. a volcanic island arc can be marked out by the geosynclinal type of the Batang Group. In the central part, the platform type Jieza Group illustrates the existence of a rather wide and stable epicontinental sea, including the comb-embayment-lagoon area about Zaqing, and the minor local upwarping of the old basement to the east. The southern part is the old landmass comprising Carboniferous and Permian sediments.

The whole trough seems to be wide in the east and narrow in the west and deep in the north and shallow in the south, laterally becoming diversified into a mobile belt and a relatively stable belt. If the epicontinental geosyncline with the consequent continental accretion is an acceptable concept,it may be suggested as well that the Carboniferous and Permian might have undergone a similar development.

## THE INTERLACED NETWORK OF THE THREE PRINCIPAL STRUCTURAL SYSTEMS

The main structures of this region are the NW trending faults and folds, of which the major fracture is Xijinwulan-Jiegu fault, a lithospheric fault zone, bounding and controlling the subordinate structures and being active over a long period of time by right-handed torsion. This structural system was responsible for the formation of the Late Triassic fault trough, the linear stretching, and the lateral variation of geological processes in this region and demarcates the following structural subdivisions (Fig. 3): (1) Dangjiang-Zhimenda fault block which might be the lower part of an "upturned" continental block; (2) Duocai-Batang mobile belt, i.e. the major fault zone itself, where NW trending structures concentrate; (3) Yage River-Dasumang subsided belt, its extent corresponds to the above-mentioned epicontinental sea; (4) Tuotuo River-Nangqen uplifted belt, corresponding to the above-mentioned old continental block.

On the background of NW structures, this region has been also subjected to intersecting NE struçtures and a superimposing of E-W ones, resulting in a rather complicated interlaced network.

There are 2 important NE structural systems, diagonally dissecting the Qinghai-Xizang Plateau: (1) Chawo La-Zonggeyong NE zone, and (2) Cegne La-Dongjiaoyong NE zone together constructing the Zaqing area where there exist the intensely developed NE structures. This area intermittently stretches northward farther beyond Dulan, and southward to join with Xigaze-Nagqu fault zone. In addition, NE structures also exist along the line of Geladandong-Wuli and the line of Xiaosumang-Zhaokalong. The principal active period of the NE structures of left-handed torsion type was from Late Mesozoic to Early Cenozoic.

The complicated E-W structures consist of not only Late Palaeozoic structures, but also Mesozoic to Cenozoic, Neogene and recent ones, predominantly attributed to the intense activity of Bangong Lake-Nujiang River fault in Zhuerkenwula-North Xizang. E-W structures are mainly distributed along four lines; from north to south, they are: (1) Wuli-Fandi ; (2) Moqu River-Zonggeyong; (3) Zaqing-Ziqudukou; (4)Soloyong-Batang.

The above-mentioned comb-embayments at the southern margin of the Late Triassic fault trough and widespread Cretaceous-Quaternary basins in the western part of this region are all genetically related to the E-W structures.

Additionally, there are also traces of NNW structure, disturbed by the N-S structure, and differences of geologic structure and of geological history between the east and west parts in this region. All of these should be worthy of notice.

## TIME AND SPACE DISTRIBUTION OF MAGMATIC ACTIVITES AND THEIR RELATIONS TO GEOLOGICAL STRUCTURE

In the region, intrusive activity is known in five periods: Late Palaeozoic, Early, Middle, and Late Mesozoic and Cenozoic. The main intrusive rocks are as follows : basic-ultrabasic rock bodies in Dangjiang-Zhimenda fault block ; NW trending quartzdiorite in Duocai-Batang mobile belt ; Zamuqu syenite, Zonggeyong-Narikongma and Jiangma granite groups in Yage River-Dasumang subsided belt. The three granite groups fall in three separate areas, repespectively, and there are some granites and alkaline rocks in Tuotuo River-Nangqen uplifted belt.

Eruptive activity is mainly divided into three periods: (1) The widespread, late Early Permian intermediate-basic eruption, among which the eruptive rocks are intercalated with manganese siliceous slate and many gabbro dykes. Information about contradictions of fossil ages in these rocks in Dangjiang-Zhimenda belt are worthy of notice. (2) The Late Triassic intermediate-basic to acid eruptives, which constitutes the Batang Group characterized by volcanic island arc sediments. (3) The Late Cretaceous intermediate-acid to alkaline volcanic and subvolcanic activities occurring in Xialaxiu-Nangqen area. In addition, beyond this region, there are also Neogene continental basaltic eruptions at Zuerkenwula.

Apparently, magmatic activity here is strictly controlled by geological structures. Chronologically, they are coincident with the intense activity of the respective main structural systems, while spatially, they vary with different structural positions. Since Mesozoic, the magmatic activity has displayed the following tendencies from north to south: (1) From old to young ; (2) From strong to weak ; (3) From uniformly controlled by NW structures to compositely controlled by several structural systems, especially by NE structures, forming the granite zone from Zonggeyong-Narikongma to Nam Lake in Xizang ; (4) From basic ultrabasic, intermediate-acid to acid, alkaline, and from relatively rich in sodium to relatively rich in potassium.

All of these are congruous with the features of the asymmetrical fault trough and the structural subdivisions.

## THE POSITION OF THIS REGION IN THE EVOLUTION OF QINGHAI-XIZANG PLATEAU

According to the main points described above, the following may be concluded:

1. As there are certain continuous and transitional relations in stratigraphy, geological structure, magmatic activity and so on between this region and Zuerkenwula-North Xizang, it may be considered that these two regions together constitute one of the "steps" of Qinghai-Xizang Plateau.

2. In this region, the Late Triassic fault trough shows a completeness and a continuity in nature, and is also consistent with the major fault zone, in which the NW trending main structures are of right-handed torsion type. To its north, the Bayan Har Mountains region, in which the main structures are of left-handed torsion type, appears to be dissected and incomplete. Therefore, it should be admitted that these 2 regions belong to 2 different land masses and form a tectonic contact along Xijinwulan-Jiegu fault. This region has rotated counter-clockwise against Bayan Har. If there was any subduction, then it was the Bayan Har region which should have been subducted.

3. In this region the NE structures, actually a part of the NE structural system throughout the Qinghai-Xizang Plateau, play a rather important role in controlling the magmatic activity and the related orogenesis. To this it is necessary to pay more attention.

We are very grateful to the following units and scientists for their help and information : Qinghai Institute of Geology ; Sencod Regional Geological Survey Team, Geological Bureau of Qinghai; Geophysical Exploration Team, Geological Bureau of Qinghai; Institute of Geology, Academia Sinica; Feng Zu-jun and Wang Er-Kang of Nanking University.

FIG.1

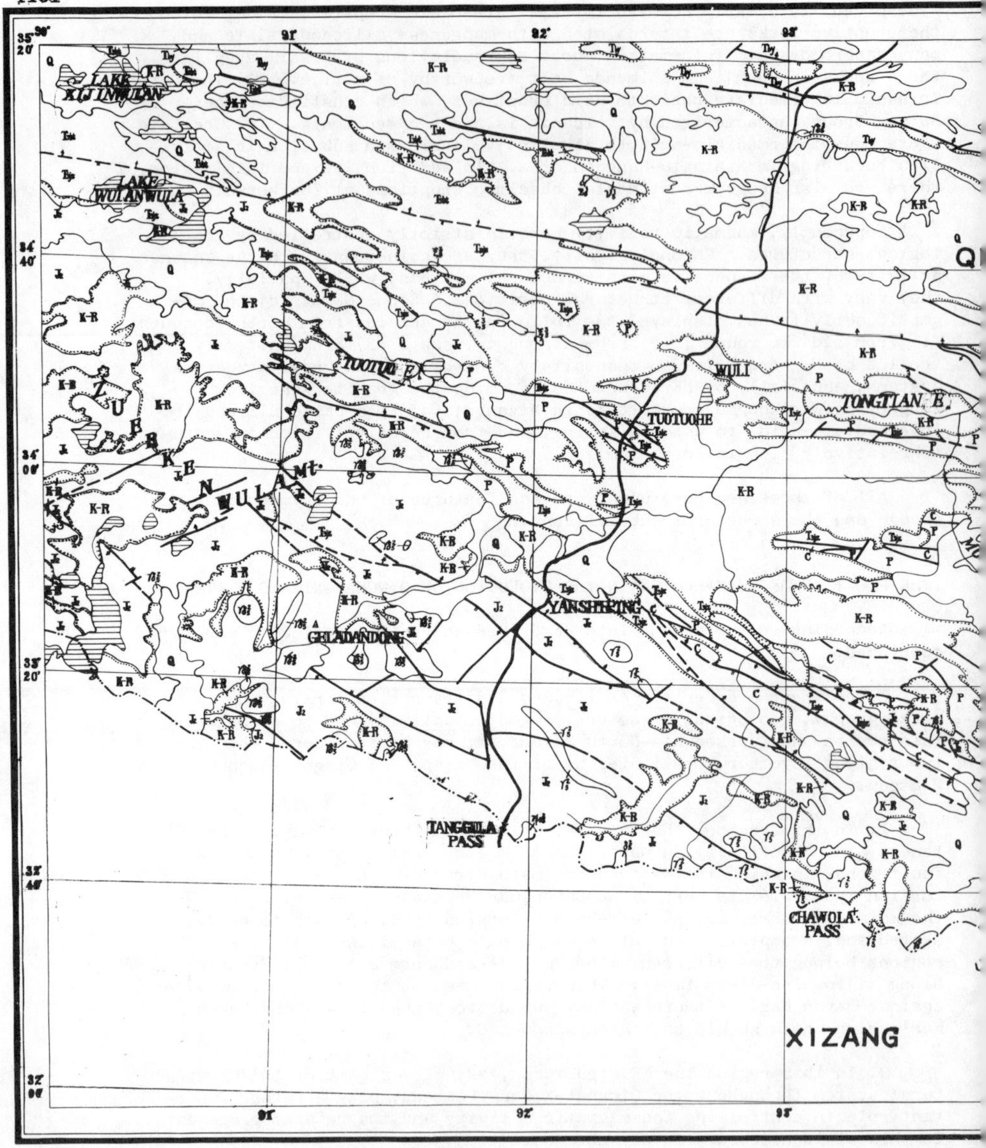


- Q — Quaternary
- K-R — Cretaceous-Tertiary
- J2 — M. Jurassic
- T3bt — U. Triassic Batang Gr.
- T3jz — U. Triassic Jieza Gr.
- T2 — M. Triassic
- Ty — Triassic and BayanHar Gr.
- P — Permian
- C — Carboniferous
- D3 — U. Devonian
- Pz1 — L. Palaeozoic
- Syenite (Cenozoic)
- Granitite (Cenozoic)
- Granite-Porphry (Cenozoic)
- Granite (Cenozoic)
- Trachyte (L.M
- Syenite (L.Mes
- Rhyolite (L.Mes
- Qtz-Monzoni (M. Mesozo
- Granite (M. Me

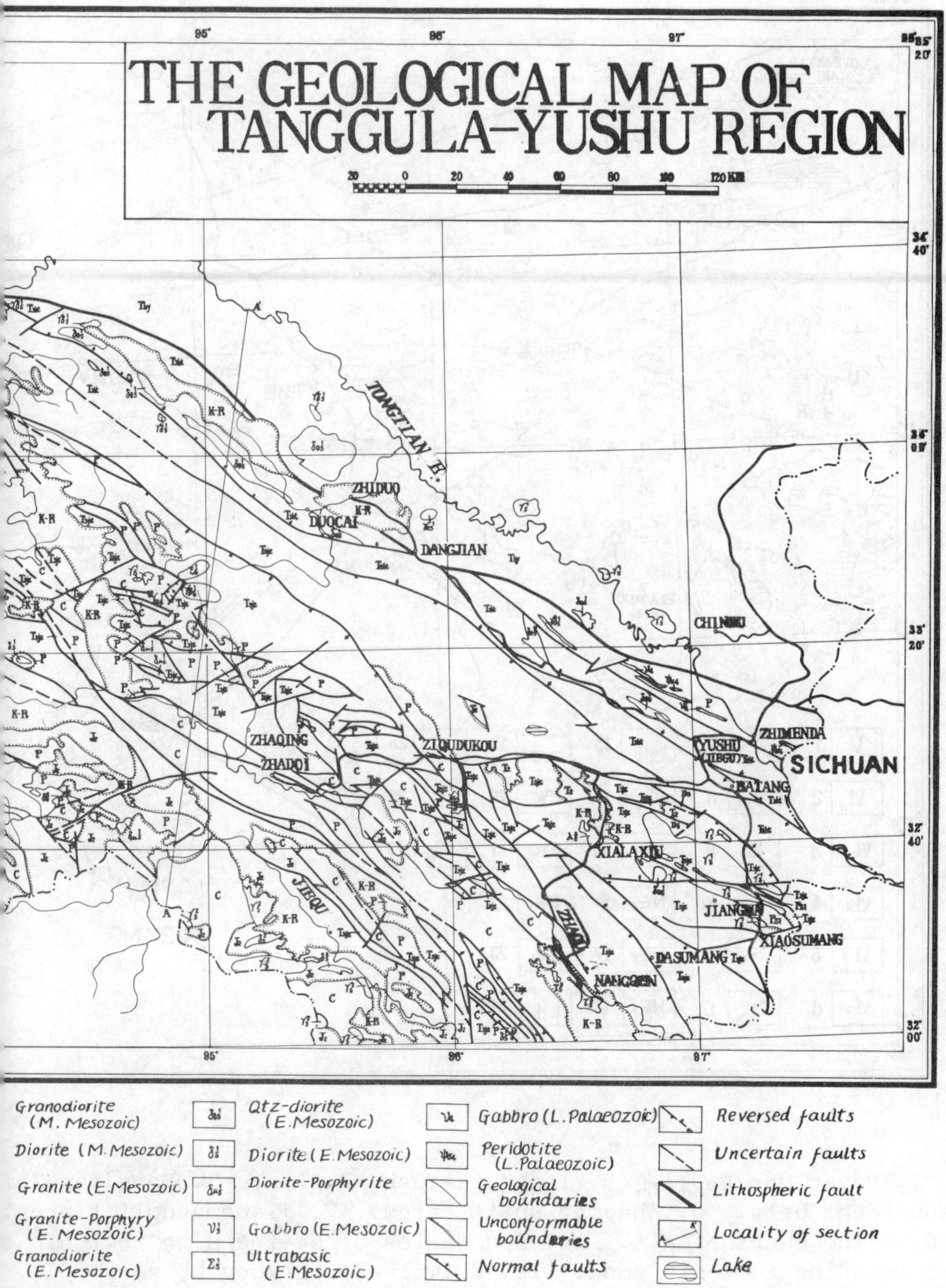
THE GEOLOGICAL MAP OF
TANGGULA-YUSHU REGION
20 0 20 40 60 80 100 120 KM
TONGTIAN H.
ZHIDUO
DUOCAI
DANGJIAN
CHINDU
ZHIMENDA
SICHUAN
YUSHU (JIEGU)
BATANG
ZIQUDUKOU
ZHAQING
ZHADOI
XIALAXIU
JIEQU
JIANGXI
XIAOSUMANG
DASUMANG
NANGQEN
Granodiorite (M. Mesozoic)
Diorite (M. Mesozoic)
Granite (E. Mesozoic)
Granite-Porphyry (E. Mesozoic)
Granodiorite (E. Mesozoic)
Qtz-diorite (E. Mesozoic)
Diorite (E. Mesozoic)
Diorite-Porphyrite
Gabbro (E. Mesozoic)
Ultrabasic (E. Mesozoic)
Gabbro (L. Palaeozoic)
Peridotite (L. Palaeozoic)
Geological boundaries
Unconformable boundaries
Normal faults
Reversed faults
Uncertain faults
Lithospheric fault
Locality of section
Lake

FIG.3

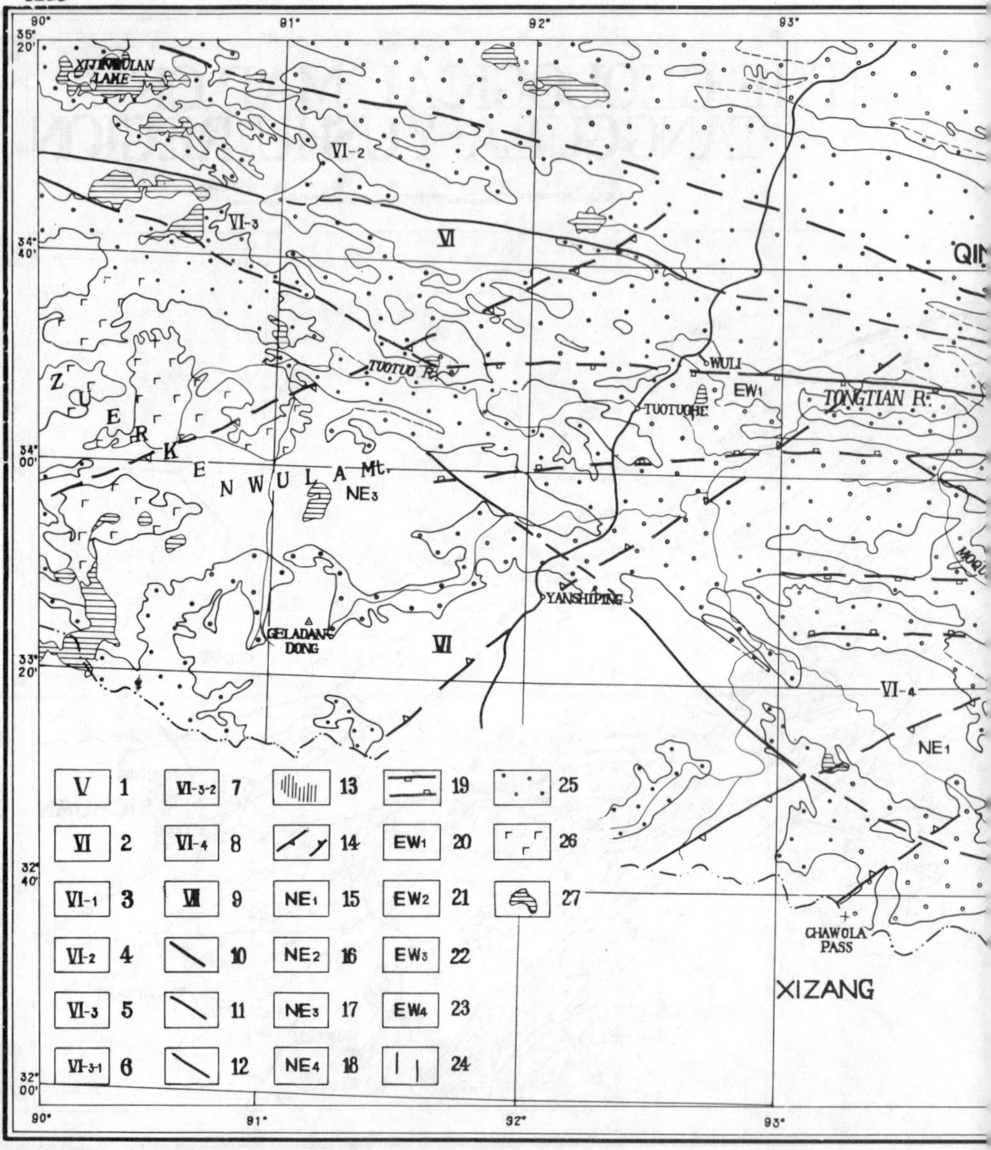


1. Bayan Har Region. 2. Tanggula-Yushu Region. 3. Dangjian-Zhi subsident belt. 6. Zhaqing uplift area. 7. Jiangma uplift area. 10. Lithospheric fault. 11. Boundaries of NW-trending tectonic d NE-trending tectonic zone. 15. Chawola-Zonggeyong NE zone. 16. Xiaosumang-Zhaokalong NE zone. 19. EW-trend tectonic zone. 20. Ziqudukou EW zone. 23. Soloyong-Batang EW zone. 24. N-S trend s 27. Lake.

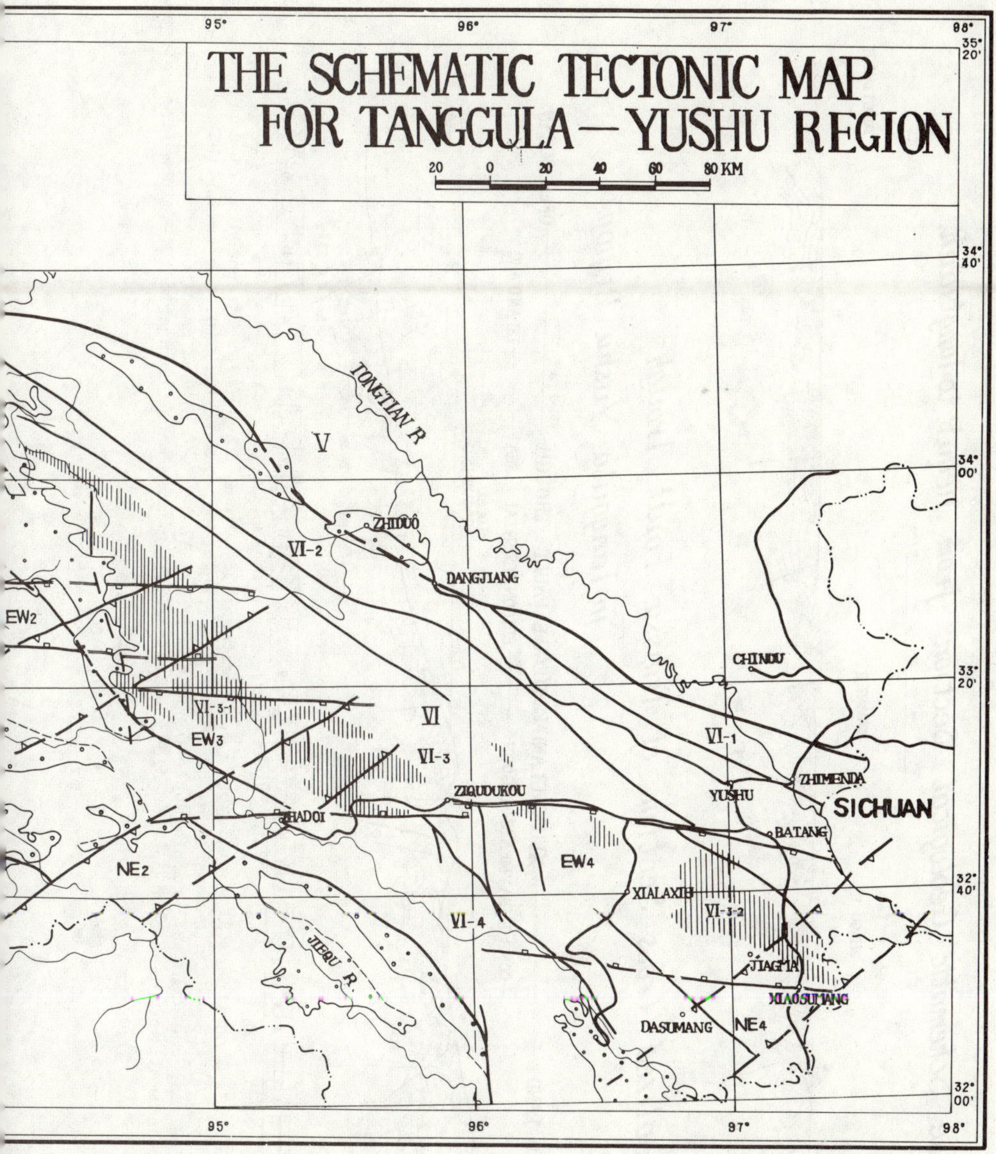


fault block. 4. Duocai-Batang mobile belt. 5. Yage R.-Dasumang
tuoiR.-Nangqen uplifted belt. 9. Zuerkenwula-North Xizang Region.
ns. 12. Subsident center. 13. Uplifts in subsident belt. 14.
ila-Dongjiaoyong NE zone. 17. Geladandong-Wuli NE zone. 18.
adidaduei EW zone. 21.Moqu R.-Zonggeyong EW zone. 22. Zhaqing-
res. 25. Creteceous-Quaternary basins. 26. Pliocene basalt.

FIG2

## The Schematic Geological Section from JiequR. to TongtianR.

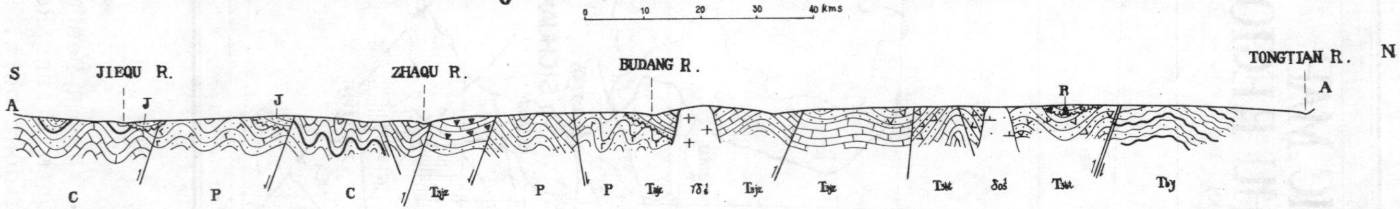


## The Idealized Cross Section of Triassic Fault Trough in Tanggula-Yushu Region

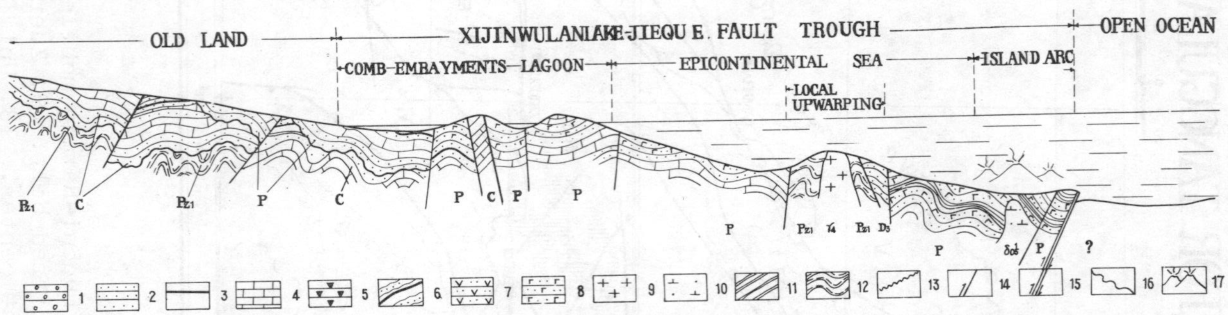


1. Conglomerate; 2. Sandstone; 3. Coal layer; 4. Limestone; 5. Gypsum layer; 6. Interbedding of sandstone and slate; 7. Intermediate volcanic rocks; 8. Basic volcanic rocks; 9. Granitite; 10. Quartzdiorite; 11. Siliceous slate; 12. Paragneiss; 13. Unconformable boundaries; 14. Faults and direction of upper side; 15. Lithospheric fault and motional direction on each side; 16. Tectonic unconformable boundaries; 17. Submarine volcanoes.

# THE SEISMICITY AND RECENT CRUSTAL MOVEMENT IN THE HIMALAYAN REGION

Ye Hong, Zhang Wen-yu, Yu Zhi-shui, Xia Qin
(Institute of Geology, State Seismological Bureau)

## Abstract

We present a study on a dynamic model of the recent tectonics in the Himalayan Region based on earthquake data recorded in the past 80 years (from 1897 — 1976). The characteristics of the avtive fault zones are studied using fault plane solutions. The directions of the recent tectonic movement in various parts of the Himalayas are calculated from the null vector data showing that the recent convergence between the NNE moving Indian Plate and the Eurasian Plate is primarily a collision process rather than underthrusting. This dynamic process includes overthrusting of the upper part of the lithosphere of the western Himalayas over the Ganges Plain, with a low dipping angle and the NE-ward movement of the Assam wedge of the India Plate along a left-lateral strike slip fault zone which is oriented NE. As a first approximation, the active fault slip rates are estimated from the seismic moment data and the rates of shortening of the Himalayas and the Tibetan (Qinghai-Xizang) Plateau north of the Yarlung Zangbo River are inferred to be 3 - 4 cm/yr and 1 - 2 cm/yr, respectively.

## EARTHQUAKE MECHANISM AND CHARACTERISTICS OF THE ACTIVE FAULT ZONES

Fig. 1 and Fig. 2 show fault plane solutions from the Himalayas. From the figures it can be seen that the characteristics of the active fault zones in various parts of the Himalayas are different.

In the outer part of the western section of the Himalayan arc, most of the active faults are thrusts with rather low dipping angles of about $20^0 - 30^0$ or less, to the NE, while, in the inner part of the arc, in addition to strike slip faults there are some normal faults, reflecting the complexity of the local stress field which maybe caused by secondary tension in the back-arc region(1).

In the eastern section of the Himalayas which trend NE, earthquake mechanism shows another pattern of recent tectonics: most of the earthquake faults are left lateral strike slip faults trending NE, with a few cases of NW dipping thrust faults.

In the NW trending Bomi-Zayu section, according to Ben-Menahem, the strongest earthquake of the area ( M = 8.6 in 1950) was caused by large-scale right lateral strike slip along the northern extension of the Miju fault(2-3), as was another earthquake occurred in 1959 in this region, which has been described by Kuo(4).

## DIRECTION OF TECTONIC MOVEMENT CALCULATED FROM NULL VECTOR DATA

As shown in Fig. 3, the line of intersection of the two nodal planes, called N, is perpendicular to the fault slip direction S. Hodgeson has called this axis the null vector. Generally S is the projection of the displacement vector V on the earthquake fault plane, so it can be shown that the null vector N should be orthogonal to the displacement vector V, which roughly represents the tectonic movement direction.

Naturally, it is impossible to determine V only from a single null vector, however, if we have a group of null vector data from earthquakes occurring in a region with a common tectonic pattern, we can obtain a plane, called the null vector plane, which consists of these null vectors, and should be the normal plane of the vector V. Thus, it ought to be possible to learn about the tectonic movement direction from the null vector data.

Let us denote the displacement vector by V = (x,y,z) and the direction cosines of the i-th null axis by

$$n_i = ( n_{xi},\ n_{yi},\ n_{zi} ).$$

Following Fara and Scheidagger(5), we set up the statistical procedure by requiring that the expression

$$F ( x, y, z ) = \Sigma\{Xn_{xi} + Yn_{yi} + Zn_{zi}\}^2 \qquad (1)$$

be a minimum subject to the auxiliary condition:

$$x^2 + y^2 + z^2 = 1 \qquad (2)$$

Expanding expression (1) for F and making notation changes, we have

$$\alpha_{11} = \Sigma n^2_{xi}$$

$$\alpha_{22} = \Sigma n^2_{yi}$$

$$\alpha_{33} = \Sigma n^2_{z}$$

$$\alpha_{21} = \alpha_{12} = \Sigma n_{xi}\ n_{yi}$$

$$\alpha_{31} = \alpha_{13} = \Sigma n_{xi}\ n_{zi}$$

$$\alpha_{32} = \alpha_{23} = \Sigma n_{y}\ n_{zi}$$

therefore:

$$F = \alpha_{11}x^2 + \alpha_{22}y^2 + \alpha_{33}z^2 + 2\alpha_{12}xy + 2\alpha_{13}xz + 2\alpha_{23}yz \qquad (3)$$

Now, the problem may be reduced to a calculation of the eigenvalues and eigenvectors of a matrix A.

$$A = \begin{vmatrix} \alpha_{11} & \alpha_{12} & \alpha_{13} \\ \alpha_{21} & \alpha_{22} & \alpha_{23} \\ \alpha_{31} & \alpha_{32} & \alpha_{33} \end{vmatrix} \qquad (4)$$

The smallest eigenvalue of the matrix A will be the minimum of F, the corresponding eigenvector is the vector $V = (x,y,z)$.

Then the tectonic movement direction angle $\alpha$ and the dip angle $\varphi$ can be easily obtained by the following two expressions:

$$\alpha = \arc\tan \frac{y}{x} \tag{5}$$

$$\varphi = \arc\sin z \tag{6}$$

At the same time, we can estimate the error by the root-mean-square and the scattering angle $\Psi$ :

$$rms = \sqrt{\frac{\lambda_{min}}{m}} \tag{7}$$

$$\Psi = \arc\sin \sqrt{\frac{\lambda_{min}}{m}} \tag{8}$$

Where $\lambda_{min}$ is the smallest eigenvalue of the matrix A and m is the number of samples.

The calculated results are shown in Table 3.

## ACTIVE FAULT SLIP RATES ESTIMATED FROM SEISMIC MOMENT DATA

According to the dislocation theory of earthquake mechanism(6), we have the Brune formula(7) :

$$\Sigma u = \frac{1}{\mu_A} \Sigma M. \tag{9}$$

Where $\Sigma u$ is the summation of the seismic displacements of a fault zone in a certain period, $\Sigma M_0$ is the summation of the seismic moments, A is the area of the entire fault zone, and $\mu$ is the elastic rigidity of the medium. Dividing $\Sigma u$ by time t, we get the mean fault slip rate for the whole active fault zone in the period,

$$V = \frac{\Sigma u}{t} \tag{10}$$

As a first approximation, the seismic moments can be estimated from magnitude Ms using an empirical formula(8). An empirical formula suitable to the Himalayas has been derived from data on the earthquakes in Table 4. which is :

$$\log M_\circ = 1.17\, Ms + 18.6 \tag{11}$$

The above method was applied to various parts of the Himalayas using the data from 1897-1976. The total amounts of displacement were then obtained for this period, as well as the mean fault slip rates (Table 5).

## DISCUSSION ON THE DYNAMIC MODEL OF RECENT TECTONICS IN THE HIMALAYAS

According to our calculation (Table 3), for the lower lithosphere, i.e., the upmost mantle of the Himalayas, the tectonic movement direction is NNE (Fig. 4). This is in accordance with the direction of motion of the India plate determined from the orientation of mid-ocean ridges and their associated transform faults, as well as magnetic anomalies(10). But it is more important to note that the dip angle of this vector is only two degrees. It means that the tectonic movement of the lower lithosphere in the Himalayas is nearly horizontal. This pattern is quite different from that of Benioff zones and this seems to support the viewpoint that the convergence between two continental plates, such as the Indian Plate and the Eurasian Plate, is a collision process rather than an underthrusting process.

For the upper lithosphere, i.e. the crust of the Himalayas, according to our calculation, the direction angles for movement of the western and eastern sections are different from each other.

In the western section, the crustal movement direction is NNE, simillar to the mantle. But the dip angle is about 30°, to the NE, much larger than that of the mantle. It indicates that there should be some thrusting movement in the crust of the Himalayas (Fig. 5). Considering the intensive collision and compression occurring here, we hold that the thrusting is upthrusting of the Himalayas over the Ganges Plain. This could also be a reasonable explanation for the strong uplifting and the significant positive isostacy gravity anomaly in this region.

In the eastern section of the Himalayas the calculated crustal movement direction is nearly north-south with a slight deviation to the west. This differs from the mantle movement direction, which is NNE as mentioned above. Considering the aforementioned characteristics of the active fault zones and that this section of the Himalayas is oriented NE, we suggest that the Assam wedge of the Indian Plate is moving northeastward along this plate boundary belt and therefore has left lateral strike slip motion relative to the Eurasian Plate. The thrust motion component in the upper lithosphere determined from earthquake mechanism is probably caused by the bedding plane slip occurring during the creation of the Himalayan arc structure or due to the counter-clockwise rotation effect of the Indian Plate proposed by some authors. In the eastern section of the Himalayas, the active fault slip rate from seismicity is estimated to be 5.6 cm/yr. Taking into account the deviation angle between the crustal movement direction and the trend of the plate boundary, it can be deduced that along this NE trending plate boundary the mean active fault slip rate in the past 80 years has been about 3-4 cm/yr. This observation corresponds with the rate of the NE-ward movement of the Indian Plate -- 5 cm/yr, determined from the linear magnetic anomalies associated with ocean ridges(11).

Assuming that there are no systematic errors in the calculations, we might infer that of the total amount of 5 cm/yr, 3-4 cm is distributed in Himalayas, resulting in the intensive compression, obduction and uplifting in this region. As a first approximation, the shortening rate of the Himalayas might be 3-4 cm/yr., The rest of the shortening are distributed throughout the whole Qinghai-Xizang Plateau north of the Yarlung Zangbo River. This results in the thickening of the lithosphere and uplifting of Qinghai-Xizang so that the shortening rate of the whole of Qinghai-Xizang might be 1-2 cm/yr.

This dynamic process could be a key to the understanding of the recent tectonics both inside and outside of Qinghai-Xizang including much of eastern Asia ; it causes an east-west extension and large-scale east-ward flow and creeping of the lithospereic material of Qinghai-Xizang, applying great pressure on South China and North China. This leads to the right lateral shifting along NW trending faults such as Red River fault in the South, and the right lateral shifting along NE trending faults such as Tan-Lu fault in the North.

Moreover, according to our estimation, the active fault slip rate from seismicity in the western Himalayas is only 0.49 cm/yr., much less than that in the eastern Himalayas. We consider two possible reasons for this: either the strain energy is released chiefly by aseismic creep there, or due to a locking condition somewhere in this region, one large or several large earthquakes are developing, causing an apparent decrease in seismicity in the last few decades. Of course, it is also possible that both of the above reasons are true, but, in our opinion, maybe the former is more acceptable. According to Hubbert(12) aseismic creep develops easily in regions, such as this, where crustal movement is predominantly upthrusting(13). According to our estimation it could also be assumed that at the present time the outer part of the Himalayan arc is more active than the inner part, which agrees well with the geological field observations(14-17) and can be interpreted as the collision zone having migrated from North to South.

References

(1) Yeh Hung et al., The analysis of the recent tectonic stress of the Himalaya Mountains Arc and its vicinities. Scientia Geologica Sinica, (1) (1975) (in Chinese).
(2) Valdiya, K. S., Himalayan transverse fault and their parallelism with subsurface structures of North Indian Plains. Tectonophys., 32 (3/4), 353-386 (1976).
(3) Ben-Menahem, A. et al., The source of the great Assam earthquake -- An interplate wedge motion, Phys. Earth Planet Interiors, 9 (4), 265-289 (1974).
(4) Kuo Tseng-Chien, Chang Cheng, Investigation of the mechanism of 22 Chinese earthquakes occurring during 1949-1960, Acta Geophysica Sinica, (1) (1962) (in Chinese).
(5) Fara, H. D., Scheidegger, A. E., An eigenvalue method for the statistical evaluation of fault plane solutions of earthquakes, B.S.S.A., 53 (4), 811-816 (1963).
(6) Aki, K., Earthquake mechanism. Tectonophys., 13 (1-4), 423-446 (1972).
(7) Brune, J. N., Seismic moment, seismicity and rate of slip along major fault Zones. J.G.R., 73 (2), 777-784 (1968).
(8) Davies, G. F., Brune, J. N., Regional and global fault slip rate from seismicity. Nature, 229 (4) 101-106 (1971).
(9) Chen, W. P., Molnar, P., Seismic movement of major earthquakes and the average rate of seismic slip in Central Asia. J.G.R., 82 (20). 2954-2969 (1977).
(10) Lepichon, X., Plate tectonics, Elsevier Scientific Publishing Company, (1973).
(11) Molnar, P., Tapponnier, P., Cenozoic tectonics of Asia effects of a continental collision. Science, 189 (4201) (1975).
(12) Hubbert, M. K., Rubey, W. W., Role of fluid pressure in mechanics of overthrust faulting. B.G.S.A., 70 (2) (1959).
(13) Chang Cheng-fa, Cheng Hsi-lan, Some tectonic features of the Mt. Jolmo Lungma area, Southern Tibet, China. Scientia Sinica, (2) (1973)
(14) Zheng Xilan, Chang Chenfa, A preliminary note on the tectonic.

features of the Lower Yalu-Tsangpo River Region. Scientia Geologica Sinica (2), 116-126 (1979) (in Chinese).

(15) Yang Li-hua, Liu Tung-sheng, On the neotectonic movements in the Mt. Jolmo Lungma Region, Scientia Geologica Sinica (2), 220-225 (1974) (in Chinese).

(16) Zhao Xitao, On the recent uplift of the Himalayas, Scientia Geologica Sinica, 3, 243-252 (1975) (in Chinese).

(17) Li Ji-jun et al., On the time, elevation and formation of the uplifting of the Qinghai-Xizang High Plateau. Scientia Sinica, (6), (1979).

(18) Ichikawa, M., Focal mechanism of earthquakes occurring in and around the Himalayan and Burmese Mountain Belts. Papers in Meteorology and Geophysics 23 (3), 149-162 (1972).

(19) Rastogi, B. K. et al., Earthquake mechanisms and tectonics in the Assam-Burma Region. Tectonophy s., 18 (3/4), (1973).

(20) Fitch, T. J., Earthquake mechanisms in the Himalayan, Burmen and Andeman Regions and continental tectonics in Central Asia. J.G.R., 75, 2699-2709 (1970).

(21) Molnar, P. et al., Fault plane solutions of shallow earthquakes and contemporary tectonic in Asia. Earth Plane Sci. Lett., 16, 101-112 (1973).

(22) Rastogi , B. K. (Hyterabad, India), Earthquake mechanisms and plate tectonics in the Himalayan Region. Tectonophys., 21 (1-2), 47-56 (1974).

Table 1 Fault plane solutions for some crustal earthquakes in Himalayas

| Number | Date and time (G.M.T.) | lat (°N) | long(°E) | magni-tude | depth(km) | Plane A | | | Plane B | | |
|---|---|---|---|---|---|---|---|---|---|---|---|
| | | | | | | strike | dip | dip angle | strike | dip | dip angle |
| 1. | 1950. 8.15. 14-09-34 | 28.4 | 96.7 | 8.5 | within crust. | 65 | SE | 81 | 150 | NE | 60 |
| 2. | 1959. 2.14. 14-25-50 | 28 | 97 | 5.6 | " | 71 | NW | 75 | 357 | NE | 40 |
| 3. | 1963. 9. 2. .01-34-30 | 34 | 74.7 | 5.4 | 23 | 43 | NW | 82 | 132 | SE | 81 |
| 4. | 1964. 9. 1 13-22-37 | 27.1 | 92.3 | 5.8 | 33 | 33 | NW | 67 | 318 | NE | 60 |
| 5. | 1964. 9.26 .00-46-028 | 30.1 | 80.7 | 6.2 | 50 | 94 | NE | 35 | 130 | SW | 60 |
| 6. | 1964 10.21 .23-09-18.8 | 28.1 | 93.8 | 6.6 | 37 | 20 | NW | 71 | 298 | NE | 68 |
| 7. | 1965. 1.12 13-32-25 | 27.5 | 87.9 | 5.6 | within crust. | 90 | N | 15 | 90 | S | 75 |
| 8. | 1965. 5.31 .10-04-43 | 32.8 | 77.9 | 5.2 | 8 | 77 | NW | 69 | 162 | SW | 77 |
| 9. | 1965. 6.1 .07-52-25 | 28.6 | 83.1 | 5.2 | 10 | 87 | NW | 40 | 123 | SW | 56 |
| 10. | 1966. 3.6 .02-15-57.2 | 31.5 | 80.5 | 6 | 50 | 324 | NE | 48 | 327 | SW | 42 |
| 11. | 1966. 6.27 .10-41-09 | 29.7 | 80.9 | 6.1 | 37 | 106 | NE | 41 | 126 | SW | 49 |
| 12. | 1966. 6.27 .10-59-18 | 29.7 | 80.9 | 6 | 36 | 117 | NE | 2 | 117 | SW | 88 |
| 13. | 1966. 6.27 .11-21-43 | 29.7 | 80.8 | 5.5 | 26 | 123 | | 90 | 123 | | 0 |
| 14. | 1966. 8.5 .01-02-59 | 32.6 | 79.5 | 5.1 | 31 | 253 | NW | 80 | 340 | SW | 68 |
| 15. | 1966.10.19 .16-53-38 | 33.5 | 78.7 | 5.1 | 25 | 9 | NW | 63 | 106 | NE | 169 |
| 16. | 1966.11.6 .02-53-04 | 28.3 | 84.0 | 5.25 | 33 | 50 | | 90 | 320 | NE | 66 |
| 17. | 1966.12.16 .20-53-14 | 29.9 | 80.8 | 6.25 | 19 | 100 | NE | 24 | 100 | SW | 66 |
| 18. | 1967. 2.20 .15-18-39.1 | 33.6 | 75.3 | 5.5 | 20 | 336 | SW | 60 | 336 | NE | 30 |
| 19. | 1967. 2.21 .12-37-43 | 33.7 | 75.4 | 5.0 | 20 | 100 | SW | 60 | 100 | NE | 30 |
| 20. | 1967. 3.11 .16-56-49 | 28.4 | 94.4 | 5.2 | 15 | 37 | NW | 65 | 134 | NE | 73 |
| 21. | 1967. 5.14 .06-58-32 | 28.4 | 94.3 | 6.2 | 12 | 34 | NW | 31 | 76 | SE | 66 |
| 22. | 1967. 9.15 .10-32-40 | 27.6 | 92.0 | 5.8 | 19 | 80 | N | 87 | 354 | E | 44 |
| 23. | 1970. 2.12 .01-51-48.4 | 29.2 | 81.6 | 5.3 | 25 | 134 | NE | 50 | 143 | SW | 41 |
| 24. | 1972. 9.4 .13-42-20 | 35.91 | 73.35 | 5.7 | 57 | 112 | SW | 70 | 153 | NE | 26 |
| 25. | 1973.10.16 .09-50-43.7 | 28.36 | 82.99 | 5.0 | 34 | 110 | NE | 58 | 110 | SW | 32 |
| 26. | 1974.12.28 .22-28-18.8 | 35.02 | 73.05 | 5.0 | 47 | 140 | NE | 24 | 140 | SW | 66 |
| 27. | 1975. 1.19 08-01 | 32.39 | 78.5 | 6.2 | 1 | 24 | NW | 50 | 174 | NE | 44 |
| 28. | 1975. 7.29 02-40 | 32.57 | 78.49 | 5.5 | 5.5 | 146 | NE | 24 | 146 | SW | 66 |

Table 1

Continued

| Number | P-axis | | T-axis | | N-axis | | Reference |
|---|---|---|---|---|---|---|---|
| | az. | pl. | az. | pl. | az. | pl. | |
| 1. | 15 | 14 | 111 | 26 | 258 | 58 | Ben-Menahem, A. et al., 1974 |
| 2. | 20 | 46 | 132 | 21 | 240 | 37 | Kuo et al., 1962 |
| 3. | 266 | 12 | 357 | 1 | 88 | 76 | Ye et al., 1979 |
| 4. | 352 | 40 | 87 | 5 | 182 | 51 | Ye et al., 1975 |
| 5. | 26 | 13 | 259 | 70 | 121 | 17 | Fitch, T. J., 1970 |
| 6. | 339 | 30 | 70 | 2 | 163 | 60 | Fitch, T. J., 1970 |
| 7. | 0 | 30 | 180 | 60 | 90 | 0 | Molnar, P. et al., 1973 |
| 8. | 298 | 25 | 35 | 6 | 130 | 65 | Ye et al., 1975 |
| 9. | 17 | 9 | 264 | 70 | 111 | 19 | Rastogi, B. K., 1974 |
| 10. | 20 | 87 | 236 | 3 | 145 | 1 | Ye et al., 1979 |
| 11. | 26 | 6 | 273 | 80 | 117 | 10 | Fitch, T. J., 1970 |
| 12. | 27 | 43 | 207 | 47 | 117 | 0 | Rastogi, B. K., 1974 |
| 13. | 33 | 45 | 213 | 45 | 123 | 0 | Ye et al., 1975 |
| 14. | 299 | 22 | 205 | 8 | 92.5 | 66 | Ye et al., 1975 |
| 15. | 147 | 35 | 320 | 55 | 55 | 3 | Ye et al., 1975 |
| 16. | 97 | 17 | 2 | 16 | 230 | 66 | Ye et al., 1979 |
| 17. | 10 | 21 | 190 | 69 | 100 | 0 | Molnar, P. et al., 1973 |
| 18. | 66 | 15 | 246 | 75 | 156 | 0 | Ye et al., 1979 |
| 19. | 10 | 15 | 190 | 75 | 100 | 0 | Rastogi, B. K., 1974 |
| 20. | 359 | 30 | 266 | 6 | 164 | 60 | Ye et al., 1975 |
| 21. | 333 | 19 | 186 | 66 | 67 | 18 | Ye et al., 1975 |
| 22. | 26 | 33 | 137 | 28 | 258 | 44 | S. S. B. China, 1973 |
| 23. | 229 | 5 | 0 | 84 | 137 | 4 | Ye et al., 1975 |
| 24. | 35 | 24 | 180 | 60 | 295 | 16 | Ye et al., 1979 |
| 25. | 200 | 11 | 20 | 79 | 110 | 0 | Ye et al., 1979 |
| 26. | 50 | 21 | 230 | 69 | 140 | 0 | Ye et al., 1979 |
| 27. | 0 | 75 | 90 | 5 | 190 | 15 | Ye et al., 1979 |
| 28. | 236 | 69 | 56 | 21 | 146 | 0 | Ye et al., 1979 |

Table 2 Fault plane solutions for some mantle earthquakes in Himalayas

| Number | Date and time (G.M.T.) | lat (°N) | long(°E) | magmi-tude | depth(km) | plane A | | | plane B | | |
|---|---|---|---|---|---|---|---|---|---|---|---|
| | | | | | | strike | dip | dip angle | strike | dip | dip angle |
| 1. | 1962. 6.17 .04-39 | 33.7 | 75.8 | | 88 | 64 | NW | 60 | 155 | NE | 89 |
| 2. | 1963. 6.26 .14-09-19.0 | 36.4 | 76.6 | | 89 | 115 | NE | 88 | 115 | SW | 2 |
| 3. | 1965. 4.11 .22-33 | 26.5 | 92.2 | | 70 | 14 | NW | 89 | 104 | NE | 79 |
| 4. | 1965. 6.22 .05-49-27.7 | 36.2 | 77.6 | 5.7 | 107 | 138 | NE | 56 | 138 | SW | 34 |
| 5. | 1966. 6.27 .18-49-50 | 29.8 | 80.7 | 5.8 | 72 | 78 | NW | 65 | 160 | SW | 75 |
| 6. | 1970. 2.26 .19-30-14 | 27.6 | 85.7 | 5.0 | 96 | 35 | NW | 52 | 122 | SW | 86 |
| 7. | 1976. 9.14 .06-43-51.4 | 29.8 | 89.6 | 5.3 | 107 | 49 | NW | 78 | 142 | NE | 76 |

Table 2 Continued

| Number | P-axis | | T-axis | | N-axis | | Reference |
|---|---|---|---|---|---|---|---|
| | az. | pl. | az. | pl. | az. | pl. | |
| 1. | 23 | 21 | 285 | 20 | 156 | 60 | Ichikawa, M., 1972 |
| 2. | 205 | 43 | 25 | 47 | 115 | 0 | Rastogi, B. K. et al., 1973 |
| 3. | 60 | 7 | 329 | 8 | 189 | 79 | Ichikawa, M., 1972 |
| 4. | 228 | 11 | 48 | 79 | 138 | 0 | Rastogi, B. K. et al., 1973 |
| 5. | 31 | 7 | 297 | 30 | 131 | 60 | Ye et al., 1975 |
| 6. | 356 | 24 | 249 | 30 | 115 | 50 | Ye et al., 1975 |
| 7. | 6 | 18 | 97 | 2 | 188 | 72 | Ye et al., 1979 |

Table 3 The directions of the tectonic movements in various parts of the Himalayas

<table>
<tr><th colspan="2" rowspan="2">parameters</th><th rowspan="2">upmost mantle</th><th colspan="2">crust</th></tr>
<tr><th>western section</th><th>eastern section</th></tr>
<tr><td colspan="2">smallest eigenvalue</td><td>0.252</td><td>1.010</td><td>0.230</td></tr>
<tr><td rowspan="3">eigen-vector</td><td>X</td><td>0.790</td><td>0.723</td><td>0.852</td></tr>
<tr><td>Y</td><td>0.612</td><td>0.432</td><td>- 0.237</td></tr>
<tr><td>Z</td><td>0.036</td><td>- 0.537</td><td>0.467</td></tr>
<tr><td rowspan="4">direction of the tectonic movement</td><td>direction angle</td><td>37</td><td>31</td><td>345</td></tr>
<tr><td>dip angle</td><td>2</td><td>-32</td><td>27</td></tr>
<tr><td>root-mean angle</td><td>0.19</td><td>0.268</td><td>0.214</td></tr>
<tr><td>scaterring angle</td><td>11</td><td>15</td><td>12</td></tr>
</table>

Table 4 Seismic moments of the earthquakes in the Himalayas and adjacent regions

<table>
<tr><th rowspan="2">Date</th><th colspan="2">Location</th><th rowspan="2">Ms</th><th rowspan="2">Moment (dyne cm)</th><th rowspan="2">References</th></tr>
<tr><th>Lat.</th><th>long.</th></tr>
<tr><td>1934. 1.15</td><td>26.5</td><td>86.5</td><td>8.5</td><td>$1.1 \cdot 10^{28}$</td><td>Chen, W.P. et al.,1977</td></tr>
<tr><td>1947. 7.29</td><td>29.5</td><td>94</td><td>7.7</td><td>$9.8 \cdot 10^{26}$</td><td>"</td></tr>
<tr><td>1950. 8.15</td><td>28.5</td><td>96.5</td><td>8.6</td><td>$2.5 \cdot 10^{29}$</td><td>Ben-Menahem,A.,1974</td></tr>
<tr><td>1951.11.18</td><td>30.5</td><td>91</td><td>8.0</td><td>$4.6 \cdot 10^{27}$</td><td>Chen, W.P.et al.,1977</td></tr>
<tr><td>1964. 8.30</td><td>27.5</td><td>88.4</td><td>5.1</td><td>$9.24 \cdot 10^{24}$</td><td>S.S.B. China, 1973</td></tr>
<tr><td>1964. 9.26</td><td>30.0</td><td>80.6</td><td>6.2</td><td>$2.30 \cdot 10^{25}$</td><td>"</td></tr>
<tr><td>1965. 1.21</td><td>35.1</td><td>87.0</td><td>5.5</td><td>$9.9 \cdot 10^{24}$</td><td>"</td></tr>
<tr><td>1965. 6.18</td><td>32.1</td><td>87.6</td><td>5.0</td><td>$2.68 \cdot 10^{24}$</td><td>"</td></tr>
<tr><td>1966. 9.26</td><td>27.5</td><td>92.7</td><td>5.7</td><td>$2.03 \cdot 10^{25}$</td><td>"</td></tr>
<tr><td>1967. 3.12</td><td>28.4</td><td>94.4</td><td>5.2</td><td>$3.17 \cdot 10^{24}$</td><td>"</td></tr>
</table>

Table 5 The summation of the seismic displacement and the average slip rate in Himalayas

| parameters | Western Section | | | | Eastern section |
|---|---|---|---|---|---|
| | Main boundary f. and Ganges boundary f. | Main central f. | Back-arc region | whole section | |
| summation of the seismic displacement | 33.11cm | 4.41cm | 1.98cm | 39.5cm | 518.6cm |
| average slip rate | 0.41cm/yr. | 0.05cm/yr. | 0.02cm/yr. | 0.49cm/yr. | 5.46cm/yr. |

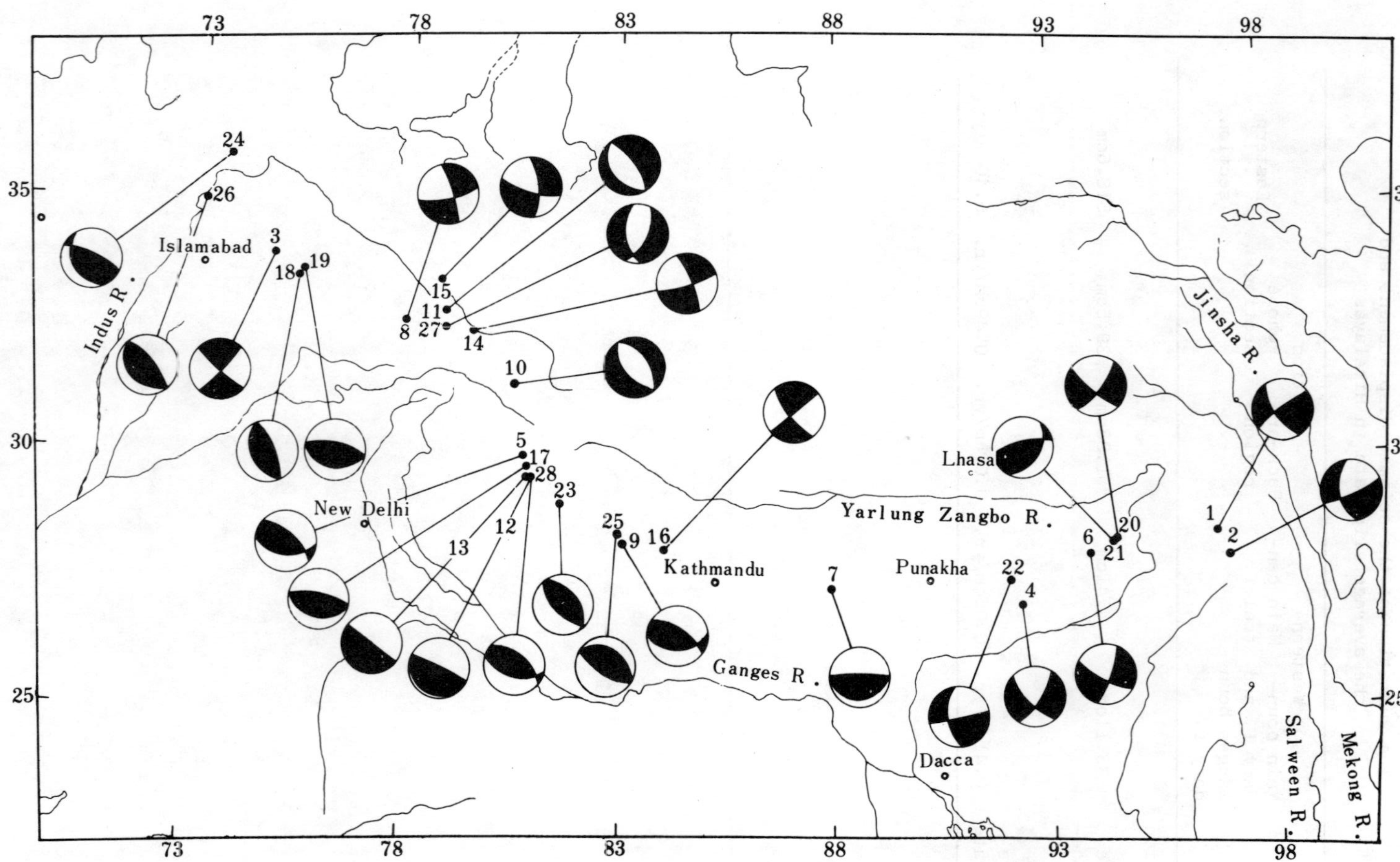


Fig. 1 Fault plane solutions of the crust earthquakes in the Himalayas (Upper hemisphere projection, black area indicates compressional first motions of P waves, the numbers of the solutions are after Table. 1)

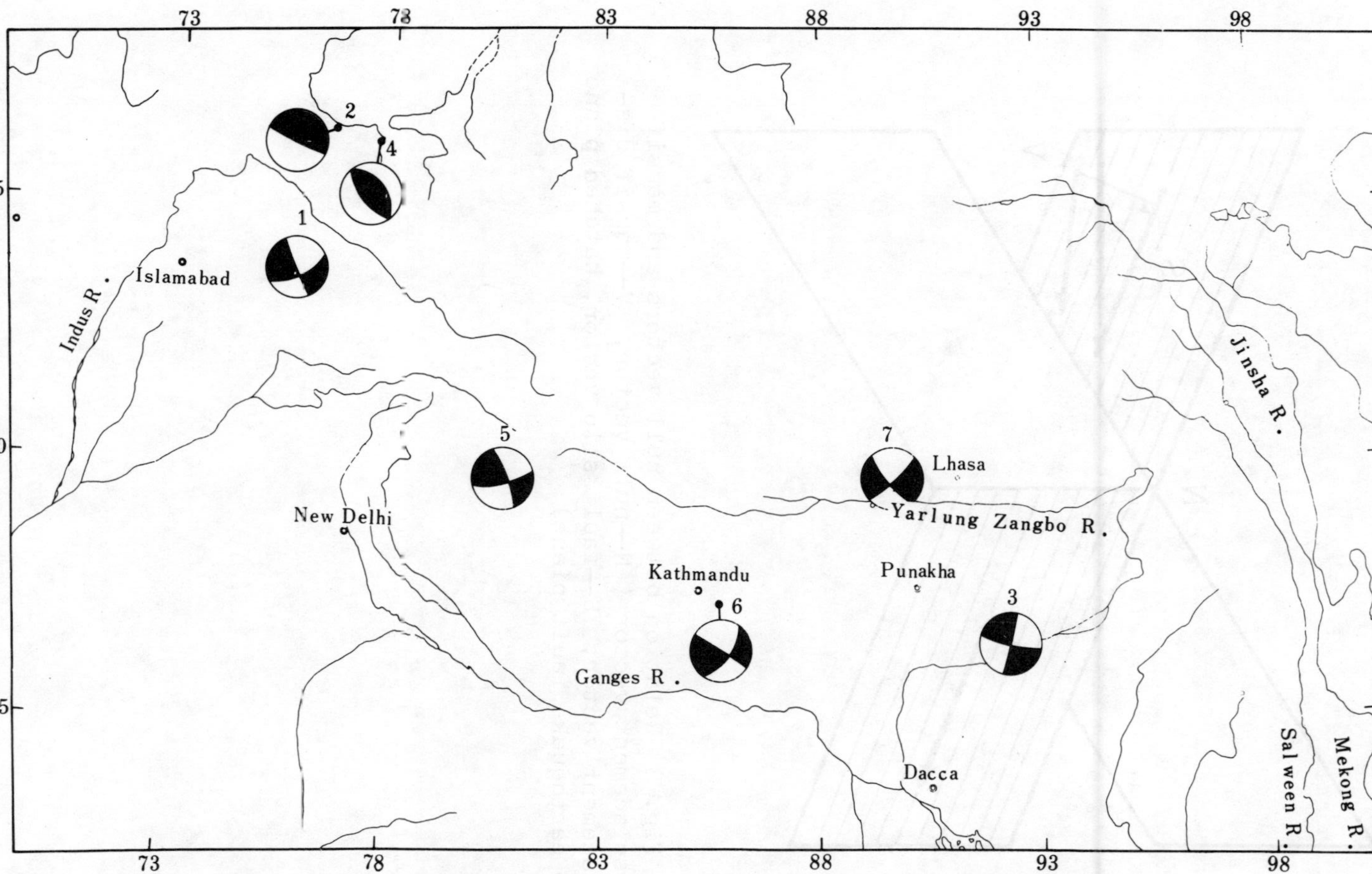


Fig. 2 Fault plane solutions of the mantle earthquakes in the Himalayas (Upper hemisphere projection, black area indicates compressional first motions of P waves, the numbers of the solutions are after Table 2)

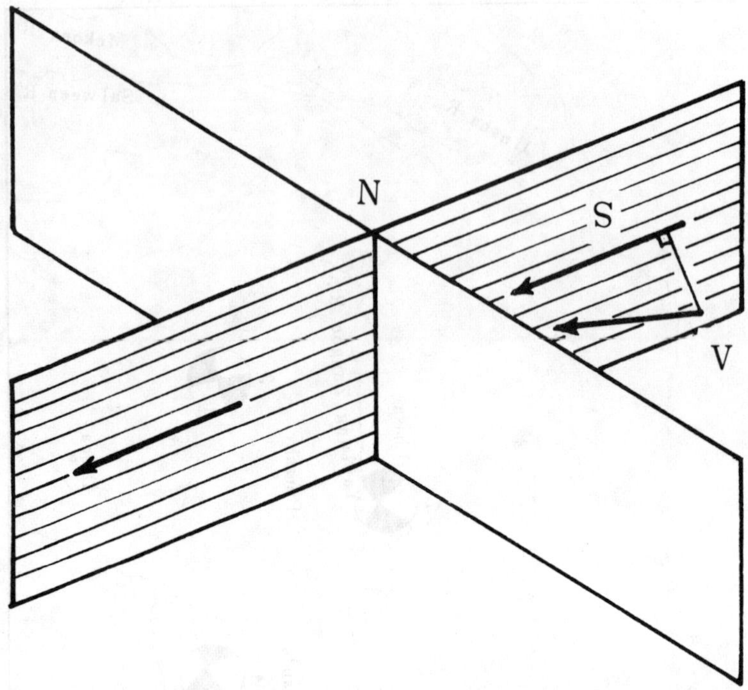


Fig. 3 Relation between null vectors and massif displacement vector ( N—null vector, V—massif displacement vector, S—fault slip vector, hatched plane — earthquake fault plane)

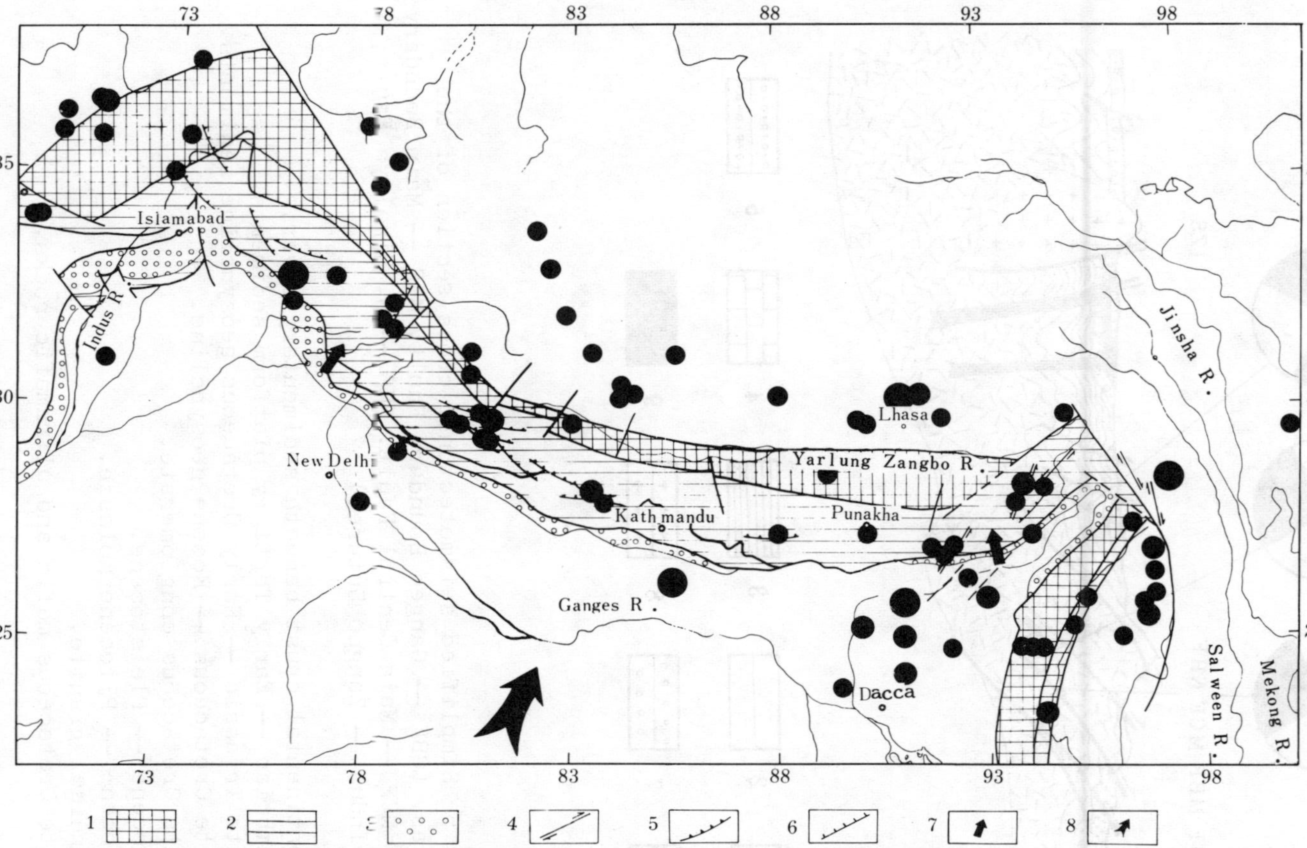


Fig. 4 Seismotectonic map of the Himalayas

1. Northern Himalayan geosycline fold belt. 2. Higher and lower Himalayan nappe belt.
3. Siwalik marginal depression belt. 4. Strike — Slip earthquake fault.
5. Thrust earthquake fault. 6. Normal earthquake fault.
7. Displacement vector of the upper lithosphere. 8. Displacement vector of the lower lithosphere.

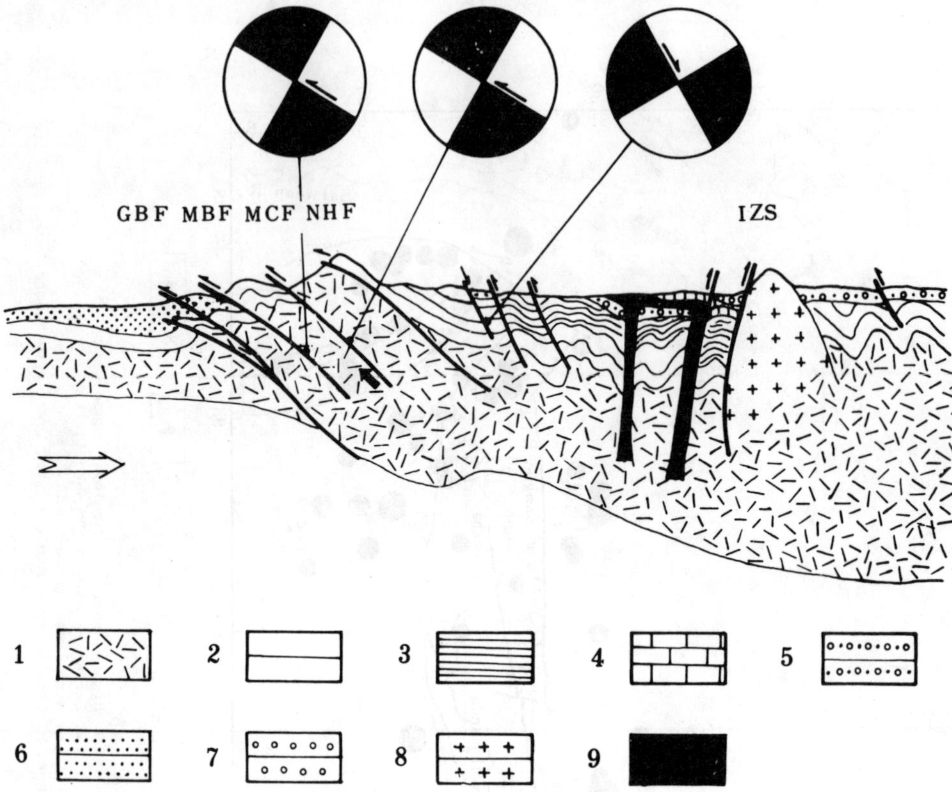


Fig. 5 Simplified seismotectonic cross section of the Himalayas. (GBF — Ganges Boundary Fault, MBF — Main Boundary Fault, MCF — Main Central Fault, NHF — N. Himalaya Fault, IZS — Indus — Zangpo Suture.)

1. Continental crust beneath sedimentary cover.
2. Cambrian — Early Tertiary platform sediments.
3. Late Triassic — Early Cretaceous geosyncline sediments.
4. Late Cretaceous — Eocene geosyncline.
5. Late Cretaceous conglomerate.
6. Miocene — Pleistocene.
7. Miocene — Pliocene molasse.
8. Gandise granite.
9. Late Cretaceous mafic and ultramafic rocks.

## TECTONIC EVOLUTION OF THE NORTH-WESTERN HIMALAYAS; IMPLICATIONS FOR SOUTHERN TIBET

Andrieux Jean, Arthaud François, Brunel Maurice, Sauniac Sophie,
Sciences Physiques de la Terre, Université of Paris 7
Dépt. Géologie Structurale, University of Montpellier

The Himalayan Range was built mainly during the last 40 millions years (Gansser, 1964; Molnar and Tapponnier, 1975) after the collision between the continental India and the Plutonic Volcanic of Ladakh. Since this time the northern border of the Indian continent deformed by reverse shear which is responsible of the classical Main Central Thrust (M.C.T.) and Main Boundary Thrust (M.B.T.). The later thrust is known to be younger than the first one and appears as a high angle fault. We are mainly concerned in this paper with the thrusting associated with the M.C.T. in the Main Himalayas.

### MAIN CENTRAL NAPPES IN THE NORTH-WESTERN HIMALAYAS

In the Central Himalayas (Nepal) it is stated that the M.C.T. is easy to trace at the base of a thick pile of gneisses in the high-amphibolite facies of Barrovian type (Le Fort, 1975, Pecher, 1977); these gneisses form the Tibetan Slab and, in the whole Himalayas, they wear a thick cover of Phanerozoïc Rocks (the Tethyan sediments).

In the Eastern and Western parts of the Himalayas the M.C.T. is not so sharply defined (Acharyya, 1977; Fuchs, 1975; Valdiya, 1978) : under the gneisses which obviously belong to the Tibetan Slab, we are able to define a number of Tectonic Units separated by thrusts. They constitute the Lower Nappes of the Himalayas. On the basis of lithological differences, three types of Units can be separated : the first type consists of quartzites, metabasalts, limestones and carbon-rich schists; the second type is made up of greywackes; and the last type, the less metamorphic one, comprises pelites and quartzites in the green-schist facies. These rock assemblages are upper-proterozoïc in age and are often intruded by late-proterozoïc the early-paleozoïc Al-rich granites, later transformed into ortho-gneisses. Underneath the Lower Nappes, the Autochtonous shows proterozoïc rocks and also sediments of upper-paleozoïc to mesozoïc age (the Blaini - Krol - Tal formations).

### TECTONICS OF THE LOWER NAPPES OF THE HIMALAYAS

A study on the tectonic style of these Nappes (Andrieux et al. 1980) shows that they have suffered the same orogenic events as the Tibetan Slab itself. They display a very conspicuous X Lineation (stretching and mineralogic lineation) parallel to the direction of thrusting, i.e. perpendicular to the general trend of the mountain belt in the orthogneisses. The shape-fabrics indicate that the bulk strain ellipsoïd, of which the X Lineation is the longest axis, is of K = 1 type; this favours a simple-shear mechanism for the strain.

From the study of the planar structural elements (schistosity, flow-cleavage, shear planes) the main steps of the deformational history can be stated as follows :

The phase 1-2 was responsible of the most conspicuous schistosity, labelled S 1-2, with a gentle dip to the North-East and formed during the stronger metamorphic climax. This S1-2 schistosity is strongly expressed in the Tibetan Slab gneisses as well as in the higher part of the Lower Nappes. In some places we observed an early schistosity folded by a second generation of folds : the second folds are associated with a new schistosity which presents the same geometrical characters as the previous one (same metamorphic paragenesis) and the earliest folds are, like the later ones, closely parallel to the X Lineation. Due to the rarity of synmetamorphic folds, the above relationships are not always available; so we prefer to speak of a phase 1-2 for all the early deformations associated with the higher metamorphism.

The phase 3 is, on the contrary, more heterogeneous; the striking structural elements are shear planes well developped close to the phase 3 thrust planes separating the various Units of the Lower Nappes; but they are also displayed in the lower part of the Tibetan Slab gneisses in such a way that we can speak of a shear-zone concerning both this slab and the Lower Nappes. During this third phase, in the upper and rear parts of the shear-zone, the shear planes are mainly parallel to the first S1-2 schistosity and then they are marked by a new cristallization of phyllosilicates and the development of quartz veins; but in the lower and frontal parts of the nappes pile, the shear planes use to be steeper or flatter than the early schistosity; the shearing strain produces locally a new cleavage, with a steeper dip than the corresponding shearing planes; the stretching lineation has always the same bearing that it had during the phase 1-2, the third phase produced folds parallel to the general trend of the Himalayas. The heterogeneous character of this phase 3 appears in the fact that the related structures develop only to the main tectonic contacts various Units, and spare the main body of these Units. Moreover, the structures related to this phase are clearly later than the highest metamorphic assemblages, and are contemporaneous of a retrogressive metamorphism.

The phase 4 is not restricted to the reverse shear-zone. Reverse faults and strike slip faults cut across the nappes contacts : the most prominent of these faults is the M.B.T. ; many other faults can be observed, not only in the Sub-Himalaya, but also in the Lower Nappes Zone and at the northern border of the Himalayas. The folding of the nappes pile by large folds with subvertical axial planes is also attribued to this phase 4; these folds are in particular well expressed as gneiss domes like the Rupshu anticline just south of the Indus Suture or like the Central Crystalline Axis.

## A MODEL FOR THE FORMATION OF THE MAIN HIMALAYAN THRUSTS

From the evidence that major thrusting intervened after the phase 1-2 deformation and metamorphism, we propose the following model of tectonic evolution for the Main Central Himalayan Shear-Zone (Brunel and Andrieux, 1980). During the first step (phase 1-2) the reverse shear was active as a homogeneous Shear-Zone, thus producing a penetrative S1-2 schistosity (fig. 2 A). Later on, as consequence of isotatic compensation and correlative erosion the S1-2 became flatter and reached a higher lever in the southern part of the Shear-Zone (fig. 2 B). Phase 3 shear movements took place along the previous S1-2 planes; new shear planes appeared in the frontal part, some steeper, some flatter than the S1-2 schistosity (fig. 2 C). As result of these new shear-planes, highly metamorphic rocks were transported farther to the south : these late thrusts explain in Kumaon, the occurrence of kyanite gneiss above greenschist facies formations . They may be responsible of the apparent "reverse"metamorphic sequences elsewhere in Himalaya .

## IMPLICATIONS FOR THE STRUCTURE OF SOUTHERN TIBET

As a consequence,the shortening in the Himalayas is not the result of a only one major shear plane (the M.C.T.) ,but of a great number of them distributed across a wide shear-zone 15 km thick. The upper part of this shear-zone consists mainly of high grade gneisses, while the lower part is made up with less metamorphic nappes. Thanks to the youngest deformations, the gneisses belonging to the Main Himalayan Shear Zone are exposed close to the Indus Suture Line (Rupshu Gneiss Dome, Nun-Kun and Suru valley gneisses) . Thus, this shear-zone has to be extended more to the North,under the Suture Line, and probably beyond, under the Trans Himalaya of Tibet. Moreover, some of the later (phase 3) thrusts gave rise to tectonic transports of about 50 km extent. For this reason the underthrusting of Indian Continent under Tibet surely plays an important part in the thickening of the Tibetan crust .

## REFERENCES


(1) Gansser A., Geology of the Himalayas, Interscience, 289p. (1964)

(2) Molnar P. and Tapponnier P, Cenozoïc Tectonics of Asia : Effects of a continental Collision. Science, 189, p. 419-426 (1975).

(3) Le Fort P., Himalays : the Collided Range. Amer. Journ. Sci., 275 A, p. 1-44 (1975).

(4) Pecher A., Geology of Nepal Himalaya : Deformation and Petrography in the Main Central Thrust Zone. Coll. Intern. C.N.R.S., Paris, n° 268, p. 301-318 (1977).

(5) Acharrya S.K., Paleogeography and Orogenic Evolution of the Eastern Himalayas. Coll. Intern. C.N.R.S. Paris, n° 268, p. 21-30 (1977).

(6) Fuchs G., Contribution to the Geology of the North-Western Himalayas. Abhard. Geol. Bund., 32, p. 3-59, (1975).

(7) Valdiya S.K., Outline of the Structure of the Kumaun Himalaya in Tectonic Geology of the Himalaya, Saklani P.S. ed., New Delhi, p. 1-14 (1978).

(8) Andrieux J., Arthaud F., Brunel M., Sauniac S., Le caractère post-métamorphique des grands chevauchements himalayens..., C.R. Acad. Sc. Paris, 291, p. 525-528 (1980)

(9) Brunel M. et Andrieux J., Sur un modèle tectonique explicatif du métamorphisme "inverse" himalayen. C.R. Acad. Sc. Paris, 291, p. 609-612 (1980)

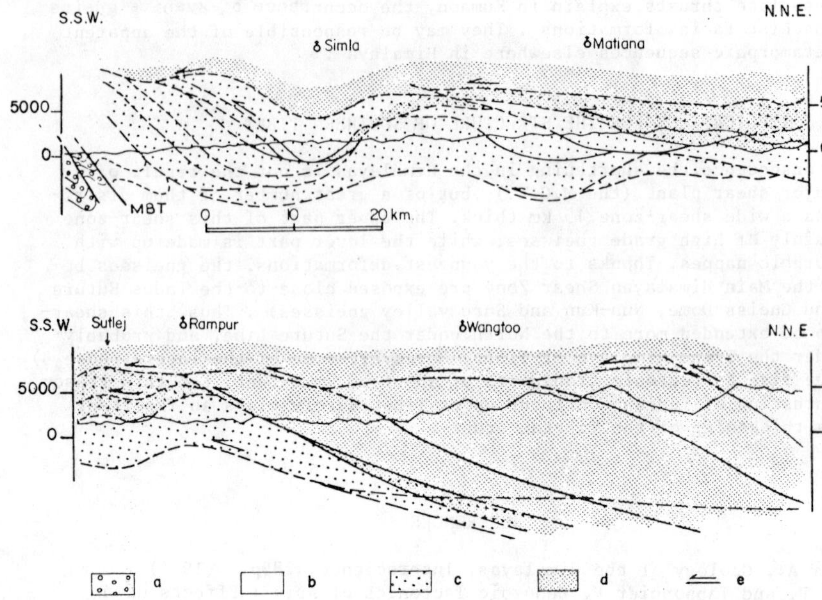


Fig. 1 : Synthetic cross-section in the North Western Himalayas. a : Sub-Himalayas; b : autochtones zone; c : Lower Nappes d : High-grade gneisses of the Tibetan Slab corresponding to the zone on strong schistosity development during phase 1-2; c+d indicates reworking during phase 3 of the base of the Tibetan Slab; e : main phase 3 shear planes (Thrusts).

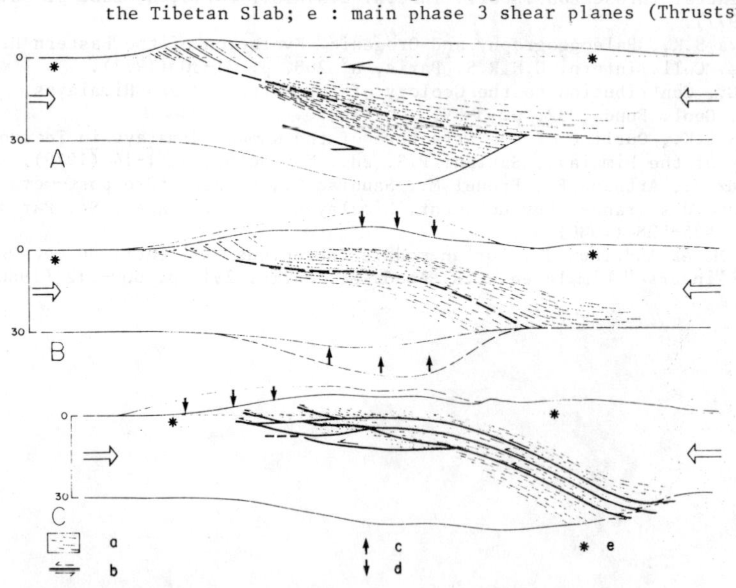


Fig. 2 : Model of the tectonic evolution of the Main Himalayan Shear Zone. a : early shear zone with S1-2 schistosity; b : phase 3 shear-plane; c : isostatic uplift; d : erosion; e : marks to appreciate the shortening.

# FORMATION AND DEVELOPMENT OF BLOCK STRUCTURES IN HIMALAYAS AND ADJACENT REGION

Li Yin-huai, Ma Fu-chen, Zhong Jia-you
(Institute of Geology, Academia Sinica)

## INTRODUCTION

Based upon the results of field investigations in Karakorum, Sulaiman and Himalayan region, and the geological and geophysical data of North Xizang of China and North India, the authors attempt to divide and describe the tectonic belts and features in the Himalayas and adjacent region, and also discuss the nature of Yarlung Zangbo River-Indus River ophiolitic belt and the deep structural features in this region. A new model of formation and development of block structures in the Himalayas and adjacent region is presented.

## DIVISION OF THECTONIC BELTS

The geological structure of the Karakorum, the Himalayas and the Xizang Plateau shows a clear zonation. According to the stratigraphic development, sedimentary formations, magmatic activities and metamorphic processes, as well as geological history, the studied region may be divided into three tectonic belts from north to south (Fig. 1).

### Northern Karakorum-Nyainqentanglha Tectonic Belt

Stratigraphically the Precambrian Nyainqentanglha group is distributed in Nyainqentanglha area. It represents a metamorphic series and consists of crystalline schists, amphibolites and gneisses.

The recent investigation in Xainza area, northern Xizang, has found a comparatively complete Palaeozoic stratigraphic profile (Fig 2), including sediments from Ordovician system to Permian ( Comprehensive Geological Team, Xizang Geological Bureau, 1980 ). This complex of sediments is a marine carbonate-clastic formation about 3,500-4,000 m in thickness. From lithological and paleontological associations it can be correlated to those in South China, and similar to that in Mt. Qomolangma area. And the Palaeozoic sediments exposed in Zhongba area on the southern side of the Yarlung Zangbo River have a similar characteristics.

In the Mesozoic era, a tectonic framework of alternate occurrence of eugeosyncline and miogeosyncline has appeared in this belt. In Lhasa area the Triassic and the Lower Cretaceous consist mainly of marine sediments, carbonate rocks with interbeds of sandstones and shales in the lower part, and clastic rocks with marls and coal series in the upper part(6). Lithological characteristics of this sedimentary series are given in Fig. 3.

The Middle Jurassic marine flysch beds are developed along the Bangong Lake-Gerze-Dongqiao deep fault. They consist of volcanic and siliceous rocks subjected then to regional metamorphism. On the southeastern slope of the Main Peak of Muggarkangri Mt., northwest of Nyima, this complex of sediments is called Muggarkangri group (Wen Shixuan, 1979). In the vicinity of Dongqiao this group has been intruded by ultrabasic rocks, which is covered by Late Jurassic Dongqiao formation in probably unconformable contact. The Dongqiao formation consists of chromspinel sandstones pebbly sandstones with interbeds of carbonaceous shales and coal seams. The upper part of this group is limestones with foraminiferas, and in basal conglomerates,gravels of ultrabasic rocks have been found.

In this belt, the Late Cretaceous-Early Tertiary Gulu formation consists of a continental purple-red sandstones-conglomerates with intermediate volcanic series, and Early Tertiary marine sediments are probably included at its top. This formation lies in angular unconformable contact upon underlying sediments and Gandise granites. This tectonic unconformity represents an intensive orogenic episode. Researchers engaged in this work call it Gandise orogenic movement. This movement accompanied with extensive intrusions of Gandise granites which formed a granitic belt up to 2,000 km in length. The isotopic dating of the main massive gives the age of 80-130 m.y. (K-Ar Dating Group of the Institute of Geology, Academia Sinica, 1979).

### Southern and Middle Karakorum-Northern Himalayan Tectonic Belt

According to materials from Comprehensive Geological Team of the Xizang Geological Bureau, the Upper Triassic Longgexie group lies discomformably or unconformably on Permian system in Kangmar and Ngamring areas. The Langexie group represents a type flysch deposits up to 10 km in thickness. In its lower part the radiolarian siliceous rocks and spilitic keratophyre are the lowest horizons in this belt.

The Langexie group extends westerly to Lhase area where it is called Jilong group. This group in Dras area is called Lamayuru flysch(2).

The Upper Jurassic-Lower Cretaceous Sanri group represents a complex of shallow-sea basic and intermediate-acidic volcanic rocks, clastic rocks and biogenic limestones with spilitic keratophyre in its lower part. Its thickness reaches 7,000 m.

The Upper Cretaceous Xigaze group is an epicontinental sedimentary formation to the north of Main Yarlung Zangbo River fault. Conglomerates about 300-1,000 m in thickness occur in its basal part. Upward there lies a sequence of interbedded sandstones and shales with conglomerates and volcanic rock intercalations about 1,500 m in thickness. In the Yajue and Qiuwu areas, this group occurs in distinctly unconformable angular contact with underlying Early Cretaceous volcanic rocks and intruded by the Gandise granites (Wu Hao-ruo et al., 1977). This unconformity is a result of Gandise orogeny.

In the south of Main Yarlung Zangbo River fault, the strata being similar to the Xigaze group, are a pelagic sedimentary formation, consisting mainly of sandstones, slates, limestones,volcanic rocks, pillow lavas and radiolarian siliceous rocks.

The Xigaze group extends westerly to the Kailas area, where it is called Kailas flysch by Gansser (1964). In Ladakh area, the Late

Cretaceous ophiolites are separated into two branches. The northern branch is distributed along the northern border of Ladakh granitic belt and on the Shyok River(2), and its extension in Skardu area joins with Bomaher group, which consists of basic and ultrabasic rocks, metatuffs and sediments with Cretaceous spores and pollens(1). The southern branch extends along the southern border of Ladakh granitic belt, and it is known as the Indus flysch and Dras volcanic rocks. The latter consists of 2,000 m - thick volcanic and clastic rocks and slates, radiolarian flinty rocks, limestones with *Orbitoides*, as well as diabase and serpentinites. As Frank and his colleagues (1977) have shown, the Dras volcanic rocks are gradually replaced by Indus flysch.

The Xigaze group around Nanjiabarbat massive has been more deeply metamorphized. This group in Southern and Middle Karakorum is divided by authors into two zones. The northern zone extends on the south of granitic belt along the mountain axis. In Hunza area, it is called Chalt formation (1). This formation joins easterly with Bomaher group, and westerly in Yasin area, it is composed of interbedded limestones, sandstones, basic volcanic flow, agglomerates and tuffs. The Chalt formation extends from Yasin area southwesterly to Udur, northwest of Kalam, and Dross and Dir, southeast of Dir. Jan and Mian (1971) called it Udur volcanic series which consists of andesites, dacites, associated with ignimbrites, tuffs, agglomerates and siliceous rocks. They form a typical calc-alkaline series. In the vicinity of Roder area, Hayden (1915) found some *Orbitoides* in this series.

The southern zone extends along the northwestern border of Nanjiabarbat massive, and consists of a complex of ňorites, gabbro, diorites, ultrabasics and migmatites without sedimentary interbeds. In Middle Indus River area, Desio (1974) called it Middle Indus noritic group. It extends southwesterly to Upper Swart area where it is called Upper Swart amphibolitic group (Martin et al., 1962). Jan and Kempe (1973) determined the age of these rocks to be the end of Cretaceous period.

The piedmont molasse formation of Miocene and Pliocene distributed intermittently along the southern border of Gandise granitic belt, Kailas granitic belt and Ladakh granitic belt, and named respectively as Luobusa group, Kailas conglomerates and Indus molasse. They overlie unconformably the above-mentioned granitic belts, Xigaze group, Kailas flysch, Indus flysch and the ultrabasic rocks intruded in. This tectonic unconformity is the result of the Shannan orogenic movement.

The Shannan movement was accompanied by granitic intrusion along the axis of 900 km in length. The isotopic dating gives its age of 8.6 m.y. (Desio, 1972). Some minor two-mica granitic intrusions are exposed in the northern Himalayan geosyncline, with an isotopic age varying between 11.3 - 32.2 m.y. (K-Ar Dating Group, Institute of Geology, Academia Sinica, 1979).

The above mentioned data show that this belt was a mobile geosyncline, as shown by the characteristics of the sedimentary formations. From Late Triassic to Late Cretaceous there were three formations of spilitic keratophyre and volcanic siliceous rocks. The most extensive one is the Late Cretaceous formation along the Yarlung Zangbo River-Indus River fault. The later intrusion represents basic and ultrabasic rocks. This petrological association together with Late Cretaceous flysch forms a disorder ophiolitic suite characterized by composition of neogenic oceanic crust.

## Thrust-Overthrust-Nappe Tectonic Belt along the Sulaiman-High and Low Himalayas Platform Border

The Precambrian Nyalam group in the Mt. Qomolangma area consists of 10 km-thick schists, gneisses and migmatites.

Overlying metamorphic rocks is an unmetamorphized series. According to the present available data fossiliferous Cambrian rocks are exposed only in Kashmir and Salt Range areas. A more complete marine sedimentary series with aboundant fauna from Ordovician to Eocene are exposed on the northern slope of Mt. Qomolangma (Fig. 4)(6).

In the Kirana area, south of Salt Range, there is a sequence of closely folded clastic rocks with rhyolites. The isotopic age of these rocks is 870 m.y. Here the platformal sedimentary cover begins at Malani rhyolites of 745 m.y.(5). The Jodhpur sandstones occur conformably upon it and change to the Eocambrian salt-bearing system and Cambrian system with trilobites, which are then unconformably overlain by erratic Late Palaeozoic Tobula formation.

The Triassic and Eocene deposits developed in Sulaiman area are also marine sedimentary series. From lithological characteristics and paleontological associations this series is generally similar to that in High Himalayan area.

There are also several thrusts and overthrusts of different sizes in the Sulaiman region. At the vicinity of Quetta,the Quetta fault led the Triassic limestones to overthrust the Eocene shales with Nummulites. In the Jiangeto area the Quetta fault led Late Cretaceous Gewar melanges to overthrust the Siwalik group. In Muslimberg area there are a number of imbricate thrusts in the Cretaceous and Jurassic terrains. All these thrusts have a slip from northwest to southeast.

The above-mentioned facts show that the Precambrian rocks in this belt is characterized by miogeosynclinal sedimentary formations. They were subjected to intensive folding and metamorphism during intervals of 660 - 700 m.y. and of 800 - 900 m.y. in the High and Low Himalayas and Sulaiman region. From Cambrian to Eocene, most of the deposits are marine. They are continuous in the High and Low Himalayan areas and lack of Palaeozoic sediments in Sulaiman region. Their lithological constitution is generally simple, but their integrated thickness is large. It shows that their sedimentary environment were platform margin. The thrusts, overthrusts and nappes are the important tectonic features in Sulaiman and High and Low Himalayan regions. Therefore, the authors call this belt a thrust-overthrust-nappe tectonic belt.

## THE NATURE OF THE OPHIOLITES

It has been suggested that the Yarlung Zangbo River-Indus River ophiolitic belt is a relict of the squeezed-out oceanic crust, and resulted from "cold" intrusion.

Authors have found that in the lower basin of Hunza River, the sequence of ophiolitic association of the northern subbelt of Chalt formation is : conglomerates and sandy conglomerates in basal part lie unconformably or disconformably upon the Dumodor formation, upward occur the interbeds of phyllites and thin bedded quartzites, chlorite schists and metamorphozed sandstones with metatuffs. They are interbedded with flysch, siliceous rocks, pillow lavas and gabbrodiabase (Fig. 5).

In the Gunar area, along the Indus River, there is a number of peridotite massives intruded into norites. The peripheries of these intrusions possess several apophysises and veins cut into the country rocks. Here ophioloites do not form the basal unit of the profile, but occur in its upper part instead. In the ophiolitic association, norites were formed earlier than ultrabasic rocks. This ophiolitic association differs distinctly from typical ophiolitic sequence (Gass, 1963).

The Late Cretaceous ophiolitic suite in southern Xizang consists of ultrabasic rocks, gabbro, spilites and andesites, as well as radiolarian siliceous rocks and flysch. From the petrographic association this suite is characterized by composition of oceanic crust. However, the geological data from Xizang show the Robusha and Zetang massives were intruded in the Late Cretaceous and Late Triassic sediments. In the south of Jiachajiesa area, a thermo-metamorphic zone of several meters in width is found around an exposed massive. And granites and vesuvianites are found in this zone. This massive contains xenoliths of phyllites and crystalline limestones. The gravity data show a gravitational high of sublatitudinal direction on Robusha massive. It reflects the extension of massive downward to a great depth, which indicates that the Yarlung Zangbo River ultrabasic belt does not form the lower part of these ophiolites, but constitutes the upper part of volcanic and sedimentary rocks as a consequence of late intrusion. It follows that this ophiolitic sequence is different from typical ophiolitic profile.

The above mentioned data indicate that the Yarlung Zangbo River-Indus ophiolitic belt is not a relict of the squeezed initial oceanic crust, but is a complex of effusive and intrusive products of melted material arosed from the upper mantle to the upper crust through the extension of deep faults that formed the neogenic oceanic crust. Since then, the compression of Cenozoic orogenesis led the ophiolitic belt undergoing repeated tectonic reconstruction and new emplacement of different extent. So we can not consider this secondary tectonic contact as a basis for the generation of cold intrusion.

## DEEP STRUCTURAL FEATURES

In the Pamir, Karakorum and Himalayas regions, a survey of deep seismic sounding of 1,500 km in length was completed by Indian, Pakistani and Italian scientists in cooperation in 1976. According to this result and in connection with gravity and aeromagnetic data, Beloussov (1979) compiled a structural profile of deep crust in the Tianshan-Pamir-Karakorum-Himalayas region (Fig. 6). He suggested that the Moho surface beneath Pamir, Karakorum and Himalayas Mountains occurs in the form of a great depression at depth about 75 km in Pamir area, and 70-73 km beneath Indus flysch and ophiolitic belt. This Moho surface arises gradually from south to north. He also suggested a complex, multi-layering and blocky structure of the vertical section of the lithosphere, consisting mainly of layered crust, upper mantle and asthenosphere. This layered structure has been cut by great deep faults into several blocks. These faults dip steeply to south or north. The Indus ophiolitic belt has a dip to south at high angle and down to the depth of 70-75 km.

Field investigations have also proved that the Indus ophiolitic belt dips to south at high angle. It is reflected in the steep southward plunge of fold axis in volcanic rocks and sediments among the ophiolites (Fig. 5). The eastward extension of the Indus ophiolitic belt expresses a series of high-angle thrusts with their dip to south in Ladakh area (Fig. 7) (Gansser, 1976).

The investigation on the Yarlung Zangbo River fault by the Geological Team of the Wuhan Geological College has shown that the plane of Yarlung Zangbo River fault dips to south at least in the east of Zhongba area. Analysis of the data on the depth of earthquake sources in the Qinghai-Xizang Plateau and adjacent region by Huan Wen-lin (1980) indicates that the plane of Yarlung Zangbo River fault dips to south and reaches a depth of 280 km.

The above-mentioned facts indicate the absence of double layered crust in the Middle Asian mountains. The Yarlung Zangbo River-Indus fault does not dip to north, but to south. Therefore, the formation and development of geological structures of the Middle Asian mountains may not be considered as the result of underthrusting of Indian Plate to north, but might be a result of extension and compression of both Indian block and Tarim block.

## THE MODEL OF THE FORMATION AND DEVELOPMENT OF BLOCKS

According to the regional geological features, paleontological associations and the distribution of Late Palaeozoic moraine deposits, as well as data on deep structure, the history of development of blocks in this region can be divided into four main stages.

### Stage of Formation of the Precambrian Basement

The Precambrian metamorphic rocks are widely distributed in the High and Low Himalayas area. The lower part of these rocks is more metamorphozed and consists mainly of various schists and gneisses, and the upper part is less metamorphozed and consists of slates, phyllites with intercalations of quartzites and stromatolitic dolomitic limestones.

On the north of Yarlung Zangbo River the Nyainqentanglha group consists also of metamorphic schists and gneisses. In the adjacent area unmetamorphozed Ordovician deposits of the platform-type are developed. Its lithological characteristics are quite similar to those of the Himalayas area. Therefore, this metamorphic group together with Nyalam group and North Col formation represents the Precambrian basement in this region.

### Stage of Development of Palaeozoic United Platform

As mentioned above, Palaeozoic deposits at both southern and northern sides of Yarlung Zangbo River do not lack the sedimentary nature in general, though they are different in lithological associations and extent of development. The finding of Late Carboniferous-Early Permian marine moraine conglomerates on the north of Yarlung Zangbo River has been reported continuously in recent years (Fig. 8).

The moraine sediments and paleontological associations show that the South Asian hinterland in that time was covered by Tacher moraine sediments of continental glacial facies (Sarbadhikari, 1973). The Bulani moraine sediments of glaciomarine facies accumulate in the High and Low Himalayas areas (Puscoe and Oldham, 1959), and the marine ice-floe sediments distribute in the Northern Karakorum, Lhasa and Qomdo and Horbaco region (Norin, 1964).

The distribution of moraine materials may prove that the Indian continent did not drift from a distance of several thousands kilometers since Late Cretaceous, though it is not a tectonic indicator.

The development of Palaeozoic sediments at both southern and northern sides of Yarlung Zangbo River, sedimentary features and distribution of moraine materials show that the studied region in that time was a united stable platform (Fig. 9).

Stage of Development of Geosyncline in Late Cretaceous- Eogene

In Late Triassic, the platform began to break up. The terrains in Southern and Middle Karakorum and Himalayan regions in the platfrom became a geosyncline. In the whole period of its development, this geosyncline subjected to two tensions and two compressions and then came to a closure (Fig. 10).

The first tension occurred in Late Triassic, Jurassic and Cretaceous. It represents the formation of spilitic keratophyre, siliceous rocks and flysch formation.

The first compression is the Gandise orogeny, It caused the whole reversion of Northern Karakorum-Nyainqentanglha tectonic belt, and the partial reversion of Northern Himalayan geosyncline accompanied with extensive intrusions of Gandise and Ladakh granites.

The second tension occurred in Late Cretaceous, and represents the formation of disorder ophiolitic suite of the neogenic oceanic crust.

The second compression is the Shannan orogenic movement occurred after the Eocene and before the deposition of Neogene Robushi group, and corresponds to the Pliocene. It accompanied with the intrusion of granites along the axis of Karakorum Mts. and two-mica granites in the Northern Himalayan area.

The Northern Himalayan geosyncline subjected to partial reversion in the Gandise orogeny, and completed reversion in the Shannan movement, and then completed the development of geosyncline.

It follows as mentioned above , there was a complete tectonic reversion in this belt, which differs from the previously named "Himalayan tectonic cycle" in broad sense, and we call it the Early Himalayan orogenic cycle.

Recent Orogenic Stage

The compression of Northern Himalayan geosyncline in the Early Himalayan orogenic cycle led the High and Low Himalayas and Siwalik terrains to be extended. It was very intensive especially in Siwalik area, where the thick Siwalik group was developed (Fig. 11).

After Early Pleistocene the High and Low Himalayas and Siwalik terrains were in the process from tension to compression, so that the Main Central Fault and Main Boundary Fault thrust and overthrust , and the latter led the deposits to overthrust the Siwalik group.

The Siwalik group is a complex of thick continental deposits, though it is not a result of geosynclinal deposition. In the group, folds, thrusts and nappes are quite extensive, forming a complete orogenic cycle of nongeosynclinal sedimentation, which differs from the previously named "Himalayan cycle" in broad sense, and is called Late Himalayan orogenic cycle.

It follows as mentioned above that since the Indosinian orogeny, the united Palaeozoic platform began to break up, and a Mesozoic geosyncline developed in the Northern Himalayan area. As a result of Yenshanian tectonic cycle, Early Himalayan tectonic cycle and Late Himalayan tectonic cycle, several regularly aligned orogenic belts were formed successively in this region from north to south.

The High and Low Himalayas block overthrust the Indian block and the Northern Karakorum-Nyainqentanglha block. Dip angle of the Main Boundary Fault in the south of this block is smaller than that of Indus-Yarlung Zangbo River fault in the north. Both eastern and western ends of this fault are bounded by the Miju fault and Kabul--Peshawar fault respectively. Therefore, here an independent but asymmetric wedge-block was formed.

A model of the development of blocks in the studied region can be summarized briefly as follows: In the early stage of the development of geosyncline, two adjoined blocks were departed away by the tension at a large distance. The crust became thin and the upper mantle up welled, so that the neogenic oceanic crust was formed. At the same time, the Moho surface arched, and its top was subjected to further tension, but two adjoined blocks were subjected to compression. In the late stage of the development of geosyncline, two adjoined blocks were compressed oppositely, so that the neogenic oceanic crust was shortened and thickened. During this period the Moho surface turned from the previous uplift to the depression, its depressed part was subjected to compression, and two adjoined blocks subjected to tension. Therefore, the authors suggest that during the tension period of the blocks the oceanic crust (geosyncline) was formed, and during the compression the continental crust (orogene) was formed, which further proves as Prof. Zhang Wen-you has repeatedly mentioned in recent years that the continental crust can be broken up in the tension to form the grabens and to create the microoceans or oceans, and the oceanic grabens can be squeezed in the compression to form the island arcs or marginal mountain ranges which later become continents. Thus, the continent and the ocean can be transformed into each other. There is a complete geotectonic cycle from depression and accumulation to folding and reversion of geosyncline. The termination of this cycle means the initiation of the development of other geotectonic cycle. Therefore, the blocky structure of this region was formed and developed as shown in this evolutional model.

## ACKNOWLEDGEMENTS

The authors are very grateful to Prof.Zhang Wen-you for his warm concern and concrete guide to the completion of this work. The authors thank also Zhao Xi-tao for his warm help in preparation of this manuscript.

## References


(1) Desio, A. et al., On the geological age of some granites of the Karakorum, Hindu Kush and Badakhshan (Central Asia). 22nd Intern. Geol. Congr., New Dehli, 11, 479-496 (1964).

(2) Frank, W. et al., Geological observations in the Ladakh area (Himalayas). A preliminary report. Schweiz Min. Petr. Mitt., 57, 89-213, 18 figs., 1 coloured geol. map. (1977).

(3) Gansser, A., The great zone between Himalaya and Tibet. A preliminary account. in: Himalaya, Colloque Intern. 268, CNRS ed. (Paris), 181-191 (1977).

(4) Marting, N. R. et al., A geological reconnaissance of the region between the Lower Swat and the Indus River of Pakistan. Geol. Bull. Panjab Univ., 2, 1-13, Lahore (1962).

(5) Stocklin, J., Structural correlation of the Alpine ranges between Iran and Central Asia. Soc. Geol. France, 12 (8), 333-353 (1977).

(6) Organizing Committee, Symposium on Qinghai-Xizang (Tibet) Plateau. Academia Sinica, A scientific guidebook to South Xizang ( Tibet ). Beijing (1980).

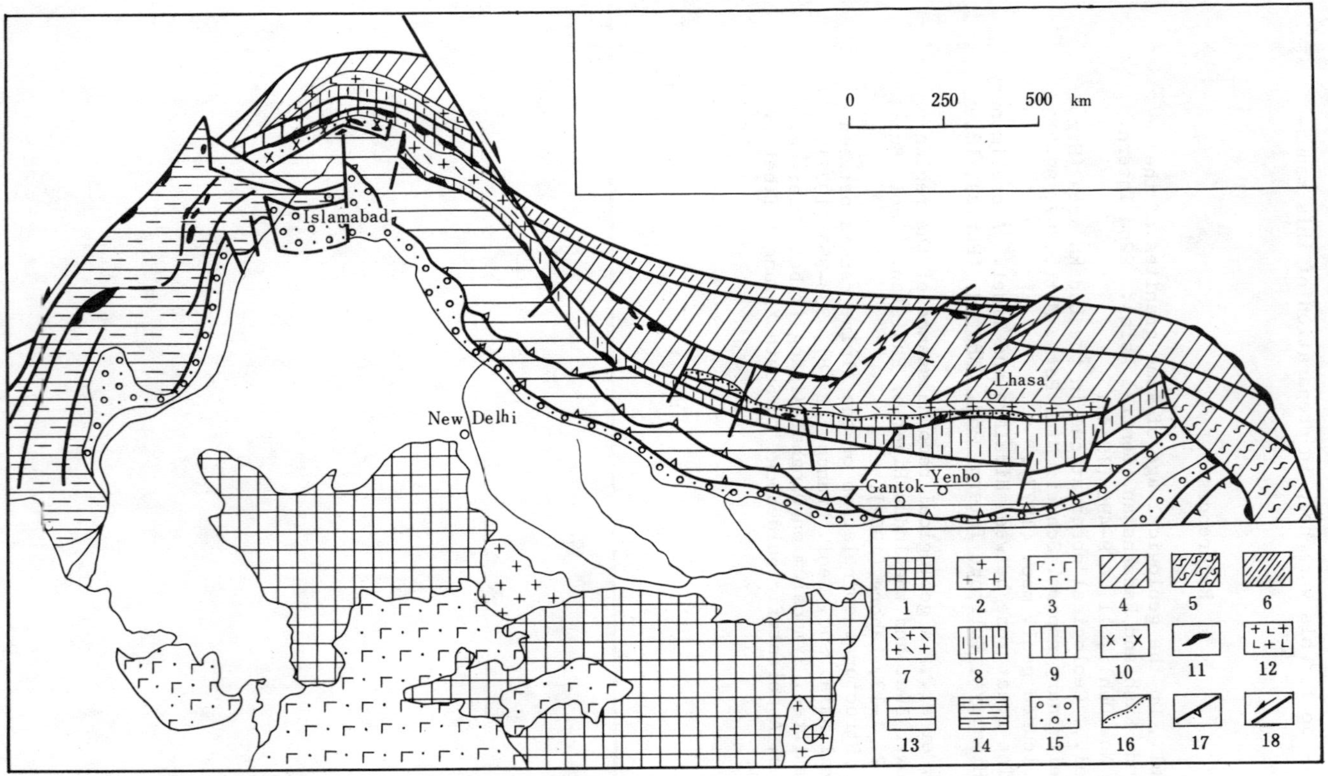


Fig. 1 Map of geotectonic outline of the Himalayas and adjacent region
1. Tectonic belt of Pre-Sinian basement; 2. Pre-Sinian granites; 3. Late Cretaceous-Eocene Deccan trap rocks; 4. Palaeozoic-Cretaceous miogeosynclinal sediments developed on the Precambrian metamorphic basement; 5. Tectonic belt reworked intensively by the Yenshanian movement; 6. Mesozoic eugeosynclinal sediments; 7. Yenshanian granites; 8. Late Triassic-Late Cretaceous eugeosynclinal sediments; 9. Late Cretaceous-Eocene sedimentary belt of marginal sea; 10. Late Cretaceous Mid-Indus River norite group; 11. Ultrabasic rocks; 12. Himalayan granites; 13. Cambrian-Eocene sediments along the platform margin in the High and Low Himalayas area; 14. Eocambrian-Eocene sediments along the platform margin in Sulaiman region; 15. Siwalik group; 16. Uncomformable boundary; 17. Thrust, overthrust and nappe; 18. Strike-slip fault

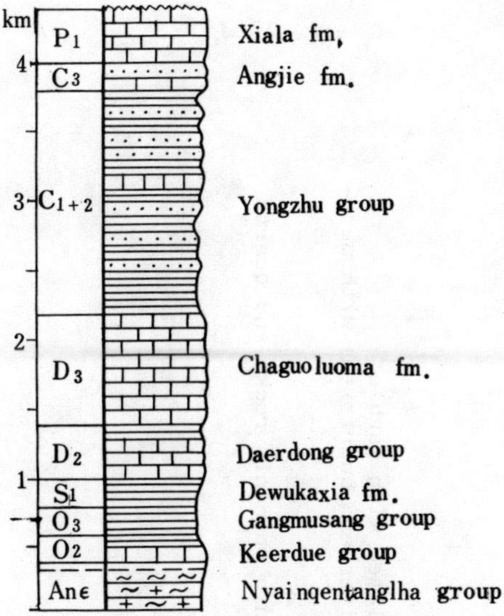


Fig. 2 Stratigraphic column of Palaeozoic sediments in Xainza area, Northern Xizang

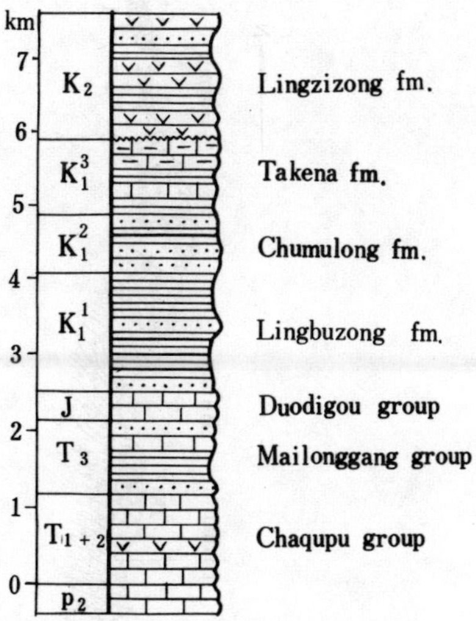


Fig. 3 Stratigraphical columnar profile of Mesozoic sediments in Lhasa area

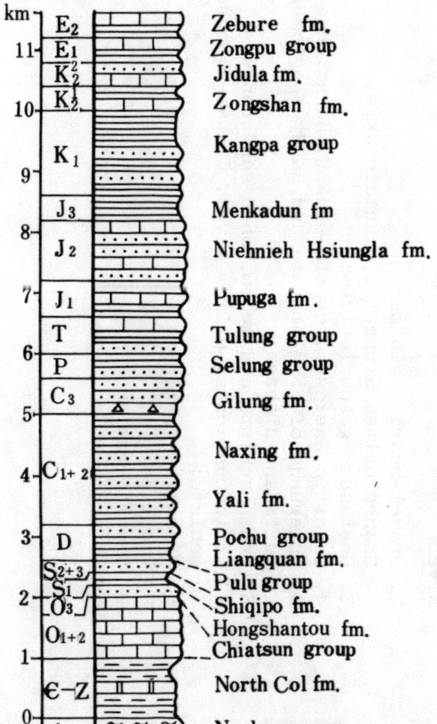


Fig. 4 Stratigraphical columnar profile of the Ordovician - Eocene deposits on the northern slope of Mt. Qomolangma

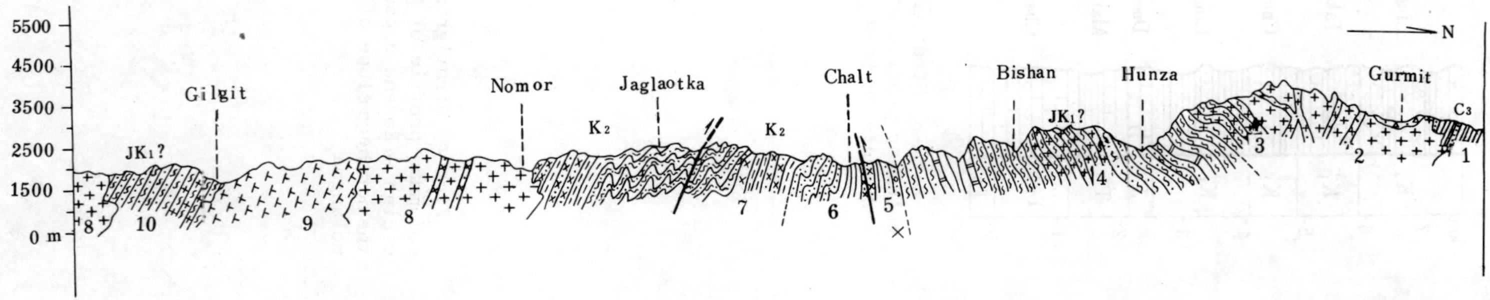


Fig. 5 Geological profile of ophiolites of Chalt formation in the lower basin of Hunza River

1. Gravel-bearing slates;
2. Granodiorites along the mountain axis;
3. Gneissose granodiorites;
4. Granitic Gneisses, quartz-schist, hornblende gneiss intercalated with marble;
5. Interbeds of conglomerate and sandy conglomerate and phyllites and quartzites with meta-tuffs and diabase;
6. Green schists and metamorphic sandstones;
7. Siliceous rocks, pillow lavas, diabase-gabbro with metamorphic volcanic rocks and quartz schists;
8. Migmatite-granites;
9. Diorites;
10. Granite-gneisses

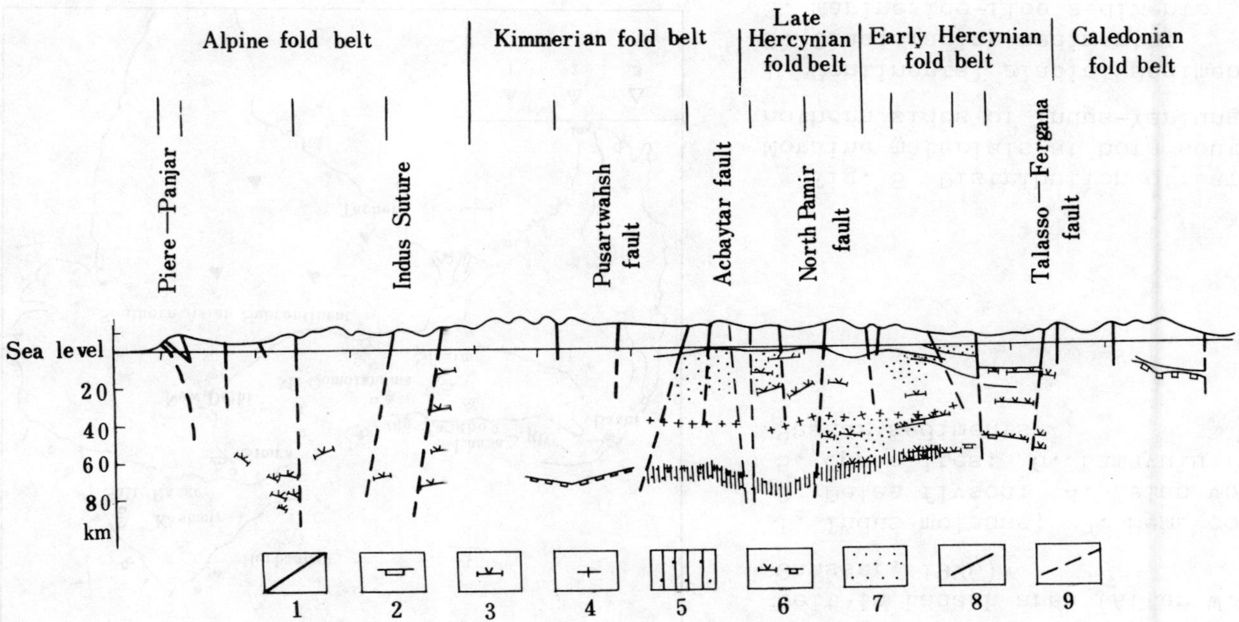


Fig. 6 Seismic profile of the crust along the Toktokur-Srinaga line (After S. A. Aliev, 1976) Isovelocities:

1. Determined from hodographs of refraction and reflection waves;
2. Determined from hodograph of refraction waves by explosion;
3. Determined from hodograph of reflection waves by explosion;
4. Determined from hodograph of inverted waves by earthquake;
5. Determined from seismic data of deep-focus earthquakes;
6. Moho discontinuity;
7. Relatively high density zone at the points of inversion;
8. Fault determined from geological data;
9. Fault determined from seismological data

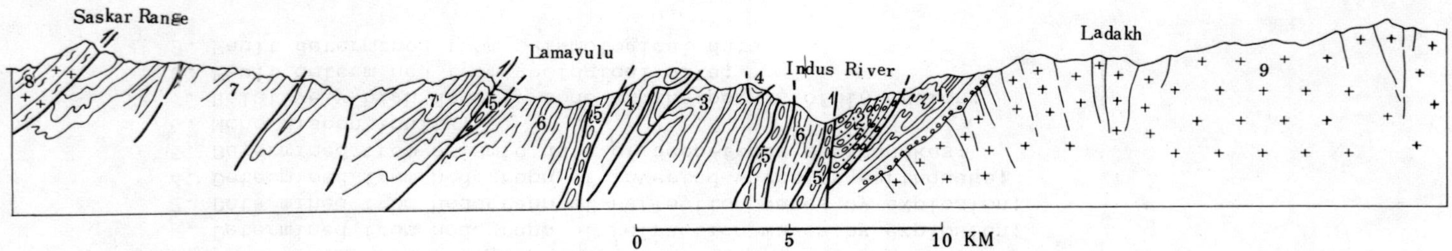


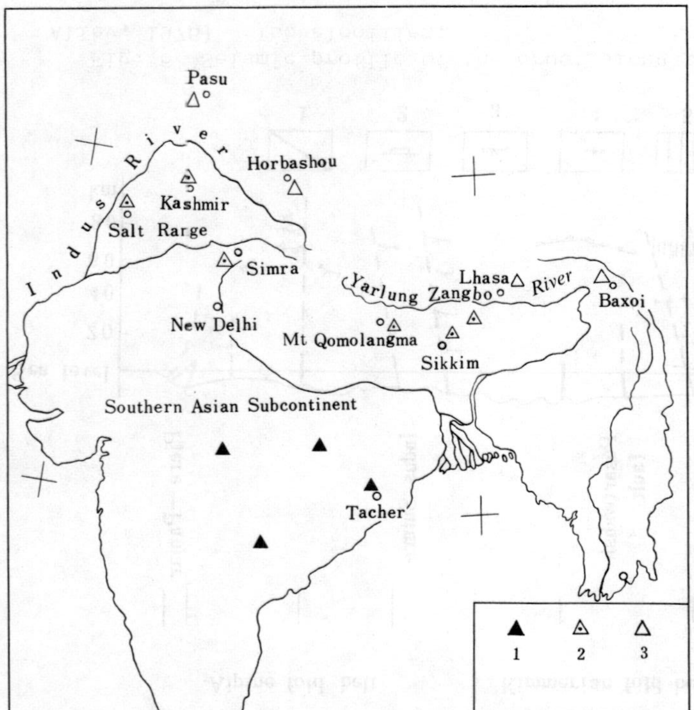


Fig. 7 Structural profile of ophiolitic belt in Ladakh area (After W. Frank and A. Gansser, 1976)

1. Indus molasse; 2. Hems conglomerates; 3. Delas flysch; 4. Delas volcanic rocks; 5. Ophiolites; 6. Lamayulu flysch; 7. Saskar sediments

Fig. 8 Distribution of Late Palaeozoic Moraine materials at both southern and nothern sides of Indus-Yarlung Zangbo Rivers

1. Continental glacial sediments;
2. Glaciomarine sediments;
3. Marine ice-floe sediments

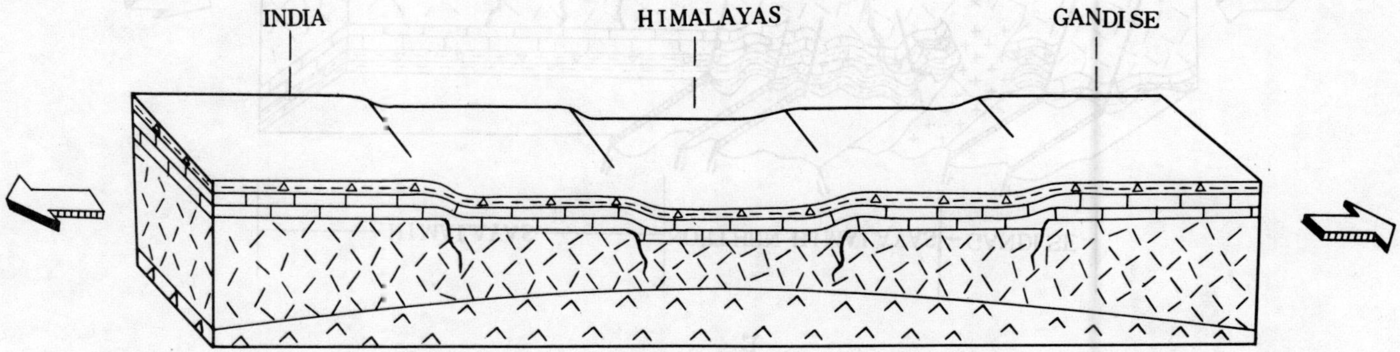


Fig. 9 Tectonic regime in the stage of development of platform in the Himalayas and adjacent region ( Index is the same as in Fig.11 )

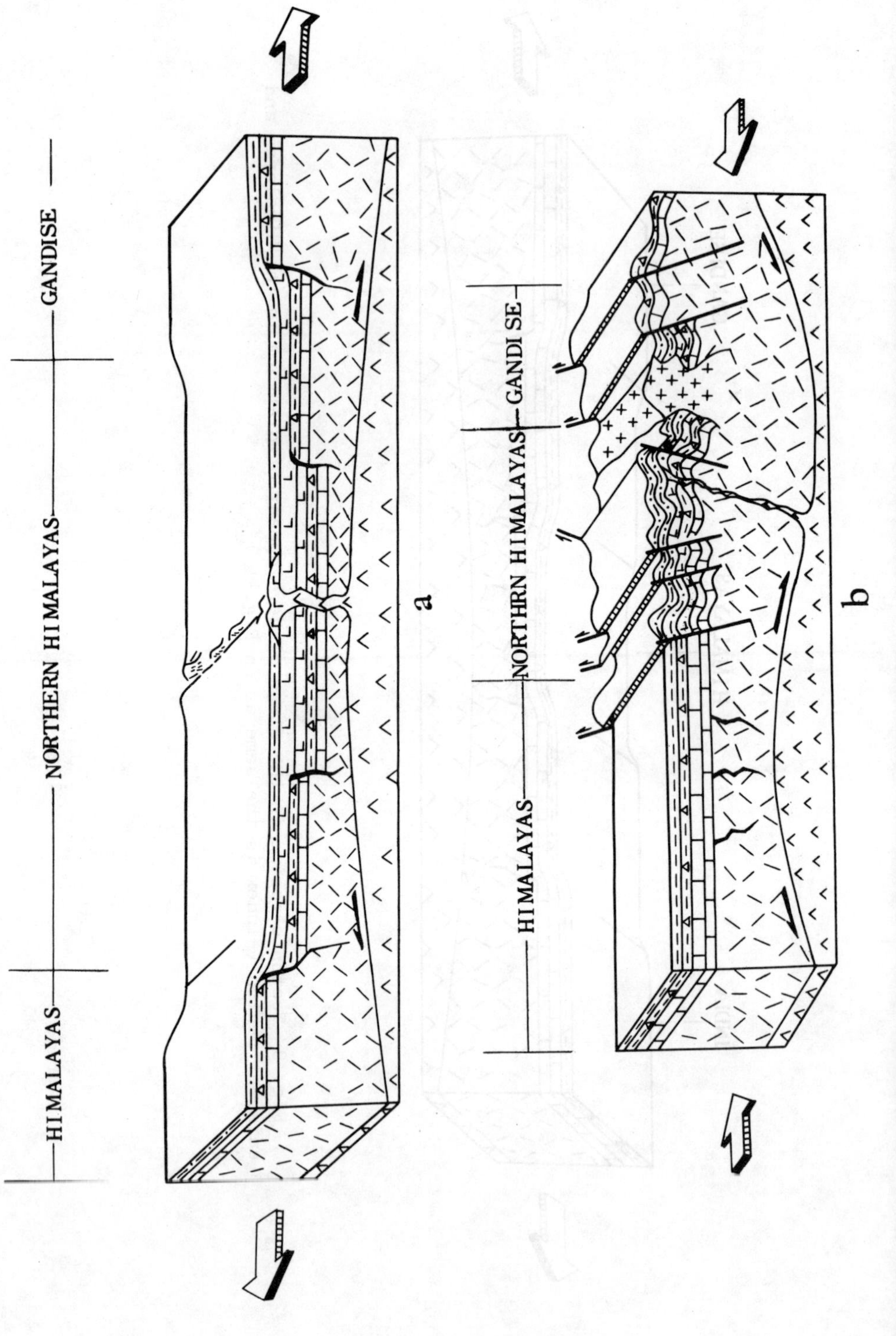
HIMALAYAS
NORTHERN HIMALAYAS
GANDISE
a
HIMALAYAS
NORTHRN HIMALAYAS
GANDI SE
b

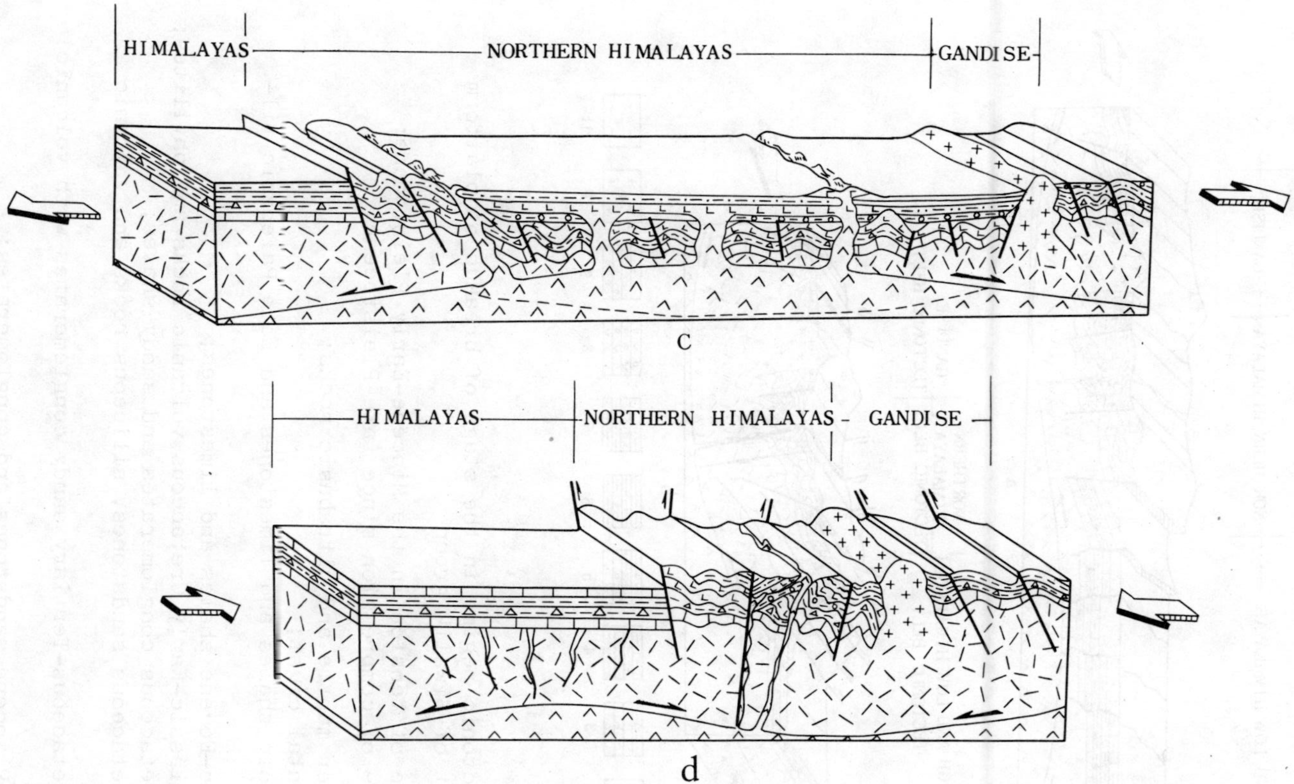


Fig. 10 Geotectonic regime in the stage of breakup of platform and the development of Northern Himalayan geosyncline (Index is the same as in Fig. 11)
(a) Stage of tension in Late Triassic–Early Cretaceous;
(b) Stage of compression in the end of Early Cretaceous;
(c) Stage of tension in Late Cretaceous;
(d) Stage of compression at the end of Late Cretaceous

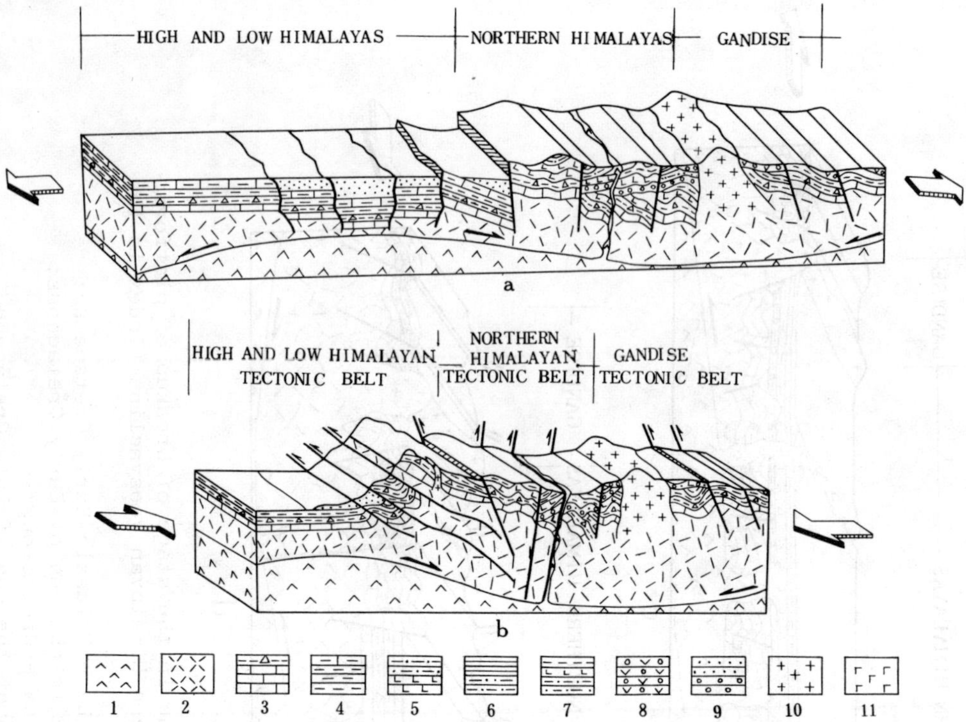


Fig. 11 Tectonic regime in the stage of breakup of platform and development of Siwalik group

(a) Substage of tension in the Miocene-Early Pleistocene;
(b) Substage of compression since Late Pleistocene.

1. The upper mantle and ultrabasic rocks;
2. Continental crust;
3. Palaeozoic shales and limestones and Late Paleozoic tillites;
4. Mesozoic-Eocene shales and limestones;
5. Late Triassic-Early Cretaceous volcanic rocks and phyllites;
6. Late Cretaceous conglomerates and sandy shales;
7. Late Cretaceous sandstones, siliceous rocks and volcanic rocks;
8. Late Cretaceous-Tertiary sandy conglomerates with volcanic rocks;
9. Middle Pliocene sandstones and conglomerates;
10. Gandise granites

# BASIC CHARACTERISTICS OF NEOTECTONIC MOVEMENT OF QINGHAI-XIZANG PLATEAU

Zhang Qing-song, Li Bing-yuan, Yang Yi-chou, Yin Ze-sheng
(Institute of Geography, Academia Sinica)

Wang Fu-bao
(Nanjing University)

The formation of the Qinghai-Xizang Plateau which covers an area of 2,300,000 km$^2$ with an average elevation over 4,000 m is indeed an extraordinary geological event. During the last one hundred years or so, people have tried to find out the answer , but differing opinions still remain. New data indicate neotectonic movement plays a decisive role on the formation of present landforms of the Plateau. It is certainly helpful, by clarifying the process of the formation of the Plateau, to the solution of the endless debate on the initial period of the neotectonic movement. Since the founding of the People's Republic of China, Chinese scientists Shi Ya-feng, Liu Dong-sheng (1), Yang Li-hua (2), Zhao Xi-tao(3), and others have made further detailed discussions on the neotectonic movement of the Himalayas. Recently, Li Ji-jun and co-authors (4) made a discussion on the period,amplitude and type of the uplift of the Plateau. In addition to the further clarification of geologic background of neotectonic movement and period of uplift of the Plateau, we shall mainly deal with the basic characteristics of the neotectonic movement as well as the cause of the uplift of the Plateau.

## GEOLOGIC BACKGROUND OF NEOTECTONIC MOVEMENT AND PERIOD OF INTENSE UPLIFT OF THE PLATEAU

The available data on stratigraphy and paleontology indicate that the Plateau was once a part of the Tethys sea which occupied the southern part of Eurasia. From the end of early Permian, the continent has constantly extended southward and the sea retreated correspondingly until middle Eocene when the whole Plateau has entirely emerged from the sea, thus forming the continent as a result of the influence of the first episode of Himalayan movement during middle Eocene. Based on it, E. Norin considered the period of the uplift of the Plateau to be late Eocene (5). This transformation from sea to land is a mark of the upheaval of the crust, however, many evidences demonstrate that no such a thing as intense uplift of the Plateau had ever occurred following this process. Sediments of Murree series of Oligocene to early Miocene at southern piedmont of the Himalayas do not chiefly come from the himalayas but from the Indian shield (6). This shows that the Himalayas had not been formed until middle Miocene. Although the deposits (i.e. Wuqia group) of the same period at southern fringe of the Tarim Basin are as thick as 2000 m or more, yet the lithology is rather simple, mainly brown-red siltpelite, demonstrating a small relief difference existed between the Kunlun Mts. and Tarim Basin then. But as for the interior of the Plateau, early Tertiary (Eocene) to Miocene is just the time for extensive development of Tanggula planation surface. The second episode of strong Himalayan movement during middle Miocene made the already peneplained surface more complicated and the occurrence of numerous faulted basins and block mountains accelerated the undulating relief, the Plateau has also gradually uplifted. The Pliocene deposits at the southern piedmont of the Himalayas and northern piedmont of the Kunlun Mts. are the manifestation of the uplift of the Plateau. Pliocene (middle Siwalik and Shule He Group) represented by these are mainly siltstone and pelite, with few coarse conglomerate, so they are not the typical molasse deposits. It shows that the rate and amplitude of the plateau's uplift were still small then. In the interior of the Plateau, many faulted basins were filled by lacustrine and fluvial clay and siltstone. Up to the end of Pliocene, ground surface was peneplained once again, newly formed broad gentle planation surface emerged which is referred to as the "primitive plateau surface". These Lacustrine siltstone are just deposits correlated with planation surface. At the same time most of the region were in humid subtropical climate with mixed forest on the higher mountains, and forest-steppe on the planation surface proper. On the forest-steppe, the Hipparion fauna was the principal inhabitants. The altitude of the planation surface is estimated about 1,000 m, and the mountains above it, such as the Himalayas, Gandise Range, and the Kunlun Mts. etc. reached 2,000 m or more. Thus the great altitude of the Plateau is mainly attained in Quaternary.

The crust movement at the end of Pliocene (the third episode of Himalayan movement) is manifested by the unconformity between Pliocene and Lower Pleistocene at the margin of the Plateau sand by the deposition of coarse conglomerate of lower Pleistocene. The unconformities are apparently displayed at Burang and Yagru Xongla (formerly Nyanyaxungla)(Fig. 1 and Fig. 2).

The maximum thickness of giant conglomerates of lower Pleistocene (really molasse-deposits) is 1,800 m on the southern slope of the Himalayas (upper Siwalik), and 2,000-3,000 m on the northern slope of the Kunlun Mts. and Altun Mts. The Gongba conglomerate of lower Pleistocene in the interior of the Plateau may reach 200-300 m, the maximum about 500 m in thickness. This shows that intense uplift of the Plateau began after the third episode of Himalayan movement and great differences in relief have been formed between the Plateau and the surrounding low lands, as well as mountain terrains and basins in the interior of it at the same time.

The facts cited above indicate that the Plateau entered a new stage of intense uplift with an unprecedented rate and amplitude at the end of Tertiary and beginning of Quaternary.

## BASIC CHARACTERISTICS OF NEOTECTONIC MOVEMENT.

The movement is characterized by three aspects: the totality, the differentiation, and the stadiality.

### The Totality

The totality of the movement is manifested by large scale uplift of the whole Plateau. As early as 1930s, Professor Li Si-guang pointed out that the Qinghai-Xizang Plateau with an average altitude of 4,000-5,000 m, is the highest of the several huge geomorphological staircases in China (7). Stairs at northern margin (Kunlun Mts.-Altun Mts. - Qilian Mt.) drop swiftly to an altitude of about 1,000 m, with a difference of 4,000 m at Tarim Basin and Hexi Corridor, while at eastern (the front of Longmen Mt. of western Sichuan ) it drops swiftly to an altitude of only about 400-500 m at the Sichuan basin, with a difference of 3,000 m. But at the southern margin, the Himalayas with a highest difference over 6,000 m drops down to the Indian Ganges plain, which is only dozens of meters a.s.l. Although the Plateau shows a tendency of gradual declining from northwest to southeast (altitude from 5,000 m declines to 3,000-4000 m), local differences of elevation and subsidence remain in its interior, numerous block mountains and faulted basins appear. But on the whole, it may yet be regarded as a great unified plateau (Fig. 3).

It is the massive block movement that has determined the totality of the plateau's uplift. Viewing the Plateau as a whole it is a rhombus faulted block. Several huge faults (such as Longmen Mt. fault, main boundary fault, and Kunlun-Qilian fault) separate its east, south and north res respectively from Sichuan basin, the Indian Ganges plain and Tarim basin(also with Hexi Corridor). These faults are all declining towards the interior of the Plateau. The entire Plateau is just like a huge wedge compressed from north and south by two continents.

## The Differentiation

Compared with the totality of the neotectonic movement, its differentiation is a subordinate feature. In addition to the declination of the Plateau from northwest to southeast mentioned before, the major manifestation is the block uplift, subsidence and the tilting. Qaidam basin is a most apparent faulted basin in the interior. Based on the available information of stratigraphy, paleontology and geomorphology, the difference on elevation and subsidence with the Kunlun Mts. has been over 2,000 m since Quaternary. Yangbajian basin is also a Quaternary faulted basin, it is only owing to the filling up of Quaternary deposits that make them retaining a flat terrain. The deposits are more than 500 m thick (Fig. 4).

Block uplift is always accompanied by subsidence. It is merely the faulted uplift and subsidence that enlarge the terrain difference within the Plateau. The Mt. Qomolangma is a typical block mountain. Viewing from the development of Quaternary glaciation, it became the highest peak in the world only after middle Pleistocene. Estimation shows that Mt. Qomolangma has uplifted 700 m more than the northern piedmont of Himalayas since late Pleistocene glaciation (8).

Tilting caused by fracturing is another manifestation of differentiation. The massive uplift of the Himalayas has become a focus of attention (Yang Li-hua and Prof. Liu Dong-sheng refer it as tilting)(2). This phenomenon is commonly seen in the interior of the Plateau though the scale is much smaller than the Himalayas. Mountains such as the Gandise Range and the Hol Xil Range etc.all appeared in an asymmetrical mountain lands, steep on the southern slope, gentle on the northern.

## The Stadiality

As for the uplift of the Plateau, in addition to totality and differentiation in space, there are distinctive stages in time. As shown before, the Plateau has uplifted totally 3,500-4,000 m since the end of Pliocene, but the rate and amplitude vary in different stages. According to Li Jijun and co-authors the magnitude of uplift in different stages of Quaternary are as follows (4): early Pleistocene, 1,000 m; middle Pleistocene, 1,000 m and since late Pleistocene, 1,700 m. These figures, though rather rough, still reflect to certain extent the stadiality of the uplift, as well as the tendency of acceleration in late period. Deduced from available data, there are three intense active periods for neotectonic movement, namely, 1) the end of Pliocene - the beginning of Quaternary; 2) the close of early Pleistocene and 3) after middle Pleistocene. The meaning is similar to "tectonic episode". Quaternary faulting and Quaternary volcanism in Kunlun Mts. area, as well as the formation of geothermal fields in Yangbajain basin are all closely correlated with their corresponding "tectonic episodes". We infer that the uplift of the Plateau might be in the style of jumping. The stadiality is a result of alternation of fast and slow uplifts.

The successive uplifts are shown by corresponding knick points almost in every river of the Plateau. For instance, three knick points of Yarlung Zangbo River are located respectively at the west of Xoka (4,500 m), Gyaca gorge (3,500 m), and Great Bend gorge (Daguaiwan Gorge) (2,800 m). Viewing from the history of the devlopment of Yarlung Zangbo River, the formation of these three knick points might be correlated to the three intense uplifts of the Plateau. Although the height between these knick points can not directly represent the magnitude of uplift in each stage, yet it qualitatively reflects that the rate of uplift of the Plateau has been the highest after the last knick point was formed.

## DISCUSSION ON DYNAMICS OF NEOTECTONIC MOVEMENT OF THE PLATEAU

Based on the formation of the Plateau and characteristics of neotectonic movement discussed before, we put forward the following points on the dynamics of the plateau's uplift.

(1) The intense uplift of the Plateau since the end of Tertiary is a result of violent NS-trending compression. The rising movement is derived from horizontal compression. This N-S horizontal compression is first reflected on the border of the Plateau. The Main Boundary Fault (MBT) along the southern slope of the Himalayas reflects that an intense activity occurred in Quaternary. The palaeostratigraphy north of it overthrusts towards south, until covering over Siwalik series, middle Pleistocene, even Holocene, ranging 20 km (9). In the faulted belt at northern piedmont of Kunlun Mts.-Altun Mts.-Qilian Mt. composed of a series of high-angle thrust fault declining towards south, horizontal compression is also very strong. The ancient strata at northern piedmont or Qilian Mt. thrust reversely toward north over Pliocene (Shule He group), lower Pleistocene (Xiyu conglomerate) and middle Pleistocene (Jiuquan conglomerate). Fault distance may reach 100-400 m (10, 11). Obviously, the strong earthquake belt of western Kunlun Mts. - Hexi Corridor is closely correlated with the recent activities of this fault zone.

(2) The south-north compression relates mainly with north-ward drift of Indian plate. On the Plateau, the Himalayas is the latest to emerge from the sea and become land mass. Since late Tertiary, especially from the end of Pliocene to the beginning of Quaternary, the rate and amplitude of the uplift of the Plateau are much higher than those of other areas, even surpassing the former ones, and became the highest peak in the world. The scale of compression is reflected by the fact that the main boundary fault is much larger than the fault at northern piedmont of Kunlun-Qilian Mountains. Investigations on mechanism of earthquake focus of the Himalayas also reveal that there is a massive compressive stress directing NS at southern slope of the Himalayas (12,13). It just coincides with MBT in position. These phenomena indicate that horizontal compressive

stress mainly comes from the Indian continent of the south. So it can be considered that the massive compressive stress produced by northward drift of the Indian plate transmits toward N through Qinghai-Xizang block, thus causing the block to overthrust northward at northern piedmont of the Kunlun-Qilian Mountains, and Tarim Basin block to underthrust passively southward. Since the Qinghai-Xizang block itself is a joint connection of numerous geologic masses, in the process of violent compression and intense uplift, the original fault belt should adjust correspondingly, either glide horizontally (such as the fault at southern piedmont of Nyainqentanglha Range), or elevate and subside vertically (such as the fault at southern fringe of Qaidam basin). So it displays the variation in the process of integrate uplift. On the other hand, the velocity of northward drift of the Indian plate might be sometimes fast, sometimes slow. This may be the reason responsible for the uplift of the Plateau in stages.

(3) The boundary of northward underthrust for present Indian plate is not situated at the fractured zone of the Yarlung Zangbo River, but at MBT of the southern piedmont of the Himalayas. Though many geologic evidences have proved that the fractured zone of Yarlung Zangbo River is a "suture" of two plates (14,15), yet its main activities are concentrated in Cretaceous and weakened considerably up to Tertiary. But there is no apparent difference in drainage patterns, planation surface distribution and mountaenous terrain morphology between the north and south sides of the fractured zone, nor there is apparent tectonic fracture in Quaternary strata of the Yarlung Zangbo suture. The evidences described above show that the northward overthrusting zone of present Indian plate should be located at southern piedmont of the Himalayas. Geologically, the boundary is just the MBT (Fig. 5).

## References


(1) Shi Ya-feng & Liu Dong-sheng, A preliminary report of scientific expedition in the Mt. Xixabangma region. Kexue Tongbao, (10) 928-938 (1964). (in Chinese).

(2) Yang Li-hua & Liu Dong-sheng, The neotectonic movement in the Mt. Qomolangma region. Scientia Geologica Sinica, (3) 207-220 (1974). (in Chinese).

(3) Zhao Xi-tao, On the recent uplift of the Himalayas, Scientia Geologica Sinica, (3) 243-252, (1975). (in Chinese).

(4) Li Ji-jun et al., A discussion on the period, amplitude and type of uplift of the Qinghai-Xizang Plateau. Scientia Sinica, 12 (11) 1314-1328 (1979). (in Chinese).

(5) Norin, E., Tertiary of the Tarim basin. Bull. Geol. Soc. China, 14 (3) (1935).

(6) Wadia, D. N. Geology of India. Mac & Co. Lim., London, (1953).

(7) Li Si-guang, Geology of China (Translated by Zhang Wen-you), Zheng feng Press, 1952.(in Chinese)

(8) Zheng Ben-xing & Shi Ya-feng, A review on Quaternary glacial periods in the Mt. Qomolangma region. Reports of scientific expedition in the Mt. Qomolangma region (1966-1968), Quaternary, Science Press (1976). (in Chinese).

(9) Gansser, A., Geology of the Himalayas. Interscience Publishers, London, New york (1964).

(10) Huang Ji-qing, Several types of neotectonic movement of China. Symposium on I Neotectonic Movement, Academia Sinica, Science Press (1956). (in Chinese)

(11) Yu Bai-liang, The characteristics and significance of neotectonic movement in the Corridor of Gansu (ditto).

(12) Shi Zheng-liang et al., Intense seismic activities of China and their correlation with plate tectonic. Scientia Geologica Sinica, (4) 281-293 (1973). (in Chinese).
(13) Ye Hong et al., A stress analysis of recent structure of Himalayas and its adjacent areas. Scientia Geologica Sinica, (1) (1975). (in Chinese).
(14) Chang Cheng-fa, Zheng Xi-lan, Tectonic characteristics of the Mt. Qomolangma region, South Tibet, China. Scientia Geologica Sinica, (1) (1973). (in Chinese).
(15) Huang Ji-Qing et al., Some new points on the tectonic characteristics of China. Acta Geologica Sinica, (1) (1974). (in Chinese).

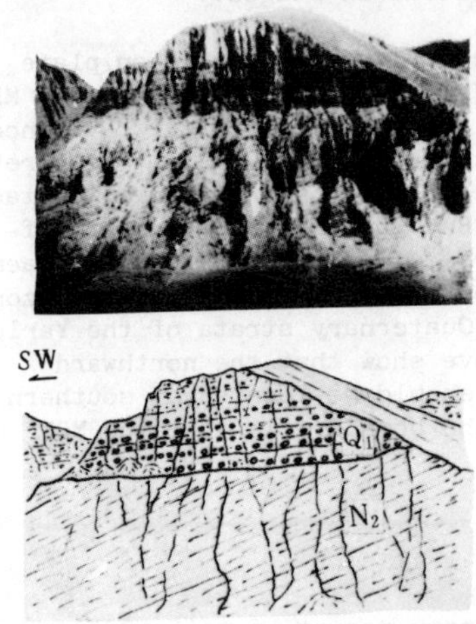


Fig. 1 The unconformity between lower Pleistocene and Pliocene near Burang

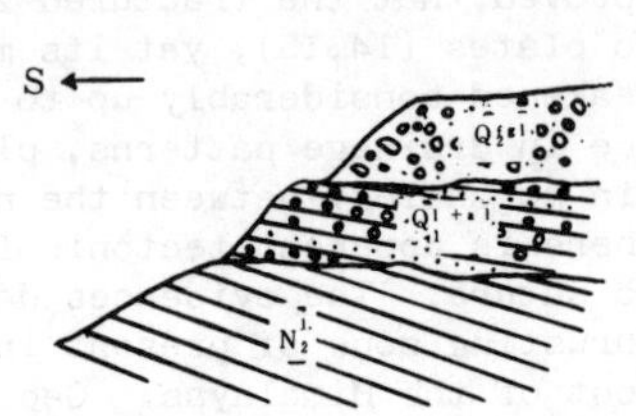


Fig. 2 The section of Pliocene and lower and middle Pleistocene on the southern slope of Yagru Xongla

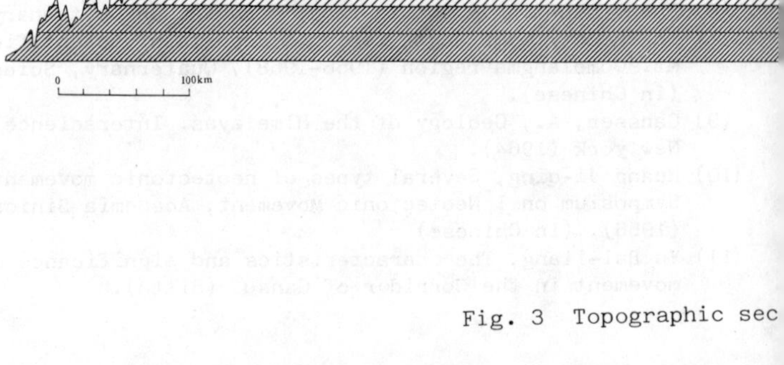


Fig. 3 Topographic sec

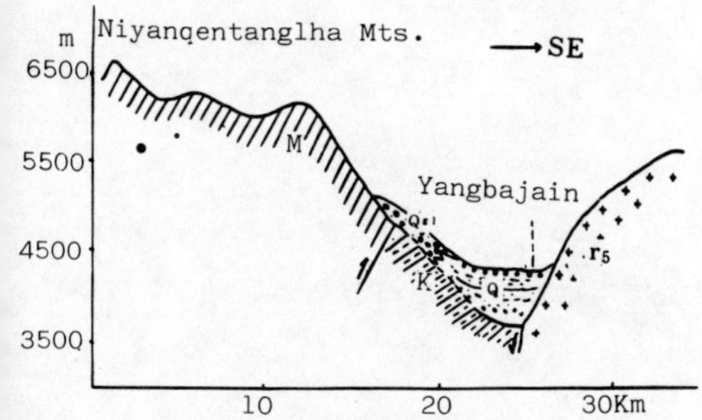


Fig. 4 The morphological and Quaternary depositional section, Yangbajain

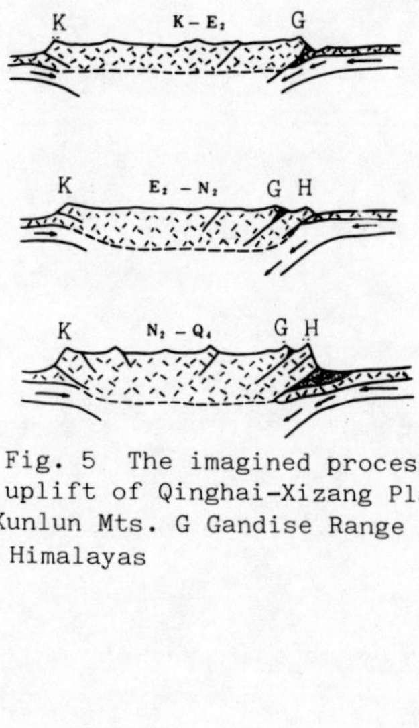


Fig. 5 The imagined processes of uplift of Qinghai-Xizang Plateau. K Kunlun Mts. G Gandise Range H Himalayas

inghai-Xizang Plateau (S – N)

# THE PROCESS OF THE UPLIFT OF THE QINGHAI-XIZANG PLATEAU

Li Ji-jun Li Bing-yuan
Wang Fu-bao Zhang Qing-song
Wen Shi-xuan Zheng Ben-xing
(The Comprehensive Scientific Expedition to the
Qinghai-Xizang Plateau, Academia Sinica)

The uplift of the Qinghai-Xizang Plateau has been studied by many scholars at different angles, creating however disagreement among them on such questions as time, process and amplitude of the upheaval. E. Norin (1936) held that a total magnitude of uplifting 6,000 m or more had been accumulated since the end of the Mesozoic era(1). But V. M. Synetzen (1962) considered the Plateau already very high at the end of the Pliocene, only 1,000-1,500 m lower than its present altitude(2).

According to Shi Ya-feng and Liu Dong-sheng (1964) who inferred from the fossil of Quercus semicarpifalia, the northern slope of Mount Xixabangma had uplifted some 3,000 m since the Pliocene(3). Basing on the data from investigation in the Qomolangma region in 1966-68, Zhao Xi-tao concluded that the Plateau had reached a height of 2,000-3,000 m in the Pliocene(4). But Cui Zhi-jiu and Zheng Ben-xing thought only a height of about 1,000 m could be reached by an estimation from the remnants of ancient karst landscape(5).

## THE QINGHAI-XIZANG REGION AFTER THE COLLISION OF THE CONTINENTS IN EOCENE

It has been made clear that the horizons of the latest marine beds and the ages of the mountain-making movement of a series of the mountains stretching roughly east-westward on the Qinghai-Xizang Plateau are getting younger from north to south; and so are the ages of their main magmatic activities. The ages are shown in Table 1 reflecting the southward extension of the Asian Continent and the gradual closing-up of the Tethys Sea when the Indian plate was drifting northward. The uplift of each mountain range has experienced a three-fold process of regression, orogeny and magmatic intrusion. The last one is marked by the fact that the Qinghai-Xizang Plateau did not suffer any transgression ever since then, and that the Himalayas, the most magnificent mountains on the globe, were formed and eventually the Qinghai-Xizang Plateau as a whole uplifted violently.

In spite of the fact that the Jurassic planation surface has already been found in Indian Peninsula(6), no conditions for the development of such surface ever existed in the Qinghai-Xizang region. Because the transgression was widespread, the Jurassic marine beds could be found almost everywhere in this region. Moreover, it was impossible for the Mesozoic era planation surface to survive on the Qinghai-Xizang Plateau after active tectonic movements of Yenshan orogeny.

Informations now available show that no immediate violent uplift occurred in this region after the gulf of Tethys Sea retreated from the Qinghai-Xizang region in the middle of the Eocene epoch. An erosional surface covered with laterite and bauxite has been found beneath the nummulitic limestone in Kashmir(7). However, the surface associated with the nummulitic limestone are subjected to strong deformation, so it is doubtful that the planation surface older than Tertiary may be preserved on the Qinghai-Xizang Plateau. It seems that there had been a long period characterized by secular movement and perhaps stable condition until a movement of greater intensity began about the middle of the Miocene. We know that the Murree group of early Miocene have originated from the sediments source of which was the iron-bearing Purana formations of the Indian Shield to the South. Accordingly, no Himalayas existed at that time. In this case, an extensive planation surface was formed at the end of the early Tertiary after a long period of denudation and planation. That the Qinghai-Xizang region was a low and flat landscape can also be deduced from the palaeontological data. The microfossil florae (pollens and spores) of early Tertiary deposits in Lumpola basin to the west of Nagqu were found to be rich in *Sequoia*, *Taxodium* and so on(8). In Qaidam basin and Gansu Corridor we have found *Magnolia*, *Kingo* and Proteaceae in the strata of the Oligocene(9). In other places the pollens of Myrtaceae and *Eucalyptus* were also discovered. Those species mentioned above all grow at low land and are warm-loving, representing a tropical environment. Figure 1 shows the main sites in the Qinghai-Xizang region and its adjacent areas where fossils of the Giant *Rhinoceros* fauna and vegetation of tropical type in the early Tertiary have been found. It is known that the Giant *Rhinoceros* was the largest terrestrial animal in the Cenozoic era and their remains were found time and again in the marginal areas of the Plateau, such as Pakistan, China's Yunnan, Gansu and Xinjiang. That kind of Giant *Rhinoceros* discovered in Yunnan is similar in many respects to that found in the Bugti strata, South Asia(10). As Grabau (1927) pointed out that "this wide distribution in Asia of an animal of this size and bulk, clearly indicates that during its existence barriers between these different regions were wanting. The animal could not have made its wide migrations after the elevation of the Himalayan and other mountain chains ..."(11) Actually, the period in which the Giant *Rhinoceros* was widely distributed may be the same time in which the planation surface was highly developed. Such a low and gently sloping landscape has had prevailing warm climate and a tropical vegetation growing on it. At present its remnants are widely preserved on the top of the mountains on the Qinghai-Xizang Plateau, with ice cap or flattopped glacier covered on its higher part (Photos 1 and 2). On the western sector of Tanggula Range, it has been discovered that basalt and andesite lava of one hundred metres in thickness overlie the surface. In other places this kind of lava also covers unconformably the Fenghuoshan group of the lower Tertiary, in turn, with the Chamabo group of upper Tertiary seen above. Therefore, it is evident that this surface must be formed at the turn of early Tertiary and Neogene at the latest. L. C. King (1976) pointed out that the planation surfaces formed in that period, i.e. Moreland Surface, are often found on the summits of the mountains in "all" the continents, with the exception of the mobile zone of the Alpine-Himalayan orogeny(12). In the case of Qinghai-Xizang Plateau, the surface mentioned above seems to be equivalent to the " Moreland Surface " which also called " African Surface " in his early articles.

## THE RISING OF THE HIMALAYAS AND THE FORMATION OF THE LOWER SURFACE IN THE NEOGENE

The tectonic movement became active again in the middle of the Miocene. The Indian plate pushing northward was obstructed and had to go on with its movement in two ways : plunging its basement beneath the Eurasia continent and making its cover slices off into a series of nappes gliding southward under great pressure. The main central thrust was formed at that time, and synorogenic tourmaline-granites invaded, with K — Ar dating of a scatter of ages from 20 to 10 m.y.old. The southern end of the nappes was fashioned into the primary Himalayas, and due to the rising mountains the molasse-like Siwaliks began to accumulate. Being different from Murree group, Siwaliks are derived wholly from the rising Himalayas, hence, presenting itself as a genuine record of the uplifting of the Himalayas. However, if we inspect the lithofacies in detail, it can be found that the piedmont breccia and fanglomerate representing the intense uplift of the mountains did not begin to appear until the Quaternary period, and on the whole the sediments of the Neogene are composed mainly of clay-stone and clayey sandstone. From Figure 2 we can see that the two graphs are similar to each other, though the distance between them is about 2,000 km. The upper graph shows the changes of the lithofacies in the Siwalik group of Haritalyangar, Himachal Pradesh in India, and the lower graph shows that of the strata of the same age (Neogene) near Yumen city in the Gansu Corridor. Both the Siwaliks in Haritalyangar and the Neogene strata in Yumen consist of many fluvial cycles, being the products of the well-developed meanders of rivers when the basin was subsiding slowly. This goes to show that the relative rise of the mountains took place also very slow, as a result, clayey component are therefore predominant (in the Baiyang He group it occupies 61 %, and in Shulehe group 45 %). Against the background of the slow uplift of the Qinghai-Xizang Plateau, the planation surface of the early Tertiary was dissected and the lower one was developed. It truncated all the ancient strata including those of the lower Tertiary, and for the case of its height it coincides with the top surface of the Pliocene strata in the intermontane basin. At present time this surface is well-preserved and generally takes the form of low and gentle sloping ridges or tabular uplands, the dissected remnants of ancient lake basin and broad valleys, which in the marginal regions of the Plateau have become high interfluves owing to deep incision of large rivers, as in the case of the Hengduan Mts.

The difference in elevation between the above mentioned two planation surfaces in the interior of the Plateau is generally about 300-700 m and more than this in the outer margins. For instance, the difference on the northern slope of the Qilian Mt. may reach 1,000 m.

In our previous paper we devoted to a consideration of the elevation of the lower planation surface at the time of its formation(13), i.e., the general altitude of the Qinghai-Xizang region at the end of the Pliocene. We concluded that the general altitude of the Qinghai-Xizang surface was around 1,000 m at that time. Subtropical plants were growing on this surface, the pollens of cedars, palms and oaks have been found on many occasions in the strata of this region. And relics of ancient karst Landforms similar to those in South China at present have been found in recent years in various parts of the Plateau. These are guides in keeping with the environment required by the ecological habits and characteristics of <u>Hipparion</u> fauna, especially <u>Rhinoceros</u>, <u>Giraffe</u> and so on. Figure 3 indicates the main places where <u>Hipparion</u> faunas were distributed in the Pliocene Asia. The species of <u>Hipparion</u> fauna discovered in recent years in Xizang and Yunnan have the same appearance with these in South Asia, indicating an environment with warm and humid climate, forest and grassland of tropical to subtropical type growing everywhere(14). The geological age of the fauna may be also the same to that of Nagri fauna of South Asia. And then occurred the change of climate over the world, the tropical forest retreated southward and the grassland expanded far away. Therefore, the *Hipparion* of the forest-living type was replaced by that of known as grassland type which roamed about almost the whole Asia. The *Hipparion* faunas found in Orma basin of Gyirong County are of the grassland type and comparable with the species often seen in North China, belonging to the middle Pliocene, reflecting a great change of environment which had occurred at that time. In South Asia, the Dhok Pathan fauna, next above the Nagri, is characterized by the sudden disappearance of the arboreal animals and the influx of steppe elements which tell us the same story. From the foregoing description it appears that during the Pliocene epoch the *Hipparion* fauna could migrate freely over the whole Asia, which proves that the ground environments were basically alike and no barriers of high mountains existed at that time. And so the Qinghai-Xizang region was not a highland. However, it is estimated by some people that the altitude of the Qinghai-Xizang region was then rather high, even as high as 2,000-3,000 m on the basis of the fact that sporo-pollen of dark coniferous forests such as spruce and fir has been found in the Pliocene strata of the Plateau. In our opinion, this estimation is misplaced for two reasons. First, there were mountains on the Qinghai-Xizang Plateau in the Pliocene and the sporo-pollen of plants on high mountains could be carried down to lower places. Second, the climate on the globe was turned cold abruptly in the late Pliocene (ice sheets formed in high latitude region), so that it is not at all surprising that dark coniferous forests grew in the mountain areas of the Qinghai-Xizang region. Moreover, sporo-pollen of dark coniferous forests has also been discovered recently in the Pliocene strata of some places in North China, sometimes occupying very high percentage (67 %). Yet no one holds the opinion that North China ever reached a height of 2,000-3,000 m in the Pliocene.

## THE VIOLENT UPLIFT OF THE QINGHAI-XIZANG PLATEAU IN THE QUATERNARY AND ITS MECHANISM

We cannot spare space to deal with the subject of the uplift events of the Qinghai-Xizang Plateau in the Quaternary. We would like to refer your attention to our paper in Scientia Sinica (Vol. 22, No. 11, 1979) for more details. In short, the plateau surface has already risen from 1,000 m in altitude at the end of the Pliocene to the present average of 4,500-5,000 m, which is remarkable. The Plateau is rising in block uplift fashion, together with which there are south-north and east-west undulating movements which can be judged from the deformation of the planation surfaces. The highest wave crest is that of the Himalayas and the lowest trough is the great fault zone of Bangong Lake to Nujiang Valley (Fig. 4). Most of the plateau lakes are controlled by the structure lines reflecting the geotectonic stress field. Among them a row of fault-guided or graben lakes on the top of the Gandise Range are magnificent, with an average height more than 5,000 m above sea level. It shows that there is an undulating movement induced by east-west tensile stress.

As for the cause of the uplift of the Plateau, our view-point is that the essential causes were the breaking-up of the Indian plate within the plate and underthrusting. The latest event of this nature took place in the middle Quaternary. The Indian plate pushed northward along the main boundary thrust so that the Siwalik group was folded into mountains. The Plateau was therefore elevated like a wooden plate raised by many wedges driven beneath it. Fig. 5 provides a brief explanation of the mechanism and process involved in the elevation of the plateau.

## References


(1) Norin, E., Tertiary of the Tarim Basin. Bulletin of the Geological Society of China, Vol. 14 (1935).

(2) Synetzen, V. M., Palaeogeography of Asia. Science press of SSSR p. 187 (1962). (in Russian).

(3) Shi Ya-feng, Liu Tung-sheng, Preliminary Report on the Mt. Xixabangma scientific expedition. Contributions at the 1964 Peking Symposium Gen. 117 (1964). (in Chinese).

(4) Zhao Xi-tao, On the recent uplift of the Himalayas. Scientia Geologica Sinica, (3) 243-252 (1975). (in Chinese).

(5) Cui Zhi-jiu, Zheng Ben-xing, The palaeokarst in Mt. Qomolangma region. Report on the scientific investigations (1966-1968) in Mt. Qomolangma district: Glaciology and Geomorphology, Science Press, (1975). (in Chinese).

(6) Vaidyanadhan, R., Recent advances in geomorphic studies of Peninsular India: A review. India Journal Earth Sciences, pp. 13-15 (1971).

(7) Wadia, D. N., Geology of India. Macmillan and Co. Limited, p. 335 (1953).

(8) Wang Kai-fa et al., On the Tertiary sporo-pollen assemblages from Lumpola Basin of Xizang, China, and their palaeogeographic significance. Scientia Geologica Sinica, (4) (1975). (in Chinese).

(9) Xu Ren (Hsü Jen) et al., Sporo-pollen assemblages from the Tertiary deposits of the Qaidam Basin and their geological significance. Acta Palaeontologica Sinica, 6 (4), (1958). (in Chinese).

(10) Zhou Ming-zhen, Some Oligocene mammal fossils from Lunan, Yunnan. Vertebrata Palasiatica, 2 (4), (1958). (in Chinese).

(11) Grabau, A. W., Summary of Cenozoic and Psychozoic. Bulletin of the Geological Society of China, Vol. 6 (1927).

(12) King, L. C., Planation remnants upon high lands. Zeit, Greom., Band 20, Heft 2, (1976).

(13) Li Ji-jun et al., A discussion on the period, amplitude and type of the uplift of the Qinghai-Xizang Plateau. Scientia Sinica, 22 (11), (1979). (in Chinese).

(14) Qi Guo-qin, Pliocene mammalian fauna of Lufeng, Yunnan. Vertebrata Palasiatica, 17 (1), (1979). (in Chinese).


Table 1 Ages of the mountains on the Qinghai-Xizang Plateau

| Mountain ranges | Ages of the latest marine beds | Orogeny | K-Ar date of main magma intrusion ($10^6$--- yr) |
|---|---|---|---|
| Altun | Devonian | Caledonian | 344-554 |
| Kunlun | Late early Permian | Hercynian | 240-260 |
| Tanggula | Mid-Jurassic | Indo-sinian | 207-210 |
| Gandise | Mid Cretaceous | Yenshan | 30-79 |
| Himalayas | Mid-Eocene | Himalayan | 10-20 |

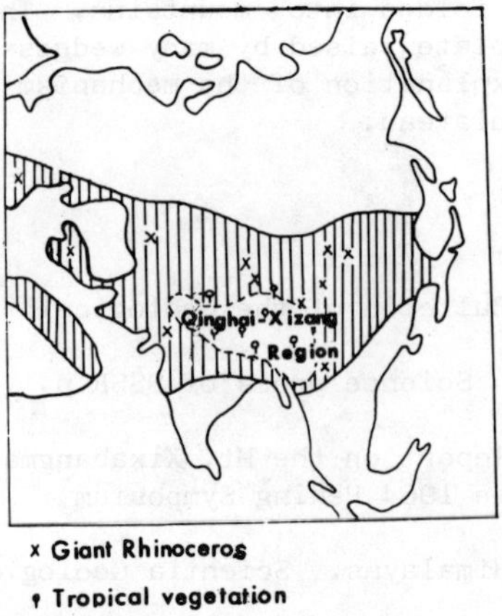


Fig. 1 Sketch map of distribution of Giant Rhinoceros fauna in Eurasia in Oligocene

Fig. 3 Sketch map of distribution of Hipparion fauna in Eurasia

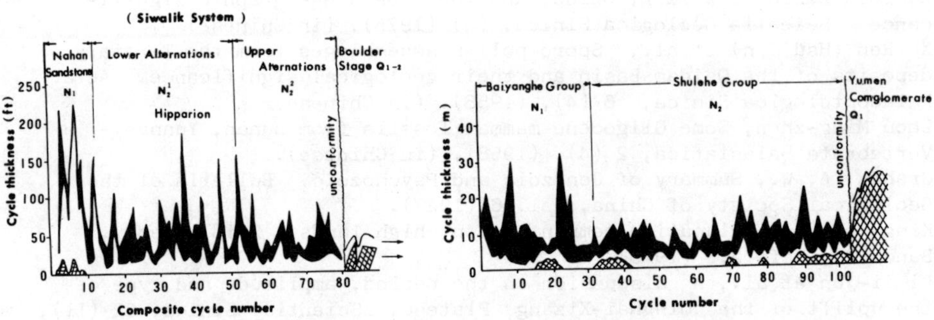


Fig. 2 Fluvial fining-upward compositecycles occured in Late Cenozoic deposits in south and north sides of Qinghai-Xizang Plateau

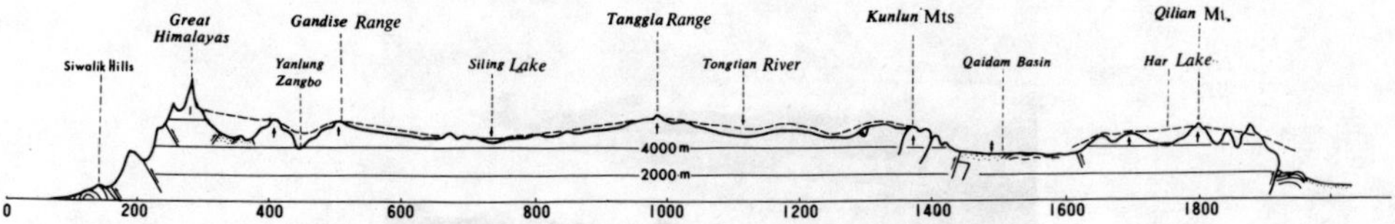


Fig. 4  Undulation of the planation surface on Qinghai-Xizang Plateau

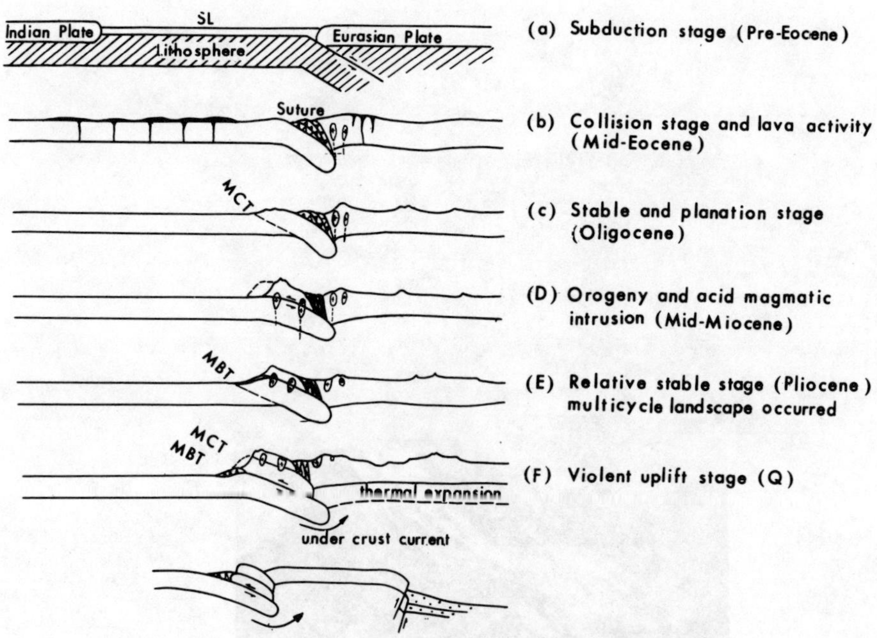


Fig. 5  Hypothetical model of the mechanism of the uplift of Qinghai-Xizang Plateau

Photo 1 Planation surface and flat-topped glacier in the Southwest Qilian Mt.

Photo 2 Dissected planation surface in the North Qilian Mt.

## SEDIMENTARY DEVELOPMENT AND FORMATION OF STRATIGRAPHIC REGION IN XIZANG

Wen Shi-xuan Zhang Bing-gao Wang Yi-gang
Sun Dong-li Wang Yu-jing Chen Chu-zhen
Dong De-yuan Liao Wei-hua Chen Ting-en
He Guo-xiong Mu Xi-nan
(Nanjing Institute of Geology and Palaeontology, Academia Sinica)

Yin Ji-xiang Wu Hao-ruo
(Institute of Geology, Academia Sinica)

In the past 15 years the stratigraphic data of Xizang have increasingly been documented (1,2,3), which make it possible to discuss the problems on the formation and development of the stratigraphic regions in Xizang.

### OUTLINE OF STRATIGRAPHIC REGIONS OF XIZANG

The strata of Xizang are not uniform in development. Generally, beds from the Ordovician to the Lower Tertiary are fairly well exposed at the northern slope of the Himalayas (1,2,4); but in other areas of Xizang the marine strata are relatively incomplete.

The sedimentary nature are various in different places and in different ages, mainly consisting of active and stable types.

The fossils are very rich in Xizang, which embrace plants, invertebrates and vertebrates, totally 30 in kind. By a rough estimation, they amount to about 3,800 species. Certainly, these materials provide a sound basis for further studies on stratigraphy in Xizang.

Judging from the stratigraphic development, sedimentary type and biotic aspects, 4 stratigraphic regions and 11 subregions may be distinguished in Xizang (see Table 1, Fig. 1).

The stratigraphic regions in Xizang are separated respectively by 3 deep fault belts running parallelly with the latitudes; but the patterns of the regions are contracted slightly tilting northwestward to the west, and contracted curving southeastward to the east. The boundaries between the stratigraphic regions are quite distinct; but those between the subregions are relatively inconspicuous, such as the boundary between the Tanggula and the Qambo subregions can be only marked out roughly. The stratigraphic division and correlation of stratigraphic regions and subregions are shown in Tables 2 and 3.

Due to the crustal mobilization which had an important bearing on the sedimentary types and biotic aspects, 6 sedimentary belts were formed in Xizang since the Palaeozoic era. Alternately, 3 are of relatively stable type and 3 of relatively active type, which are nearly identical in distribution with the stratigraphic regions listed below (from south to north): 1) Southern subregion of Tethys-Himalayas, sedimentary belt of relatively stable type; 2) Northern subregion of Tethys-Himalayas, sedimentary belt of relatively active type; 3) Shiquanhe-Xainza subregion and Lhasa-Bomi subregion, sedimentary belt of relatively stable type; 4) Biru subregion and Muggarkangri subregion, sedimentary belt of relatively active type; 5) Karakorum subregion, Tanggula subregion and Qamdo subregion, sedimentary belt of relatively stable type; 6) Hoh Xil subregion and Kunlun subregion, sedimentary belt of relatively active type.

In the sedimentary belt of relatively active type, the deposits, yielding rare fossils, consist of graywacke, flysch, and siliceous rocks, all of which are more or less metamorphosed. But in the sedimentary belt of relatively stable type, the deposits are of shallow-sea clastic rocks, quartz sandstone or other clastic rocks, which contain rich fossils.

## A BRIEF HISTORY OF SEDIMENTARY DEVELOPMENT

The Pre-Ordovician fossils have not been reportedly known in Xizang. However, it is likely that the Pre-Ordovician deposits exist in the Nyalam group of the Himalayan region or in the metamorphic rocks elsewhere of Xizang. The Nyalam group, exceeding 20,000 m in thickness, underlies the fossiliferous Ordovician rocks, and its original rocks are largely composed of graywacke, feldspathic sandstone, clay siltstone, etc., which strikingly exhibit a rhythmic structure. All these characteristics reveal that in the sedimentary process the crustal activities were probably much stronger and the rate of sedimentation was very rapid.

From Ordovician to Tertiary, 3 periods may be distinguished in the sedimentary development of Xizang.

### Ordovician-Devonian

In Ordovician-Devonian time, Xizang as a whole was in a stable shallow-sea, where plenty of benthonic, nectonic and planktonic organisms dwelled, and consequently the deposits of relatively stable type were developed extensively.

The Ordovician deposits are known to outcrop in every region of Xizang with the exception of the Hoh Xil-Kunlun region. Lithologically, they are composed of thick-bedded limestone, dolomite, bioclastic limestone and calcareous shale. The Ordovician rocks contain rich faunas (1). For instance, the Lower Series yields nautiloids *Ordosoceras*, *Manchuroceras*, *Wutinoceras*, *Ormoceras*, *Paradnatoceras*; brachiopods *Aporthophyla*, *Glyptomena*; gastropods *Maclurites*, *Tropidodiscus*, *Cymbularia* and trilobite *Pseudocalymene*. These fossils bear similarities to those of North China type. The Middle and Upper Series carry nautiloids *Sinoceras*, *Eosomichelinoceras*, *Beloitoceras*; brachiopod *Hirnantia*; trilobite *Dalmanitina*, which are closely related to those of the Central China type. But it is noteworthy that in the eastern part of Qambo subregion the Lower Ordovician, quite different from that in other regions, is formed by variegated slate and sandstone, which carries graptolites *Isograptus*, *Didymograptus*, *Dichograptus*, *Tetragraptus* and *Phyllograptus* (5). Evidently, these deposits are of the relatively active one, belonging to the South China type.

The Silurian deposits are only known to crop out in a few localities in Xizang. The Lower Silurian yields graptolites *Streptograptus*,

Climacograptus, Oktavites, Demirastrites (1); brachiopod Hindella; corals Palaeofavosites, Solenihalysites, Amplexeoides, Orthophyllum, etc. (6). The Middle and Upper Silurian contain graptolites Pristiograptus, Monoclimacis; nautiloid Michelinoceras (Kopaninoceras) jucundum (Barrande) and bivalve Cardiola cf. docens Barrande etc. These faunas bear a strong resemblance to those from western Yunnan and western Sichuan.

The Devonian rocks are less widely exposed in Xizang. Lithologically, they consist of limestone, shale, quartz sandstone, bioclastic limestone and dolomite. In southern Xizang, the Lower Series yields tentaculites Nowakia acuaria (Richter), Guerichina xizangensis Mu; graptolites Neomonograptus atopus rigidus Mu et Ni, Monograptus thomasi Jaeger, M. cf. yukonensis Jackson et Lenz. etc. (1), which bear a striking likeness to those from the coeval deposits of western Yunnan and Sichuan. In northeastern Xizang the Middle and Upper Series all contain rich animal fossils. The main fossils are brachiopods Athyrisina, Stringocephalus, Cyrtospirifer, Hypothyridina, and corals Thamnopora, Hexagonaria, Nalivkinella, Cladopora, etc., which strongly resemble those from the contemporaneous deposits of South China.

## Carboniferous and Permian

The Carboniferous and Permian strata in Xizang show great differences. For instance, northwards in the Hoh Xil-Kunlun region there exist variegated sandstones and flysch intercalated with argillaceous shales of a great thickness, which represent the relatively active type. But in the rest of Xizang the deposits are of the relatively stable type, consisting of quartz sandstone, shally limestone and shale. Furthermore, the continental intercalations with flora are developed at some places.

So far as the Carboniferous faunas are concerned, two types of marine faunas may be recognized. In the Himalayan and Gandise-Nyainqentanglha regions (7), the fusulinids are rarely found; the Lower Carboniferous yields ammonoid Gattendorfia and brachiopods Fusella, Ovatia, Marginirugus, Syringothyris, Balakhonia (1); and the Upper Carboniferous contains the forms of the Gondwana-Tethyan type, including brachiopods Stepanoviella, Attenuatella, Trigonotreta and coral Empodesma, etc. (1). In contrast, the fusulinids are very plentiful in other regions of Xizang, which bear a close similarity to those of South China. The Lower Carboniferous yields fusulinid Eostaffella; corals Cystophrentis, Aulina, Kueichouphyllum, Yuanophyllum; brachiopods Gigantoproductus, Schuchertella, Praewaagenoconcha, etc. The Upper Carboniferous bears fusulinids Profusulinella, Fusulinella, Triticites, Pseudoschwagerina; brachiopods Choristites, Natannella and corals Kepingophyllum, Axolithophyllum, Caninia, etc., which are apparently of the Cathaysia-Tethyan type. Noticeably, in the Tanggula-Hengduan Mts. region the Early Carboniferous continental intercalations carry fossil plants, such as Archaeocalamites scrobiculatus (Schloth.), Sphenophyllum tenerrinum Ett., Cardiopteridium spitsbergense Nath, which are the same species as those in the contemporaneous flora of South China.

The Early Permian deposits have a wide distribution in Xizang; but the Upper Permian was less widely distributed, which is probably due to the withdrawal of sea water from vast areas of Xizang at that time. In Permian the marine faunas like the Carboniferous ones, may be grouped into 2 types (see Table 4). In the early Early Permian the faunas of the Gondwana-Tethyan type are seen in the Himalayan region, the Gandise-Nyainqentanglha region (8,9), and those of the Cathaysia-Tethyan type are visible in the Tanggula-Hengduan Mts. region. In the late Early Permian the faunas of the Gondwana-Tethyan type (not including those from the exotic blocks) are observed in the Himalayan region and those of the Cathaysia-Tethyan type are present in the rest of Xizang (10). Near Lhasa and Xainza the upper Lower Permian lies directly on the lower Lower Permian. The former carries the faunas of the Cathaysia-Tethyan type and the latter yield those of the Gondwana-Tethyan type. In the Tanggula and Qamdo subregions and in the vicinity of Lhasa, the Late Permian faunas are entirely of the Cathaysia-Tethyan type.

On discussing the Permian fusulinid biogeographical province, Wang Yu-jing et al. first proposed the names of the Gondwana-Tethyan type and Cathaysia-Tethyan type. Here we apply the two terms to all the Permian and Late Carboniferous marine faunas. The faunas of the Gondwana-Tethyan type mainly include solitary corals and large-sized, thick-shelled brachiopods, but no or rare fusulinids are involved. It seems that they were living in the sea water of lower temperature. But the faunas of the Cathaysia-Tethyan type contain the most abundant fusulinids in addition to some reef-building corals and some brachiopods. They were likely to live in warmer marine water.

In Permian time the continental floras, same as the marine faunas, may be divided into two types. In Early Permian the plants of the Gondwanaland type, such as Glossopteris, Raniganjia, Dizeugotheca and Dichotomopteris occurred in the Himalayan region (8,11); and in Late Permian forms of the Cathaysia flora belonging to the Eurasia, namely Gigantopteris, Gigantonoclea, Lobatannularia and Rhizomopsis, etc. were present in the Tanggula-Hengduan Mts. region (3.10).

## Triassic-Tertiary

Comparatively the Triassic-Tertiary deposits in Xizang are much more complicated. Nevertheless, all the marine faunas obviously belong to the Tethyan type. Subsequent to the regression, the vast area of Xizang has uplifted into land ever since.

The Triassic deposits have hitherto not been found in the Gandise-Nyainqentanglha region, apart from its southern marginal area (vicinity of Lhasa) and northern marginal area (Biru subregion). However, the Triassic deposits are extensively distributed in other stratigraphic regions. From the aspect of sedimentation, in the southern part of Hoh Xil-Kunlun region, succeeding the Permian the Triassic is still a flysch formation, which belongs to the relatively active type (3). In Muggarkangri subregion, Biru subregion and the Northern subregion of Tethys-Himalayas, two sedimentary belts of relatively active type successively exist, but in the rest of Xizang the Triassic is composed of shallow-sea carbonate rocks and quartz sandstones, belonging to the relatively stable type. In fact, the configuration of the present 6 sedimentary belts was roughly shaped out at that time already. From the palaeontological viewpoint, the faunas flourished much more fully in Triassic than in other geological periods (1,12). The Early Triassic fauna includes ammonoids Otoceras, Ophiceras, Gyronites, Procarnites and bivalves Claraia, Eumorphotis, etc. The Middle Triassic fauna comprises ammonoids Japanites, Anacrochordiceras, Ptychites, Protrachyceras, Hoplotropites, Parahaurites and bivalve Daonella, etc. The Late Triassic is prolific in fossils, containing ammonoids Nodotibetites, Griesbachites, Indojuvavites, Cyrtopleurites, Pinacoceras and bivalves Halobia, Lilangina, Burmesia, Prorotrigonia, Indopecten, Pergamidia and the Ichthyosaurus Himalayasaurus. In the Tanggula-Hengduan Mts. region, the Upper Triassic continental coal-bearing beds carry fossil plants Equisetites arenaceus (Jaeger), Hyrcanopteris sevanensis Krysh. et Prgn., Pterophyllum minutum Li et Tsao, Nilssoniopteris jourdyi (Zeiller), which share the same forms as those from the contemporaneous sediments of Southwest China that can be referred to the flora of Eurasian continent.

The Jurassic, same as the Triassic, is known to crop out in the same regions of Xizang with the exception of the Hoh Xil-Kunlun region where the marine deposition came to a close at that time. Comparatively, the Lower Jurassic has a less wide distribution; whereas the Middle and Upper Jurassic are outcropped in a vast area of Xizang, which is probably due to the large-scale transgression at that time. It is interesting to note that in the Gandise-Nyainqentanglha region the Jurassic deposits, just like the Triassic, are only developed in the marginal areas, which look like a continental barrier extending across the central part of Xizang, causing the distinctive difference in faunal and sedimentary characteristics between the Himalayan region in the south and the Tanggula-Hengduan Mts. region in the north. Here special mention should be made of the Himalayan region where the Jurassic is developed rather completely and contains the commonest forms of the Tethyan type together with some endemic elements of the Himalayas (1). In the Tanggula-Hengduan Mts. region the Lower and Upper Jurassic are poorly exposed, but the Middle Jurassic is best developed (3, 13). The main fossils are brachiopods Burmerhynchia, Holcothyris, Kallirhynchia; and bivalves Chlamys (Radulopecten), Meleagrinella, Pseudotrapezium, Ceratomya. This fauna bear remarkable endemic features, though it strongly resembles those from the contemporaneous deposits in western Yunnan of China and in Shan State of Burma. Furthermore, it also includes Psilunio, Lamprotula (Eolamprotula), Undulatula, Cuneopsis, Margaritifera, which are all of the Asian fresh-water type found in the continental intercalations.

From Early to Late Cretaceous the marine deposits were gradually diminishing, while the continental deposits were expanding. These marine deposits with the Tethyan fauna are distributed from west to east in 4 unequal banded areas as follows: 1) Southern subregion of Tethys-Himalayas (where the marine Cretaceous is exposed most completely, containing very abundant fossils) (1, 14). 2) Northern subregion of Tethys-Himalayas (where the Cretaceous is composed of radiolarian siliceous rocks and flysch formation with intercalations of intermediate and acid volcanic rocks, which are known to be of relatively active type) (15,16). 3) Shiquanhe-Xainza subregion and the western part of Lhasa-Bomi subregion, and 4) Karakorum subregion (an extension from southwestern Xinjiang southeastwards to Xizang).

In Xizang the Cretaceous continental deposits vary, in ascending order, from dark-colored deposits in the west to red beds in the east, that is to say, the Early Cretaceous continental strata are mainly of coal-bearing deposits, and the Late Cretaceous largely belongs to red beds. In these deposits there occur bivalves Trigonioides (Trigonioides) sinensis Gu et Ma, T. (Diversitrigonioides) xizangensis Gu, the reptilia Homalocephale cf. calathocercos (Maryanska) and fossil plants Weichselia reticulate (Stokes et Webb), Onychiopsis elongata (Geyler), cf. Klukia browniana (Dunker), Elatides curvifolia (Dunker) etc., all of which belong apparently to the Eurasian type.

On the north of the Yarlung Zangbo River, the Upper Cretaceous is composed of red coarse clastic rocks. Generally, between the Upper Cretaceous and the underlying beds there exists an unconformity. The underlying bed is of the Cenomanian age, consisting of marine limestones with foraminifera Orbitolina bangoinica Zhang, which indicates that a tectonic movement most probably took place in Post-Cenomanian.

In Tertiary time the marine water retreated from vast areas of Xizang, except the southernmost part of southern subregion of Tethys-Himalayas (1), Zhongba and the vicinity of Lhasa (17), where the marine deposits were still in existence. In the southern subregion of Tethys-Himalayas, these deposits are composed of shallow-sea carbonate rocks, which are apparently of relatively stable type. The Palaeocene contains foraminifers Keramosphaera tergestina (Stache), Miscellanea miscella (d'Archiac et Haime) and gastropods Campanile ganesha (Noetling), Confusiscala indica Douvillé, etc. The highest marine bed in Xizang yields larger foraminifers Orbitolites, Fasciolites, Nummulites, Assilina, Discocyclina, etc., which are of the Lutetian age (Middle Eocene).

The Tertiary continental deposits are mostly of lacustrine or, sometimes, of fluviatile facies and of piedmont sediments. In northern Xizang the Tertiary rocks are red clastic sediments, reflecting an arid sedimentary environment. In the northernmost part they carry a large quantity of thin-bedded gypsum. Contrastingly in southern Xizang the Tertiary is composed of dark-colored coal-bearing beds or coal bands which were deposited in a humid climate.

## THE FORMATION OF STRATIGRAPHIC REGION

As stated above, from Ordovician to Devonian, the whole of Xizang was in a sedimentary area of relatively stable type. Later in Carboniferous and Permian, a sedimentary belt of relatively active type came into existence in the Hoh Xil-Kunlun region; whereas in other regions the sedimentary belts of relatively stable type remained unchanged. The sedimentary belt of relatively active type ceased to exist in the Hoh Xil-Kunlun region at the end of Triassic. But, it is interesting to note, in Triassic there appeared successively two sedimentary belts of relatively active type. One was in the Muggarkangri and Biru subregions, and the other in the northern subregion of the Tethys-Himalayas. Although the two sedimentary belts continuously developed in Jurassic, the former belt came to a close in Late Jurassic. In Cretaceous the sedimentary belt of relatively active type occurred only in the Northern subregion of the Tethys-Himalayas, but it was out of sight at the end of Cretaceous. It seems that the sedimentary belts of relatively active type appeared successively from north to south. Furthermore, the differentiation existed in sedimentary development between different geological ages, so that various stratigraphic regions and subregions were gradually formed in Xizang.

On a whole, the Palaeozoic sediments, covering a vast area of Xizang, were mainly of relatively stable type. From the beginning of the Mesozoic, the sedimentary area of relatively stable type became diminishing; whereas that of relatively active type came into existence and expanded its scope successively and increasingly. This process indicates that the original sedimentary area of relatively stable type were cut up into pieces.

## References


(1) Xizang Scientific Expedition, Academia Sinica, A Report of Scientific Expedition in the Mount Jolmo Lungma Region (1966-1968), Geology, 1-222, Science Press (1974) (in Chinese).

(2) Mu En-tze, Yin Chi-hsiang, Wen Shih-hsuan, Wang Yi-kang, Chang Ping-kao, Stratigraphy of the Mount Jolmo Langma Region in Southern Tibet, China. Scientia Sinica, 16 (1), 96-111 (1973).

(3) Wen Shi-xuan, New materials of biostratigraphy in the Northern Qiangtang Plateau. Acta Stratigr. Sinica, 3 (2), 150-156 (1979) ( in Chinese).

(4) Wang Yi-gang, Sun Dong-li, He Guo-xiong, New advances in the stratigraphy of the Himalayas (China's side). Journ. Stratigr., 4 (1), 55-59 (1980)(in Chinese).

(5) Dong De-yuan, Mu Xi-nan, Sun Dong-li, Chen Xu, Lin Yao-kun, Lower Ordovician strata in the vicinity of Qunyido, Jomda County of Eastern Xizang (Tibet). Acta Stratigr. Sinica, 3 (3), 230 (1979) (in Chinese).

(6) Liang Shou-sheng, Xia Jin-bao, Discovery of Silurian strata in North Xizang. Acta Palaeont. Sinica, 17 (1), 99-100 (1978) (in Chinese).

(7) Li Pu, A preliminary study of the geology in Eastern Tibet. Kexue Tongbao, (7), 62-71 (1955).

(8) Yin Ji-xiang, Guo Shi-zeng, On the discovery of the stratigraphy of Gondwana facies in northern slope of the Qomolangma Feng in Southern Xizang, China Scientia Geol. Sinica, (4), 291-322 (1976) (in Chinese).

(9) Ching Yu-kan, Liang Xi-luo, Wen Shih-hsuan, Additional material of animal fossils from the Permian deposits on the northern slope of Mount Qomolangma Feng. Ibid., 1977 (3), 236-249, 2pls. (1977) (in Chinese).

(10) Zhang Lin-xin, Mu Xi-nan, Sun Dong-li, Dong De-yuan, New observation on the "Toba Coal Series" of Qamdo, Xizang (Tibet). Acta Stratigr. Sinica, 3 (3), 231-232 (1979) (in Chinese).

(11) Hsü Jen, On the discovery of a Glossopteris flora in Southern Xizang and its significance in geology and palaeogeography. Scientia Geol. Sinica, 1976 (4), 323-331 (1976) (in Chinese).

(12) Gu Qing-ge, Triassic marine strata in Tingri area, Xizang. Geol. Review, 23 (4), 256-262 (1965) (in Chinese).

(13) Sun Dong-li, Zhang Bing-gao, Some problems of the marine Jurassic strata of the Karakorum-Tanggula area. Acta Stratigr. Sinica, 3 (4), 317-321 (1979) (in Chinese).

(14) Hayden, H., The geology of the provinces Tsang and U in Tibet. Mem. Geol. Surv. India, 36 (2), 122-201 (1907).

(15) Wang Yi-gang, Wang Yu-jing, Wu Hao-ruo, Discussion on the Jiabula formation and the discovery of Longzi Lower Jurassic in Southern Xizang, China. Scientia Geol. Sinica, 1976 (2), 149-156 (1976) (in Chinese).

(16) Wu Hao-ruo, Wang Dong-an, Wang Lian-chen, The Cretaceous of Laze-Jiangze district, Southern Xizang. Ibid., 1977 (3), 250-262 (1977) (in Chinese).

(17) Zhang Bing-gao, Mu Xi-nan, Discovery of the marine Tertiary rock from the north of the Yarlung Zangbo Jiang, Tibet. Acta Stratigr. Sinica, 3 (1), 65-66 (1979) (in Chinese).

Table 1 Division of stratigraphic regions in Xizang

| Stratigraphic region | Stratigraphic subregion |
|---|---|
| Himalayan Region (I) | Southern Subregion of Tethys Himalayas (I)<br>Northern Subregion of Tethys Himalayas ($I_2$) |
| Gandise-Nyainqentanglha Region (II) | Shiquanhe-Xainza Subregion ($II_1$)<br>Lhasa-Bome Subregion ($II_2$)<br>Biru Subregion ($II_3$) |
| Tanggula-Hengduan Mts. Region (III) | Muggarkangri Subregion ($III_1$)<br>Karakorum Subregion ($III_2$)<br>Tanggula Subregion ($III_3$)<br>Qamdo Subregion ($III_4$) |
| HohXil-Kunlun Region (IV) | HohXil Subregion ($IV_1$)<br>Kunlun Subregion ($IV_2$) |

Table 2 Division and Correlation of Palaeozoic Strata in Xizang

| Age | | Himalayan Region (I) | | Gandise–Nyainqentanglha Region (II) | | | Tanggula–Hengduan Mts Region (III) | | | | Hohxil–Kunlun Region (IV) | |
|---|---|---|---|---|---|---|---|---|---|---|---|---|
| | | Southern Subr. of Tethys Himalayas (I₁) | Northern Subr. of Tethys Himalayas (I₂) | Shiquanhe–Xainza Subr. (II₁) | Lhasa–Bome Subr. (II₂) | Biru Subr. (II) | Muggar-kangri Subr (III₁) | Karakorum Subr. (III₂) | Tanggula Subr. (III₃) | Qamdo Subr. (III₄) | Hohxil Subr. (IV₁) | Kunlun Subr. (IV) |
| P | $P_2$ | | | | Lielonggou Fm. | | | | Raggyorcaka Fm. | Kaxamda Fm.<br>Toba Fm. | | |
| | $P_1$ | Selung Gr.: Quburega Fm.<br>Qubu Fm. | Baidinpu Fm.<br>Kangmar Fm. | Xiala Fm. | Lobadoi Fm.<br>Urulung Fm. | | | Minzhocaka Ls.<br>Toinlungkongba Fm. | ? Kamrbo Fm. | Tyokar Fm.<br>Manco Fm. | Ninggongqu–Jiulong Fm. | Epimetamorphic sandstone & shale |
| C | $C_2$ | Jilong Gr.: Quartz sandstone Mem.<br>Siltstone Mem.<br>Diamictite Mem. | Bilong Fm.<br>Kewoxiga Fm. | Angjie Fm. | Pondo Gr. | | | Lungmuco Gr. | | Lica Gr.<br>Aoqu Gr. | | Variegated sandstone intercalated with marlite |
| | $C_1$ | ?<br>Naxing Fm.<br>Yali Fm.: Upper part | Quartz sandstone<br>Sandy shale | Yongzhuqiao Gr. | ?<br>Nuoco Fm. | | | Sandstone, Conglomerate & Limestone | ?<br>Macala Fm.<br>Limestone & sandy shale | Macala Fm.<br>Wuqingna Fm. | | Sandstone intercalated with limestone |
| D | $D_3$ | Yali Fm.: Lower part | | Chagguoluoma Fm. | Yarze Fm.<br>Kongboshan Fm. | | | Lazhulong Fm. | Jianhu Gr. | Qiangge Fm.<br>Zhuogedong Fm. | | |
| | $D_2$ | Pochu Gr. | Quartz sandstone<br>Marlite | Dzerdong Gr. | Niantang Fm. | | | Yaxier Gr. | | Dingzonglong Fm.<br>Haitong Fm. | | |
| | $D_1$ | Liangquan Fm. | | | | | | | | | | |
| S | $S_{2+3}$ | Pulu Gr. | | Phyllites & sandstone | | | | | | | | |
| | $S_1$ | Shiqipo Fm. | | ? Dewulaxia Fm. ? Dongka Fm. | | | | Feldspathic–Quartz sandstone | | | | |
| O | $O_3$ | Hongshantou Fm. | | Gangmusang Fm. | | | | | | | | |
| | $O_2$ | Chiatsun Gr.: Upper Fm. | Marlite | Keerduo Fm. | | | | | | | | |
| | $O_1$ | Chiatsun Gr.: Lower Fm. | | Mendeema Fm. | Goqen Fm. | | | | | Qunyido Fm. | | |
| Є | | Rouqiecun Gr.<br>—Fault—<br>Nyalam Gr. | | Nyainqentanglha Gr. | | —Fault—<br>Jiayuqiao Gr. | | | —Fault—<br>Various schists | | | |

Table 3 Division and Correlation of Mesoozoic and Cainozoic Strata in Xizang

| Region / Age | | Himalayan Region (Ⅰ): S. Subregion of Tethys Himalayas (Ⅰ₁) | Himalayan Region (Ⅰ): N. Subregion of Tethys Himalayas (Ⅰ₂) | Gandise–Nyainqentanglha Region (Ⅱ): Shiquanhe–Xainza Subregion (Ⅱ₁) | Gandise–Nyainqentanglha Region (Ⅱ): Lhasa–Bome Subregion (Ⅱ₂) | Gandise–Nyainqentanglha Region (Ⅱ): Biru Subregion (Ⅱ₃) | Tanggula–Hengduan Mts Region (Ⅲ): Muggarkangri Subregion (Ⅲ₁) | Tanggula–Hengduan Mts Region (Ⅲ): Karakorum Subregion (Ⅲ₂) | Tanggula–Hengduan Mts Region (Ⅲ): Tanggula Subregion (Ⅲ₃) | Tanggula–Hengduan Mts Region (Ⅲ): Qamdo Subregion (Ⅲ₄) | HohXil–Kunlun Region (Ⅳ): HohXil Sub region (Ⅳ₁) | HohXil–Kunlun Region (Ⅳ): Kunlun Sub region (Ⅳ₂) |
|---|---|---|---|---|---|---|---|---|---|---|---|---|
| N | | Woma Fm. | Luobusha Gr. | Lunpola Fm. | Oiyug Fm.<br>Manxiang Fm. | Bilung Fm. | Brown–red sandstone & siltpelite | | Purple–red sandstone | Lawula Gr. | Brick–red sandstone | Brick–red sandstone |
| E | | Zhepure Fm.<br>Zongpu Gr. | ?<br>Maila Congl. Fm. | Niubao Fm. | Rigongla Fm.<br>Limestone | ? | | | Dengqen Fm.<br>Kyangngoin Fm. | ? | | |
| K | K₂ | Jidula Fm.<br>Zongshan Fm.<br>Kangba Gr.: Kangbacun-kou Fm. | Xigaze Gr. / Zong-zhuo Fm. | ?<br>Jingzhushan Fm. | ?<br>Lingzizong Fm. | Zonggei Gr. | | ?<br>Red bed intercalated with marlite | Purple–red sandstone & conglomerate | Konjo Gr. | | |
| | K₁ | Kangba Gr.: Chaqiala Fm.<br>Kangba Dongshan Fm. | Jiabula Fm. | Langzhan Fm.<br>Doiba Fm.<br>Chuanba Fm. | Takna Fm.<br>Chumulong Fm.<br>Leibuzong Fm. | ?<br>Duoni Fm. | | | | ?<br>Qamdun Gr. | | |
| J | J₃ | Menkatun Fm. | Weimei Fm. | Volcano–clastic rocks | Lime-stone<br>? / Lagong-tang Fm. | Lagong-tang Fm. | ?<br>Dongqo Fm. | ?<br>Jieshandaban Fm. | | | | |
| | J₂ | Niehnieh Hsiungla Fm. | Gyangze Gr. | Daheishan Fm. | Liuwun Fm. | Guoqu Gr. / Liuwun Fm. | Muggarkangri Gr. | | ?<br>Yanshiping Gr.<br>Sewa Fm. | Chag-yab Gr. / Liuwun Fm.<br>?<br>Lime-stone | | |
| | J₁ | Pupuga Fm. | Ritang Fm. | ? | | | Fault<br>Genalongba Gr. | | | | | |
| T | T₃ | Derirong Fm.<br>Qulonggongba Fm.<br>Tulung Gr.: Dashalong Fm.<br>Zhamure Fm. | Gyilong Fm.<br>Langjiexue Gr. | | Mailonggang Gr. | Sandstone and Slates | | | ?<br>TumaingelaFm.<br>Gyiza Gr. | ?<br>Bagong Fm.<br>Bolila Fm.<br>Jiapila Fm. | Upper Bayanhar Shan Gr. | |
| | T₂ | Tulung Gr.: Laibuxi Fm. | Shale intercalated with thin–bedded limestone | | Chaqupu Gr. | | | | ?<br>Kangnan Fm. | Congla Fm.<br>Walasi Fm. | | |
| | T₁ | Kanshare Fm. | Sandy shale intercalated with limestone | | | | | | Ningshuiquan Fm.<br>Kanglu Fm. | Serongsi Fm.<br>Pushuiqiao Fm. | | |

Table 4 Different type of fauna in the Permian marine deposists of Xizang

| age | Cathaysia —Tethyan type | | Gondwana — Tethyan type | |
|---|---|---|---|---|
| $P_2$ | fusulinids<br>brachiopods | *Reichelina, Palaeofusulins*<br>*Peltichia zigzag* (Huang),<br>*Cathaysia chonetoidea* (Chao) | | |
| $P_1^2$ | fusulinids<br><br><br>corals | *Neoschwagerina, Sumatrina,*<br>*Verbeekina, Lantschichites,*<br>*Yabeina*<br>*Iranophyllum, Wentzelella,*<br>*Ipciphyllum* | brachiopods | *Costiferina,*<br>*Marginifera* |
| $P_1^1$ | fusulinids | *Misellina, Cancellina,*<br>*Szechuanophyllum, Yatsengia* | Corals<br>brachiopods<br><br><br>ammonoids<br>bryozoans | *Lytvolasma, Wannerophyllum*<br>*Taeniothaerus. Calliomarginatia,*<br>*Chonetella,*<br>*Neospirifer kubeiensis* Ting<br>*Uraloceras*<br>*Streblotrypa marmionensis* Etheridge |

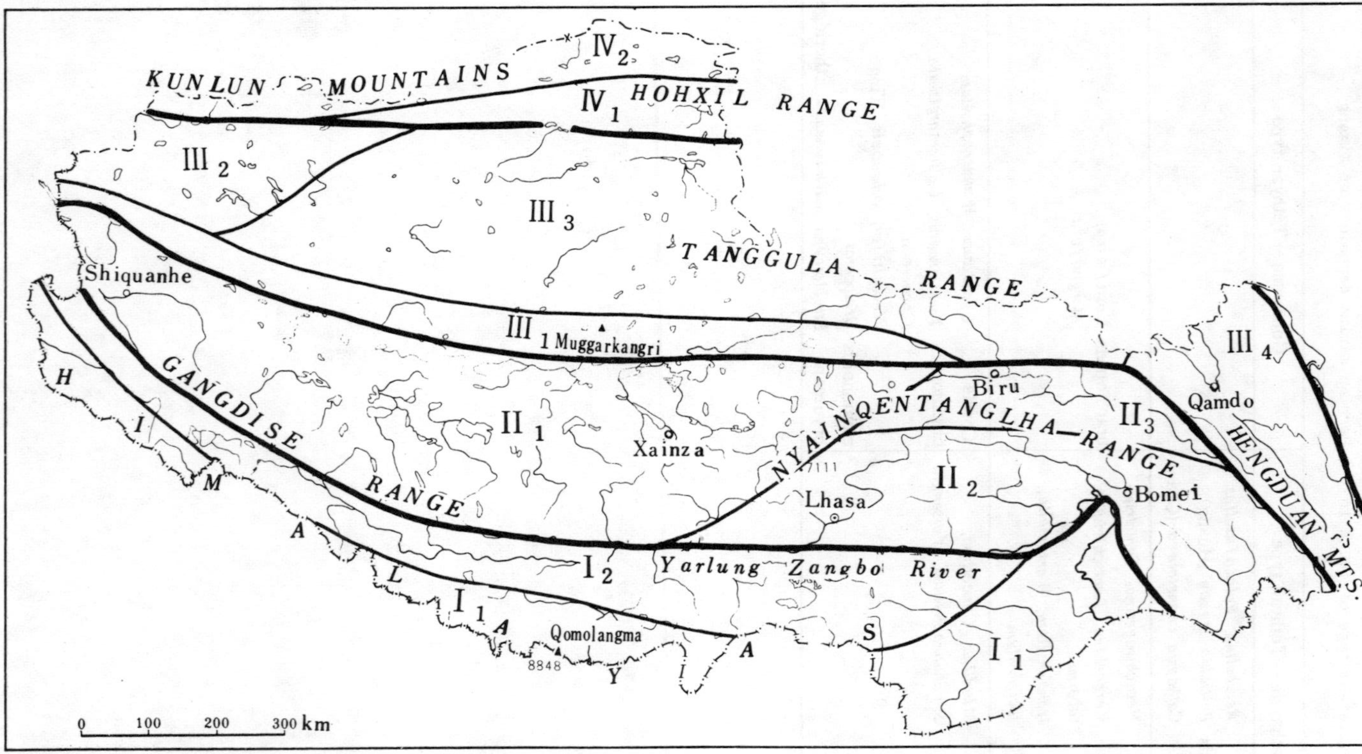


Fig. 1 Map showing the division of stratigraphic region in Xizang

# ALTITUDES OF TERTIARY FORESTS ESTIMATED FROM PALEOTEMPERATURE*

Daniel I. Axelrod
Dept. Botany, Univ. California, Davis, Calif. 95616 USA

To estimate altitude of a Tertiary flora, an adequate sample is required because each vegetation zone has a range in elevation. Some species may be widely distributed in it, or extend into other zones, but some are more restricted. A good sample indicates whether the fossil flora is similar to the upper, lower or middle part of related modern forests.

Altitude can be inferred from the temperature requirements of derived modern forests. Mean annual temperature (T) alone is not sufficient. The annual range (A) must also be estimated because it provides the annual wave of temperature that influences the modern flora similar to the fossil. Plotting the data on a Bailey nomogram (Fig. 1) provides an indication of a) warmth of climate (W) by radii of specified temperature that define the duration of the growing season as the number of days (d) with mean temperature warmer than W, and b) the temperateness or equability of climate (M) shown by a field of arcs nearly normal to the W radii that provide an index to thermal extremes that increase in all directions from an ideal centered at T 14°C and A 0°. The nomogram also shows frost frequency lines significant in determining distribution today, as well as the major biotherms of treeline (W 10°C) and the edge of the tropics (W 18°C). For any point defined by the coordinates T and A, the mean monthly temperatures can be recovered if it is assumed that the march of temperature lies under a sine wave. Since all factors are interrelated, the nomogram can be used to assess any temperature field, either today or in the past (1-6).

To estimate altitude of a Tertiary flora, reference is made to a flora at sealevel. The difference in mean temperature, when multiplied by a normal terrestrial lapse rate (183 m / -°C) gives an approximation of elevation. The following examples illustrate some key points in constructing the nomogram (1) and its use in estimating altitude (3-6).

Fig. 2 We start with a point representing a meterological station with an adequate record, preferably 20 years or more. Temperature is plotted on a graph in which the scale of mean annual temperature (T) is half that of mean annual range of temperature (A). In Fig. 2, mean annual temperature is 15°C, the range is 20°C. The mean temperature of the cold month (tc) is 5°C and mean temperature of the warm month (tw) is 25°C. Note that tc 5° + tw 25°C = 30°C, which when divided by 2 gives the mean temperature (T) of 15°. Also note that tw 25° - tc 5° = 20°C, which is the mean annual range (A). These relations are true for any station in which the ratio of T to A is 1:2. Further, any line that diverges upward at 45° represents tw, whereas any line diverging downward at 45° from the T scale represents tc (see 1, fig. 1).**

---

*Research supported by National Science Foundation, U.S.A.

**The relations in Fig. 2, presented initially by (1), were modified to a 1:1 ratio for T and A in order to show climatic parameters of forests using tw and tc only (7,8). The method does not add to an understanding of either present or past climate, because neither the duration of the growing season (W) nor the temperature extremes (M) are measurable. Further, system can not provide a method for estimating altitude.

Fig. 3 Each station defined by T and A obviously includes the mean temperatures of all 12 months. If the march of temperature falls under a sine wave, which is normal for most areas, the mean monthly temperatures can be recovered. The chief exceptions are monsoonal areas where tw falls earlier than July and fog-bound coasts with tw delayed to late summer. The normal sine wave can be divided into 12 equal parts by the ratio of 0.0, 0.5, 0.866 and 1.0, using a year of 12 months of 30 days each. In Fig. 3, the solid lines diverging to the T axis show the mean temperatures of the paired months.

Fig. 4 shows a comparison of the actual and the simulated temperatures for Beijing and Kunming. The latter were determined from the T and A of each station which define a point for centering the template (Fig. 3). The curves for Beijing are quite similar, with only a slight difference in mean temperatures of the warm and cold months. For Kunming, the chief difference is the flattened curve for summer owing to the heavy cloud cover during the rainy season. Clearly, if we can make a reasonable estimate of T and A of a fossil flora, monthly temperatures can be inferred.

Fig. 5 illustrates the temperature requirements of the major forests of California in terms of tw and tc only. The dots represent actual stations in the forests. At low elevations, the upper limit of the oak-pine woodland (Quercus-Pinus) and mixed conifer forest (Abies, Pinus, Pseudotsuga, Calocedrus) is controlled by tc 5.6° and -1°C, respectively. By contrast, the distribution of the fir (Abies) and subalpine (Abies, Tsuga, Pinus) forests in the higher mountains is limited by tw of 17° and 14°C, respectively. Note that the inland distribution of the coastal Sequoia forest is limited by a frost frequency of about 1% of the hours of the year. The oak-madrone-tanoak (Quercus, Arbutus, Lithocarpus) woodland has its inner distribution in areas where tw is about 21°C.

Fig. 6 shows the major biotherms of the earth. Treeline is largely determined by tw 10°C, either in high mountains or in polar regions. This biotherm has 0 days with a mean temperature warmer than 10°C. The outer margin of the tropics is defined by tc 18°C, where all days (365) of the year have a mean temperature warmer than 18°. Summerless climates lie above the line of 10°C, within the lines of 18°C are winterless or tropical climates, and between them in the middle latitudes are temperate climates, ranging from warm to cold temperate.

Fig. 7 shows the major biotherms plotted on a graph in which the ratio of T and A is 1:2. Midway between them at T 14° there are 183 days warmer than 14°C. The graph can be divided further for each degree or half degree according to the equation by Bailey (see Fig. 1; 1,4).

Fig. 8 depicts the lines of warmth (W) diverging from the T axis. Note that in areas with a mean temperature less than 14°C we move into areas of colder, longer winters and shorter, cooler summers. Conversely, in areas with T greater than 14°C, climates have shorter, warmer winters and longer, hotter summers. And increased A brings greater ranges of temperature (I, II, III).

The arcs spreading out from the centrum T 14° and A 0° provide a measure of climatic extremes expressed in terms of equability (see Fig. 1). The centrum is midway between polar and tropical climates and is approximately the mean temperature of the earth's surface. At the center, the rating is M 100 and it decreases outward in all directions falling to 0 on Antarctica.

Equability at Kunming is M 64, at Beijing it is M 42. Kunming has only 0.4% of the hours of the year with frost, but Beijing as 10%.***

The arrangement of the thermal scale corresponds with the real world. Warm temperatures are at the bottom (equatorial), cooler ones at higher levels (polar). It also depicts a rise in elevation from a low or moderate level to a higher, cooler zone. Note also that mean annual temperature (T) alone has no climatic or biologic significance, because the lines of warmth (W) increasingly transgress the T scale as one moves away from 14°C.

Fig. 9 charts the forests of California (see Fig. 5) in terms of warmth (W) and equability (M). The parameters of tw and tc are still there (Fig. 2), as are the mean monthly temperatures for each station (Fig. 3). Emphasis is placed on the fact that since the lines of warmth (W) delineate the duration of growing seasons, their significance greatly exceeds that of tw or tc which can only indicate the thermal limits of a particular vegetation zone, and can not provide a measure of either the duration of the growing season or the thermal extremes (cf. 7). Note that forests of similar warmth (i.e. W 13.5, or 168 days warmer than 13.5°C) in California (redwood, oak madrone, mixed conifer) are separated by their equability rating, a critical fact not determinable from a mere plot of tc and tw of the stations as has been supposed (7).

Fig. 10 shows similar relations for the major forests of eastern Asia. Stations of comparable warmth are widely separated in terms of equability. The mixed mesophytic forest of Japan has a similar warmth (W) as the mixed evergreen forest of the Himalayan mid slope. Both Darjeeling 1 and Fukushima have similar warmth (W 13.5°C) but the differences in vegetation (more evergreens at Darjeeling) reflect the difference in equability (M 64 vs M 47). Note also that the simple deciduous forest and mixed conifer forest, as well as the mixed mesophytic and mixed northern hardwood forests are separated chiefly by their equability (M) rating for they largely have similar warmth (W).

To evaluate the elevation of a Tertiary flora, we must remember that climatic conditions differed from those of today. There were no major ice sheets until late in the Tertiary, deserts had not yet spread to subcontinental extent, continuous mountain chains and high plateaus were lower or not yet in existence, and ocean temperatures were warmer in middle and high latitudes. Since the global thermal gradient was lower, conditions were more equable everywhere so differences in lapse rate could not have been pronounced, contrary to (7). Therefore, in estimating the elevation of a Tertiary forest we must take into account the fact that the range of temperature (A) was less than in most modern forests.

Fig. 11 illustrates this principle by comparing 2 floras in the lowlands with 2 in the mountains that live under similar warmth (W). In transect A-A' a difference in T of 9°C separates the floras, but in B-B' the difference is 19°C. B' is 1830 m higher than A' even though they have the same warmth (W).

---

***Contrary to one investigator (7,8) a low range of temperature does not define an equable climate. Singapore (A 1.7°C) and Bogota (A 1.4°C) both have a very low range but can scarcely be considered similar in equability; Singapore (T 26.7°C) has a rating of M 45 but Bogota (T 14.2°C) has a rating of M 93! Even Campbell I. (Lat. 52°S), with a greater range (4.9°C) than Singapore, is more equable (M 55), as is San Francisco (M 74) with a range of 6.3°C.

Thus, in estimating the altitude of a Tertiary flora we must infer an A less than that in modern forests. In general, it seems safe to assume that fossil forests lived under conditions no more extreme than those in the mildest parts of the related, derived modern forests, and certainly were more equable in the Miocene to Eocene.

Fig. 10 provides evidence to estimate the distance from the Miocene lower Namling flora (9), Tibet, to treeline. The flora is reported to be a deciduous hardwood forest. If we assume that Miocene temperatures were like those in the upper part of the forest (Fig. 10), and that the range of temperature (A) was no greater than that in the mildest part of the modern forest (it certainly was less), than 13°C is the difference between the Namling flora and treeline, or about 2350 m (13° x 183 m / °C). What was the elevation of the lower Namling flora? The coastal strip at this latitude supported a rich evergreen sclerophyll forest now under study by Dr. Li Hao-min, and above it was a mixed mesophytic forest as shown by the Shangwang flora which resembles the moderm forest in the central Yangste basin (10). Comparison with thermal conditions estimated for the Namling flora indicates the difference in mean temperature between it and sealevel was about 8°C. This implies that the Namling flora had an elevation near 1500 m as compared with 3800 m today, suggesting about 2400 m uplift since the Miocene. Modification of this estimate would depend on a larger, more adequate sample of the flora, and on a better knowledge of the modern Asian forests than I now possess. Nonetheless, this method of estimating altitude has given acceptable results (3-6), and can do so in eastern Asia.

## References


(1) Bailey, H. P., A method of determining the warmth and temperateness of climate. Geografisker Ann. 47, 1-16, 1960.

(2) Bailey, H. P., Toward a unified concept of the temperate climate. Geogr. Review 54 (4), 516-545, 1964.

(3) Axelrod, D. I., A method for determining the altitudes of Tertiary floras. The Paleobotanist 14, 144-171, 1965.

(4) Axelrod, D. I. and H. P. Bailey, Tertiary vegetation, climate and altitude of the Rio Grande depression, New Mexico-Colorado. Paleobiology 2, 235-254, 1976.

(5) Axelrod, D. I., Tertiary floras and topographic history of the Snake River Basin, Idaho. Geol. Soc. Amer. Bull, 79, 713-734, 1968.

(6) Axelrod, D. I., Contributions to the Neogene Paleobotany of central California. Univ. Calif. Pub. Geol. Sci. 121, 1-212, 1980.

(7) Wolfe, J., Temperature parameters of humid to mesic forests of eastern Asia and relation to forests of other regions of the Northern Hemisphere and Australasia. U. S. Geol. Surv. Prof. Paper 1106. 37 p.

(8) Wolfe, J. A., Tertiary climates and floristic relationships at high latitudes in the Northern Hemisphere. Paleo. Paleo. Paleo. 30, 313-323, 1980.

(9) Li Haomin and Guo Shuang-xing, The Miocene flora from Namling of Xizang. Acta Palaeontologica Sinica 15 (1), 7-18, 1976.

(10) Hu, H. H. and R. W. Chaney, A Miocene flora from Shantung Province, China. Carnegie Inst. Wash. Pub. 507. 147 pp.

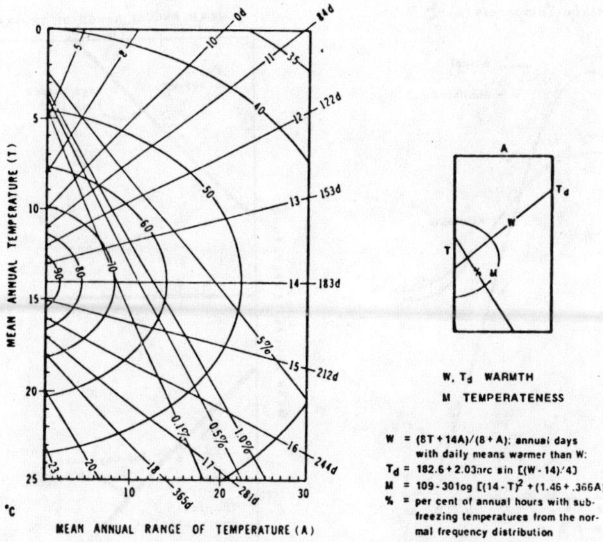


Fig. 1. Bailey nomogram (1, 2) showing warmth (W) and equability (M) of climate, and frost-frequency lines.

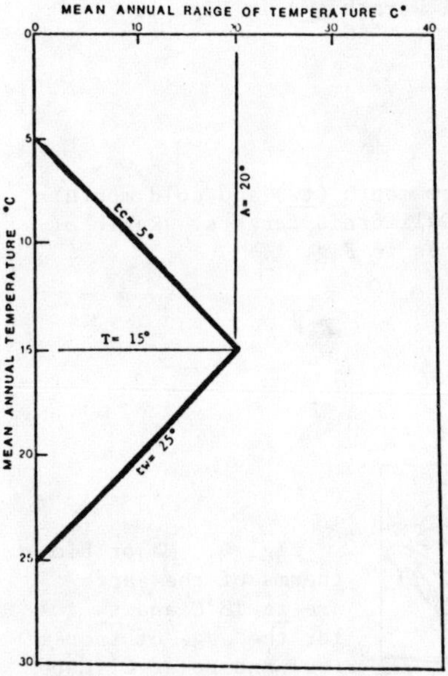


Fig. 2. Relations of mean annual temperature (T) and mean annual range of temperature (A) on which nomogram (Fig. 1) is based. Ratio to T to A is 1:2.

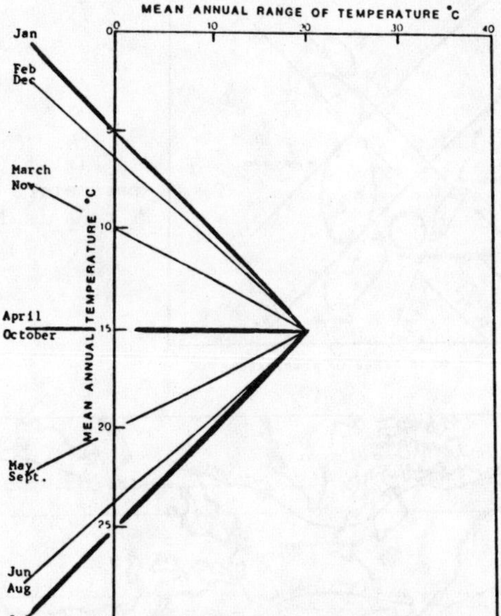


Fig. 3. Template for simulating mean monthly temperatures, either today or in the past.

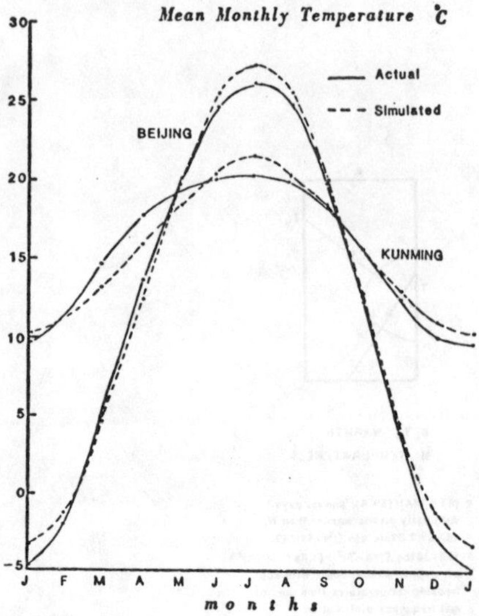


Fig. 4. Comparison of actual with simulated temperatures using Fig. 3 as a template.

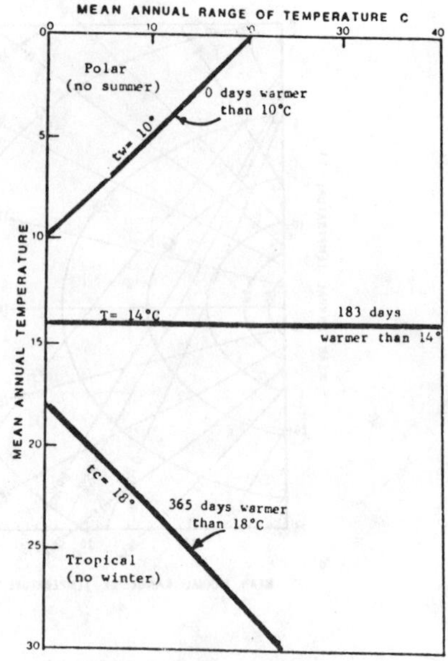


Fig. 7. Biotherms of Fig. 6 plotted to show relations to similar lines of warmth (W) in Fig. 1.

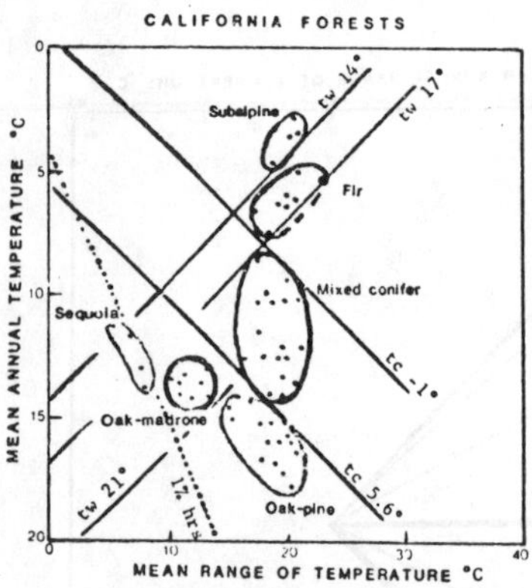


Fig. 5. Warm month (tw) and cold month parameters of California forests. Ratio of T to A is 1:2, as in Fig. 1.

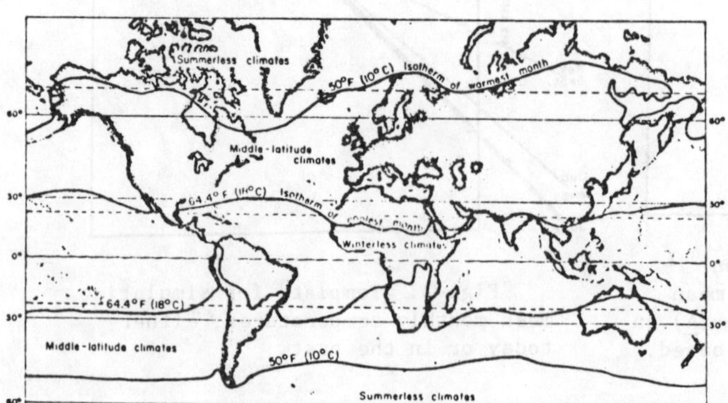


Fig. 6. Major biotherms of the earth are tc 18°C and tw 10° for the edge of tropical and polar climate.

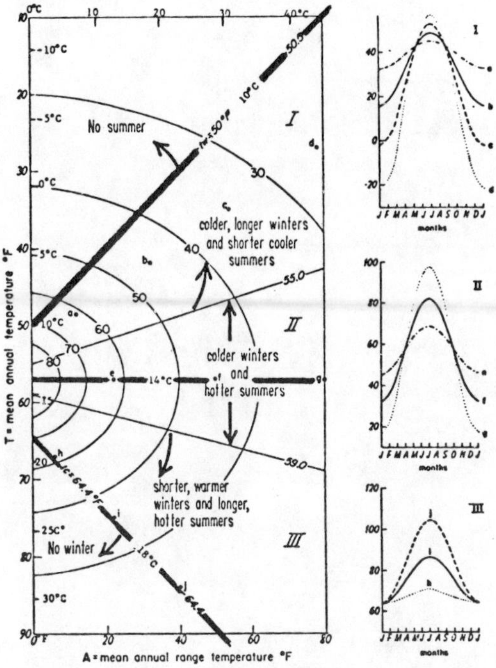


Fig. 8. Greater ranges and extremes of temperature occur with increasing (A).

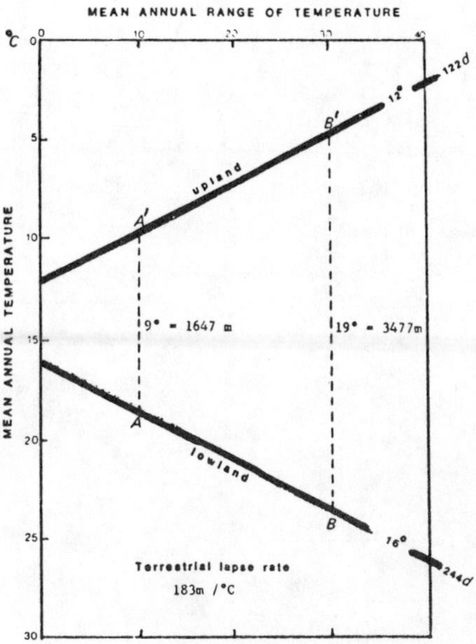


Fig. 11. To estimate altitude, range of temperature (A) must be known or estimated. The higher the (A), the greater the difference in elevation.

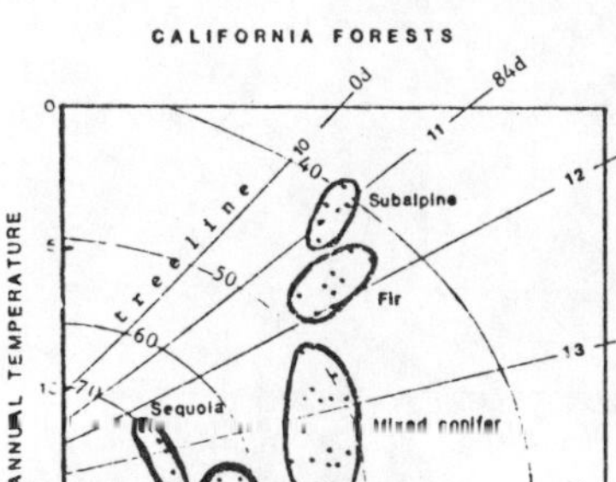


Fig. 9. Warmth & equability of California forests. Compare Fig. 5 which shows only warm and cold month temperatures.

Fig. 10. Thermal parameters of east Asian forests. Lower range (A) must be inferred for Tertiary forests.

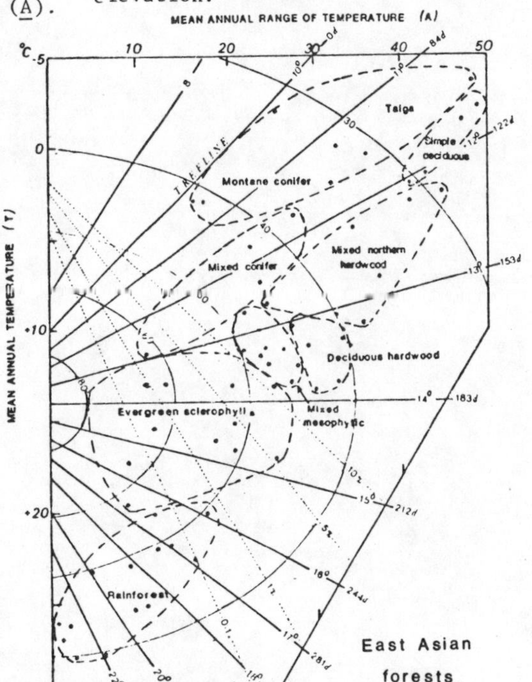

# VEGETATIONAL CHANGES IN THE PAST AND THE UPLIFT OF QINGHAI-XIZANG PLATEAU

Xu Ren*
(Institute of Botany, Academia Sinica)

## INTRODUCTION

One of the most fascinating events happened in the Cenozoic period is the uplift of the Qinghai-Xizang Plateau. This plateau at present is the most vast and highest plateau in the world, known as the "World's Roof". It covers nearly three million sq. km in extent, and a large proportion of it is about 4,500-5,500m in elevation, except that the northeastern part of the plateau, such as, the Qaidam Basin is about 3,000m and Xining is more than 2,000m above the sea level. The vegetation is mostly of alpine steppe with some coniferous forests scattering in some deep valleys. The northern border of the plateau is defined by some high mountain ranges, the Kunlun Mountains, the connecting Altun Mountains and the Qilian Mountain. To the south of the plateau is the Himalayas, the loftiest mountain range on the earth. This range curves into the northwestern part of Yunnan and connects with some mountain ranges of Yunnan. Between the Kunlun and the Himalayan ranges within Xizang there are some other mountain ranges running nearly parallel with each other, notably the Tanggula and the Gandise Ranges. How the plateau was formed, how the vegetation changed and what is the cause of the uplift of the plateau and why all the mountain ranges are parallel with each other, are the problems which have continuously interested all scientists.

In the past one and a half decades, Academia Sinica has sent several scientific expeditions to Xizang to work on these problems. A few years ago these expeditions found the Qubu Formation bearing the Gondwana _Glossopteris_ flora of Permian age north to the Himalayas. The Qubu Formation is mainly composed of quartzite sandstone and carbonaceous shales of continental phase and located only about 300-400 km north to the Raniganj coal field of West Bengal of India. After critical examination I have suggested that it is probably the northern extention of the Raniganj Series in southern Xizang. Therefore, I further suggested that the uplift of Qinghai-Xizang Plateau is actually caused by the collision of the drifting India plate against the Eurasia plate and the suture between these two plates would be Zangbo thrust, parallel to and located at the south of the Yarlung Zangbo River as some geologists predicated. It is surprised to note that along this thrust, there lies a belt of basic to ultrabasic rocks that leads me to believe that the southern part of Xizang did not belong to the ancient Asia but was a part of the Gondwanaland (Hsü, 1973, 1976a, 1976b). This strongly supports the view of Wagener (1928), Sahni (1936) and Du Toit (1937) that the India block drifted in Cretaceous from the southeastern corner of Africa and later on in Eocene joined up with Asia to become its subcontinent, though the suture does not lie in the Himalayas

* Hsü Jen

as predicated.

At this occasion it is no need to repeat the findings of the Glossopteris flora. What I will say is something about the past vegetation of Xizang and Qinghai in order to show how the uplift of Qinghai-Xizang Plateau is formed.

Here I may say a few words about the age of the Qubu Formation. According to the stratigraphical sequence in India, the Raniganj Series conformably underlies the Panchat Series of lower Triassic. But some Chinese geologists believe that the Qubu Formation conformably underlies the Shailong Group which is considered equivalent to the Mao-Kou Section of the Yangxin Series of early Permian. Anyway, it is still an open question and has to be worked out.

Recently Li (1981) has made some notes on three new species of this Glossopteris flora and thrown some doubts about the identification of the fossils collected from Qubu. He thinks that Raniganjia qubuensis might belong to the genus Stellotheca, Dizeugotheca qubuensis and Dichotomopteris qubuensis found only on sterile fragments, all the definitely known species of Dizeugotheca have been collected in the early Permian and the described species of the Glossopteris flora of the Qubu Formation are all with wide vertical range. However, I should point out that the leaves of Stellotheca are free up to the base (Maithy and Mandal, 1976), while the leaves of Raniganjia qubuensis are united at the base for about 1/3 - 2/3 of their length. So, Raniganjia qubuensis has nothing to do with Stellotheca. Although these specimens of Dizeugotheca and Dichotomopteris are sterile, yet in India both genera are restricted in the Raniganj Series. Again, the venation of our specimens is of the same type as described by Maithy (1972, 1973). There is no reason to name them as Pecopteris qubuensis and Merianapteris qubuensis until the fertile specimens are found. Moreover, Glossopteris communis and G. angustifolia have not been recorded in India under the Raniganj Series. So, at present, I have not found any reason to assign the Qubu Formation to be Lower Permian as Li suggested.

## THE PAST VEGETATION OF XIZANG AND QINGHAI FROM LATE CARBONIFEROUS TO LATE CRETACEOUS

By the time of late Carboniferous the Lepidodendropsis flora was prosperous in northern Xizang. The Gigantopteris flora extended from Yunnan and Sichuan to northwestern Xizang during late Permian. The upper Triassic flora of northern Xizang is closely related to that of Yunnan, but the upper Triassic flora of eastern Xizang and Jurassic flora of the Qaidam Basin are related to those of northern China respectively. All these indicate that the Kunlun Mountains had already extended to eastern Xizang by that time. The early Cretaceous flora found in Lhasa and the late Cretaceous flora of Xigaze show close relationship with those of northern hemisphere and have nothing to do with those of southern hemisphere. All the floras are no doubt tropical in nature. The late Cretaceous flora was mainly composed of Eucalyptus, especially with the tropical plants Phrynium, Dianella, Ficus and Cassia. In fact, by those periods Xizang was immersed in the Tethys, only left a part of the land occasionally elevated above the sea as islands.

Judging from these facts above-mentioned, the climate of most parts of Xizang was hot and moist but that of eastern Xizang and Qinghai was rather dry and warm. That means the Kunlun Mountains had already uplifted, being the barrier to prevent the seasonal winds of the Indian Ocean passing over the Kunlun Mountains. Actually there was a chain of mountains running from the northern boundary of Xizang up to eastern China, such as Qilian Mountain and Qinling Mountains. These ranges since then have separated China into two different floristic provinces: the northern and the southern ones. These mountains I believe were caused by plate tectonics of both the Indian and the Pacific plates against Eurasia. As a result of these movements some interior parts of Eurasia folded, thrusted, and uplifted into mountain ranges.

## VEGETATIONAL CHANGES DURING TERTIARY AND UPLIFT OF QINGHAI-XIZANG PLATEAU

Now, let us turn to the vegetational changes during the Tertiary in order to estimate the rate of uplift of Qinghai-Xizang Plateau.

In early Eocene, about 50 m.y. ago, evergreen broad-leaved forests were flourishing in Xizang. A flora found in Moincer of the Moincer Series of Ngari region in western Xizang was mainly composed of *Eucalyptus* and *Ficus*, with *Eucalyptus geinitzi*, *E. oblongifolia*, *Cassia marshalensis*, *Rhamnus menchiensis*, *Phrynium tibetica*, and *Cyperacites haydenii*. As *Eucalyptus* is at present widespread in Australia and *Phrynium* and *Ficus* have never lived beyond tropic, the climate was then hot and moist as before. Hence, by that time Xizang was no doubt less than 500 m above the sea level. The isotopic determination of the rocks taken from the Gandise gives an age of 30-79 m.y., so the orogeny of Gandise could have started much earlier than that of the Himalayas.

At the time of Miocene, about 20 m.y. ago, the vegetation gradually changed into coniferous and alpine evergreen and deciduous broad-leaved mixed forests. In central Xizang it was first dominated by evergreen oaks and deciduous forests of *Populus latior*, *P. balsamoides*, *P. glandulifera*, *Juglans*, *Salix*, along with *Cedrus*, *Tsuga* and *Sabina* cf. *caridaium*. Later on, this flora was replaced by forests of *Sophora macrocarpa* and *S. xizangensis* with underground *Cyperacites namlingensis*. Spores of Cyatheaceae and pollen of *Podocarpus* and *Cycas* were preserved in the rocks. That means the topography of central Xizang was rather complex. Some spots might be over 2,000 m in elevation. But some deep valley less than 1,000 m in elevation would have existed. In general, the elevation of central Xizang was lower than 1,500 m. Later on, in late Miocene the elevation was higher than 1,500 m while the deciduous broad-leaved forests of *Populus balsamoides*, *P. glandulifera*, *P. latior*, *Albizzia*, *Sophora* and *Salix* were flourishing in the hills of western Xizang.

By that time, only *Abies*, *Picea*, *Cedrus*, *Tsuga*, *Carpinus*, *Betula*, *Castanea* and *Fagus* were grown scatteredly somewhat in the Qaidam Basin. It is interesting to note that pollen of some xerophytic shrubs and herbs, such as, *Ephedra*, Chenopodiaceae, Compositae, Zygophyllaceae, Leguminosae and Gramineae are preserved in the rocks of the Qaidam Basin. That means the Tanggula Range had already uplifted at least more than 2,000 m in elevation.

Passing to Pliocene, about 7 m.y. ago, coniferous and evergreen and deciduous broad-leaved forests of *Lithocarpus*, *Quercus*, *Tsuga yunnanensis*, *Picea*, *Abies*, *Pinus*, *Cedrus*, *Fagus*, *Salix*, *Carya*, *Melia* and

Ilex were widespread in northern Xizang. It is interesting to note that some pollen of tropical elements, such as Podocarpus and Araliaceae are still found in the rocks. Moreover, spores of the epiphytic ferns Drynaria were preserved in the same sediments. This means that some deep valleys might still have existed, so the topography of northern Xizang was also complex even by that time.

Up to late Pliocene, some subtropical alpine forests were still widespread in the northern slope of the Himalayas, among which evergreen oaks Quercus semecarpifolia, Q. pannosa, and Q. senecens and Cedrus deodara were predominant. Fossils were collected from a spot at present being 5,700-5,900 m above sea level which was only 2,500-3,000 m in altitude by that time. It is worthwhile to note that in the Ngari region of western Himalayas, Picea became quite dominant, along with Betula and Abies representing a timber line vegetation. At this time Ulmus was flourishing in northeastern Xizang. The climate became cooler and drier than before. That means the elevation of Ngari region would be between 3,000 m and 4,000 m.

By that time, the Qaidam Basin became drier and cooler; only few trees of Abies, Picea and Betula lived on the hills. The xerophytic plants Ephedra, Tamarix and Nitraria became more dominant. A continental climate was prevailing there. That means the Tanggula Range had folded and uplifted to certain extent, say, more than 3,000 m in elevation.

## VEGETATIONAL CHANGES DURING QUATERNARY AND THE UPLIFT OF XIZANG PLATEAU

The most active period in the uplift of Xizang Plateau and the mountain ranges is the Quaternary. Due to the influence of climatic changes of the whole world during the Pleistocene period, flora in Xizang changed very rapidly. Alternation of floras occurred in Xizang regularly during the glacial and the interglacial periods. During the glacial periods no tree could live on the hills. At the interglacial periods the trees returned to the hills. But during the early Quaternary upheaval of Xizang Plateau the environmental conditions were more or less profoundly modified. Changing in ecosystem also influenced the composition of plant communities. Anyway, the alpine mesophile flora was then replaced by an alpine xerophilous flora. So the flora of Xizang changed to alpine plant cushions and alpine steppe, even to alpine desert.

During early Pleistocene forests of Cedrus, Pinus, Betula, Quercus, Carpinus and Alnus were widespread in the central Himalayas. But Larix, Abies, Picea, Pinus and Cupressaceae were flourishing in the western Himalayas and forests of Picea and Cupressaceae were better developed in the northern Xizang. The xerophilous Tamarix and Nitraria which had come from Qinghai lived with Artemisia and Chenopodiaceae. That means the Himalayas and other mountain ranges arrived at a higher level. Judging from the remains of water plants Nelumbo, Typha, Potamogeton and Sparganium and the plankton algae Pediastrum boryanum and P. duplex, lakes and marshes would still have existed in the northern Xizang. So the climate there was not too dry.

During middle Pleistocene mixed forest of Pinus, Betula, Quercus, Carpinus and Alnus lived in the central Himalayas. Remains of Picea spinulosa, Sabina recurva, Cedrus deodora, and Quercus semecarpifolia were found in Jiabula.

By that time the xerophytic elements Ephedra, Tamarix, Nitraria, Artemisia and Chenopodiaceae became dominant in the northern and northeastern Xizang. That means the elevation of northern Xizang was about 3,000-3,600 m above the sea level.

Turning to late Quaternary, trees were scanty in northern Xizang. Remains of Picea and Abies have only been observed. An alpine steppe was developing. The elevation would be over 4,000 m.

Passing to Post Glacial age the vegetation of most parts of eastern Xizang became alpine steppes.

But some remains of Lonicera, Viburnum, Rosa, Rhamnus, and Salix along with palaeolithic implements have been found in Yali on the southern slope of the Himalayas. This kind of vegetation at present grows at an altitude of 3,400-3,700 m in the southern slope of central Xizang, whereas Yali is now 4,300 m high. That means the Himalayas has uplifted by 600-900 m since 7,000-9,000 years ago.

## CONCLUSION

Judging from the past vegetation, the uplift of Qinghai-Xizang Plateau probably began from Upper Cretaceous and went on in Eocene, Miocene and Pliocene. The Quaternary is the most active period of upheaval.

1. During early Eocene, the flora was tropical-subtropical forest. Xizang and Qinghai were less than 500 m in elevation.

2. In Miocene, Xizang was situated in subtropical zone, but the vegetation was temperate in nature. Some spots might uplift over 2,000 m; others were less than 1,500 m in elevation.

3. Up to late Pliocene Cedrus-evergreen oak forests were prevailing in the Himalayas. The Himalayas was 2,500-3,000 m in elevation.

4. During early Pleistocene alpine coniferous-deciduous broadleaved forests were predominant. The Xizang Plateau was about 3,000 m in elevation.

5. In middle Pleistocene, an alpine coniferous forest was dominant. The Xizang Plateau was 3,000-3,500 m in elevation.

6. During late Pleistocene, Abies-Picea forests were grown scatteredly in the valleys. The elevation was about 4,000 m above sea level and alpine steppe was predominant in Xizang.

7. At the time of the post glacial age an alpine steppe was prevailing. Xizang was about 4,500 m in elevation.

8. During Holocene, the vegetation of Xizang was alpine steppe. Many high mountain ranges uplifted. The plateau upheaved to 4,500-5,500 m in elevation.

9. As a whole, the uplift rate goes on in acceleration. If we plot them in figure, it looks like a parabola curve (see the Fig. 1).

References


(1) Archangelsky, S., de la Sata, Elias, Contribution al conacimierto de las filices Permicas de Potagonia extraandina. Acta Geol. Lilloana. 3, 85-126, Pls. 1-7 (1960).

(2) Du Toit, A.L., Our wondering continents. Oliver and Royd., Edinburgh (1937).

(3) Maheshwari, H.K.,Floristics of the Permian and Triassic Gondwanas of India. The Palaeobotanist, 23 (2), 145-160 (1974).

(4) Hsü Jen, On the discovery of some plant fossils from the Mt. Jolmo Lungma Region, Southern Tibet and its significance. Acta Bot. Sin., 15 (2), 254-258, Pls. 1-2 (1973) (in Chinese).

(5) Hsü Jen, On the discovery of a Glossopteris flora in southern Xizang and its significance in geology and palaeogeography. Scientia Geologica Sin., 4, 320-335, Pls. 1-4 (1976a) (in Chinese).

(6) Hsü Jen, On the palaeobotanical evidence for continental drift and the Himalayan uplift. The palaeobotanist, 25, 131-145 (1976b)

(7) Li Xing-xue, Notes on the three new species of Glossopteris flora from the Qubu formation, South Xizang, with a discussion on the age at the formation. In this Proceedings of Symposium on Qinghai-Xizang (Tibet) Plateau, Academia Sinica (1981).

(8) Kong Zhao-chen et al., Preliminary studies on the vegetation of the Qinghai-Xizang Plateau during Neogene and Quaternary. Ibid. (1981).

(9) Maithy, P.K., New type of fern frond from Lower Gondwana of India. The Palaeobotanist, 21 (3), 365-367, Pl. I (1972).

(10) Maithy, P.K., Some contribution to the knowledge of Indian Gondwana ferns. Ibid., 22 (1), 29-33, Pls. 1-2 (1973).

(11) Maithy, P.K., Mandal, J., Further observations of Stellotheca robusta Surange and Prakash. The Palaeobotanist, 25, 279-289 (1976).

(12) Sahni, B., Wegener's theory of continental drift in the light of Palaeobotanical evidence. Jour. Indian Bot. Soc., 15 (5), 319-332 (1936).

(13) Wagener, A., In: Theory of continental drift, Symposium Amer. Ass. Petrol. Geol. Tusla (1928).


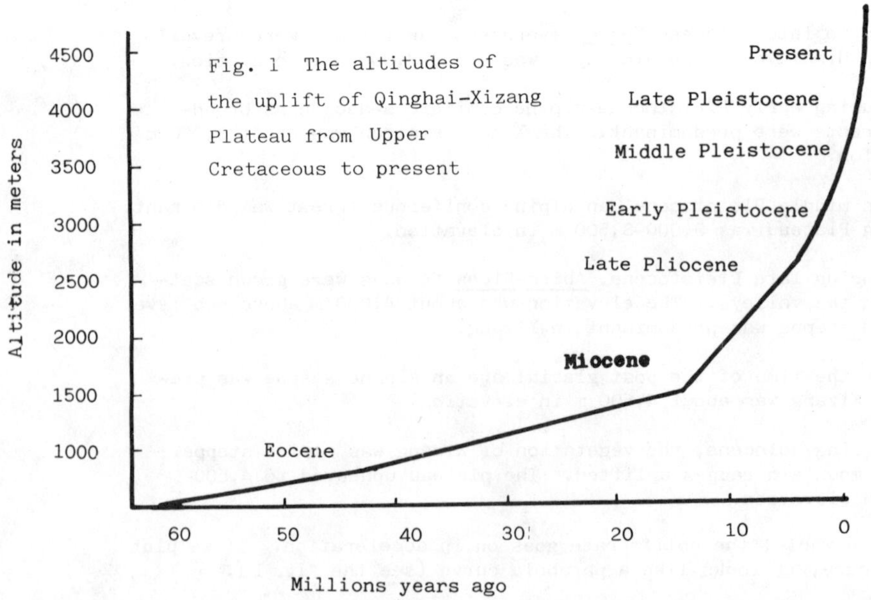


Fig. 1 The altitudes of the uplift of Qinghai-Xizang Plateau from Upper Cretaceous to present

# DISCOVERY OF CATHAYSIA FLORA IN THE QINGHAI-XIZANG PLATEAU WITH SPECIAL REFERENCE TO ITS PERMIAN PHYTOGEOGRAPHICAL PROVINCES

Li Xing-xue    Yao Zhao-qi
(Nanjing Institute of Geology and Palaeontology, Academia Sinica)

The finding by the Swedish geologist, Bexell, of a Late Permian Angara flora overlying directly the Permo-Carboniferous floras of Cathaysia-type in the Nanshan section of West Qilian Mountain has been well-known since 1935 (1,2).

In recent years, owing to the occurrence of a Permian Glossopteris flora in the Qubu formation of South Xizang (3,4), the northern border of the Gondwanaland is considered to reach at least the northern slope of the Himalayas. Since then, some of the following questions may be naturally arisen: (1) Did the Permian Angara flora extend further southwards from Qilian Mt.? (2) Where should the northern border of Gondwanaland really be? and (3) What kind of Permian flora would be found in the Qinghai-Xizang Plateau proper? All these have attracted more and more attention among authors in the circle of Earth Sciences.

## THE CATHAYSIA FLORAS IN THE PLATEAU

Early in the fifties, there were only two Chinese articles(5, 6), recording a few fossil plants of Cathaysia-type in Qamdo of East Xizang and in the Wuli coalfield of South Qinghai. However, in the last decade, rather rich materials of Late Permian Cathaysia flora were available at our disposal. The following are those of the most important:

### The Toba Flora from Qamdo of East Xizang

The Toba formation in Qamdo is a rich fossiliferous coal series, from which at least 35 species of plant fossils were obtained (7). The plants found in its upper portion are: *Sphenophyllum koboense* Kob., *Annularia pingloensis* (Sze), *Paracalamites stenocostatus* Gu et Zhi, *Rajahia* (*Pecopteris*) *calceiformis* Li et Yao, *R.* (*P.*) *bifurcata* Końno, *Fascipteris stena* Gu et Zhi, *Gigantopteris dictyophylloides* Gu et Zhi, etc.; whereas in its lower portion, besides those recorded in the upper, are *Lepidodendron oculus-felis* (Abbado), "*Lep.*" *acutangulum* (Halle), *Sphenophyllum sinocoreanum* Yabe, *Lobatannularia multifolia* Końno et Asama, *Schizoneura manchuriensis* Końno, *Pecopteris anderssonii* Halle, *Rajahia* (*P.*) mirabillis Gu et Zhi, *R.* (*P.*) *qamdoensis* Li, Yao et Deng, *Ptychocarpus* (*Fascipteris*) *densatus* Gu et Zhi, *Gigantopteris* cf. *nicotianaefolia* Schenk, *G. dictyophylloides* Gu et Zhi, *Gigantonoclea* spp. and *Rhipidopsis panii* Chow, etc.

The plant fossil association suggests a close relationship with that of the early stage of the wide-spread *Gigantopteris* coal series of South China, but differs from it in the absence of such forms as *Tingia car-*

*bonica* (Schenk), *Cordaites principalis* (Germ.) of a little earlier age. Besides, marine fossil invertebrates from both the overlying and underlying beds as well as the marine intercalations of the coal-bearing deposits, also indicate that the Toba formation is of an age neither earlier than late Early Permian, nor younger than the latest Permian or the Changhsingian stage. The flora is thus regarded as a typical Cathaysia flora of early Late Permian belonging to the type of its Southern subprovince (8). In other words, the Toba flora is equivalent to the Lungtan flora in strict sense of South China.

## The Shuanghu Flora of Raggyorcaka Formation in North Xizang

There are 13 definitely or somewhat doubtfully determined species recorded from the upper part of Raggyorcaka formation in Shuanghu area of North Xizang (9). All the plant fossils came from the same horizon and locality. Among them the following 9 are more significant than the other: *Selaginellites tibeticus* Li et Zhu, *Sphenophyllum* aff. *speciosum* (Royle), *Annularia pingloensis* (Sze), *Rajahia* (*Pecopteris*) *calceiformis* Li et Yao, *Pecopteris shuanghuensis* Li et Yao, *Gigantonoclea guizhouensis* Gu et Zhi, *G. meridionalis* Li et Yao, *Compsopteris contracta* Gu et Zhi and *Lobatannularia* sp.

At the first glance, the Shuanghu flora has rather an appearance close to the Lungtan flora of South China, but is distinguished from it by containing many new forms. The flora is definitely a little younger in age, namely the Changhsingian stage. This is based also on the evidence of marine fossil invertebrates. For instance, the formation preceding the plant-bearing horizon, i.e. the Lower Raggyorcaka formation, is characterized by *Palaeofusulina sinensis* Sheng, *P. nana* Likharev, etc. and the plant-bearing horizon is immediately followed by the Kanglu formation with *Claraia stachei* (Bittner), *Cl. aurita* (Hauer) of Scythian age. It is obvious that the discovery of this flora in North Xizang, adds much to our knowledge for the first time of the lastest stage of the Cathaysia flora. The Shuanghu flora might be regarded as the uppermost plant-assemblage of the Cathaysia flora, representing probably the latest phase of the Late Permian flora in the world.

In composition, the present flora bears a strong resemblance to the Late Permian flora or the Lungtan-Changhsingian flora of South China rather than that of the Upper Shihhotse formation of North China. It is also worthy to be noted that none of the described plants gives any indication of affinity of the Angara flora, nor do they show a close relationship with *Glossopteris* flora, although the genus *Glossopteris* is known to occur in the Qubu formation of South Xizang, about 500 km. to the south of Shuanghu area.

The presence of *Selaginellites tibeticus* in the flora is also of interest, because this is the first record of the genus *Selaginellites* in the Palaeozoic floras of East Asia.

## The Permian Plants from South Qinghai

In the southeastern part of Yüshu area of Qinghai, there is a florule characterized by some elements of the Permian Cathaysia flora, such as *Lobatannularia multifolia* Kon'no et Asama, *Annularia* cf. *pingloensis* (Sze), *Rajahia* (*Pecopteris*) *calceiformis* Li et Yao, *Comsopteris* cf. *contracta* Gu et Zhi and some indeterminable plant remains. Judging from the floral composition and the geographical location, the flora of Yüshu is obviously close to Toba flora of Qamdo, East Xizang, belonging to the early Late Permian Cathaysia flora of South China.

### A Mixed Permian Flora from the West Qilian Mountain

In 1978 and 1979, some Chinese geologists and palaeontologists made a series of investigations of the well-known Nanshan section surveyed by Bexell in West Qilian Mt. The new results of these trips have been served to emphasize at least the following two points: (1) Bexell's Nanshan section, consisting of 8 rock-divisions and amounting to a thickness of almost 2,000 meters, with 4 plant-bearing zones named ascendingly A, B, C, D is established not upon a natural but a generalized profile, although the stratigraphical succession of these plant-bearing zones is correct. (2) The Late Permian Angara flora in the Nanshan area was previously considered by Bexell (1) and Halle (2) to be younger in age than the Permian Cathaysia flora. However, in a much well-exposed geological section in the Sunan county, some 100 km east to the Nanshan area, with an unbroken sequence of the 4 plant-bearing horizons surveyed by Bexell, they have surprisingly found that the Permian Cathaysia flora is mixed with some typical elements of the Late Permian Angara flora in an equivalent of the plant-bearing zone C of Bexell. The Cathaysia flora is here characterized by such forms as *Annularia gracilescens* Halle, *Lobatannularia lingulata* (Halle), *Pecopteris lativenosa* Halle, *P. tenuicostata* Halle, *Compsopteris wongii* (Halle) and *Rhipidopsis* cf. *lobata* Halle. These plants are almost all of common and index forms of the Upper Shihhotse formation of North China, while their association is mainly represented by such typical Permian Angara forms as *Zamiopteris glossopteroides* *Schmalh*.; *Z. lanceolata* Neub., *Phyllotheca* sp., *Callipteris* ex gr. *zeilleri* (Zal.) and some coniferous remains of *Sashinmia*-type.

## THE PERMIAN PHYTOGEOGRAPHICAL PROVINCES IN THE PLATEAU

To sum up the discoveries mentioned above and take into consideration other new palaeozoological and geological information available, it seems appropriate to outline the boundaries of the Permian phytogeographical provinces as follows (Fig. 1): (1) The boundary between the *Glossopteris* and the Cathaysia floras may be considered to coincide with the plate suture or the deep fault zone running along the valley of the Yarlung Zangbo River; (2) The Tienshan Mts.-Hinggan Mts. deep fault zone, lying outside to the north of the plateau, has generally been regarded as a natural boundary between the Angara and Cathaysia floras. However, as the sea-trough in this area disappeared at the end of Early Permian, the Late Permian Angara flora spread southward into Qilian Mt. and made a mixture with the Cathaysia flora, so that no boundary between these two floras ever exists in the plateau; (3) There is, however, a subdividing line for the Permian Cathaysia floral province itself, i.e. the boundary between the Northern and Southern Cathaysia floral subprovinces in this plateau to extend north-westerly from the northern border of Qinghai Lake to the adjacent area of Lenghu district.

The new findings of the Cathaysia flora in the Qinghai-Xizang Plateau might also afford rather a justifiable interpretation of a strong support to the occurrence of Late Permian Cathaysia plants in Turkey (10) and Iraq (11) in the Near East.

References

(1) Bexell, G., On the Stratigraphy of the plant-bearing deposits of Late Palaeozoic and Mesozoic age in the Nanshan region (Kansu). Geografiska Annuler, Stockholm, p. 63 (1935).

(2) Halle, T G., The relation between the Late Palaeozoic floras of Eastern and Northern Asia. C.R. 2e Congr. Strat. Carbon., Heerlen, 1935, 1, 237-245 (1937).

(3) Hsu Jen, On the discovery of some plant fossils from the Mt. Jolmo Lungma region, southern Tibet, and its significance. Acta Bot. Sin., 14 (1), 50-69 (1973).

(4) Hsu Jen, On the palaeobotanical evidence for continental drift and Himalayan uplift. Palaeobotanist, 25, 131-142 (1978).

(5) Li Pu, A preliminary study of the geology in eastern Tibet. Kexue Tongbao, (7), 62-71 (1955).

(6) Pan Kuang, On the occurrence of Gigantopteris flora in the Tsinghai-Tibet platform. Ibid., (1), 341 (1957).

(7) Li Xing-xue and Yao Zhao-qi, Carboniferous and Permian floral provinces in Eastern Asia. Paper read at the 9th Congr. Carbon. Strat. Geol., Urbana, 1979 (in press).

(8) Li Xing-xue, Yao Zhao-qi, Deng Long-hua, Early Late Permian flora from Toba, Qamdo of eastern Xizang. The Series of the Scientific Expedition to the Qinghai-Xizang Plateau, Palaeontology of Xizang, Book V (in press).

(9) Li Xing-xue, Yao Zhao-qi, Zhu Jia-nan, Duan Shu-ying, Hu Yu-fan, Late Permian plants from Northern Xizang. Ibid. (in press).

(10) Wagner, R.H., On a mixed Cathaysia and Gondwana flora from SE Anatolia (Turkey). C.R. 4e Congr. Strat. Geol. Carbon., Paris, 3, 745-752 (1962).

(11) Ctyroky, P., Permian flora from the Ga'ara region (Western Iraq). N. Jb. Geol. Palaeont. Mh., H. 7, 43-47 (1973).

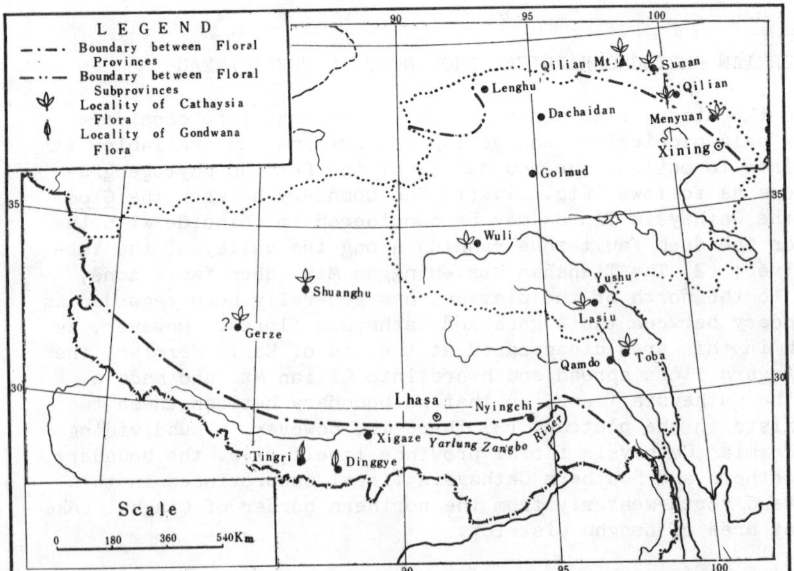


Fig. 1 Showing main localities of Permian plants in Qinghai-Xizang Plateau

# PALAEOBIOGEOGRAPHY OF QINGHAI-XIZANG PLATEAU, EVIDENCE FOR CONTINENTAL DRIFT

Wen Shi-xuan
(Nanjing Institute of Geology and Palaeontology, Academia Sinica)

In recent years large-scale scientific investigations and regional geological surveys have been made in the Qinghai-Xizang Plateau and subsequently a great number of fossils have been described. In this paper an attempt is made, by using fossil evidences, to explain the palaeobiogeography of the plateau.

## EVIDENCE FOR PANGAEA

The Pre-Carboniferous fossils from the Qinghai-Xizang Plateau afford substantial arguments in favor of the Pangaea advocated by Wegener. Here for convenience, an introduction is made, to begin with, of the materials obtained from the Himalayas, which is situated on the southern margin of the plateau.

In Mt. Qomolangma region no Cambrian fossils have hitherto been found, but in its adjacent areas, for example, the Salt Range of Pakistan and Spiti of India, there occur Cambrian trilobites *Redlichia* and *Chittidilla*, etc. which bear close similarities to those from Southwest China and the Tarim Basin of Northwest China.

From Ordovician, fossils are known to occur in the Himalayas at different ages (1,2). For instance, in the Mt. Qomolangma region, the Lower Ordovician fauna include nautiloids, gastropods and brachiopods. *Ordosoceras*, *Manchuroceras* and *Wutinoceras* are known to be the commonest forms in North China and the southern part of Northeast China. *Maclurites* cf. *acuminatus* (Billing) may be compared with that from Newfoundland of Canada, Vermont of U.S.A. and from the Zhuozi Mountain of Nei Monggol of North China as well; *M.* cf. *kotoi* (Kobayashi) was originally recorded in the southern part of Northeast China. *Orthambonites minusculus* Phleger is known to appear in Nevada of U.S.A.; *O.* cf. *rotundiformis* Cooper and *Glyptomena* cf. *sculpturata* Cooper are the important elements of North America; *Aporthophyla* is usually seen in North America; *Hemipronites* is known to be spread in Europe.

The Middle Ordovician fauna contain nautiloids, gastropods and brachiopods. *Sinoceras* is known to be a characteristic form of the Changjiang River region and *Eosomichelinoceras* coming next over there Especially the species *S. chinense* (Foord), *S. densum* (Yü), *E. huananense* Chen, *Michelinoceras xuanxianense* Chang, *M. yui* Lai, *M. paraelongatum* Chang, etc. are all commonly seen in the Changjiang River region.

It should be pointed out that fossils have been rarely recorded from the Upper Ordovician of the Himalayas with the exception of *Michelinoceras* which, as a cosmopolitan form, has recently been known to occur in Mt. Qomolangma region.

In Mt. Qomolangma region the Early Silurian fauna contains nautiloids and graptolites. *Harisoceras* and *Columenoceras* are known to be distributed in East Europe, the former has also been seen in North America. *Geisonoceras*, *Michelinoceras* are evidently ubiquitous in the world. In addition, the Early Silurian graptolites *Climacograptus scalaris* (Hisinger), *Streptograptus lobiferus* (M'Coy), *S. becki* (Barrande), *Monograptus priodon* (Brunn), *M. balli* (Barrande), *M. marri* Perner, *Oktavites contortus* (Perner), *O. spiralis* (Geinitz), *O. proteus* (Barrande) bear similarities to those of Nepal and have affinities to those of South China and Southwest China.

The Middle and Upper Silurian Faunas include the nautiloid *Michelinoceras* (*Kopaninoceras*) *jucundum* (Barrande), which has been seen in Europe, and the graptolites *Pristiograptus dubius* (Suess), *P. dubius latus* (Boucek), *Monoclimacis vomerina* cf. *subgracilis* Pribyl, which have been recorded in Europe and South China.

The Early Devonian graptolites *Neomonograptus atopus* (Boucek) *rigidus* Me et Ni, *Monograptus thomasi* Jaeger, *M.* cf. *yukonensis* Jackson et Lenz have been found from the Mt. Qomolangma region. Obviously, this graptolite fauna includes the elements of North America, Australia and Europe. Furthermore there occur tentaculites including *Nowakia acuaria* (Richter) and *Guerichina xizangensis* Mu. Both species very closely resemble those of Czechoslovakia but, notably, the first one has a wide distribution.

In the central part of the Himalayas the Early and Middle Devonian corals *Favosites goldfussi* Orbigny, *Stringophyllum isactis* (Frech), *Sociophyllum longiseptatum* (Bulvanker), *Neostringophyllum* aff. *concacum* Walther and *Cyathophyllum dianthus* Goldfuss have been found in Dolpo of Nepal (3). These corals are commonly known in South China and are widely spread in Europe and U.S.S.R.

In the Mt. Qomolangma region no fossils have hitherto been found in the Middle Devonian. However, in the Upper Devonian there occurs only one species of bivalves *Paracyclas* cf. *dubia* Beus., which may be compared with that from Rhine of Germany.

From the above statement, the Cambrian-Devonian faunas of the Himalayas, though showing some differences, bear similarities to those of North China and the southern part of Northeast China or, sometimes, to those of South China, Southwest China and Northwest China. On the whole, they all fall into the same biogeographical province.

Geographically, North and East Xizang, and South Qinghai are situated between the Himalayas and the above-mentioned regions of China. It is interesting to note that the Ordovician-Devonian faunas from North and East Xizang and South Qinghai bear a striking likeness to those from both regions. Therefore, we firmly believe that in Pre-Carboniferous the whole Qinghai-Xizang Plateau (including the Himalayas) and the above-mentioned regions of China were comprised in one palaeobiogeographical unit, which may be a part of larger palaeobiogeographic unit by reason that the faunas from there have a close relation to those from North America, Europe and Australia. Certainly, this leads a support to the hypothesis of Pangaea proposed by Wegener.

## REMARK ON THE BOUNDARY BETWEEN EURASIA AND INDIAN SUBCONTINENT

According to the hypothesis of plate tectonics, the Qinghai-Xizang Plateau was formed by the collision of the Indian subcontinent and Eurasia. Where lies the boundary between the two continents has long attracted much attention of the geologists concerned. Some discussions will be made on the basis of the Post-Devonian palaeobiogeographic materials as given below.

### Analysis of Marine Faunas

The Carboniferous faunas are strikingly differentiated on the southern and northern sides of the Bangong Lake-Nujiang River fault. On the southern side of the fault the Lower Carboniferous carrying no fusulinids, is predominated by the brachiopods *Fusella*, *Ovatia*, *Marginirugus*, *Syringothyris*, *Balakhonia* (4), which are allied to those of the Himalayas and Australia, but dissimilar to those of South China; while on the northern side of the fault, the Lower Carboniferous bears abundant fusulinids, corals and brachiopods, such as *Eostaffella*, *Cystophrentis*, *Aulina*, *Kueichouphyllum*, *Yuanophyllum*, *Gigantoproductus*, *Schuchertella* and *Praewaagenoconcha*, which are greatly different from those on the southern side of the fault but, instead, are closely similar to those of South China.

On the south of the fault the Late Carboniferous faunas are apparently of the Gondwana-Tethys type,represented by the brachiopods *Stepanoviella*, *Attenuatella*, *Trigonotreta* and the coral *Empodesma*. It is to be noted that the deposits contained the diamictites of glacial origin (5). But on the north of the fault the Late Carboniferous faunas belong to the Cathaysia-Tethys type, including copious fusulinids, brachiopods and corals, such as *Profusulinella*, *Fusulinella*, *Triticites*, *Pseudoschwagerina*, *Choristites*, *Natannella*, *Kepingophyllum*, *Xillia*, *Axolithophyllum*, *Caninia*, which are similar to those of South China but quite unlike to those on the south of the fault.

The Permian faunas (4,6,7,8), same as the Late Carboniferous ones may be distinguished into two types: the Gondwana-Tethys type and Cathaysia-Tethys type (see Table 1). The former fauna may be compared with those of the Himalayas, the Timor Island and the western part of Australia; and the latter is nearly the same as that of South China.

As a matter of fact, the Gondwana-Tethys and the Cathaysia-Tethys are different parts of the Tethys. But in contrast with the Cathaysia-Tethys fauna, the Gondwana-Tethys fauna was apparently living in the marine water of low temperature (4,6,7). The reason that the water temperature became lower may be due to high latitudes or palaeoglaciers.

From the above statement, in Late Carboniferous and early Early Permian the boundary between the Gondwana-Tethys fauna and the Cathaysia-Tethys fauna should be drawn along the Bangong Lake-Nujiang River fault; but in late Early and Late Permian it moved southwards to the Yarlung Zangbo River fault. The removal of distributional limits of the two faunas in different ages is probably due to the glacial growth and decline in the Gondwanaland.

## Analysis of Continental Biotas

Although the removal of the distribution of different faunas in the same Tethys between two continents was due to the environmental changes, the continental biotas can provide more convincing evidences for the continental limitation. Therefore,we must pay more attention to the distribution of biotas in the Gondwanaland and Eurasia, which were separated from each other by the Tethys.

Beginning from the Carboniferous, the continental biotas were developed in the Qinghai-Xizang Plateau (see Table 2). As clearly seen, in the Early Permian the Gondwana flora occurred on the south of the Yarlung Zangbo River fault (5,9,10). Before Cretaceous the flora of Eurasian type was confined to the north of the Bangong Lake-Nujiang River fault (8), whereas in Early Cretaceous it transplanted southwards across the fault (11), but not exceeded the Yarlung Zangbo River.

## Discussion on the Boundary Line

As stated above, from Carboniferous there existed two boundaries, the Bangong Lake-Nujiang River fault and the Yarlung Zangbo River fault, both of which placed restrictions on the distribution of different biotas. On the north of the Bangong Lake-Nujiang River fault, all the marine or continental biotas are nearly entirely of the Eurasia or Cathaysia-Tethys type. Apparently, this area belongs to the Eurasia. On the south of the Yarlung Zangbo River fault the marine faunas are not entirely of Gondwana-Tethys type; but the Permian continental flora is obviously of the Gondwana type. Therefore,this area belongs to the Gondwanaland.

In the area between the two faults, the marine faunas are not of uniform type, and the Eurasia biotas did not appear until the Cretaceous. It was in the Early Cretaceous that the Eurasia flora transplanted southwards to the Gandise Range. At the end of Cretaceous they expanded as far as to the Yarlung Zangbo River fault. It seems that the Bangong Lake-Nujiang River fault may be the boundary of the two continents which closed up before Cretaceous, and the Yarlung Zangbo River fault is another one which closed up at the end of Cretaceous. The area between the two boundaries was probably a part of the Gondwanaland or a small land within the Tethys between the two continents.

# DRIFT OF INDIAN SUBCONTINENT

The drifting of the Indian Subcontinent can be traced from the distribution of the Permian faunas. According to Wang Yu-jing's and others' views, the Permian faunas comprise, in addition to the Gondwana-Tethys and the Cathaysia-Tethys types, the Angara-Tethys type which was distributed on the margin of the Angara land. This fauna, similar to the Gondwana-Tethys fauna, is known only in the Early Permian. It contains the coral *Lytvolasma*, the ammonoid *Uraloceras* and the brachiopods Strophalosiidae and Spiriferidae, which are the identical forms in the Gondwana-Tethys fauna. Furthermore, this fauna also bears some endemic forms.

Judging from the faunal aspects, the Angara-Tethys and Gondwana-Tethys faunas were most probably living in the marine water of low temperature. The two faunas were distributed respectively on the northern and southern sides of the Cathaysia-Tethys fauna which were living in the warm sea. A long time ago, some geologist called the intermediate part as the "equatorial province" (E. Haug, 1908). However, Wang et al. held that the Angara-Tethys was in the north temperate zone and the Gondwana-Tethys in the south temperate zone and the Cathaysia-Tethys in the tropics and subtropics. Nevertheless, Sheng Jin-zhang and Jin Yu-gan believed that the similarity of the Angara-Tethys fauna to the Gondwana-Tethys fauna may be compared with that of the modern marine faunas between the north temperate zone and the south temperate zone.

Recent palaeontological records indicate that in the early Lower Permian the boundary between the Angara-Tethys and the Cathaysia-Tethys faunas (or, for short, the northern boundary) extends from the northern part of Honshu of Japan, via the northern part of the Korean Peninsula, Yinshan Mt., Beishan Mts., Tianshan Mts., the southern extremity of the Aral Sea, the northern extremity of the Black Sea to the northern part of the Balkan Peninsula. The boundary between the Gondwana-Tethys and Cathaysia-Tethys faunas (or, for short, the southern boundary) extends from North Iran through Kashmir, the Bangong Lake-Nujiang River fault, as far as to Baxoi of eastern Xizang, and there it turns straight southwards along E. Long. 99° to the northern part of the Sumatra Island of Indonesia and then stretches out towards the east along S. Lat. 7° (see Fig. 1).

Judging from the two boundaries, it seems that the northern boundary is a normal and unchangeable one as proved collaterally by the fact that in eastern Jilin two types of faunas are found to be mixed together (4), and that the northern boundary was a latitudinal line at that time (12). Noticeably the southern boundary runs parellel with the latitudinal line in the western and eastern parts, but turned southwards in the central part. Strange to say, in the western part it stands not far from the northern boundary as seen from the upper reaches of the Nujiang River to the Beishan Mountain of Gansu where the distance between the two boundaries is 1,200 km (nearly equal to an extent of 11° latitude degrees). Between the two boundaries one may draw a presumed equatorial line, which separates the area into two with latitude 5°30 on each side. In that case, it is inconceivable why the tropics and subtropics where the Cathaysia-Tethys fauna was then living are so narrow in scope.

The writer suggests that the present state of southern boundary is related to the drifting of the Indian Subcontinent to the north along E. Long. 99°. However, during the drifting the Indian Subcontinent was Obstructed by Eurasia and thus diminished its scope greatly as clearly observed from the western sectors of the boundaries. Besides, on the west of the Indian Subcontinent and the Arabian Peninsula, the distance between the two boundaries is not wide.

The present day's latitudinal extent in the eastern part between the two lines is about 48°. Supposing there lies an equator between them, it has Lat. 24° on both sides, a scope approximately within the tropics and subtropics, which is an ideal environment requisite for the Cathaysia-Tethys fauna. It seems that the eastern sector of the southern boundary was unaltered. Besides, the present day's latitudinal difference between the eastern and western sectors of the southern boundary is about 40° a figure indicating that the Indian Subcontinent drifted northwards to a distance of about 4,400 km.

## EVOLUTION OF EAST TETHYS

Evolutionarily, the East Tethys may be divided into several stages as follows:

From Cambrian to Devonian the Indian Subcontinent and the Eurasia were connected together as a part of the Pangaea. On the Pangaea, the shallow-sea faunas were largely developed. It seems that the shallow sea in the east of the Pangaea is likely to be the embryonic stage of the East Tethys.

In Carboniferous time, the eastern part of the Pangaea began to split by the drifting of the Indian Subcontinent towards south, and consequently the East Tethys came into existence. In Permian time the East Tethys expanded more widely. According to the latitudinal position, it may be separated from north to south into the Angara-Tethys, the Cathaysia-Tethys and the Gondwana-Tethys, each with its own distinct fauna.

Beginning from the Mesozoic era, the East Tethys gradually reduced its scope because of the drifting of the Indian Subcontinent once again northwards which brought about the result that the faunas are nearly identical of the Tethys type. Finally, in Middle Eocene, the sedimentary history of the East Tethys came to an end.

## References


(1) Xizang Scientific Expedition, Academia Sinica, A Report of Scientific Expedition in the Mount Jolmo Lungma Region (1966-1968), Geology. Science Press, 1-222 (1974) (in Chinese).

(2) Mu An-tze, Yin Chi-hsiang, Wen Shih-hsuan, Wang Yi-kang, Chang Ping-kao, Stratigraphy of the Mount Jolmo Lungma Region in southern Tibet, China, Scientia Sinica, 16 (1), 96-111 (1973).

(3) Flügel, H., Paläozoische Korallen aus Tibetischen Zone von Dolpo (Nepal), JB. Geol., B.A., Sonderband, 12, 101-120 (1966).

(4) Zhang Shou-xin, Ching Yu-kan, Late Palaeozoic Brachiopods from the Mount Jolmo Lungma Region, A Report Sc. Exp. Mt. Jolmo Lungma Region (1966-1968), Palaeont, Fasc. 2, 159-242, 19 pls. (1976) (in Chinese).

(5) Yin Ji-xiang, Guo Shi-zeng, On the discovery of the stratigraphy of Gondwana facies in northern slope of the Qomolangma Feng in southern Xizang, China. Scientia Geol. Sinica, 1976 (4), 291-322 (1976) (in Chinese).

(6) Wu Wang-shi, The coral Fossils from the Mount Jolmo Lungma Region, A Report Sc. Exp. Mt. Jolmo Lungma Region (1966-1968), Palaeont. Fasc. 1, 83-128, 10 pls. (1975) (in Chinese).

(7) Ching Yu-kan, Liang Xi-luo, Wen Shih-hsuan, Additional material of animal fossils from the Permian deposits on the northern slope of Mount Qomolangma Feng, Scientia Geol. Sinica, 1977 (3), 236-249, 2 pls. (1977) (in Chinese).

(8) Wen Shi-xuan, New materials of biostratigraphy in the northern Qiangtang Plateau, Acta Stratigr. Sinica, 3 (2), 150-156 (1979) (in Chinese).

(9) Hsü Jen, On the discovery of a Glossopteris flora in southern Xizang and its significance in geology and palaeogeography, Scientia Geol. Sinica, (4), 323-331 (1976) (in Chinese).

(10) Hsü Jen, On the palaeobotanical evidence for continental drift and Himalayan uplift, Palaeobotanist 25, 131-142 (1978).

(11) Tuan Shu-yin, Chen Yeh, Keng Kuo-chang, Some Early Cretaceous plants from Lhasa, Tibetan Autonomous Region, China, Acta Botanica Sinica, 19 (2), 114-119, 3 pls. (1977) (in Chinese).
(12) Stehli, F.G., Possible Permian climatic zonation and its implication, America Jour. Science, 255 (9), 607-618 (1957).

Table 1 Representatives of Permian marine faunas and their distribution

| Age | Cathaysia–Tethys type | Gondwana — Tethys type |
|---|---|---|
| $P_2$ | fusulinids *Reichelina, Palaeofusulina*<br>brachiopods *Peltichia zigzag* (Huang), *Cathaysia chonetoidea* (Chao)<br>On the north of the Yarlung Zangbo River fault | |
| $P_1^2$ | fusulinids *Neoschwagerina, Sumatrina, Verbeekina, Lantschichites, Yabeina*<br>corals *Iranophyllum, Wentzelella, Ipciphyllum*<br>On the north of the Yarlung Zangbo River fault, and in the exotic block on the south of the Yarlung Zangbo River fault | brachiopods *Costiferina, Marginifera*<br>On the south of the Yarlung Zangbo River fault |
| $P_1^1$ | fusulinids *Misellina, Cancellina*<br>corals *Szechuanophyllum, Yatsengia*<br>On the north of the Bangong Lake–Nujiang River fault | corals *Lytvolasma, Wannerophyllum*<br>brachiopods *Taeniothaerus, Calliomarginatia, Chonetella, Neospirifer kubeiensis* Ting<br>ammonoid *Uraloceras*<br>bryozoan *Streblotrypa marmionensis* Etheridge<br>On the south of the Bangong Lake — Nujiang River fault |

Table 2 Representatives of Carboniferous Cretaceous Continental biotas and Their Distribution

| Age | Boita and Its Distribution | Characteristic forms |
|---|---|---|
| $K_2$ | flora of North America type in the Yarlung Zangbo River fault | *Cercidiphyllum ellipticum* (Newberry), *Juglandites sinuatus* Lesq., *Ficus daphnagenoides* (Heer) |
| $K_1$ | molluscan fauna of Asia type on the north of the Yarlung Zangbo River fault | *Trigonioides (Trigonioides)*, *T. (Diversitrigonioides)* |
|  | flora of Eurasia type on the north of the Yarlung Zangbo River fault | *Welchselia reticulata* (Stokes et Webb), *Onychiopsis elongata* (Geyler), cf. *Klukia browniana* (Dunker) |
| $J_2$ | molluscan fauna of Asia type on the north of the Bangong Lake–Nujiang River fault | *Psilunio, Lamprotula* (Eolamprotula), *Undulatula, Cuneopsis, Margaritifera* |
| $T_3$ | flora of Asia type on the north of the Bangong Lake–Nujiang River fault | *Equisetites arenaceus* (Jaeger), *Hyrcanopteris sevanensis* Krysh. et Pryn., *Pterophyllum minutum* Li et Tsao, *Nilssoniopteris jourdyi* (Zeiller) |
| $P_2$ | Cathaysia flora on the north of the Bangong Lake–Nujiang River fault | *Gigantopteris, Gigantonociea, Lobatannularia, Rhizomopsis, Sphenophyllum* |
| $P_1$ | Gondwana flora on the south of the Yarlung Zangbo River fault | *Glossopteris, Raniganjia, Dizeugotheca, Dichotomopteris* |
| $C_2$ | flora of South China type on the north of the Bangong Lake–Nujiang River fault | *Archaeocalamites scrobiculatus* (Schloth.), *Sphenophyllum tenerrinum* Ett., *Cardiopteridium spitsbergense* Nath. |

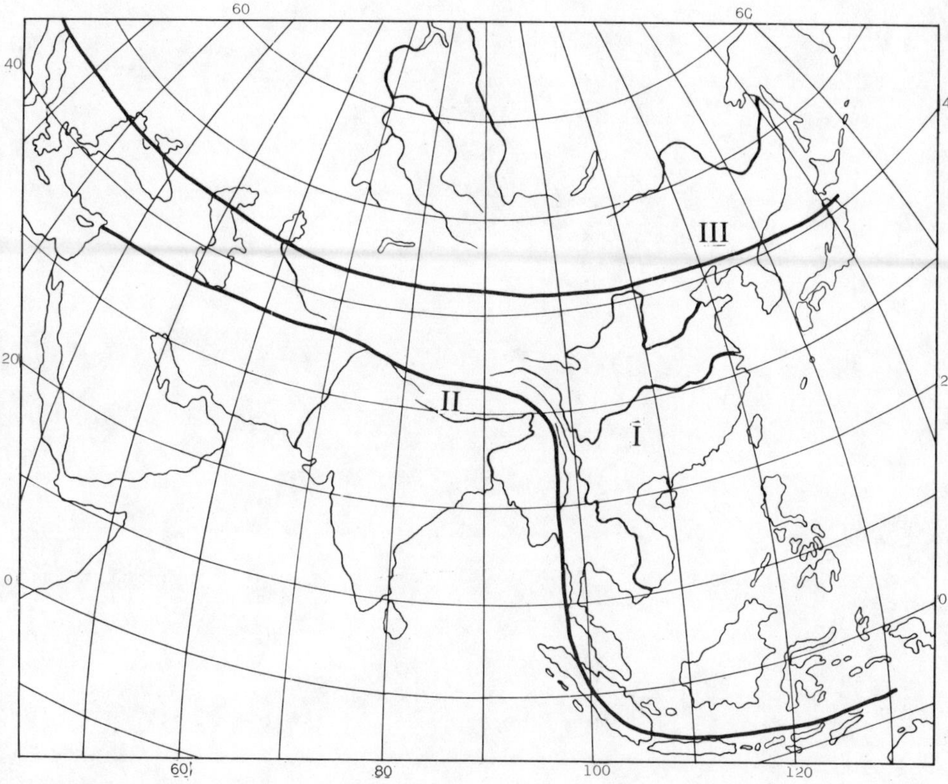


Fig. 1 Boundary line of various marine faunas in early Early Permian.

I. Cathaysia-Tethys type;
II. Gondwana-Tethys type;
III. Angara-Tethys type

# ORDOVICIAN AND SILURIAN CEPHALOPOD PROVINCIALISM IN XIZANG WITH COMMENTS ON THE CONTINENTAL DRIFT

Chen Ting-en
(Nanjing Institute of Geology and Palaeontology, Academia Sinica)

The Ordovician and Silurian strata containing very rich fossil cephalopods in Xizang are spread over the following 2 belts from east to west.

1) The northern slope of the Himalayas, south of the Yarlung Zangbo River, covering some counties as Gyirong, Nyalam, Tingri (including the northern slope of Mt. Qomolangma), Dinggye and Yadong.

2) Near Xainza county and northeast of Baingoin county (northern Xizang).

In the Early Ordovician, there are carbonate sediments, in which some sandstones are intercalated in South and North Xizang.

Ordosoceras (Pomphoceras) - Wutinoceras is known as the leading member of fossil cephalopods, including 22 species belonging to 12 genera: Ordosoceras (Pomphoceras) nyalamense (Chen), O. (Pomphoceras) chiatsunense (Tsou), O. (Pomphoceras) contractum (Chen), O. (Pomphoceras) yaliense (Chen), Wutinoceras remotum Chen, W. sp., Ormoceras nyalamense Chen, O. sp., Manchuroceras spp., Bassleroceras sp., Paradnatoceras yaliense Chen, Ancistroceras yaliense (Tsou), Dideroceras cf. hsichi (Yu), D. sp., Tarphyceras sp., Hopeioceras sp., ?Stereoplasmoceras tofanggoense (Kobayashi), Mysterioceras sp., Georgina taylori Wade. Among them, actinoceroids make up 70 per cent of the total amount of fossils. In our country, the early Ordovician cephalopods have been divided into actinoceroid fauna of north China type and orthoceroid fauna of south China type. In Xizang, which is a part of southwestern China, the caphalopod fauna is very similar to that of North China.

Pomphoceras, the subgenus of Ordosoceras containing 6 species, is an endemic element belonging to actinoceroids in the North China platform. Four species of them have been found in South Xizang. Among them, Ordosoceras (pomphoceras) chiatsurense (Tsou) and O. (Pomphoceras) yaliense (Chen) were also found in Nei Monggol (Inner Mongolia). The other three genera, Ordosoceras, Polydesmia and Meitanoceras, found in Nyalam last year, are typical ones of the family Polydesmiidae (Kobayashi, 1940). This family is an endemic one and widely distributed in Nei Monggol, the North China platform in the Early Ordovician. Wutinoceras and Ormoceras are also widely spread in the North China platform, Northeast China, Nei Monggol, Qinghai and Xinjiang. But Manchuroceras and Hopeioceras are widely presented not only in Liangjiashan Formation of Northeast China but also in Honghuayuan Formation of South China. Georgina found in Nyalam is a very peculiar actinoceroid and has been discovered in Xinjiang. The characters of this genus collected from Xizang may be compared with those of Georgina Basin in Northwest Australia. Georgina and actinoceroid fossils as mentioned above were all typical benthic organisms. It indicates that the depositional environments belong to the lower latitudinal and tropical climatic condition and during Early Ordovician, Xizang, Xinjiang and Australia were not far apart from each other.

At the beginning of the Middle Ordovician, the strata exceeding 80-

100 meters in thickness are limestone intercalated with mud-bands, just like Pagoda limestone in South Xizang. In North Xizang, they are 280 meters thick, mainly purple mud-banded limestone intercalated with pure-limestone.

The Middle Ordovician cephalopods in Xizang are characterized by Sinoceras fauna, and 19 species belonging to 12 genera are recognized: Sinoceras chinense (Foord), S. densum (Yu) Michelinoceras xuanxianense Chang, M. yui Lai, M. paraelogatum Chang, M. yaliense Chen, M. pararegulare Chen, Didero ceras sp., Curtoceras sp., Eosomichelinoceras huananense Chen, E. sp., Beloi toceras xizangense Chen, B. regulare T. E. Chen, Paramoceras pyriforme Chen, Actinomorpha simplex Chen, Trocholites sp., Ormoceras sp., Anaspyroceras sp.. Sinoceras is contained in the Middle Ordovician Pagoda limestone in Southwest China, as well as in Gansu, Xinjiang and western Yunnan. Neither Sinoceras nor actinoceroids mentioned above has been found in the Indian shield yet.

Eosomichelinoceras huananense and Michalinoceras xuanxianense, M. paraelogatum, M. pararegulare present widely in Gansu, Xinjiang, western Yunnan and the Changjiang River Valley Region. Now, Sinoceras and Eosomichelinoceras have been also discovered in North Xizang. In addition, the nautiloid in coil type, such as Curtoceras, Trocholites and Eurasiaticoceras have been found here. Obviously in the Middle Ordovician, the cephalopod fauna of South Xizang is in accord with that of North Xizang.

Accompanying to the transgression in the Early Silurian, a graptolite shale deposits which is similar to the Lungmachi Formation in the Changjiang River Valley Region occurred in South and North Xizang. Above the graptolite bearing beds, the limestone of Dongka Formation lies, yeilding cephalopod fossils (6 species belonging to 3 genera). Among them, Harrisoceras is a common element in the Middle Silurian cephalopod fauna of North America and Europe (Bohemia). The others, such as Allanoceras and Ederoceras are the important elements of Upper Silurian cephalopod fauna in Central Asia (Table 4).

Druing the Middle and Late Silurian, the deposits in Nyalam and Dinggye counties were quite similar to the "reticulated limestone" of western Yunnan, attaining a thickness of 40-80 meters. In this period, Michelinoceras together with its subgenus Michelinoceras (Kopaninoceras) was overwhelmingly flourishing in Xizang. Totally 17 species of fossil cephalopods have been collected. They are Michelinoceras michelini (Barrande), M. simiale (Barrande) M. chiatsunense Chen, M. transiens (Barrande), M. extenuatum (Barrande), Michelinoceras (Kopaninoceras) jucundum (Barramde), M. (Kopaninoceras) capax (Barrande), M. (Kopaninoceras) dorsatum (Barrande), M. (Kopaninoceras) fergan ense Blashov Geisonoceras robustum Chen, G. severum (Barrande), ?Geisonoceras squamatulum (Barrande), Geisonocerina sp. A. Harrisoceras simplex Chen, Dawsonoceras sp., Columenoceras priscum Chen, Virgoceras polemon (Barrandee).

In this assemblage as mentioned above 9 species of Orthoceroids are numerous in individuals covering 70 per cent of the total specimens, while Pseudorthiceratidae is less. All of them are distributed far and wide in Europe (Bohemia), central Asia and western Yunnan of China. Many of the European species are typical ones for example, Geisonoceras severum (Barrande), Virgoceras polemon (Barrande), etc. They are quite different from those in Southwest China.

Judging from the facts mentioned above, it may be concluded that: 1. In Xizang, both sides of the Yarlung Zangbo River were a combined continent in the Early Paleozoic era. The Ordovician and Silurian nautiloids of South Xizang and North Xizang are fairly similar to each other. 2. In the late Early Paleozoic era from North China up to the continent, the Tethyan Sea was gradually formed and the Pangaea was split into two large landmasses. Then, in Mesozoic, they reattached again to each other. 3. The Early Paleozoic strata extending as two belts may be controlled by great faults.

## References


(1) Barrande, J., Systeme Silurian du centre la Boheme, Pt. 1,2, (1865-1877).

(2) Chang Ri-tung, Early Ordovician nautiloids of Zhuozishang region, Yikezhaomeng Nei-Megu, Acta Palaeont. Sinica, 7 (4), 1-71 (1959) (in Chinese).

(3) Chang Ri-tung, Polydesmia occurrence and its relation with Ordosoceras. Ibid., 8 (1), 41-53 (1960) (in Chinese).

(4) Chang Ri-tung, Several species of Middle Ordovician cephalopod, Huanxian of Gansu. Ibid., 10 (4), 514-521 (1962) (in Chinese).

(5) Chen Jun-yuan, The fossil nautiloids from the Mount Jolmo Lungma Region, A report of scientific expedition in the Mount Jolmo Lungma Region (1966-1968), Palaeont., Fasc. I, 267-308 (1975) (in Chinese).

(6) Endo, R., The Canadian and Ordovician Formations and fossils of South Manchuria, U. S. Nat., Bull. 164, 1-152 (1932).

(7) Flower, R. H., Studies of Paleozoic Nautiloidea I-VII, Pt. 4, Bull. Amer. Palaeont., 28 (169), 30-67 (1943).

(8) Flower, R. H., Ordovician cephalopods of the Cincinnati region Amer. Palaeont., 29 (116), 1-656 (1946).

(9) Gansser, A., Geology of the Himalayas. Interscience publ. John Wiley and S. Londres, 1-289 (1964).

(10) Hayden, H. H., The Geology of Spiti. Men. Geol. Surv. India, 36 (1) 1-129 (1904).

(11) Heim. A. Gansser, A., Central. Himalaya, Geological (1939).

(12) Mu An-tze et al, Stratigraphy of the Mount Jolmo Lungma Region in Southern Tibet. Scientia Sinica, 16 (1), 96-111 (1973).

(13) Wade, M., Georginidae, new family of actinoceratoid cephalopods Ordovician, Australia. Mem. Queensland Mus. 18 (L), 1-15 (1977).

(14) Wang Yi-gang, The stratigraphy in Mount Jolmo Lungma Region. A report of scientific expedition in the Mount Jolmo Lungma Region (1966-1968), Geology, 24-47 (1975) (in Chinese).

(15) Yin Ji-xiang, Guo Shi-zeng, Stratigraphy in the Mount Jolmo Lungma and its north slope. Scientia Sinica. 21 (5) 629-644 (1978).

Table1 DISTRIBUTION OF CEPHALOPODS IN EARLY ORDOVICIAN

| | S. Xizang | N. Xizang | S. China | Xinjiang | Gansu | W. Yunnan |
|---|---|---|---|---|---|---|
| O₂ | *Sinoceras* (2 sp) | ● | ● | ● | ● | ● |
| | *Michelinoceras paraelogatum* | | ● | | ● | ● |
| | *M. pararegulare* | | ● | | ● | ● |
| | *M. xuanxianense* | | ● | | ● | |
| | ect. spp. | ● | | ● | ● | ● |
| | *Eosomichelinoceras* 2sp. | ● | ● | | | ● |
| | *Paramoceras* 1sp. | | | | | ● |
| | *Ormoceras* 2sp. | | ● | | | ● |
| | *Trocholites* 1sp. | ● | ● | ? ● | | ● |
| | *Beloitoceras* 2sp. | | ● | | | |
| | *Anaspyroceras* 2sp. | | | | | ? ● |
| | *Actinomorpha* 1sp. | | ● | | | |
| | *Curtoceras* 1sp. | | ● | | | |
| | *Dideroceras* 1sp. | | ● | ● | | ● |

Table 2 DISTRIBUTION OF CEPHALOPODS IN MIDDLE ORDOVICIAN

| | S. Xizang | N. Xizang | N. China | NeiMonggol | S. China | Xinjiang | S. Qinghai | Australia |
|---|---|---|---|---|---|---|---|---|
| $O_1$ | *Ordosoceras* (*Pomphoceras*) (5sp.) | | ● | ● | | | | |
| | *Wutinoceras* (2sp.) | | ● | ● | | | ● | ● |
| | *Ormoceras* (2sp.) | | ● | ● | | ● | ● | |
| | *Meitanoceras* (1sp.) | Cephalopods and *Crinoids* | | ● | | | | |
| | *Georgina* (1sp.) | | | | | ● | | ● |
| | *Manchuroceras* (2sp.) | | ● | ● | ● | | | ● |
| | *Hopeioceras* (1sp.) | | ● | ● | ? ● | | | |
| | *Paradnatoceras* (2sp.) | | | ● | ● | | | |
| | *Ancistroceras* (1sp.) | | | ● | ● | | | |
| | *Mysterioceras* (2sp.) | | | | | | | ● |
| | *Tarphyceras* (1sp.) | | | | | | | |
| | *Dideroceras* (2sp.) | | ● | ● | ● | | ● | ● |

Table 3 DISTRIBUTION OF CEPHALOPODS IN MIDDLE AND LATE SILURIAN

| | S. Xizang | N. Xizang | W. Yunnan | Europe | S. W. China |
|---|---|---|---|---|---|
| S $_{2-3}$ | *Michelinoceras* | | | | |
| | *michelini* | | ● | ● | |
| | *M. simiale* | | ● | ● | |
| | *M. transiense* | | ● | ● | ● |
| | et spp. | | ● | ● | ● |
| | *M. (Kopaninoceras)* | | ● | ● | |
| | *jucundum* | | ● | ● | |
| | *M. (K.) capax* | | ● | ● | |
| | *M. (K.) dorsatum* | | ● | ● | |
| | *M. (K.) ferganense* | | ● | ● | |
| | et spp. | | ● | ● | |
| | *Harrisoceras* 2sp. | ● | ● | ● | ● |
| | *Geisonoceras* 3sp. | | ● | ● | |
| | *Dawsonoceras* 1sp. | | ● | ● | |
| | *Geisonocerina* 1sp. | | ● | ● | |
| | *Columenoceras* 1sp. | | ? | ● | |
| | *Virgoceras* 1sp. | | ● | ● | |
| | | *Allanoceras* | | ● | |
| | | *Ederoceras* | | ● | |

Table 4 DISTRIBUTION OF CEPHALOPODS IN ORDOVICIAN AND SILURIAN

| | | South Xizang | North Xizang: Yongzhuqiao, Xainza | North Xizang: Dongka Lake Bangoin | North China | South China | West Yunnan |
|---|---|---|---|---|---|---|---|
| $S_3$ | Pulu Gr. | *Michilinoceras (Kopaninoceras) jucundum* Assemblage | | | | *Yunnanoceras*<br>*Enthocycloceras* zone | Camarocrinus beds<br>*Michelinoceras (Kopaninoceras) jucundum* |
| $S_2$ | Pulu Gr. | *Michelinoceras michelini* Assemblage | | | | *Sichuanoceras* zone<br>*Michelinoceras* | *Michelinoceras michelini* |
| $S_1$ | Shiqipo Fm. | *Streptogr. lobiferus.*<br>*Climacogr. normalis* | Dewukaxia Fm.<br>*Climacogr. medius*<br>*Cl. normalis* | Dongka Fm.<br>*Harrisoceras*<br>*Allanoceras* | | *Songkanoceras* zone<br>*Yichangoceras* zone | *Streptogr. crispus* zone<br>↑<br>*Diplogr.modestus* zone |
| $O_3$ | | Hongshantou Fm.<br>*Michelinoceras* | Gangmusang Fm.<br>*Cl. normalis*<br>*Glyptogr* cf. *persculptus*<br>*Dalmanitina* | | | *Hirrantia—Dalmanitina*<br>*Dicellgr. szechuanensis* | *Michelinoceras* |
| $O_2$ | Chiatsum Gr. | Upper Fm.<br>*Sinoceras chinese* zone<br>*Curtoceras*<br>*Eurasiaticoceras* | Keerduo Fm.<br>*Sinoceras*<br>*Curtoceras*<br>*Eurasiaticoceras* | | *Gonioceras badouense* zone | *Sinoceras chinense* zone<br>*Eurasiaticoceras* | *Sinoceras chinense* zone<br>*Eurasiaticoceras* |
| $O_1$ | Chiatsum Gr. | Lower Fm.<br>*Ordosoceras (Pomphoceras)—Wutinoceras* zone<br>*Georgina* | Mendeema Fm.<br>Cephalopods<br>Crinoids | | *Stereoplasmoceras Pseudoseptatum* zone<br>*Ordosoceras—Polydesmia* zone<br>*Manchuroceras—Coreonoceras* zone | *Dideroceras Protocycloceras* zone<br>*Manchuroceras—Coreonoceras* | *Dideroceras*<br>*Protocycloceras*<br>*Didymograptus* |

# STRATIGRAPHIC DISTRIBUTION OF FOSSIL CORALS AND THEIR PALAEOBIOGEOGRAPHIC PROVINCES IN XIZANG

Wu Wang-shi Liao Wei-hua

(Nanjing Institute of Geology and Palaeontology, Academia Sinica)

The Palaeozoic, Mesozoic and early Cenozoic marine strata are well developed in Xizang and rich in fossil corals. As far as our knowledge goes, the oldest coral-bearing beds in Xizang belong to Lower Silurian and the highest strata to Middle Eocene.

Up to now, the Early Silurian corals in Xizang have been discovered in only two localities: the one in the Nyalam district of southern Xizang while the other in the Baingoin district of northern Xizang. In the former, *Triplophyllum nyalamense* and *Favosites* sp. were found, while in the latter some tabulata such as *Palaeofavosites*, *Catenipora*, *Halysites* and a few solitary rugose corals somewhat similar to the elements of the Lojoping Formation in western Hubei were contained.

Devonian corals are mainly distributed in the Qamdo region of eastern Xizang and Rutog district of northwestern Xizang. Obviously they bear a much closer relationship to those in South China, the Kuznetsk basin, the Urals, the Mt. Holy-Cross, the Harz, Torquay and the Mackenzie River. However, no Devonian corals have yet been discovered in the area to the south of Yalung Zangbo River in Xizang, where there are merely some kinds of pelagic organism, such as graptolite and tentaculites, with only a few coral specimens from the Dolpo of Nepal being reported lately (1).

Carboniferous rocks are mainly exposed in both banks of the Lancang River, where they may be differentiated into two divisions: the Lower Carboniferous and the Upper Carboniferous. The former can be subdivided into two Formations: the Wuqingna Formation below and the Macala Formation above. The former formation contains *Cystophrentis kolaohoensis* Yü, an important element of the Kolaoho member of Tournaisian stage in southern Guizhou. In the Upper part of the latter Formation *Yuanophyllum*, *Clisiophyllum*, *Aulina*, *Dibunophyllum* and *Palaeosmilia* occur, which are the representative genera of the Zhaojiashan Formation of Lower Namurian stage in western Guizhou.

The Upper Carboniferous rocks of Riwoqe, Gonjo and Markam of Qamdo region contain the following fossils: *Sestrophyllum*, *Lithostrotionella*, *Bothrophyllum* and *Kepingophyllum* which were also recorded from the Upper Carboniferous Weining Formation and the Maping Formation in western Guizhou. Judging from the fossil aspects, the Carboniferous coral faunas from eastern Xizang may belong to the South China type. However, the Jilong Formation in southern Xizang, probably Late Carboniferous in age, yields the cold-water type coral *Empodesma* which has hitherto been known only from the lower part of Pennsylvanian Group in Oklahoma and Texas in U.S.A.

It is now generally recognized that there were two different biogeographic provinces of coral faunas during the Early Permian (2). The difference between the faunas of Lytvolasma and Iranophyllum is very celarly shown in Fig. 1. In southern Xizang the so-called Lytvolasma fauna is only some solitary corals, often associated with thick-shelled brachiopods. Doubtless to say, this fauna should belong to the cold-water type, distributing along the borders of the Gondwana land, such as Australia, Timor, Nepal and Pakistan. Just as was known to us, there existed the continental or shore glaciers on the Gondwana land during the late Palaeozoic. The Irannophyllum fauna discovered in northern Xizang is characterized by hermstypic corals which were supposedly living in a warm-water environment and may easily be compared with those of the other Tetyan regions such as Afghanistan, Iran and Turkey.

According to the data available from Iran and Greenland, the geological range of the Lytvolasma fauna is from the Parafusulina zone to the Neoschwagerina zone of fusulinid zones. However, the Iranophyllum fauna is only confined to the Neoschwagerina zone of fusulinid zones. With regard to the biostratigraphical sequence of lower Permian coral faunas in Xizang, it is possible that the Lytvolasma fauna is slighty lower than the Iranophyllum fauna, with a few elements of the former extending to the latter.

On the contrary, beginning from the Mesozoic, the distinction between the two biogeographic realms in the Permian gradually died away. Both seem to belong to the Tethyan type.

The Late Triassic coral-bearing strata are widespread in Xizang especially in the Hengduan Mts. and the Himalayas. Some genera and species such as Distichophyllum norica (Frech), Paradistichophyllum weberi (de Regny), P. aff. clathrata (Cooper Reed), Stylophyllosis mosjvari Frech, Thamnostropis dronori (Melnikova), T. loretzi (Volz), Omphalophylia baletiformis (Munster), Craspedophyllia cristata Volz and Retiophyllia cf. paraclathrata Roniewicz, previously reported in foreign countries, were also found in Xizang. Among them, the former three species occur simultaneously on both sides of the Yalung Zangbo River.

The Jurassic marine strata of Xizang (excluding on the east of the Lancang River) have been well-known (3). The Middle Jurassic contains the following corals: Epistreptophyllum duncani (Gregory), E. cornutiformis (Gregory), Comoseris jumarense radiata Gregory, Rhipidogyra flabellum Michelin, Complexastrea kachensis Gregory, Cladophyllia turbinata (Gregory) and Trocharea sp., which were originally reported from Middle Jurassic in India; while the Upper Jurassic Scleractinia bear a close resemblance to those faunas of Portugal, France, F.R. Germany, Switzerland, Poland as well as the Caucasus of U.S.S.R. Among them, there exist a lot of identical species, for instance, Isastrea goldfussi (Koby), Isastrocoenia kachensis Gregory, Thamnastrea confluens (Quenstedt), Goniocora annulata Roniewicz, Schizosmilia excelsa Koby, Plesiosmilia truncata Koby, Pseudocoenia maxima (Beauvais), Epistreptophyllum cylindratum Milashevitsch and Fungiastrea multicinata (Koby), of which the former two are to be seen everywhere in Xizang.

The Early Cretaceous corals have been found to the south of the Nagqu-Ngari Highway and to the west of the Qinghai-Xizang Highway. Among these corals, there are a few species, such as Opisthopyllum vesiculare Ogilvie, Actinastraea pseudominina major Morycowa, Fungiastraea turbinata (de Fromental), Dermosmilia laxata Etallon. Microsmilia oppeli (Remeš), Danacosmilia cf. corallina de Fromentel, Aulastraeopora deangelisi Prever, Hydnophora insignis (Duncnn), etc., which are common in the corresponding beds in Portugal, France, Italy, Tunisia, Bulgaria, the Crimea of U.S.S.R. as well as South India. It should be pointed out that the Early Cretaceous fauna was still kept in a Late Jurassic aspect.

The Upper Cretaceous marine beds are only distributed in the Himalayas and along the Yarlung Zangbo River. From the upper part of the Zongshan Formation in Gamba some corals such as Paracyathus, Cyathocera and Trochocyathus occur, with a curly spongy columella which is a characteristic axial structure in Tertiary Scleractinia. However, according to the geological ranges of other kinds of fossils, such as foraminiferas, echinoidea, algae, lamellibranchia and gastropods, the upper part of this formation is believed to belong to the Upper Cretaceous Maestrichtian stage.

Up to the present, the highest marine strata in Xizang is known to be of Middle Eocene. It seems that in Xizang, the history of marine sedimentation came to an end in the period of post-Eocene. The Early Tertiary marine deposits are distributed only in some partial areas, showing the characteristics of a residual sea basin. The lower part of Zongpu Group (Palaeocene) and the Zhepure Formation (Eocene) contain Astrocoenia and Bantamia which may be compared with the corals from Pakistan, Indonesia as well as North America.

After Middle Eocene, the sea withdrew completely from the Himalayas, yet a tremendous uplifting movement in the formation of the Xizang Plateau took place only after the Pliocene epoch.

Why is it that there was such a great difference in the aspects of coral fossils between the south and the north of the Yarlung Zangbo River during the late Palaeozoic and that from the beginning of Mesozoic such difference became more and more indistinct and finally disappeared?

Two possibilities may be involved in. The first one is that during the late Palaeozoic, there was still a very far distance between the Gondwana land and the Eurasian land. The northern Xizang within the Eurasian land was situated in the tropics or subtropics and, therefore, plenty of reef-building corals enjoyed their living. On the other hand, southern Xizang, India as well as the other portions of the Gondwana land were located in the frigid zones, where continental and mountain glaciers might even have happened. It is for that reason that only a few species of cold-water solitary corals could exist.

From the beginning of Mesozoic, the Indian Plate drifted rapidly northward and gradually got closer to the Eurasian continent until it finally reached the temperate zone. Therefore, the characteristics of the faunas in both realms gradually tended to be quite similar to each other.

The second possibility is that the climatic changes of the earth can be explained by the variations in the obliquity of the earth's ecliptic. So far as we know, the difference of the obliquity of the earth's ecliptic may have an influence on the radiative heating power from the sun, and then further affect the extent of climate zones, both warm and cold as well as the average annual temperatures on earth's surface (4).

In Triassic, the vertebrate fossils (mainly Amphibia and Reptilia) were wide spread on land from Antarctica to the Spitsbergen Island in the Arctic Circle (5), while the northern limits of the distribution of reef-building corals in Jurassic were approximately Lat. 60°N. (6) (7) (Fig.2). Such case mentioned above can be interpreted that the warm climate zone in Mesozoic was more extensive than that in late Palaeozoic or at present. Therefore, both parts of Xizang belong to the same, climate zone. In addition, corals were spread far and wide with the help of planulae, and similar species could be found in both the north and south of Xizang's sea.

It seems true that the continent might have drifted and that the obliquity of the earth's ecliptic might also have changed during the past long geological history. Accordingly these two possibilities mentioned above are likely to be acceptable. In late Palaeozoic, it was still a very far distance between northern Xizang and southern Xizang. Moreover, the warm climatic zone became narrower because the obliquity of the ecliptic was comparative bigger. At that time, northern Xizang was within the confines of tropics or subtropics, where there were plenty of reef-building colonial corals. On the other hand, being located in the frigid zone, southern Xizang witnessed only a few solitary corals belonging to cold-water type.

From the beginning of Mesozoic, southern Xizang as a portion of the Indian Plate drifted rapidly northward and gradually got closer to northern Xizang. On the other hand, as the obliquity of the ecliptic became smaller, the warm climatic zone was in turn wider and wider. Now that the weather environments in both southern and northern Xizang were quite the same, similar coral fauna multiplied to a great extent.

## References


(1) Flügel, H., Paläozoische Korallen aus der Tibetischen zone von Dolpo (Nepal). Jb. Geol. B.A. Sonderband, 12, 101-120 (1966).

(2) Wu Wang-shi, The coral fossils from the Mt. Jomo Lungma Region. A Report of Scientific Expedition in Mt. Jomo Lungma Region (1966-1968), Palaeontology, Fasc. 1, 83-128 (1975). (in Chinese)

(3) Liao Wei-hua and Li Zhang-rong, Jurassic Scleractinia from Amdo, Northern Xizang (Tibet). Acta Palaeontologica Sinica, 19 (3), 228-238 (1980) (in Chinese).

(4) Xu Qin-qi, On the causes of ice ages. Scientia Geologica Sinica, (3), 252-263 (1979) (in Chinese).

(5) Liu Shi-fan, Obliquity of the ecliptic in the Geological time. Acta Astronomica Sinica, 18 (2), 259-263 (1977). (in Chinese).

(6) Beauvais, L., Upper Jurassic hermatypic corals. In Hallam, A. (Edit.): Atlas of Palaeobiogeography, 317-328 (1973).

(7) Beauvais,L., Beauvais, M., Studies on the world distribution of the Upper Cretaceous corals. Proceedings of the Second International Coral Reef Symposium, 1, 475-494 (1974).

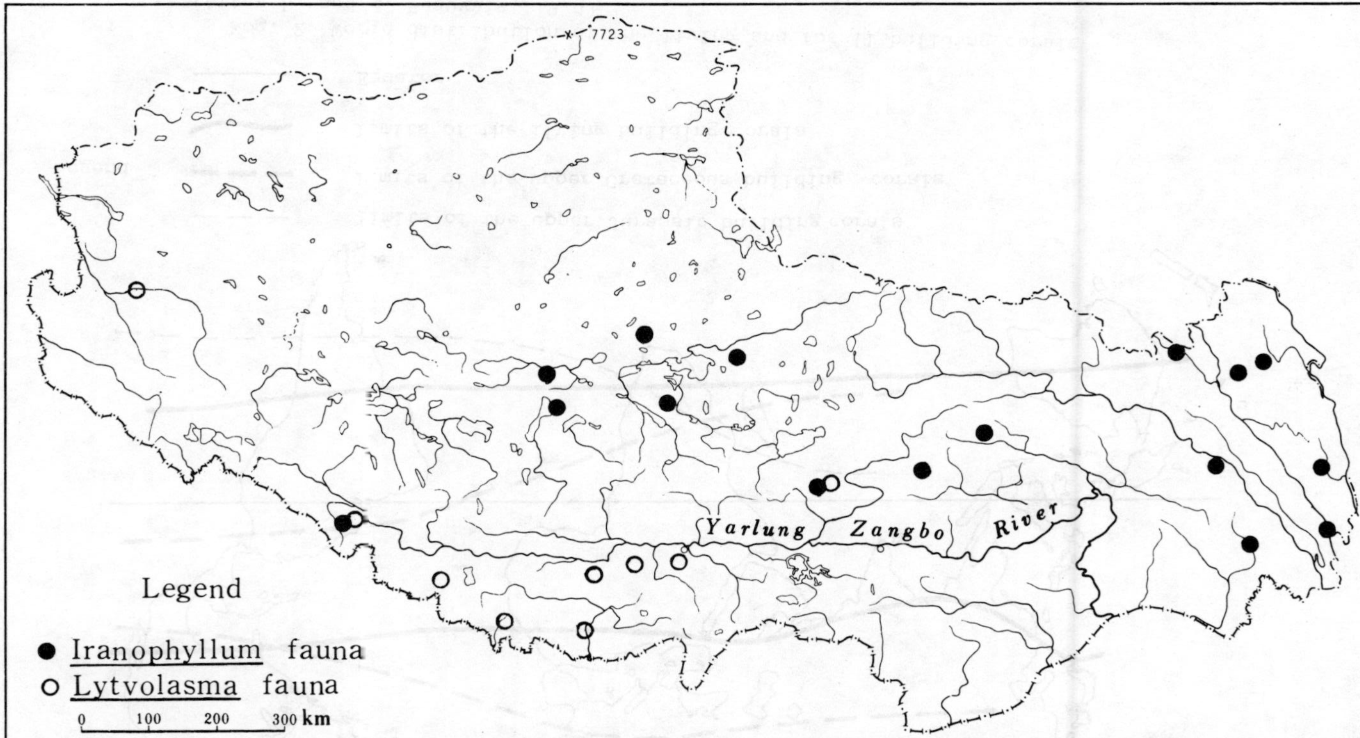


Fig. 1 Sketch map showing the distribution of the Iranophyllum fauna and Lytyolasma fauna in Xizang during Early Permian period

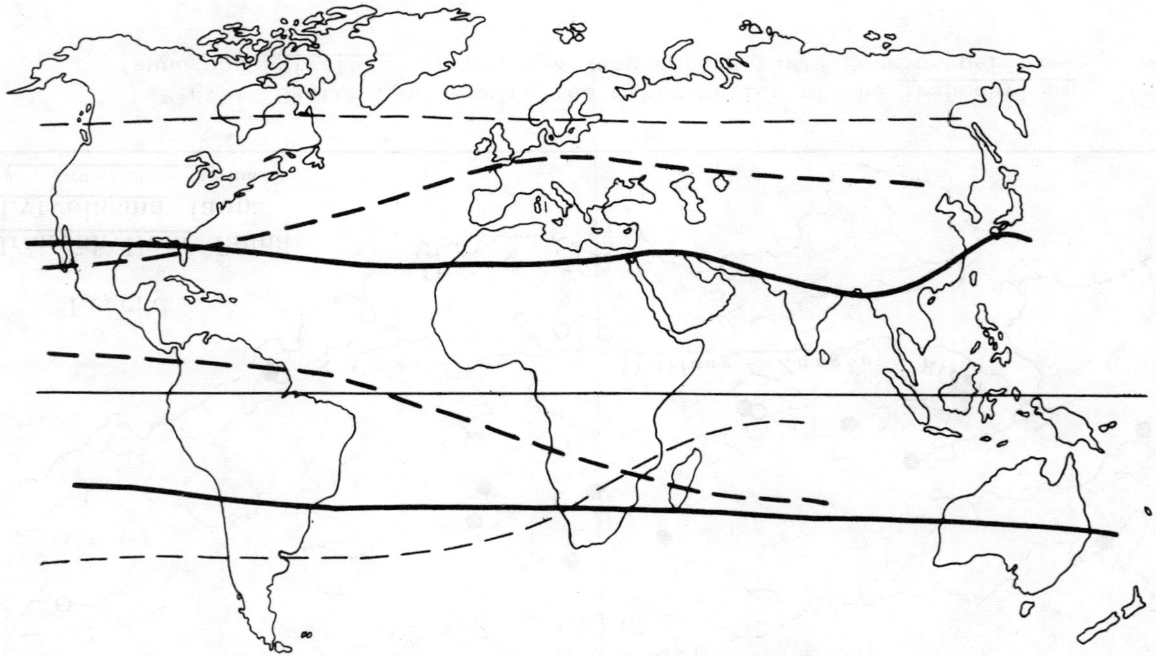

Legend

- limits of the upper Jurassic building corals
- limits of the upper Cretaceous building corals
- limits of the living building corals
- Equator

Fog. 2 World distribution of the living and fossil building corals (After L. and M. Beauvais, 1974)

# ON THE PALEOECOLOGICAL RELATION BETWEEN GONDWANA AND TETHYS FAUNAS IN THE PERMIAN OF XIZANG

Jin Yu-gan
(Nanjing Institute of Geology and Palaeontology, Academia Sinica)

Using the lithological characters of the fossiliferous sequences, Diener (5) divided the Permian marine faunas of the Central Himalayas into two facies, the Himalayan and the Tibetan facies. The former is characterized by a dominance of endemic elements and terrigenous clastic rocks, and the latter contains a fauna similar to that from the Middle and the Upper Productus Limestones in the Salt Range and usually occurs in thick bedded limestone. Recently, a series of stratigraphic and palaeontologic studies of the Permian in Xizang and other areas of the Central Himalayas have provided a great abundance of fundamental information regarding the distribution of marine faunas. Faunas broadly corresponding to those of Diener's two facies were interpreted as different faunal provinces, namely the Gondwana or cool water fauna, and the Tethys or warm water fauna. Some palaeontologists suggested that the Permian rocks bearing these faunas were deposited in widely separated marine areas during the Permian and were brought together later. But, primarily based upon the distribution patterns of brachiopods, three latitudinal faunal regions in China have been identified (8,16). The southern region is approximately restricted to the south of $32^{o}$ north latitude and delimited to the east by the Nujiang River; the faunas are closely related to those of the northern marginal sea of Gondwana. The northern region, with its southern boundary at about $42^{o}$ north latitude, is a Paleozoic eugeosynclinal region that communicated with the Boreal Realm. Other areas stretching from the Kunlun Mountains in the west to Taiwan in the east are all included in the central region, and are characterized by flourishing, typical faunas of the Tethys Realm. The distribution pattern of these faunal regions may be largely due to differences in climatic zones. Boundaries between these faunal regions were actually delineated by ancient topographic barriers, though they broadly parallel latitudes. Furthermore, each region is distinguished by a unique suite of communities from onshore to offshore. It is of interest that the faunas previously classified into the Tethys fauna and non-Tethys fauna, such as the Gondwana fauna and the Boreal fauna, are possibly contemporaneous communities in different positions in the same region. In order to further discuss this problem,a brief description of ecological characters of the Permian faunas in the southern region and related areas will be given below. Although many of the conclusion presented in this paper are speculative and will undoubtedly be modified by more detailed study in the future, the writer presents them in the hope of emphasizing a different line of study and shedding some new light on a solution of major problems in regional geology.

## FAUNAL ASSOCIATIONS AND LITHOLOGICAL RELATIONSHIPS

In the southern region, the lower boundary of the Permian was arbitrarily placed either above or below the pebble slate group with the

*Stepano viella* fauna. This group extends from the Mount Qomolangma region northwards through Kangmar into Lhasa and Bomi and as far north as the Orba Lake (10). Typically, it consists, from bottom to top, of pebble mudstone, sandstone and shale interbedded with thin argillaceous limestone. This unit probably accumulated in a trough developed between Gondwana to the south and the carbonate platform to the north. As a whole, the amount of terrigenous material and the size of pebbles in this unit tends to decrease upward and northward and it is presumed to be replaced by argillaceous limestone in the areas north of this region such as Markam and Gonjo. Marine fossils found in the fine-grained sandstone are of high density but always low diversity in Lhasa and are closely similar to those from the Umaria marine bed which is overwhelmingly dominated by *Stepanoviella*. Specimens of *Chaoiella*, *Syringothyris*, *Leiorhynchoidea*, *Spiriferellina* and *Deltopecten* were also collected but the percentage of each taxon in total number of specimens never exceeds 2%. The fossiliferous bed of the Jilong Formation consists of well sorted, fine-grained, bioclastic sandstone, and yields a more diverse fauna including pelecypods, brachiopods, trilobites and solitary corals dominated by *Stepanoviella* and *Trigonotreta*; the latter is usually small in size and characteristically thin shelled. In a transgressive sequence, the *Stepanoviella* fauna has been discovered below the bed with a fauna containing foraminifers, bryozoans and brachiopods, for example, the *Callimarginatia* fauna of the Urulung formation in Lhasa (Jin and Sun, in press), and above the bed yielding the pelecypod dominated *Eurydesma* fauna in Kashmir (2). Following the model of late Carboniferous marine invertebrate communities in clastic sediments proposed by Bretsky (3), evidence mentioned above suggests that the latest Carboniferous communities from onshore to offshore in this region and related areas are (1): the intertidal to subtidal *Eurydesma* community (2), the subtidal *Stepanoviella* community and (3) the *Liraplecta* shelf community which includes some schwagerinids and always appears in the central region.

Permian rocks distributed in southern Xizang may be grouped into three subparallel major depositional belts formed in the shelf sea of the northern margin of the ancient Indian continent. They are inner shelf, bank and outer shelf. Farther northward, in eastern and northern Xizang, the Permian sequences are closely comparable to those of South China, but with volcanic rocks in various horizons, and are considered to have been deposited in a setting sloping from the carbonate platform into a foreland trough south of the Tarim uplift. This trough extends along the Kunlun Mountains towards the east of the Nujiang River where there is a well developed flysch sequence. On the other hand, in the foothills of the Eastern Himalayas, a long narrow belt of terrestrial deposits extends from eastern Nepal to the Abor Hills. The terrestrial deposits are interbedded with marine beds at several places which contain *Lingula* (1). The deposits seem to have accumulated in a supratidal to shallow subtidal depositional belt which grades northward into an inner shelf depositional belt.

The inner shelf depositional belt was wide, occupying most areas south of the Yarlung Zangbo River. The lower part of the Permian sequence in this province consists of fine-grained sandstone, siltstone and slate bearing plant fossils in the Mount Qomolangma region. The Qubu Formation, which may be a northward extension of the Damuda Group developed in the Eastern Himalayas, grades further northwards into black slate with a limestone band in Kangmar and in Karganglha Pass. Its equivalents in the other two depositional belts are composed of thin bedded biospararenite limestone with bryozoans and brachiopods as its main biogenic components, for example, the Urulung Formation in Lhasa and the limestone found in Lhaze. A marine fauna dominated by *Calliomarginatia*, *Krotovia* and fenestellid bryozoans was collected in Karganglha Pass, Lhaze and Lhasa. These terrigenous sediments were presumably deposited in shallow, nearshore environments and the carbonate rocks may have been deposited in shallow to moderately deep water. The upper part of the Permian sequence in this depositional belt, such as the Quburega Formation, include calcareous sandstone, siltstone and shale as well as bioclastic limestone which includes coquina and biosparrudite with brachiopod debris and crinoidal stems as its main biogenic components. The marine fossils are dominated by a large, thick shelled brachiopod fauna with low diversity. *Spiriferella* and *Neospirifer* are often found with a great abundance of pedicle valves, but brachial valves are rare and were possibly destroyed by strong wave action. Solitary corals and fusulines have not been discovered so far. This fauna and sediments may be assigned to Diener's Himalaya facies and were considered as a representative of the cool water or Gondwana fauna, but are interpreted here to be formed in inner shelf, high energy environments. Corresponding units of the Quburega Formation in the Kangmar area, a little farther north, but still in this depositional province are more calcareous and somewhat thicker.

Some isolated Permian rocks outcrop in the belt extending from Lhaze to Chitichum along the Yarlung Zangbo River are included in the bank depositional belt. They are all composed of carbonate deposits ranging from biosparrudite to biospararenite and primarily consisting of bryozoan and brachiopod fragments. Brachiopods from the 'exotic block' in Zhongba are very diverse and contain some reef or bioherm elements such as *Richthofenia*, *Enteletes* and *Leptodus* (7). Species of *Comuquina*, *Marginifera* and some pediculate brachiopods are common and usually form locally abundant populations. Associated organisms include abundant colonial corals, ramose bryozoans, calcareous algae, sponges and so on. The fauna from the Chitichun Limestone, the type of the Tibetan facies, is closely comparable to that from Zhongba and is thought to have been formed in this westward, discontinuous bank belt.

Rocks of the outer shelf depositional belt distributed in the areas north of the Yarlung Zangbo River such as Lhasa, Xainza and Bomi, which are lithologically characterized by a dominance of foraminifers and algae biosparite to foraminiferal biomicrite with bryozoans and brachiopods biospararenite bands in the upper part. The faunas are remarkable for a dominance of colonial corals, such as *Iranophyllum* and fusulines with subsphaeroid shell and thinner spirotheca such as some species of *Neoschwagerina*, *Verbeekina*, which chiefly inhabited gentle, near wave base environments.

*Nankinella* is abundant in some horizons in Xainza and Bomi. According to Choi (4), this genus seems to be best adapted to deep water and slightly muddy environments. In addition, a characteristic productid species of inner shelf depositional province, *Costiferina alata*, was also collected from the bryozoan and brachiopod biospararenite band in Lhasa, indicating that this band is possibly a northward extension from the south.

To conclude, the lithofacies change of the Permian deposits in the southern region is characterized by decreasing terrigenous clastic rocks and shifting carbonate rocks from coarser grained to finer grained northward. The distribution pattern of depositional belts is lithologically and geographically similar to those in the present shelf sea of the northeastern Gulf of Mexico (6). Corresponding to the Permian depositional provinces in the southern region and related areas, from strand line to offshore are (1) quartz-shell sand (quartz sand 50%), quartz-shell sand (quartz sand 50%), (2) shell sand, (3) oolitic sand and foraminiferal sand and silt belts. Analogy between them strongly suggests that the Permian rocks of the southern region have accumulated in a single shelf sea deepening northward, and that no other sea basins have ever been developed between these depositional belts during Permian time.

## A COMPARISON WITH THE NORTHERN REGIONS OF CHINA AND THE WESTERN UNITED STATES

A comparison of the Permian ecological distribution of the southern region of Xizang with that of the northern region of China seems to be particularly useful as faunas from both regions share so many genera and species and their communities are so structurally similar that the inferred community model of the southern region may be tested by the data from the other one. In the northern region the early Permian deposits consist of a flysch-like sequence. Thus the change of communities is relatively rapid.

These flysch-like series are mainly composed of rhythmical sequences of clastic rocks intercalated with bioclastic limestone lenses. Generally, they largely consist of coarser-grained sandstone in northern Gansu, fine grained sandstone in northern Nei Monggol and siltstone in Northeast China. There is a tendency to decrease in grain size eastward. Correspondingly, the limestone lenses are characterized by a large numbers of coquina and crinoidal biosparrudits bryozoan and fusulines biospararenite in northern Gansu, by crinoidal biosparrudite, bryozoan and fusuline biospararenite in northern Nei Monggol, and by biomicrite in Northeast China.

Faunas from northern Gansu are similar to those from the inner shelf depositional belts of the southern region. Both regions contain a large, thick shelled brachiopod fauna with low diversity, abundant pedicle valves of <u>Spiriferella</u> and <u>Neospirifer</u>, solitary corals but no fusulines or colonial corals were found. From northern Nei Monggol, in addition to the former fauna, a bioherm fauna or bank fauna is also developed and contains diverse brachiopods and colonial corals which are similar to those of the Tibetan facies. Fusulines of this bioherm fauna are often dominated by monotonous schwagerinids but include hardly any neoschwagerinids. Neoschwagerinids, verbeekinids and Tethys elements among the brachiopods, corals and other groups are abundant in Northeast China, and may be comparable to those from the outer shelf depositional belts of the southern region, to some extent.

It is worthy of notice that the faunas respectively similar to those of the Tibetan facies and the Himalayan facies have been found repeatedly in different portions of the same bioherm in Jisu Honger district, northern Nei Monggol. As a rule, the succession of biohermal communities begins with crinoidal and brachiopod debris and a community characterized by flourishing solitary corals, <u>Spiriferella</u>, <u>Neospirifer</u> and other large, thick shelled brachiopods, which may be comparable to those in the Himalayan facies. Then, the frame builder, sponges, bryozoans, algae, colonial corals and reef brachiopods crowd on the shell debris and form a community similar to the Tibetan facies. It is also worthy to note that some bioherms in Northeast China commonly begin with monotonous schwagerinids and then form a diverse community with neoschwagerinids and verbeekinids.

In addition to the foregoing region, the distribution of the Tethyan and the non-Tethyan faunas in Western North America also can be cited as an example for comparison. The Tethyan fauna is marked by the dominance of neoschwagerinids and verbeekinids and was inferred as what developed in the Palaeopacific Ocean far from the non-Tethyan fauna which is characterized by dominance of schwagerinids or brachiopods with Boreal affinities.

According to Stevens (1), Permian rocks in western United States can be grouped into eight depositional provinces. Deposits formed in inner and outer shelf contain considerable amounts of terrigenous clastic sediments and shallow water marine fossils. The shelf margin deposits are distinguished by coarser grained and better sorted calcarenite and abundant colonial corals. These three depositional provinces can be compared with inner shelf and bank depositional belts in the southern region. Rocks accumulated in the interior sea and uplift belt composed of clastic sediments and no counterparts of them have been found in the southern region. Sequences consisting of andesitic volcanic, volcanic clastic rocks, shale and limestone lenses are developed in the remaining depositional provinces to the west of the uplift belt, in which Tethyan elements were always reported from the rocks overlying those with non-Tethyan elements. Those provinces were inferred as back-arc basin and volcanic arc, and the overlapping pattern was interpreted (11) as a result of collison between plates far apart from each other. In light of a comparision with the ecological distribution of the Permian fauna in the southern and the northern regions of China, a tentative explanation may be suggested here that the Tethyan fauna and the non-Tethyan fauna are different communities of the same community succession formed in the open sea, similar to faunas in the outer shelf depositional belts of South and Northeast China in the northern region.

## CONCLUSION

The result of this tentative comparative study shows that: 1. Changes of sediments and faunas of the Permian progressing from onshore to offshore in southern Xizang, northern China and western United States are roughly comparable. 2. The so-called Gondwana Fauna dominated by large, thick-shelled brachiopods and solitary corals and devoid of fusulines, is a ubiquitous faunal association and commonly developed in terrigenous clastic rock and coarser grained calcarenite formed in nearshore environments. 3. In general, the Tethyan fauna is dominated by fine grained calcarenite formed in the open sea far from shorelines. 4. An intermediate faunal association, containing a bank fauna distinguished by monotonous schwagerinids and/or bioherm faunas, is also formed in the position between them.

Therefore, it is difficult to see any reliable faunal or lithological evidence for the suggestion that the areas on both sides of the Yarlung Zangbo suture were widely separated during Permian time. We believe that apart from climatic conditions, distribution of marine animals is also dependent on other ecological factors. In delimitation of the faunal distribution and using it in estimating the configuration of continents, paleoecological analysis should not be neglected.

## References


(1) Acharyya, S.K., Shah, S.C., Biostratigraphy of the marine fauna associated with the diamicites of the Himalayas. Bull. Ind. Geol. Assoc., 8 (2), 9-24 (1976).

(2) Acharyya, S.K., Ghosh, S.C., Ghosh, R.N., Shah, S.C., The continental Gondwana Group and associated marine sequences of Arunachal Pradesh (Nefa). Himalayan Geology, 5, 60-82 (1975).

(3) Bretsky, P.W. Jr., Evolution of Paleozoic benthic marine invertebrate communities. Palaeogeogr. Paleoclimat. Paleoecol., 6, 45-59 (1969).

(4) Choi, D.R., Distribution of the Upper Permian fusulinids with relation to limestone lithofacies in the Southern Kitakami Mountains, N.E. Japan. Journ. Geol. Soc. Japan, 82 (2), 113-125 (1976).

(5) Diener, C., The Permo-Carboniferous fauna of Chitichun, No. 1, Palaeontologia Indica, ser. 15, Himalayan Fossils, 1, 1-105 (1897).

(6) Gould, H.R., Stewart, R.H., Continental terrace sediments in northeastern Gulf of Mexico. In Finding Ancient Shorelines, Soc. Econ. Paleontol. and Mineralog., New Haven, Conn., 1-19 (1955).

(7) Grant, R.E., Brachiopods in the Permian reef environments of West Texas. North Amer. Paleont. Convention, Pt. J, 1444-1481 (1969).

(8) Jin Yu-gan, Liang Xi-luo, Wen Shi-hsuan, Additional material of animal fossils from the Permian deposits on the northern slope of Mount Qomolangma Feng. Sci. Geol. Sinica, (3), 237-249 (1977) (in Chinese).

(9) Monger, J.W., Ross, C.A., Distribution of Fusulinaceans in the Western Canadian Cordillera. Canadian Jour. Earth Sci., B, 259-278 (1971).

(10) Norin, E., Geological expeditions in Western Tibet. In Sine-Swedish Expedition report. Stockholm, 29, 1-24 (1946).

(11) Stevens, H., Permian depositional provinces and tectonics, Western United States. Paleozoic Paleogeography of the Western United States. Pacific Sect., Soc. Econ. Pal. Min., Los Angeles, California, U.S.A., 113-135 (1977).

(12) Wang Yu-jin, Mu Xi-nan, Some new observations on the Permian biostratigraphy of the Himalayan Province in southern Xizang. Jour. Stratigraphy, 4 (2), 145-152 (1980) (in Chinese).

(13) Waterhouse, J.B., Gupta, V.T., Early Permian fossils from southern Tibet, like faunas from Peninsular India and Lesser Himalayas of Garhwal. Jour. Geol. Soc. India, 20, 461-464 (1979).

(14) Zhang Bing-gao, Permian, In the Stratigraphy in Mt. Jolmo Lungma Region. A report of sci. exped. in Mt. Jolmo Lungma region (1966-1968), Geology, Sci. Press, 58-80 (1975) (in Chinese).

(15) Zhang Shou-xin, Jin Yu-gan, Late Palaeozoic brachiopods from Mt. Jolmo Lungma Region. A report of Sci. exped. in Mt. Jolmo Lungma Region (1966-1968), Palaeont., fasc. 2, Sci. Press, 159-242 (1976).

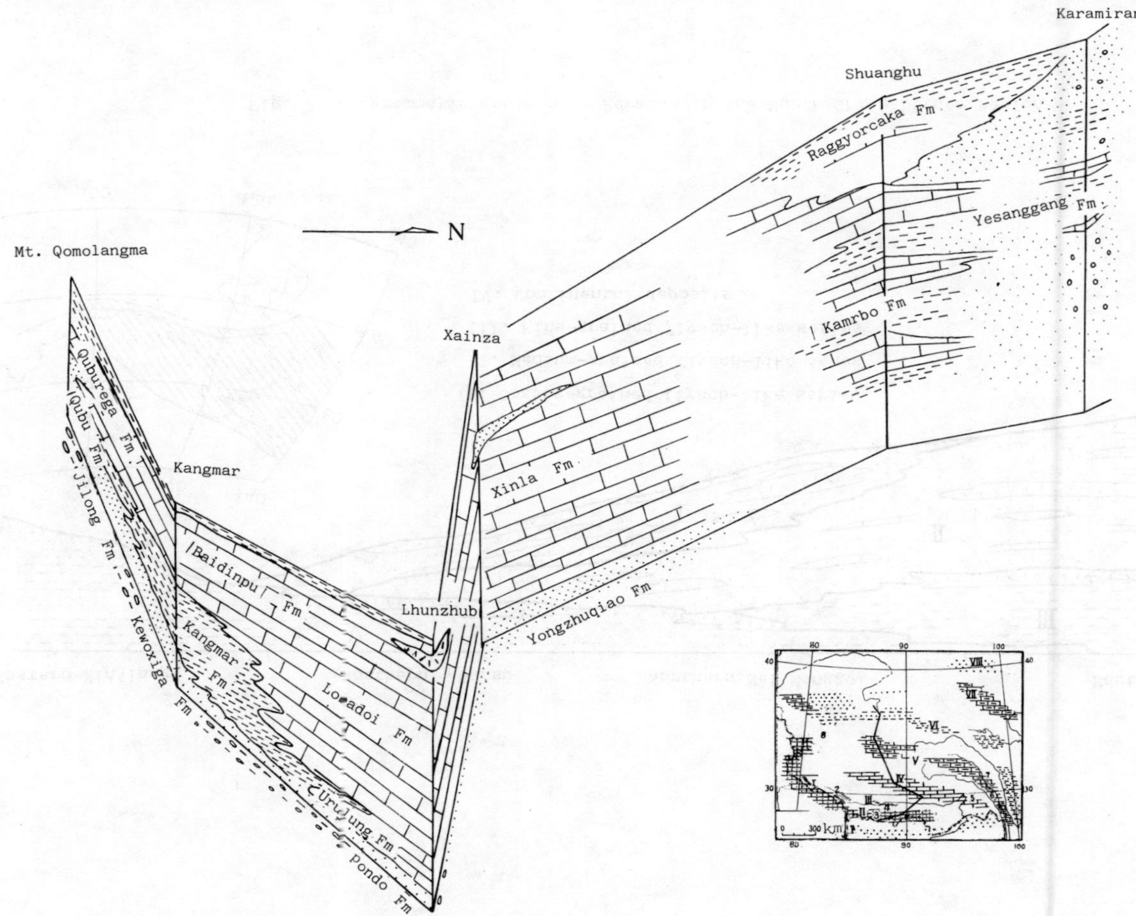


Fig. 1 Fence diagram showing thickness and facies relationships. List of localities 1. Chitichum, 2. Zhongba, 3. Karganglha Pass, 4. Lhaze, 5. Bomi, 6. Markam, 7. Gonjo, 8. Orba Lake. Depositional belts I. supratidal, II. inner shelf, III. bank, IV. outer shelf, V. shelf basin, VI. flysh belt, VII. shelf, VIII. supratidal

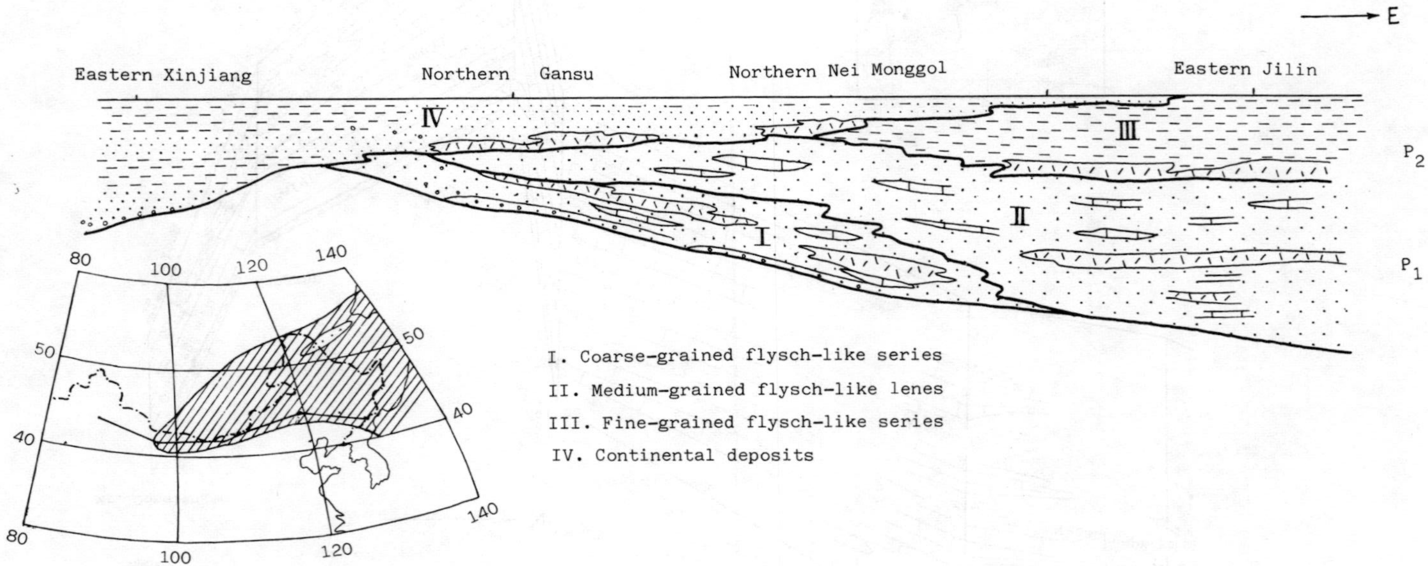


Fig. 2 Diagrammatic section of Permian in the North China border region

# NATURE OF THE PERMIAN BIOTAS IN XIZANG AND THE NORTHERN BOUNDARY OF THE INDIAN PLATE

Wang Yu-jing, Mu Xi-nan
(Nanjing Institute of Geology and Palaeontology,
Academia Sinica)

The Xizang Permian system is widely distributed. On the basis of the regional geological structure, litho-facies and characteristics of the biotas, it can be easily divided into two provinces, the Himalayas and Northern Xizang. The latter can be further subdivided into five subprovinces, known as Gandise—Nyainqentanglha, Karakorum, Tanggula, Hengduan Mts. and Hoh Xil-Kunlun (Fig. 1). The stratigraphic sequences, lithology and biotas in each province and subprovince are given in Table 1.

There are two types in Xizang Permian biotas, the Gondwana-Tethyan and the Cathaysia-Tethyan. The Gondwana-Tethyan type is chiefly distributed in the northern marginal sea of the Gondwana land including Himalayas and Gandise-Nyainqentanglha subprovinces in Xizang. Its characteristic is that the cool-water biotas developed in early Early Permian and the warm-water ones in late Early Permian. No or rare real marine fossils of Late Permian have been found yet. The cool-water biotas contain the Glossopteris flora and the faunas of Lytvolasma, Taeniothaerus, Monodiexodina and Uraloceras. Whereas the warm-water ones include Iranophyllum, Neoschwagerina faunas and the calcareous algal flora. The Cathaysia-Tethyan type distributes in Tanggula, Hoh Xil-Kunlun and Hengduan Mts. subprovinces in Xizang. This type is nearly similar to the warm-water biotas including the faunas of Iranophyllum, Misellina-Neoschwagerina and the calcareous algal flora in Early Permian and the Cathaysia flora as well as the fusulinid Palaeofusulina fauna in Late Permian.

## THE NATURE OF THE PERMIAN BIOTAS IN XIZANG AND PERMIAN BIOGEOGRAPHICAL PROVINCES

We have divided the Xizang Permian biotas above-cited into the cool-water and warm-water types. According to the law that the present is the key to the past, we will give out the following evidences.

### The Evidence of the Cool-Water Biota

1. The Gondwana Flora (The Glossopteris Flora). This flora consists mainly of Glossopteris, Gangamopteris, Neoggrathiopsis and so on. They have woods with annual rings which indicate the mark of a cool climate.

2. The Lytvolasma Fauna. This fauna is composed of some solitary individual corals with thicker septa but without dissepiments. They are similar to the ahermatypic corals in present Scheractinia living in the sea water with a temperature range of 4°-10°C, or even further below 0°. They mostly

live near the shore of Norway and are referred to members of the cold-water faunas. It can be inferred that the Lytvolasma fauna may also be a sort of cool-water fauna.

3. The distribution of Uraloceras The ammonite genus Uraloceras contained in the Quburega formation is chiefly distributed in Arctic Canada, Greenland, Ural and northern Siberia, East Australia and is an important fossil of the Arctic and Antarctic Realm.

In addition, no calcareous algal flora is found in the rocks yielding the cool-water fauna.

## The Evidence of the Warm-Water Biota

1. The Cathaysia Flora. This flora includes such elements as the genera Gigantopteris, Gigantonoclea, Lobatannularia and some Lepidodendron and Calamites. Generally speaking, the ferny plant is the mark of the warm climate, whereas the Lepidodendron and Calamites are the main elements of the Euamerican flora which lack woods with annual rings and are considered to be of warm humid climate.

2. The Calcareous Algal Flora. The Xizang Permian fossil algae contain the genera Mizzia, Eogoniolina, Clavaporella of the family Dasycladaceae, Anchicodium, Succodium of Codiaceae and Gymnocodium, Permocalculus of Gymnocodiaceae. The members of recent Dasycladaceae and Codiaceae are a sort of typical thermophilic algae chiefly living in the tropical and subtropical seas. The family Gymnocodiaceae is an extinct algae, but it structurally bears resemblance to the members of the recent red algae Chaetangiaceae which also lives in present tropical and subtropical seas. Judging by this fact, the members of Gymnocodiaceae might also be a sort of thermophilic marine algae.

3. The Iranophyllum Fauna. This fauna is composed of the colonial corals with Tertiary septa which are gathered in the reef limestone. Judging from modern reef corals commonly living in clear shallow sea water 30 degrees from the equator in latitude and with water temperatures from 25°-30° C, the Iranophyllum fauna might also occur in the Tethyan sea with similar environments and was a warm-water fauna.

The cool-water biotas mentioned above were chiefly distributed in two branches along the northern and southern margins of the Tethys respectively, namely the southern marginal sea of Angaraland and the northern one of Gondwanaland, whereas the warm-water biotas were developed in the Cathaysia-Tethyan sea between them (Fig. 2).

According to the hypothesis of the continental drift and plate tectonics, the factors controlling this distribution pattern of the biotas might probably result from the temperature in relation to the effect of the glaciation on the land masses, the latitude position and the ocean current.

Judging from the above fact, we have obtained some new views on the biogeographical provinces of the Permian Tethyan biotas.

The Permian marine biotas may generally fall into three biogeographical Realms: the Boreal Realm, the Antarctic Realm or Australia Realm and the Tethyan Realm. The first two Realms were situated at the south and north poles of that time and a cold-water fauna developed. The Tethyan Realm is subdivided into three provinces, known as the Gondwana-Tethyan, Angara-Tethyan and Cathaysia-Tethyan provinces.

The Gondwana-Tethyan province. This province was located at the northern margin of the Gondwanaland (including New Zealand, Timor, Himalayas, Karakorum, Salt Range, Afghanistan and Iran) and situated in the southern temperate zone. It is characterized both by the cool-water faunas represented by Lytvolasma, Taeniothaerus, Uraloceras and Monodiexodina in early Early Permian and by the warm-water Tethyan biotes such as Iranophyllum, Neoschwagerina and Dasycladaceae, Gymnocodiaceae.

The Angara-Tethyan province. This province was situated in the southern margin of the Angaraland (including Beishan Mt. of Gansu, northern Nei Monggol, northern Heilongjiang, central Jilin in China and East Siberia in U.S.S.R. and Kitakami Massif of Japan) and located in the northern temperate zone. The aspect of their biotas is compared to those of the Gondwana-Tethyan province in Early Permian.

The Cathaysia-Tethyan province. This province developed between the former two (including westerly from Tunisia via South Europe, West, Central, Southeast Asia, China to West America) and occupied a tropical and subtropical terrain. It was characterized by the typical warm-water biotas in Permian.

## ON THE NORTHERN BOUNDARY OF THE INDIAN PLATE

There are three different opinions on the northern boundary of the Indian Plate in Paleozoic up to the present. (1) It is located on the southern slope of the Himalayas; (2) Along the Yarlung Zangbo River Valley; (3) Along the line of Bangong Lake-Dongqiao-Nujiang River.

We agree to the second opinion, due to the recent discovery of the Glossopteris flora and the strata of the Gondwana facies from the northern slope of Mt. Qomolangma in Xizang, but we support also the third boundary in view of the following two reasons.

1. Judging from the fact that the cool-water Lytvolasma fauna and lower "Stepanoviella" fauna distributed in West Australia, Timor, Himalayas, Kashmir, India and Iran are also found in the Gandise-Nyainqentanglha subprovince, north of Yarlung Zangbo River Valley, it is suggested that this region might be a part of the northern marginal sea of the Indian Plate.

2. Moreover, according to Jin et al. (1977), the conglomerate slates of the Pondo group in the Gandise-Nyainqentanglha region are approximately corresponding to the Chataje diamictite member in the northern slope of Mt. Qomolangma and to the diamictite of the Kewoxiga formation in the area north of Kangmar both lithologically and stratigraphically (Table 2). It seems suggestive that the above-cited regions might share a common continental shelf for sedimentation. This means that the northern boundary of the Indian Plate would at least reach the Gandise-Lhasa-Bomi fold zone and might reach the Bangong Lake-Dongqiao-Nujiang River line.

Table 1 Stratigraphical Sequence, Rocks and Biotas of Xizang Permian System

| System | | Permian | | | | |
|---|---|---|---|---|---|---|
| Series | | Lower Permian | | | Upper Permian | |
| Stage | | Chihsian | Tingrian | Maokouan | Wuchiapingian | Changhsingian |
| Himalayan Province (I) | Kangmar | Kangmar Fm.<br>black slate 100 m | Baidinpu Fm.<br>crystalline limestone 250 m<br>*Lytvolasma* fauna<br>*Taeniothaerus* fauna | | | |
| | Tingri | Selung Group<br>Qubu Fm.<br>sandstone, shale 20 m<br>*Glossopteris* flora | Quburega Fm.<br>siltstone intercalated with shale and bioclastic limestone 375 m<br>*Lytvolasma* fauna<br>*Taeniothaerus* fauna<br>*Uraloceras* | | | |
| | Zhongba–Burang | Selung Group<br>calcareous, shales intercalated with argillaceous limestone 300–400 m<br>*Lytvolasma* fauna<br>*Taeniothaerus* fauna | | Lasaila Ls. (Chitichun Ls.)<br>limestone, cherts 100 m<br>*Iranophyllum* fauna<br>*Neoschwagerina* fauna<br>*Cyclolobus*<br>calcareous algal flora | | |
| Northern Xizang Province (II) | Gandise–Nyainqentanglha Subprov. ($II_1$) | Urulung Fm.<br>bryozoan–bearing limestone interchanged with blackish slates 40 m<br>*Lytvolasma* fauna | | Lobadoi Fm.<br>Majula Mem.: limestone 400 m<br>*Iranophyllum* fauna<br>*Neoschwagerina* fauna<br>Lobadoi reservoir Mem.: massive and breccia limestones intercalated with neutral-basic volcanics 100 m<br>*Yabeina* fauna | | |
| | Karakorum Subprov. ($II_2$) | Toinlungkongba Fm.<br>sandstone, slate, massive limestone 300 m<br>*Monodiexodina* fauna | | Minzhocaka Ls.<br>massive limestone, marble 200 m<br>*Neoschwagerina* fauna | | |
| | Tanggula Subprov. ($II_3$) | | | Kamrbo Group<br>limestone, shale >1900 m<br>*Neoschwagerina* fauna | | Raggyorcaka Fm.<br>sandstone, shale, mudstone, limestone intercalated with thin coal bed 200 m<br>*Palaeofusulina* fauna<br>Cathaysian flora |
| | Hoh Xil–Kunlun Subprov. ($II_5$) | | | Ninggongqujiulong Fm.<br>flysch composed of light metamorphic sandstone, shale and limestone 900 m<br>few members of *Neoschwagerina* fauna | | |
| | Hengduan Mts Subprov. ($II_4$) | Manco Fm.<br>limestone, tuff lava 300 m<br>*Misellina* fauna<br>Calcareous algal flora | | Tyokar Fm.<br>limestone, calcareous shale, argillaceous sandstone, fine–grained sandstone 200 m<br>*Neoschwagerina* fauna<br>Calcareous algal flora | Toba Fm.<br>sandstone, shale, coal bed intercalated with limestone 880 m<br>Cathaysian flora<br>Brachiopods | Kaxamda Fm.<br>mudstone, sandstone, shale 550 m<br>*Palaeofusulina* fauna<br>Xiyacun andesite<br>andesite pyroclastic rock, hornblande andesite 270–710 m |

References

(1) Gupta, V.J., Indian Palaeozoic Stratigraphy, Hindustan Pub. Corporation, Delhi, 1-204 (1973).

(2) Hsü Jen, On the discovery of some plant fossils from the Mt. Jolmo Lungma Region, southern Tibet, and its significance. Acta Botanica Sinica, 15 (2), 254-260 (1973) (in Chinese).

(3) Hsü Jen, Discorery of Glossopteris flora in South Xizang and its significance in geology and palaeogeography. Scientia Geol. Sinica, (4), 223-339 (1976) (in Chinese).

(4) Ching Yu-kan, Liang Xi-luo, Wen Shih-hsuan, New data on Permian fossil fauna at the northern slope of Mt. Qomolangma. Scientia Geol. Sinica, (3), 236-249 (1977) (in Chinese).

(5) Nakazava, K., Kapoor, H. M., Correlation of the marine Permian in the Tethys and Gondwana. Four. Intern. Gondw. Sym. (1977).

(6) Plumatead, E.P., The Late Palaeozoic Glossopteris Flora. In: A. Hallam (Editor), Atlas of Palaeobiogeography, 187-205 (1973).

(7) Wang Yu-jing, Mu Xi-nan, Some new observation on the Permian biostratigraphy of the Himalayan province in southern Xizang. Journ. Stratigr., 4 (2), 145-151 (1980) (in Chinese).

(8) Waterhouse, J.B., Gupta, V.J. Permian faunal zones and correlation of the Himalayas. Bull. India Geol. Assoc., 10 (2) (1977).

(9) Wu Wang-shi, The Coral fossils from the Mount Jolmo Lungma Region, A Report Sc. Exp. Mt. Jolmo Lungma Region (1966-1968), Palaeont., Fasc. 1, 83-128 (1975) (in Chinese).

(10) Yin Ji-xiang, Guo Shi-zeng, Dicovery of beds of Gondwana facies at the northern slope of Mt. Qomolangma. Scientia Geol. Sinica, (4), 291-32 (1976) (in Chinese)

(11) Yin Ji-xiang Guo Shi-zeng, Mt. Qomolangma and the beds at its northern slope. Scientia Sinica, 21 (5), 629-644 (1978).

(12) Zhang Shou-xin, Ching Yu-Kan, Late Palaeozoic Brachiopods from the Mount Jolmo Lungma Region, A Report Sc. Exp. Mt. Jolmo Lungma Region (1966-1968), Palaeont., Fasc. 2, 159-242 (1976) (in Chinese).

Table 2

| Himalayan Province | | | | Northern Xizang Province | |
|---|---|---|---|---|---|
| Kangmar | | Tingri | | Gandise-Nyainqentangla Subporov. | |
| Lower Permian | Baidingpu Fm.<br>Kangmar Fm. | Selung Group | Quburega Fm.<br>Qubu Fm. | | Urulung Fm. |
| Upper Carboniferous | Bilong Fm.<br>quartzite<br>Kewoxiga Fm.<br>dark gray diamictite | Jilong Fm. | Chaya quartzite sandstone Member<br>*Stepanoviella* bed<br>Chataje diamictite Member | Pondo Group | *Stepanoviella* fauna<br>mudstone sandy slate with some breccias |

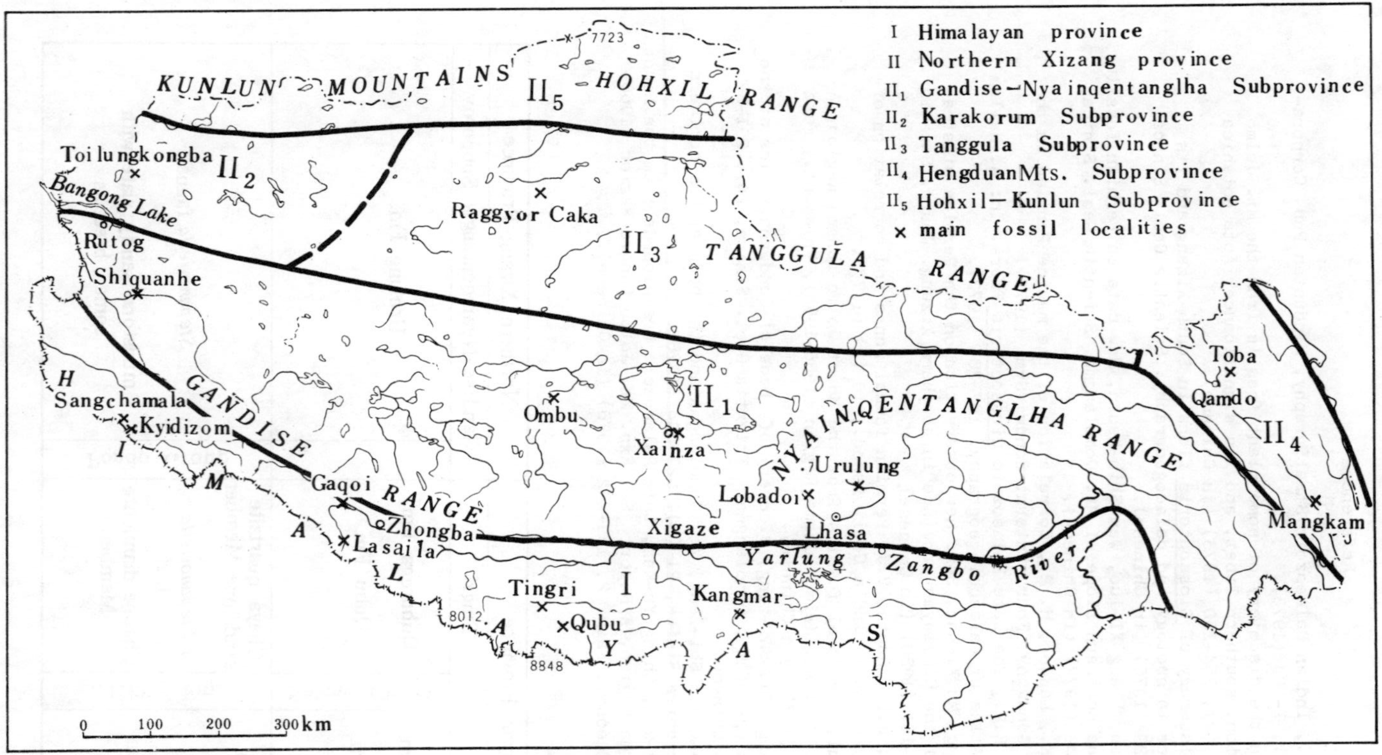


Fig. 1 Map showing provincialism of Xizang Permian System and main fossil localities

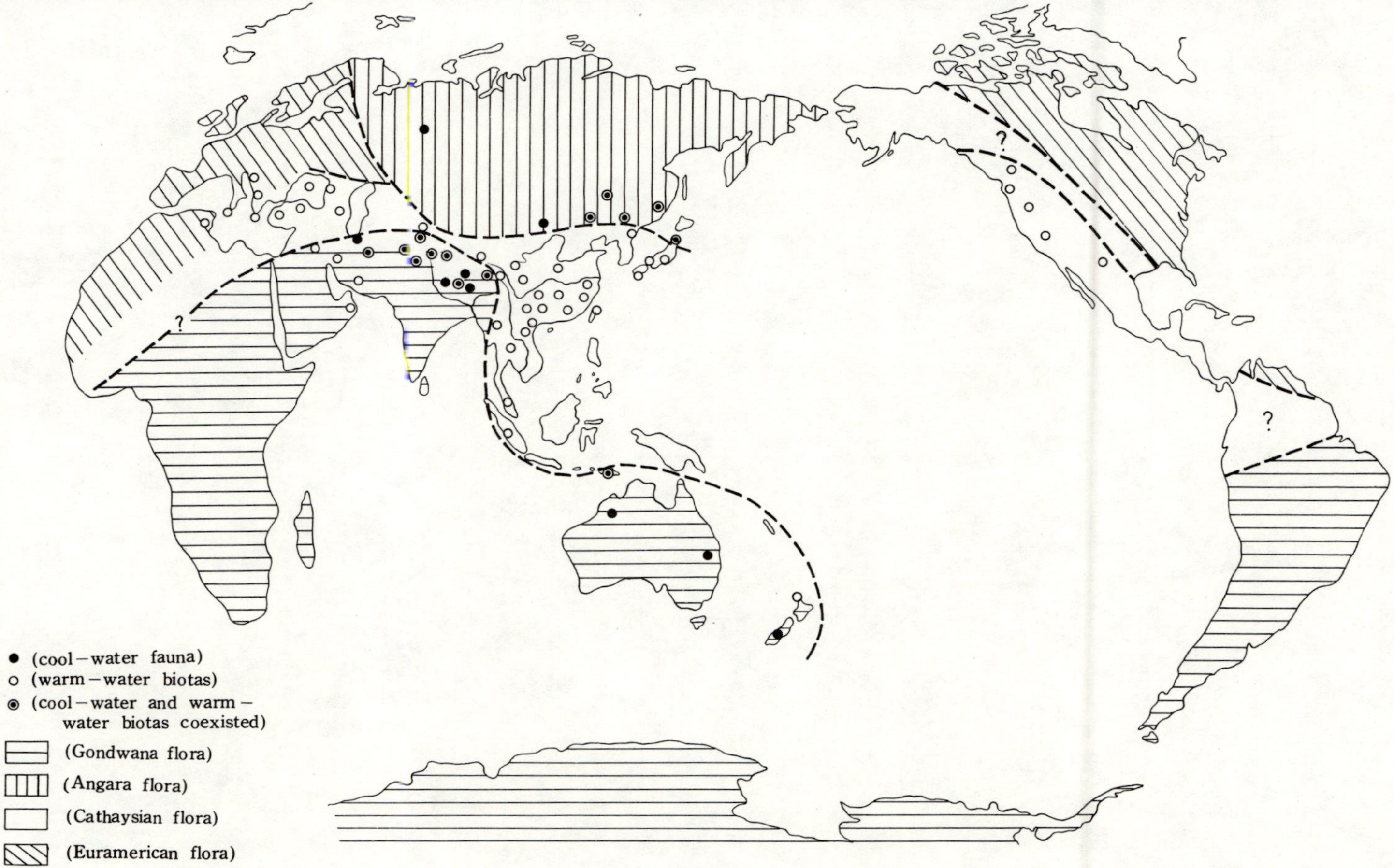


Fig. 2 Map showing distribution of Permian cool - water and warm - water biotas

# THE CARBONIFEROUS AND PERMIAN FUSULINID SUCCESSIONS IN EASTERN QINGHAI-XIZANG PLATEAU

Zhang Lin-xin
(Nanjing Institute of Geology and Palaeontology, Academia Sinica)

Carboniferous and Permian sediments are widely distributed in eastern Qinghai-Xizang Plateau (between 28°-32° N and 96°-102° E) and are developed in other parts of Xizang, Qinghai and Sichuan provinces(Fig. 1). They are extensively exposed in simple structure and represented by beds of more than two thousands metres in thickness, most of which are limestones containing abundant fusuline foraminifers. The fusulinids described and illustrated comprise 267 species and subspecies belonging to 53 genera and subgenera, of which one genus and 52 species and subspecies are new.

The Carboniferous and Permian strata are exposed near Gemazhapu, a village of Zhongzan, south of Batang county, western Sichuan. In Gemazhapu section, the strata exceeding 1,400-1,500 metres in thickness are composed almost all of thick-bedded to massive limestones, and are extensively exposed in simple structure. The fusulinid succession is quite clear. The fusulinids are well distributed from the top of the Lower Carboniferous to the Upper Permian. The divisions and details of the succession reading from top downward are as follows:

Super-formation: Lower Triassic

---------- disconformity----------

Upper Permian

Beds 27-21 Light grey, thick-bedded, dolomitic limestones with cherty limestones in the lower part (Beds 21-22), containing the following fusulinids (587, 586, 578): Reichelina cribroseptata Erk, R. changhsingensis Sheng et Chang, Eoverbeekina batangica Zhang sp. nov., Nankinella quasihunanensis Sheng, Sphaerulina zisongzhengensis Sheng

...............465.01.550 33 m

Upper part of Lower Permian

Beds 20-14 Light grey and whitish grey massive limestones yielding following fusulinids (575, 573-567): Sumatrina longissima Deprat, Paraverbeekina ellipsoidalis Chen, Verbeekina heimi Thompson, V. verbeeki Geinitz, Neoschwagerina cheni Sheng, N. sphaerica nana Sheng, N. sp., Chusenella schwagerinaeformis Sheng, Pseudodoliolina sp.

......... 181.71 m

Lower part of Lower Permian

Bed 13 Light grey, massive finely crystalline limestones. This bed may be subdivided into two parts:

The Upper part contains the following forms (566): Nankinella hunanensis Chen, Pisolina excessa Lee.

The lower part contains the following fusulinids (565, 563, 562): Misellina claudiae Deprat, M. ovalis Deprat, M. (Brevaxina) compressa Deprat, M. (B.) zhongzanica Zhang sp. nov.

............75.22 m

Upper part of Upper Carboniferous (Mapingian)

Bed 12 Grey and light grey, massive subcrystalline limestones, with Staffella batangica Zhang sp. nov ................20.0 m

Beds 11-10 Light grey and whitish grey, massive fine crystalline limestones, yielding the following forms (560-577, 554-552): Eoparafusulina shengi Chang, Schwagerina sp. B, S. Zhongzanica Zhang sp. nov., Zellia crassialveola Chang, Z. heritschi Kahler et Kahler, Pseudoschwagerina zhongzanica Zhang sp. nov., P. batangica Zhang sp. nov., Triticites pauper Dunbar et Henbest, T. flabellus Zhang sp. nov., T. sp. C, Schubertella kingi exilis Suleimanov, Pseudofusulina batangica Zhang sp. nov., P. cylindrica Skinner et Wilde ..............121.88 m

Beds 9-8 Light grey, massive subcrystalline oölitic limestones, the following fusulinids have been found (551, 549, 547, 546): Triticites bashkirius Rosovskaya, T. sp. D, T. subrhomboides Chen, T. springvillensis Thompson, Verville et Bissell, T. dictyopharus Rosovskaya, T. flubellus Zhang sp. nov., Schubertella ex. gr. pseudoglobulosa Safonova ..............54.58 m

Lower part of Upper Carboniferous (Weiningian)

Beds 7-6 Whitish grey and light grey, massive oölitic limestones, containing the following forms (544-539): Fusulinella colaniae Lee et Chen, F. zhongzanica Zhang sp. nov., F. provecta Sheng, F. praebocki Rauser, Profusulinella zhongzanica Zhang sp. nov., P. pseudoaljutovica Rauser, P. deprati Beede et Kniker, P. paraaljutovica Safonova, Fusulina pseudokonnoi Sheng, Pseudostaffella sphaeroidea (Ehrenberg), Millerella pura Malakova ..............204.1 m

Beds 5-4 Light grey, thick-bedded and massive oölitic limestones, with deep grey limestones. The following fusulinids have been found (538-530, 527): Profusulinella paraaljutovica Safonova, P. batangica Zhang sp. nov., P. ovata Rauser, Pseudostaffella larionovae Rauser et Safonova, Millerella pura Malakova, Schubertella polymorpha Safonova, S. subkingi Putrja, Fusiella sp. ................131.51 m

Lower Carboniferous

Bed 3 Reddish to whitish grey, thin-bedded pure limestones, the following forms have been obtained (519, 517). Fusulinids: Eostaffella irenae Ganelina, E. sp.; Corals: Dibunophyllum sp., Arachnolasma sinense Yabe et Hanzawa ................13.36 m

Bed 2 Grey and whitish grey, subcrystalline limestones with Kueichouphyllum sp. ............92.61 m

Bed 1 Light grey, massive subcrystalline limestones with red clastic limestones in the lower part ...........33.32 m

------------disconformity-------------

Sub-formation: Upper Devonian

The rapid evolution of fusulinids in the Gemazhapu section permits biostratigraphic zonation of the Carboniferous and Permian Systems. Ten fusulinid zones are established in descending order as follows:

10. Zone of Reichelina cribroseptata-Eoverbeekina batangica (Bed 22 to Bed 20)
9. Zone of Neoschwagerina cheni (Bed 19 to Bed 14)
8. Zone of Nankinella hunanensis-Pisolina excessa (Upper part of Bed 13)

7. Zone of Misellina claudiae (Lower part of Bed 13)
6. Zone of Staffella batangica (Bed 12)
5. Zone of Zellia-Pseudoschwagerina (Bed 11 and Bed 10)
4. Zone of Triticites (Bed 9 and Bed 8)
3. Zone of Fusulinella-Fusulina (Bed 7 and Bed 6)
2. Zone of Profusulinella (Bed 5 and Bed 4)
1. Zone of Eostaffella (Bed 3)

The Eostaffella zone belongs to the latest Lower Carboniferous age and is regarded as the lowermost fusuline zone of the Carboniferous system. In the Riwoqe County of East Xizang, it is known as the Macala formation. The Eostaffella zone corresponds probably to the Zhaojiashan formation of Western Guizhou or to the Hochow formation of Anhui (1).

The Profusulinella zone and the Fusulinella-Fusulina zone are belonging to the Weiningian of the early Upper Carboniferous. The Profusulinella zone consists of four species of the genus profusulinella and Pseudostaffell larionovae Rauser et Safonova. This zone has also been found in Markam and Riwoqe counties of Eastern Xizang. The appearance of Fusulinella is drawn as the boundary between the Profusulinella zone and the Fusulinella-Fusulina zone, but the elements of Profusulinella still exist in the latter zone. The lower part of Upper Carboniferous (the Aoqu group) is found in Markam county of Eastern Xizang. Two fusulinid zones are present in the Aoqu group, the lower Profusulinella zone consisting of Profusulinella pseudoaljutovica Rauser, P. parva (Lee et Chen) and Pseudostaffella larionovae Rauser et Safonova; the upper Fusulinella-Fusulina zone containing Fusulinella obesa Sheng, F. vozhgalensis Safonova, Fusulina schellwieni Staff and Neostaffella sphaeroidea (Ehrengberg). The fusulinid zones of the Aoqu group bear a strong resemblance to those of the Huanglung formation (2) or the Weining fromation of South China.

The Triticites zone,the Zellia-Pseudoschwagerina zone and the Staffella batangica zone belong to the Mapingian of the late Upper Carboniferous age. The strata containing them are widely distributed. In Qinghai-Xizang Plateau, the fusulinids belonging to those three zones have been found in more than 12 counties, such as Markam, Qamdo, Riwoqe, Deqen, etc. The Mapingian fusulinid bearing formation is named the Lica group of Eastern Xizang. It can be correlated with the Chuanshan formation or the Maping limestone (3) of South China.

The Misellina claudiae zone is characterized by the presence of the Misellina claudiae, and represents the lowest fusuline horizon of the Permian system in Eastern Qinghai-Xizang Plateau. In this zone, the elements of Misellina are very rich, Triticites, Quasifusulina, Rugosofusulina and Pseudoschwagerina have not been found. The zone of Nankinella hunanensis-Pisolina excessa is belonging to the lower part of Lower Permian. The main elements all belong to Staffellidae, the others appear rarely or fail to find. The upper part of Lower Permian is characterized by the presence of the genus Neoschwagerina, and is referred to the zone of Neoschwagerina cheni. The associates of Neoschwagerina are: Verbeekina, Paraverbeekina, Pseudodoliolina, Chusenella and Sumatrina. In Markam county, the Lower Permian includes two formations, the lower Manco formation and the upper Jiaoga formation. The Manco formation cab be correlated with the Chihsia formation, and the Tyokar formation is probably equivalent to the Maokou formation (4) of South China.

The zone of Reichelina cribroseptata-Eoverbeekina batangica is belonging to the Upper Permian. This zone contains Reichelina cribroseptata

Erk, R. changhsingensis Sheng et Chang, Eoverbeekina batangica Zhang sp. nov., Nankinella quasihunanensis Sheng and Sphaerulina zisongzhengensis Sheng. It may be correlated with the Wuchiaping formation of South China.

No Palaeofusulina has been found in the Gemazhapu section, but it is found near the Toba village of the Qamdo county and is found in the vicinity of Lhandun of the Markam county. Rocks bearing Palaeofusulina in Toba are named the Kaxamda formation. In this formation, the following forms have been found: Palaeofusulina subcylindrica Sheng. P. fusiformis Sheng, P. tobensis Zhang and Gallowaiinella jiaogensis Zhang. The Kaxamda formation can be correlated with the Palaeofusulina zone of South China.

To sum up what has been stated above, a table showing the relations of the Carboniferous and Permian fusulinid zones in Eastern Qinghai-Xizang Plateau and South China is given below.

References


(1) Chang Lin-hsin, Fusulinids from the Hochow limestone, Hohsien, Anhui. Acta Palaeont. Sinica, 10 (4), 432-441 (1962) (in Chinese).

(2) Sheng Jin-zhang, Min Qing-kui, Wang Li-li, Fusulinid zones of the Middle Carboniferous Huanglong limestone at Jinsigang near Nanjing. Acta Palaeont. Sinica, 15 (2), 176-212 (1976) (in Chinese).

(3) Wu Wang-shi, Zhang Lin-xin et all. Upper Carboniferous rocks of the Pu'an Xian, Guizhou with comments on its upper boundary. Science Press (1979) (in Chinese).

(4) Sheng Jin-chang, Permian Fusulinids of Kwangsi, Kueichow and Szechuan, Palaeontologia Sinica, New Series B, No. 10, 1-247 (1963).


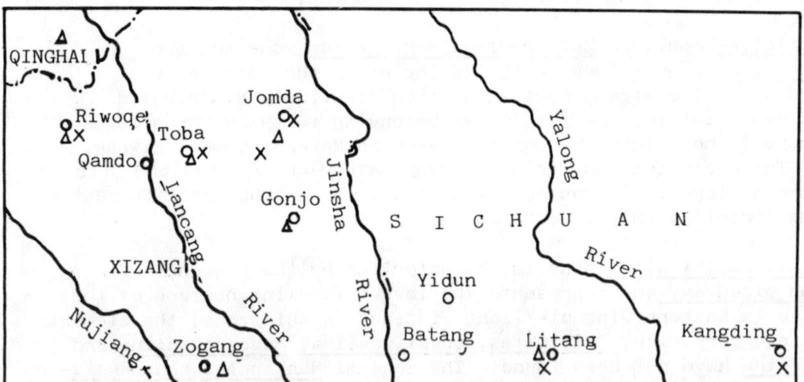


Fig. 1 Carboniferous and Permian fusulinid locations of eastern Qinghai-Xizang Plateau

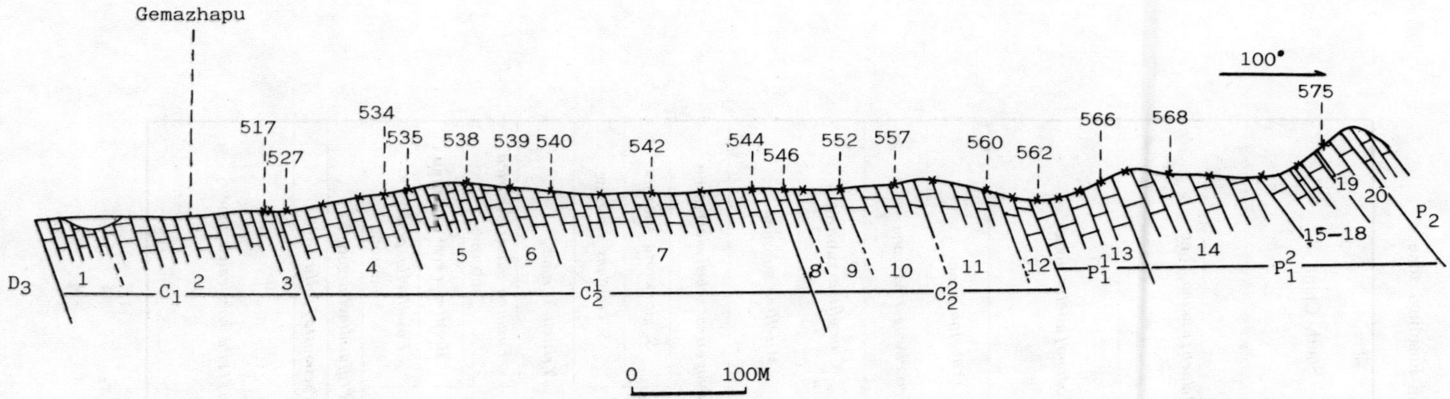


Fig. 2 Section of the Carboniferous and Permian strata in Gamazhapu-Chidantan (I)

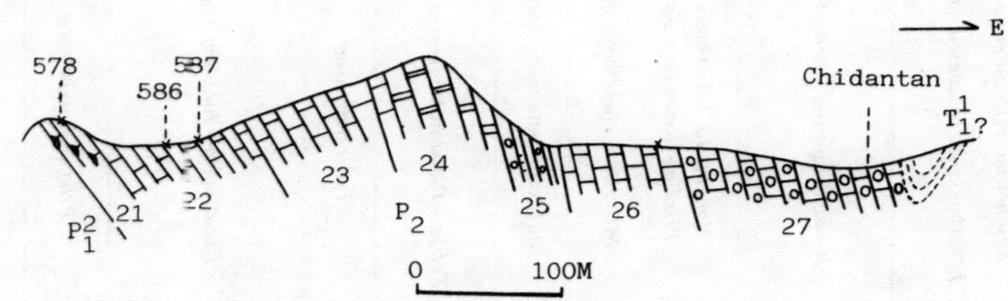


Fig. 3 Section of the Carboniferous and Permian strata in Gamazhapu-Chidantan (II)

Table 1 Correlation chart of the Carboniferous and Permian Fusulinid zone

<table>
<tr><th>System</th><th>Series</th><th>Formation</th><th>Eastern Qinghai—Xizang Plateau</th><th colspan="2">South China</th></tr>
<tr><td rowspan="6">Permian</td><td rowspan="2">Upper Series</td><td>Changhsing</td><td>Palaeofusulina zone</td><td colspan="2">Palaeofusulina zone</td></tr>
<tr><td>Wuchiaping</td><td>Reichelina cribroseptata—Eoverbeekina batangica zone</td><td colspan="2">Codonofusiella zone</td></tr>
<tr><td rowspan="4">Lower Series</td><td rowspan="2">Maokou</td><td rowspan="2">Neoschwagerina cheni zone</td><td colspan="2">Yabeina zone</td></tr>
<tr><td colspan="2">Neoschwagerina zone</td></tr>
<tr><td rowspan="2">Chihsia</td><td rowspan="2">Nankinella hunanensis—Pisolina excessa zone<br>Misellina claudiae zone</td><td rowspan="2">Parafusulina zone</td><td>Cancellina subzone</td></tr>
<tr><td>Misellina subzone</td></tr>
<tr><td rowspan="10">Carboniferous</td><td rowspan="9">Upper Series</td><td rowspan="4">Maping</td><td>Staffella batangica zone</td><td colspan="2">Robustoschwagerina schellwieni zone</td></tr>
<tr><td rowspan="2">Zellia– Pseudoschwagerina zone</td><td rowspan="2">Sphaeroschwagerina zone</td><td>S. glomerosa subzone</td></tr>
<tr><td>S. constans subzone</td></tr>
<tr><td>Triticites zone</td><td colspan="2">Triticites zone</td></tr>
<tr><td rowspan="5">Weining</td><td rowspan="3">Fusulinella– Fusulina zone</td><td rowspan="3">Fusulinella – Fusulina zone</td><td>Fusulina quasicylindrica subzone</td></tr>
<tr><td>Beedina cheni subzone</td></tr>
<tr><td>Fusulinella praebocki subzone</td></tr>
<tr><td rowspan="2">Profusulinella zone</td><td colspan="2">Profusulinella zone</td></tr>
<tr><td colspan="2">Pseudostaffella zone</td></tr>
<tr><td>Lower Series</td><td>Hochow</td><td>Eostaffella zone</td><td colspan="2">Eostaffella hohsienica zone</td></tr>
</table>

# LATE TRIASSIC PLANTS AND THEIR GEOGRAPHICAL PROVINCES IN QINGHAI-XIZANG PLATEAU

Li Pei-juan, Wu Xiang-wu
(Nanjing Institute of Geology and Palaeontology, Academia Sinica)

Over the last two decades, great progress was made in the study on the Late Triassic floras of China. From the phytogeographical point of view,the widespread Late Triassic floras in China can be classified into two types, i. e. the Dictyophyllum-Clathropteris flora in the littoral floristic province and the Danaeopsis-Bernoullia flora in the continental floristic province. They are separated by a boundary extending from Tianshan Mts. eastward through the southern part of Qilian Mt. South Qinling Mts.,Dabei Mt. to eastern Jilin (Fig. 1).

The Dictyophyllum-Clathropteris flora (including such floras in Daqiaudi and Hsuchiaho formations (1,2) of Sichuan, Ipinglang and Xiangyun formation (3) of Yunnan, Anyuan formation (4) of Hunan and Jiangxi Siaoping formation (5) of Guangdong, Wuzau formation (6) of Zhejiang, Fuanjiatan formation of Jiangsu, Daken and Wenbinshan Formation (7) of South Fujian etc. (8,9,10)) is distributed in the area to the south of the approximate boundary of the two floras. The general characteristics of the flora are mentioned below:

1. Filicales is regarded as a dominant class of this flora, yielding Dipteridaceae, Marattiaceae, Angiopteridaceae, Asterotheceae, Matoniaceae Osmundaceae etc., among which Dipteridaceae is one of the most remarkable element.

2. Pteridospermae is represented by Ptilozamites chinensis, Hyrcanopteris sevanensis, H. sinensis, Lepidopteris, Thinnfeldia, Ctenozamites etc.. Among these forms, Pt. chinensis and H. sevanensis are the distinctive species in South China.

3. Cycadophyta occupies the lagrest amount among this flora, they are: Pterophyllum jaegeri, Pt. ptilum, Pt. sinense, Sinoctenis calophylla, Anomozamites loczyi, Anthrophyopsis crassinervis, A. venulosa, Nilssoniopteris jourdyi, Drepanozamites nilssoni, etc. These species are restricted to the Late Triassic period in South China and may be regarded as the typical forms of this flora.

4. Only a few elements of Ginkgoales and Coniferales have been recognized, such as Baiera, Glossophyllum, Podozamites, Cycadocarpidium, Glyptolepis and Hirmerella.

The Danaeopsis-Bernoullia flora is represented by the flora of the Yenchang Group (11,12) of Shaanxi, Gansu and Ningxia, and the Xiaoquango Group of Xinjiang. The Yenchang flora of northern Shaanxi was under detailed study by Prof. H. C. Sze (1956), characterized by the following features:

1. Filicales is the dominant class, among which Asterotheca szeiana, *Bernoullia zeilleri, *Danaeopsis fecunda, *Todites shensiensis, Cladophlebis gracilis, Cl. kaoiana and so on are characteristic forms, but the Dipteridaceae member is nearly absent.

2. Equisetales is also superior in numbers of the flora, all of them are much closely related to the older Mesozoic known forms.

3. Pteridospermae and Ginkgoales, such as Thinnfeldia, Sphenobaiera, Ginkgoites and Glossophyllum hold a prominant position.

4. Of the Cycadophyta, only a few forms e.g. Sinozamites, Sphenozamites, ?Drepanozamites have been described.

5. Coniferales is rather scarce.

The following are some Late Triassic plant-bearing strata which have been recently discovered in Qinghai-Xizang Plateau: (1) the Bagang formation in eastern Xizang; (2) the Tumaingela formation in northern Xizang; (3) the Lamaya formation in western Sichuan and northern Yunnan; (4) the Gyiza group in southern Qinghai; (5) the Babaoshan group in Burhanbudai Mt. area of Qinghai; (6) the Nanyinger and Mole groups in Qilian Mts. of Qinghai. (A sketch map showing geographical distribution is given in Fig. 2)

The Bagong formation is of continental and marine alternated deposits with coal seams, containing fossil plants: Hysterites friesi, Equisetites platyodon, E. rogersi, Neocalamites cf. hoerensis, Danaeopsis fecunda, Marattia asiatica, M. munsteri, Asterotheca szeiana, Todites shensiensis, Reteophlebis simplex, Gleichenites yipinglangensis, Thaumatopteris contracta, T. remauryi, Dictyophyllum nathorsti, Clathropteris meniscioides, Cl. mongugaica, Cl. platyphylla, Goeppertella microloba, Hyrcanopteris sevanensis, Pterophyllum aequale, Pt. astartense, Pt. minutum, Pt. schenki, Sinoctenis calophylla, Nilssoniopteris jourdyi, Drepanozamites nilssoni, Glossophyllum sp., Cycadocarpidium sp. and cf. Hirmerella munsteri (13,14).

The lithological characters of the Tumaingela formation are almost alike to the Bagong formation. It comprises the following plant fossils: Equisetites arenaceus, E. longidens, E. sarrani, Neoclamites carrerei, Taeniocladopsis rhizomoides, Danaeopsis cf. marantacea, Todites crenatum, Thaumatopteris fuchsi Clathropteris meniscioides, Cl. tenuinervis, Allophyton denqinensis, Hyrcanopteris sevanensis, Pterophyllum haneianum, Pt. minutum, Pt. schenki, Pt. cf. zygotacticum, Dictyozamites baqenensis, Otozamites cf. indosinensis, Nilssoniopteris jourdyi, Baiera cf. elegans, cf. Hirmerella munsteri etc. (15).

The Lamaya formation is mainly of continental origin, yielding fossil plants: Selaginellites yunnanensis, Equisetites cf. takahashi, Marattia asiatica, M. hörensis, Todites denticulatus, T. goeppertianus, T. recurvatus, Gleichenites yipinglangensis, Dictyophyllum nathorsti, D. muensteri, Clathropteris meniscioides, Cl. mongugaica, Goeppertella xiangchengensis, Cladophlebis scariosa, Cl. cf. raciborskii, Hyrcanopteris sinensis, H. magnifolia, Pterophyllum aequale, Pt. jaegeri, Pt. ptilum, Pt. sinense,

*They are also found in the Dictyophyllum-Clathropteris flora, but not so flourishing as those in the Danaeopsis-Bernoullia flora.

Nilssoniopteris jourdyi, Sinoctenis calophylla, Nilssonia cf. muensteri, Ctenozamites cycadea, Ginkgoites qamdoensis, Glossophyllum sp., Pityophyllum longifolium, Podozamites lanceolatus and Cycadocarpidium sp. (16)

The upper part of Gyiza group is composed of the marine and continental alternated deposits, with a floral aspect similar to that of the Bagong and Tumaingela formations, carrying essentially elements: Equisetites arenaceus, Equisetostachys sp., Phlebopteris xiangyunensis, Dictyophyllum nilssoni, Clathropteris meniscioides, Hyrcanopteris cf. sinensis, Pterophyllum cf. jaegeri, Pt. sinense, Pt. minutum and Anthrophyopsis venulosa (17)

The upper part of Babaoshan group consists of continental beds intercalated with marine deposits. The fossil flora of this formation comprises the following species: Neocalamites carrerei, Asterotheca cottoni, A. szeiana, Dictyophyllum nathorsti, Clathropteris meniscioides, Areolatophyllum superbiense, Gleichenites? sp., Cladophlebis haiburensis, Lepidopteris ottonis, Pseudodanaeopsis sinensis, Ptilozamites chinensis, Protoblechnum sp., Scoresbya furcata, Ctenozamites cycadea, Pterophyllum ctenoides, Glossophyllum shensiense, Taeniopteris cf. tenuinervis,etc. (17).

The above-mentioned formations are either generally associated with marine intercalations or underlain and/or overlain by marine sediments, which contain many typical Norian lamellibranchia such as Unionites emeiensis, Trigonodus keuperinus, Yunnanophorus gracilis, Y. boulei, Halobia superbescens, Myophoria (Costatoria) napenensis and Permophorus sp..

The Nanyinger group is composed chiefly of continental beds, yielding such fossil plants as Equisetites cf. sthenodon, Neocalamites carcinoides N. carrerei, N. meriani, N. rugosus, Danaeopsis fecunda, Asterotheca cottoni, Bernoullia zeilleri, Todites shensiensis, Sphenopteris chowkiawanensis, Cladophlebis gigantea, Cl. gracilis, Thinnfeldia rigida, Sagenopteris lanceolata and S. cf. spatulata (17).

The sediments of Mole group are almost all of continental origin, interbedded occasionally with marine beds, carrying the same plants as those of the Nanyinger group, such as Equisetites sarrani, E. sthenodon, Neocalamites carcinoides, N. carrerei, N. meriani, N. rugosus, Danaeopsis fecunda, Asterotheca szeiana. Bernoullia zeilleri, Cladophlebis gigantea Cl. gracilis, Sphenopteris chowkiawanensis, Thinnfeldia rigida, Protoblechnum hughesi, Pseudodanaeopsis sinensis, Lepidopteris sp., Sagenopteris spatulata, Sphenobaiera crassinervis, Glossophyllum shensiense, Dictyophyllum? sp., Clathropteris sp., Pterophyllum sp., Anomozamites sp. and Nilssonia sp. (17). Some marine lamellibranchia like Unionites griesbachi, Plagiostoma nuitoensis, Nuculana cf. alpina, Palaeonucula strigilata etc. have been obtained from the upper part of Mole group.

As remarked above, the composition of the floras in Bagang, Tumaingela, Lamaya, Gyiza and Babaoshan formations is most likely corresponding to that of the Noric (or Noric-Rhaetic) floras of Ipinglang, Hsuchiaho, Anyuan etc. in South China. The floras of Nanyinger and Mole groups resemble closely that of the Yenchang formation in North China, which, according to Pan (1936) and Sze (1956), was designated to the age of Keuper-Rhaetic.

Judging from the floral and lithological evidences, we come to the conclusion that the Late Triassic floras of Qinghai-Xizang Plateau may also be divided into two different types. The floras of such formations as Bagong, Tumaingela Lamaya, Gyiza and Babaoshan belong to the type of Dictyophyllum-Clathropteris flora under the effect of oceanic climate, reflecting humid and thermal tropical or tropical-subtropical conditions. On the other hand the floras of Nanyinger and Mole groups may be included within the type of Danaeopsis-Bernoullia flora under the influence of continental climate, showing slightly warm and moist subtropical-temperate conditions. The boundary of the two floras in Qinghai-Xizang Plateau lies approximately between South Qilian Mt. and Qinling Mts. (Lat. 34° 30'-38°30').

During the period of early to early-middle Late Triassic, the Qinghai-Xizang Plateau was mostly covered by marine invasions, so that plant remains have not been found. As a result of the Indo-China Movement as well as the transgressions and regressions of Tethys, the littoral marshland was widely developed in Xizang and the southern part of Qinghai at the middle-late Late Triassic time. The tropical or tropical-subtropical plants were in full flourish in this area and provided a favourable condition for the formation of coal-bearing deposits. While the Qilian Mt. area of Qinghai belong chiefly to the interior basin because of the control of the continental climate, the composition of the flora is obviously distinct from that of the former area, though the southern alope of Qilian Mt. was invaded by marine water for a short period of time. The Galedesi formation (upper part of the Mole group) has a mixed flora mainly of the elements of Danaeopsis-Bernoullia flora with some elements of Dictyophyllum Clathropteris flora, such as Dictyophyllum, Clatropteris, Pterophyllum and Anomozamites, suggesting that the climate then was moister and warmer than that of the floras in the continental province.

## References


(1) Xu Ren (Hsü Jen) et al., Late Triassic Baoding flora, SW Sichuan of China. Sci. Press, Beijing (1979) (in Chinese).

(2) Lee Pei Chuan, Fossil plants from the Hsuchiaho series of Kwangyuan, northern Szechuan. Mem. Inst. Geol. & Palaeont. Acad. Sin., (3), 101-178 (1964) (in Chinese).

(3) Li Pei-juan (Lee Pei Chuan), Cao Zheng-yao (Tsao Cheng-yao), Wu Shun-qing (Wu Shun Ching), Mesozoic plants from Yunnan. Mesozoic fossils from Yunnan I, Science Press, 87-150 (1976) (in Chinese).

(4) Feng Shao-nan et al., Mesozoic plants. Fossil Atlas of Middle-south China III, Geol. Press, Beijing, 199-150 (1978) (in Chinese)

(5) Tsao Cheng-yao, Fossil plants from the Siaoping series in Kaoming, Kwangtung. Acta Palaeont. Sin., 13 (3), 510-528 (1965) (in Chinese

(6) Chen Qi-shi et al.,, Age of the coal-bearing Wuzao formation in Yiwu of Zhejiang. Acta Stratigr. Sin., 2 (1), 74-76 (1978) (in Chinese).

(7) Zhou Tong-shun; On the Mesozoic coal-bearing strata and fossil plants from Fujian Province. Prof. Pap. Strata Palaeont.IV, Geol. Press Beijing, 88-134 (1978) (in Chinese).

(8) Wu Shun-ching, Notes on some Upper Triassic plants from Anlung, Kweichow. Acta Palaeont. Sin., 14 (2), 233-241 (1966) (in Chinese).

(9) Chen Ye, Duan Shu-ying, Zhang Yu-cheng, New species of the Late Triassic plants from Yanbian, Szechuan I-III. Acta Bot. Sin., 21 (1), 57-63 (1979a); 21 (2), 186-190 (1979b); 21 (3), 269-273 (1979c) (in Chinese).

(10) Sun Ge, On the discovery of Cycadocarpidium from the Upper Triassic of eastern Jilin. Acta Palaeont. Sin., 18 (3), 312-325 (1979) (in Chinese).

(11) Pan C. H., Older Mesozoic plants from North Shensi. Palaeont. Sin. Ser. A., 4 (2), 1-49 (1936).
(12) Sze H. C., Older Mesozoic plants from the Yenchang formation, northern Shensi. Palaeont. Sin. N. S. A. (Whole Ser. no. 139) 5, 1-217 (1956).
(13) Li Pu, A preliminary study of the geology in Eastern Tibet. Kexue Tongbao, (7), 62-71 (1955).
(14) Wu Xiang-wu, Late Triassic plants from eastern Xizang. The Series of the Scientific Expedition to the Qinghai-Xizang Plateau, Palaeontology of Xizang, Book V (in press) (in Chinese).
(15) Wu Xiang-wu, Fossil plants from Triassic Tumaingela formation in Anduo-Baqen Area, northern Xizang. The Series of the Scientific Expedition to the Qinghai-Xizan Plateau, Palaeontology of Xizang, Book V (in press) (in Chinese).
(16) Li Pei-juan, Wu Xiang-wu, Fossil plants from the Upper Triassic Lamaya formation of western Sichuan. Stratigraphy and Palaeontology in West Sichuan and East Xizang, China, II (in press) (in Chinese).
(17) He Yuan-liang, Li Pei-juang, Wu Xiang-wu, Mesozoic plants from Qinghai. Fossil Atlas of Northwest China, Qinghai volume (II), Geol. Press Beijing, pp. 129-158 (1979) (in Chinese).
(18) Sze, H. C., Li Xing-xue (Lee Hsing Hsueh) et al.: Mesozoic plants from China. Science Press, (1963) (in Chinese).

Tabla 1 Approximate correlation of Late Triassic formations from Qinghai-Xizang Plateau

| Region / Formation / Stage | E. Xizang | N. Xizang | W. Sichuan | S. Qinghai | Kunlun area Qinghai | Qilian Mt. area Qinghai | | |
|---|---|---|---|---|---|---|---|---|
| Rhaetian | Bagong Fm. | Tumengela Fm. | Lamaya Fm. | Gyiza Gr. | Babaoshan Gr. | Mole Gr. | Galedesi Fm. | Nanyinger Gr. |
| | | | Lanashan Fm. | | | | | |
| Norian | Bolila Fm.* | | Tumgou Fm. * | | | | Atasi Fm. | |
| Carnian | Jiapila Fm. | | Qugasi Fm. * | | | | | |

* marine bed

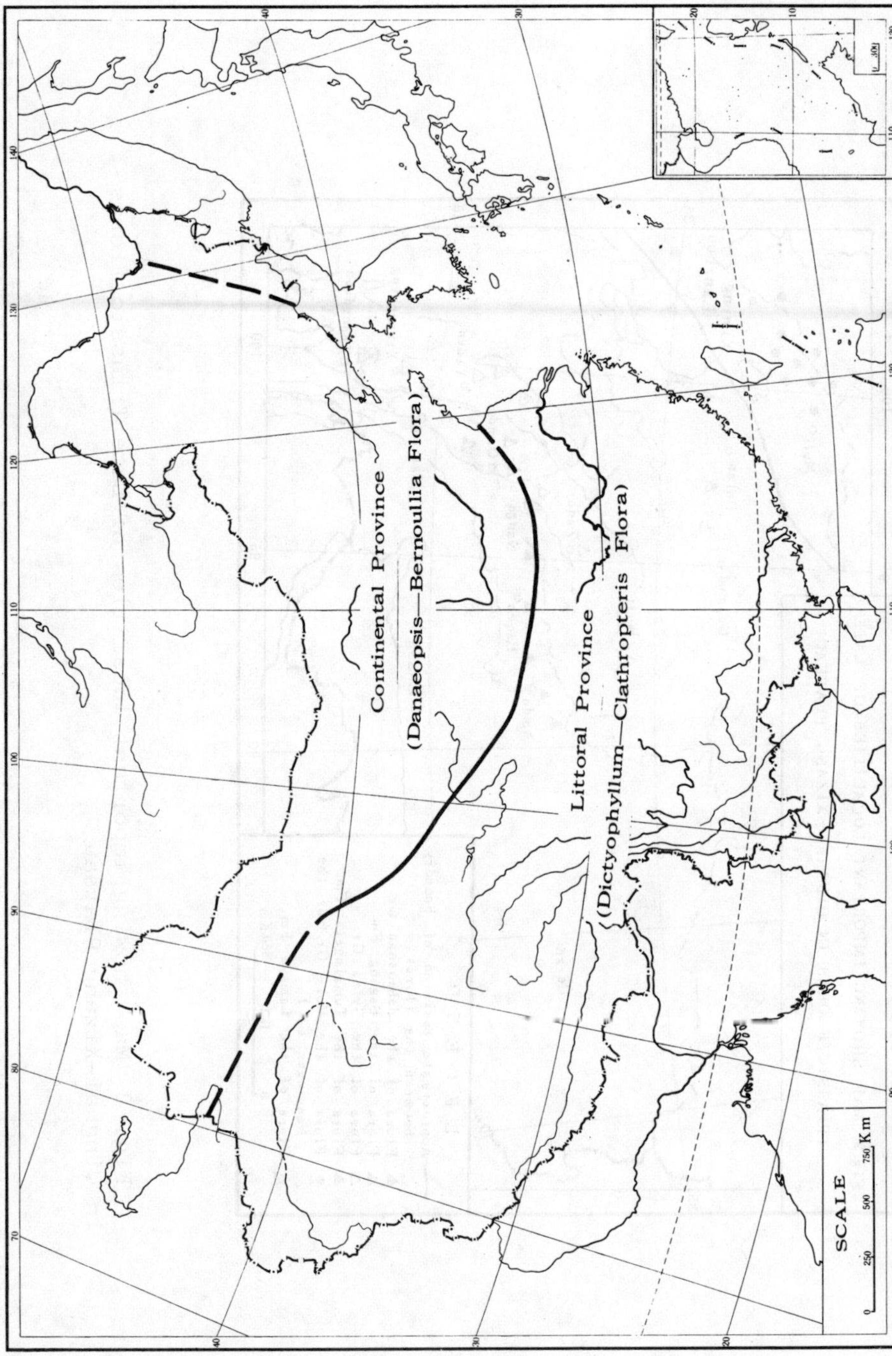


Fig. 1 Map showing the distribution of the continental and the littoral provinces in China

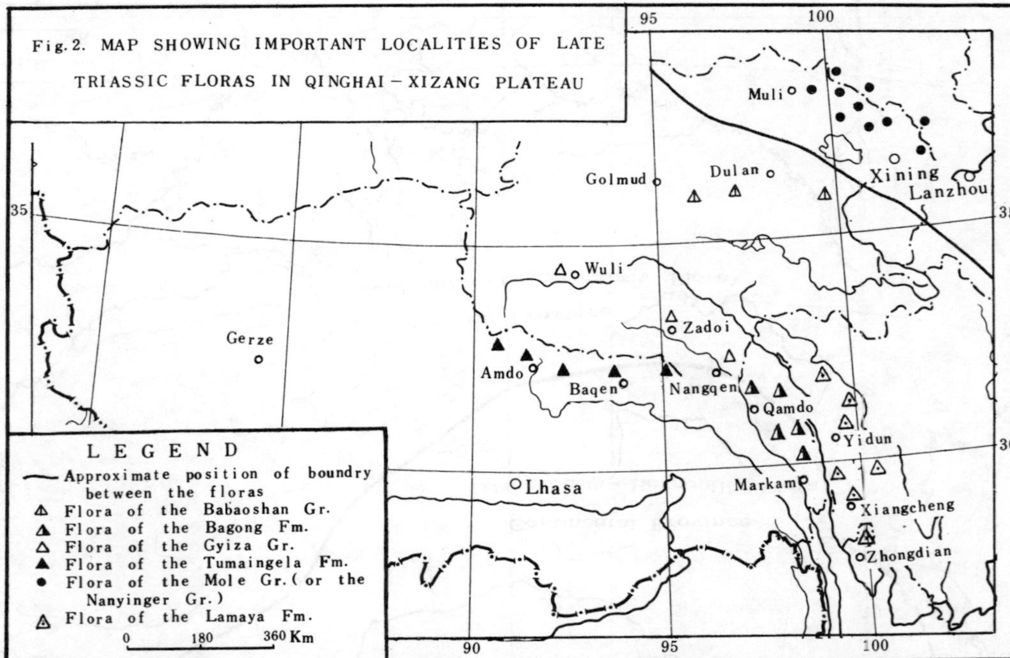


Fig. 2 Map showing important localities of Late Triassic floras in Qinghai-Xizang Plateau

## ON THE ELEVATION AND CLIMATIC CHANGES OF THE QINGHAI-XIZANG PLATEAU BASED ON FOSSIL ANGIOSPERMS

Guo Shuang-xing
(Nanjing Institute of Geology and Palaeontology, Academia Sinica)

Before the founding of New China, only one species, Typha latissima, of the fossil angiosperm on the Qinghai-Xizang Plateau was recorded (1). Over the last ten years some Late Cretaceous and Tertiary floras on the plateau have been discovered and collected by the Geological Prospecting Team of the Xizang Bureau of Geology and the Comprehensive Scientific Expedition to the Qinghai-Xizang Plateau, Academia Sinica. These floras were described (2-4,6-8) and their localities are as follows: Late Cretaceous flora from the Xigaze Group in Xigaze, Ngamring and Sagya counties on the Middle and Upper Yarlung Zangbo River Valley; Eocene and Late Miocene floras from the lower and upper parts of the Moincer Group in Gar county on the southern slopes of Gandise Range; The Middle to Late Miocene flora from Wulong formation in Namling county, central Xizang and, in addition, two Late Pliocene floras from the Sanying formation in Eryuan county, Yunnan and from the Sigeda formation in Dechang county, Sichuan, both situated on the eastern plateau and a Miocene flora in Zekog county, Qinghai on the northeastern plateau (See Fig. 1). On the basis of these floras from the plateau, the present paper aims to make a new approach of inquiry on its elevation, climatic changes and floristic successions.

### ELEVATION AND CLIMATE DURING LATE CRETACEOUS

The Late Cretaceous flora (1) from the Xigaze Group is mostly composed of angiosperms with just a few fragments of fern and gymnosperm. The flora consists mainly of subtropical and warm-temperate broadleaved deciduous forests with several evergreen trees, such as Quercus orbicularis, Cercidiphyllum ellipticum, Rhamnites eminens, Celastrus minor, Ficus daphnogenoides, F. stephensonii, Laurophyllum sp., Cassia fayetensis, Eucalyptus angusta, E. geintzi and Aralia firma. There are also some temperate deciduous plants and swampy monocots, for example, Populus latoir, Salix cf. meeki, Juglandites sinuatus, Viburnum asperm, Cyperacites haydenii and Typhs sp.. Among them lots of species are commonly known in the northern hemisphere, especially in Late Cretaceous deposits of North America. The flora contains 64% of deciduous species, 36% of evergreen ones, with their leaves being 29% of entire margins and 71% of non-entire margins. There has been no montane plant, but there exist alternate strata of shallow marine and nonmarine deposits. These facts make us believe that the Middle and Upper Yarlung Zangbo River Valley might be scores of to one or two hundred meters in altituds above the sea level and a subtropical and subhumid climate during Late Cretaceous.

### ELEVATION AND CLIMATE DURING EOCENE

The Eocene flora from the lower part of the Moincer Group in Gar was given a list by Hsu Jen (3). This flora is composed of all angiosperms, containing broadleaved evergreen trees with a couple of deciduous species, such as Myrica parvifolia, Ficua mytifolia, Eucalyptus angustifolia, E. geinitzi, E. attenuata, Eugenia cf. hilgardiana, Cassia marshallensis, Rhamnus menshiensis, Cyperacites haydenii and Phrynium paleocapitatum, among them these species of Eucalyptus are dominated. Many species of the flora have been found in Eocene deposits in North America. The leaves of fossil plants are all of entire margins. Their characteristics are quite similar to those of native and cultivated living species distributed along the coast of Southeast China. Therefore it is probable that the altitude in Gar area at that time was not quite high, about 100 or 300 m in height, but not exceeding 300 m above the sea level. However, the fossil locality is situated at an altitude of 5,140 m today. It is evident that since Eocene the southern slopes of Gandise Range have been uplifted nearly 5,000 m regardless the factors of denudation. The Eocene flora represents a tropical or paratropical and wet climate, reflecting a much warmer and wetter one than that of Late Cretaceous.

## ELEVATION, CLIMATE AND FLORISTIC SUCCESSIONS DURING NEOGENE

Some Neogene floras are distributed in southern and central Xizang, northwestern Yunnan, western Sichuan and eastern Qinghai respectively.

1. A Late Miocene flora (3) from the upper part of the Moincer Group in Gar is at the same locality as above mentioned Eocene flora. This small flora consists of all temperate broadleaved deciduous forests, such as Populus latior, P. balsamoides, P. glandulifera and Salix viminalis with a gymnosperm, Sabina sp.. They are quite similar to the living species distributed widely in the hills of North, Northeast and Northwest China. This flora can be easily distinguished from the above stated Eocene flora by its composition and characteristics. It was represented by typically temperate elements, showing that the southern slopes of Gandise Range during Late Miocene already uplifted very high, proximately over 1,500 m above the sea level in altitude. But the fossil locality today occurred at an altitude of 5,140 m. Therefore, we have sound reason to consider that from Late Miocene up to now the southern slopes of Gandise Range has uplifted over 3,000 m once again. The rising speed is much faster after Miocene than before.

2. A Middle to Late Miocene flora (4) from Wulong formation in Namling county is mostly composed of subtropical and warm-temperate broadleaved deciduous forests and some subalpine evergreen forms with a few temperate members (See Table 1).

These fossil remains are closely similar to the living species growing in Hengduan Mountains at present-day. The fossil flora is characterized by 70% of deciduous species and 30% of evergreen ones. Its leaves are 22% of entire margins and 78% of non-entire margins. It must be emphasized that some montane plants, such as the alpine species of Quercus, Rhododendron and microthermic Betula have been found in the flora. The composition of the fossil plants bears a close resemblance to that of the present plants in Hengduan Mountains. The distributive range of the living species similar to the fossil remains is from 2,644 to 3,456 m in altitude on an average, indicating Namling area already reached an altitude of at least 2,644 m during Miocene.

Contemporary studies on phytogeography and paleobotany, according to Li Ji-jun (5) show that the mean annual temperature today in southeastern

Xizang is 8°-9°C higher than that in East China of the same latitude. The upper limit of subalpine coniferous forests on the plateau is about 4,300-4,400 m, but around 3,000 m in East China under the same latitude. These facts show that the uplift of the plateau has caused the temperature to rise higher, and as a consequence the distributive limits of the forest zones move upward. It is known as the Masserhebung effect. On the basis of the Masserhebung effect, the actual altitude of Namling area during Middle to Late Miocene should be determined at about 1,000 m from estimation based on the distributive limits of the living forests. That is to say the Miocene plants in Namling area grew at about 1,644 m in altitude, i.e. over 1,500 m. But Namling area lies in an altitude from 3,700 to 4,400 m nowadays. So obviously from then on Namling area has uplifted nearly 2,000 m once more. From here we learn that the rising speed in central Xizang is also faster after Miocene than before. The climate represented by the fossil flora shows the subtropical and subhumid environments.

3. In the northeastern part of the plateau there is a Miocene flora (6) in Zekog county of eastern Qinghai. This flora is made up of all temperate broadleaved deciduous forests, except one evergreen gymnosperm *Taxus qinghaiensis*, such as *Salix* sp., *Cercis miochinensis*, *Podogonium oehningense*, *Acer pseudocarrpinifolium*, *A. subginnala*, *Carpites chakaensis*, with a few herbs, such as *Ranunlus anoseleratus*, *Typha latissima*, *Phragmites oeningensis* and *Cyperacites* sp. The fossil assemblage is similar to the living vegetation of Northwest China, the Huanghe River Valley and to its north. The fossil forests are characterized by coriaceous or semi-coriaceous leaves with dentate margins, while the herbaceous plants grew in shallow lakes, ponds and swamps. The fossil flora reflects a temperate and subarid climatic conditions. Judging from the components of the fossil flora, the altitude shown by Zekog flora during the Miocene may be estimated at about 1,500 m above the sea level. Since the present altitude of the fossil locality lies in 3,200 m, it is apparent that from Miocene up to now the Zekog area has uplifted over 1,700 m again.

4. A Late Pliocene flora (7) from the Sanying formation in Eryuan county, Yunnan is composed mainly of subtropical and warm-temperate broadleaved evergreen forests and several deciduous elements with a gymnosperm (See Table 2).

The Eryuan flora is characterized by sclerophyllous leaves with non-entire margins. It includes 60% of evergreen species and 40% of deciduous ones. The alpine species of *Quercus* is dominant. The fossil assemblage bears striking resemblance in characters to the living species of Eryuan where the fossil remains occurred. From Table 2 we learn that the similar living species of fossil plants today are distributed in a range from 1,925 m to 3,162 m on an average. At present, the fossil locality is situated at an altitude of 1,000 m. If we take 1,925 m as a criterion, then since Late Pliocene there was a little uplift in this area. The Late Pliocene flora represents a subtropical and subhumid climate accompanied, however, by seasonal dorought.

5. A Late Pliocene flora (8) from Sigeda formation in Dechang county, Sichuan is composed chiefly of subtropical broadleaved evergreen trees and shrubs with a few deciduous members (See Table 3). This flora is dominated by 71% of the evergreen species but only occupying 29% of the deciduous ones, all leaves with dentate margins. The Fagaceae containing four species is the largest in the flora. The similar species of the fossil remains are distributed within a range between 1,100 and 2,180 m above sea level to the south of the Changjiang River Valley, especially rich in Hengduan Mountains, Southwest China. At present-day, the fossil locality lies at an altitude of 1,600 m. Therefore, it is estimated that since Late Pliocene the altitude in

Dechang area has uplifted nearly 500 m once more without calculating denudation. The climate represented by the fossil flora indicates a subtropical and subhumid environments.

## CONCLUSION

1. Since Late Cretaceous the uplift of the plateau has been unbalanced and asymmetrical. Before Miocene its upheaval was very slow but obviously turned fast at the beginning of Miocene. At that time the whole plateau might be situated at about 1,500 m above the sea level in altitude. From then on its rising speed has been increasing continuously.

2. During Late Cretaceous along the Yarlung Zangbo River Valley there was a subtropical and subhumid climate. From the region to the south of Gandise Range there seems to be a tropical or paratropical and wet climate, reflecting it was much warmer and wetter than that in Late Cretaceous. During Miocene in central Xizang there was a subtropical and subhumid climate, which is quite similar to that of Hengduan Mountains today. On the eastern plateau the Late Pliocene climate is near to that of Hengduan Mountains at present, being a subtropical and subhumid one. On the northern plateau the Miocene climate was temperate and subarid.

3. Since Late Cretaceous the fossil plant remains of the plateau, except *Eucalyptus*, are all commonly known in northern hemisphere, especially in eastern Asia and North America. The Late Cretaceous flora in southern Xizang is dominated by fossils of subtropical and warm-temperate broadleaved deciduous forests, containing a few evergreen trees. The Eocene flora is composed of fossils of tropical and paratropical broadleaved evergreen forests with a few deciduous plants. Comparing with Late Cretaceous plants, the evergreen components of Eocene flora increase in percentage of species, or deciduous ones decrease. The Middle to Late Miocene flora is made up of fossils of subtropical broadleaved deciduous plants containing some evergreen forms. On the eastern plateau the Late Pliocene floras consist mainly of subtropical broadleaved evergreen forests with a few deciduous elements. On the northern plateau the Miocene flora is composed entirely of temperate broadleaved deciduous plants.

## References


(1) Chaney, R. W., The Kucha flora in relation to the physical condition in central Asia during the Late Tertiary. Sven Hedin Geogr. Ann., 12, 75-195, pls. 1-3 (1935).

(2) Guo Shuang-xing, The plant fossils of the Xigaze Group from Mount Jolmo Lungma Region. A report of Scientific Expedition in the Mount Jolmo Lungma (1966-1968), Palaeontology, Fasc. I. 411-423, 3 pls. (1975) (in Chinese).

(3) Guo Shuang-xing, Pliocene floras of western Sichuan. Acta Palaeontologica Sinica, 17 (3), 343-350, 2 pls. (1978) (in Chinese).

(4) Li Hao-min, Guo Shuang-xing, The Miocene flora from Namling of Xizang. Acta Palaeontologica Sinica, 15 (1), 7-18, 3 pls. (1976) (in Chinese).

(5) Li Ji-jun et al., A discussion on the period, amplitude and type of the uplift of the Qinghai-Xizang Plateau. Scientia Sinica, 22 (11), 1314-1328 (1979)

(6) Guo Shuang-xing, A Miocene flora in Zekog, Qinghai. Acta Palaeontologica Sinica, 19 (5): 406-411, pl. 1 (1980) (in Chinese).

(7) Tao Jun-rong, Kong Zhao-chen, The fossil florule and sporo-pollen assemblage of the Shang-in coal series in Erhyuan, Yunnan. Acta

Botanica Sinica, 15 (1), 120-126, 4 pls. (1973) (in Chinese)
(8) Hsü Jen, On the palaeobotanical evidence for continental drift and Himalayan uplift. Palaeobotanist, 25, 131-141 (1978).

Table 1 A Miocene Flora from Wulong Formation in Namling County, Xizang

| FOSSIL SPECIES | SIMILAR LIVING SPECIES | | | | | | | | | | | |
|---|---|---|---|---|---|---|---|---|---|---|---|---|
| | LIVING SPECIES | HABIT | | LOCALITY | | | | | | | ALTITUDE (m) | |
| | | TREE | HERB | 1 | 2 | 3 | 4 | 5 | 6 | 7 | LOWER | UPPER |
| *Populus latior* | *Populus* sp. | D | | | | | | | | | | |
| *Betula parautilis* | *B. utilis* | D | | | + | | + | + | | | 3800 | 4500 |
| *Carpinus grandis* | *Carpinus* sp. | D | | | | | | | | | | |
| *Carpinus wulongensis* | *C. fargesiana* | D | | | + | | | | | | 1000 | 2000 |
| *Quercus wulongensis* | *Q. pannesa* | E | | | | | | + | | | 2600 | 3900 |
| *Quercus prespathulata* | *Q. spathulata* | E | | | | + | + | + | | | 1300 | 2400 |
| *Quercus namlingensis* | *Q. gilliana* | E | | | | | | + | | | 2400 | 3100 |
| *Quercus semicarpifolia* | *Q. semicarpifolia* | E | | | | | | + | | | 2400 | 3200 |
| *Ulmus hedini* | *U. glaucescens* | D | | | + | + | | | | | | |
| *Thermopsis prebarbata* | *T. barbata* | | | | | | | + | | | 4300 | 4800 |
| *Sophora xizangensis* | *S. moocroftiana* | D | | | | | | + | | | 3000 | 4200 |
| *Rhododendron sanzugawaense* | *R. metternichii* | E | | | | | | | | + | | |
| *Rhododendron namlinegense* | *R. lepidotum* | E | | | | | | + | | | 3000 | 3000 |
| *Phragmites* sp. | *Phragmites* sp. | | D | | | | | | | | | |
| *Cyperacites* sp. | *Cyperus* sp. | | D | | | | | | | | | |
| Average altitude | | | | | | | | | | | 2644 | 3456 |

Table 2 A Late Pliocene Flora from Sanying Formation in Eryuan County, Yunnan Province

| FOSSIL SPECIES | HABIT | | LOCALITY | | | | | | | ALTITUDE (m) | |
|---|---|---|---|---|---|---|---|---|---|---|---|
| | TREE | SHRUB | 1 | 2 | 3 | 4 | 5 | 6 | 7 | LOWER | UPPER |
| *Pinus yunnanensis* | E | | | | | | + | + | | 1600 | 2800 |
| *Populus* sp. | D | | | | | | | | | | |
| *Quercus semicarpifolia* | E | | | | | | + | | | 2300 | 3200 |
| *Quercus spathiulata* | E | | | | + | + | + | | | 1300 | 2400 |
| *Quercus pannosa* | | E | | | | | + | | | 2600 | 4300 |
| *Quercus monimotricha* | | E | | | + | + | + | | + | 2400 | 3900 |
| *Quercus gilliana* | | E | | | | | + | | | 2400 | 3100 |
| *Celtis bungeana* | D | | + | + | + | + | + | | | | |
| *Acer paxii* | D | | | | | | + | | | 1500 | 2500 |
| *Viburnum ovacifolium* | | D | | | | | + | | | 1300 | 3100 |
| Average altituae | | | | | | | | | | 1925 | 3162 |

Table 3 A Late Pliocene Flora from Sigeda Formation in Dechang County, Sichuan Province.

| FOSSIL SPECIES | SIMILAR LIVING SPECIES | | | | | | | |
|---|---|---|---|---|---|---|---|---|
| | LIVING SPECIES | HABIT | LOCALITY | | | | ALTITUDE (m) | |
| | | | 1 | 2 | 3 | 4 | LOWER | UPPER |
| *Castanopsis anningensis* ........... | *C. orthacantha* | E | | | + | | 1800 | 3200 |
| *Quercus consimilla* ............... | *Q. nubium* | E | | | + | + | 600 | 1400 |
| *Quercus protospathulata* ........... | *Q. spathulata* | E | | + | + | | 1300 | 2400 |
| *Quercus preglauca* ............... | *Q. glauca* | E | | + | + | + | 500 | 1500 |
| *Actinodaphne dolichophylla* ........ | *A. reticulata* | E | | | + | | | |
| *Sorbus gracilinervosus* ............ | *S. wilsoniana* | D | | + | + | + | 1300 | 2500 |
| *Acer florinii* ................... | *A. pictum* var. *parviflorum* | D | + | | + | | | |
| Average altitude | | | | | | | 1100 | 2180 |

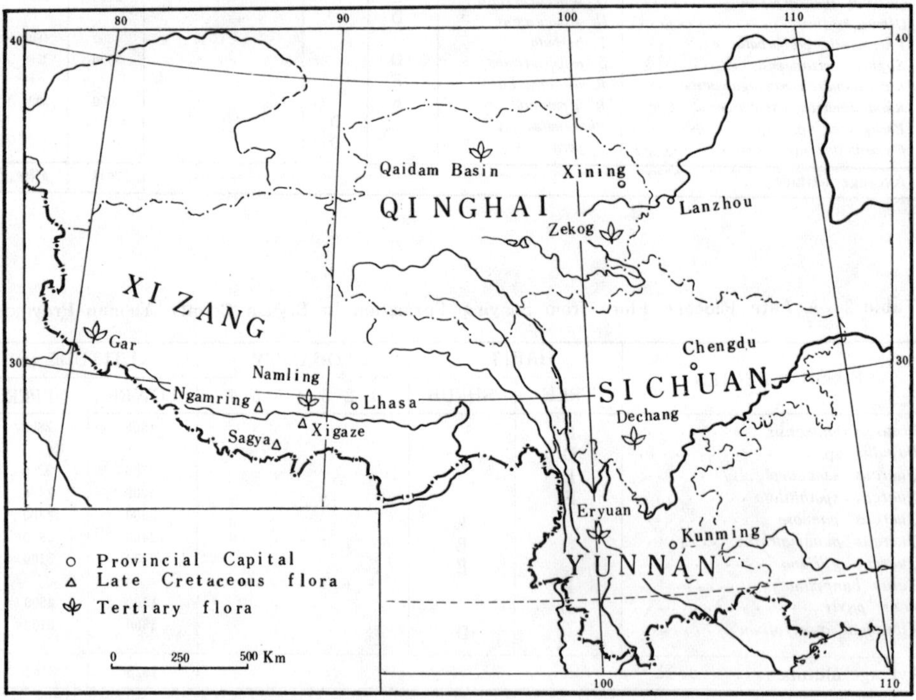


Fig. 1 The Localities of Late Cretaceous and Tertiary floras on the Qinghai-Xizang Plateau

# TERTIARY PALYNOLOGICAL ASSEMBLAGES FROM XIZANG WITH REFERENCE TO THEIR PALEOGEOGRAPHICAL SIGNIFICANCE

Song Zhi-chen Liu Geng-wu
(Nanjing Institute of Geology and Palaeontology, Academia Sinica)

Standing towering in all its splendour in the southern part of the Eurasian continent is the Qinghai-Xizang Plateau. Since it has an average altitude of over 4,000 m as well as a unique natural landscape, it has an enormous influence on the local physical environment and great attention has been aroused among scientists both at home and abroad. Thanks to the effective efforts of many scientists, a number of works dealing with geology, geography and paleontology of this plateau were published in the past twenty years, and especially in the past decade. In the present paper the Tertiary spore-pollen sequence is briefly introduced, with a discussion on Tertiary floral provincialism and the uplift of the plateau from the viewpoint of palynology.

The main Tertiary spore and pollen-bearing stratigraphic units are given in Table 1. Most of them are in the Loinbo La Basin and the Namling area. The Lower Tertiary deposits in the Loinbo La Basin, the Niubao Group are composed of purple and grayish green sandstones, mudstones and conglomerates with a thickness of about 2,200 m; they lie unconformably on Cretaceous sediments. In the northeastern part of Xizang, they often bear gypsum and salt. The Neogene in this basin, known as the Lunpola Group, is mainly lacustrine facies, and consists of yellowish green, grey and dark gray sandy shales with a thickness of about 1,100 m. The Tertiary deposits of Namling in southern Xizang are, in ascending order, the Rigongla, Wulong, Laiqing, Dangjintang and Caidou formations. The first formation is Paleogene age, while the rest are Neogene. They are mostly lacustrine or bog deposits often with coal beds. Red beds are absent.

Tertiary palynomorphs have been frequently obtained from some localities in Xizang (Fig. 1). Palynological evidence suggests that during the Tertiary, especially the Paleogene, there were two floristic provinces in Xizang area, the Northern Province and the Southern Province, with boundary extending approximately along the Gandise Range. This is also supported by lithological evidence. Tertiary sediments, especially those of Lower Tertiary age in the Northern Province are composed of continental clastic rocks often with red beds. So far it seems that no coal has been found. In the Southern Province, however, coal-bearing beds are common but no red beds. In the Neogene, because of probable climatic changes in the Northern Province, differences between the two provinces seem less distinct than those in the Paleogene, with the eastern end of the boundary shifted more or less northward.

Spore-pollen assemblages of the Loinbo La Basin may be taken as representative of the Northern Province (Fig. 2) (1). The earliest one, assigned to the Middle and Late Eocene was obtained from the lower part of the Niubao Group. It is characterized by abundant angiosperm pollen, including <u>Fraxi-</u>

noipollenites, Euphorbiacides, Meliaceoidites, Rutaceoipollis, Retitricolpites, Retitricolporites, Triporopollenites, Subtriporopollenites, Plicapollis, Ulmipollenites, etc. Gymnosperm pollen is mainly Ephedripites, representing about 10% of the total. Spores are chiefly represented by Pterisisporites in addition to minor amounts of Schizaeoisporites, Deltoidospora and Leptolepidites. This palynoflora seems comparable with those from the Ganjo Group, eastern Xizang (2), the Jinsha Formation (according to data of the Paleontologic Group of the Geological Laboratory of the Jianghan Petroleum Bureau) and the gypsum- and salt-bearing formation of Hubei Province (3). Judging from the palynological and lithological characteristics, there could have been a plain or low mountains with an arid subtropical climate, possibly belonging to the Eocene arid subtropical zone of central China. The spore-pollen assemblage from the upper part of the Niubao Group, ranging in age from Late Eocene to Middle Oligocene, resembles the previous one except for the appearance of a number of saccate pollen types in place of Ephedripites and Pterisisporites, The Oligocene assemblage of the Northern Province has been found from the lowest part of the Lunpola Group of the Loinbo La Basin (1,4) and from the Dengqen Formation of Dengqen County (1). The warmth-loving elements, such as members of Myrtaceae, Meliaceae, Carya and Taxodiaceae are present in minor amounts in the Early or Middle Oligocene palynofloras (1,4,10). During the Late Oligocene, possibly including the Early Miocene, conifer pollen increased rapidly,making up almost 90% of the total, with Abiespollenites and Piceaepollenites representing over 20%. Such a striking change in composition may be considered to indicate a drop in temperature caused mainly by crustal uplift in this area, although an extensive temperature drop occurred over most of the Northern Hemisphere in this period. The temperature drop in the Loinbo La area of the Northern Province is believed to be only a regional one related to crustal uplift because at nearly the same paleolatitude in the present Jianghan Basin, some subtropical to warm temperate Oligocene assemblages are found (data from the Jianghan Petroleum Bureau, 1976).

No typical thermophilous elements are found to persist into the Late Tertiary. Corresponding to the stratigraphic division (Table 1), two Neogene assemblages may be recognized from the Loinbo La Basin. Both assemblages are dominated by angiosperm pollen. In the first assemblage, Quercus represents 18-27.5%. Other types are Salix, Juglans, Ulmus,Celtis,Rhus and pollen of the herbaceous Chenopodiaceae, Compositae, etc. Gymnosperm pollen, amounting to 34-48 % of the total, is mainly Pinus, in addition to Picea, Abies and Keteleeria. A few spores belonging chiefly to the Polypodiaceae have been found. The assemblage of the upper Lunpola Group closely resembles the previous one except for more Salix and pollen of herbs. Based on this fact, it seems probable that the assemblage of the upper Lunpola Group belongs to the Miocene period.

Although no Eocene palynological information is available at present from the Southern Province, we gain a general impression of the vegetation from megafossils of the Eocene flora of Gar County recorded by the Scientific Expedition to the Qinghai-Xizang Plateau. This flora contains such forms as Ficus myrtifolia, Eucalyptus genitzi,E. attenuata, E. latifolia, cf. Eugenia hilgardiana, Myrica beatoensis, Cyperacites heydenii and Phrynium sp. These taxa, mostly belonging to the Myrtaceae, provide some idea of the temperature requirements of this flora. In addition, the Moincer Group bears 7-12 coal seams that are 10 km. long. It seems to us that during the Eocene the altitude of southern Xizang might not be too high, and the climate must have been rather warm and humid.

In the Southern Province, only a samll Oligocene pollen-spore assemblage has been obtained from the upper part of the Rigongla Formation (5) (Fig. 3). This assemblage has a number of warmth-loving members, such as _Castanea_ and _Engelhardtia_, that constitute more than 40% of the total, in association with 20% or so of gymnosperm pollen and over 20% of spores. It is very similar, paleogeographically, to the earlier megaflora.

A Miocene assemblage in the Southern Province is represented by pollen and spores of the Wulong Formation of Namling. This palynoflora is somewhat similar to that of the Rigongla Formation which is dominated by angiosperm pollen. Noteworthy is the fact that in the upper part of this formation the percentage of saccate pollen, including a number of species of _Picea_ and _Abies_ sometimes amounts to 50% of the assemblage. Therefore, during the Miocene period, the climate in the area south of the Nyainqentanglha Range was warmer and more humid than that to the north of the range, comparable probably to that of the present Changjiang River Valley. It was not until the Late Miocene, that a marked drop in temperature resulted from elevation near the Namling area. The assemblages of the Dangjintang and Caidou formations, assigned to the Pliocene, are characterized by abundant conifer pollen (5,6). That the pollen of _Picea_ and _Abies_ represents about 25 percent of the palynoflora implies a climate similar to that in Late Miocene. Palynological assemblage from the Himalayan area described by Xu et al. (1973), Zheng (in press), and Cao (in press) are believed to be Pliocene age (7,8,9). The assemblage of the Yebokangala Formation, dating from Middle to Late Pliocene, is dominated by alpine trees and shrub plants, including a number of species of _Cedrus_, _Picea_, _Abies_, _Quercus_ and the like (7). Therefore, from present botanical evidence, it is apparent that major elevation of the Himalayas the highest mountain in the world occurred mainly in Neogene and Quaternary, and especially in Quaternary.

Regardless of its specific age, the palynoflora suggests certain climatic conditions, and thus affords a means of reconstructing the ancient topography. The climatic conditions under which modern _Picea_ and _Abies_ forests live provides a basis for estimating the elevations at which the fossil _Picea_ and _Abies_ floras existed. In Fig. 4 seven selected modern _Picea_ and _Abies_ forests in different parts of China are shown (12,13). The seven localities, the dominating species and the altitude of each locality are shown in the square frames on the map.

Spruce and fir forest mixed with _Larix olgensis_ and some birches is found from 1,100 m up to 1,500 m on the northern slope of the Changbai Mountains.

_Picea heycri_ and _P. wilsonii_ are important members of the subalpine conifer forest between 1,800-2,700 m on the Wutai and Taihang Mountains.

On the northern slope of the Taibai Mountain, _Abies sutchenensis_ occurs above 2,500 m (2,500-3,500 m), forming an _Abies_ needle-leaved forest.

_Abies fabri_ and _Tsuga_ forest is found on the western slope of the Erlang Mountain at an altitude from 2,700-2,800 m .

A subalpine needle-leaved forest of _Abies fabri_ ranges from 3,000 m up to 3,800 m on the western slope of the Gaoligong Mountain.

On the southern slope of Mount Qomolangma a subalpine needle-leaved forest, in which *Abies spectabilis* is predominant, exists from 3,100 m to 3,800 m (or even 4,000 m).

*Pices morrisonicola* is found between 2,300-3,000 m on the Central Range of Taiwan, forming a pure needle-leaved forest.

With two exceptions, the Changbai and Taibai Mountains, the lower limits of modern *Picea* and *Abies* forests in China are above 2,000 m It is, therefore, reasonable to assume that the altitudes in the areas where a large amount of *Abiespollenites* and / or *Piceaepollenites* have been found were over 2,000 m.

In summary, we may come to the following preliminary conclusions:

1. During the Early Tertiary, there were two floristic provinces in Xizang area, namely the Northern Province and the Southern Province with a boundary line extending approximately along the Gandise Range. These provinces differ from each other in both botanical and lithological features. They existed until Late Tertiary. Probably because of the climatic change in the Northern Province in Neogene, differences between them are less pronounced in Neogene than in Palaeogene, and their boundary is shifted northward at its eastern end, extending possibly along the Nyainqentanglha Range.

2. Since the climate in the Southern Province was moister and warmer than that in the Northern Province, it was favourable to the formation of Tertiary coal beds in southern Xizang.

3. The elevation of the entire Xizang area did not rise at the same t time. By the end of the Oligocene, the altitude of the northern part of the plateau is estimated to be about 2,000 m. It was not until the end of the Miocene, however, that the same altitude occurred in the Namling area south of the Nyainqentanglha Range. As far as the Himalayan area is concerned, this altitude came into being only in the Middle-Late Pliocene.

## References


(1) Song Zhi-chen, Liu Geng-wu, Early Tertiary Palynoflora and its significance of paleogeography from northern and eastern Xizang. The Series of the Comprehensive Scientific Expedition to the Qinghai-Xizang Plateau, Paleontology of Xizang, Book V. Science Press (in Chinese) (in press).

(2) Song Zhi-chen, Li Man-ying, Eocene palynological assemblages from the Gonjo group in eastern Tibet Autonomous Region (Xizang Zizhiqu), Stratigraphy and Paleontology of the Eastern Qinghai-Xizang Plateau. Sichuan People's Press (in Chinese) (in press).

(3) Li Man-ying, Song Zhi-chen, Li Zai-ping, Some Cretaceous-Tertiary palynological assemblages from the Yangtze-Han River Plain. Memoris of Nanjing Institute of Geology and Paleontology, Academia Sinica, 9, 1-44, pl. 1-12 (1978) (in Chinese).

(4) Wang Kai-fa, Yang Jiao-wen, Li Zhi, Li Zeng-rui, On the Tertiary sporo-pollen assemblages from Lunpola Basin of Xizang, China and their paleogeographic significance. Scientia Geol. Sinica, 366-374, pl.1-4 (1975) (in Chinese).

(5) Song Zhi-chen, Liu Jing-lin, The Tertiary Sporo-pollen assemblages from Namling of Xizang, The Series of the Comprehensive Scientific Expedition to the Qinghai-Xizang Plateau, Paleontology of Xizang, Book V. Science Press (in Chinese) (in press).
(6) Li Hao-ming, Guo Shuang-xing, The Miocene flora from Namling of Xizang. Acta Palaeontologia Sinica, 15 (1), 7-18, pl. 1-3 (1976) (in Chinese).
(7) Hsu Jen, Tao Jun-rong, Sun Xing-jun, On the discovery of a Quercus semicarpifolia bed in Mount Shisha Pangma and its significance in botany and geology. Acta Botanica Sinica, 15, 103-109, pl. 1-4 (1973) (in Chinese).
(8) Zheng Ya-hui, Sporo-pollen assemblages of the Woma Formation of Gyirong Basin in Xizang, The Series of the Comprehensive Scientific Expedition to the Qinghai-Xizang Palteau, Quaternary of Xizang, Science Press (in Chinese) (in press).
(9) Cao Liu, Pliocene Palynological flora from Disong, Pulan district, Xizang. Ibid. (in Chinese) (in press).
(10) Hsu Jen, Sung Tzi-chen, Chow Ho-i, Sporo-pollen assemblages from the Tertiary deposits of the Tsaidam Basin and their geological significance. Acta Palaeont. Sinica, 6, 429-440, pl. 1-6 (1958) (in Chinese).
(11) Li Gi-jun, Wen Shi-xian, Zhang Qing-song, Wang Fu-bao, Zheng Ben-xing, Li Bing-yuan, A discussion on the period, amplitude and type of the uplift of the Qinghai-Xizang plateau. Scientia Sinica, 22, 1325-1328 (1979).
(12) Hou Xue-yu, On the characteristics of the vertical mountainously vegetational belts of the different floristic regions of China, ed. Botanic Society of China. Scientific Information Press, 254-258 (1963) (in Chinese).
(13) The Editorial Committee on Chinese Vegetation, The Vegetation of China, Science Press, 1373p. (1980).

Table 1

| | Northern Province | | | Southern Province | | |
|---|---|---|---|---|---|---|
| | Loinbo La | Dengqen | Gonjo | Namling | Mount Xixabangma | Moincer |
| Neogene | | | | Caiduo Formation | Yebokangg-ala Group | |
| | | | | Dangjingtang Fm. | | |
| | Lunpola Group: Upper part | | | Laiqing Formation | | |
| | Lunpola Group: Lower part: Upper part part | | | Wulong Fm.: Upper member | | |
| | | | | Wulong Fm.: Lower member | | |
| | Lunpola Group: Lower part: lower lower part | | | Rigongla Formation | | |
| Paleogene | Niubao Group: Upper part | Dengqen Formation | | | | Moincer Group |
| | Niubao Group: Lower part | | Gonji Group | | | |

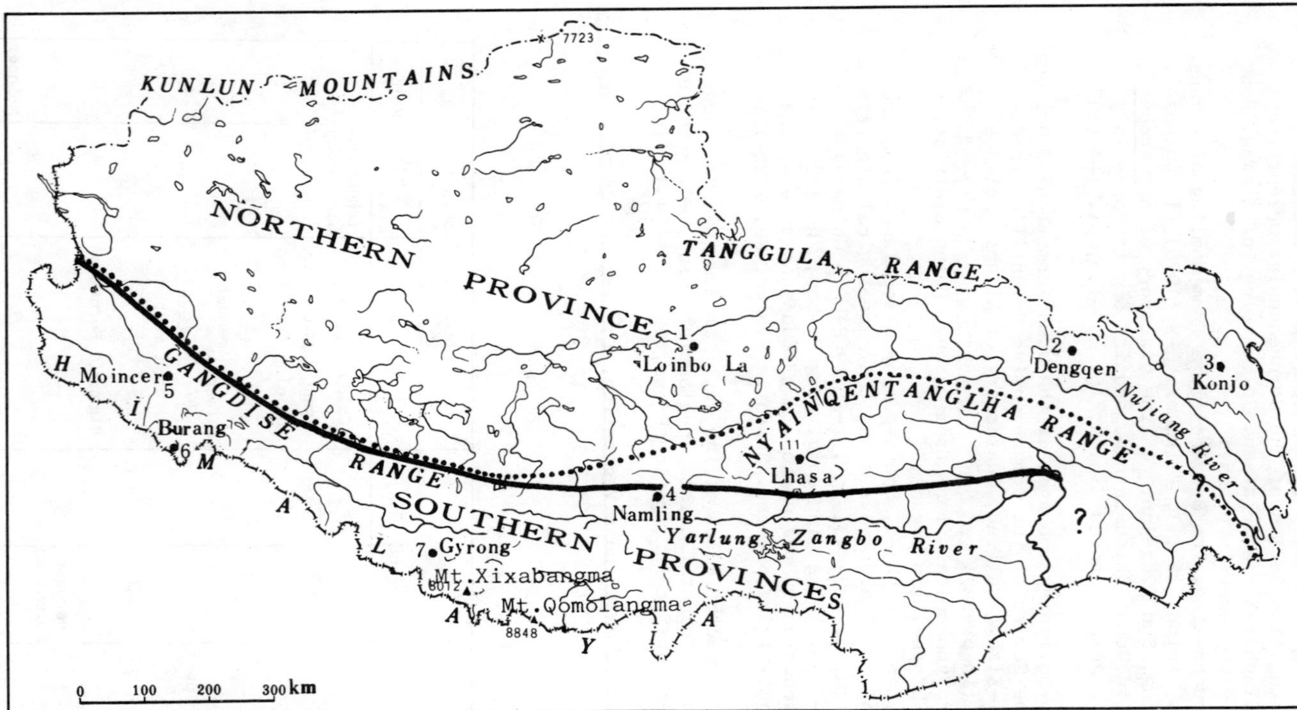


Fig. 1 Sketch map showing the Tertiary floristic provinces and sampling localities. The black line indicates the boundary between the two Paleogene provinces, the dotted line indicates that between the two Neogene provinces.

1. The Loinbo La Basin; 2. Dengqen; 3. Gonjo; 4. Wulong; 5. Moincer (megafossil, data from the Scientific Expedition to the Qinghai-Xizang Plateau); 6. Jiaxing (microfossil, Cao, in press); 7. Woma (microfossil, Zheng, in press); 8. Mount Xixapangma (megafossil and microfossil, Xu et al., 1973)

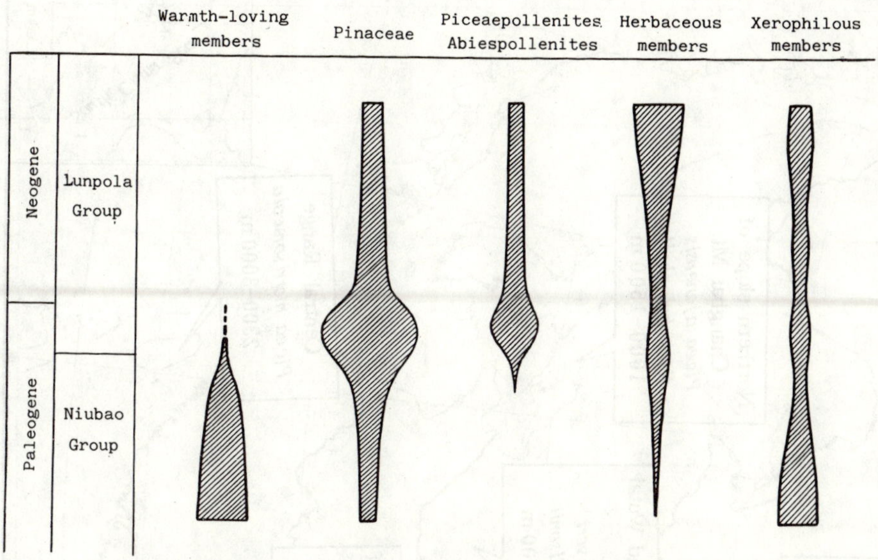


Fig. 2 Stratigraphic distribution of selected taxa from N. Province (mainly according to the data from Loinbo La)

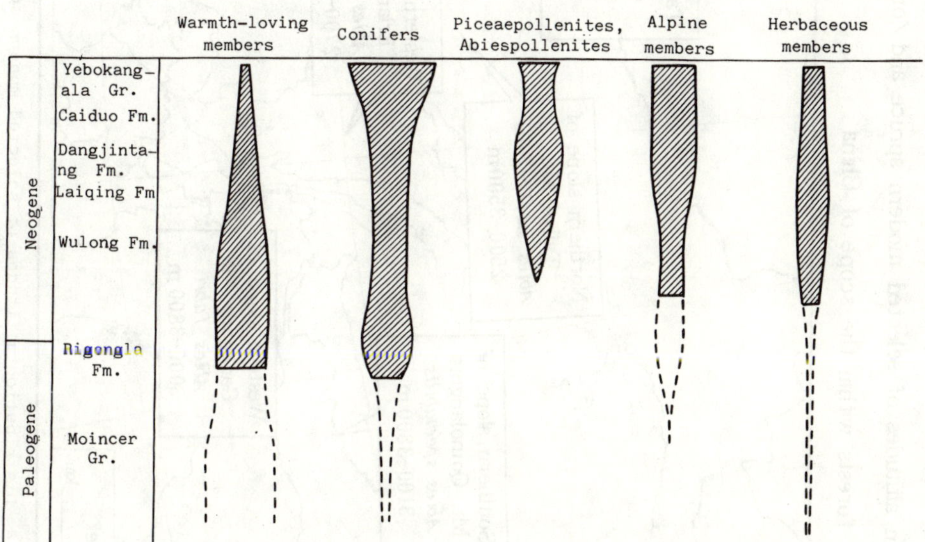


Fig. 3 Stratigraphic distribution of selected taxa from S. Province (mainly according to the data from Namling)

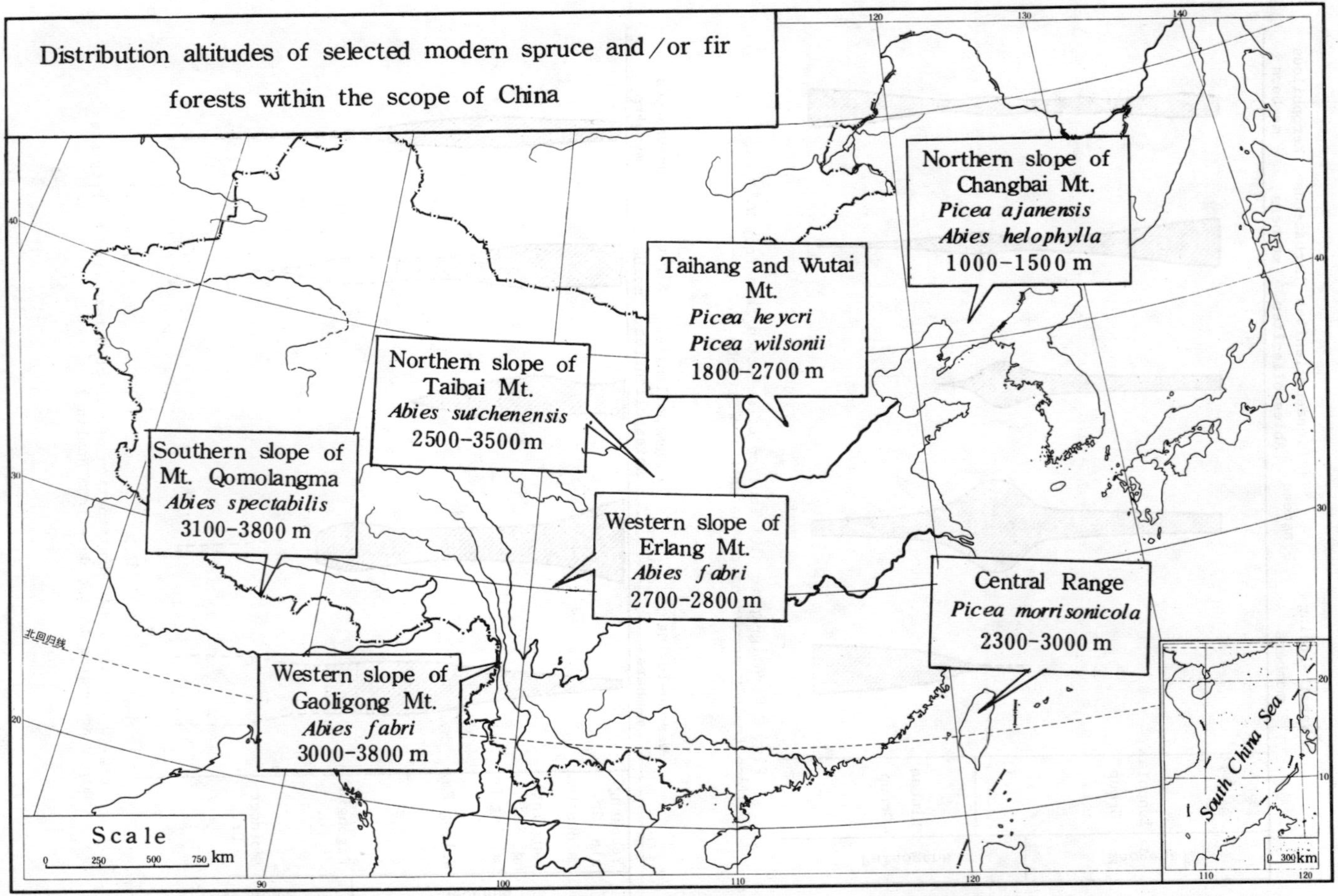


Fig. 4 Distribution altitudes of selected modern spruce and/or fir forests within the scope of China

# BASED UPON PALYNOLOGICAL STUDY TO DISCUSS THE NATURAL ENVIRONMENT OF THE CENTRAL AND SOUTHERN QINGHAI-XIZANG PLATEAU OF HOLOCENE*

Huang Ci-xuan Liang Yu-lian
(Institute of Geography, Academia Sinica)

In recent years, we carried out pollen analysis on some sediments of the Holocene of the Qinghai-Xizang Plateau. We analysed the exposed sections of Nara Yumco Lake, Yamzhog Yumco Lake and Ngamring in South Xizang, and that of Coqen and drill holes of Chagcam Caka (Lake II and Lake III) in North Xizang (Fig. 1).

The Plateau possesses wide distribution of Holocene sediments which are principally alluvial, diluvial, glacial, fluvio-glacial deposits, aeolian accumulation, as well as lacustrine and shore deposits in nature. All samples were taken from deposits of lacustrine and shore facies 4,000-5,000 m in elevation. Its natural zones include 1) Montane shrubby steppe in South Xizang (warm and cool, semi-arid) and 2) Steppe in Qiangtang Plateau (cold and semi-arid) (1). Both temperature and precipitation decrease progressively from southeast to northwest on the plateau surface, and aridity increases from southeast to northwest. Since the vegetation distribution of the plateau is controlled by climate and the flora components are closely correlated with the neighbouring areas, the frutex Rhododendron, some species of herbaceous Compositae, Saxifragaceae, Gentianaceae, etc. widely distribute in China's southwest mountainous area and the east Himalayas, while Ephedra, Tamaricaceae and Chenopodiaceae distribute more commonly in Central Asia. Vegetation on Qinghai-Xizang Plateau is bound up in different degree with that in East and Central Asia. This feature has also been embodied in sporo-pollen of plant association of the Holocene as well.

According to sporo-pollen analysis and data on dating of absolute chronology with radio carbons, this paper primarily reconstructs the actual environment in which the plant association of the plateau have developed, and deals with the division of Holocene climatic periods as well as the discussion on evolution of natural environment of the same period (Holocene).

## RESULTS OF SPORO-POLLEN ANALYSIS

The main feature of sporo-pollen groups of the Holocene deposited in

---

*Sporo-pollens were collected in the field by scientific workers of Qinghai Institute of Saline Lake, Academia Sinica, Li Bing-yuan of the Institute of Geography, Academia Sinica, and Wang Fu-bao of Nanjing University; the $C^{14}$ dating of absolute chronology was finished by Qinghai Institute of Saline Lake, and Institute of Archaeology, the Chinese Academy of Social Sciences; and animal fossils were determined by the division of abdominal foot, Nanjing Institute of Geology and Palaeontology, Academia Sinica.

central Xizang is that herbaceous pollen within the temperate plant species occupies a dominant position, being about 50-90 % of the group. The Xylophyta pollen makes up less than 50 %. The spore in Pteridophyta makes up the least with only a few percentage.

In Xylophyt pollen, the amount of Pinus of coniferous trees makes up the maximum. Besides, there are pollens of Picea, Abies, Cupressaceae and Tsuga, which are subtropical alpine plants. Pollens of Betulaceae occupy the largest amount in the broad-leaved trees, including those of Betula, Alnus, Carpinus and Corylus, and the amount of Quercus pollen comes secondly. In the pollens of frutice species, there are Rosaceae, Ericaceae, Tamaricaceae, Ephedra etc. In the pollens of herbaceous, Artemisia and Cyperaceae are most common, besides, there are also Polygonaceae, Ranunculaceae, Thalictrum and Compositae. Both the species and quantity of pteridophyta pollens are few, they are mainly Polypodiaceae, Pteridaceae and Davalliaceae, etc.

## EVOLUTION OF PLANT ASSOCIATION AND DIVISION OF HOLOCENE CLIMATIC PERIODS

It is possible for us to reconstruct the actual environment of Palaeophytes by means of determination and analysis of sporo-pollen fossils deposited in different periods. The section of Nara Yumco Lake and drill holes of Chagcam Caka (Lake III) are the case.

Sporo-pollen analysis of Nara Yumco Lake (Fig. 2, Table 1) reflects that vegetation development have experienced the following three stages. The first stage is Cyperaceae meadow of scattered trees of birch family. The second is plant association composed of birch family, Quercus, Tsuga and Mesophyte herbs, as well as Alisma. In the third stage, mesophyte herbs were predominant while mixed with some Pinus and shruby plants, Rosa and Rhododendron.

$C^{14}$ dating of two peat samples situated at the profile of the second stage reveals that the age for one at 3.80 m deep is 6380±100 years, and the one at 0.57 m deep is 3625±100 years. This shows that the three plant evolution stages correspond to early, middle and late periods of the Holocene respectively, and it is 3,000-7,500 years from the middle to the present.

Based on the profile of Lake III of Chagcam Caka (Fig. 3), two stages for plant development may be distinguished. The first one is of Artemisia, Pinus, Tamaricaceae and Ephedra, with a few Picea and Abies. The second one is similar to the first, but almost without Picea and Abies, with broad leaf trees of Alnus and Betula in high percentage and Cyperaceae in greater amount. Climate was warmer and more humid than the previous period. The $C^{14}$ dating of peat at 4.5 m deep is 4780±185 years B. P. and the bottom mud (13.2 m deep) for the first stage is 13,357±165 years. Below it, lithology changes greatly, only a few pollens of Artemisia have been found. $C^{14}$ dating of two of them reveals that one is 15,369±165 years (14m deep ), and another is 19,963±351 years (15.7 m deep). So it can be said that the two stages of plant evolution within the profile may correspond to early and middle Holocene periods and the strata below 13.2 m may correspond to late Pleistocene.

A grey-clay and sub-clay strata of 3-9 m deep are commonly found in middle sections of Yamzhog Yumco, Ngamring and Coqên. They consist of peat, plant residues, or fresh-water snail fossils. Deposits of this strata contain maximum amount of sporo-pollen, while the quantity of Xylophyta reaches the highest within the entire strata. $C^{14}$ dating shows that the

Table 1 Comparison of Holocene's Sporo-Pollen Groups

| Profile / Comparison | Nara Yumco altitude 4750 m | | Cogen altitude 4675 m | | Chagcam Caka (Lake III) altitude 4400 m | |
|---|---|---|---|---|---|---|
| | Sediment $C^{14}$ | Pollen group | Sediment $C^{14}$ | Pollen group | Sediment $C^{14}$ | Pollen group |
| Late Holocene 0-3000 B.P. | yellow silty | Pinus Rosaceae Rhorodendron medium life herb. 88.7% | gray-yellow fine sand | Ephedra Artemisia | | |
| Middle Holocene 3000-7500 B.P. | gray brown sub - clay Peat 3625±100 Peat 6380±100 | Betulaceae Quercus Tsuga Alisma medium life herb. 56-60% | gray-yellow silty sub-clay relic plant fossil 7010±150 | Pinus Betula Cyperaceae Artemisia herb. 56-73% | black mud (include halic) 4780±185 | Ephedra Tamaricaceae Alnus Pinus Artemisia Cyperaceae herb. 90% |
| Early Holocene 7500-13,000 B.P.? | sub-clay | Betulaceae Cyperaceae herb. 80% | | | black mud 13357±165 | Pinus Tamaricaceae Ephedra Artemisia herb. 90% |

age of peat of the strata 3.5 m deep of Ngamring section is 4,490±195 years, and the same strata where freshwater snail fossils (Radix auricularia Linneans) have been found in Coqên (8.5 m deep) are 7010±150 years old.

So it is beyond doubt that the deposits of middle Holocene respond to a period when it was climatically optimum with a warmer and wetter environment and vegetation grew best.

To sum up, three divisions can be grouped climatically in connection with plant association development during Holocene period (Table 2): 1) The early division ranges from 7,500-13,000 B. P. with wide distribution of sparse shrubsteppe over plateau surface. 2) The middle division from 3,000-7,500 years B. P., is the most flourishing period for plant association of the plateau during middle Holocene. The southern part reflects a forest-shrub-steppe natural landscape. Further westward variations occur and further north-west, sparse shrub-steppe predominates the Chagcam Caka area. On the whole it reflects that physical environment of the plateau has changed gradually better, but there were still some differences, because it was warm-humid in South Xizang, cold and semiarid in the North. 3) The late Holocene is 0-3000 years B. P. The landscape is typical of small shrub-steppe or steppe, similar to the present one.

Scholars of differnt countries have adopted different ways to classify climatic periods of the Holocene (Table 3). Based on the study of peat deposits in North Europe, A. Blytt and R. Sernander divided Holocene into five climatic periods,taking the Atlantic period as climatic optimum. M. E. Neustadt has studied the peat and lacustrine facies sediments in Soviet Union and divided Holocene into four climatic periods in which the middle Holocene (7,700-2,500 years B. P.) was taken as the climatic optimum (2). Based on the investigations in central Honshu, Japan, K. Sohma has divided Holocene into four climatic periods and he suggested that the temperature of the warmest period might have been 2°C higher than it is today (3); some scholars of North America have studied the deposits of lacustrine and marsh-facies and divided Holocene into five climatic periods. For example, Richard N. Mack et al. have studied the sporo-pollen assemblages of Holocene lacustrine facies in North America. By using data of $C^{14}$ dating they divided Holocene into five climatic periods in which the third was the warmest and wettest one (7,600-3,000 years B. P.)(4).

The work of sporo-pollen analysis engaged by Chinese scientists is mainly concentrated in North and Northeast China. Most of them have divided Holocene into three climatic periods while the others divided it into four or five periods. Chen Cheng-hui analysed the sporo-pollen assemblages of Holocene sediments in Southern part of Liaoning Province, based on the dating of absolute chronology of radio carbon, he divided Holocene into three climatic periods in which the middle Holocene was considered to be the warmest and wettest, being 8,000-2,500 years B. P. (5).

In the past, the division of Holocene strata in Qinghai-Xizang plateau was mainly based on lithology of sediments. During the scientific expedition to the region of Mt. Qomolangma in 1966-1968, travertine sediments and various species of shrub plant fossils were found in Holocene sediments in Yali region, South Xizang, and it was divided into two climatic periods: early Holocene was the climatic optimum period (i. e. Yali period ) when it was warm and humid; late Holocene was Rongpude little ice age when it was cold (6).

All the above mentioned criteria for the division of Holocene suggest that there existed the climatic optimum periods which are comparable with

Table 2 Comparison of Holocene Physical Environment

| Period \ Physical Environment | | Fresh lake of S. Xizang | | Salt water lake of N.Xizang | | Saline lake of N. Xizang | |
|---|---|---|---|---|---|---|---|
| | | vegetation | climate | vegetation | climate | vegetation | climate |
| Present | | fruticeta steppe | warm and cool semi-arid | steppe | cold semi-arid | desert steppe | cold and dry |
| Holocene | $Hl_3$ | fruticeta steppe | warm and cool semi-arid | little fruticeta steppe | cold semi-arid | | |
| | $Hl_2$ | forest fruticeta steppe | warm semi-humid | forest fruticeta steppe | warm and cool semi-arid | sparse fruticeta | warm and cool semi-arid |
| | $Hl_1$ | sparse fruticeta steppe | warm and cool semi-arid | | | sparse fruticeta steppe | turn into warmer semi-arid |

Table 3 Division and Comparison of the Holocene

| Region | C. and N. Xizang | Mt. Qomolangma (Guo Xu-dong) | S. Liaoning (Chen Cheng-hui) | N. and N-E. China | U. S. S. R (Neustadt M.E.) | N. America (Richard N. Mack) | N. Europe (A. Blytt-R. Sernander) |
|---|---|---|---|---|---|---|---|
| Division | Late Holocene<br>$Hl_3$<br>0-3000 B.P. | Rongpude Little Ice Age (cold) | Zhuanghe Period<br>0-2500 B.P. | Late Holocene<br>0-2500 B.P. | Late Holocene<br>0-2500 B.P. | ZONE V<br>0-1500 B.P.<br>ZONE IV<br>1500-3000B.P. | Sub-Atlantic (rather cold and humid)<br>0-2500 B.P. |
| | Middle Holocene<br>$Hl_2$<br>3000-7500 B.P. | Yali Period<br>(warm-humid) | Dagushan Period<br>2500-8000 B.P. | Middle Holocene<br>2500-7500 B.P. | Middle Holocene<br>2500-7700 B.P. | ZONE III<br>3000-7600 B.P. | Sub-Boreal (rather warm and dry)<br>2500-5000 B.P.<br>Atlantic (rather warm and humid)<br>5000-7500 B.P. |
| | Early Holocene<br>$Hl_1$<br>7500-13000 B.P. ? | | Pulandian Period<br>8000-10000 B.P. | Early Holocene<br>7500-10000 B.P. ? | Early Holocene<br>7700-9800 B.P.<br>Oldest Holocene<br>9800-12000B.P. | ZONE II<br>7600-8300 B.P.<br>ZONE I<br>8300-9500 B.P. | Boreal (rather warm and dry)<br>7500-9500 B.P.<br>Pre-Boreal (climate became better)<br>9500-10300 B.P |

our division of middle Holocene.

## CHANGES OF HOLOCENE PHYSICAL ENVIRONMENT

The distribution and development of plant association on the plateau surface during Holocene period are closely correlated with the global climatic changes of the same period, and also clearly correlated with the uplift of the plateau. These characteristics of changes of physical environment during Holocene may be summarized as following three aspects:

(1) 3,000 years B. P. was a limit of greater change both in climate and in vegetation. The landscape was typical of small shrub-steppe or steppe of simple species, but before this period there were trees and more species of plants.

(2) Natural environment progressively changed from southeast to northwest. Controlled by plateau climate, present-day vegetation in southeast is principally shrub-meadow or steppe with a high vegetation coverage, while in northwest semi-desert and desert landscape prevails, with an extermely low vegetation coverage or even without vegetation cover. A trend of change from southeast to northwest also appeared during Holocene period. For example, there were more Xylophyta components and higher proportion of broad-leaved trees in southeast, but further northwestward, Xylophyta decreased, so did the broad-leaf trees. On the other hand,Xerophytic component increased. Particularly some plants relatively in favour of warm and wet such as Quercus and Tsuga existed in vegetations in southeast during early and middle Holocene periods. They decreased further westward and disappeared up to northwest, but Xerophyte such as Ephedra and Tamaricaceae appeared instead, and the number of plant species was much lesser.

(3) Some specific characteristics of changes of Plateau physical environment existed during Holocene. It is of general significance for trend of climatic changes from cold turning warm, to warm and humid, then to dry and cold in Northern Hemisphere during Holocene period. If compared with North China Plain which is similar in latitude, plant association was of Pinus of needle-leaf trees dominating in early Holocene, while in the middle broad-leaf forests composed of mainly Quercus family of broad-leaf trees, or first mixed needle-and-broad leaf forests, then Pinus of needle-leaf trees in late Holocene. Therefore, it was warm and humid in middle period, and colder in early and late periods (7,8). Holocene climate of the plateau roughly coincides with this trend, but it also has its specific characteristics. For instance, no fundamental changes have ever occurred so far as plateau plant change concerned from early to middle periods. It is only a quantitative change, because cold adaptable components decreased and Thermophilous and Hygrophilous increased. Up to late Holocene, dramatic changes took place on plant components because Xylophyta components decreased greatly, even disappeared, while Xerophyte components and herbaceous plant increased rapidly. This demonstrates that it is far more inferior to be obvious in warmness and humidity in middle period here than that in other areas of North China, and the trend to be drier and colder in late period is much distinct. Cold and aridity governed the environment of the entire Holocene period. Generally speaking, it was much drier and colder than plain areas of the same latitude with herbal plants well developed. The altitude of the plateau was constantly rising because the continuous uplift of the plateau during Holocene period, as a result the climate became dry and cold. This factor partly compensated the increasing temperature of the period and

reduced the magnitude of warm and humidity. The uplift, on the other hand, accelerated the dry and cold process during later period, consequently the plateau is not as distinctly warm and humid as that of plain areas in the middle Holocene, and it was even drier and colder in later period. From this one can see that the change of plateau physical environment in Holocene is closely correlated with the uplift of the plateau, thus it possesses both the general situation within global scope and specific characteristics of the plateau proper.

References


(1) Zheng Du et al., Physico-geographical differentiation of the Qinghai-Xizang Plateau, Acta Geographica Sinica. 34 (1), (1979). (in Chinese)

(2) Neustadt M. E. and Gujeleak V. K., Problems of Holocene V. A. S. "Problems of Holocene", (1961). (in Russian)

(3) Neustadt M. E., Some problems of Holocene study in some places of Asia and Australia. V. A. S. "Problems of Holocene", (1961). (in Russian)

(4) Richard N. Mack, N. W. Rutter, Vaughn M. Bryant, Jr., and S. Valastro, Reexamination of postglacial vegetation history in northern Idaho: Hager pond, Bonner Co., Quaternary research, 10 (2), (September, 1978).

(5) Chen Cheng-hui et al., Evolution of natural environment in southern part of Liaoning Province during the last 10,000 years. Scientia Sinica,6 (1977). (in Chinese)

(6) Guo Xu-dong, Quaternary interglacial period and palaoclimatology in Mt. Qomolangma Region. Report of investigations on Mt. Qomolangma region (1966-1968), Quaternary Geology, Science Press, (1976). (in Chinese)

(7) Liu Jin-ling et al., Sporo-pollen assemblages of peat in the southern piedmont of Yanshan Mts. Quaternary Research in China, 4 (1), (1965). (in Chinese)

(8) Zhou Kun-shu, Investigations on two buried peats in the vicinity of Beijing and their sporo-pollen analysis. Quaternary Research in China, 4 (1), (1965). (in Chinese)


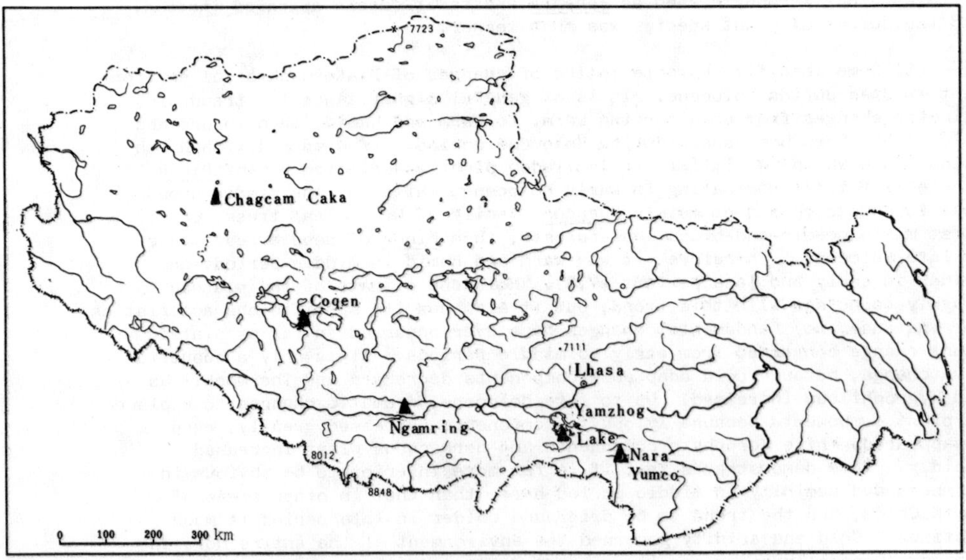


Fig. 1 Profile's positions of Holocene sporo-pollen analysis.

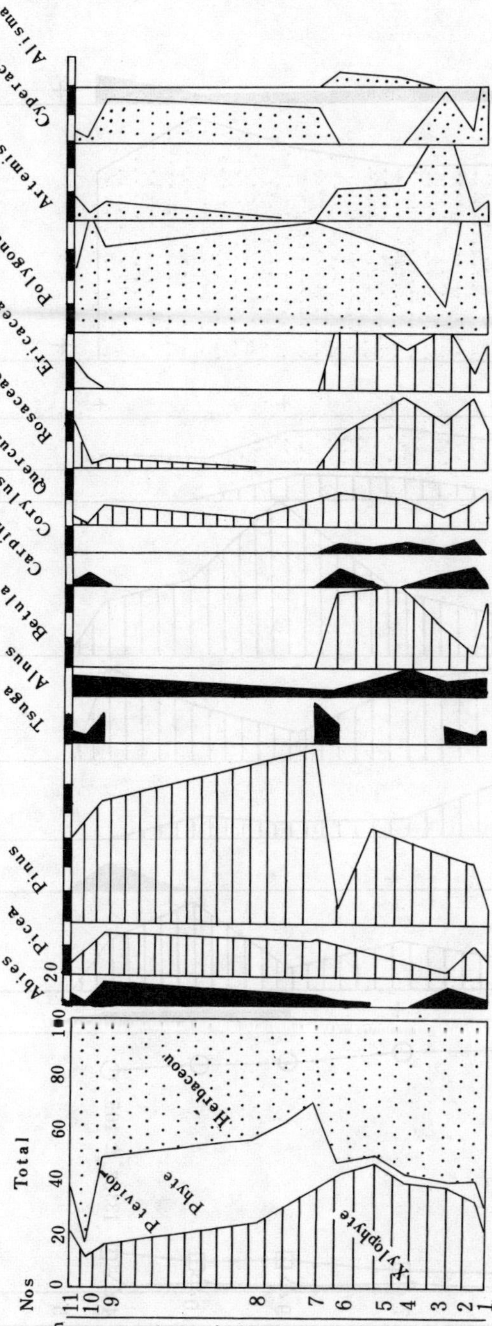


Fig. 2 Pollen diagram of sediments of Lake Maro Yumco

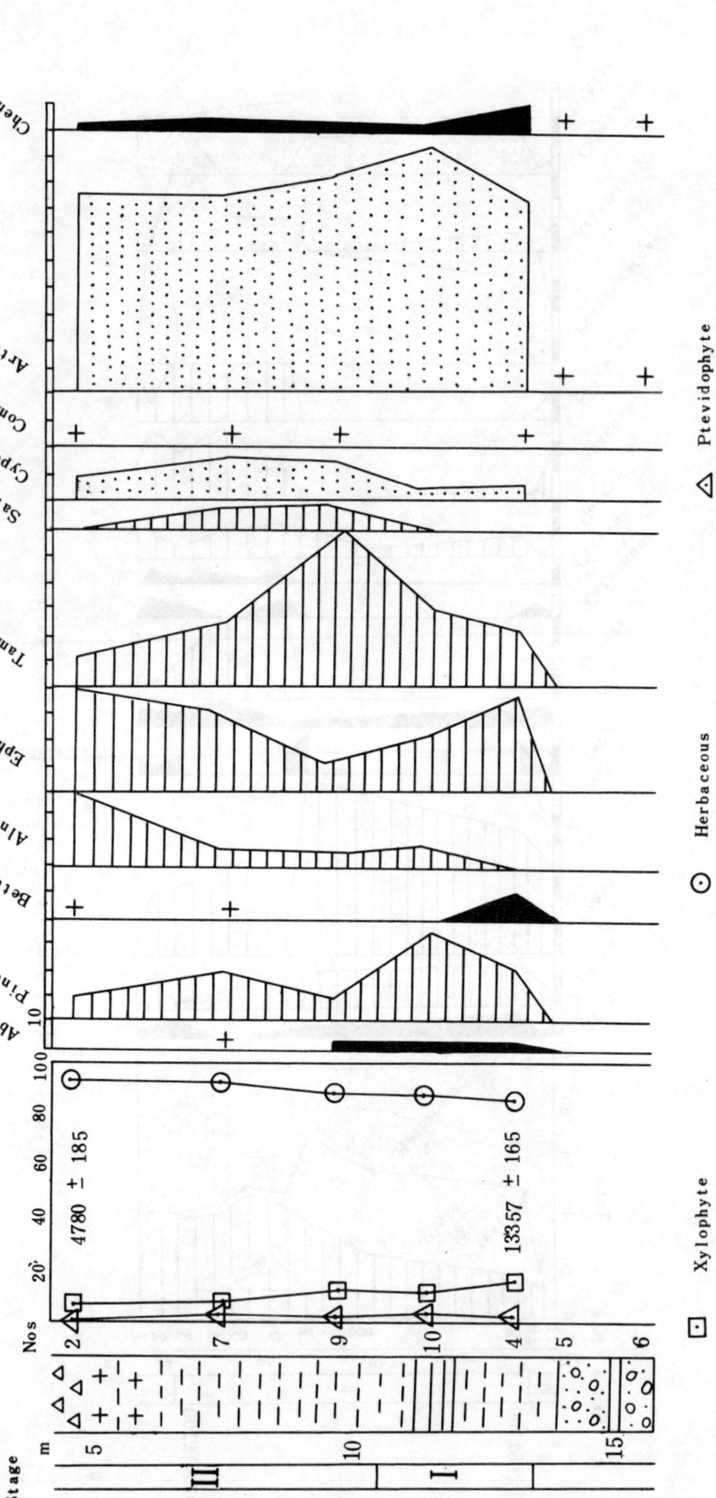


Fig. 3 Pollen diagram of sediments of Chagcam Caka (Lake III)

THE CLIMATE AND UPLIFT OF THE QINGHAI-XIZANG (TIBET) PLATEAU IN THE LATE PLEISTOCENE AND HOLOCENE

Huang Wan-po Ji Hong-xiang

(Institute of Vertebrate Palaeontology and Paleoanthropology, Academia Sinica)

From 1975 to 1977, the Scientific Expedition to the Qinghai-Xizang Plateau investigated several Late Pleistocene and Holocene sites in the Xizang region, where some mammalian fossils and microlithic were discovered (Fig. 1).

In this paper, the climate and uplift of the Qinghai-Xizang Plateau during the Late Pleistocene and Holocene are discussed in the light of these fossils of mammalian remains and archaeological finds.

## FOSSILS OF BIBOS SP. AND LIANA OF LATE PLEISTOCENE FOUND IN THE NYINGCHI BASIN

In the Nyingchi Basin, deposits of lakefacies about 140 metres thick, composed chiefly of pale-green, fine sand-stone, have been found, and fossils of Bibos and liana unearthed from among them. The chronologic age has been dated by $C^{14}$ method as 36,000 y. or even older.

The existent species of genus of Bibos are now living only in the tropical or subtropical areas south of the Changjiang River. And the discovery of Bibos and liana in Late Pleistocene deposit of Nyingchi Basin shows that the climate there was then warm and moist.

As analysed by Chen Wan-yong by means of electronmicroscopic slices, the pale-green, fine sand-stones bearing fossils are made up of hydromica-kaolinite-ancudite. This composition could be produced only under subtropical wet and hot weather conditions now prevailing in areas south of the Changjiang River. At present, the altitude of the Nyingchi Basin containing the Bibos and liana is about 2,900 metres a.s.l. yet, it might have been 1500-2000 metres a.s.l. in Late Pleistocene. Therefore, the climate in the Nyingchi Basin was probably not so dry at that time as it is today.

## PERIGLACIAL CLIMATIC MAMMALIAN FOSSILS FROM ABA AREA OF XIZANG

In early 1954, the Laboratory of Vertebrate Paleontology received some mammalian fossils, collected from the vicinity of Sogzang Si in the Aba Autonomous Prefecture of Xizang on the southern bank of the upper reaches of the Huanghe River.

These fossils include a rhinoceros skull, two upper molars belonging to the same animal and a few molars of Equus.

After a study of them in 1959, Professor Zhou Ming-zhen (Chow Min-chen) pointed out, "The rhinoceros materials belong to the *Coelodonta antiquitatis*, and the *Equus* materials to the *Equus* (Hemoinus) sp. The geological age of the fossil bearing beds is regarded as Late Pleistocene" (1)

The discovery is particularly significant because the locality where the fossils were found is situated over 3,000 metres a.s.l. at the southernmost point of the area where the species were distributed and where the climate was then cold.

Up to the present, *Coelodonta*-*mammuthus* fauna-bearing strata have been only found in areas 50-500 metres above present sea level in north China. If we presume that the strata bearing *Coelodonta*-*mammuthus* fauna in north China are located 50-500 metres a.s.l. at present, then the altitude of the Aba area with such strata has been raised 500-1,000 metres since Late Pleistocene.

## FIRST DISCOVERY OF THE FOSSIL *BOS GRUNNIENS* FROM QINGHAI-XIZANG PLATEAU

Some *Bos* and *Equus* fossils were discovered in the sands and gravels near Yangbajain in Xizang in 1975. The geological age of the fossil-bearing beds belongs to Late Pleistocene. The fossil *Bos* is evidently quite close to the recent species, *Bos grunniens* of Xizang, as described by Linnaeus in 1766. Therefore it is very probable that it was the ancestor of living *Bos grunniens* of the present day.

The fossils, though insufficient, were discovered for the first time in Xizang. Most *Bos grunniens* now live in cold areas with high elevations. So from ecology of living *Bos grunniens* we might conclude that the fossils of *Bos* remains indicate that the climate on Qinghai-Xizang Plateau was cold in Late Pleistocene.

## THE CLIMATE AT THE NEOLITHIC SITE OF KARUB IN XIZANG

A Neolithic site at Karub near Qamdo dated 4,690±130 y.b.p., was found by the archaeological team of Xizang Autonomous Region in 1977. The animal bones found there have been identified preliminarily as follows:

- *Macaca*
- *Lepus* sp.
- *Ochotona* sp.
- *Rattus* sp.
- *Marmota himalayana*
- *Vulpes* sp.
- *Sus* sp.
- *Hydropotes inermis* Sm.
- *Cervus elaphus* L.
- *Capreolus capreolus* L.
- Bovidae indet.
- *Procapra picticaudata* Hodgson
- *Naemarhedus garal* Hardwicke
- *Capricornis* sp.

From the list we can see that most of them were animals of artiodactyla.

An element of this fauna, the existent species of _Capreolus capreolus_ is now living mainly in areas in the north-eastern, northern and north-western parts of China, but is not mentioned so far in the chronicles of the Xizang region. The discovery of _Capreolus_ at this Neolithic site might offer proof that the climate there then was the same as that of northern part of China.

Living species of _Hydropotes inermis_ and the _Macaca_ are now found distributed in provinces in the Changjiang River basin, especially in the low-altitude areas of these provinces. The _Macaca_ is found in the tropics and subtropics; _Hydropotes inermis_ live by rivers and lakes where bulruchs flourish; _Macaca speciosa_ dwell in dense forests or on rocky cliffs. The discovery of bones of these animals at the Karub site suggests that there might have existed rivers, lakes or swamps and the climate there then might have been warmer and wetter than it is at present.

_Capricornis_ and _Cervus elaphus_ also like to live in a rather warm environment.

There are _Falco_ and _Anser_ on the list of the unearthed animals. Together with _Hydropotes inermis_ that prefer waterside as their habitat, they give further proof that there were then rivers or lakes around Karub site.

The spores and pollens found in this site are as follows:

- _Polypodium nipponicum_ Mett
- _Pteris yunnanensis_ Christ
- Chenopodiaceae
- Artemisia
- Compositae
- Gramineae
- Selaginellaceae

Some of these, such as _Polypodium nipponicum_ and _Pteris yunnanensis_ etc. grow today only in the tropics or subtropics.

After an intensive study of the climatic changes in China over a period of 5,000 years, Zhu Ke-zhen (Chu Ko-chen) said that 5,000 to 3,000 years ago, in other words, from the Age of Yangshao Culture to that of the Yin Dynasty at Anyang, the climate was warm in China. The discovery of the Karub site, Qamdo county, further proves that this warm stage (Thermal Maximum) was not accidental but a common phenomenon in prehistorical China.

## DISCOVERY OF MICROLITHIC FROM THE NORTHERN XIZANG PLATEAU

While making a geological survey of the Northern Xizang Plateau, the Scientific Expedition to the Qinghai-Xizang Plateau discovered 12 sites bearing some flaked stone microlithic. All the stone microlithic were collected from the beaches of the Siling Lake. As neither pot-shard nor polished stone implements was found with them, it is certain that they are microlithic. What is interesting is that the nature of the culture is different from Indian "Madras Culture", but is similar to that in northern China.

At the present day Siling Lake, where the microlithic was found, lies at an elevation of some 4,800 meters a.s.l., a height that is difficult for human beings to live on It seems that the Northern Xizang Plateau was uplifted after the microlithic epoch, or tremendous changes had taken place in its climate. The warm climate had been growing very cold and people began to move from Siling Lake to the low land.

This is the first time that microlithic has been discovered on such a high plateau in western China, and this opens the way for researches into human fossils on the Qinghai-Xizang Plateau.

## EFFECT OF QINGHAI-XIZANG PLATEAU UPLIFT ON MAMMALIAN EVOLUTION

Qinghai-Xizang possesses a peculiar landscape. The mammalians living here are mainly of the alpine-prairie type. According to the data supplied by Feng Zuo-jian in 1979, there are such special kinds of mammalian on the plateau as:

*Pantholops hodgsoni*
*Marmota himalayana*
*Cervus albirostris*
*Ochotona ladacensis*
*Lepus oiostolus*
*Cricetulus kamensis*
*Pitymys leucurus*

The last, *Pitymys*, are typical. Most scholars were of the opinion that *pitymys* might be divided into three subgenuses:

*Pitymys*
*Neodon*
*Phaiomys*

*Pitymys* subgenus is distributed mainly in Europe and north America; *Neodon* subgenus, in Kazakhstan of USSR, Nepal, Sikkim and the Hengduan Mountains of China; *Phaiomys* subgenus, on the Qinghai-Xizang Plateau and nearby Kashmir. *Pitymys* which is considered a subgenus endemic of the Qinghai-Xizang Plateau, is widely distributed over the meadow or plateau-prairie, with an elevation of 3,500-4,500 metres. They adapted themselves well to high altitudes and cold areas, but their ancestors had used to live in a warm and damp climate. Along with the uplift of the Plateau during the Quaternary, the climate was getting from warm and damp to cold and dry, and the forests turned into grasslands. They gradually adapted and acclimatized themselves to the circumstances and, through natural selection, they formed their own system different from their ancesters, and became new members in the region.

## CONCLUSION

Let us make a general review of the above-mentioned points. On the Qinghai-Xizang Plateau, animal fossils of the Late Pleistocene Age include several species. One of them, *Bibos* sp. is now found only in south China. When *Bibos* lived there, the climate of the Qinghai-Xizang Plateau was probably not so dry and cold as it is now.

On the other hand, the archaeological finds in the Northern Xizang Plateau and the Aba Area prove that the climate there was warm. However, the Plateau has now become an uninhabited place for humans with its cold climate.

Both mammalian remains and archaeological finds also show that the climate of the Qinghai-Xizang Plateau since the Late Pleistocene has changed from warm to cold several times. This is in conformity with the history of uplift and climatic fluctuation of the great plateau during the same period.

## References


(1) Chow Min-chen, Pleistocene mammalian fossils from the eastern border of the Xizang Plateau. Quaternaria Sinica, Vol. I, 8-13 (1959). (in Chinese).

(2) Falconer, H., On the fossil rhinoceros of Central Xizang. Paleontological Memoirs and Notes, 173-185 (1868).

(3) Huang Wan-po, Note on Holocene Gazo Site of Qamdo, Xizang. Vertebrate Palasiatica 18 (2) 163-168 (1980). (in Chinese)

(4) Huang Wan-po, The Quaternary mammalian fossil localities from Xizang. Palaeontology of Xizang, Book I, 48-53 (1980). (in Chinese)

(5) Zhang Shen-shui, New discovery of the Microlithic materials from North Xizang. Palaeontology of Xizang, Book I, 70-76 (1980). (in Chinese)

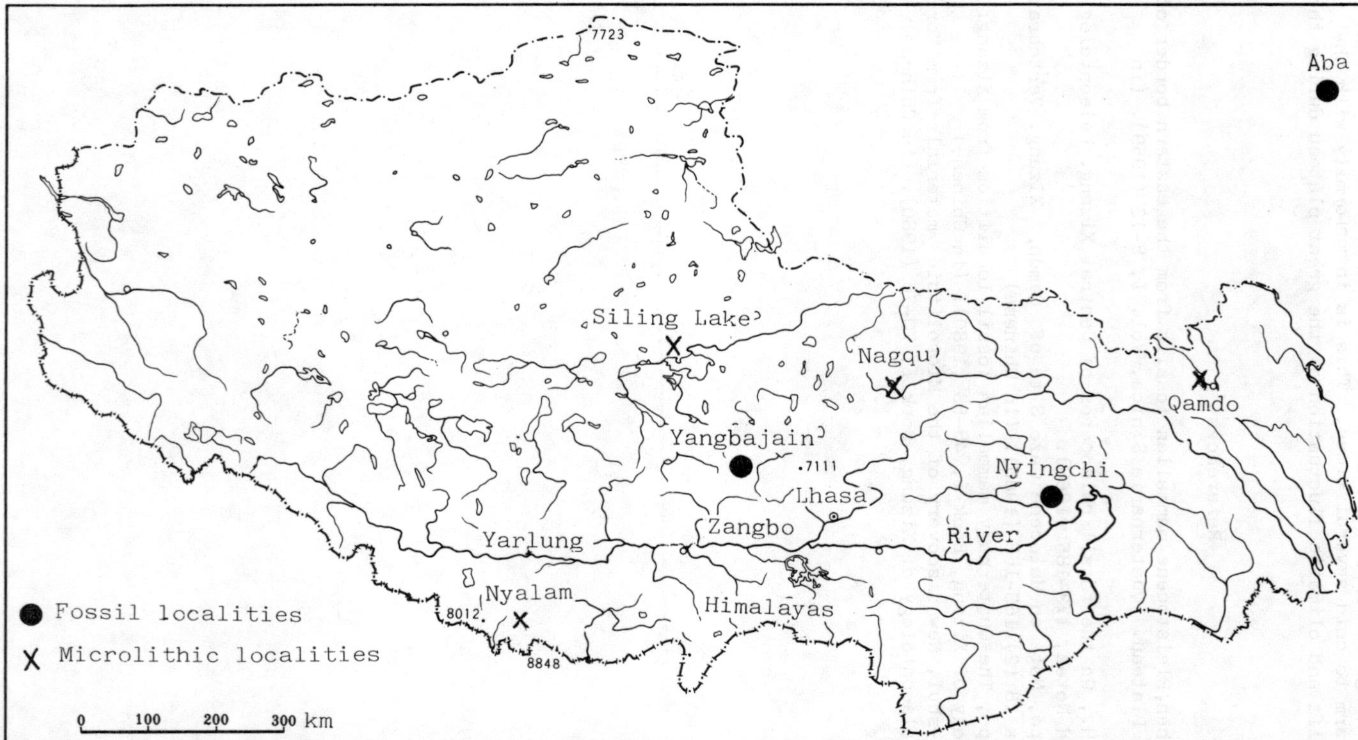


Fig. 1 Mammalian fossil and microlithic localities of Late Pleistocene and Holocene in Qinghai-Xizang Plateau

# THE PLIOCENE AND QUATERNARY ENVIRONMENT ON THE QINGHAI-XIZANG PLATEAU

Wang Fu-bao
(Department of Geography, Nanjing University)

Li Bing-yuan Zhang Qing-song
(Institute of Geography, Academia Sinica)

This paper is based on the data which were collected in the last few years from South and East Xizang. The authors attempt to divide more detailedly the strata of Pliocene and Quaternary between Himalayas and Kunlun Mountains, and make a preliminary analysis on the evolution of geographical environment during this period.

## DESCRIPTION OF PLIOCENE AND QUATERNARY STRATA

### The Pliocene Series

Distribution of the Pliocene system may be divided into two areas, i.e. the south and north. In the south, the Pliocene series are well-developed in the areas of Qusum, Pagri, Tingri, Yagru Xongla, Xixabangma, Gyirong, Burang and Zanda; while in the north they are well-exposed in the areas of Nagqu, Bulong, Qeling-Loinbo La and Quguo, etc.

The Jilong group and Zada group of the Pliocene epoch are best exposed in the south, and both of them consist mainly of sediments of lake facies. The Jilong group may be divided into three parts (Fig. 1). The lower part known as Zong Ga formation consists of grey, greyish yellow conglomerates, intercalated with thin layers of brown coal containing several kinds of pollen including *Riccia*, *Onychium*, *Cedrus*, *Pinus*, etc. This part is over 50 meters thick and is unconformably overlying the Jurassic series.

The middle part known as Woma formation (Fig. 2, Photo 1) consists of greyish yellow mudstone, sandstone and red ferrous sandstone. The middle section of this part is about 60-90 m thick and contains the *Hipparion* fauna (1) and Chenopodiaceae, Polypodiaceae, *Ephedra* pollen.

The upper part known as Gongba Nanmulin formation consists of greyish yellow mudstone and sandstone, which contains at the lower section the pollen mainly of *Cedrus*. More Chenopodiaceae, Polypodiaceae, *Pinus*, *Picea* were found in the sporo-pollen assemblages at the top of the upper part (Zheng Yahui, 1979). This layer is about 150 meters thick.

The Zada group which was found in Zanda basin, is about 800 meters thick, (Photo 2). This layer consisting of greyish yellow mudstone, sandstone and conglomerates contains several kinds of fauna fossils of the *Paleotragus microdon* at its alower section.

In the north of the Plateau, the lower part of the Pliocene system is known as the Bulong formation, whereas the middle and upper parts as the Qugu formation. The Bulong formation about 50 m thick, being similar to Chinji formation of Siwalik group in south Asia, also contains the Hipparion fauna (2). The Qugu formation over 370 m in thickness consists alternately of brown-greyish clay and siltstone, sandstone. Its paleomagnetic date belongs to the Gilbert and Gauss epoch (Qian Fang 1978).

## The Pleistocene Series

1. The lower limit of Quaternary. According to the facies, lithological characters and fossils of Dati formation and Gongba conglomerate in the southern slope of the Yagru Xongla are all different (Fig. 3). The fossils in Dati formation contain Hipparion and others. The age belongs to the upper Pliocene, the palaeomagnetic date Gauss normal epoch. As the palaeomagnetic date of Gongba conglomerate that covers Dati formation belongs to Matuyama reversed epoch, the age belongs to the lower Quaternary. Because the contact between Dati formation and Gongba conglomerate is uncomformable, we take the lower limit of the Matuyama epoch as the base of the Quaternary, its age is about 2.43 million years.

2. The early Pleistocene system series. On the northern slope of the Himalayas and in the lake basins of the Pliocene, there often exposed are the layers of conglomerate, as thick as 150-500 m, known as the Gongba conglomerate (Photo 3). This is a set of conglomerate strata consisting mainly of limestone and quartz sandstone with calcium cementation. In the north Xizang the early Pleistocene sediments known as Qiangtang formation consists mainly of lacustrine deposits, such as clay and silt sandstone, with a total thickness of about 200-500 m. According to the pollen analysis made by Kong Zhao-chen, the strata from bottom upwards can be divided into three cycles: Pinus, Picea, Betula, Ulmus—Chenopodiaceae, Ephedra to Picea, Pinus—Chenopodiaceae to Betula, Quercus, Picea—no sporo-pollen or Chenopodiaceae (3,4). The Gongba conglomerate and Qiangtang formation belong to Matuyama reversed epoch, so they are included in the early Pleistocene.

3. The middle Pleistocene system. The differences between the early Pleistocene and the middle Pleistocene are marked in lithological characters and stratum distribution. At the mountain terrain, the middle Pleistocene consists mainly of glacial and lacustrine deposits. The lacustrine deposits in North Xizang are often found in the present lake basins. The glacial deposits in South Xizang were formed during two former glacial periods, and during the interglacial ages the weathering crust of red earth and paleosol were formed (Fig. 4).

4. The late Pleistocene series. According to the glacial tills, fluvioglacial deposits and lacustrine deposits the late Pleistocene can be subdivided into 2 or 3 glacial periods. Loess deposits formed during Würm glacial period can be seen at the piedmont.

## The Holocene Series

The lower limit of the Holocene was formed about ten thousand years ago. In the south, thick layars of diatom earth deposits can be found in many lake basins.

In addition, in the old broad valleys and lake basins of the south terrain Holocene part usually developed 9,970-3,050 years ago (Fig. 5).

## DISCUSSION ON THE ENVIRONMENTAL CHANGES SINCE PLIOCENE

Owing to the uplift of the Qinghai-Xizang Plateau, the effect of the global climatic changes on the environments of the Plateau has been strengthened. As a result, the range of the climatic changes on the Plateau is greater than that on the surrounding plain. The difference between them become even more obvious since the Holocene.

The climatic change in the Pliocene may be divided into three stages. In the early Pliocene, the climate in all parts of the Plateau is represented by tropical forest. In the valleys of the Himalayas, for example, there were *Quercus*, *Castanopsis*, *Platanus*, *Podocarpus*, *Humata trigoliata*, *Pseudodrynaria* (Cao Liu, 1979) etc. The *Hipparion* fauna of forest type dwelt there too. In the middle Pliocene, there was obvious difference of the climate in all parts of the plateau. For example, in the south Xizang from early to late Pliocene the vegetation turned from subtropical forests to tropical sparse steppe successively, while in the west and north Xizang there existed the vegetation of forests and steppes of warm temperate zone. In the late Pliocene, the climate of subtropical forests appeared again at the northern slope of middle Himalayas. Later on, however, it became the climate of steppe successively. In the west and north there appeared alternately the climate of theropeucedrynion consisted of *Picea*, *Abies* and *Betula*, and the climate of steppe.

Since Quaternary, the climate differentiation in space and the cyclic changes from cold or dry to warm or humid were even more apparent than that in the Pliocene. Moreover, the climate was constantly becoming colder and drier.

In the Pleistocene, large areas of northern part of the plateau were full of water. The climate underwent thrice alternation of the forest-steppe and wilderness. But it was still warm on the whole.

In the southern part of the plateau, rivers began to cut down. The Pliocene lake basins disappeared. It seems that climate in the south was drier than in the north.

After the dissection and uplift in late middle Pleistocene, the plateau formed in the middle Pleistocene was similar to the outline of the present one. The mountainous region with a height above the zone of maximum precipitation, received yet a lot of precipitation. As a result, the glaciers reached its maximum scale and occurred twice since Quaternary. In the interglacial age, however, the temperature of the plateau increased at great speed. In the valleys of the Nyainqentanglha Range and the southern Gandise Range there developed fully red earth weathering crust. In the late Pleistocene, the situation of atmospheric circulation changed as a result of the strong uplift of the plateau. During the glacial period with the decreasing of precipitation, glaciation prevailed, while the cryopedological process developed widely. The plateau, thus, presented in a severe climate similar to that of polar regions. The climatic changes were very noticeable since the Holocene. According to the data obtained from Chagcam Caka Lake, wet period appeared thrice and dry period four times in the past 10 thousand years. The wet periods, appeared respectively 10,000, 8,000-7,000 and 6,000-5,000 years ago. Between two wet periods was a dry period. The climate changed rapidly since 3,800 B. P. The water level of Zhari Nam Lake in 7,000 ± 150 years B. P. was about 22 m higher than it is today. It was a freshwater lake. The lakeside was covered with forests consisted mainly of *Pinus* and *Betula*. But it is a salt lake now (Huang Ci-xuan, 1979).

On the northern slope of the Himalayas in southern Xizang there were many places along the lakesides covered with forests 3,000-4,000 years ago, the Yamzho Lake was still a lake with outlet. From that time the climate became colder and drier rapidly, cold period appeared thrice on the plateau. In $C^{14}$ 2,980±150, 1,540±85 , 1,920±110 years B. P. and 17th-19th centuries it is the New Ice-age. There are evidences left by the water level showing that the lakes lowered rapidly. Many lakes for example, the Yamzho Lake, became inland lakes. In many broad valleys, for example, the Yarlung Zangbo River Valley, desert landscapes developed extensively. Forests disappeared from inner part of the plateau. It is estimated that the temperature in the middle Holocene was over 3-4°C higher than that of today. The temperature in the New Ice-age, however, reduced 4-5°C.

## References


(1) Ji Hung-xiang et al. The Hipparion fauna from Gyirong Basin, Xizang, (Palaeontology of Xizang) (Book I). The Series of the Scientific Expedition to Qinghai-Xizang Plateau, Science Press (1980).

(2) Zheng Shao-hua, The Hipparion fauna of Bulong Basin, Biru, Xizang, (Palaeontology of Xizang) (Book I). The Series of the Scientific Expedition to Qinghai-Xizang Plateau, Science Press (1980).

(3) Tang Ling-yu et al, Spore-pollen assemblage and significance of the 203 m borehole in the Qingshui River across Qinghai-Xizang Highway. Acta of Lanzhou University (Natural Sciences) No. 2 (1976).

(4) Tang Ling-yu et al., Spore-pollen assemblage and significance of the Quaternary lacustrine deposits in the Kunlun Mountain Pass Basin, Qinghai. The Series of Lanzhou Institute of Glaciology and Cryopedology, Academia Sinica, Science Press, No. 1 (1976).

Note: All references are in Chinese.

Table 1 Pliocene and Pleistocene stratigraphic environmental changes on the Qinghai-Xizang Plateau

| Stage | Strata: South Xizang | Strata: North Xizang | $C^{14}$(B.P.) Pollen and Paleovertebrates | Plaeoclimatic types | Temperature fluctuation (°C) Low | Temperature fluctuation (°C) High |
|---|---|---|---|---|---|---|
| Holocene | Present deposits | 4th Evaporite | | | | |
| Holocene | Rongbudesi moraine | 3rd Evaporite | 1540±85-1920±110 2980±150 | Steppe-Desert | 4-5 | |
| Holocene | Nari Lake lacustrine deposits | 1st and 2nd Evaporite | 3050±100-9970±150 | Forest-Steppe Steppe | | 3-4 |
| Late Pliestocene | Sipangoer diatom earth 20 m | | 11700±180 23500±1200 | Steppe-Desert | 9-10 | |
| Late Pliestocene | 3rd Palaeosoil | | | Steppe, Forest | | 7-9 |
| Late Pliestocene | Pengou lacustrine layer 100 m | | *Moschus moschifekus* *Equus hemoionus* | Steppe-Desert | 7-9 | |
| Middle Pliestocene | 2nd palaeosoil | | | Forest, Forest-Steppe | | 10-13 |
| Middle Pliestocene | Fuqu moraine | | | Steppe, Forest-Steppe | | |
| Middle Pliestocene | 1st palaeosoil | | *Picea*, *Quercus* | Forest | | 10-13 |
| Middle Pliestocene | Nyalam moraine 200 m | Kunlun gap Moraine 20-100 m | | Forest-Steppe Steppe | | |
| | Unconformity | Unconformity | | | | |
| Lower Pliestocene | Gongba conglomerate 150-500 m | Qiangtang formation 200-500 m | Chenopodiaceae-*Betula*, *Quercus*, *Picea* Chenopodiaceae-*Picea*, *Pinus* Chenopodiaceae, *Ephedra*-*Picea*, *Pinus*, *Betula* | Forest-Steppe alternation | | |
| | Unconformity | Unconformity | | | | |
| Pliocene (Jilong group) | Congbananmulin formation 150 m | Quguo formation 3,700 m | *Picea*, *Pinus*, Polypodiaceae *Cedrus* | Subtropical forest | | 16-18 |
| Pliocene (Jilong group) | Woma formation 60-90 m | | Woma *Hipparion* fauna Chenopodiaceae, *Ephedra* | Subtropical forest steppe | | |
| Pliocene (Jilong group) | Zongga formation 50 m | Bulong formation 50 m | Bulong *Hipparion* fauna *Riccia*, *Onychium*, *Cedrus* | Tropical or Subtropical forest | | 18-20 |

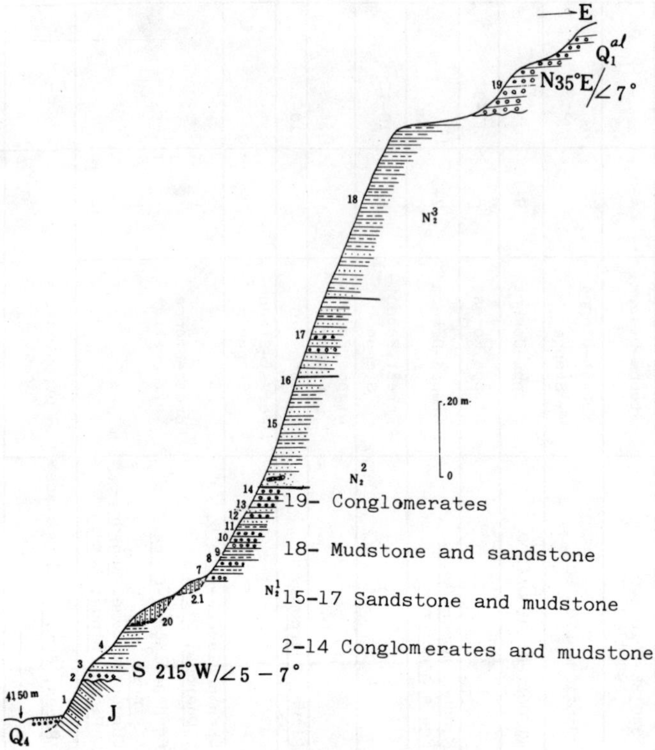


Fig. 1 The section of the Jilong group in the eastern part of the Gyirong county

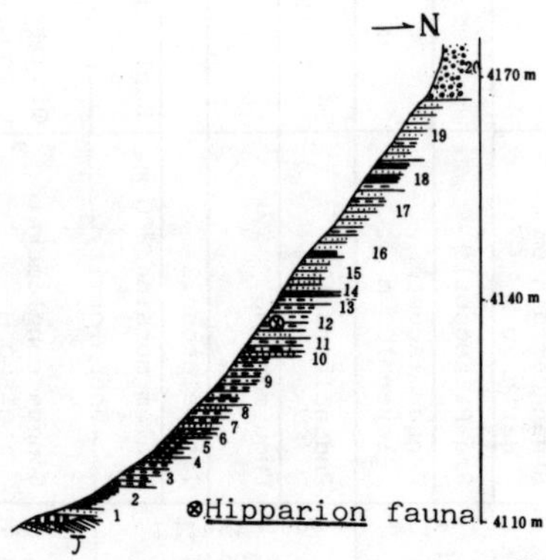


Fig. 2 The section of Woma formation in the southern Gyirong basin

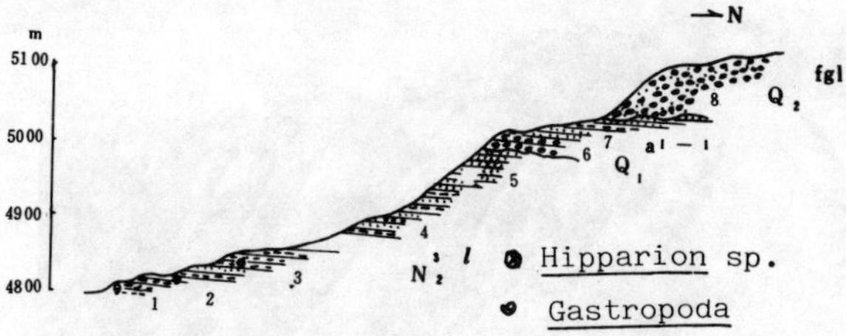


Fig. 3 Schematic profile of the southern slope of Yagru Xongla

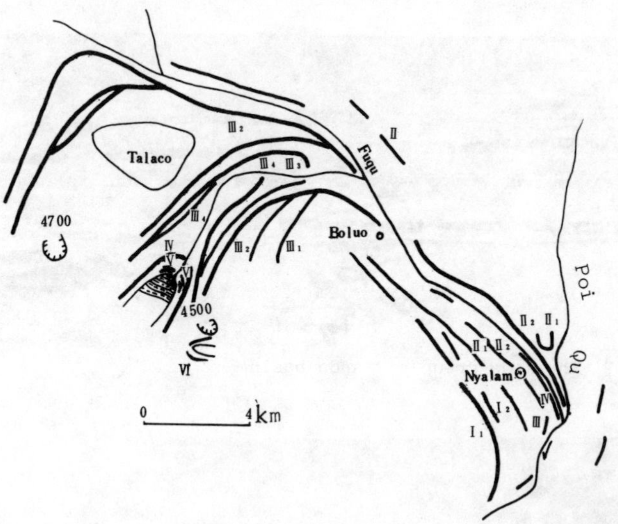


Fig. 4 The sketch map of the Quaternary glaciation of the low reaches of the Fuqu River at the vicinity of Nyalam county

I--Nyalam glacial age ($Q_2^1$)

II--Fuqu glacial age ($Q_2^2$)

$III_{1-2}$--Boluo glacial age ($Q_3^1$)

$III_2$--Jilongsi glacial age ($Q_3^2$)

$III_{3-4}$--Rongbusi glacial age ($Q_3^3$)

IV--New ice-age

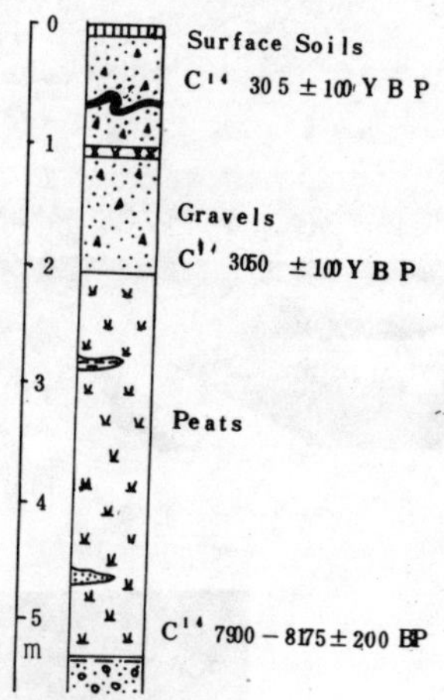


Fig. 5 Holocene section of Qilongduo at Yangbajain

Photo 1 The Woma formation in the southern Gyirong basin

Photo 2 The Zada group at Zanda basin

Photo 3 The Gongba conglomerate at the southern slope of Yagru Xongla

# PRELIMINARY STUDY ON THE VEGETATION OF THE QINGHAI-XIZANG PLATEAU DURING NEOGENE AND QUATERNARY PERIODS*

Kong Zhao-chen, Du Nai-qiu
(Institute of Botany, Academia Sinica)

Modernization of the flora and vegetation in Qinghai-Xizang Plateau is not only determined by the evolutionary succession of plants themselves, but also effected by the change of environment conditions as the result of collision of the drifting India Plate against the Eurasia Plate, which has made the mountain ranges of Gandise, Himalayas, Tanggula, Kunlun and the Qinghai-Xizang Plateau itself uplifted, the Tethys Sea disappeared and the climatic and soil conditions altogether changed since the beginning of Neogene.

In the south-western Xizang at the foot of the southern slope of the Gandise Range, there a coal series in the locality Moincêr at an altitude of 5,140 m in Zanda of Ngari Prefecture has been found. From the shales of the coal series Eucalyptus spp. and Phrynium sp. were obtained. The flora probably represents a tropical flora of lower altitude about 1,000 m in elevation. This flora is tentatively assigned by paleobotanists to late Cretaceous-early Eocene in age.

In the Lumpola basin of the northern Xizang and Qaidam basin in Qinghai during Oligocene, subtropical-temperate floras were distributed(1).

That how the vegetation on the Qinghai-Xizang Plateau has been developed since Miocene can be seen in the sketch map of the distribution of plant fossil localities on Qinghai-Xizang Plateau during Neogene and Quaternary (Fig. 1).

Almost all the sites where fossil plants were found are in the alpine steppe meadow and cold desert at an elevation of 4,000-6,000 m. Fig. 2 shows the geological section of the Kunlun Mts. during late Miocene or early Pliocene. In the palynological investigations, pollen or spores of Tsuga, Melia, Carya, Ilex, Quercus, Lithocarpus and Proteacites and ferns have been found in the sediments. (Plate I).

Sporo-pollen assemblages of late Miocene or early Pliocene obtained from the Kunlun Mts. show some subtropical broad-leaved deciduous forests and needle to broad-leaved mixed forests were flourishing there(2). The climate was wet and warm then. Ferns, such as Drynaria and Lepisorus, were epiphytic or growing under trees. At that time, the vertical distribution of vegetational belts was not clear. Judging from the vegetational changes from late Miocene to early Pliocene, due to the uplift the altitude of the Qinghai-Xizang Plateau at that time was not less than 2,000 m in elevation(2, 3).

---

* The authors wish to express deep thanks to Prof. Xu Ren(Hsü Jen) under whose guidance and encouragement the present paper is prepared.

At the time of late Pliocene Cedrus (C. deodara) and Quercus (Q. semecarpifolia) forests were widespread in the Himalayas(4). During late Pliocene Abies-Picea forests were well developed in the Kunlun Mts. Tanggula Range, Qaidam Basin, Ngari Prefecture, and northern Xizang. By that time, the Plateau might have uplifted up to 3,000 m above sea level, along with falling in temperature, which were effected by the atmospheric circulation and topographic changes. As for floristic changes, deciduous forests were gradually decreasing and xerophytic plants (Ephedra, Nitraria, etc. ) were gradually increasing in the Kunlun Mts. Plate II shows some typical xerophytic plants, Nitralia, Ephedra, etc. On the southern slope of the Himalayas, Quercus, Cedrus forests were widespread with undergrowing herbs mainly of Caryophyllaceae. In the northern Xizang and Qaidam Basin there were deciduous forests, mainly of Betula, Corylus, etc. By that time the horizontal and vertical zonation of plants were better distinguished.

As a result of uplift of the Plateau and falling in temperature during the Quaternary period, vegetation changed rapidly in the glaciation and interglaciation periods of the early Pleistocene. Alpine steppe and forests alternately developed.

During the glaciation period of early Pleistocene, only herbs and shrubs could have grown in the Qaidam Basin, Ngari Prefecture and northern Xizang, though some Abies-Picea forests might have migrated down to the valleys. Plate III shows some typical plants in the glaciation period of the early Pleistocene, such as Artemisia, Chenopodium, Compositae etc.

The flora of the interglaciation period of early Pleistocene was mainly composed of Pinus, Betula, Carpinus, Quercus, Alnus, and some subtropical alpine trees, such as Cedrus. Meanwhile the aquatic plants such as Potamogeton, Sparganium, Typha, Nelumbo and Green alga, Pediastrum boryanum, P. duplex grew in some shallow, stagnant ponds, lakes or marshes of the Qaidam Basin, the Kunlun Mts. and northern Xizang (Plate IV).

At that stage, the vegetation was represented by the Zanda flora found in the district Zanda, at an altitude of 4,200 m a.s.l. Now, the Zanda basin is situated at 31.4°N, 79.8°E. The climate is cold and dry. Annual mean temperature is about 1°C, and the annual precipitation is 100 mm. The vegetational zone belongs to typical alpine desert steppe. It is mainly composed of Compositae, Gramineae, Leguminosae, Cruciferae, Cyperaceae, Chenopodiaceae, Ranunculaceae and Rosaceae. Gymnospermae are very poorly represented, only Ephedra and Sabina present. Ferns are even scanty. Based on the sporo-pollen assemblages we have seen that during the interglaciation period of the early Pleistocene alpine coniferous deciduous broad-leaved forests were predominant(5). Meantime, fern plants such as Pteris, Polypodium, and Lepisorus grew under the forests. By that time the climate was warm and moist. Their correlation has been made with pollen zones by other authors who have published pollen diagrams of the Kashmir and Pamirs (Hsü Jen 1958, Pahomov, M. M., 1971). According to the information mentioned above, we consider that either Himalayas or Kunlun Mts. in the early Pleistocene did not obstruct air current coming from the Indian Ocean and their floras could be compared between one another.

During the interglaciation period of the middle Pleistocene and the late Pleistocene, although some Picea forests began to decay in Ngari Prefecture, northern Xizang and the Qaidam Basin, yet the Picea forests were still flourishing in Gyarag of the southern Xizang.

Up to the late Pleistocene and Holocene, the vegetation changed into alpine steppe or alpine meadow just as those found in Kunlun Mts.,Ngari Prefecture and northern Xizang at present. But in the eastern and southern Xizang the forests have been preserved even up to the present time. Some broad-leaved forests still existed in the eastern Xizang, and bushes of *Rhododendron*, *Viburnum* and *Lonicera* were flourishing in Yali on southern slope of the Himalayas. It is estimated during post-glacial the elevation of Yali was about 3,700-3,900 m whereas it is 4,300 m(6) at present.

According to the result of analysis of the fossil plants and spore-pollen assemblages, some conclusions can be drawn:

(1) The Miocene or early Pliocene flora was mainly composed of *Quercus*, *Lithocarpus*, *Melia*, *Ilex*, *Tsuga* and some ferns. At that time the climate was wet and warm. The vegetation was of subtropical and warm-temperate forests. The annual mean temperature was about 10°C and the annual meen precipitation was about 700 mm. The Kunlun Mts. was above 2,000 m in elevation.

(2) The flora of late Pliocene on the southern slope of the Himalayas was mainly composed of *Quercus* and *Cedrus* with some subtropical to temperate, needle to broad-leaved mixed forests. The climate was wet and cool, but in the later stage was drier than before. The xerophyic plants were gradually becoming predominant in the Kunlun Mts. region. It is estimated this area was then about 3,000 m in elevation.

(3) The floras and climate of the glaciation and interglaciation periods of the Quaternary were well-distinguished. Floras in glaciation of early Pleistocene were represented by alpine-steppe. The climate was dry and cold. Floras in the interglaciation periods of early Pleistocene were mainly composed of Amentiferae, but some herbaceous plants, such as Chenopodiaceae, Caryophyllaceae and Compositae were present. Meanwhile, many lakes and marshes might have existed somewhere on Qinghai-Xizang Plateau.

(4) During the Post Glacial the annual mean temperature rose again, shrubs of *Rhododendron*, *Lonicera* and *Viburnum* were flourishing in the Yali region in southern Xizang, but in the Kunlun Mts. the flora from Holocene to the present time is quite similar to that of the Post Glacial.

As a result of comparison of palaeobotanical materials obtained on the Qinghai-Xizang Plateau with those from its adjacent areas, we come to know the rate of uplift of the Plateau and its adjacent areas, as well as the development of floras and the variation of climate during the Neogene and the Quaternary.

References

(1) Xu Ren(Hsü Jen), On the palaeobotanical evidence for continental drift and the Himalayan uplift. The Palaeobotanist,25, 131-145 (1976).
(2) Kong Zhao-chen et al., Neogene-Quaternary palynoflora from Kunlun Mts. to Tanggula Range and the uplift of Qinghai-Xizang Plateau. Problems of Era, Amplitude and Mode of the Uplift of Qinghai-Xizang Plateau, Science Press, 78-87 (1981) (in Chinese).
(3) Li Hao-min and Guo Shuang-xing, The Miocene flora from Namling of Xizang. Acta Palaeont. Sin., 15 (1), 7-18 (1976) (in Chinese).
(4) Xu Ren (Hsü Jen) et al., On the discovery of *Quercus semecarpifolia*

bed in Mount Xixabangma and its significance in botany and geology. Acta Bot. Sin., 15 (1), 103-119, Pls. 1-4 (1973) (in Chinese).
(5) Wang Rui et al., Preliminary study on the vegetational development and climatic changes in Ari prefecture since Pliocene. Proceedings of Symposium on Glacialogy and Cryopedology, Society of Geography of China (1981) (In press) (in Chinese).
(6) Xu Ren (Hsü Jen) et al., Quaternary palaeobotanical studies on the Mount Qomolangma region and Himalayan uplift, Kexue Tongbao. 18 (6), 274-279 (1973) (in Chinese).

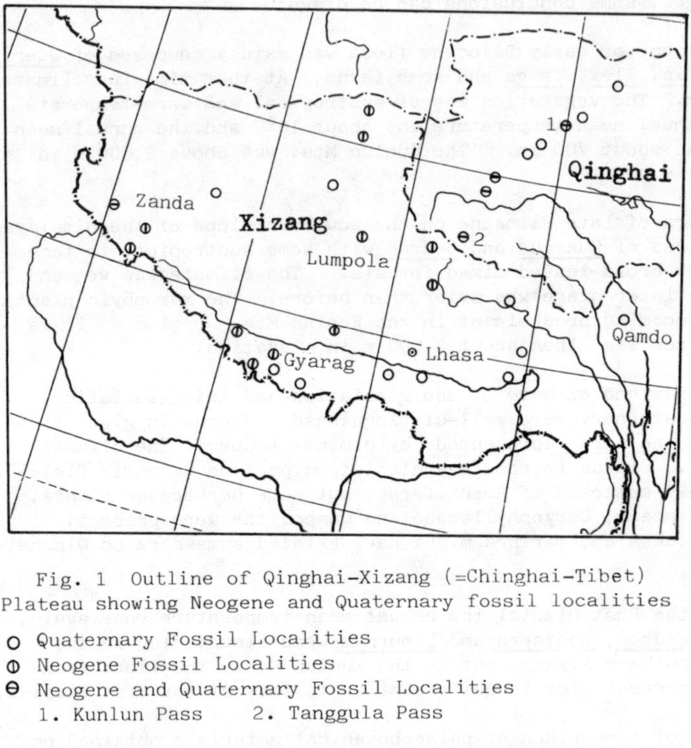


Fig. 1 Outline of Qinghai-Xizang (=Chinghai-Tibet) Plateau showing Neogene and Quaternary fossil localities

○ Quaternary Fossil Localities
⦶ Neogene Fossil Localities
⊖ Neogene and Quaternary Fossil Localities
1. Kunlun Pass 2. Tanggula Pass

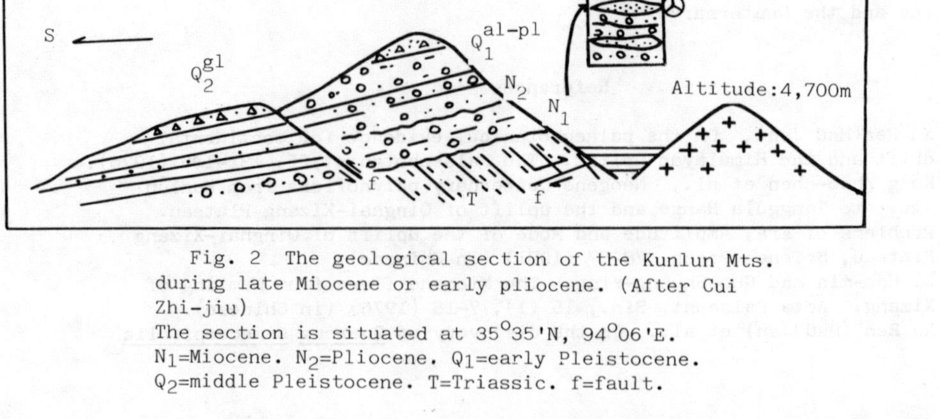


Fig. 2 The geological section of the Kunlun Mts. during late Miocene or early pliocene. (After Cui Zhi-jiu)
The section is situated at 35°35'N, 94°06'E.
$N_1$=Miocene. $N_2$=Pliocene. $Q_1$=early Pleistocene.
$Q_2$=middle Pleistocene. T=Triassic. f=fault.

Plate I

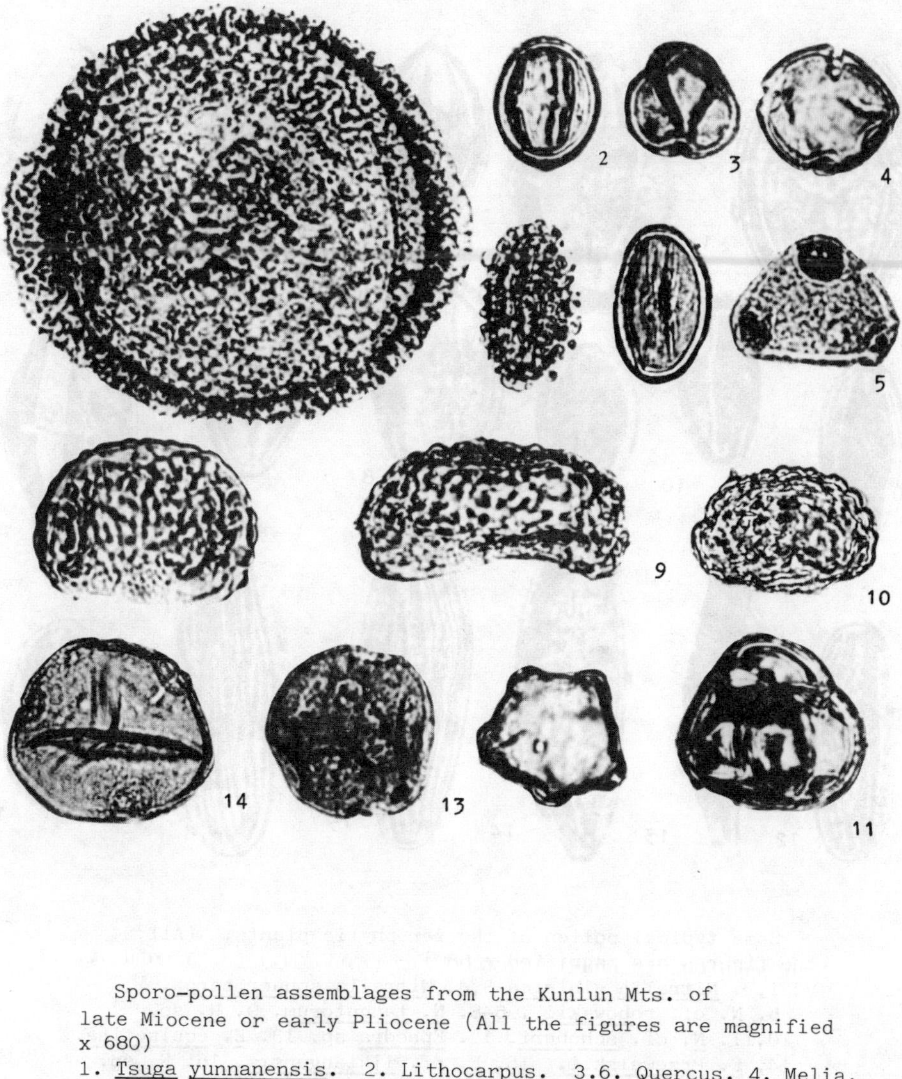


Sporo-pollen assemblages from the Kunlun Mts. of late Miocene or early Pliocene (All the figures are magnified x 680)
1. Tsuga yunnanensis. 2. Lithocarpus. 3.6. Quercus. 4. Melia. 4. Proteacidites. 7. Ilex. 8.10. Lepisorus. 9. Drynaria. 11. Rhododendron. 12. Pterocarya cf. insignis.13. Altingia cf. excelsa. 14. Fagus longipetiolata.

Plate II

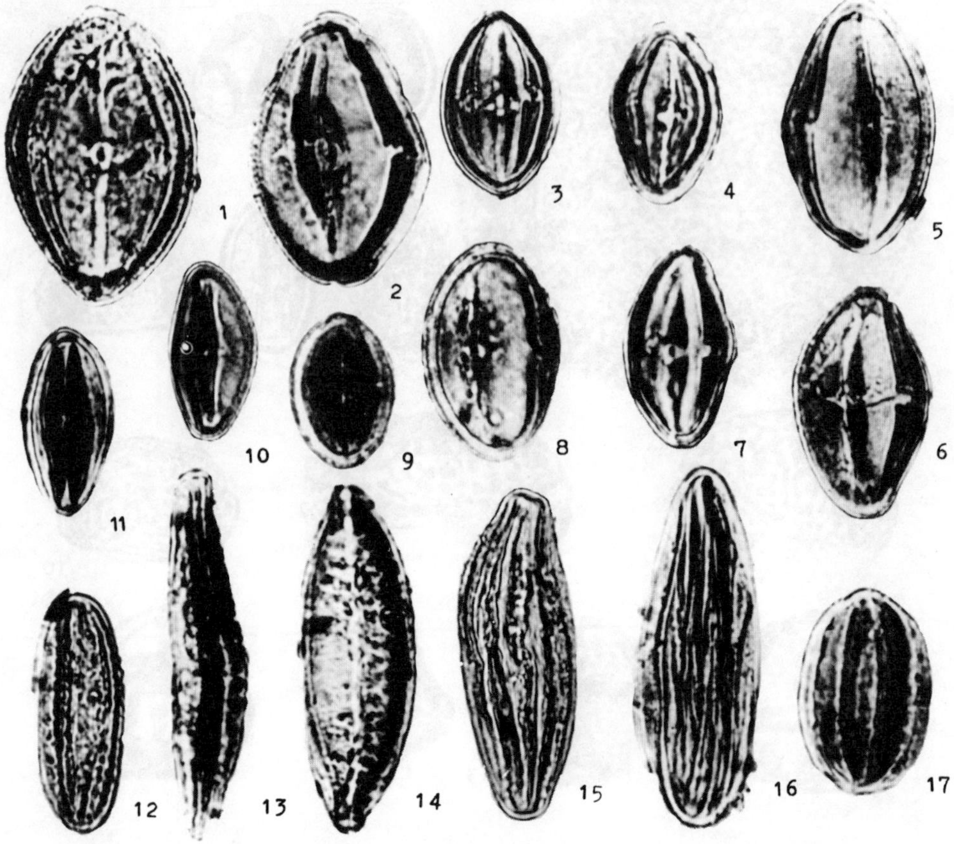

Some typical pollen of the xerophytic plants. (All the figures are magnified x 680)

1,2. Nitralia sibirica 3,4. Nitralia sphaerocarpa.
5. N. cf. robowskii . 6–8. N. tangutorum. 9. N. sp.
10,11. N. cf. schoberi. 12. Ephedra sp. 13. E. equisetina.
14. E. Przewalskii. 15. E. cf. likiangensis. 16. E. sp.
17. E. minuta.

Plate III

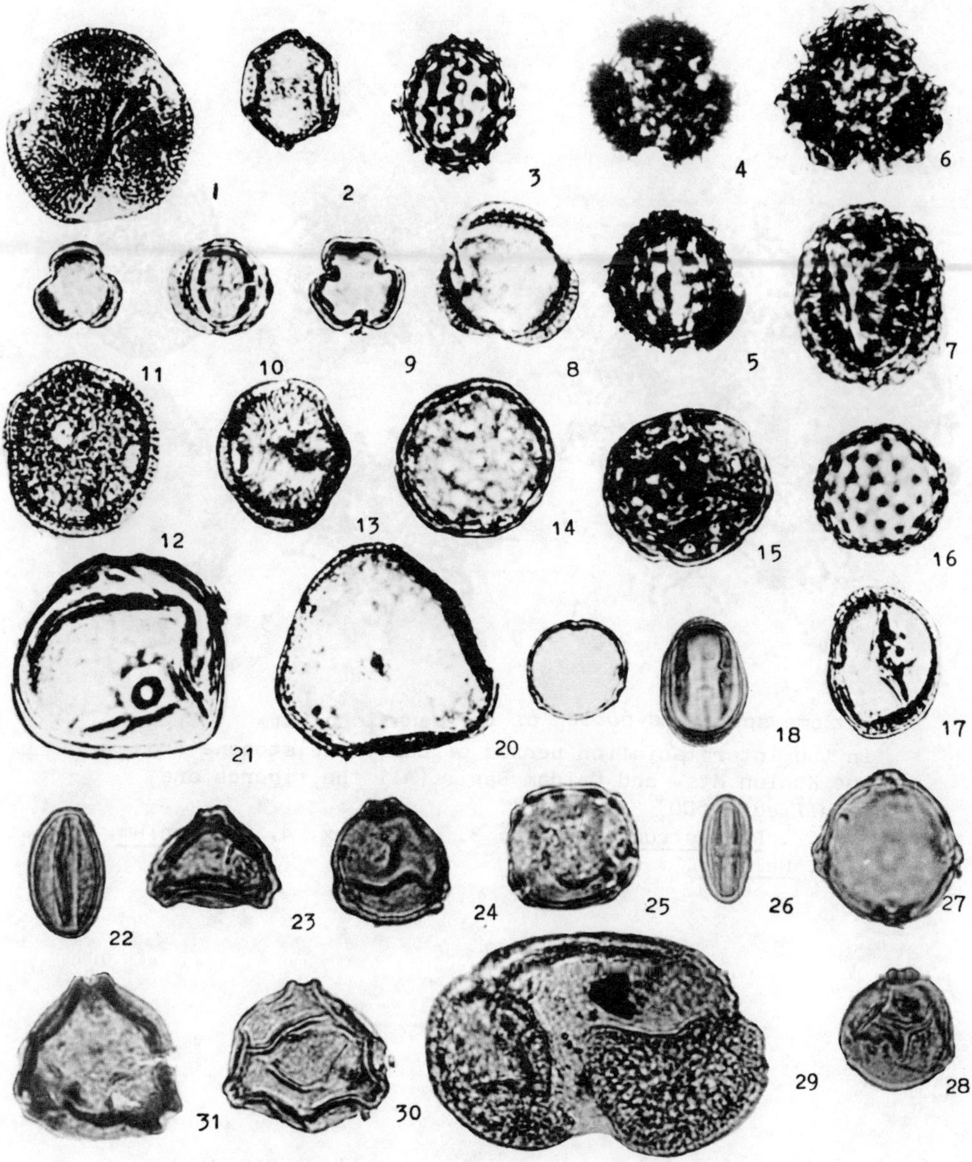


Some typical pollen of the Glaciation and Interglaciation periods from the Kunlun Mts., Tanggula Range and Himalayas. (All the figures are magnified x 680)

1. Gentiana. 2-8. Compostae. 9-11. Artemisia.
12,13. Caryophynaceae. 14-16 Chenopodiaceae.
17,18. Polygonum. 19. Humulus. 20. Cyperaceae. 21. Gramineae.
22. Quercus. 23,24,28,30 31, Betula. 25. Alnus 26. Castanea.
27. Carpinus. 29. Pinus.

Plate IV

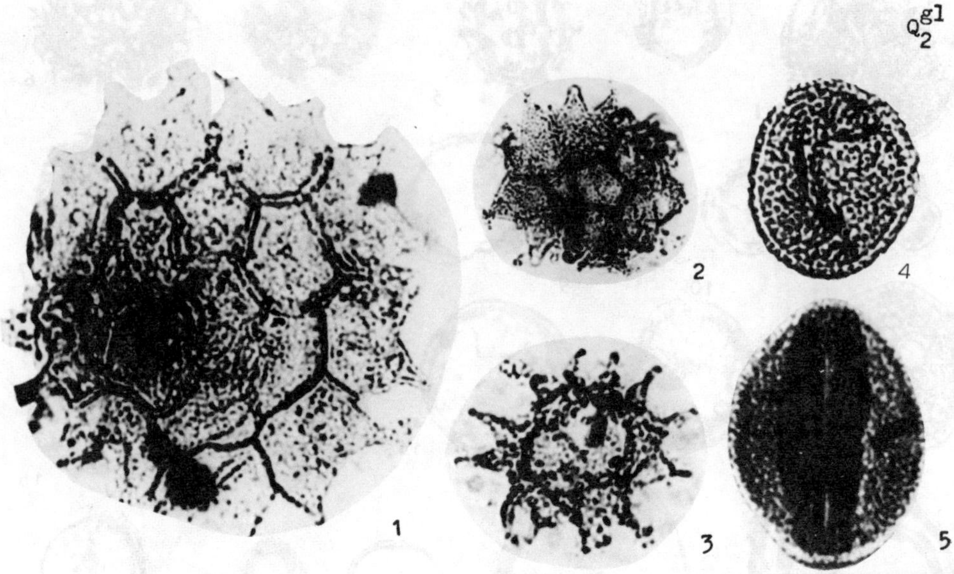


Some sporo and pollen of the aquatic plants in the Interglaciation period of early Pleistocene from the Kunlun Mts. and Qaidam Basin (All the figures are magnified x 680)

1,2. Pediastrum boryanum. 3. P. duplex. 4. Sparganium. 5. Nelumbo.

# THE EVOLUTION OF THE PALAEOGEOGRAPHIC ENVIRONMENTS IN THE TANGGULA MOUNTAINS IN THE PLIOCENE—QUATERNARY

Xu Shu-ying
(Lanzhou University)

## THE SUBTROPICAL FOREST-STEPPE ENVIRONMENT IN THE PLIOCENE

During the early Tertiary period, the Tanggula Range suffered upift and dissection with great quantities of deposits accumulating on the down-faulting flanks of the mountains. These sediments were terrestrial facies of purple-red clastics. They show fining-upward characteristics possibly because of weakening of the tectonic movements and the consequent reduction in relief. By the Oligocene, the mountain region had became a planation surface.

The Himalayan movement reached a new high tide in the Miocene, and this caused the dismemberment of the integral planation surface and a series of new mountains and fault basins formed, thus a new stage in the geological development of the Tanggula Range began.

The widespread Pliocene strata both within and beyond the Tanggula Range provide indications of changes in the palaeogeographical environment. A series of narrow Pliocene basins aligned E-W were located on the southern piedmont of the Tanggula Range from Amdo to Baqen, northern Xizang. In the Bulong Basin, located at Xagqü Qu, one of upper reaches of Nujiang River, Hipparion fauna were discovered in Bulong formation, thus confirming the Pliocene age of the beds (1). Similar fault basins, also aligned E-W were formed on the northern piedmont. Boreholes in a fault basin on the Tuotuohe revealed purple-red and grey-green sandy mudstone and mudstone with gypsum, 70 m thick of which could be seen, of late Pliocene age beneath lacustrine beds of the early Pleistocene. There is also a gypsum layer at the top of the Pliocene lacustrine beds in the fault basin of the Tongtian River. Sporo-pollen analysis of these two fault basins reveals a subtropical forest-steppe assemblage in the Pliocene beds. One borehole (No. 18) on the Tongtian River shows a dominance of arboreal pollen in the lower part (60-100 m), e.g., Picea, Pinus, Ulmus, Abies, Quercus and Cedrus etc., while herbaceous pollens are mainly of Chenopodiaceae and Ephedra spp. Another borehole (No.17) on the Tuotuohe also shows a predominance of arboreal pollen towards the bottom (287-297 m), e.g., Ulmus, Betula, Pinus, Picea, Cedrus and Salix etc. * The lithofacies and the contained faunal and floral remains of Pliocene beds in the piedmont areas of the Tanggula Range indicate moderate tectonism throughout the Pliocene with fine grain lacustrine sediments predominating in the piedmont areas. The warm and humid climate at the beginning of the Pliocene became dry in the late Pliocene. This concentrated gypsum in the inland lacustrine sediments. The lakes may not have been linked by rivers until the beginning of the Quaternary.

*Kong Zhao-chen et al., Unpublished work, 1977.

The Pliocene beds are also often found within the fault basins of the interior mountain belt, and have a thickness of 700 m in the upper reaches of the Quguo River on the western side of the Tanggula pass. It is named the Tangquangou formation of conglomerate with sandy or clay lenses. Clasts in the conglomerate do not exceed 5-6 cm in diameter and are subrounded. The main lithologies are the limestone and sandstone of Yanshiping group (mid-Jurassic). It contains none of the acidic intrusive rocks of the Mesozoic age, now exposed far and wide on both sides of the Tanggula pass. The calcareous cement of red-yellow conglomerate in this fluviatile indicates an arid and hot subtropical environment. These deposits are well developed in the conditions of low relief and the streams did not yet cut down into the intrusive rocks beneath the sedimentary cover. At present the upper limit of the conglomerate reaches an altitude of more than 5,200 m, almost as high as the widespread Pliocene erosion surface making up a part of the main ridge of the mountain.

In addition to the evidence of the Pliocene strata, the palaeogeomorphology of the dissected area is also a useful guide to the geographical environments at that time. Palaeokarst forms, such as stone walls, Fenglin (needle karst), and stone forests, occur in the limestone all over the mountain tops, except the glaciated area. Their height is comparable with the Pliocene planation surface. The karst surfaces on both flanks of the Tanggula Range, especially those on the southern slope where the palaeokarst is particularly well developed from No. 114 Highway Squad to Bei Shan of Amdo and from Xagquka along Heihe-Qamdo Highway to northern slope of Baqen, escaped modification by the Quaternary glaciation. Although these karst features are small in scale (50 m in height at most) and do not match the typical Fenglin morphology of the present tropical monsoon and equatorial rainforest zone, they nevertheless indicate a warm and humid climate in contrast to the present cold one. Clearly, relief was low and favoured the development of karst in these mid-latitude ($32^{o}$ — $34^{o}$ N) mountains, and the subtropical palaeokarst was thus exposed.

## THE WARM AND HUMID ENVIRONMENT IN THE EARLY PLEISTOCENE

Orogenic movement has been characterized by extensive and broad amplitude arching since the end of the Pliocene. Because of the uplift of the whole mountain body at the beginning of the Early Pleistocene, the difference in height between the mountains and peidmonts is small. Subsidence ceased in the Pliocene in both the southern piedmont basins such as, Xagquka, Bulong, Sog County and the northern ones such as the Tongtian River. Here the Quaternary beds are thin and the river terraces of bedrock are formed. In the mountain interior around Quguo River Pliocene beds tectonism produced southeasterly dips of between $10^{o}$ and $20^{o}$.

The thin strata of early Pliocene age in the Tanggula Range and their piedmont consist mainly of fluviatile and lacustrine fine lutites. Boreholes in the early Pleistocene piedmont fault basin on the southern side show a weathered mantle of grey-white granite, 30 m thick, covered by 80 m of lacustrine and paludal blue-grey muddy silts and light-brown sandy clays. In the Chamucuo Basin which stretches east-west along the northern piedmont, the early Pleistocene lacustrine muddy silts is over 160 m thick. In the Quguo River fault valley, the Quguo formation of lacustrine origin with a thickness of 360 m, rests uncomformably on the Tangquangou formation conglomerate (Fig. 1). The Quguo formation consists of clay and silt with palaeomagnetic age of 4.30-2.20 million years, i.e. late Pliocene to early Pleistocene (2).

The lithofacies and lithology of the early Pleistocene sediments in the Tanggula Range and its piedmont areas contrast with the sediments on both the southern and northern margin of the Qinghai-Xizang Plateau. To the south of the Himalayas we found over 1,000 m of Siwalik conglomerate of the early Pleistocene age, while the Xiyu conglomerate of the same age on the northern piedmont of the Kunlun Mountains is 2,000 m or more thick. This indicates that on the mountains bordering the Plateau in the early Pleistocene intense and rapid tilting occurred, while in the interior of the Plateau, including the Tanggula Range, tilting was strong but steady and structual difference of motion was small, so the Pliocene planation surface was fundamently preserved. Only in the subsiding piedmont and intermontane basins there was fine substantial deposition of fluviolacustrine facies.

Sporo-pollen assemblages in the early Pleistocene lacustrine beds indicate a period of temperature decline with herbs predominant. For example, sporo-pollen taken from early Quaternary travertine deposits in a limestone cave (4900 m) at the Bei Shan of Amdo consists mainly of meadow species including Cyperaceae, *Chrysanthemum* sp., *Artemisa* sp., Gramineae, etc., as well as some *Picea*, *Alnus cremastogyne* etc. Boreholes in beds of the same age at Tuotuohe Valley show 69-74% of pollen consists of herbs, mainly Chenopodiaceae and *Ephedra* spp. * The decline in temperature did not result in development of glaciers in the Tanggula Range, because summit altitudes were lower than 2,500 m. During a large part of the early Pleistocene, the climate in the mountains was warm and humid favouring the growth of warm temperature mixed broadleaf and coniferous forests. Sporo-pollen in the early Pleistocene Quguo He lacustrine beds (e.g. the third segment of the Quguo formation dated palaeomagnetically as 2.90-2.20 million years old) shows a predominance of *Quercus* sp., *Betula* sp., and *Corylus* sp. etc. It is evident from the boreholes along Tongtain River Valley and Tuotuohe Valley, that a vast area of the piedmont in the early Pleistocene was covered by mixed broadleaf forest, whereas on the mountains forest-steppe and coniferous forest predominated. The great extent of fluviatile and lacustrine deposits represents warm and humid climate at that time with the exception of the northern slope where the climate was a little arid, for the deposits are richer in calcium, indicating that the Tanggula Range had already been uplifted so as to constitute a barrier to the Indian monsoon.

Many lakes had endorheic characteristics during the Pliocene but, because of uplift and a change to more humid conditions in the early Pleistocene, they became linked by a dense drainage, so network of lacustrine deposits was replaced by fluviatile gravels both horizontally and vertically. Because of mountain gradients, rivers were not deeply incised. Deposits are mainly of well-rounded gravels of rather small size (less than 10 cm in diameter), and cross sections of river valley are broad and gentle. In the rift valley, because of slow depression, fluviatile gravels accumulated to thickness of several hundred meters, e.g. the Buqu River Valley on the northern slope contains 400-500 m of the early Pleistocene sediments. On some sections of the wide valley where limestone occurs, low Fenglin(20-30 m high), stone forests and corrosion caves are common, especially on the eastern and southern slopes which have a higher monsoon rainfall (Fig. 2).

---

* Kong Zhao-chen et al., Unpublished work, 1977.

## ENVIRONMENTS OF THE GLACIAL AND INTERGLACIAL STAGES DURING THE MIDDLE AND LATE PLEISTOCENE

At the end of the early Pleistocene tectonic movements caused 3,000-3,500 m uplift of the Tanggula Range. The elevation of the broad and well-preserved planation surface of the Pliocene into the rich snow zone favored the development of glaciers in large scale. This was the Tanggula glaciation which produced widespread glacial deposits on the planation surface at altitudes of 5,300-5,400 m on both flanks of the range. Till is common on the southern piedmont around No. 110 Highway Squad south to Za'gya Zangbu River. On the northern piedmont till attains a maximum thickness of over 30 m and occurs as far as the slope near No. 101 Highway Squad. Till of the same age in the Buqu fault valley is buried under 200 m from the surface ground and is over 216 m thick*. It is estimated, therefore, that the Tanggula glaciation spread from To'gyü Mt. in the south to Buqu River gorge in the north, forming an ice cap 80-90 km wide covering the whole mountain axis and flowing down to the piedmonts along wide valleys dating from the early Pleistocene.

Middle Pleistocene outwash consisting of 100 m light-yellow gravels with clay occurs in the Amdo fault basin in the southern piedmont. This covers 80 m of fluviatile and lacustrine sediments also of the early pleistocene age, as mentioned above. On the northern piedmont, as far north as the Buqu River fault basin,a 170-200 m thick outwash fan is formed.

Following the Tanggula glaciation, the mountain terrain was again uplifted. As the Buqu River flowed out the fault basin, it cut deeply into the early Pleistocene glacial fluvial sediments. On the southern piedmont, the Za'gya Zangbu River cut down into the hummock till plain and flowed into a wide but shallow lake as indicated by the thick lacustrine sediments (Fig. 3). Sporo-pollen assemblages of the later part of the middle Pleistocene interglacial period show a predominance of tree pollens, mainly of *Picea*, *Pinus*, *Corylus* and *Betula* etc., ** indicating a warm and humid climate. In many places on both the southern and northern mountain slopes brown forest soils developed on the glacial deposits of the Tanggula glacial period.

The Qinghai-Xizang Plateau and its marginal mountains were elevated to a still greater height during the late Pleistocene. This attenuated the effects of Indian monsoon in the interior of the Plateau and precipitation declined in the Tanggula Range region. This had the effect of reducing the scale of the glaciation in this area. On the basis of the glacial erosional and depositional evidence, the late Pleistocene age may be divided into two glacial periods and one interglacial separating them. The former glaciation is called the Za'gya Zangbu glacial period and the latter glaciation is called the Basicuo glacial period.

During the Za'gya Zangbu glacial period, extensive areas of tills 20-30 m thick were formed on the southern piedmont in the Za'gya Zangbu Valley at an altitude of between 4,950-5,100 m. On the southern side of the Tanggula Pass, end moraines in the Bu'nyog River Valley developed mainly along the Tumengela-Qinghai Highway north to Za'gya Zangbu. On the northern piedmont tills of the same age also occur in the valleys,

---

* Hu Hai-tao, Unpublished work, 1976

** Kong Zhao-chen et al., Unpublished work, 1977

such as the low spread of moraine 50-60 m above the valley floor along the Qinghai-Xizang Highway and the till plain 40-50 m above the river level in the valley of the Baiduo River, a branch of the Buqu River. Till deposits of the same age are well developed along the piedmont east of the Buqu Basin. The till distribution indicates deposition by piedmont glaciers and dendritic valley glaciers with a total area of 22,700 $km^2$, which is about 11 times that of the modern glaciers. Valley glaciers could have extended as far as 20 km or more.

The relics of the Basicuo glacial period have very distinctive morphology. North of the Za'gya Zangbu River and east of the Qinghai-Xizang Highway there are 3-4 end moraines at the outlet of the glacial trough. Among these end moraines about 10-30 m high, there are small lakes or ponds. There are two lateral moraines on the right side of the trough 50-80 m and 100-120 m above the valley floor respectively. Similar condition appears at the eastern side of the Pass (Fig. 4). Here several end moraines occur on the northern side of the Basicuo Lake, the most complete one running very close to the lake shoreline. The Basicuo Lake is the source of the Zhile River which runs northwards cutting through the moraine ridges. East of the pass on the eastern side of the Menjiari glacial trough two lateral moraines occur, 70 m and 120 m respectively above the valley floor. Well preserved roche moutonnees occur in groups along the valley axis.

The brevity of the late Pleistocene interglacial period, together with the uplift of the Tanggula Range to an evelation of 4,000 m or more promoted the growth of coniferous forests in the subalpine mountains and on the plateau surfaces. Spore-pollen analysis indicates a predominance of _Picea_ and _Pinus_ etc. with alpine bush-steppe.

As the mountain area was uplifted and the climate became colder and more arid, the extent of glaciated area was reduced. At the same time, the periglacial zone became enlarged. Periglacial phenomena dated from this time included abundant freeze-thaw debris, large polygen ground (100-150 m in diameter) on the southern side of the mountain and enormous stone circles, 14-25 m in diameter near the source of the Changjiang River on the northern side of the Geladandong snow-capped mountains. These stone circles are similar to those found in high latitudes and are clearly not the product of the present climate, but dated as the product of the Late Pleistocene.

Permafrost is extensively developed in the Tanggula Range. The southern limit of the continuous permafrost zone runs through their southern peidmont, and the section of their main ridges is still the thickest part of the permafrost on the Qinghai-Xizang Plateau at present. A borehole in the Early Pleitocene lacustrine beds of the Quguo River fault basin showed a permafrost thickness of 128 m (3).

## PERIGLACIAL ENVIRONMENTS IN THE HOLOCENE

Further tectonic uplift occurred at the end of the Late Pleistocene. Significant subsidence and elevation occurred mainly along north-south faults, which cut across the west north-west regional fault trend. At points of fault intersection geothermal phenomena, such as hot springs, occurred and permaforst was melted in these areas, creating unfrozen lacunnae in the hinterland of the Plateau. At present hot springs and relic sinter terraces and cones covering the outwash fans of the Basicuo glaciation occurred along the margins of the Buqu fault basin.

Although tectonic movement continued in the Holocene, the climate changed. During the Basicuo glacial period, the periglacial aeolian loess driven from theperiglacial zone, together with large amounts of limestone breccial produced by the periglacial action of the Late Pleistocene, became mixed during the extensive solifluction which occurred during the mid-Holocene warm period. The material has taken the form of terraces of 10 m thick on the slopes and fans in the valleys covering the outwash fans and terraces of the Basicuo glaciation. Unlike those of the present day, these periglacial forms and deposits are large and stable. In addition, the periglacial aeolian sand which accumulated during the Basicuo glaciation became fixed or partly fixed by plants during the mid-Holocene warm period. This is very common in the piedmont areas of the Tanggula Range. During this warm period rivers became incised and lakes were formed and the No. 1 Terrace of the Buqu River and the Za'gya Zangbu River were formed. Glaciers near the pass melted and produced such a large quantity of meltwater that mountain lakes, such as the Basicuo Lake, were formed.

The glaciers began to advance again during the Neoglaciation, for example, glaciers east of the pass advanced as far as the southern shore of the Basicuo Lake, the tongue was 100 m lower and the ice tongue 4-5 km longer than the present glaciers (Fig. 5). These Neoglacial tills now have a soil profile and grass vegetation on them. It is clear that the climate changed several times during the Neoglaciation. For several retreats moraines have been recognized dating from this period. With the uplift of the summit surface to 5,000 m a.s.l. and more, periglacial activity remained active during the Neoglacial. The turf of the grassland steppe vegetation was locally varied by solifluction. One such turf layer, 1.5 m thick, was dated by $C^{14}$ assay as 1350±150 years B. P.*This clearly dates the Neoglacial on the southern shore of the Basicuo Lake. Freeze-thaw deformation of the flood plain deposits (near the No. 100 Highway Squad) on the west bank of the Buqu River, the stone circles around the fluvial fans and icewedge polygons on the outwash fans of the Basicuo glaciation in the Buqu fault basin, also date from the Neoglacial period.

The present glaciers in the Tanggula Range occupy altitudes between 5,400-6,100 m. Snowlines lie between 5,500-5,800 m and a little higher on the southern slope than on the northern. The glaciers occupy an area of 2,082 $km^2$ and include valley glaciers, cirque glaciers and hanging glaciers. The largest valley glacier is the Jinggaindiru glacier on the western end of the range which is 12.8 km in length.

The whole mountain area suffers a periglacial climate with an average temperature in most of the area lying between -4°C and -6°C. The diurnal temperature range is great, while the annual temperature range is relatively small. This is conducive to strong periglacial action and maintenance and development of permafrost. The periglacial zone has a vertical range of 1,000 m (4,500-5,400 m a.s.l.) under the alpine freezing-desert and alpine meadow-steppe environments.

## References

(1) Huang Wan-po, Ji Hong-Xiang , The first discovery of Xizang Hipparion fauna and its significance to the plateau rise. Kexue Tongbao, 24 (19), 885-888 (1979). (in Chinese)

* Cui Zhi-jiu, Unpublished work, 1979.

(2) Duan Wan-ti, et al., The research of Quaternary glaciation along the Qinghai-Xizang Highway. Kexue Tongbao, 24 (10), 455-458 (1979). (in Chinese)
(3) Wang Jia-cheng, et al., Permafrost along the Qinghai-Xizang Highway. Acta Geographica Sinica, 34 (1), 18-32 (1979). (in Chinese)

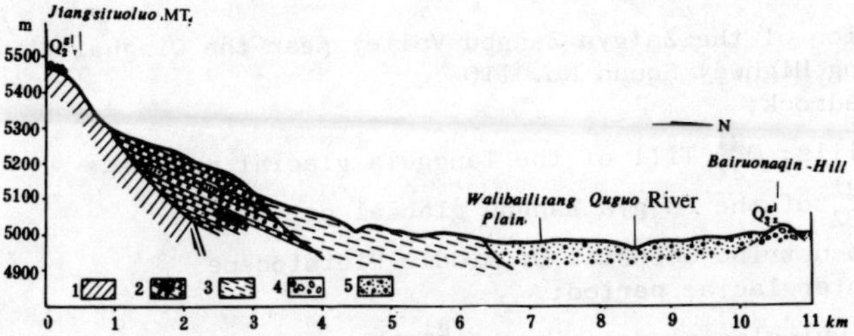


Fig. 1 Section of the Quguo River Valley in the Tanggula Range
1. Bedrock;
2. Tangquangou formation;
3. Quguo formation;
4. Tills: the $Q_{2T}^{gl}$ Middle Pleistocene till of the Tanggula glacial period; the $Q_{3Z}^{gl}$ Late pleistocene till of the Za'gya Zangbu glacial period;
5. Late Pleistocene fluvioglacial deposits

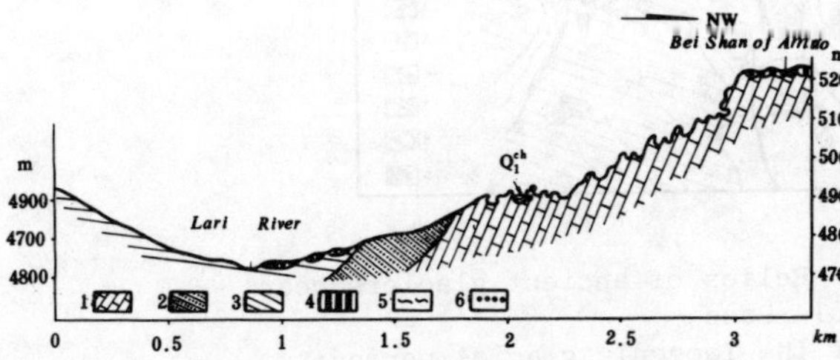


Fig. 2 Section of the Lari River-Bei Shan of Amdo
1. Triassic limestone;
2. Jurassic sandstone;
3. Early Tertiary conglomerate;
4. Pliocene weathered mantle;
5. Early Pleistocene cave deposits;
6. Pleistocene fluviatile deposits

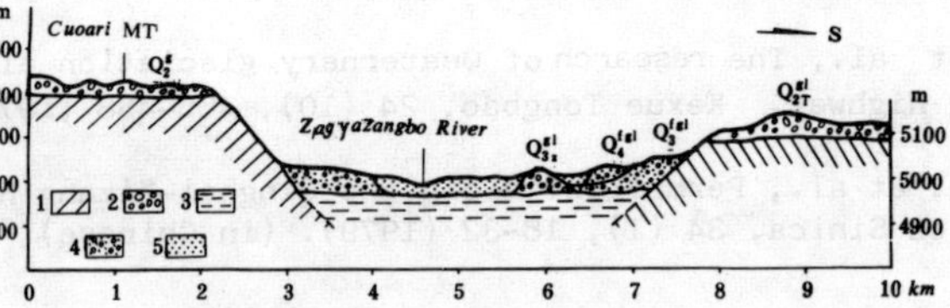


Fig. 3 Section of the Za'gya Zangbu Valley near the Qinghai-Xizang Highway Squad No. 110
1. Bedrock;
2. Tills: $Q^{gl}_{2T}$ Till of the Tanggula glacial period; $Q^{gl}_{3Z}$ of the Za'gya Zangbu glacial period;
3. Lacustrine deposits of middle Pleistocene interglacial period;
4. Fluvioglacial deposits: $Q^{fgl}_{3}$ Late Pleistocene fluvioglacial gravels; $Q^{fgl}_{4}$ Holocene fluvioglacial gravels;
5. Holocene alluvial deposits

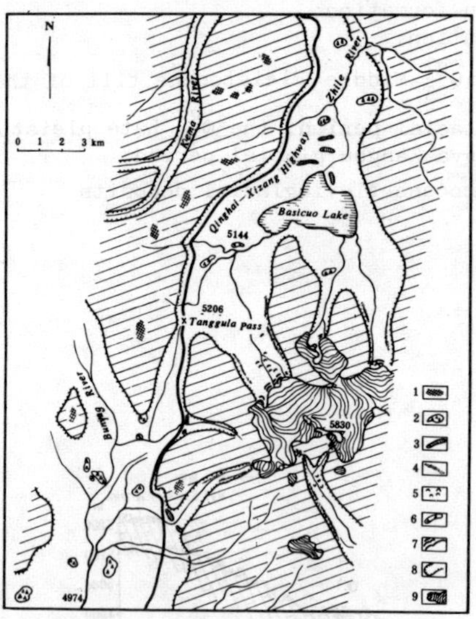


Fig. 4 Relics of ancient glaciers near the Tanggula Pass
1. Till of the Tanggula glacial period;
2. Till of the Za'gya Zangbu glacial period;
3. End moraine of the Basicuo glacial period;
4. Lateral moraine of the Basicuo glacial period;
5. Roches moutonnees of the Basicuo glacial period;
6. End moraine of Neoglacial period;
7. End moraine of modern glacier;
8. Late Pleistocene trough valley;
9. Existing glaciers

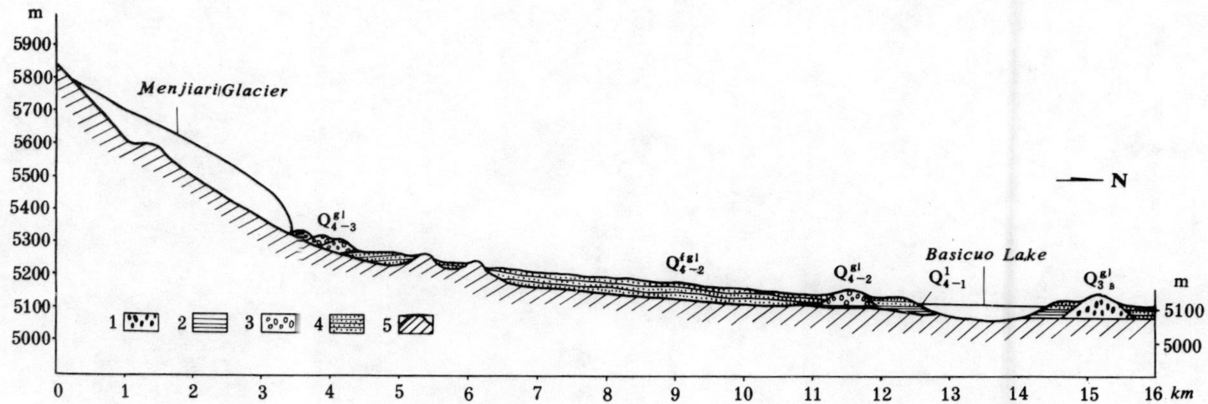


Fig. 5 Section of the Menjiari Glacier-Basicuo Lake

1. End moraine of the Basicuo glacial period;
2. Early Holocene lacustrine deposits;
3. Tills: $Q^{gl}_{4-2}$ End moraine of the Neoglacial period;
   $Q^{gl}_{4-3}$ Modern end moraine;
4. Fluvioglacial deposits of the Neoglacial period;
5. Roches moutonnees of the Basicuo glacial period

# TETHY'S ZONE OF NEPAL HIMALAYAS

M.N. RANA, Y.L. SINGH
(Department of Mines and Geology, Nepal)

## ABSTRACT

The Tethys Himalayan Zone is distributed widely in the Higher Himalayas of Nepal and neighbouring Tibetan Plateau. For convenience this zone is subdivided into two groups - the lower Himal group and the upper Tethys group. The Himal group is represented by high grade metamorphic rocks of various kinds such as Kyanite gneiss, biotite gneiss and migmatites belonging to the Precambrian age. The Tethys group is represented by sequences of fossiliferous sedimentary rocks of different ages ranging from Cambrian to Paleogene. The lower part of this group consists of unfossiliferous low grade metamorphic rocks.

## INTRODUCTION

According to the tectonic positions and lithocharacters, the Nepal Himalayas are subdivided into three major zones. They are from north to south as follow:

1. The Tethys Himalayan Zone,
2. The Midland Meta Sediment Zone,and
3. The Siwalik Zone.

The Siwalik zone is represented by a thick sequence of fluviatile sedimentary rocks the age of which ranges from Miocene to Pleistocene. This zone is seperated from the overlying pre-Tertiary rocks of the Midland meta-Sediment zone by a thrust called the Main Boundary Fault. The Midland meta Sediment zone is again seperated from high grade metamorphic rocks of the Tehtys Himalayan zone to the north by another major thrust by the name of the Main Central Thrust.

These two major thrusts run parallel to the main trend of the Himalayas throughout the kingdom of Nepal.

Attempt is made here to deal with the Tethys Himalayan zone alone. Unfortunately much work has not yet been done in this zone of Nepalese territory as the Higher Himalayas are not easily accessible.

## The Tethys Himalayan Zone

The rocks of the Tethys Himalayan zone overlies the Midland metasediment zone along a thrust plane, the Main central thrust. It extends from the Higher Himalayas of Nepal to north into Tibetan Plateau. According to the nature of lithostratigraphy this zone is sub-divided into two groups - the lower Himal group and the upper Tethys Group.

### The Tethys Group

The Tethys group lies unconformably over the Himal group. This group consists of limestone, dolomites, sandstones, shales, conglomerates and low grade crystalling schists. In Nepal Himalayas, the rocks of this group are distributed in the southern most margin of the huge Tibetan Tethys basin in three regions - the Sagarmatha region, the Langu Manang region and the Saipal-Amlang region. The main part of this group lies to the north in the Tibetal Plateau.

The lower part of this group is represented by low grade metamorphic schists such as fine grained biotite schist, calcareous biotite quartz schist and so on. The upper part of this group consists of limestones, dolomites, marbles, sandstones, shales, conglomerates and metamorphosed shales.

The sediments of this group are supposed to be deposited in a narrow but shallow basin (miogeosyncline) developed along the Indus-Tsanpo zone. The sediments of this group are highly fossiliferous and hence the stratigraphy is well established, from Cambrian to Paleogene in regular sequences. The best known fossils are trilobites, craptolites, brachiopods, corals, crinoids, ammonoids and beleminites.

Generally the Cretaceous succession is characterised by glauconitic sandstone. The Jurassic sequences are arenaceous characterized by metamorphosed shales, conglomerates, oolitic limestones and sandstones. The Triassic succession are mostly calcareous shales and limestones where as the Permian succession is mainly metamorphosed shales. The Carboniferous period is represented by crionoidal limestones. Silurian Devonian sequences are made of mainly shales and quartzites, and the Cambro-Ordovician period consists of mainly dolomitic limestones, argillaceous limestone and calcareous schists.

Distribution:

1. In far-western Nepal in Saipal-Amlang region they are well exposed west of Mugu Karnali; from Saipal to Khojar and northwestern border areas.
2. In western Nepal in Langu Manang region, the widely distributed areas are Kangibora Himal; Churen Himal, Annapurna Himal and Mustang and Manasalu areas.
3. In eastern Nepal in the Sagarmatha region, they are distributed from the Sagarmatha area to the extreme north of Dudh Kosi river.

Age:- Cambrian to Paleogene

The Himal Group:

This group is made up of various kinds of high grade metamorphic gneisses and migmatites often intruded by tourmaline granites. It occurs immediately north of a thrust zone over the autocthonous Midland metasediments and generally known as Basement complex or Himalayan gneiss. The thrust is known by the name of Main central thrust.

The upper part of this group is generally represented by granitic migmatites and the lower part by banded gneiss of various kinds. The mineral contents in this group of rocks are mainly kyanite, staurolite, garnet, sillimanite, muscovites and biotites. Minerals such as epidote, hornblende and cordeirite also do occur in some places. According to the variation of mineral contents in the rocks they are differently named as mica gneisses, kyanite gneisses, sillimanite gneisses and kyanite garnet biotite gneisses and so on.

The gneisses and migmatites are often intercalated by layers of marbles, slate and amphibolites. They are generally leucocratic, but sometimes melanocratic types are also met with. As a rule the gneisses are coarse grained sometimes with Augen structures and pyroblasts made of feldspars.

Distribution:

1. In the east they are well exposed in the southern slope of Mt. Chamlang, Namche Bazar, the upper reaches of Tama Kosi, Khimti Khola and the Dudh Kosi.
2. In the central Nepal they occur around Helambu region, Langtang Himal, Ganesh Himal and Gosaikunda.
3. In the western Nepal they are distributed in the southern slopes of Mt. Machhapuchare, the Annapurna Himal, Simikot and Ninal Pass and others.

Age:- Precambrian

## EVOLUTION OF THE TETHYS SEDIMENTS

In Nepal Tethys sediments are distributed along the southern margin of Tibetan Tethys basin along Sagarmatha region, the Langu Manang region and the Saipal Amlang region. This basin is known as the southern marginal basin in order to differentiate it from the Main Tethys basin of Tibet in the north.

Lower Paleozoic sediments are observed directly deposited upon the highly metamorphosed crystalline rocks of the Precambrian age. This means that the area might have been exposed to the surface from Precambrian up to early Paleozoic. The area started sinking to form the Tethys basin in the latest Eocambrian. Huge amounts of argillaceous sediments then started pouring in the basin from Paleozoic in the period older than Ordovician. Sediments in proper sequences were deposited in this basin from Paleozoic to Paleogene. The whole area then started uplifting during the upper (or after) Cretaceous time during which the northern fore deep, along the so-called Indus suture line was developed. This trough seperate the northern marginal mobile zone of Hindusthan shield from the Karakorum upland. Here the sediments are found intermixed with basic to ultrabasic volcanic materials thus representing an ophiolitic belt. The Nepalese Himalayas as a whole are the marginal parts of the Hindusthan shield. No ophiolites are reported from Nepal yet.

## REFERENCE


(1) Gansser, A. 1964 ; Geology of the Nepal Himalayas Interscience Publishers, a division of John Wiley & Sons Ltd., London, New York, Sydney.

(2) Gerhard, 1967 ; Jum Bau Des Himalayas An Austrian Dhaula Himal Expedition, an unpublished report.

(3) Hagen, T. 1969 ; Report on the Geological Survey of Nepal Volume I, Preliminary Reconnaissance, Druck Von Art. Institut Orell Fussli AG, Zurich. Kommissions verlag von Gebruder, Fretz Ag Zurich.

(4) Geology of the Nepal Himalayas, 1973 ; Geology of the Nepal, Edited by Yoshihide Ohta, Supervised by Seiji Hashimoto, Himalayan Committee of Kokkaido University, Sapporo, Japan.

(5) Tater, J.M. 1980 ; The Geological Setting of Nepal Himalayas, IGCP National Committee, Nepal National Commission for Unesco, Department of Mines and Geology, Lainchaur, Kathmandu, Nepal.

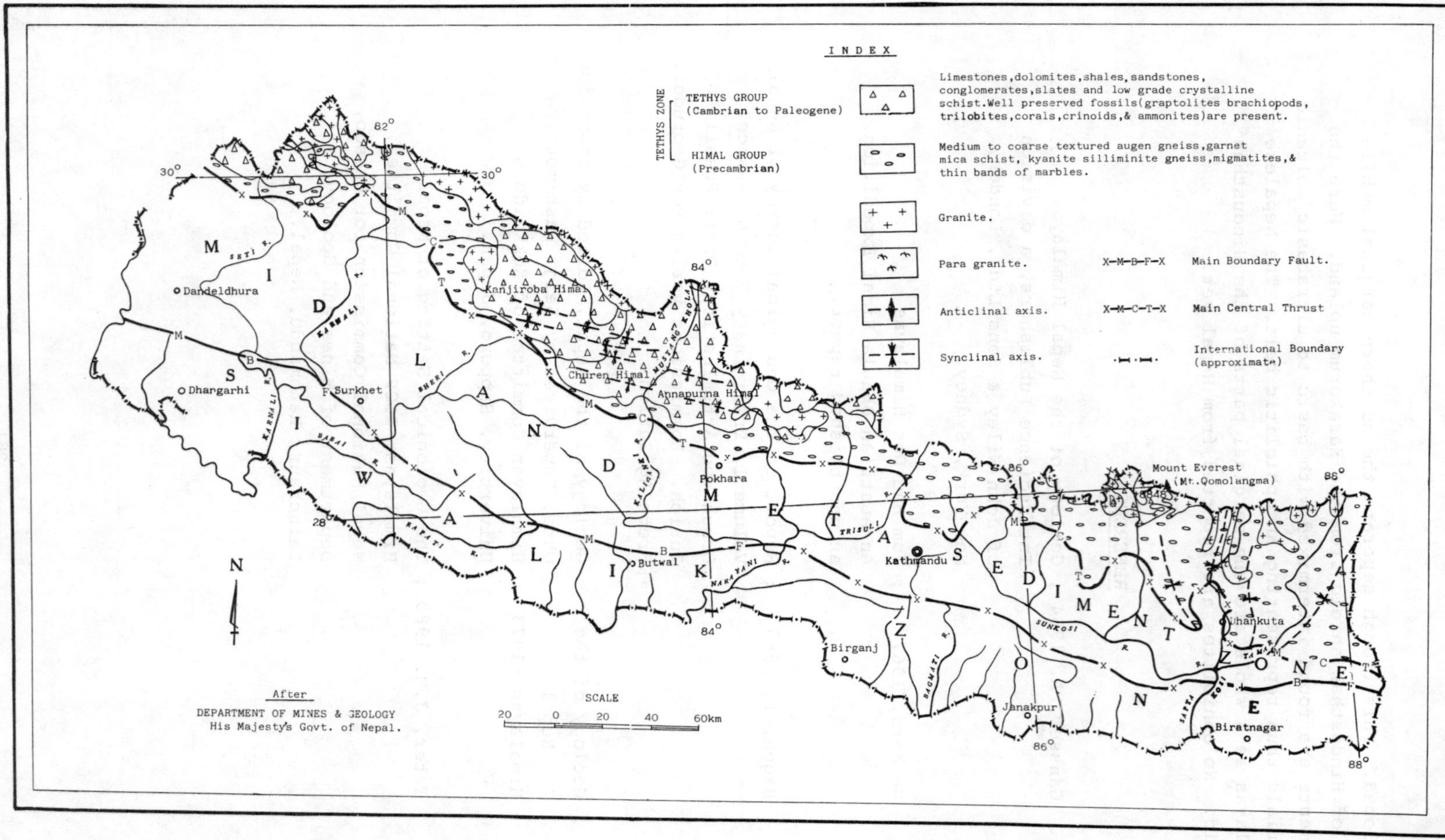


Fig. 1 Geological map of Tethys zone of Nepal Himalayas

# STRATIGRAPHIC AND GEOLOGIC CHARACTERISTICS OF PRE-ORDOVICIAN ROCKS IN THE CENTRAL HIMALAYAS

Yin Ji-xiang, Zhang Qi, Li Da-zhou, Li Shao-hua
(Institute of Geology, Academia Sinica)

The Himalayas are an arcuate range with mountains protruding southwards. The southernmost convex part includes Nepal, Sikkim and Xigaze in Xizang at the southwestern frontier of China, which is referred to the middle segment of the Himalayas or the Central Himalayas. As far as we know, there are 14 peaks above 8,000 m in Asia, and 9 of which are situated in the Central Himalayas. Among them Mt. Qomolangma ( 8,848 m), the world's highest peak, is right on the Sino-Nepalese border. These high peaks extending in a row from NWW to SEE clearly display the unique geologic and structural charateristics in the region. The main bodies of these peaks and their southern slopes consist chiefly of pre-Ordovician rocks. At present there is debate about the stratigraphic succession, the age and the metamorphism due to complicated geologic and structural settings in the region concerned (Fig. 1).

Based on the differences in the stratigraphic development and structural deformation, the Central Himalayas may be subdivided into four structural lithofacies belts from north to south : the Tethys Himalayan belt, the High Himalayan belt, the Lower Himalayan belt and Sub-Himalayan belt. The Sub-Himalayan belt and the Lower Himalayan belt of the Central Himalayas are situated entirely in Nepal and Sikkim, while the former two belts can be seen in the Xigaze district in southern Xizang. Except the Sub-Himalayan belt where pre-Ordovician rocks are absent, the other belts are quite different from each other in their stratigraphic features of pre-Ordovician rocks.

## THE TETHYS HIMALAYAN BELT

The Tethyan sediments in the middle section of the Tethys Himalayan belt outcrop mainly in the Xigaze district of southern Xizang, and in northwestern Nepal, and can also be seen subordinately in northern Sikkim. The Tethyan sediments of Ordovician to Early Tertiary age in the region concerned are characterized by fossiliferous and nearly unmetamorphosed rocks. Their sedimentary nature is readily determined. Before the Ordovician the Tethyan sediments in the same region are generally characterized by unfossiliferous and slightly metamorphosed rocks. Their stratigraphic and geologic features are difficult to be determined owing to erasing of a great amount of geologic records by faulting.

In the Xigaze district the slightly metamorphosed rocks conformably underlying the Rouqicun formation and the lower formation of the Chiatsun group of Early Ordovician are represented by the North Col formation at the northern slope of Mt. Qomolangma(1,2). At Mt. Qomolangma the North Col formation occurs in a section below 8,200 m in altitude with a visible thickness of 1,270 m. It consists chiefly of biotite-quartz schist intercalated with quartz schist and thin marbles in the lower part and biotite-quartz phyllite with quartz schist in the upper part. It is separated from the underlying augen and banded migmatites and gneiss by yellowish cataclastic muscovite granite gneiss (or granite mylonites) (Fig. 2). In Nyalam county the biotite-quartz schists of the North Col formation are intruded by many granite sills and dykes, and are separated from underlying migmatites of the Nyalam group by the Jangling granite. The metamorphic grade of the same formation, which contains the chlorite-biotite zone in its lower part and the chlorite zone in upper part(3), belongs to green schist facies.

In the Tethyan belt of the Central Himalayas, the slightly metamorphosed rocks conformably underlying the fossiliferous Ordovician succession are referred to as the North Col formation at the northern slope of Mt. Qomolangma and the Larjung formation of Thakkhola region in northwestern Nepal(4). They can be correlated, in lithology and position, with the Parahio and the Haimanta series and their equivalents in Spiti region ( see Table 1). So these rocks may possibly be Sinian-Cambrian in age.

## THE HIGH HIMALAYAN BELT

The High Himalayan belt extending in correspondence with the High Himalayas of the Central Himalayas is situated at the high mountains region of Sino-Nepalese border. Metamorphosed rocks of the High Himalayan belt are represented by the geologic section of the High Himalayas from Chiatsun to the Friendship Bridge, Nyalam county, Xizang (Fig. 3). This section has been surveyed by Yin Si-huai in 1966, who suggested that this metamorphosed succession could be regarded as a series of continuous strata and might be subdivided into four formations and five metamorphic zones from kyanite to muscovite zones showing the gradual decrease of regional metamorphism from bottom to top(5). Recently, the section has been reexamined and a new classification in stratigraphic units and metamorphic zones, which differs from the previous one, has been given by the authors engaged in this study.

According to the difference in lithology, the Nyalam group, a temporarily named unit, formerly called the Jolmo Lungma group, may be subdivided into two parts. The lower part contains beds 1-5 and the upper part contains beds 6-7. The two parts are separated by the Quxam thrust sheet.

The stratigraphic sequence of the lower part, with a visible thickness of 13,000 m can be briefly described in ascending order as follows:

1. Kyanite-sillimanite gneiss, with its basal part extending into Nepal. It may be separated from the underlying Midland meta-sediments group in the vicinity of Kodari, Nepal by the Biqu thrust or the Main Central Thrust.
2. Kyanite schist.
3. Sillimanite gneiss, intercalated with almandine-biotite schist, diopside marble and quartzite, which is intruded by gneissose tourmaline granite in the upper part.
4. Staurolite-kyanite-sillimanite gneiss, intercalated with quartz schist.
5. Sillimanite gneiss, intercalated with almandine-biotite schist

and quartz schist.

The lower part of the Nyalam group, which contains the kyanite zone, the sillimanite zone I and the sillimanite zone II from bottom to top, may provide an example of inversed metamorphism.

The upper part of the Nyalam group with thickness of 9,000 m, i.e. beds 6-7 of the section, is composed chiefly of augen and banded migmatites, intercalated in the upper horizons with sillimanite gneiss, kyanite-sillimanite gneiss, biotite schist and marble. The metamorphosed zone in the upper part consists of sillimanite zone II to sillimanite zone I from bottom to top with a normal stratigraphic succession.

The Himalayan gneiss, which unconformably underlies the slightly metamorphosed pelitic rocks of the Tethys Himalayan belt in the region between the southern slope of Mt. Qomolangma and the Gosain Kund in eastern Central Nepal, is composed mainly of calcareous and siliceous gneiss with basic dykes in the lower part and granitic gneiss in the upper part(6). The authors, however, believe that the contact between the Himalayan gneiss and its overlying slightly metamorphosed rocks might be a thrust sheet as shown in Fig. 1. Metamorphosed rocks in accordance with the Himalayan gneiss also were named the Tibetan Slab (La Dalle du Tibet) by French scientists. The stratigraphic succession of the Tibetan Slab, exceeding 5,000-10,000 m in thickness, can be briefly described in ascending order as given below: 1. Biotite- kyanite (sillimanite) gneiss; 2. Pyroxene-horblende gneiss, intercalated with amphibolite and marble ; 3. Dark banded fine hornblende gneiss; 4. Augen gneiss and embrechites(7,8). In northwestern Nepal the stratigraphic sequence of the Dhumpu gneiss, conformably underlying the Larjung formation, may be described in ascending order as given below : 1. Quartz-mica schist; 2. Grey gneiss (1,510m); 3. Calcareous gneiss (1,960 m); 4. Augen gneiss (3,900 m)(4).

As mentioned above, the metamorphosed rocks of the High Himalayan belt cropping out both in the Xigaze district in Xizang and in Nepal may be compared, in lithology and position, with those of High Himalayan belt in other regions of the Himalayas (see Table 1).

## THE LOWER HIMALAYAN BELT

The Lower Himalayan belt in Nepal and Sikkim, which extends in accordance with the Lower Himalayas, contains chiefly of pre-Ordovician rocks, an outlier derived from the High Himalayan belt, a small amount of Gondwana sediments of Permo-Carboniferous, and marine strata of Cretaceous to Eocene. The stratigraphic succession of the pre-Ordovician in Central Eastern Nepal may be generalized in descending order as given below (Modified from Hashimoto, S. et al. eds., 1973):

——— Thrust ———

Kathmandu group (Early to Middle Palaeozoic, > 7,200 m)

An upper calcareous and arenaceous subgroup (2,900 m) consists chiefly of sandstone and shale with four limestone horizons; the lowest limestone, named as Godavari limestone, yielded fossils of crinoids, trilobites and brachiopods etc., Ordovician to Devonian in age.

A middle arenaceous subgroup (1,300 m) is composed mainly of medium grained sandstone and quartzite with ripple marks.

A lower calcareous and arenaceous subgroup ( > 3,600 m) consists chiefly of graywacke and quartzite interbeded with phyllite, intercalated with limestone only at the top and in the middle part, and with graded bedding only in the lower part.

~~~~~~~~~~~~ ? ~~~~~~~~~~~~

Midland meta-sediments group (Z, > 15,000 m)

An upper subgroup (3,500 m) is composed mainly of stromatolite -bearing limestone, intercalated with slate, phyllite and metamorphosed sandstone.

A middle subgroup (1,500 m) consists chiefly of quartzite with ripple marks, intercalated with phyllite and slate. There is quartzose conglomerate in the basal part.

— — — — — — — — Hiatus — — — — — — — — —

A lower subgroup ( > 10,000 m) consists chiefly of medium-coarse grained sandstone, intercalated with chlorite phyllite. There is a thin conglomerate in the basal part, pebbles of which are composed mainly of quartz and a few of granite.

(The basal subgroup consists mainly of slate, for which we do not know whether the stratigraphic position is accurate)

~~~~~~~~~~~~ ? ~~~~~~~~~~~~

Basement gneiss

The middle and lower part of the Kathmandu group or Phulchauki group and the Midland meta-sediments group in Nepal are representative of the pre-Ordovician lightly metamorphosed rocks of Lower Himalayan belt. The mid-upper parts of the Midland meta-sediments group were probably equivalent in lithology and position to the Bhimphedi group of Late Precambrian or Sinian age in Phulchauki district(9,11), and can be roughly correlated with the North Col formation and its equivalents in Tethys Himalayan belt. The Kathmandu group distributed around Kathmandu refers only to fossiliferous, Early and Middle Palaeozoic sediments in the Midland of Nepal. It is probable that the Tethyan transgression reached there during the Early Palaeozoic, and the palaeogeographic connection between the Lower Himalayas and the Tethyan Himalayas is not clear. The upper part of the Kathmandu group contains fossils probably of Ordovician-Devonian, so the underlying thick calcareous and arenaceous succession may be considered as Cambrian in age.

The Phulchauki group (? Є-D) belonging to the upper part of the "Kathmandu Complex", outcropping in the vicinity of Kathmandu and the middle section of the Mahabharad Range, seems to be equivalent to the Kathmandu group in stratigraphic position. It is worthy of note that the Tistung formation, situated at the top of the Bhimphedi group which belongs to the lower part of the "Kathmandu Complex", rests unconformably on the Markhu formation of Late Precambrian in age(11).

## A FEW SPECIAL REMARKS ON STRATIGRAPHIC GEOLOGY OF PRE-ORDOVICIAN IN THE CENTRAL HIMALAYAS

In light of foregoing statements, generally speaking, the stratigraphic and geologic characteristics of pre-Ordovician rocks in Central Himalayas may be briefly generalized as follows :

1. Migration of sedimentary basins in Central Himalayas. The time boundary of the early Early Ordovician separates the sedimentary successions before and after it. During the Sinian to Cambrian period, these

sedimentary basins, in which flysch-like clayey and arenaceous rocks, intercalated with graywake, were deposited and situated mainly in the present Lower Himalayas. These basins seem to be of the miogeosyncline type. After early Early Ordovician, these basins being of a platform type, migrated northwards to the present Tethyan Himalayas (Fig. 4).

2. Nappe structures of the High Himalayan belt in eastern Central Himalayas. The High Himalayan belt of the Central Himalayas as a whole is characterized by flat structures with a northward dip and is subjected to the effects of the NS-trending undulatory folds. This structure might wholly thrust southward along the Main Central Thrust and cover on the slightly metamorphosed succession of the Lower Himalayan belt. It is possible that the High Himalayan belt in the Central Himalayas may be the largest nappe of the world. It can be also subdivided into two subordinate nappes by the Quxam thrust sheet, named respectively as the Kangshan nappe and the Dingringbo nappe by the authors (Fig. 5).

3. Taking into consideration of the absence of pebbles, originating from the underlying crystalline basement, in the marine Tethyan deposits with Gondwana elements(12) and the presence of Tethyan fossils in the Lower Himalayas(9), it is suggested that the High Himalayan belt or the Central Crystalline Belt must have been buried deeply under the Tethyan sediments during the Sinian to Middle Tertiary period. Therefore, it is unreasonable to call the High Himalayan belt as "Central Crystalline Axis "," Ancient Island Arc " or "Himalayan Microcontinent".

## References


(1) Yin Chi-hsiang, Kuo Shih-tseng, Stratigraphy of the Mount Jolmo Lungma and its North slope, Scientia Sinica, 21(5), 629-644 (1978).

(2) Yin Jixiang, Guo Shizeng, Stratigraphy of the Mt. Qomolangma and its northern slope, and a special discussing on Cambro-Sinian, Carboniferous and Permian sequences in comparison with those of the adjacent regions, in Report on the Scientific Expedition to Mt. Qomolangma (1975), Geology, Science Press, 1-71 (1979)(in Chinese).

(3) Liu Bingguang, Zhang Hongbo, The metamorphic rocks, migmatite and magmatic rocks in the northern slope of the Mt. Qomolangma,in Report on the Scientific Expedition to Mt. Qomolangma (1975), Geology, Science Press, 123-181 (1979) (in Chinese).

(4) Bordet, P. et al., Recherche géologiques dans l'Himalaya du Népal région de la Thakkhola. (Eds., Centre Nat. Rec. Scient.) 15, Paris, 7e, 1-279 (1971)

(5) Yin Sihuai, Magmatic rocks, metamorphic rocks and migmatite of the Mt. Qomolangma region, in Report on the Scientific Expedition to Mt. Qomolangma region (1966-1968), Geology, Science Press, 233-261 (1974) (in Chinese).

(6) Hashimoto, S. et al. (Eds.), Geology of the Nepal Himalayas. Himalayan committee of Hokkaido Univ., Sapporo, Japan, 1-286 (1973).

(7) Bordet, P., On the position of the Himalayan Main Central Thrust within Nepal Himalaya, in Seminar on geodynamics of the Himalayan region, edited by H. K. Gupta, Nat. Geophys. Research Institute, Hyderabad, India, 148-155 (1973).

(8) Bordet, P., Colchen, M., Le Fort, P., Some features of the geology of the Annapurna range, Nepal Himalaya, Himalayan Geology, 2, 537-563 (1972).

(9) Gupta, V. J., Virdi, N. S., Fossil records from the lesser Himalaya and their stratigraphical and tectonic implications, in Tectonic geology of the Himalaya, Edited by P. S. Saklani, Today and Tomorrow's Printers and Publishers, 113-118 (1978).

(10) Gansser, A., Geology of the Himalayas. Intersciences Publ., 1-289 (1964).

(11) Kumar, R., Shah , A. N., Bingham, D. K., Positive evidence of a Precambrian tectonic phase in Central Nepal Himalaya, Geol. Soc. India, J., 19(11), 519-522 (1978).

(12) Yin Jixiang, Guo Shizeng, On the discovery of the stratigraphy of Gondwana facies in northern slope of the Qomolangma Feng in southern Xizang, China, Scientia Geologica Sinica, 4, 291-322 (1976) (in Chinese).

|  | Kashmir | Spiti (India) | Khumaon (India) | Nepal | Xigaze (China) | Sikkim | Bhutan | Nanjagbarwa Mt and Siang (China) |  |
|---|---|---|---|---|---|---|---|---|---|
| € — Z | Trilobite shale and slate<br>Dogra Slate | Parahio Series<br>Haimanta | Mic. and Low. Garbyang Series<br>Martoli Series | Larjung Formation | North Col Formation |  |  |  | N |
|  |  |  |  |  | MNT |  |  |  |  |
| AnZ | Salkhala Series | Vaikrita Series | Vaikrita Series | Dhumpu Gneiss (Himalayan Gneiss) | Nyalam Group | Darjeeling Gneiss | Thimpu Series | Namche Bava Group | ↓ |
|  |  |  |  |  | MCT |  |  |  |  |
| Z | (?) | (?) | Chhiplakot Group (Simla Slate) | Midland meta-sediments Group (Bhimphedi Group) |  | Daling Schist | Shumar Series | Siang Group | S |

Table 1  Correlation of pre-Ordovician teotonic sequences in Himalayas

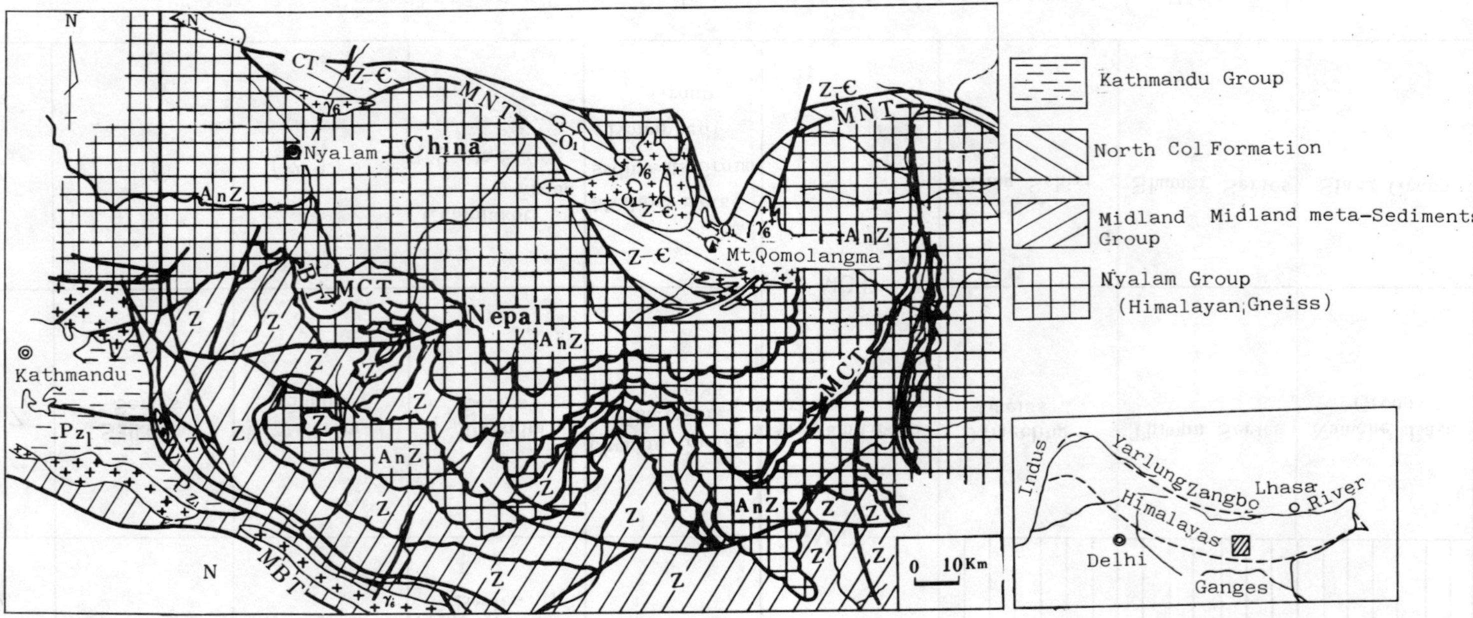


Fig. 1 Geological sketch map of the Mt. Qomolangma Region, showing the distribution of the pre-Ordovician rocks (MBT-The Main Boundary Thrust; MCT-The Main Central Thrust; MNT-Main Northern Thrust; CT- Chiatsun Thrust, BT- Biqu Thrust; N—Yebokanggale group; $O_1$-Jolmo Lungma group)

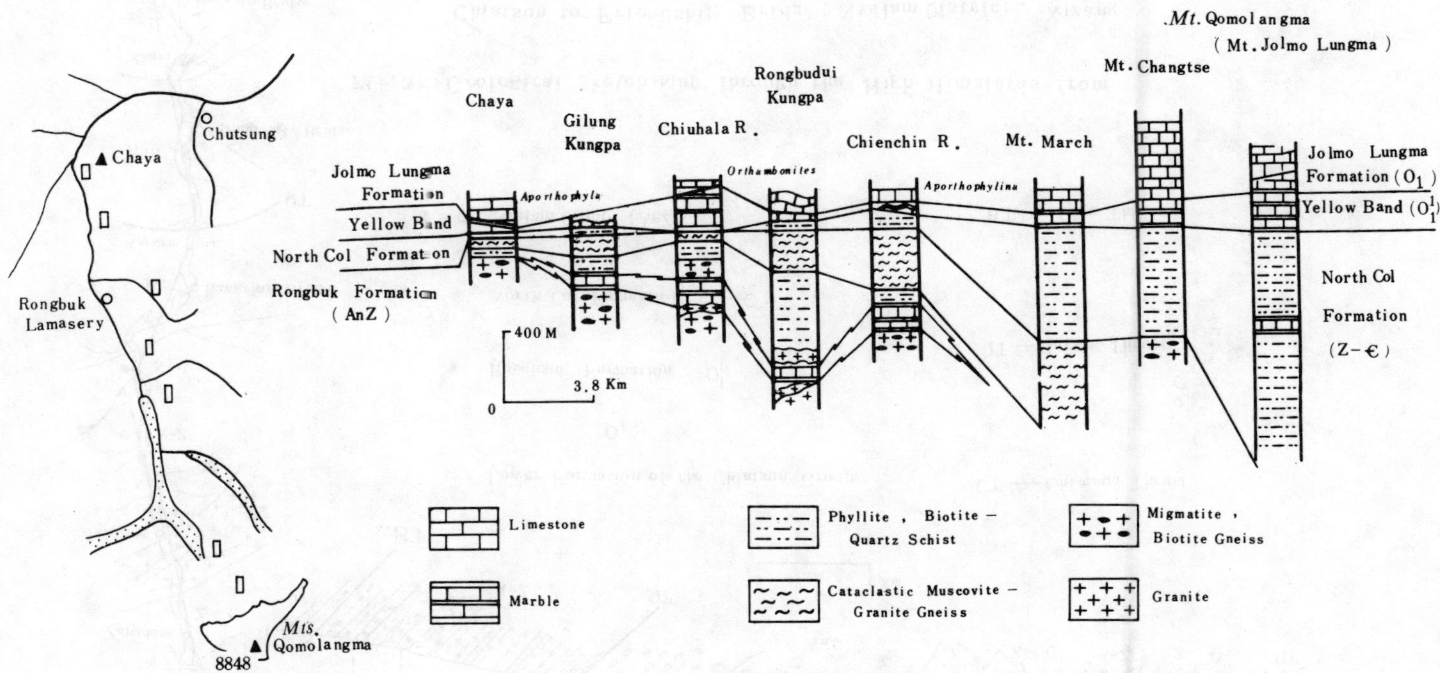


Fig. 2 Stratigraphic columns of the northern slope of Mt. Qomolangma and their correlation (Simplified from Yin Chih-hsiang and Guo Shih - tseng, 1978)

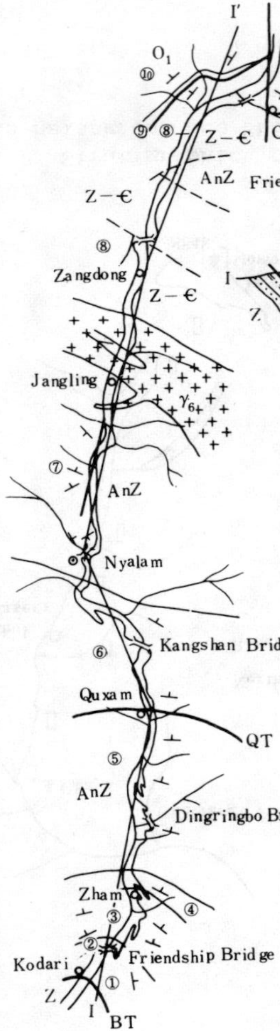


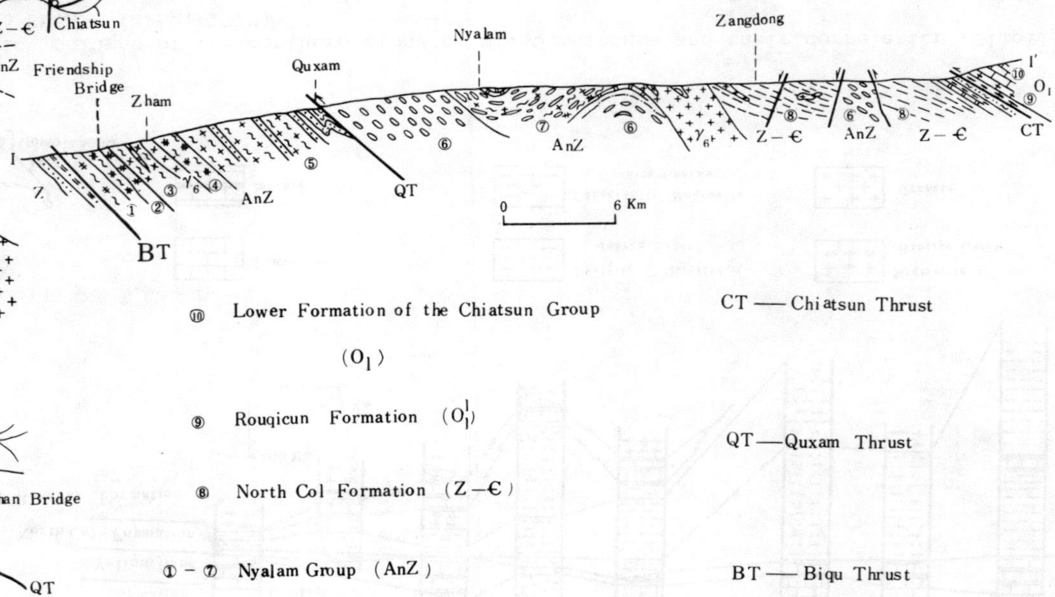


Fig. 3 Geological Sketch Map Through the High Himalayas from Chiatsun to Friendship Bridge, Nyalam District, Xizang

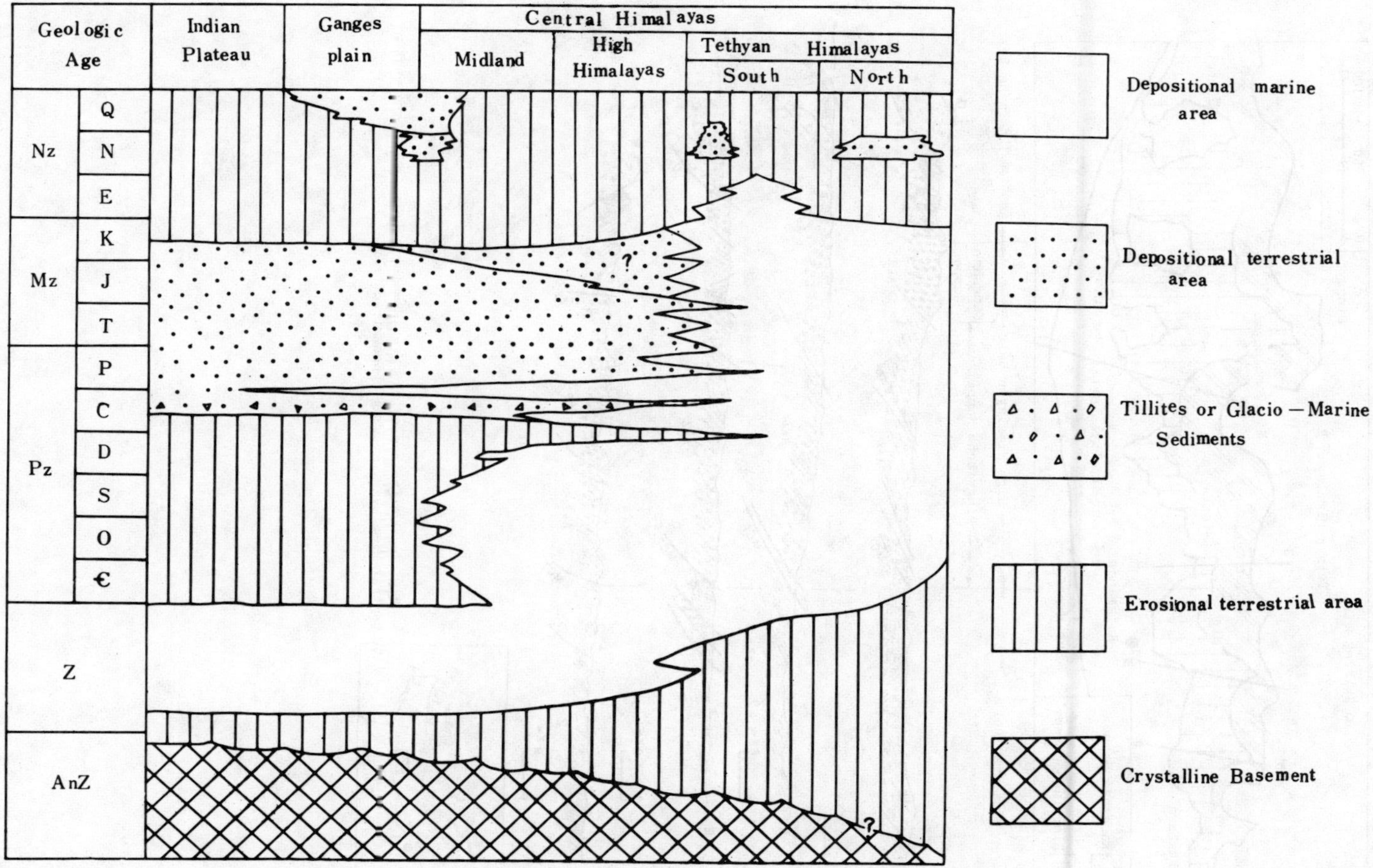


Fig. 4 Migration of sedimentary basins (Modified from Hashimoto, S. et al., ed. 1973)

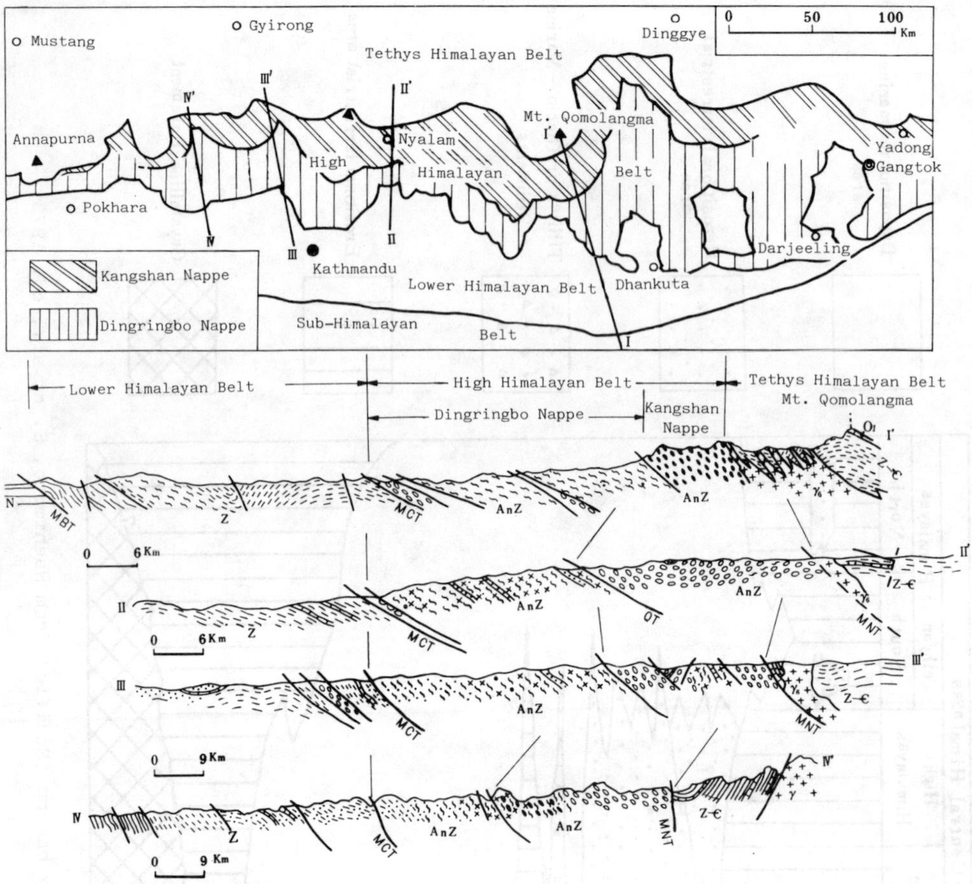


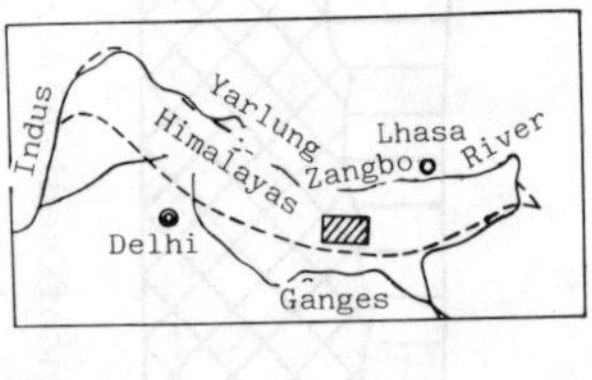


Fig. 5 Outline map of the High Himalayan Belt in eastern part of Central Himalayas, showing nappe structures

## DEVONIAN FOSSILS FROM TANGCHU, CENTRAL BHUTAN, HIMALAYA

V. J. Gupta
(Centre of Advanced Study in Geology,
Panjab University, Chandigarh-160014, India)

Genevieve Termier
(Department of Structural Geology,
University of Paris VI, 75230, Paris, France)

The tectonic slices of fossiliferous Lower Palaeozoic rocks belonging to Tethys Himalaya are known to occur in the Phulchauki-Godavari section of Nepal(1, 5, 6) and Tangchu basin of Central Bhutan(3, 7). The fossils from these two localities have important bearing on the palaeobiogeographic, geosynclinal and tectonic history of the Himalaya. The fauna from the Phulchauki-Godavari section has been described by several workers during the last few years and mention may be made of contributions by Bordet and Cavet(1), Gupta(4) and Gupta and Stocklin(5). On the basis of these publications it is now generally accepted that a complete sequence of rocks from Cambrian to Devonian is exposed in the Phulchauki-Godavari section. The Cambrian rocks are believed to have unconformable contact with the underlying Precambrian sequence.

The occurrence of Lower Palaeozoic fossils from the black slates, dark grey calcareous siltstones and limestones from the Tangchu basin of Central Bhutan was recorded for the first time by Gansser(2). Gupta(3) referred to the occurrence of *Atrypa reticularis*, *Euryspirifer* (?), *Fenestella*, and parts of crinoid stems from the Tangchu basin and assigned Middle Devonian age to the fossiliferous beds. Termier and Gansser(7) illustrated and described the following fossils from the Tangchu basin of Central Bhutan: *Atrypa* cf. *reticularis*, Elythidae gr. *Elita-Torynifer*, *Fenestella* ? sp., ? Phacopidae, *Hexacrinites* ? sp. Termier and Gansser(7) assigned Middle to Upper Devonian age (probably Frasnian) to the fossiliferous beds.

The present note records the following forms in addition to those described by Termier and Gansser(7) from the same locality but from a slight lower stratigraphic horizon: *Spinocyrtia* sp., *Tylothyris* sp. or *Eleutherokomna* sp., *Megachonetes* ? sp., Proetid pygidium, crinoid stems (probably *Hexacrinites* ? sp. ), internal casts of solitary corals.

The fossils referred to above are in a very poor state of preservation and indicate Givetian-Fransnian age supporting the stratigraphic conclusions arrived at earlier by Termier and Gansser(7).

The nearest known locality yielding Devonian fossils are the metasediments belonging to the upper part of the Chekha Group in the Black Mountain Section of Central Bhutan. These sediments have yielded poorly preserved brachiopods, bivalves, corals, trilobites, crinoids, bryozoans, etc. In view of the poor preservation, it is very difficult to make generic or specific identifications of the fossils from the Chekha Group. However, the fauna in general has close similarity with the Devonian fauna from the

Tangchu Basin. The possibility of the lower units of the Chekha Group as belonging to Ordovician and Silurian cannot be ruled out.

## References

(1) Bordet, P., Cavet, J., Sur L'existence d'une faune d'age silurien dans la region de Kathmandu (Himalaya du Nepal). C. R. Ac. Sc., 248, 1547-1549 (1959).

(2) Gansser, A., Geology of the Himalaya. Interscience Publishers. John Wiley & Sons, London, 289 p. (1964).

(3) Gupta, V. J., Devonian fossils from the Lesser Himalayan Zone of Central Bhutan. Curr. Sci., 40(18), 490 (1971).

(4) Gupta, V. J., Upper Devonian conodonts from Phulchauki, Nepal. Him. Geol., 5, 153-168 (1975).

(5) Gupta, V. J., Stocklin, J., Stratigraphy and structure of the Phulchauki-Chandragiri area, Nepal. Recent Researches in Geology, 7, 263-275 (1978).

(6) Gupta, V. J., Termier, G., Lower Palaeozoic Echinodermata from Chandragiri Pass, Near Kathmandu, Nepal. Recent Researches in Geology, 7, 472-480 (1978).

(7) Termier, G., Gansser, A., Les Series devoniennes du Tang Chu (Himalaya du Bhoutan). Eclogae Geologicae Helvetiae, 67(3), 587-596 (1974).

# PERMIAN BRACHIOPOD AND BIVALVE ZONES IN THE HIMALAYA OF INDIA AND NEPAL

J. B. Waterhouse
(Department of Geology,
University of Queensland, Brisbane, Australia)

V. J. Gupta
(Centre of Advanced Study in Geology,
Panjab University, Chandigarh-160014, INDIA)

## INTRODUCTION

The present contribution supplements our paper (Waterhouse and Gupta, 1977) on Permian faunal zones of the Himalaya, drawing attention to further advances in Permian biostratigraphy of the Himalaya.

### *Eurydesma cordatum* Zone

This zone is well developed in the basal part of the Nagmarg Formation of the Agglomeratic Slate Succession of Kashmir at Bren Spur, just above the Golabgarh Boulder Formation. *Eurydesma cordatum* Morris is the most characteristic form and is associated with the bivalves *Deltopecten*, and *Myonia*, and brachiopod *Tomiopsis*. The distribution in the Himalayas is summarized by Waterhouse and Gupta (1977), supplemented by a report of *E. cordatum* from the Malung Shales of Ladakh (Gupta and Waterhouse, 1978).

The *Eurydesma cordatum* Zone is of late Asselian age and is correlative with the extensive *Eurydesma* faunas of Gondwana, ranging through Afghanistan, Argentina, South Africa and Australia.

### *Brachythyrinella narsahensis* and *Stepanoviella umarionsis* Zone

The *Eurydesma* beds are followed by a variety of faunas in various parts of the Himalayas and elsewhere, that appear to be correlative, insofar as they have similar relationships to underlying and overlying zones. *Brachythyrinella narsahensis* (Reed) of the Boulder Slate sequence in the Bijni tectonic unit of Garwhal Himalaya (Waterhouse and Gupta, 1979a) with Chonetid, *Cancrinella*, *Anidanthus* and other species is found also in the Ji-long or Chi-Lung Group of the Tibetan Domain (Waterhouse and Gupta, 1979 b), and as the Umaria fauna of Peninsular India. *Stepanoviella* is found in Tibet and Umaria, and in correlative faunas of Afghanistan (Termier *et al*., 1974). The faunas from the Bomte beds of the Siang district in Arunachal Pradesh are possibly correlative, with rare *Eurydesma*, chonetid, *Cancrinella*, *Subansiria*, *Ambikella* (=*Tomiopsis*), *Platyteichum* and *Uraloceras* (Singh, 1978 a, b, c) although there are few direct links. Conceivably the chonetid-conulariid beds of the Salt Range are similar, and we suggest that the lower Tiverton Formation of the Bowen Basin, Queensland, with *Uraloceras* and with *Attenuatella* as in Tibet is correlative. The Beckett Shale of Western Australia has *Stepanoviella*. The faunas are not well clearly developed in Kashmir, where beds of nondescript bivalves above the *Eurydesma* zone, in the lower Nagmarg Formation of Bren

Spur, might correlative. The fauna is classed in the Tastubian Substage, Sakmarian Stage.

## Reedoconcha permixtus Zone

The upper units of the Nagmarg Formation exposed at Bren Spur, Kashmir, have yielded a diverse fauna represented by *Cyrtella* (or *Punctocyrtella*) *nagmargensis*, *Aperispirifer*, *Tomiopsis*, and abundant *Reedoconcha permixtus* (Reed), and rare *Eurydesma*, also occurring in the Marbal valley and other areas of Kashmir. Similar faunas are in the Tramwala Formation of Bhallesh Group in Kishtwar and Katargali Formation. *Reedoconcha* and allied faunas extend into Afghanistan (see Termier *et al*., 1974) and Iran in Geirud Member D. Allied genera including *Taeniothaerus* are now found to be particularly abundant in the Bowen Basin of Queensland, in the Upper Tiverton Formation, Riverstone Sandstone and Pindari beds and *Eurydesma* band of the Cattle Creek Formation. The faunas are regarded as Sterlitamakian.

## Retimarginifera Zone

Limestone intercalations within the Ralakung Volcanic Succession of the Luneak valley at Zanskar, Ladakh, have yielded well preserved *Derbyia*, *Plicatifera*, *Juresania*. *Retimarginifera*, *Reticularia*, and *Cleiothyridina semiconcava* (Waagen), the latter shared with the lower Amb Formation of Salt Range, Pakistan and suggestive of a Baigendzinian age. No other fauna of the Himalayas is similar (Gupta and Waterhouse, 1979).

## Lamnimargus himalayensis Zone

This zone is typified by faunas of the lower Zewan Formation of Kashmir, with *Waagenoconcha gangeticus* (Diener), *Lamnimargus himalayensis* (Diener), *Costiferina alatus* Waterhouse, *Anidanthus fusiformis* Waterhouse, *Spiriferella rajah* (Salter), *Neospirifer moosakhailensis* (Davidson) var. *ambiensis* (waagen), *Fusispirifer nitiensis* (Diener) and *Hoskingia latouchei* (Diener) prominent. Similar faunas are also known to occur in the Talai Formation of Kishtwar; Salooni Formation of Chamba; Kuling Shales of Kalapani-Kuti sections of northeastern Kumaun; Lachi Group of Sikkim; Productus Shales of Black Mountain area of Bhutan; Selung Group of South Tibet and the Nangung Formation of Namlang Group of west Nepal. The exotic blocks of Chitichun No. 1 and Malla Sangcha in south Tibet may possibly belong to this zone (Waterhouse and Gupta, 1978a, 1979c).

The *Lamnimargus himalayensis* Zone is correlated with the lower Punjabian stage, especially Kalabaghian Substage.

## Pyramus silicius Zone

This zone was hesitantly recognised by Waterhouse and Gupta (1977; Waterhouse, 1978a) on the basis of principally bivalve faunas with *Pyramus*, *Schizodus*, and *Permophorus* in the Popa Member of the Senja Formation in northwest Nepal, matched with a more diverse member in Zewan Member C (lower Fauna 111) of Nakazawa *et al*. (1975) in Kashmir, with bivalves, conodonts, and gastropods including *Bellerophon blandfordianus* (Waagen) and *Retispira ornatissima* (Waagen). Foraminifera were compared with

Urustenian - Abadehian forms of Armenia. These faunas are now recognised beyond the Himalayas. Similar molluscan genera and species are found in the "topmost" or bivalve bed of Waagen (1891), or upper Chhidruan, of the Salt Range, and similar bivalves, with Urushtenian brachiopods, were recorded from the Karakorum by Merla (1934). The age is Urushtenian, of the Djulfian Stage.

Krotovia arcuata Zone

This zone is represented in the Pija Shale Member of the Senja Formation in northwest Nepal and is characterised by Krotovia and other brachiopod species. There are no perceptible links with the Kashmir sequence, but the "white sand" near the top of the Permian beds in the Salt Range might be correlative. The zone is assigned to the Baisalian Substage, Djulfian Stage.

Marginalosia kalikotei Zone

This zone is represented in the Nisal, Nambdo and Luri Members of the Senja Formation in northwest Nepal, with a suite of fossils including the named species, and species of Cancrinella, Neospirifer, Spiriferella, Hoskingia, etc. The zone has been traced east to the Kali Gandaki Valley of Nepal by Waterhouse (1978 b) into thin beds previously mapped as Triassic. No such fauna has been found in Kashmir. The zone is of lower Dorashamian age and belongs to the Vedian Substage.

Atomodesma variable Zone or Subzone

Atomodesma variabile Wanner dominates a small fauna with Glabrichonetes in Senja Formation in northwest Nepal. Similar faunas are also possibly found with the faunal division IV of Nakazawa et al. (1975) within the Zewan Formation of Kashmir.

This zone is of lower Dorashamian age and may belong to Ogbinan Substage. It might match the Paratirolites beds of Armenia and Madagascar, or be slightly older.

Claraia-Otoceras-Ophiceras Zonal Complex

The faunas of this zone have traditionally been placed at the base of the Triassic Period due to the complete absence of the Palaeozoic brachiopods Diener (1909). However, Permian-type brachiopods are now widely known with Ophiceras and especially Otoceras (Waterhouse, 1976, 1978 b). Some authorities rely on the ammonoids, Otoceras, Ophiceras but historically, in establishing the boundary, prime stress was laid on the presence or absence of Strophomenida and Productida. Others now stress certain bivalve genera, such as Eumorphotis, but these have no significance in the original definition of the Triassic Period, and the one form that might have been important, Claraia, is known in Permian rocks (Nakazawa et al., 1975).

The Otoceras-Ophiceras beds are widely represented in the Himalayas, in Nepal, Kashmir, etc. as discussed by Waterhouse and Gupta (1977) and Waterhouse (1978 a,b).

Summary

There is at least one major gap in the Himalayan sequences between the Retimarginifera and Lamnimargus zones, of early Middle Permian age. Waterhouse and Gupta (1977, p. 15) suggested that this may be partly closed by a Gondolella rosenkrantzi conodont assemblage in Ladakh. Early Permian Fusulinacea are also found in this region, perhaps close to the Eurydesma zone in age, or a little younger. It will be observed that Fusulinacea, Ammonoidea, and conodonts, though favoured by some authorities, are too rare, and too isolated to be of practical use in correlation: the burden falls on brachiopods especially, and to lesser extent on bivalves, as is the case for most Permian sequences of the world outside of the palaeotropical realm (Waterhouse, 1976).

A particularly interesting find has been the discovery of Linoproductus in the Kroll Limestone near Nainital, in the Lesser Himalayas (Valdiya, 1980). We point out that Valdyia was misinformed about the age implications of this genus: it extends from Middle Carboniferous (if indeed not Visean) into the very late Permian - and has for instance been described from the late Permian of Nepal. Ahluwalia (1978) has also recorded upper Palaeozoic foraminifera Porifera from the overlying lower Tal Formation and Valdiya (1980) has stated that a Permian age was established for the lower Tal through moravamminid fossils.

## References

(1) Ahluwalia, A. D., Discovery of Upper Palaeozoic fossils (Foraminifera or Porifera ?) from Mussoorie Phosphorite, lower Tal Formation, Kumaun Lower Himalaya, India. Contrib. Himal. Geol. 1, 20-24 (1978).

(2) Gupta, V. J., Waterhouse, J. B., Eurydesma from Lahaul and Laddakh. Bull. Ind. Geol. Assoc. 11 (2), 141-152 (1978).

(3) Gupta, V. J., Waterhouse, J. B., Permian invertebrate fauna from the Ralakung Volcanics, Ladakh Himalayas, India. Rec. Resear. Geol. 5, 31-49 (1979).

(4) Diener, C., Lower Triassic Cephalopoda from Spiti, Malla Johar, and Byans. Palaeont. Ind. Ser. 15 (2), 1-186 (1909).

(5) Merla G., Fossils anthracolitici del Caracorum. Spedizione italiana De Filippi nell'H malaia Caracorum c Turchestan Cinese (1913-1914) Ser. 25, 101-319 (1934) Bologna.

(6) Nakazawa, K., Kapoor, H. M., Ishii, K., Bando, Y., Okomura, U., Tokuoka, I., Murata, M., Nakamura, K., Nogami, Y., Sakagami, S. and Shimizu, D., The Upper Permian and the Lower Triassic in Kashmir, India. Mem. Fac. Sci., Kyoto Univ. Ser. Geol. Mineral. 42 (1), 106 p. (1975).

(7) Singh, Tricholan., Lower Permian Cephalopods from eastern Himalaya, India. Himal. Geol. 8, 178-193 (1978a).

(8) Singh, Tricholan.,Lower Permian gastropods and bivalves from eastern Himalaya, India. Rec. Res. Geol. 7, 276-310 (1978b).

(9) Singh, Tricholan., Brachiopods from Permian Formation of Siang district, Arunachal Pradesh. Contrib. Himal. Geol. 1, 171-188 (1978c).

(10) Termier, G., Termier, H., De Lapparent, A. F. and Martin, P., Monographie du Permo-Carbonifere de Wardak (Afghanistan central). Doc. Labor. Geol. Fac. Sci. Lyon H. S. 2, 135 p. (1974).

(11) Valdiya, K. S., Discovery of late Palaeozoic brachiopod in the Upper Krol of the Nainital Hills, Kumaun Himalaya. J. Geol. Soc. Ind. 21, 97-101. (1980).

(12) Waagen, W., Salt Range Fossils. Palaeont. Ind. ser. 13 4 (2), 89-242 (1891).

(13) Waterhouse, J. B., World correlations for marine Permian faunas. Pap. Dep. Geol. Univ. Qd, 7 (2), 1-232 (1976) Brisbane.
(14) Waterhouse, J. B., Permian Brachiopoda and Mollusca from northwest Nepal. Palaeontographica A 160, 1-178 (1978a).
(15) Waterhouse, J. B., Permian rocks of the Kali Gandaki area, (Thakkola), north central Nepal. Contr. Him. Geol. 1, 195-213 (1978b).
(16) Waterhouse, J. B., Gupta, V. J., Permian faunal zones and correlations of the Himalayas. Bull. Ind. Geol. Assoc. 10 (2), 1-19 (1977).
(17) Waterhouse, J. B., Gupta, V. J., Permian invertebrate faunas from the *Lamnimargus himalayensis* Zone of Spiti and Ladakh regions, northwest India. Contrib. Himal. Geol. 1, 5-19 (1978).
(18) Waterhouse, J. B., Gupta, V. J., Early Permian fossils from the Bijni Tectonic Unit, Garhwal, Himalaya. Rec. Res. Geol. 4, 410-437 (1979a).
(19) Waterhouse, J. B., Gupta, V. J., Early Permian fossils from Southern Tibet like faunas from Peninsular India and Lesser Himalayas of Garhwal. J. Geol. Soc. Ind. 20, 461-464 (1979b).
(20) Waterhouse, J. B., Gupta, V. J., Late middle Permian brachiopods from Marbal Paaa, Kashmir, India. Bull. Ind. Geol. Assoc. 12 (1), 1-42 (1979c).

Table 1 Permian brachiopod and bivalve zones of the Himalaya in India and Nepal. Note that "typical genera" refers to species and generic abundance, not to restricted generic ranges

| International Stage | Substage | Himalayan Zone | Typical genera | Typical area |
|---|---|---|---|---|
| Dorashamian | Ellesmerian - Gangetian | Ophiceras & Otoceras | Claraia, Ophiceras, Otoceras, Glyptophiceras | Panjang Khola, NW Nepal Guryul Ravine, Kashmir |
| | Ogbinan | Atomodesma variabile | Glabrichonetes Atomodesma | NW Nepal, Kashmir |
| | Vedian | Marginalosia kalikotei | Rugaria Megasteges Cancrinella Pondospirifer | NW Nepal |
| Djulfian | Baisalian | Krotovia arcuata | | NW Nepal |
| | Urustenian | Pyramus silicius | Permophorus, Schizodus, Bellerophon, Retispira spp. | Kashmir, NW Nepal |
| Punjabian | low-mid.Chhidruan Kalabaghian | Lamnimargus himalayensis | Waagenoconcha, Costiferina, Anidanthus, Spiriferella Fusispirifer spp. | lower Zewan Ftm Kashmir |
| Kazanian | Sosnovian | | | |
| | Kalinovian | | | |
| Kungurian | Ufimian | | | |
| | Irenian | | | |
| | Filippovian | | | |
| Baigendzinian | Krasnoufimian | Retimarginifera | Derbyia, Juresania | Ladakh |
| | Sarginian | | | |
| Sakmarian | Aktastinian | | | |
| | Sterlitamakian | Reedoconcha permixtus | Taeniothaerus, Punctocyrtella, Tomiopsis | Bren Spur Kashmir |
| | Tastubian | Stepanoviella Brachythyrinella | Neochonetes, Subansiria, Tomiopsis | Bijni, Garwhal, Umaria |
| Asselian | Kurmaian | Eurydesma cordatum | Myonia, Tomiopsis | Bren Spur, Kashmir |
| | Uskalikian | | | |
| | Surenan | | | |

# PRECAMBRIAN TECTONICS IN AUSTRALIA, VIEWED FROM THE HIMALAYAS AND THE QINGHAI-XIZANG PLATEAU *

K.A. Plumb
(Bureau of Mineral Resources, Canberra, Australia)

## INTRODUCTION

The origin of Proterozoic polymetamorphic belts of reactivation, of which the Australian Shield is a typical example, remains enigmatic[1]. Ensialic models prevail[2,3,4,5,6,7]. Precambrian palaeomagnetic data[2] are still indecisive[8], but tectonic patterns suggest that major crustal blocks maintained their relative positions throughout the Proterozoic[3,4,5]. Despite apparent differences in Proterozoic and Phanerozoic tectonic styles, there are also significant parallels in tectonic cycles of all ages[5]. Over all Proterozoic mobile belts, younger cover may conceal critical zones.

This paper describes three ancient intracontinental mobile belts - the Arunta-Musgrave Blocks, Albany-Fraser Province, and Capricorn Orogen (Fig. 1) - for comparison with the modern Himalayas and Qinghai-Xizang Plateau.

## REGIONAL SETTING

Figure 1 is a simplified and modified[5,6] Tectonic Map of Australia[9]. The belts under consideration are parts of the interconnected Central Australian Mobile Belts[5] - linear belts of reactivation adjacent to or between ancient cratons.

## REGIONAL DESCRIPTIONS

### Central Australia

The Arunta and Musgrave Blocks lie between the North Australian and Gawler Cratons (Fig. 1); these cratons have distinctly different histories and tectonic styles[5]. Proterozoic palaeomagnetic data[2], from North Australia in particular, are inadequate to confidently define the relative positions of these cratons. Thrusting correlates with northward strike-slip faulting in the North Australian Craton[6]. Central Australia is characterised by the largest Bouguer anomalies in Australia.

The <u>Arunta Block</u> is divided into three tectonic zones (Fig. 3)[6]; major steep faults separate them. Three stratigraphic divisions, separated by unconformities, are identified in each. These units, at least in the Northern Zone, are continuous with Early Proterozoic sequences of the North Australian Orogenic Province. The Northern Zone is characterised by mostly low-grade metasediments of Division 2 and abundant "S-type" granites. The Central Zone

---

* Published with permission of Director, Bureau of Mineral Resources, Geology and Geophysics.

is the main area of Division 1, low-pressure granulites, metamorphosed during the 1800-1750 m.y. Strangways Event. The Southern Zone is characterised by intense young (1050 m.y.) migmatisation and reactivation (Ormiston Event), superimposed on 1700-1550 m.y. amphibolite metamorphism (Chewings Event); granites are mostly "I-type".

The Musgrave Block mirrors the Arunta Block (Fig. 2). A southern zone comprises 1650 m.y. gneisses, apparently continuous with the Gawler Craton[10], and intruded by 1150-1100 m.y. granite. A central zone of major northward-directed thrusts contain high-pressure granulites[11], 1200 m.y. old with relict 1550 and 1330 m.y. ages[12]; the granulites may be a new, younger sequence[12,13]. Emplacement of the Giles Complex[14,15] overlapped a period of major thrusting[10,15]. The northern zone[16] resembles the Arunta Southern Zone - retrogression, migmatisation and granite emplacement - the 1150-1100 m.y. Kulgeran Event.

Late or post-tectonic bimodal volcanics and granites were emplaced into the Musgrave Block about 1040 m.y.,[15] and post-tectonic granites into the Arunta Block[6] at 1000 m.y. Platform cover (Amadeus, Ngalia, Officer Basins) was then deposited over the whole belt. Intracontinental thrusts and nappes deformed basement and cover at the end of the Proterozoic (Petermann Ranges Orogeny)[16] and during the Early Carboniferous (Alice Springs Orogeny)[6].

The most popular crustal models[7] (Fig. 4) postulate major thrusts extending to the mantle, and major relief at the base of the crust. Major Bouguer anomaly highs correlate with upthrust deep-crustal granulite blocks.

## Albany-Fraser Province

The Albany-Fraser Province resembles the Grenville Province. It truncates Archaean trends in the Yilgarn Block (Fig. 1) at a high angle. The belt is marked by a major gravity gradient. Five tectonic zones are identified[4] (Fig. 5). Archaean rocks are traced into the Province, where several episodes of amphibolite or granulite metamorphism and granite emplacement occurred between 1850 and 1100 m.y. The western boundary (Fraser Front) is gradational. The Fraser Complex is a floored body of supracrustal mafic granulite. The only Proterozoic sediments are metamorphosed platform deposits. Preferred gravity models[4,7] show major faults extending into the mantle (Fig. 6); the nature of the faulting is uncertain, but sinistral strike-slip displacement is probable[4].

## Capricorn Orogen

The Capricorn Orogen[4], between the Archaean Pilbara and Yilgarn Blocks, is largely covered by the Bangemall Basin (Fig. 1). The Pilbara and Yilgarn Blocks have distinctly different tectonic histories and styles[4,5], but are supposed to have behaved as one unit since the end of the Archaean[2,3]. Shelf sequences on both cratons pass into turbidites in the central orogen[4] (Fig. 7); olistostromes and pillow basalts are recorded[3]. The orogen was multiply deformed, metamorphosed, and intruded by granites 1700-1500 m.y. ago. Explosion-seismic profiles[17] (Fig. 8) show different crustal structures and thickness and a sharp boundary between the Yilgarn and Pilbara Blocks. Lateral changes in structure correlate with faults. A gravity high over the velocity gradient has been explained as an intracontinental rift[3,7].

## CONCLUSION

These belts are all zones of repeated reactivation between major cratons. They have sialic basements. Major structures extend to the mantle, delineate distinctive tectonic zones, and produce major relief of the Moho. Deep crustal rocks are exposed. Platform sediments are metamorphosed.

Such features are common in Tibet. Perhaps Proterozoic ensialic models are inappropriate. Students of Precambrian tectonics require comparative data on the deep-seated rocks and geochemical signatures of collision belts, which the great uplift of the Tibet Plateau and Himalayas can provide.

Dr A.J. Stewart reviewed the manuscript, and the figures were prepared in the BMR Drawing Office.

## REFERENCES

(1) Windley, B.F., The Evolving Continents. John Wiley, 385 pp. (1977).
(2) McElhinny, M.W., Embleton, B.J., Precambrian and early Palaeozoic palaeomagnetism in Australia. Phil. Trans. R. Soc. Lond. A, 280, 417-431 (1976).
(3) Horwitz, R.C., Smith, R.E., Bridging the Yilgarn and Pilbara Blocks, Western Australia. Precambrian Res., 6, 293-322 (1978).
(4) Gee, R.D., Structure and tectonic style of the Western Australian Shield. Tectonophysics, 58, 327-369 (1979).
(5) Plumb, K.A., The tectonic evolution of Australia. Earth-Sci. Rev., 14, 205-249 (1979).
(6) Plumb, K.A., Derrick, G.M., Needham, R.S., Shaw, R.D., The Proterozoic of northern Australia. In, Precambrian of the Southern Hemisphere, Developments in Precambrian Geology 2, (Edit. D.R. Hunter), Elsevier (in press).
(7) Mathur, S.P., Shaw, R.D., Geophysics and geology of the Australian geosutures - a review. Precambrian Res. (in press).
(8) Burke, K., Dewey, J.F., Kidd, W.S.F., Precambrian palaeomagnetic results compatible with contemporary operation of the Wilson Cycle, Tectonophysics, 33, 287-299 (1976).
(9) GSA (Geological Society of Australia), Tectonic Map of Australia and New Guinea 1:5 000 000, Sydney (1971).
(10) Thomson, B.P., Musgrave Block - regional geology. In, Economic Geology of Australia and Papua New Guinea, 1. Metals. Aust. Inst. Min. Metall. Mon., 5, 451-454 (1975).
(11) Moore, A.C., Goode, A.D.T., Petrography and origin of granulite-facies rocks in the western Musgrave Block, central Australia. J. Geol.Soc. Aust., 25, 341-358 (1978).
(12) Gray, C.M., Geochronology of granulite-facies gneisses in the western Musgrave Block, Central Australia. J. geol. Soc. Aust., 25, 403-414 (1978).
(13) Collerson, K.D., Oliver, R.L., Rutland, R.W.R., An example of structural and metamorphic relationships in the Musgrave Orogenic Belt, central Australia, J. geol. Soc. Aust., 18, 379-393 (1972).
(14) Nesbitt, R.W., Goode, A.D.T., Moore, A.C., Hopwood, T.P., The Giles Complex central Australia: a stratified sequence of mafic and ultramafic intrusions. Spec. Publs. geol. Soc. S. Afr., 1, 547-564 (1970).
(15) Daniels, J.L., The geology of the Blackstone region, Western Australia. Geol. Surv. W. Aust. Bull., 123 (1974).
(16) Forman, D.J., Regional geology of the southwest margin, Amadeus Basin, central Australia. Bur. Miner. Resour. Aust. Bull., 144 (1966).
(17) Drummond, B.J., A crustal profile across the Archaean Pilbara and northern Yilgarn cratons, northwest Australia. BMR J. Aust. Geol. Geophys., 4, 171-180 (1979).

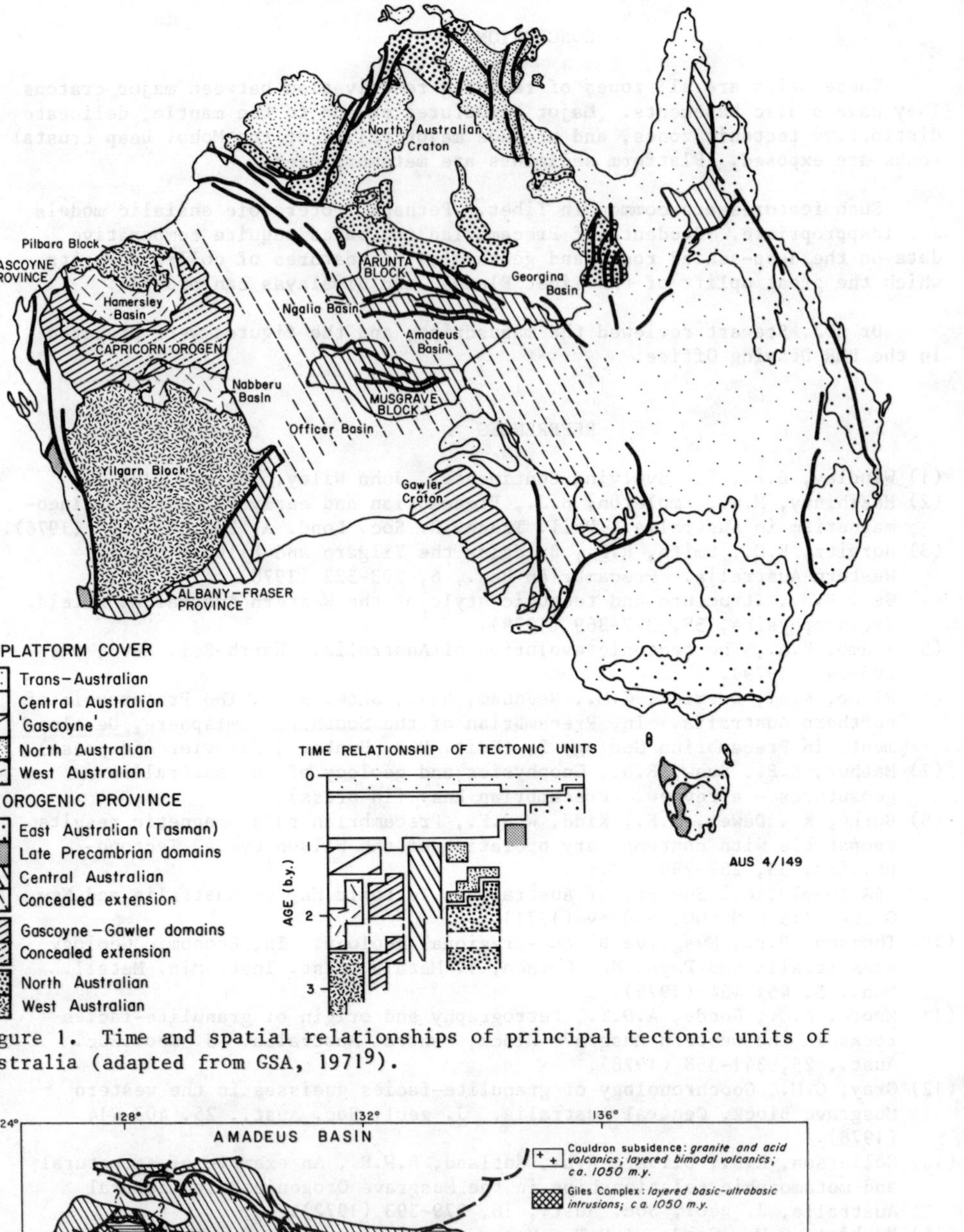


Figure 1. Time and spatial relationships of principal tectonic units of Australia (adapted from GSA, 1971[9]).

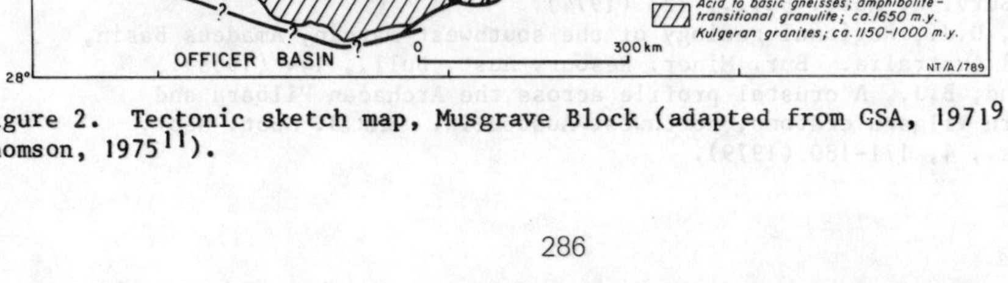


Figure 2. Tectonic sketch map, Musgrave Block (adapted from GSA, 1971[9]; Thomson, 1975[11]).

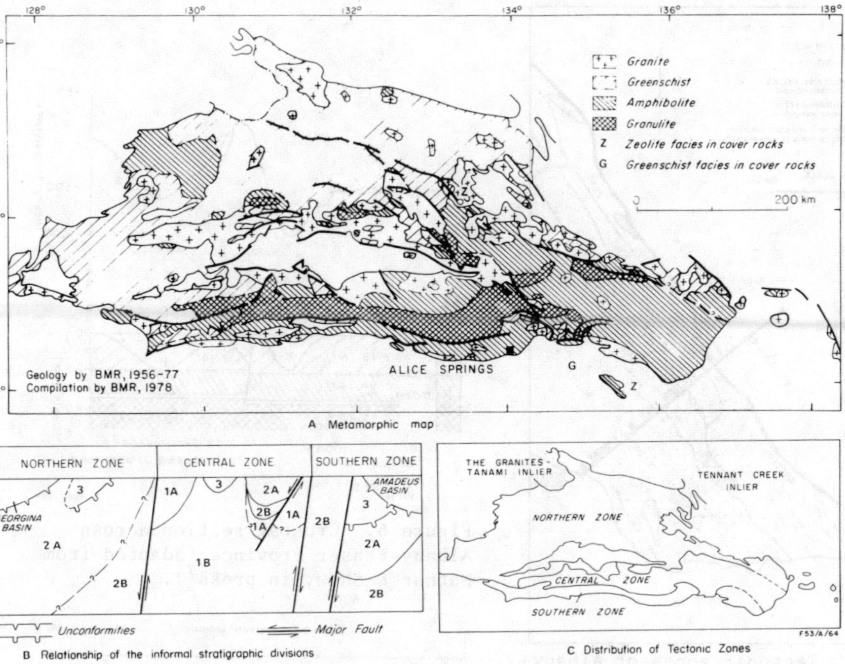


Figure 3. Metamorphic rocks, stratigraphic relationships, and tectonic zones, Arunta Block (from Plumb et al., in press[6]).

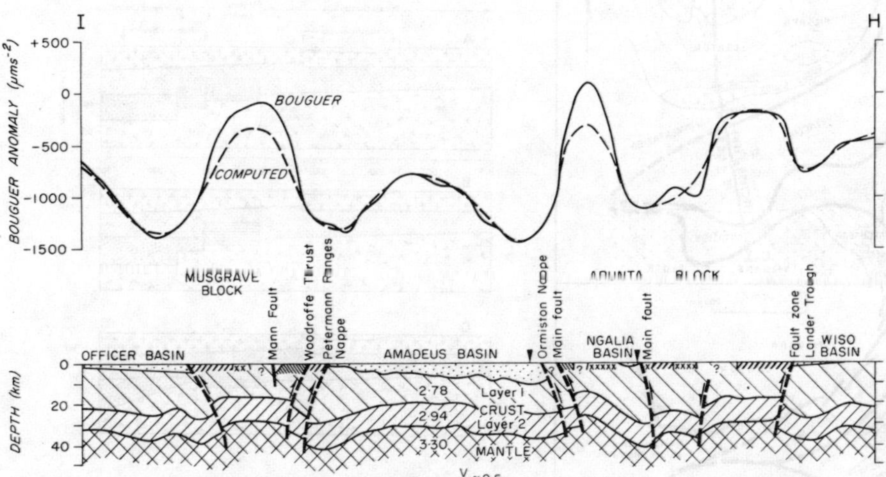


Figure 4. North-south crustal section across central Australia (from Mathur & Shaw, in press[7]).

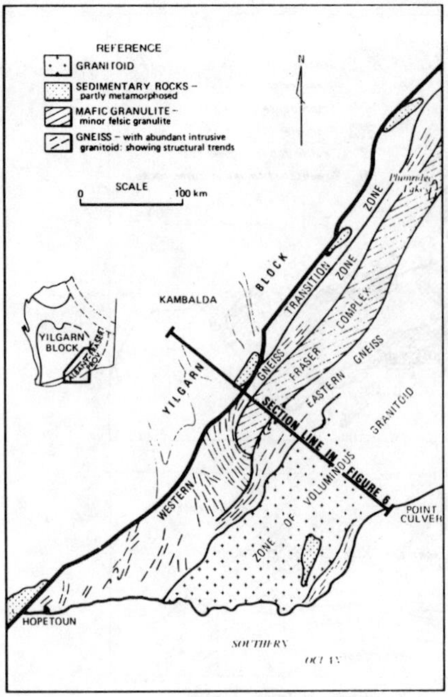


Figure 5. Tectonic zones of Albany-Fraser Province (from Gee, 1974[4]; with permission of Elsevier).

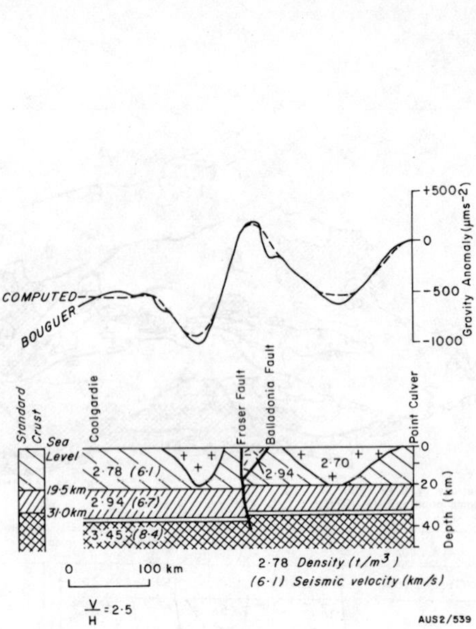


Figure 6. Crustal section across Albany-Fraser Province (adapted from Mathur & Shaw, in press[7]).

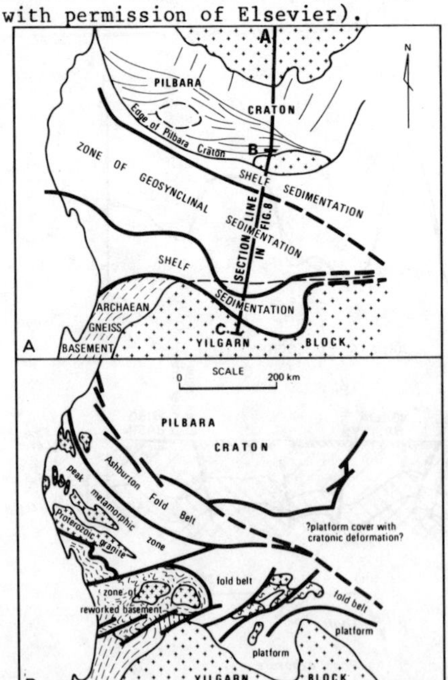


Figure 7. Evolution of Capricorn Orogen. A. Sedimentation, 2000 my BP. B. Orogenesis, 1600 my BP (from Gee; 1974[4], with permission of Elsevier).

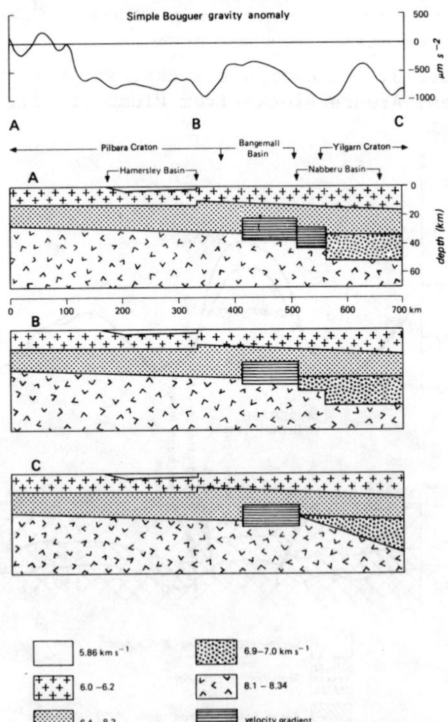


Figure 8. Alternative crustal profiles across Capricorn Orogen (from Drummond, 1979[17]).

# THE CRETACEOUS - TERTIARY BOUNDARY IN ZANSKAR RANGE (WESTERN HIMALAYA)

Maurizio Gaetani
(Istituto di Paleontologia, Università di Milano, Italia)

In a recent paper A. Nicora, I. Premoli Silva and myself described a sequence through the Cretaceous-Tertiary boundary in Zanskar Range, Ladakh, which is the most complete for the Western Himalaya, so far known (Gaetani et al., 1980). Here I will summarize the sequence and results (Fig. 1).

## THE SEQUENCE

From bottom to top:

### Kangi La Flysch

Dark grey, burrowed, rusty weathered sandstones to silty slates; calcareous cement. Thickness about 400 m; age: no own data; according to Fuchs (1979) ? Campanian - Maastrichtian.

### Grey dark limestones and marls

1. Grey silty marls in poorly defined beds, densely burrowed. 40-80 m

2. Omphalocyclus Beds. Dark grey wackestone to packstone, more ralely grainstone. Richly fossiliferous, with two fossil assemblages, separated or mixed up in some beds. Assemblage A is dominated by Omphalocyclus macroporus (Lamarck) both A and B forms; subordinate are small Miliolids and Texturaliids and fragmented algae. Assemblage B is dominated by Dasycladaceae ( Acroporella, Cymopolia, Trinocladus ) and subordinate Corallinaceae ( Archaeolithothamnium) ; Forams are rare with O. macroporus, Iraqia cf. complanata Henson, Pseudorbitolina sp., Kilianina sp., Minouxia sp. 65.3 m

3. Grey silty marls, poorly exposed. about 15 m

4. Very mature whitish quartzarenites. 9.6 m

5. Grey silty marls and grey marly packstone very rich in Siderolites calcitrapoides Lamarck and with rare O. macroporus and Gavelinella sp. about 40 m

6. Dark grey wackestone to packstone in poorly defined beds. At 10 m from the base abundant Daviesina danieli Smout and rarer D. khatiyahi Smout. Lockhartia sp., Sphaerogypsina sp., Rotalia sp., algae in fragments as Archaeolithothamnium, Dystichoplax, Ethelia and Peyssonelia are also present. 35 m

7. Dark marls. 8 m

8. Dark grey coarse grainstone to packstone, in poorly defined beds, extremely rich in Daviesina langhami Smout and with rarer D. khathiyahi and very rare Ranikothalia sp. 5 m

9. The sequence follows up for more than 100 m with dark grey fossiliferous wackestone to packstone alternating with dolomitized wackestone.

Purple and green litharenites and shales

This fissile, varicoloured unit represents the top of the sequence in the area of the Zanskar Synclinorium.

The Cretaceous-Tertiary boundary thus lies between assemblages with Siderolites calcitrapoides and with Daviesina danieli, corresponding to levels (5) and (6) of the sequence. Late Maastrichtian and Middle Palaeocene are the ages detected. The entirely shallow marine sequence shows a smooth trend from restricted very shallow platform to a somewhat more open marine environment.

## CORRELATIONS

No obvious correlations can be made with other series outcropping in Ladakh and in Western Himalaya, where or flyscioid-molassic sequences developed in this interval, or carbonate shallow marine sequences are not preserved or described till now.

The best correlations can be made with Southern Xizang (Tibet) (Mu En Zhi et al., 1973; Ho et al., 1976; Zhang Bing-gao, 1979). Levels (2) to part of (5) can be correlated with the topmost part of the Zongsham Formation and/or the Jidula Formation. Daviesina assemblages are possibly correlatable with Unit 2 and may be with the Unit 3 (as suggested me by Dr. Wang during the Symposium) of the Zongpu Group.

Moreover, the occurrence of Lower-Middle Eocene in Ladakh is long known (Stolickza, 1866; Lydekker, 1880; La Touche, 1888; Fossa Mancini, 1928; Metzeltin and Nicora, 1977; Fuchs, 1977,1979). Also in Southern Xizang the Zebu Ri Formation, locally ending the marine sequence, contains abundant fossils of Lower-Middle Eocene age (A Scientific Guidebook to South Xizang (Tibet), 1980).

Consequently, a continuous connection during Maastrichtian and Middle Palaeocene to Middle Eocene between Ladakh and Southern Xizang, seems to be sufficiently demonstrated.

Similar, somewhat correlatable successions outcrop in Pakistan, namely in the Northern Indus area and in the Baluchistan-Sind.

## References

(1) Fossa Mancini,E., In Fossa Mancini,E., Parona,C.F., Stefanini, G., Fossili del Secondario e del Terziario. Spedizione italiana De Filippi nell'Himalaia, Caracorum e Turchestan Cinese (1913-14) 2 (6),1-334,Zanichelli Bologna (1928).

(2) Fuchs,G.,Traverse of Zanskar from the Indus to the Valley of Kashmir - a preliminary note. Jb. Geol.B-A, 120 (2), 219-229, Wien (1977).

(3) Fuchs, G., On the Geology of Western Ladakh. Jb. Geol. B-A, 122 (2), 513-540, Wien (1979).

(4) Gaetani,M., Nicora, A., Premoli Silva, I., Uppermost Cretaceous and Paleocene in the Zanskar Range (Ladakh-Himalaya). Riv. It. Pal. Strat., 86 (1),127-166, Milano (1980).

(5) Ho Yen, Zhang Ping-kao, Hu Lang-ying, Sheng Jing-chang, Mesozoic and Cenozoic Foraminifera from the Mount Jolmo Lungma Region. A Report Sc. Exp. Mount Jolmo Lungma Region (1966-1968), Palaeontology, (2),1-125, Beijing (1976).

(6) La Touche, T.H.D., Re-discovery of Nummulites in Zanskar. Rec. Geol. Survey India, 21 (4), Calcutta (1888).
(7) Lydekker, R., Geology of Ladakh and neighbouring districts, being fourth notice of geology of Kashmir and neighbouring territories. Rec. Geol. Surv. India, 13 (1), 26-58, Calcutta (1880).
(8) Metzeltin, S., Nicora, A., A new Eocene outcrop from Zanskar Range, Himalaya. Riv. It. Pal. Strat.,83 (4),803-824, Milano (1977).
(9) Mu En-zhi, Yin Ji-xiang, Wen Shi-xuan, Wang Yi-gang, Zhang Binggao, Stratigraphy of the Mount Jolmo Lungma Region in Southern Tibet, China. Scientia Geol. Sinica, 1,3-36, Beijing (1973).
(10) Organizing Committee, A Scientific Guidebook to South Xizang (Tibet). 1-104, Beijing (1980).
(11) Stoliczka, F., Summary of geological observations during a visit to the Provinces Rupshu, Karnag, South Ladakh, Zanskar, Suroo and Dras of Western Tibet, 1865. Mem. Geol. Surv. India, 5(3), 337-354, Calcutta (1866).
(12) Zhang Bing-gao, The Cretaceous-Tertiary Boundary of Southern Tibet, China. In Christensen, W, Birkelund, T., (Eds.) Cretaceous-Tertiary Boundary Events Symposium, Proceedings-Appendix,258-262, Copenhagen (1979).

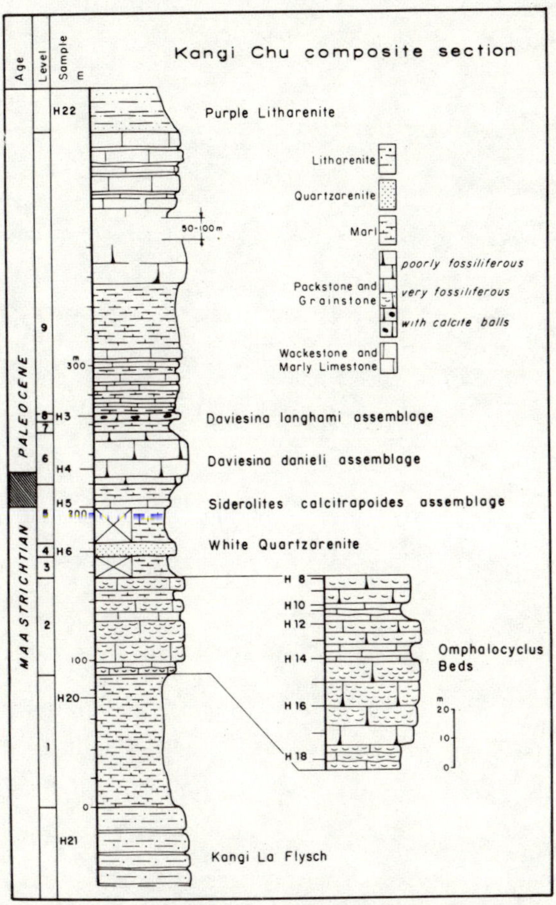


Fig 1 - Kangi Chu composite section. The Cretaceous-Tertiary boundary should be placed between Siderolites and Daviesina assemblages.

## PALAEOZOIC BIOSTRATIGRAPHY AND PALAEOGEOGRAPHY OF THE HIMALAYA

V. J. Gupta
(Centre of Advanced Study in Geology,
Panjab University, Chandigarh-160014, India)

Well developed sequences of fossiliferous Palaeozoic rocks are exposed in different parts of Kashmir, Ladakh and Spiti. Incomplete sequences of these rocks are also exposed in parts of Nepal, Bhutan and Lesser Himalayas. A general review of the data published till 1976 on the Palaeozoic biostratigraphy of the Himalaya has already been published by Gupta(8). The present paper deals with the contributions made by the author and other colleagues from India and abroad on the Palaeozoic biostratigraphy between the years 1977 and 1980.

### KASHMIR-KISHTWAR-CHAMBA REGIONS

The Cambrian trilobites from the Hundwara basin, Baramula District, Kashmir, have been redescribed and some of the identifications revised on the basis of latest nomenclature. The find of graptolites by the author from the Ordovician and Silurian rocks of Kashmir has been supported by the officers of the Geological Survey of India(4) as they have also found the occurrence of Late Silurian graptolites in the Harpatnar Valley near Aishmuqam and Banihal area, Kashmir. Gupta and Janvier(11) have described the Devonian fishes from Margan Pass, Lutherwan and Kishtwar areas. The stratigraphic position of the Tanawal Formation has been reviewed on the basis of additional palaeontological and stratigraphical data collected in recent years(2). The Syringothyris Limestone of Kotsu Hill, Kashmir, has yielded fairly well preserved conodont assemblage of Tournaisian to Visean age(9).

The brachiopod and bivalve fauna from the Fenestella Shales exposed in different parts of Kashmir has been revised and mid-Visean to Bashkirian age has been assigned to these shales. The fauna from the Fenestella Shales is approximately correlative with the similar faunas from the middle Naxing Formation of South Tibet(23).

The Agglomeratic Slate Succession of Kashmir has been investigated in detail and it has been observed that the volcanic influence within this succession was not as strong as might be expected from the name of the formation. Nonvolcanic rocks predominate this succession and the presence of definite tuffaceous, pyroclastic, glaciated horizons and interbedded traps has been encountered within this unit.

The so-called "Agglomeratic Slate Succession" has been classified as part of the Panjal Group which has been further subdivided into Golabgarh Boulder Formation, Nagmarg Formation and Panjal Formation. The Panjal Traps exposed in the Guryul Ravine section show pillow structures(23).

The Lower Permian succession exposed north of Sinthan Pass has yielded well preserved specimens of *Deltoblastus batheri*(19). The find of this blastoid is significant from palaeogeographic and plate tectonic point of view as identical forms are known to occur in Timor(19). The *Gangamopteris* Beds exposed at Risin Spur, near Zewan have yielded a palaeonisciform actinopterygian fish *Gardinerichthys tewarii*(10). The presence of fauna belonging to *Eurydesma cordatum* Zone (Late Asselian), *Taeniothaerus permixtus* Zone (Sakmarian - possibly Sterlitamakian) and *Lamnimargus himalayensis* Zone (Punjabian) has been recorded in different parts of Kashmir (23). A well preserved late Middle Permian fauna belonging to the *Costeferina alatus* fossil community, in the *Lamnimargus himalayensis* Zone has been described by Waterhouse and Gupta(25) from the Marbal area of Kashmir.

## LADAKH LAHAUL SPITI KINNAUR BEGIONS

Of the significant contributions made on the Palaeozoic stratigraphy of this region mention may be made of the establishment and classification of the Palaeozoic rocks in the Luneak valley of Zanskar area, Ladakh. Gupta(9) has classified the Palaeozoic rocks of this area into Phe Formation (Cambrian to Lower Ordovician), Karsha Formation (Upper Ordovician to early Late Silurian), Tanze Formation (Late Silurian to early Middle Devonian), Muth Quartzite (Middle to Upper Devonian), Luneak Formation (Tournaisian to early Visean) and Ralakung Volcanic Succession ( Upper Carboniferous to Permian). The ages assigned to different units are based on faunal assemblages collected from different stratigraphic horizons(11, 19). The Tanze Formation has yielded well preserved specimens of Osteolepiformes fish *Indolepis zanskarensis*(13). The Lower Carboniferous limestone exposed southeast of Surichun La near Zanskar has yielded crinoids(18) and conodonts(9).

The marine intercalations associated with the Ralakung Volcanic Succession exposed in the Tanze Section of Luneak Valley have yielded Atkansian (and possibly Kungurian) brachiopods(25). Similarly, thin sedimentary band at the top of the Ralakung Volcanic Succession has yielded Punjabian ammonoids. Pillow structures are frequently found within the Ralakung Volcanics of Ladakh.

The Kuling Formation exposed near Baralacha Ban area in the Lahaul valley has yielded representatives of *Eurydesma manendergarhensis* and *Deltopecten* cf. *mitchelli* whereas Malung Shales (Asselian) of Ladakh have yielded *Eurydesma cordatum*(6, 20, 24). The Kuling Shales exposed near Losar in the Upper Spiti Valley have yielded brachiopod fauna belonging to the *Lamnimargus himalayensis* Zone(22). The Lamayuru Flysch in Ladakh contains large exotic blocks of limestone. These Limestones exposed near Mulbekh have yielded Upper Palaeozoic fauna.

The Lower Devonian limestone exposed near Sumdo in the Lower Spiti Valley has yielded 14 ostracod taxa which show remarkable similarity with the Lower Devonian ostracods from the Hargan Formation of U.S.A.(1) Fragmentary fish scales and teeth and conodonts have been recovered from the sequence overlying the Muth Quartzite in the Yulang River Section of Kinnaur. The faunal assemblage is suggestive of uppermost Givetian to lowermost Frasnian age for the fossiliferous beds(12).

## BHUTAN

The occurrence of fossiliferous Devonian rocks was recorded for the first time by Gansser(3) and this fauna was subsequently described by Termier and Gansser(21). Gupta visited this area and made fresh collection of fossils from the Tangchu basin of Central West Nepal. The stratigraphical and palaeontological observations based on fresh fossil collections are discussed by Gupta and Termier(15) in a separate paper being published in the present volume.

## NEPAL

The Palaeozoic succession exposed in the Phulchauki-Godavari and Chandrasiri areas, near Kathmandu, Nepal has been worked out in detail and fresh fossil collections have been made from different stratigraphic levels(7, 14, 15). The fauna collected from different stratigraphic levels clearly indicates the presence of Cambrian to Upper Devonian succession in this area. The Permian stratigraphy of the Thakkola region of Central Nepal has been worked out in detail by Waterhouse(22).

## LESSER HIMALAYAS

The find of marine fossils (brachiopods, pelecypods, gastropods, bryozoans, etc. ) from the Boulder Slate Sequence of the Lower Bijni Tectonic Unit of Garhwal has thrown important light on the Palaeogeography of this region during the Upper Palaeozoic times. The fossils from this region in addition to others include *Coronalosia blijniensis*, *Brachythrinella* cf. *narsarhensis*, *Cancrinella farleyensis*, *Deltopecten* cf. *mitchella*, *Bicarinella* sp. suggesting Sakmarian age for the fossiliferous beds. The fauna is correlative with the upper units of the Nagmarg Formation of Kashmir and Umaria Marine Bed of Peninsular India. The early Permian fauna recorded by the Chinese geologists from the Chi-Lung or Jilong Group (section exposed at Mt. Chi-Lung-Kun-hsi of Tingem country) from the northern slope of the Mount Everest also show remarkable similarity with those of Umaria and Garhwal. The fauna from Garhwal, Tibet and Umaria belongs to one biozone dominated and linked by *Brachythrinella narsarhensis*(25).

## References


(1) Bhatia, S. B., Jain, S. P., Gupta, V. J., Lower Devonian Ostracods from Spiti and its palaeogeographical significance. Proc. Him. Geology Seminar, Geol. Surv. Ind. In press, Calcutta (1976).

(2) Fuchs, G., Gupta, V. J., On the stratigraphic position of the Tanawal Formation, Kashmir. Cont. Him. Geol., 1, New Delhi (1978).

(3) Gansser, A., Geology of the Himalaya, Interscience Publishers. John Wiley and Sons, London, 289 p. (1964).

(4) Geological Survey of India, The geology of parts of the Kashmir Himalaya. Guide to Excursion, Himalayan Geology Seminar, Geological Survey of India, Calcutta, 22 (1976).

(5) Gupta, V. J., Devonian fossils from the Lesser Himalayan Zone of Central Bhutan. Current Science, Bangalore, 40 (18), 490 (1971).

(6) Gupta, V. J., Indian Palaeozoic Stratigraphy, Hindustan Publishing Corporation (I), New Delhi, 207 p. (1973).

(7) Gupta, V. J., Upper Devonian conodonts from Phulchauki, Nepal.Himalayan Geology, Dehradun, 5,153-168 (1975).

(8) Gupta, V. J., Palaeozoic biostratigraphy and palaeogeography of the Himalaya. Colloq. Intern., C.N.R. S., Paris, 268, 197-202 (1976).

(9) Gupta, V. J., A note on the Palaeozoic Stratigraphy of Luneak Valley with a description of a new species of *Retichonetes* from Surichun La, Ladakh. Recent Researches in Geology, New Delhi, 4, 37-45 (1977).

(10) Gupta, V. J., Heyler, D., Janvier, Ph., A new Actinopterygian fish from the Lower Permian of Kashmir, India. Bull. Ind. Geol. Assoc., Chandigarh, 11(1), 1-10 (1978).
(11) Gupta, V. J., Janvier, Ph., A review of the Devonian vertebrate localities of the Indian Himalayas (Kashmir, Ladakh and Kumaun), with remarks on their stratigraphical and palaeobiogeographical significance. Contributions Himalayan Geology, New Delhi, 1, 78-83 (1978).
(12) Gupta,V. J., Janvier, Ph., Late Devonian vertebrate remains from Western Himalayas, Himachal Pradesh, India. Bull. Ind. Geol. Assoc., 161-179 (197 ).
(13) Gupta, V. J., Janvier, Ph., *Indolepis zanskarensis* gen. n., sp. n. from the Tanze Formation of Luneak Valley, Zanskar, Ladakh. Contributions Himalayan Geology, 2, In press (1980).
(14) Gupta, V. J., Stocklin, J., Stratigraphy and Structure of the Phulchauki-Chandragiriarea, Nepal. Recent Researches in Geology, New Delhi, 7, 263-275 (1978).
(15) Gupta, V. J., Termier, G., Lower Palaeozoic Echinodermata from Chandragiri Pass, near Kathmandu, Nepal. Recent Researches in Geology, New Delhi, 7, 472-480 (1978).
(16) Gupta, V. J., Waterhouse, J. B., Permian invertebrate faunas from the Ralakung Volcanics, Ladakh Himalaya, India. Recent Researches in Geology, New Delhi, 5, 31-49 (1978).
(17) Gupta, V. J., Waterhouse, J. B., *Eurydesma* from Lahaul and Ladakh. Bull. Ind. Geol. Assoc., Chandigarh, 11 (2), 141-152 (1979).
(18) Gupta, V. J., Webster, G. D., *Deltoblastus batheri* from the Kashmir Himalaya. Rivista Italiana Paleont., 82 (2), 279-284 (1976).
(19) Gupta, V. J., Webster, G. D., *Deltoblastus batheri* from the Permian of Himalaya. Palaeontological data and its implications on plate tectonics. Journ. Geol. Soc. Ind., Bangalore, 21 (7) (1980).
(20) Srikantia, S. V., Bhargava, O. N., Kapoor, H. M., A note on the occurrence of *Eurydesma* and *Deltopecten* assemblage from the Kuling Formation of Baralacha Ban area, Lahaul, Himachal Himalaya. Journ. Geol. Soc. Ind., Bangalore, 19(2), 77-78 (1978).
(21) Termier, G., Gansser, A., Les Series devoniennes du Tangchu (Himalaya du Bhoutan). Eclogae Geologicae Helvetiae Zurich, 67 (3), 587-596 (1974).
(22) Waterhouse, J. B., Permian rocks of the Kali Gandaki area, Thakkola, north central Nepal. Contr. Himalayan Geology, 1, 195-213 (1978).
(23) Waterhouse, J. B., Gupta, V. J., Permian faunal zones and Correlations of the Himalayas. Bull. Ind. Geol. Assoc., Chandigarh, 10 (2), 1-19 (1979).
(24) Waterhouse, J. B., Gupta, V. J., Brachiopods from the Carboniferous Fenestella Shales of Kashmir. Becent Researches in Geology, New Delhi, 3, 145-187 (1977).
(25) Waterhouse, J. B., Gupta, V. J., Lower Permian fossils from the Bijni Tectonic Unit of Garhwal, Lesser Himalayas. Recent Researches in Geology, New Delhi, 4, 410-437 (1979).

# SPECIAL REMARKS ON GEOLOGICAL HISTORY OF THE HIMALAYAS

Wang Yi-gang
(Nanjing Institute of Geology and Palaeontology, Academia Sinica)

Based on the data collected from scientific expedition to Mt. Qomolangma during 1966-1968 and the scientific expedition in wide areas in Xizang during 1973-1976, both organized by Academia Sinica, as well as the adjacent regions, this paper discusses some problems concerning the geological history of the Himalayas.

## BASEMENT OF THE HIMALAYAN SEDIMENTARY REGION

The geological history of the Himalayas may be traced as far back as early Proterozoic or Late Archaeozoic. The deposits are roughly shown in Table 1.

From the composition of metamorphic rocks comprising the basement of the Himalayan sedimentary region, and, especially, on the basis of a careful study on rocks of Nyalam Group (1), the same as 'Jolmo Lungma Group' (2), it is inferred that the original rocks may be mainly composed of greywacke in excess of $Al_2O_3$ with a few intercalated beds of argillaceous and feldspathic sandstones, i.e., a series of argillaceous clastic sediments. Only the upper layers in the sequence are relatively stable sediments of limestone from solvent calcium carbonates, and furthermore there are no volcanic rocks. This inference (2) indicates the characteristics of a relatively active deposition.

The basement of the Himalayan sedimentary region is probably correlated with Aravalli Group of Indian shield. The original rocks of the latter ought to have been a series of argillaceous clastic sediments with basic volcanic rocks, hence representing the deposition of violently active type. An inference may be made from the above comparison that there exist certain differences between the basement of the Himalayan sedimentary region and that of the Indian shield.

## BOUNDARY BETWEEN THE METAMORPHIC (OR 'BASEMENT') AND NORMAL SEDIMENTARY STRATA, AND THE DATE OF METAMORPHISM

Gansser (1974) once expounded: "Due to the gradually decreasing metamorphism it is often difficult to fix the top of the metamorphics or 'basement' and the beginning of the normal sediments. We noted how this boundary rises within the sediments along the Himalayas from the west to east" (3).

We have noticed that the shifting of the boundary between the basement and normal sediments is gradually upward from south to north, from west to east, but very limited.

In accordance with the data derived from 12 samples in K-Ar method,

6 samples in Rb-Sr method, and 5 samples in U-Pb method (4), the Nyalam Group, at least, underwent two major metamorphoses. The first one is about 640-660 m.y. in age, and the second one about 10-20 m.y. in age. The basement, Nyalam Group was mainly derived from the first metamorphism and the age is coincident with the beginning of normal sediments. Eastward from the Kashmir region, the beginning of the normal sediments is shown in Table 2.

We also further noticed that along the Lower Himalayas, as viewed from south to north metamorphism occurred about 700-800 m.y. ago (5). This is coincident with the fact that the normal sediments began from Sinian.

As far as the above-mentioned matter goes, we consider that the metamorphism of the basement of the Himalayan sedimentary region began from Sinian, gradually migrated from south to north and from west to east, and ended in Early Caledonian.

## ABOUT THE SO-CALLED PALEOHIMALAYAN RIDGE

Fuch (1968) emphasized that "the uniform geosynclinal sedimentation in the Himalayas was ended by the Caledonian orogeny. This early Paleozoic orogeny is of greatest important because since then a ridge exists in the Himalayas separating the fossiliferous marine sequence of the Tethys zone from the fossiliferous deposits southwest its" (6).

We hold that Caledonian orogeny certainly had influence over the Indian shield. The deposits from Ordovician to Devonian were often interrupted along the marginal regions of the Indian shield, from where northward it locally expanded to land, and provided great clastic matter for the Himalayan basin. But, Caledonian orogeny had no considerable influence over the Himalayas. The metamorphic rocks composing the basement were continuously overlain by normal sediments, and persisted to Ordovician,Silurian, Devonian and Lower Carboniferous (7,8). On the diagrammatically restored sections of the Ordovician to Devonian of the northern Tethyan Himalayas, it may be seen: the Himalayan basin with Kumaon as a center of depression was in a process of regression (see Fig. 1). The Muth Quartzite Group, being mainly consisted of the sea shore deposits originated from the marginal region of Indian shield, was the result of the regression during the stage of the Caledonian epeirogeny in the Himalayas. The subsequent deposits, lacking Middle Carboniferous (?) and Upper Permian, were evidently interrupted by local epeirogenic movements. Besides, there is no reliable evidence of unconformities yet detected. Recently, more and more information obtained clearly demonstrates that south of the so-called Paleohimalayan ridge was not a unfossiliferous sedimentary region. Many marine fossils including brachiopods, bryozoa, corals, fusulinids, algae, gastropods, etc., have been found there from Carboniferous to Mesozoic strata (9, 10, 11, 12, 13), while at many places in the northern and southern parts of the Tethyan Himalayas including Kashmir, North Nepal, North Sikkim and three counties Tingri, Dinggye, and Kangmar of China (1, 14, 15), sediments and faunas possessed of the Gondwana facial characteristics have been found.

All above-mentioned facts serve as a convincing explanation that the southern and the northern flanks of the so-called Paleohimalayan ridge were interlinked with each other and there was no presence of barrier in between. Features of their faunas do not differ widely. As to some minor differences, local environment may sufficiently account for. Therefore, judging by the data and information available, the presence of the Paleohimalayan ridge can not be verified for lack of appropriate geological conditions and ample evidences.

## SEDIMENTARY DIFFERENTIATION OBSERVED IN MESOZOIC STRATA

The Paleozoic strata of the northern Tethyan Himalayas crop out in odds and ends. Their nature is very similar to or coincident with those of the southern Tethyan Himalayas. Thence, the sedimentary environment of the northern Tethyan Himalayas was on the whole the same as that of the southern Tethyan Himalayas during Paleozoic.

During Mesozoic, the Himalayan sedimentary region was mainly in a depressed stage. But, obvious sedimentary differentiation occurred in the northern Tethyan Himalayas and the southern Tethyan Himalayas. The former may be called as a violently active deposition, and the latter, a relatively active deposition.

The sedimentary differentiation began from Triassic, became weakened in Jurassic owing to the extension of transgression, and reached the summit during Cretaceous. However, the sedimentary facial changes a rather rapid in vertical and horizontal directions in the Mesozoic, especially in the Cretaceous and in the northern Tethyan Himalayas.

The northern Tethyan Himalayas occurred as a region of violently active deposition under the environment of deep-sea or subdeep-sea, commonly consisting of great thick clastic rocks with flysch rocks, basic volcanic rocks and radiolarian siliceous rocks, with only some pelagic faunas. But, in the southern Tethyan Himalayas, it was a relatively active deposition under the shallow-sea environment, comprising mainly carbonate-clastic rocks, without such distinctive sediments as those of the northern Tethyan Himalayas, but containing abundantly various kinds of marine fossils such as benthonic-pelagic faunal assemblage.

It should be mentioned that during Cretaceous at the northern marginal region of the Himalayas, the sedimentary differentiation derived from the rapid changes of palaeogeographic environments was more obvious than ever. It seems clear that from south to north, a suite of marine Cretaceous sediments with flysch and siliceous rocks existed along the east of Gyangze; at the southern side of the ultrabasic belt, there was a series of volcanic-siliceous rocks under deep-sea environment; at the other (northern) side of the ultrabasic rock belt mostly continental sediments interbedded with marine sediments in the lower sequence and intercalated with minor marine beds in the upper sequence (16) occurred.

## SOME PROBLEMS CONCERNING UPLIFTING OF THE HIMALAYAS

### About the Highest Marine Strata

The large-scale episodes of the Himalayan movement started right after the end of marine sedimentary history. Therefore, the end of marine sedimentary history is a momentous sign during the development of the Himalayan geological history. The Early Tertiary strata of the Himalayas are the deposits of a relic of sea-basin. As far as our knowledge is concerned, the Himalayan sea ended after the middle Eocene at the latest. But, the marine sediments locally ended as early as Late Cretaceous along the Kumaon. Thence, it is deduced that the Himalayan sea retracted at first from the Kumaon region; then the sea water was rapidly evacuated from the whole Himalayas, eastward and westward, through either flank of the Kumaon region.

### The Source of Strata of Xizang's (Tibetan) Facies

The so-called strata of Xizang's facies are a synthethic sequence of the 'exotic blocks', which discontinuously and latitudinally distribute along the northern Tethyan Himalayas from Ladakh region through Ngari region, Zhongba of China at least to Gyangze, China eastward. The strata of Xizang's facies mainly include some red limestones and siliceous rocks from Permian to Cretaceous. Based on the characters of strata and features of faunas, they are often interpreted as the deposition of deep-water environment and considered to have a source from the farther north of the Himalayas.

As far as we know, the Permian of Xizang's facies contains some warmth-liking brachiopods, corals and fusulinids, etc., formerly recognized as a characteristic of the region north of the Himalayas and different from the Himalayas itself. The recent discovery of warmth-liking faunas including brachiopods, corals, fusulinids, algae, etc., in Tal formation of the Lower Himalayas in Kumaon region (13) is a weighty evidence that they are not limited to the north of the Himalayas in distribution. We interpret this contradictory fact as an indication that the horizon of the warmth-liking faunas might overlie that of the faunas fit for cooler water, as a consequence of the re-rising of temperature after the Himalayan region underwent a stage of cooler-water environment.

The Triassic of Xizang's facies contains some early Triassic, Anisian and Carnian ammonoids. It is plain for the present author that the features of these ammonoid faunas are in the main similarities to those from the Himalayas.

The Jurassic includes some Early Jurassic ammonoids and Late Jurassic Calipionellids. The nature of the former bears some relations with Late Hettangian and Sinemurian ammonoids from the Adneth limestone in Austria. In recent years, the author et al. have obtained some Liassic (from Late Hettangian to Pleisbachian) ammomoids from normal sequence at Tulong, Nyalam in central Himalayas and at Ritang, Lhünze in Eastern Himalayas (7, 17). Their features are also very close to those of the "exotic blocks". The latter including _Calpionella_ _alpina_ and _C_. _elliptica_ are commonly warmth-liking faunas of the Tethyan type, but they have also been found in western Australia, namely they are not confined to typical Tethyan regions.

For reasons given above, the author presumes that the north of the present Himalayas was a possible source of the so-called strata of Xizang's facies, which were in the same sea-basin with those of the Himalayas and mostly disappeared under the subducted belt.

About Molasse Deposits of the Northern Margin of the Himalayas

The Siwalik Group occurring at the southern margin of the Himalayas, is a suite of the famous molasse deposits. Frank and Gansser et al. (1977) (18) reported the presence of molasse deposits such as those in Ladakh region, Indus molasse and Hemis conglomerate, with intercalated beds of Nummulites limestone in the lower sequence of the latter. At Gandise Range, the equivalent strata are called the 'Kailas conglomerate'. Recently, in the vicinity of Mai La Zhongba, China, the author et al. have found a suite of the so-called 'Mai La conglomerate', likewise intercalated with Nummulites beds. Besides, there has also been found a suite of conglomerate over 1,000 m in thickness at Ruobusa, Zetang at the northern margin of East Himalayas. It contains some lamellibranchia, gastropods, charopyta, etc., indicating its age from Late Oligocene to Miocene.

These conglomerates of the northern and southern margins of the Himalayas are alike, but the conglomerates of the northern margin appear a little earlier than those of the southern margin and ought to have been the result of the uplifting of the Gandise Range. This leads to a presumption that the uplifting of northern margin of the Himalayas might have been a little earlier than that of the southern margin; and that, in the course of uplifting, the superficial configuration of the northern margin underwent a violent undulation due to a diversity of the uplifting time in different localities.

## CONCLUDING REMARKS: SUBDIVISION AND EVOLUTION OF HIMALAYAN SEDIMENTARY REGION

Strata of the Himalayan sedimentary region are latitudinally well distributed. Recently, according to the development of strata, sedimentary environment, the assemblages of distinctive rocks and the features of biota. The author recognize that the Himalayan sedimentary region may be divided into five or six sedimentary zones. They are from north to south as follows:

### Zone 1, The Northern Tethyan Himalayan Zone

This zone is seated between the lower watershed of South Xizang and the suture-belt of Yarlung Zangbo-Indus. It mainly contains locally a series of very thick Mesozoic strata, low-grade metamorphic rocks, with a few radiolarian cherts, intermediate and basic volcanic rocks and flysch deposits.

The Paleozoic strata are distributed only at several localities in odds and ends. The Mesozoic strata of Zone 1 are apparently different from those of Zone 2. The molasse deposits are intermittently distributed along the northern margin from Ladakh in west Himalayas eastward to Zetang in East Himalayas. In addition, there is a latitudinally metamorphic belt in this zone, but its composition, sequence and age are not clear yet. The general features of this zone are of violently active deposit.

### Zone 2, The Southern Tethyan Himalayan Zone

This zone is confined to the southern part of the lower watershed of South Xizang and the northern part of the High Himalayas and its farther northern regions. The strata of this zone are almost all of marine deposits, about 10,000 m in thickness, ranging from Cambrian to Tertiary, deposited continuously with local interruptions of Middle Carboniferous and Late Permian. They overlie the metamorphic Precambrian or Preordovician, containing various kinds of abundant marine fossils, and have some sedimentary characters of Gondwana facies. This zone is of relatively active deposit.

## Zone 3, The Crystalline Rock Zone

This zone is mainly situated along the high Himalayas. It consists of a series of 'old metamorphic rocks' from high-to low-grade composing the basement of the Himalayan sedimentary region and was divided into 6 mineral crystalline zones by Ying (2). It is necessary to point out that there are also a few low-grade metamorphic or normal sedimentary rocks locally covering the 'old metamorphic rocks' in this zone.

## Zone 4, The Lower Himalayan Sedimentary Zone

The deposit of this zone is relatively stable. The 'old metamorphic rocks' are mainly overlain by the deposits of Gondwana facies, and are similar to the deposit of the northern margin of Indian shield.

## Zone 5, The Tertiary Zone

This zone crops out at the subhimalayas along the southern margin of the Himalayas and comprises a suite of the molasse clastic deposits of Siwalik and other Tertiary deposits.

The strata of Xizang's facies, originating possibly from the north of the present Himalayas, should have been sediments in an off-shore shallow-sea. These strata and those of Zone 1 and Zone 2 should have been the sediments of different environments in the same sea-basin. Therefore, another sedimentary zone of Xizang's facies representing relatively active type might have been originally present to the north of violently active deposit of Zone 1.

The distribution of these sedimentary zones in space mentioned above reflects generally the sedimentary character of the Himalayan sedimentary regions as the margin of the 'Southern Continent'. From the Indian shield outward to the Himalayas, a succession of sedimentary types from south to north may be recognized as follows: relatively stable type (Zone 4) - relatively active type (Zone 2) - violently active type (Zone 1) - relatively active type (Zone of Xizang's facies).

The difference of the various sedimentary zones mentioned above reflects veritably that they were the products in different stages or under different sedimentary environments and conditions in the same stage. There were three stages in the evolution of the Himalayan sedimentary regions. In Pre-Paleozoic stage, the whole Himalayan sedimentary region as primarily deposited in the sedimentary environment of relatively active type, while the Indian shield in sedimentary environment of violently active type. In Paleozoic stage, zone 4 was first evolved into relatively stable environment from relatively active environment, but the north of Zone 4 was still in relatively active environment. In Mesozoic, the Himalayan sedimentary region was in a very active stage, while the sedimentary differentiation of the Himalayas was remarkable; Zone 1 was isolated to evolve into violently active environment from the relatively active environments of both Zone 2 and the Zone of Xizang's facies (see Fig. 2).

## References


(1) Yin Ji-xiang, Guo Shi-zeng, On the discovery of the stratigraphy of Gondwana facies in northern slope of the Qolmolungma Feng in Southern Xizang, China. Scientia Geologica Sinica, (4), 291-322 (1976) (in Chinese).

(2) Ying Si-huai, Magmatic, metamorphic and migmatic rocks of the Mount Jolmo Lungma Region in Southern Tibet, China. A Report of Scientific Expedition in the Jolmo Lungma Region (1966-1968), - Geology, Science Press, 233-261 (1974) (in Chinese).

(3) Gansser, A., The Himalayan Tethys: Rivista Italiena di Paleontogia e Stratigrafia, Men. 14, 393-411 (1974)

(4) Laboratory of Isotopic Geology, Kweiyang Institute of Geochemistry, Academia Sinica, Isotopic dating of the metamorphic rocks from the Mount Jolmo Lungma Area, China. Scientia Sinica, 16 (3), 385-395 (1973)

(5) Krummenacher, D., Determinations d'age isotopique faites sur quelques roches de l'Himalaya, Schweiz. Mineral. und Petrogr. 41 (2), 273-283 (1961); Nepal Central: geochronometrie des series de l'Himalaya, ibid., 46 (1), 43-54 (1966).

(6) Fuchs, G., The geological history of the Himalayas. International Geological Congress Report of the Twenty-Third Session, Sec. 3, 161-174 (1968).

(7) Wang Yi-gang, Wen Shi-xuan, Yin Ji-xiang, Zhang Bing-gao, Zhang Ming-liang, Stratigraphy of the Jolmo Lungma Region (A detailed Report). A Report of Scientific Expedition in the Mount Jolmo Lungma Region (1966-1968), - Geology, Science Press 1-213 (1974) (in Chinese).

(8) Mu An-tze et al., Stratigraphy of the Mount Jolmo Lungma Region in Southern Tibet, China. Scientia Sinica, 16 (1), 96-111 (1973)

(9) Maithani, J.H.P., A study of the molluscan fauna from the Tal formation (Garhwal) and its significance in correlation. Himalayan Geology, 2, 239-251 (1972).

(10) Ganesan, T.M., Fenestellid bryozoa from the Boulder Slate Sequence of Garhwal. Himalayan Geology, 2, 431-451 (1972).

(11) Singh, P., A note on the fossiliferous formations in Lesser Himalaya of Nepal and Bhutan. Himalayan Geology, 3, 372-380 (1973).

(12) Singh, T., Note on the Upper Palaeozoic fauna from Subansiri District, Arunachal Pradesh. Himalayan Geology, 401-410 (1973).

(13) Valdiya, Lithology and age of the Tal formation in Garhwal, and implication on stratigraphic scheme of Krol Belt. Journal Geol. Soc. India, 16 (2), 119-134 (1975).

(14) Gupta, V.J., Indian Paleozoic Stratigraphy, Indian Mesozoic Stratigraphy (1973), (1975).

(15) Muir-Wood, H.M. et al., Upper Paleozoic faunas of North Sikkim. Paleont. Indica, N.S., 31 (1), (1941).

(16) Wu Hao-ruo, Wang Dong-an, Wang Lian-chen, The Cretaceous of Lazi-Jiangze district, Southern Xizang. Scientia Geologica Sinica, (3), 250-262 (1977).

(17) Wang Yi-gang, Wang Yu-jing, Wu Hao-ruo, Discussion on the Jiabula formation and the discovery of Longzi Lower Jurassic in Southern Xizang, China. Scientia Geologica Sinica, (2), 149-156 (1976) (in Chinese).

(18) Frank, W, Gansser, A. et al., Geological observations in the Ladakh Area (Himalayas). Schweiz. Mineral Petrogr. Mitt., 57 (1), 89-113 (1977)

Table 1

| Kashmir | Spiti & W. Kumaon | E. Kumaon | Nepal | Qomolangma region |
|---|---|---|---|---|
| Salkhala Series | Yaikrita Series | Jutogh Series | Dhunpu Series | Nyalam Group |
| W ← | | | | → E |

Table 2

| | | Kashmir | Spiti | Kumaon | Nepal | Qomolangma region | E. Himalayas |
|---|---|---|---|---|---|---|---|
| | | W ← | | | | | → E |
| Gansser, 1974 Beginning of Normal Sediments | | $Є_1$ | Є | U.part of Є | Doulagiri (Nilgiri) | Middle Paleozoic | raising to east |
| This Paper Beginning of Normal Sediments | South ← North; L. Himalayas / Tethyan Himalayas | $Є_1$ | $Є_2$ | U.part of Є | O1 | O1 | not clear |
| | | | | | Sn | | |

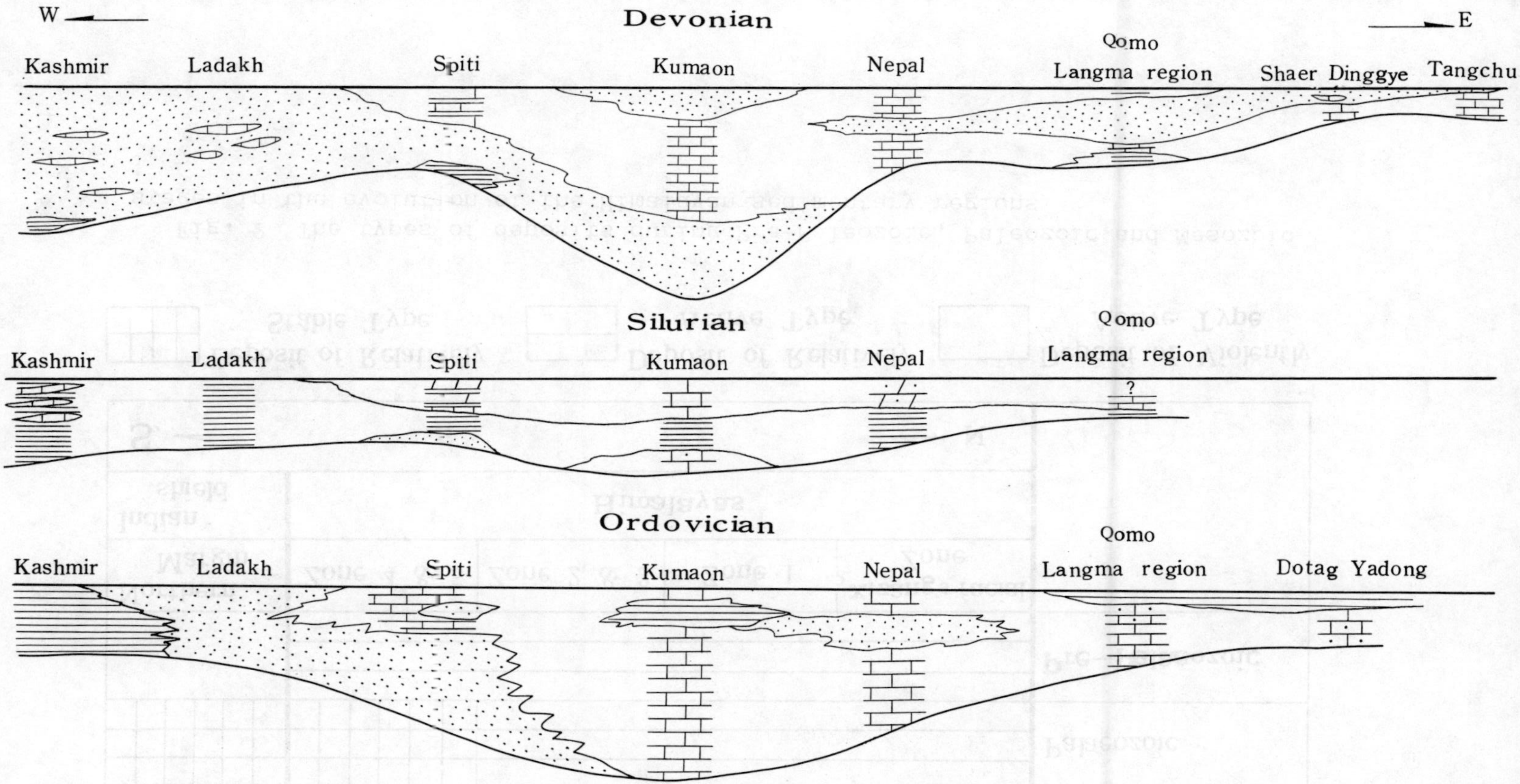


Fig. 1 The diagrammatically restored section of Ordovician to Devonian of the Northern Tethyan Himalayas

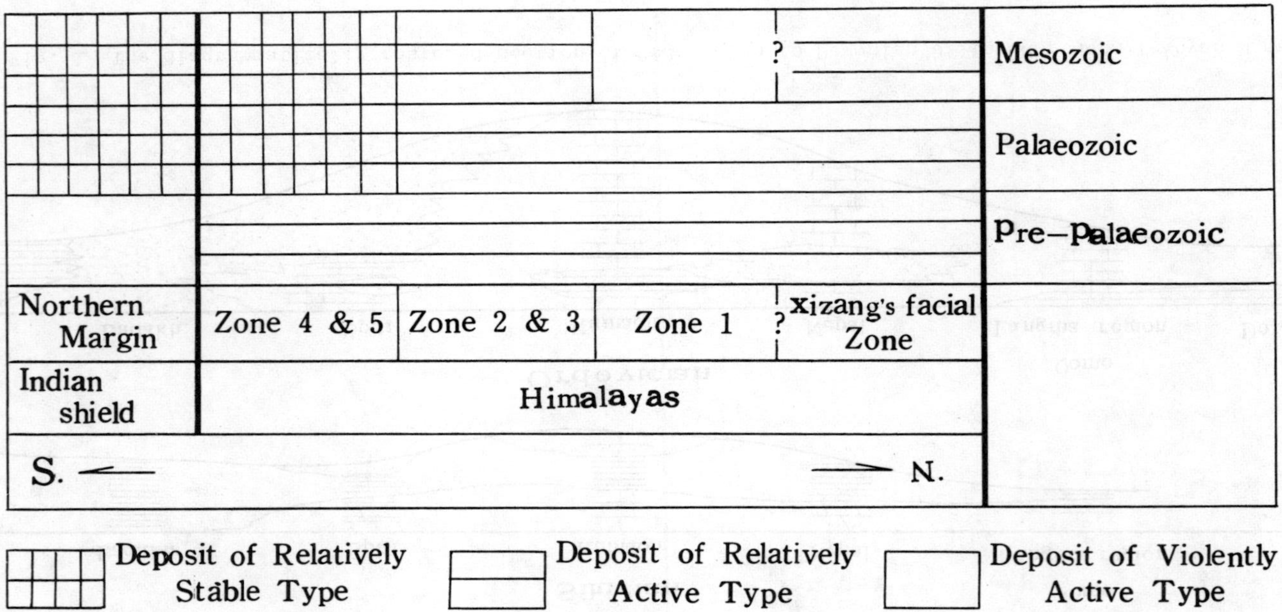


Fig. 2 The types of deposits during Pre-Paleozoic, Paleozoic and Mesozoic stages in the evolution of the Himalayan sedimentary regions

# THE STRUCTURAL AND SEDIMENTOLOGICAL EVOLUTION OF THE THAKKOLA-MUSTANG GRABEN (NEPAL HIMALAYA) IN RELATION TO THE UPLIFT OF THE HIMALAYAN RANGE

Monique FORT,
(Lab. Géographie Physique, Université Paris 7)
Pierre FREYTET,
(Lab. de Géologie structurale, Université Paris-Sud)
Michel COLCHEN,
(Lab. de Géologie stratigraphique et Structurale,
Université de Poitiers, France)
(GRECO, C.N.R.S. 1300 12, HIMALAYA-KARAKORUM)

The Thakkhola-Mustang Graben (Nepal) together with the Yagru-Xixabangma area in Tibet (4, 27), and possibly the Upper Sutlej Valley in India (25), is one of the very few areas on the northern side of the Himalayan Range where Late Cenozoic sedimentation and recent tectonic evolution may be observed.

The Thakkhola-Mustang region is a high intramontaneous basin, altitudes varying between 3-4000 m at the bottom to more than 6000 m on the surrounding summits (24). Located at the northern foot of the *Dhaulagiri and* Annapurna Ranges (Fig. 1), framed by the Paleozoic and Mesozoic sediments of Tibetan series (3) or Tethysian series (21), it was filled with a thick series of probably Plio-Pleistocenous age, the Tetang-Thakkhola Formations (2), formerly called Thakmar Series (23). This deposition was strongly influenced by transverse-fault system of the Thakkhola. Although absolute datations are lacking, the recent paleogeographic and structural evolution of the area can be reconstructed from the continuous outcrops of the present Kali Gandaki valley and its tributaries.

**The Graben.** In the structural history of the Himalaya, the individualisation of the Graben corresponds to a modern distensional phase (5), very different from the previous phases, produced by the compression interpreted as resulting from the collision between Indian and Tibetan-Asiatic plates.

The new structural pattern shows a series of transversal faults and cleavages oriented N20-N40° (7). The characteristics are : - its transvers orientation, - the predominance of vertical faults with a horizontal senestral sliped component, maximal throws being 3,000 m (vertical) and 4,500 m (lateral slip) ; - mylonites outcroping to the south, between *Dhaulagiri* and Annapurna are in direct prolongation with the Thakkhola System (11), to which they probably belong.

The Thakkhola system is accompanied by numerous other minor faults, oriented N115° and N150-160°.

**Lithostratigraphic units.** Two main groups of formations have been recognized :

1) filling formations : Tetang and Thakkhola formations (2) :

- Tetang Formation (0 - 180 m) outcrops in the southern part of the Graben. It has buried a dissected paleotopography. Four units of fluviatile and lacustrine - palustrine facies have been recognised (17) ;

- Thakkhola Formation (maximum 700 m) lies directly and unconformably upon Tetang Formation and the "Tethysian" substrate. It has a wider extension, and three main fluvio-lacustrine terms have been individualised (17) ;

- the Tetang and Thakkhola Formations are affected by a late stage of deformation.

**2) entrenched formations : Sammargaon breccia (2) and terraces levels :**

- Sammargaon Formation is a sedimentary breccia of unsorted and very poorly layered Paleozoic detritus, cemented by a red matrix. It is strictly limited to the western faulted ridge of the graben, and is itself faulted ;

- more recents deposits are found below terraces levels, situated more and more deeper throughout the Graben filling. Two types are known (13) : fluvio-glacial deposits, and thin alluvial cover of terraces (3 to 5 levels).

## Sedimentological data and their paleogeographical implications.

The recognition of **several detrital discharges** depends on geomorphic, petrographic and granulometric data. Lower Tetang sediments had a local origin, but Upper Tetang materials came from the north-east (granitic pebbles). Thakkhola sediments had a south-west origin (metamorphosed Paleozoic pebbles) (2), Sammargaon breccia also had a western origin too (Paleozoic rocks). The more recent deposits comprising terraces are related to a north-south transport (Kali Gandaki).

In the graben, the conditions of deposition were alternatively fluviatile, lacustrine and palustrine (17, 18), as it is known in southern France (16). Fluviatile deposits are represented by lensoid conglomerates (channels fillings); and siltstones representing flood plain materials, having suffered pedological modifications, because they contain typical pedological features as marmorizations, carbonates nodules, root traces. Lacustrine and palustrine deposits are basically limestones, occurring in thin but extensive lenses, including Cyanophyts and Charophyts remains (shallow lakes). Palustrine deposits result from the emersion of lacustrine muds and exhibit desiccation cracks, pedological features (root traces with clay coatings, nodules, recrystallizations). A tuffaceous facies is also well developped, containing vegetal remains encrusted by Cyanophyts, and consolidated by an alternation of phreatic and vadose cements (17, 18).

## The history of the Thakkhola-Mustang basin.

Following the Oligo-Miocene emersion of the Himalayan Range (21), we consider that the more recent history may be subdivided into three periods :

**The pre-Thakkhola period.** As already noticed (24), the summits surrounding the basin have a subconstant altitude (6,000 m), suggests the existence of an old erosional surface, of pediment type, shaped under rather semi-arid conditions ; similar surfaces are know in Tibet (36). It is probably the northern equivalent of the transit piedmont surface of the Siwaliks series southward. Ultimately, this surface suffered a tropical, rubefacient weathering, under more humid conditions ; comparable soils have been described in Tibet (20). Thus, the general low altitudes and warm climates characterised the pre-Thakkhola period (maximum : 1,500 m).

**The Tetang-Thakkhola period.** (Figure 2). Although deformations seems to have been continuous during the whole period, but four stages are visible

(6) : 1, individualisation of the Tetang Graben by normal faults F1 ; 2, normal faults F2 which affect the whole area, initiate the dissymetry of the Graben and give a *weak* dip to the Tetang layers ; 3, F3 fault occur throughout the Thakkhola depositional period, determining an important syntectonic sedimentation ; 4, F4 stage is the last, but important refaulting of the Thakkhola system (1,500 m), synclinal bending (N20° axis), general tilting (5-10°) towards the north of the topographic surface of the basin. This period corresponds to a major paroxysm related *to the uplift of the Himalayan Range*. We also consider that the Thakkhola fault system is older than the Plio-Pleistocene movements, as reported for other Himalayan transverse fractures (34, 22) ; it could explain the setting of the primitive, probably antecedent, hydrographic pattern.

**The Kali Gandaki period.** It is characterised by three major features.

1. The installation of the Kali Gandaki River began at the end of the Thakknola sedimentation (Fig. 2). The new hydrographic network is responsible for major dissection (600 m) of the Graben filling and its *hinterland*. The inversion of drainage might be due to several factors (14) : - the southward tilting of the Graben itself, - the uplift of the Higher Range, - a later adjustment of the Thakkola faulted system (faulted Sammargaon breccia), minor faults and cleavages in Upper Pleistocene lacustrine sediments in Lower Thakkhola (11, 12).

2. The cooling of the climate and the appearence of glacial processes are a direct consequences of uplift of the Himalayan Range. Fluvio-glacial and glacial deposits fossilized the previous Kali Gandaki gorges at the confluences with major affluents. The Lower Thakkhola was flooded by the paleolake of Marpha (11), and possibly partly filled by ice which descended from *Dhaulagiri*, Nilgiri and Annapurna peaks down to Larjung, Dhumpu, Ghasa (8, 9, 12).

3. The latter plus the more recent evolution is characterised by a new period of vertical dissection, interrupted from time to time by short phases of alluvial terraced modelation, as in all the Range (33).

## Discussion, comparisons and chronology.

The relative chronology may be summarized as follows: 1, surrection of the first Himalayan Range ; 2, shapping of the pediment surface ; 3, tropical weathering throughout the area ; 4, first movement of faults and Tetang Graben ; 5, Tetang sedimentation ; 6, second movement of fault system, short erosional stage ; 7, Thakkhola sedimentation, closely connected to the activity of the Thakkhola faulted system (northward flowing) , 8, last recurrent faulting of Thakkhola Graben immediately followed by -9, deep vertical dissection (600 m) by Kali Gandaki, southern flowing ; 10, important glacial stage ; 11, more recent staged dissection with 3-5 levels.

Concerning litho-stratigraphical chronology, previous authors have suggested Middle Tertiary to Pleistocene (24), "probably Pleistocene" (21), equivalent to the Mio-Pliocene Molasses of the Siwaliks (3). We consider a Pliocene age highly probable for the Tetang formation, because of : a) predominance of the Coniferous pollens comparable to the flora of Middle Siwaliks zones II and III (31) ; and also to the Lower Pliocene pollens of Yagru Zongla (1) dated by *Hipparion* ; b) occurrence of Ostracod *Illyocypris gibba* (RAMDORH) within both Tetang (14) and Lower Yagru Zangla deposits.

Chinese authors have observed two planation surfaces in Tibet : the upper was edified during the Oligo-Miocene, and the lower during the Pliocene (36). The red soil episode ranges from the end of the Pliocene to early Pleistocene (20).

The thick and rather widespread Thakkhola fluvio-glacial sediments may be correlated to the Tibetan Nynyxunglaha Glaciation (37) attributed to the Middle Pleistocene (32). The Late Pleistocene and Holocene Thakkhola Glaciation (8, 9, 14) can be directly correlated with those of other parts of Himalaya (10, 15, 19, 26, 28, 29, 30) and of Tibet (32, 35).

## Conclusion.

The paleogeographic and geodynamic importance of the Thakkhola-Mustang area may be understood by stressing following points :

- the original history of the area, initiated after the northern pediment was disrupted, began with the development of a thick detrital sedimentation related to the functioning of major transversal faulted tectonics. The sedimentation has always compensated the graben subsidence. The deposition of alternating fluvial, lacustrine, and palustrine layers indicates that the bottom of the graben has never been occupied by a deep lake, but by a rather flat piedmont plain with torrential fans and small lakes. These tectonic and sedimentological events coincided with the beginning of Himalayan Range uplift ;

- at the end of the very history of the Thakkhola-Mustang area, a new geomorphological pattern appeared as a consequence of the recent vertical tectonic, and general upheavel of the whole area, including Tibet and Higher Range. These marked changes provoked fluvial dissection of the Kali Gandaki towards Ganga, to the probably first appearence of glaciation on the side of the Range, and subsequently to development of terraces, due both to Himalayan uplift and climatic changes (increasing of cold and aridity). Both factors still influence the geomorphology of the northern side of the Higher Himalayan Range.

## Références


(1) ACADEMIA SINICA, Symposium on Quinghai-Xizang (Tibet) Plateau. *A Scientific Guidebook to south Xizang (Tibet)*, June 2-14, Beijing, 104 p. (1980).

(2) BASSOULET J.P. and COLCHEN C., Les formations tardi-orogéniques de la Thakkhola, Himalaya du Népal, *2ème Réun. Sc. Terre, France*, p. 34 (1974).

(3) BORDET P., COLCHEN M., KRUMMENAKER D., LE FORT P., MOUTERDE R. and REMY J.M., *Recherches géologiques dans l'Himalaya du Népal*, CNRS, Paris, 275 p. (1971).

(4) CHANG CHEN FA, ZHENG XI LANG and PAN YU SHENG, The geological history, tectonic zonation and origin uf uplifting of the Himalaya, *Acta Sinica Geologica*, Beijing 1-15 (1977).

(5) COLCHEN M., On the paleogeographic and structural evolution of the Tibetan area of the Nepalese Himalaya *Himalayan Geology*, 5, 83-103 (1974).

(6) COLCHEN M., FORT M. and FREYTET P., Sédimentation et tectonique plio-quaternaire dans le Haut Himalaya : l'exemple du fossé de la Thakkhola (Himalaya du Népal), *7ème Réunion Sciences de la Terre*, Lyon, France, p. 121- (1979).

(7) COLCHEN M., FORT M. and FREYTET P., Evolution paléogéographique et structurale du fossé de la Thakkhola-Mustang (Himalaya du Népal) : implications sur l'histoire récente de la chaîne himalayenne, *C.R. Ac. Sc.*, Paris, 290, D, 311-314 (1980).

(8) DOLLFUS O. and USSELMANN P., Recherches géomorphologiques dans le centre-ouest du Népal, *Cahiers Népalais*, CNRS, Paris, 56 p. (1972).

(9) FORT M., Quaternary deposits of the Middle Kali Gandaki Valley (Central Nepal), *Himalayan Geology*, 6, 499-507 (1976).

(10) FORT M., Etudes sur le Quaternaire en Himalaya, la haute vallée de la Buri Gandaki, Népal. *Cahiers Népalais*, CNRS, Paris, 236 p., (1979).
(11) FORT M., Les formations quaternaires lacustres de la Basse Thakkhola (Himalaya du Népal) : intérêt paléogéographique, néotectonique et chronologique, *C.R. Ac. Sc.*, Paris, 280, D, 171-174 (1980 a).
(12) FORT M., Néotectonique et géomorphologie dans le Haut-Himalaya, *8ème Réunion Sc. Terre*, Marseille, France, p. 152 (1980 b).
(13) FORT M., La formation de la Kali Gandaki : état des connaissances et implications sur l'histoire récente de la Haute Chaîne himalayenne, *Bull. Ass. Géogr. Français*, Paris, n° 470-471 (1980 c).
(14) FORT M., BASSOULET J.P., COLCHEN M., FREYTET, Sedimentological and structural evolution of the Thakkhola-Mustang Graben (Nepal Himalaya) during Late Neogene and Pleistocene, *Field Conference on Neogene/Quaternary Boundary*, I.G.C.P., Project 41, Chandigarh, India (in press) (1979).
(15) FORT M. and FREYTET P., L'évolution sédimentaire récente du bassin intra -montagnard de Pokhara (Himalaya, Centre-Ouest Népal), *C.R. Ac. Sc.*, D, 289, 1195-1198 (1979).
(16) FREYTET P., Petrography and paleoenvironments of continental carbonated deposits, with a particular reference to Upper Cretaceous and Lower Eocene of Languedoc, Southern France, *Sedimentary Geology*, 10, 25-60 (1973).
(17) FREYTET P., BASSOULET J.P., COLCHEN M., Preliminary sedimentological data on the continental Thakkhola formations (Higher Himalaya, *Himal. Geol.* (in press) (1977).
(18) FREYTET P. and FORT M., Figures diagénétiques, traces d'émersion, paléosols et figures de sédimentation lacustre dans les formations récentes (Plio-Quaternaire) de la Kali Gandaki et du bassin de Pokhara (Himalaya du Népal), *Bull. Assoc. Géogr. Français*, Paris, n° 470-471 (1980).
(19) FUSHIMI H., Glaciations in the Khumbu Himal, SEPPYO, 40, 71-79, (1978).
(20) GAO YIXIN, CHEN HONGZHAD and WU ZHIDONG, Paleosols and the upheaval of the Xizang Plateau, *Symposium on Qinghai-Xizang Plateau*, Beijing (abstr.) p. 274 (1980).
(21) GANSSER A., *Geology of the Himalaya*, Interscience Publ., London, 289 p. (1964).
(22) GANSSER A., Structural map of Himalaya and South Tibet, 1/3,500,00, *in Coll. Intern. CNRS*, n° 268, *Himalaya, Sciences de la Terre*, Paris (1977).
(23) HAGEN T., Géologie des Thakkhola (Népal), *Eclog. Geol. Helv.*, 52/2, 709-719 (1959).
(24) HAGEN T., Report on the Geological Survey of Nepal, vol. 2 : Geology of the Thakkhola, including adjacent areas. *Denkschriften der Schweizerischen Naturforschenden Gesellschaft*, 86, 2, 160 p. (1968).
(25) HEIM A., GANSSER A., Central Himalaya, Geological observations of the Swiss Expedition 1936, *Mem. Soc. Helv. Sc. Nat.*, 73, 1, 1-245 (1939).
(26) HIGUCHI K., Glaciers and climates of Nepal Himalaya. The data report of Glaciological expedition to Nepal, 1973-74. SEPPYO, 38 (1976).
(27) HUANG WAMPO, JI. HONGXIANG, The first discovery of Xizang *Hipparion* fauna and its significance to the Plateau rise, *Kexue Tongbao*, 19, 885-888 (1979)
(28) IWATA S., Late Pleistocene and Holocene moraines in Sagarmatha (Everest) region, Khumbu Himal, SEPPYO, 38, 109-114 (1976).
(29) KALVODA J., The relief of the Himalaya and its recent modelation. *Rozpravy CSAV*, 86, Praha, I, 52 p. (1976).
(30) KALVODA J., The Quaternary history of the Barun Glaciers, Nepal Himalaya, *Vestnik Ustredniho Ustavu Geologickeho*, 54, 1, 11-23 (1979).
(31) NANDI B., Palynostratigraphy of the Siwalik Group of Panjab, *Himalayan Geology*, 5, 411-424 (1975).
(32) SHI YAFENG, LI *JIJUN*, A glaciological research of the Qinghai-Xizang

Plateau, *in Proc. Symp. Qinghai-Xizang (Tibet)* (abstracts), Beijing, 221-222 (1980).
(33) THOURET J.C., Contribution à l'étude de l'évolution des grands versants de la montagne himalayenne. Exemple du bassin versant de l'Ankhu Khola, *Thèse 3ème cycle*, Paris, 419 p. (unpublished) (1977).
(34) VALDIYA K.S., Himalayan transverse faults and folds and their parallelism with subsurface structures of the northern Indian Plain, *Tectonophysics*, 32, 353-386.
(35) XIZANG SCIENTIFIC INVESTIGATION TEAM, Academia Sinica, Report on the scientific investigations (1966-1968) in the Mt. Qomolangma district. *Quaternary Geology* Science Press, 112 p.
(36) ZANG QINGSONG, LI BINGYUAN, YANG YICHOU and YIN ZESHENG, Planation surfaces on the Qinghai-Xizang Plateau (abstract), *Symp. on Qinghai-Xizang Plateau*, p. 259 (1980).
(37) ZHENG BENXING, LI JIJUN, Quaternary glaciation on the Qinghai-Xizang plateau, *Proc. of Symposium on Qinghai-Xizang Plateau* (abstracts), Beijing, p. 229 (1980).

Explanation of plate :

Fig.1a: m.c.t.= Main Central Thrust, m.b.t.= Main Border Thrust, s.w. = Siwaliks.

Fig.1b: Lo Mg = Lo Mantang, Sm= Sammargaon, KG= Kali Gandaki, Te= Tetang, Kg= Kagbeni, MK= Muktinath, Th n. and Th s.= Thorungse Nord and Thorungse Sud, Jm= Jomosom, Ma= Marpha, Lj= Larjung, Tg= Taglung, Le = Lete , Gz= Gaza.

: 1= Pleistocene and recent formation, 2= Thakkhola formation, 3= Tetang formation, 4= Mesozoic, 5= Paleozoic, 6= Tibetan Slab , 7= Mustang granite, 8= Mylonite, 9= fault, 10= crest line.

Fig.2 : I= Mesozoic, II= Tetang formation, III= Thakkhola formation, IV= Pleistocene and recent formation, F1,F2, F3 and F4= faults.

# THAKKHOLA-MUSTANG
(sketch map)

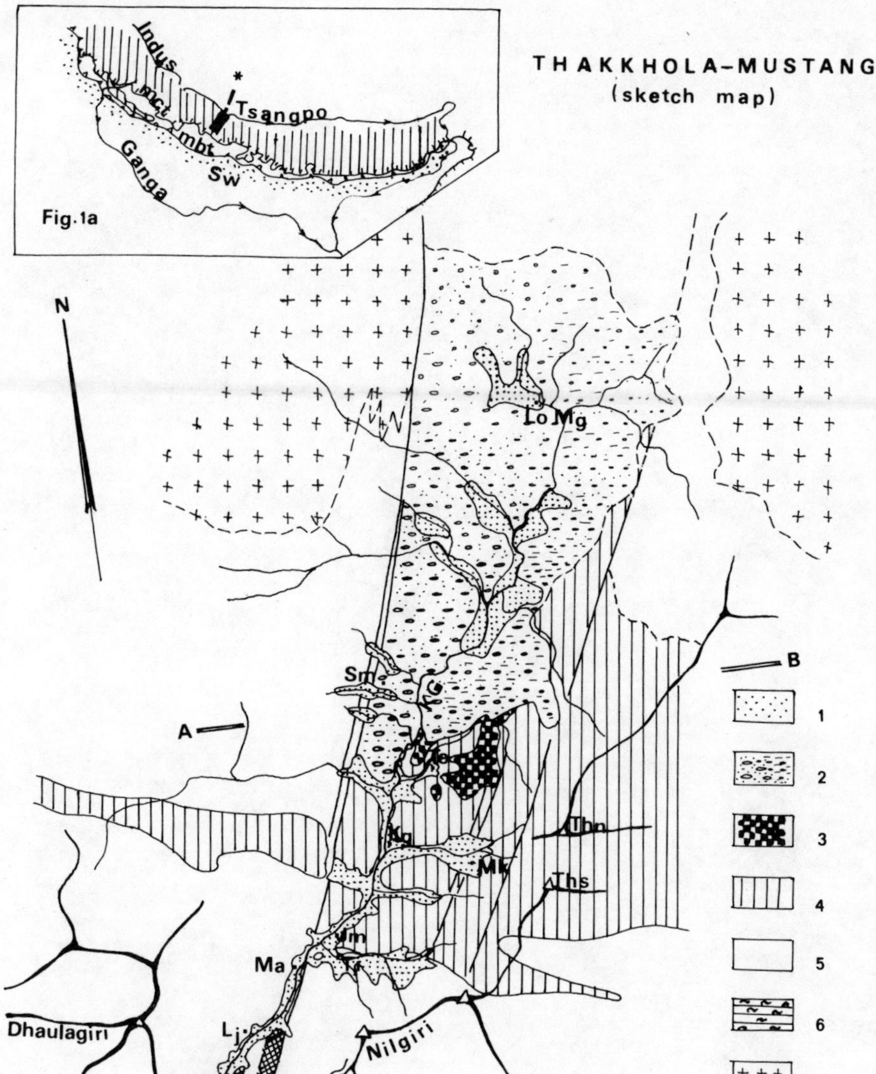


Fig.1a

Fig.1b

Fig. 2

## SOME ANALOGIES BETWEEN THE TECTONIC HISTORIES OF AFGHANISTAN AND TIBET

M. Mattauer*, P. Tapponnier**, F. Proust*

*Département des Sciences de la Terre, U.S.T.L. 34060 Montpellier, Frane

**Institut de Physique du Globe, 4 Place Jussieu, 75230 Paris Cédex 05 France

Abstract

The tectonic histories of Tibet and Afghanistan may have been similar in many ways since the end of the Paleozoic. Both appear to be the result of successive accretion of fragments of Gondwana to the active margin of Laurasia. This margin, which, in Afghanistan, appears to lie along the present Herat and Panjchir faults, South of Hindu Kuch, swings around the central Pamirs and can presumably be traced along the present western Altyn Tagh and Kun-Lun faults, South of Kun-Lun. North of this boundary, Paleozoic rocks have been deformed in the upper Paleozoic, whereas South of it, there is no trace of the Hercynian orogeny. The first collision of Gondwanian fragments with Laurasia probably occurred in the early Mesozoic along the Hindu Kuch and Kun-Lun. To the South, ophiolites along the Panjao and Pang Gong-Nu Chiang sutures (respectively in central Afghanistan and central Tibet) testify for another suturing event in the upper Jurassic or lower Cretaceous. The Indus-Tsang-Po suture between India and Tibet corresponds, in eastern Afghanistan, to two ophiolite subbelts, near Kabul and Khost. Both ophiolite complexes have been emplaced between Maestrichtian and lower Eocene, presumably by choking of two northward dipping subduction zones. In Pakistan, the Kohistan mafic sequence probably represents the crust of an intervening Island Arc obducted on top of the Indian margin. Blueschist metamorphism was associated with the obduction. After complete contact between the Indian and Asian continents was achieved, presumably in the end of Eocene, the penetration of India into Asia caused large scale intracontinental shortening. A large part of the shortening was accomodated by strike slip faulting along the Mesozoic and more ancient sutures. In particular, both Tibet and Afghanistan were extruded to the East and West respectively along the Herat and Altyn Tagh-Kun Lun sutures.

We here summarize the work of different French groups on the geology of Afghanistan (Mennessier, 1972, Lapparent, 1972, Boulin and Bouyx, 1977, Andrieux and Brunel, 1979, Blaise and Al, 1978, Beun and Al, 1979, Cassaigneau, 1979 etc..) We try to interpret major facts in the framework of plate tectonics, and also propose a comparison between the tectonic evolutions of Afghanistan and Tibet (Tapponnier et Al, 1981).

Afghanistan is a particularly suitable place to study the Alpine collision belt because the several ranges of the belt merge into a narrow zone. One can therefore make a complete cross section from the Indian shield to the Eurasian platform over a rather short distance. It is clear that results of geological investigations in Afghanistan are of great interest to understand some aspects of the structure and history of Tibet, since both regions are almost adjacent. In fact, the similarities are so striking that Afghanistan may be nicknamed "little Tibet".

Geology and structure of Afghanistan. The major features of the geology of Afghanistan are shown on the simplified structural map of Figure 1 and the schematic cross section of Figure 2.

At least four sutures can be identified in Afghanistan between India and the Turan platform. The three southern most sutures separate blocks of Gondwanian origin (India and the Kabul block, the Helmand block and the Farah block). The fourth suture (along the Herat fault) corresponds to the southern margin of Laurasia. All blocks south of it have paleozoic marine platform sediments, faunas and floras characteristic of Gondwana and do not bear the imprint of the upper Paleozoic orogeny which has been pervasive to the north, in Laurasia (Stocklin, 1977).
The ages of the sutures increase from south to north : overthrusting of the ophiolites in the Khost suture, locally associated with blueschist metamorphism occurs before the upper paleocene (Mattauer et Al, 1978, Tapponnier et Al, 1981). For the Kabul suture, the age is less certain but it is likely that the ultramafic nappes were overthrust before the Eocene. After that time, it appears that India and the Kabul block were still separated by a rather deep and large basin where up to 10,000 m of tertiary flysch and molasses have piled up (Arthaud et Al, 1977, Cassaigneau, 1979).
These sediments have been tightly folded between the Eocene and the present. To the west, the Kabul and Khost sutures join the Chaman fault zone. To the east, their lateral equivalents are found North and South of the Kohistan Island Arc, (Khan Tahirkheli et Al, 1979) which continues in Ladakh and in the Kangdese mountains.

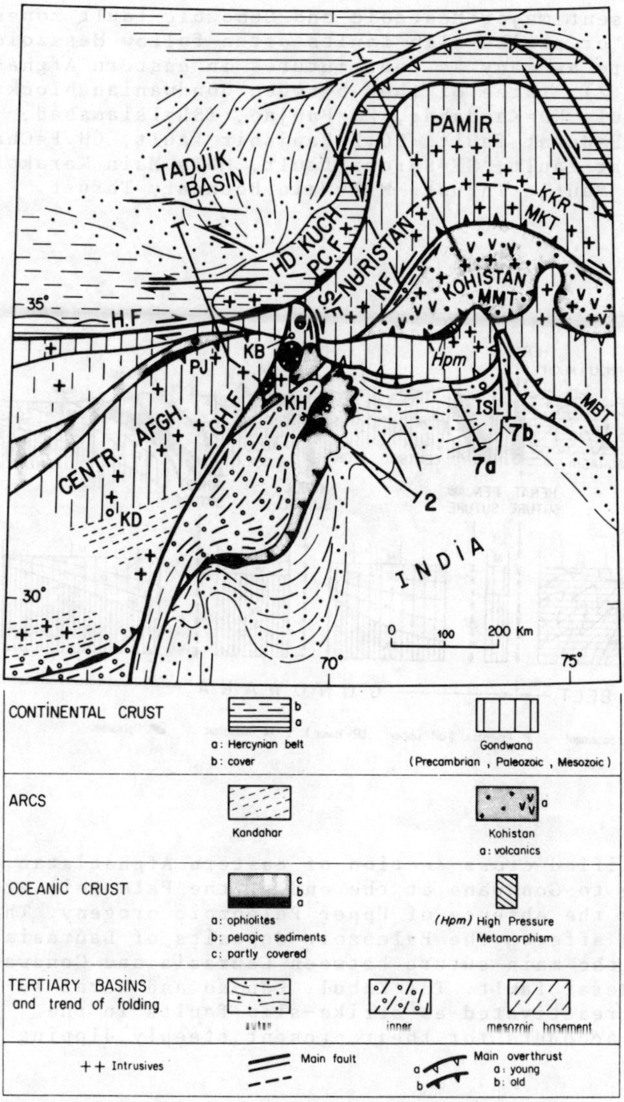


KH: Khost _ KB: Kabul _ KD: Kandahar _ PJ: Penjao _ ISL: Islamabad _ KKR: Karakorum
Fault: HF: Herat _ PC.F: Panjchir _ CHF: Chaman _ KF: Kunar _ SF: Sarobi
Thrust: MKT: Main Karakorum _ MMT: Main Mantle _ MBT: Main Boundary

Figure 1: Simplified structural map of Afghanistan and western Himalayan syntaxis (From Tapponnier et Al, 1981). Cross section number 2 refers to Figure 2. Vertical bars indicate Gondwanian basement, and horizontal bars Laurasian basements.

Bold lines represent major Mesozoic and Cenozoic fault zones; Cenozoic thrusts or strike-slip faults often follow Mesozoic sutures. There are as many as four sutures in eastern Afghanistan (see Fig. 2), three of them between Gondwanian blocks. KH=Khost, KB=Kabul, KD=Kandahar, PJ=Panjao, ISL=Islamabad, KKR=Karakoram, HF=Herat fault, PC.F=Panjshir fault, CH.F=Chaman fault, K.F.= Kunar fault, SF=Sarobi fault, MKT= Main Karakoram Thrust, MMT=Main Mantle Thrust, MBT=Main Boundary Thrust.

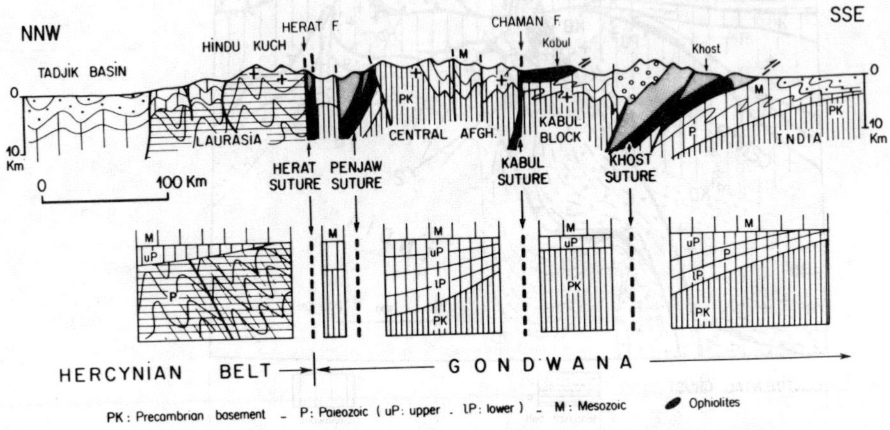


Figure 2 : Simplified cross-section of eastern Afghanistan. Blocks belonging to Gondwana at the end of the Paleozoic are characterized by the absence of Upper Paleozoic orogeny. This orogeny strongly affects the Paleozoic deposits of Laurasia. Correspondingly the main suture between Laurasia and Gondwana lies along the Herat fault. The Kabul, Panjao and Herat sutures have been reactivated as strike-slip faults in the Tertiary, which accounts for their present steeply dipping geometry.

Central Afghanistan is separated into two blocks by the Panjao-Farah suture. We do not know how the ophiolitic melange and the upper Triassic to Jurassic flysch have been emplaced in this zone because it has been reactivated as a strike-slip fault, which makes the present structure more complex. Nevertheless, P. Bordet, C. Montenat, and J.P. Carbonnel (1978) have found evidence for high pressure metamorphism along it, and shown that the emplacement occurred before the middle Cretaceous. Only splinters of serpentine outcrop along the Herat fault (whose present structure is also that of tertiary strike slip fault), but it probably represents the major paleogeographic discontinuity of Afghanistan.

For the sake of simplicity, intrusives have not been emphasized on Figure 1 and 2, but there are two main belts of granites and granodiorites, whose petrology, chemistry, and ages have been studied by A. Desio (1975), P. Lefort and F. Debon (1978), and others. One belt extends along the Hindu-Kush north of the Herat and Panjschir faults : its calco-alkaline plutons have yielded ages of about 210 M.y. Another belt is located in the central Afghan mountains. Also calco-alkaline in nature, it appears to be quite younger : radiometric ages cluster around 100-110 M.y but some are tertiary. It is likely that these granite belts are a result of subduction of oceanic crust, at different times, under different blocks now part of of Asia.

<u>Opening and closure of the Tethyan basins</u>. There is little doubt that the tectonics of Afghanistan can be interpreted as a result of subduction of different parts of the Tethyan ocean and successive collision with Asia of blocks previously detached from Gondwana. India is only the last and the biggest of these blocks. This large scale history appears to be strikingly similar to that described for Tibet by Chang Cheng Fa and Cheng Hsi Lan (1973). Despite some uncertainties in the timing, it is possible to reconstruct the main stages in this history. Our views are presented in Figure 3 and 4 (From Tapponnier et Al, 1981).

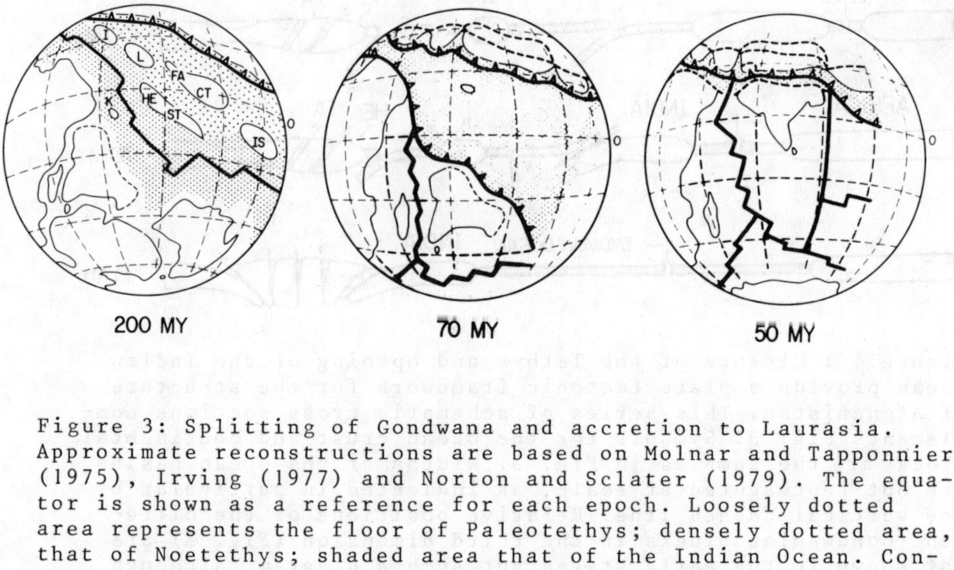


Figure 3: Splitting of Gondwana and accretion to Laurasia. Approximate reconstructions are based on Molnar and Tapponnier (1975), Irving (1977) and Norton and Sclater (1979). The equator is shown as a reference for each epoch. Loosely dotted area represents the floor of Paleotethys; densely dotted area, that of Noetethys; shaded area that of the Indian Ocean. Contours and positions of small continental blocks of Gondwanian origin at 200m.y. are hypothetical: I=Iran, L=Lut, FA and CT=Farah block and central Tibet, HE and ST=Helmand block and southern Tibet, IS=Indonesian block. Bold lines represent

major plate boundaries; normal lines: ridge and transform faults, lines with open triangles on one side; subduction zone. Position of subduction zones can be inferred with some confidence but location and sometimes existence of ridges within Paleotethys and Neotethys is, of course, impossible to specify since the corresponding ocean floor has essentially disappeared. Dashed lines represent suture zones. If the reconstruction at 70 m.y. is approximately correct, the geologic record and timing in Afghanistan, Pakistan and Ladakh, requires two active subduction zones between India and Eurasia in the Upper Cretaceous.

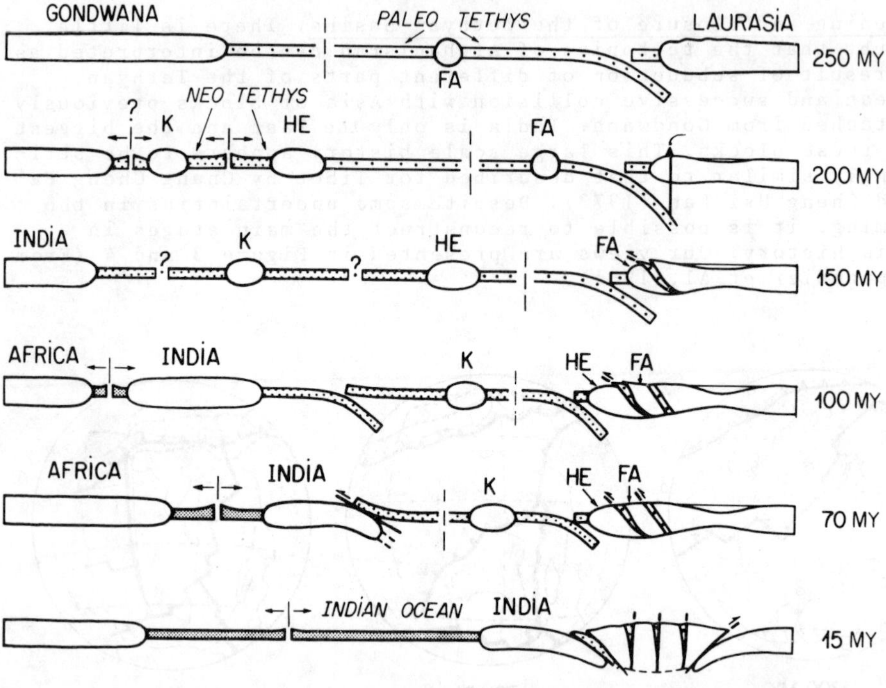


Figure 4 : Closure of the Tethys and opening of the Indian Ocean provide a plate tectonic framework for the structure of Afghanistan. This series of schematic cross sections complements Fig. 3. Symbols for the ocean crust and continental blocks are the same as in Fig. 3. Widths of the ocean basins are not represented at scale, as indicated in particular by the vertical dashed line. Relative positions of the different Gondwanian blocks in the third dimension (Fig. 3) are not known in the early stages but such a diagram, although two dimensional, is essential to understand the large-scale kinematics. The greatest uncertainties in this evolution scheme concern the age of the ocean floor in the Panjao basin (between FA and HE) and the dip of the subduction zone that consumed it.

In the Permian, the paleotethyan ocean separated Gondwana from Laurasia. Intrusion of the Hindu-Kuch granites during the Triassic indicate that the floor of this ocean was then subducting under Laurasia. In the beginning of the Jurassic, large realms of Paleotethys has essentially disappeared, while to the south the basins of Neotethys were opening between Gondwana and continental blocks detached from it. By the middle of the Cretaceous, the central Afghan blocks (FA and HE) had collided with Laurasia, forming first the Herat, then the Panjao sutures. The principal remnants of Paleotethys are to be found along the Herat-Panjschir-Akbaytal suture (Stöcklin, 1977) whose equivalent in Tibet can be traced south of the Altyn Tagh, and along the Kun Lun fault south of the western Kun Lun. It is also tempting to correlate the upper Jurassic-lower Cretaceous Panjao-Farah suture in central Afghanistan with the Pang Gong-Nu Chiang suture (PGNC), of comparable age, between southern Tibet and North Central Tibet (ST and CT on figure 3; N. Tibet and S. Tibet on Figure 5 )(Chang Chen Fa and Cheng Hsi Lan, 1973)). In the Cretaceous, subduction of the Neotethyan ocean floor is responsible for the Calco-Alkaline magmatism in central Afghanistan and the "Kohistan Island Arc" (Khan Tahikheli et Al, 1979).

Ophiolites in the Khost suture, which is a lateral equivalent of the Oman-Muslimbagh suture appear to be thrust onto the Indian shield in the uppermost cretaceous or paleocene. The Kabul suture is perhaps of upper paleocene age. The emplacement of the ultramafic thrustsheets in both sutures appears to result from choking of two northward-dipping subduction zones by buoyant continental crust (Davies, 1971, Mattauer et Al, 1977).

However, Figure 3 draws attention on a major discrepancy between inferences from the reconstructions(i.e.: Molnar and Tapponnier, 1975) and the geological record. If we accept both the idea that complete contact between the Indian and Asian continents was achieved only at the end of the Eocene, and the approximate age of emplacement of the Khost and Kabul ophiolites, then the ophiolites were not emplaced at the time of the India-Eurasia collision and the two continents were still separated by a rather wide ocean basin in the Eocene. This basin is readily identified with the Katawaz basin, where shortening may have amounted to several hundred Km in the upper Tertiary. A present day equivalent of this situation may be found in the Gulf of Oman : whereas ultramafic nappes have been thrust onto the Arabian shield to the South, the collision between Arabia and Iran is not yet complete and subduction is still active under the thick Makran flysch prism in the north (Jacob and Quittmeyer, 1979).

This is not without implications for the tectonics of the Yarlung-Tsang Po suture zone : if the ages of ophiolite emplacement along strike prove to be similar to those in Afghanistan, the reconstructions may require significant departures from accepted models. Perhaps the unique and simple suture line envisaged between the passive northern margin of India and the "Andean" margin of Cretaceous Asia is instead composed, as in Afghanistan, of two or more sutures with intervening blocks between them.

Intracontinental deformation and the impingement of the "Pamir Wedge". After complete collision between India and Eurasia, slip line field theory (Tapponnier and Molnar, 1976) may help predict both the resemblance and the assymetry between the Tectonics of central Afghanistan and Tibet, on either side of the protruding "Pamir Wedge".

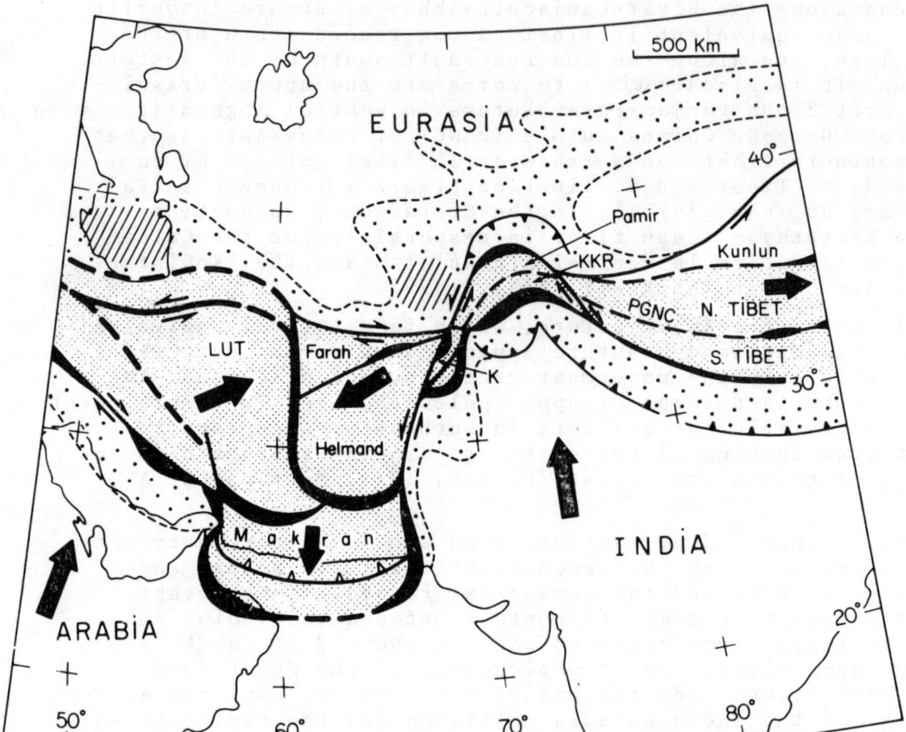


Figure 5 : Geometry of the principal alpine sutures between Eurasia, Arabia and India. Intervening blocks between the three major continental masses are shaded. Ophiolitic belts are in black. Dotted areas correspond to zones of alpine deformation. Hatched areas correspond to residual oceanic crust. Bold lines are major Cenozoic faults. Arrows indicate the approximate motions of intervening blocks in response to the penetration of India and Arabia into Eurasia. As in Afghanistan and Kohistan, the Tsang Po suture might be subdivided into two sutures. The Panjao suture corresponds to the Pang Gong Nu Chiang suture and the Herat suture, to the Kun Lun suture.

Much as central Afghanistan, extruded to the west between the Chaman and Herat faults, Tibet may have been squeezed out to the East between the Karakorum and Altyn Tagh faults (Tapponnier and Molnar, 1977, 1981) (Figure 5). Both regions display the same roughly triangular shape. The process has taken place on a grander scale for Tibet, however, since it was located in a more frontal position with respect to the Indian indenter.

As in Afghanistan, segments of the large strike slip faults in Tibet and central China appear to be reactivated sutures (Tapponnier and Molnar, 1977). For instance, the western part of the Altyn Tagh fault, and the Kun Lun fault may be equivalents of the Herat fault, along Laurasia's ancient boundary. Yet, only sections adequately oriented with respect to the collision's stress field have been reactivated. For instance, sutures in central Tibet, almost perpendicular to the push of India, did not allow large strike-slip displacements in the Cenozoic. The fact that ancient sutures never heal completely and can be reactivated as large strike-slip faults, must be an important feature of many other areas of intracontinental deformation.

## References


Andrieux, J. and M. Brunel, 1977: L'évolution des chaînes occidentales du Pakistan, Mém.h.sér.Soc.Géol.Fr., 8, 189-207.

Arthaud, F., F. Mégard and M. Seguret, 1977: Cadre tectonique de quelques bassins sédimentaires, Bull. Centre. Rech. Elf-Aquitaine, 1-1, 147-188.

Beun, N., P. Bordet and J.P. Carbonnel, 1979: Premières données quantitatives relatives au coulissage du décrochements de Chaman (Afghanistan du Sud-Est) C.R. Acad. Sci. Paris, Ser. D, 288, 931-935.

Blaise, J., P. Bordet, J.P. Carbonnel and C. Montenat, 1978 : Flyschs et ophiolites dans la région de Panjaw : une suture Eocimmérienne en Afghanistan Central, Bull.Soc.Géol.Fr.7, XX, N°5, 210-204.

Boulin, J., and E. Bouyx, 1977: Introduction à la geologie de l'Hindu Kouch occidental, Mém.h.sér., Soc.Géol.Fr., 8, 87-105.

Cassaigneau, C., 1979 : Contribution à l'étude des sutures Inde-Eurasie : la zone de suture de Khost (S.E. Afghanistan), Thèse spécialité, Montpellier.

Chan Cheng-fa and Cheng Hsi-lan, 1973: Some tectonic features of the Mt. Jolmo Lungma area, southern Tibet, China, Sci. Sinica 16, 257-265.

Davies, H.C., 1971 : Peridotite-gabbro-basalt complex in eastern Papua : an overthrust plate of oceanic mantle and crust, Dep.Nat.Dev.Bur.Miner.Resour., Australia, Bull.128.

Debon, F., P. Le Fort and J. Sonet, 1978: Des caractères géochimiques de deux provinces plutoniques d'Afghanistan : Hindou-Kouch et "Montagnes Centrales", 6ème réun. Ann. Sci.Terre, Orsay, 135.

Desio, A.,1975 : Geology of central Badakshan (North East Afghanistan) and surrounding countries, in : Italian Expedition to Karakorum and Hindu-Kush, Rep.III, 3 (Brill, Leiden).
Irving, E., 1977 : Drift of the major continental blocks since the Devonian, Nature, 270, 304-309.
Jacob, K.H., and R.C. Quittmeyer, 1979 : The Makran region of Pakistan and Iran : trench arc system with active plate subduction, in : Geodynamics of Pakistan, Geol. Surv. Pakistan, Spec.Mem, 305-371.
Khan Tahirkheli, R.A., M. Mattauer, F. Proust, and P. Tapponnier, 1979: The India Eurasia suture zone in Northern Pakistan : synthesis and interpretation of recent data at Plate Scale, in : Geodynamics of Pakistan, Geol. Surv. Pakistan, Spec. Mem, 125-130.
De Lapparent, A.F., 1972: L'Afghanistan et la dérive du continent indien, Rev. Géogr. Phys. Géol.Dyn. XIV, 4, 449-455.
Mattauer, M., F. Proust and P. Tapponnier, 1977: Tectonic mechanism of obduction in relation with high pressure metamorphism, Coll. Int. C.N.R.S., 272, Orogenic Mafic and Ultramafic Association, Grenoble (1977) 197-201.
Mattauer, M., F. Proust, P. Tapponnier and C. Cassaigneau, 1978: Ophiolites, obductions et tectonique globale dans l'Est de l'Afghanistan, C.R. Acad. Sci. Paris, Sér. D, 287, 983-985.
Mennessier, G., 1972: Géologie de la chaîne d'Altimour (Afghanistan Oriental), Rev.Géogr.Phys.Géol.Dyn.XIV, 4, 345-356.
Molnar, P. and P.Tapponnier, 1975 : Cenozoic tectonics of Asia : effects of a continental collision, Science 189 , 419-426.
Norton, I.O. and J.G. Sclater,1979 : A model for the evolution of the Indian ocean and the break up of Gondwanaland, J. Geophys. Res. 84, 6803-6830.
Stöklin, J., 1977: Structural correlation of the Alpine ranges between Iran and Central Asia, Mém.h.sér.,Soc.Géol. Fr.8, 333-353.
Tapponnier, P., and P. Molnar, 1976: Slip-line field theory and large-scale continental tectonics, Nature 264, 319-324.
Tapponnier, P., and P. Molnar, 1977 : Active faulting and Cenozoic tectonics of China, J. Geophys. Res. 82, 2905-2930.
Tapponnier, P., M. Mattauer, F. Proust, and C. Cassaigneau, 1981 : Mesozoic ophiolites, sutures, and large-scale tectonic movements in Afghanistan, Earth Planet. Sci. Lett., 52, in the press.
Tapponnier, P., P. Molnar and Wang Ping Chen, 1981 : Active normal faulting in Tibet, this volume.

# ANALOGIES AND DIFFERENCES ALONG THE INDUS-TSANGPO SUTURE ZONE

Peter Schindler
Swiss Federal Institute of Technology Zürich

In a conference or symposium like the present one we participate in general not only to report on our work and to present new information, but also to learn about the work and the discoveries of other scientists. As for this particular symposium I know that a number of scientists have come also to find out what and where the most serious gaps in our knowledge of the Qinghai-Xizang plateau are. I count myself among these people not only because the work I will report on has been done by others but because it seems particularly appropriate at this first conference of its kind to carefully define open problems and fields of future common work.

Among earth scientists there is apparently a consensus on both the approximate ages and the significance of the separate ophiolite belts of the Qinghai-Xizang region. The Kokoxili-Kunlun range is reported to have the oldest ages, even though recent, quaternary volcanic activity has also been observed by Deng Wanming (1). Further towards the south, ages become progressively younger, with estimates of some 120 to 70 million years in the Tsangpo region. Himalayan orogeny is still much more recent and in fact presumably continuing at this time. A consensus is also evident on the interpretation of the Indus-Tsangpo ophiolite zone as the remnant of the collision between the Indian plate and the Asian continent. Marked disagreement has however become manifest on the questions of the existence and the vergence of subductions. Seismic data seem to allow interpretation of both northward and southward dip of such subductions, if they exist at all.

In the Ladakh area the ophiolite zone is apparently doubled (2). The northern Shyok area has not been generally accessible, but information gathered by Gansser (2) seems to point to the existence of two separate suture zones with different ages. Whether or not the northern suture zone continues towards the east is not known. From satellite photos it seems possible, but this hypothesis needs to be proven or disproven in the field. Of major interest should also be the connecting faults.

Frank et al. (3) distinguish at least six subzones within the Indus suture zone. The northernmost zone is in the form of a large batholith, the Ladakh granites. Preliminary age determinations by Frank communicated to Gansser (2) indicate some 60 m.y. for these rocks. They are interpreted as accumulation of intrusives after the consumption of the oceanic crust between the Indian and Asian plates. If this interpretation were correct, then one should expect relatively low initial Sr - isotope ratios. For the Ladakh granites this mass spectrometer determination has not yet been performed. However, as Professor Liu Tung-sheng has pointed out in the opening speech to this symposium, an exactly analogous situation pevails in the Tsangpo region of the suture. Whereas in the High Himalayas the $Sr^{87}/Sr^{86}$ - ratio is 0.747, it is only 0.704 in the Kangdese mountains. This would suggest that there is a close relationship between the Ladakh granites and the Kangdese granites. It is tempting to conclude that the same kind of orogeny, at about the same period, is responsible for both mountain ranges.

The next zone south of the Ladakh intrusives within the Indus suture zone are the molasse series. In eastern Ladakh they are quite spectacular and show indisturbed thicknesses of over 3000 meters. The conglomerates contains mostly granites and other plutonic rocks, but also some limestone pebbles with alveolinas allowing a lower Eocene dating. These comglomerates have their counterpart in the Kailas conglomerates, containing similar pebbles (Heim und Gansser 4) In the map published for the excursion after this symposium, some smaller conglomerate outcrops are marked. The relationship between these and the Kailas and Indus conglomerates should also be of very high interest.

The complications of the flysch zone and the ophiolitic mélanges are such that it is difficult to define which parameters should be investigated first. Again different is the case of the Dras volcanics. Situated between the molasse and the flysch and separated from those by the ophiolitic mélange, they seem to have no analogy further to the east. These are older than the Ladakh granites: fossils found in limestone bands (orbitolinas) give the Dras a cretaceous age. Geochemical analysis has not yet been done, so the genesis of these rocks is unknown.

As a smaller equivalent to the huge Amlang La ophiolite nappe, a klippe (Spongtang) is reported within the sedimentary section in the south of the suture. No very convincing explanation is known for the emplacement of such huge "exotic blocks" on the southern sediments.

These sediments, named "Zaskar sediments" by Frank et al. (3), are of paleozoic to triassic age. They rest on an old sialic crust. Although not quite as complicated as the flysch and ophiolite belts, the correspondence to the sediments in Xizang is unknown and merits clarification.

On the north end of the north-south profile just described, the situation in Ladakh and in the Lhasa area in Xizang seems to be fairly similar. If the conditions along the whole length of the Indus-Tsangpo suture have been the same for the orogeny of the Ladakh granites and the Kangdese mountains, then one conclusion implicitly drawn by many people is that the continental collision happened over the whole length at the same period. Although not impossible, this is not necessarily plausible. The existence of the Dras series only in the west may point to a different development of the suture at both ends, but this is not really understood.

Further to the south, complexity of the problems increases enormously. It is probably fair to say that we are just at the beginning of the understanding of the Indus-Tsangpo suture zone. Now that information from regions until today inaccesible becomes available, chances for fast progress in the understanding are great. The Academia Sinica deserves praise from all scientists interested in the Himalayas and the Xizang plateau for letting us participate in the exploration of this fascinating and important part of the world.

## References

(1) Deng Wanming. A preliminary study on the petrology and petrochemistry of the quaternary rocks of northern Tibet Autonomous region:Institute of Geology, Academia Sinica, Beijing, page 149 - 163.
(2) Gansser,Augusto. Ophiolitic belts of the Himalayan and Tibetan region. Geological Society of America Map and Chart series, MC-33, 1979, page 12 - 14.
(3) Frank, Wolfgang, Gansser,A., Trommsdorff,V., Schweiz. mineral. petrogr. Mitt. (57), 89 - 113 (1977).
(4) Heim,A. Gansser,A., Central Himalaya, Geological Observations of the Swiss Expedition 1936. Denkschriften der Schweizerischen Naturforschenden Gesellschaft, (73), page 1 - 245 (1939).

# LATEST MARINE SEDIMENTS IN XIZANG AND PROCESS OF EARLY TERTIARY REGRESSION

Zhang Bing-gao
(Nanjing Institute of Geology and Palaeontology,
Academia Sinica)

Up to now the Lower Tertiary deposits are the latest marine sediments in Xizang. They chiefly distribute in Tingri-Gamba region to the south of Yarlung Zangbo River, and they are also found in Mai La and Lhünzhub to the north of Yarlung Zangbo River.

The Lower Tertiary sediments to the south of Yarlung Zangbo River may be divided into the Zongpu Group in the lower part and the Zhepure Formation in the upper. The problem concerning the uppermost marine sediments in Xizang is the age of the Zhepure Formation, exposed chiefly in the Zhepure Mountain of Tingri.

The upper part of Zhepure Formation is composed of limestones, containing the following large foraminifers (Xia, 1962): *Orbitolites complanatus* Lamarck, *Fasciolites oblongus* (d'Orbigny), *F. oviculus* (Nuttall), *Nummulites lucasi* Douvillé, *N. incrassatus* de la Harpe, *N. paronai* (Prever), *N. afghanicus* Cizancout, *Assilina placentula* (Deshayes), *A. rota* Davies, *A. orientalis* Douvillé, *Suscocyclina nummulitica* (Guembal), *D. clarki* (Cushman), *D. partti*, *D. sowerbyi* Nuttall. These fossils commonly occur in the Middle Eocene. Only a few of them (for example *Nummulites incrassatus* and *Discocyclina sowerbyi*)range into the Late Eocene. Wen (1974) referred this part to the lower part of the Upper Eocene (Lutetian).

According to Ho et al, (1976) the foraminifers in the Zhepure Formation are chiefly *Orbitolites cotentinensis* Lehmann, *Fasciolites ellipsoidalis* (Schwager), *F. nuttalli* (Davies), *F. oviculus* (Nuttall), *Nummulites laevigatus* (Bruguiere), *N. rotularius* Deshayes, *N. atacicus* Davies, *Assilina dandotica* Davies, *A. subspinosa* Davies, *A. granulosa* (d'Archiac) and *Discocyclina sowerbyi* Nuttall, etc. All of these fossils are common in the middle stage of Eocene and may be compared with those of South Asia (in the Assam region and Baluchistan Basin). In the Upper Eocene, the characteristic fossil is *Pellatispira*. In the middle and lower part of the Eocene the assemblage includes a large number of *Assilina*, *Fasciolites*, *Dictyoccooides*, *Lockhartia* and *Nummulites*, etc. but no *Pellatispira* has been found in the Zhepure Formation. The faunal aspect of this formation closely related to that of the Middle Eocene in South Asia. It is therefore that the age of the upper part of the Zhepure Formation, should be referred to the Middle Eocene (Lutetian).

Based on the present material, the sea was still united in the late Late Cretaceous of Himalayan region (Fig. 1).

The marine sediments covered an area including Kashmir, Salt Range, Spiti, Tingri-Gamba region, Assam and the western part of Gandise Range, etc. The ancient sea was distributed nearly east-west in direction in the Himalayan region.

The marine Tertiary sediments are very different from those of Cretaceous. They have not been found in the western part of the Gandise Range. The Kumaon region may have also been uplifted. From the beginning of the Tertiary the western part of the Gandise Range was uplifted and united with the Kumaon region. Based on the material illustrated of the uppermost marine sediments in Xizang and adjacent regions during the Tertiary we may discuss the course of the Tertiary regression. (Fig. 2).

Fig. 2 shows that the age of the marine sediments is the oldest from north at the sides of Gandise Range and in the Kumaon region, and gradually becomes younger away from this area. This change indicates that two separated and disconnected seas were formed perhaps in the Himalayan region and the sea regressed from both sides of the Gandise Range and the Kumaon region. The ancient sea in Xizang regressed from west to east from the western parts of Nepal passing through the Tingri-Gamba region to Assam and the western part of Burma during Early Tertiary. In Palaeocene the marine area included Lhünzhub to the north of Lhasa and in Middle Eocene it even extended to Mai La, at the middle part of Gandise Range.

According to some authors, the Indus-Yarlung Zangbo River is the suture line connecting the Indian plate with the Eurasian plate. But the contact age is quite different as deduced by various authors. Because the two disconnected seas were formed in the Himalayan region during early Tertiary, it appears that the Eurasian and Indian continents were then connected, so that the collision of the Indian plate with the Eurasian plate probably occurred prior to Tertiary period.

## References

(1) Eames, F.E., Discussion of the faunas of certain standard sections, and their bearing on the classification and correlation of the Eocene in Western Pakistan. Q.J.G.S. London, 107, 173-196 (1952).

(2) Hashimoto, S. et al., Geology of the Nepal Himalayan. Himalayan Committee of Hokkaido Univ., Sappora, Japan, 1-286 (1973).

(3) Ho Yen, Zhang Bing-gao, Hu Lan-ying, Sheng Jing-chang, Mesozoic and Cenozoic foraminifera from the Mount Jolmo Lungma Region. A Report of the Scientific Expedition in Mount Jolmo Lungma Region - (1966-1968), Palaeontology, Fasc. II, 1-124, Science Press (1976) (in Chinese).

(4) Kureshy, A.A., Eocene biostratigraphy of Pakistan B.R.G.M., (69), 219-224 (1969).

(5) Nagappa, Y., Foraminifera biostratigraphy of Cretaceous-Eocene succession in the India-Pakistan-Burma region. Micropal. 5 (2), 145-192 (1959).

(6) Rao, L.R., The problem of the Cretaceous-Tertiary boundary with special reference to India and adjacent countries. Bangalore, India. 1-66 (1964).

(7) Samanta, B.K., The Eocene beds of Garo Hills, Assam, India. Geol. Mag., 105 (2), 124-135 (1968).

(8) Wen Shi-xuan, The stratigraphy in Mount Jolmo Lungma Region -- Tertiary system. A Report of the Scientific Expedition in Mount Jolmo Lungma Region - (1966-1968), Geology, 184-212, Science Press (1974)

(in Chinese).
(9) Zhang Bing-gao, Mu Xi-nan, Discovery of marine Tertiary to the north of Yarlung Zangbo River, Xizang. Jour. Stratigraphy, 3 (1), 65-66 (1979) (in Chinese).
(10) Zheng Xi-lan, Chang Chen-fa, A preliminary note on the tectonic features of the lower Yalu-Tsangpo River region. Scientia Geologica Sinica, (2), 116-126 (1979) (in Chinese).

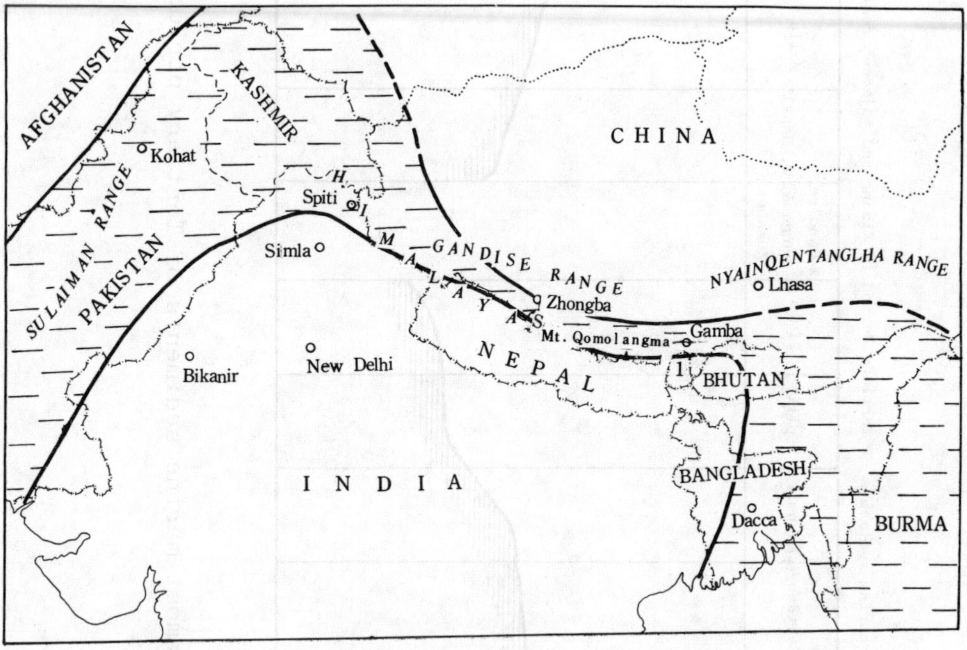


Fig. 1 Map showing the distribution of late Late Cretaceous sea in Xizang and adjacent regions

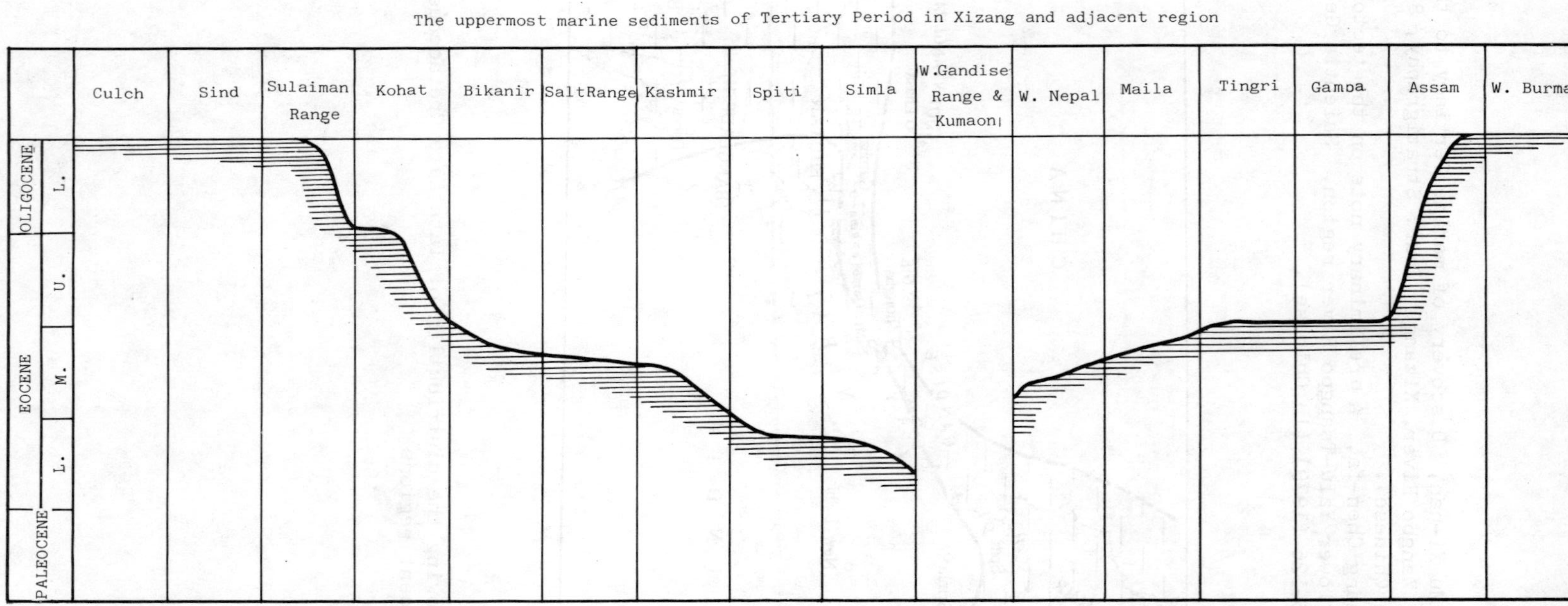


Fig. 2 The uppermost marine sediments of Tertiary period in Xizang and adjacent region.

# ON JURASSIC SYSTEM AND JURASSIC-CRETACEOUS BOUNDARY IN XIZANG

Sun Dong-li
(Nanjing Institute of Geology and Palaeontology, Academia Sinica)

In Jurassic period, Xizang was covered with sea water extensively. The "Tibetan Sea" as a part of Tethys, almost occupied the whole territory of Xizang except its northern extremity. The banded Jurassic rocks are widely exposed in a east-west direction and the strike turns to a north-south direction in the Hengduan Mountains region.

As being one of the most developed marine sediments in China, the Jurassic strata distributed in this realm are here divided into seven zones in sedimentary formation according to the characteristics in lithology, tectonics, biotas and the different modes of deposition.

Recent studies on Jurassic ammonoids, belemnoids, brachiopods and microfossils have led to the subdivision of Jurassic strata and the establishment of biostratigraphic sequence. The present paper aims to deal with the palaeogeographic position of "Tibetan Sea" at that time and the interchanges of transgression and regression. In this article the boundary between Jurassic and Cretaceous is discussed in obedience to the point of view that Phanerozoic system represent evolutionary stages in life history. From this point of view the author tends to place the boundary between Berriasian and Valanginian.

## SUBDIVISION OF ZONES IN SEDIMENTARY FORMATION AND CORRELATION OF JURASSIC STRATA IN VARIOUS ZONES

(The distribution of various zones is given in Fig. 1)

1. <u>The zone of sedimentary formation of stable littoral neritic pelitic carbonatites and clastic carbonatites in Southern Tethyan Himalayan Region.</u>

This sedimentary formation is mostly composed of limestones, quartz sandstones, shales, organic limestones, marls with plentiful faunas.

Lower Jurassic: The Pupuga formation is composed of shallow-sea pelitic carbonate formation containing ammonoids, brachiopods, scleractinia, bivalves, foraminifera, etc.

Middle Jurassic: The Niehnieh Hsiungla formation is composed of limestones, oolitic limestones, quartz sandstones with ammonoids, belemnoids, brachiopods, bivalves and hexacorals. The ammonoids are mainly found from the upper member (Callovian stage) and the lower member (Bajocian stage), while there are only brachiopods and lamellibrachiata in the middle member (Bathonian stage).

Upper Jurassic: The Menkatun formation consists chiefly of shales and marls intercalated with sandstones and limestones. The principal fauna is ammonoids. Brachiopods and bivalves are rare.

2. <u>The zone of sedimentary formation in Northern Tethyan Himalayan Region is composed of comparatively active epicontinental bathyal siliceous-argillaceous carbonatites, associated with neutral and basic volcanic rocks.</u>

Lower Jurassic: The Ritang formation is characterized by grey-black shales and siliceous-argillaceous limestones with plentiful ammonoids. They constitute a typical planktonic and nectonic community.

Middle Jurassic: The Gyangze group is composed of flysch in its lower part and flaggy limestones, marls intercalated with tuff of stightly basic volcanic rocks, in the upper. Fossils are very rare, merely in the vicinity of Kangmar county <u>Delecticeras</u> sp. of Bathonian age is found.

Upper Jurassic: The Weimei formation contains, in ascending order, pebbly sandstones, grey-dark shales, siliceous shales, but in the upper part intercalated with andesites, carrying mainly ammonoids and belemnoids, and constituting a complete cycle of sedimentation.

3. <u>The Lhasa zone of the stable littoral carbonate sedimentary formation.</u>

This zone generally covers the southern part of Gandise-Nyainqentanglha region, though the lithological characters of the Jurassic in the area west of Lhasa are still not clear. In Lhasa, the upper strata are represented by the Leibuzong coal-bearing formation, the Duodigou group in the lower is composed of pale-grey, grey-black, thick-bedded limestones, bioclastic limestones with hexacorals and gastropods. A little slates are found at the base, while the zone merges successively eastwards into the comparatively active neritic clastic carbonate deposits with intermediate and basic volcanic rocks. The Upper-Middle Jurassic Lagongtang formation is formed by grey-dark shales and quartz sandstones, resembling those of the east-member of the following fourth zone.

4. <u>The sedimentary formation zone of comparatively active, littoral, neritic, argillaceous, fragmental, carbonate on the southern side of Bangong Lake-Nujiang River abyssal fault.</u>

This zone occupies the northern part of Gandise-Nyainqentanglha region. In Shiquanhe-Xainza area, the upper part, Chuanba formation, is composed of coal-bearing sediments. The Daheishan formation in the lower is composed of limestones and mudstones, containing bivalves and gastropods. But in the Middle-Upper Jurassic strata in Baingoin-Nagqu area, there is grey-green volcanic clastic rocks and limestones partially intercalated with volcanic rocksbearing ammonoids, bivalves and gastropods. In the Biru-Banbar area, the Upper Jurassic, namely Guoqu group, consists mostly of striped sandy slates intercalated with tuffaceous sandstone, liparite volcanic agglomerates with ammonoids. The Shading group below is composed of sandstones and slates with no fossil record. To the east of Lhorong, the Douni formation known as the top of the Upper Jurassic, constitutes an alternate marine and continental coal-bearing deposits with plants. The Middle-Upper Jurassic Lagongtang formation is composed of grey-black, grey-green shales, arenaceous shales in association with ammonoids, bivalves and salenioids. The Shading group lying immediately below the formation is composed of limestone, marls, quartz sandstones, with brachiopods and bivalves, corresponding seemingly to the Liuwun formation.

5. <u>The zone of comparatively active, littoral, neritic, argillaceous,</u>

clastic sedimentary formation on the north side of Bangong Lake-Nujiang River abyssal fault.

This zone is located in a narrow and long terrain on the northern side of Bangong Lake-Nujiang River abyssal fault. The Upper Jurassic series-Dongqo formation (or Butuo formation in Dengqen county) is composed of limestones, dolomitic limestones, calcareous slates, with hexacoralla, hydroid polyp and stromatoporoids. The Middle Jurassic Muggarkangri group consists of epimetamorphic grey-green sandstones and shales. The Lower Jurassic Genalongba group is composed of dark sandstones, shales and carbonaceous shales with ammonoids and plants.

6. The Karakorum-Tanggula zone of stable shore epicontinental, detrital carbonate sedimentary formation intercalated with beds of terrestrial deposits.

In the west of the zone, the Upper Jurassic Hongqilapu and Jieshandaban formations are mainly composed of limestones with ammonoids, brachiopods, bivalves and hexacoralla. The Middle-Lower Jurassic Bagongbulansha group mostly consists of grey sandstones and shales intercalated with continental sediments, yielding such faunas as ammonoids, bryozoans, belemnoids, bivalves, hexacoralls and plants.

The eastern part of the zone is characterized by the vast distribution of Middle Jurassic deposits represented by the Yanshiping group and the Sewa formation, the latter consisting mainly of dark shales intercalated with mudstones and marls with ammonoids and bivalves. In Dengqen district, there is no deposit corresponding with the Sewa formation of a considerable thickness. The Yanshiping group is representative of neritic fragmental sediments intercalated with terrestrial clastic deposits becoming thinner towards east and west. In its southern part, it is mainly composed of marine deposits characterized by Burmirhynchia-Holcothyris fauna, while towards north it contains many intercalated beds of continental deposits.

7. The Qamdo zone of comparatively active, littoral, clastic, sedimentary formation and argillaceous microclastic sedimentary formation with lacustrine facies.

Lying mainly on the eastern bank of the Nujiang River, this zone is composed of red and variegated clastic rocks known as Chagyab group. In Riwoqe, west of Qamdo and Markam, the dominant faunas are brachiopods, pelecypods and foraminiferas, while in Darma La Mountains the biota contains plants and dinosaurian.

The lithological characters of each zone of sedimentary formation mentioned above seem to coincide with the tectonic position. On both sides of "slab suture" or abyssal fault, the deposits reflect correspondingly their mobility, while in the Southern Tethyan Himalayan Region and the Karakorum-Tanggula Region, known as the continental marginal shelf, comparatively stable littoral neritic clastic carbonate deposits called as platform type of deposits occur.

## THE PROPERTY OF BIOLOGICAL GROUP AND THE SEQUENCE OF MAIN FOSSILS

In Jurassic period, the Xizang sea belonged to the eastern province of the Tethyan realm in which the faunas, particularly nectonic ammonoids and belemnoids etc. are among the commonest type, while the benthonic brachiopods, bivalves, corals, stromatoporoids and gastropods contain numerous endemic genera and species. The fossil-bearing sequence is given in Table 1.

## THE PALEOGEOGRAPHICAL POSITION AND THE INTERCHANGE OF TRANSGRESSION AND REGRESSION

On the basis of studies on the Jurassic of Xizang and some neighbouring areas including Pamir, Xinjiang, Qinghai. West Yunnan and Burma, the arc margin of Laurasia in this region can be roughly outlined. The Xizang sea lay to the south of Taxkorgan, Quanshuigou, Bangdag Lake, Margai Caka, Hoh Xil Range, Wudaoliang and Yushu and to the west of Riwoqe,Qamdo, Markam, Weixi and Yongde. D. V. Ager, (1975) and A. Hallam, (1975) made an effort to reconstruct a map of distribution of land and sea on the basis of geological, paleomagnetic and other geophysical evidences. The coast line sketched in the region was located only at Lat. 28° N. However, under our investigation, the real coast line should stand further north, about Lat. 35° N.

In the wake of Indo-China orogeny in Early Jurassic, the sea water extended eastward to Markam and northward to Hongqilapu in southwest Xinjiang and to the south of Baingoin and Amdo, i.e., the fifth and sixth sedimentary zones and the southern part of the seventh zone.

In Middle Jurassic, transgression was so widespread almost over the whole region and expanded northward to the northern side of Tanggula Range that the sediments in this epoch overlay the older Triassic strata. Among them, the largest transgression took place in Bathonian and Callovian ages.

In Late Jurassic, due to the influence of Yenshan Movement, the sea water began to recede from north-east to south-west. The Oxfordian and early Kimmeredgian marine deposits merely reached the sixth sedimentary zone. So,the Cretaceous terrestrial purple clastic rocks overlie unconformably the Upper or the Middle Jurassic. No Upper Jurassic marine strata are found from the seventh sedimentary zone. During the Tithonian-Berriasian ages, the sea water receded to as far as the fourth sedimentury zone.

## THE BOUNDARY BETWEEN JURASSIC AND CRETACEOUS

This is one of the most outstanding problem in the study on systematic stratigraphy and has been a subject in hot debate for more than a century, but no unanimous conclusion has been reached up to now.

The following views on this problem are well known to us at present.
1. J. Wiedmann holds that Berriasian enters into Jurassic as a

substage of the Tithonian stage.

2. V. L. Yegoyan and others suggest that Berriasian should be attributed to Cretaceous as an independent stage.

3. G. Le Hegarat and others uphold the traditional boundary of the Jurassic and Cretaceous systems in Berrias area, south-east of France, i. e. to draw a boundary between the *Berriasella jacobi* zone and the *B. grandis* zone.

The writer deems it right that the evolutionary stages in a life history, i.e. the discontinuity, must be considered in demarcating the boundaries of a system. Of course, we cannot follow the viewpoint of the so-called "catastrophists", and are looking forward to a "natural break". However, we should find out a reasonable boundary on the basis of the evolutionary history of the whole biota and the developmental rule of various organisms including not only ammonoids but also ostracods, brachiopods, foraminifers, tintinnids, mammalia, etc. Indeed, no major step has been taken on the traditional boundary in regard to brachiopods and ammonoids. D. V. Ager, (1975) drew the boundary between the Berriasian stage and the Valanginian stage as evidenced by the extinction of Pygopids and the rapid development of Cyclothyrids which seems to be suitable according to the current development.

The marine Jurassic-Cretaceous continuous sediments of Xizang are mainly developed in the Himalayan Region, especially in Gyangze county and Nagarze county, where the sections of continuous sedimentation are well exposed. More detailed surveyed sections are those from Jiabula and Weimei villages, Gyangze county as well as those from Kadong bridge and Sherangre, Gongboxue in Nagarze county.

The author holds that the Weimei formation and the upper part of the Menkatun formation with *Himalayites*-*Blanfordiceras* assemblage including *Haplophylloceras*, *Berriasella*, *Spiticeras*, *Aulacosphinctes*, etc. may be approximately corresponding to Berriasian stage and the uppermost part of Tithonian stage. Moreover, the lower part of the Menkatun formation bearing *Virgatosphinctes*-*Aulacosphinctoides* assemblage is possibly equivalent to Tithonian stage and may be compared to *Virgatosphinctes transitorious* zone. The base of the Jiabula formation and the Kangba-Dongshan formation carrying ammonoids (*Calliptychoceras*, *Neohaploceras*, *Killiangella*, Olco*stephanus*, *Pterolytoceras*) and *brachiopods* (*Peregrinella*) is considered as the base of the Cretaceous -Valanginian stage.

Wang Yi-gang and others (1976) surveyed the Jiabula section again. The writer agrees with them to draw the boundary between Layer 3 with *Calliptychoceras* sp., *Neohaploceras* sp., *Pterolytoceras exotium* (Oppel) and Layer 2 with *Himalayites* sp., *Berriasella* sp., *Haplophylloceras strigile* (Blanford).

The section from Weimei village may be compared layer by layer with that from Jiabula village. The Jurassic and Cretaceous boundary is placed between Layer 6 with *Himalayites* sp., *Berriasella* sp. and Layer 7 with *Calliptychoceras* sp.

The sections from Kadong bridge and Sherangre outcrop very well. They may be approximately correlated layer by layer with those from Jiabula and Weimei villages. But there is a grey macrocrystalline-microcrystalline andesite and volcanic breccia layer at the base of Jiabula formation, in the lower part of which the ingredient of clastic rock increases. In the section of Kadong bridge, *Spiticeras* sp. and *Blanfordiceras* sp. (79LY2) were found from the Weimei formation (i.e. Layers 1-2) in 1979. The Geological Bureau of Xizang also collected Himalayites sp. and Berriasella sp. from these layers. In Layers 3-6, there are Neocomites sp. (79LY3,4), Pseudomelonia arginensis Yang et Wu, Hibolithes jiabulensis minor Chen (79LY5), belonging to the lower part of Jiabula formation. Killianella sp., Calliptychoceras sp., and Olcostephanus sp. were also collected from the equivalent layers by the same Bureau.

In the section from Sherangre, Gongboxue, Layer 1 with Haplophylloceras sp. and Spiticeras sp. is assigned to the Weimei formation; while Layer 2-4 to the Jiabula formation. The ammonoids Olcostephanus sp., Killianella sp., Neocomites sp. and the brachiopod Peregrinella langarzensis Sun (sp. nov.) was found from the lower part of this formation.

Generally speaking, it is preferable to refer the extinction of Berriasellinae, Spiticeratinae and Himalayitinae, the expansion of Neocomitidae and the appearance of Olcostephanidae belonging to Desmoceraticea and Peregrinella of Dimerellidae to the beginning of Valanginian stage as the base of Cretaceous system.

## References


(1) Ager, D. V., The Jurassic world ocean, Jurassic Northern North Sea Symposium, published by the Norwegian Petroleum Society, Stavanger (1975).

(2) Ager, D. V., Brachiopods at the Jurassic-Cretaceous boundary. Memoirs du B.R.G.M., (86), 150-162 (1975).

(3) Bassoullet, J. P., Mouterde, P., Les formation sedimentaires Mesozoiques du domaine Tibetain de l'Himalayan du Nepal. R.C.R. (253), 362-372, Ecologie et Geologie de l'Himalaya central (1977).

(4) Dyuschits, V.V., The Borriasian of the Crimia and its stratigraphical relations. Memoires du B.R.G.M., (86), 337-341 (1975).

(5) Fatmi, A. N., Stratigraphy of the Jurassic and Lower Cretaceous rocks and Jurassic ammonites from northern areas of west Pakistan. Bull. Brit. Mus. Nat. Hist. (Geol.), 20 (7), 299-380 (1972).

(6) Desio, A., Scientific Reports, In Italian Expedition to the Karakorum ($K^2$) and Hindu Kush, 3 (1975).

(7) Gupta, V. J., Indian Mesozoic Stratigraphy, Hindustan Publishing Corporation, Delhi (India) (1975).

(8) Hallam, A., Jurassic Environments. Cambridge University Press (1975).

(9) Wang Yi-gang, Wang Yu-jing, Wu Hao-ruo, Discussion on the Jiabula Formation and the discovery of Longzi Lower Jurassic in southern Xizang, China. Scientia Geologica Sinica, (2) 149-156 (1976) (in Chinese).

(10) Wang Yi-gang, Zhang Ming-liang, The stratigraphy of the Mount Jolmo Lungma Region, Jurassic. A Report of Scientific Expedition in the Mount Jolmo Lungma Region (1966-1968), Geology, Science Press, 127-146 (1974) (in Chinese)

(11) Wiedmann, J. Jurassic-Cretaceous boundary as one of the Mesozoic system boundaries. Memoires du B.R.M., (86), 358-362 (1975).

(12) Yegoyan, V. L., Tithonian and Berriasian boundary is the boundary between the Jurassic and Cretaceous system. Memoires du B.R.G.M., (86), 363-369 (1975).

(13) Yin Ji-xiang, Fang Zhong-jing, Marine Jurassic in western Yunnan. Scientific Geologica Sinica, (3), 217-237 (1973) (in Chinese).
(14) Zhao Jin-ko, Jurassic and Cretaceous ammonoids from the Mount Jolmo Lungma region, A Report of Scientific Expedition in the Mount Jolmo Lungma Region (1966-1968), Palaeontology, Fasc. 3, Science Press, 503-545 (1976) (in Chinese).

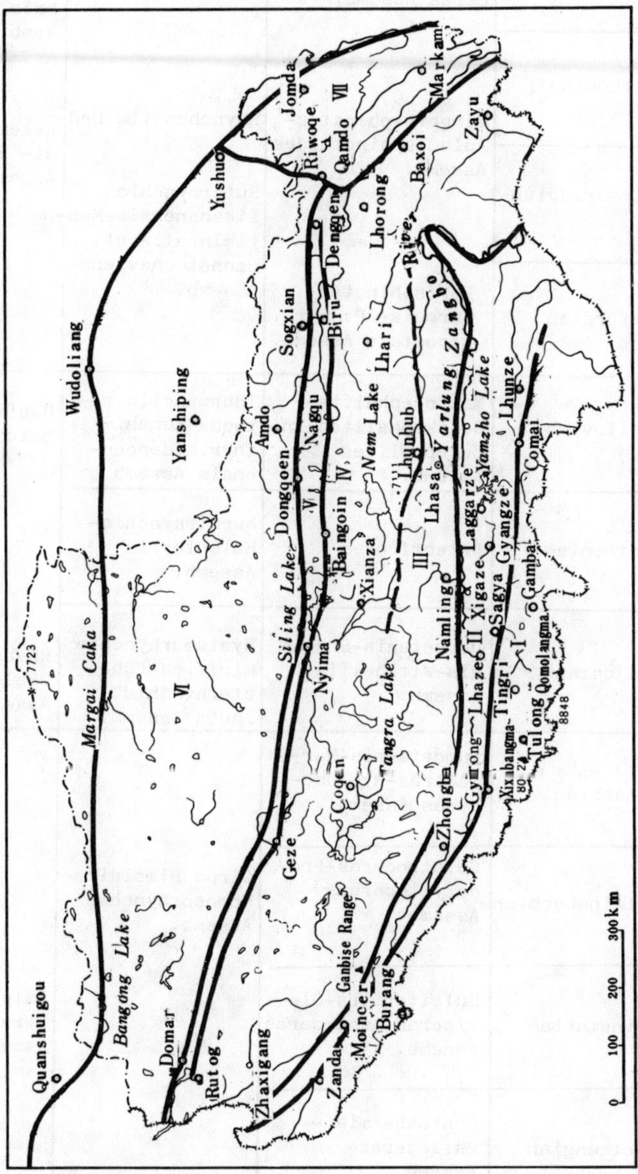


Fig. 1 Sketch Map of zones of Jurassic sedimentary formations in Xizang

Table 1 Jurassic fossil assemblages of Xizang

| Series | Stage | Ammonoids | Brachiopods | Belemnoids |
|---|---|---|---|---|
| Upper Jurassic | Berriasian | Spiticeras - Himalayites-Blanfordiceras Assemb. | | Hibolithes semisulcata-Belemnopsis uhligi Assemb. |
| | Tithonian | Virgatosphinctes-Aulacosphinctoides Assemb. | Rhynchonella Bed | Belemnopsis gera di- B. alfuraca Assemb. |
| | Kimmeridgian | | Rutorhynchia iieshanensis-Monticlarella cf. czenstochaviensis Assemb. | |
| | Oxfordian | Perisphinctes furcula- P. curvicostata Assemb. | | |
| Middle Jurassic | Callovian | Macrocephalites-Dolikephalites-Indocephalites Assemb. | Thurmanella rotunda-Kutchithyris denqenensis Assemb. | Hastites - Holco belus Assemb. |
| | Bathonian | Delecticeras | Burmirhynchia-Holcothyris Assemb. | |
| | Bajocian | Dorsetenia-Sonninia-Witchellia Assemb. | Nyalamurhynchia mirifica-"Rhactorhynchia" lauta Assemb. | Rhabdobelus-?Belemnoteuthis Assemb. |
| Lower Jurassic | Toarcian | Dumortieria-Harpoceras-Nyalamoceras Assemb. | | |
| | Pleinsbachian | Galatioceras-Prodatylioceras Assemb. | Cirpa himalaica-Homoeorhynchia Assemb. | |
| | Sinemurian | Sulciferites-Gleviceras-Arnioceras Assemb. | | Salpigoteuthes-Atractites Assemb. |
| | Hettangian | Schlotheimia - Psiloceras- Assemb. | | |

| Pelecypods | Foraminifera | Hexacoralla | Gastropods |
| --- | --- | --- | --- |
| uchia sptiensis-<br>aevitrigonia car-<br>iniformis - As-<br>retoides ding-<br>iensis Assemb. | | Phacellastrea ruto-<br>gensis Thecosmil-<br>lia tibetana Dungu-<br>lia simplex Isastrea<br>goldfussi Colligno-<br>natrea jumarensis<br>Epistreptophyllum<br>cf. diatritum Com-<br>plexastrea cf.<br>magnifica | Aptyxiella sp.<br>Cossmanea sp.<br>Polyptyxis sp.<br>Procerithium sp. |
| seudotrapezium<br>ordiforme-chla-<br>ys (Radulopec-<br>en) vagens-Go-<br>iomya sulcata-<br>eiostrea birma-<br>ica Assemb. | Haurania shuang-<br>huensis Recto-<br>cyclammina sp.<br>Pseudocyclammina<br>simplex | | |
| uenstedtia tibe-<br>ica-Entolium de-<br>issum Assemb. | | | Nerinea tulongen-<br>sis N. nyalamensis<br>Onustus acuminatus |
| starte cf. val-<br>zii-Protocardia<br>f. africana-Weyla<br>bongoensis<br>ssemb. | Haurania deserta<br>Pseudocyclammina<br>aff. liasica<br>Rhapydionina<br>urensis Labyrin-<br>thina recoarensis<br>Orbitopsella<br>praecursor | Enallhelia venus-<br>ta Isastrocoenia<br>kachensis | Monoplocus? niti-<br>dus Sulcoactaeon<br>bellireticulatus |

# NATURAL ENVIRONMENT OF THE PLIOCENE BASIN IN GYIRONG, XIZANG

Chen Wan-yong
(Institute of Vertebrate Palaeontology and Palaeoanthropology, Academia Sinica)

Gyirong basin, situated in the central part of the folded region of the Himalayas and at the northern slope of Mt. Xixabangma, is an intermont basin formed in the Miocene fold-fault system. It has an area of about 200 $km^2$. A *Hipparion* fauna (Pl. 1) was known in this basin in 1975 (Huang Wan-Po and Ji Hung-Xiang, 1979). The fossil locality is 4,100-4,300 meters above sea level.

## LITHOSTRATIGRAPHY

Based on the data of fossils and lithofacies and paleogeography, the Pliocene deposits of the basin, named Woma formation, about 450 m. thick, are divided into three members (Fig. 1):

1. Upper Member. Sediments attributed to the Upper Member of Woma formation consist of yellow, light purple marls and fine sandstones, underlying unconformably the Quaternary Gongba conglomerates. The total thickness is about 40-80 m. The marls yielded *Gangetia* ex gr. *rissoides* and *Tectochara* cf. *minutissima*.

2. Middle Member. The Middle Member comprises alternations of gray conglomerates, yellow and gray sandstones, which is separated from the Lower Member by an obvious erosional surface. Thickness 20-244 m.

3. Lower Member. Composed of sanstone, mudstone and laminated lignites, about 13-130 m. and containing *Hipparion gyizhongensis*, *Chilotherium* sp., *Hyaena* sp., *Palaeotragus microdon* and *Ochotona gyizhongensis*. This member overlies the limestones and slates of Jurassic in an angular unconformity.

## LITHOFACIES AND PALEOGEOGRAPHY

Analysis of the sediments and fossils of the basin shows that four different sedimentation stages and paleogeographic environments existed in the basin during the Pliocene time.

### Fluvial Facies-Conglomerate Stage (Fig. 2)

Before the ancient Gyirong basin came into being, a Pliocene intermontane river, which flowed from south to north, had been present along the fold-fault system formed in the Miocene, in which a series of cemented, sandy and muddy conglomerates were deposited. The components vary gradually to gray and yellow sandstones toward the northern part of the basin, which shows that the lake-bog began to appear during this period.

Lacustrine-Bog Facies-Mudstone-Sandstone Stage (Fig. 3)

The ancient Gyirong river was barricaded by the Mala Mountain in the north, so that the Gyirong lake was formed. Sediments with fine stratifications (reflecting seasonal changes) containing <u>Gangetia</u> ex gr. <u>rissordes</u>, <u>Ilyocypris</u> <u>bradyi</u> were deposited in this basin. It is thought that water in this basin was quite deep and calm, and the water volume was not greatly changeable. Marsh land could have been present along the bank of the lake at that time.

Fluvio-Lacustrine Facies-Sandstone-Conglomerate stage (Fig. 4)

Sediments with abundant fossils of <u>Picea</u> sp., <u>Abies</u> sp., suggest that the climate was cool and rainy in that period. The appearance of interbedding of conglomerates, sandstones and cross-bedding might indicate that influx of the water into the basin increased, water was disturbed and its volume was changeable

Lacustrine Facies-Mudstone Stage (Fig. 5)

After the disturbing period, the lake became stable and a series of marls with yellow and gray micro-stratification, containing <u>Gangetia</u> ex gr. <u>rissordes</u> and <u>Tectochara</u> cf. <u>minutissima</u> were deposited. It seems that water in this lake was clear and calm again, but the climate at that time was rather dry and parched, and the tendency of continuous vaporization might have taken place during the later sedimentation.

PALEOCLIMATIC CHARACTERISTICS

Clay minerals, heavy minerals and organic matter and chemical analytical data, as well as the evidences of the fauna and flora make it possible to interpret the paleoclimatic characteristics of Gyirong basin during Pliocene.

1. <u>Hipparion</u> fauna of the basin seems to reveal that the climate, during the early stage of sedimentation was rather moist and hot.

2. Sporomorph associations exhibit the existence of different climates in different sedimentation intervals <u>Loropetalum</u> <u>chinense</u>, <u>Palmae</u>, <u>Podocarpaceae</u> sp., and <u>Pteris</u> sp., <u>Quercus</u> sp. in the Lower Member show the same climate condition as the <u>Hipparion</u> fauna does. <u>Picea</u> sp., <u>Abies</u> sp., and <u>Tsuga</u> <u>chinensis</u> in the Middle Member is indicative of coolness. Few spores found in the Upper Member might display the dry weather.

3. As <u>Gangetia</u> ex gr. <u>rissordes</u> (Pl. I, Fig. 1) is now still living in the fresh lakes in South Asia, there is no reason to suspect that it lived in the same climate at that time in Gyirong basin as that in South Asia today. Besides, <u>Ilyocypris</u> <u>bradyi</u> exists in some subtropic lakes, such as Yichang area, Hubei, at the bottom of which the annual average temperature of the water remains at 17°C.

4. Sediments in the lower part, incorporated in gray sandstones and

marls suggest that they were deposited in a moist climate; interbedding of gray and yellow sandstones and mudstones in the middle part demonstrates an alternative climate of wet and dry; the Upper Member consisting of yellow or yellow purple mudstone shows that the climate was already dry.

5. The composition of heavy minerals dominated by stable or extreme stable minerals also indicates the hot and wet climate (Fig. 6)

6. Smaller quantity of $CaCO_3$, and larger quantity of organic matter are contained in the Lower Member, which may reflect that sedimentation in the early stage took place in a moist climate. The quantity of the oxides in the Mid-Upper Member was in the contrary condition (Table 1).

7. Composition of clay minerals varies. The kaoline-hydromica, reflecting the hot and wet climate in the Lower Member; the montmorilonite-kaolinite-hydromica, represented the alternation of dry and wet climate in the Middle Member and the montmorillonite-hydromica in the Upper Member is thought to have been dry.

The climate change which took place in the ancient lake is shown in Fig. 7: hot and wet in the early period, alternative of dry and wet, of cool and warm in the middle period, and in the late period, it was getting dry and parched.

## THE NATURAL PALEOENVIRONMENT OF *HIPPARION* FAUNA

During the early period of sedimentation, when the *Hipparion* fauna lived, the Gyirong lake was about 700 m above sea level, in a irregular rectangular shape with wider in the south and narrower in the north, of about 200 $km^2$ (Fig. 8). Clear and calm lake water and sufficient sunshine were suitable for *Gangetia* ex gr. *rissoides*, *Ilyocypris bradyi* and etc. to live in; and other plants, such as *Potamogeton* sp. lived along the margin; about 1,000 m above sea level, there were *Loropetalum chinense*, *Trachycarpus fortunei* and etc; and also the conisilvae zone, such as, *Picea* sp. and *Abies* sp. The *Hipparion* fauna might have existed on the lake-rim plain and hilly area. At that time, the moist and hot seasonal wind from the Indian Ocean made Gyirong basin an environment of subtropic all the year round.

## ON THE HIMALAYAN UPLIFT

After Pliocene, the Himalayan Barrier was high enough to prevent the warm and damp Indian Ocean seasonal wind from passing by. Today, the annual average temperature in the Gyirong basin is about 2°-10°C, and annual rainfall is only 300-400 cm. Such an environment is occupied by some special animal forms, *Marmota himalayana*, *Procaptra picticaudata*, *Vulpes montan*, and *Panthera uncis* for example, and by some paramos plants, such as *Stipa purpurea*, *Orinus thoroldii*, and *Pennisetum flaccidum* and etc.

Table 1 Organic matter and carbonate contents in the Pliocene deposits of Gyirong Basin

| Horizon | | No. of specimens | Organic matter quantity | Carbonatite quantity |
|---|---|---|---|---|
| pliocene | upper member | D 7 | 0.44 | 37.00 |
| | | B34 | 0.17 | 21.50 |
| | | B32 | 0.25 | |
| | | A23 | 0.33 | 22.80 |
| | | A 24 | 0.20 | |
| | | C 15 | 0.40 | 38.50 |
| | | E 9 | 0.27 | |
| | middle member | A 16 | 0.55 | |
| | | A 20 | 0.25 | |
| | | B 19 | 0.22 | 7.15 |
| | | C 13 | 1.60 | 26.90 |
| | lower member | A 6 | 3.55 | |
| | | A 7 | 0.25 | 4.10 |
| | | C 4 | 4.53 | 27.40 |
| | | B 4 | 0.55 | |
| | | D 4 | 0.47 | |

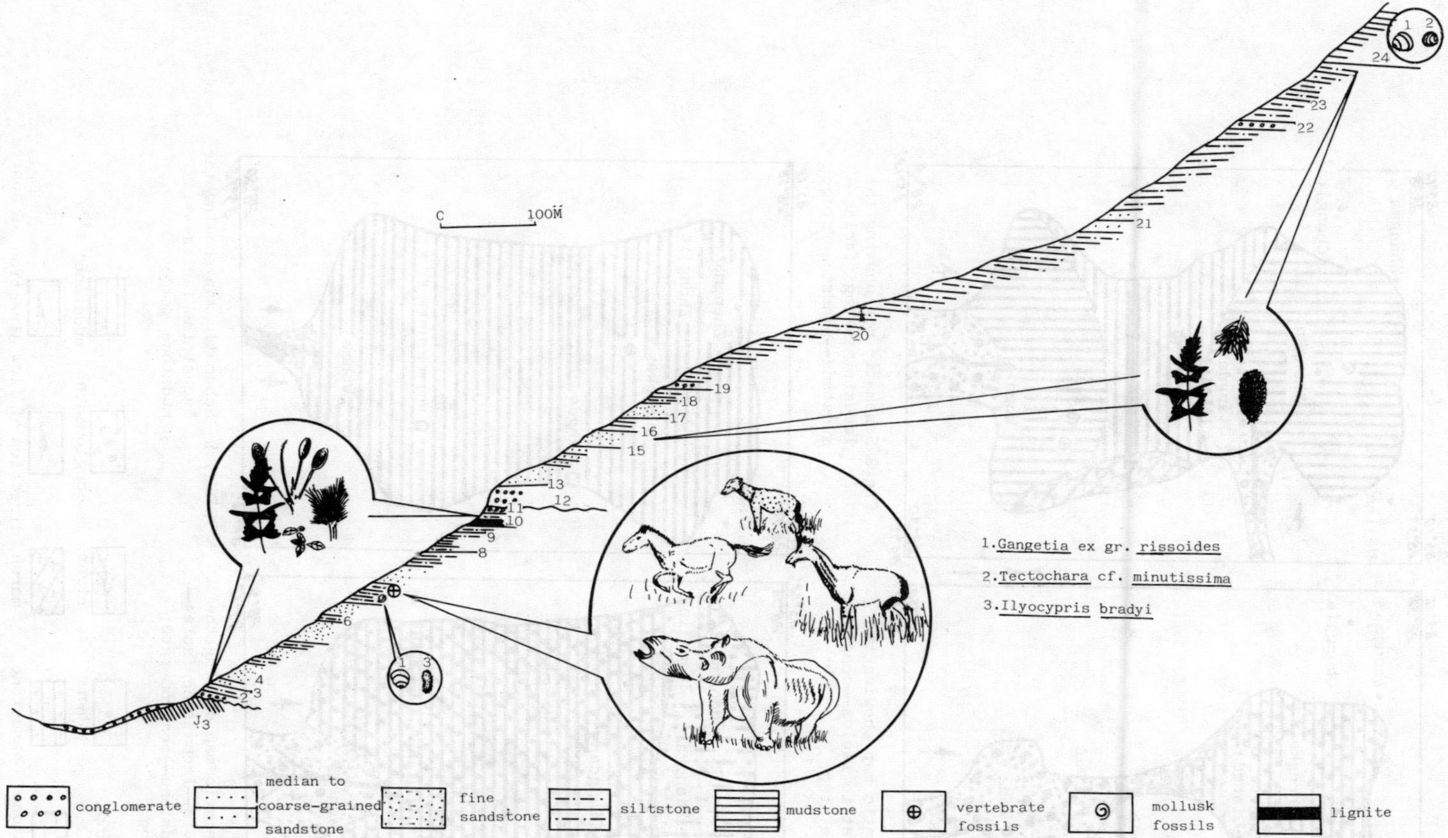


Fig. 1 Geological section of Pliocene Woma formation of Gyirong basin, Xizang

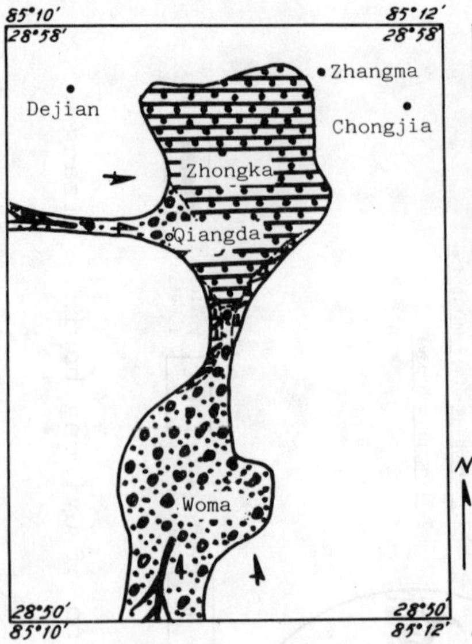


Fig. 2 Schematic map showing the fluviatile facies—conglomerate stage

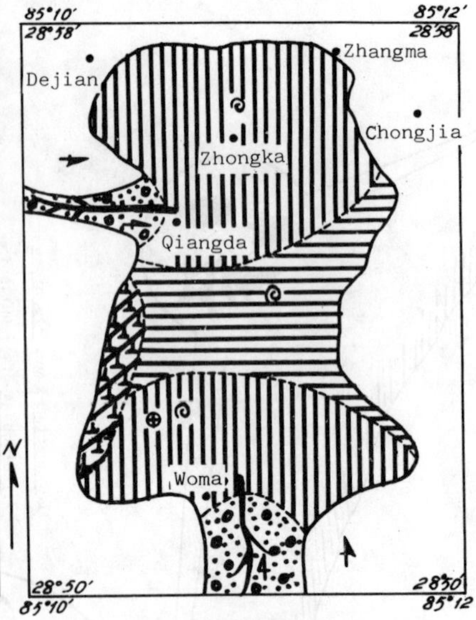


Fig.3 Schematic map showing the lacustrine—bog facies—mudstone—sandstone stage

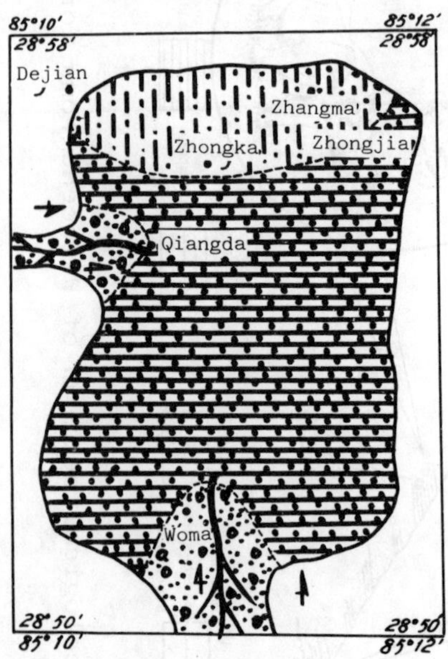


Fig. 4 Schematic map showing the fluvio—lacustrine facies—sandstone—conglomerate stage

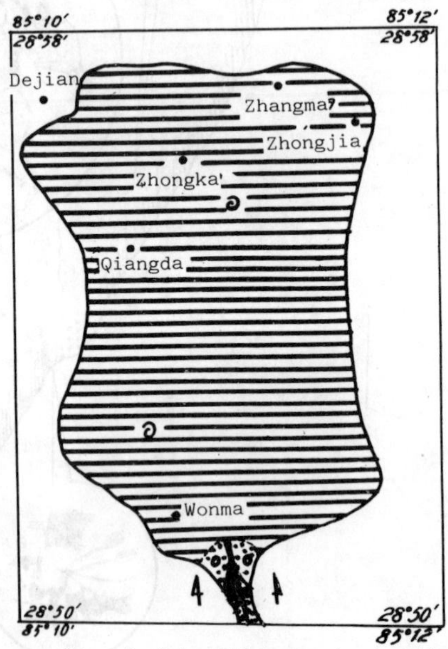


Fig. 5 Schematic map showing the lacustrine facies—mudstone stage

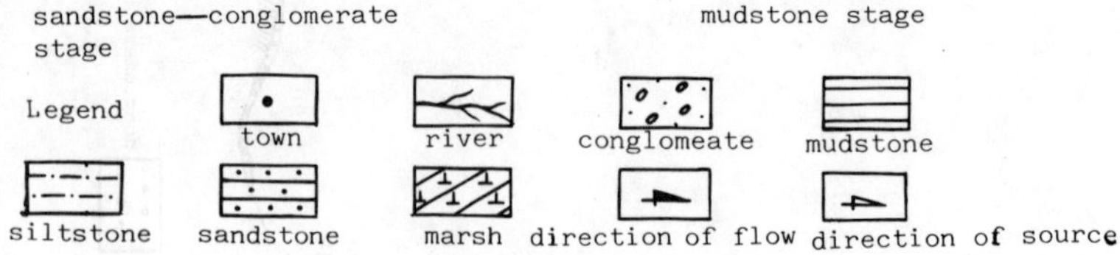

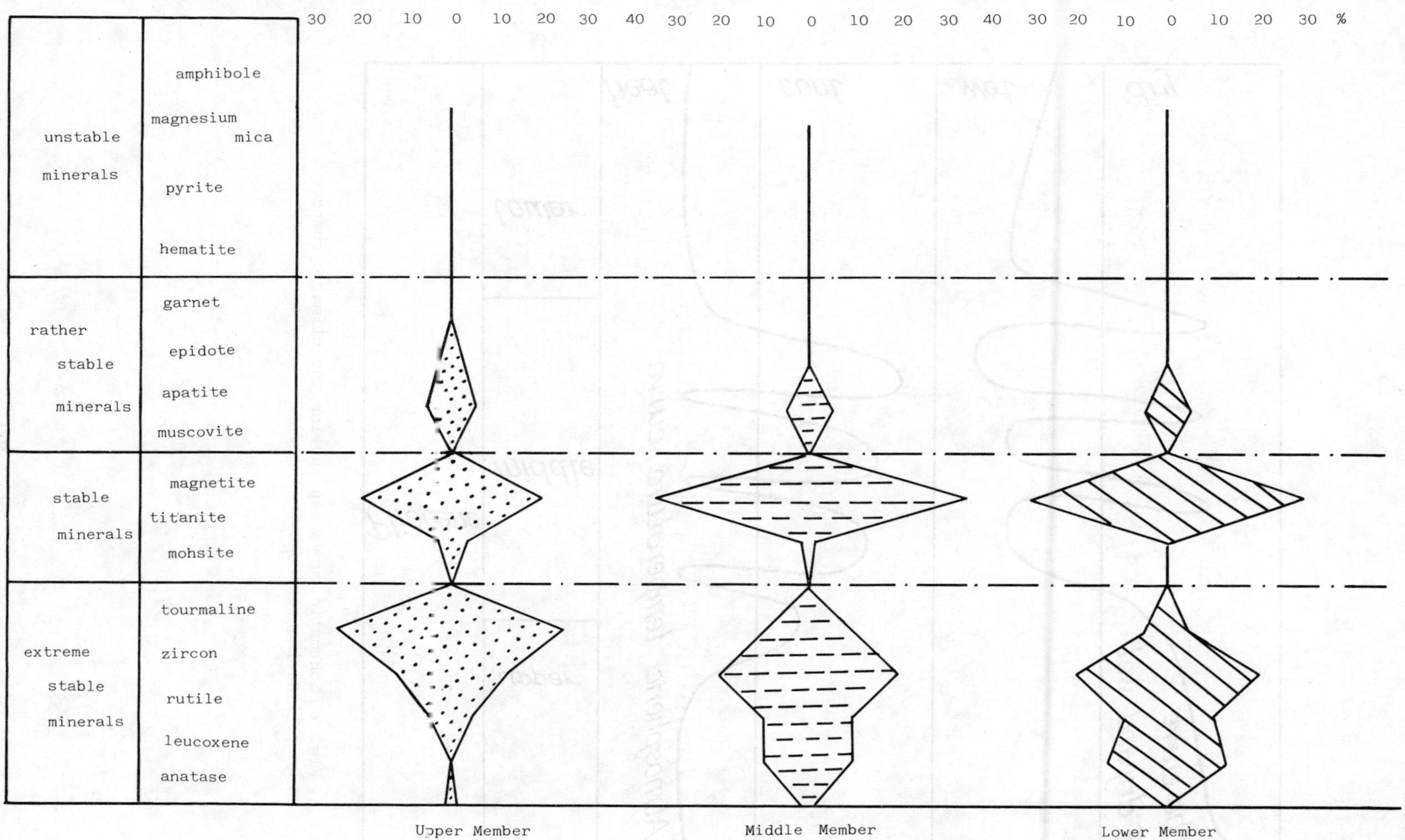


Fig. 6 Mean value of the composition of heavy minerals in Pliocene deposits of Gyirong basin

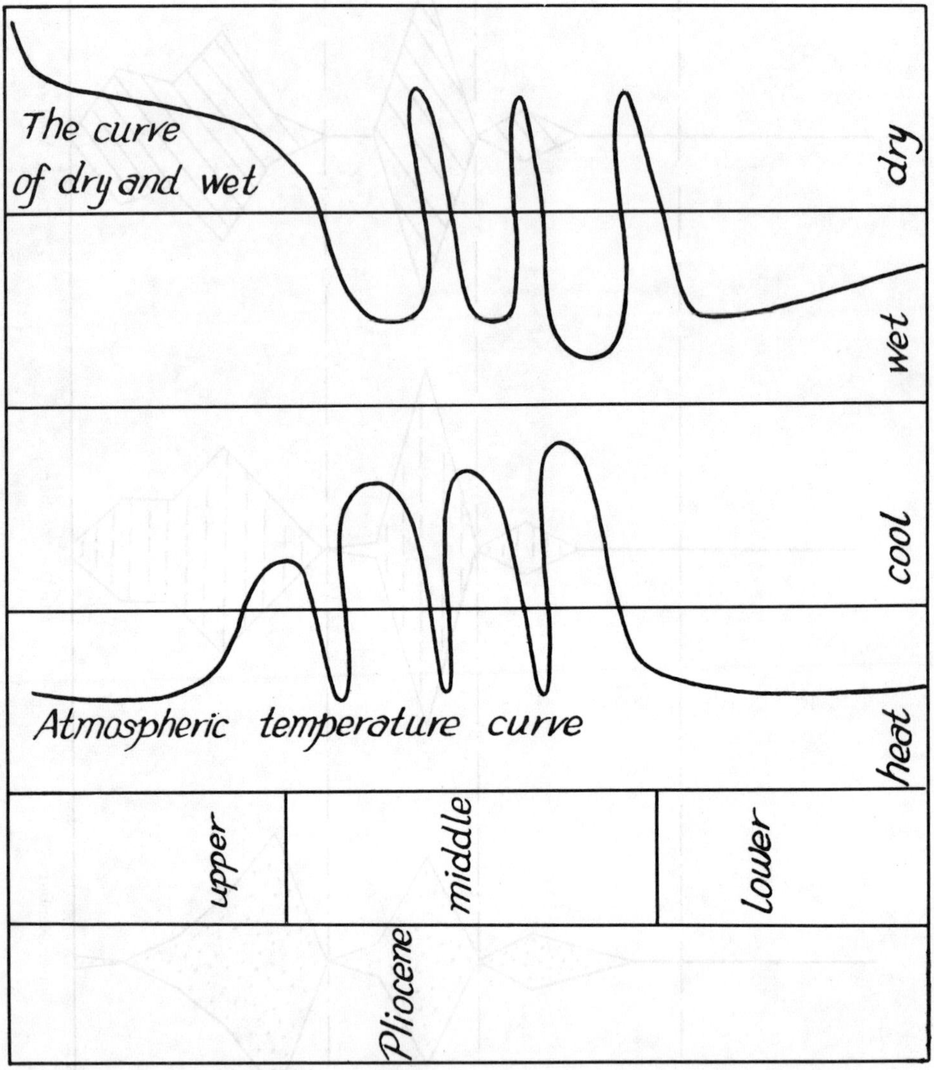


Fig. 7 Schematic curve map showing the climatic changes during Pliocone

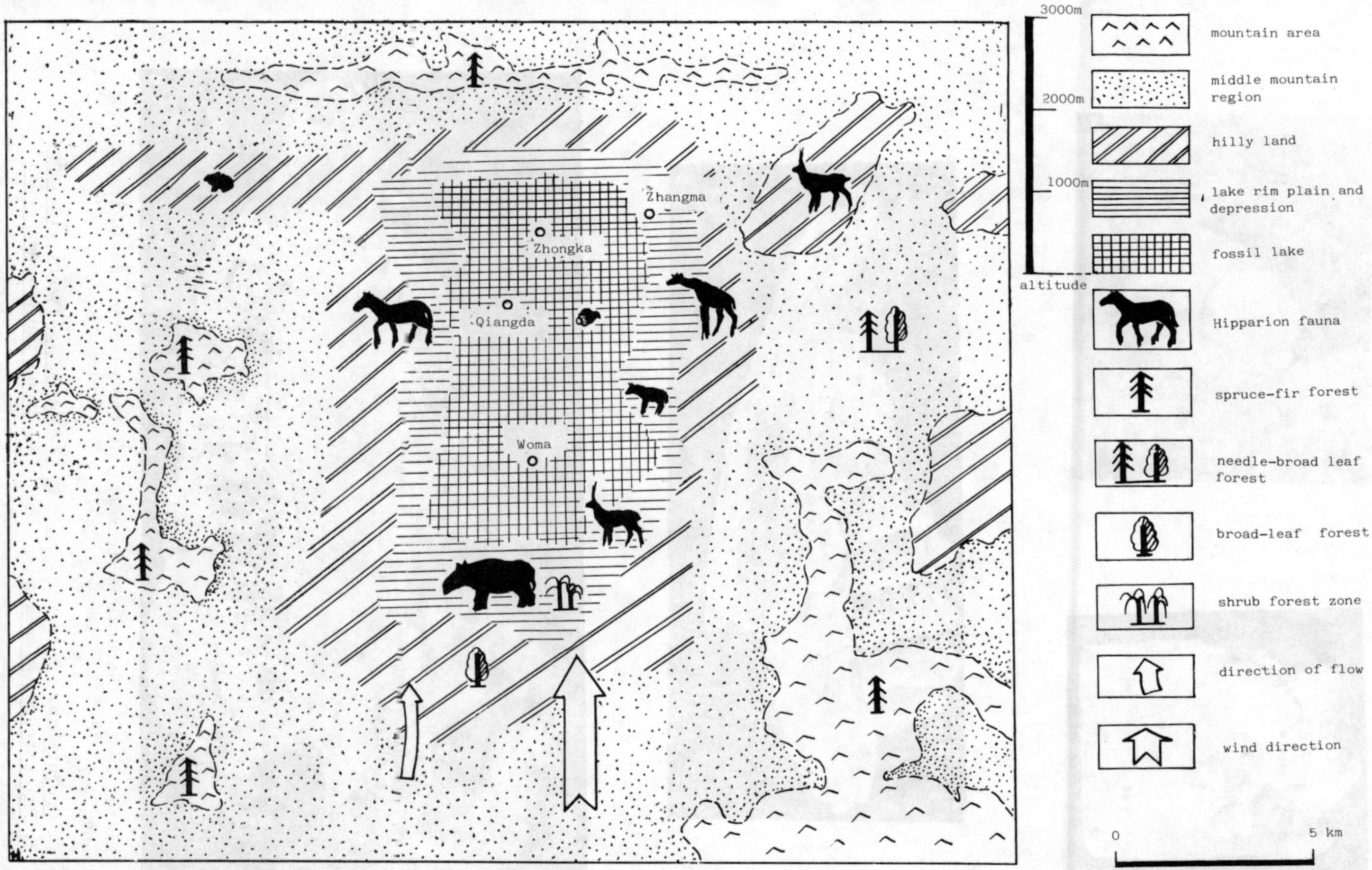


Fig. 8 A schematic map of natural enivronment of the Pliocene basin of Gyirong

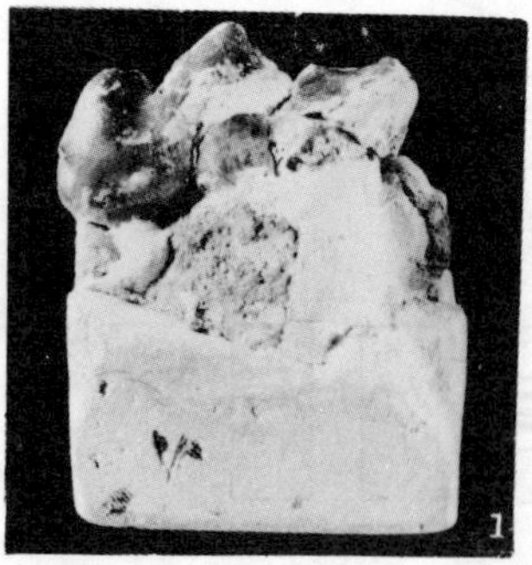

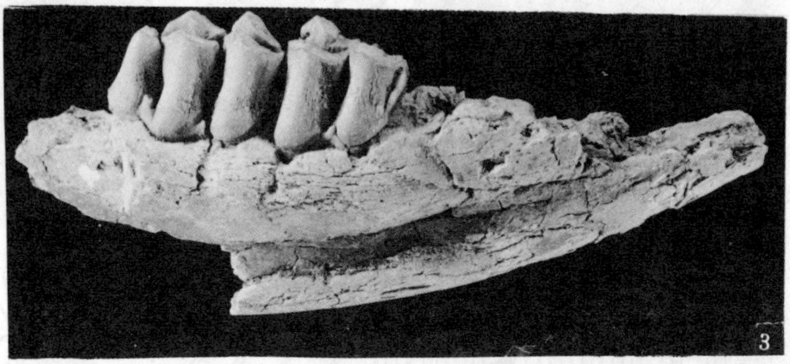

Plate I Natural Environment of the Pliocene Basin in Gyirong Xizang

1. *Crocuta* sp.
2. *Gangetia* ex gr. *rissiodes*
3. *Palaeotragus microdon*
4. *Hipparion gyizhongensis*

## CHARACTERISTICS AND EVOLUTION OF GRANITOIDS OF SOUTHERN XIZANG

Tu Guang-chi*, Zhang Yu-quan,
Zhao Zhen-hua, Wang Zhong-gang
( Institute of Geochemistry, Academia Sinica)

Granitoids are widespread in southern Xizang, covering an area of 100,000 $km^2$, and occurring in three parallel belts, each with its own petrographic characteristics. These are the Gandise Belt to the north, the Lhagoi Kangri Belt in the middle, and the Himalayan Belt to the south. Investigation of petrography and geochemistry of the granitoids was deemed necessary for the correct interpretation of the geological evolution of the Qinghai-Xizang Plateau.

### GENERAL PETROGRAPHY AND GEOCHEMISTRY OF GRANITOIDS

The plutons of these three belts all occur in the upthrust sides of the great thrust zones. The Gandise granitoids are located in the upthrust side of the Yarlung Zangbo thrust zone which proves to be the suture between the Indian Plate and the Eurasian Plate. The Lhagoi Kangri and the Himalayan granites are situated to the north, or the upthrust sides of the Main Central Thrust and the Main Boundary Thrust respectively.

The Gandise Belt is characterized by the clustering of great batholiths with composite plutons. The main rock types are diorite, granodiorite and tonalite. Norite, gabbro, hypersthene diorite, porphyritic biotie granite and biotite granite are less important. Both the Lhagoi Kangri and the Himalayan Belts, on the other hand, have only small-sized widely-separated plutons with simple petrography. The main rock types are gneissic two-mica granite for the former, and tourmaline muscovite granite for the latter.

The granitoids of the three petrographic belts exhibit much difference in their alkali feldspar coefficient (K). The peak values of K are 35, 55, 65 and 75 % for diorite and granodiorite of the Gandise Belt, granites of the Lhagoi Kangri and the Himalayan Belts respectively (see Xie Ying-wen et al., Fig. 1).

The An values for the plagioclases of the granitoids also vary in the same fashion. For most of the plagioclase of the Gandise granitoid, An is larger than 30. 28-21 is the An value for Lhagoi Kangri granites while for the Himalayan granites, it is only 10.

The characteristic accessory minerals are magnetite, sphene, apatite and allanite for the Gandise granitoids, and ilmenite, tourmaline, and monazite for the Lhagoi Kangri and the Himalayan granites (see Pan Jin-ming et al., Table 1).

On the basis of a total of 213 whole rock chemical analyses, it is

* Tu Kuang-chih

found that the average chemical composition of Gandise granitoids approaches very closely that of a granodiorite, with low differentiation index, low content of Si, K and high content of Ti, Fe, Mg and Ca. The average chemical composition of both the Lhagoi Kangri and the Himalayan granites is that of a granite, with high differentiation index, high Si, K and low Ti, Fe, Mg, Ca content (see Wang Zhong-gang et al., Table 1).

The granitoids of the three belts also show difference in REE distribution pattern. The REE model for the diorite and granodiorite of the Gandise Belt is characterized by a smooth curve, with a very indistinct Eu anomaly. Biotite granite and porphyritic biotite granite of the same belt show clear Eu anomaly, while granites of both the Lhagoi Kangri and the Himalayan Belts give very strong negative Eu anomaly (see Zhao Zhenhua et al., Fig. 3-6).

The initial $^{87}Sr/^{86}Sr$ value are quite different for granitoids of the three belts, probably indicating different source material (Table 1).

The initial $^{87}Sr/^{86}Sr$ values for granodiorite of Quxu pluton (0.7033-0.7036) implies a deeper source, probably of upper mantle, or an oceanic crust origin. A definite crustal origin is assigned to Gubug granite, due to its high initial $^{87}Sr/^{86}Sr$ value (0.7478). For the Kangmar two-mica granite, the source problem is puzzling, it might have diversified sources.

## TEMPORAL AND SPATIAL DISTRIBUTION OF GRANITOIDS

It is interesting to note that granitoids of the three above-mentioned petrographic belts, together with those of the Tanggula and the Kunlung Mountains farther to the north, show obvious tendency of southward decreasing geological ages. Granitoids of the Kunlung Mountains are largely of Hercynian in age, while those of the Tanggula Range have Early Mesozoic datings. The principal granitoids of the Gandise Belt have isotopic ages 120-70 m.y. with subordinate 50-10m.y. (Fig. 1). The Lhagoi Kangri granites have most datings in 50-30 m.y. while the ages of the Himalayan granites fall in the range from 20-10 m.y. .

The general picture of migration of younger granitoids toward the continental margin can also be observed in southeastern China. Here the Pre-Cambrian and Palaeozoic granites are distributed mostly in the inland part, while the Late Yenshanian granites are much more developed in the coastal regions with Early Yenshanian granites lying between them (Fig.2).

However, superimposed on the general pattern of migration of younger granitoids toward the continental margin, we also observe the opposite trend of granitoid formation. This is the case of some of the Gandise granitoids.

For example, in the Quxu-Lhasa batholith, the early gabbro is met to the south of Yarlung Zangbo River, later diorite occurs in the vicinity of Quxu, still later granodiorite crops out widely close to the Lhasa Bridge, while the latest granite is seen to the north of Lhasa. This shows a very clear northward migration of granitoids (Fig. 3).

A similar picture of temporal and spatial distribution of granitoids in the Rutog batholith is observed. Here a northward migration of early diorite thru granodiorite and porphyritic hornblend biotite granite, to late equigranular biotite granite, is very obvious (Fig.4).

In southeastern China we can observe the same phenomenon, i. e., superposition of opposite trend of granitoid formation on the general picture of granitoids migration toward the continental margin. The granites of Yunkai Mt. - Darong Mt. -Shiwan Mts. offer a good example. According to the recent finding of Mo Zu-sun*, these granites show similar petrography and are interrelated in their evolution. It is interesting to note that the granites in Yunkai Mt. to the southeast were formed in Early Palaeozoic while the granites in Darong Mt.- Shiwan Mts. to the northwest were formed in Late Palaeozoic and Triassic. The direction of granite formation in this particular composite batholith is thus opposite to the general direction of migration toward the continental margin (Fig. 5).

A west to east migration of granitoid along the strike is also manifested to some extent in the Gandise Belt. The early diorite and granodiorite which comprise two thirds of the total granitoids, occur mostly in the west and the central portion of the Gandise Range, while the more acidic and younger granitoids such as the biotite granite and porphyritic biotite granite, are usually seen in the east Range (Fig.6).

## CORRELATION BETWEEN GRANITOIDS OF SOUTHERN XIZANG AND CIRCUM-PACIFIC BELT

In recent years Circum-Pacific petrologists tend to make distinctions between I-type and S-type granitoids. We find this distinction useful in correlation between granitoids of Southern Xizang and Circum-Pacific Belt.

In general, granitoids of Gandise Belt are very similar to the I-type granitoids of the western North and South America in their modes of occurrence, ages of intrusion, petrographic characteristics, constituents of main and accessory minerals, as well as in their trend of evolution, close association with basic to intermediate volcanic rocks and Cu mineralization of the porphyry type. Granites of the Lhagoi Kangri and the Himalayan Belts correspond with the S-type granitoids of southeastern Asia.

In correlating the Gandise granitoids with those of western North and South America, we find the following common features:

1. They form great linear petrographic belts with numerous batholiths. Some of the petrographic belts in western North and South America match in size with the Gandise Belt, thousand of kilometers in length.

2. Their ages of formation are close to one another, with main phases of intrusion in Mid-to Late Cretaceous.

3. They intruded into eugeosynclines, preceded by intensive basic to intermediate volcanic eruptions.

4. They show similar trend in magmatic evolution, beginning with basic or intermediate-basic and ending in acidic. In most cases diorite, tonalite and granodiorite are the dominant rock types with subordinate granites.

5. The characteristic accessory minerals are very alike, with magnetite, sphene, apatite and allanite as the chief components. In dark-colored

* "Genesis and evolution of Nanling granitoids", 1979.

minerals, hornblende is an important constituent.

6. An inlandward migration of granitoid formation is observed in many places. Such is the case in Gandise, in south-central Alaska, the Sierra Nevada of California, the Coast Mountains of British Columbia and in Northern Chile.

7. A close genetic and spatial relationship is observed for granitoids to porphyry Cu mineralization both in the western North and South America, as well as in the Gandise Belt and its eastern continuation.

In petrological and mineralogical characteristics, granites of the Lhagoi Kangri and the Himalayan Belts are similar to the S-type granitoids of Southeast Asia, with simple rock types, mostly granites of various kinds. The distinguishing accessory minerals are muscovite, ilmenite with monazite as the chief REE mineral. However, there is a great difference in their modes of occurrence. The granitoids in both the Lhagoi Kangri and the Himalayan Belts occur only in small-sized, widely separated plutons, while in southeastern China, as well as in Malaysia, in Thailand, besides small plutons, granitoids usually form large multi-stage composite batholiths, with individual batholith covering an area larger than 1,000 $km^2$.

## DISCUSSION ON SOME PROBLEMS PERTINENT TO GRANITOIDS OF SOUTHERN XIZANG

The time and space distribution of granitoids of the five petrographic belts in Qinghai-Xizang Plateau, the petrological and geochemical characteristics of granitoids of the three southern belts, as well as their correlation with the Circum-Pacific granitoids can find their explanation in plate tectonics theory.

According to recent geophysical findings, the Indian Plate, having located in a geographical position farther to the south started moving northward at about 120 m.y. ago with a rate of about 8-9 cm/year. The northward movement of the Indian Plate, or the subduction of the Indian oceanic crust underneath the Eurasian continental crust persisted for about 50 m.y. until the collision and compression of the two plates took place. It is interesting to note that the duration of major granitoid formation in the Gandise Belt coincided with the period of subduction (120-70 m.y.).

For the other two southern Xizang granitoid belts, it is necessary to state that ages of granite formation fit in quite well with the appearance of the Main Central Thrust (50-30 m.y.) and the Main Boundary Thrust (20-10 m.y.).

It is postulated that in the course of subduction of Indian Plate underneath the Eurasian Plate, magmatic substances were generated along plate margins. Partial melting probably took place in the oceanic crust and its overlying sediments. The resulting intermediate to basic magmas gave rise to gabbro, diorite, granodiorite intrusions and andesite and basalt flows of the Gandise Belt.

When the two plates started colliding against each other about 50 m.y. ago, intensive compression made part of the sialitic material remelt, which resulted in the more acidic granite formation in the Lhagoi Kangri and the Himalayan Belts. The compression of plates also led to the appearance of the low-angle thrust zones.

The southward migration of later granitoid formation of the five petrographic belts of Qinghai-Xizang Plateau reflects the course of continental growth in this particular part of Eurasian Plate since Late Palaeozoic. A similar but less distinct mode of continental growth is also observed in southeastern China.

The multi-stage continental growth in these two large areas on the earth, Qinghai-Xizang Plateau and Southeast China, with accompanying granitoid formation is tentatively explained by microplate accretion hypothesis. In both areas, there were landmasses to start with, the Tarim and Qaidam Massives and the Jiangnan Old Landmass. From these old landmasses outward, microplates started to grow one after the other at sequential plate movements, giving rise to migration of granitoid formation toward continental margins.

As for the opposite direction of granitoid formation observed in certain batholiths, it is confined only in one plate movement. It is probably due to the subduction of an oceanic crust underneath a continental one, as was the case in Gandise and in the west coast of North and South America.

The last but not the least important problem to be tackled is the interrelationship between the crust thickening and the late large-scale granitoid formation. From geophysical studies, the crust in the Gandise Belt reaches a thickness of 70-72 km., the thickest for the whole Qinghai-Xizang Plateau. Recent investigations show the same order of crust thickness for the Peru-North Chile Batholiths. The well-known Sierra Nevada Batholiths have crust thickness of about 50 km. In all these three cases subduction of an oceanic crust underneath a continental one took place in comparatively recent geologic time, accompanied by large-scale I-type granitoid intrusion. It is worth while to mention that when S-type granitoids occur extensively, crust never attains such a thickness. For example, in southeastern China, where Mesozoic granites are widespread, crust thickness is only around 35 kilometers. There seems to be a mutual dependence between the largest crust thickening and recent large-scale I-type granitoid formation (Table 2).

## References


(1) Chang Cheng-fa et al., A brief discussion on the tectonic evolution of Qinghai-Xizang (Tibet) Plateau, In this Proceedings of Symposium on Qinghai-Xizang (Tibet) Plateau, Academia Sinica (1981).

(2) Laboratory of Isotopic Geology, Institute of Geochemistry, Academia Sinica and Laboratory of Isotopic Geology, Institute of Geological Sciences of Hupeh Province, Geochronological study of the granites in Nanling and adjacent regions in Southern China, Scientia Sinica, 19(3), 362-383 (1976).

(3) Chappell, B. W., White, A. J.R., Two contrasting granite types, Pacific Geology, (8), 173-174 (1974).

(4) Nanking University, Granitic rocks of different geological periods of southeastern China and their genetic relations to certain metallic deposits, Scientia Sinica, (17), 55-72 (1974).

(5) Takahashi, M., Aramaki, S., Ishihara, S, Magnetite-series / Ilmenite-series U. S. I-type/S-type granitoids, Mining Geology Special Issue, (8), 13-28 (1980).

Table 1 Initial $^{87}Sr/^{86}Sr$ of different granitoids of Southern Xizang

| | |
|---|---|
| Quxu granodiorite, Gandise Belt | 0.7033 - 0.7036 |
| Kangmar gneissic two-mica granite, Lhagoi Kangri Belt | 0.7186 |
| Gubug tourmaline muscovite granite, Himalayan Belt | 0.7478 |

( After Wang Jun-wen et al.)

Table 2 Interrelationship between crust thickness, plate movement and granitoid types

| locality | granitoid type | crust thickness (km) | plate movement |
|---|---|---|---|
| Gandise | I-type | 70-72 | Convergent, subduction of oceanic crust underneath a continental one |
| Peru - North Chile | I-type | 70 | |
| Sierra Nevada, U.S.A. | I-type | 50 | |
| Himalayas | S-type | 50 | Collision of two plates |
| SE China | S-type | 35 | ? |

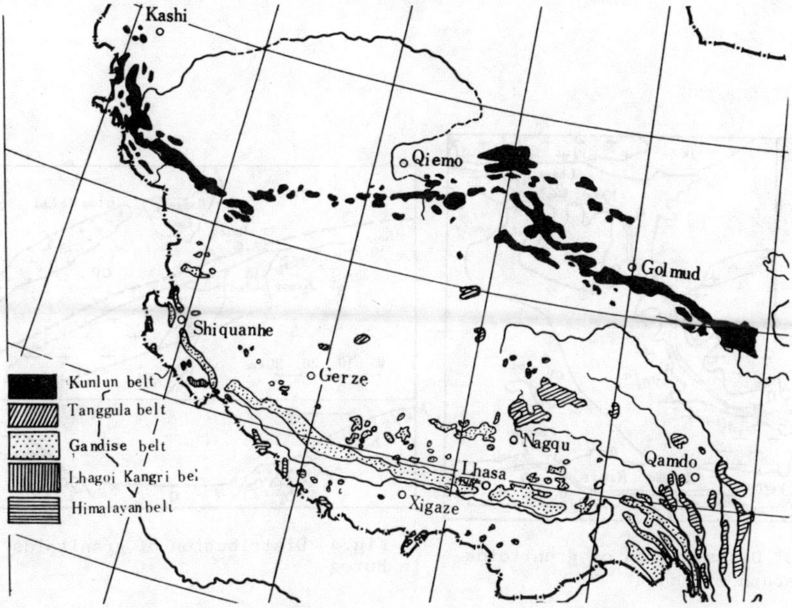


Fig. 1 The five petrographic belts of Qinghai-Xizang Plateau

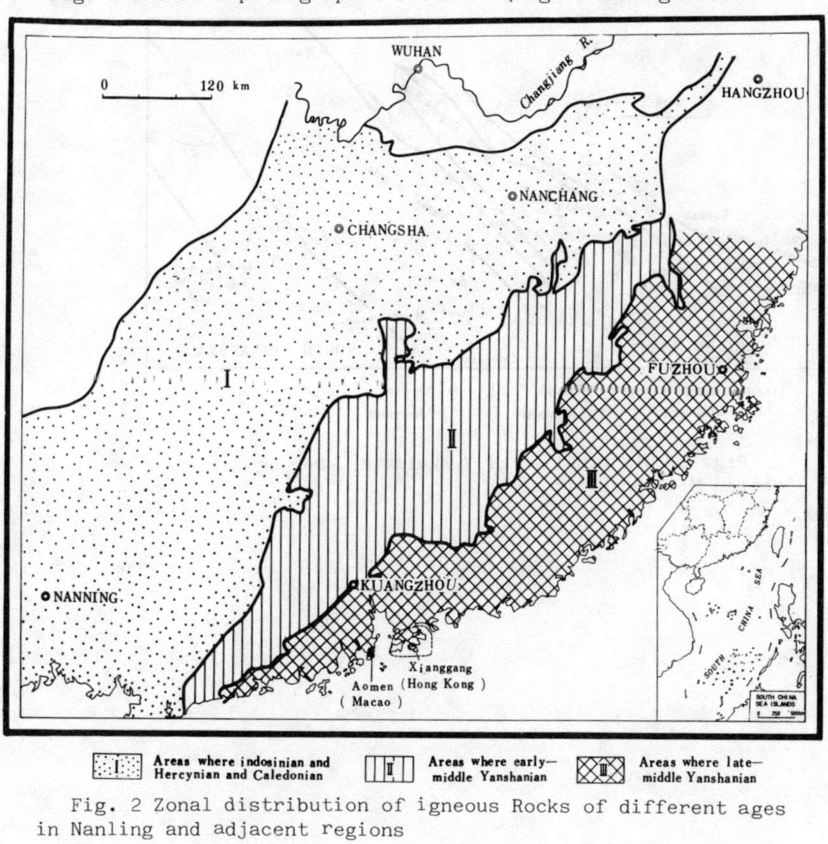


Fig. 2 Zonal distribution of igneous Rocks of different ages in Nanling and adjacent regions

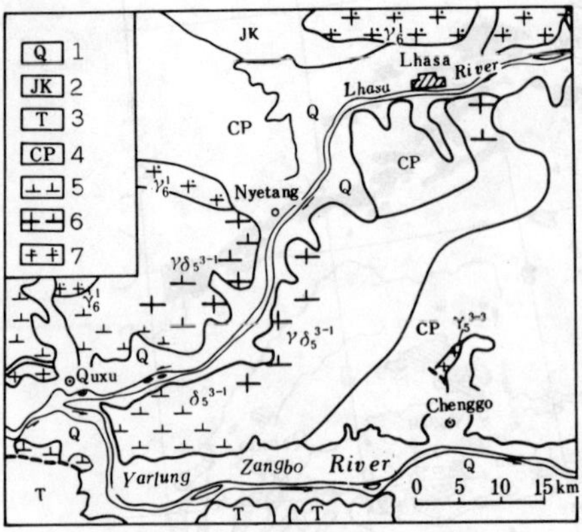


Fig.3 Distribution of granitoids to the south of Lhasa

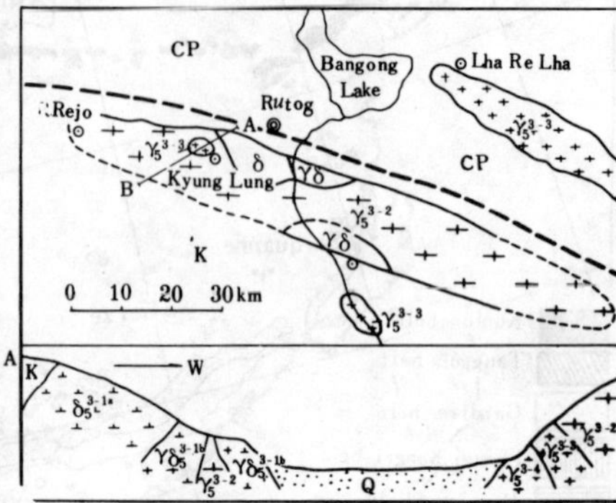


Fig.4 Distribution of granitoids in Rutog

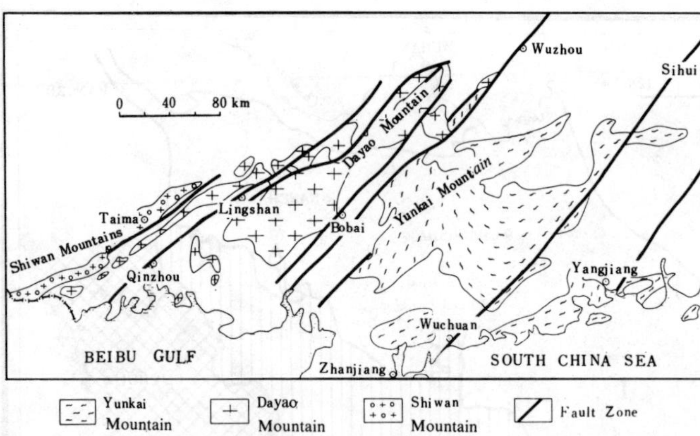


Fig. 5 Distribution of Yunkaida Mt.-Darong Mt.-Shiwan Mts. granitic rocks

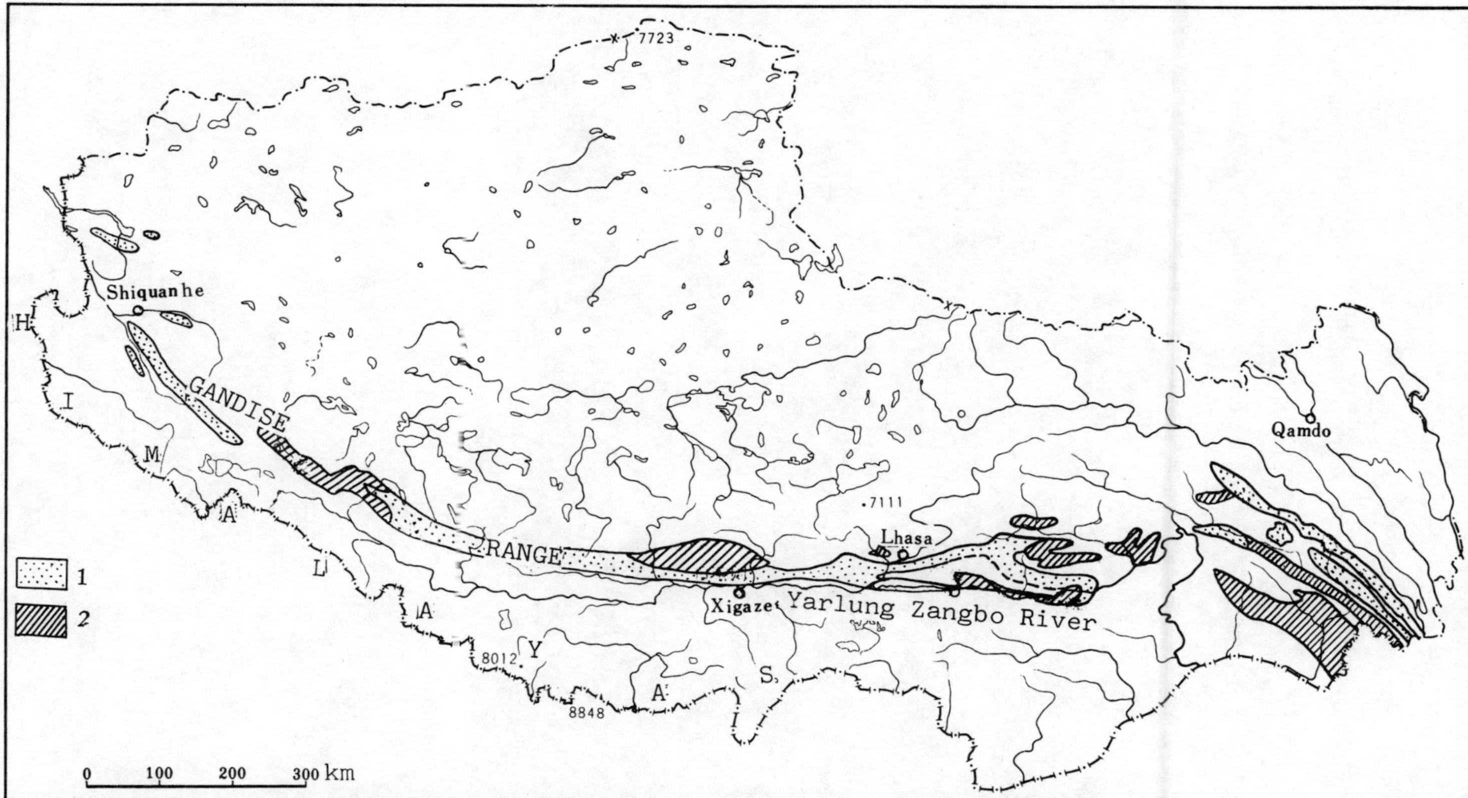


Fig. 6 Distribution of Gandise granitoid rocks

1. Diorite and granodiorite
2. Biotite granite

# THE MIGRATION AND EVOLUTION OF MAGMATISM AND METAMORPHISM IN XIZANG SINCE CRETACEOUS AND THEIR RELATION TO THE INDIAN PLATE MOTION A POSSIBLE MODEL FOR THE UPLIFT OF QINGHAI-XIZANG PLATEAU

Zhou Yun-sheng, Zhang Qi, Jin Cheng-wei, Deng Wan-ming
(Institute of Geology, Academia Sinica)

## INTRODUCTION

We have attempted to interpret the formation of some magmatic and metamorphic rock belts in Xizang on the basis of plate tectonics in some of our previous papers(1,2,3). Lately,a summarization of the information on magmatic and metamorphic rocks in Xizang is presented in a monograph entitled "Magmatism and metamorphism in Xizang" (Science Press, in press). On the basis of these works and the recently published information regarding the motion of the Indian plate, we try to discuss the migration and evolution of the magmatism and metamorphism since Cretaceous in Xizang by a threefold model of continental convergence-continental collision - intracontinental convergence. A tentative view on the large-scale uplift of the Qinghai-Xizang Plateau since Pliocene will also be discussed. As for the details of the regional geology, petrology and petrochemistry relevant to magmatic and metamorphic rocks, readers may refer to other articles(4,5,6,7,8) of this symposium.

## DISTRIBUTION OF THE MAGMATIC AND METAMORPHIC ROCKS

In Xizang, the magmatic and metamorphic rocks formed since Cretaceous are widespread. They distribute mainly in the Nyainqentanglha, the Gandise and the Himalayan regions to the south of the Bangong Lake-Dongqiao-Nujaing River faults zone. To the north of this zone, there are only the Quaternary volcanic rocks which occur at the margin of Kunlun Mts. of northern Xizang.

These rocks make up a series of belts of magmatic and metamorphic rocks. According to their tectonic setting, associations of rocks and ages of formation,the magmatic rocks can be divided into five belts as follows (Fig. 1): (1) the Nyainqentanglha belt of magmatic rocks, (2) the Gandise belt of magmatic rocks, (3) the Yarlung Zangbo belt of ophiolitic rocks, (4) the Himalayan belt of granitic rocks, that can be subdivided into the Lhagoi Kangri zone of gneissic granites and the High Himalayan zone of tourmaline granites, and (5) the North Xizang belt of Quaternary volcanic rocks. According to the facies series, pressure types and ages of the regional metamorphic rocks and the tectonic settings in which they formed, the regional metamorphic rocks are divided into five belts (Fig. 2) : (1) the Nyainqentanglha metamorphic belt, (2) the Gandise metamorphic belt, (3) the Yarlung Zangbo metamorphic belt, (4) the Lhagoi Kangri metamorphic belt , and (5) the High Himalayan metamorphic belt.

All of the belts mentioned above align parallely with each other, and extend in approximately east-west direction like a series of arcs. Their trends are consistent with those of Himalayas, Gandise and

Nyainqentanglha Ranges. In these belts, volcanism, plutonism and regional metamorphism show a similar pattern of migration in space and time, while each of them displays a special tendency of evolution, which will be discussed in the following.

## MIGRATION AND EVOLUTION OF VOLCANISM

Since Cretaceous, the spacial migration of volcanism with time has shown the following tendency. The Early Cretaceous volcanic activities occurred in a wide area of the Nyainqentanglha-Gandise region. During Late Cretaceous, volcanic activities shifted to the southern part of the region and the Yarlung Zangbo belt. From Tertiary to Quaternary, the activities developed again northward and migrated to the southern margin of Kunlun Mts. The characeristics of volcanism are different from one region to another.

In the Nyainqentanglha-Gandise region, the Early Cretaceous volcanic rocks are associated with shallow-sea deposits and the Late Cretaceous-Tertiary volcanics are in company with terrestrial deposits. Therefore, the environment of volcanism underwent a change from shallow sea, island arc into inland basin during Cretaceous-Tertiary. The volcanic rocks are characterized by abundant tuffs and volcanic agglomerates,which show apparently some eruptive cycles together with lavas. They are mostly basaltic, andesitic, dacitic and rhyolitic rocks which belong to calc-alkaline series in chemical composition. However, there are also some alkaline rocks such as leucite basalt. Of these rocks, andesitic and dacitic rocks are predominant. When the age of volcanic rocks become younger, the acid members of associated volcanic rocks tend to increase abundantly. Espcially in the Gandise region, composition of the volcanic rocks tends to become more acidic and more alkaline from south to north as their geological ages become younger. (9,5).

In the Yarlung Zangbo belt, the volcanic rocks formed probably from Middle to Late Cretaceous are associated with abyssal deposits. Lavas are of great abundance and have the well-developed pillow structure, whereas the pyroclastic rocks such as tuffs are lacking. Therefore, the volcanism occurred in a deep-sea environment. These rocks are mainly pillow, massive and sheeted basaltic and spilitic lavas. Most of them are oceanic tholeiitic in chemical composition and some are island-arc tholeiitic(6,10). They are associated with ultramafic and mafic plutonics, sheeted basic dyke swarm, radiolarian chert and abyssal sediments, and they compose together a suite of sequence-type ophiolites representing probably the fragments of oceanic lithosphere in Tethys(11).

In the southern margin of Kunlun Mts. the Quaternary volcanic rocks occur as lava sheet and lava flows overlying the Tertiary lacustrine or fluvial deposits and other older sedimentary rocks. Therefore,the volcanism took place in a continental environment, and showed effusive eruption. The volcanics have complex types of association, including dacite, pyroxene andesite, alkalic basalts and alkaline rocks. However, their composition changes apparently southward from calc-alkaline into alkaline(2).

As to the petrochemistry of volcanics in the said regions, all the rocks from the Yarlung Zangbo belt can be referred to sodic subseries, whether they are alkalic or subalkalic, whereas all the rocks from Nyainqentanglha-Gandise region and from the southern margin of Kunlun Mts. in northern Xizang can be referred to potassic subseries (Fig. 3).

## MIGRATION AND EVOLUTION OF PLUTONISM

Since Cretaceous, the plutonism in Xizang has been characterized by a series of large - scale intrusions of intermediate-acidic magma. Mafic-ultramafic plutonic rocks occurred as a tectonic emplacement along the Yarlung Zangbo belt.

The K-Ar ages of the intermediate-acidic intrusive rocks formed since Cretaceous are shown in Fig. 4.They may be divided into four groups, i.e. (1) 130 — 90 m.y. in Nyainqentanglha and Gandise regions, (2) 90 — 40m.y. in Gandise region, (3) 40 — 20 m.y. in Gandise and Lhagoi Kangri regions, and (4) 20 — 10 m.y. in Himalayas.

In general, the activities of the intermediate-acidic magma show a southward migration from Nyainqentanglha to Himalayas with the lapse of time. Their intensity and scale were most intense and magnificent in the Nyainqentanglha and Gandise regions, and were relatively weaker and less in Himalayan region. However, the activities were completely lack or very weak in the Yarlung Zangbo belt. The intrusions in various regions display different features.

In the Nyainqentanglha and Gandise regions, a series of multi-period and multi-phase intrusions of intermediate-acidic magma continued from Early Cretaceous to Eocene. On the basis of the distribution of K-Ar ages of intermediate-acidic intrusive rocks and their geological relationship, three intrusive periods may be distinguished, i.e. (1) Early to Late Cretaceous intrusive period, (2) Late Cretaceous to Eocene intrusive period and (3) Eocene to Miocene intrusive period.

The intrusive activities in the (1) period occurred mainly in the Nyainqentanglha region, and those in the (2) period migrated southward to the Gandise region. The intrusions of both periods began with the emplacement of quartz-diorite or diorites, which was followed by those of quartz--monzonite and granodiorite, and ended with that of monzogranite or granite. Among them the activities of granodiorite and monzogranite were most intensive. All of these activities brought about two huge complex batholthic zones, namely the Nyainqentanglha batholthic zone and the Gandise batholthic zone(12,13). The activities in the (3) period occurred within the Gandise zone and at its northern side, and were characterized by the emplacement of some small granite bodies, including two mica- and muscovite-granite, microcline granite and tourmaline-bearing granite(12).

It should be pointed out specially that the ages of the intrusions of intermediate- acidic magma in the Gandise region apparently tend to become younger and younger from south to north (Fig. 4), and their composition tends to increase in $Na_2O + K_2O$ and $K_2O$ contents(12).

In contrast with the plutonism in both regions mentioned above, the plutonic activities in the Himalayan region are much younger in age and much smaller in scale, and are characterized by the emplacement of smaller bodies which consist of the granitic rocks containing high $SiO_2$ and $K_2O + Na_2O$ content in composition(14). According to the ages of the granitic rocks and their tectonic setting occurred, two intrusive periods of granitic magma are distinguished, i.e. (1) Eocene to Miocene intrusive period and (2) Miocene to Pliocene intrusive period.

In the (1) intrusive period, the emplacement of granitic magma occurred along the Lhagoi Kangri Mt. of northern Himalayas, and formed a series of the diapiric domes which consist of gneissic two-mica granite or muscovite granite. In the rocks, gneissosity is well-developed and shows the same orientation as the schistosity of metamorphic rocks surrounding them. This implies that the rocks were compressed by regional structural stress in their emplacing and crystallizing process. Similarly, this phenomenon has also been observed on the northern side of the Gandise batholithic zone. In the (2) intrusive period, the emplacement of granitic magma occurred along the High Himalayas and formed tourmaline-muscovite granite and two-mica granite without or with weak gneissose structure, which are associated with migmatites. These rocks are the youngest plutonic rocks found in Xizang at present. Therefore, the plutonism in the Himalayan region shows a southern migration with the lapse of time.

In Fig. 5, composition of the intermediate- acidic intrusive rocks from all the above- mentioned regions are compared using Q-Or-Ab diagrams. The points of the rocks from the Nyainqentanglha and the Gandise regions scatter in a wide area in Q-Or-Ab diagram and show the tendency shifting from the central part of the compositional field of granitic rocks toward Ab corner (Fig. 5c and 5d), whereas those of the rocks from the Himalayan region concentrate in the central part of the compositional field of granitic rocks and do not show such shifting tendency (Fig. 5a and 5b). This implies that generating of the magma in the Nyainqentanglha and the Gandise region is at a greater depth of the crust than that of the magma in the Himalayan region.

## MIGRATION AND EVOLUTION OF REGIONAL METAMORPHISM

In Xizang, the regional metamorphism since Cretaceous shows the same tendency of migration in space and time as that of the plutonism. The main metamorphic periods are Cretaceous-Tertiary and Tertiary. The regional metamorphism resulted in the formation of five metamorphic belts as shown in Fig. 2.(15)

In the Nyainqentanglha and the Gandise metamorphic belts, the metamorphism might take place during Cretaceous to Tertiary. The metamorphic facies series change from low greenschist facies to amphibolite facies. The mineral assemblages containing andalusite and sillimanite are common in relatively high temperature zones. The average $b_0$ value of potassic white mica in relatively low temperature zones is 8.989Å and 9.012(2). The regional metamorphic rocks associate with abundant granitic rocks and migmatites. Temperature of metamorphism increases toward both Nyainqentanglha and Gandise batholithic zones which could represent the positions of two parallel thermal axes of regional metamorphism. As shown by above features, the metamorphism is of the low-medium pressure type.

In the Yarlung Zangbo metamorphic belt, the metamorphism took place probably during Late Cretaceous-Tertiary(15). The metamorphic facies series is from low greenschist to epidote amphibolite facies. Almandine in metapelites is present in greenschist facies and associates with albite. The potassic white mica in metapelites has a higher average $b_0$ value up to 9.043Å (2). The metamorphic rocks are associated with mafic-ultramafic plutonic rocks and mafic volcanics. Although no high pressure metamorphic mineral assemblage has been found up to the present, the above features strongly indicate that the metamorphism is probably of a medium-high pressure type. The belt probably constitutes together with the Gandise metamorphic belt as a pair of metamorphic belts, whose pressure types are obviously different from each other(7,15).

In the Lhangoi Kangri metamorphic belt, the uppermost metamorphosed strata are of the Cretaceous. The metamorphic grade is generally low, from low greenschist facies to greenshist facies. But it may reach amphibolite facies in the surrounding of some gneissic granitic domes, where the mineral assemblages containing staurolite + almandine and kyanite + almandine or andalusite + almandine often occur. The average $b_o$ value of white mica in metapelites is 8.987Å. Therefore, the metamorphism is probably of the low-medium pressure type.

The High Himalayan basal metamorphic belt is a Precambrian Barrovian-type belt(15,16,17). However, its young K-Ar ages (Fig. 6) indicate that it was influenced by a thermal event during 20-10 m.y., which might have caused a degraded metamorphism. This metamorphism is closely related to the formation of the High Himalayan tourmaline granites. The low-pressure mineral assemblages containing cordierite are present in gneisses around some of the granitic bodies, suggesting that the degraded metamorphism is probably of the low-pressure type.

To sum up, the magmatism and metamorphism in Xizang since Cretaceous have brought about four magmatic and metamorphic belts which reflect different tectonic settings during their formation:

1. the Early Cretaceous to Tertiary Nyainqentanglha-Gandise magmatic and metamorphic belt, which is characterized by the close association of calc-alkaline volcanic rocks, intermediate-acidic plutonic rocks and low-pressure regional metamorphic rocks, indicating tectonic settings of active continental margin and island arc at that time;

2. the Late Cretaceous to Eocene Yarlung Zangbo ophiolitic and medium-high pressure metamorphic belt, which consists of a series of ophiolitic blocks representing fragments of the Tethyan oceanic lithosphere and some medium-high pressure metamorphic terrains, indicating that the belt bears the character of the suture between two plates and represents the subducting position of the Mesozoic Tethyan oceanic lithosphere;

3. the Tertiary Himalayan granitic and degraded metamorphic belt, in which the young granitic plutonism, low-pressure degraded metamorphism and intensive migmatization in the Precambrian basal rocks associate with a series of large-scale thrusting, indicating the remobilization of a stabilized continental crust block;

4. the Quaternary Northern Xizang volcanic belt, in which the terrestrial volcanics tend to increase more and more in alkalic content from north to south, implying probably a tectonic setting of interaction between blocks within a continent.

The regular arrangement of these belts in space, their order of formation in time and respective tectonic settings provide important evidences for investigating the history of the Indian Plate motion since Cretaceous and its effects on the geological evolution of the Qinghai-Xizang Plateau.

## THE RELATION OF THE MAGMATISM AND METAMORPHISM TO THE MOTION OF THE INDIAN PLATE

On the basis of the results of the above discussion and using the data of some studies on the Indian Plate motion (18,19,20), we propose here a threefold model of continental convergence - continental collision -

intracontinental convergence to explain the migration and evolution of magmatism and metamorphism in Xizang since Cretaceous (Fig. 7). A brief account of the model is given as follows.

1. The slow convergence stage of the Indian continent toward the Eurasian continent during Early Cretaceous. It is suggested that the Indian continental block was separated from Antarctic-Australian continental block at about 130 m.y. ago and moved slowly northward in a velocity of 3-5 cm/y.(18) This coincides with the distribution of Early Cretaceous calc-alkaline volcanic rocks and occurrence of intermediate-acidic plutonic rocks with K-Ar age of 130-90 m.y. in the Nyainqentanglha-Gandise belt. This coincidence implies that the oceanic lithosphere of the Mesozoic Tethys intervening between the two continents began underthrusting slowly beneath the Eurasian Plate along the Yarlung Zangbo trench in Early Cretaceous due to the northward motion of the Indian continent. The subduction of the oceanic lithosphere along the underthrusting zone with northern dip led to the Early Cretaceous volcanism, plutonism and associated low-pressure regional metamorphism in the Nyainqentanglha and the Gandise regions. According to the available information(13,9), the front of magmatic activity in that time reached at least to the northern side of the Nyainqentanglha. Judging from this, the slope of the underthrusting zone was rather gentle.

2. The rapid convergence stage of the Indian continent toward the Eurasian continent from Late Cretaceous to Eocene. The northward motion of the Indian continent began to accelerate about 85 m. y. ago, and its average velocity amounted to 15 cm/y.( 18, 19 ). As discussed previously, the magmatism of the Nyainqentanglha-Gandise belt showed clearly southward migration at the beginning of Late Cretaceous, i.e. about 90 m.y. It appears that the migration of magmatism and the velocity-change of Indian Plate motion are synchronous. The southward migration of magmatism implies that the slope of the underthrusting zone of Tethyan oceanic lithosphere was probably steepened at that time. Due to the rapid drift of the Indian continent, the Tethyan oceanic lithosphere underthrusted and subducted rapidly in a steep slope along the Yarlung Zangbo trench zone, resulting in the large-scale volcanism and plutonism of calic-alkaline magma and the regional metamorphism of low pressure-high temperature type in the Gandise region, as well as medium-high pressure metamorphism at the upper part of the underthrusting zone.

3. The collison stage of the two continents at the beginning of Eocene. The rate of motion of the Indian continent turned slow from rapid at about 55 m.y. ago, and the average velocity was 4-6 cm/y.(18). At the beginning of Eocene, in the Yarlung Zangbo belt the terrestrial deposits began to occur, and the previous abyssal deposits were subjected to intensive folding and large-scale thrusting(21,22). These imply that the Indian continent could collide with the Eurasian continent along the Yarlung Zangbo trench zone, leading to the disappearance of Misozoic Tethys and the structural uplift of residual oceanic crust and medium-high pressure metamorphic rocks at the upper part of underthrusting zone.

4. The intracontinental convergence stage since Eocene. After the collision of the continents, the Indian continent continuously pushed slowly northward(18,20). In such a case, the following scene can be deduced easily. Under the strong horizontal compression resulted from the continuous pushing of the Indian continent,a grand scale of thrusting movements between continental crustal blocks probably occurred within these two continents, resulting in the shortening and thickening of continental crust. Thus, a partial melting of crustal materials might be

generated within the thickened continental crust, leading to new magmatic activities and associated regional metamorphism. We call such tectonic process between crustal blocks within continent the intracontinental convergence.

Since Eocene, the great thrusting structures in continental crust have been well developed in Xizang. Especially in the Himalayan region, a series of great thrust nappe structures was successively formed southward(22). Such structures have also been found in area to the north of the Yarlung Zangbo belt that marks the collision of the Indian with the Eurasian continent(23). The magmatic activities since that time have migrated regularly successively from the belt toward south and north respectively, which is consistent with the development of the grand scale thrusting structures. The phenomenon probably reflects the intracontinental convergences from opposite directions, which occurred to the south and the north of the collided belt.

The Eocene-Miocene granitic plutonic activities in the Lhagoi Kangri region and at the northern side of the Gandise batholithic belt occurred symmetrically in spcial position relative to the Yarlung Zangbo belt. Their tectonic positions are to the north of the main northern thrust of Himalayas and to the south of the southern marginal fault of Nyainqentanglha respectively. The petrochemical characteristics of the granites indicate that they are probably the products of the partial melting in the upper crust(24,8). Therefore, the plutonism and associated metamorphism reflect probably the intracontinental convergences from opposite directions along the above two thrusts, as shown in Fig. 7d.

The Miocene-Pliocene granitic activities in the High Himalayan region occurred to the north of the main boundary thrust of Himalayas i.e. the southern boundary fault of the Qinghai-Xizang Plateau. The petrochemical characteristics of the granites are similar to those of granites in the Lhagoi Kangri region, and show a similar genesis(24,8). The Quaternary volcanism in northern Xizang occurred to the south of the marginal fault of Kunlun Mts., i.e. the northern boundary fault of the plateau. The compositional change of volcanic rocks in space is attributed to the "wedging" of the Tarim block beneath the Qinghai-Xizang Plateau(2). Therefore, the granitic plutonism and associated metamorphism and the volcanism reflect probably the intracontinental convergences from opposite directions along the southern and northern boundary faults of the plateau. As for the difference of magmatism in the two districts, it may be due to the different slopes and descending depths of the two intracontinental convergence belt.

## A TENTATIVE MODEL FOR THE LARGE-SCALE UPLIFT OF THE QINGHAI-XIZANG PLATEAU

It is pointed out that the Himalayas began to rise in Miocene, and that the violent uplifting of the Qinghai-Xizang Plateau chiefly took place in Quaternary, and up to now the accumulative magnitude of the uplift amounts to 3,000-4,000 m(25). Based on this fact and the above recognized development of magmatism and metamorphism since Cretaceous, we further propose a tentative model for the large-scale uplifting of the Plateau, which is shown in Fig. 8.

In this model, we suggest that the large-scale uplifting of the Qinghai-Xizang Plateau is not a direct result of the collision between the

Indian continent and the Eurasian continent, but is caused by the intracontinental convergences from opposite directions resulting from the continuous northward pushing of the Indian continent after the collision. Particularly, the intracontinental convergences from opposite directions along the southern and the northern boundary faults of the plateau since Pliocene have played an important role in the uplift of the plateau as a whole. These intracontinental convergences, on the one hand, have caused the thickening of continental crust and the facies transformation of materials within continental crust, and on the other hand, transformed the horizontal movement of the Indian continent into the vertical movement within the plateau. The combination of these processes probably led to the violent uplifting of the plateau as a whole.

## References


(1) Jin Cheng-wei,Zhou Yun-sheng,Igneous rocks belts in the Himalayas and the Gandise arc and their genetic model. Scientia Geologica Sinica, (4), 297-311 (1978) (in Chinese).

(2) Deng Wan-ming, A preliminary study on the petrology and petrochemistry of the Quaternary volcanic rocks of northern Xizang Autonomous Region. Acta Geologica Sinica, (2), 148-161 (1978) (in Chinese).

(3) Zhang Qi et al., The $b_o$ value of potassic white mica in the metamorphic rocks from the metamorphic belts of Xizang and its geological significance. Kexue Tongbao, (3), 120-123 (1979) (in Chinese).

(4) Chang Cheng-fa, Pan Yu-sheng, The history of tectonic evolution of the Qinghai-Xizang Plateau. In this Proceedings of Symposium on Qinghai-Xizang (Tibet) Plateau, Academia Sinica (1981).

(5) Jin Cheng-wei, The Mesozoic-Cenozoic volcanic rocks in Xizang and their bearings on geology. In this Proceedings of Symposium on Qinghai-Xizang (Tibet) Plateau, Academia Sinica (1981).

(6) Deng Wan-ming, Wu Hao-ruo, Chang Cheng-fa, Liu Guan-jian, The sequence, formation and emplacement of the Yarlung Zangbo ophiolite belt, Xizang. In this Proceedings of Symposium on Qinghai-Xizang (Tibet) Plateau, Academia Sinica (1981).

(7) Zhang Qi, Li Da-zhou, Li Shao-hua, A pair of metamorphic belts in southern Xizang. In this Proceedings of Symposium on Qinghai-Xizang (Tibet) Plateau, Academia Sinica (1981).

(8) Liu Guan-jien, Zhang Kui-wu, Petrochemistry of granites in Xizang and an approach to their origin. In this Proceedings of Symposium on Qinghai-Xizang (Tibet) Plateau, Academia Sinica (1981).

(9) Jin Chengwei, Volcanic rocks of Xizang. In Magmatism and metamorphism in Xizang. Science Press, 212-270 (1981) (in press) (in Chinese).

(10) Mei Hou-jun,Lin Xue-nong,Ophiolite of Xizang. In Magmatism and metamorphism in Xizang. Science Press, 147-211 (1981) (in press) (in Chinese).

(11) Deng Wan-ming, Studies on the Igneous Petrology of the Yarlung Zangbo ophiolite. In this Proceedings of Symposium on Qinghai-Xizang (Tibet) Plateau, Academia Sinica (1981).

(12) Zhou Yun-sheng, Zhang Kui-wu, Gandise intermediate-acidic intrusive rocks belt. In Magmatism and metamorphism in Xizang, Science Press, 50-98 (1981) (in press) (in Chinese).

(13) Jin Cheng-wei, Northern Xizang intermediate-acidic intrusive rock belt. In Magmatism and metamorphism in Xizang. Science Press, 99-111 (1981) (in press) (in Chinese).

(14) Zhou Yun-sheng,Zhang Kui-wu,Himalaya granite belt. In magmatism and metamorphism in Xizang. Science Press, 1-49 (1981) (in press) (in Chinese).

(15) Zhang Qi, Li Shao-hua,Metamorphic belt and metamorphism of Xizang.

In Magmatism and metamorphism in Xizang, Science Press (1981) (in press) (in Chinese).
(16) Liu Bing-guang,Zhang Hong-bo,Metamorphic, migmatic and magmatic rocks at northern slope of the Qomolangma Feng in southern Xizang, China. In Report on the Scientific Expedition to Mt. Qomolangma Region (1975), Geology, Science Press (1979) (in Chinese).
(17) Ying Si-huai,Magmatic, metamorphic and migmatic rocks of the Mount Qomolangma region in southern Xizang, China. Scientia Geologica Sinica, (2), 104-128 (1973) (in Chinese).
(18) Powell, C. McA., A speculative tectonic history of Pekistan and surroundings : some constraints from the Indian Ocean. in Geodynamics of Pakistan, Geological Survey of Pakistan, 5-24, (1979).
(19) McKenzie, D. P., Sclater, J. G., The evolution of the Indian Ocean since the Late Cretaceous. Royal Astron. Soc. Geophys. Jour., 25, 437-528 (1971).
(20) Molnar, P., Tapponnier, P., Cenozoic tectonics of Asia : Effects of a continental collision. Science, 189, 419-426 (1975).
(21) Zhang Bing-gao, The latest marine sediments in Xizang and the process of the Early Tertiary regression. In this Proceedings of Symposium on Qinghai-Xizang (Tibet) Plateau, Academia Sinica (1981).
(22) Chang Cheng-fa,Cheng Hsi-lan,Some tectonic features of the Mt. Jolmo Lungma area, southern Tibet, China. Scientia Sinica, 6 (2), (1973).
(23) Pan Yu-sheng,The nappe structure of Xizang and its geological significance. Scientia Geologica Sinica, (1), 11-17 (1980) ( in Chinese).
(24) Tu Kuang-chi et al., Characteristics and evolution of granitoids of South Xizang. In this Proceedings of Symposium on Qinghai-Xizang (Tibet) Plateau, Academia Sinica (1981).
(25) Li Ji-jun et al., The process of the upheaval of the Qinghai-Xizang Plateau. In this Proceedings of Symposium on Qinghai-Xizang (Tibet) Plateau, Academia Sinica (1981).
(26) Irvine, T. N., Baragar, W. R. A., A guide to the chemical classification of the common volcanic rocks. Canadian journal of earth sciences, 8, 524-548 (1971).
(27) Winkler, H. G. F., Petrogenesis of metamorphic rocks. Springer-Verlag, New York Inc, 278-323 (1976).
(28) Group of K-Ar geochronology, Institute of Geology, Academia Sinica, K-Ar dating and division of the Himalayan movement in southern Xizang. Scientia Geologica Sinica, (1), 13-21 (1979) (in Chinese).
(29) Li Pu et al., Report on the age results of rocks and minerals dated by K-Ar method. Scientia Geologica Sinica, (1), 24-34 (1964) (in Chinese).
(30) Li Pu et al., K-Ar dating of rocks from Mount Shisha Pangma area, Tibet, China. Kexue Tongbao (10), 925-926 (1965) (in Chinese).
(31) Division of Isotope Geology, Institute of Geochemistry, Academia Sinica, Isotope dating of metamorphic rocks from the Mt. Jolmo Lungma area, southern Tibet, China. Scientia Sinica, (3), 280-288 (1973) (in Chinese).

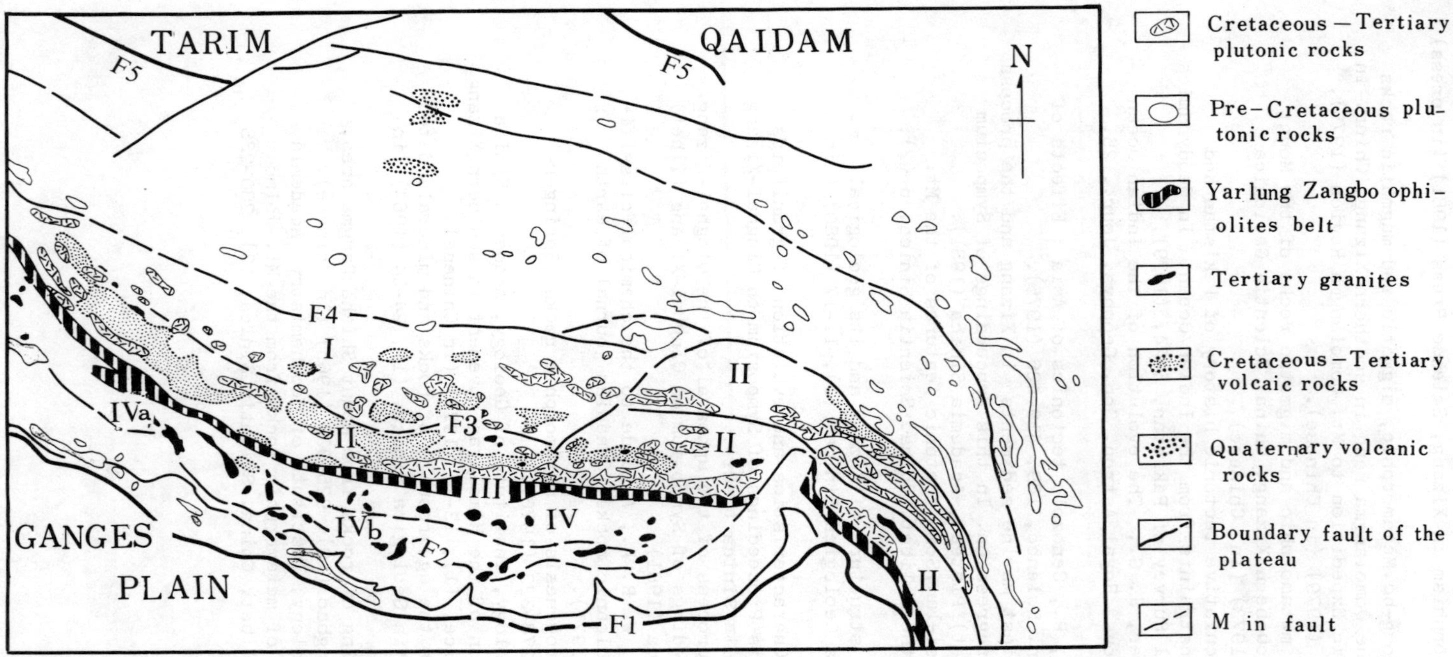


Fig. 1 Distribution of the magmatic rocks formed since Cretaceous and the magmatic belts in Xizang and its neighbourings

1. The Nyainqentanglha magmatic belt;
2. The Gandise magmatic belt;
3. The Yarlung Zangbo ophiolitic belt;

4-1. The Lhagoi Kangri granitic belt, 4-2. The High Himalayan granitic belt;

5. The Northern Xizang Quaternary volcanic belt. $F_1$ = the main boudary thrust, $F_2$ = the main northern thrust, $F_3$ = the southern marginal fault of Nyainqentanglha, $F_4$ = Bangong Lake-Dongqiao-Nujaing River fault and $F_5$ = the northern boundary fault of Qinghai-Xizang Plateau

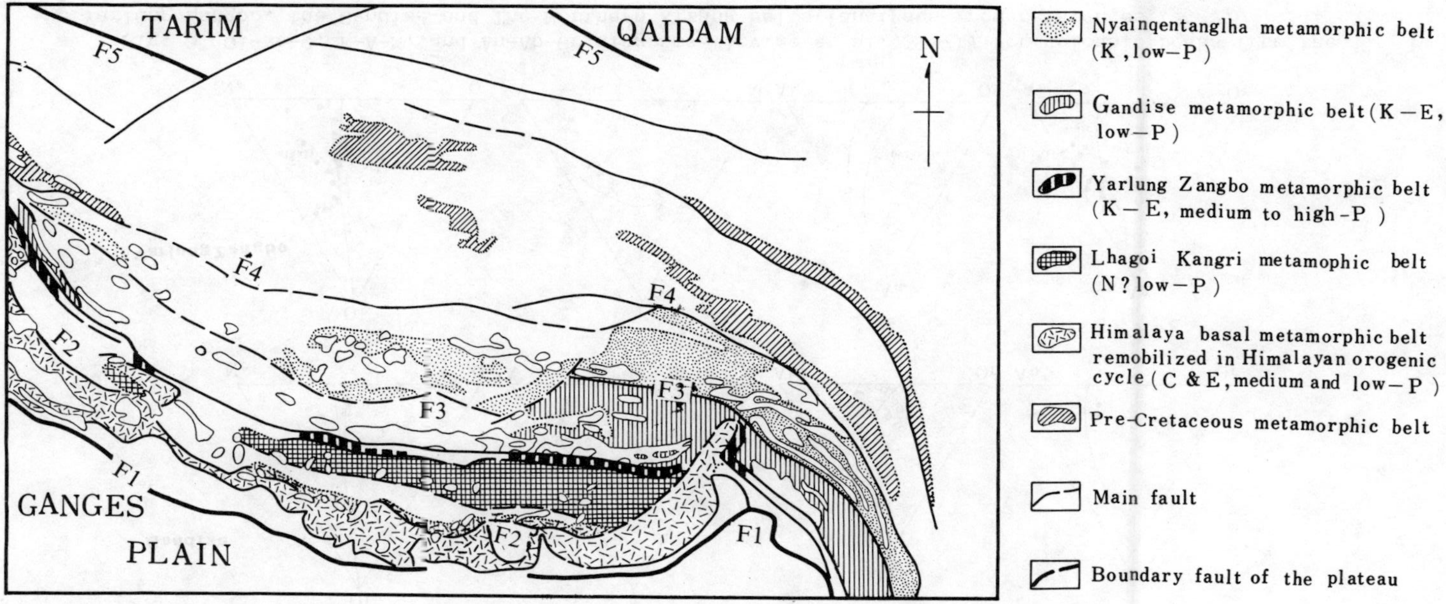


Fig. 2 Distributin of the metamorphic rocks formed since Cretaceous and the metamorphic belts in Xizang and its neighbourings. $F_1$, $F_2$, $F_3$, $F_4$ and $F_5$ are same as those of Fig. 1

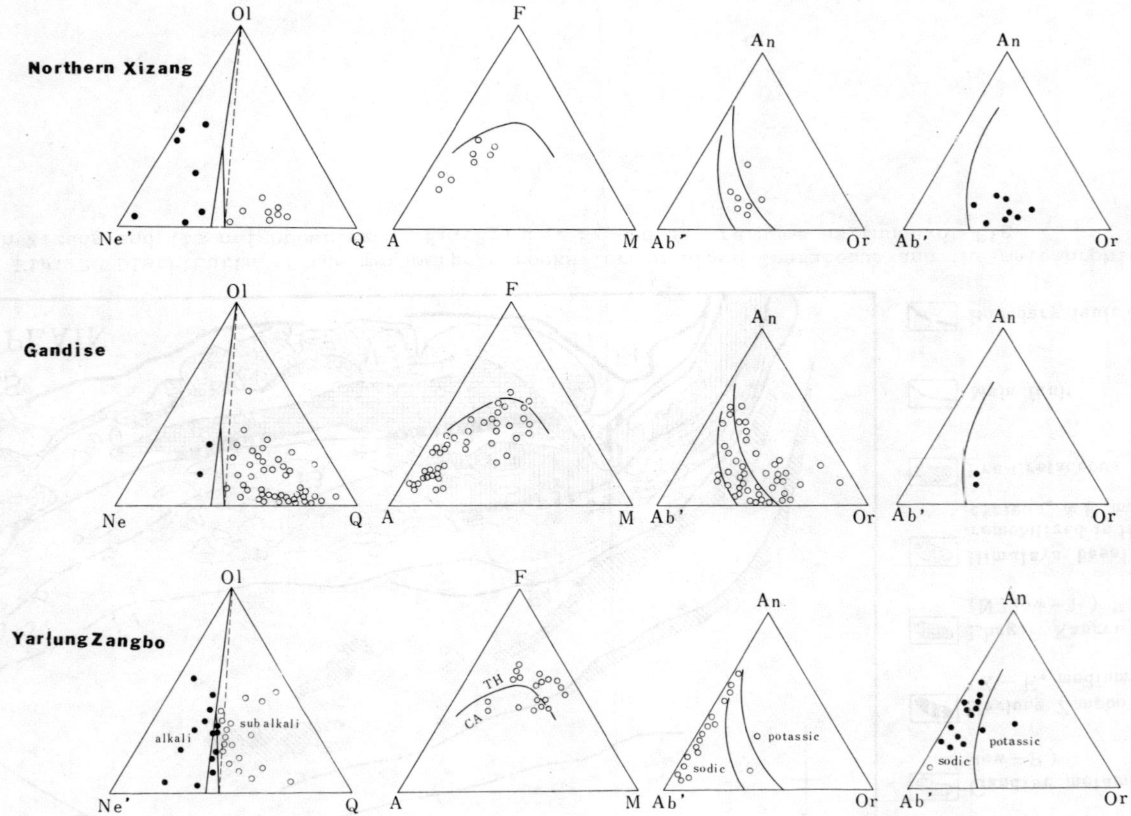


Fig. 3 Ol-Ne-Q, F-A-M, and An-Ab-Or diagrams (Irvine et al., 1971) for volcanic rocks from the Yarlung Zangbo, the Gandise and The Northern Xizang belts (analyses from references (2, 9))

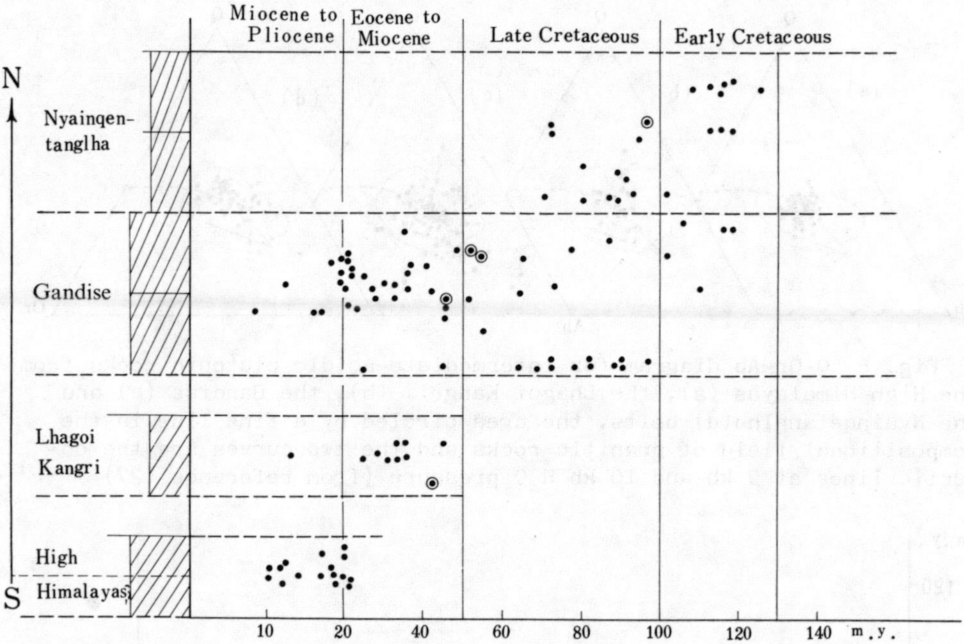


Fig. 4 Diagram showing distribution of K-Ar age of magmatic rocks in the Nyainqentanglha, the Gandise, the Lhagoi Kangri and the High Himalayan belts. The black points are representative of the ages of granitic rocks, and the circles are those of volcanic rocks (data from references (9, 12, 13, 14, 24 and 28))

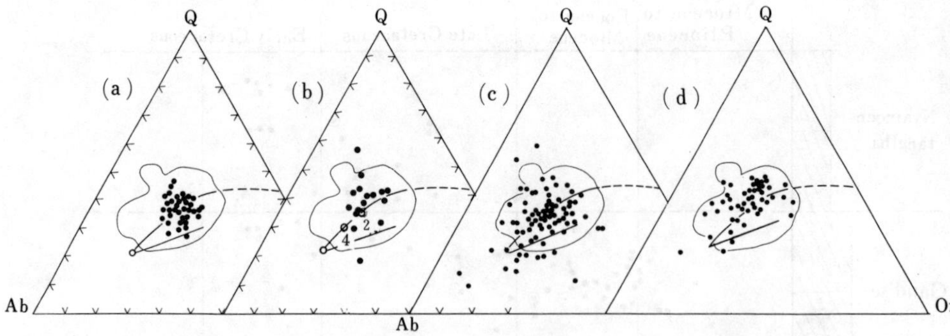


Fig. 5  Q-Or-Ab diagram for intermediate-acidic plutonic rocks from the High Himalayas (a), the Lhagoi Kangri  (b), the Gandise (c) and the Nyainqetanglha(d) belts. the area circled by a fine line is the compositional field of granitic rocks and the two curves are the co-tectic lines at 2 kb and 10 kb $H_2O$ pressure (from reference (27) )

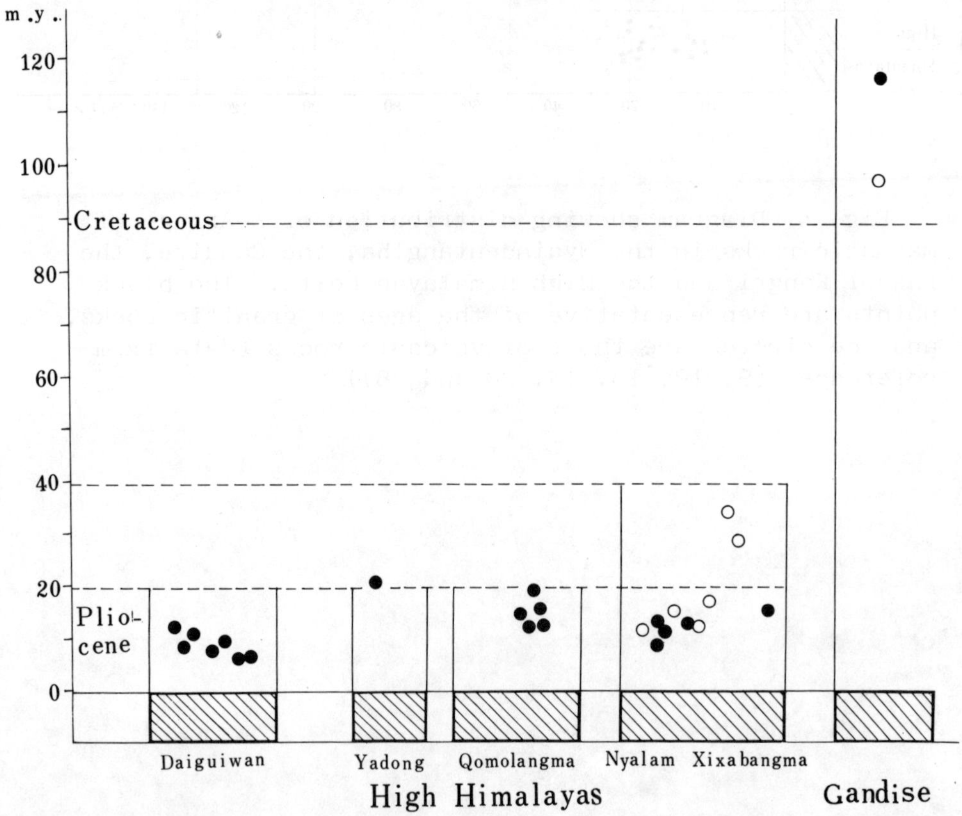


Fig. 6  Diagram showing distribution of K-Ar age of metamorphic rocks (black points) and migmatite (circles) in the High Himalayas and the Gandise metamorphic belts (data from references (28), (29), (30), (31) ).

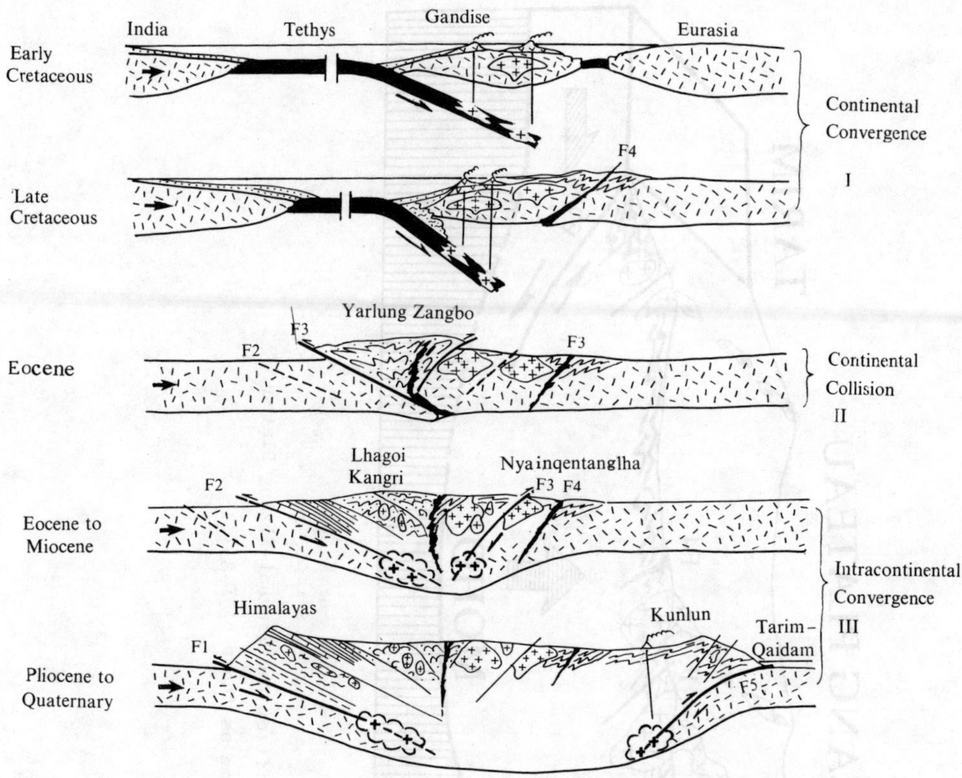


Fig. 7 A threefold model for migration and evolution of magmatism and metamorphism in Xizang since Cretaceous

1. Continental convergence stage;
2. Continental collision stage;
3. Intracontinental convergence

$F_1$, $F_2$, $F_3$, $F_4$, and $F_5$ are same as those of Fig. 1

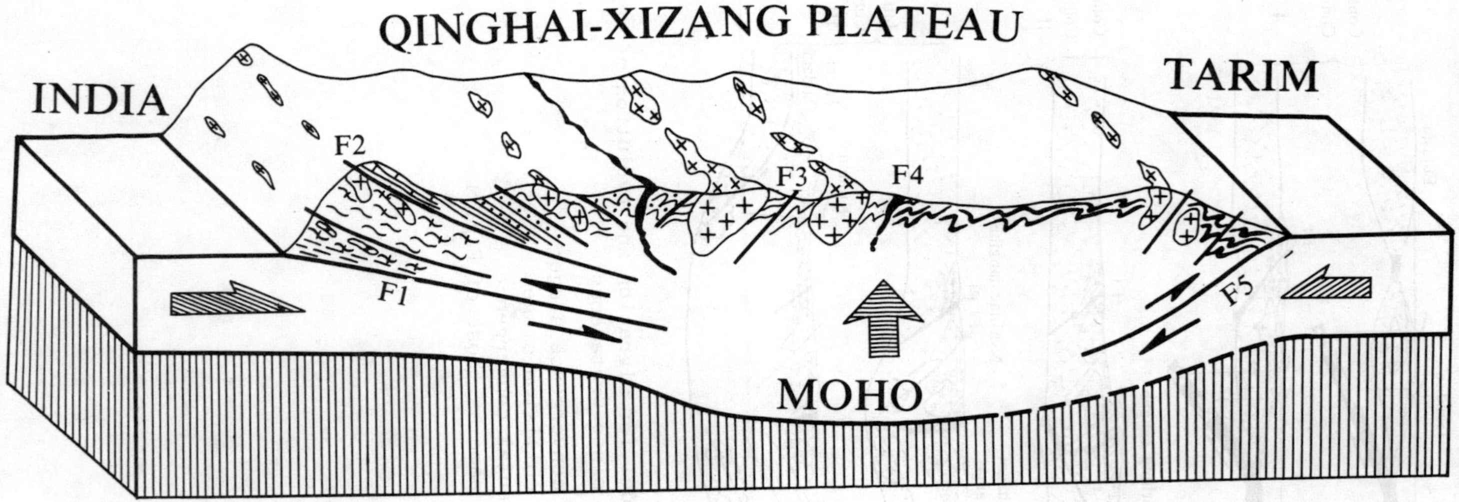


Fig. 8 A tentative model for the uplift of the Qinghai-Xizang Plateau
$F_1$, $F_2$, $F_3$, $F_4$ and $F_5$ are same as those of Fig.7

# ON THE CRUSTAL MOVEMENT CHARACTERISTICS OF THE XIZANG PLATEAU WITH SPECIAL REFERENCE TO INTERMEDIATE-ACID MAGMATIC ACTIVITIES

Zhang Yu-quan, Wang Zhong-gang, Zhao Zhen-hua
Xu Yong-hua, Lin Xue-nong, Xie Ying-wen
(Institute of Geochemistry, Academia Sinica)

## GEOLOGICAL BACKGROUND

Tectonic framework : The regional structural line of South Xizang is NWW in strike, essentially consistent with the strike of the ranges there.

Stratigraphy : In the areas of Nyingchi and Bomi in the east, Devonian strata are widely exposed while in the areas of Shiquanhe and Rutog in the west, Jurassic-Cretaceous strata are largely distributed. In the south, i.e. to the south of the Yarlung Zangbo River, Palaeozoic-Cainozoic strata, especially Triassic-Jurassic-Cretaceous strata, are largely exposed.

Metamorphic rocks : In terms of the degree of metamorphism, two metamorphic belts are divided:(1) higher-grade metamorphic belt and (2) lower-grade metamorphic belt. Higher-grade metamorphic belt lies along the national boundary of China while the latter is distributed in the east of the Lhagoi Kangri Mts. and the Gandise Range. The metamorphic degree is greater for some metamorphic rocks in the local parts, for example, on the both sides at the bend of the Yarlung Zangbo River, there occur gneisses associated with migmatites.

Volcanic rocks : Distributed on a large scale along the Gandise Range are Cretaceous basalts, andesites and rhyolites.

## THE CHARACTERISTICS OF INTERMEDIATE-ACID MAGMATIC ACTIVITIES IN THE SOUTHERN PART OF XIZANG

Magmatic activities in the southern part of Xizang are characterized by their high frequency, large scale, extensive distribution, NWW zonal distribution pattern and decreasing geological age from north to south, as exemplified by the Gandise belt, the Lhagoi Kangri belt and the Himalayan belt (Fig. 1).

1. The Gandise belt. This petrographical belt stretches along the Gar River — Yarlung Zangbo River and their northern bank, southeastward to the Bomi and Zayu areas, with a total length of 2,000 km, covering an area of about 70,000 km$^2$. The main rock types include diorite, granodiorite, porphyritic biotite granite , biotite granite and two-mica granite, etc. The Gandise belt is approximately at the palaeo-suture line between the Indian Plate and Eurasian Plate(1,2,3,4,5). Along this palaeo-suture line, a giant intermediate-acid magmatic arc is recognized. The granitoids range in age from late Yenshanian to Himalayan, but the main parts are late Yenshanian in age.

2. The Lhagoi Kangri belt. It lies along the Lhagoi Kangri Mountains, characterized by small and disperse intrusive bodies. The major rock types include two-mica monzonitic granite. The age of the granite is basically Eocene corresponding to the formation of the Main Central Thrust zone.

3. The Himalayan belt. The belt stretches along the Himalayas, also with small and disperse intrusive bodies. The chief rock types are tourmaline-muscovite granite and two-mica granite. Its age closely corresponds to Miocene during which the Main Boundary Thrust Zone was formed.

As shown in Table 1, the Gandise belt, largely composed of composite and complex plutons, is characterized by multicyclic and multiepisodic intrutions on a large scale, with the intruding sequence beginning with intermediate rocks, followed by acid ones. Although basic, intermediate, acid and alkaline rocks are all developed in the belt, the major part is composed of Yenshanian intermediate rocks (e.g. diorite, granodiorite), accounting for 2/3 of the total area of the whole belt, and distributed along the Yarlung Zangbo River and its northern bank. The small Himalayan acid granites are mainly distributed in the middle-east segments of the belt, such as Yangbajain, Nangxian and Nyingchi granite bodies. The Lhagoi Kangri and Himalayan belts are mostly composed of acid rocks of single intrusion.

The Gandise belt shows a tendency varying in chemical composition from intermediate to acid as one goes from south to north. The calculated average chemical composition for granitoids of the Gandise belt, on the basis of the chemical analyses of 213 rock samples by area weight method shows that this petrographical belt is rich in Ti, Fe, Mg, Ca, whereas the Lhagoi Kangri and the Himalayan belts are poor in these elements.

From the projected points in the diagram (Fig. 2), we can see that the chemical composition of the Gandise belt seems to be characteristic of calc-alkaline magmatic series. In addition, the curve plotted on the basis of REE distribution patterns is a smooth one, indicating indistinct Eu anomaly. Moreover, trace elements, such as V, Cr, Ni, have higher Clark values as compared with those of acid rocks in the crust. The initial $^{87}Sr/^{86}Sr$ ratio is found to be relatively low. The other two belts show an opposite tendency. All these facts strongly suggest that the source of the granitoids of the Gandise belt can be traced to the upper mantle while the other two belts display the characteristics of crustal partial melting.

## EFFECT OF THERMODYNAMIC PROCESS

The Gandise belt and the Lhagoi Kangri belt were thermodynamically affected after their formation, thus leading to the distorsion, fragmentation and recrystallization of rock-forming minerals in the granitoids. As a result, gneissose structures are commonly observed ( 7, Photos. 1, 3, 4, 5).

## INTRUSION AGES OF THE THREE PETROGRAPHICAL BELTS AND TIME OF ROCK-FORMING MINERAL RECRYSTALLIZATION

1. The Gandise belt. The major portion of the Gandise belt, mainly composed of diorite, tonalite, granodiorite, porphyritic biotite, monzonitic

granite, etc., give an age ranging from 120 to 70 m.y., belonging to late Yenshanian, equivalent to Cretaceous. Subordinately, medium-fine grained biotite granite and two-mica granite in the Gandise belt yield an age in the range of 50 to 10 m.y., belonging to Himalayan, corresponding to the time interval from Eocene to Miocene.

2. The Lhagoi Kangri belt. Two-mica monzonitic granite from the Lhagoi Kangri belt has an age ranging from 50 to 30 m.y., and the two-mica K-feldspar granite is as old as 13 m.y.

3. The Himalayan belt. Tourmaline-muscovite and two-mica K-feldspar granites from the belt range from 20 to 10 m.y. in age, belonging to late Himalayan, corresponding to Miocene. The time of biotite recrystallization for the Lhagoi Kangri and the Himalayan belts stands between 30 and 20 m.y.

## THE CHARACTERISTICS OF CRUSTAL MOVEMENT AND THE UPLIFTING HISTORY OF THE XIZANG PLATEAU

During the Mesozoic era before the plate subduction, the southern Xizang being the Tethyan Sea was still a part of the Indian Plate until the plate subduction and collision took place at the late Mesozoic time (Fig. 3).

Subduction Stage : Due to the India Ocean sea-bottom spreading, the Indian Plate was forced to drift northwardly. At the beginning, the front slide of the oceanic crust — the northern part of the Tethyan Sea, underthrust towards the Eurasian Plate, and remelted at deeper levels, thus giving rise to the major part of the Gandise rock belt, geochronologically corresponding to Cretaceous (120-70 m.y.). It is postulated that the plate subduction began at early Cretaceous and continued for about 50 m.y. During this long period of time, the predominant horizontal movement was accompanied by the vertical movement, which subsequently led to the occurrence of lower-Cretaceous limestone formation and upper-Cretaceous flysch formations.

Collision Stage(6) : The continuous spread of the Indian Ocean resulted in the northward drift of the Indian Subcontinent underlying the Eurasian Plate. At 50 m.y., the two continents began to come closer. Because of their similar specific gravity, the Indian Plate could no longer underthrust the Eurasian Plate. They started collide against each other. The plate collision and compression led to crust thickening. Due to the increasing uplifting rate of the crust, a series of thrust faults was resulted . The Main Central Thrust and the Main Boundary Thrusts at the southern slope of the Himalayas are the major ones resulted from the plate collision and compression. At the same time, the Lhagoi Kangri belt and the Himalayan belt came into being. Recrystallization phenomena show that the plate collision and compression have been lasting even up to the present.

## References

(1) Fu Cheng-yi, Continental drift, sea-floor spreading and plate tectonics. Science Press (1972) (in Chinese).
(2) Yin Tsan-hsun, Plate tectonics: A review. Scientia Geologica Sinica, (1), 60 (1973) (in Chinese).

(3) Chang Chengfa et al., Tectonic features of the Mount Jolmo Lungma Region in Southern Tibet, China. Scientia Geologica Sinica, (1), 7 (1973) (In Chinese).
(4) Li Chunyu, Tectonic evolutions of some mountain ranges in China, as tentatively interpreted on the concept of plate tectonics. Acta Geophysica Sinica, 18(1), 66 (1975) (in Chinese).
(5) Huang Jiqing et al., An outline of the tectonic characteristics of China. Acta Geologica Sinica, (2), 125 (1977) (in Chinese).
(6) Molnar, P., Tapponniner, P., The collision between India and Eurasia. Scientific American, 236(4), 30-41 (1977).
(7) Zhang Yu-quan et al., Recrystallized biotites in granitoid rocks in southern Xizang and their bearing on plate collision. In this Proceedings of Symposium on Qinghai-Xizang (Tibet) Plateau, Academia Sinica (1981) (in Chinese).

Table 1

Intrusive stages of some plutons

| Rock Belts | Names of plutons | Late Yenshanian | | | | | | Himalayan | |
|---|---|---|---|---|---|---|---|---|---|
| | | I | II | III | IV | V | VI | Early | Late |
| Gandise | Rutog | | Tonalite | Granodiorite | Porphyritic biotite granite | Biotite granite | Alkaline granite | | |
| | Quxu | Belugite | Diorite | Granodiorite | | " | | Monzonitic granite | |
| | Gyabge | Norite | Hypersthene diorite | " | Porphyritic two-mica granite | " | | | |
| | Azhag Kanglha | | Diorite | " | " | " | Alkaline granite | | |
| | Demola | | " | " | Porphyritic biotite granite | " | | | |
| Lhagoi Kangri | Dag La | | | | | | | Two-mica monzonitic granite | |
| | Marla Mt. | | | | | | | " | |
| | Naimonanyi | | | | | | | " | |
| | Gyacola | | | | | | | | Two-mica granite |
| Himalayas | Cona | | | | | | | | " |
| | Lhozhag | | | | | | | | Tourmaline--muscovite granite |
| | Gabug | | | | | | | | " |
| | Todam | | | | | | | | " |

Table 2 Granitoid rocks in southern Xizang and their recrystallized biotite isotopic ages

| Rock belt | Major rock type | Sample num. | Mineral | Age range (m.y.) |
|---|---|---|---|---|
| Gandise | diorite, granodiorite | | biotite | |
| | porphyritic biotite granite | 35 | zircon | 120 - 70 |
| | " " | 6 | biotite* | 30 - 20 |
| Lhagoi Kongri | gneissose two-mica | 10 | biotite | 50 - 13 |
| | monzonitic granite | | zircon | |
| | " " | 4 | biotite* | 30 |
| Himalayas | tourmaline-muscovite | | biotite | |
| | granite | 10 | | 20 - 6 |
| | two-mica granite | | muscovite | |

* recrystallized biotite

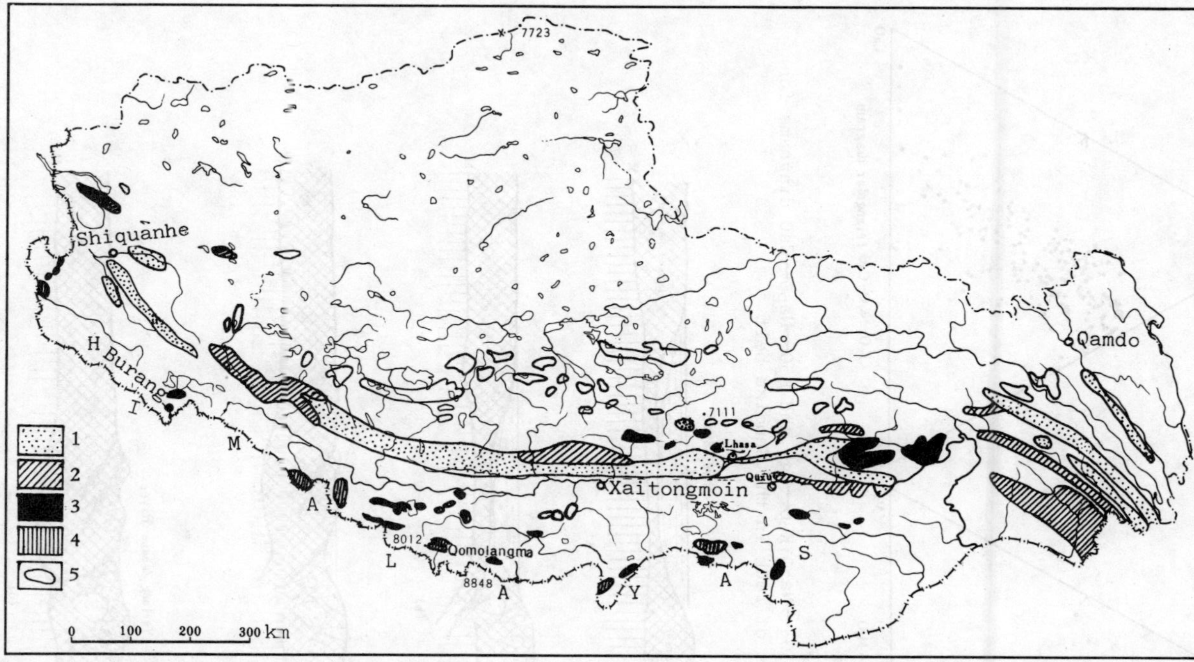


Fig. 1 The schematic map showing the distribution of intermediate-acid magmatic rocks in the southern part of Xizang

1. Gandise belt: Diorite granodiorite and porphyritic biotite granite;
2. Gandise belt: Biotite granite and two-mica granite;
3. Lhagoi Kangri belt: Gneissose two-mica granite;
4. Himalayan belt: Tourmaline-muscovite granite and two-mica granite;
5. Age-undivided granite.

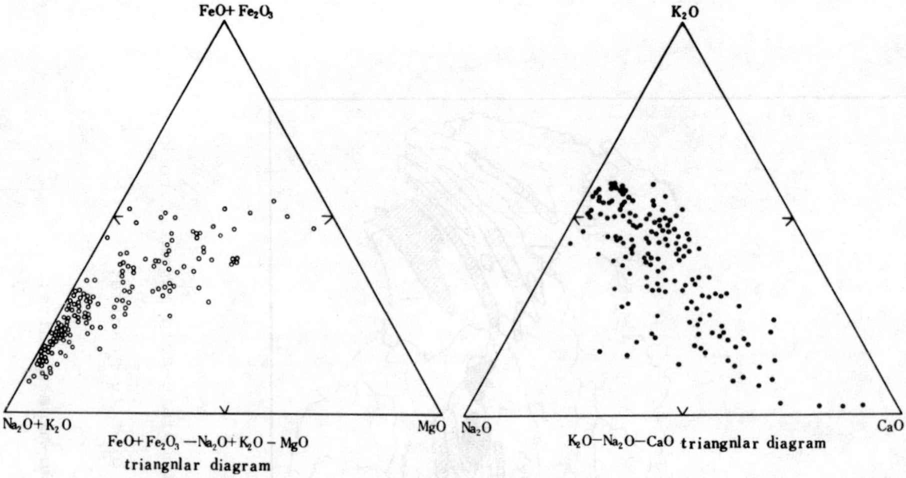


Fig. 2 $FeO+Fe_2O_3$—$Na_2O+K_2O$—$MgO$ and $K_2O$—$Na_2O$—$CaO$ diagrams for various main types of rock in the Gandise belt

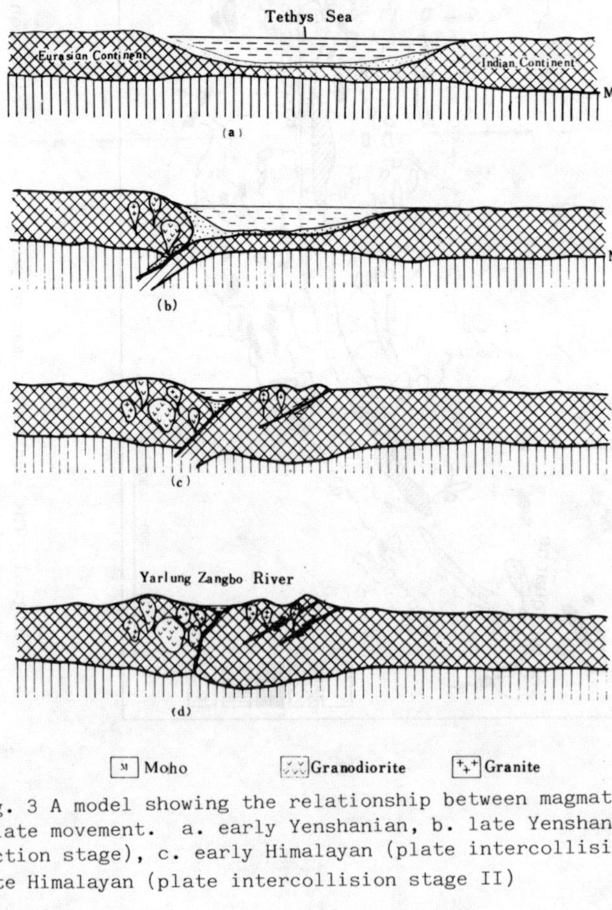


Fig. 3 A model showing the relationship between magmatic activities and plate movement. a. early Yenshanian, b. late Yenshanian (plate subduction stage), c. early Himalayan (plate intercollision stage I), d. late Himalayan (plate intercollision stage II)

# ON THE EVOLUTION OF YARLUNG ZANGBO RIVER ABYSSAL FAULT AS VIEWED FROM THE SPACE-TIME DISTRIBUTION PATTERN OF THE GANDISE MAGMATIC COMPLEX BELT

Chen Fu-min
(Institute of Geochemistry, Academia Sinica)

It has been widely accepted by world geological circles that the Yarlung Zangbo River abyssal fault is a boundary between the Indian and Eurasian Plates. The continuous northward push of the Indian Plate since Mesozoic and the consequent subduction underneath the Eurasian Plate and collision with it resulted in the formation of the Gandise intermediate-acid magmatic complex belt and the Yarlung Zangbo River ophiolite belt. In this paper the author attempts to explain the genesis and development of the Yarlung Zangbo River abyssal fault and the direction of the relative motion between the two plates in terms of the space-time distribution pattern of the Gandise magmatic complex belt and its primary and secondary structures.

## THE GEOLOGICAL SETTING OF THE ABYSSAL FAULT AND ITS FEATURES

The Yarlung Zangbo River abyssal fault, a giant fault zone with the total length of more than 2,000 km, lies approximately along the Gar River and the southern bank of the Yarlung Zangbo River, striking EW from the east of Saga and NW from the west. The belt is considered as a boundary, i.e.the suture line between the Indian and Eurasian Plates. The suture line is westwardly connected with the Indus River suture(1) and eastwardly with the suture line in Burma.

The Yarlung Zangbo River abyssal fault is characteristically marked by the occurrence of ophiolite belts which are composed of basic, ultrabasic, intermediate-basic volcanic rocks and radiolarian cherts (Fig.1). Most of basic and ultrabasic rocks occur as lenses and lentioles. As for these rock bodies, there may be two origins. One is the tectonic emplacement, namely the pre-existing rocks were pushed upward by compressive forces to the upper part of the crust through the abyssal fault. The rocks formed by tectonic emplacement are characterized by strong fragmentation and remarkable foliation and brecciation. The other is "achors" intrusion(2), which is characterized by noticeable alteration of country rocks and dendritic occurrence of rock bodies in the host rocks. Facies belts can even be distinguished in some larger rock bodies emplaced by this mechanism. The majority of these rock bodies extend and dip southwards. On the northern side of the abyssal fault are flysch deposits and the Gandise magmatic complex belt while on the south are Triassic lower-grade metamorphic rocks. On both flanks of the abyssal fault pair metamorphic belts are developed ; one is the high-temperature low-pressure metamorphic belt on the northern flank, and the other is the low-temperature high-pressure metamorphic belt on the southern flank.

The Yarlung Zangbo River abyssal fault shows different dipping directions in the different periods of geological history. During the Mesozoic, the abyssal fault dipped to the north when the Indian Plate underthrust beneath the Eurasian Plate(3). In the Cainozoic after the collision of the Indian continental slab with the Eurasian, it became dipping to the south.

## THE CHARACTERISTICS OF THE SPACE-TIME DISTRIBUTION PATTERN OF THE GANDISE MAGMATIC COMPLEX BELT

This belt is located in the northern side of the Yarlung Zangbo River abyssal fault (Fig.1), stretching approximately along the Gandise, running parallel to the abyssal fault. It is about 1,500 km in length from Shiquanhe in the west to Nyingchi in the east. The exposure of the belt is relatively continuous in the east to Saga while intermittent in the west. The belt is composed of diorite, granodiorite and granite. The intrusion relations between the rock bodies of different ages are distinct. And the intrusion sequence is such that diorite was emplaced first, granodiorite next and granite last. Therefore, from the intrusion sequence, it can be seen that the rocks show a tendency to vary in their chemical composition from intermediate through intermediate-acid to acid.

The space-time distribution pattern of the belt displays an obvious regular variation. Exposed on the southern side of the belt are a number of small diorite plutons, and northward, widely exposed are granodiorites, both of which are the major members of the Gandise belt. More northward granites(4) are predominantly exposed. For example, gabbro, diorite, granodiorite and biotite granite are found exposed in Quxu-Lhasa section from south to north (Fig. 2). $SiO_2$ and $K_2O$ also increase progressively from south to north, indicating that the granitoids gradually become more acid and younger in age from south to north. It is estimated that granodiorite was emplaced during the late Yenshanian period (70-120 m.y.), whereas granites in the north during the Himalayan period (20-70 m.y.). This consideration indicates that the Gandise magmatic complex belt was resulted from the subduction of the northern slide of oceanic crust of the Indian Plate underneath the Eurasian Plate along the Yarlung Zangbo River abyssal fault, which gave rise to the anatexis and upward intrusion of anatectic magmas, followed by remelting of continental crusts of different thickness and mixing of the materials.

Similarly, the belt has a regular space-time variation from west to east. The western part of the belt (i.e. the rock bodies around Shiquanhe) (Fig. 1) is mainly composed of diorite and granodiorite (in which diorite is predominant). Two diorite bodies gave an age of 160 m.y., equivalent to early Yenshanian.

The middle part of the belt (i.e. the rock bodies from Saga to Quxu) mainly consists of granodiorite, ranging in age from 70 to 120 m.y.(5), corresponding to late Yenshanian.

The eastern part of the belt (i.e. all the rock bodies from the east of Quxu to Nyingchi) is chiefly made up of granite, its isotopic ages ranging from 20 to 70 m.y.(6), equivalent to Himalayan.

The above mentioned facts show that from west to east the belt has a regular variation in petrology from intermediate through intermediate-acid to acid, and as to their formation ages from early Yenshanian through late Yenshanian to Himalayan.

## THE GENESIS AND DEVELOPMENT OF THE YARLUNG ZANGBO RIVER ABYSSAL FAULT AND ITS RELATIONSHIP WITH THE INDIAN PLATE

The regular variation of the Gandise magmatic complex belt in time and space shows that the Yarlung Zangbo River abyssal fault initiated, first of all, in the west and was gradually developing and expanding to the east. That is to say, the northwestern part of the Indian Plate began to underthrust towards the Eurasian Plate in the early Yenshanian period (Fig. 3)(7), later on, it underthrust on a larger scale eastward during the late Yenshanian and Himalayan periods. This process controlled the formation and variation of the Gandise magmatic complex belt.

As the Indian Plate continuously underthrust northward and the two plates gradually came closer to each other, the northwestern part of the Indian Plate first collided with the Eurasian Plate (Fig. 3). Due to the counteraction of the Eurasian Plate, the northeastern part of the Indian Plate was forced to turn northwards, thus leading to the complete collision between the two plates. Subsequently, the Yarlung Zangbo River abyssal fault turned to incline drastically to the south (Fig. 4), resulting in another abyssal fault inclining to the north at the foot of southern Himalayas. Under the strong compression stress produced by the northward movement of the Indian Plate, the Himalayas between these two abyssal faults promptly elevated, thus bringing forth the highest peak in the world (Fig. 4).

## THE DIRECTION OF RELATIVE MOTION BETWEEN THE TWO PLATES

Studies on the primary and secondary structures of the middle belt enable us to re-establish the stress field of regional structure during and after the belt formation.

The orientational arrangement of feldspars, hornblendes, mica and other minerals, as well as segregates and inclusions is NNW in strike (Fig. 5) on the southern side of the belt, close to the abyssal fault. However, on the north, far from the abyssal fault, the orientational arrangement is close to EW in strike. As for the relevant primary joints (Fig. 6), on the southern side of the belt, the torsion joints are directed to NEE and NNW, while the tension joints are close to NW; northwards, the torsion joints are NNE and NNW in direction while the tension joints close to SN. From the stress analyses of primary structures (Fig. 6), it can be deduced that the belt was formed under the compression in the direction from south to north, and at the same time the two plates were dextrorotated.

## CONCLUSION

It is concluded from the studies on regular variation of the Gandise magmatic complex belt in space and time both from south to north and from west to east, and on primary structures in the belt that: (1) the Gandise magmatic complex belt is the product of anatexis caused by the subduction of the Indian Plate underneath the Eurasian Plate; (2) during the northward shift of the Indian Plate, its northwestern part first underthrust towards the Eurasian Plate, followed by a overall subduction from west to east. The Yarlung Zangbo River abyssal fault initiated from the west and gradually developed and expanded to the east; (3) the northwestern slide of the Indian Plate first collided with the Eurasian Plate and this forced the northeastern part of the Indian Plate to turn northwards, thus leading to the complete collision with the Eurasian Plate ; (4) during the subduction, the faulting planes dipped to the north at great angle, and after the complete collision between these two plates faulting planes became southward dipping at great angle; and (5) during the subduction of the Indian Plate underneath the Eurasian Plate, horizontal compression from south to north predominated and at the same time the two plates were dextrorotated.

## References


(1) The Chinese Academy of Geological Sciences, Geological map of Asia (1:5,000,000), Geology Press (1975) (in Chinese).

(2) Mei Houjun, Lin Xuenong, The ophiolite system in Xizang (Tibet). The Series of the Scientific Expedition to Qinghai-Xizang Plateau, Magmatic and metamorphic rocks in Xizang (Tibet), Science Press (in press) ( in Chinese).

(3) Chang Chenfa, Zheng Xilan, Tectonic features of the Mount Jolmo Lungma Region in Southern Tibet, China. Scientia Geologica Sinica, (1), 1-11 (1973) (in Chinese).

(4) Jin Chengwei, Zhou Yunsheng, Igneous rock belts in the Himalayas and the Gandise arc and their genetic model. Scientia Geologica Sinica, (4), 297-311 (1978) (in Chinese).

(5) Zhang Yuquan, Lin Xuelong, Chen Fumin, Xu Yonghua, A report on isotopic geological ages from Southern Tibet, China. Procedings of Symposium on Isotopic Geology, China, Fasc. 2, (1979) (in Chinese).

(6) Institute of Geology, Academia Sinica, K-Ar dating and division of the Himalayan movement in Southern Xizang. Scientia Geologica Sinica, (1), 13-20 (1979) (in Chinese).

(7) Molnar, P., Tapponniner, P., The collision between India and Eurasia. Scientific American, 236(4), 30-41 (1977).

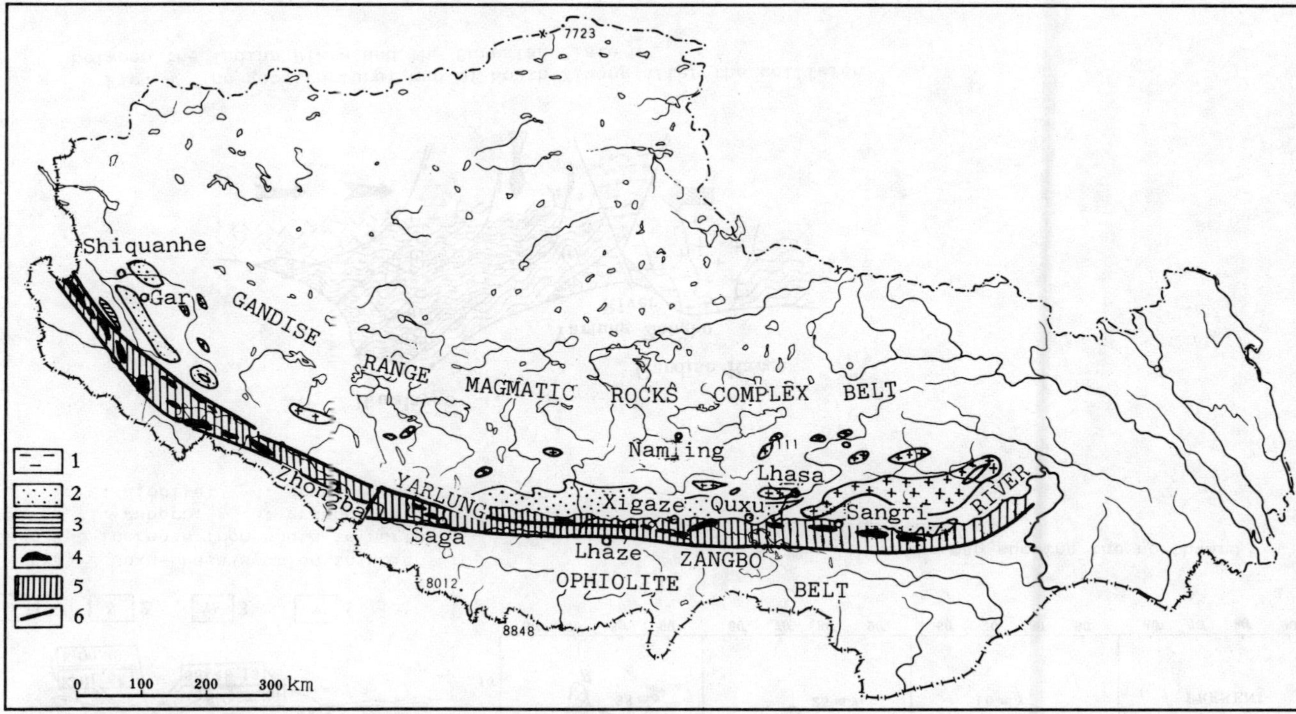


Fig. 1 The schematic map showing the locations of the Gandise intermediate-acid magmatic complex belt and the Yarlung Zangbo River ophiolite belt

1. granite; 4. basic and ultrabasic rocks;
2. granodiorite; 5. ophiolite belt;
3. diorite; 6. fault

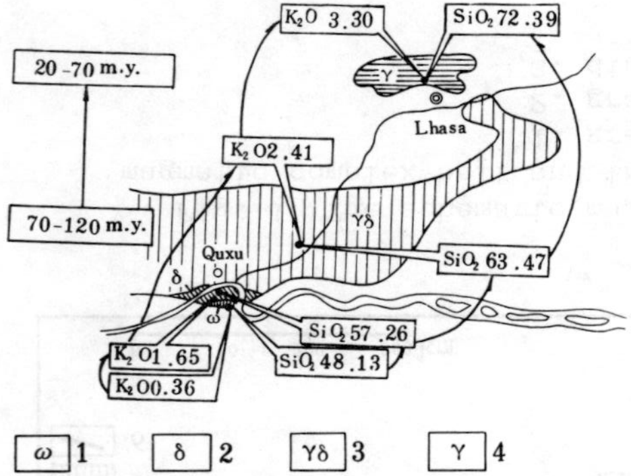


Fig. 2 Quxu-Lhasa section in which $SiO_2$ and $K_2O$ increase from south to north
1. gabbro; 3. granodiorite;
2. diorite; 4. granite

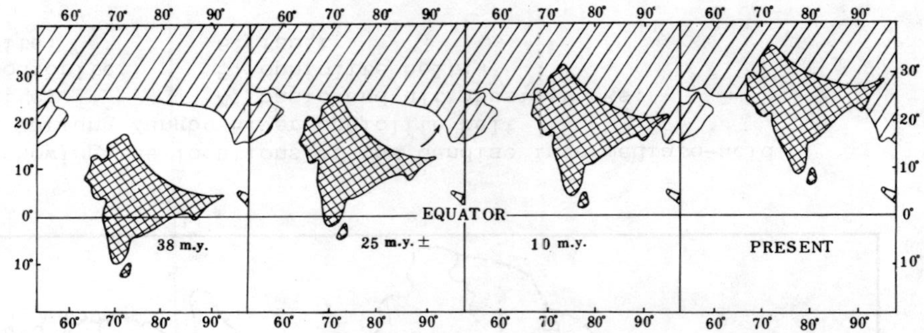


Fig. 3 The schematic map showing the northward shift of the Indian Plate

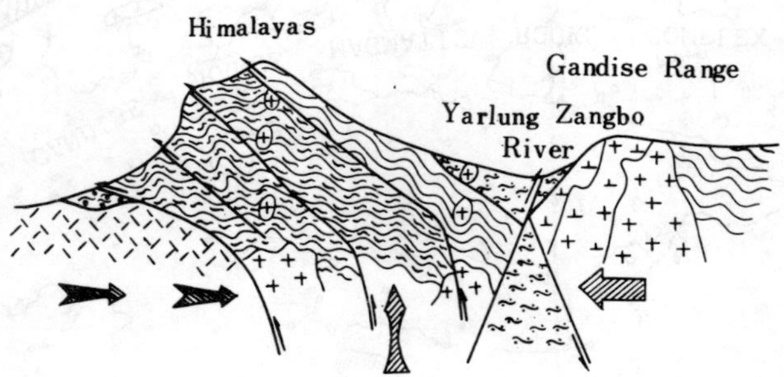


Fig. 4 The geostructural map of South Xizang after the collision between the Indian Plate and the Eurasian Plate

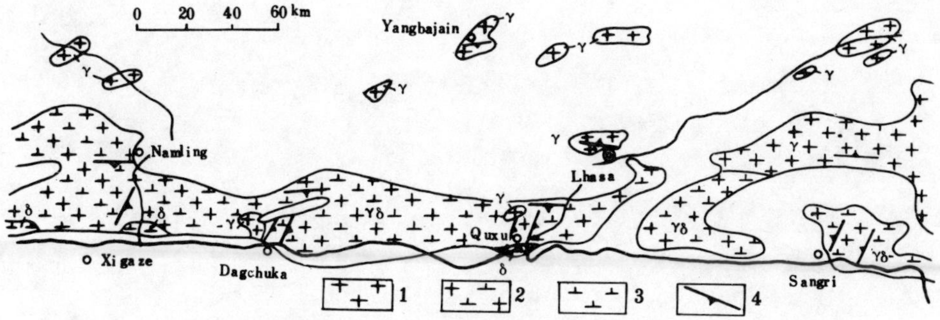


Fig. 5 The distribution of primary structures in the middle part of the Gandise magmatic complex belt

1. granite ; 3. diorite;
2. granodiorite; 4. primary orientational structures

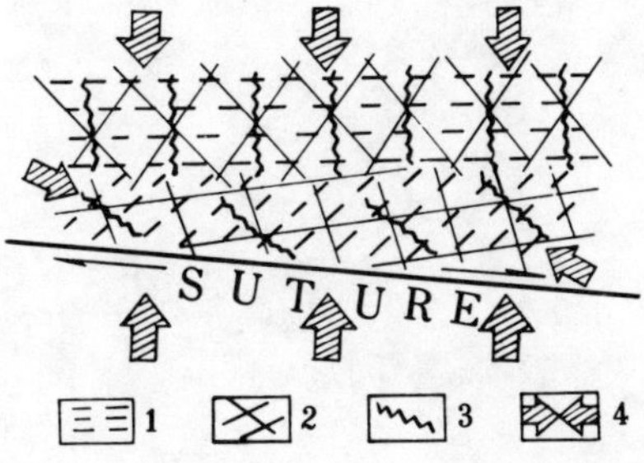


Fig. 6 The relationship between the primary structures in the Gandise complex belt and regional stress

1. primary orientational structures;
2. torsion joints;
3. tension joints;
4. stress direction

# GRANITOID BELTS WEST AND SOUTH OF TIBET
# ABOUT THEIR GEOCHEMICAL TRENDS AND Rb-Sr ISOTOPIC STUDIES

François DEBON, Patrick Le FORT and Jacques SONET
(Centre de Recherches Pétrographiques et Géochimiques,
case officielle N°1, 54500 Vandoeuvre-lès-Nancy, France)

A great number and variety of granitoid belts or alignments is known West and South of Tibet. Figure 1 schematically shows ten of them, numbered from I to X and referred here after accordingly to their number :

I - the Western Hindu-Kush - Western Badakhshan belt forms an arcuate ridge for some 500 km before it enters the Pamir . It intrudes into Precambrian to Lower Paleozoic or even sometimes Carboniferous and Triassic series (1),(2).

II - the Helmand belt of the Central Mountains of Afghanistan extends for 350 km. Most of its plutons intrude into Precambrian but some are intrusive in Paleozoic series and one of them, the Behsud granodiorite, in Lower Cretaceous (3).

III - the Arghandab belt follows for 450 km the South-Eastern border of the Central Mountains of Afghanistan. It intrudes Precambrian formations and series of Paleozoic to Lower Cretaceous age for one of them (4).

IV - the Eastern Badakhshan ensemble of plutons outcrops in Afghanistan on some 150 km SSW-NNE. It intrudes into Precambrian formations and sometimes into sedimentaries up to Upper Jurassic (1),(5).

V - the Nuristan ensemble occurs as a wide band extending for 350 km more or less parallel to the Afghanistan - Pakistan border. In this poorly known region, some plutons intrude metamorphic Precambrian formations, others intrude sedimentary series attributed to Paleozoic and Triassic (1),(6).

VI - the Karakorum belt extends East-West for some 800 km. The central batholith intrudes sedimentary and metamorphic formations of Paleozoic to possible Mesozoic age (7),(8).

VII - the Transhimalayan belt, one of the largest in the world, follows the Northern edge of the Himalaya for some 2000 km. It corresponds to the Kangdese belt of the Chinese geologists. It can be followed towards the South-West crossing Northern Pakistan (the Kohistan belt) and ending in Afghanistan in Eastern Nuristan (the Konar valley). Altogether the length approximates 2600 km. It intrudes Upper Cretaceous limestones in Northern Kashmir (9) whilst according to (10) it intrudes Lower Cretaceous and is covered by Upper Cretaceous in South central Tibet. It is definitely older than the Tertiary Kailas conglomerate (11), (12).

VIII - the North Himalaya alignment is made of some fifteen granitic domes South of the Indus-Tsangpo suture, clearly shown for the first time by (13) on his structural map of Himalaya and South-Tibet. It can be followed for some 900 km more or less in the central portion of the Himalaya. It seems to intrude the Tibetan sedimentary series at least up to the Upper Paleozoic (10).

IX - the Higher Himalayan belt is made up of some ten plutons, extends for 1900 km and covers an area around 8000 km$^2$ (14),(15),(16). It intrudes the Paleozoic and Mesozoic (up to Cretaceous) Tibetan sedimentary series of the Higher Himalaya.

X - the Lesser Himalayan belt, made up of some fifteen different plutons, more or less follows for 1600 km the Southern edge of the Himalaya (17). These plutons have intrusive contacts with unfossiliferous metamorphic (mainly epizonal) formations of debated age : Upper Precambrian to Lower Paleozoic according to most authors. They are often gneissified particularly towards their border. Their occurrence only in the MCT thrusted limb is characteristic.

GEOCHEMICAL TRENDS (fig.2 and 3)

From the 900 geochemical samples collected by us in the ten belts and ensembles, from the work we have already published (14 to 21) and from the results which are not yet published (Debon and Le Fort, in preparation) we may give the main geochemical characteristics of the granitoid belts listed above. All analysis have been performed on the "quantometer" of the Centre de Recherches Pétrographiques et Géochimiques in Nancy (K. Govindaraju, analyst); thus, the set of data is very homogeneous and really comparable.

I - Western Hindu Kush and Badakhshan granitoids globally correspond to a typical cafemic and calc-alkaline association. This association is very diversified from gabbro-diorites to granites with a predominance of granodiorites. The plutons taken one by one are also generally cafemic and calc-alkaline excepted for a number of light-coloured plutons (Chaïdan, Salang pass, Warsaj) presenting an alumino-cafemic trend and a distinct sub-alkaline tendancy.

II - two different types are to be distinguished in the Helmand belt. The first one (IIa) is made up of biotite or two mica granites s.s., generally porphyritic, corresponding to an aluminous association. The second one (IIb), heterogeneous, mainly corresponds to a cafemic granitoid association with predominating light-coloured members and an intermediate nature between the sub-alkaline and calc-alkaline types (plutons of Nili, Sharestan, ENE Gezab, Behsud).

III - the Arghandab belt also is heterogeneous. Most of it corresponds to a cafemic and calc-alkaline batholith surrounded by a swarm of small plutons having also a cafemic character but with a sub-alkaline and even sometimes peralkaline or saturated alkaline tendancy type (Zarkachan pluton for example). North of the quaternary Dacht-e Nawar volcanoes, the belt is made up of separated plutons, some cafemic and calc-alkaline (as Alishah), the others aluminous with biotite and/or two-mica granitoids.

IV - in the Eastern Badakhshan region, the plutonism mainly occurs as alumino-cafemic or cafemic light-coloured associations. Their type may be intermediate between calc-alkaline and sub-alkaline or readily sub-alkaline. Two-mica aluminous associations also exist but they have a very subordinate volume abundance.

V - the granitoids of Southern Nuristan may be divided into three groups. The first one (Va) largely predominating is made up of light-coloured granitoids that form generally biotite or/and two-mica aluminous associations, or, sometimes, alumino-cafemic with a sub-alkaline to calc-alkaline nature. The second group (Vb) is exclusively made up of dark-coloured typically cafemic granitoids with an intermediate type between sub-alkaline and calc-alkaline. Finally the third group (Vc) has a very small volume and is represented by a nepheline syenite. At first sight, these three groups are genetically independent.

VI - the Karakorum axial batholith has been described as an association of biotite ± hornblende granites and granodiorites (7). The analysis of a few samples collected by one of us during a short excursion, shows that the axial batholith is made up of a granodioritic series typically cafemic and calc-alkaline. The occurrence of syenitic rocks reported by (7) as the presence of two-mica granites and leucogranites is noteworthy and altogether with the calc-alkaline batholith recalls the Arghandab association (III).

VII - the enormous Transhimalaya - Kohistan - Konar belt that follows the peri-indian suture is a very composite ensemble made up of cafemic associations of variegated nature : calc-alkaline, sub-alkaline, trondhjemitic calc-alkaline. There is a definite predominance of dark coloured members. It is to be noted that in the Afghan and Pakistan areas there is no gap in the field between the suture ophiolites and the granitoid belt whilst they are clearly separated in the Ladakh.

In Ladakh, the predominant association is neither calc-alkaline nor tholeiitic but sub-alkaline with gabbro-diorites and monzodiorites. Calc-alkaline quartz-diorites and tonalites occur but are not the dominating type of rocks as suggested by (22). Granites s.s. also occur. The calc-alkaline character is largely predominating in Kohistan as already stated by (23),(24),(25). Sub-alkaline trends may sometimes clearly appear as with the high K-diorites (24) and with the quartz-syenites and granites North of Kalam. We came also across several occurrences of trondhjemites. In Eastern Nuristan (Konar valley), the sub-alkaline or even saturated alkaline associations largely occur. In addition calc-alkaline associations, with a sub-alkaline tendancy in general, as well as a few trondhjemites also occur.

IX - as shown by our study based on the Manaslu pluton, the Higher Himalayan belt is made of two-mica (muscovite > biotite) leucogranites with a very constant composition. The granites are aluminium, feldspar and alkali rich (sodium rich in general). The main variations occur in the relative proportion of sodium and potassium with a reverse correlation for the two oxydes. The leucocratic character is extreme.

X - the Lesser Himalayan granites were described as a collection of separate plutons of hornblende granodiorite to leucogranite composition and with a variable but generally alpine age. (17) showed that they formed a single belt of two-mica cordierite bearing granites with typical microgranular inclusions. Most of the Lesser Himalayan granitoids belong to a typically aluminous association, with biotite and two mica ± cordierite (and even andalusite). This association is feldspar depleted (actually sodium depleted) but quartz rich. These characteristics relate the Lesser Himalayan granitoids to the S - type of (26).

## Rb-Sr ISOTOPIC DATING

The ten granitoid belts described above spread over a considerable span of time. Most of them have been reworked by subsequent movements or orogenies. Rb-Sr whole rock dating is therefore a method better adapted than K-Ar dating in order to obtain the age of setting and crystallization of the plutons. This is why we have essentially used the Rb-Sr method and why we limit to it the following lines. Rb-Sr whole rock analysis show a large variety of compositions and when possible a wide range of isochron ages and initial Sr ratios.

I - up to now only separated results have been published concerning Rb-Sr dating of the granitoids of Western Hindu-Kush and Western Badakhshan; they have given conventional ages on minerals and rocks as well as one internal isochron (27),(28). The results are more scattered in Badakhshan (12 to 213 m.y.) than in Western Hindu-Kush (203 to 216 m.y.). We have completed a 8 points whole rock isochron (fig.4a) giving a 234 ± 6 m.y. age (Upper most Permian) with a Sr initial ratio of 0.7044 ± 0.0002.

II - in the Helmand belt, the two types geochemically distinguished are met again by Rb-Sr isotopic analysis :

- the aluminous association (IIa) is of Lower Paleozoic age as shown by the Behsud NE porphyritic granite six points whole rock isochron at 502 ± 24 m.y. with a Sr initial ratio of 0.7093 ± 0.0018 (29);
- the mainly cafemic association (IIb) has been paid little attention up to now. We have completed a seven points whole rock isochron on the light-coloured granitoids of the Nili and Sharestan plutons giving a 82 ± 4 m.y. age with a Sr initial ratio of 0.7065 ± 0.0004 (fig.4b).

III - the Arghandab belt has been paid more attention. An old and poor three points whole rock isochron including a sample of cross cutting pegmatitic dyke had given a 153 ± 20 m.y. age with a 0.7111 ± 0.0015 Sr initial ratio for the Alishah pluton of the Northern part (30). Recently whole rock isochrons have been obtained for the main batholith and for the monzo-syenitic pluton of Zarkachan. Samples from the Southern part of the batholith have yielded a 110 ± 4 m.y. age (31). The eight point isochron gives a low Sr initial ratio around 0.705 (S. Deutsch, personal communication). The Zarkachan pluton has given a 106 ± 8 m.y. age for a six whole rock, one biotite, one K - feldspar isochron with a Sr initial ratio of 0.7056 ± 0.0002. Both isochrons point out to the "Middle Cretaceous" age of emplacement of the belt.

IV - Only (27) have published Rb-Sr determinations for a biotite of the Baharak adamellite pluton and another for the Abu Abdal granite. An Oligocene age of 32 m.y. and a Miocene age of 17 m.y. have been obtained. Our preliminary results on Abu Abdal give a provisional age around 27 m.y. with a 0.708 initial ratio.

V - There are two Rb-Sr determinations made on biotites of a biotite-hornblende cataclastic granodiorite of Northern Nuristan (27). Upper Cretaceous ages have been obtained (93 and 86 m.y.). A measure on the Tirich Mir granodiorite, a probable prolongation of the Western Nuristan granitoids in Pakistan, has given a 115 ± 4 m.y. (Lower Cretaceous) age (32). Our preliminary results on a two-mica leucogranite from Poshal valley give a provisional age around 26 m.y. with a high 0.722 Sr initial ratio.

VI - Very young ages have been obtained (24, 13, 9 m.y.) for biotites of deformed and recrystallized samples of the axial batholith of Karakorum (27).

VII - No satisfactory results have yet been obtained with Rb-Sr dating of the Transhimalayan - Kohistan belt. Difficulties seem to arise from the too homogeneous or too heterogeneous character of the granitoids. Several intrusions seem also to have successively taken place. Quite long ago (27) published a 48 m.y. age for a rock of Deosai plateau (Northern Pakistan). Preliminary results give an average age around 60 m.y. in Ladakh (Frank quoted by (22)) and around 18 m.y. for the Quxu pluton of central Tibet (33). The only clear point reached is the characteristically low Sr initial ratio around 0.703 to 0.704.

VIII - One of the pluton of the North Himalayan alignment, the pluton of Kangmar has been recently analyzed for Rb-Sr isotopes. A six points whole rock isochron has been obtained giving 485 ± 6 m.y. age with a high Sr initial ratio of 0.7186 ± 0.0018 (33). These results compare very well with those obtained for the Lesser Himalayan granites (see below, X).

IX - The Manaslu leucogranite, one of the Higher Himalayan belt, presents no isochron (34 and fig.4c). However, taking into account the recent age given by the stratigraphic observations (probably post-Eocene and pre-Pliocene), the initial Sr ratios of the different samples analyzed are very high, in the range 0.735 to 0.760.

X - The Lesser Himalayan granites give Lower Paleozoic isochron ages. The Manserah pluton of Pakistan gives a well defined seven point isochron with a Cambrian age of 516 ± 16 m.y. and a high 0.7189 ± 0.0006 Sr initial ratio (21) very similar to the 518 ± 100 m.y. and 0.718 Sr initial ratio proposed by (35) for the Mandi pluton of India that we consider to be part of the same belt. The Simchar pluton, South-West of Kathmandu (Nepal) also gives a Lower Paleozoic age of 509 ± 56 with a slightly lower Sr initial ratio of 0.7087 ± 0.0049 (Le Fort, Debon, Sonet, in preparation). It is to be noted that similar ages have been obtained independently for intrusive granites in the Lahul crystalline nappes (512 ± 16 m.y.; $Sr_0$ = 0.720 ± 2) (36).

## CONCLUSION

Tibet is surrounded by granitoid belts scattered over some 500 m.y. Three main types of geodynamic pattern seem to take place successively.

The Lower Paleozoic plutonism found from Afghanistan (IIa) to Nepal (X) seems also to have affected other parts of Gondwanaland (29). It can be compared to similar events in Australo Antarctica (37) as well as to the late pan-African events or to the Ross orogeny of Antarctica (38). Actually such events do not seem to pertain to a true orogeny but to be distributed along lineaments of continental extension. Deformation and metamorphism may be present but they seem to play a minor part. The fact that they are still more or less parallel to alpine trends speaks of their deep crustal setting that has guided the following weakness lines of the Earth.

During Upper Paleozoic, Mesozoic and maybe Paleogene most of the plutonism is of cafemic type and of calc-alkaline or sub-alkaline nature (I, IIb, III, Vb,

VI, VII). Minor amounts of a predominating alumino-cafemic and sub-alkaline plutonism (IV) as well as an aluminous plutonism (Va) also occur although both plutonisms have an unsettled Meso- or Cenozoic age. The change from a cafemic type to an aluminous type, through an alumino-cafemic type, associated with increasing Sr initial ratios may be considered to express the increasing contribution of the continental crust versus the mantle in the generation of these associations (Debon and Le Fort, in preparation). As suggested by different workers for different belts (39),(40),(18),(41),(42),(etc.) cafemic associations mainly derive from oceanic subductions of island arc or of Andean type. The presence of a strike-slip component in a subduction zone may have a direct repercussion on the nature and the diversity of the plutonism emplaced in the overriding plate. Thus such strike-slip would be responsible for the coexistence in the Arghandab belt (III) of two different and contemporaneous magmatisms : one typically calc-alkaline, the other sub-alkaline to alkaline (18).

Cenozoic plutonism seems to be mainly restricted to the Higher Himalayan belt (IX). The generation of these leucogranites during the Himalayan continental collision results in a typical aluminous character, a remarkably constant anatectic composition and high $^{87}Sr/^{86}Sr$ values (14 to 16), (34). The Rb-Sr inhomogeneity is probably linked to the intense fluid circulation (Cocherie et al., in preparation). In Nuristan the granitoids are partly similar (Va pro parte) and may have resulted from similar processes.

## References


(1) Chmyriov, V.M., Mirzad, S.H., Geological map of Afghanistan. 1/1,000.000. Depart. of Geol. and Mines, Kabul (1972).

(2) Boulin, J., Bouyx, E, Introduction à la géologie de l'Hindqu Kouch occidental. Mém. h. sér. Soc. géol. France, 8, 87-105 (1977).

(3) Bordet, P., Blaise, J., Fourcade, E., Montenat, C., Sur la présence d'une microfaune du valanginien à la base de la série volcano-sédimentaire de Kadjao, dans la région de Behsud (Hazarajat, Afghanistan central). C.R. Acad. Sci.(Paris), D, 282, 1245-1248 (1976).

(4) Lapparent, A.F. de, Colleau, A., Marie, P., Moullade, M., Données stratigraphiques sur l'âge du granite de Zarkachan, près de Mokur (Afghanistan). C.R. Acad. Sci.(Paris), D, 270, 588-591 (1970).

(5) Desio, A., Geology of central Badakhshan (North-East Afghanistan) and surrounding countries. In : Italian expeditions to the Karakorum ($K_2$) and Hindu-Kush. Scientific reports, III, Geology-Petrology, 3rd vol., 628 p., E.J. Brill, Leiden (1975).

(6) Fuchs, G., Matura, A., The geology of the Nilaw area in central Nurestan, Afghanistan. Jb. Geol., B.-A., 119 (2), 97-128 (1976).

(7) Desio, A., Martina, E., Geology of the upper Hunza valley, Karakorum (W. Pakistan). Bull. Soc. Geol. It., 91, 283-314 (1972).

(8) Desio, A., Corrélation entre les structures des chaînes du Nord-Est de l'Afghanistan et du Nord-Ouest du Pakistan. Mém. h. sér. Soc. géol. France, 8, 179-188 (1977).

(9) Wadia, D.N., The Cretaceous volcanic series of Astor-Deosai, Kashmir, and its intrusions. Rec. geol. Surv. India, 72 (2), 151-161 (1937).

(10) Academia Sinica, A scientific guidebook to South Xizang (Tibet), June 2-14, 1980. Organizing Committee Symp. Qinghai-Xizang (Tibet) Plateau, 104 p., one map 1/1.500,000 (1980).

(11) Heim, A., Gansser, A., Central Himalaya. Geological observations of the Swiss expedition 1936. Reprinted Hindustan publ. Corp., Delhi, 245 p. (1939-1975).

(12) Frank, W., Gansser, A., Trommsdorff, V., Geological observations in the Ladakh area (Himalayas). Bull. suisse Minéral. Pétrogr., 57, 89-113 (1977a).

(13) Gansser, A., The great suture zone between Himalaya and Tibet. A preliminary account. Colloque Internat. n°268, Himalaya, Editions Cent. Natl. Rech. Sci., Paris, 1976, vol. Sci. de la Terre, 181-191 (1977).

(14) Le Fort, P., Les leucogranites à tourmaline de l'Himalaya sur l'exemple du granite du Manaslu (Népal central). Bull. Soc. géol. France, (7), XV, 555-561 (1973).
(15) Le Fort, P., The anatectic himalayan leucogranites with emphasis on the Manaslu tourmaline granite. In "Recent researches in geology", vol.II, Hindustan publ. Corp., Delhi-110007, 76-90 (1975a).
(16) Le Fort, P., Le granite du Manaslu. In : Recherches géologiques dans l'Himalaya du Népal, région du Nyi-Shang , Editions Cent. Natl. Rech. Sci., Paris, ch.III, 49-65 (1975b).
(17) Le Fort, P., Debon, F., Stebbins, J., Mise en évidence d'une ceinture de granites à cordiérite et enclaves "microgrenues" en Bas-Himalaya (Népal, Indes, Pakistan). 6è réunion annuelle Sci. de la Terre, Orsay, Soc. géol. France édit., 243 (1978).
(18) Afzali, H., Debon, F., Le Fort, P., Sonet, J., Le massif monzo-syénitique de Zarkachan (Afghanistan central) : caractères, âge Rb-Sr et implications tecto-orogéniques. C.R. Acad. Sci.(Paris), D, 288, 287-290 (1979).
(19) Debon, F., Le Fort, P., Sonet, J., Des caractères géochimiques de deux provinces plutoniques d'Afghanistan : Hindou-Kouch et "Montagnes Centrales". 6è réunion annuelle Sci. de la Terre, Orsay, Soc. géol. France édit., 135 (1978).
(20) Debon, F., Le Fort, P., Sonet, J., Typologie chimique et chronologie des roches plutoniques d'Afghanistan. 8è réunion annuelle Sci. de la Terre, Marseille, Soc. géol. France édit., 122 (1980).
(21) Le Fort, P., Debon, F., Sonet, J., The "Lesser Himalayan" cordierite granite belt. Typology and age of the pluton of Manserah (Pakistan). In : Proc. of the Int. Geodynamic Conference, Peshawar, nov-dec.1979, Geol. Bull. Univ. Peshawar (Pakistan), 13, 51-61 (1980).
(22) Gansser, A., The peri-indian suture zone. Mém. Bureau Rech. Géol. et Minières, 115, 140-148 (1980).
(23) Jan, M.Q., Petrology of the obducted mafic and ultramafic metamorphites from the Southern part of the Kohistan island arc sequence. In : Proc. of the Int. Commit. Geodynamics. Spec. issue, Geol. Bull. Univ. Peshawar (Pakistan), 13, 95-107 (1980).
(24) Majid, M., Paracha, F.A., Calc-alkaline magmatism at destructive plate margin in Kohistan, Northern Pakistan. Geol. Bull. Univ. Peshawar (Pakistan), 13, 109-120 (1980).
(25) Bard, J.P., Maluski, H., Matte, Ph., Proust, F., The Kohistan sequence : crust and mantle of an obducted island arc. Proc. of the Int. Commit. Geodynamics. Spec. issue, Geol. Bull. Univ. Peshawar (Pakistan), 13, 87-94 (1980).
(26) Chappell, B.W., White, A.J.R., Two contrasting granite types. Pacific Geol., 8, 173-174 (1974).
(27) Desio, A., Tongiorgi, E., Ferrara, G., On the geological age of some granites of the Karakorum, Hindu Kush and Badakhshan (central Asia). Himalayan and alpine orogeny, XXII Int. Geol. Congress India, 479-496 (1964).
(28) Blaise, J., Bordet, P., Lang, J., Lapparent, A.F.de, Leutwein, F., Sonet, J., Mesures géochronologiques de quelques roches cristallines d'Afghanistan central. C.R. Acad. Sci.(Paris), D, 270, 2772-2775 (1970).
(29) Montenat, C., Blaise, J., Bordet, P., Debon, F., Deutsch, S., Le Fort, P., Sonet, J., Métamorphisme et plutonisme au Paléozoïque ancien en domaine gondwan sur la marge nord-ouest des Montagnes Centrales d'Afghanistan. Bull. Soc. géol. France, in press (1981).
(30) Blaise, J., Leutwein, F., Sonet, J., Sur la datation à 140-150 M.A. de la granodiorite d'Alishah et sur ses conséquences pour la géologie des montagnes à l'Ouest de Kaboul (Afghanistan central). C.R. Acad. Sci.(Paris), D, 274, 1457-1460 (1972).
(31) Carbonnel, J.P., Deutsch, S., Montenat, C., Post Paleozoic granodioritic axes in the Afghan Central mountains and peri-indian sutures. VI Europ. Colloq. on geochronol., cosmochronol. and isotope geol., Lillehammer, 3-6 sept. 1979, p.11 (1979).

(32) Desio, A., Guj, P., Pasquarè, G., Notes on the geology of Wakhan (North-East Afghanistan). Atti della Accad. Naz. dei Lincei, Roma, Mem., ser.8, 9, n°2, 37-52 (1968).
(33) Wang, J., Gui, X., Chen, Z., Xu, R., Rb-Sr isotopic studies of some intermediate-acidic plutons in Southern Xizang. Symposium on Qinghai-Xizang (Tibet) plateau, Abstract, Acad. Sinica, Pekin may 1980.
(34) Vidal, Ph., Rb-Sr systematics in granite from central Nepal (Manaslu) : Significance of the Oligocene age and high $^{87}Sr/^{86}Sr$ ratio in Himalayan orogeny : Comment. Geology, vol.6, n°4, p.196 (1978).
(35) Jäger, E., Bhandari, A.K., Bhanot, V.B., Rb-Sr age determination on biotites and whole rock samples from the Mandi and Chor granites, Himachal Pradesh, India. Eclog. geol. Helvetica, vol.64, n°3, 521-528 (1971).
(36) Frank, W., Thöni, M., Purtscheller, F., Geology and petrography of Kulu - South-Lahul area. Colloque Internat. n°268 Himalaya, Editions Cent. Natl. Rech. Sci., Paris, 1976, vol. Sci. de la Terre, 147-172 (1977b).
(37) Crawford, A.R., Campbell, K.S.W., Large scale horizontal displacement within Australo-Antarctica in the Ordovician. Nature, 241, 11-14 (1973).
(38) Elliott, D.H., Tectonics of Antarctica : a review. Amer. J. Sci., 275 A, 45-106 (1975).
(39) Le Fort, P., La chaîne himalayenne et la dérive des continents. Revue Géogr. phys. Géol. dynam., 2, vol.13, fasc.1, 5-12 (1971).
(40) Tahirkheli, R.A.Khan, Mattauer, M., Proust, F., Données nouvelles sur la suture Inde-Eurasie au Pakistan. Colloque Internat. n°268 Himalaya, Editions Cent. Natl. Rech. Sci., Paris, 1976, vol. Sci. de la Terre, 209-212 (1977).
(41) Bordet, P., Carbonnel, J.P., Desmet, A., Montenat, Ch., Vachard, D., Une marge continentale active d'âge mésozoïque dans la région de Kandahar (S. Afghanistan). 7è réunion annuelle Sci. de la Terre, Lyon, Soc. géol. France édit., p.70 (1979).
(42) Boulin, J., Bouyx, E., Les évènements éocimmériens en Afghanistan. 7è réunion annuelle Sci. de la Terre, Lyon, Soc. géol. France édit., p.78 (1979).
(43) La Roche, H.de, Leterrier, J., Grandclaude, Ph., Marchal, M., A classification of volcanic and plutonic rocks using $R_1R_2$ diagram and major elements analyses. Its relationships with current nomenclature. Chemical geol., 29, 183-210 (1980).
(44) Hine, R., Williams, I.S., Chappell, B.W., White, A.J.R., Contrasts between I- and S- type granitoids of the Kosciusko batholith. J. geol. Soc. Australia, 25, n°4, 219-234 (1978).
(45) Couturié, J.P., Le massif granitique de la Margeride (Massif central français). Ann. sci. de l'Univ. de Clermont, n°62, 319 p. (1977).
(46) Leterrier, J., Etude pétrographique et géochimique du massif granitique de Quérigut (Ariège). Sci. de la Terre, Nancy, Mém.23, 320 p. (1972).
(47) Debon, F., Les massifs granitoïdes à structure concentrique de Cauterets-Panticosa (Pyrénées occidentales) et leurs enclaves. Une étude pétrographique et géochimique. Sci. de la Terre, Nancy, Mém.33, 420 p. (1975).
(48) Pagel, M., Leterrier, J., The subalkaline potassic magmatism of the Ballons massif (Southern Vosges, France) : shoshonitic affinity. Lithos, 13, 1-10 (1980).
(49) Barrière, M., Le complexe de Ploumanac'h, Massif Armoricain. Essai sur la mise en place et l'évolution pétrologique d'une association plutonique subalcaline tardi-orogénique. Thèse ès Sciences Univ. de Brest, 291 p. (1977).
(50) Leterrier, J., Maury, R.C., Représentation graphique de la cristallisation fractionnée des séries volcaniques. Application aux séries alcalines. Geol. Rdsch., 67, n°3, 943-952 (1978).
(51) La Roche, H.de, Sur l'expression graphique des relations entre la composition chimique et la composition minéralogique quantitative des roches plutoniques. Présentation d'un diagramme destiné à l'étude chimico-minéralogique des massifs granitiques ou granodioritiques. Application aux Vosges cristallines. Sci. de la Terre, Nancy, 9, 293-337 (1964).
(52) Hamet, J., Allègre, C.J., Rb-Sr systematics in granite from central Nepal (Manaslu) : significance of the Oligocene age and high $^{87}Sr/^{86}Sr$ ratio in Himalayan orogeny. Geology, 4, 470-472 (1976).

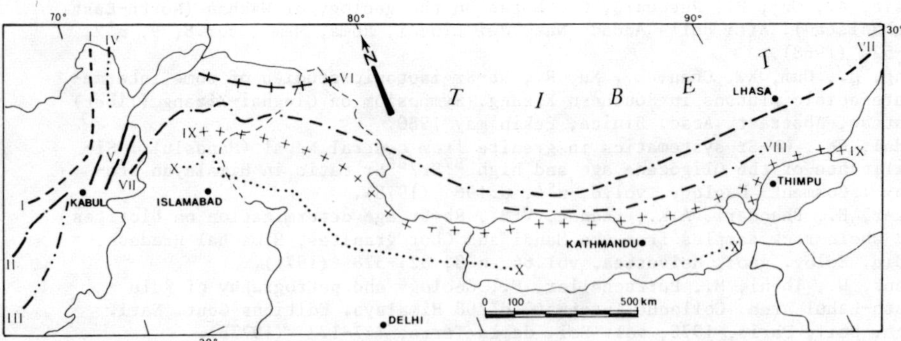


Fig. 1 - Schematic localization of ten granitoid belts and ensembles surrounding Tibet from West to South. They are numbered with roman numerals and referred with this number in the text as in figures 2 and 3.

I = West Hindu-Kush and Badakhshan (205 chemical analyses) , II = Helmand (IIa : 8 anal., IIb : 27 anal.) , III = Arghandab (51 anal.) , IV = East Badakhshan (46 anal.) ; V = Nuristan (Va : 146 anal., Vb : 15 anal, Vc : 1 anal.);VI = Karakorum (8 anal.) ; VII = Transhimalaya, Kohistan and Konar (22 + 61 + 42 = 125 anal.) ; VIII = North Himalaya ; IX = Higher Himalaya (60 anal.) , X = Lesser Himalaya (207 anal.).

Fig. 2 -"Characteristic minerals" diagram (21).
It enables to distinguish three main types of association : 1) cafemic associations mainly localized in the cafemic area, trends having a negative slope ; 2) alumino-cafemic associations mainly localized in the aluminous area, trends having a negative slope ; 3) aluminous associations entirely localized in the aluminous area, trends having generally a positive or slightly negative slope. This classification would reflect the mantellic nature and/or crustal nature of the material from which the granitoid associations derive.

Real and theoretical mineral compositions have been plotted on the diagram 2a : And : andalusite, Ap = apatite, Bi = biotite (hatched domain), Cd = cordierite, Chl = chlorite, Ct = calcite, Di = diopside, Ep = epidote, Fk = K feldspar, Foïd = leucite or/and nepheline, Gr = pyralspite, Hb = hornblende, Hy = hypersthene, Ilm = ilmenite, Mt = magnetite, Mu = muscovite, Ol= olivine, Pl = plagioclase, Q = quartz, Sill = sillimanite, Sph = sphene, Tr = tourmaline. GR : granite, AD : adamellite, GD : granodiorite, TO : tonalite, D : diorite, G : gabbro, MZ : monzonite, S : syenite = mean chemical compositions as given in (43). All parameters as millications-g in 100 g of rock or mineral.

Reference trends : ALUM for aluminous associations with (K) for the Kosciusko batholith, S-type (44) and (M) for the Margeride pluton (45) ; CALK for calc-alkaline associations (46-47) ; SALK for sub-alkaline associations with (B) for the Ballons pluton (48) and (P) for the Ploumanac'h pluton (49) ; ALKS for saturated alkaline associations (50).

The locations of the sedimentary and metamorphic rock types correspond to average ranges of usual composition (from C.R.P.G. data files).

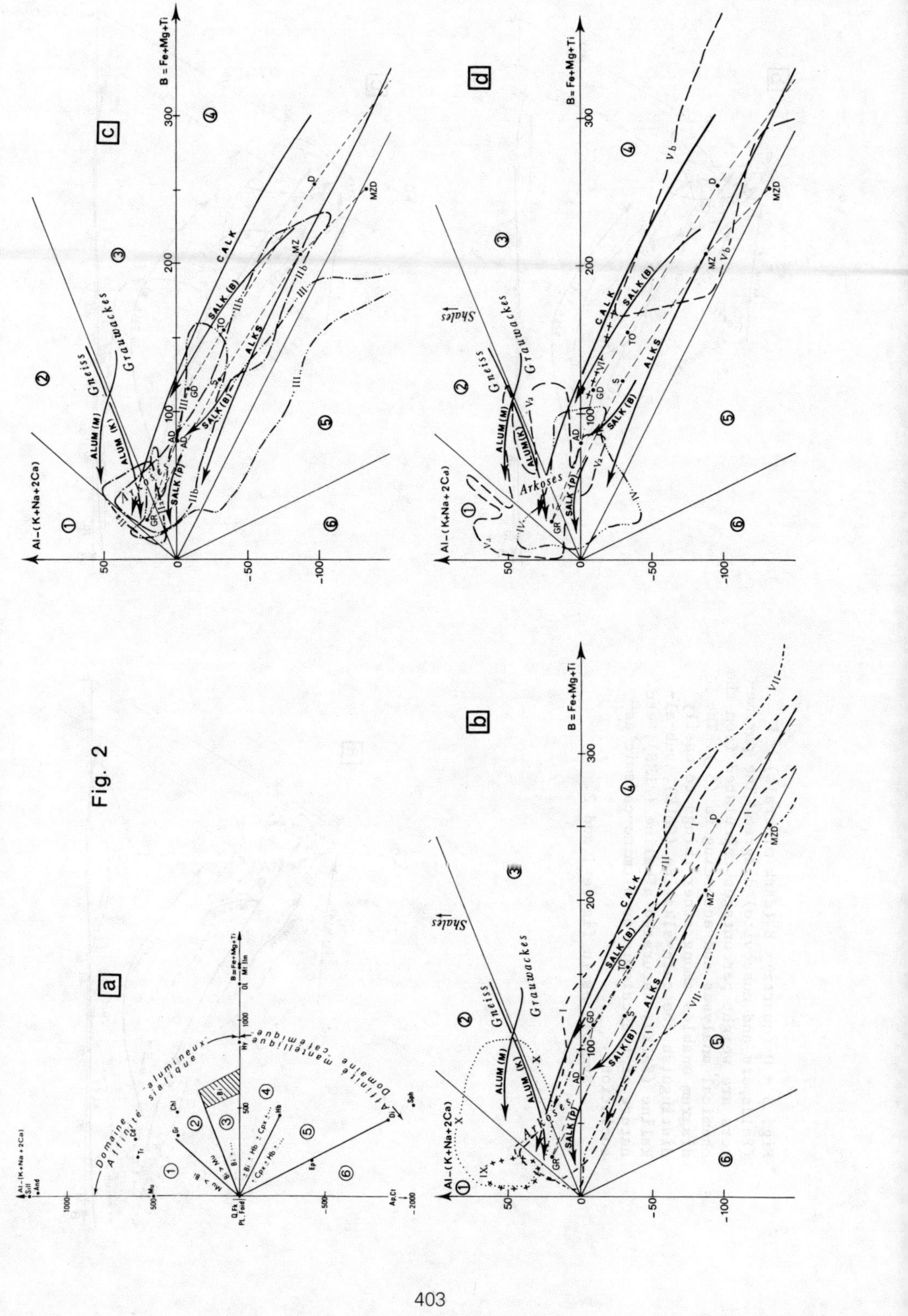
a
b
c
d
Al-(K+Na+2Ca)
B = Fe+Mg+Ti
Domaine alumineux
Affinité sialique
Domaine calcomagnésien
Affinité mantellique
Gneiss
Grauwackes
Arkoses
Shales
ALUM (M)
ALUM (K)
SALK (P)
SALK (B)
CALK
ALKS
GR
AD
GD
TO
S
MZ
D
MZD

Fig. 2

Fig. 3 - Q (quartz)- B (dark minerals) - F (feldspars and muscovite) diagram. The parameters are weight percentages calculated from the chemical analyses and according to (51). The diagram enables among other things to easily distinguish the calc-alkaline (CALK), sub-alkaline (SALK), saturated alkaline (ALKS), etc. nature of the cafemic or alumino-cafemic associations.

Symbols as in figure 1 and 2.

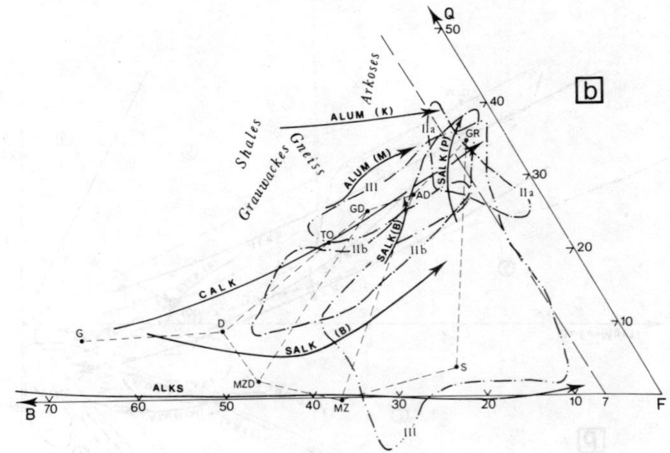


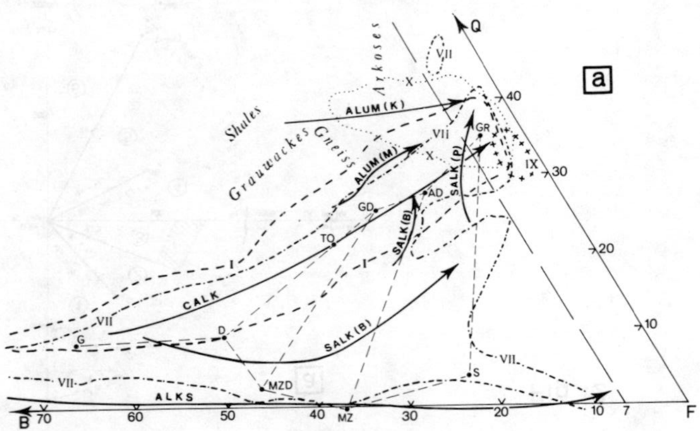


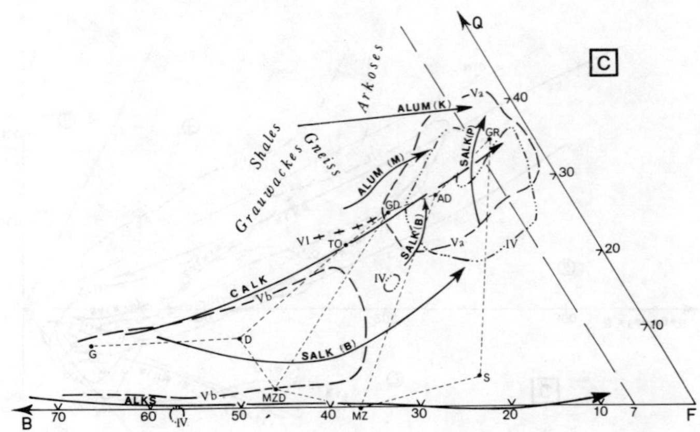

Fig. 4 - New Rb-Sr whole rock data concerning three of the granitoid belts.

4a : eight points isochron for two plutons (Chaïdan and Warsaj) distant of 250 km in the Western Hundu-Kush and Badakhshan (I) ; they are of light-coloured type. Border facies do not lie on the isochron but two previous measurements including one of a third pluton (Salang Pass) (28) and taking into account the lower value of A 21 fall on the line. Three whole rock-biotite measurements give slightly younger ages (193, 203, 212 m.y.).

4b : the light-couloured samples of the contiguous Nili and Sharestan plutons of the Helmand belt (IIb) give a rather good alignment at 82 m.y. On the contrary, dark-coloured samples prove to be isotopically inhomogeneous.

4c : Nine samples of the Manaslu leucogranite and dykes of it have been analyzed (34 and Vidal unpublished). The alleged isochron of (52) does not exist ; four samples at least being forged. Only a whole rock-muscovite measurement gives a Miocene age (34).

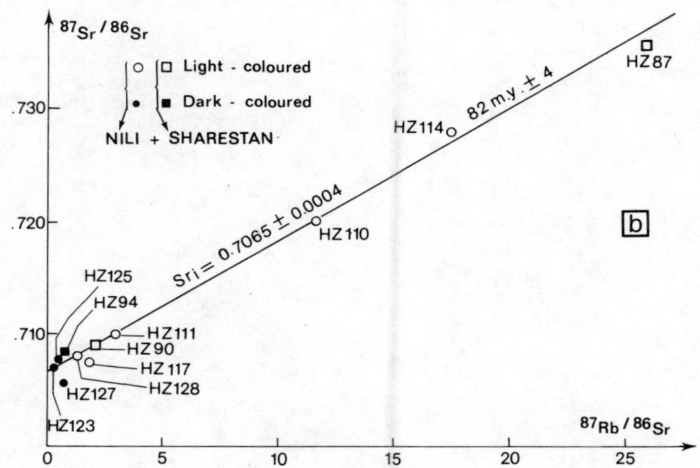


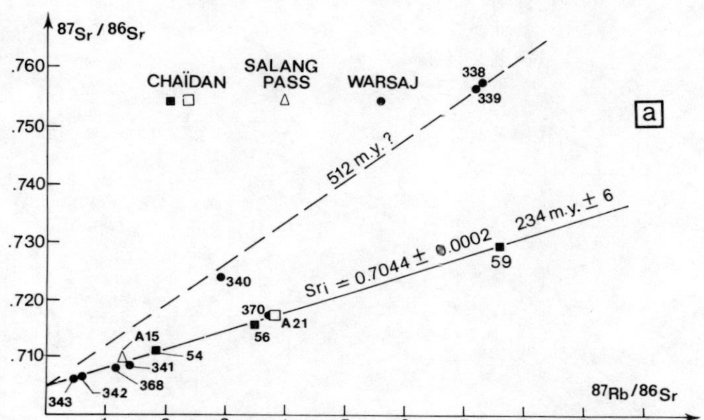


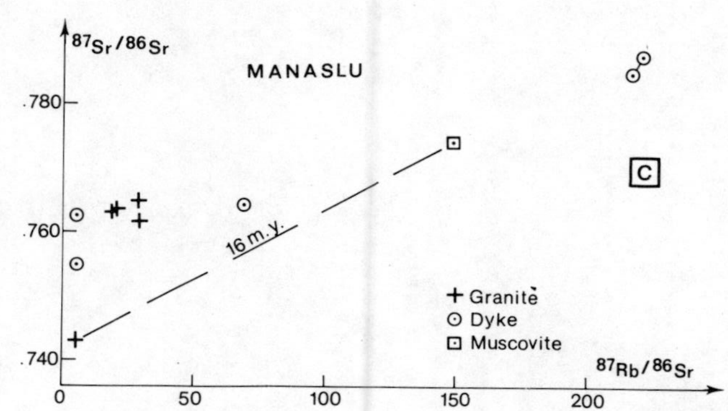

# SOME GEOCHEMICAL CHARACTERISTICS OF TRACE ELEMENTS IN GRANITOID ROCKS IN SOUTHERN XIZANG

Wang Yi-xian, Zhao Zhen-hua, Wang Zhong-gang
(Institute of Geochemistry, Academia Sinica)

This paper includes two parts: 1. Multi-component statistical analyses on trace elements, and 2. Distribution features of some trace elements.

## MULTI-COMPONENT STATISTICAL ANALYSES ON TRACE ELEMENTS

Multi-component statistical analyses were carried out for 29 elements ($SiO_2$, $TiO_2$, $Al_2O_3$, $Fe_2O_3$, FeO, MnO, MgO, CaO, $Na_2O$, $K_2O$, $P_2O_5$, F, $RE_2O_3$, $Nb_2O_5$, $Ta_2O_5$, $Li_2O$, $Rb_2O$, U, Th, Be, Sr, Ba, Cu, Pb, V, Cr, Zr, Co, and Ga) in 94 samples from granitoid rocks in Southern Xizang. The sample descriptions are shown in Table 1.

### Petrographic Characters and Ages of Instrusion

1. Discriminatory analyses. Using the method of discriminatory analysis we have classified the petrographic characters and ages of the granitoid rocks. The calculations were conducted for six different cases. The discriminant functions have been established for four types of granitoid rocks, including diorite-granodiorite, porphyritic biotite granite, biotite granite and two-mica tourmaline muscovite granite. The U-values are all very small, indicating that the classification is satisfactory.

It is worthy of note that between any two of the four types of granitoid rocks above, the $D^2_{ef}$-values are all great, indicating great differences in petrographic characters exist between them. In addition, the discriminant matrixes for classification of petrographic characters are also obtained.

By utilizing discriminatory analysis, the rocks of Late Yenshanian Period and those of Himalayan Period are clearly distinguished. In the discriminant functions for ages of intrusion the U-values are all small, indicating satisfactory discrimination. Also we obtained $D^2_{ef}$-and FHG-values and the discriminant matrixes ( not shown here ). It is clear that there exist noticeable differences in times of formation among Gandise, Lhagoi Kangri, and Himalayan belts.

2. Cluster Analyses. Cluster analyses were carried out by two different methods. 1) Correlation coefficient method: The correlation coefficients calculated by this method for various samples are all very great, showing that there are only small differences among these samples; 2) Systematic cluster analysis: The systematic cluster analyses were carried out using three groups of variables and eight different calculation approaches. Shown in Fig. 1 are the results by intermediate distance approach for all elements of interest. It can be seen from Fig. 1 that the results calculated clearly divided the samples into five groups, i.e.,

diorite-granodiorite, biotite granite — porphyritic biotite granite, tourmaline muscovite granite, biotite tourmaline granite and two-mica granite.

## Correlations between Trace Elements and Major Elements

Utilizing the stepwise regression method, the correlations between trace and major elements for all 94 samples have been clarified. The calculations were conducted for six different cases, and all regression equations have been established. It is evident that between the regression equations involving fluorine and those without fluorine taken into account no noticeable difference can be found, indicating that the variable of fluorine exerts little effect on the results.

In order to check up the results of regression analyses, a variance analysis was conducted on them. For all elements except for lithium oxide the regression analysis results all explicitly occur at the level of $\alpha = 0.01$ (namely,with 99 % confidence), whereas for lithium oxide at the level of $\alpha=0.10$ (namely, with 90 % confidence), demonstrating that the regression treatment is satisfactory.

For the sake of exploring the sequential order of the influence of independent varialbles on the dependent variables, the regression coefficients are transformed into standard ones, leading to a series of new regression equations as below:

$$Y_{RE_2O_3} = 0.00088 + 0.53TiO_2 + 0.09MnO - 0.19Na_2O + 0.68K_2O$$

$$Y_{Nb_2O_5} = 0.10 - 2.19SiO_2 - 1.14Al_2O_3 - 0.44Fe_2O_3 - 0.15MgO - 1.27CaO - 0.40K_2O - 0.19P_2O_5 + 0.13F$$

$$Y_{Rb_2O} = -0.073 + 0.31SiO_2 - 0.14TiO_2 + 0.18MgO + 0.18Na_2O + 0.45K_2O - 0.07P_2O_5 + 0.15F$$

$$Y_{Th} = 0.0011 + 0.21TiO_2 - 0.17Al_2O_3 + 0.21MnO - 0.08Na_2O + 0.48K_2O$$

$$Y_{Be} = -0.0027 + 0.34SiO_2 - 0.20TiO_2 + 0.56Al_2O_3 - 0.18Fe_2O_3 + 0.10P_2O_5 + 0.12F$$

$$Y_{Ba} = -0.018 + 0.48Al_2O_3 + 0.31Fe_2O_3 - 0.32CaO - 0.27Na_2O - 0.10F$$

Concerning the correlations between trace elements and F, Nb, Ta, Rb and Be are all positively correlated with it, whereas Ba is negatively correlated with it. This fact indicates that the migration and enrichment of Nb, Ta, Rb and Be is related with the activities of fluorine. The correlation coefficients between them are, however, all very small, implying that the role of fluorine in their migration and enrichment is not important. In addition, there exist a positive correlation between F and CaO and between F and $K_2O$ in this area. The $K_2O$ contents increase whereas CaO contents decrease in the following order: diorite - granodiorite, porphyritic biotite granite, biotite granite, two-mica granite and tourmaline muscovite granite. This trend, therefore, is not in favour of the enrichments of neither fluorine nor the trace elements correlated with it.

## DISTRIBUTION FEATURES OF SOME TRACE ELEMENTS

Shown in Table 2 are the abundances of trace elements in granitoid rocks in Southern Xizang. The abundances of Li, Rb, U, Th, Nb and Ta are close to or somewhat lower than their average abundances in the acidic rocks of the crust, whereas the abundances of V, Cr, and Ni are higher than those in crustal acidic rocks.

Shown in Fig. 2 are concentration frequency histograms for Li, Rb, REE, Nb, Ta, U, Th, and F. It shows an exclusively positive declination for all elements concerned, implying a fluctuated variation of the concentrations of all these elements and the trend to disperse.

From diorite-granodiorite thru biotite granite, porphyritic biotite granite, two-mica granite and finally to tourmaline muscovite granite, the Li, Rb, Be, and F contents gradually increase. In other words, they gradually increase in their contents from Gandise belt to Lhagoi Kangri belt, to Himalayan belt, and from Late Yenshanian to Himalayan Period.

Although Be has only a very low abundance in general, its content in tourmaline muscovite granite, which possesses a higher differentiation index, is noticeable, thus leads to the formation of beryl in several parts of the pluton. In one or two plutons chrysoberyl was also found.

The highest Sr contents were observed in diorite and granodiorite, then, the Sr content gradually decreases from porphyritic biotite granite to biotite granite and two-mica granite. In Early Himalayan two-mica granite, the Sr content increases slightly, but it decreases again in Late Himalayan tourmaline muscovite granite. The variation of Ba contents shows a pattern most similar to Sr.

In tourmaline muscovite granite the U contents increase noticeably, leading to the formation of independent uranium minerals such as uranothorite and uraninite.

The variation pattern of the contents of Th and REE is closely similar. They have their highest contents in porphyritic biotite granite of Late Yenshanian Period, and monazite and uranothorite are often observed.

As for the differentiation indices of rocks, they obviously increase from Late Yenshanian Period to Himalayan Period. With intensifying magmatic differentiation, the ratios of Mg/Li, K/Rb, and Ba/Rb all gradually lower. These ratios obviously decrease from diorite, granodiorite to biotite granite, porphyritic biotite granite, two-mica granite and to tourmaline muscovite granite (Table 2).

## SUMMARY

Among the granitoid rocks of Southern Xizang, only porphyritic biotite granite shows some characteristics of differentiation, namely its Li content and Rb, Be, F, and REE contents are higher, whereas the ratios of Mg/Li, K/Rb and Ba/Rb are relatively low. In addition, its REE distribution pattern shows a more typical pattern for fractional crystallization, while, other granitoid rocks have no evident characteristics of differentiation, namely the correlations between F and Nb, Ta, Rb, and Be are all positive with very small correlation coefficients. In general, the behaviours of Sr and Ba are quite different during the crystallization of magma, i.e., Sr was enriched in plagioclase which crystallized first, while, in contrast, Ba remained in the melt, leading to a progressively enrichment towards the late stages of the magmatic differentiation. However, no evidence enables us to say so for this area. As mentioned above, in our instances, Sr and Ba show similar distribution. Moreover, the mineralization of trace elements is not remarkable in this area. All the above mentioned facts imply that the granitoid rocks of this area were formed by partial melting of the crustal materials in different parts and to various extents, rather than cognate differentiation or cognate remelting differentiation in general.

Thus, it can be deduced that the granodiorite in Southern Xizang stemed from the lower crust whereas the granites stemed from the upper crust. Therefore, they show obscure successive relations in the distribution and evolution of trace elements.

## References


(1) Institute of Geochemistry, Academia Sinica, Geochemistry of Granitoids in South China, 225-267, 333-349, Science Press, (1979) (in Chinese).

(2) Wedepohl, K. H., Handbook of Geochemistry, II(2), 3-E-1——3-0-1, 37-E-8——37-E-9, Springer-Verlag (1970).


Table 1 Descriptions for samples

| Sample No. | Age | Rock type grouping No*. | Age grouping No. |
|---|---|---|---|
| 1 - 20 | Late Yensh. Period | 1 | 1 |
| 21 - 30 | Late Yensh. Period | 2 | 1 |
| 31 - 51 | Late Yensh. Period | 3 | 1 |
| 52 - 67 | Late Yensh. Period | 4 | 1 |
| 68 -94 | Himalayan Period | 4 | 2 |

* 1. Diorite and granodiorite; 2. Porphyritic biotite granite; 3. Biotite granite; 4. Two-mica granite and tourmaline muscovite granite.

Table 2 Contents of trace elements and ratios of Mg/Li, K/Rb and Ba/Rb in granitoid rocks of different types (ppm)

| Rock Type | Li | Rb | Be | Sr | Ba | U | Th | F | Mg/Li | K/Rb | Ba/Rb |
|---|---|---|---|---|---|---|---|---|---|---|---|
| Late Yenshanian Period | | | | | | | | | | | |
| Diorite and granodiorite | 30 | 124 | 1.2 | 450 | 250 | 2.6 | 15.9 | 925 | 573 | 211 | 2.02 |
| Porphyritic biotite-granite | 50 | 251 | 1.9 | 310 | 210 | 4.0 | 30.9 | 1221 | 105 | 172 | 0.48 |
| Biotite-granite | 42 | 239 | 1.4 | 240 | 170 | 3.3 | 25.0 | 967 | 125 | 179 | 0.71 |
| Two-mica granite | | | 1.8 | 230 | 190 | 2.0 | 12.0 | | | | |
| Early Himalayan Period | | | | | | | | | | | |
| Two-mica granite | 72* | 282* | 3.3 | 390 | 250 | 3.6 | 9.0 | 1137* | 60 | 149 | 1.30 |
| Late Himalayan Period | | | | | | | | | | | |
| Tourmaline-muscovite granite | | | 3.9 | 250 | 210 | 6.3 | 13.6 | | | | |
| Average content | 39 | 183 | 1.6 | 380 | 230 | 3.1 | 19.0 | 1002 | | | |

* Average for two-mica granites of Late Yenshanian and Early Himalayan Period and tourmaline-muscovite granite of Late Himalayan Period.

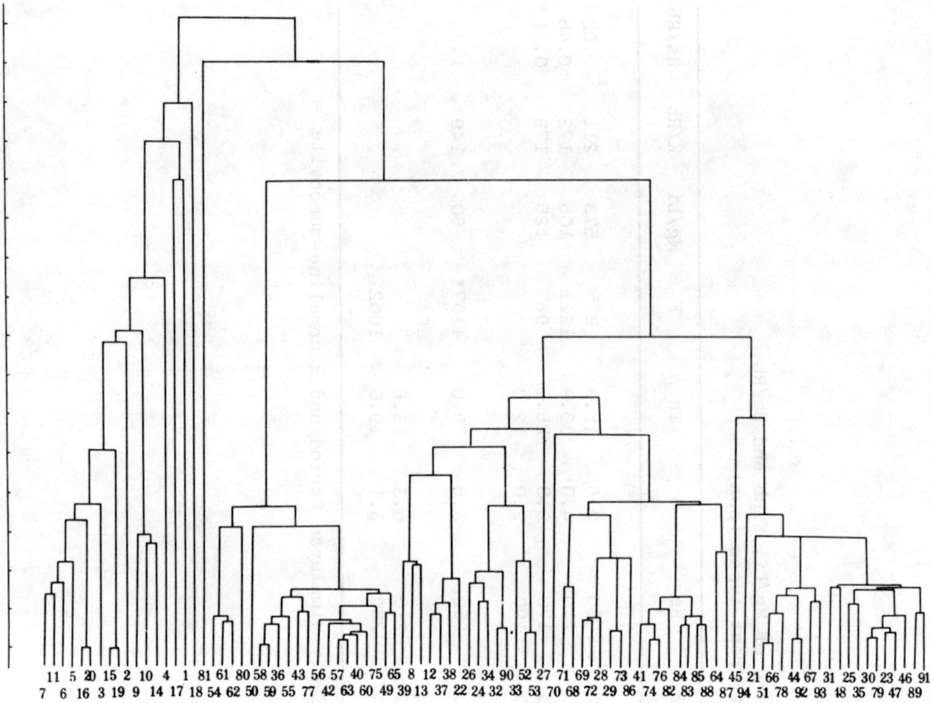

Fig. 1 Diagram showing the results of systematic cluster analyses for granitoid rocks in southern Xizang

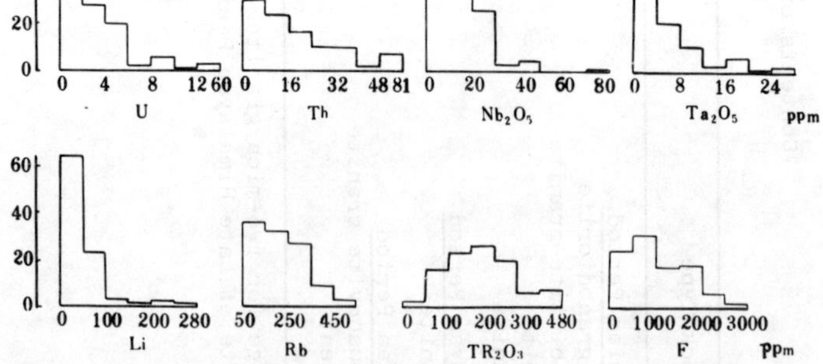

Fig. 2 Concentration distribution frequency histograms for various trace elements

# PLATE TECTONICS AND GLOBAL SEALEVEL CHANGES: A COMPARISON OF THE GLOBAL CRETACEOUS RECORD WITH THAT OF TIBET

Erle G. Kauffman

(Department of Paleobiology, U.S. National Museum, Smithsonian Institution, Washington, D.C., and Department of Geological Sciences, University of Colorado, Boulder, CO, United States of America)

During the Mesozoic-Early Cenozoic eras, intense global tectonics initiated a series of plate movements that led to the present configurations of continents and ocean basins. The Cretaceous Period was a particularly active interval, and serves as an example of the complex interactions between global tectonics and the sedimentologic, biologic, and oceanographic histories of the marine realm. The greatest Mesozoic changes involved the position, size, and volume of ocean basins. New oceans and seas formed (e.g. the Atlantic, Caribbean) and others were diminished (e.g. the Pacific, Mediterranean) in size. Large topographic features were intermittently built and destroyed on the ocean floor. These events should have caused major volumetric changes in the world ocean system, and changes in relative sealevel attaining hundreds of meters during the Cretaceous. Such eustatic changes would have a profound effect on Cretaceous paleoceanography, sedimentation patterns, and biologic history.

Five plate tectonic processes have been proposed, acting individually or collectively, to account for worldwide changes of sealevel (Tectono-eustacy) (Fig. 1): (1) The building of oceanic ridges and island arcs during spreading episodes, and subsequently their destruction during periods of relative plate quiescence; (2) block uplift and subsidence of large segments of the seafloor; (3) broad doming of oceanic crust over hot spots, and subsequent collapse of the crust as hot spots subsided or as continental crust migrated over them, (4) change in the volume of ocean basins due to size changes of the earth, or due to changes in the ratio of oceanic to continental crust; and (5) simultaneous vertical movements of whole continental and/or oceanic crustal blocks (reviewed in Kauffman, 1979). All five processes would have to be reversible to account for alternating rise and fall of sealevel. This constraint, and available geophysical evidence, favor hypotheses 1-3 over 4 and 5. There were no Cretaceous glacio-eustatic events to confuse the tectono-eustatic picture.

To test the tectono-eustatic hypothesis, we must prove from the sedimentary record that the times of initial and peak transgression, and of peak regression of epicontinental seas, were essentially synchronous worldwide. Such a test requires well preserved exposures of transgressive-regressive sequences (marine megacyclothems), a system of precise global correlation normally based on biostratigraphy, and sufficient radiometric data to assist in correlation and allow measurement of rates of sealevel change. Detection of sealevel changes is most obvious in the marine stratigraphic record of epicontinental seas. Two kinds of evidence are used to detect sealevel changes in cratonic and continental margin facies: (1) Correlative regional migrations of the strand and marginal marine facies, depicting transgressions (marine advance) and regressions (strand retreat); these migrations are the most sensitive indicators of sealevel change; (2) Vertical deepening-shallowing stratigraphic sequences offshore from the strand line. Both phenomena can also be produced by local tectonic and sedimentologic factors; but the proof that such events represent eustatic fluctuations lies in the precise correlation of transgressive and regressive peaks in diverse parts of the world, utilizing biostratigraphic zones and geochronology. Different tectonic settings produce different global patterns of eustatically generated transgressions and regressions so that such correlations in proof of eustacy must be made between tectonically similar settings. In most cases, eustatic changes of the magnitude predicted by Cretaceous plate tectonic events would have been large enough to overprint the effects of local and regional tectonics, their resultant sedimentation patterns (especially in epicontinental sites), and geoidal eustacy. Tectono-eustatic events should be clearly distinguishable in the Mesozoic stratigraphic record all over the world.

The best example of regionally correlative megacyclothems on a large craton is that developed for the Western Interior Cretaceous Seaway of North America (Kauffman, 1977, 1979). Here, five major transgressive-regressive pulses are represented by well documented strandline migrations (e.g. Gill and Cobban, 1973), by regional migrations of nearshore vs. offshore facies (e.g. McGookey, et al, 1972), and by definition of marine cyclothems based on vertical stratigraphic sequences offshore from the strandline (e.g. Hattin, 1964; Kauffman, 1969, 1977). These cyclothems may have distinct facies expression in: (a) stable platform environments (asymmetrical, with little or no preserved transgressive sediments: e.g. Hattin, 1975, Fig. 5); (b) subsiding troughs (subsymmetrical, with similar transgressive-regressive facies suites in opposing sequences; Kauffman, 1969); and (c) tectonically active margins of rapid sedimentation (shoreward vs. seaward migration of successive progradational and transgression events: e.g. Ryer, 1977). But their transgressive peaks are correlative within one faunal zone (0.25-0.5 MY interval) throughout the basin, strongly suggesting a predominant eustatic control on their development.

The epicratonic eustatic theory developed from the American Interior has been subsequently tested against Cretaceous sequences in many parts of the world: coastal North America, Central America, northern South America, Africa, Europe, Japan and Australia. Nearly 75 percent of the sections tested in these regions show evidence for essentially synchronous transgressive peaks which broadly overprint local tectonic and sedimentologic effects. Epicontinental transgressive peaks which are broadly correlative in some or all of these areas occur in the Upper Valanginian, Upper Hauterivian, Middle Barremian, Upper Aptian, Upper Albian, Lower Turonian, Coniacian-Santonian, Lower Campanian, Upper Campanian, and Middle to Upper Maastrichtian. Of these, the Aptian, Albian, Turonian, Coniacian-Santonian, Upper Campanian, and Maastrichtian events are the most widespread. The periodicity and duration of these eustatically generated events is irregular, with transgressions generally longer than regressions, but shorter than would be predicted from single geophysical causes. They were produced by complex interaction of numerous plate tectonic events, in some cases enhanced by large scale continental tectonics. Evidence strongly suggests that similar tectono-eustatic events dominate earlier Mesozoic and Paleogene marine sedimentation patterns on a smaller scale.

Epicratonic regressive peaks are only broadly correlative since transgressive and regressive marine-nonmarine boundaries are diachronous from the shelf to the cratonic interior in response to 4-5 MY eustatic events (rise or fall). Marine sedimentation begins later and ends earlier in intracratonic basins than it does in continental margins for any event. Correlation of regressive peaks in support of a tectono-eustatic hypothesis must therefore be done in continental shelf and margin sites; this has been extensively done by Vail, et al (1977). Global eustatic curves must therefore be drawn from two sources, representing different tectonic settings. True transgressive peaks can only be obtained from inner shelf and intracratonic sites near the strand margin; major regressive peaks are best obtained from outer shelf and continental margin sites, normally through seismic stratigraphic analysis.

Global Cretaceous sealevel curves have been constructed from these data which prove the hypothesis of tectono-eustatic control (Kauffman, 1977, 1979; Vail, et al, 1977). Eustatic rise and widespread transgression broadly correlate with active periods of plate movement, seafloor spreading, building of oceanic ridges and island arcs, and regional continental tectonics. Eustatic fall and marine regression reflect times of plate and regional tectonic quiescence, collapse or destruction of topographic highs on the seafloor, and filling of epicontinental basins. This theory, developed for the Cretaceous Period, predicts that sealevel fluctuations should leave a diagnostic sedimentary record during any period of large scale plate tectonic activity, at any time in the history of the earth.

Definition of Cretaceous sealevel curves has not yet been attempted in China. Since this large area is somewhat unique among Cretaceous continental masses in both its topographically positive nature and the intensity of Cretaceous tectonic events associated with subduction of Gondwanan plates below Eurasia, we might expect it to have a complex history of relative sealevel change reflecting both regional and tectono-eustatic control. In a sense, China presents the ultimate test of the Cretaceous eustatic model; if a eustatic overprint exists on the sedimentary history of the Chinese Cretaceous, it will greatly affect the debate over whether regional tectonic or global tectono-eustatic events exercise primary control on cratonic sealevel history.

## Cretaceous Sealevel Fluctuations in China

With accretion of Gondwanan continental masses beginning in the Late Paleozoic, the Asian subcontinent was enlarged and progressively uplifted as a positive tectonic block. This is clearly reflected in the pattern of Permian through Cenozoic marine transgressions, which generally decrease from north to south in size and extent of epicontinental incursion through the Mesozoic and Paleogene. Epicontinental seas did not invade the interior of the Asian Craton after the Eocene. This large scale tectonic influence created anomalies in the Chinese marine record, as compared to other major continental areas, and especially during the Cretaceous Period.

Epicontinental Cretaceous seas were unusually restricted in Asia, and mainly occupied narrow coastal zones and, on the craton, deep narrow tectonic basins trending east-southeast to west-northwest across Tibet and northwest China (Fig. 2). These were parallel to the principal subduction and continental suture zones. Unlike North America and Europe, deep basinal facies dominated over shallow platform marine facies in China, and the extent of marine transgression successively decreased through the Cretaceous. Northern Cretaceous basins of China have the shortest, most discontinuous, and geographically least extensive marine Cretaceous record; southern basins of the Himalayan Tectonic Zone have the most complete marine record (Figs. 3,4).

Marine Cretaceous facies occur intermittently in each of the major tectonic basins (Fig. 3), being separated by disconformities and/or thick continental facies; the latter dominate over most of China during the Cretaceous. These fluctuations suggest changes in relative sealevel which could be dominantly controlled by regional tectonic or global eustatic processes. Dominance of eustatic control would be reflected by close correlation of marine incursions between environmentally and tectonically distinct Cretaceous basins, and collectively between the Chinese record and the contemporaneous marine history in other parts of the world (e.g. Kauffman, 1979, Fig. 2, for cratonic regions). Dominance of regional tectonic control on relative sealevel would produce low levels of correlation between transgressions in different basins. The marine stratigraphic record of the Chinese Cretaceous is complex, and seems to show both regional tectonic and eustatic overprints (Figs. 3,4); interpretation of this Cretaceous record is further complicated by: (1) the difficulty in deciphering eustatic fluctuations and transgressive-regressive marine history from basinal facies (black shales, marls, flysche) where deep water, fine-grained sediments continue to be deposited during early to middle regressive history, and where mass flow events (turbidites) producing coarser clastics do not necessarily reflect shallowing phases; (2) lack of sufficient lithostratigraphic detail in many areas, especially in regard to identification of smaller eustatic pulses and more subtle marine disconformities or paraconformities which may reflect marine erosion during smaller regressive phases; and (3) the rudimentary nature of Cretaceous biostratigraphy and sedimentary geochronology at present, which precludes detailed correlation of the Chinese sequence with other areas of the world. Factors (2) and (3) largely reflect the immense size and inaccessible nature of many Cretaceous outcrop areas, especially in Tibet, and focus on broad stratigraphic reconnaissance by Chinese scientists as a logical first step in geological interpretation of this complex region.

Nevertheless, analyses of published reports on the Chinese Cretaceous, including those of the Qinghai-Xizang (Tibet) Plateau symposium (1980), of unpublished information graciously provided by Chinese colleagues during May and June, 1980, and personal observations (1980) of the Tibetan Cretaceous have allowed preliminary interpretations of Cretaceous marine history in seven principal areas (Figs. 3,4) spanning all four tectonic regions of China (Fig. 2). Figure 3 summarizes these data in stratigraphic columns which are generalized from more detailed field data (several sections) in each region. Attempts have been made to show varying opinions of different workers in the same areas. Marine transgressions ($T_1$-$T_8$) are assigned ages on the basis of limited biostratigraphic data now available for most regions; they are, at best, generalized hypotheses regarding the timing and duration of transgressive-regressive events and possible eustatic fluctuations. Much additional work needs to be done. As a preliminary test of eustatic control on these marine incursions, relative sealevel curves derived from each region have been abstracted from the stratigraphic data (Fig. 4) compared between regions, and correlated with Kauffmans' (1979) global model for eustatic fluctuations in cratonic areas (Fig. 4).

Nine discrete peak levels of Cretaceous marine transgression into China are apparent from these data (Figs. 3,4). There is a general correlation of transgressive peaks between some or all tectonic basins at seven different levels (Fig. 4): (1) The Middle-Upper Valanginian boundary zone in the Kangpa Basin, North Xigaze region, and in northeastern China and the USSR west of Vladivostok; (2) the upper Middle Aptian of northern Tibet, the Lhasa area, and possibly the area south and west of Xigaze (reworked conglomerate pebbles only); (3) a middle Lower to early Upper Albian interval in the Kangpa Basin, Lhasa area, and in North Tibet; this incursion is possibly represented by reworked conglomerate pebbles (limestone, chert) south and west of Xigaze, and by large olistostrome blocks northeast of Xigaze; (4) a Late Cenomanian-Early Turonian interval in all areas sampled (Fig. 4); (5) a latest Turonian through Santonian interval in southern basins around Kangpa and Xigaze, but not to the north; (6) a late Lower to early Upper Campanian peak in the Kangpa Basin, and all around Xigaze; this is also part of a broad peak in northwestern Chinese basins; and (7) a middle Late Campanian peak near Kangpa, north of Xigaze, and in northwestern China.

These correlations suggest a dominantly eustatic origin for Cretaceous transgressions and regressions in China. Further, there is close correlation of each of these transgressive peaks with those found widely on other cratons of the world and which Kauffman (1979; see Figs. 3,4 herein) considers to be of eustatic origin ($T_1$, $T_4$, $T_5$, $T_6$, $T_7$, $T_8$, $T_9$ of Kauffmans' terminology). These are consistently the strongest transgressive pulses of the 10 described by Kauffman for the Cretaceous. Considering the complex tectonic history of China during this time (including strong tectonic influences on the history of Cretaceous marine basins), and the generally positive nature of the Asian subcontinent, this apparent eustatic overprint is remarkable. The great magnitude of Cretaceous eustatic fluctuations (up to 650m) predicted by Vail, et al (1979) from seismic stratigraphy, and by Hancock and Kauffman (1979) from epicontinental stratigraphic sequences, seems to be reflected in the

Chinese Cretaceous record, where eustatic sealevel fluctuations broadly overprint both the tectonically positive nature of the Asian block, and the complex regional tectonic controls on marine basins associated with Asian-Gondwanan collision events.

There are, of course, major discrepancies in the Chinese stratigraphic record when compared to the eustatic record of more stable cratons such as the North American Interior (Kauffman, 1977). Correlations of Chinese transgressive and regressive peaks are not always as precise as they are between other continental areas (e.g. Europe vs. North America), or are only precise in certain areas and not in others. For example, both the Late Albian ($T_5$) and the Cenomanian-Turonian ($T_6$) boundary zone transgressive peaks show offsets of approximately one substage from section to section in China, but are characteristically closely correlative within a single biostratigraphic zone elsewhere. The two Campanian transgressive peaks ($T_8$-$T_9$) which are well defined in North America, Europe, and in the development of distinct rudist-bearing limestone sequences in the Kangpa Basin, are only broadly developed around Xigaze and in northern China as a single marine event, insofar as is known. They are not developed at all near Lhasa and in North Tibet. There appear to be two discordant marine incursions found in China but not commonly elsewhere in the world; the Berriasian transgression near Lhasa, and the small Middle Maastrichtian pulse in the Kangpa Basin.

These levels of discrepancy between the well studied North American-European-African Cretaceous eustatic record, and a tectonically active region like China are not unexpected. In part, apparent temporal variations in transgressive peaks as we can presently plot them are only reflective of a poorly known marine paleobiologic and geochronologic record in China, and our inability to precisely correlate large intervals of the stratigraphic column. Detailed marine biostratigraphic studies are just now under way for large portions of China. Further, we can legitimately ask the question of whether the Middle Maastrichtian transgression in the Kangpa Basin is well dated, or whether it possibly represents miscorrelation of the slightly younger global Maastrichtian peak ($T_{10}$; Kauffman, 1979)? The unusual but well dated Berriasian marine peak around Lhasa could reflect local tectonic uplift which overwhelmed the eustatic rise of sealevel producing a Middle Valanginian transgressive peak elsewhere in China (Figs. 3,4).

Two interesting tectonic patterns emerge from the Cretaceous marine stratigraphic record (Figs. 3,4). It is obvious that during some time intervals, regional tectonic uplift affecting all, or large portions of, the cratonic interior of China, has outpaced eustatic rise and thus denied access of epicontinental seas to the continental interior. The two most obvious intervals reflecting this are: (1) The Hauterivian-Barremian stages, which seem to be represented widely by an erosional interval or only by continental sediments, even though small global marine transgressive peaks in the latest Hauterivian and Middle Barremian (Fig. 4, left column; Kauffman, 1979) are widely recorded elsewhere; and (2) the post-Middle Turonian interval from central Tibet (Lhasa region) to the north; with the exception of a small southward Campanian incursion into northwestern China, widespread erosional intervals and continental deposits predominate throughout this interval in North China. At the same time, widespread marine transgression similar to that found elsewhere in the world (Hancock and Kauffman, 1979, Fig. 5; Kauffman, 1979) characterized the basins of southern Tibet (Figs. 3,4). This discrepancy reflects the widely noted north-to-south retreat of epicontinental seas, and the migration of marine basins associated with episodes of subduction and plate accretion, during Mesozoic and Early Cenozoic time. Secondly, to the extent that they can be dated, the preserved Cretaceous marine incursions into China are less extensively developed and more short-lived than the same transgressive episodes on stable cratons or on broad continental coastal plains throughout the world. This sedimentation pattern is typical of the more central regions of cratonic blocks that are undergoing broad uplift and regional tectonism (Kauffman, 1973), and that are far enough removed from ocean basins so that only the periods of maximum eustatic rise are sufficient to inundate them. Finally, the Barremian through Cenomanian (middle Cretaceous) marine record in areas closest to the Jarlung-Zangbo ophiolite belt (suture zone) is erratically developed (offset or only partial preservation of transgressive events), or else it consists only of reworked clasts (conglomerate pebbles, olistostromes) in younger Cretaceous sediments (Fig. 3). These observations suggest a major period of tectonism, including compression, uplift, and erosion of sediments formed in deep arc-related trenches along the suture zone, during pre-Turonian middle Cretaceous time.

In conclusion, the Cretaceous marine record in China seems to predominantly reflect global eustatic history despite the positive nature of the Asian Subcontinent, and the strong influence of regional tectonics associated with Asian-Gondwanan plate collision events during this Period. Figure 4 is an hypothesis of this relationship which must be tested against much more detailed field work, and more precise biostratigraphic-geochronologic time matrix, than currently exists in China. Hopefully, this will be one focus of prospective international cooperation in geological and paleobiological studies of China.

## References Cited


(1) Gill, J.R., and Cobban, W.A., Stratigraphy and geologic history of the Montana Group and equivalent rocks, Montana, Wyoming, and North and South Dakota. U.S. Geol. Surv., Prof. Paper 776, 1-37 (1973).

(2) Hancock, J.M. and Kauffman, E.G., The great transgressions of the Late Cretaceous. Jour. Geol. Soc. of London, 136 (2), 175-186 (1979).

(3) Hattin, D.E., Cyclic sedimentation in the Colorado Group of west-central Kansas; in Merriam, D.F. (ed.), Symposium on cyclic sedimentation. Kansas Geol. Surv. Bull., 169, 205-217 (1964).

(4) ______, Stratigraphic study of the Carlile - Niobrara (Upper Cretaceous) unconformity in Kansas and northeastern Nebraska; in Caldwell, W.G.E. (ed.), The Cretaceous System in the Western Interior of North America. Geol. Assoc. Canada, Spec. Paper 13, 31-54 (1975).

(5) Kauffman, E.G., Cretaceous marine cycles of the Western Interior. Mountain Geol. 6 (4), 227-245 (1969).

(6) ______, Stratigraphic evidence for Cretaceous eustatic changes. Geol. Soc. America, Abstr. Vol., 1973 Ann. Meeting, 687 (1973).

(7) ______, Geological and biological overview: Western Interior Cretaceous basin; in Kauffman, E.G. (spec. ed.), Cretaceous facies, faunas, and paleoenvironments across the Western Interior basin. Mountain Geol. 14 (3,4), 75-99 (1977).

(8) ______, Cretaceous; in Robinson, R.A., and Teichert, C. (eds.), Treatise on invertebrate paleontology; Pt. A, Introduction; fossilization (taphonomy), biogeography and biostratigraphy. A418-A487 (1979).

(9) McGookey, D.P., Haun, J.D., Hale, L.A., Goodell, H.G., McCubbin, D.G., Weimer, R.J., and Wulf, G.R., Cretaceous System; in Mallory, W.W. (ed.), Geologic Atlas of the Rocky Mountain Region. Rocky Mountain Assoc. Geol., Denver, CO, 190-232 (1972).

(10) Organizing Committee, Symposium on Qinghai-Xizang (Tibet) Plateau, Academia Sinica, Proceedings of Symposium on Qinghai-Xizang (Tibet) Plateau, May 25 - June 1, 1980 (Abstracts), 1-323 (1980).

(11) ______, A scientific guidebook to south Xizang (Tibet), June 2 - 14, 1980, 1-104 (1980).

(12) Ryer, T.A., First day: Coalville and Rockport areas, Utah; in Kauffman, E.G. (spec. ed.), Cretaceous facies, faunas, and paleoenvironments across the Western Interior basin. Mountain Geol. 14 (3,4), 105-128 (1977).

(13) Vail, P.R., Mitchum, Jr., R.M., Todd, R.G., Widmier, J.M., Thompson III, S., Sangree, J.B., Budd, J.N., and Hatlelid, W.G., Seismic stratigraphy and global changes of sea level; in Seismic stratigraphy - applications to hydrocarbon exploration. American Assoc. Petrol. Geol. Mem. 26, 49-212 (1977).

(14) Vail, P.R., and Hardenbol, J., Sea-level changes during the Tertiary. Oceanus, 22 (3), 71-79 (1979).

(15) Wen, S., Zhang, B., Wang, Y., Sun, D., Wang, Y., Chen, C., Dong, D., Liao, W., Chen, T., He, G., Mu, X., Yin, J. and Wu, H., Development of sediments and formation of stratigraphical regions in Xizang. Proc. Sympos. Qinghai-Xizang (Tibet) Plateau (Abstracts); Peking, Academia Sinica, 1 (1980).

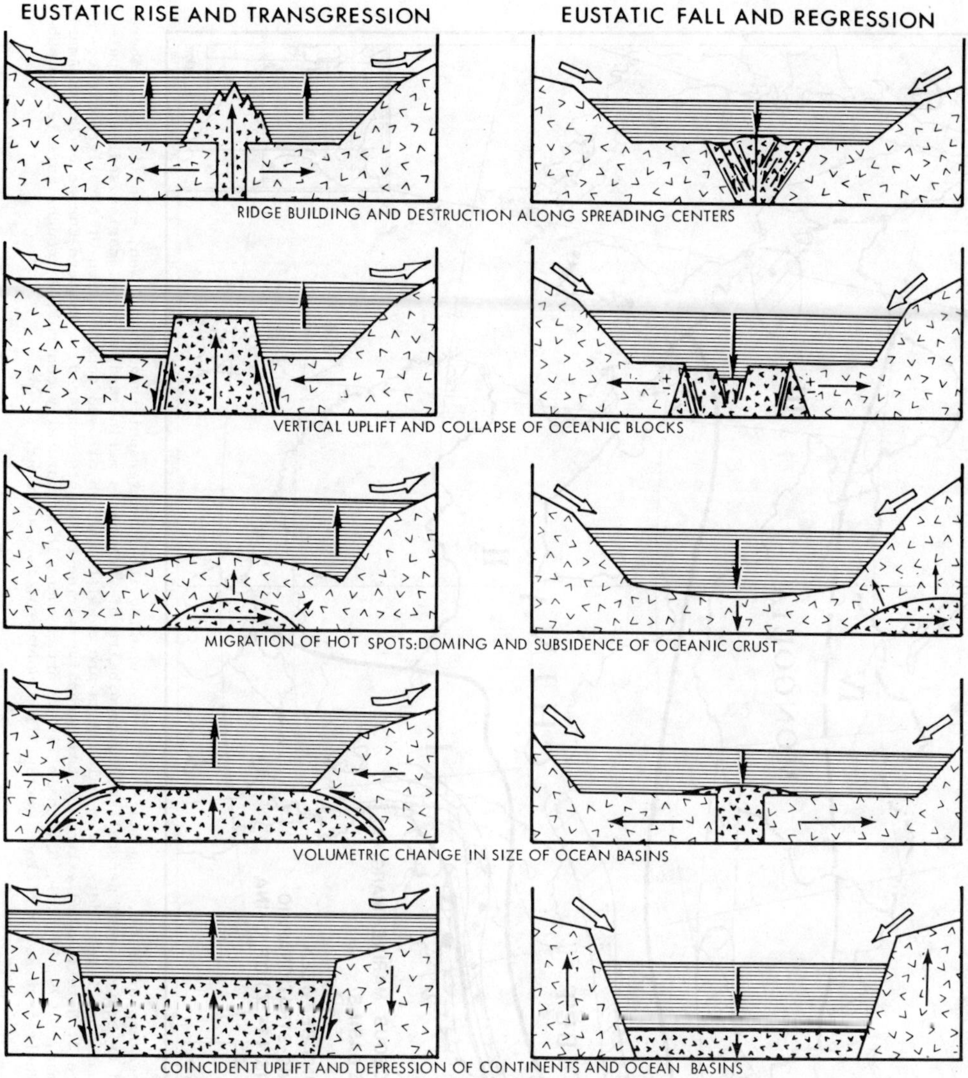


FIGURE 1. HYPOTHESES FOR TECTONO-EUSTATIC FLUCTUATIONS OF SEALEVEL.

Schematic diagrams of hypotheses proposed for tectono-eustatic rise and fall of global sealevel, causing synchronous worldwide epicratonic transgressions and regressions. Top three hypotheses are most widely accepted. Marine Cretaceous record implies that more than one process may have acted to produce global sealevel changes of several hundred meters magnitude. These hypotheses are summarized and referenced in Kauffman (1979).

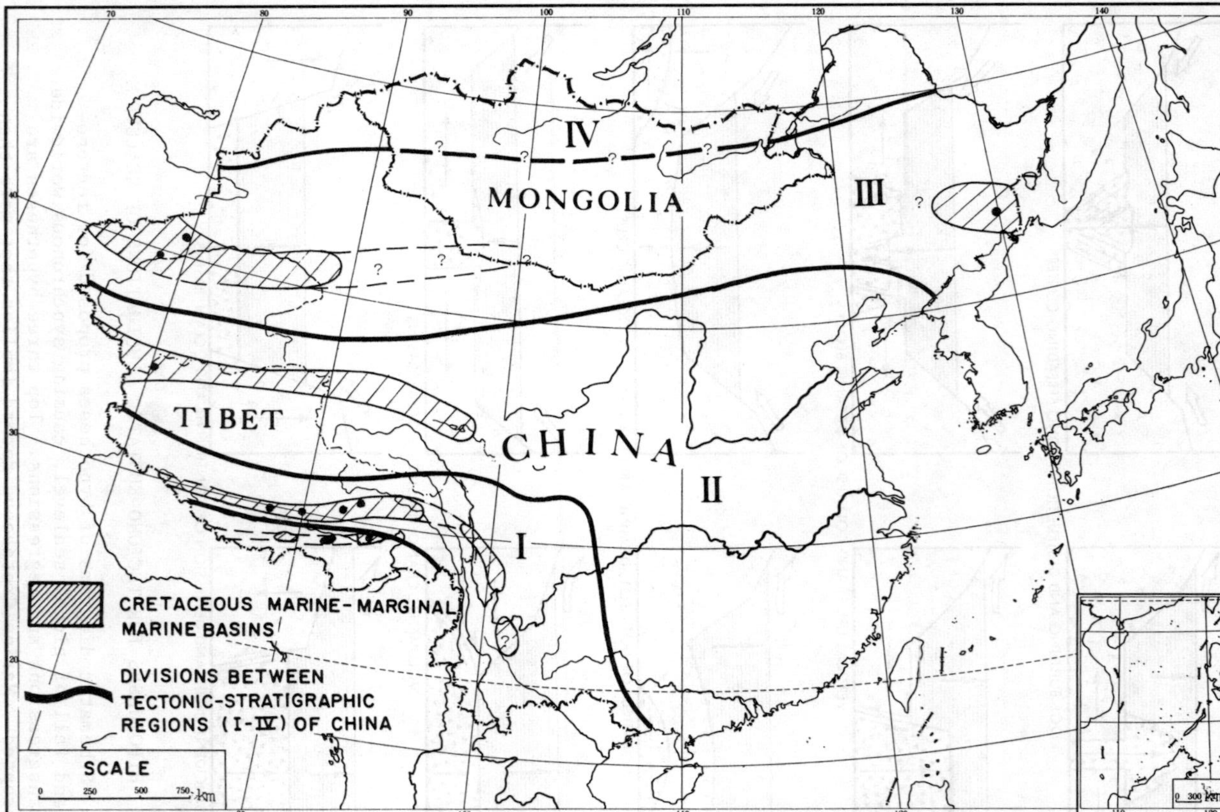


Fig. 2 Generalized map of China, Mongolia showing the stratigraphic regicns of Wen, et al (1980) numbered I – IV and separated by heavy lines, generalized outcrop belts of Cretaceous marine and marginal marine sediments (patterned), and principal localities (dots) from which data are available which allowed construction of stratigraphic columns shown in Figure 3. Note limited extent of transgressive marine sediments compared to Western Europe or North America at this time (Kauffman, 1979, Figs. 1, 5), their linear arrangement in structurally controlled troughs parallel to subduction zones, and their increasing regional extent from northwest to southwest.

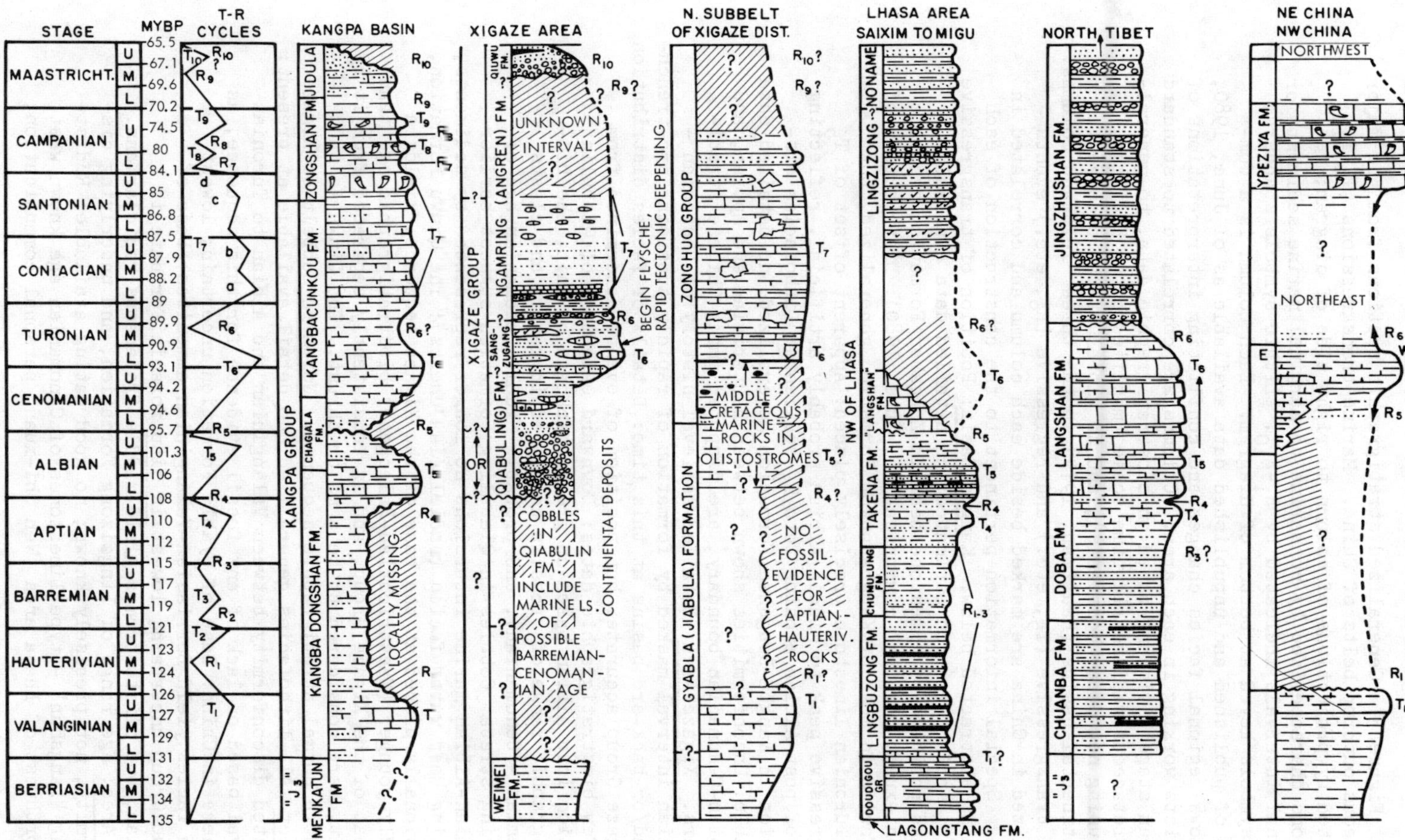


FIGURE 3. GENERALIZED STRATIGRAPHIC SECTIONS FROM MAJOR CRETACEOUS OUTCROP BELTS OF CHINA.

Figure 3. Generalized stratigraphic sections from the major Cretaceous outcrop belts of China. Marine transgressions ($T_1$-$T_{10}$) indicated by projections of column to right side of diagram; regressions characterized by nonmarine to marginal marine sediments or erosional intervals indicated by $R_1$-$R_{10}$. Lithic symbols standard; other symbols keyed at bottom of diagram. Each column is a composite of published and unpublished data available as of June, 1980, and shows regional facies changes and conflicting interpretations of scientists working in each area. Columns are correlated to standard European stage divisions based on biostratigraphic data, to generalized Cretaceous radiometric time scale (modified after Kauffman, 1979, using new decay constants), and to global eustatic curve for epicontinental basins (Kauffman, 1979) in the three lefthand columns. Global transgressive ($T_5$, etc.) and regressive ($R_6$, etc.) events recognized in China are marked beside each column and correlated in Fig. 4. Special information pertinent to the construction of each column is abstracted below. Kangpa Basin: Position of transgressive peak in Jidula Formation estimated; no detailed data available: Ages and position of limestone beds in Zongshan Formation accurate (Wen Shi-xuan, personal communication, June, 1980): Distribution of lithofacies in most of Kangbacunkou Formation generalized, but Lower Turonian limestone precisely placed: Apparent offset of $T_5$ transgressive peak to Middle Albian probably artificial, reflecting depth of post-Albian erosion into $T_5$ sediments: Lithofacies distribution in Kangba-Dongshan Formation generalized, no detailed data available: Unconformities shown between Hauterivian and Aptian, and at Albian-Cenomanian boundary, are disputed among Chinese stratigraphers. Xigaze Area: Detailed sealevel history of Coniacian - Campanian interval masked by formation of rapidly subsiding arc-trench-gap and/or back-arc basins at this time: Details of facies distribution in Xigaze Group accurate, but position of events determined from very few biostratigraphic dates: Downward extension of Qiabulin Formation into Albian based on great thickness of unit, the biostratigraphic range of fossils contained in matrix and clasts (Barremian - Cenomanian), and probable Middle Cenomanian age of overlying strata: Cobbles in Qiabuling Formation suggest earlier Barremian-Albian marine incursions no longer represented by sediments, including deep water facies (radiolarian cherts): The Qiuwu Formation, traditionally placed at the base of the Xigaze sequence, yields a markedly younger (Maastrichtian-Paleogene) biostratigraphic age than the rest of the Xigaze Group, and is consistently separated from it by a fault, here considered to be a major thrust. The Qiuwu is thus placed in normal stratigraphic sequence. Northern Subbelt of Xigaze District: All lithofacies generalized, no detail available at present: Suggested disconformity between Valanginian and Albian to Turonian interval based on lack of any fossil evidence for this sequence, as elsewhere in China: Base of Gyabla Formation uncertain; lowest marine facies yield Berriasian-Valanginian ammonites: Base of Zonghuo Group uncertain; lowest *in situ* foraminifer date is Turonian, but olistostromes yield middle Cretaceous radiolarians. Lhasa Area: Age range of Lingzizong Formation, and underlying disconformity, not precisely known; no good dating available; Rudist-bearing "Langshan" - type limestones of Cenomanian age known west and northwest of Lhasa area (Wen Shi-xuan, personal communication,

June, 1980), missing near Lhasa below erosion surface: A regressive, largely nonmarine sequence separates Cenomanian limestones from well dated $T_5$ marine sequence of Takena Formation near Lhasa: All lithofacies, except for Takena Formation, generalized from available data: Position of $T_4$ transgression based on occurrence of Upper Aptian marine fossils west of Lhasa area, separated from $T_5$ by nonmarine redbeds. North Tibet: All lithofacies generalized from available data; no detailed placement of individual units possible at present: $T_4$ marine transgression and $R_4$ regression well dated: Insufficient detail available to differentiate $T_5$ and $T_6$ transgressions from limestone-marl sequences of Langshan Formation. Northeast and Northwest China: Ypeziya Formation contains Senonian rudists and could extend lower than shown; no data available at present: $T_6$ transgressive event based on single Middle Cenomanian-Lower Turonian brackish water biota interbedded with freshwater facies near Chanling, Kirin Province: Eastward expansion of $T_6$ interval reflects fully marine sequence containing Albian - Senonian rudists near Vladivostok, USSR: The Albian - Valanginian unconformity is based on a reported gap near Vladivostok; debated among Chinese workers. $T_1$ transgressive pulse based on occurrence of Berriasian - Valanginian ammonites near Vladivostok, just across Sino-Soviet border, and equivalent age brackish water beds in northeast China.

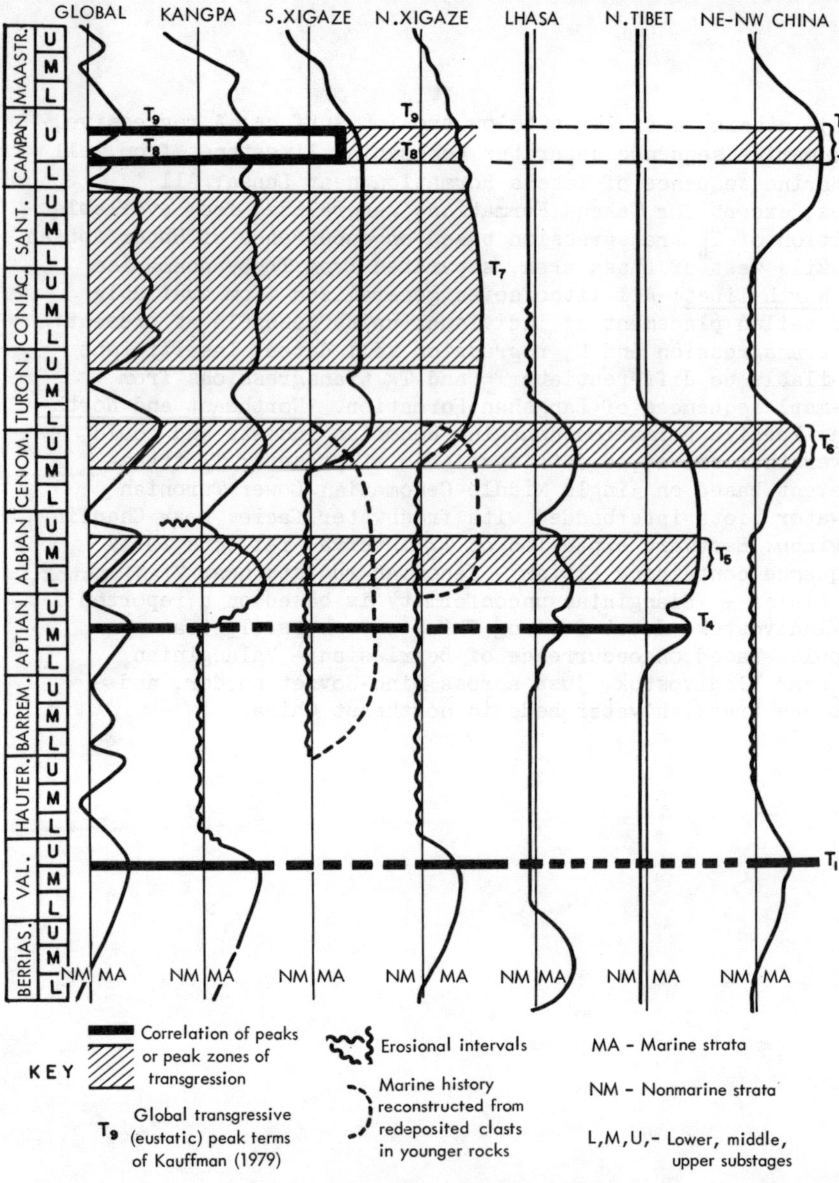


FIGURE 4. SUMMARY OF CRETACEOUS TRANSGRESSIVE-REGRESSIVE HISTORY IN KEY OUTCROP AREAS, CHINA.

Figure 4. Summary of the transgressive-regressive history of Cretaceous marine and marginal marine strata in key outcrop areas of China (labeled at top), compared with the global eustatic sealevel curve of Kauffman (1979; left column) for epicratonic sequences. Chinese sealevel curves taken directly from Figure 3. Key to symbols at bottom of diagram. Note seven main levels or intervals at which Chinese sealevel curves correlate between some or all areas, and generally with the global epicratonic eustatic curves (left column); this strongly suggests a tectono-eustatic overprint on the marine stratigraphic history of China, even in the most active areas of regional tectonics. Discussion of offsets in curves, and aberrant Berriasian and Maastrichtian transgressions, found in text. Note longer history of marine incursions in southern Tibetan sequences than in north, reflecting southward migration of tectonic troughs with uplift of plateau and continental accretion.

# PETROCHEMICAL STUDY OF GRANITOID ROCKS IN SOUTHERN XIZANG

Wang Zhong-gang, Zhang Yu-quan, Zhao Hui-lan
(Institute of Geochemistry, Academia Sinica)

It has been well known that granitoid rocks are widely distributed throughout the whole area of southern Xizang, covering more than one hundred thousand square kilometers. They can be assigned to three rock belts of Gandise, Lhagoi Kangri, and Himalayas (see Tu, Fig. 1, this Proceedings)(1) and constitute a continuous series from granodiorite to alkali-feldspar granite in petrographic characters. This paper is confined to the discussion on the problems of their petrochemical characteristics.

## AVERAGE CHEMICAL COMPOSITIONS OF GRANITOID ROCKS

200 chemical bulk analyses have been carried out on samples from granitoid rocks of different types and intrusive periods in this area and then the average chemical compositions of the three rock belts and the whole area have been calculated by the area weight method. All results are listed in Table 1 and Table 2.

It is clearly seen from Table 1 that there exist some explicit regularities in the evolution of main components from earlier granodiorite to later granitic rocks. For example, the $SiO_2$ content increases from 62.38 % to 72.61 %. Other components which increase in company with $SiO_2$ mainly include $K_2O$. On contrast, however, $TiO_2$, $Fe_2O_3$, FeO, MgO, CaO and $P_2O_5$ contents all decrease remarkably, whereas the $Na_2O$ content remains almost unchanged. Fig. 1 was drawn after the method of Harker, from this figure the trends above mentioned can be seen more clearly (Fig. 1).

Table 2 shows the average chemical compositions of granitoid rocks for individual rock belts and for the whole area. Obviously, the average chemical composition of the granitoid rocks of the whole area is closely similar to that of Daly's granodiorite. A comparison of it with the granitoid rocks in China, leads to the conclusion that the average composition of granitoid rocks in South Xizang is evidently more basic, and is close to the composition of the large batholith in California, USA.

Moreover, it can be also seen that, there exist certain differences among the average chemical compositions of individual rock belts. The acidity of the rocks increases gradually from North to South. The average compositions of Lhagoi Kangri belt and Himalayan belt are evidently different from that of Gandise belt, implying that they have different magma sources and different genetic characters.

Some information about the differentiation indices and solidification indices for the granitoid rocks of different intrusive periods and different types are listed also in Table 1. Diorite-granodiorite formed at the first stage of Late Yenshanian is closely similar to the average granodiorite over the world. However, other rock types fall into the range between the average granite and alkli-granite over the world. Generally speaking, the magma stemed from the mantle have the solidification indices more than or about 40. But, in our instances, they are all less than 40.

According to the statistical analysis results obtained by Prof. Yang Wei-hua of our institute, there exists a linear correlation between Ti contents and Al contents for those granitoid rocks which are believed to be metasomatic in origin, but this linear correlation does not hold true for magmatic origin. So a Ti-Al content diagram was drawn for the granitoid rocks formed at different stages in this area. From the diagram it can hardly say that there exists such a linear correlation between Ti and Al, and it implies again that the granitoid rocks of this area have magmatic origin.

## TEMPERATURE AND PRESSURE CONDITIONS DURING THE FORMATION OF GRANITOID ROCKS

The compositions of various standard minerals of various granitoid rocks are plotted into the Ab-Or-Q-$H_2O$ diagram (Fig. 2). The distribution of the dots shows two different types. The first type is characterized by closely accumulated dots around the eutectic point, and these dots represent porphyritic biotite granite, biotite granite, two-mica granite, gneissose two-mica granite. The other type is characterized by dispersive dots at the positions far from the eutectic point. These dots represent granodiorite which was formed at the earliest stage.

Connecting the dots of the first type into isopleths (Fig. 3), the location of these isopleths is close to that of acidic extrusive rocks. This immediately leads to magmatic origin of these rocks.

After transforming the average chemical compositions of various rocks into standard minerals and then plotting the compositions of the standard minerals obtained into the Ab-Or-Q-$H_2O$ diagram after Tuttle and Bowen (1958)(Fig. 4) (2) it can be clearly seen that the temperature-pressure conditions for all types of granitoid rocks seem to be similar except for granodiorite, namely, the temperatures are about 700 -750°C, pressures about 1 - 2 kb. Granodiorite is not uniform in composition showing a great variation in petrographic characters, and it extends along the Yarlung Zangbo River (the suture line of the two plates) as a large batholith. It has, obviously, a deeper source and an emplacment at greater depth. According to this diagram, the temperature and pressure at which this type of rocks was formed are about 750°C and 5 kb, respectively, corresponding to a depth of 15 -25 km.

The main conclusions can be summarized as follows. The granodiorite of the first stage of Late Yenshanian Period was formed at the stage when the Indian Plate underthrusted towards the Eurasian Plate (see Zhang Yu-quan, Fig. 3, this Proceedings)(3). At that time the front part of the Indian Plate — the part of oceanic crust, together with their continental materials which were washed and then sedimented on the oceanic crust, was dragged into the subduction zone and then was fused to give rise to the granodiorite and diorite magma. We agree with Armstrong's viewpoint that the sial materials which were disintegrated from the continent were transported to the oceanic basin, and then, were dragged into the subduction zone, re-fused together with the oceanic basaltic crust to give rise to a mother magma similar to that of the batholith in North America. Furthermore, we also hold that this kind of magma assimilated and replaced the materials in the conduits during its intrusion upwards from the depth. As the result it became non-uniform in composition. However, the granitic rocks generated at later stages were formed at the time when the inter-collision of the two plates proceeded. Their mother magma has a sial source, and was generated from the remelting of the upper crust, and at that time, this part of crust had become very thick already.

References


(1) Tu Guang-chi et al., Characteristics and evolution of granitoids of Southern Xizang. In this Proceedings of Symposium on Qinghai-Xizang (Tibet) Plateau, Academia Sinica, (1981).

(2) Tuttle, O. F., Bowen, N. L., Origin of granite in the light of experimental studies in the system $NaAlSi_3O_8$- $KAlSi_3O_8$-$SiO_2$-$H_2O$. Geol. Soc. Amer. Mem, 74, 65-75 (1958).

(3) Zhang Yu-quan et al., On the crustal movement characteristics of the Xizang plateau with special reference to intermediate-acid magmatic activities. In this Proceedings of Symposium on Qinghai-Xizang (Tibet) Plateau, Academia Sinica, (1981).


Table 1 Average chemical compositions of the granitoid rocks for different types and intrusive periods in South Xizang

| Intrusive periods and types of rock | $\gamma_5^{3-1}$ | $\gamma_5^{3-2}$ | $\gamma_5^{3-3}$ | $\gamma_5^{3-4}$ | $\gamma_6^{1}$ | $\gamma_6^{2}$ |
|---|---|---|---|---|---|---|
| Number of samples | 54 | 20 | 44 | 25 | 27 | 30 |
| $SiO_2$ | 62.38 | 70.56 | 72.33 | 73.38 | 72.41 | 72.61 |
| $TiO_2$ | 0.63 | 0.33 | 0.23 | 0.21 | 0.19 | 0.15 |
| $Al_2O_3$ | 16.24 | 14.29 | 13.68 | 13.70 | 14.77 | 14.79 |
| $Fe_2O_3$ | 2.75 | 1.32 | 1.09 | 1.12 | 0.76 | 0.60 |
| FeO | 3.01 | 1.74 | 1.44 | 1.22 | 1.09 | 1.08 |
| MnO | 0.08 | 0.06 | 0.04 | 0.04 | 0.05 | 0.04 |
| MgO | 2.25 | 0.62 | 0.46 | 0.28 | 0.55 | 0.32 |
| CaO | 4.86 | 2.22 | 1.95 | 1.66 | 1.70 | 1.40 |
| $Na_2O$ | 3.68 | 3.20 | 3.60 | 3.50 | 3.71 | 3.83 |
| $K_2O$ | 2.85 | 4.56 | 4.36 | 4.18 | 3.80 | 4.46 |
| $P_2O_5$ | 0.19 | 0.13 | 0.07 | 0.06 | 0.10 | 0.09 |
| Ignition loss | 0.88 | 0.69 | 0.56 | 0.77 | 0.70 | 0.71 |
| Sum | 99.80 | 99.72 | 99.81 | 100.12 | 99.87 | 100.08 |
| DI | 65 | 83 | 86 | 87 | 86 | 88 |
| SI | 15.5 | 5.4 | 4.2 | 2.7 | 5.6 | 3.1 |

$\gamma_5^{3-1}$ : Diorite-granodiorite (Late Yenshanian Period )

$\gamma_5^{3-2}$ : Porphyritic-biotite granite (Late Yenshanian)

$\gamma_5^{3-3}$ : Biotite granite (Late Yenshanian)

$\gamma_5^{3-4}$ : Two-mica granite (Late Yenshanian)

$\gamma_6^{1}$ : Two-mica granite (Early Himalayan Period)

$\gamma_6^{2}$ : Muscovite granite (Late Himalayan Period)

Table 2 Comparison between average chemical compositions of granitoids in South Xizang and those in other areas

| Locality* | 1 | 2 | 3 | 4 | 5 | 6 | 7 |
|---|---|---|---|---|---|---|---|
| Area (x 100 $km^2$) | 693.4 | 42.7 | 27.5 | 763.6 | | | |
| $SiO_2$ | 65.66 | 72.41 | 72.61 | 66.57 | 70.18 | 70.40 | 72.05 |
| $TiO_2$ | 0.49 | 0.19 | 0.15 | 0.46 | 0.39 | 0.31 | 0.28 |
| $Al_2O_3$ | 15.37 | 14.77 | 14.79 | 15.32 | 14.47 | 14.48 | 13.72 |
| $Fe_2O_3$ | 2.15 | 0.76 | 0.60 | 2.02 | 1.57 | 1.38 | 0.78 |
| FeO | 2.44 | 1.09 | 1.08 | 2.32 | 1.78 | 1.77 | 1.96 |
| MnO | 0.06 | 0.05 | 0.04 | 0.07 | 0.12 | 0.08 | 0.08 |
| MgO | 1.57 | 0.55 | 0.32 | 1.47 | 0.88 | 0.94 | 0.66 |
| CaO | 4.11 | 1.70 | 1.40 | 3.56 | 1.99 | 1.93 | 1.38 |
| $Na_2O$ | 3.55 | 3.71 | 3.83 | 3.57 | 3.48 | 3.77 | 3.22 |
| $K_2O$ | 3.47 | 3.80 | 4.46 | 3.53 | 4.11 | 3.79 | 4.54 |
| $P_2O_5$ | 0.16 | 0.10 | 0.09 | 0.15 | 0.19 | 0.18 | |
| Ignition loss | 0.79 | 0.73 | 0.71 | 0.78 | 0.84 | 0.65 | |
| Sum | 99.82 | 99.87 | 100.08 | 99.82 | | | |

* 1, Gandise rock belt ; 2, Lhagoi Kangri rock belt; 3, Himalayan rock belt ; 4, Whole area; 5, Mean values of granitoids, after R. A. Daly; 6, Mean values of granitoids in China, after Li Tong ; 7, Mean values of granitoids in South China, after Institute of Geochemistry, Academia Sinica.

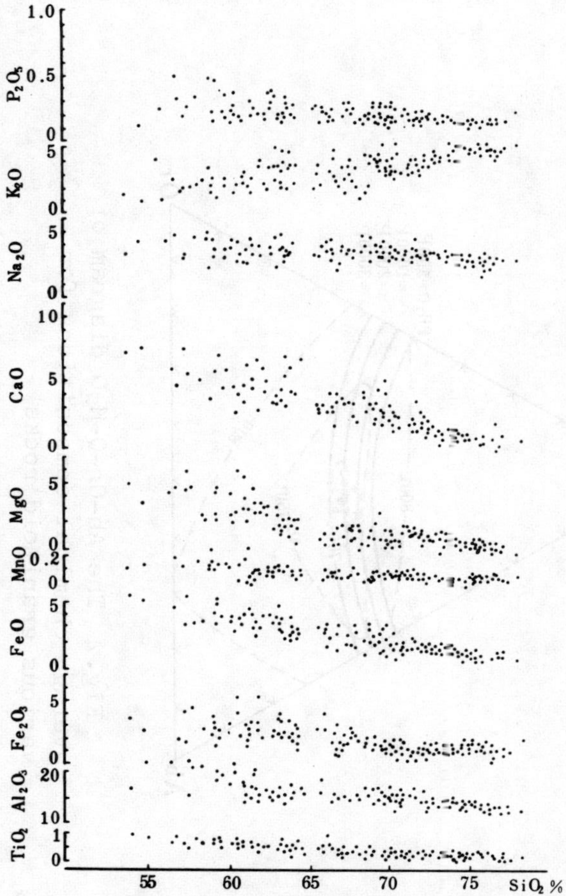


Fig.1 Changes of the main components versus $SiO_2$ for granitoid rocks in this area

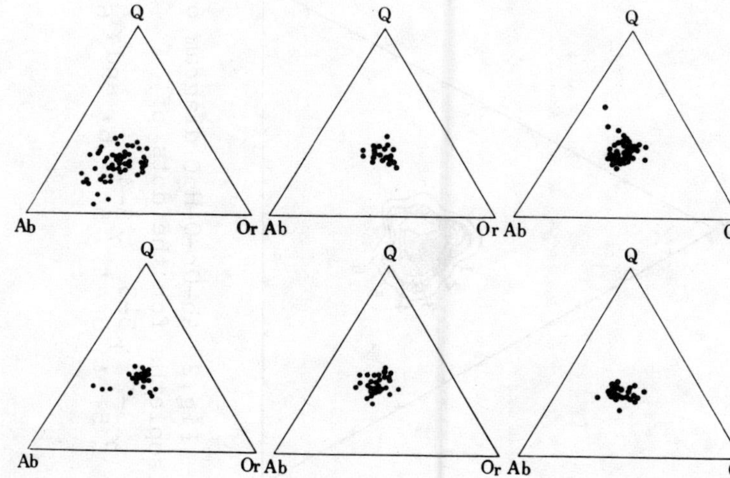


Fig.2 Ab-Or-Q-$H_2O$ diagrams

a-$\gamma_5^{3-1}$; b-$\gamma_5^{3-2}$; c-$\gamma_5^{3-3}$;

d-$\gamma_5^{3-4}$; e-$\gamma_6^1$; f-$\gamma_6^2$

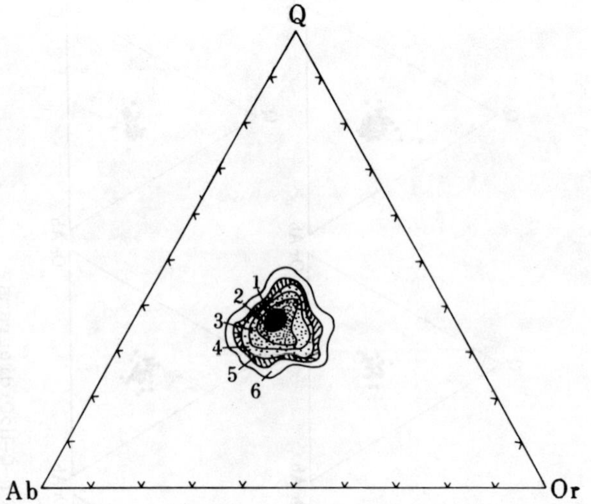


Fig.3 Ab-Or-Q-$H_2O$ diagram of isopleths for the dots of $\gamma_5^{3-2}$, $\gamma_5^{3-3}$, $\gamma_5^{3-4}$, $\gamma_6^1$, and $\gamma_6^2$

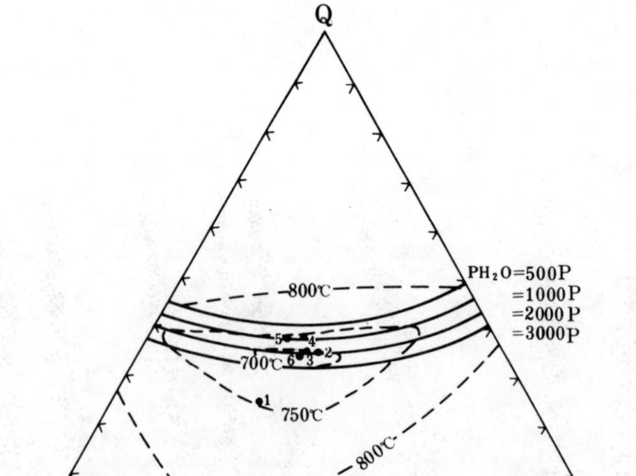


Fig.4 The Ab-Or-Q-$H_2O$ diagram of average chemical compositions for various granitoid rocks

# THE MESOZOIC-CENOZOIC VOLCANIC ROCKS IN XIZANG AND THEIR BEARING ON GEOLOGY

Jin Cheng-wei
(Institute of Geology, Academia Sinica)

Mesozoic and Cenozoic volcanic activity occurred extensively in Xizang. Among them the Late Cretaceous-Early Tertiary is the most magnificent, but the Quaternary also arouses the interest of all (Fig. 1).

Triassic volcanic activity occurred mainly in the Jinsha River and Lancang River regions of Qamdo and in the Yarlung Zangbo region. In the Jinsha River region Triassic metamorphosed volcanic rocks of geosynclinal facies are a series of alkali basalts and tholeiites. The Late Triassic volcanic rocks of Gyapeila formation are mainly a dacitic - andesitic pyroclastic volcanics that belongs to sodic and ordinary subseries of the calc-alkali series. The eruptions in the cycle were tranquil at the beginning and became vigorous in later periods. In the Yarlung Zangbo region, for example, near southern Chachala and Mainzangbamala, basalt, basaltic andesite and diabase are intercalated in sedimentary formations.

Jurassic volcanic rocks are distributed along the Nujiang River-Bangong Lake fault zone and its neighbouring lake district of northern Xizang (Fig. 1). The rocks consist mainly of intermediate to basic lavas and pyroclastic rocks. In Nyainrong, pyroclastic rocks are intercalated with variegated sandstones and shales. The Jurassic rock series are rather complex. Alkali basaltic and tholeiitic series occur mainly in fault zones and their northern sides, while the calc-alkali series occurs on their southern sides (Fig. 2).

The Early Cretaceous volcanic rocks crop out in a vast region from north of Yarlung Zangbo to the lake district of northern Xizang, but most of them are distributed north of Yarlung Zangbo ophiolite belt and south of the Gandise granite batholith, i.e. Sangsang, Ngamring, Xaitongmoin, Geding and Sangri etc. (Fig. 1). They are mainly andesitic - dacitic lavas and pyroclastic rocks. They are unconformably covered by the Late Cretaceous Xigaze group. In some areas, the ratio between the lava and pyroclastic is variable, but the latter are usually predominant. In Xaitongmoin the pyroclastic formation (200m in thickness) are conformable with marble and siliceous shale and is intruded by a granodiorite whose age is 70-90 m.y. The pyroclastic formation shows a cyclothem consisting of tuff breccia, crystallinoclastic and rock debris tuff, incompetent ignimbrite and rhyolite (Fig. 3). The Early Cretaceous volcanic rocks belong mostly to potassic and common subseries of calc - alkali series (Fig. 4).

In Xizang, the most intense volcanic activity occurred from Late Cretaceous to Early Tertiary and scattered mainly over two regions: The Yarlung Zangbo suture zone and north of the Gandise granite batholith. In the Yarlung Zangbo zone spilite and tholeiite, associated with basic and ultrabasic rocks, radiolarian chert and flysch, exhibit eruptive features of under-water facies, such as the pillow structure of spilite in Rinbung and Congdu or a quenching texture.

The petrochemistry of volcanic rocks show high $Na_2O$ and low $K_2O/SiO_2$ (0.02). The Gandise zone extends nearly a thousand km eastwards from Ngari with a width of about 100 km at its widest section (Fig. 1). The main rock types are andesitic-dacitic lava and pyroclastic rocks intercalated with a little basalt. The acidity or clastic ratio increases from west to east, while in Ngari intermediate lava is predominant. In the middle and eastern section intermediate-acidic rocks prevail and pyroclastic rocks increase too. The geological section from Dagzhuka to Yangbajain may be divided into four formations (Fig. 5). (1) Intermediate pyroclastic rocks and a little trachyte andesite. The pyroclastic rocks show a cyclothem consisting of volcanic breccia and crystalloclastic tuff. (2) Acidic crystalloclastic tuff and tuffaceous sandstone. This formation lies unconformbaly on the former formation. (3) Rhyolite and andesitic basalt. Their radiometric ages are 40 and 52 m.y.,respectively. (4) Intermediate-acidic pyroclastic rocks and tuffaceous sandstone intruded by Xoggu La two-mica granite of 23-35 m.y. in age. These rock series belong mainly to potassic and common subseries of calc-alkalies (Fig. 6). The alkali series have only been found to the north of Margyang and the piedmont of Nyainqentanglha Range, belonging also to the potassic subseries (Fig. 6). This character differs markedly from the Late Cretaceous volcanic rocks of sodic subseries in Yarlung Zangbo ophiolite. Geographically, the volcanics tend to increase in $K_2O/SiO_2$ from south (0.02) to north (0.1) in Gandise. Along the line of Lunggar-Damxung-Poindo, potasium-rich rocks of potasium-rich trachyandesite ($K_2O$ = 5.60-6.80) and leucite basalt outcrop.

The scale of the Early Tertiary volcanic activity is rather small. In Lhunzhub, Mar and Dongo La acidic pyroclastic rocks and lavas lie unconformbly on the Late Cretaceous red beds.In addition, the acidic pyroclastics have been discovered in the Tertiary basins of northern Xizang, Gandise, Qamdo and Ngari, which also belong to the calc-alkali series. Tertiary rhyolite with Norm An/Or less than 0.11 is the equivalent of two-feldspar granite (after Streckeisens classification). Moreover, $K_2$-E rhyolite is the eqivalent of alkali-feldspar granite, whose An/Or is more than 0.11.

Quaternary volcanic eruptions occurred in the Qiangbaqiun, Yongbo Lake and Bamaoqiongzong of Qiangtang (Fig. 1). They are effusive eruptions along certain tectonic fissures. Augite dacite, trachy-andesite. augite andesite and rhyolite are distributed at Yongbo Lake and Qiangbaqu in lying on the southern side of Kunlun Mountains. They belong to the calc-alkali series. Bamaoqiongzong volcanic rocks are characterized by feldspathoid and alkali mafic mineral, for example, shihlunite, leucite tephrite, noselite phonolite and pseudoleucite porphry. They belong to the potassic subseries of alkali series (Fig. 7). The volcanic rocks lie unconformably on the Quaternary lacustrine facies the stratigraphy of which appears to have been subjected thermal metamorphsm.

The diagram of Solidification Index ($SI = \frac{100MgO}{MgO+Fe_2O_3+FeO+Na_2O+K_2O}$) versus $Na_2O + K_2O$ from Triassic to Quaternary shows (Fig. 8): (1) The evolutionary trends of volcanic rocks in Yarlung Zangbo ophiolite belt are different from those of $K_2$ - E volcanic rocks in Gandise. (2) The evolutionary trends of the Early Cretaceous calc-alkali volcanic rocks on the north of Yarlung Zangbo is quite similar to those of the ophiolite belt of Yarlung Zangbo. (3) The magma of $K_2$ - E calc-alkali volcanic rocks in Gandise may be dacitic - andesitic in the range of SI = 25-30.

In Xizang the Cenozoic - Mesozoic volcanic activities are intimately related to the northward pushing of the Indian Plate. Since Cretaceous, India separated from Australia and drifted towards the north. Thus, calc-alkali andesitic magma and island arc tholeiitic magma were produced during the subduction of the oceanic lithosphere. This process of magma generation has been interpreted by many petrologists. Nevetheless, another kind of magmatism must have existed in the Gandise island arc region. During the period from 70 to 40 m.y., large-scale volcanic activities occurred along a narrow area in Gandise, extending nearly a thousand kilometers in an E-W direction. A reasonable explanation is that the dip angle of the Benioff zone at that time must have been great (Fig. 9). The drifting velocity of the Indian Plate was thus accelerated considerably and amounted probably to about 15-20 cm/yr. The rapid subduction of the Indian Plate, in turn, led to the acceleration of mantle convection, resulting in the generation of another kind of magma through the partial melting of Gandise crust. Other evidence supporting this speculation is that the $Sr^{87}/Sr^{86}$ ratio is 0.708 for the $K_2$ -E volcanic rocks in Gandise.

The Tethyan sea was closed about 40 m.y. ago. At the same time, activity of the Yarlung Zangbo convergence belt stopped. A series of intracontinental convergence belts were formed. Quaternary volcanic eruptions in the northern Qiangtang Plateau were caused by the southward downthrusting of Tarim block along a new intracontinental, continental convergence belt or the resurgent Kunlun fault.

## References


(1) Yin Jixiang, Wang Yigang, Zhang Minliang, Stratigraphy of the Mt. Qomolangma region ——Triassic. In Report on the Scientific Expedition to Mt. Qomolangma Region (1966-1968), Geology, Science Press, 81-127 (1974) (in Chinese ).

(2) Deng Wanming, A preliminary study on the petrology and petrochemistry of the Quaternary volcanic rocks of Northern Tibet Autonomous region. Acta Geological Sinica, (2) 148-162 (1978) (in Chinese).

(3) Jin Chengwei, Zhou Yunsheng, Igneous rocks belt in the Himalayas and the Gandes arc and their genetic model. Scientia Geologica Sinica, (4), 287-312 (1978) (in Chinese).

(4) Jin Chengwei, Volcanic Rocks of Xizang. In Magmatism and metamorphism in Xizang, China ( in press )(in Chinese).

(5) Wu Haoruo et al., The Cretaceous of Laze-Jiangze district, Southern Xizang, Scientia Geologia Sinica, (3), 250-262 (1977) (in Chinese).

(6) Boettcher, A. L., Volcanism and orogenic belts —— the origin of andesites. Tectonophysics, 17, 223-240 (1973).

(7) Green, T. H., Ringwood, A. E., Genesis of the calc-alkaline igneous rocks suite. Contr. Miner. Petrol., 18, 105-162 (1968).

(8) Green , T. H., Crystallization of calc-alkaline andesites under controlled high-pressure hydrous condition. Contr. Miner. Petrol., 34, 150-166 (1972).

(9) Hutchison, C. S., Laboratory handbook of petrographic techniques. A Wiley-Intersince Publication (1974).

(10) Irvine, T. N., Baragar, W. R. A., A guide to the chemical classification of the common volcanic rocks. Can. Jour. Earth Sci., 8 (5), 523-548 (1971).

(11) Kushiro, I., Origin of some magmas in oceanic and circumoceanic regions. Tectonophysics 17, 211-222 (1973).

(12) Middlemost, E. A. K., A simple classification of volcanic rocks. Bull. Volcanol., 36 (2), 382-397 (1972).

(13) Poldervaart, A., Parker, A. B., The crystallization index as a parameter of igneous differentiation in binary variation diagrams. Amer. Jour. Sci., 262, 281-289 (1964).
(14) Powell, C. McA., Geodynamics of Pakistan. (Abul Farah, Editor; et al) Geol. Surv. Pakistan. Quetta Pakistan 5-24 (1979).
(15) Streckeisen, A. N., Classification of the common igneous rocks by means of their chemical composition, a provisional attempt. Neues Jahrb. Mineral., Monatsh., 1, 1-15 (1976).
(16) Wyllie, P. J., Experimental petrology and global tectonics —— A Preview. Tectonophysics, 17, 189-209 (1973).

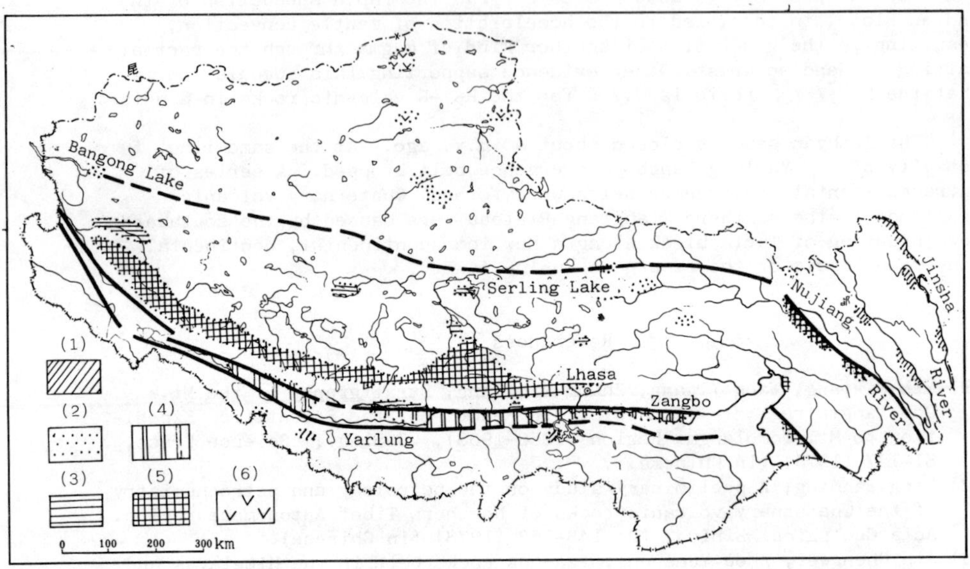


(1) Triassic (2) Jurassic (3) Early Cretaceous
(4) Late Cretaceous (5) Late Cretaceous-Early Tertiary (6) Quaternary

Fig. 1 Simplified diagram showing the distribution of the Mesozoic and Cenozoic volcanic rocks in Xizang

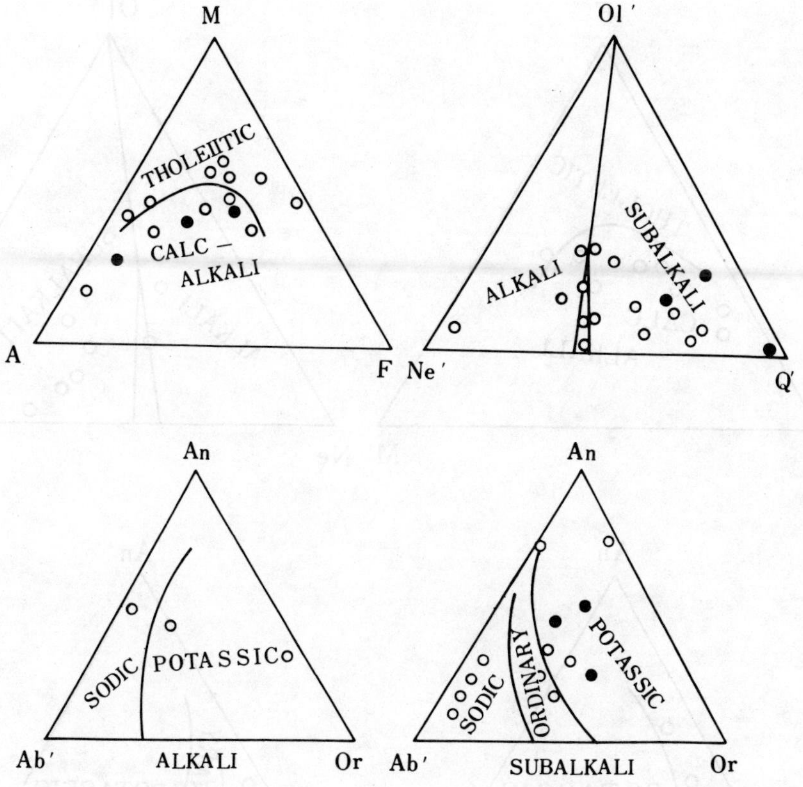


Fig. 2 AFM, Ne'-Ol'-Q' and An-Ab'-Or diagram from Jurassic volcanic rocks
The solid circles represent the sample collected from south of Bangong Lake-Nujiang River zone

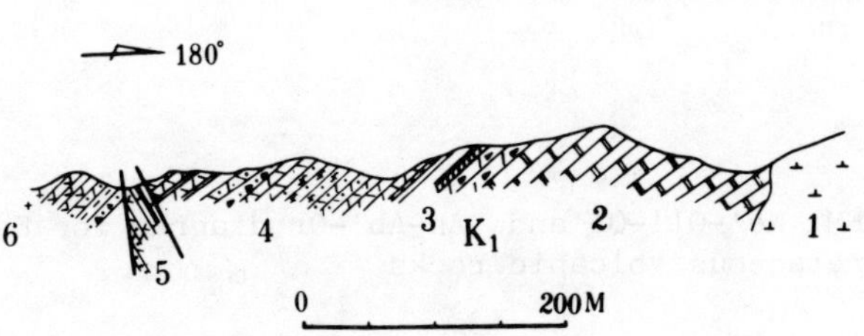


Fig. 3 Geological section in Nagunju, Xaitongmoin,

1. diorite
2. marble
3. siliceous shale and ore bed
4. intermediate acidic volcanic clastic rocks
5. marble and intermediate—acidic volcanic clastic rocks
6. granite

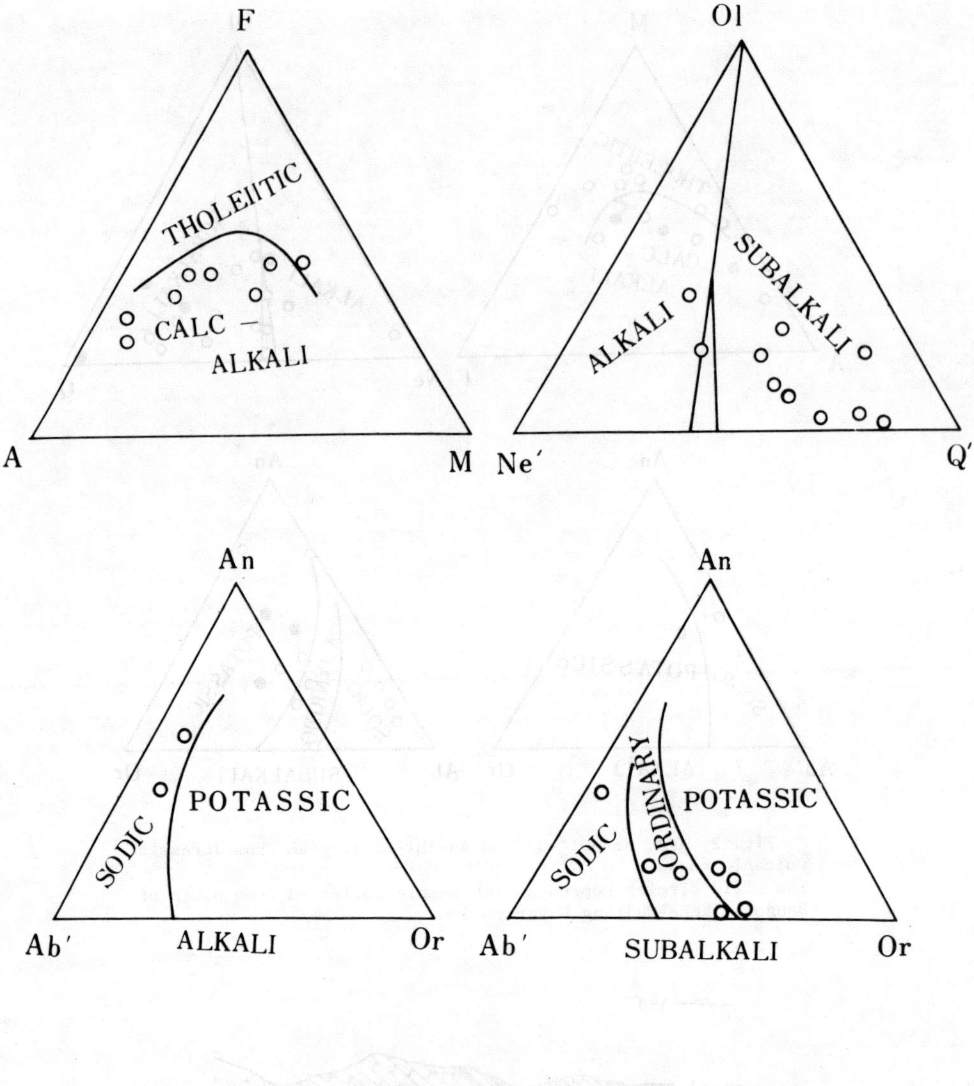


Fig. 4 AFM, Ne'-Ol'-Q' and An-Ab'-Or diagram for Early Cretaceous volcanic rocks

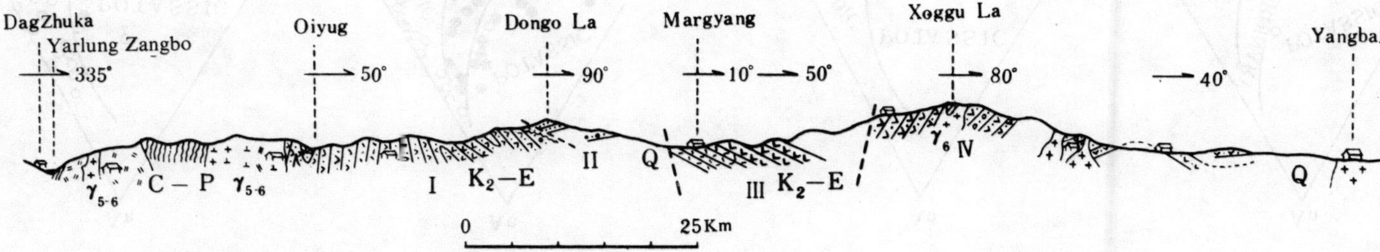


Fig. 5 Geological section between Dag Zhuka and Yangbajain

I. Andesitic clastic rocks and trachyte andesite;
II. Rhyolitic crystalloclastic tuff and tuffaceous sandstone;
III. Rhyolite and andesite basalt (40-52 m. y.);
IV. Andesitic-dacitic clastic rocks and tuffaceous sandstone

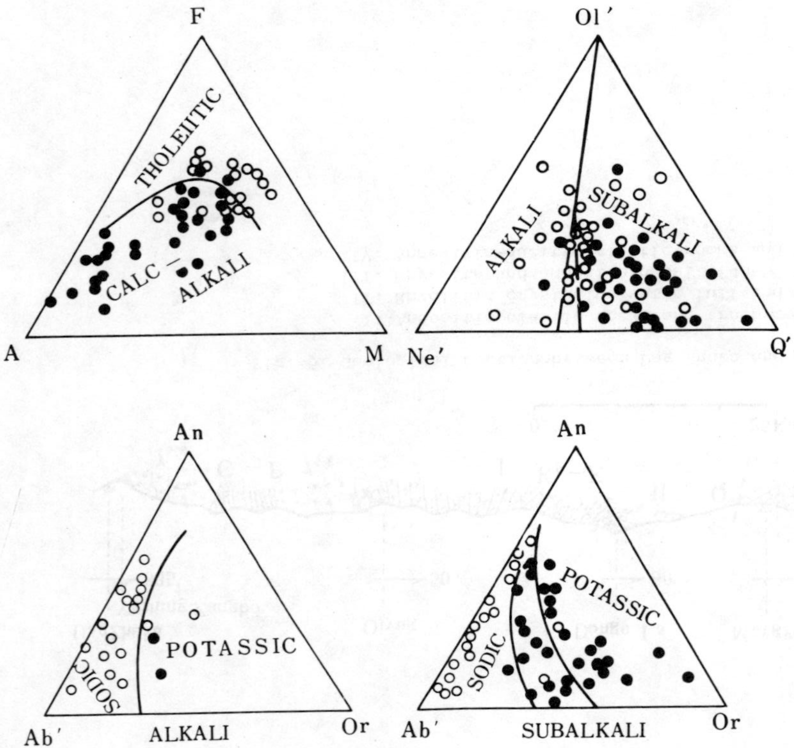


Fig. 6 AFM, Ne'-Ol'-Q' and An-Ab'-Or diagram for Late Cretaceous to Early Tertiary volcanic rocks Yarlung Zangbo o Gandise ●

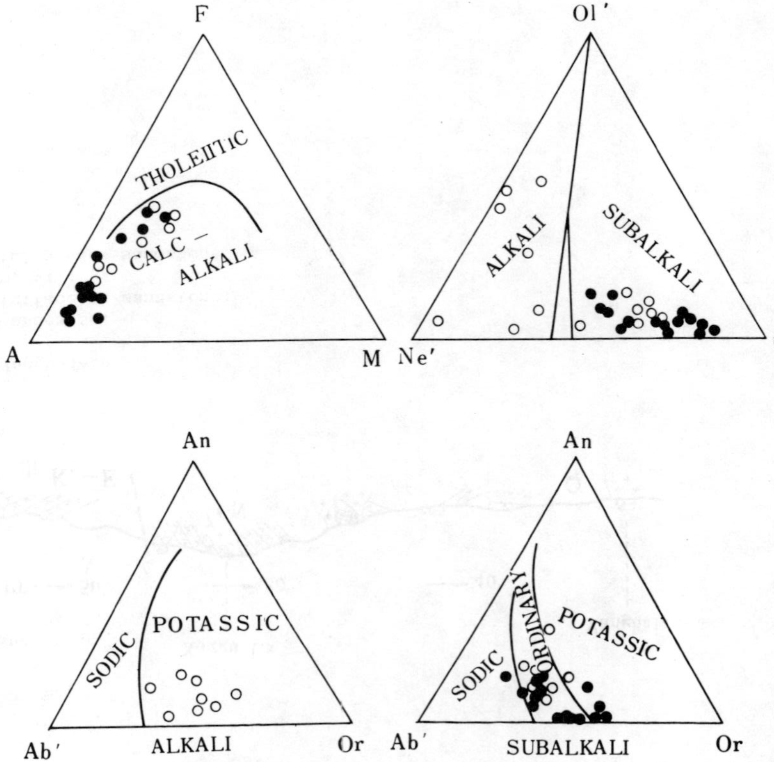


Fir. 7 AFM, Ne'-Ol'-Q' and An-Ab'-Or diagram for Tertiary (●) and Quaternary (o) volcanic cocks

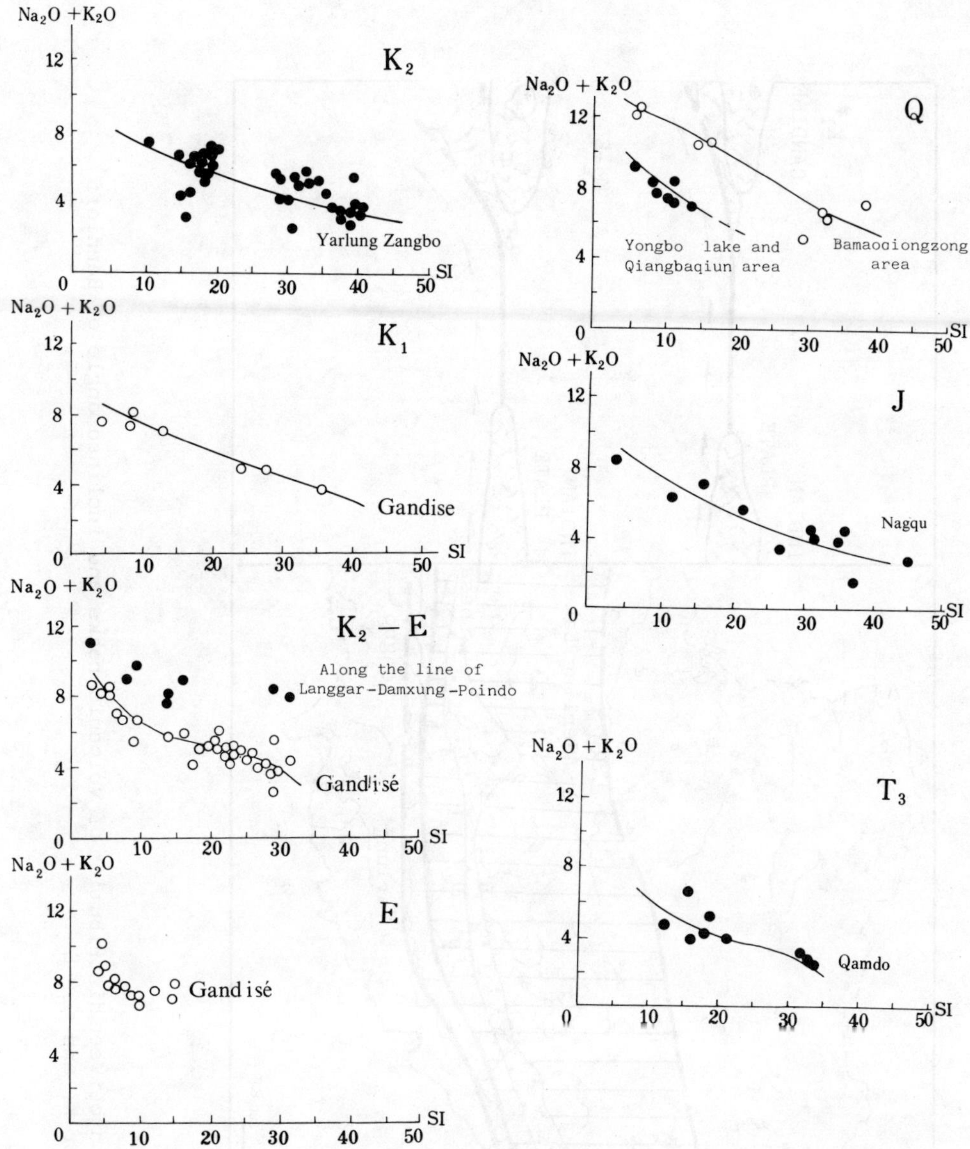


Fig. 8 SI versus $Na_2O + K_2O$ diagram for volcanic rocks from T to Q

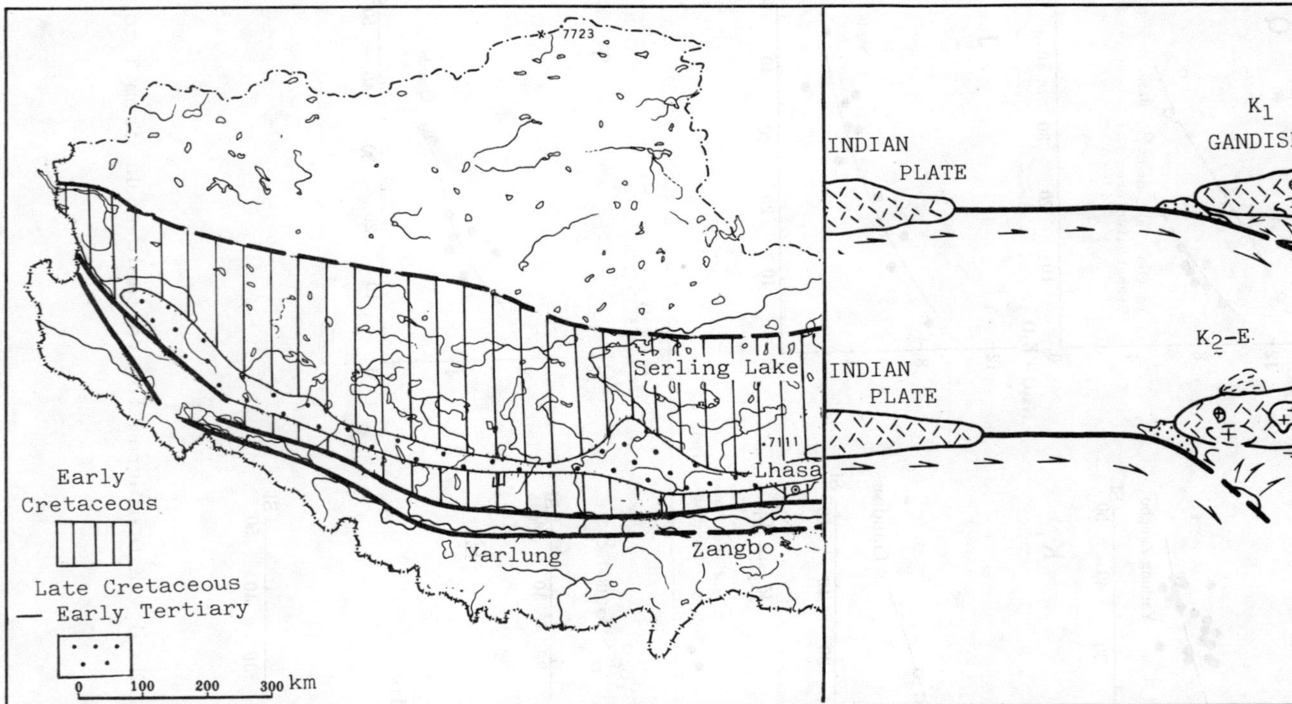


Fig. 9 Relation between distribution of volcanic rocks and inclined angle of Bennioff belt for $K_1$ and $K_2$-E

# A PAIR OF METAMORPHIC BELTS IN SOUTH XIZANG

Zhang Qi, Li Da-zhou, Li Shao-hua
(Institute of Geology, Academia Sinica)

Recently, we have made a preliminary investigation of the metamorphic belts in Xizang and found two entirely different metamorphic belts. In the following, a description of the characteristics of these two belts is given and a tentative interpretation for their origin and interrelation is proposeed.

Fig. 1 shows the distribution of the two metamorphic belts in South Xizang. The Gandise metamorphic belt lies in the north and Yarlung Zangbo River metamorphic belt lies in the south. The former consists of greenschist facies and amphibolite facies. There are andalusite schist, sillimanite gneiss, and staurolite-andalusite schist in this belt, and the latter consists of greenschist facies and possibly, prehnite-pumpellyite facies.

As shown in Fig. 2, it is a cross section through the Gandise metamorphic belt from Chongsar of Gongbogyamda to Nangxian, granite and migmatite are wide spread in vast areas, while metamorphic rocks, mostly of amphibolite facies, occur as relic blocks. Therefore, the contact metamorphism of country rocks is fairly well developed, too. This can hardly be distinguished from regional metamorphism in some cases, and belongs to "the contact-low pressure type of regional metamorphism"(16).

The Yarlung Zangbo River metamorphic belt differs greatly from the former. First, it is narrowly exposed and breaks up now and then along its trend. Second, all of the rocks in it have undergone only low-grade metamorphism, and these are associated exclusively with ophiolites. Sinha Roy in 1977(15) reported glaucophane schist exposed in the Lower Zayu (i.e.Lohit River) district, which is situated at the eastern end of the Yaulung Zangbo River metamorphic belt. He named them high pressure greenschists (glaucophane schist), prehnite-pumpellyite and laumontite facies.

There is a suite of greenschists and pelitic schists in Mt. Namjagbarwa district. It is named the Wa'nyai Qiao formation. Its bottom is composed of metapelites. The upper part is mainly composed of greenschists such as epidote-chlorite schist, biotite-actinolite schist, epidote-biotite-chlorite schist, actinolite-talc schist etc.. Probably, the primary rocks were mainly basic volcanic rocks and serpentine. They are accompanied by ultrabasic rocks. This belt can be subdivided into the biotite zone and almandine zone.

The Lhaze district is situated in the middle of the Yarlung Zangbo River belt. So far, a stratigraphic division has not yet been established unequivocally in this area. In our opinion, it belongs to Late Cretaceous. The conglomerate above the unconformity is likely to be Tertiary (Fig. 3). Most of the rocks have undergone slight regional metamorphism. The metamorphic minerals in siliceous rocks are actinolite, chlorite, stilpnomelane etc. Epidote, chlorite, albite, calcite, prehnite, quartz etc. are frequently present in the volcanic rocks.

What pressure type does Yarlung Zangbo River metamorphic belt belong to? This is a very interesting problem. In order to make a valid judgment about the metamorphic pressure type of this belt, we are presenting here some available information regarding metamorphic pressure.

First, according to Banno (1964), the early appearance of almandine and hornblende indicates that the metamorphic pressure is rather high(16). The earliest almandine associates with albite in Wa'nyai Qiao formation, Mt. Namjagbarwa district. This implies that the estimated P-T gradient of Yarlung Zangbo River belt is intermediate between that of East Otago and that of Barrow's zone in Dalradian.

Second, we notice that the average $b_o$ value of muscovites in Gandise belt is only 9.003 Å, which is rather small, whereas that of Yarlung Zangbo River belt is 9.043Å, which lies between the $b_o$ values of East Otago(9.038Å, after Sassi et al., 1974) and Sanbagawa (9.045Å, after Iwasaki et al., 1978),and falls in the high pressure field.

Third, when we studied the muscovites in metamorphic rocks in Wa'nyai Qiao formation, Mt. Namjagbarwa district, we obtained a magnesia-RM* diagram of muscovite-phengite series in greenschist-blueschist facies (Fig. 4). Three areas can be distinguished in this diagram. The first one is called area I of high pressure facies series. The muscovites falling into this area are exclusively associated with glaucophane and lowsonite . Actually, the latter are widely found in metabasites and some metagraywackes.

The muscovites falling into area II of the high pressure facies series are evidently different from those in area I. Generally speaking, in this area there are almost no typical high pressure mineral assemblages like those in area I. For example, all the muscovites from the metapelites of Sanbagawa belt and Sangun belt in Japan and Xingyang metamorphic belt in Honan, China,belong to the high pressure type and fall into area II. The muscovites from the metapelites in Wa'nyai Qiao formation can hardly fall into the same area by mere coincidence. Perhaps they belong to the high pressure type, too. Area III represents the field of medium pressure facies series in metapelites. But, here we shall not discuss it further.

To sum up, the features of metamorphism in Gandise belt and Yarlung Zangbo River belt are totally different, as shown in Table 1.

As shown above, the two metamorphic belts show essentially different characteristics and appearances, but they are closely associated with each other, extend in parallel direction, and have similar ages of metamorphism. These features strongly suggest that they probably constitute a pair of metamorphic belts: The Gandise belt is the low-pressure type, and the Yarlung Zangbo River belt is the high-medium pressure type. Conceivably, they were formed by the rapid underthrusting of the Indian Plate beneath Eurasian Plate during the period from Late Cretaceous to Tertiary.

So far as we know, this pair of metamorphic belts extends far ayay as shown in Fig. 1. Towards the north-west it extends to Kashmir and Pakistan, and towards south-east it passes through India, Burma, Malaya and arrives in Sumatra and Java. It is clearly shown in the map that

* RM= $\frac{1}{2}Fe_2O_3 + FeO + MgO$ (molecular number).

the Nanga Parbat and Namjagbawa districts lie at the north-west and north-east corners of the Indian Plate, respectively. Both of them wedge deeply into the Eurasian Plate, and have been strongly squeezed. Consequently, in the low-pressure belt the metamorphic grade is higher, whereas in the high-medium pressure belt the metamorphic pressure is higher. So far all the glaucophanes that have been found are distributed almost exclusively over the areas close to the two corners, for example, Lower Zayu district(15), Dras volcanic rocks in Kashmir(8), and Lower Swart-Buner schist group in Pakistan( 5).

## References


(1) Black, P. M., Mineralogy of New Caledonian metamorphic rocks.Contr. Mineral. Petrol., 49, 269-284 (1975).

(2) Brown, E. H., The greenschist facies in part of eastern Otago, New Zealand. Contr. Mineral. Petrol. 14, 259-292 (1967).

(3) Butler, B. C. M., Chemical study of minerals from the Moine schists of the Ardnamurchan area, Argyllshire, Scotland. J. Petrol. 8, 233-267 (1967).

(4) Cipriani, C., Sassi, F. P., Scolari, A., Metamorphic white micas: Definition of paragenetic fields. Schweiz. Mineral. Petrog. Mitt., 51, 259-302 (1971).

(5) Desio, S. A., The occurrence of blueschists between the middle Indus and the Swat Valleys as an evidence of subduction (Northern Pakistan). Accad. Naz. Lincei, Roma, S. 8, VLXII, 1-9 (1977).

(6) Ernst, W. G., Significance of phengitic micas from low-grade schists. Am. Mineral., 48, 1357-1373 (1963).

(7) Ernst, W. G., Seki, Y., Onuki, H. et al., Comparative stuuy of low-grade metamorphism in the California Coast Ranges and the outer metamorphic belt of Japan. Geol. Soc. Amer., Mem., 124, 276p (1970).

(8) Frank. W., Gansser, A., Trommsdorff, V., Geological observations in the Ladakh area (Himalayas). Schweiz. Mineral. Petrog. Mitt., 57, 89-113 (1977).

(9) Iwasaki, M., Sassi, F. P., Zirpoli, G., New data on the K-white micas from the Sanbagawa metamorphic belt and their petrologic significance. Jour. Japan. Associa. Mineral. Petrol., 73, 274-280 (1978).

(10) Lambelt, R. St. J., The mineralogy and metamorphism of the Maine schists of the Morar and Knoydart districts of Inverness-shire. Trans. Rog. Soc. Edin., 63, 553-588 (1959).

(11) McNamara, M., The lower greenschist facies in the Scottish Highlands. Geol. Foren. Stockholm Forh., 87, 347-389 (1965).

(12) Sassi, F. P., Scolari, A., The b₀ value of the potassic white micas as a barometric indicator in low-grade metamorphism of pelitic schists. Contr. Mineral. Petrol. 45, 143-152 (1974).

(13) Seki, Y., Basal spacing of metamorphic white micas and type of metamorphism. Geol. Soc. Jap. Jour., 79, 611-620 (1973).

(14) Shibata, S. (ed.) Petrologic manual of Japan, Part IV, Metamorphic rocks, Asakura Shoten,(1978)(in Japanese).

(15) Sinha, Roy S., Metamorphism and tectonics of the Himalayas, as an example of Eastern Himalaya. Geotectonics, 2, 66-74 (1977) (in Russian).

(16) Turner, F. J., Metamorphic petrology, McGraw-Hill, New York, 403p 1968.

(17) Velde, B.,$Si^{+4}$content of natural phengites. Contr. Mineral. Petrol., 14, 250-258 (1974).

(18) Yeh Da-nian, Li Da-zhou, Dong Gung-fu, Jiu Xiu-wen, 3T-phengite and C-type eclogite in Xingyang metamorphic belt and their tectonic significance, 24 (5), 217-220 (1979),(in Chinese).

Table 1 Comparison of some characteristics of Gandise metamorphic belt with those of Yarlung Zangbo River metamorphic belt

| | Gandise metamorphic belt | Yarlung Zangbo River metamorphic belt |
|---|---|---|
| Metamorphic rocks | Slate, Phyllite, Schist, Gneiss, Amphibolite and Migmatite | Slate, Phyllite, Schist, Metavolcanic Rock |
| Metamorphic facies | Greenschist and Amphibolite | Greenschist and Prehnite-Pumpellyite |
| Index minerals | Andalusite and Sillimanite | |
| Accompanied rocks | Granitic rock | Ophiolite |
| Average bo value of muscovites | 9.003 (Å) | 9.043 (Å) |
| Strata | $O_1$ - K | $K_2$, $T_3$ (?) |
| Metamorphic age | K - R | The End of $K_2$ - R |

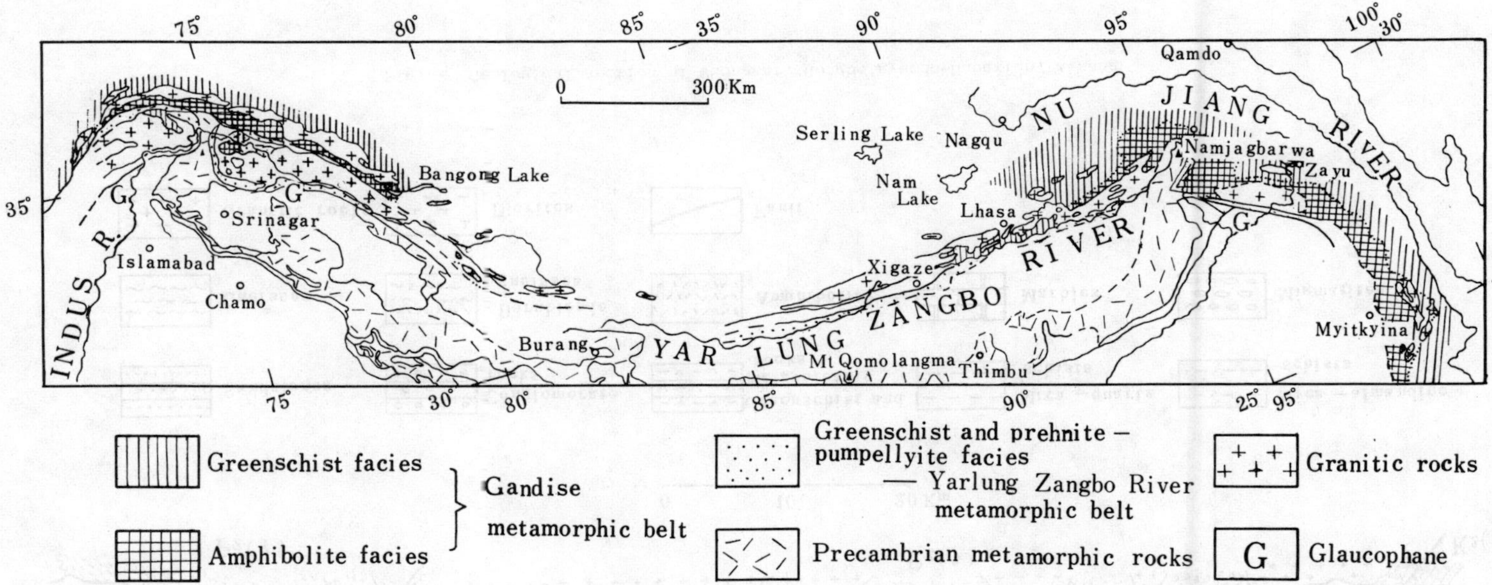


Fig.1 Simplified map showing distribution of metamorphic belts in South Xizang and it's neighbouring districts

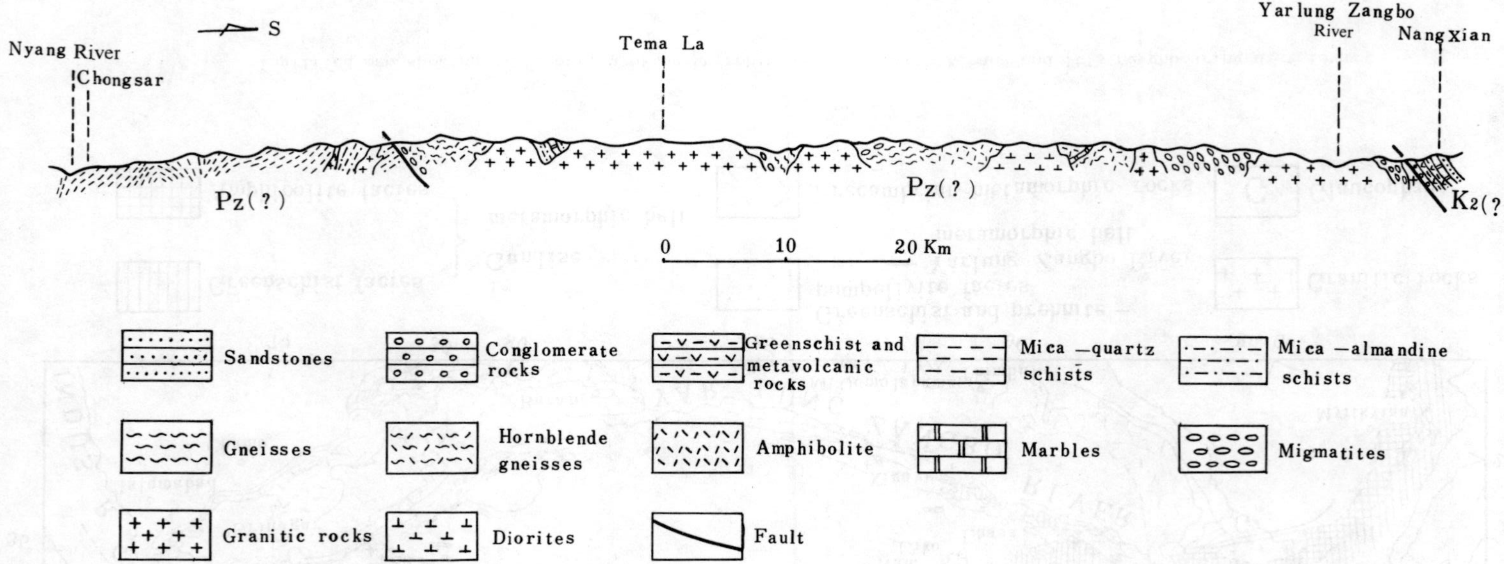


Fig. 2 Geological section of Chongsar, Gongbo'gyamda-Nangxian, Xizang

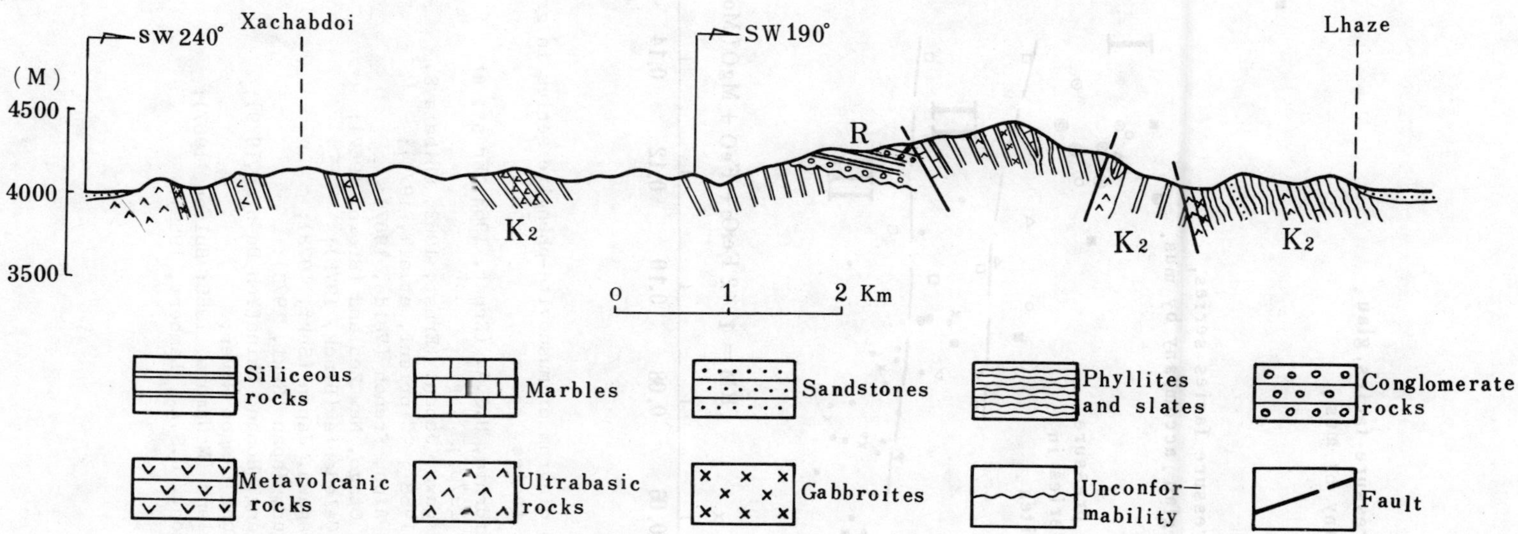


Fig. 3 Geological section of Xachabdoi-Lhaze, Xizang

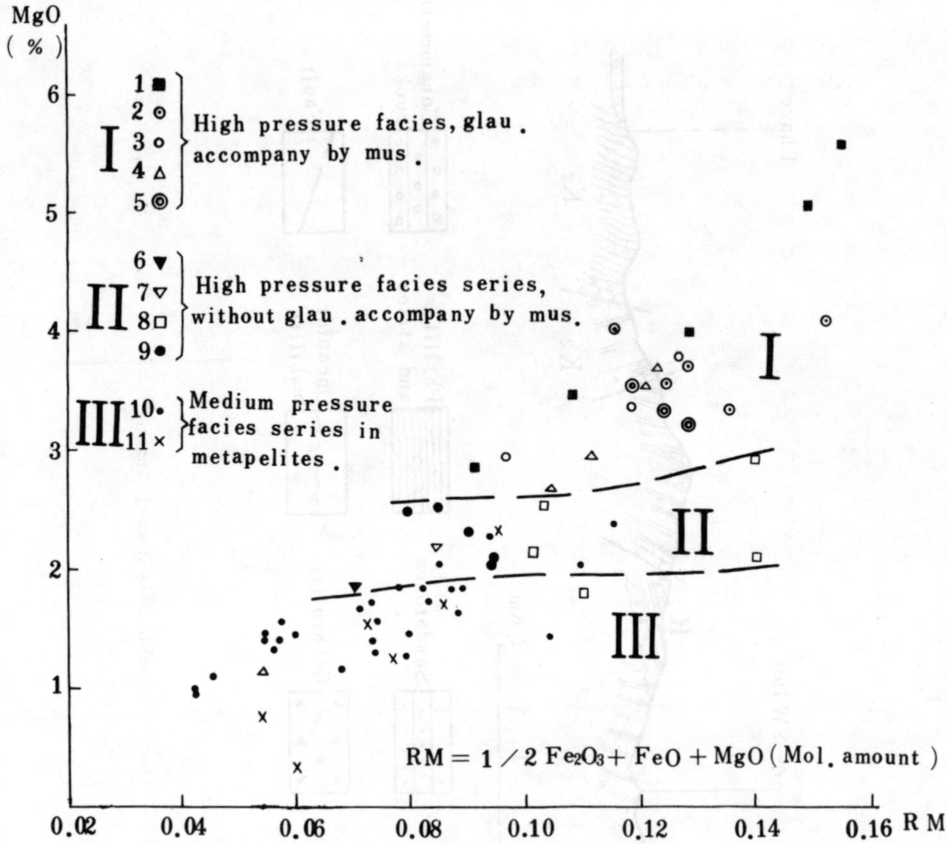


Fig. 4 MgO—RM diagram of muscovite-phengite series in greenschist-glaucophane schist facies.

1. California, U.S.A. (Ernst, 1963; Ernst, et al., 1970);
2. Sanbagawa, Japan. (Ernst, 1963; Shibata S. ed. 1968 ; Cipriani, et al., 1971);
3. West Alps, French (Velde, 1967);
4. East Otago, New Zealand (Brown, 1967);
5. New Caledonia (Black, 1975);
6. Sanbagawa, Japan (Seki, 1973);
7. Sangun, Japan (Seki, 1973);
8. Xinyang, Honan, China(Yeh Da-nian,1979);
9. Yarlung Zangbo River;
10. Scotland (McNamara, 1955; Bulter, 1967);
11. Vermont, U.S.A. (Lambert, 1959).

# PETROCHEMISTRY OF GRANITES IN XIZANG AND AN APPROACH TO THEIR ORIGIN

Liu Guan-jian, Zhang Kui-wu
(Institute of Geology, Academia Sinica)

## INTRODUCTION

Various granitic rocks of different ages are distributed extensively along the trend of mountains in Xizang. According to the geographical distribution and the rock type, from south to north, they can be roughly divided into following five belts: the tourmaline-muscovite granite belt in High Himalayas; the gneissic two-mica granite belt in Lhagoi Kangri in the lower watershed of Southern Xizang; the Gandise granite belt; the Northern Xizang granite belt; and the granite belt in Qangtang-Qamdo (Fig. 1). In this paper, the authors give a brief discussion of the petrochemical features of major elements in granites from the five belts, the differences among them and their different origins.

The intrusive bodies of the tourmaline-muscovite granite belt in the High Himalayas stretch along the border of Xizang, China. Their position is obviously controlled by the Main Northern Thrust. Except for a small amount of medium-sized stocks, they are mostly sills and dikes, and are emplaced in the crystalline schist and the northern Lower Palaeozoic sequence. The emplacement age is 20-10 m.y.

The intrusive bodies of the gneissic two-mica granite belt in Lhagoi Kangri are emplaced in a suite of low-grade metamorphic rocks, and are obviously controlled by buried faults of E-W trend and some minor faults with N-S trends in the lower watershed in South Xizang. Their isotopic ages are 40-20 m.y. Their dimensions are similar to those of the tourmaline-muscovite granite in the High Himalayas.

The Gandise granite belt extends more than 2,000 km from east to west within Xizang with a width from dozens to more than a hundred km from south to north. Rocks in this belt are products of the intermediate to acidic magmas emplaced in different episodes. The principal rock types are gabbro, diorite, tonalite, granodiorite, and granite, etc. Granodiorite is dominant among them. All bodies lie north of Yarlung Zangbo ultrabasic rock belt as huge batholiths and large stocks. From south to north, the rocks change from basic to acidic. The hypogene action in Gandise belt is accompanied by strong basic to acidic volcanism whose ages range from 110 to 12 m.y., but concentrate mainly within 100-50 m.y.

The rock types, body dimensions and distribution pattern of the Northern Xizang granite belt are similar to those of the Gandise belt. Their isotopic ages are from 120 to 70 m.y.

The dimensions of the granite belt in Qangtang-Qamdo are much smaller than those of the Gandise granite belt. This belt is characterized by small stocks which crop out discontinuously. Stocks are products of the

Hercynian to Early Yenshanian magmatism. The rock type is similar to that of Gandise granite belt. However, the variation trend of the petrochemical features is opposite to that of the Gandise granite belt, i.e., its acidity increases gradually from north to south.

Detailed descriptions of the petrological and geological features of the five belts mentioned above can be found in a monograph entitled " Magmatism and Metamorphism in Xizang, China. "(11).

## PETROCHEMICAL FEATURES

The data on the chemical composition of these granitic rocks have already been given and classified in "Magmatism and Metamorphism in Xizang, China". The average chemical compositions of the five belts are listed in Table 1 of this paper.

Petrochemical vector diagrams of Zavarickii were constructed respectively for different petrographical belts. The granitic rocks in Xizang may be divided into two types, a Himalayan type and a Gandise type : (1) the tourmaline-muscovite granites in High Himalayas and the gneissic two-mica granites in Lhagoi Kangri are called Himalayan type ; and (2) the granitic rocks of Gandise, Northern Xizang and Qangtang-Qamdo are called Gandise type.

The Himalayan Type. The rocks of this type are mostly acidic and high alkali and aluminium over-saturated.Only a few of them belong to a normal series. On the alkali plane, however, points of tourmaline-muscovite granite in High Himalayas are slightly away from sb axis, and the inclinations of vectors are rather steep (Fig. 2a).

The Gandise Type. The granitic rocks of this belt may be similarly sub-divided into the following two chemical suites. (1) The normal suite. Its points distribute in an inclined zone on the alkali plane. From bottom to top, the points migrate to the right, and the slopes of vectors become smaller and smaller, reflecting a continuous variation from intermediate-acidic to acid. (2) The aluminium-oversaturated suite. Points are distributed also in an inclined zone on the alkali face, and scatter in the upper left of the normal suite. The inclinations of the vectors are similar to those of gneissic two-mica granites in Lhagoi Kangri, or steeper (Fig. 2b and 2c).

The difference between areas of distribution of the two suites described above is most pronounced in Gandise granite belt. Rocks of the normal suite are mainly exposed in the southern flank of Gandise Range and are relatively concentrated in the western and middle parts along E-W direction. Rocks of the aluminium-oversaturated suite distribute dominantly on the crest and the northern flank of the Gandise Range and are relatively concentrated in the eastern part along the E-W direction.

Diagrams of $SiO_2$ versus several principal oxides show that rocks of various belts in Xizang can also be divided into two groups as mentioned above.

The Himalayan Type. The contents of principal oxides in various rocks are similar (Fig. 3).

The Gandise Type. Contents of other principal oxides of granitic rocks of this type exhibit a variation similar to that of $SiO_2$. The contents of principal oxides vary in the following way : (1) contents of MgO, FeO + $Fe_2O_3$, CaO and $Al_2O_3$ decrease with an increase of $SiO_2$; (2) $Na_2O$ content does not vary with the increase of $SiO_2$ ; and (3) $K_2O$ content increases markedly with the increase of $SiO_2$ content (Fig. 3).

In Q-Ab-Or and An-Ab-Or diagrams, points of the tourmaline-muscovite granite in High Himalayas and the gneissic two-mica granite in Lhagoi Kangri fall into similar areas. According to their geographical distributions, granitic rocks in the Gandise, Northern Xizang and Qangtang-Qamdo can be divided into two or three groups and are plotted separately, i.e., from south to north, the southern and northern parts of the Gandise granite belt, and the Temo La, Azhag Kanglha and Zhaxize sub-belts of the North Xizang. Toward this direction, contents of $SiO_2$ and $K_2O$ increase. However, $Na_2O$ content does not vary from south to north. So the trend mentioned above reflects only a relative variation for contents of $K_2O$ and $Na_2O$. Contrary to that in the Gandise belt, the contents of $SiO_2$ and $K_2O$ decrease clearly from south to north in the Qangtang-Qamdo granite belt (Fig.4).

Granitic rocks in each belt are classified on the basis of their norms. Average normal quartz, albite and orthoclase are plotted in Q-Ab-Or diagram (Fig. 5). All of the points of granites fall near the eutectic point, whereas those with higher basicity from Gandise, Northern Xizang and Qangtang-Qamdo belts fall (to a greater extent) near the albite corner. Fig. 5 shows that confining pressures are different for different suites during the formation of magmas (Fig. 5).

The results, mentioned above, are consistent with those of the granitic rocks from Circum-Pacific areas(1,2,3,5,6,7,8).

## DISCUSSION

According to the results mentioned above, we believe that except for the Qangtang-Qamdo granite belt, the formation of granitic rocks of other four belts was related to the northward motion of the Indian Plate, the subduction of the front margin of Tethys beneath the Eurasian Plate and finally the collision between the Indian and Eurasian plates. The interaction between the two plates during subduction was friction and deformation and resulted in the generation of a large amount of heat. Besides, the initial temperature in the upper mantle was very high then. For the partial melting of pyrolite and the possible underthrusting oceanic crust, water was indispensible, which was released through dehydration of the hydrous minerals in the underthrusting oceanic lithosphere. In mid-Mesozoic (130 m.y.) India was separated from the Antarctic continent and moved northward. The northern margin of Tethys subducted beneath Eurasian Plate. The rate of the Indian Plate towards Eurasia was very small, about 4-6 cm/y(4). Since the underthrusting rate was also rather slow, sufficient heat in oceanic lithosphere could be generated only when it was subducted to great depths. For a similar reason, it could introduce enough hydrous minerals into the mantle in order to dehydrate them only when it had been subducted for quite a long time. As a result, emplacement of the granitic magmas occurred in the north, and the Northern Xizang granite belt was formed. During Late Cretaceous or Early Tertiary (80 m.y.), the Indian Plate was again separated from Madagascar, and the average northward rate increased greater than ten cm/y(4). It is conceivable that adequate heat and water could be produced at shallower depths in the subduction zone. Consequently, the resulting granitic magmas were emplaced to the south of Northern Xizang granite belt, and the

Gandise granite belt was thus formed.

The Indian Plate collided with the Eurasian Plate in Early Eocene (40 m.y.). The latter was an impediment for the former's northward motion. The buried fault parallel to the lower watershed of South Xizang (about 40 m.y.), and the main northern thrust parallel to the Himalayas (about 20 m.y.) were then formed. At the same time, partial melting of the crust occurred and magmas were emplaced upward. Subsequently, the two belts, namely, the gneissic two-mica granite belt in Lhagoi Kangri (40-20 m.y.) and tourmaline-muscovite granite belt in High Himalayas (20-10 m.y.) were finally formed.

## References


(1) Batema, P. C., Dodge, F. C. W., Variations of major chemical constitutent across the Central Sierra Nevada Batholith. Geol. Soc. Amer., Bull., 81 (2), 409-420 (1970).

(2) Dickinson, W. R., Petrogenetic significance of geosynclinal andesite volcanism along the Pacific margin of North America. Geol. Soc. Amer. Bull., 73, 1241-1256 (1962).

(3) Kuno, H., Origin of Cenozoic petrographic provinces of Japan and surrounding areas. Bull. Volcan.,Ser. 2, 20, 37-76 (1959).

(4) Molnar, P., Tapponnier, P., Cenozoic tectonics of Asia; Effects of a continental collision. Science, 189 (4201), 419-426 (1975).

(5) Moore, J. G., The quartz diorite boundary line in the Western United States. Jour. Geol., 67, 198-200 (1959).

(6) Moore, J. G., Grantz, A., Blake, M. C., The quartz diorite line in northwestern North America. Geol. Sur. Res., 1962 U.S. Geol. Survey Prof. 450E, E89-93 (1963).

(7) Oba, N., Petrochemical studies of Kyushu Oquter zone granites, Japan. Jour. Geol. Soc. Japan, 68 (789), 162-171(1962).

(8) Smith, T. E., Riddle, C., Jackson, T. A., Chemical variation within the coast plutonic complex of British Columbia between lat.$53^{o}$ and $55^{o}$N. Geol. Soc. Amer. Bull., 90 (4), 346-359 (1979).

(9) Tuttle, O. F., Bowen, N. L., Origin of granite in light of experimental study in the system $NaAlSi_3O_8$-$KAlSi_3O_8$-$SiO_2$-$H_2C$. Geol. Soc. Amer. Mem., 74, 153 (1958).

(10) Zavarickii,A. H., Introduction for petrochemistry of igneous rocks. Izd. Akad. Nauk., SSSR, (1950) (in Russian).

(11) Zhou Yun-sheng et al., Magmatism and Metamorphism in Xizang, China. Science Press (in press)(in Chinese).

Table 1 The average chemical composition of granitic rocks from different granitic belts in Xizang

| Belt | Rock type | $SiO_2$ | $TiO_2$ | $Al_2O_3$ | $Fe_2O_3$ | FeO | MnO | CaO | MgO | $Na_2O$ | $K_2O$ | $P_2O_5$ | Number of analyses |
|---|---|---|---|---|---|---|---|---|---|---|---|---|---|
| High Himalayas | Tourmaline-muscovite granite | 73.73 | 0.15 | 13.85 | 1.22 | 0.56 | 0.01 | 0.77 | 0.52 | 3.89 | 4.28 | 0.14 | 36 |
| Lhagoi Kangri | Gneissic two-mica granite | 73.63 | 0.14 | 13.75 | 1.54 | 0.85 | 0.006 | 1.63 | 0.76 | 3.51 | 3.77 | 0.07 | 16 |
| Gandise | Gabbro | 51.33 | 0.72 | 19.17 | 5.78 | 4.40 | 0.12 | 7.86 | 5.00 | 3.50 | 0.55 | 0.34 | 1 |
| | Quartz-diorite | 58.30 | 0.65 | 16.73 | 5.00 | 3.68 | 0.21 | 6.17 | 3.21 | 3.78 | 1.89 | 0.28 | 10 |
| | Quartz-monzodiorite | 63.97 | 0.65 | 16.14 | 3.72 | 2.05 | 0.07 | 3.63 | 1.95 | 3.65 | 3.58 | 0.17 | 12 |
| | Granodiorite | 68.22 | 0.67 | 14.44 | 2.91 | 1.77 | 0.06 | 3.09 | 1.65 | 3.62 | 2.98 | 0.14 | 18 |
| | Biotite monzonitic granite | 73.05 | 0.22 | 13.28 | 1.68 | 1.04 | 0.02 | 2.73 | 0.74 | 3.69 | 3.71 | 0.10 | 25 |
| | Granite | 74.24 | 0.19 | 12.68 | 1.12 | 1.12 | 0.008 | 1.01 | 0.76 | 3.17 | 4.97 | 0.08 | 8 |
| | Average | 67.56 | 0.48 | 14.66 | 2.89 | 1.81 | 0.07 | 3.33 | 1.66 | 3.58 | 3.43 | 0.16 | |
| Northern Xizang | Quartz -diorite | 61.85 | 0.82 | 15.39 | 2.95 | 4.08 | 0.10 | 5.09 | 3.28 | 2.91 | 2.39 | 0.25 | 9 |
| | Quartz-monzodiorite | 66.78 | 0.63 | 15.06 | 2.31 | 2.42 | 0.05 | 4.18 | 1.75 | 3.55 | 2.68 | 0.19 | 5 |
| | Granodiorite | 67.54 | 0.38 | 14.81 | 2.96 | 2.30 | 0.08 | 3.39 | 1.63 | 3.24 | 3.93 | 0.17 | 7 |
| | Biotite granite | 72.93 | 0.35 | 12.09 | 1.11 | 1.74 | 0.02 | 1.75 | 0.94 | 3.09 | 4.10 | 0.16 | 17 |
| | Granite | 73.32 | 0.27 | 13.25 | 1.41 | 1.12 | 0.02 | 1.37 | 0.74 | 2.92 | 4.48 | 0.15 | 9 |
| | Average | 69.59 | 0.43 | 13.56 | 1.99 | 2.10 | 0.05 | 2.87 | 1.46 | 3.10 | 3.50 | 0.19 | |
| Qangtang-Qamdo | Quartz-diorite | 61.34 | 0.73 | 15.85 | 1.63 | 4.37 | 0.09 | 4.76 | 2.61 | 3.10 | 2.09 | 0.09 | 9 |
| | Granodiorite | 68.24 | 0.53 | 14.73 | 1.31 | 2.15 | 0.05 | 2.37 | 1.50 | 3.03 | 4.18 | 0.08 | 8 |
| | Biotite hornhlende granite | 73.35 | 0.34 | 13.13 | 1.09 | 1.21 | 0.03 | 1.56 | 0.64 | 3.35 | 4.88 | 0.09 | 10 |
| | Granite | 72.45 | 0.25 | 13.78 | 0.78 | 1.14 | 0.04 | 1.43 | 0.49 | 2.52 | 5.32 | 0.10 | 4 |
| | Average | 68.30 | 0.52 | 14.24 | 1.43 | 2.53 | 0.06 | 2.95 | 1.55 | 2.97 | 3.68 | 0.09 | |

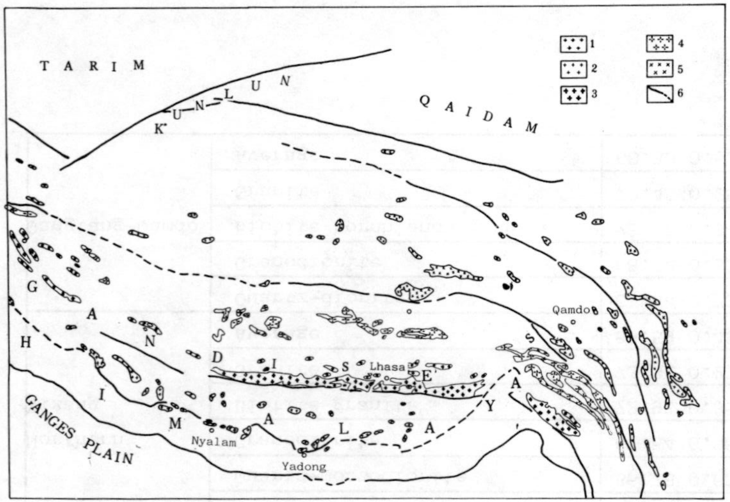


Fig. 1 Distribution of granitic rocks in Xizang

1. Tourmaline-muscovite granite belt in High Himalayas;
2. Gneissic two-mica granite belt in Lhagoi Kangri;
3. Gandise granite belt;
4. Northern Xizang granite belt;
5. Granite belt in Qangtang-Qamdo;
6. Fault

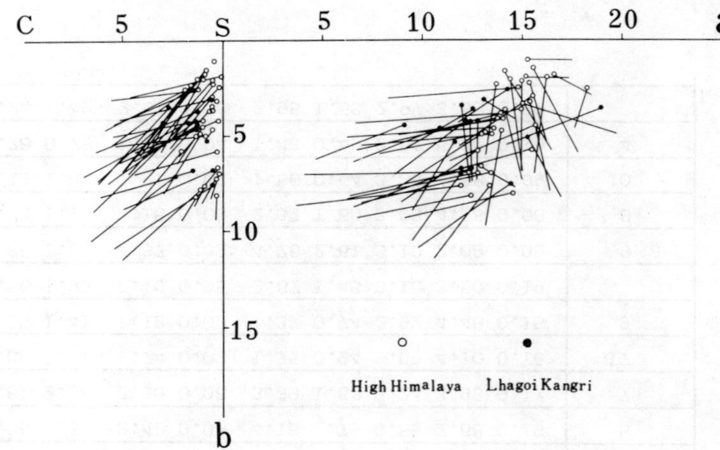


Fig. 2 - a Zavarickii's petrochemical vector diagram of granitic rocks from Xizang

High Himalayas and Lhagoi Kangri

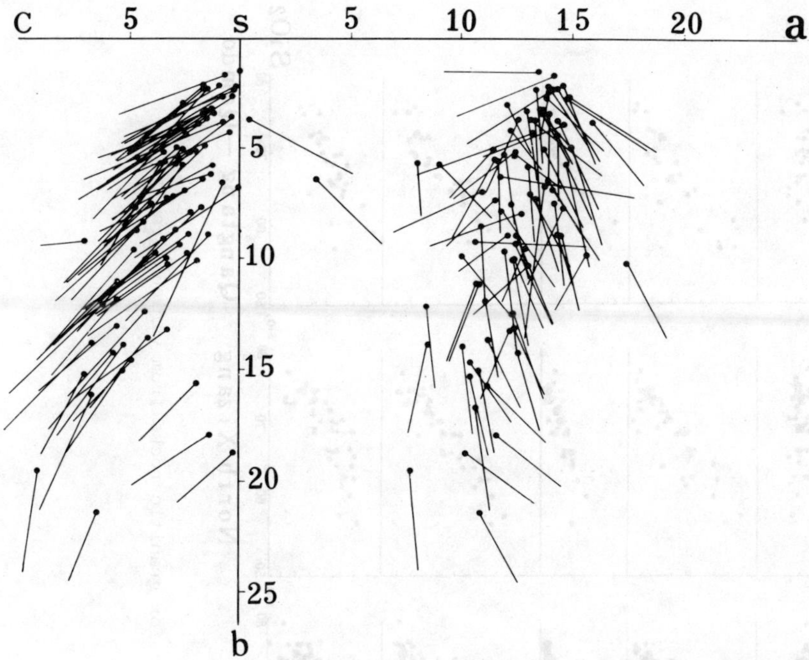


Fig. 2 - b Zavarickiis petrochemical vector diagram of granitic rocks from Xizang

Gandise

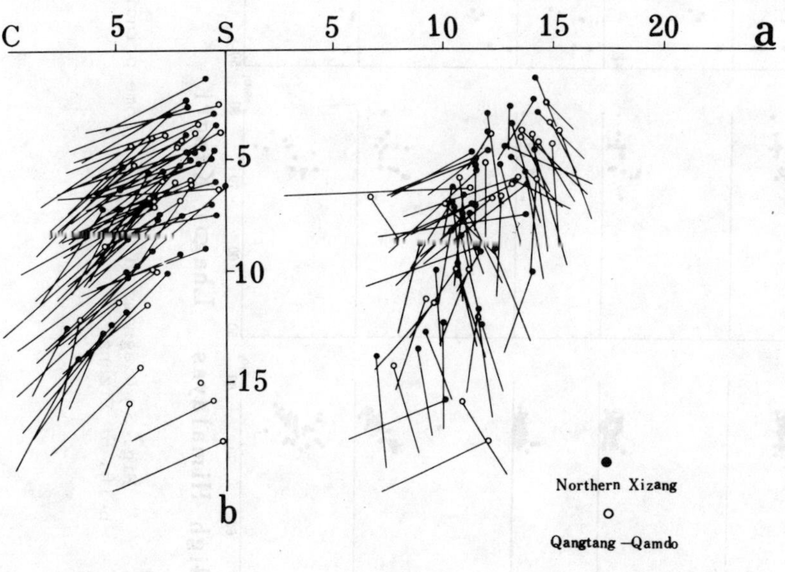


Fig. 2 - c Zavarickii's petrochemical vector diagram of granitic rocks from Xizang

Northern Xizang and Qangtang-Qamdo

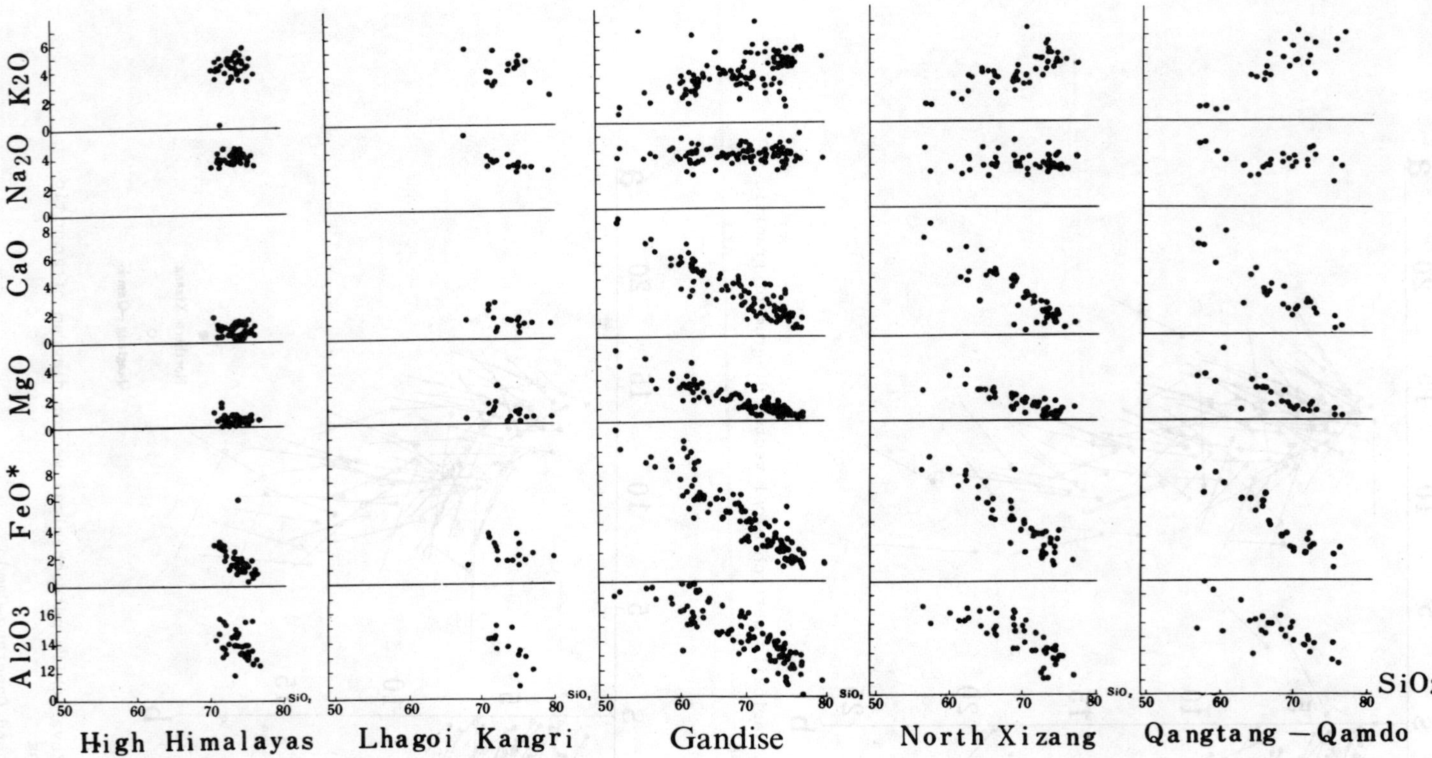


Fig. 3 Diagrams of $SiO_2$ vs. some principal oxides for granitic rocks from various granite belts of Xizang

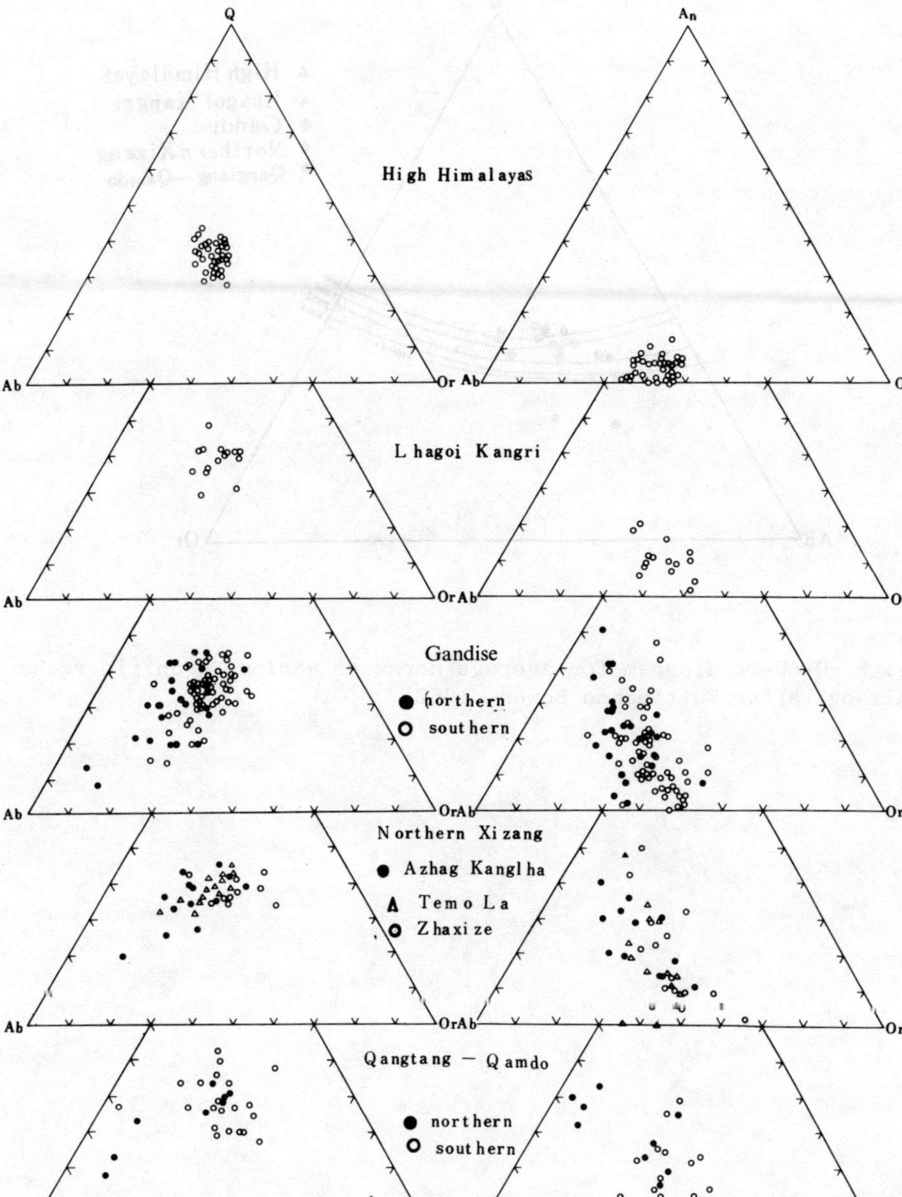


Fig. 4 Q-Ab-Or and An-Ab-Or diagrams for granitic rocks from Xizang

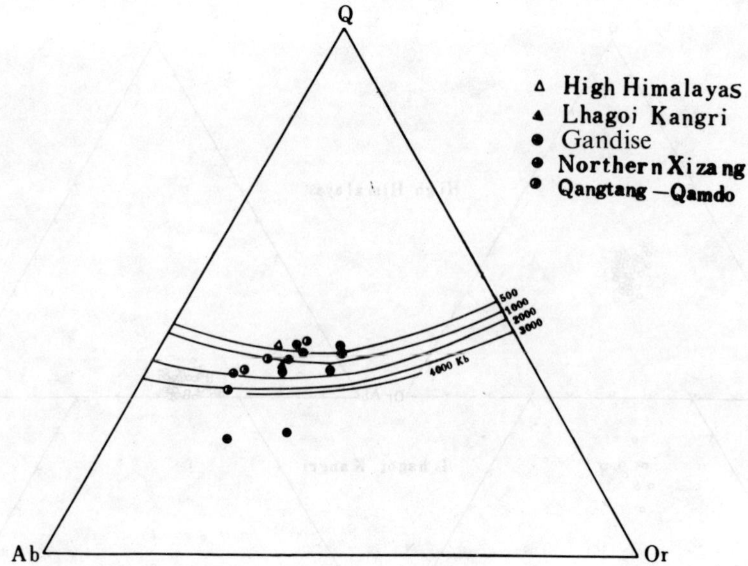


Fig. 5 Q-Ab-Or diagram for average norms of various granitic rocks from Xizang (after Tuttle and Bowen, 1958)

# CHARACTERISTICS OF FELDSPAR MINERALS AND THE EVOLUTION OF XIZANG GRANITES

Xie Ying-wen, Zhang Yu-quan, Deng Bing-jun
(Institute of Geochemistry, Academia Sinica)

The study on feldspars in granites will not only give us a better understanding of granite genesis and evolution but also serve as an aid in identifing the ages of rock bodies.

Samples studied in this paper were collected from Gandise rock belt, Lhagoi Kangri rock belt and Himalayan rock belt in southern Xizang. These rocks, including gabbro, diorite, granodiorite, porphyritic granite, two-mica granite and tourmaline-muscovite granite, range in age from late Yenshanian to Himalayan.

Feldspar minerals in this region consist of plagioclases and alkali feldspars.

From optical measurement, chemical analysis and X-ray powder diffraction of feldspar minerals in granitic rocks, the following characteristics can be outlined.

## ALKALI FELDSPAR COEFFICIENT (K) AND ITS FREQUENCY DISTRIBUTION IN GRANITES INCREASING WITH YOUNGER AGE

Alkali coefficient of southern Xizang granites ranges from 0 to 90 percent. However, the frequency distribution of K is different for granites of different ages and stages (Fig.1)

Fig. 1 shows the regular variation of frequency distribution of K with respect to different ages and different stages. They increase with increasingly younger ages from late Yenshanian to Himalayan as well as from the 1st stage to the 4th stage within the Yenshanian.This indicates that alkali concentrated in magma during its evolution.

## COMPOSITIONAL VARIATION OF FELDSPARS IN GRANITES

### An Content of Plagioclases Decreases toward Younger Ages.

An and Ab contents of plagioclases are calculated from refractive indices and optic axial angle measurements on spindle stage for 126 specimens collected from the 24 rock bodies in the three rock belts in southern Xizang. A few chemical data have also been taken into consideration for this purpose (Fig.2). It can be seen from Fig. 2 that An contents in plagioclase of late Yenshanian granites are higher, mostly exceed 30, while that of Himalayan granites are all less than 30.

Fig. 3 shows the frequency distribution of An contents for granites of different ages. It is obvious that An content of late Yenshanian varies within a wider range, from 2 to 66 percent, averaging 27.15. In comparison, An contents of the Himalayan ranges from 8 to 36 percent with an average of 17.34. So, it is clear that An content of late Yenshanian granites is higher than that of the Himalayan.

The frequency distribution of An content in plagioclase for different intrusion stages during the same period is presented in Fig. 4. From this diagram we can see that An content decreases continuously from 38.1 for the first stage of late Yenshanian to 9.43 for the fourth stage within the same period.

Difference in An content is also recognized between the phenocryst and groundmass within the same specimen. For example, An content of plagioclase phenocryst is 33 percent in coarse-grained porphyritic granodiorite of late Yenshanian, in contrast of 30 percent in the groundmass. In two-mica granite of Himalayan age, An is 34 percent in plagioclase phenocryst, but 28 percent in groundmass. Therefore, An content is lower in groundmass than in phenocryst, which is in agreement with the general sequence of crystallization.

As can be seen from Tables 1 and 2, the same trend of An variation is noticed in composite as well as complex rock masses.

From the above description it appears that plagioclase becomes more acid in composition during the process of evolution.

## Compositional Characteriatics of Alkali Feldspars

On the basis of chemical analyses, the following variation trend can be noticed:

1. Ab contents in alkali feldspar is different among granitic rocks of different ages. It is lower in Yenshanian granites than in Himalayan ones (Fig.5).

2. With respect to the same period, Ab is lower in earlier stages than in later stages (Fig. 6). This is probably due to the difference in physical conditions during rock genesis (at shallower depth) and the relative enrichment of Na in later magmas.

3. Ab content of alkali feldspars is lower in earlier stages than in later stages within the same rock body. In Demola composite rock mass, as an example, Ab content of alkali feldspar is 17.2 in the coarse-grained porphyritic biotite granite of the 2nd stage of later Yenshanian, in comparison with 20.5 percent in the two-mica granite of the 4th stage of late Yenshanian. Same phenomenon is observed in Quxu mass. This can probably be accounted for by the concentration of Na in granites formed at later stages as a result of magma differentiation at greater depth that gave rise to the emplacement of multistage composite mass.

4. An content in alkali feldspars tends to be lowered from Yenshanian to Himalayan. Perhaps this may be related to the general tendency of magmatic evolution in Xizang, i.e., it becomes more and more acid with time.

## STRUCTURAL STATE OF FELDSPARS IN GRANITES

As is well known that structural state study of feldspars can be expected to afford a deeper insight into the environment and physical factors involved in their formation as well as other aspects of genetic interest. X-ray powder diffraction is used as the major method in our studies for this purpose.

1. The structural state of plagioclase belongs to low-temperature ordered type. More than 30 feldspar specimens are measured by X-ray diffraction technique for $2\theta_{131}$, $2\theta_{1\bar{3}1}$, $2\theta_{\bar{2}41}$, and $2\theta_{2\bar{4}1}$. The An contents are determined with spindle stage. Degree of ordering are calculated based on the relation of An with $\Delta\theta_1=2\theta_{131}-2\theta_{1\bar{3}1}$ and with $\Delta\theta_2=2\theta_{\bar{2}41}-2\theta_{2\bar{4}1}$.

A glance of Fig. 7 show that plagioclase ordering varies within a limited range (60-100) and most of them are ordered type with only a few exceptions of transitional ordering. It is also noticed that Yenshanian plagioclase is of higher degree of ordering than that of the Himalayan.

2. The structural state of alkali feldspars is more complex, with all gradations from disordered orthoclase to the most fully-ordered microcline. Based upon X-ray powder diffraction data on more than one hundred alkali feldspars, the degree of ordering are estimated from (060) and (204) by using the formula proposed by Ragland (1970). The results show that degree of ordering of alkali feldspars in Xizang granite falls within very wide range. All gradations have been noticed from disordered orthoclase to the most fully-ordered microcline. Additionally, feldspar of late Yenshanian is of higher degree of ordering than that of the Himalayan (Fig.8).

3. Plagioclase always exhibits higher degree of ordering in comparison with alkali feldspars in the same rock mass.

4. Within the same rock body, ordering of alkali feldspar in phenocryst and groundmass is basically of the same degree. Table 3 presents the comparison of alkali feldspar ordering between phenocryst and groundmass from nine rock masses. As may be seen from this table, the ordering of alkali feldspars is similar between phenocrysts and groundmass in most cases. Observations suggest that these phenocrysts are not metacrysts but are resulted from magma crystallization. It must be pointed out, however, that in the two rock bodies of Da La and Ma La, alkali feldspar phenocrysts are obviously of higher degree of ordering than the groundmass. Field and microscopic evidence shows that potash feldspar phenocrysts in these two masses are euhedral tabular in shape, and increases in size with increasing distance toward the interior of the rock mass. This is just the phenomenon characteristic of an origin through magma crystallization. So, further study is needed to explain why alkali feldspar phenocrysts in these two masses are of higher degree of ordering.

Generally speaking, the structural state of feldspar may be affected by many factors, such as the age of granite formation, conditions involved in petrogenesis as well as volatile components. But recent data indicate that, due to the intensive plate activity in southern Xizang, rock-forming minerals in granitic rocks are subjected to obvious reformation, resulting in a remarkable increase in the degree of ordering of feldspars. This fact leads to the believe that plate tectonic is the major factor that affects feldspar ordering in southern Xizang.

In summary, abundant data on feldspar composition and structural state provide evidence of magmatic origin for southern Xizang granites.

## References


(1) Institute of Geochemistry, Academia Sinica, Geochemistry of granites in Southern China. Science Press, 87-128 (1979) (in Chinese).

(2) Troger, W. E., Optische Bestimmung der gesteinsbildenden Minerale. STUTTGART, 127 (1971).

(3) Smith, J. V., Feldspar minerals, 2. Springer-Verlay Berlin Heidelberg, New York, 15-126 (1974).

(4) Valui, G. A., Evolution of composition and structure state of feldspars in process of crystallization of small depth granitoids. Proceedings of Soviet Society of Mineralogy, (5), 524-534 (1974) (in Russian).

(5) Kamentsev, I. V. et al., Determination of Al-Si ordering and feldspar composition by powder analysis method. Proceedings of Soviet Society of Mineralogy, (4), 476-481 (1977) (in Russian).


Table 1 The compositional variation of plagioclase in Rutog composite rock massif

| Period | | Rock type | Refractive index Ng | Nm | Np | Plagioclase An | Axial angle |
|---|---|---|---|---|---|---|---|
| Late Yenshanian | I | Diorite | 1.553 | 1.550 | 1.547 | 34 | 84 |
| | | Granodiorite | 1.551 | 1.547 | 1.544 | 30 | -88 |
| | II | Porphyritic hornblende biotite granite | 1.550 | 1.547 | 1.543 | 28 | -83 |
| | III | Biotite granite | 1.543 | 1.540 | 1.534 | 12 | 83 |
| | IV | Granite | 1.542 | 1.538 | 1.533 | 10 | 90 |

Table 2 Compositional variation of plagioclase in Gyabge complex

| Period | | Rock type | Refractive index Ng | Nm | Np | Plagioclase An | Axial angle |
|---|---|---|---|---|---|---|---|
| Late Yenshanian | I | Norite | 1.571 | 1.565 | 1.563 | 65-67 | 88 |
| | | Hypersthene diorite | 1.556 | 1.551 | 1.548 | 37-38 | 90 |
| | II | Granodiorite | 1.550 | 1.548 | 1.544 | 28-32 | -85 |
| | III | Porphyritic two-mica granite | 1.547 | 1.543 | 1.540 | 21-25 | -86 |
| | IV | Biotite granite | 1.546 | 1.542 | 1.539 | 17-20 | -88 |

Table 3 Comparison of ordering of phenocrystals with that of groundmass in alkali feldspars

| Period | No. | Pluton | Rock type | Ordering of phenocrystal | Ordering of groundmass |
|---|---|---|---|---|---|
| Yenshanian | 1 | Zayu | Gneissose porphyritic biotite granite | 0.66 | 0.71 |
| | 2 | Demola | Porphyritic biotite granite | 0.66 | 0.63 |
| | 3 | Pyikang | " | 0.98 | 0.98 |
| | 4 | " | " | 0.95 | 0.89 |
| | 5 | Yangbajain | " | 0.86 | 0.98 |
| | 6 | Kangrinboqe | Porphyritic granodiorite | 0.45 | 0.45 |
| | 7 | Demola | Porphyritic two mica granite | 0.80 | 0.80 |
| | 8 | Gyabge | " | 0.45 | 0.86 |
| | 9 | Ma La | Porphyritic biotite granite | 0.66 | 0.33 |
| Himalayan | 10 | Ladakai | Porphyritic granodiorite | 0.89 | 0.80 |
| | 11 | Zayu | Two-mica granite | 0.83 | 0.98 |
| | 12 | Da La | " | 0.80 | 0.36 |

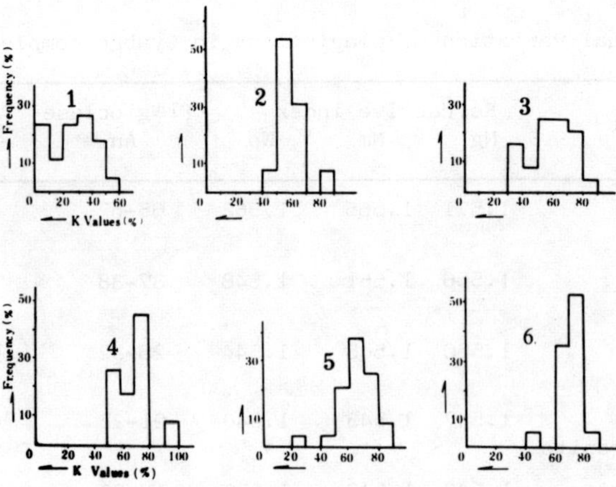


Fig.1 K frequency for the granites of different ages

1. Yenshanian I; 4. Yenshanian IV;
2. Yenshanian II; 5. Early Himalayan;
3. Yenshanian III; 6. Late Himalayan

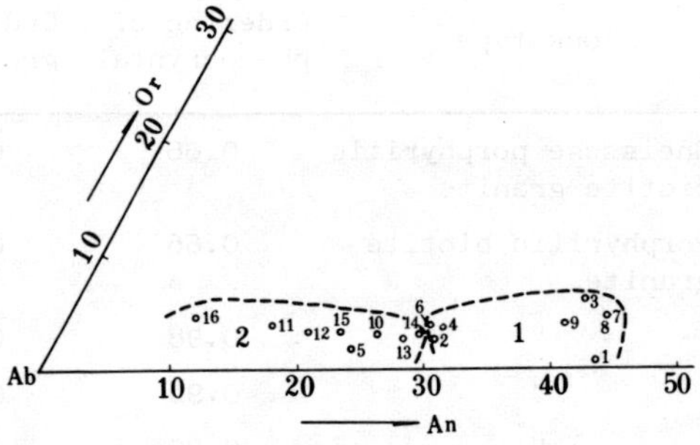


Fig. 2 Or–Ab–An triangular diagram for plagioclases

1. Yenshanian; 2. Himalayan

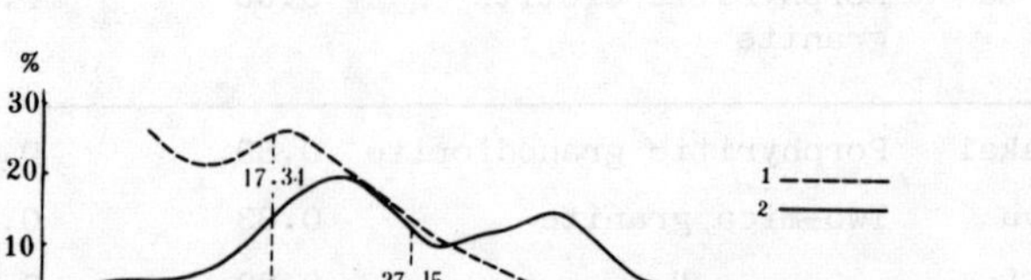


Fig. 3 An frequency of plagioclase in granites of different ages

1. Himalayan; 2. Yenshanian

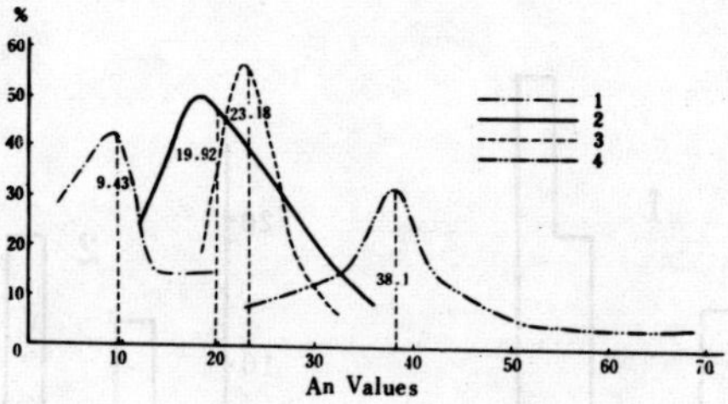


Fig.4 An frequency of plagioclase in granites of different intrusion stages during late Yenshanian

1.$\gamma_5^{3-1}$; 2.$\gamma_5^{3-2}$; 3.$\gamma_5^{3-3}$; 4.$\gamma_5^{3-4}$

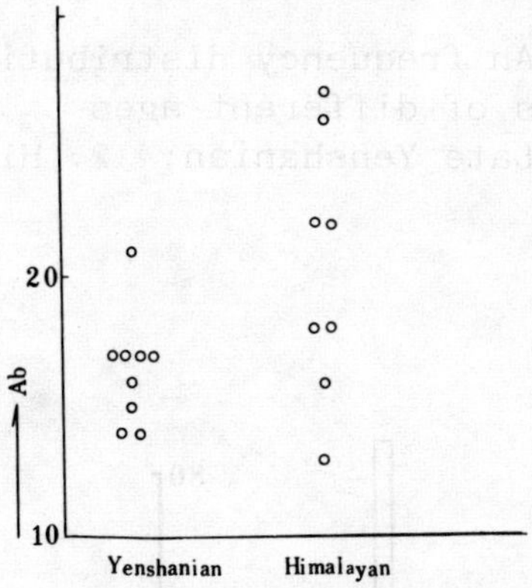


Fig.5 Ab variations of alkali feldspars in granites of different ages

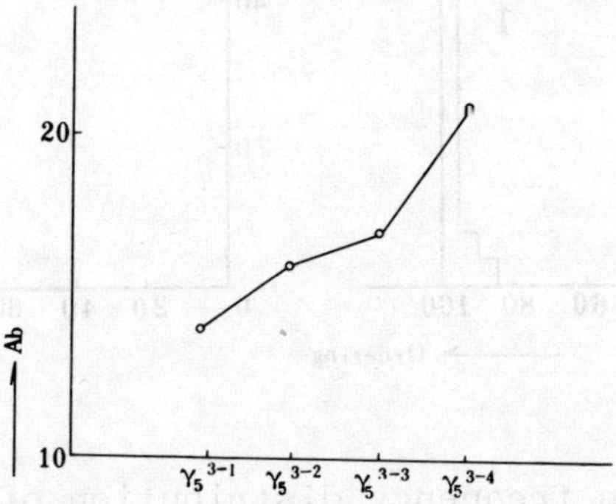


Fig.6 Ab variations of alkali feldspars in granites of different intrusion stages during late Yenshanian

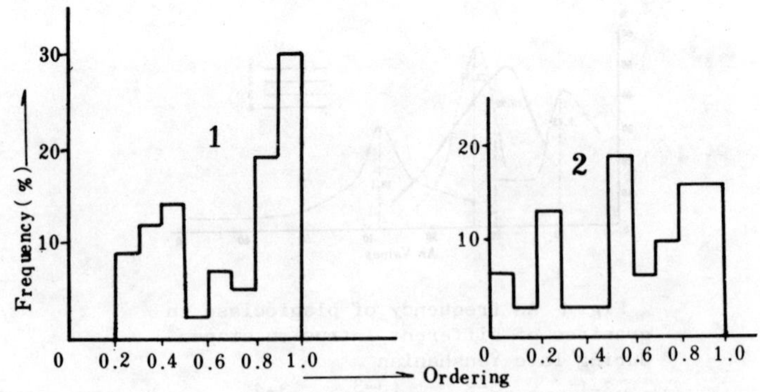


Fig. 7 An frequency distribution of ordering of plagioclases of different ages
1. Late Yenshanian; 2. Himalayan

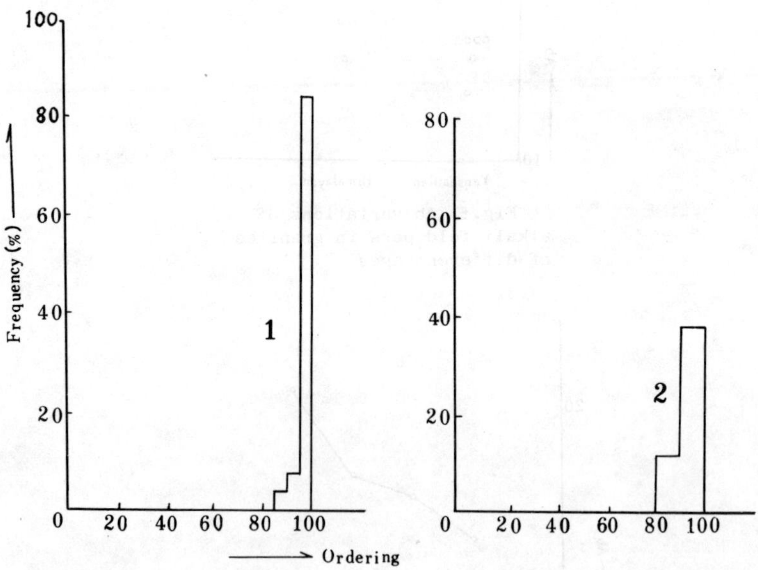


Fig.8 An frequency distribution of ordering of alkali feldspars of different ages
1. Late Yenshanian; 2. Himalayan

# FEATURES OF ACCESSORY MINERALS IN GRANITOID ROCKS IN SOUTHERN XIZANG

Pan Jing-ming, Zhang Yu-quan  Hong Wen-xing
(Institute of Geochemistry, Academia Sinica)

The authors have identified 42 artificial concentrations of heavy minerals which were collected from 13 plutons along Gandise, Lhagoi Kangri and Himalayan belts. Each concentration sample weighed ten to twenty kilograms. The plutons along Gandise belt mainly belong to Yenshanian Period, and most of them are identified as grandiorite, biotite-granite, porphyritic biotite-granite and so on, which are referred to as relatively melanocratic granitoid rocks in this paper. Meanwhile, the plutons along Himalayan and Lhagoi Kangri belts belong to Himalayan Period, and they consist of two-mica granite, porphyritic two-mica granite, tourmaline muscovite granite and are referred to as relatively leucocratic granitoid rocks.

The samples used are secondary samples, i.e., zircon and apatite were separated in advance from the samples for isotopic dating. Therefore, the contents of these two minerals have not been reckoned. After dressing, every concentration sample was separated into ten to sixteen different magnetic fractions, and the identifications of every fraction, including determinations of physical and optical properties of various minerals, were conducted under microscope, in conjunction with microchemical analyses and more than sixty X-ray powder diffraction determinations. Finally, the content for each accessory mineral except for zircon and apatite was calculated.

## FEATURES OF ACCESSORY MINERALS FROM GRANITOID ROCKS OF DIFFERENT TYPES AND DIFFERENT INTRUSION TIMES

More than 50 species of accessory minerals have been found in these 42 concentrations, including rare element minerals (zircon, zirconolite, euxenite, columbite-tantalite, chrysoberyl, monazite, xenotime, bastnasite and allanite), radioactive minerals (uraninite, thorite, uranothorite, thorianite, cheralite and torbernite) tungsten-tin-molybdenum minerals (scheelite, cassiterite, molybdenite), polymetallic minerals (native copper, pyrite, chalcopyrite, bornite, malachite, cinnabar, arsenopyrite and cerusite), iron and titanium minerals (magnetite, limonite, ilmenite, rutile, anatase, sphene and leucoxene), volatile minerals (apatite, tourmaline, fluorite, topaz) and other minerals ( barite, calcite, magnesite, rhodochrosite and siderite garnet, staurolite, vesuvianite, epidote, zoisite, clinozoisite, $\alpha$ - zoisite, chlorite, corundum and calamine). In individual plutons one could often find 7 to 26 species of accessory minerals, among which zircon, apatite, magnetite and allanite possess highest occurrence frequencies (usually exceeds 90 % ).

Contents of accessory minerals from granitoid rocks of Yenshanian Period (i.e., the relatively melanocratic rocks) are generally higher than those of Himalayan Period (the relatively leucocratic rocks). For example, among 31 samples of Yenshanian Period, the contents of accessory minerals in 11 samples exceed 1 kg/t, and only one sample give a content below 50 g/t. The average contents of total 31 samples are calculated to be 2,203 g/t. At the same time, 40 species of accessory minerals have been recognized in Yenshanian relatively melanocratic granitoid rocks, with allanite, uranothorite, scheelite, magnetite, sphene as the dominant minerals, and zirconolite, bastnasite, thorianite, cheralite and molybdenite as the characteristic ones.

On contrast, among 11 concentrations from Himalayan relatively leucocratic granitoid rocks, only three have the contents exceeding 1 kg/t of accessory minerals, while up to 4 of the 11 samples have the contents below 50 g/t. The average contents are estimated to be 1,431 g/t. 43 species of accessory minerals have been found in total. Among these species, monazite, xenotime, uraninite, thorite, copper minerals and tourmaline predominate. While, chrysoberyl, columbite-tantalite, euxenite, cassiterite and cinnabar are characteristic minerals.

## FEATURES OF CRYSTAL FORMS OF ZIRCON FROM DIFFERENT BELTS

The crystal forms of zircon are quite different for different rock belts (Fig. 1). For example, in Gandise belt, the crystal forms of zircon are more complex, showing both prismatic faces of {110} and {100}, at the same time, the pyramidal face {131} also occurred. As more as seven types of crystal forms have been recognized. In addition to various prismatic crystals, the tabular crystals with well developed prismatic face {110} and the rhombic tabular crystals and distorted crystals have also been discovered. The distorted crystals only occurred at the eastern part of Gandise belt.

On contrast, along Himalayan belt, the crystal forms of zircon are more simple and uniform, only the prismatic crystal with {100} has been found.

## POLYMETALLIC SULPHIDE MINERALS

A number of polymetallic sulphide minerals have been observed in granitoid rocks in southern Xizang. In particular, a variety of sulphide minerals with noticeable contents have been found in the plutons around Lhasa. Thus, the region around Lhasa can be called as "Higher Sulphide Content Region". In addition, if ore minerals other than sulphide are also included, the calculated content ratios of metallic to non-metallic minerals will be higher also in the plutons around Lhasa.

## AVERAGE SPECIFIC GRAVITIES OF ACCESSORY MINERALS AND THE VARIATION OF RELATIVE AMOUNT OF VOLATILE MINERALS

We calculated the content in percentage of every accessory mineral for every concentration. Then we distinguished seven main types of accessory mineral associations and named them after the mineral predominating in corresponding associations. For example, we have magnetite type, allanite type, garnet type and sphene type and so on (Fig.2). Magnetite type and garnet type mainly occur in Gandise belt, while tourmaline type in Himalayan belt. Allanite type and sphene type are more characteristic in Gandise belt.

The average specific gravity for each sample of accessory minerals is calculated in terms of weighted average method. It can be seen from Fig. 2 that the ranges of average specific gravities of accessory minerals and the weighted means are different for different associations. From lower to higher, we have tourmaline type — 3.33, sphene type 3.58, allanite type 3.74, garnet type 3.91, magnetite type 4.52 and pyrite type 5.0. These numerical figures show a close relation existing between average specific gravities and the accessory mineral associations.

We plotted average means of specific gravities of various samples into Fig. 3 in reference to the sample localities, and then connected those dots of the same belt into curves. In addition, the content ratios of volatile-bearing to non-volatile minerals calculated were plotted also into the same figure. This figure shows clearly that the average specific gravities of accessory minerals from Gandise belt are noticeably higher than those from Himalayan belt, and that the average means become progressively higher from west to east for both Gandise and Himalayan belts.

As for the content ratios of volatile-bearing to non-volatile minerals, they show a contrast tendency to become progressively lower from west to east for Himalayan belt. Because little volatile-bearing minerals are found in Gandise belt, no similar curves are shown in the figure for them.

## SUMMARY

Generally speaking, the average specific gravities of accessory minerals in granitoid rocks and the content ratios of volatile-bearing to non-volatile minerals can be correlated with the depths at which these rocks were formed. We think, those plutons that are characterized by higher average specific gravities of accessory minerals and lower content ratios of volatile-bearing to non-volatile minerals might be formed at some greater depths. Under such an assumption, we may deduce that the formation depth of the plutons along Gandise belt might probably be greater than that for Himalayan belt. And there is a tendency to become deeper from west to east for both belts.

Based on the comparison between features of accessory minerals from Gandise belt and those from Himalayan belt (Table 1), it can be concluded that the accessory minerals in granitoid rocks in southern Xizang are characterized by zonal distribution in space and by distinct stages of development in time.

Table 1 Comparison of accessory minerals from Gandise Belt with those from Himalayan Belt

| | Gandise Rock Belt | Himalayan Rock Belt |
|---|---|---|
| Time | Yenshanian Period | Himalayan Period |
| Depth | Deep | Shallow |
| Species of acces. min. | Less in variety | More in variety |
| Contents of acces. min. | Higher | Lower |
| Main and characteristic acces. minerals | Magnetite, allanite, sphene and W, Mo, Th-minerals | Tourmaline, chrysoberyl, columbite-tantalite and Cu, U, Sn-minerals |
| Main types of acces. mineral associations | Magnetite type, allanite type and sphene type | Tourmaline type |
| Average specific gravities of acces. minerals | Higher | Lower |
| Proportions of volatile-bearing minerals | Higher | Lower |

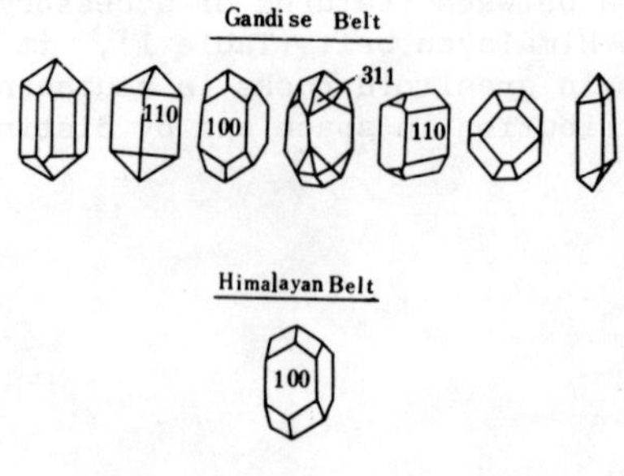


Fig. 1 Crystal forms of Zircon

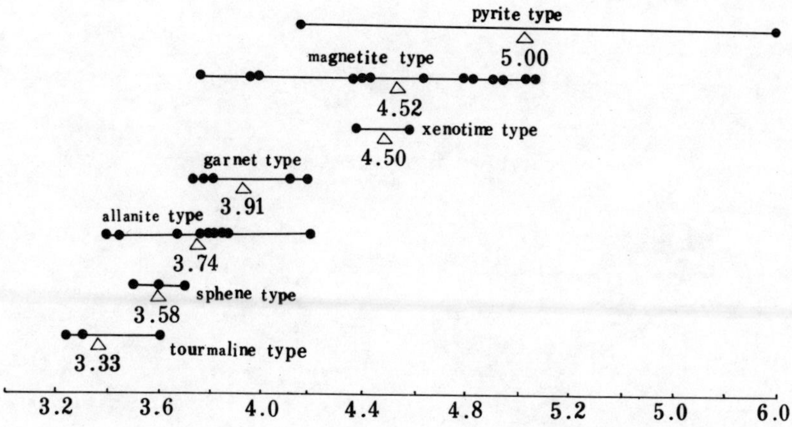


Fig. 2 The ranges of average specific gravities for various types of accessory mineral associations

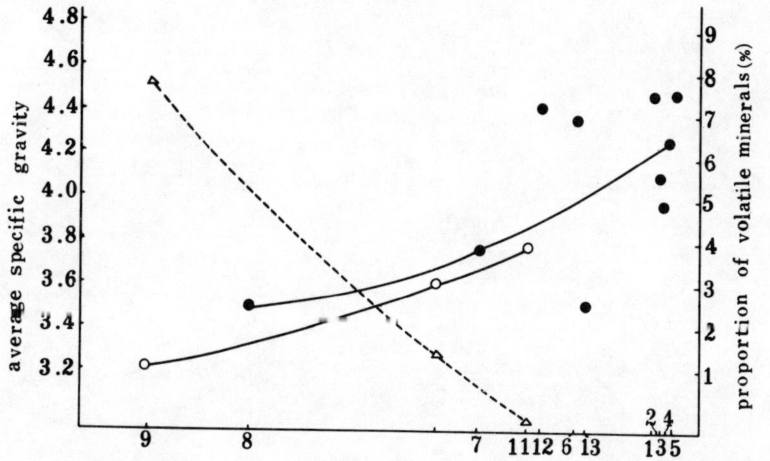


Fig. 3 Variations of the average specific gravities of accessory minerals and of the proportions of volatile-bearing minerals in various rock belts

The dots of plutons along the abscissa correspond to the projection of their positions onto the abscissa. The solid circles represent the average specific gravities for Gandise belt, open circles the average specific gravities for Himalayan belt and the open triangles the proportions of volatile-bearing minerals for Himalayan belt.

# GENESIS OF CERTAIN CHROMITE DEPOSITS IN XIZANG

Chi San-chuan, Shi Qi-zheng
(Wuhan Geological College)

A certain ultrabasic igneous rock body is situated in the eastern segment of the South Xizang ultrabasic igneous rock belt. It extends in a NWW-SEE direction and plunges due south. Its trend is nearly the same as that of the Yarlung Zangbo deep fault. It is a composite igneous rock body of multiple intrusions with a marked intrusive contact relation to its country rocks of late Late Triassic in age. The magmatic differentiation is rather complete. The state when the magmas were intruded into the deep-lying magmatic chamber is considered to have been still molten or in the state of "crystal mush" and could hardly have been in cold solid state. The chromite-bearing body, having a high Mg/Fe ratio and being rich in chromium, is a type of magnesian ultrabasic rock.

The southern flank of the country rocks of the igneous rock body is a thick series of shallow marine flyschoid clastic sediments of Noric age. The sediments are argillaceous, sandy, and calcareous, intercalated with eruptive rocks (spilite, keratophyre and volcanic tuff), with radiolarian chert in the upper part. On the northern flank is a continental molasse formation of a graben type of Oligo-Miocene age. There are some basic and ultrabasic pebbles in the multicomponental conglomerates in its upper part, indicating that the former contact of the igneous rock body with the Tertiary was a sedimentary contact, but appears now as a fault contact (Fig. 1).

The contact of the igneous rock body with the Triassic rocks is apparently an intrusive one. The evidence is the following : (1) There are apophyses intruded into the upper Triassic at the border of the igneous rock body (Fig. 2);(2) There are many xenoliths including phyllite, volcanics, limestone and siliceous rocks at the border of the igneous rock body, and the modes of occurrence of the xenoliths are consistent with the upper Triassic being near the contact. No hypogene allochthonous xenoliths however, have been found; (3) contact thermal metamorphism superimposed on the regional metamorphism in the contact rocks and xenoliths is common. Metamorphic grades are uneven, for instance, spilitic volcanic tuff near the igneous rock body in the central part of Xinka Mt. has been metamorphosed to vesuvianite-garnet-chlorite schist outwards (Fig. 3). Vesuvianite and grossularite diminish gradually, and tremolite occurs. The metamorphic grade is of amphibolite-pyroxenite facies. The contact rock near Lakangba has been metamorphosed to tremolite-zoisite hornfels and its metamorphic grade is of albite-epidote facies. The limestone xenoliths have been metamorphosed to anthophyllite marble ( Fig. 2b) ; (4) the primary flow structure is well developed, and its direction is generally consistent with the zones of rock facies.

These above facts show that the contact of the igneous rock with the Noric strata is an intrusive contact, and the magma intruded in a state of a high temperature molten melt or "crystal mush", and not as a "solid cold intrusion".

Chromite ore bodies occur mainly in the middle and lower parts of the dunite-harzburgite zones with an approximately E-W direction and are concentrated segmentally in groups. Chromite ore bodies are mostly in veins and lenses. The ores are of high quality, and most are of metallurgically rich ores.

The general characteristics are analogous to the Alpine podiform chromite ores, but the ores have properties of their own in so far as the genetic classification and the controlling factors are considered.

According to the main features, three types of ore-deposits are recognized as follows: (1) Ores by crystal-fractionation in the late orthomagmatic stage. Two subtypes are differentiated by modes of formation : A. Ores of segregation in situ ; B. Ores by filter pressure. (2) Ores by cumulative deposits in the late orthomagmatic stage. (3) Ores by injection of residual chromite-rich magma at the latest stage.

The main properties of cumulative ores are : (1) Ore-forming chromites are magnesio-chromite and aluminochromite (Fig. 4, Points 3, 7, 14) ; (2) The ore-forming chromites are in the form of medium to coarse aggregates of pisolitic ores ; (3) The pisolitic ores occur mainly in the dunite-harzburgite zone, but the nearby country rocks are exclusively of dunite ; (4) The arrangement of the pisoliths commonly has no definite pattern, but in some ore bodies , it shows an orientation along the border of the footwall and "bedding planes" parallel to the base of the ore body ( Figs. 5, 6). In the olivine matrix among the pisoliths, there are fine to medium grained disseminated chromites showing gradations from light-medium-dense disseminations to pisolitic ores. "Bedding" and "rhythmic" structures analogous to sedimentary rocks are shown. Although such an igneous accumulative texture is not as typical as the layered intrusion of Bushweld, the pisolitic arrangement is parallel to the base of the ore bodies. Hence, we assume that they accumulated in the crystal mush ; (5) The interior structure of the pisolith shows that an individual pisolith is composed of aggregates of tens of grains of chromite crystals, forming a three axis ellipsoidal body. The outer margin is often surrounded by a thin layer of serpentine (Fig. 7). Pisoliths are often compressed to form elongated and flattened shapes welded to each other gradually passing into massive ores. These are the results of remelting, deformation, overgrowth and serpentinization of a later stage after the original accumulation.

When we examine the pisoliths of the ore body No. 5, some embryo forms of pisoliths of loosely connected grains of chromite were found. Some have just gotten together (pisoliths only in outlines) and some had already gotten together to form irregularily shaped pisolith. Some are rounded and ellipsoidal, whereas some show embryo forms of a big pisolith and aggregates of several small pisoliths. Disseminated chromites and pisoliths are arranged in bands, like flow structures (Fig. 8).

Because of the viscosity of the magma, the rate of the setting crystal is rather slow. The tectonic environment of the igneous rock body is unstable so the magma has been compressed and has migrated, especially near the channel. There are eddy currents in the crystal mush, causing the accumulative crystals to be aligned in a definite pattern (Fig. 9). These allowed chromites to be aggregated to form pisoliths around the particles by flow of magma and to form other igneous cumulative textures.

Ores by injection of residual chromite-rich magma in the latest stage show the main properties (1) Ore-forming chromites are magnesiochromites (Fig. 4, Points 5-11, 13), which are quite different from the composition of accessory chromites ; (2) Most are massive, and some are densely disseminated ; (3) There are vein-and lens-shaped ore bodies (Fig. 10) having zigzag edges (Fig.11) which are caused by pre-ore forming fractures; (4) The contact between the ores and the country rocks is sharp. The directions of the ore bodies are oblique to the flow structure of the igneous rock body. The nearby country rocks often show weak to intermediate serpentinization

The altered and discolored borders are a few centimeters to a meter wide.

The three types of mineralization mentioned above can occur separately or in combination.

## CONCLUSIONS

1. This igneous rock body must have been derived from the upper mantle. The magma intruded as a high temperature molten melt or "crystal mush". Therefore,it is obviously inconsistent with the model of " solid cold intrusion".

2. The chromite deposits were formed in an unstable tectonic environment. Structural deformation occurred in each mineralization stage. Fractures that were derived from the plastic deformation of the late magmatic stage are the main ore-controlling factors.

3. The genesis of the "Alpine-type" ultrabasic igneous rocks and the podiform chromite in geosynclinal terrains, especially the genesis of pisolitic ores, have long been argued. H. B. Pavloff suggested that the pisolitic ores had been formed by magmatic immiscibility ; and J. Dickey suggested that they had been formed as "snow balls tumbling in the magmatic segregation zones at the margin of the plates. The separation of chromite by magmatic immiscibility in a magnesian melt seems hardly possible, in light of experimental data. Therefore, the present authors suggest that the pisolitic ore formed as culmulative products of gravitational settling through continuing differentiation.

## References


(1) Wilson, H. D. B. (ed), Magmatic mineral deposits (collected treatises); A Chinese translation by Wuhan Geological College, Geology Press (1977).

(2) Dickey, J., A hypothesis of origin for podiform chromite deposits, Geochimica et Cosmochimica Acta, 39 (6-7), 1061-1074 (1975).

(3) Pavlov, N. V. et al., Chromite nodules as indicators of liquation in magmatic melt, Bull. Acad. Sci. SSSR, Sec. Geology, (11), 29-45 (1975).

(4) Academy of Science and Ministry of Geology of SSSR, The origin of endogenic mineral deposits, Science Press, 31-41, 100-119, Moscow (1976) (in Russian).

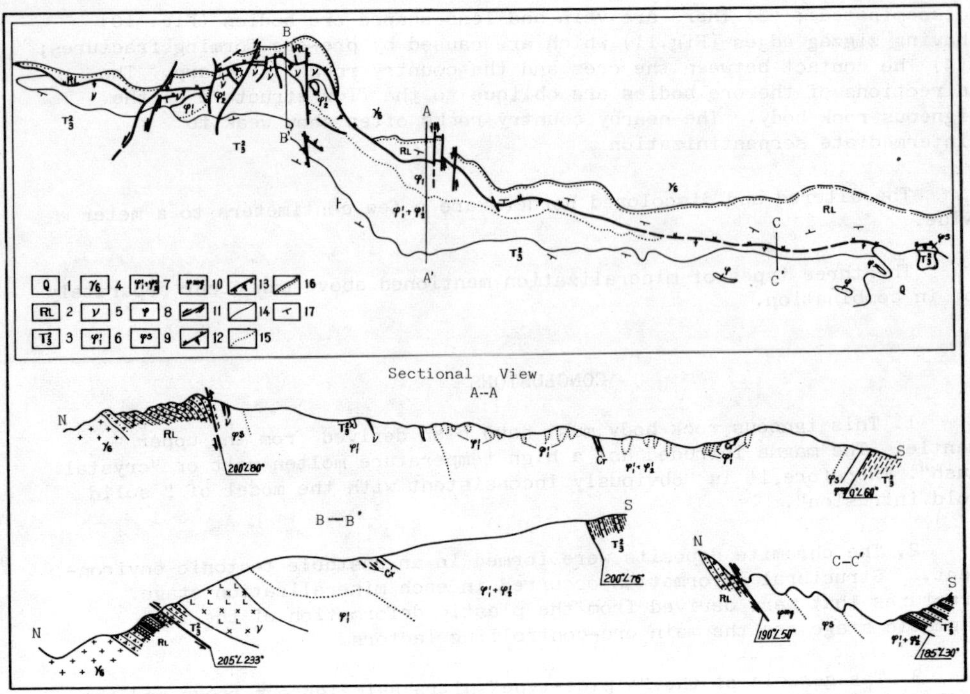


Fig. 1 Sketch map of the ultrabasic igneous rock bodies in Xizang

1. Quaternary system;
2. Tertiary;
3. Upper Triassic;
4. Biotite-granite, Quartz-diorite;
5. Peridotite-gabbro;
6. Dunite zone;
7. Dunite-Harzburgite zone;
8. Ultrabsic undifferentiated;
9. Serpentinite;
10. Talc-Magnesitite;
11. Tear-faults;
12. Reverse-faults;
13. Normal-faults;
14. Geological-boundaries;
15. Facies-boundaries;
16. Unconformity;
17. Mode of occurrence

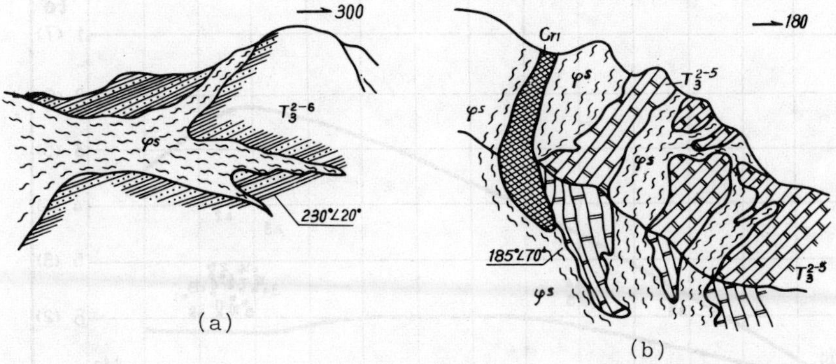


Fig. 2 The intrusive contact relationship between igneous rock body and countryrocks (Sectional view).

(a) Northern contact belt;

(b) Southern contact belt.

$T_3^{2-6}$ — Sandstone, slate;

$T_3^{2-5}$ — Anthophyllite marble;

$\varphi$ s — Serpentinite

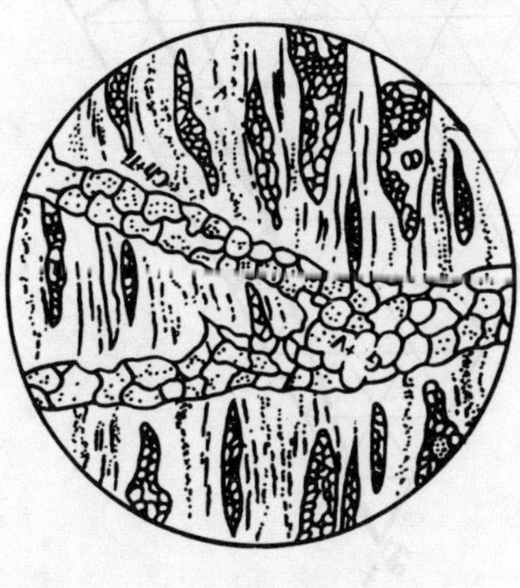

Fig. 3 Vesuvianite-garnet-chlorite schist

G-garnet;
V-vesuvianite;
Ch-chlorite;
Ti-leucoxene

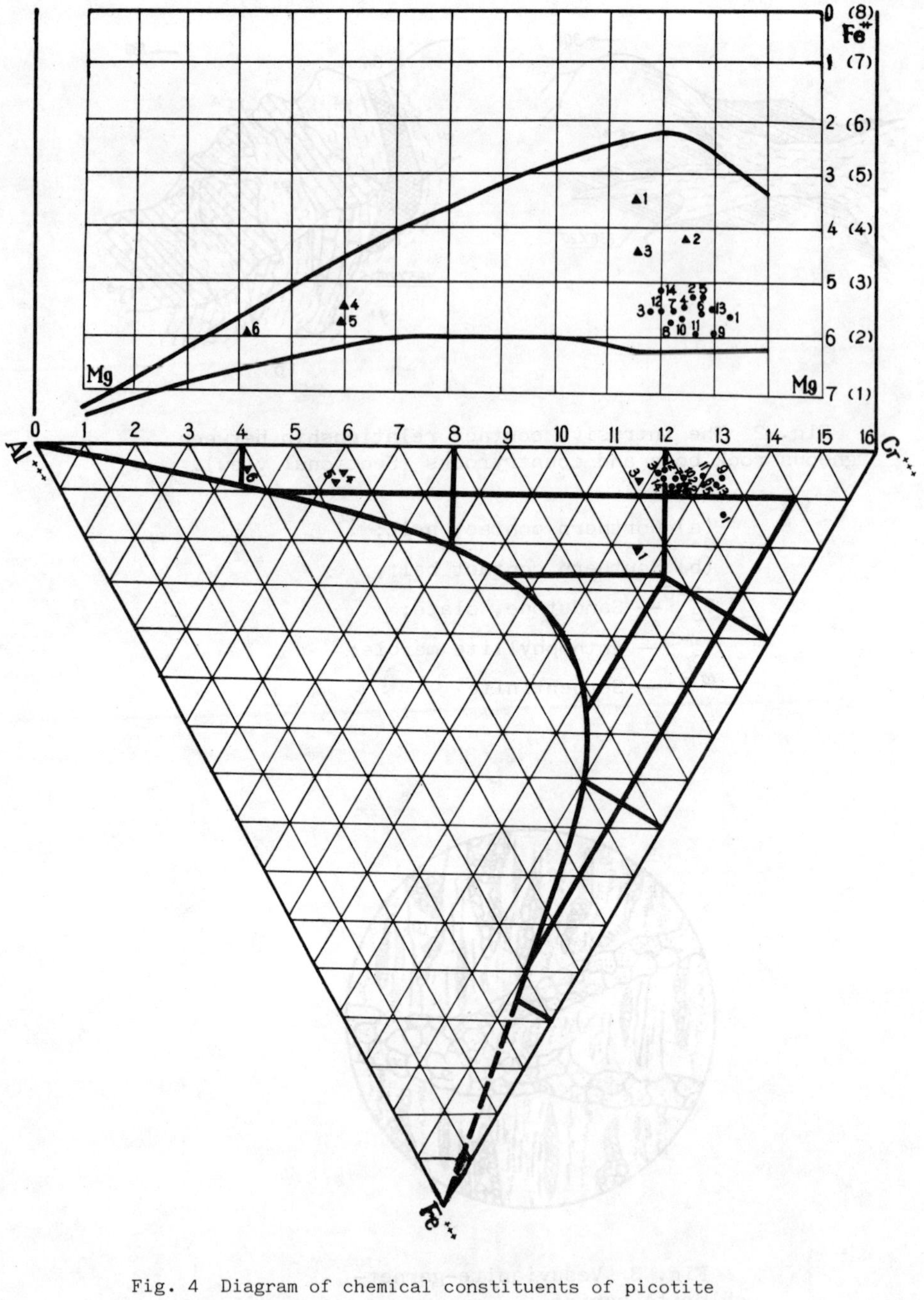

Fig. 4 Diagram of chemical constituents of picotite

▲ Accessory picotite; ● Ore-forming picotite

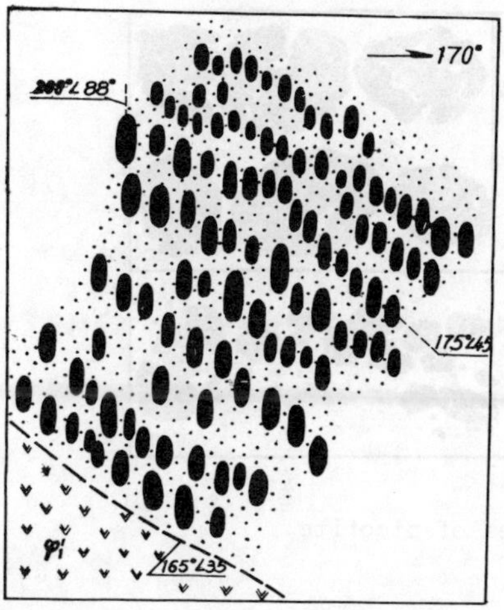


Fig. 5 Orientation of the pisolites in pisolitic ores (Sectional view)

Fig. 6 Pisolitic ores

Fig. 7 Pisolitic ores

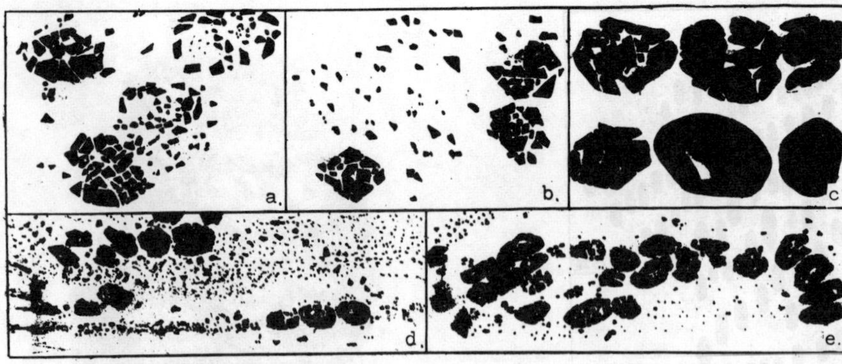


Fig. 8 Cumulative form of aggregates of picotite

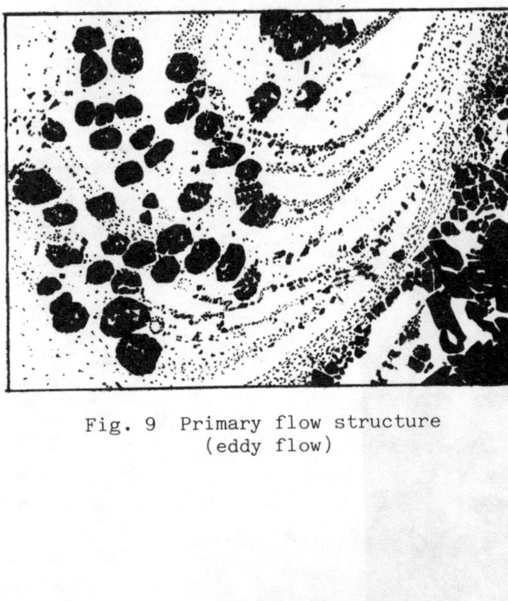

Fig. 9 Primary flow structure (eddy flow)

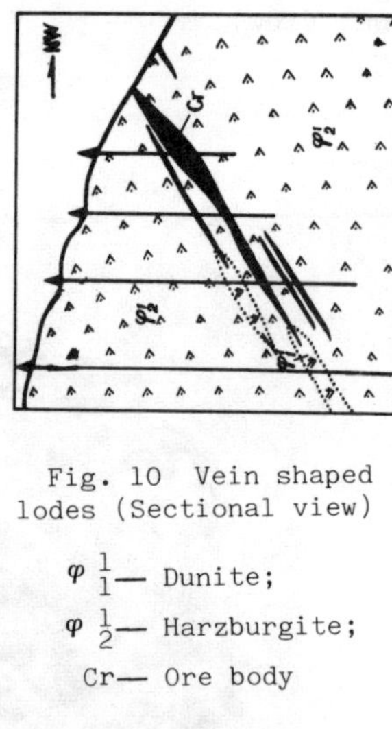


Fig. 10 Vein shaped lodes (Sectional view)

$\varphi\ \frac{1}{1}$— Dunite;

$\varphi\ \frac{1}{2}$— Harzburgite;

Cr— Ore body

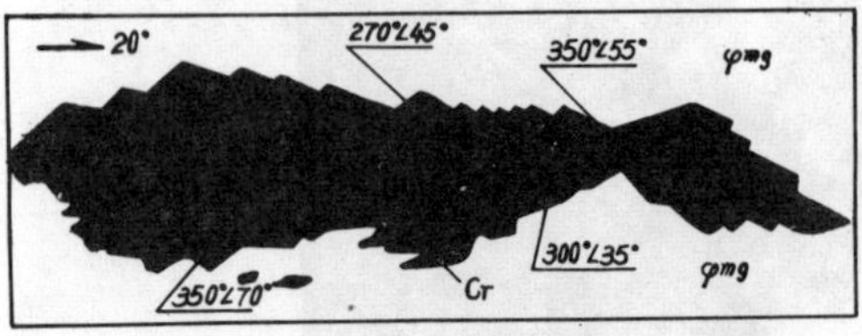


Fig. 11 Ore body with zigzag boundary

# ISOTOPIC GEOCHRONOLOGY OF GRANITOID ROCKS IN SOUTHERN XIZANG PLATEAU

Zhang Yu-quan, Dai Tong-mo, Hong A-shi
(Institute of Geochemistry, Academia Sinica)

Widespread in the southern part of the Xizang Plateau are granitoid rocks, characterized by their fairly intensive intrusion and distinct NWW extension. With decreasing geological ages from north to south, these granitoid rocks can be divided into three petrographical belts as the Gandise belt, the Lhagoi Kangri belt and the Himalayan belt.

Age determinations were conducted by K-Ar and U-Pb methods on various types of rock from the three belts mentioned above. For both methods, minerals used were collected from the same rock sample. It is found from the isotopic ages, in conjunction with the research results on biotites in granitoid rocks that the biotites in granitoid rocks in the southern part of Xizang once underwent recrystallization. Therefore, biotite K-Ar ages are partially representative of the time of biotite recrystallization. Some individual anomalous zircon U-Pb ages will be discussed in another paper.

The decay constants proved at the 25th International Geological Congress held in 1976 were adopted in the calculation of geological isotopic ages as follows:

$$\lambda_{^{40}K}(\beta) = 4.962 \times 10^{-10}\ yr.^{-1}$$

$$\lambda_{^{40}K}(e) = 0.5811 \times 10^{-10}\ yr.^{-1}$$

$$^{40}K/K\ (\text{atomic ratio}) = 1.667 \times 10^{-4}$$

$$\lambda_{^{238}U} = 1.5513 \times 10^{-10}\ yr.^{-1}$$

$$\lambda_{^{235}U} = 9.8485 \times 10^{-10}\ yr.^{-1}$$

$$^{238}U/^{235}U = 137.88$$

## RELIABILITY OF THE DATA OBTAINED

In order to check the reliability of the isotopic data, we made the following experiments apart from the blank assay of argon extraction system and the age determinations of standard samples.

### Repeated Experiments

The results of repeated tests are given in Table 1, from which we can see the ages determined twice are fairly in agreement (Table 1).

## The Comparison of Isotopic Data Obtained by Volumetric and Dilution Methods

In Table 2 is the comparison of isotopic data determined by volumetric method with those by isotopic dilution method. It can be seen from this table that the isotopic data determined either on different minerals from a same rock body, or on a same kind of mineral by dilution method and volumetric method are fairly consistent.

## The Comparison of Isotopic Ages Determined by K-Ar and U-Pb Methods

In Table 3 is the comparison of K-Ar ages with U-Pb ages. We can see biotite K-Ar ages and zircon U-Pb ages for a same type of rock are close to each other as well, but zircon U-Pb ages are slightly higher. In certain cases, the zircon age data are much lower than the biotite ones. This anomaly is to be discussed elsewhere in this book.

## RECRYSTALLIZED BIOTITE-BEARING ROCKS AND THEIR AGE DETERMINATIONS

Biotite recrystallization phenomena are commonly observed in some of the plutons in the Gandise belt, such as Quxu, Nangxian, Nyingchi, Azhag Kanglha and Zayu granite bodies; and in the Lhagoi Kangri belt, such as Ka La, Kangmar and Mala Mt. granite bodies (Table 4). The age determinations are given in Table 4 for zircons and recrystallized biotites present in most of the granitoid bodies. Table 4 shows the comparison of biotite K-Ar ages with zircon U-Pb ages. From this table we can see that the zircon ages are in coincidence with the geological setting of rocks, representing the time of crystallization of rocks during the emplacement. However, the biotite ages are much younger as compared with the ages of host rocks, which is in consistency with the later biotite recrystallization indicated by their microscopic examinations. Thus, recrystallized biotite ages can only represent the time of biotite recrystallization.

## RESULTS AND DISCUSSION

In order to determine the ages of granitoid rocks in the southern part of Xizang, samples were collected from more than 30 plutons during 1973 to 1976, yielding 126 age determinations. Among these 30 plutons, 17 were sampled in the Gandise belt, yielding 69 age determinations, and the rest from the Lhagoi Kangri and Himalayan belts. The host rocks from which minerals were separated for isotopic dating include diorite, granodiorite, porphyritic biotite granite, biotite granite, two-mica granite and tourmaline-muscovite granite. The results are given in Table 5 and Fig. 1.

From this table, we can see that the isotopic geological ages from the Gandise belt are in the range of 128 to 8 m.y., corresponding to late Yenshanian to Himalayan(1,2,3,4). Some of the plutons in the Gandise belt give a biotite age ranging from 30 to 20 m.y. The isotopic geological ages from the Lhagoi Kangri belt range from 31 to 6 m.y., in which the isotopic ages yielded by biotites from gneissose two-mica granites are mostly in the range of 30 m.y. As for the Himalayan belt, 16 isotopic geological ages ( of which 6 had been published before) were determined, mostly within the range of 20 to 12 m.y.

From the foregoing isotopic geological data determined in recent years, it is evident that the granitoid rocks widespread in the southern part of Xizang Plateau have an age ranging from 128 to 8 m.y., corresponding to late Yenshanian to Himalayan(1,2,3,4,5).

Table 6 shows that the isotopic geological ages reflect, to a certain extent, the intrusion sequence of late Yenshanian composite plutons. Early and late intrusive bodies usually have distinct ages, but those emplaced during the middle stage frequently yield crossover or very close isotopic geological ages. This may be attributable to the time-close emplacement of intrusive plutons in late Yenshanian composite rocks. Although their intrusion sequence and isotopic ages seem undistinguishable, their age range is fairly distinct. As for the composite rocks composed of late Yenshanian and Himalayan intrusives, their isotopic ages are highly distinguishable.

The isotopic geological ages of three petrographical belts discussed are given as follows:

1. The Gandise belt. The main part of the Gandise belt, composed of diorite, granodiorite and porphyritic biotite granite, give 40 mica and zircon isotopic ages ranging from 120 to 70 m.y., belonging to late Yenshanian, corresponding to Cretaceous. Additionally, 31 isotopic ages were determined on medium-fine grained biotite granite and two-mica granite samples from the Gandise belt, which fall into the interval of 50 to 8 m.y., belonging to Himalayan, corresponding to Oligocene to Miocene.

2. The Lhagoi Kangri belt. It is characterized by relatively small intrusive bodies and simple petrography, mainly composed of gneissose two-mica monzonitic granites. Mica ages are mostly in the range of about 30 m.y., which only represent the time of mica recrystallization at that time. Geochronologically, the plutons, such as Da La, Ma La Mt., etc. emplaced during early Himalayan are similar to Nangxian County, Yangbajain pluton, etc. of the same type in this region, within a range of 55-50 m.y. Moreover, two-mica granites in the belt yielded the isotopic ages in the range of 13-6 m.y.

3. The Himalayan belt. This belt is characterized by small intrusive bodies, such as Gabug, Toda, Korqag tourmaline-muscovite granite bodies, and Oona and Dodo Mt. two-mica granite bodies. Their isotopic geological ages are in the range of 20-10 m.y., belonging to late Himalayan, equivalent to Miocene.

Based upon the mica and zircon isotopic ages of various types of granitoid rock, in conjunction with their geological setting, magmatic intrusive activities responsible for granitoid rocks in the southern part of Xizang can be divided into three major stages: the first stage magmatic intrusion, which took place at late Yenshanian with an age ranging from 120 to 70 m.y., corresponding to Cretaceous; the second-stage magmatic intrusion, which occurred at early Himalayan with an age in the range of 50-30 m.y., equivalent to Palaeocene to Oligocene; the third stage magmatic intrusion, which took place at late Himalayan, with an age of 20-10 m.y., corresponding to Miocene.

The geological characteristics of granitoid rocks mentioned above and their isotopic geological ages show regular variations from north to south, which is well coincident with both the plate subduction and collision in the southern part of Xizang. From the foregoing discussion, it is suggested that the formation of the Gandise belt has close connection with the subduction of the Indian Plate underneath the Eurasian Continental Plate 120-70 m.y. ago while the formation of the Lhagoi Kangri belt and the Himalayan belt has a close bearing on the inter-collision between the Indian Subcontinent overlying the Indian Plate and the Eurasian Plate 50-10 m.y. age.

## References


(1) Li Pu et al., A report on the absolute ages of rocks and minerals determined by K-Ar method. Scientia Geologica Sinica, (1), 24 (1964) (in Chinese).

(2) Li Pu et al., Absolute age determination of rocks from the Xixabangma Region, Tibet, China. Kexue Tongbao, (10), 925 (1965) (in Chinese).

(3) Laboratory of Isotopic Geology, Institute of Geochemistry, Academia Sinica, Isotopic dating of the metamorphic rocks from the Mt. Jolmo Lungma area, China. Scientia Sinica, (3), 280 (1973).

(4) Chang Yuquan et al., A report on isotopic geological ages from Southern Tibet, China. Proceedings of Symposium on Isotopic Geology, China, Geology Press, 2, 185 (1979) (In Chinese).

(5) Group of K-Ar Geochronology, Institute of Geology, Academia Sinica, K-Ar dating and division of the Himalayan movement in Southern Xizang (Tibet). Scientia Geologica Sinica, (1). 13 (1979) (in Chinese).

Table 1 The results of repeated tests

| Name of pluton | Sample No. | Rock type | Mineral | Age (m.y.) I | Age (m.y.) II | Relative error (%) | Method |
|---|---|---|---|---|---|---|---|
| Da La | 74-256 | Gneissose two-mica granite | Muscovite | 29.5 | 29.5 | 0 | Dilution |
| | 74-256 | " " | Biotite | 31.7 | 28.7 | 10.0 | Volumetric |
| Zhaxize | 73-558 | Biotite granite | " | 115.1 | 112.5 | 2.3 | " |
| Yangbajain | 75-159 | Two-mica granite | " | 46.7 | 50.9 | 8.6 | Dilution |
| Zayu | 73-190 | " " | " | 8.2 | 8.8 | 7.0 | Volumetric |

Table 2 The comparison of isotopic ages determined by volumetric method with those by dilution method

| Name of pluton | Sample No. | Rock type | Mineral | Age (m.y.) Volumetric | Age (m.y.) Dilution | Relative error (%) |
|---|---|---|---|---|---|---|
| Da La | 74-256 | Gneissose two-mica granite | Biotite | 30.2 | | 2.3 |
| | 74-256 | " " | Muscovite | | 29.5 | |
| Pyikang | 75-612 | Porphyritic biotite granite | Biotite | 116.9 | 111.2 | 5.0 |
| Quxu | 74-382 | Porphyritic granodiorite | Hornblende | 46.8 | 44.2 | 5.7 |

Table 3 The comparison of K-Ar ages with U-Pb ages

| Name of pluton | Sample No. | Rock type | Age (m.y.) K-Ar (Biotite) | Age (m.y.) U-Pb (Zircon) | Relative error (%) |
|---|---|---|---|---|---|
| Pyikang | 75-612 | Porphyritic biotite granite | 114.1 | 134 | 8.0 |
| Demola | 73-390 | Granodiorite | 120.3 | 128 | 6.2 |
| Demola | 75-586 | Porphyritic biotite granite | 75.0 | 77 | 2.6 |

Table 4 The comparison of recrystallized biotite K-Ar ages with zircon U-Pb ages

| Name of pluton | Sample No. | Rock type | Age (m.y.) Biotite (K-Ar) | Age (m.y.) Zircon (U-Pb) | Age error (m.y.) |
|---|---|---|---|---|---|
| Zayu | 73-97 | Porphyritic biotite granite | 19.6 | 87 | 67.4 |
| Azhag Kanglha | 73-756 | Granodiorite | 35.2 | 72 | 36.8 |
| Nyingchi | 74-848 | Gneissose two-mica granite | 18.7 | 73.5 | 54.8 |
| Kangmar | 74-92 | " | 31.8 | 266 | 234.2 |

Table 5.1 Isotopic age data from the Gandise belt

| Location | No. | Sample No. | Rock type | Mineral | K% | Ar ($10^{-6}$G/G) | $^{40}Ar/^{40}K$ | Age (m.y.) | Method |
|---|---|---|---|---|---|---|---|---|---|
| Zhaxize | 1 | 73-820 | Granodiorite | Biotite | 7.65 | 0.0701 | 0.00766 | 127.2 | Volumetric |
| | 2 | 73-637 | Two-mica granite | " | 7.78 | 0.0705 | 0.00757 | 125.8 | Dilution |
| | 3 | 73-639 | " " | " | 7.61 | 0.0653 | 0.00717 | 119.4 | Volumetric |
| | 4 | 73-588 | Biotite granite | " | 5.96 | 0.0493 | 0.00691 | 115.1 | " |
| | 5 | 73-558 | " " | " | | | 0.00675 | 112.5 | " |
| | 6 | 73-650 | Porphyritic biotite granite | Zircon | | | | 85 | U-Pb |
| | 7 | 73-642 | Alkali granite | Biotite | 5.93 | 0.0330 | 0.00465 | 78.3 | Volumetric |
| Demola | 8 | 73-390 | Granodiorite | Biotite | 6.71 | 0.0580 | 0.00723 | 120.3 | Volumetric |
| | 9 | 73-390 | " | Zircon | | | | 128 | U-Pb |
| | 10 | 73-247 | Porphyritic two-mica granite | Biotite | 8.07 | 0.0563 | 0.00583 | 97.7 | Dilution |
| | 11 | 73-586 | Porphyritic biotite granite | " | 7.01 | 0.0341 | 0.00406 | 68.6 | Volumetric |
| | 12 | 73-586 | " | Zircon | | | | 77 | U-Pb |
| | 13 | 73-586 | " | " | | | | 72 | " |
| | 14 | 73-588 | " | Biotite | 7.41 | 0.0395 | 0.00445 | 75 | Volumetric |
| | 15 | 73-589 | Two mica granite | Muscovite | 8.88 | 0.0499 | 0.00469 | 79 | Dilution |
| Azhag Kanglha | 16 | 73-723 | Diorite | Biotite* | 7.76 | 0.021 | 0.00227 | 38.6 | Volumetric |
| | 17 | 73-724 | " | " | 6.75 | 0.00935 | 0.0116 | 19.8 | " |
| | 18 | 73-724 | " | Zircon | | | | 65 | U-Pb |
| | 19 | 73-725 | Granodiorite | Biotite* | 7.80 | 0.00111 | 0.00119 | 20.3 | Volumetric |
| | 20 | 73-756 | " | " | 7.83 | 0.0193 | 0.00206 | 35.2 | " |
| | 21 | 73-756 | " | Zircon | | | | 72 | U-Pb |
| | 22 | 73-751 | Biotite granite | Biotite* | 7.06 | 0.0121 | 0.00140 | 24.5 | Dilution |
| | 23 | 73-25 | Porphyritic biotite granite | Biotite | 6.15 | 0.0558 | 0.00758 | 125.9 | Volumetric |
| | 24 | 73-26 | Biotite granite | " | | | 0.00527 | 88.6 | " |
| | 25 | 73-26 | " " | Zircon | | | | 74 | U-Pb |

* Recrystallized biotite

Table 5.1 Continued

| Location | No. | Sample No. | Rock type | Mineral | K% | Ar ($10^{-6}$G/G) | $^{40}Ar/^{40}K$ | Age (m.y.) | Method |
|---|---|---|---|---|---|---|---|---|---|
| Zayu | 26 | 73-97 | Porphyritic biotite granite | Zircon | | | | 87 | U-Pb |
| | 27 | 73-97 | " | Biotite* | 7.19 | 0.00984 | 0.00114 | 19.6 | Volumetric |
| | 28 | 73-512 | " | " | 7.38 | 0.00897 | 0.00102 | 17.4 | Dilution |
| | 29 | 73-106 | Gneissose lamprophyre | " | 6.98 | 0.00631 | 0.000754 | 12.9 | Volumetric |
| | 30 | 73-73 | Two-mica granite | Biotite | 7.04 | 0.0165 | 0.00190 | 33.5 | " |
| | 31 | 73-73 | " " | Zircon | | | | 51 | U-Pb |
| | 32 | 73-212 | Gneissose two-mica granite | Biotite* | 7.8 | 0.0110 | 0.00118 | 20.2 | Volumetric |
| | 33 | 73-521 | Biotite granite | Biotite | 7.38 | 0.0176 | 0.00151 | 26.3 | Dilution |
| | 34 | 73-521 | " " | " | | | 0.00172 | 29.4 | Volumetric |
| | 35 | 73-144 | Two Mica granite | Biotite | 7.23 | 0.0163 | 0.00093 | 17.7 | Volumetric |
| | 36 | 73-508 | " | " | 8.9 | 0.0149 | 0.00140 | 24 | " |
| | 37 | 73-508 | " | " | | | 0.00112 | 19.2 | Dilution |
| | 38 | 73-192 | " | " | 7.80 | 0.00451 | 0.000511 | 8.8 | Volumetric |
| | 39 | 73-190 | " | " | | | 0.00048 | 8.2 | " |
| Nyingchi | 40 | 74-847 | Gneissose biotite granite | Biotite* | 8.23 | 0.0121 | 0.000123 | 21 | " |
| | 41 | 74-848 | Gneissose two-mica granite | " | 8.05 | 0.0105 | 0.00109 | 18.7 | Dilution |
| | 42 | 74-848 | " | Zircon | | | | 73.5 | U-Pb |
| | 43 | 74-829 | two-mica granite | Biotite | 7.81 | 0.0115 | 0.00123 | 21 | Volumetric |
| | 44 | 74-854 | " | " | 7.92 | 0.0104 | 0.00110 | 18.8 | Dilution |
| Gyabge | 45 | 75-611 | Granodiorite | " | 7.85 | 0.0496 | 0.00529 | 88.7 | Volumetric |
| | 46 | 75-611 | " | " | | | 0.00555 | 93.1 | Dilution |
| | 47 | 75-608 | Porphyritic two-mica granite | " | 8.38 | 0.0525 | 0.00514 | 88 | Volumetric |
| | 48 | 74-779 | Biotite granite | " | | | 0.00529 | 88.8 | " |
| Nangxian | 49 | 75-577 | Two-mica granite | " | 8.08 | 0.0318 | 0.00329 | 55.7 | Dilution |
| Mijisumdo | 50 | 75-555 | Biotite granite | " | 7.33 | 0.0178 | 0.00203 | 34.6 | " |

Table 5.1 Continued

| Location | No. | Sample No. | Rock type | Mineral | K% | Ar $(10^{-6}G/G)$ | $^{40}Ar/^{40}K$ | Age (m.y.) | Method |
|---|---|---|---|---|---|---|---|---|---|
| Pyikang | 51 | 75-612 | Porphyritic biotite granite | Biotite | 7.79 | 0.0664 | 0.00702 | 116.9 | Volumetric |
| | 52 | 75-612 | " | " | 7.79 | 0.0612 | 0.00666 | 111.2 | Dilution |
| | 53 | 75-612 | " | Zircon | | | | 136 | U-Pb |
| | 54 | 75-613 | Biotite granite | Biotite | 7.11 | 0.0161 | 0.00189 | 32.2 | Volumetric |
| Yangbajain | 55 | 75-165 | Porphyritic biotite granite | " | 7.88 | | 0.00448 | 75.5 | Dilution |
| | 56 | 75-165 | " | Zircon | | | | 96.5 | U-Pb |
| | 57 | 75-168 | Biotite granite | Biotite | 7.1 | 0.0189 | 0.00297 | 50.4 | Dilution |
| | 58 | $75\text{-}159_3$ | Two-mica granite | " | 5.83 | 0.0209 | 0.00300 | 50.9 | " |
| | 59 | $75\text{-}159_3$ | " | " | | | 0.00275 | 46.7 | " |
| Lhasa | 60 | 74-319 | Biotite granite | " | | | 0.0058 | 97.2 | Volumetric |
| | 61 | 74-1 | Granodiorite | " | 7.52 | 0.0385 | 0.00428 | 72.2 | " |
| Quxu | 62 | 74-382 | Porphyritic granodiorite | Hornblende | 0.82 | 0.0270 | 0.00276 | 46.8 | " |
| | 63 | 74-382 | " | " | 0.82 | 0.0255 | 0.00260 | 44.2 | Dilution |
| | 64 | 74-382 | " | Biotite | 8.28 | 0.0160 | 0.00161 | 27.5 | Volumetric |
| Dagzhuka | 65 | 74-160 | Granodiorite | " | 7.4 | 0.0476 | 0.00538 | 90.2 | " |
| Xaitongmoin | 66 | 75-37 | " | " | 6.95 | 0.0365 | 0.00438 | 73.9 | Dilution |
| Mai La | 67 | 75-510 | Porphyritic biotite granite | " | 6.45 | 0.0541 | 0.00687 | 116.7 | Volumetric |
| | 68 | $75\text{-}510_a$ | Biotite granite | " | 6.60 | 0.0370 | 0.00459 | 75.8 | " |
| Shiquanhe | 69 | 76-110 | Tonalite | Zircon | | | | 307.5 | U-Pb |
| Shiquanhe | 70 | 76-112 | Biotite granite | Zircon | | | | 104 | U-Pb |
| | 71 | 76-112 | " | Biotite | 7.66 | 0.0329 | 0.00359 | 60.7 | Dilution |
| Luma | 72 | 76-52 | Granodiorite | Zircon | | | | 67 | U-Pb |
| Rutog | 73 | 76-87 | Diorite | Biotite | 6.66 | 0.0385 | 0.00473 | 81.2 | Dilution |
| | 74 | 76-91 | Granodiorite | " | 7.31 | 0.0356 | 0.00407 | 68.7 | Volumetric |
| | 75 | 76-78 | Porphyritic biotite granite | " | 7.86 | 0.0507 | 0.00537 | 90.5 | " |

Table 5.1 Continued

| Location | No. | Sample No. | Rock type | Mineral | K% | Ar ($10^{-6}$G/G) | $^{40}Ar/^{40}K$ | Age (m.y.) | Method |
|---|---|---|---|---|---|---|---|---|---|
| | 76 | 76-80 | Biotite granite | Biotite | 5.87 | 0.00351 | 0.00499 | 83.9 | Dilution |
| | 77 | 76-68 | " | " | 5.98 | 0.0353 | 0.00493 | 82.9 | Volumetric |
| Gar | 78 | 76-2 | Quartz-monzonite | " | 7.00 | 0.0206 | 0.00246 | 41.8 | Dilution |
| | 79 | 76-3 | " | Zircon | | | | 60 | U-Pb |
| | 80 | 76-4 | Alkali granite | " | | | | 55 | " |
| | 81 | 76-8 | " | " | | | | 58 | " |

Table 5.2 Isotopic age data from the Lhagoi Kangri belt

| Location | No. | Sample No. | Rock type | Mineral | K% | Ar ($10^{-6}$G/G) | $^{40}Ar/^{40}K$ | Age (m.y.) | Method |
|---|---|---|---|---|---|---|---|---|---|
| Da La | 82 | 74-256 | Gneissose two-mica granite | Muscovite | 8.54 | | 0.00173 | 29.5 | Dilution |
| | 83 | 74-256 | " | " | 8.54 | | 0.00173 | 29.5 | " |
| | 84 | 74-256 | " | Biotite | | | 0.00186 | 31.7 | Volumetric |
| | 85 | 74-256 | " | " | | | 0.00168 | 28.7 | " |
| Kangmar | 86 | 74- 94 | " | " | | | 0.00181 | 31.8 | " |
| | 87 | 74- 94 | " | Zircon | | | | 266 | U-Pb |
| Ma La | 88 | 75-48 | " | Biotite | 7.95 | | 0.00187 | 31.8 | Volumetric |
| Naimonanyi | 89 | 76-23 | " | " | | | 0.00095 | 16.3 | " |
| | 90 | 76-31 | Two-mica granite | Muscovite | | | 0.00040 | 6.8 | " |
| | 91 | 75-56 | " | Biotite | | | 0.00079 | 13.6 | " |

Table 5.3 Isotopic age data from the Himalayan belt

| Location | No. | Sample No. | Rock type | Mineral | K% | Ar ($10^{-6}$G/G) | $^{40}Ar/^{40}K$ | Age (m.y.) | Method |
|---|---|---|---|---|---|---|---|---|---|
| Cona | 92 | 74-202 | Two-mica granite | Biotite | | | 0.00118 | 20.1 | Dilution |
| | 93 | 74- 33 | Tourmaline muscovite granite | Muscovite | 9.02 | 0.00924 | 0.000856 | 14.7 | Volumetric |
| | 94 | 74- 33 | " | " | | | 0.00118 | 20.1 | " |
| | 95 | 74- 33 | " | " | | | 0.00108 | 18.4 | " |
| Gyirong | 96 | 75- 39 | " | " | 8.40 | 0.0124 | 0.00121 | 21.1 | Dilution |
| | 97 | 75- 39 | " | Biotite | | | 0.00119 | 20.4 | Volumetric |
| Pedo La | 98 | 75-532 | Two mica granite | " | 7.89 | 0.0199 | 0.00207 | 36 | Dilution |
| | 99 | 75-532 | " | " | | | 0.00117 | 20 | Volumetric |
| Qusum | 100 | 76- 58 | Tourmaline muscovite granite | Muscovite | 8.64 | | 0.00187 | 31.9 | Dilution |
| Korgag | 101 | 76- 22 | Pegmatite | " | 8.59 | 0.00335 | 0.000326 | 5.6 | Volumetric |
| Nyalam | 102 | M33- 1 | " | " | 8.44 | 0.0077 | 0.00073 | 12.5 | Dilution |
| | 103 | M33- 1 | " | " | 8.44 | 0.0072 | 0.00070 | 12 | Volumetric |
| Xixabangma | 104 | X-1 | Granite | Biotite | 7.40 | 0.0124 | 0.00070 | 12 | Volumetric |
| | 105 | X-2 | Pegmatite | " | 7.45 | 0.0129 | 0.00070 | 12 | " |
| | 106 | X-3 | " | Muscovite | 8.40 | 0.0170 | 0.00080 | 13.7 | " |
| | 107 | X-4 | Granite | " | 8.95 | 0.0154 | 0.00070 | 12 | " |

Table 6 The intrusion sequence of Gandise composite rocks and their isotopic ages (m.y.)

| Period | Name of pluton | Intrusion stage | | | | |
|---|---|---|---|---|---|---|
| | | I | II | III | IV | V |
| Late Yenshanian | Demola | Diorite Granodiorite (120) | Porphyritic biotite granite (77) | Porphyritic two-mica granite (97) | Biotite granite | Two-mica granite (68) |
| | Rutog | Diorite Tonalite (93.81) | Granodiorite (86) | Porphyritic biotite granite (90) | Biotite granite (82) | Alkaline granite |
| Late Yenshanian to Himalayan | Pyikang | Porphyritic biotite granite (116) | Fine-grained biotite granite (32) | | | |
| | Yangbajain | Porphyritic biotite granite (96) | Two-mica granite (50) | | | |

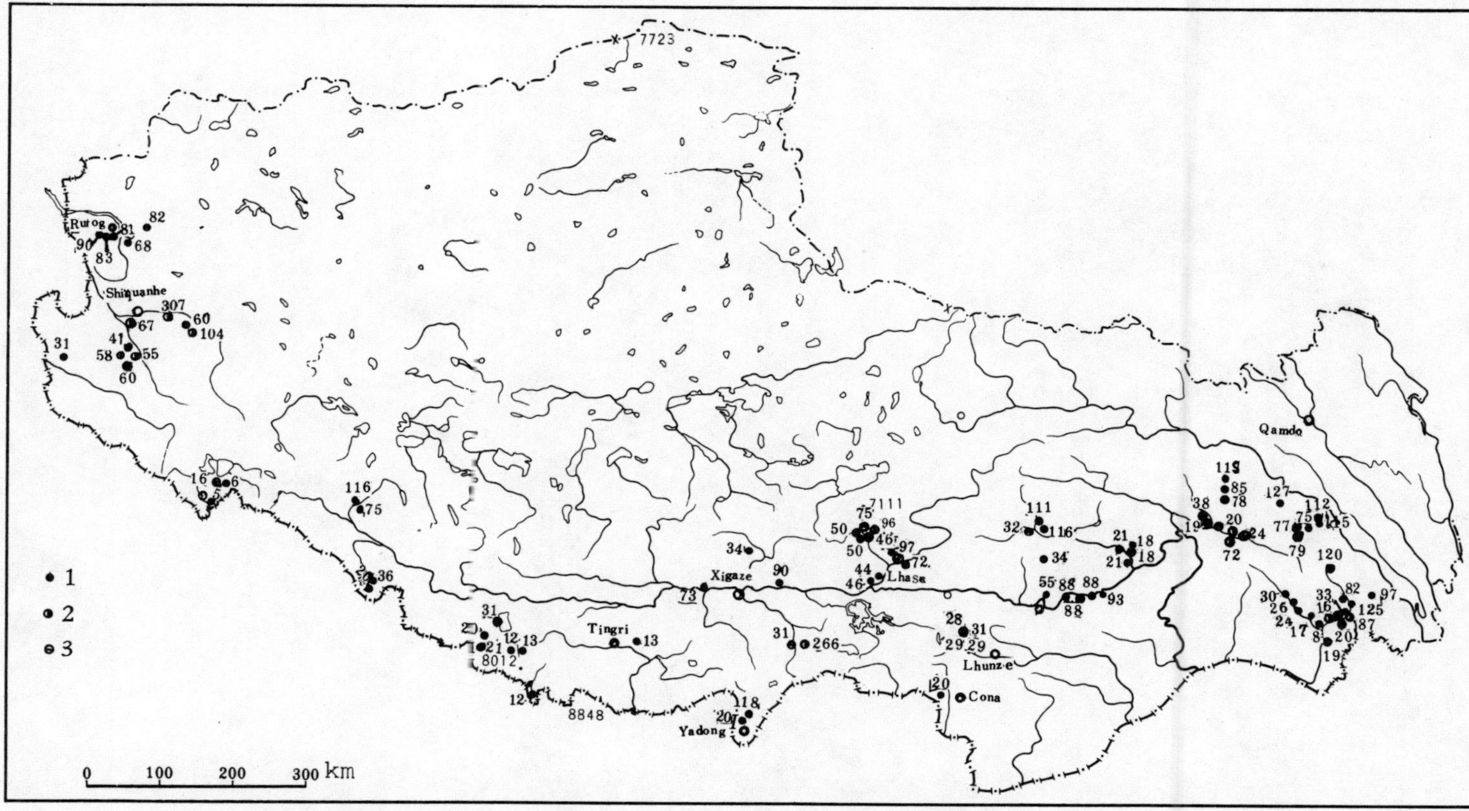


Fig. 1 A map showing the sampling localities and the isotopic geological ages for granitoid rocks in Southern Xizang
1. Mica ages 2. Zircon ages 3. Recrystallized biotite ages

# A DISCUSSION ON ZIRCON U-Pb DATING OF INTERMEDIATE-ACID ROCKS IN SOUTHERN XIZANG

U-Pb Dating Group, Laboratory of Isotopic Geology,
Institute of Geochemistry, Academia Sinica

In the southern part of Xizang are widely exposed intermediate-acid rocks, characterized by their distinct regularity in petrographical features, formation ages and geographical distribution. From north to south, three major petrographical belts can be divided as the Gandise belt (including the Bomi-Zayu belt in the east), which is the oldest in age, corresponding to late Yenshanian, and mainly composed of diorite, granodiorite, porphyritic biotite monzonitic granite and biotite-potash-feldspar granite; the Lhagoi Kangri belt which was formed later, corresponding to early-middle Himalayan, and chiefly consists of gneissose two-mica monzonitic granites; and the Himalayan belt which is the latest, corresponding to late Himalayan, and substantially made up of tourmaline-muscovite potash-feldspar granite and two-mica potash-feldspar granite. The most extensive exposure of plutons is found in the Gandise belt and its eastern Bomi-Zayu area, and the plutons are largely the composite rocks of multi-episode and multi-stage magmatic intrusion origin. These plutons usually contain significant amounts of zircon, but the plutons in the rest two belts show opposite tendency. Therefore, our emphasis in the present paper is placed on zircon U-Pb dating of the Gandise belt and its eastern Bomi-Zayu area.

U-Pb isotopic age determinations were carried out on 34 zircons, 2 feldspars, 2 sphenes, 2 apatites and 2 whole rock samples from the Gandise belt and its eastern Bomi-Zayu area. The origin of common lead and its modes of occurrence in zircons were studied and the intrusion ages were established for the intermaediate-acid rocks in the Gandise belt and the Bomi-Zayu area. A possibility was proposed for the granitoid rocks in this region having two different sources. Moreover, possible reasons for the fact that the U-Pb isotopic ages are lower than the K-Ar isotopic ages from the same pluton were also studied in the paper.

## ORIGIN OF COMMON LEAD AND ITS MODES OF OCCURRENCE IN ZIRCONS

Isotopic analyses indicate that common lead content in the zircons of granitoid rocks in this region is very high, averaging 70 %. It is well-known that the existence of large amounts of common lead must introduce great errors into the isotopic analyses of radiogenic lead(1), and thus, the reliability of determined isotopic ages directly depends on how to reasonably estimate zircon-containing common lead in terms of isotopic composition. The study of the origin of common lead and its modes of occurrence in zircons is particularly needed as for such younger geological samples as intermediate-acid rocks in the soutern part of Xizang Plateau.

The measurements of Pb content and Pb isotopic composition of environmental contaminants in our laboratory show that Pb isotopic composition of dust is close to that of galena in the given area which is lower in Pb content. From our experiments in which the background was controlled below 10 % of total Pb, it is evident that the significant amounts of common lead in zircons are inherent in the minerals themselves.

Little has been reported concerning the mode ccurrence of common lead in zircons. S. A. Magomedov (1970) found that common lead evaperized out when zircon was vacuum-heated to 1,000°C, and he suggested that the common lead could be introduced into the crystal lattice of zircon by substituting zircunium, or accumulated on the crystal surface.

Microscopic examinations show that there exists a large number of irregular fractures and solid inclusions within the crystals of zircons, and common lead seems most likely to have been introduced into zircons along with late hydrothermal solution which filled the fractures of zircons. In the heating experiment of zircon ( in $H_2$ stream ), we obtained two Pb volatilization peaks at 600° and 800°C, respectively (Fig. 1), which were produced by common lead as viewed from their isotopic composition.

Two zircon water loss temperatures were determined, using an inorganic gas chromatograph (200° and 400°C ). From the gas chromatograms of zircons, it is indicated that zircons began to vary in their structure when the temperature reached 400°C, thus providing confident evidence that there exists common lead within zircon crystals.

It was found in the heating experiment of zircon (73-724) that Pb evaporation was simultaneously accompanied by the escape of $H_2S$ at the exit of hydrogen gas, indicatin; the d composition of sulphide phase in the sample at the same time. The temperature curves of Pb volatilization of galena from this region are given in Fig. 2. It may be concluded that there are at least two modes f occurrence for common lead in zircons : one occurring within the crystals and the other as sulphide form.

Common lead in zircons from this region is estimated in terms of Pb isotopic composition of galena from the same region, and the calculated isotopic ages are basically in agreement with those obtained by internal isochron method (Table 2). This further demonstrates that the significant amounts of common lead in zircons are inherent in the minerals themselves, rather than introduced by extraneous contaminants.

## A DISCUSSION ON THE ISOTOPIC GEOLOGICAL AGES FROM THE GANDISE BELT AND ITS EASTERN BOMI-ZAYU AREA

Based upon the determined isotopic data, we are going to discuss the geology of intermediate-acid rocks in this region and their material sources.

On account of the geological modes of occurrence of the plutons and their biotite K-Ar ages, three groups of zircon U-Pb ages are roughly defined (see Table 3).

The first group is approximately consistent with biotite K-Ar ages, or slightly higher ; the second, older than biotite K-Ar ages ; and the third, younger than biotite K-Ar ages.

The first group of age data were mostly yielded by rock samples taken from the eastern Gandise belt (Fig. 2 ), such as Demola, Azhag Kanglha plutons. In the light of the age values obtained, the intrusion age of the eastern Gandise belt should be assigned to late Yenshanian and early Himalayan. The second group was mainly from Bomi-Zayu fold zone to the east of Nyingchi. As confirmed by zircon U-Pb ages, this terrain should have an age corresponding to late Yenshanian and Himalayan. However, its younger biotite K-Ar ages can be explained by biotite recrystallization caused by later thermodynamic process in the rocks. The above two groups of age values are in coincidence with the geological modes of occurrence. As a special case in our study, the third group of age data — anomalous age values which account for the majority of the obtained age values — were mostly given by rock samples taken from the middle-western segment of the eastern Gandise belt, mostly close to the ultrabasic rock belt or along the plate suture zone. The U-Pb ages there are within the range of 17-30 m.y., precisely in disaggreement with the geological modes of occurrence (overlain by Early Tertiary strata).

As to these anomalous zircon U-Pb age values, we are surely to ask "Why?" Attempts were made to trace the answer from our experimental errors, but in vain. For example, the zircon U-Pb ages from coarse-grained porphyritic hornblende biotite granites in Mai La pluton are half lower than the biotite K-Ar ages, although they are consistent with sphene- and apatite-isochron ages.

In the study on the influence of the accumulated amount of intermediate members of the radioactive U-decay series in those extremely young minerals on the isotopic age values, D. York suggested in accordance to his calculations that it would take about several hundred million years to decrease the concentration of intermediate members down to 0.1 % of $^{206}$Pb, if minerals contain no $^{234}$U at the time of their formation. From the data on inclusions, petrochemistry and REE distribution patterns from this region, it is shown that the widely exposed diorites and granodiorites in the Gandise belt are of deeper derivation. Again, from the fact that zircon crystals are irregular in form and coated with melting crusts, it is inferred that $^{234}$U depletion in the magma of this kind is possible, thus leading to the later start for zircon U-Pb dating clock as compared with K-Ar dating clock. Other geological factors for lower U-Pb ages are awaiting for further study.

## POSSIBLE SOURCES FOR INTERMEDIATE-ACID ROCKS IN THE REGION

It is found that there exist remarkable differences in the amounts of common lead in zircons from the western segment of the Gandise belt and its eastern Bomi-Zayu area, as well as in lead isotopic composition. As for the Bomi-Zayu area, the content of common lead in zircons ( from 13 intermediate-acid rocks) counts up to 65 % in average, and for the Gandise belt, the content of common lead in 30 zircons averages 80 %, suggesting that these two kinds of common lead are derived from different sources (Fig. 3). According to the radiogenic $^{207}$Pb/$^{206}$Pb in zircons, the plutons in the Bomi-Zayu area and the western segment of the Gandise belt possess the characteristics of crust-derived intrusive rocks, whereas the plutons in the middle segment and those with anomalus isotopic ages may have more diverse material sources, indicating that these rocks are plutonic in origin from different depths. For example, granodiorites distributed in the areas along the Yarlung Zangbo River may have derived from a magmatic source, 1,000 m.y. old in age.

## CONCLUSIONS

From the foregoing statements, several conclusions are drawn as follows:

1. Zircons in the intermediate-acid rocks of Southern Xizang contain significant amounts of common lead which occurs within the crystals of zircons or as sulphide phases associated with zircons, but absolutely not introduced by contaminants during analytical procedure.

2. The content of common lead in zircons and Pb isotopic composition have a close bearing on the source, intrusion age and hydrothermal activity of granitoid rocks ; for example, zircons in granites emplaced at the later stage have a higher content of common lead.

3. Apart from the isotopic ages of the plutons, the content of common lead and Pb isotopic composition also provide strong evidence that the eastern-western segment and the middle part of the Gandise belt, as well as those plutons with anomalous isotopic ages are different from one another in their sources.

4. Further study is needed for the phenomena in which zircon U-Pb ages are lower than biotite K-Ar ages from this region.

## Refernces

(1) Chernyshev, I. V., Analyzing the errors in the method of absolute age determination. Geokhimia, (1), 73-81 (1962) (in Russian).
(2) Magamedov, Sh. A., Study on the mechanism of radiogenic product migration in zircon. Geokhimia, (2), 263-267 (1970) (in Russian).
(3) York, D., Farquhar, R. M., The Earth's Age and Geochronology 73 (1972).

Table 1 Isotopic composition of lead volatilizing out of zircon at 600°C

| Zircon sample | $^{204}Pb\%$ | $^{206}Pb\%$ | $^{207}Pb\%$ | $^{208}Pb\%$ |
|---|---|---|---|---|
| NM | 1.154 | 35.39 | 18.99 | 18.99 |
| 73-724 | 1.38 | 25.21 | 21.10 | 21.10 |
| 74-843 | 1.36 | 26.05 | 21.03 | 21.03 |

Table 2 The model ages and internal isochron ages of zircons

| Name of pluton (Sample No.) | Sample | U% | Th% | Pb% | $^{204}Pb$% | $^{206}Pb$% | $^{207}Pb$% | $^{208}Pb$% |
|---|---|---|---|---|---|---|---|---|
| Zayu pluton 73-97 | Zircon | 0.202 | 0.557 | $6.27x10^{-3}$ | 0.41 | 45.18 | 7.86 | 46.55 |
| " | Feldspar | $2.11x10^{-3}$ | $2.9x10^{-4}$ | $1.05x10^{-3}$ | 1.30 | 25.31 | 20.92 | 52.48 |
| " | Whole rock | $2.7x10^{-4}$ | $1.9x10^{-3}$ | $1.97x10^{-3}$ | 1.32 | 25.00 | 20.45 | 53.23 |
| Demola pluton 73-586 | Zircon | 0.12 | 0.315 | $4.6x10^{-3}$ | 0.789 | 40.02 | 12.76 | 46.43 |
| " | Feldspar | $5x10^{-5}$ | $2.1x10^{-4}$ | $5.9x10^{-4}$ | 1.32 | 25.05 | 20.99 | 52.64 |
| " | Whole rock | $7.3x10^{-4}$ | $1.09x10^{-3}$ | $2.1x10^{-3}$ | 1.28 | 29.90 | 19.92 | 48.90 |
| Kangrinboqe pluton 76-40 | Zircon | 0.101 | 0.295 | $3.08x10^{-3}$ | 1.106 | 31.234 | 17.402 | 50.207 |
| " | Sphene | $7.83x10^{-3}$ | $2.12x10^{-2}$ | $5.67x10^{-4}$ | 1.220 | 28.29 | 18.349 | 52.141 |
| " | Phosphorite | $7.58x10^{-3}$ | $1.81x10^{-2}$ | $4.30x10^{-4}$ | 1.234 | 28.309 | 19.056 | 51.402 |

Table 2 (Continued)

| Name of pluton (Sample No.) | Sample | $\alpha$ | $\gamma$ | $\mu_0$ | $\omega$ | $^{206}Pb/^{238}U$ | $^{208}Pb/^{232}Th$ | Isochron ages $^{206}Pb-^{238}U$ | Isochron ages $^{208}Pb-^{232}Th$ |
|---|---|---|---|---|---|---|---|---|---|
| Zayu pluton 73-97 | Zircon | 110.20 | 113.54 | 6790.38 | 1934.56 | 87 | 77 | | |
| " | Feldspar | 19.47 | 40.37 | 133.58 | 18.971 | m.y. | | 87 | 77 |
| " | Whole rock | 18.94 | 44.33 | 25.59 | 65.241 | | | m.y. | |
| Demola pluton 73-586 | Zircon | 50.73 | 58.85 | 2857.20 | 7743.67 | 72 | 50 | | |
| " | Feldspar | 18.98 | 39.88 | 5.548 | 24.08 | m.y. | | 67 | 53 |
| " | Whole rock | 23.36 | 38.20 | 23.469 | 36.21 | | | m.y. | |
| Kangrinboqe pluton 76-40 | Zircon | | 44.57 | 25.35 | 7626 | 23 | 12 | | |
| " | Sphene | 22.96 | 41.90 | 963.1 | 2697 | m.y. | | 22 | |
| " | Phosphorite | 22.72 | 40.88 | 1214 | 3004 | | | m.y. | |

Table 3 U-Pb ages from the Gandise belt and its eastern part (Bomi-Zayu area)

| Sample No. | Location | Rock type | U % | Th % | Pb % | $^{204}Pb\%$ | $^{206}Pb\%$ | $^{207}Pb\%$ | $^{208}Pb\%$ |
|---|---|---|---|---|---|---|---|---|---|
| | | | | | ( Type I ) | | | | |
| 73-390. | Demola | Diorite | 0.114 | 0.043 | 0.0094 | 0.996 | 39.60 | 16.06 | 43.34 |
| $73\text{-}586_I$ | " | Porphyritic biotite granite | 0.094 | 0.045 | 0.0037 | 0.874 | 42.76 | 14.87 | 41.50 |
| $73\text{-}586_{II}$ | " | " | 0.12 | 0.315 | 0.0046 | 0.789 | 40.02 | 12.76 | 46.43 |
| 73-26 | Azhag Kanglha | Biotite granite | 0.229 | 0.163 | 0.017 | 1.03 | 37.24 | 17.54 | 44.18 |
| 75-612 | Pyikan | Porphyritic biotite granite | 0.165 | 0.054 | 0.041 | 1.22 | 30.46 | 19.54 | 48.78 |
| 75-165 | Yangbajain | " | 0.231 | 0.078 | 0.016 | 1.021 | 38.41 | 17.09 | 43.47 |

Table 3 ( Continued )

| Sample No. | Location | Rock type | Common Pb % | $T(^{206}/_{238})$ (m.y.) | $T(^{207}/_{235})$ (m.y.) | $T(^{208}/_{232})$ (m.y.) | $T(^{207}/_{206})$ (m.y.) | Accepted age Value (m.y.) | K-Ar (m.y.) |
|---|---|---|---|---|---|---|---|---|---|
| | | | | | | ( Type I ) | | | |
| 73-390. | Demola | Diorite | 74.29 | 128 | 34 | 230 | – | 128 | 120.3 |
| $73\text{-}586_I$ | " | Porphyritic biotite granite | 65.18 | 77 | 63 | 139 | – | 77 | 77 |
| $73\text{-}586_{II}$ | " | " | 59.66 | 72 | 22 | 50 | – | 72 | 77 |
| 73-26 | Azhag Kanglha | Biotite granite | 77.84 | 99 | 148 | 28 | 1108 | 99 | 90 |
| 75-612 | Pyikan | Porphyritic biotite granite | 92.30 | 136 | 134 | – | 139 | 136 | 116.3 |
| 75-165 | Yangbajain | " | 78.23 | 96.5 | 107 | 75 | 383 | 107 | 78.8 |

Table 3 ( Continued )

| Sample No. | Location | Rock type | U % | Th % | Pb % | $^{204}Pb\%$ | $^{206}Pb\%$ | $^{207}Pb\%$ | $^{208}Pb\%$ |
|---|---|---|---|---|---|---|---|---|---|
| ( Type II ) | | | | | | | | | |
| 73-724 | Azhag Kanglha | Diorite | 0.107 | 0.05 | 0.003 | 0.901 | 38.61 | 15.63 | 44.75 |
| 73-756 | " | Granodiorite | 0.088 | 0.175 | 0.003 | 0.691 | 43.75 | 12.21 | 43.34 |
| 73-25 | " | Porphyrite biotite granite | 0.102 | 0.034 | 0.004 | 0.936 | 41.08 | 16.08 | 39.91 |
| 73-97 | Zayu | " | 0.202 | 0.557 | 0.006 | 0.41 | 45.18 | 7.86 | 46.55 |
| 73-73 | " | Two-mica granite | 0.170 | 0.096 | 0.016 | 1.21 | 30.34 | 19.55 | 48.90 |
| 74-848 | Nyingchi | " | 0.172 | 0.0413 | 0.004 | 0.533 | 60.34 | 11.07 | 28.04 |
| 74-852 | " | " | 0.111 | 0.0771 | 0.002 | 1.15 | 36.13 | 18.08 | 44.62 |
| 76-68 | Rutog | Biotite granite | 0.29 | 0.117 | 0.009 | 0.510 | 61.40 | 10.36 | 27.73 |
| 76-112 | Shiquanhe | " | 0.058 | 0.073 | 0.004 | 1.227 | 32.18 | 18.61 | 47.98 |



Table 3 ( Continued )

| Sample No. | Location | Rock type | Common Pb % | T(206/238) (m.y.) | T(207/235) (m.y.) | T(208/232) (m.y.) | T(207/206) (m.y.) | Accepted age Value (m.y.) | K-Ar (m.y.) |
|---|---|---|---|---|---|---|---|---|---|
| ( Type II) | | | | | | | | | |
| 73-724 | Azhag Kanglha | Diorite | 67.31 | 39 | 50 | 110 | 668 | 39 | |
| 73-756 | " | Granodiorite | 51.54 | 72 | 62 | 59 | – | 72 | 19.8 |
| 73-25 | " | Porphyrite biotite granite | 69.81 | 60 | 80 | 104 | 793 | 60 | |
| 73-97 | Zayu | " | 31.01 | 87 | 68 | 77 | – | 87 | 19.6 |
| 73-73 | " | Two-mica granite | 91.5 | 51 | 75 | 22 | – | 51 | 33.5 |
| 74-848 | Nyingchi | " | 41.48 | 73.5 | 81.5 | 110 | 1056 | 73.5 | 18.7 |
| 74-852 | " | " | 85.79 | 154 | – | – | – | 154 | 17.7 |
| 76-68 | Rutog | Biotite granite | 38.89 | 121 | 109 | 123 | – | 121 | 82.9 |
| 76-112 | Shiquanhe | " | 93.57 | 104 | – | – | – | 104 | 60.7 |

Table 3 ( Continued )

| Sample No. | Location | Rock type | U % | Th % | Pb % | $^{204}Pb\%$ | $^{206}Pb\%$ | $^{207}Pb\%$ | $^{208}Pb\%$ |
|---|---|---|---|---|---|---|---|---|---|
| | | | | | ( Type III ) | | | | |
| 73-650 | Zhaxize | Porphyritic biotite granite | 0.193 | 0.200 | 0.007 | 0.693 | 46.42 | 12.22 | 40.62 |
| 75-606 | Nangxian County | Granodiorite | 0.0454 | 0.0113 | 0.0009 | 1.085 | 31.46 | 17.50 | 49.78 |
| 74-779 | Guabge | Biotite granite | 0.088 | | 0.0018 | 1.02 | 38.07 | 17.25 | 43.66 |
| 74-382 | Quxu | Porphyritic granodiorite | 0.10 | | 0.0016 | 0.96 | 33.87 | 15.67 | 49.55 |
| 74-391 | Lhasa | Biotite granite | 0.164 | | 0.0037 | 1.13 | 34.19 | 18.19 | 46.25 |
| 75-37 | Dongqu | Granodiorite | 0.093 | 0.245 | 0.003 | 1.194 | 27.81 | 18.88 | 52.12 |
| $75\text{-}510_a$ | Mai La | Biotite granite | 0.255 | 0.082 | 0.005 | 1.023 | 38.79 | 17.08 | 43.10 |
| 76-4 | Gar | Granite | 0.113 | 0.048 | 0.006 | 1.123 | 34.23 | 18.10 | 46.55 |
| 76-8 | " | " | 0.111 | 0.143 | 0.008 | 1.143 | 31.29 | 18.31 | 49.26 |
| 76-40 | " | Porphyritic granodiorite | 0.102 | 0.295 | 0.003 | 1.106 | 31.23 | 17.40 | 50.26 |
| 76-52 | Luma | Granodiorite | 0.162 | 0.259 | 0.0213 | 1.221 | 29.68 | 19.02 | 50.08 |
| 76-96 | Rutog | " | 0.169 | 0.205 | 0.007 | 0.965 | 39.93 | 15.55 | 46.54 |
| 76-87 | " | Diorite | 0.061 | 0.087 | 0.002 | 0.883 | 38.57 | 15.21 | 45.33 |
| 76-91 | " | Granodiorite | 0.088 | 0.096 | 0.002 | 1.031 | 37.13 | 16.41 | 45.42 |
| 76-78 | " | Porphyritic biotite granite | 0.181 | 0.104 | 0.004 | 0.965 | 38.50 | 16.54 | 43.99 |
| 76-80 | " | Biotite granite | 0.215 | 0.231 | 0.006 | 0.947 | 38.98 | 15.88 | 44.18 |

Table 3 ( Continued ) * Overlain by Early Tertiary strata

| Sample No. | Location | Rock type | Common Pb % | T(206/238) (m.y.) | T(207/235) (m.y.) | T(208/232) (m.y.) | T(207/206) (m.y.) | Accepted age Value (m.y.) | K-Ar (m.y.) |
|---|---|---|---|---|---|---|---|---|---|
| | | | | | ( Type III ) | | | | |
| 73-650 | Zhaxize | Porphyritic biotite granite | 51.68 | 85 | 68 | 102 | | 85 | 101.1 |
| 75-606 | Nangxian County | Granodiorite | 78.46 | 19 | 30 | 140 | 1052 | 19 | 37.52 |
| 74-779 | Guabge | Biotite granite | | 26 | 35 | | 492 | 26 | 88.8 |
| 74-382 | Quxu | Porphyritic granodiorite | 69.58 | 28.5 | 25.6 | | 576 | 28.5 | 46.81 |
| 74-391 | Lhasa | Biotite granite | 81.89 | 24 | 33 | | 778 | 24 | 97.2 |
| 75-37 | Dongqu | Granodiorite | 86.52 | 17 | 35 | 21 | 1576 | 17 | 73.9 |
| $75\text{-}510_a$ | Mai La | Biotite granite | 77.31 | 31 | 34 | 33 | 248 | 31 | 40.7 |
| 76-4 | Gar | Granite | 85.64 | 55 | 28 | 27 | | 55 | |
| 76-8 | " | " | 87.16 | 58 | 22 | 38 | | 58 | |
| 76-40 | " | Porphyritic granodiorite | 84.34 | 23 | | 12 | | 23 | * |
| 76-52 | Luma | Granodiorite | 93.23 | 67 | | | | 67 | |
| 76-96 | Rutog | " | 71.63 | 56.8 | 16.5 | 63.6 | | 56.8 | 80-90 |
| 76-87 | " | Diorite | 66.46 | 44 | 51 | 40 | 418 | 44 | 94.2 |
| 76-91 | " | Granodiorite | 78.70 | 33 | | 19 | | 33 | 68.73 |
| 76-78 | " | Porphyritic biotite granite | 73.59 | 32 | 41 | 39 | 632 | 32 | 90.5 |
| 76-80 | " | Biotite granite | 72.22 | 45 | 39 | 34 | | 45 | 83.9 |

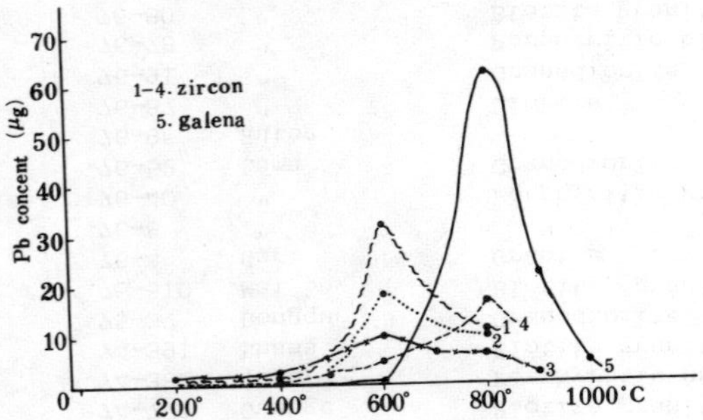


Fig. 1  Pb volatilization curves at different temperatures in the heating experiment of zircon

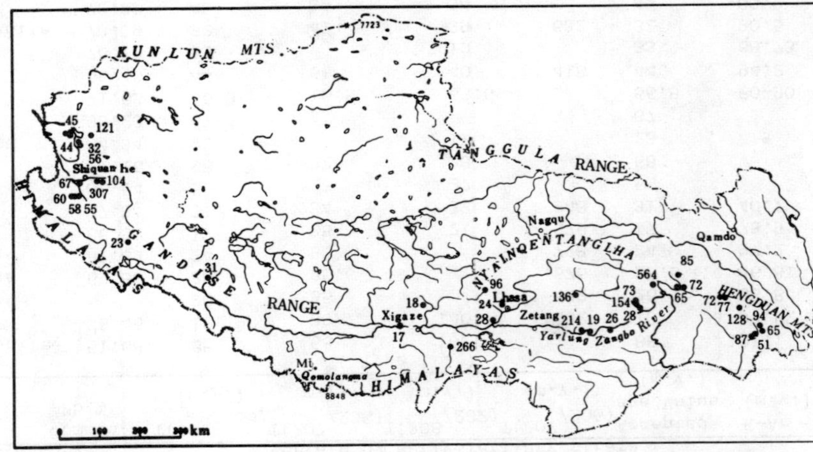


Fig. 2  A map showing the distribution of zircon U-Pb age data for intermediate-acid rocks in the Gandise belt and its eastern Bomi-Zayu area in Southern Xizang

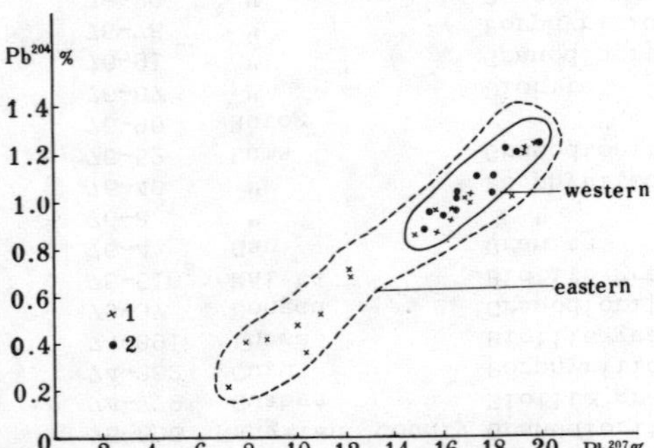


Fig. 3  Zircon $^{204}Pb$—$^{207}Pb$ plot for granitoid rocks in the Gandise belt and its eastern Bomi-Zayu area.

1. sample from the eastern part
2. sample from the western segment of the Gandise belt

# GEOCHEMISTRY OF REE IN INTERMEDIATE-ACIDIC INTRUSIVE ROCKS IN SOUTHERN XIZANG

Zhao Zhen-hua, Wang Yi-xian, Qian Zhi-xin
Zhang Yu-quan, Wang Zhong-gang
(Institute of Geochemistry, Academia Sinica)

The geochemical characteristics of REE in intermediate-acidic intrusive rocks in southern Xizang are described in this paper with an emphasis on the abundances and the distribution patterns and on their variance with the petrographic types. At the same time, a preliminary discussion is also made on the origin of the intrusive rocks.

Using cation resin for separation and X-ray fluorescence spectrometry for analysis, we determined the contents of individual REE in diorite, granodiorite, biotite granite, two-mica granite and tourmaline muscovite granite. The determination uncertainties for this method are better than 30 %. The total REE contents were determined with the uncertainties of about 10-20 % by colorimetry using extraction technique for separation.

## ABUNDANCES OF REE AND THEIR VARIANCE

The localities and petrographic features of samples are listed in Table 1. The abundances of REE were calculated by area weight method (Table 2). It is clearly seen from the table that the abundances of REE drastically decrease from Late Yenshanian to Early and Late Himalayan periods. The REE abundances in this area are lower than the average values for acidic rocks of the crust. This feature is also reflected by the petrographic types (Table 2). The contents of REE increase in conformity with the following sequence: tourmaline muscovite granite and two-mica granite—diorite and granodiorite— biotite granite— porphyritic biotite granite. Accompanying with the variance of REE abundances in this area, the associations of accessory minerals change regularly. For Late Yenshanian period, allanite-sphene type and xenotime type are typical accessory mineral associations, but for Late Himalayan period, they are replaced by monazite type association.

The study on the regularities of REE abundance variance leads to find out the control function of alkaline metals. REE concentrations increase with increasing potassium content, and decrease with increasing aluminum content. The K/Al ratios are indicators of this regularity. The REE contents will be higher than 200 ppm when the atomic ratio of K/Al is greater than 0.35, while the REE contents will be lower than 100 ppm when this ratio becomes less than 0.35 (Fig. 1). Moreover, the ratio of Na/K controls the REE contents also. When the atomic ratio of Na/K closes to unity, the REE contents will be higher than 200 ppm, but when this ratio becomes greater than 1.2, the REE contents will drastically decrease (lower than 100 ppm) (Fig. 2).

## DISTRIBUTION PATTERNS OF REE

Intermediate-acidic rocks of this area can be divided into four types, based on the obtained chondrite-normalized REE patterns of samples and the features of Eu anomaly:

1. The pattern declines to the right and almost in smooth curve, showing little anomaly of Eu. $\delta$Eu values range from 0.78 to 0.95, and heavy REE are more abundant. The ratios $\Sigma$ La-Eu/$\Sigma$ Gd-Lu+Y for this type range from 2.35 to 4.0. The rocks that can be assigned to this type include diorite and granodiorite ( Fig. 3).

2. The pattern has a V-shaped curve, showing intermediate Eu anomaly, $\delta$Eu values range from 0.50 to 0.70, and the light REE are relatively enriched. Biotite granite can be assigned to this pattern (Fig. 4).

3. This type declines to the right in V-shaped curve with noticeable Eu anomaly, $\delta$ Eu values fall into the range of 0.36 - 0.39. Two-mica and tourmaline muscovite granite are good representatives of this type. This type is identicaly similar to that for leucogranite of Manaslu in Central Nepal (Fig. 5).

4. This type declines to the right in V-shaped curve showing very strong Eu anomaly, $\delta$ Eu values fall into the range of 0.27 - 0.36. Porphyritic biotite granite belongs to this type (Fig. 6).

From the above descriptions it is evident that the variation of REE patterns is closely related to the rock types. In the sequential order: diorite-granodiorite — biotite granite — porphyritic biotite granite — two-mica granite and tourmaline muscovite granite, the rocks vary from intermediate-acidic to acidic, and from smooth REE patterns to V-shaped curves, from obscure Eu anomaly ( $\delta$ Eu about 0.90) to very strong Eu anomaly ( $\delta$ Eu about 0.30). Therefore, the REE distribution patterns demonstrate clearly the characteristics of magmatic evolution. It can be seen from the diagram showing the relation between differentiation index(DI) and the Eu anomaly that the Eu anomaly becomes stronger when the differentiation index becomes greater (Fig.7). This feature is closely similar to that of Californian batholith in the United States ( Table 3). From the table, the Eu anomaly becomes more evident in the following sequential order: gabbro — tonalite — granodiorite — leucogranite.

## DISCUSSION ON THE MECHANISM OF FORMATION FOR INTERMEDIATE-ACIDIC ROCKS IN SOUTHERN XIZANG

Plotting the REE parameters of intermediate-acidic intrusive rocks into the Treuil and Joron's diagram of partial melting and fractional crystallization patterns for igneous rocks (Fig. 8), it is clearly seen that the dots are conformed with the partial melting and fractional crystallization patterns. So it can be concluded that the intermediate-acidic magmatic rocks of this area were derived from the partial melting of certain mother materials. It should be pointed out that the porphyritic biotite granite displays a peculiar feature, i.e., its dots are conformed with the fractional crystallization pattern. In addition, its Ce/Yb ratio and the differentiation index and its Li, Rb, Cs, and F contents all show more evident fractional crystallization features than biotite granite (Table 4). This is similar to the Hercynian young porphyritic biotite granite in Northern Portugal, which shows more evident fractional characters than other granite in the same belt.

According to the features of REE patterns, the intrusive rocks of this area can be divided into two types,namely abyssal type and hypabyssal type. The former includes diorite and granodiorite which constitute the principal part of Gandise belt, and characterized by smooth REE pattern and little Eu anomaly. Moreover, Gandise belt is composed of complex variety of plutons, including diorite, granodiorite, biotite granite and two-mica granite with a large number of andesite, basalt and intermediate-acidic volcanic rocks. It is characterized by a low $^{87}Sr/^{86}Sr$ ratio. In mineralogy, hornblend is the typical melanocratic mineral, and magnetite, allanite and sphene are typical accessory minerals. All of these features suggest that their parent materials should be stemed from some deeper sources, that is to say, from lower part of the crust and/or upper mantle. As we know these sources are rich in Eu but depleted in light REE. Based on the isotopic ages, this type of intrusive rocks should be formed by the partial melting of the materials in lower crust and upper mantle during the underthrust stage of the Indian Plate towards the Eurasian Plate (70 - 120 m.y.). Two-mica granite and tourmaline muscovite granite, constituting the principal part of Lhagoi Kangri and Himalayan belts, belong to the hypabyssal type. They are uniform and simple in petrographic types. Their REE patterns are drastically declined and bended with very strong Eu anomaly. This type of patterns is similar to that of gneiss. Moreover, they gave rise to a high $^{87}Sr/^{86}Sr$ ratio and low content of melanocratic minerals. Monazite and garnet are their typical accessory minerals. These features suggest that their parent materials should be stemed from the sial layer of the crust, which is relatively rich in light REE but depleted in Eu. According to isotopic ages, this type should be formed by partial melting of the materials in sial of the upper crust during the intercollision stage of the Indian and Eurasian Plates.

## References


(1) Albuguerque, C. A. R., Rare earth elements in "younger" granite, Northern Portugal. Lithos, 11(3), 220-228 (1978).

(2) Allegre, C. L., Minster, J. F., Quantitative models of trace element behaviour inrmagmatic processes. Earth and Planet. Sci. Lett. 38(1), 16-17 (1978).

(3) Cocherie Alain, Donnees Preliminaries sur la Geochimie des Terres Rares dans le Massif Leucogranitioue du Manaslu ( Nepal Central ). Himalaya Sciences de la Terres, Editions du CNRS, Paris, 93-110 (1977).

(4) Heskin, L. A., Frei, F. A. Distribution of rare earth elements in lithosphere and cosmos. Publisher : "Mir", Moscow, 87(1968).

Table 1 Localities and petrographic features of samples

| Rock type | Sample No. | Pluton | Contents of rock-forming minerals ( % ) | | | | | | | |
|---|---|---|---|---|---|---|---|---|---|---|
| | | | plag. | horn. | qz. | bi. | K-fel. | microc. | mus. | tour. |
| Diorite | 73-724 | Azhag Kanglha | 66.9 | 22.9 | 9.0 | 1.3 | | | | |
| Granodiorite | 74-349<br>73-725 | Quxu<br>Azhag Kanglha | 34-36 | 3-12 | 18-20 | 10 | 21-32 | | | |
| Biotite granite | 73-26<br>73-554<br>73-630<br>73-781 | Azhag Kanglha<br>Azhag Kanglha<br>Zhaxize<br>Lhasa | 26-31 | | 26-29 | 6-8 | (27) | 36 | | |
| Porphyritic biotite granite | 73-25<br>73-650<br>73-97 | Azhag Kanglha<br>Zhaxize<br>Zayu | 26-33 | | 20-25 | 6-18 | 20-30 | 34-41 | | |
| Two-mica granite | 73-73<br>73-143<br>74-94<br>73-637 | Zayu<br>Zayu<br>Kangmar<br>Zhaxize | 7-27 | | 28-35 | 3-8 | | 33-50 | 2-4 | |
| Tourmaline muscovite granite | 74-35 | Gabug | 23.5 | | 26.9 | tr. | 44.0 | | 4.9 | 1.5 |

Table 2 REE contents in various rocks

| Periods | $RE_2O_3$ (ppm) | Rock type | $RE_2O_3$ (ppm) |
|---|---|---|---|
| Late Yenshanian (96*) | 200 | Diorite, Granodiorite (31) | 190 |
| Early Himalayan (18) | 138 | Porphyritic biotite-granite(15) | 231 |
| Late Himalayan (24) | 136 | Biotite-granite (34) | 215 |
| Average (138) | 194 | Two-mica and tourmaline-muscovite Granite (58) | 151 |

* number for samples.

Table 4 A comparison between porphyritic biotite-granite and biotite-granite

| | Ce/Yb | Eu | Li (ppm) | Rb (ppm) | $RE_2O_3$ (ppm) | F (ppm) | DI | $K_2O$+$Na_2O$ (%) |
|---|---|---|---|---|---|---|---|---|
| Porphyritic biotite-granite | 9 | 0.32 | 50 | 251 | 231 | 1187 | 88 | 8.5 |
| Biotite-granite | 6 | 0.60 | 42 | 239 | 215 | 976 | 82 | 7.4 |

DI: differentiation index

Table 3 REE Distribution of Southern Californian batholith in USA

| Rock type | La | Ce | Pr | Nd | Sm | Eu | Tb | Dy | Ho | Tm | Yb | Lu | Y | ΣRE | $\frac{\Sigma Ce}{\Sigma Y}$ | δEu* | $\frac{Ce^{**}}{Yb}$ |
|---|---|---|---|---|---|---|---|---|---|---|---|---|---|---|---|---|---|
| gabbro | 4.0 | 14.5 | 2.2 | 7.7 | 2.2 | 1.05 | 0.48 | 2.8 | 0.57 | 0.25 | 1.7 | 0.26 | 16 | 59 | 1.4 | 1.37 | 1.7 |
| tonalite | 13.4 | 23 | 5.7 | 17 | 3.7 | 1.06 | 0.84 | 2.6 | 0.75 | 0.39 | 2.0 | 0.30 | 22 | 97 | 2.2 | 0.81 | 2.3 |
| granodiorite | 25 | 34 | 7.0 | 22 | 4.0 | 0.80 | 0.83 | 3.0 | 0.90 | 0.37 | 2.9 | 0.39 | 25 | 121 | 2.7 | 0.60 | 2.4 |
| leucogranite | 25 | 88 | 5.6 | 24 | 3.8 | 0.63 | 0.63 | 3.1 | 0.74 | 0.26 | 1.30 | 0.20 | 17 | 180 | 6.7 | 0.53 | 13.6 |

* calculation by authors of this paper
** ratio of chondrite normalized

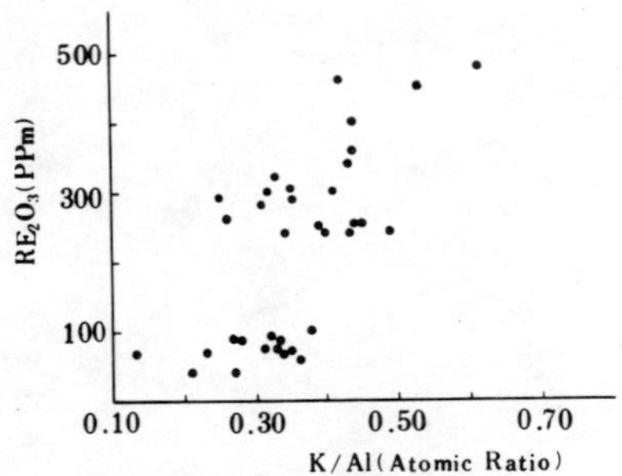

Fig.1 Relation between REE contents and K/Al atomic ratios

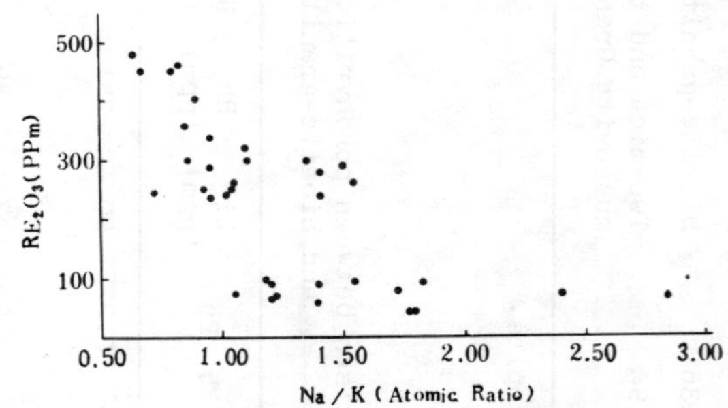

Fig. 2 Relation between REE contents and Na/K atomic ratios

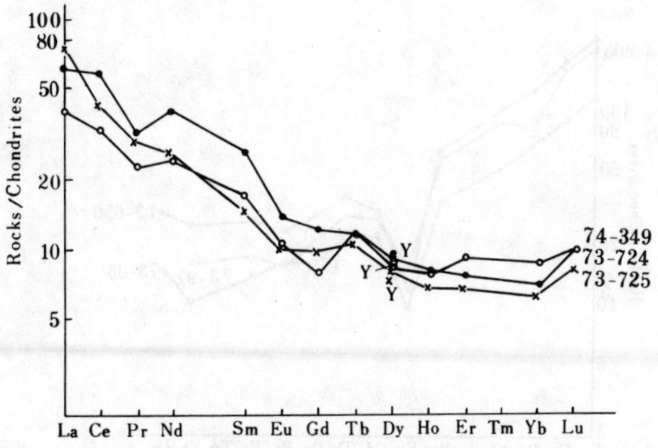


Fig. 3 REE pattern of diorite and granodiorite

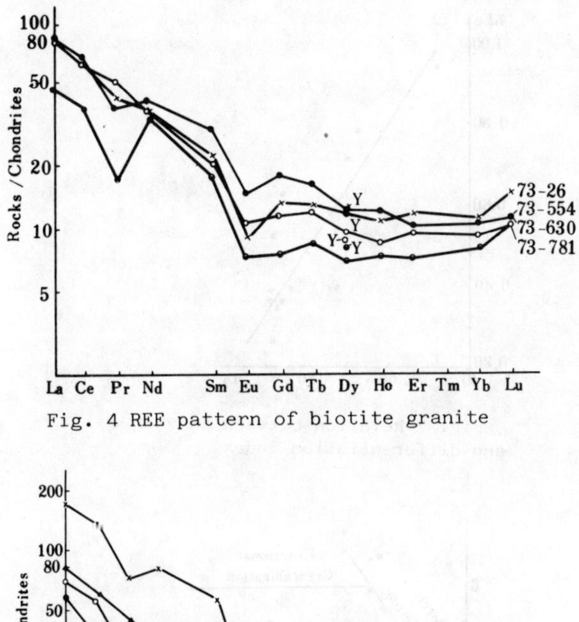


Fig. 4 REE pattern of biotite granite

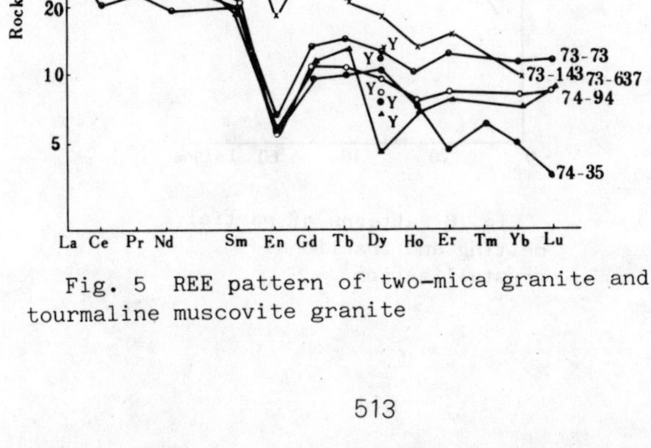


Fig. 5 REE pattern of two-mica granite and tourmaline muscovite granite

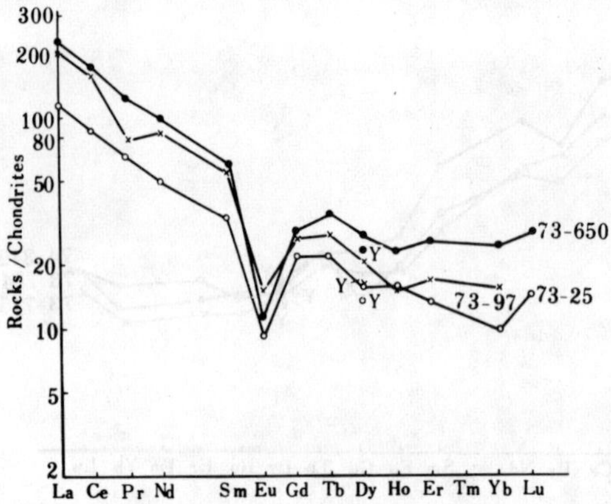


Fig. 6 REE pattern of porphyritic biotite granite

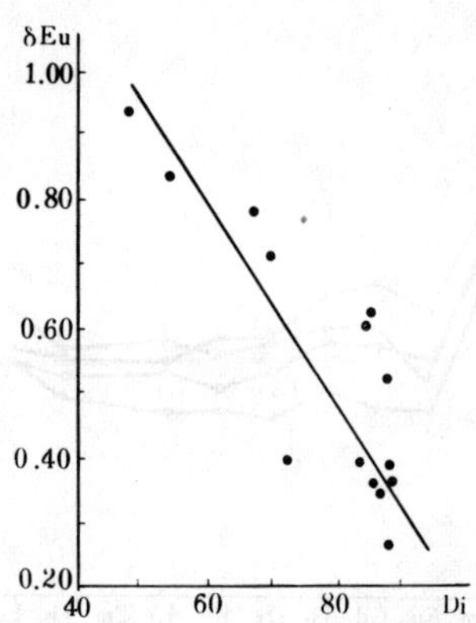


Fig. 7 Relation between Eu and differentiation index

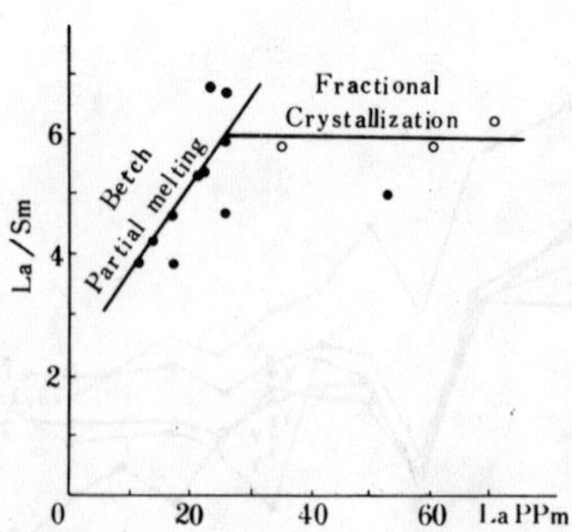


Fig. 8 Patterns of partial melting and fractional crystallization

## Rb-Sr ISOTOPIC STUDIES ON SOME INTERMEDIATE-ACID PLUTONS IN SOUTHERN XIZANG

Wang Jun-wen, Chen Zhong-li, Gui Xun-tang
Xu Rong-hua, Zhang Yu-quan
(Institute of Geochemistry, Academia Sinica)

Widespread in Southern Xizang are intermediate-acid rocks characterized by their regular distribution in space and time(1). The isotopic data determined by K-Ar method(2) for the three petrographical belts are as follows:

The major parts of granitoids of the Gandise belt (the North belt) give an age of 120-70 m.y.; the age of the Lhagoi Kangri belt (the Middle belt) is in the range of 50-30 m.y. and the Himalayan belt (the South belt) has an age ranging from 20 to 10 m.y. As for their petrography, these three belts show a noticeable tendency from complicated to simple. In the North belt, the sequence from granodiorite to alkaline granite is well developed ; the Middle belt is mainly composed of gneissose two-mica monzonitic granite and the South belt consists essentially of tourmaline - muscovite K-feldspar granite.

Zircon U-Pb ages of these petrographical belts appear more intricate(3). In order to explore their formation time and genesis, we have determined the Rb-Sr whole rock isochron ages on granodiorite from the Quxu pluton in the Gandise belt, gneissose two-mica granite from the Kangmar pluton in the Lhagoi Kangri belt, and tourmaline-muscovite granite from the Gabug pluton in the Himalayan belt. Meanwhile, their $^{87}Sr/^{86}Sr$ initial ratios have been also worked out, respectively.

### EXPERIMENTAL PROCEDURE

0.05-0.3 gram of pulverized whole rock sample was taken and spikes $^{87}Rb$ and $^{84}Sr$ were added at the same time. Then, $HF + HClO_4$, a mixed acid, was used to dissolve the rock sample. The separation of Rb from Sr was carried out on Dowex-50 x 8 strongly acid cation exchange columns. The sample in the form of chloride was analyzed on $CH_5$ Type mass spectrometer by Yunnan Institute of Geology. The contents of Rb and Sr were measured by means of isotopic dilution method. Another portion of sample was determined for the ratio of $^{87}Sr/^{86}Sr$ which was then corrected using 0.1194. Finally, the data processing was performed by least-square method, using the constant $\lambda = 1.42 \times 10^{-11}\ yr^{-1}$. The decomposition of sample and the separation of Rb from Sr were all carried out in platinum and quartz glass wares. The reagents used were repeatedly purified.

### GEOLOGICAL SETTING, RESULTS AND DISCUSSION

#### Quxu Pluton

The Quxu pluton stretches on both sides of the Lhasa River, winding through the city of Lhasa to Quxu county. The pluton is composed of gabbro, diorite, granodiorite and granite with granodiorite predominant. The contact relationship characterized by the intrusion of granodiorite into diorite and granite into granodiorite can be observed in the field. Its country rock consist of Carboniferous-Permian low-grade metamorphic slate and sandstone, as well as Jurassic metamorphic sandstone and limestone. Closely-spaced inclusions are found in granodiorite here and there, aligning orientationally and running parallel to the strike of the country rocks.

The granodiorite is generally grey in colour and has equigranular texture with clinopyroxene, K-feldspar, quartz, biotite and hornblende as their major rock-forming minerals, and magnetite and sphene as their predominant accessory minerals. In the local parts of the pluton some prophyritic granodiorites are recognized.

Table 1 shows the Rb-Sr isotopic data for five whole rock specimens from Quxu granodiorite ( Q-74-181, Q-74-340, Q-74-349, Q-74-376 and Q-74-338). The results calculated on the basis of the five points plotted on the Rb-Sr isochron are as follow:

r (correlation coefficient) = 0.9687
a (intercept) = 0.7036
b (slope) = 0.0002671
t (age value) = 17.3 m.y.

The results calculated on the basis of the four points representing Q-74-181, Q-74-340, Q-74-376 and Q-74-338 are:

r = 0.9940 a = 0.7033
b = 0.0002671 t = 18.3 m.y.

The isotopic ages of granitoids determined by K-Ar method from the whole Gandise belt are in the range of 120 - 70 m.y. The K-Ar ages for the Quxu pluton are 90 m.y. and 73 m.y., whereas the hornblende K-Ar ages for porphyritic granodiorite are 46.8 m.y.(volumetric method) and 44.2 m.y.(dilution method); the biotite K-Ar age — 27.5 m.y.; the zircon U-Pb age — 28 m.y., all lower than the age of the Gandise belt. The noticeable difference between these isotopic ages may imply that Himalayan orogenetic movement might have exercised great influence on the Quxu pluton as exemplified by the well-developed K replacement at the later stage observed under microscope in granodiorite. It is postulated that this K replacement during the later stage may have associated with the addition of rubidium isotopes, consequently giving rise to a younger isotopic age for the pluton.

Its lower $^{87}Sr/^{86}Sr$ initial ratio (about 0.7036) indicates that the granodiorite in the Quxu pluton seems to have been derived from the partial remelting of Tethyan oceanic crust, or may be the product of the primary differentiation of basaltic magmas from the upper mantle.

## Kangmar Pluton

The Kangmar pluton lies to the north of Kangmar county, occurring as a small stock in Palaeozoic strata. It is 15-16 km long in SN direction and 6-7 km wide in EW direction, constituting the core of a short-axis anticline. Gneissose structures are commonly developed in the pluton. The mode of occurrence of gneissose lamination is basically in accordance with that of the contact zone, and varies with the latter. This pluton is mainly composed of light-grey, medium-grained gneissose two-mica granite which consists of plagioclase, K-feldspar, quartz, biotite and muscovite, as well as some accessory minerals, such as zircon, apatite, garnet, etc.

Listed in Table 2 and Fig. 2 are the Rb-Sr isotopic data for six whole rock samples from the pluton. Calculated by least-square method, the age value, t = 485 ± 6 m.y., $^{87}Sr/^{86}Sr$ initial ratio = 0.7186 ± 0.0018 and the correlation coeficient, r = 0.9987. The Kangmar pluton yielded a biotite K-Ar age of 31.8 m.y. and gave a zircon U-Pb age of 266 m.y.

The $^{87}Sr/^{86}Sr$ initial ratio indicates that the Kangmar pluton is the granitic rock of transitional type with extremely diverse source, which may be derived from the partial melting of ancient crust. However, no matter what source it may be, it can be said with certainty that gneissose two-mica granite must contain significant amounts of sialic material during its emplacement.

The age values calculated by three different methods are badly in disagreement, and thus further studies are needed for their geological implications.

It is known from the literature(4,5) that there are three groups of isotopic age data so far dated on Rb-Sr isochrons, which represent Precambrian (1,800, 1,200, 650 m.y.), lower Palaezoic (500 to 400 m.y.) and Cainozoic (Himalayan) (22,28 m.y.) rocks widespread in diferent parts of the Himalayan Region. Obviously, thermal events once took place extensively during lower Palaezoic-Caledonian time before the Indian plate began pushing northward. This is confirmed by the present Rb-Sr whole rock isochron age value of 484 m.y., as well as by $^{207}Pb/^{206}Pb$ ratio of zircon from the Kangmar pluton which gave an age of 451 m.y. Another thermal event was recorded during the late Hercynian (Permian) (Zircon $^{206}Pb/^{238}U$ ratio gave an age of 266 m.y.). During the Himalayan, tectonic compression resulted in gneissose lamination and led to recrystallization of mica minerals. As a result, we have obtained a younger biotite K-Ar age of 31.8 m.y. The objectivity of our knowledge still needs further confirmation.

## Gabug Pluton

The Gabug pluton is located to the north of Yadong county and extends from east to west, in fault contact with Ordovician limestone to the north and in intrusive contact with lower Palaeozoic metamorphic schists to the south. Petrographically, it is mainly composed of medium - and coarse-grained, tourmaline - muscovite leucogranites. Well-developed in the pluton are a number of aplitic and pegmatitic dykes, which are characterized by abundant tourmaline.

The Gabug pluton gave a muscovite K-Ar age of 14.7 m.y. The Rb-Sr whole rock ages dated on seven samples from the pluton are given in Table 3 and Fig. 3. The correlation coeficient, r = 0.9910, $^{87}Sr/^{86}Sr$ initial ratio = 0.7478 ± 0.00097, age value obtained by least-square method, t = 43 ± 3 m.y.

Its $^{87}Sr/^{86}Sr$ initial ratio shows that the Gabug pluton is the typical granite body of crustal origin, although its Rb-Sr whole rock isochron age is obviously higher than its K-Ar ages.

It is postulated from the available geological data that the collision of the Indian plate with the Eurasian plate might have begun at Eocene, and as a consequence, the Main Central Thrust Zone was set in activity at the same time, thus leading to the remelting of sialic material within the deep-seated shear zone, followed by the emplacement of widespread leucogranites with abundant tourmaline. Moreover, later tectonic movement seems to have certain influence on the pluton, too, reflected by its muscovite K-Ar ages.

## CONCLUSION

1. Rb-Sr whole rock isochron age values and $^{87}Sr/^{86}Sr$ initial ratios for some intermediate-acid rocks in southern Xizang are given as follows:

Granodiorite in the Quxu pluton of the North belt: t=17-18 m.y. ; a = 0.7036 - 0.7033

Gneissose two-mica granite in the Kangmar pluton of the Middle belt : t = 485 ± 6 m.y. ; a = 0.7186 ± 0.0018

Tourmaline-muscovite granite in the Gabug pluton of the South belt : t = 43 ± 3 m.y. ; a = 0.7478 ± 0.0010 .

2. The comparison of the above isotopic data with the K-Ar age data disclose great difference. The age data from the North belt are much lower than most of the K-Ar ages, but close to the zircon U-Pb ages; the age values for the Middle belt are greatly higher than the K-Ar ages and the zircon U-Pb ages; and the ages for the South belt are slightly higher as compared with the K-Ar ages.

3. The $^{87}Sr/^{86}Sr$ initial ratios show that the granodiorite in the North belt is most likely to be derived from the partial remelting of Tethyan oceanic crust, or the product of primary differentiation of magmas from the upper mantle ; the gneissose two-mica granite in the Middle belt may contain a significant amount of sialic material during its emplacement; and the tourmaline-muscovite granite in the South belt is typical of crustal origin.

## References

(1) Tu Guang-chi, et al., Characteristics and evolution of granitoids of Southern Xizang. In this Proceedings of Symposium on Qinghai-Xizang (Tibet) Plateau, Academia Sinica (1981).

(2) Chang Yu-quan et al., Isotopic geochronology of granitoid rocks in Southern Xizang Plateau. In this Proceedings of Symposium on Qinghai-Xizang (Tibet) Plateau, Academia Sinica (1981).

(3) U-Pb Dating Group, Laboratory of Isotopic Geology, Institute of Geochemistry, Academia Sinica, A discussion on zircon U-Pb dating of intermediate-acid rocks in Southern Xizang.In this Proceedings of Symposium on Qinghai-Xizang (Tibet) Plateau, Academia Sinica (1981).

(4) Bhanot, V. B. et al., Rb-Sr ages for granites and gneisses from Himalaya. Short Papers of Fourth International Conference, Geochronology, Cosmochronology, Isotope Geology, 32-33 (1978).

(5) Hamet, J., Claude-Jean Alligre. Rb-Sr systematic in granite from central Nepal (Manaslu), significance of the Oligocene, age and high $^{87}Sr/^{86}Sr$ ratio in Himalaya Orogeny. Geology, 4(8), 470-472 (1976).

Table 1 Rb and Sr isotopic data for porphyritic granodiorite from Quxu pluton in Xizang

| Sample No. | $^{87}Rb$ ($\mu m/g$) | $^{86}Sr$ ($\mu m/g$) | $^{87}Sr/^{86}Sr$ | $^{87}Rb/^{86}Sr$ |
|---|---|---|---|---|
| Q-74-181 | 16.1866 | 0.7121 | 0.7092 | 22.7308 |
| Q-74-338 | 0.2255 | 0.5541 | 0.7037 | 0.4070 |
| Q-74-340 | 0.3685 | 0.5699 | 0.7036 | 0.6466 |
| Q-74-349 | 0.2230 | 0.5287 | 0.7047 | 0.4218 |
| Q-74-376 | 0.1474 | 0.7387 | 0.7029 | 0.1995 |

Table 2 Rb and Sr isotopic data for gneissose two-mica granite from the Kangmar pluton in Xizang

| Sample No. | $^{87}Rb$ ($\mu m/g$) | $^{86}Sr$ ($\mu m/g$) | $^{87}Sr/^{86}Sr$ | $^{87}Rb/^{86}Sr$ |
|---|---|---|---|---|
| K-74-90 | 0.8669 | 0.08116 | 0.7839 | 10.6814 |
| K-74-91 | 0.8788 | 0.02078 | 1.0098 | 42.2907 |
| K-74-92 | 0.9064 | 0.09936 | 0.7804 | 9.1224 |
| K-74-94 | 0.8411 | 0.09490 | 0.7830 | 8.8630 |
| K-74-99 | 0.7191 | 0.09698 | 0.7725 | 7.4149 |
| K-74-100 | 0.8693 | 0.04592 | 0.8540 | 18.9307 |

Table 3 Rb and Sr isotopic data for tourmaline-muscovite granite from the Gabug pluton in Xizang

| Sample No. | $^{87}Rb$ ($\mu m/g$) | $^{86}Sr$ ($\mu m/g$) | $^{87}Sr/^{86}Sr$ | $^{87}Rb/^{86}Sr$ |
|---|---|---|---|---|
| I-74-22 | 1.08584 | 0.05401 | 0.7605 | 20.1044 |
| I-74-29 | 1.1706 | 0.08849 | 0.7567 | 13.2286 |
| I-74-30 | 2.2580 | 0.06721 | 0.7678 | 33.5962 |
| I-74-33 | 1.3262 | 0.05951 | 0.7603 | 22.2853 |
| I-74-36 | 1.3523 | 0.08017 | 0.7574 | 16.8079 |
| I-74-41 | 1.4416 | 0.04516 | 0.7673 | 31.9221 |
| I-74-42 | 1.1479 | 0.03199 | 0.7704 | 35.8831 |

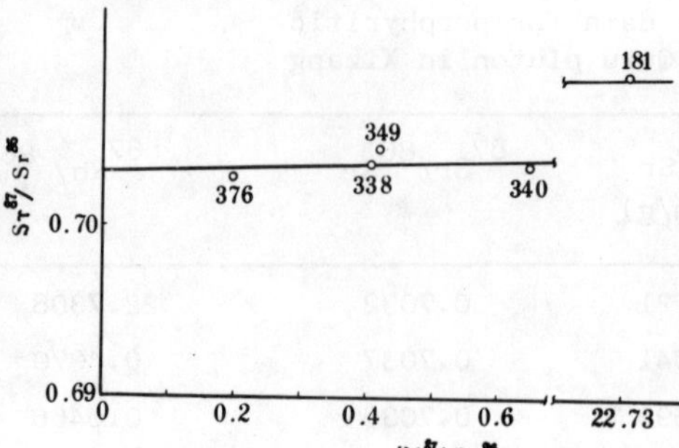


Fig. 1 Rb-Sr isochron for Quxu granodiorite of Xizang

t(4) = 18.3 m.y.
t(5) = 17.3 m.y.
a = 0.7033 a = 0.7036
r = 0.9940 r = 0.9687

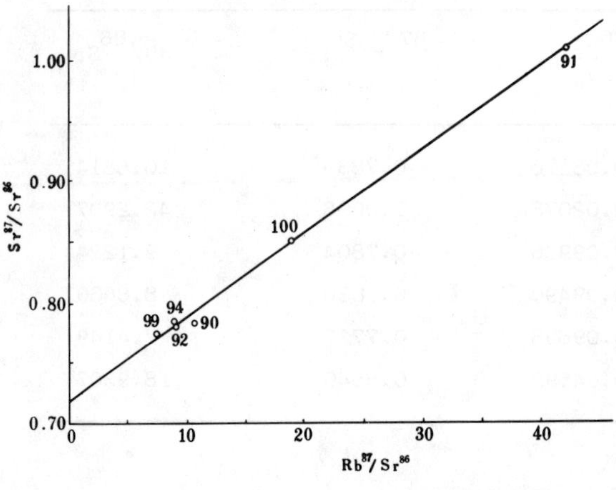


Fig. 2 Rb-Sr isochron for gneissose granite from the Kangmar pluton in Xizang

$t = 485 \pm 6$ m.y.
$a = 0.7168 \pm 0.0018$
$b = 0.006904 \pm 0.00010$

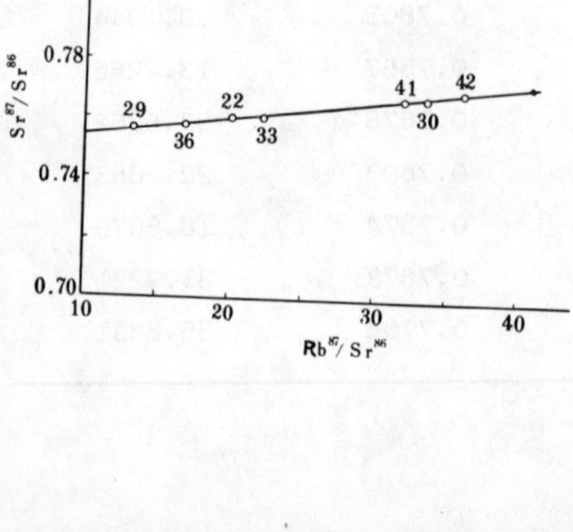


Fig. 3 Rb-Sr isochron for tourmaline-muscovite granite from the Gabug pluton in Xizang

$t = 43 \pm 3$ m.y.
$a = 0.7478 \pm 0.00097$
$b = 0.0006101 \pm 0.000369$

# RECRYSTALLIZED BIOTITES IN GRANITOID ROCKS IN SOUTHERN XIZANG AND THEIR BEARING ON PLATE COLLISION

Zhang Yu-quan, Xie Ying-wen, Xu Yong-hua
(Institute of Geochemistry, Academia Sinica)

Mica minerals are the important rock-forming minerals of granitoid rocks, and the main objects applicable to K-Ar isotopic age dating. Studies on mica minerals can offer much insight into the genesis and evolution of magmatic rocks and provide grounds for the establishment of intrusion ages of them and their geological history. Biotites to be discussed in the present paper are referred to those recrystallization under later thermodynamic effect, following petrogenetic crystallization in granitoid rocks, but not to those crystallization in sedimentary rocks during metamorphism. Based upon the chemical analyses and X-ray powder analyses, it is evident that these recrystallized mica are still biotites in nature.

## MODES OF OCCURRENCE OF RECRYSTALLIZED BIOTITES

Biotite recrystallization phenomena are commonly observed in some of the plutons, such as Quxu, Nangxian, Nyingchi, Azhag Kanglha, and Zayu pluton in the Gandise belt, and Da La, Kangmar and Mala Mt. pluton in the Lhagoi Kangri belt. Petrographically, biotite recrystallization can be detected in diorite, granodiorite, porphyritic biotite monzonitic granite, two-mica granite and two-mica monzonitic granite.

The above mentioned plutons all display gneissose structures, however, the degrees of gneissose structure development are different due to the different tectonic positions of the plutons. Nevertheless, it can be said with certainty, that gneissose structures can be recognized wherever recrystallized biotites are found. Additionally, the distorsion and fragmentation of rock-forming minerals and well-developed orientational microfractures can also be observed in the rocks where recrystallized biotites occur. At the same time, early crystallized biotites were subjected to distorsion, fragmentation and Fe extraction, together with the appearance of recrystallized biotites (see Photos 1-12).

## CHARACTERISTICS OF RECRYSTALLIZED BIOTITES

### General Features

Macroscopically, recrystallized biotites are usually in orientational arrangement, thus constituting gneissose structure, while microscopically, they are distributed along the orientational microfractures. Some recrystallized biotite aggregates diffused outwards along the surrounding fractures, frequently in association with irregular sphenes. Recrystallized biotites usually occur as laths with the ratio of length to width 3:1-9:1. A few inclusions are found in them. Their polychroism (Np- light yellow, Ng- light brown or brown), lower refractive indexes (Nm = 1.645 - 1.650 ), and distribution along the orientational fractures are indicative of their later recrystallization origin. This is in aggreement with the younger ages given by K-Ar method on recrystallized biotites.

### Characteristics of Crystal Axis ($b_o$)

The comparison of X-ray powder diffraction data between recrystallized biotites and those without recrystallization in various types of rock indicates remarkable difference in their $b_o$ (Table 1). The size of $b_o$ for biotites is related to the properties and chemical composition of original rocks. As shown in Table 1, $b_o$ for recrystallized biotites in various types of rock are all small, as compared with that of biotites without recrystallization, and similar to biotites crystallizing in metasedimentary rocks, indicating their crystallization under similar conditions.

### Characteristics of Chemical Composition

Fig. 1 shows the comparison of chemical compositions between recrystallized biotites and those without recrystallization in the same type of rock.

Recrystallized biotites show a tendency to decrease in their FeO, $Fe_2O_3$, $TiO_2$ and CaO, and increase in their $Al_2O_3$, MgO, $K_2O$ contents. This is in consistency with the fact that Fe leaching took place during biotite recrystallization, as indicated by thin-section observation.

### Characteristics of Trace Elements

Semi-quantitative spectrometric analyses of recrystallized biotite and those without recrystallization in various types or rock show that recrystallized biotites display a tendency to decrease of become depleted in Be, Co, Cr, Cu, Sc, Sn and Zr. This phenomenon is in aggreement with the fact that the outward migration of these elements occurred during the biotite recrystallization.

### Time of Biotite Recrystallization

On the basis of K-Ar age determinations, it can be seen that biotite recrystallization took place at 30 m.y. or more for the Lhagoi Kangri belt. However, geochronologically, the plutons include not only early Himalayan intrusives, but also pre-Himalayan intrusive rocks. In the middle Gandise belt, biotites from late Yenshanian intrusive rocks have a recrystallization age ranging from 30 to 20 m.y., with an zircon U-Pb age ranging from 100 to 70 m. y. It all goes to prove that the time of biotite recrystallization is much later than that of the solidification and crystallization of plutons.

## A POSSIBLE MECHANISM FOR THE RECRYSTALLIZATION OF BIOTITES

The plutons in which biotites recrystallized are mainly distributed in the areas along the Yarlung Zangbo faulting belt and the Lhagoi Kangri Range (Fig. 2). If the subduction and collision did take place in South Xizang, the former would be the suture line between the Indian Plate and the Eurasian Plate(1,2,3,4,5). The age of biotite recrystallization ( 35 to 20 m.y.), corresponds to the middle stage of the collision between the Indian Plate and the Eurasian Plate.

From the tempero-spatial distribution of recrystallized biotites, in conjunction with the EW and NW striking gneissose structures of the plutons in which biotites recrystallized, it may be considered that the recrystallization of biotites is attributed to the following hypothesis that at 30-20 m.y. the north-northeastward drift of the Indian subcontinental plate was counteracted by the Gandise belt formed during the early subduction stage, thus resulting in the rise of pressure and, consequently, temprerature at the suture line along the Yarlung Zangbo River which led to biotite recrystallization. Moreover, the Lhagoi Kangri region was under compression at that time (Fig. 3), thus giving rise to the recrystallization of biotites in the plutons produced during the early and pre-collision stages. With time passing by, the terrain under compression expanded southwardly, thus bringing about newly recrystallized biotite belt.

## SUMMARY

1. From the foregoing discussion on phenomenon of biotite recrystallization and recrystallized biotites themselves, it is indicated that in the southern Xizang Plateau there did take place continental plate inter-collision, and much light has been thrown to the time of dramatic elevation of the Xizang Plateau.

2. Studies on these recrystallized biotites are of great help to the investigation of the origin of granitoid rocks. It is usually considered that the granites with gneissose structure in particular are the products of syntectonism. Although it can not be said that there is no existence of such granites with gneissose structure elsewhere, in the granitoid rocks in the southern part of Xizang Plateau, gneissose structures were generated by biotite recrystallization during post- petrogenetic stage.

3. The recrystallization of biotites is of great significance in the mineralogical studies on biotite minerals themselves. Particularly, biotites are the main objects for isotopic geological dating by K-Ar method, and therefore, recrystallization of biotites is of some help to the explanation of biotite K-Ar ages and to the discussion on the conditions of Ar retention.

## References

(1) Fu Chengyi, Continental Drift, Sea-floor Spreading and Plate Tectonics. Science Press, 17 (1972) (in Chinese).

(2) Yin Tsan-hsun, Plate tectonics : A review. Scientia Geologica Sinica, (1), 60 (1973) (in Chinese).

(3) Chang Chengfa et al., Tectonic features of the Mount Jolmo Lungma

Region in Southern Tibet, China. Scientia Geologica Sinica, (1), 7(1973) (in Chinese).

(4) Li Chunyu, Tectonic evolutions of some mountain ranges in China, as tentatively interpreted on the concept of plate tectonics. Acta Geophysica Sinica, 18 (1), 66 (1975) (in Chinese).

(5) Huang Jiqing et al.,An outline of the tectonic characteristics of China. Acta Geologica Sinica, (2), 125 (1977) (in Chinese).

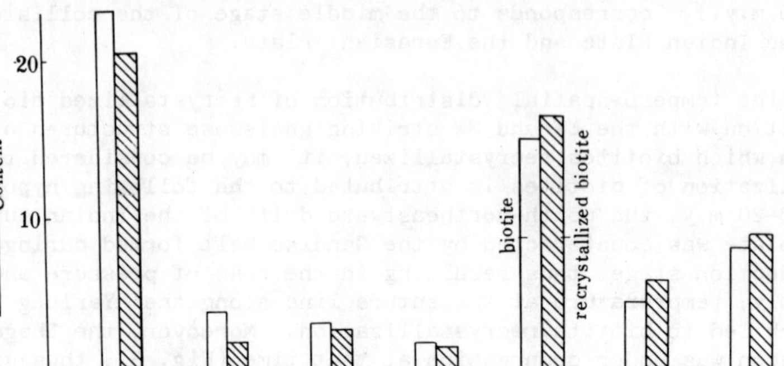


Fig. 1 The comparison of chemical compositions between recrystallized biotites and those without recyrstallization in the same type of rock

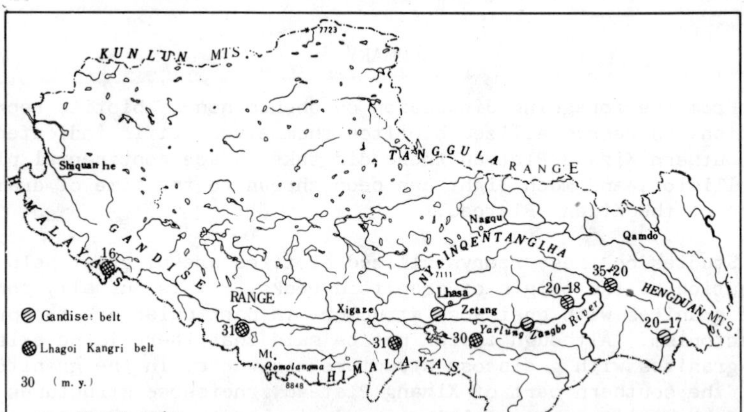


Fig. 2 The schematic map showing the distribution of recrystallized biotites in granitoid rocks in Southern Xizang

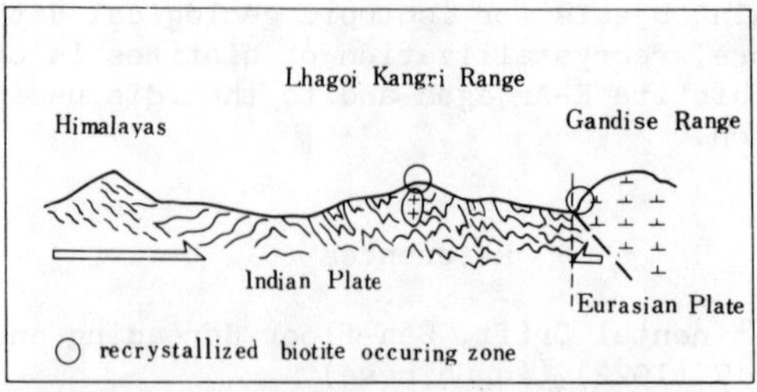


Fig. 3 Schematic representation of the relationship between plate collision and biotite recrystallization

Table 1 The comparison of $b_o$ for biotites in various types of rock

| Rock type | Sample No. | Biotite | $d_{060}$ | $b_o$ | Nm |
|---|---|---|---|---|---|
| Porphyritic biotite granite | 73-25 | Biotite | 1.547 | 9.282 | 1.650 |
| | 73-586 | " | 1.549 | 9.294 | 1.645 |
| Gneissose porphyritic biotite granite | 73-97 | Biotite* | 1.541 | 9.246 | 1.650 |
| | 73-97 | " | 1.544 | 9.264 | 1.645 |
| | 75-165 | " | 1.542 | 9.250 | 1.649 |
| Two-mica granite | 73-247 | Biotite | 1.544 | 9.264 | 1.649 |
| Gneissose two-mica granite | 74-256 | Biotite* | 1.540 | 9.240 | 1.645 |
| Gneiss | 73-758 | Biotite | 1.542 | 9.250 | |

* Recrystallized biotite

Table 3 Recrystallized biotite K-Ar ages

| Rock belt | Name of pluton | Sample No. | Rock type | Age (m.y.) |
|---|---|---|---|---|
| Gandise | Zayu | 73-97 | Gneissose porphyritic biotite granite | 19.6 ± 1.7 |
| | | 73-512 | " " | 17.4 ± 0.8 |
| | | 73-212 | Gneissose porphyritic two-mica granite | 20.2 ± 1.9 |
| | Azhag | 73-723 | Diorite | 38.6 ± 2.6 |
| | Kanglha | 73-724 | " | 19.8 ± 1.6 |
| | | 73-756 | Granodiorite | 35.2 ± 2.0 |
| | Nyingchi | 74-847 | Gneissose biotite granite | 21.0 ± 0.6 |
| | | 74-848 | Gneissose two-mica granite | 18.7 ± 1.3 |
| Lhagoi Kangri | Kangmar | 74-92 | Gneissose two-mica monzonitic granite | 31.8 |
| | Da La | 74-256 | " " | 30.2 |
| | Ma La Mt. | 75-48 | " " | 31.8 |

Table 2 The comparison of trace elements between normal biotites and recrystallized biotites in various types of rock

| Rock type | Sample No. | Mica | Ba | Be | Co | Cr | Ni | Cu | Ga | Li | Sc | Sn | V | Zn | Zr |
|---|---|---|---|---|---|---|---|---|---|---|---|---|---|---|---|
| Porphyritic biotite granite | 75-612 | Biotite | 300 | 2 | 30 | 500 | 30 | 20 | 50 | 1000 | 30 | 50 | 300 | 500 | 100 |
| | 73-653 | " | 300 | 2 | 10 | 30 | 20 | 20 | 30 | 200 | | 70 | 100 | 500 | 100 |
| Gneissose porphyritic biotite granite | 73-79 | Biotite* | 100 | – | 10 | 50 | 10 | 1 | 10 | 300 | – | – | 100 | – | 70 |
| | 73-97 | " | 200 | < 1 | 10 | 50 | 10 | 3 | 10 | 3000 | – | 30 | 100 | – | < 100 |
| Gneissose biotite granite | 74-843 | Biotite* | 2000 | | – | – | 30 | 5 | 10 | – | – | – | 200 | – | < 100 |
| Two-mica granite | 73-247 | Biotite | 200 | 3 | 10 | 70 | 20 | 3 | 20 | 1000 | 30 | 100 | 100 | – | 400 |
| | 73-242 | " | 200 | 1 | 10 | 70 | 30 | 5 | 30 | 1000 | 50 | 100 | 200 | – | 700 |
| Gneissose two-mica granite | 74-256 | Biotite* | 500 | – | – | 50 | 50 | 3 | 30 | 700 | – | – | 100 | – | 50 |
| | 74-92 | " | 200 | – | – | 50 | 50 | 3 | 30 | – | – | 100 | 100 | – | 50 |

* Recrystallized biotite

Photo 1 Distorted early biotite ($Bi_1$) with Fe exsolution
Porphyritic biotite monzonitic granite 73-97 x 21 (+)

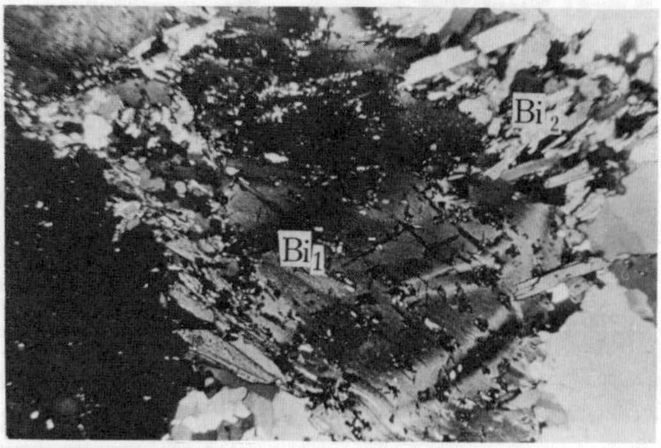


Photo 2 $Bi_1$ represents distorted early biotite showing undulatory extinction; $Bi_2$ indicates recrystallized biotite
Porphyritic biotite monzonitic granite 73-91 x 21 (+)

Photo 3 Recrystallized biotite ($Bi_2$) filling in plagioclase (Pl) microfractures gabbro-diorite 80-100 x 80 (+)

Photo 4 Recrystallized biotite ($Bi_2$) filling in microcline (Mi) fractures
Porphyritic biotite monzonitic granite 73-91 x 21 (+)

Photo 5 Recrystallized biotite ($Bi_2$) commonly associated with secondary sphene ($Sp_2$) and zircon (Zr)
Porphyritic biotite monzonitic granite 73-91 x 80 (+)

Photo 6 Recrystallized biotite aggregates tending to fill in the surrounding fractures (Q)
Porphyritic biotite monzonitic granite 73-82 x 33 (+)

Photo 7 Distorted plagioclase (Pl) due to compression, associated with recrystallized biotite ($Bi_2$)

Porphyritic biotite monzonitic granite 73-91 x 33 (+)

Photo 8 Sub-boundaries caused by compression in plagioclase (Pl) showing overlapping extinction

Tourmaline-muscovite granite 74-34x30 (+)

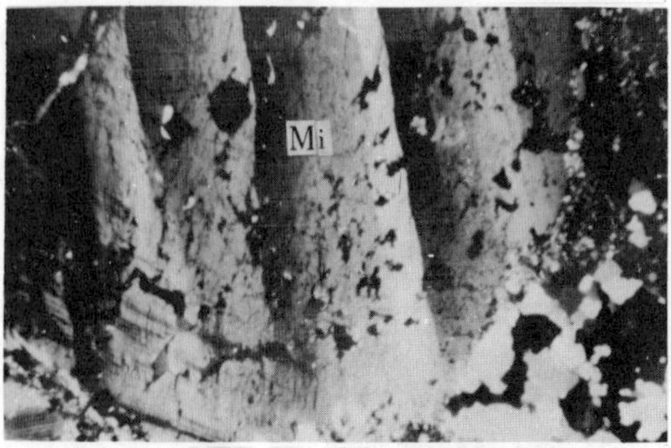


Photo 9 Microcline (Mi) with subboundaries, showing overlapping extinction

Porphyritic biotite monzonitic granite 73-79 x 12 (+)

Photo 10 Distorted plagioclase (Pl) with internal microfractures and marginal dissolution

Diorite 74-361 x 80(+)

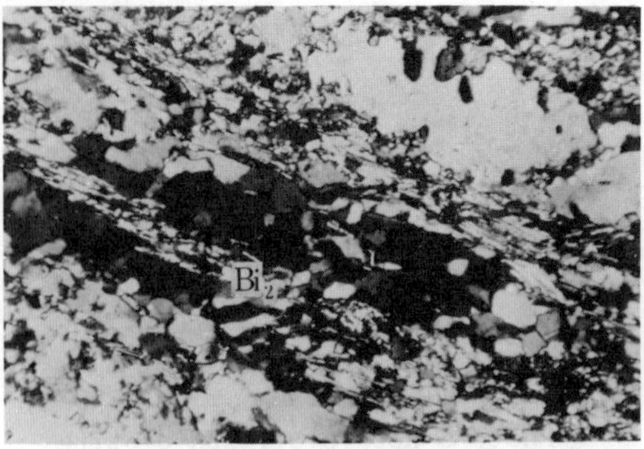


Photo 11 Recrystallized minerals distributed along the orientational fractures

Diorite 74-361 x 80 (+)

Photo 12 Orientationally arranged recrystallized biotites ($Bi_2$) filling in plagioclase (Pl) fractures

Two-mica monzonitic granite x 44 (+)

# A PRELIMINARY STUDY ON THE PETROLOGY AND GENESIS OF THE YARLUNG ZANGBO OPHIOLITE BELT

Deng Wan-ming
(Institute of Geology, Academia Sinica)

## INTRODUCTION

The Yarlung Zangbo ophiolite belt extends along southern Xizang in an E-W direction. It extends westward into Ladakh and connects with "Indus ophiolite belt" (Fig. 1). In the east it encompasses the great bend of the Yalung Zangbo River, and extends discontinuously along Naga Hills to the India-Burma boundary. The large Gandise and Ladakh granitoid belts occur on the northern side of the ophiolite. On the south it is separated from the Mesozoic low-grade metamorphics by faults. Its total length within China exceeds 2,000 km and with a width of several km to more than 10 km from north to south.

In the Xigaze-Lhaze district, on the north of the ophiolite belt there is the Xigaze group, i.e. a flysch zone of Late Cretaceous age. The ophiolite and flysch are in fault contact. Within the belt, the spatial relations of various ophiolite rocks from north to south are : mafic volcanics,ultramafic rocks and radiolarian rocks (Fig. 2). The members mentioned above are all separated from each other by thrusts.

## A POSSIBLE ORIGINAL SEQUENCE

Fig. 3 and Fig 4 are two sections in the Yarlung Zangbo ophiolite belt. The first section is situated 18 km south of Xigaze. Blocks of volcanics and rodingite in the ultramafic rock body are found. At the northern side of the body, gabbro and gabbro-diabase, mafic massive lava, and mafic pillow lava crop out in sequence. In the pillow lava, there are several massive sulfide bodies (Fig. 3).

The second section is situated south of Tangga in Lhaze area. In the section a large number of erupted sheeted volcanic flows occur as comformably intercalated beds in the siliceous shale. In addition, two zones of ophiolitic melange are found (Fig. 4).

In Bainang and other districts, it is found that the conglomerate overlies uncomformably upon the ultramafic body, in which a great deal of dike swarms of gabbro occur.

A possible original stratigraphical sequence of the Yarlung Zangbo ophiolite can be established as follows (from bottom up) : 1. basal ultramafic complexes ; 2. gabbros and gabbro-dibases ; 3. mafic volcanic rocks (Fig. 5).

This sequence agrees roughly with that of the typical sections over the world and is similar to those of recent oceanic crust. But the ultramafic and mafic cumulates and leucocratic rocks are dismembered or absent in most of the districts of the ophiolite belt. Only certain rock bodies situated in the east show magmatic differentiation fairly well and these rocks show cumulate structure. Under the volcanics in the sequence there are generally no diabasic dike swarms. These characteristics are different from other ophiolites.

## THE PETROLOGICAL AND PETROCHEMICAL FEATURES

The basal ultramafic rock bodies occur usually as lenses, generally exhibit extensive foliated and brecciated structures and contain tectonic blocks of siliceous rock, volcanics and limestones. The ultramafic rocks have undergone advanced serpetinization and as a result, the original texture has become obscure. A little fresh or unaltered ultramafics remain within some bodies. Harzburgite is the most important original rock type and consists of 80-85 % olivine ($F_o$ 90-95) and 6-8 % orthopyroxene ( En 90-93 ) with accessory magnetite and chromite. A limited amount of dunite and lherzolite is present in the harzburgite. Olivine grains invariably show kink banding, which is also present in the orthopyroxene. Petrochemically, the average MgO/(FeO) ratio for harzburgite ranges from 8 to 10.

The basal ultramafic rock of the Yarlung Zangbo ophiolite belt seems to belong to the metamorphic peridotite of Coleman (Coleman, 1977). They differ in occurrence, petrological characteristics and petrochemistry from the high-temperature peridotites. It is considered that the metamorphic peridotite is a partial melting residue formed within the upper mantle where its deformation was developed at subsolidus temperatures.

Within certain basal ultramafic rock bodies there occur gabbroic dike swarms, but they are not very well developed in the ophiolite. Sometimes they occur at the margin of an ultramafic body and have the typical association of clinopyroxene + calc plagioclase (An 55-65) and fine gabbroic structure. The petrochemical composition of the gabbros agrees with that of the tholeiite.

The volcanic rocks of the ophiolite occupy the upper part of the possible sequence and are subdivided into three kinds. From the lower to the upper part, they are mafic massive lava, mafic pillow lava and mafic sheeted lavas successively. The massive lavas are characterized by massive structure, bearing ophitic or trachybasaltic texture. Plagioclase and clinopyroxene are the predominant minerals of the massive lavas and glassy material is rare and very susceptible to alteration to chlorite.

The diameter of the individual pillows is about 1 m in the Congdu section of Xigaze. There are many dense amygdaloidal or vesicular structures on the surface of pillows. The total thickness of the pillow lava is about 1 km. The diameter of the pillows is only 20-30 cm in Bainang district and no amygdaloidal or vesicular structures are found there. The textures of pillows are rather variable. For petrography, the pillow lavas are mostly spilite, basalt, trachybasalt or diabase. Plagioclase and clinopyroxene are the most common minerals in pillow lava, and within the margin of pillows glassy material predominates. The pillow lavas have clearly undergone hydrothermal metamorphism.

The sheeted lavas occurring in siliceous rocks are developed very well in the Lhaze-Tangga section (Fig. 4). Within this section there are

15 beds of sheeted lava from north to south. Rock types of the sheets are mainly spilite and basalt. Basaltic glass in the erupted sheeted lava amounts to 40-50 % and the rock sometimes contains many titaniferous augite grains.

The average values of major elements of the various igneous rock from the Yarlung Zangbo ophiolite belt are presented in Table 1.

In Fig. 6, massive and pillow lava concentrate mainly in the area of the oceanic ridge TH, whereas sheeted lavas are in the area of the island-arc TH. For the massive and pillow lavas, the ratio of $Al_2O_3/TiO_2$ ranges from 10 to 20. $TiO_2$ content is low. The area distribution of the plotted points agrees well with that occupied by mid-ocean ridge basalt. However, the $Al_2O_3/TiO_2$ ratio of sheeted lava is always lower than 6 and the $TiO_2$ content is usually high (Fig. 7).

In the AMF figure (Fig. 8), the distribution plotted points of both massive and pillow lavas and sheeted lavas are distinctly different from each other. This suggests that volcanic activities occurred in different tectonic environments.

It appears more likely that both massive and pillow lavas are related to the mafic magma originating from the differentiation of pyrolite, and they erupted from a spreading ridge in the ocean. The sheeted lavas are considered to be genetically different from the underlying pillow and massive lavas and are thought to have formed at some later time in the island arc or marginal sea basins.

## FORMATION AND EMPLACEMENT OF THE OPHIOLITE

It is generally agreed that from Jurassic to Cretaceous Tethys was formed by sea floor spreading. The mafic magma formed by partial melting of the upper mantle directly overlay on the newly formed oceanic basement. A small amount of this magma perhaps intruded into ultramafic rock which existed previously within the upper mantle.

At the oceanic ridge, the massive and pillow lavas were formed by submarine extrusions. The width of the Tethys became smaller and smaller due to the continuous northward drifting of the Indian plate. Later sheeted lava was formed under the environment away from the spreading axis and overlapped on the massive and pillow lavas. During the Indian plate subduction beneath the Eurasian plate, the ophiolite as a part of fragments of the Tethyan oceanic crust was emplaced upward and in Eocene was squeezed into the upper crust by compression during the collision of the two continental crusts (Fig. 9).

As a direct result of the above mentioned events, the original section of the Tethyan oceanic crust was dismembered or a part of sequence was lost. Along the contact zones between adjacent units in the original section tectonic sliding occurred during the tectonic emplacement of the ophiolite into Gandise continental margin. Thus, tectonic relationships between the members of the ophiolite are exhibited. On the other hand, in the contact zone between the ophiolitic thrusting slabs and their surrounding rocks there usually are no high-temperature contact metamorphic rocks which are characteristic of the ultramafic magmatic intrusion. The basal ultramafic body may contain some of the tectonic blocks of overlying members, which do not show the features of melting.

## CONCLUSIONS

The Yarlung Zangbo ophiolite assemblages consist of polygenetic igneous rocks. The author suggests that the ophiolite was formed on the Tethyan oceanic floor in Mesozoic, and represents partial fragments of the Tethyan oceanic crust.

The ophiolite belt together with its accompanying melange and exotic blocks constitutes the Yarlung Zangbo tecto-lithofacies belt, which obviously represents the suture zone between the Indian and Eurasian plates.

## References


(1) Gansser, A. et al., Geological observation in Ladakh area (Himalayas). A preliminary report. SMPM 57, 89-113 (1977).

(2) Coleman, R. G., Ophiolite Ancient Oceanic Lithosphere ? Springer-Verlag Berlin Heidelberg. New York (1977).

(3) Cong Bo-lin et al., The study on the Cenozoic basalts in North China Fault Block. Acta Geologica Sinica, 53(2), 112-124 (1979) (in Chinese).

(4) Sun Shen-Su, Nesbitt, R. N., Geochemical regularity and genetic significance of ophiolitic basalts. Geology, 6, 689-693 (1978).

Table 1 Average percentages of major elements of the various igneous rocks from the Yarlung Zangbo ophiolite belt, Xizang

| Rocks | $SiO_2$ | $TiO_2$ | $Al_2O_3$ | $Fe_2O_3$ | FeO | MnO | MgO | CaO | $Na_2O$ | $K_2O$ | $Cr_2O_3$ | NiO | $H_2O^+$ | $H_2O^-$ | $CO_2$ | $P_2O_5$ | Total | Analytic Values |
|---|---|---|---|---|---|---|---|---|---|---|---|---|---|---|---|---|---|---|
| Sheeted lavas | 48.33 | 3.14 | 13.63 | 8.33 | 6.32 | 0.32 | 4.43 | 5.77 | 3.90 | 0.23 | | | 4.43 | 0.48 | 0.39 | 0.45 | 100.19 | 17 |
| Pillow lavas | 49.08 | 1.38 | 16.86 | 6.07 | 3.69 | 0.15 | 5.68 | 6.72 | 4.49 | 1.32 | | | 3.72 | 0.58 | 0.14 | 0.14 | 100.02 | 9 |
| Massive lavas | 51.36 | 1.04 | 14.98 | 4.98 | 3.94 | 0.77 | 6.43 | 6.88 | 4.44 | 0.26 | | | 4.20 | 0.57 | 0.13 | 0.16 | 100.14 | 11 |
| Gabbro-diabases | 48.36 | 1.07 | 15.18 | 4.35 | 5.32 | 0.20 | 7.74 | 9.54 | 3.33 | 0.25 | | | 3.53 | 0.51 | 0.50 | 0.19 | 100.17 | 8 |
| Harzburgites | 41.28 | 0.08 | 1.10 | 3.04 | 5.98 | 0.22 | 41.79 | 1.18 | 0.17 | 0.03 | 0.38 | 0.28 | 4.08 | 0.31 | 0.41 | | 100.33 | 24 |

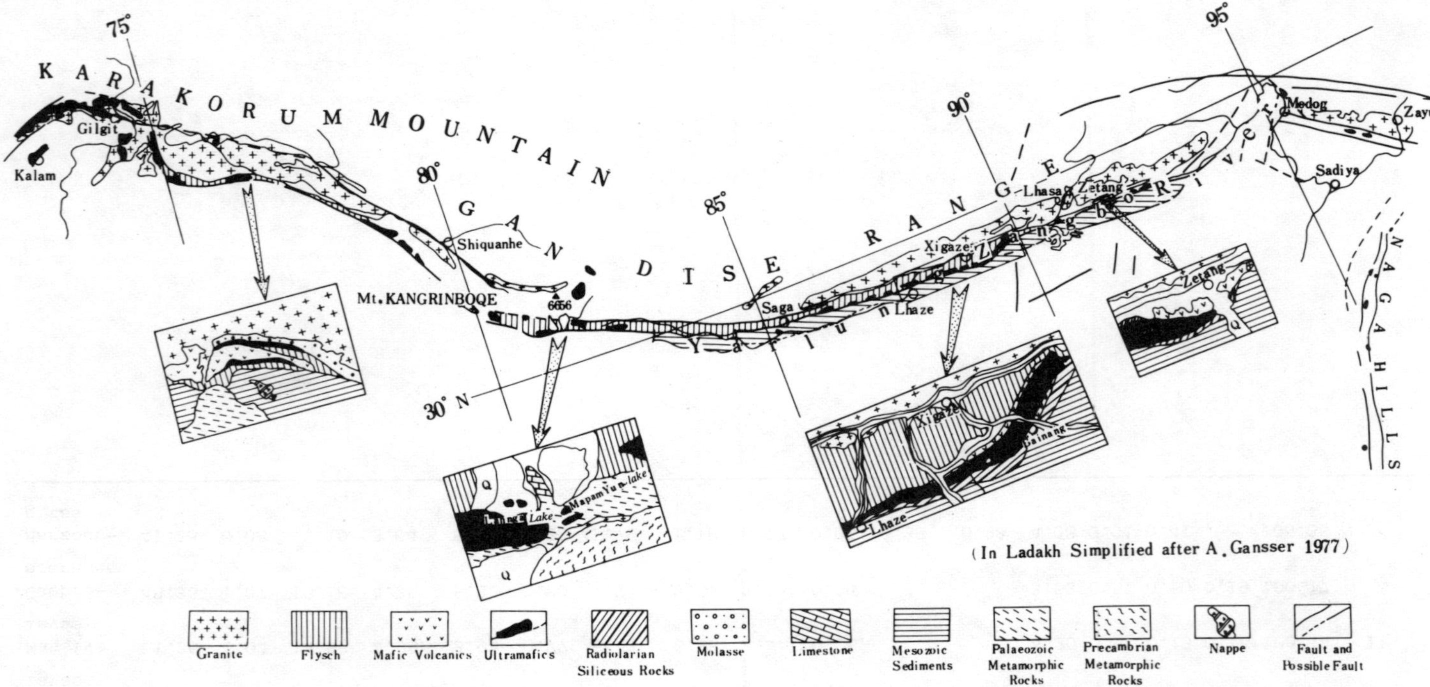


Fig. 1 Geological sketch map of the Indus— Yarlung Zangbo ophiolite belt

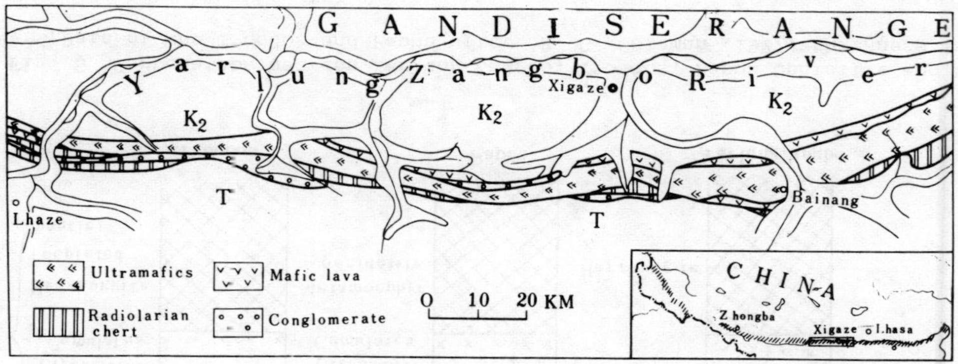


Fig. 2 The Yarlung Zangbo ophiolite belt in the Lhaze — Xigaze area

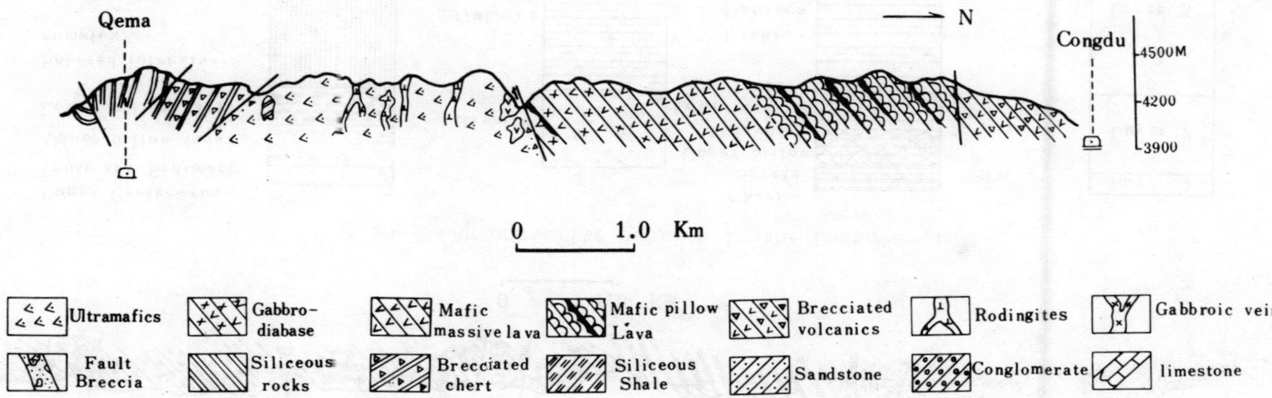


Fig. 3 An ophiolite section in the Xigaze area

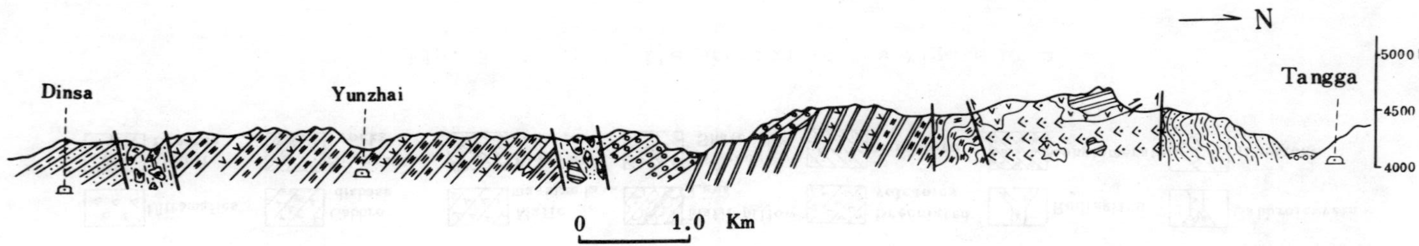


Fig. 4 An ophiolite section in the Lhaze area

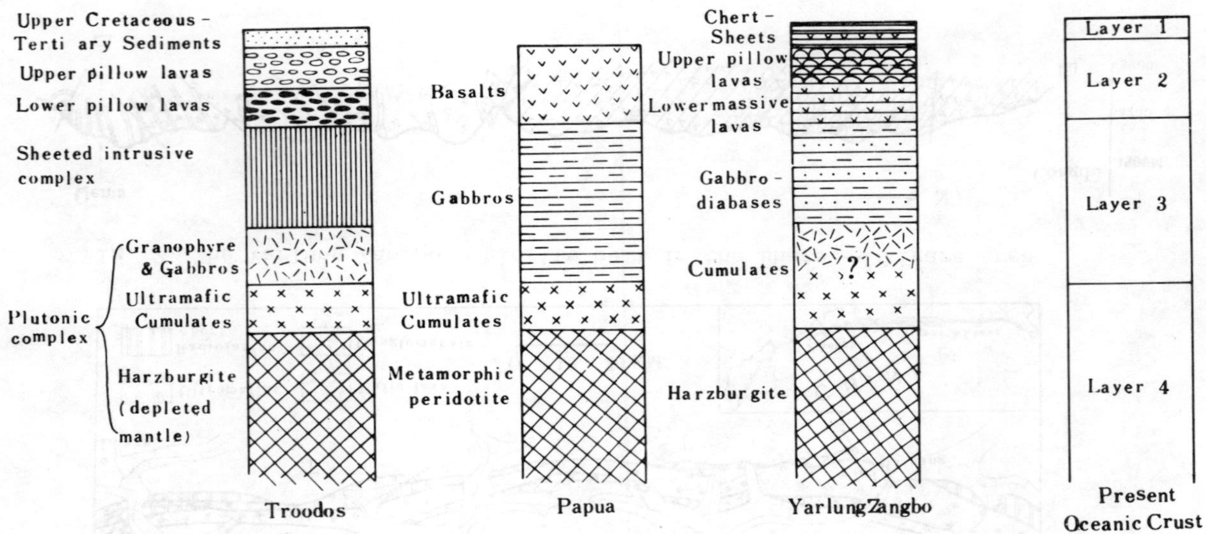


Fig. 5 Correlation between Yarlung Zangbo, Troodos, Papua Ophiolite and oceanic layers, Data of the Troodos and Papua after R. G. Coleman (1977) thickness is schematic

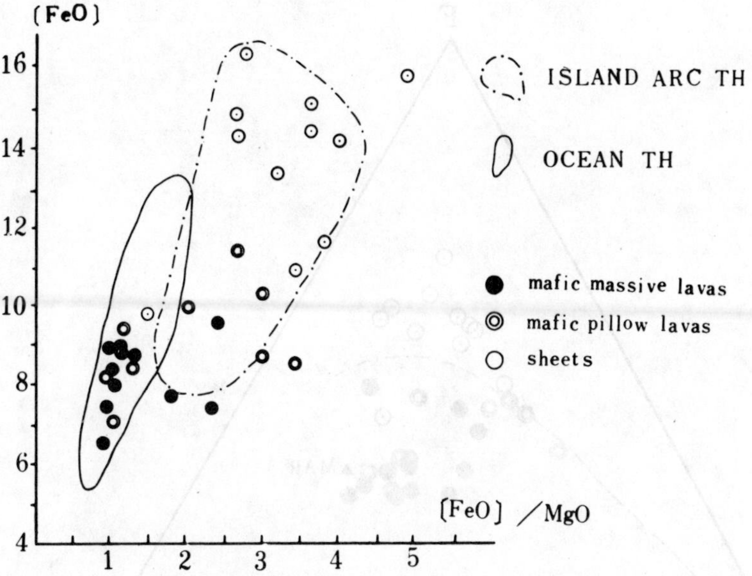


Fig. 6 (FeO) versus (FeO) / MgO diagram of volcanic rocks from the ophiolite (Modified and simplified after Cong Bolin,1979)

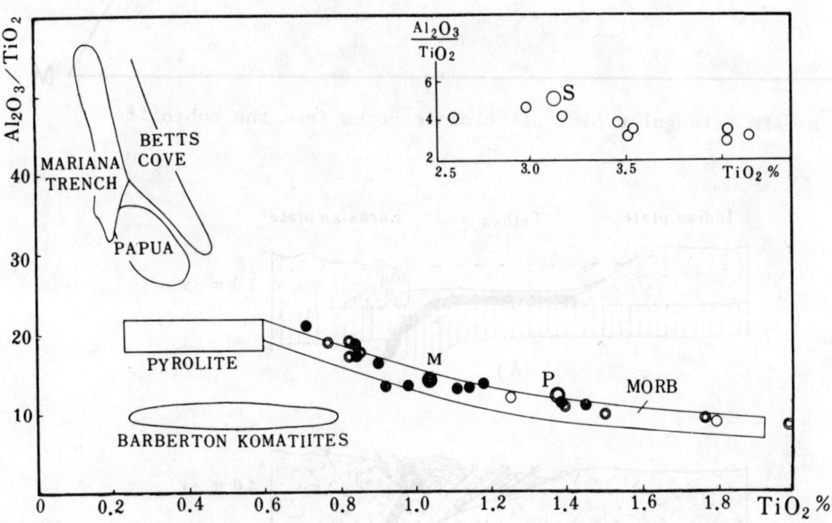


Fig. 7 A plot of $TiO_2$ abundance against $Al_2O_3$/ $TiO_2$ of volcanic rocks from the ophiolite (After Shen—su Sun and R. N. Nesbitt (1978)

MORB — MID-OCEAN RIDGE BASALT

M. P. S.— Average massive, pillow and sheeted lava respectively.

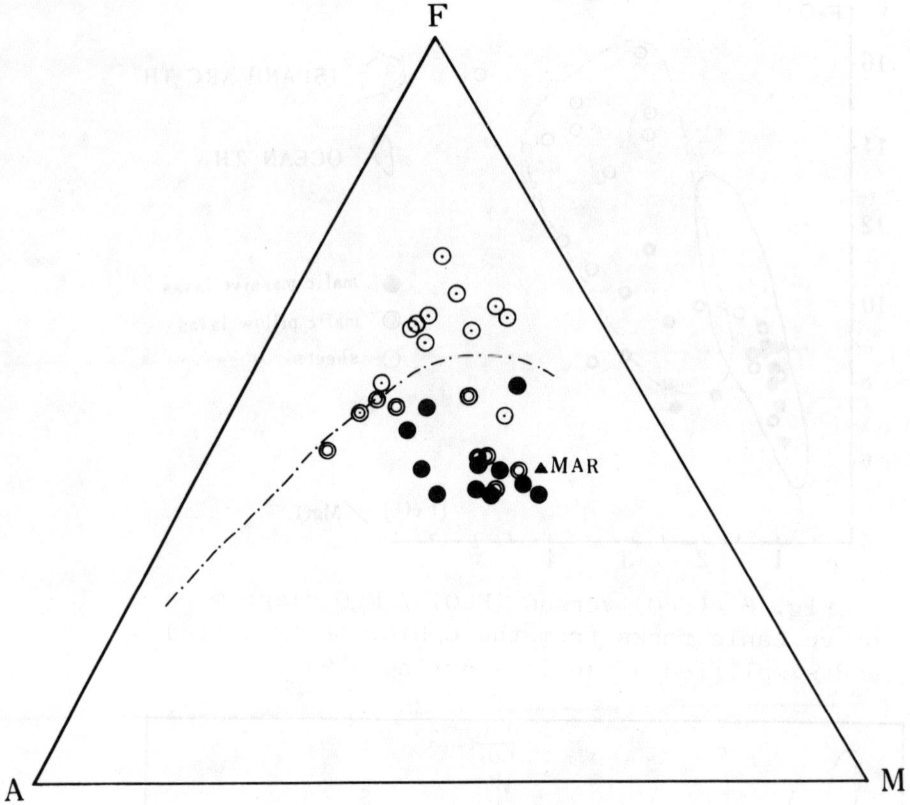


Fig. 8 AFM triangular plot of volcanic rocks from the ophiolite

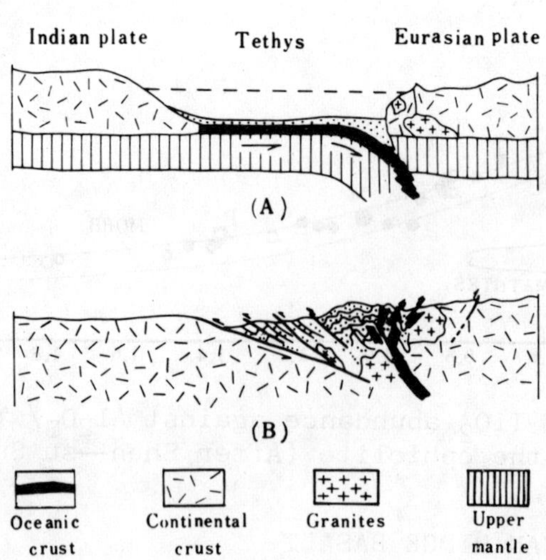


Fig. 9 Schematic sections of the formation and emplacement of the Yarlung Zangbo ophiolite

# ULTRABASIC ROCKS AND MANTLE EVOLUTION

Asish R. Basu
(Department of Geological Sciences, University of Rochester
Rochester, New York, U.S.A.)

The most direct source of information on the earth's upper mantle comes from peridotites believed to be derived from the upper mantle. The peridotite xenoliths found in alkali olivine basalts and in kimberlites are generally accepted to be broken fragments of the solid upper mantle included accidentally by the invading magma during its ascent from the mantle to the crust. A second group of ultrabasic rocks, known as the "Alpine-type", occurs along the central axis or "the suture" of folded mountain belts, and are believed to be mantle-exposures emplaced during plate-tectonic processes. The ultrabasic rocks associated with the ophiolites of the Indus-Tsangpo suture zone in Qinghai-Xizang are excellent examples of this latter group.

In this article, some of the pertinent data recently acquired on these ultrabasic, mantle-derived rocks are reviewed with emphasis of their implication on the nature and evolution of the earth's upper mantle. These data include Sm-Nd systematics, a recently developed tool in cosmochronology and the data on the isotopic composition and the concentration of the five noble gases, He, Ne, Ar, Kr and Xe measured on these samples. The word "ultrabasic" is used here to include rocks and minerals which are ultramafic such as peridotites, as well as rocks low in silica such as carbonatite, alnoite and kimberlite.

## KIMBERLITES, INCLUSIONS AND THEIR DEPTHS OF ORIGIN

Phase equilibria studies, conducted during the past twenty years or so, on basic and ultrabasic rock compositions, have greatly improved our understanding of the nature of the upper mantle and have established the fact that the upper mantle is predominantly of ultrabasic composition. The ultrabasic rock kimberlite is almost always found in stable continental regions and are reported from southern Africa, Siberia, North America, Brazil, India, Australia and in the eastern part of China [1-2]. They most commonly occur as vertical, narrow pipelike bodies or as thin dikes, but rarely as sills. This rock has an inequigranular, porphyritic texture; the porphyritic texture is due to relatively large crystals of olivine, enstatite, chrome-diopside, pyrope, ilmenite and phlogopite set in a fine-grained matrix consisting mostly of serpentine, carbonate, phlogopite, magnetite and perovskite. Many of the large crystals are derived from the disaggregation of the many inclusions that the kimberlites incorporate from the overlying mantle during their journey through the mantle. Chemically, the kimberlites are characterized by high Mg/Fe, K/Na ratios and low $SiO_2$ content. High carbon dioxide and water contents are considered to be major factors in controlling the extremely rapid ascent of kimberlite from depth.

Garnet lherzolites are the most common ultrabasic inclusions picked up by the intruding kimberlite. Such inclusions in southern African kimberlites are grouped into two types based on mineralogy, texture and bulk compositions (3-4). The first group, called granular type, is well established, on the basis of the application of experimental phase equilibria studies to the natural mineral assemblages of the lherzolites, to have originated at depths of 100-150 km. The second group, called the sheared-type, is believed to have originated at 150-200 km in depth. Figure 1, taken from Boyd (3), shows a plot of the estimated equilibration conditions for a suite of lherzolite inclusions in kimberlites from northern Lesotho. This figure demonstrates that since the kimberlites enclose the two groups of inclusions, they must be derived from at least as deep as 180 km, the site of the deepest sheared xenolith. Furthermore, the relations shown in figure 1 also indicate that the kimberlites have originated in the stability field of diamond - a fact substantiated by the occurrence of diamond in some of the kimberlites. It is also clear from figure 1 that kimberlites are produced at a depth in the mantle where no hydrous mineral, such as phlogopite, is stable. It is noteworthy that no other rock type is known to have originated at greater depths that are readily available for analysis.

## SM-ND SYSTEMATICS OF ULTRABASIC ROCKS

The decay of $^{147}$Sm to $^{143}$Nd with a half-life of 1.06 x $10^{11}$ years is not only used for dating meteorites and rocks, but the variation of the $^{143}$Nd/$^{144}$Nd ratio has also been used as a natural tracer and as a genetic indicator in terrestrial rocks of crust-mantle origin. The advantage of the Sm-Nd system is that both Sm and Nd are rare earth elements and their geochemical behavior is more predictable than other radioactive parent-daughter pairs, such as U-Pb and Rb-Sr.

High precision measurements of Sm and Nd contents of kimberlites from southern Africa, India, the U.S.A. and from the U.S.S.R. are shown in figure 2 along with the chondrite-normalized complete rare earth element (REE) abundance patterns in the Weselton and Premier kimberlite pipes in South Africa. The conspicuous feature of these REE patterns is the extreme enrichment of the light REE compared to the heavy REE resulting in a high La/Yb ratio. Also shown in figure 2 is the complete REE pattern of the basaltic achondrite, Juvinas which has the same Sm/Nd ratio and age as average chrondrites and may, therefore, be considered to represent the bulk chondritic earth.

Basu and Tatsumoto (5-6) have demonstrated that the initial $^{143}$Nd/$^{144}$Nd ratios of kimberlites are identical to those of Juvinas at the time of pipe emplacement during a period of 90 to 1300 million years before present (figure 3). Figure 3 shows the initial Nd-isotopic ratios in ultrabasic rocks with respect to the Juvinas growth curve in 4.56 billion years with the following data symbols: A to I are kimberlites, A = Bultfontein pipe, B = Benfontein carbonatitic sill in kimberlite, C = kimberlite pipe at Colorado-Wyoming border in the U.S.A., D and E = kimberlites from southern India, F, G = Premier and DeBeers pipe, respectively, from South Africa, H = Avon diatreme in SE Missouri, U.S.A., I = Yakutian pipe in the U.S.S.R.; J = Magnet Cove carbonatite in Arkansas, K = Alnoite from the Solomon Islands, and L = McClure Mountain carbonatite from Colorado. The most significant conclusion that emerges from figure 3 is that kimberlites are derived in space and time from a relatively undifferentiated, primitive, chondritic mantle reservoir.

Figure 4 shows that the ultrabasic inclusions in kimberlites of South Africa have a different Nd-isotopic signature than their host kimberlites [(6)]. This is shown by a parameter $^{\varepsilon}$Nd which is defined as:

$$\varepsilon_{Nd} = \frac{(^{143}Nd/^{144}Nd) \text{ initial in sample at time T}}{(^{143}Nd/^{144}Nd) \text{ in Juvinas at time T}} - 1 \quad X10^4.$$

Thus, all kimberlites have $^{\varepsilon}$Nd = near 0 and their inclusions show both positive and negative $^{\varepsilon}$Nd values. This data clearly indicate the existence of vertically heterogeneous, and geologically complex subcontinental mantle beneath southern Africa. It is also clear from this date that all inclusions, basic and ultrabasic, in kimberlites are unrelated to their host. These host kimberlites are derived from a deeper, relatively primitive chondritic reservoir.

The above Nd-isotopic data are shown schematically in figure 5, which is reproduced here from Basu and Tatsumoto [(6)]. The source of kimberlite has, at depths greater than or equal to 200 km, a chondritic Sm/Nd ratio. This part of the mantle must be unfractionated which is overlain by a less primitive mantle with a higher Sm/Nd ratio than chondrites. The sources for the mid ocean ridge basalt is the shallower zone in the mantle and is depleted in most large ion lithophile elements with an even higher Sm/Nd ratio than chondrites. The different trace elemental ratios in various levels of the mantle result by the extraction of variable amounts of melt in the past from an initially homogeneous, chondritic mantle reservoir same as the source of kimberlites. The above processes of melt extraction also cause the mantle to become heterogeneous both vertically and laterally with time. However, the most important feature of this diagram is the increasing degree of primitiveness of the mantle with depth.

## NOBLE GASES IN MANTLE-DERIVED SAMPLES

The elemental abundance pattern and isotopic composition of the noble gases He, Ne, Ar, Kr and Xe in the earth's atmosphere are unlike those of noble gases trapped in chondritic meteorites. It has been recognized for some time that the solar system has two components of the noble gases: The "solar" component and the "planetary" component. The solar wind is the profuse source of the "solar type" noble gases, whereas the "planetary type" noble gas is characteristic of the undifferentiated chondrites. The "planetary" noble gas pattern differs from the "solar" pattern by being enormously depleted in He and Ne abundances. The abundance pattern in the earth's atmosphere is very similar to the "planetary" pattern except for Xe, which seems to be preferentially adsorped in shale and shows a depletion by a factor of 10 in the atmosphere. The recent discovery [(7)] of primordial He in glassy rims of oceanic tholeiitic basalts indicates that primordial He is still being degassed from the earth's mantle. It is increasingly becoming clear that primordial components of possibly all the noble gases are currently being degassed.

The noble gas data on mantle-derived ultrabasic xenoliths were reviewed in 1977 by Basu [(8)] and subsequent data collected mostly by the Berkeley group were reviewed by Kyser [(9)]. Some of the pertinent data are mentioned here. The $^3He/^4He$ ratio of $10^{-5}$ in the mantle-derived samples is well established now. As mentioned earlier, the only viable explanation for a high $^3He/^4He$

ratio is primordial He with a high $^3He/^4He$ ratio. Some glassy rims of submarine basalts have excesses of $^{21}Ne$ and $^{20}Ne$ and high He/Ne and $^{40}Ar/^{36}Ar$ ratios. However, ultrabasic xenoliths in alkalic basalts and in kimberlites have much lower $^{40}Ar/^{36}Ar$ ratios than the same ratio in glassy rims of submarine basalts. Sometimes this ratio is found to be as low as the atmospheric value of 296. Such a low value of $^{40}Ar/^{36}Ar$ in the mantle is, however, difficult to reconcile. Because the atmosphere has no potassium and $^{40}Ar$ is produced only in the solid earth with the $^{36}Ar$ being non-radiogenic, the $^{40}Ar/^{36}Ar$ ratio in the atmosphere can only increase with time by the degassing of the earth and that degassing removes both $^{40}Ar$ and $^{36}Ar$ from the earth. While $^{40}Ar$ continues to be produced from the decay of $^{40}$,, there is no mechanism by which $^{36}Ar$ is produced in the earth. Therefore, $^{40}Ar/^{36}Ar$ ratio in the mantle increases faster after each degassing episode. The low $^{40}Ar/^{36}Ar$ ratio in the mantle, then, indicates the existence of primordial $^{36}Ar$ which has not yet degassed. This conclusion is in concert with the inference from the He isotope data from the same set of samples. The Kr isotopic ratios in the mantle are very similar to those of the atmosphere. Excess $^{129}Xe$ is present in the mantle due to the presence of now extinct $^{129}I$ early in the history of the solar system. However, all mantle samples do not show this excess $^{129}Xe$, which is a decay product of $^{129}I$ with a half life of 17 million years. The presence of excess $^{129}Xe$ in some samples possibly indicates a heterogeneous mantle in terms of its noble gas contents and their isotopic ratios.

In conclusion, the two sources of data on Nd isotopes and the noble gases reviewed above indicate that the earth has retained within the mantle some primitive material from which it formed 4.56 billion years before present.

## References


(1) Dawson, J. B., Advances in Kimberlite Geology, Earth Sci. Rev., 7, 187-214 (1971).

(2) Zhang Pei-yuan, Ji Yu-xing and Hu Hong-qian, Geological Conditions of the Formation of the Diamond-bearing Kimberlites in Southern Liaoning Province, Geological Review (Beijing, China), vol. 26, No. 1, 29-34 (1980).

(3) Boyd, F. R., A Pyroxene Geotherm, Geochim. Cosmochim. Acta 37, 2533-2546 (1973).

(4) MacGregor, I. D., and Basu, A. R., Thermal Structure of the Lithosphere - A Petrologic Model, Science, 185, 1007-1011 (1974).

(5) Basu, A. R. and Tatsumoto, M., Nd-isotopes in Selected Mantle-derived Rocks and Minerals and their Implications for Mantle Evolution, Contributions to Mineralogy and Petrology, in Press (1980).

(6) Basu, A. R. and Tatsumoto, M., Sm-Nd Systematics in Kimberlites and in the Minerals of Garnet Lherzolite Inclusions, Science 205, 388-401 (1979).

(7) Lupton, J. E. and Craig, H., Excess $^3He$ in oceanic basalts: Evidence for Terrestrial Primordial Helium, Earth Planet. Sci. 26, 133-139 (1975).

(8) Basu, A. R., Noble Gases in the Upper Mantle, Ind. Jour. Earth Sci. S. Ray Volume, 157-169 (1977).

(9) Kyser, T. K., Stable and Rare Gas Isotopes and the Genesis of Basic Lavas and Mantle Xenoliths, Ph. D. Thesis, Univ. Calif., Berkeley, 188 p. (1980).

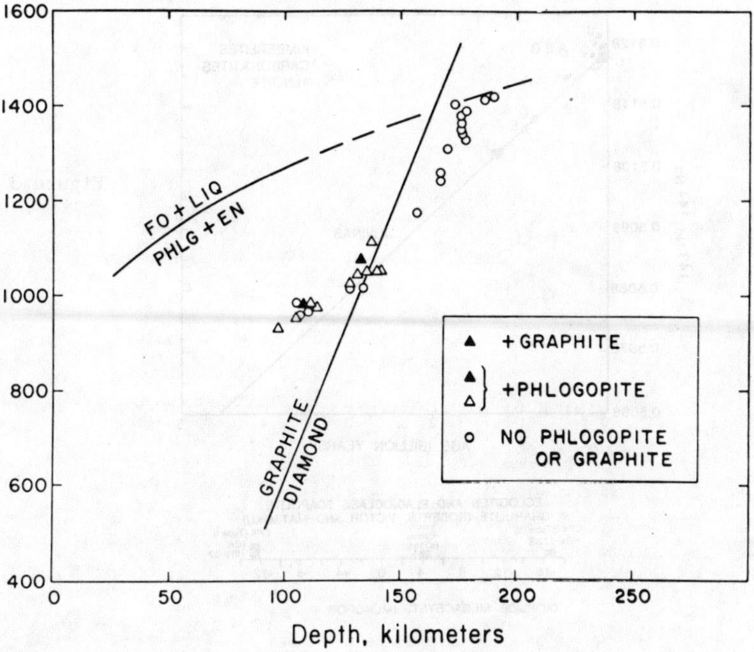


Figure 1 (after Boyd, 1973)

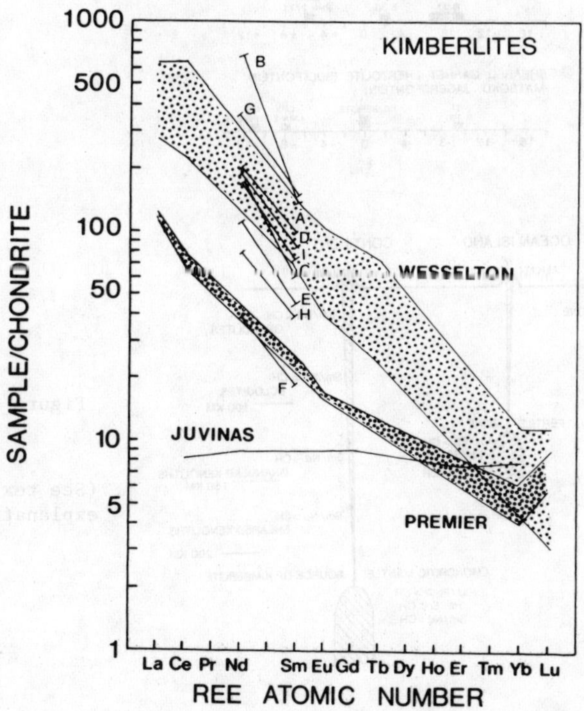


Figure 2

( See Text for explanation )

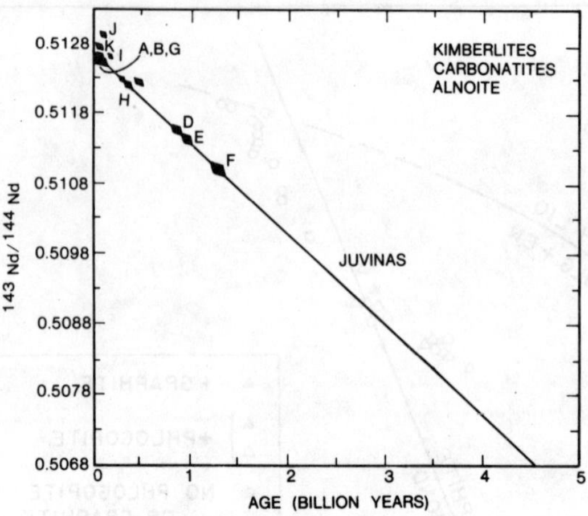


Figure 3

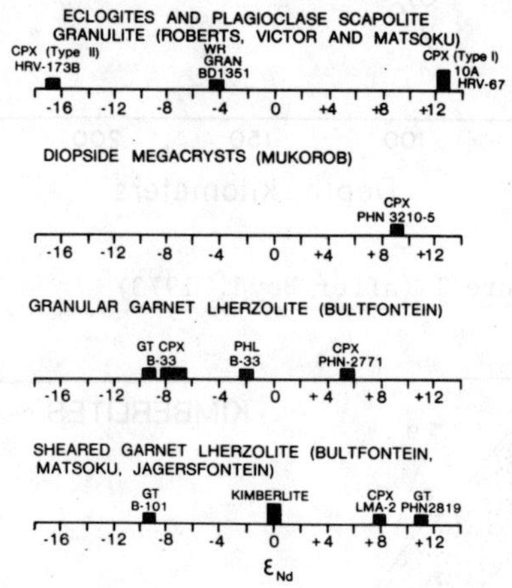


Figure 4

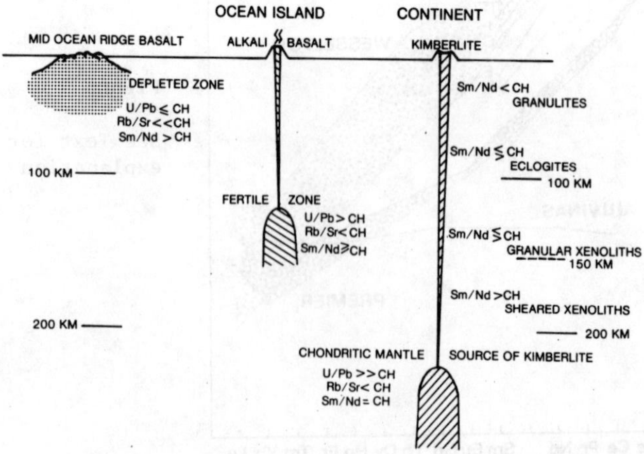


Figure 5

(See text for explanation )

# ON OPHIOLITE SYSTEM ON QINGHAI-XIZANG PLATEAU WITH PARTICULAR REFERENCE TO ITS GENESIS IN WEST XIZANG

Mei Hou-jun, Lin Xue-nong,
Chi Jia-xiang, Zhang Guang-yu
(Institute of Geochemistry, Academia Sinica)

Wu Ming-tang
(Geological Bureau, Xizang Autonomous Region)

Several ophiolite belts in Ngari, Xizang, ophiolites exposed along the upper reaches of the Yarlung Zangbo River, and a few small serpentinite bodies occurring along the Nujiang River were recognized during our investigation tour to Xizang from 1973-1976. A review of the geological literature and the unpublished reports as well as the arguments from our field studies, has shed much light to the general distribution patterns of ophiolite belts in Xizang (Fig. 1).

The Yarlung Zangbo River-Xiangquan River ophiolite belt stretches about 1,400 km long, starting from Gyemdong of Nangxian County in the east and ending near the Manam Mountain Pass at Zanda County. In the belt there occur some ultramafic massifs on the ridges with an altitude of 5,000 m above sea level, some of which extend more than ten kilometers, and the Quewuzhang ultramafic dyke in Saga County is as high as about 6,000 m (Fig. 2). The NW striking Indus River-Gar River ophiolite belt comprises several ultramafic massifs set in echelon pattern. As viewed from the fact that the volcanic rock sequences in the two belts and in the Ngang-lung Kangri Range and Bangong Mt. ophiolote belts in Ngari are Cretaceous in age, it can be considered that the four ophiolite belts were orogenetically formed at the end of Mesozoic era, whereas the two E-W striking ophiolite belts in the so-called Lake Area of northern Xizang were probably formed at early Mesozoic.

A well-developed geological section of ophiolite suite lies along the northern mountains near the old city of Saga County (Fig. 3). The outcrops are seen from north to south : 1) sandstones interbedded with shales, of the Xigaze group ; 2) the Yang La ultramafic massif ; 3) reddish cherts intercalated with spilites ; 4) fluidal gabbros and spilitic flows intercalated with concordant diabase dykes; 5) reddish cherts with compact spilitic lava flows ; 6) laminated and compact spilitic lave flows with limestone lenses ; 7) shales and phyllites with sandstones. At the south mountains near Ngamring the ultramafic massif is associated with a wider fluidal gabbro intrusive ;between them occurs a lense of ultramafic cumulate about 200 m wide and several hundred metres long. Near the lense there is an olivine meta-gabbro body about ten metres long of one metre wide within the fluidal gabbro intrusive. The geological section of ophiolite at Ngari is simpler than that at Saga County, e.g. the sections near Rutog and near Jaggang at Rutog County are mainly composed of ultramafic rocks, intermediate and basic volcanic rocks and pale cherts intercalated with limestones.

Most of the ultramafic massifs in the western half of Xizang are

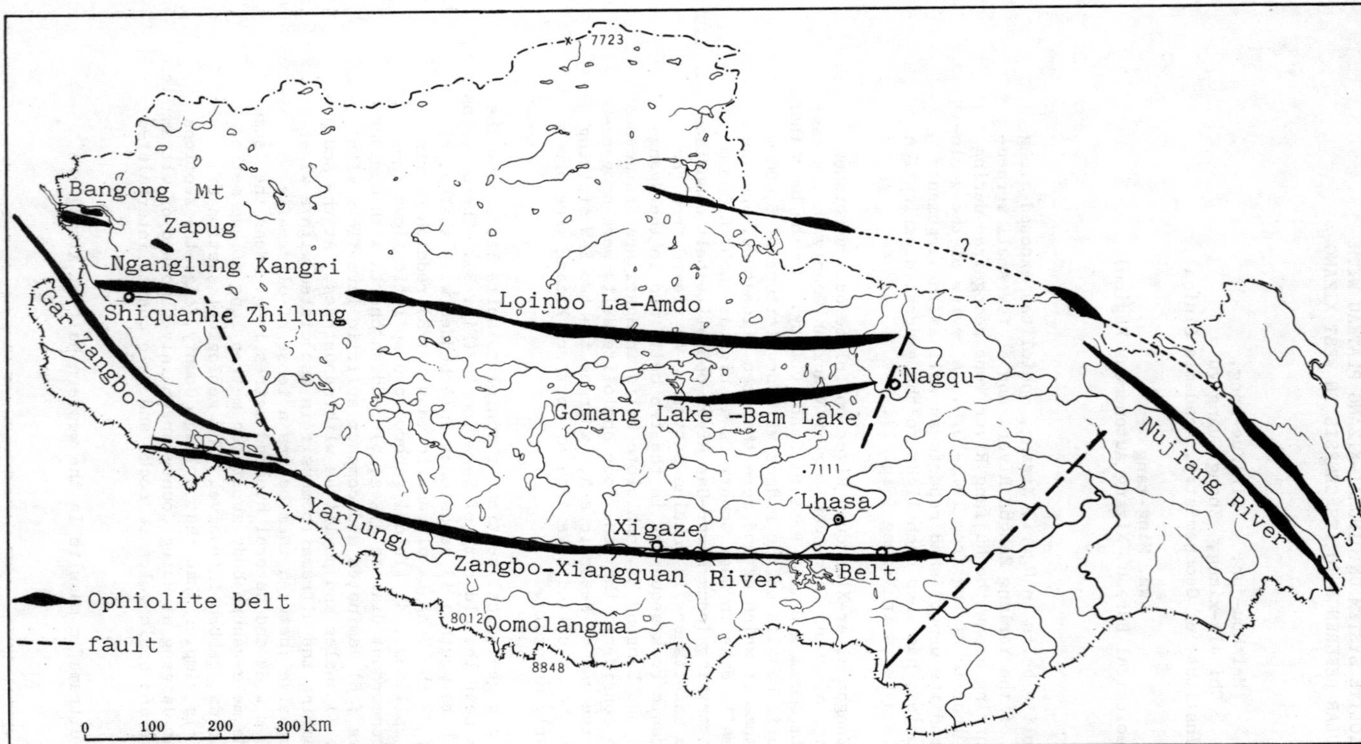


Fig. 1 Map showing the distribution of ophiolite belts in Xizang

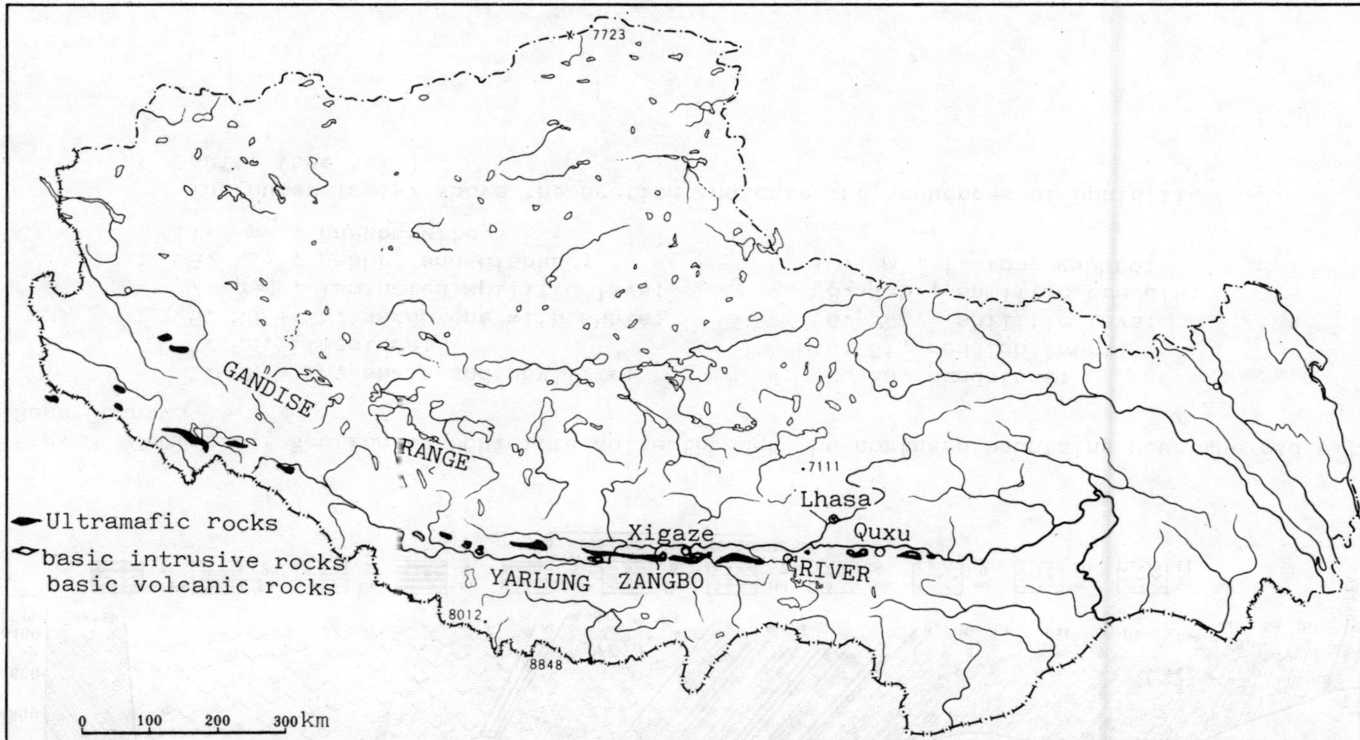


Fig. 2 Schematic map of Yarlung Zangbo River - Xiangquan River ophiolite belt, Xizang

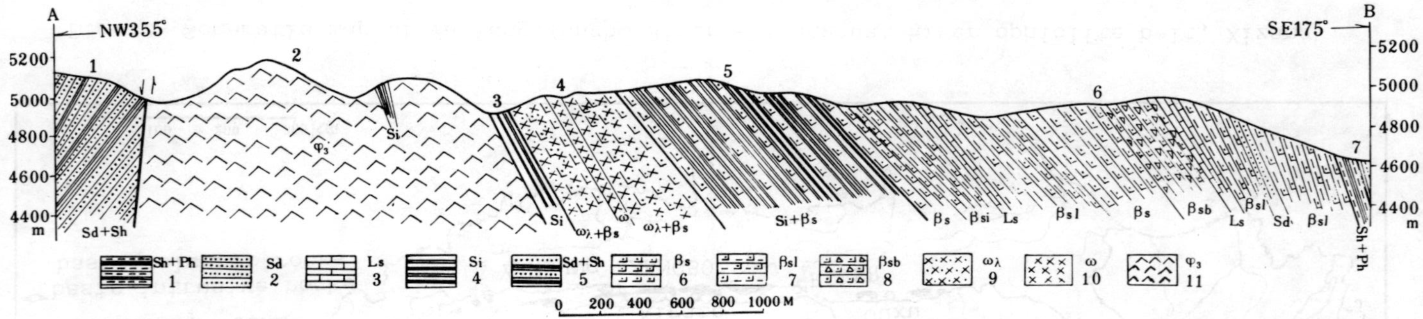


Fig. 3 Geological section of ophiolite suite through the northern mountains near the old city of Saga County

1. Sh + Ph, shale and phyllite;
2. Sd, sandstone;
3. Ls, limestone;
4. Si, reddish chert;
5. Sd + Sh, sandstone with shale;
6. $\beta s$, spilitic lava;
7. $\beta sl$, laminated spilitic lava;
8. $\beta sb$, spilitic breccia;
9. $\omega\lambda$, gabbro and diabase;
10. $\omega$, fluidal gabbro;
11. $\varphi_3$, harzburgite.

The numerals 1–7 above the section indicate the sequences of ophiolite suite (see text).

composed predominantly of harzburgites, and the rest show higher differentiation, as exemplified by the occurrence of dunite, harzburgite, lherzolite and pyroxenite. Highly differentiated massifs always contain more podform chromite ore-bodies. The major rockforming minerals in the ultramafic rocks are magnesian olivine ( $F_{o\ 88-94}$), magnesian orthopyroxene ($En_{88-92}$) and chrome diopside ($En_{48.4}$ $Wo_{46.2}$ $Fs_{5.4}$) as well as a minor amount of chrome spinel grains. The ultramafic rocks occurring in the terrain from Ngamring westward all show hypidiomorphic granular textures, and are characterized by most of orthopyroxene crystals containing a great deal of clinopyroxene lamellae which run parallel to the planes of optic axes of host crystals probably caused by exsolution. Fine lamellae run through their host crystals, while wider ones run disjointedly. The principal refractive indexes determined on a wide lamella are similar to that of chrome diopside grains. Sometimes it can be seen that olivine grains are enclosed by orthopyroxene crystals with clinopyroxene lamellae (Fig.4). Olivine crystals in the ultramafic massifs occurring in the terrain east of Dagzhuka exhibit more kink bands. In terms of the two pyroxene geothermometers from Wood and Banno(1973)(1), it can be estimated from chemical analyses of a pair of pyroxene that the ultramafic massif near the Lannga Lake at Burang County perhaps has a formation temperature of 1050°C.

The ultramafic cumulate near Ngamring corresponds to bronzite-bearing wehrlite or lherzolite, in which the contents of both pyroxenes are erratic here and there. It shows orthocumulus texture. Cumulus cyrstals are chiefly chrysolites ($Fo_{81}$), with a minor amount of chromite grains, whereas intercumulus minerals include bronzite ($En_{86}$), diopside ($En_{42.5}$ $Wo_{47.0}$ $Fs_{10.5}$) and a few hornblendes replacing pyroxenes. Many pyroxenite dykes, which comprise predominantly bronzite ($En_{86}$) and pegmatoid norite dykes, occur within the ultramafic massifs. The latter consist of bronzite ($En_{82-91}$), anorthite ($An_{97}$) and tremolite. The decrepitation temperature was determined on liquid microinclusions in anorthite to be 750 ± 5° C (after Lu Huan-zhang).

There are two or three later stage-formed serpentinized ultrabasic dykes intervening into the gabbro intrusive at the south mountains near Ngamring. The ultramafic massifs in the terrain from Ngamring to Bainang contain a great number of greyish-white fragments of rodingite. Besides, there are some ultramafic massifs in the Lake Area of northern Xizang with skarn iron-ore occurrences in their aureoles(2). The facts stated above, particularly the textural features of the ultrabasic rocks occurring in the terrain from Ngamring westwards, lead us to postulate that the Alpine ultramafic massifs might be derived from a crystal mush chemically similar to magnesian peridotite, which had first intruded from the mantle into the lower crust and formed chromite deposits during differentiation(3), and then, could be pushed into the higher level of the crust during serpentinization by the compressing force between two plates.

Cretaceous basic volcanic rocks of ophiolite belts in western Xizang are petrologically variable from one place to another, with olivine basalt, basalt with clinopyroxene phenocrysts, basalt with labrador feldspar phenocrysts, altered basalt with ophitic texture, variolite with feldspar microlites, plagioclase-rich spilite, and spilitic lava comprising microlites of epidote, albite, actinolite and chlorite. The plot of alkali oxides against silica of the volcanic rocks shows that some of them can be assigned to tholeiite suite, and the others, to alkali basalt suite (Fig. 5). According to H. Kuno's diagram (1960), in the western segment of the Gandise Range occurs high-alumina basalt. On FAM-diagram the scattering

Notes : 1. Za -94, from an ultramafic massif at Lannga Lake, Burang ; 2. Za - 188, from a dyke in an ultramafic massif near Rutog ; 3. Za - 76, from a dyke in an ultramafic massif, Lannga Lake, Burang ; 4. ZR - 14, from a dyke in an ultramafic massif, Qusum ; 5. Zc - 05, from a dyke, at the south mountains near Ngamring ; 6. Za - 95, from a dyke near Lannga Lake, Burang ; 7. Za - 313, from Cretaceous volcanic rocks, Butog ; 8. Za-47, from Cretaceous volcanic rocks, Moincer, Gar ; 9. Zc - 129, from Cretaceous, Saga ; 10. 74 - 596, from Cretaceous volcanic rocks, Zetang ; 11. ZR - 10, from Cretaceous volcanic rocks, Qusum ; 12. Za - 22, from Cretaceous volcanic rocks, in the basalt-andesite-rhyolite series, Large Wild Donkey Valley, Gar ; 13. Za - 94, from harzburgite, Lannga Lake, Burang ; 14. Za - 94, from harzburgite, Lannga Lake, Burang.

Fig. 4 Orthopyroxene crystals with clinopyroxene lamellae enclosing olivine grains, cross polarized, x 44

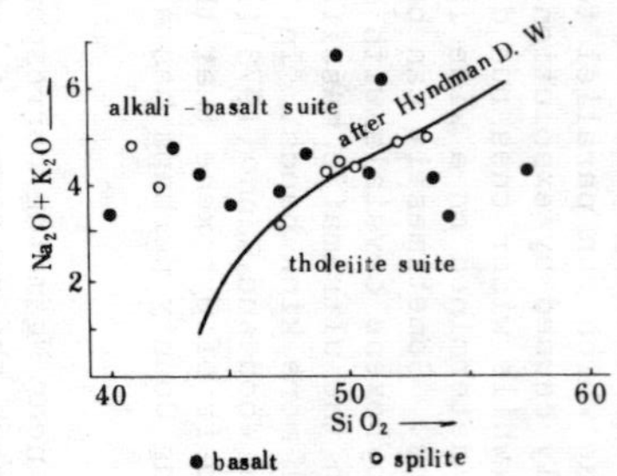


Fig. 5 The diagram of $(Na_2O + K_2O)$ vs. $SiO_2$, for the basic volcanic rocks in ophiolite belts in Western Xizang

pattern of Cretaceous volcanic rocks in western Xizang is similar to that of the Melanesian island arc(4) and that of an island arc in the Mediterranean Sea(5). All these show clearly that Cretaceous volcanic rocks in western Xizang can not be simply referred to low potasium tholeiites constituting the oceanic floor, instead, there are many types of basalt in western Xizang. Basic volcanic rocks in some segments of ophiolite belts in western Xizang should be assigned to alkali basalt suite. Basing on the study of rare elements, J. Loeschke considered Paleozoic spilite in Eastern Alps as altered alkali olivine basalt(6). In the western half of Xizang, Cretaceous volcanic rocks are characteristic of an island arc terrain.

The Ti-Zr diagram after J. A. Pearce(7) shows that most of the basic volcanic rocks in western Xizang are not probably derived from ocean-floor basalts (Fig.6). Also, the Ti-Zr-Sr diagram illustrates the same case, but the Sr data seem deficient. On the $TiO_2$-$K_2O$-$P_2O_5$ diagram(8), it is obvious that most of Cretaceous intermediate and basic volcanic rocks in western Xizang should be regarded as being of continental type, and the rest of oceanic type. However, because of their association with cherts and limestones, most of Cretaceous intermediate and basic volcanic rocks should be considered as marine basalts; whereas Cretaceous continental basalt-andesite-rhyolite formations are developed only on the Gandise Range and the Nganglung Kangri Range. Therefore, it is postulated that most of the basic volcanic rocks in western Xizang are marine basalts building up the continental crust of island arc terrain, and partly are continental volcanic rocks in the island arc terrain or on the margin of the continent. In addition, a few slices of the oceanic floor may be recognized in the island arc terrain. In a word, during the Cretaceous period the western half of Xizang was an island arc terrain above the plate subduction zone.

Some representative chemical analyses for the ophiolite igneous rocks in western Xizang are listed in Table 1.

Using a contour line 3,000 m above sea level drawn on the map, the outline of the Qinghai-Xizang Plateau can be marked. More than ten large ophiolite belts or ultramafic rock belts and some small ones formed during the last billion years are winding lengthwise and crosswise on the plateau. Considering the investigation on igneous rocks in the Qinghai-Xizang Plateau for the past 30 years conducted by many geologists, it can be said that in the central part of the highest plateau in the world almost no large traps represented by flood basalts occur, but a large number of closely-spaced ophiolite belts. In other words, the magmatic formations particular to PLATFORM do not occur within this so-called PLATEAU. On the other hand, nearly all traps in the world are generally distributed on the plateaus never exceeding 3,000 m above sea level. So the Qinghai- Xizang Plateau may not be considered as a real plateau, instead it is only an assemblage of orogenic belts formed during the last billion years.

If an Alpine ultrabasic rock belt is thought as the backbone of an ophiolite belt, it can be seen clearly that the four main mountain systems in the world —— Qinghai-Xizang Mountain System, Altay Mountain System, Alps Mountain System and Circum Pacific Mountain System —— are precisely the areas where ophiolite belts, which were formed in the last billion years, closely gather. Among them there are many landmasses on which traps and kimberlites occur (Fig. 8). Thus, the four mountain systems can be endowed with geological sense as the assemblage of orogenic belts, and termed as four OPHIOLITE SYSTEMS, i.e. Qinghai-Xizang Ophiolite System, Altay Ophiolite

Table 1 Chemical analyses for ophiolites in western Xizang

| Sample No. | 1 | 2 | 3 | 4 | 5 | 6 | 7 | 8 | 9 | 10 | 11 | 12 | 13 | 14 |
|---|---|---|---|---|---|---|---|---|---|---|---|---|---|---|
| Rock type | harz-burgite | bron-zitite | norite | dial-lagite | gabbro | dia-base | alkali basalt | tho-leiite | spi-lite | oliv. basalt | basalt | alum. basalt | ensta-tite | diopside |
| $SiO_2$ | 43.35 | 49.85 | 48.86 | 50.10 | 48.97 | 46.97 | 49.25 | 54.05 | 47.17 | 43.66 | 51.44 | 50.09 | 56.57 | 51.11 |
| $TiO_2$ | 0.16 | 0.22 | 0.06 | 0.11 | 0.96 | 1.36 | 1.09 | 0.61 | 3.76 | 0.95 | 0.85 | 1.38 | 0.04 | 0.08 |
| $Al_2O_3$ | 1.94 | 1.78 | 10.71 | 2.10 | 16.31 | 15.78 | 21.01 | 20.12 | 12.87 | 20.85 | 14.56 | 17.35 | 2.67 | 3.74 |
| $Cr_2O_3$ | 0.51 | 0.35 | 0.28 | 0.50 | 0.02 | 0.39 | 0.11 | 0.12 | 0.00 | 0.13 | 0.05 | 0.00 | 0.60 | 1.46 |
| $Fe_2O_3$ | 4.77 | 6.94 | 3.79 | 0.95 | 3.38 | 2.93 | 6.07 | 4.81 | 13.84 | 6.81 | 3.64 | 9.15 | 1.28 | 1.10 |
| FeO | 8.61 | 8.33 | 4.28 | 2.95 | 5.24 | 8.98 | 3.69 | 5.19 | 3.77 | 4.07 | 7.54 | 5.52 | 4.75 | 2.23 |
| MnO | 0.14 | 0.28 | 0.12 | 0.11 | 0.20 | 0.16 | 0.14 | 0.15 | 0.23 | 0.13 | 0.23 | 0.20 | 0.11 | 0.08 |
| NiO | 0.23 | 0.13 | 0.06 | – | 0.03 | 0.02 | 0.02 | 0.08 | 0.02 | 0.00 | 0.01 | 0.06 | 0.04 | 0.04 |
| MgO | 37.68 | 26.04 | 18.01 | 20.34 | 6.58 | 5.62 | 3.24 | 3.85 | 3.85 | 6.97 | 6.62 | 2.66 | 32.84 | 18.95 |
| CaO | 1.72 | 3.70 | 10.92 | 19.44 | 10.98 | 14.10 | 5.20 | 5.49 | 8.60 | 7.87 | 3.55 | 8.24 | 0.85 | 21.42 |
| $Na_2O$ | 0.06 | 0.10 | 0.42 | 0.07 | 3.54 | 2.63 | 5.30 | 3.28 | 2.39 | 2.52 | 5.42 | 3.48 | 0.02 | 0.20 |
| $K_2O$ | 0.03 | 0.03 | 0.03 | 0.01 | 0.24 | 0.14 | 1.52 | 0.06 | 0.76 | 1.68 | 0.89 | 0.75 | – | – |
| $H_2O^+$ | 0.61 | 2.51 | 2.16 | 2.72 | 2.53 | 0.81 | 3.22 |  | 2.90 |  | 4.30 | 1.43 | 0.06 | 0.11 |
| $H_2O^-$ |  |  |  |  | 0.07 |  |  |  | 0.03 |  |  |  |  |  |
| $P_2O_5$ | 0.06 | 0.04 | 0.03 | 0.00 | 0.18 | 0.14 | 0.24 | 0.09 | 0.49 | 0.43 | 0.15 | 0.26 | – | – |
| $CO_2$ |  |  |  | 0.24 | 0.14 |  |  | 0.04 |  |  |  |  |  |  |
| ign.loss |  |  |  |  |  |  |  | 2.82 |  | 3.66 |  |  |  |  |
| sum. | 99.87 | 100.30 | 99.73 | 99.64 | 99.37 | 100.03 | 100.10 | 100.72 | 100.72 | 99.73 | 99.25 | 100.57 | 99.83 | 100.52 |

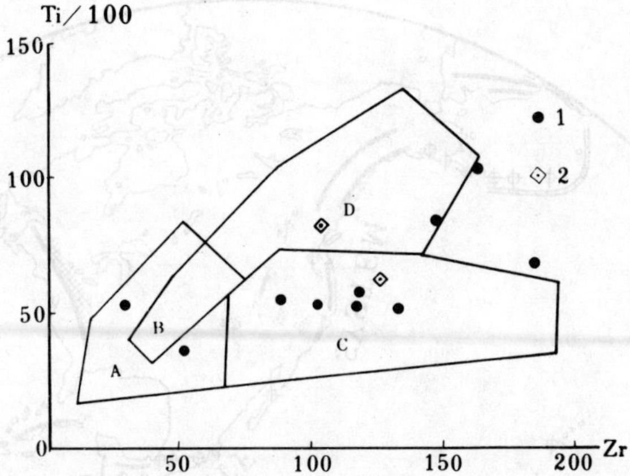


Fig. 6 Ti-Zr discriminating diagram showing the basaltic clans of ophiolite belts in western Xizang
1. basic volcanic rocks;
2. basic dykes;
D and B. oceanic floor; A. tholeiite;
C and B. calc-alkali basalt (after J. A. Pearce et al., 1973)

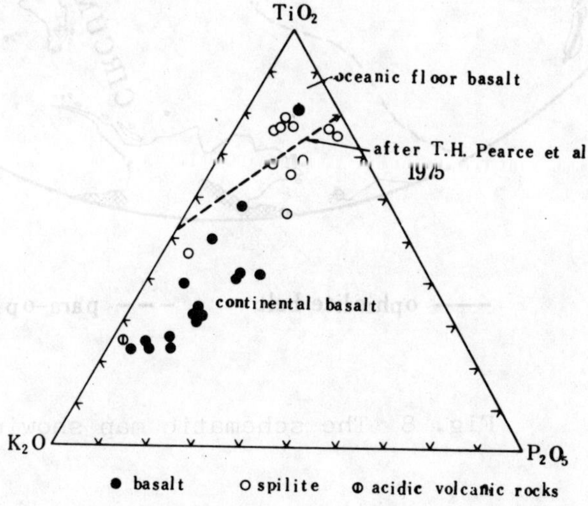


Fig. 7 $TiO_2$-$K_2O$-$P_2O_5$ diagram of Cretaceous volcanic rocks in ophiolite belts in western Xizang

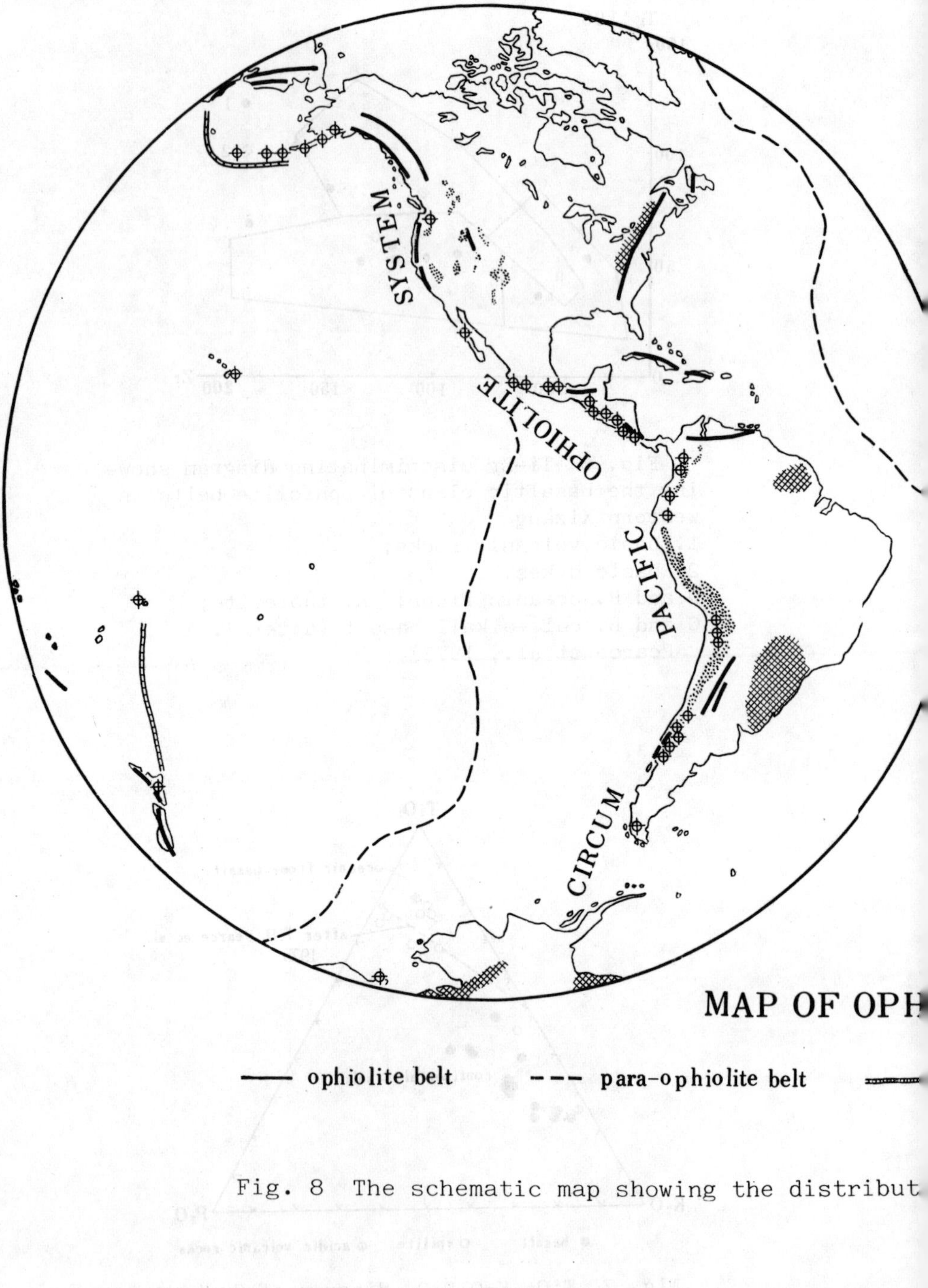


Fig. 8 The schematic map showing the distribut

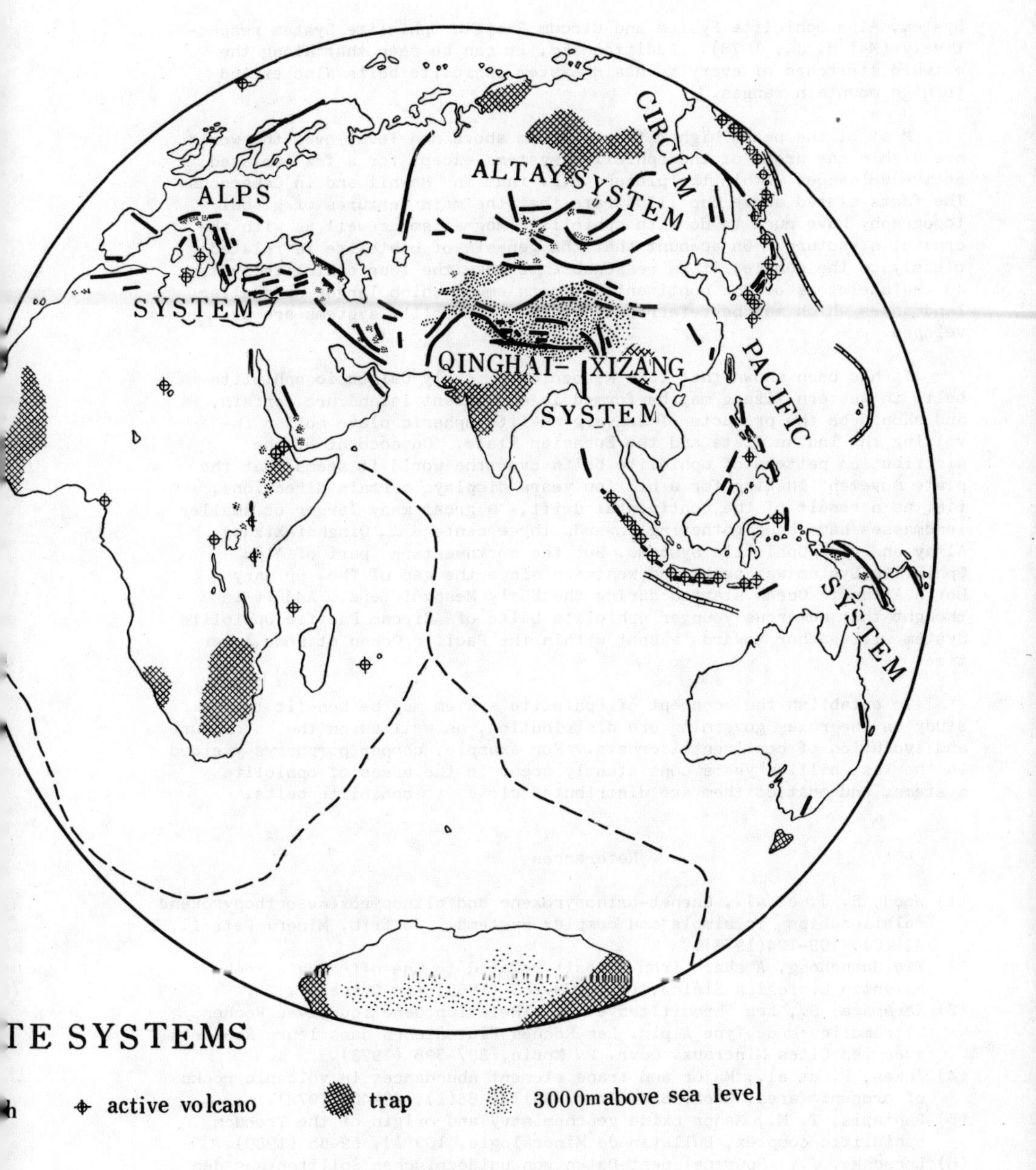


TE SYSTEMS

h ✦ active volcano trap 3000m above sea level

ophiolite systems in the world

System, Alps Ophiolite System and Circum Pacific Ophiolite System respectively (Mei H. J., 1978). Additionally, it can be seen that along the outward stretches of every mountain system ophiolite belts also extend in high mountain ranges.

Most of the peaks higher than 3,000 m above sea level over the world are within the areas of the ophiolite systems except for a few isolated active volcanoes within the plates, e.g. that in Hawaii and in Cameroons. The facts stated above can illustrate that the main features of global topography have much to do with ophiolitic magmatism as well as with the crustal structure. On account that the genesis of ophiolite is related closely to the mantle, it is reasonable to take the four ophiolite systems as the skeletons of the continental crusts among which larger or smaller landmasses which may be referred to as trap-kimberlite systems are developed.

It has been shown that late Mesozoic — early Cainozoic ophiolite belts in western Xizang may be formed in an ancient island arc terrain, and should be the products of convergent lithospheric plate motion involving the Indian Plate and the Eurasian Plate. On account of the distribution pattern of ophiolite belts over the world it seems that the plate movement lasting for a billion years displays certain directions, i.e. as a result of the continental drift, a great many larger or smaller landmasses have been gathering towards three centers — Qinghai-Xizang, Altay and Alps Ophiolite Systems. But the northwestern part of Alps Ophiolite System was separated westward since the gap of the primary North Atlantic Ocean started during the Early Mesozoic era. And it is thought that numerous younger ophiolite belts of Circum Pacific Ophiolite System will gather towards a spot within the Pacific Ocean at some later time.

To establish the concept of ophiolite system may be benefit to the study on the rules governing ore distribution, as well as on the structure and evolution of continental crusts. For example, copper porphyrys yielded in the last billion years consistently occur in the areas of ophiolite systems, and most of them are distributed close to ophiolite belts.

## References


(1) Wood, B. J. et al., Garnet-orthopyroxene and clinopyroxene-orthopyroxene relationships in simple and complex systems. Contrib. Miner. Petrol., 42(10), 109-124(1973).

(2) Xie Guanghong, A skarn iron deposit related to the ultrabasic rock. Scientia Geologica Sinica, (3), 278-283 (1965) (in Chinese).

(3) Karamata, S., Les Chromitites et leur Relation Genetique avec Roches Ultramafiques de Type Alpin. Les Roches Plutoniques dans leurs Rapports avec les Gites Mineraux. Ouvr. P. Morin, 397-398 (1973).

(4) Jakes, P. et al., Major and trace element abundances in volcanic rocks of orogenic area. Geol. Soc. Amer. Bull., 83(1), 29-40 (1972).

(5) Pantazis, T. M., Major oxide geochemistry and origin of the Troodos ophiolitic complex, Bulletin de Mineralogie, 103(1), 59-65 (1980).

(6) Loeschke, J., Spurenelement-Daten von paläozoischen Spiliten aus den Ostalpen und ihre Bedeutung fur geotektonische Interpretation, Geologische Rundschau, 64(1), 62-74 (1975).

(7) Pearce, J. A. et al., Tectonic setting of base volcanic rocks determined using trace element analysis. Earth and Planet. Sci. Lett., 19(2), 290-300 (1973).

(8) Pearce, T. H. et al., The $TiO_2$-$K_2O$-$P_2O_5$ diagram: A method of discriminating between oceanic and non-oceanic basalts. Earth and Planet. Sci. Lett., 24(3), 419-426 (1975).

# OPHIOLITIC PRODUCTS AND PROCESSES: CYPRUS AND THE OMAN

Professor I.G. Gass
(Open University, Milton Keynes, England)

The Troodos massif, Cyprus and the Semail Nappe of the Oman are of particular significance among Phanerozoic ophiolites for, being well-exposed, relatively undeformed and readily accessible, they have been subjected to detailed study over the past 30 years. In both cases a full (Penrose) ophiolite sequence is present ranging upwards from tectonized harzburgites, through layered peridotites and gabbros, to sheeted dyke complexes and pillow lavas. The processes that produced these sequences will be described and discussed. It will be concluded, on the basis of palaeogeographic, structural and compositional characteristics, that both these ophiolites are relics of oceanic lithosphere formed at constructive plate margins beneath small, back-arc basins.

## INTRODUCTION

Originally, the envisaged tectonic timing and setting for ophiolites was during the early stages of eugeosynclinal development and workers such as Brunn (1961), Dubertret (1952), and Aubouin (1965) regarded ophiolites as magmatic masses erupted at lines of tectonic weakness along the submarine flanks of developing geosynclines (Fig.1a). Within these "balloons" of basaltic melt, dykes were injected through the chilled skin to produce second-stage eruptions of pillow-lavas whilst the molten interior crystallized to produce a cumulate ultramafic-mafic sequence (Fig. 1b & c).

In 1972, subsequent to the near-universal acceptance of plate tectonics, the Geological Society of America's Penrose conference on ophiolite redefined an ophiolite as an assemblage of basic and ultrabasic rocks with a specific composition and ordering (see Table 1).

Table 1 gives a complete vertical sequence. In areas of structural complexity whole units in the sequence may be missing or transposed and so the term "fragmented" or "dismembered" ophiolite is appropriate. However, in many cases the original, pre-disruption ophiolite stratigraphy can be reconstructed and in these cases, and in those of undeformed masses, the sequence given in Table 1 is invariably present. It is this remarkable uniformity of succession, irrespective of age or geographic setting, that strongly implies a common origin.

All lines of evidence indicate that ophiolites were formed in a submarine tensional environment where a plentiful supply of near-surface basaltic magma was readily available. Geophysical evidence indicates that such requirements are met at oceanic ridges and many workers (e.g.Coleman, 1977; Gass & Smewing, 1980) identify ophiolites as land-bound fragments of oceanic lithosphere formed at fossil constructive plate margins. If this is the case, then the study of ophiolites is not only critical in its own right but it can enhance our understanding of magmatic, metamorphic and structural processes at present-day constructive plate margins. So, ophiolite studies are important and carry implications far beyond the confines of the individual area being studied. It is unfortunate therefore, that most ophiolites have suffered extensive deformation and metamorphism during their emplacement into fold-mountain belts and their investigation and interpretation is difficult and complex. This being so, many of the detailed studies have concentrated on well-exposed and relatively undeformed examples such as the Troodos massif of Cyprus and the Semail Nappe of Oman. It is from these studies that much of our understanding of ophiolites has come. The ophiolite model, as deduced from studies in Cyprus and the Oman, is now described and discussed before a concluding section on genetic implications is presented.

## THE OPHIOLITE MODEL

Figure 2 illustrates not only the disposition of the various ophiolitic rock units but their genetic relationship one to another.

It is widely accepted that the tectonized harzburgite at the base of the sequence represents a lherzolitic depleted upper mantle from which a basaltic melt has been extracted to accumulate in a high-level chamber. In the chamber crystallization takes place and a layered plutonic sequence is produced. Melts tapped from the chamber are injected as dykes and erupted as commonly pillowed lava flows. Using illustrations from Troodos and the Oman the ophiolite sequence will be described and discussed under the headings of (i) Depleted mantle (tectonized harzburgite), (ii) Magma chamber products (iii) Dykes and lavas and (iv) Metamorphic-mineralization processes.

1. Depleted mantle. Field relations and chemical and mineralogical homogeneity of the tectonized harzburgites suggest that they are the residues of a plagioclase lherzolite mantle from which basaltic liquid has been extracted (Greenbaum, 1972; Menzies and Allen, 1974). This proposal is supported by the presence of gabbroic pods within the tectonites which are best explained as batches of basaltic melt that crystallized before they could escape from the mantle and localized patches of plagioclase lherzolite which probably represent mantle that had escaped melt depletion and from which little or no basaltic magma had been extracted.

Harzburgites form about 80% of the total outcrop and enclose irregular lenticular masses of dunite, which form most of the remaining 20% subordinate bodies of gabbroic and lherzolitic composition also occur. In the harzburgites foliation is well developed and is enhanced by the metamorphic segregation of olivine and orthopyroxene into separate folia. Lineation is also present as chromite and orthopyroxene elongations. Foliation and lineation directions are coincident with the orientation of the olivine a - axis and this, in turn, equates with the fastest p-wave direction. All are perpendicular to the spreading axis and in line with the spreading direction (Christensen and Smewing, In press). Textures are xenoblastic granular and olivine, which forms over 70% of the mode, is in the compositional range $Fo_{90-92}$. Orthopyroxene never exceeds 30% in modal abundance and has a constant composition close to $En_{90-91}$. Dunitic lenses usually measure 0.5-1.0km in the foliation plane and are usually between 5-100m thick. In detail, the dunite-harzburgite contact is intricately and deeply infolded and the dunite masses themselves have been internally folded and have obviously suffered the same deformation as the host harzburgites. The dunitic olivine has the same Fo-rich compositional range ($Fo_{90-92}$) as that of the harzburgites. The chromite however has a distinctly higher Cr/Al ratio and furthermore occurs as layered euhedral grains that strongly suggest a cumulate origin for the dunitic bodies (Greenbaum, 1972 & 1977, Neary & Brown, 1979). It is within these dunitic lenses that the economic deposits of chromitite occur.

Most workers accept that the tectonized harzburgite represents depleted mantle - the refractory residue from which basaltic or picritic melts have been extracted to form the overlying layered plutonics, sheeted dyke complex and pillow lavas. The only possible alternative, a cumulate origin, is implausible on size considerations (the harzburgites of Oman, Papua and the Bay of Islands are over 7km thick) and because the harzburgite is a homogeneous assemblage of olivine and orthopyroxene with no cumulate textures, no rhythmic or cryptic layering, no bulk compositional variations and is enriched in refractory element such as Mg, Cr and Ni. The model proposed to explain these features is that at depths of ca. 25-30km (7.5-9kb), basaltic magma first separated from the mantle. The two elements, the magma and the residual harzburgite, both move upwards under the influence of mantle convection. The magma, whose initial composition will depend on the degree of partial melting, will change in composition as it rises because the olivine phase volume increases with decreasing pressure and the melt temperature does not vary significantly under adiabatic decompression. This causes the rising melt to stay within the olivine and chromite. It has been proposed from studies on Troodos (Allen, 1975), and has been confirmed by work in the Oman, that the dunitic bodies within the harzburgite result from olivine and chromite precipitation from rising melts. The melt remaining within the ascending diapir rises to feed the main high-level chamber beneath the spreading axis.

2. <u>Magma Chamber Products.</u> Depending on the degree of partial melting the basaltic melts segregated from the pristine lherzolitic mantle collected in a magma chamber (or chambers) some 1-2km below the crust - sea-water interface. With a high temperature difference between the top and bottom of this chamber, crystallization proceeded rapidly with the formation of olivine, two pyroxenes and plagioclase within the magma. At the same time melt chilled against the chamber roof underplates it with microgabbros or dolerite. From compositional considerations it seems that about half of the chamber solidified in this way before magma was tapped off, intruded as dykes and extruded as pillow lavas.

The layered rocks have a laminar fabric. At the top of the sequence phase layering and other features of igneous sedimentation are common. With depth plagioclase may disappear and gabbros are sometimes replaced as the dominant rock type by peridotites in which cyclic successions involving dunites, wehrlites, pyroxenites and troctolites occur. In the gabbros ol, opx, cpx & pl are present. Olivine is always the first phase to form, but it is varyingly followed by opx, cpx or pl. Detailed petrographic and geochemical studies of layered sequences has identified multiple repetition of ol-px precipitation in ultramafic zones and ol-px-pl in gabbros. The only realistic way of explaining these cyclic phenomena is to invoke the repeated replenishment of the main magma chamber by small batches of fresh magma. This proposal is totally compatible with the diapiric rise of melt within the mantle invoked to explain the dunitic bodies within the tectonized harzburgite.

Lastly, much recent work has concentrated on deducing the space form of the high-level chambers and on identifying whether the constructive margins, at which the ophiolites formed, were maintained by a single extensive chamber or a plexus of smaller chambers. Situations vary. For Troodos, a slow spread -ing axis (<1cm/flank/yr) a multiple model of numerous small (1-2km deep, ca 5km diameter) chambers has been proposed in field and petrological data (Allen, 1975; Spooner, 1977). For the Oman, formed at a faster ( 2cm/flank/ yr) ridge the chambers are large (3-5km deep and up to 20km across).

3. Dykes and lavas. A unique feature of many, although not all, ophiolites is the presence, between the overlying pillow lavas and the underlying gabbros, of a Sheeted Dyke Complex that is usually about 1.5km thick and which, in its central part, consists of 100% dyke with no intervening host. Individual dykes vary in width from 0.1-5m and are fine grained, generally aphyric, non-vesicular diabase with ophitic or variolitic textures. Hydrothermal metamorphism to greenschist and amphibolite facies is usual and although modifying the mineralogy has not grossly affected the igneous texture. Chilled margins are ubiquitous and the phenomenon of preferential one-way chilling has been identified on all complexes in which it has been sought. The regular aspect and often near-vertical orientation of the dykes, emphasized by jointing that is predominantly parallel to dyke margins, gives larger outcrops a distinctly sheeted appearance. At the top of the dyke complex the transition from dominantly dyke to dominantly lava is rapid and dyke abundance can fall from 70-20% within vertical distances of 25-50m. At the base, the transition from dykes to gabbros is equally rapid although the gabbros immediately below the dyke complex are cut by dykes that form 10-15% of the outcrop. Most dykes root immediately adjacent to the dyke-gabbro contact but others, cutting the gabbro, have their origin in deeper magma bodies. In all complete ophiolitic sequences the uppermost igneous unit is of basaltic lavas that are commonly pillowed. Immediately overlying the dyke complex are lavas that are compositionally equivalent to the dykes and like the dykes were formed by processes at constructive margins. Gass & Smewing (1973) invoked the term Axis Sequence to identify the dykes and lavas produced at the constructive margin. In most ophiolites, and certainly in Troodos and the Oman, the Axis lavas are overlain by other pillow lavas that differ geochemically from them. These "upper" pillow lavas show trace element abundances that Pearce & Cann (1971) among other workers, have identified as indicating an island arc tectonic environment.

4. Metamorphism & Mineralization. Coleman (1977) divided the metamorphic processes that affect ophiolites into (i) those that occur whilst the ophiolite is still an integral part of the oceanic lithosphere and (ii) those that occur during or after obduction. Here I consider only the oceanic *in situ* processes and furthermore exclude serpentinization, as this has been treated exhaustively elsewhere (see Coleman *op. cit.*).

Most metamorphic studies on ophiolites started in the early 1970's (eg. Gass and Smewing, 1973, Spooner and Fyfe, 1973). These studies indicated that the metamorphic zones were originally horizontal and that zeolite facies overlies greenschist facies which gave way, with depth, to lower amphibolite facies. No directional fabric is present and metamorphism obviously occurred *in situ* for conformably overlying sediments are unmetamorphosed.

As the metamorphic processes obviously occurred whilst the ophiolite was part of the oceanic lithosphere, and as there are no directional fabrics, it was proposed that the metamorphic agent was thermal, the heat emanating from the underlying magma, and the metamorphic processes involved the circulation of sea-water through newly-formed, still hot, oceanic crust with thermal gradient greater than $150^{o}C\ km^{-1}$. Convection is envisaged as a single-pass system with recharge continually taking place at the water/rock interface and discharge occurring above thermal highs along discrete channels. During convection, water-rock reaction in the downward part of the cycle imposed a prograde hydrothermal metamorphism, increasing with depth from zeolite through greenschist to amphibolite facies, on the pillow lavas, Sheeted Dyke Complex and uppermost gabbros. Convecting fluids leach transition metals (Fe, Mn, Cu, Zn, Pb, Ni, Co, Ag, Au) from the rocks and also redistributed silica and other elements. During their return to the surface the metal-enriched fluids were channelled along fault planes and this meant that emission of the fluid into sea-water was localized. In favourable sheltered locations, such as bathymetric depressions, reaction between the hot, metal-enriched brines and sea-water resulted in the precipitation of sulphides to form massive sulphide ore bodies and Fe and Mn to give ferromanganoan hydroxide sediments. The processes involved are diagrammatically represented in Figure 3.

## CONCLUSIONS

Most workers accept that ophiolites were formed at ancient oceanic constructive plate margins. But, although ophiolite basalts are usually comparable to oceanic basalts in their trace element content (e.g.Pearce and Cann 1973) and rare earth distribution they often show calc-alkaline tendencies and the fractionation of the plutonic rocks leads to calc-alkaline trondhjemites and tonalites. This trend can be explained by the entry of water into the magma chamber from an underlying subduction zone. This in turn, implies that most ophiolites represent fragments of oceanic lithosphere produced above subduction zones, in back-arc, marginal sea environments. This proposal is in keeping with (i) palaeogeographic reconstructions, (ii) the thinness of ophiolite crustal layers compared to those of major ocean basins and (iii) the short time gap between the age of the ophiolites and the sediments at the margin where they become obducted.

So, seemingly, ophiolites are not representatives of major ocean basins and therefore they do not represent "normal" oceanic lithosphere. Despite this, they are the best land-based evidence available on constructive margin processes and the very wide variety of phenomena they display demonstrates two things (i) that despite the general pattern, sequences vary from complex to complex and there is little doubt that the present day constructive margin processes also vary from margin to margin and (ii) the complexity of processes at constructive margins involves a variety of pene-contemporaneous episodic processes in the mantle, the magma chambers, during dyke injection and lava eruption as well as metamorphism after the oceanic crust has formed.

## References


(1) Allen, C.R., The petrology of a portion of the Troodos plutonic complex, Cyprus. Unpubl. Ph.D. thesis, Cambridge University, 315 pp. (1975).

(2) Aubouin, J., Geosynclines. Elsevier, Amsterdam, 335 pp (1965).

(3) Brunn, J.H., Mise en place et differenciation pluto-volcanique due cortege ophiolitique. Rev. Geogr. Phys. Dyn.3, 115-132. (1960)

(4) Coleman, R.G., Ophiolites. Ancient Oceanic Lithosphere? Springer-Verlag. 229 pp. (1977).

(5) Dubertret, L., Geologie des roches vertes du Nord-Quest de la Syrie et due Hatay. Notes et mem. Moyen-Orient 6, 2-179 (1955).

(6) Gass, I.G., Smewing, J.D., Intrusion, extrusion and metamorphism at constructive margins: evidence from the Troodos massif, Cyprus Nature, 242.,26-29 (1973).

(7) Gass, I.G., Smewing, J.D., Ophiolites: obducted oceanic lithosphere. In the Sea v.7, (Ed. Emiliani, C.) Wiley New York (1980).

(8) Greenbaum, D., Magmatic processes at ocean ridges: evidence from the Troodos massif, Cyprus. Nature, Phys. Sci., 238, 18-21 (1972)

(9) Greenbaum, D., The chromitiferous rocks of the Troodos ophiolite complex, Cyprus. Econ. Geol., 72, 1175-1194 (1977).

(10) Menzies, M.A., Allen, C.R., Plagioclase lherzolite-residual mantle relationships within two eastern Mediterranean ophiolites. Contrib. Mineral. Petrol., 45, 197-213 (1974).

(11) Neary, C.N., Brown, M.A., Chromites from the Al 'Ays complex, Saudi Arabia and the Semail complex, Oman. In: Evolution and mineralization of the Arabian-Nubian Shield, Bull.3. Inst. Applied Geol. Jeddah, 193-205 (1979).

(12) Pearce, J.A., Cann, J.R., Ophiolite origin investigated by discriminant analysis using Ti, Zr and Y. Earth Planet Sci. Lett. 12, 339-349 (1971).

(13) Pearce, J.A., Cann, J.R., Tectonic setting of basic volcanic rocks determined using trace element analyses. Earth Planet Sci. Lett. 19, 290-300 (1973).

(14) Spooner, E.T.C., Hyarothermal model for the origin of the opiolitic cupriferous pyrite ore deposits of Cyprus. Sp. Publ. Geol. Soc. Lond., 7, 58-71 (1977).

(15) Spooner, E.T.C., Fife, W.S., Sub-sea-floor metamorphism, heat and mass transfer. Contr. Mineral. Petrol., 42, 287-304 (1973).

Table 1. The complete ophiolite sequence as defined at the Penrose Ophiolite Conference in 1972.

Basic volcanics
(commonly pillowed)

------------------------------ Merging into ------------------------------

Basic sheeted dyke complex

---

| | | |
|---|---|---|
| Trondhjemites | ) | |
| | ) | High level intrusive |
| Gabbros | ) | |
| Olivine gabbros | ) | |
| | ) | |
| Pyroxenites | ) | Layered plutonic sequence |
| Peridotites | ) | |

---

Harzburgite (sometimes lherzolite, dunite and chromitite with a metamorphic tectonized fabric and commonly serpentinized)

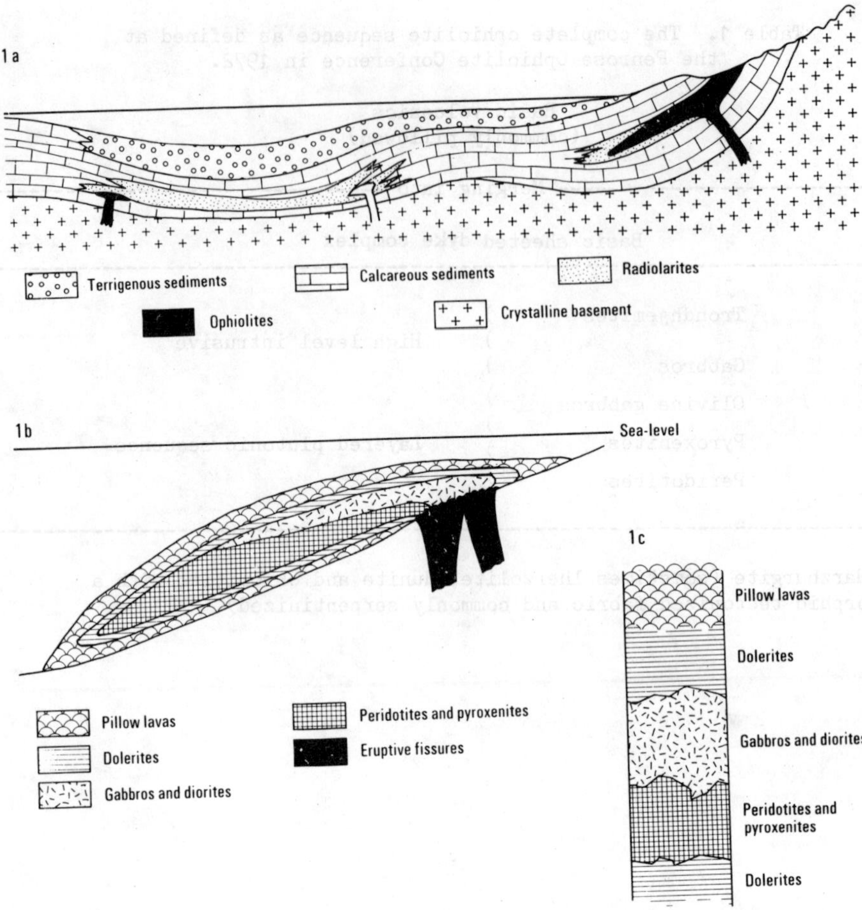


Figure 1. The geosynnclinal concept of ophiolites.

1a. The postulated position of ophiolite masses along the flanks of eugeosynclines.
1b. The internal structure of eruptive magma masses.
1c. The ophiolite sequence through the magam bodies (largely after Aubouin 1956).

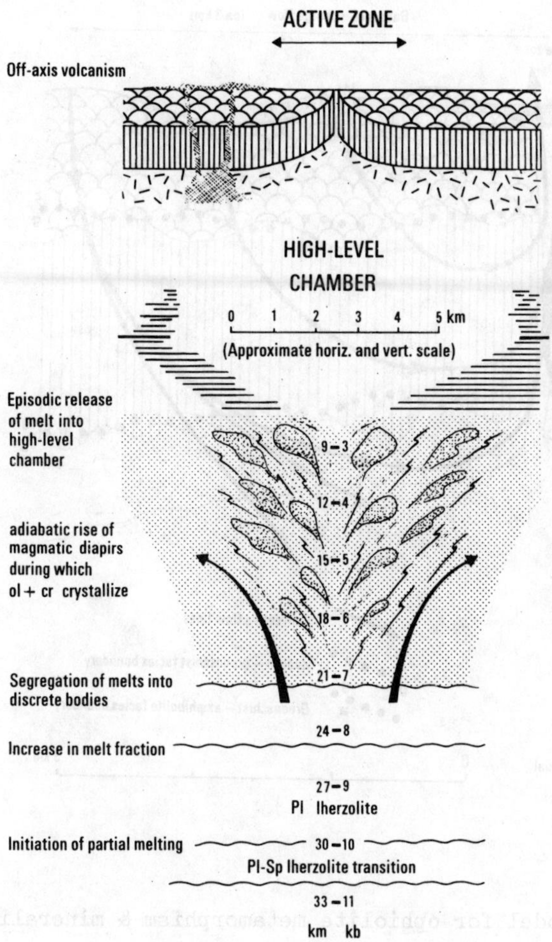


Figure 2. Space form of an ophiolite magma chamber and its relation to various rock units.

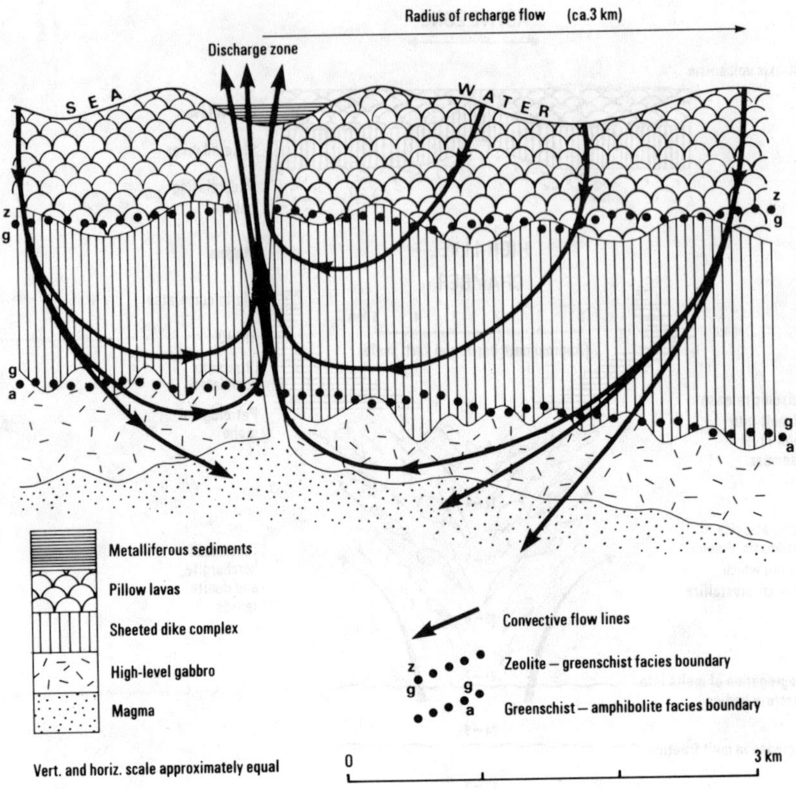


Figure 3. Schematic model for ophiolite metamorphism & mineralization

# GEOLOGICAL OUTLINE AND GEOLOGICAL HISTORY OF THE YARLUNG ZANGBO SUTURE ZONE

Wu Hao-ruo
(Institute of Geology, Academia Sinica)

During the past two decades or so, the famous Indus Suture Zone has been studied and described frequently, but up to now, information about the Yarlung Zangbo Suture Zone to the east has been very rare. Since 1974, I have worked in Yarlung Zangbo valley. In this paper I will describe briefly a geological outline of the Yarlung Zangbo Suture Zone and review its geological history. Moreover, the situation of the northern boundary of Gondwana will be discussed.

## EXTENT

The Yarlung Zangbo Suture Zone, from the Mayum La (30°34'N, 82°29'E) in the west to the great bend of the Yarlung Zangbo in the east, extends over 1,000 km.

Its northern border runs roughly along the southern slope of the Gandise, and is a major unconformity formed most probably in the mid-Cretaceous. It is a deep fault. Since the Mesozoic, particularly in its late period, the area to the north of the fault uplifted and extensive intermediate-acid magma activity developed. In contrast, the area to the south subsided and the intense basic volcanic activity occurred.

Its southern border, marked by the Lhagoi Kangri granite belt with intermittent outcrops, may be a buried fault, which roughly follows the crests of the Lhagoi Kangri (the lower watershed of southern Xizang). This fault can be called Lhagoi Kangri Fault. Because also limiting the Himalayas from the north, it also may be named the North Boundary Thrust (N.B.T.). In Mesozoic, the area to the south of the fault kept properties of a stable platform, whereas that to the north of the fault transformed into active geosyncline. Therefore, the Yarlung Zangbo valley to the north of the fault should not be designated as a part of the Himalayas geologically. It should be regarded as a independent tectonic unit (Fig. 1).

In the Xigaze region, based on the basic volcanics mainly distributed to the north of the line of Lhaze-Bainang-Degyiling, the suture zone can be subdivided into two subzones : the northern subzone - a eugeosynclinal zone and the southern subzone - a miogeosynclinal zone. A major south-dipping thrust separates them.

## THE NORTHERN SUBZONE

It is an E - W linear structural zone, which mainly consists of the Upper Triassic and the Upper Cretaceous, and the great ultramafic rock body that cuts across them.

The stratigraphy of the northern subzone is shown in Table 1. The

Pre-Permian Palaeozoic, Upper-Middle Triassic and Lower Jurassic rocks crop out only in the western Zhongba district (29°53'N, 83°10'E). The Permian limestone appears mostly as klippes except in the Zhongba district. The Middle-Upper Jurassic and Lower Cretaceous have been uncertain in the subzone up to now. The Eocene marine beds are found only in Mai La, northwest of Zhongba. The Miocene red sandstone and conglomerate are distributed sparsely on the flanks of the ultramafic rock body.

The Upper Triassic, which is widely exposed in the southern part of the northern subzone, is called the Gyirong Group. It is characterized by cyclothems of variegated siliceous shale and platy chert, interbedded with limestone, sandstone and basalts rich in titaniferous augites.

The widespread Upper Cretaceous includes five units. The basal conglomerates - Qiuwu Formation are found only at the northern boundary. The second conglomerate sequence - the Qiabulin Formation - overlaps the Triassic rocks southwards, and a similar position is occupied by the Sangzugang Formation. Upward, the upper part of the Cretaceous obviously may be subdivided into two lithofacies belts, the Angren Formation, a flysch belt in the north, and the volcanic-siliceous sedimentary belt, the Laze Group, in the south (24). West of Jangseb Village, about 60 km southwest of Xigaze, on the south of the flysch there are 40 m of purple and green siliceous shale bearing a lot of volcanic materials. Southward it transforms into a set of basic volcanics : the Jangseb Volcanics, which crops out to the south of the Angren Formation, mainly consists of pillow lava, and forms the mountain crests.

The strata in the northern subzone regionally dip southwards. The Qiuwu Formation, Qiabulin Formation, Sangzugang Formation and the Laze Group generally appear as monoclines or open folds, and the Angren Formation and Gyirong Group constitute a series of closed folds, that may have resulted from lithological difference between them.

The great ultramafic rock body extends through the southern part of the northern subzone. To the west of Rinbung, it often attains a width of 2-3 km, or more than 5 km in some places, and almost without any interruption. But to the east of Rinbung, it is markedly reduced in size and crops out only intermittently. The rock body consists mainly of harzburgite with dunite and lherzolite as a minor part, and shows intense serpentinization, schistosity and brecciation. It usually shows a tectonic contact with the surrounding rocks, but locally, phenomena relating to contact metamorphism have been found. The exotic blocks, such as chert, volcanics, limestones, conglomerate etc. vary with types of wall rocks. The upper part of the rock body are intruded by gabbro dykes.

The highest horizon which is structurally overlain by ultramafic rocks is Upper Cretaceous in age, but the highest marine bed in the northern subzone is Eocene. In eastern Qusum County (29°04'N, 92°20'E), the conglomerates, overlying unconformably on the ultramafic rocks, contain Miocene gastropods and charophyta. Therefore, the final emplacement age of the ultramafic rocks is considered to be Oligocene.

## THE SOUTHERN SUBZONE

Its stratigraphy is shown in Table 2. Differing from the northern subzone, the strata here from Permian-Carboniferous to Upper Cretaceous occur, but the Triassic crops out most widely. This subzone appears roughly as a synclinorium. Its southern limb is gently dipping, and the strata preserved are rather complete. Its northern limb was destroyed by faulting, making the strata there steeply dipping and incomplete. The Permo - Carboniferous rocks constitute a few anticlines with short axes on the southern flank of the synclinorium, and at the core, the Jurassic and Cretaceous are preserved.

The Kewoxika Formation, lying in the basal part of the Permo-Carboniferous, consists mainly of pebbly slate, in which various unsorted and sub-angular pebbles account for 5 per cent and are irregularly disseminated in a gray brown graywacke matrix. Their acute edges can cut downward through the bedding. In contrast, pebbles in the conglomerate beds, intercalated in the pebbly slate, were sorted and rounded very well. Lithologically and horizontally the Kewoxika Formation can be correlated with glaciomarine deposits of the same age in the Himalayas. Upwards, the Permo-Carboniferous sequence passes from quartz sandstone into slate and limestone constituting a sedimentary cycle. The limestone contains abundant fossils of benthonic organisms.

The Chogar Group, occurring most widely in the southern subzone, is characterized by a great thickness of slate and marl interbedded with beds of thick-bedded quartz sandstone with thicknesses of 10-30 m.

The Lower to Middle Jurassic has been found in the Gyangze and Lhunze (28°20'N, 92°30'E) districts. In Lhaze district, the Upper Jurassic quartze sandstone directly overlies the Triassic Chogar Group. The Cretaceous extends from Lhunze in the east through Chigu Lake and Yamzhog Lake to Gyangze in the west. The Lower Cretaceous Jiabula Formation can be subdivided into three parts, the lower is shale with calcareous and sandy concretions and lenses, containing abundant ammonites, belemnites and bivalves etc. The middle is siliceous shale with chert, containing a few belemnites and bivalves. The upper is a great thickness of pyrite-bearing shale, containing rare ammonites and belemnites. The Late Cretaceous Zongzhuo Formation is characterized by a large number of olistostromes,and interbedded with variegated pelagic limestone beds, which contain Campanian-Maestrichtian foraminifera *Globotruncana stuartiformis* (Dalbiez), *G. linneiana tricarinata* (Quereau) etc. They are the latest marine deposits in the southern subzone.

The magmatic activity in the southern subzone is mainly subdivided into two types. One is represented by gneissic two-mica granite, such as the Surco Granite, the Gyaco La Granite, the Mabja Granite etc., occurring along the Lhagoi Kangri Fault. The Kangmar Granite may be controlled by a N-S fault, cutting the core of the N-S elongated Kangmar anticline. Radiometric dating indicates that these granite bodies were formed in the Oligocene-Miocene.

The other type is represented by diabase sills or dykes, which develop particularly in the Chogar Group between Lhaze and Gyaco La. South of Gyaco La, the diabase dykes are distributed close to the Gyaco La Granite. However, no intrusion of diabase dyke or sill into the granite has been observed, so that the diabase may be older than the granite.

## GEOLOGICAL HISTORY

Based on the above data, the evolution of the Yarlung Zangbo Suture Zone may be generalized as follows (Fig. 2):

1. In Palaeozoic, this zone was mainly represented by an epicontinental sea environment. Only at the end of Carboniferous to the beginning of Permian did the littoral environment appear. The conglomerate intercalations of the Kewoxika Formation shows sedimentary features of the coast belt, and the pebbly slate might be derived from glaciers that entered the sea, suggesting that the climate of the land nearby was cold in this period. The Palaeozoic sedimentary sequence, faunal content and climatic conditions in this zone were similar to those in the Himalayas. Thus, they all belong to a continental shelf of the northern margin of the Gondwanaland.

2. During Early to Middle Triassic, the sea water gradually deepened and a pelitic, calcareous sedimentary sequence, bearing pyrite, but lacking in benthonic organism fossils accumulated.

3. From Triassic to Early Cretaceous, the southern subzone was generally represented by the above mentioned bathyal to abyssal environment, locally fine clastic flysch and some volcanic materials indicate some unstablity. Only in Middle to Upper Jurassic did a regression occur once. As mentioned above, the Late Jurassic quartz sandstone overlies strata with different ages, and this disconformity was reported from Hazara(5), Salt Range(6), northeastern part of Kumaun(17) and Thakkhola(1). In this period, the Himalayas still belonged to a shallow sea around the northern margin of the Gondwanaland, and the Yarlung Zangbo Suture Zone should have been the continental slope outside the shelf.

4. From Late Triassic, the differentiation between the northern subzone and the southern subzone began. The Late Triassic Gyirong Group in the northern subzone may be the representative of an island environment(23). During Jurassic to Early Cretaceous, sedimentary records in the northern subzone seem to be scarce at present. Similarly, in the Gandise zone the Lower and Middle Jurassic have not been found up to now, and the Upper Jurassic to Lower Cretaceous consist of shallow marine or lagoonal deposits. The Cenomanian *Orbitolina* limestone is widespread in the northern subzone and in the Gandise zone, so both areas could belong to the same shallow sea. Therefore, in this period, the northern subzone or its northern part was closely related to the Gandise zone.

5. Beginning from Turonian, the Gandise area uplifted and became land. The Late Cretaceous red beds lie unconformably upon the Cenomanian limestone. On the other hand, the Yarlung Zangbo northern subzone subsided greatly accumulating the flysch of Angren Formation, which represented an environment from continental slope to trench. To the south of it, a large amount of radiolarian accumulation of the Laze Group may represent a rather open oceanic basin. Therefore, I tend to consider that the Gandise probably drifted rapidly northward to split further the Yarlung Zangbo zone during this time. Perhaps, this was a cause of the Indian block drifting northward very slowly before 80 m. y. B. P.

6. Until the later stage of Late Cretaceous, the Indian block drifted rapidly northward(16),(9). By Campanian, the sea became rather narrow. In the Zongzhuo Formation of the southern subzone there are some blocks of conglomerate, basic volcanics and oolitic limestone derived from the northern subzone(25). The Jangseb Volcanics of the northern subzone often show chemical compositions of the calc-alklic series, representing an active continental margin or island arc, that may have formed in this period.

7. By Eocene, this zone passed into a coastal environment. The

Nummulite-bearing conglomerate in Zhongba district is the final remnant of the sea.

## THE NORTHERN BOUNDARY OF THE GONDWANA

The history of the Yarlung Zangbo Suture Zone indicates that it was only in Mesozoic, it developed into bathyal and deep sea from a shallow marine environment around the northern margin of Gondwana. Therefore, only at this time, due to the detachment from Gondwana and the northward drift of the Gandise, it became the northern boundary of Gondwana. Nevertheless, where was this boundary in Palaeozoic?

So far as I know, southward from the Tanggula in the wide area of Xizang, the Palaeozoic strata generally belong to those of a platform type. Particularly , the Permo-Carboniferous are all similar to Gondwanan sedimentary sequences(13),(3). Northwards, the lower-grade metamorphic Permian flysch were found in the Kunlun Mts. ( 22 ), so that the northern boundary of the Gondwana may be situated along the southern margin of the Kunlun Mts.

However, how shall we explain these palaeobiogeographical problems, such as the Gondwana Lystrosaurus fossils that occur in Tarim ( 20 ), north of the Kunlun Mts., the Cathaysian flora found in northern Xizang (22), south of the Kunlun Mts., and the mixed Cathaysian-Gondwana flora that reported in Thailand(15), New Guinea(10) and Kashmir(14)? I think, it may be necessary to reexamine the Permian - Triassic palaeogeography of the eastern Tethys.

According to Dietz and Holden(4), the Tethys at the end of Permian became wider and wider eastward, but on their map, the southern border is uncertain and shown in a dashed line. On the other hand, quite different settings of the ancient continents were shown in some recent palaeogeographical maps, such as those of Mintz(12) and Ziegler et al. (27), in which blocks of ancestral China drifted northward on a large scale since the Palaeozoic. However, their positions are speculative.

Recently, the palaeomagnetic data of the Late Cretaceous red bed near Lhasa, suggested a possible palaeolatitude near the equator(26). The plant fossils also indicate tropical climatic conditions(8), (21). Other palaeomagnetic data suggested that the Iran-Afghanistan and Malaya etc. also might have drifted northwards in the geological past (11),(2). So certain northward motion of the Asia plate in the geological past is possible, and that could be one of the prerequisites for the northward drift of thousands of km of the Indian block after Late Cretaceous. Therefore, the eastern part of the Tethys at the end of the Permian might not have been as wide as Dietz and Holden (1970) postulated. Moreover, in the interior of the sea there might have existed some microcontinents, such as Iran-Afghanistan (7), (19), Karakorum-Gandise, Tanggula, Yangtze, Southeast Asia (18) etc., by which animals and plants might have migrated from one continent to another in certain periods.

## References

(1) Bordet, P., Colchen, M., Krummenacher, D.,Le Fort, P., Mouterde, R., Remy, M., Recherches géologiques dans l'Himalaya du Népal, région de la Thakkhola. Paris, Ed. C.N.R.S., 279 P. (1971).

(2) Bunopas, S., Pitakpaivan, K., Sukroo, J., Vella, P., Preliminary palaeomagnetic results from Thailand sedimentary rocks, Proceedings of the Third Regional Conference on Geology and Mineral Resources of Southeast Asia. Ed. Prinya Nutalaya, 25-32 (1978).
(3) Ching Yukan, Liang Xiluo, Wen Shixuan, Additional material of animal fossils from the Permian deposits on the northern slope of Mount Qomolangma Feng. Scientia Geologica Sinica, 236-249 (1977) ( in Chinese ).
(4) Deitz, R. S., Holden, J. C., Reconstruction of Pangaea : breakup and dispersion of the continents, Permian to Present. Geophysic. Res. Jour., 4939-4956 (1970).
(5) Fatmi, A. N., Stratigraphy of Jurassic and Lower Cretaceous rocks and Jurassic Ammonites from northern areas of west Pakistan. Bull. Br. Mus. Nat. Hist. (Geol.) London, 20(7), 297-380 (1972).
(6) Gansser, A., Geology of the Himalayas, Interscience, Publ . 289 p. (1964).
(7) Gealey, W. K., Ophiolites obduction and geologic evolution of the Oman Mountains and adjacent areas. Bull. Geol. Soc. Amer., 88, 1183-1191 (1977).
(8) Guo Shuangxing, Some plant fossils from Rikaze Group in Mount Qomolangma region, A Report of scientific expedition in Mount Qomolangma Region (1966-1968), Palaeontology (1), 411-423 Science Press (1975) (in Chinese).
(9) Johnson, B. D., Powell, C. McA., Veeves, J. J., Spreading history of the eastern Indian Ocean and Greater India's northward flight from Antarctica and Australia. Geol. Soc. Amer. Bull., 87, 1560-1566 (1976).
(10) Konno, E., Some connections between the Cathaysian flora and the *Glossopteris* flora in India during the later Permian age. Palaeobotanist, 14, 26-35 (1966).
(11) Krumsiek, K., Zur Bewegung der Iranisch-Afghanischen platte. Geol. Rundsch. 65, 909-929 (1976).
(12) Mintz, L. W., Historical Geology, the science of dynamic earth. Charles E. Merrill Publishing Company, 588p.(1977).
(13) Norin, E., Geological expedition in western Tibet. Rep. Sino-Swedish Expedition, 29 (1946).
(14) Pal, A., Lower Carboniferous plant beds in Kashmir Himalaya. Geol. Surv. India, News, 8(4), 10 (1977).
(15) Playsin, S., Review of the Ratburi Group. Geology of Thailand, 1, Eds., Stokes, R. B., Tantisukrit, C., Campbell, K. V., 65-100 (1975).
(16) Sclater, J. G., Fisher, R. L., Evolution of the east central Indian Ocean with emphasis on the tectonic setting of the Ninetyeast Ridge. Geol. Soc. Amer. Bull., 85, 683-702 (1974).
(17) Shan, S. K., Sinha, A. K., Stratigraphy and tectonics of the 'Tethys' zone in a part of western Kumaon, Himalaya. Himalaya Geology, 4, 1-27 (1974).
(18) Stauffer, P. H., Malaya and Southeast Asia in the pattern of continental drift. Geol. Soc. Malaysia, Bulletin, 7, 89-138 (1974).
(19) Stocklin, J., Structural correlation of the Alpine ranges between Iran and Central Asia, Mem. h., ser. Soc. Geol. France, (8), 333-353 (1977).
(20) Sun, A. L., Permo-Triassic reptiles of Sinkiang. Sci. Sinica, 16, 152-156 (1973).
(21) Tuan Shuyin, Chen Yeh, Keng Kuochang, Some Early Cretaceous plant from Lhasa, Tibetan Autonomous Region, China. Acta Botanica Sinica, 19(2), 114-119 (1977) (in Chinese).
(22) Wen Si-xuan, New materials of biostratigraphy in the northern Qiangtang Plateau, Acta Stratigraphica Sinica, 3(2), 150-156 (1979).
(23) Wu Haoruo, Deng Wanming, Basic geological features of the Yarlung Zangbo Ophiolite Belt, Xizang, China, in Ophiolites, Proceedings of the International Ophiolite Symposium, Cyprus, 1979. Ed by A.

Panayiotou, 462-472 (1980).
(24) Wu Haoruo, Wang Dongan, Wang Lianchen, The Cretaceous of Laze-Jiangze district, southern Xizang, Science Geologica Sinica, 252-262 (1977) (in Chinese).
(25) Wu Hao-ruo, Wang Dong-an, Melanges of Jiangze district, southern Xizang. In Research on Sedimentology, Science Press, 120-127 (1980) (in Chinese).
(26) Zhu Xiangyuan, Liu Chun, Ye Sujuan, Lin Jinlu, Remanence of red beds from Linzhou, Xizang and the northward movement of the Indian plate. Science Geologica Sinica, 47-51 (1977) (in Chinese).
(27) Ziegler, A. M., Hansen, K. S., Johnson, M. E., Kelly,M. A., Scotese, C. R., Van der Voo, R., Silurian continental distribution, paleogeography, climatology, and biogeography. Tectonophysics, 40, 13-51 (1977).

Table 1 Stratigraphy of the northern subzone

<table>
<tr><th>Age</th><th colspan="2">Unit</th><th colspan="2">Lithology</th><th colspan="2">Chief fossils</th><th colspan="2">Thickness (m)</th></tr>
<tr><td>N</td><td colspan="2"></td><td colspan="2">Red conglomerate with sandstone and mudstone</td><td colspan="2"></td><td colspan="2">100</td></tr>
<tr><td>E</td><td colspan="2">Mai La Conglomerate</td><td colspan="2">Yellow and purple conglomerate with marl lenses</td><td colspan="2">Nummulites</td><td colspan="2">200</td></tr>
<tr><td rowspan="4">$K_2$</td><td rowspan="2">Laze Group</td><td>Jangseb Volcanics</td><td rowspan="2">Radiolarian with basic volcanics</td><td>Pillow and massive basic volcanics</td><td rowspan="2">Radiolaria</td><td>Radiolaria</td><td rowspan="2">700</td><td>500</td></tr>
<tr><td>Angren Formation</td><td>Turbidites</td><td>Mammites</td><td>1,000</td></tr>
<tr><td colspan="2">Sangzugang Formation</td><td colspan="2">Gray bioclastic limestone</td><td colspan="2">Orbitolina, Rudistes</td><td colspan="2">100-200</td></tr>
<tr><td colspan="2">Qiabulin Formation</td><td colspan="2">Purple and yellow conglomerate with sandstone</td><td colspan="2">Equisetites, Trochactaeon</td><td colspan="2">150-500</td></tr>
<tr><td>$J_1$</td><td colspan="2">Qiuwu Formation</td><td colspan="2">Purple conglomerate with coal-bearing sand-shale sequences<br>Gray limestone</td><td colspan="2">Eucalyptus, Ficus<br>Phylloceras</td><td colspan="2">200-1,000<br>50</td></tr>
<tr><td>$T_3$</td><td colspan="2">Gyirong Group</td><td colspan="2">Variegated siliceous shale and calcareous shale with radiolarian,sandstone and basic volcanics</td><td colspan="2">Monotis, Halobia</td><td colspan="2">$>$1,000</td></tr>
<tr><td>$T_{1-2}$</td><td colspan="2"></td><td colspan="2">Gray limestone and shale</td><td colspan="2">Anaflemingites, Owenites</td><td colspan="2">50</td></tr>
<tr><td>P</td><td colspan="2"></td><td colspan="2">Gray limestone and shale</td><td colspan="2">Neospirifer, Chonetella</td><td colspan="2">100-400</td></tr>
<tr><td>C</td><td colspan="2"></td><td colspan="2">Gray quartz sandstone and shale</td><td colspan="2">Eochoristites</td><td colspan="2">160</td></tr>
<tr><td>S-D</td><td colspan="2"></td><td colspan="2">Gray quartz sandstone and thin-bedded marl</td><td colspan="2">Stylion</td><td colspan="2">50</td></tr>
<tr><td>O</td><td colspan="2"></td><td colspan="2">Gray limestone and shale</td><td colspan="2">Orthoceratidae</td><td colspan="2">600</td></tr>
</table>

Table 2 Stratigraphy of the southern subzone

| Age | Unit | Lithology | Chief fossils | Thickness (m) |
|---|---|---|---|---|
| $K_2$ | Zongzhuo Formation | Olistostrome, siliceous flysch-like deposits, pelagic limestone | *Globotruncana* spp. *Heterohelix* spp. | 1,100 |
| $K_1$ | Jiabula Formation | Gray-black shale, calcareous shale, siliceous shale | *Calliptychoceras*, *Beudanticeras*, *Hibolithes* | 1,100 |
| $J_3$ | Weimei Formation | Arenaceous shale, and thick-bedded sandstone | *Himalayaites*, *Haplophylloceras*, *Inoceramus* | 700 |
| $J_2$ | Gyangze Formation | Thin-bedded limestone and shaly marl with tuff beds | *Delecticeras* | 200 |
| $J_{1-2}$ | Jiazhong Flysch | Cyclothems of fine sandstone, siltstone and shale, intercalated with dacite | | 500 |
| $J_1$ | Ritang Formation | Gray black calcareous mudstone and marl | *Psiloceras*, *Arietites*, *Prodactylioceras* | 350 |
| T | Chogar Group | Gray black siltstone, calcareous slate with quartz sandstone and marl | *Anasibirites*, *Placites*, *Claraia*, *Daonella* | 870 |
| P | Baidingpu Formation | Alternations of limestone and black slate | *Neospirifer*, *Spiriferella*, *Stenocisma*, *Tachylasma* | 250 |
| | Kangmar Formation | Black slate | | 100 |
| | Bilong Formation | Quartz sandstone | | 80 |
| $C_3$ | Kewoxika Formation | Pebbly slate with conglomerate | | 200 |

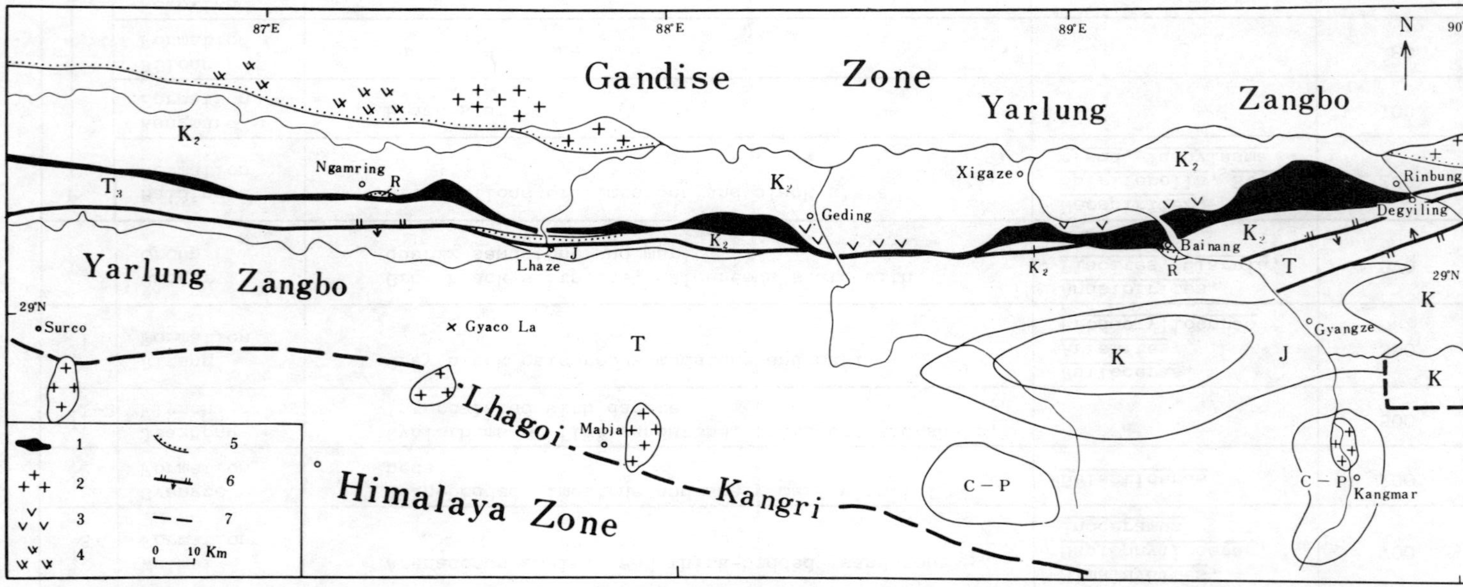


1. ultramafic rocks 2. granites 3. basic rocks 4. intermediate—acid volcanics 5. unconformity 6. thrust 7. fault

Fig. 1 The Yarlung Zangbo Suture Zone in the Xigaze region

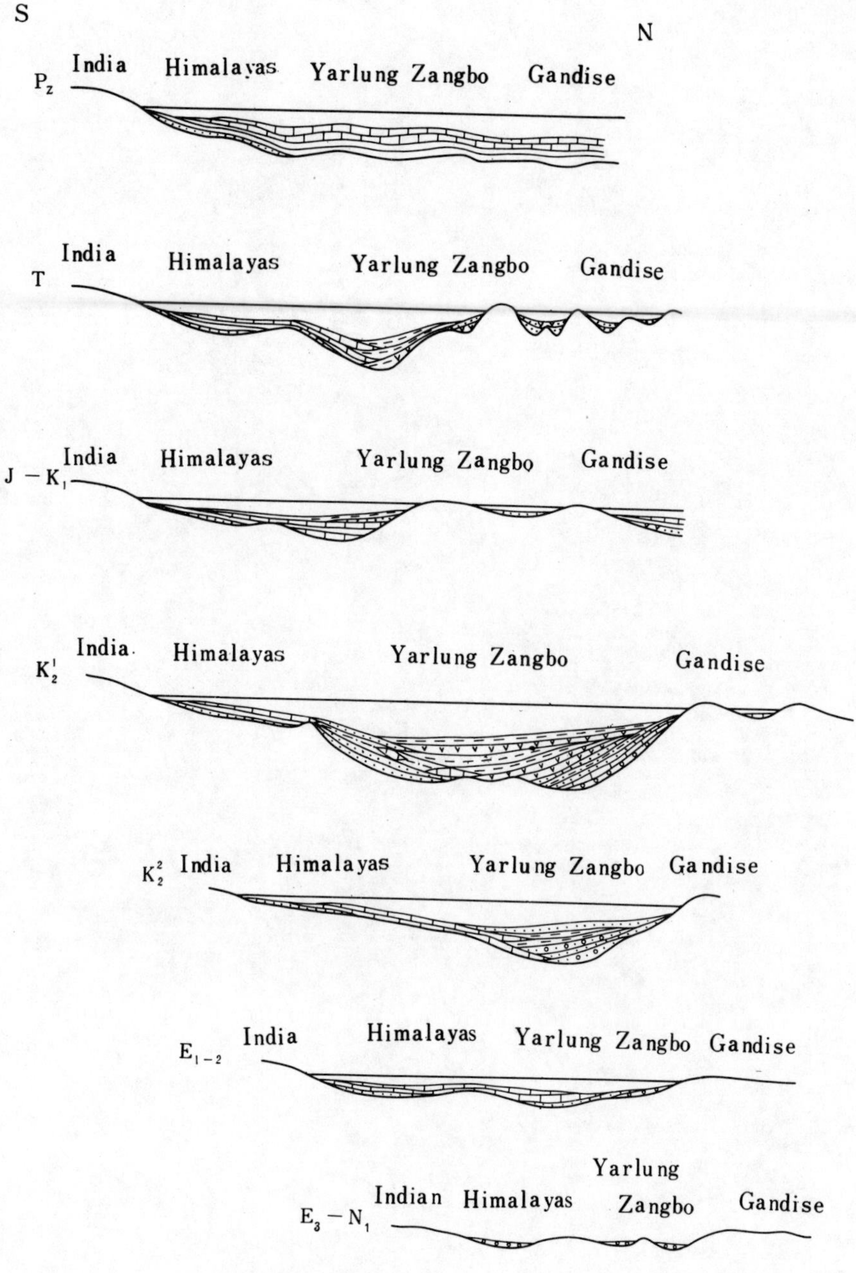


Fig. 2 Geological evolution scheme of the Yarlung Zangbo Suture Zone

# FIELD EVIDENCES FOR CONTINENTAL RIFTING IN TRIASSIC TIME IN THE LADAKH PART OF THE INDUS SUTURE ZONE

J.P. BASSOULLET, M. COLCHEN
(Laboratoire de Géologie stratigraphique et structurale, Université de Poitiers, France)
J. MARCOUX
(Laboratoire de Géologie historique, Université Paris XI, Orsay)
G. MASCLE
(Laboratoire de Géologie structurale, Université Paris VI)
GRECO, C.N.R.S. 1300 12, HIMALAYA-KARAKORUM

The Indus-Tsang-po suture zone lies North of the Himalayan ranges between the Eurasiatic and the Indian continental blocks. On a large scale geological map (fig. 1) it emphasized by a belt of discontinuous outcrops of ophiolites, approximatively lined up in North-West-South-East direction with the exception of a few massifs lying South (A. GANSSER, 1964). On a cross section the suture appears as an asymetric fan shaped structure thrusting both India and Eurasia respectively to the South and to the North.

## I. The structural sketch

The upper Indus valley separates two main ranges. To the North, the Ladakh ranges, mainly granitic and granodioritic, represent a large Cretaceous to Eocene batholith of andian type ; to the South extend the Zanskar ranges (fig. 2).

From South to North, along a cross-section from Ringdom Gompa to Lamayuru several structural units are distinguished (fig. 2 and 3) :

1. The Himalayan group (Tethys or Tibet Himalaya of Gansser, 1964). This group shows a crystalline basement (Tibetan slab) and a classic "himalayan type" mesozoïc sequence superstructure. The Ringdom Gompa amphibolites lying on the top of the basement are correlated with the Permo-Carboniferous Panjal traps ; a mainly carbonatic sequence represents the equivalent of Kioto limestones (Trias-Early Jurassic) ; upper jurassic black shales (Spiti) and detritic beds (early Cretaceous Giumal sandstones) are the main characteristic lithologies ; the upper part starts with upper-Cretaceous limestones (Chikkim), then pelites, shales, sandstones and slates (with *Omphalocyclus*) (METZELTIN S. and NICORA A. 1977 ; GAETANI M. and *al.*, 1980) and slates, limestones constitute the "so called" Kangila flysch.

Two main phases of deformation are documented, the first one is characterized by large Southward recumbent folds (with associated regional cleavage) ; in the southernmost areas this cleavage merge into the regional Tibetan slab foliation. The structures associated with the second episode mainly correspond to Northward overturned folds and associated fracture cleavage.

2. The Lamayuru nappe : this unit is characterized by flysch like series made of epimetamorphic alternances of pelites, limestones and shales (middle Triassic to middle Jurassic (J.P. BASSOULLET and *al*, 1978). The whole formations are overturned and thrusted to the South onto the Himalayan group.

3. The Nindam nappe, also contains a flysch like series made of pelites, greywackes and quartzites, including a few volcanic intrusions. In western

Lakakh this unit appears enriched in volcanites and is know as the Dras volcanics ; its age is attributed to the Cretaceous. The unit is mainly outcropping on the Southern side of the Indus valley but is also known in small klippen (Dumbur) and at the basis of the "ophiolitic nappe" South of Ladakh (J.P. BASSOULLET and *al*, 1980).

4. The Zanskar nappe, exhibits a mesozoic series including the followings : upper triassic sandy limestones and sands (with plants) ; uppermost triassic limestones (with Megalodonts) ; liassic limestones (with Algae and Lithiotis) ; upper Jurassic Ammonitiferous black shales (Spiti) ; lower Cretaceous Giumal sandstones ; upper Cretaceous varicoloured marly limestones (Fatula limestones of BASSOULLET and *al*, 1978). The epimetamorphic and strongly folded formations are thrusted over the Lamayuru nappe which crop out to the North and to the South. Four distinct phases of deformation are known. The first one relates to large Southward recumbent folds (and associated minor folds and cleavage) ; the second one is documented by Northward overturned folds and cleavage ; the third one results in a regional flat lying fracture cleavage ; the last one corresponds to a synformal deformation of the whole structure.

5. The ophiolitic nappes are outcropping in two different configurations ; South of Photaksar and Sirsir La, large ophiolitic bodies (Photaksar klippe of BASSOULLET and *al*, 1978 and 1980 or Spongtang klippe of FUCHS, 1977) exhibit various of the ophiolitic assemblage particularly harzburgites with a metamorphic tectonic fabric, vulcanic flows (pillow-lavas and agglomerates) and metamorphic rocks (metaquartzite ...).

The ophiolites are Southward thrusted onto a complex sole made of slices of Nindam flysch, serpentinites and triassic volcanites ; the sole itself tectonically rests on the Lamayuru nappe and the Himalayan Tertiary.

A few serpentinite bodies, associated with radiolarites, volcanites and permian limestones, appear also embedded within the tectonic contacts between the Zanskar or Nindam nappes and the Lamayuru nappe.

Conclusion

The observed structural pattern results of a polyphased tectonism. The two earliest phases are obviously post upper-Cretaceous ; the two last phases are post middle-late-Eocene. The earliest phases, documented in the Indus suture zone by upper Cretaceous reworking of granitic pebbles are only represented here by a gentle unconformity of the upper Cretaceous. Such an observation agrees well with the ideas that the "Zanskar ranges" represent a very intense shortening zone.

The polyphased nature of the tectonics tends to preclude a simple palinspatic restoration. Therefore, the initial paleogeographic disposition of the various structural units will not be discussed here.

## II. Field evidences for extensional tectonic processus in connection with intragondwanian continental rifting

*a) The exotic block of Lamayuru* (fig. 2 and 3)

For Geological setting and structural position see fig. 2 and 3. From bottom to the top, the following succession can be observed (fig. 4 and 5) :

1 - Pale green or pink crinoidal limestones, 20 to 30 m thick, in decimetric beds, folded in antiform. Thin sections show structures of Biosparudite or Biosparite with remains of Bryozoa, Crinoids, Brachiopoda and many Foraminifers and Algae (LYS, 1980), frequent in the mesogean Djulfian :

*Tubiphytes obscurus* Maslov, *Gymnocodium bellerophontis* (Roth), *G. nodosum* Ogilv., *Mizzia velebitana* (Schubert), *Deckerella composita* Reit., *Climacammina sphaerica* Pot., *C. sumatrana* (Voltz), *Hemigordius ovatus* Gr. *H. gr. reicheli* Lys, *Agathammina pusilla* (Gein.), *Globivalvulina vonderschmitti* Reichel, *Paraglobivalvulina mira* Reit., *Dagmarita chanakchiensis* Reit., *Frondina permica* S. de C. et Des., *Ichtyolaria permotaurica* S. de C. et DES., *Neoendothyra reicheli* (Reit.), *Reichelina cribroseptata* Erk, *Nankinella orbicularia* Lee, *Codonofusiella nana* Erk. Occuring with very characteristic forms of the mesogean Djulfian :

*Ungarella stellata* Korde, *Tuberitina collosa* Reit., *Valvulinella bykowski* (Schubert) = *Abadehella coniformis* Okim. et Ishii, *Protonodosaria globifrondina* S. de C. et Des., *P. praecursor* (R. Chern.), *Nodosaria sagitta* Mik. Maklay, *Stipulina* n. gen. Lys, *Colaniella parva* (Colani), *Parareichelina reticulata* Mik. Maklay.

The corraded irregular surface of the precedent limestones is coated by millimetric to centimetric polymetallic crusts.

2 - Limestones or dolomitized limestones, infilling many cavities and neptunians dykes and sills.

Some dykes could be followed on several meters into the permian limestone. Sometimes their thickness reach 1 dm thick. One of these cavities contains a micritic limestone with accumulation of Ammonite and also some fragments of volcanic tuff. The following species were determined (GUEX *in* BASSOULLET and *al.*, 1978). *Meekoceras lilangense* Krafft, *Clypeoceras* sp. aff. *crassus* Krafft, *Meekoceras* aff. *discus* Waagen *in* Krafft et Diener or *disciforme* Krafft. This fauna is similar to the Meekoceras beds one (Lower Scythian) of the Lilang section in the Spiti region (HAYDEN *in* GUPTA, 1975).

3 - Tuffs, tuffaceous sandstones, submarine lava flows (pillows) agglomerates are interbedded by dark radiolarites and red tuffaceous limestones.

*b) <u>The exotic blocks at the basis of the ophiolitic nappe</u>* (fig. 2, 3, 6, 7)

Within the tectonic sale of the ophiolitic nappe itself, we can observe :

1) Several small exotic blocks exhibiting primary contact between lavas (vesicular pillows) and red fossiliferous limestones (Hallstatt type) ; in some cases Conodonts indicate an upper Triassic age (close up fig. 6).

2) In the context in the upper Photang river area (fig. 2 and 3 geological map), below the main ophiolitic unit we have the following succession (fig. 7):

A - Lamayuru flysch more or less tectonized

$\emptyset_1$ – a major overthrusting contact

B - A tectonic assemblage of numerous formations :

$B_1$- highly disturbed tectonic breccian reworking elements from A and B.

$\emptyset$ - tectonic contact

$B_2$- black and grey massive limestone, algae, corals, brachiopods, etc., indicate a shallow water sedimentary environment;this limestone is highly recrystallised and now we have any biostratigraphic data. At the top of this limestone we can observe a polymetallic crust (Fe Mn ?).

$B_3$- red and brown fine argilites.

B4- green tuffs.

B5- Red, pink and white well bedded limestones ; this highly fossiliferous (Ammonites) exhibit close similarity with the classical Hallstatt facies. Conodonts assemblage suggest a Karnian age (according to Dr. HIRSCH geological survey of Israel).

B6- in primary contact above this upper Triassic limestone, a first flow of massive lava and agglomeratic tuff reworking small blocks af the same pink limestone.

B7- ≃ 100 m of lavas flows in same part exhibiting pillows structure.

Ø2 - major overthrusting contact.

C - Schistozed and highly tectonised serpentine.

## Conclusions

This observations suggest that the continental rifting processes initiated at the lower-most part of triassic is confirmed up to the end of the Triassic. This continental breakup is probably linked with the first stages of the neotethyan breakup and perhaps the oceanization itself.

## Références


ANDRIEUX J., BRUNEL M. and SHAH S.K., La suture de l'Indus au Ladakh (Inde), *C.R. Ac. Sc.*, D, 284, 2327-2330, Paris (1977).

BASSOULLET J.P., COLCHEN M., MARCOUX J. et MASCLE G., Une transversale de la zone de l'Indus de Khalsi à Photaksar, *C.R. Ac. Sc.*, D, 286, 563-566 (1978).

BASSOULLET J.P., COLCHEN M., GUEX J., LYS M., MARCOUX J. et MASCLE G., Permien terminal néritique, Scythien pélagique et volcanisme sous-marin, indices de processus tectono-sédimentaires distensifs à la limite Permien-Trias dans un bloc exotique de la suture de l'Indus (Himalaya du Ladakh), *C.R. Ac. Sc.*, D, 287, 675-678 (1978).

BASSOULLET J.P., BELLIER J.P., COLCHEN M., MARCOUX J. et MASCLE G., Découverte de Crétacé supérieur calcaire pélagique dans le Zanskar (Himalaya du Ladakh), *Bull. Soc. géol. Fr.*, 7e sér., 20, 961-964 (1978).

BASSOULLET J.P. et COLCHEN M., La limite Permien-Trias dans le domaine tibétain de l'Himalaya du Népal (Annapurnas-Ganesh Himal), *Coll. intern. CNRS*, n° 268, Ecologie et Géologie de l'Himalaya, 41-52, Paris (1979).

BASSOULLET J.P., COLCHEN M., MARCOUX J. and MASCLE G., The Indus suture zone : possible evidence for early triassic oceanization, *Publ. Comm. Intern. expl. Scient. Méditerranée*, vol. 25/26, fasc. 2a, 107-108, Monaco, (1979).

BASSOULLET J.P., COLCHEN M., JUTEAU Th., MARCOUX J. et MASCLE G., L'édifice de nappes du Zanskar (Ladakh, Himalaya), *C.R. Ac. Sc.*, D, 290, 389-392, Paris (1980).

BERNOUILLI D. and JENKYNS H.C., Alpine, Mediterranean, and central Atlantic Mesozoic facies in relation to the early evolution of the Tethys, *Soc. Econ. Pal. Miner.*, Sp. Publ., 19, 129-160 (1974).

CHANG CHENG-FA and PAN YUSHENG, The history of tectonic evolution of the Qinghai-Xizang (Tibet) plateau, *Proceeding of Symp. on Qinghai-Xizang (Tibet) Plateau*, abstracts, 58-59, Pékin (1980).

DIENER C., Notes on the geological structure of the chitichum Region, *Mem. geol. Surv. India*, 28 (1), 1-27 (1898).

FRANK W., GANSSER A. and TROMMSDORF V., Geological observations in the Ladakh Area (Himalayan). A preliminary report. *Schweiz. Petr. Mitt.*, 57, 89-113 (1977).

FUCHS G., Traverse of Zanskar from the Indus to the Valley of Kashmir - a preliminary note -, *Jahrb. Geol. B.A.*, 120, (2), 219-229, Wien, 1977.
FUCHS G., On the Geology of Western Ladakh, *Jahrb. Geol. B.A.*, Band 122, Heft 2, S. 513-540, Wien (1979)
GAETANI M., NICORA A. and PREMOLI SILVA I., Uppermost Cretaceous and Palcocene in the Zanskar Range (Ladakh-Himalaya) ; *Rev. Italiana Paleont. Strat.*, vol. 86, n. 1 (in press), Milano (1980).
GANSSER A., Geology of the Himalayas, *Interscience Publishers, John Wiley and Sons Ltd*, London, New-York, Sydney, I-289 (1964).
GUPTA V.J., Permo-Triassic boundary in the Himalaya, *Osterr. Akad. Wissensch., Schr. Reihe Erdwissensch. Kommiss.* (Stratigr. Alpin. Mediterr. Trias Symposium Wien 1973), vol. 2, 97-99 (1974).
HAYDEN H.H., *in* V.J. GUPTA, Indian Mesozoic Stratigraphy, *Hindustan Pub. Co.*, p. 53, Delhi (1975).
HEIM A. and GANSSER A., Central Himalaya, geological observations of the Swiss expedition 1936. *Mem. Soc. Helv. Sci. nat.*, 73, (1), 1-245, Zürich (1939).
KRYSZTYN L., SCHAFER G. and SCHLAGER W., *N. Jb. Geol. Pal. Abh.*, Stuttgart, 137, 284-304 (1971).
MASCLE G., *Riv. Ital. Paleont.*, Milan, 83, 389-408 (1974).
METZELTIN S. and NICORA A., A new Eocene outcrop from Zanskar Range, Himalaya. *Riv. It. Paleont. Strat.*, vol. 83, n. 4, p. 803-824, 4 pl., 6 fig., Milano (1977).
PECHER A., Déformations et métamorphisme associés à une zone de cisaillement. Exemple du grand chevauchement central himalayan (M.C.T.), transversale des Annapurnas et du Manaslu (Népal). *Thèse Sciences*, Grenoble, 354 p., (1978).
SHAH S.K., SHARMA M.L., GERGAN J.T. and TARA C.S., Stratigraphy and structure of the Western Part of the Indus suture belt, Ladakh, Northwest Himalaya, *Himalayan Geology*, 6, 534-556, Delhi (1976).
WENDT J., Stratigraphische Kondensation in triadischen und jurassischen Cephalopodenkalk der Tethys, *N. Jb. Geol. Pal. Mh.*, Stuttgart, n° 7, 433-438 (1970).

## Explanation of plate

Figure 1 : Geological sketch map of the Indus suture zone NP = Nanga Parbat, mct = main central thrust, mbt = main boundary thrust.

Figure 2 : The Indus suture zone in Ladakh : structural pattern of Zanskar. 1. Ladakh batholith, 2. Detritic formations of the Indus Valley (cf. Indus Molasse, Upper Cretaceous Eocene), 3. Nindam nappe, 4. Lamayuru nappe, 5. Zanskar-Shillakong nappe, 6. Ophiolitic sole, 7. Ophiolitic nappe, 8. Serpentinites, 9. Himalayan group, 10. Quaternary.
Kh = Khalsi, Nm = Nindam, Mb = Mulbeck, Lm = Lamayuru, Sk = Shillakong, Wa = Wakka Chu, Ya = Yapola Chu.

Figure 3 : Schematic cross-section of a Lamayuru transverse (cf A. fig. 2) ; Him = Himalaya, Oph = Ophiolite, Lm = Lamayuru, Sk = Shillakong, Nm = Nindam, In = Indus ; 1 = exotic block of Lamayuru, 2 = exotic blocks at the basis of the ophiolitic nappe.

Figure 4 : Cross section of the Lamayuru exotic block. A = Lamayuru flysch in reverse position, B = Lamayuru exotic block : 1) Neritic limestones, upper most Permian ; 2) Polymetallic crusts, neptunian dykes and pelagic lens of lower Triassic limestone ; 3) Pillow lavas and agglomerates ; 4) Tuffitic sandstone ; S) Tectonic sheet of Serpentine ; C = Nindam-Dras flysch unit : 1) Tuffs and radiolarites ; 2) Shales and Quartzites ; $\Phi_1$ and $\Phi_2$ tectonic contacts.

Figure 5 : Top of the Lamayuru exotic block ; 1) Late Djulfian neritic limestone (*Colaniella, Palaeofusulina* biozone -det. M. LYS), 2) Polymetallic crust, 3) Lower Scythian pelagic limestone (Fauna with *Meekoceras lilangense* Kraft, *Meekoceras* aff. *discus* Waagen, ... etc., similar to the fauna of the Lilang section (*Meekoceras* beds) in Spiti area - det. J. GUEX), 4) Tuffs and pillows lavas.

Figure 6 : Primary contact between ; 1 = Triassic limestone with fossils : 2 = (Ammonites, Crinoides, etc...) and 3 = basalt volcanic rocks.

Figure 7 : Cross section of the exotic block at the basis of the ophiolitic nappe (Legend in text).

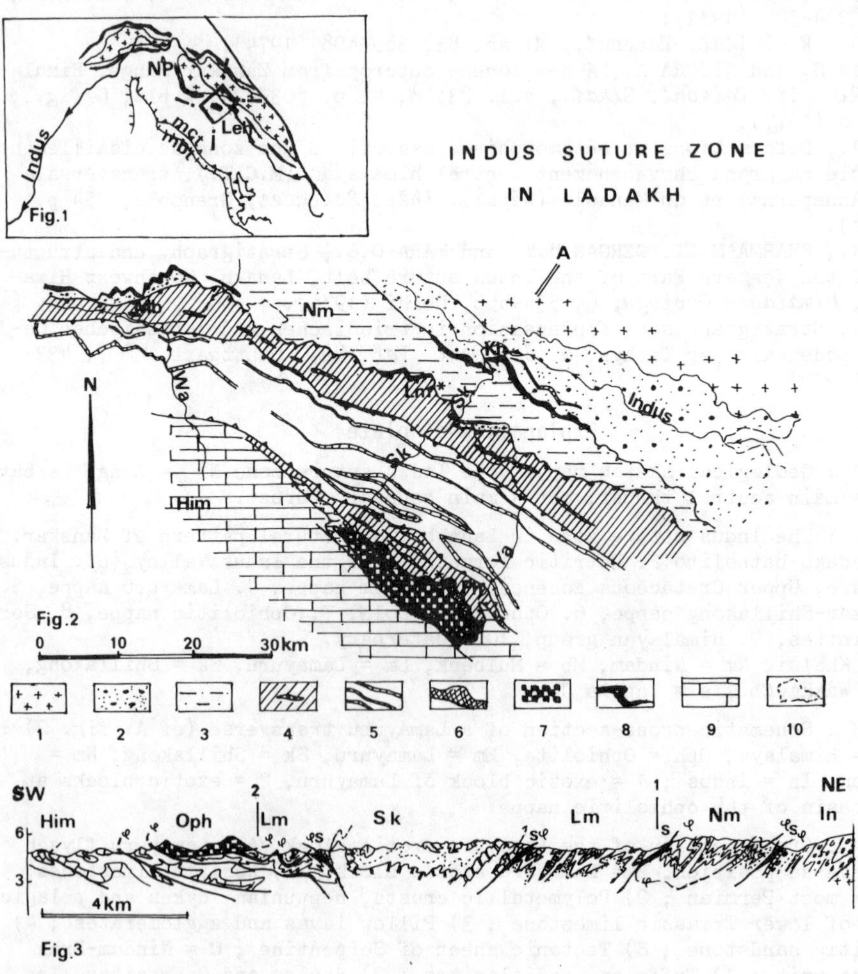

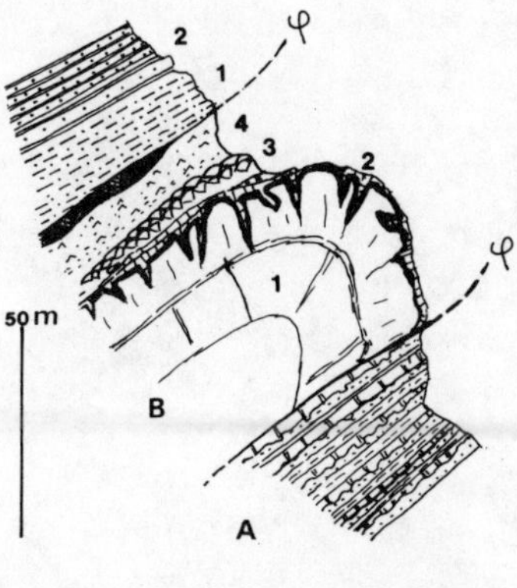

Fig. 4

Fig. 5

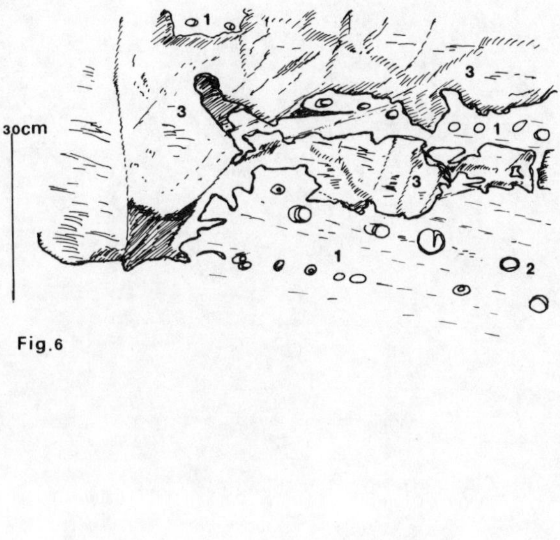

Fig. 6

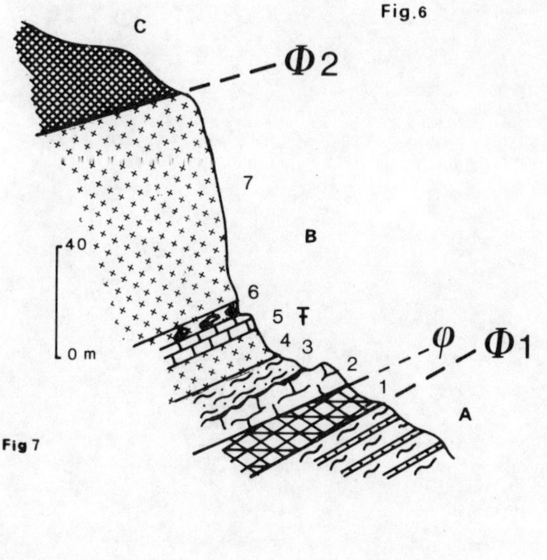

Fig 7

## TECTONIC COMPRESSION CHARACTERISTICS OF THE DAG CHUKA ULTRABASIC ROCKS

Lin Xue-nong
(Institute of Geochemistry, Academia Sinica)

Dag Chuka, a small town, is located near a ferry on the southern bank of the Yarlung Zangbo River. From Dag Chuka town southward to the mountain ridge, the ultrabasic rocks are exposed in the Dalang Pu valley, constituting the eastern part of the Rinbung-Bainang ultramafic massif, which is about 12 km wide and at an altitude of 4,000-5,500 m above sea level. The ultrabasic massif consists of harzburgite intercalated with dunite, which accounts for about 10 %. The dunite facies occurs as lenses with distinct boundaries and shows a consistent mode of occurrence with the massif, namely, they strike nearly eastwesterly and dip to south at an angle of about 45° - 85°. The northern part of the massif contacts with Cretaceous Xigaze group by a fault, while its southern boundary has not yet been traced (Fig.1).

Most of the ultramafic rocks appear rather fresh. The various rocks are mainly composed of forsterite, enstatite, diopside and chrome spinel with forsterite as the dominant mineral, which accounts for almost 90-99 volume % ( with Fo 93-97) of dunite, and 80-90 volume % (with Fo 93-94) of harzburgite. Enstatite is next in its content, coming up to 10-20 volume % (with En 92-95) of harzburgite. Diopside is a minor mineral comprising about 4 volume % of harzburgite. Chrome spinel is disseminated in various kinds of ultramafic rocks, comprising only 1-2 volume % in general, or up to 5 volume % in special cases (Table 1). The grain size and texture are quite different for these minerals, suggesting a complex mechanism of their formation.

### EVIDENCES OF TECTONIC COMPRESSION OBSERVED IN THE ROCKS

1. Undulatory extinction. The undulatory extinction is mainly observed in coarse forsterite crystals in dunite and often shows certain extinction directions. Some forsterite crystals show two or more different extinction azimuths. In harzburgite, olivine crystals also show undulatory extinction very often, while the extinction azimuths vary to a great extent (Photo 1).

2. Kink band. Kink band is also found in coarse olivine crystals and is the commonest evidence of tectonic compression. Kink bands look like multiple twin, but do not conform to any twinning law. Kink bands are often associated with undulatory extinction and often progressively transform into the latter (Photo 2). The width of kink band may be as wide as wide as about 0.1-0.5 mm (Photo 3), and may also be as narrow as about 0.001-0.005 mm. Kink bands may be developed throughout the whole grain as well as only within a local region. The difference in distribution between kink bands in dunite and in harzburgite is similar to the difference in undulatory extinction for these two rock types.

3. <u>Flattening of crystals and bending of the cleavage plane</u>. The coarse forsterite grains in dunite have been flattened with the length to width ratio often exceeding 4 to 10, whereas for fine-grained recrystallized forsterite crystals, this ratio is often less than 2 (Photo 4). In harzburgite, coarse enstatite crystals often occur as laths, and the cleavages in them are often bended or even broken together with the diopside lamellae (Photo 5). At the same time, undulatory extinction and kink bands are commonly observed in adjacent forsterite grains.

The main characteristics of deformation caused by tectonic compression can be summarized as follows:

(1) The compressed deformation is exclusively found in coarse forsterite and / or enstatite crystals (commonly greater than 1 mm), and the deformation is generally much more complex than that observed in peridotite xenoliths in basalts, indicating that these coarse forsterite and enstatite crystals were subjected to multiple and diverse compressions.

(2) In dunite, the deformation is commonly observed throughout the whole rock showing certain fixed azimuths. On contrast, in harzburgite the deformation is often observed in some separated and local regions, and no fixed azimuths or several different azimuths are discovered. It may reflect the difference in formation process between dunite and harzburgite. We suggest that harzburgite might probably experience a stage of fluid magmatic crystal mush. All these features indicate that the rock massif was affected repeatedly by complex compression.

## THE MAGMATIC STRUCTURAL FEATURES RETAINED IN THE ROCKS

1. <u>The rocks commonly possess a hypidiomorphic-inequigranular or pseudoporphyritic texture</u>. The coarse forsterite and/or enstatite crystals probably crystallized out at the earlier stages (Photo 6).

2. <u>Poikilitic texture</u>. Rounded olivine crystals are often found to be enclosed in chrome spinel and show different optical orientations. Sometimes olivine crystals enclosed in enstatite are also observed.

3. <u>Lamella texture</u>. Diopside lamellae run parallel to the cleavage plane of enstatite (Photo 7). They were products of exsolution during the crystallization of magma.

4. <u>Xenomorphic-granular texture</u>. Fine-grained enstatite, forsterite and chrome spinel crystallized with xenomorphic-granular texture in interstices between coarse grains in harzburgite (Photo 8), suggesting that they are the eutectic products.

## TECTONIC COMPRESSION AFTER SOLIDIFICATION, AND LOCAL REMELTING, METASOMATISM AND RECRYSTALLIZATION

1. In harzburgite, radiating cracks are common in forsterite crystals around the automorphic-granular chrome spinel, and sometimes even cut across adjacent grains (Photo 2), indicating that these cracks are caused by compression following solidification.

2. The veinlets of fine-grained olivine worm their way into the cracks

in enstatite crystals with distinct boundaries (Photo 5), suggesting they are resulted from the filling of the cracks during local remelting.

3. Sieve texture. Rounded olivine and chrome spinel crystals are enclosed in enstatite grains, and olivine crystals show common optical azimuth (Photo 9) indicating that olivine was replaced by enstatite.

4. Chrome spinel crystals are enclosed in olivine grains with fixed orientation. Automorphic granular chrome spinel in coarse olivine grains might be the product of local recrystallization (Photo 10).

5. Part of dunite showing automorphic granular texture might probably also be resulted from the recrystallization.

Based on the above information, we have got some preliminary views as follows:

1. The ultramafic massives have undergone intensive and repeated tectonic compression. The tectonic compression before intrusion made the coarse forsterite and enstatite crystals deformed. The tectonic compression during intrusion led the ultramafic massives to emplace into the place close to where they lie today. And under the action of compression soon after solidification of most part of the massives, relative sliding took place leading to the cold emplacement of the massives into the place where they are found, so they are generally fenced by faults and no contact aureoles could be observed. Moreover, the compression after solidification could complicate the deformation further, leading to local remelting, metasomatism, recrystallization and fissuring.

2. The compression stress is closely related to the collision between Indian and Eurasian plates. The massives are located just in the collision zone between the two plates. The immediate country rocks belong to the Xigaze group of Cretaceous. Therefore,the intrusion might have probably taken place in Early Himalayan period. And it is interesting to note that the most intensive tectonic compression took place about 70-40 m. y. ago, according to the isotopic ages of the granites in adjacent area. Therefore, the intrusion and compression coincide with each other in time.

3. The ultramafic massives probably experienced a stage of magmatic "crystal mush" during their intrusion. And there once existed in such mush a lot of large dunite fragments which have become the dunite lenses observed in present massives. The smaller dunite fragments and individual crystals together with the fluidal part all have made contributions to the formation of harzburgite and other types of ultramafic rocks in the massives.

Table 1 Mineral compositions for various ultramafic rocks (in volume %)

| Rocks \ Minerals | Forsterite | Enstatite | Diopside | Chrome spinel |
|---|---|---|---|---|
| Dunite | 94-100 (98.4)* <br> Fo 93-97 | 0-3 (0.4) | 0-0.8(0.1) | 0.2-1 ; 5.6** <br> (1.1) |
| Harzburgite | 79-90(82.4) <br> Fo 93-94 | 8-18.6(14.7) <br> En 92-95 | 0-3.9(1.7) | 0.7-1.8(1.2) |

* Average in parenthesis.

** In some local and limited regions.

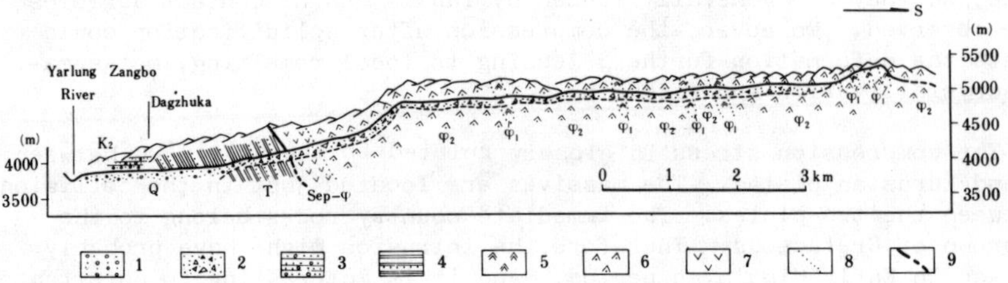


Fig. 1 Sketch profile of Dag Chuka

1. Quaternary aluvium;
2. Quaternary moraine-breccia;
3. Late-Cretaceous sandy conglomerate;
4. Triassic phyllite and sandstone;
5. Dunite;
6. Harzburgite;
7. Serpentinized ultramafic rock;
8. Facies boundary;
9. Fault

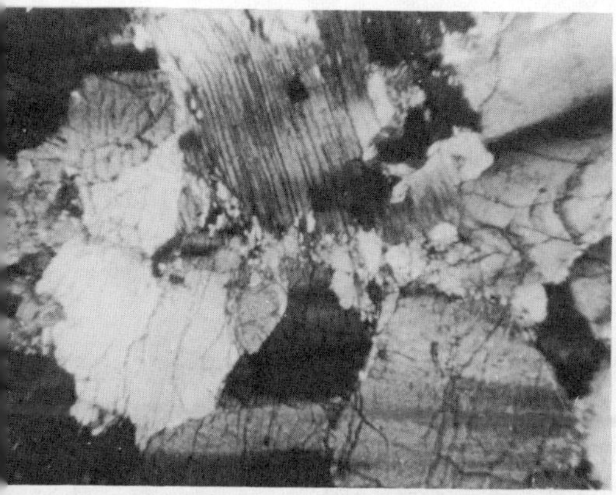

Photo 1 Harzburgite
D516 X 55

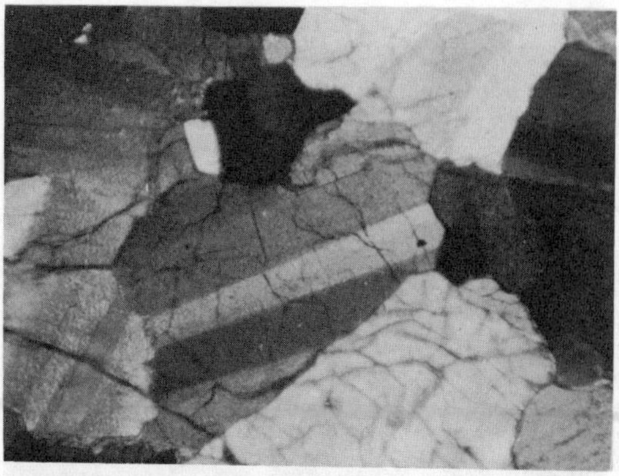

Photo 2 Harzburgite
D510 X 55

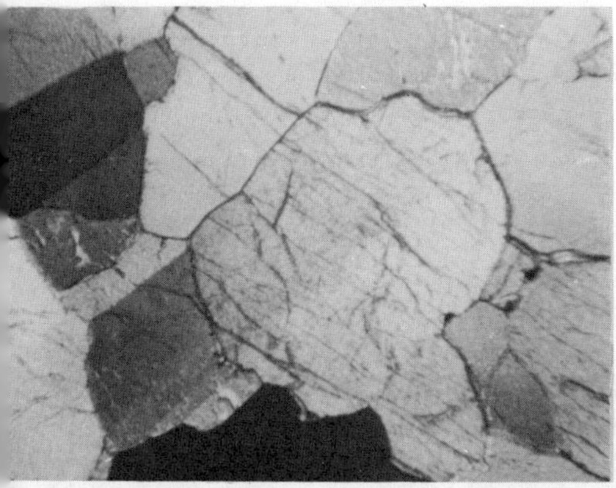

Photo 3 Dunite
D543 X 55

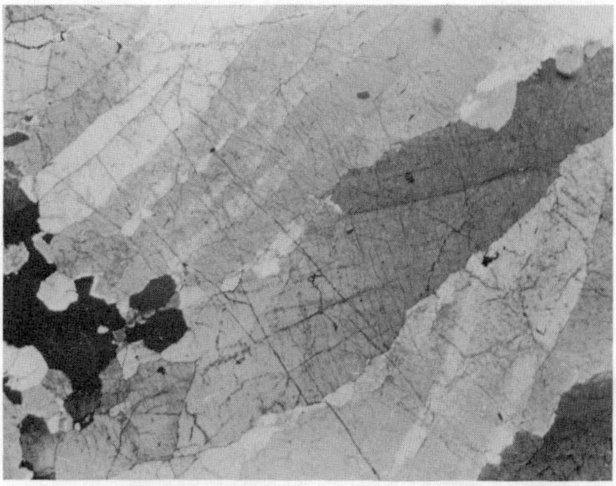

Photo 4 Dunite
D502 X 11

Photo 5 Harzburgite
D518 X 11

Photo 6 Harżburgite
D516 X 11

Photo 7 Harzburgite
D518 X 220

Photo. 8 Harzburgite
D518 X 55

Photo 9 Harzburgite
D518 X 55

Photo 10 Dunite
D542 X 55

# SOME CONTRIBUTIONS ON THE OPHIOLITES OF LADAKH AND NAGALAND

V. J. Gupta
(Centre of Advanced Study in Geology,
Panjab University, Chandigarh-160014, India)

The ophiolites of Ladakh and Nagaland represent most important tectonic feature of the Himalaya and have important bearing on the plate tectonics. The ophiolites of both these regions are associated with the glaucophane schists of blue schist facies and are indicative of their having been tectonically derived due to subduction related process.

## LADAKH

The Dras Volcanics (including strips of the intrusive basic and ultrabasic rocks of the ophiolite suite) and associated rocks exposed in a long linear belt extending across Ladakh in WNW-ESE direction upto Burzil Pass, Nanga Parbat and beyond in northwest have been given varying names by different workers, i.e. Dras Volcanics(15, 16, 19, 20, 22), Indus Ophiolites(5, 8); Dras Complex Group(14); Sumdo Formation(17); Dras Volcano-sedimentary Series(2), etc. The rocks constituting these consist of 3,000 metres thick succession of varying composition from ultrabasics to acidic nature and are represented by volcanic, pyroclastic, hypabyssal and plutonic rocks. The rocks belonging to the ophiolite suite within this succession are essentially represented by agglomerates with serpentinite dunite occurring mostly as large bodies in a melange form mixed with other sedimentaries (radiolarites, Orbitolina bearing limestones, agglomerates, ash beds, jasperites, etc.). At places exotic blocks containing pelagic limestones of varying ages (Late Palaeozoic to Mesozoic) are encountered within these ophiolites suggesting the existence of an ocean floor during the Late Palaeozoic.

The stratigraphy and tectonic implications of the ophiolite zone of Ladakh have been discussed in detail by several workers during the last few years(4, 6, 7, 8, 9, 11, 12, 13, 21). In view of the limited space available for the paper in the present volume, it would not be possible to discuss details on the ophiolites of Ladakh and it is suggested that for detailed geological maps of the ophiolite belt of Ladakh and other geological aspects reference may be made to papers referred to above.

## NAGALAND

The ophiolite belt of Nagaland is very little known in comparison to the ophiolites of Ladakh which have been investigated in detail during the last few years. The ophiolite belt of Nagaland constitutes part of the Zepuhu Formation which covers a linear tract of approximately 90 km x 12 km and includes a typical ophiolite-ultramafic-mafic suite of rocks in association with quartzite and limestone (partly radiolitic) succession. The ophiolite belt lies over the Cainozoic rocks ( Disang Group) with a thrusted contact.

The geological succession exposed in the ophiolite belt has been broadly classified by Agrawal(1) and Singh(18) into the following lithostratigraphic units:

4. Jopi Formation - Molasse sequence comprising conglomerate-greywacke-arkose lithofacies containing plant fossils. Exposed along the crestal region of Mollen-Jopi-Kamku-Pokphur ridge

_____________ Unconformity _____________

3. Disang Group - Thick sequence of flysch deposits comprising shale, sandy shale, slates with intercalations of greywacke, subgreywackes, argillaceous limestones, etc. containing gastropods, lamellibranchs and plant fossils in the upper units exposed west of Zepuhu Formation around Meluri, Kiphire, Tuensang, Noklak, etc.

_____________ Thrust _____________

2. Zepuhu Formation - Chert, quartzite, greywacke, limestone ophiolite sequence representing a typical melange deposit and pillow lavas. The ophiolite constitute basically mafic and ultramafic sequence of rock suites consisting of ultramafic harzburgite, iherzolite, dunite, pyroxenite, serpentinite, hornblendite, basalt, spilite, tuffs, pyroclastics, glaucophane schists, chlorite-sericite-prehnite-antigonite schists, etc. Ultramafics range in thickness from a few metres to a few kms (upto 10 kms). The following stratigraphic sequence can be broadly worked out Within the various rock units constituting the Zepuhu Formation(18).
   (f) Pelagic sediments;
   (e) Quartz plagioclase rocks;
   (d) Spilites;
   (c) Gabbros;
   (b) Dunite, pyroxenite (websterite and clinopyroxenite) and sericite-chlorite-tremolite, schists;
   (a) Harzburgite and serpentinites

_____________ Thrust _____________

1. Nimi Formation - Preflysch marine sediments comprising quartzites, phyllites, carbon phyllites, schists and limestone with granite and minor serpentinite bodies. Exposed east of the ophiolite complex (sensu stricto) around Phokhungri, Nimi and Saramati for a strike length of about 45 km with a maximum outcrop width of 7 km just along the Indo-Burma border

The Zepuhu Formation lies west of the Nimi Formation with a faulted/thrust contact. It consists of chert-quartzite-greywacke-limestone-ophiolite succession representing a typical 'melange' deposit. The ophiolites exposed are basically a mafic-ultramafic sequence of rock suites and consist of harzburgite, iherzolite, dunite, pyroxenite, gabbro, serpentinite, hornblendite, basalt spillite, tuff, pyroclastics, glaucophane schist, sericite prehnite-antigorite-schist, etc., with associated iron, chromium, nickel and cobalt mineralizations. There are various gradations and variations in each rock type. The ultramafics range in width from a few metres to a few kms (upto approx. 10 km) and are arranged in an en-echelon pattern in conformity with the regional trend of the enveloping pre-flysch and flysch sediments and have some preferential associations and distributions throughout the belt. The Zepuhu Formation can be traced in south from Mollen-Molite to NE of Ania Schu in the north where it enters into Burma.Agrawal(1) recorded the discovery of tectonically significant linear patches of glaucophane schist (east of Pungro, around Moke and Washelo), actinolite schist and eclogites within the ophiolite suite. According to him the Zepuhu Formation probably represents a melange indicating its formations along a benioff-zone. The rocks of this formation have been subjected to a low grade regional metamorphism of green-schist facies and high pressure-low temperature blue schist metamorphism. Kumar et al.(10) on the basis of their work on the peridotites, pyroxenites, serpentinites and gabbros of the northeastern part of the ophiolite belt show the presence of mineralization of chromium, nickel, cobalt and iron oxides. They have recorded the presence of sulphides associated with the spilitic and basaltic rocks. A sequence ranging from peridotite to spilite at the top has been maked by them through various mafic and ultramafic rocks. According to them the melange zone is well marked and consists of blue schist rocks associated with peridotite, harzburgite, dunite, pyroxenite serpentinite, hornblendite, etc.

The ophiolite belt of Nagaland is characterised by the presence of high angle reverse faults (trending between N-S to NE-SW) which have dissected the various lithological units into parallel to subparallel longitudinal segments. Among the major fractures associated with this belt mentioned may be made of Phokhungri fault, Zepuhu fault and Tizuzunki fault which run concordantly to the regional trend of the formations. The first two faults approximately mark the western limit of the Nimi and Zepuhu Formations respectively. In addition to these there exist a large number of transverse faults which are generally reverse-cum-wrench faults(1).

The refractory minerals and rocks encountered in association with ophiolite belt of Nagaland include asbestos, chromemagnetite, chromite, limestone, olivine, quartzite, minor magnetite and talc.

## CONCLUSIONS

The ophiolites of Indus Suture Zone and Nagaland are associated with the glaucophane schists of blue schist facies and these indicate their having been tectonically derived due to subduction related processes. Some workers in the past have expressed opinion in favour of their origin due to obduction but the association of blue schists with the ophiolitic melange in the Indus region and Nagaland rules out the possibility of their having been derived due to obduction. The tectonic setting in the area south of the Indus Suture Zone is indicative of marginal tectonic emplacements. The presence of tonalites in the Kargil area shows that the tonalite magma must have been derived by the partial melting of the subducted oceanic floor in this region. Similarly the presence of potash rich suite in the Nagaland is indicative of the deep seated source supporting its origin due to subduction processes.

## References

(1) Agrawal, O. P., Prospects of refractory minerals in the ophiolite belt of Nagaland. Proc. All India Seminar on Raw Materials for Glass Ceramic and Refractory Industries, Shillong (1977).

(2) Andrieux, M., Brunel, M., Shah, S. K., La Suture De L'Indus Au Ladakh (India). Comptes rendus des C. R. Acad. Sci. Paris, 284. Ser. D, 2327-2330 (1977).

(3) Casnedi, R., The ophiolites of the Indus Suture Line (Dras and Astor Zones - Kashmir). Estratto Da Ofiolite. Bull. del Group di Lavoro Sulli ofiolite Medit. Ann., 365-371 (1976).

(4) Colchen, M., Le Fort, P., Some remarks and questions concerning the

geology of the Himalaya. Colloq. Intern. Ecol. & Geol., Himalaya, C.N.R.S., 268, 131-137 (1976).

(5) Dasarthi, N., Tectonic setting of Indus Ophiolite. Proc. Upper Mantle Project, Hyderabad, 167-174 (1967).

(6) Frank, W., Gansser, A., Trommsdorff, V., Geological observations in the Ladakh area (Himalaya) - A preliminary report. Schweiz Mineral Petrog., 57, 89-113 (1977).

(7) Gansser, A., Geology of the Himalayas. Wiley-Interscience London, 289 p. (1964).

(8) Gansser, A., The great suture zone between Himalaya and Tibet - A preliminary account. Intern. Colloquium on the Geology and Ecology of the Himalaya, Paris, 268, 181-192 (1976).

(9) Gupta, V. J., Kumar, S., Geology of Ladakh, Lahaul and Spiti regions of Himalaya with special reference to the stratigraphic position of flysch deposits. Geol. Rund., 64 (2), 540-563 (1975).

(10) Kumar, R., Singh, R. N., Prabhakar, A., Environmental aspect of Fe, Cr, Co, Ni, and associated mineralisation in and around Pukhpur in the north eastern part of ophiolite belt, Nagaland. 9th Seminar on Himalayan Geology, Dehra Dun, abst. No. 76, 72-73 (1978).

(11) Powell, C. McA., Conaghan, P. J., Plate tectonics and the Himalayas. Earth and Planetary Science Letters, 20, 1-12 (1973).

(12) Rai, H., Gupta, K. R., Geology of the area between Pashkyum and Badhkharbu Ladakh, 9th Seminar on Himalayan Geology, Dehra Dun. abst. No. 104, 99-100 (1978).

(13) Rai, Hakim, Pande, I. C., Geology of the Kargil Igneous Complex, Ladakh, Jammu and Kashmir State, India. Recent Researches in Geology, 5, 190-210 (1977).

(14) Raiverman, V., Mishra, V. N., Suru tectonic axis, Kargil area, Ladakh. Geol. Min. Met. Soc. Ind. Golden Jubilee Symposium (1974).

(15) Shah, S. K., Indus ophiolite belt and the tectonic setting of the Malla Johar-Kiogad. Exotics in Himalaya. Colloq. Intern. Ecology and Geol. Him., C.N.R.S., 268, 361-368 (1976).

(16) Shah, S. K., Sharma, M. L., Gergan, J. T., Tara, C. S., Stratigraphy and structure of the western part of the Indus Suture Belt, Ladakh, Northwest Himalaya. Him. Geol., 6, 534-556 (1976).

(17) Shankar, R., Padhi, R. N., Prakash, G., Thusu, J. L., Wangdus, C., Recent geological studies in upper Indus Valley and the concept of plate tectonics. Abst. Intern. Sym. Tectonics Metallogeny of South East Asia and Far East (1974).

(18) Singh, R. N., Geology and petrography of the Naga Hills ophiolite, N. E. India. Abst. 26th Intern. Geol. Congr., 1, 91, Paris, (1980).

(19) Srikantia, S. V., Bhargava, O. N., The Indus tectonic belt of Ladakh Himalaya: Its Geology, Significance and Evolution. Tectonic Geology of the Himalaya: Saklani, P. S. (Ed.), 43-62 (1978).

(20) de Terra, H., Geological studies in Northwest Himalaya between the Kashmir and the Indus Valleys: Mem. Connecticut Acad. of Arts and Science, New Haven, 8 (2), 18-76 (1935).

(21) Virdi, N. S., Thakur, V. C., Kumar, S., Blue schist facies metamorphism in the Indus Suture Zone of Ladakh and its significance. Him. Geol., 7, 479-482 (1977).

(22) Wadia, D. N., The Cretaceous and volcanic series of Astor, Deosai, Kashmir and its intrusions. Rec. Geol. Surv. Ind., 72 (2), 151-161 (1937).

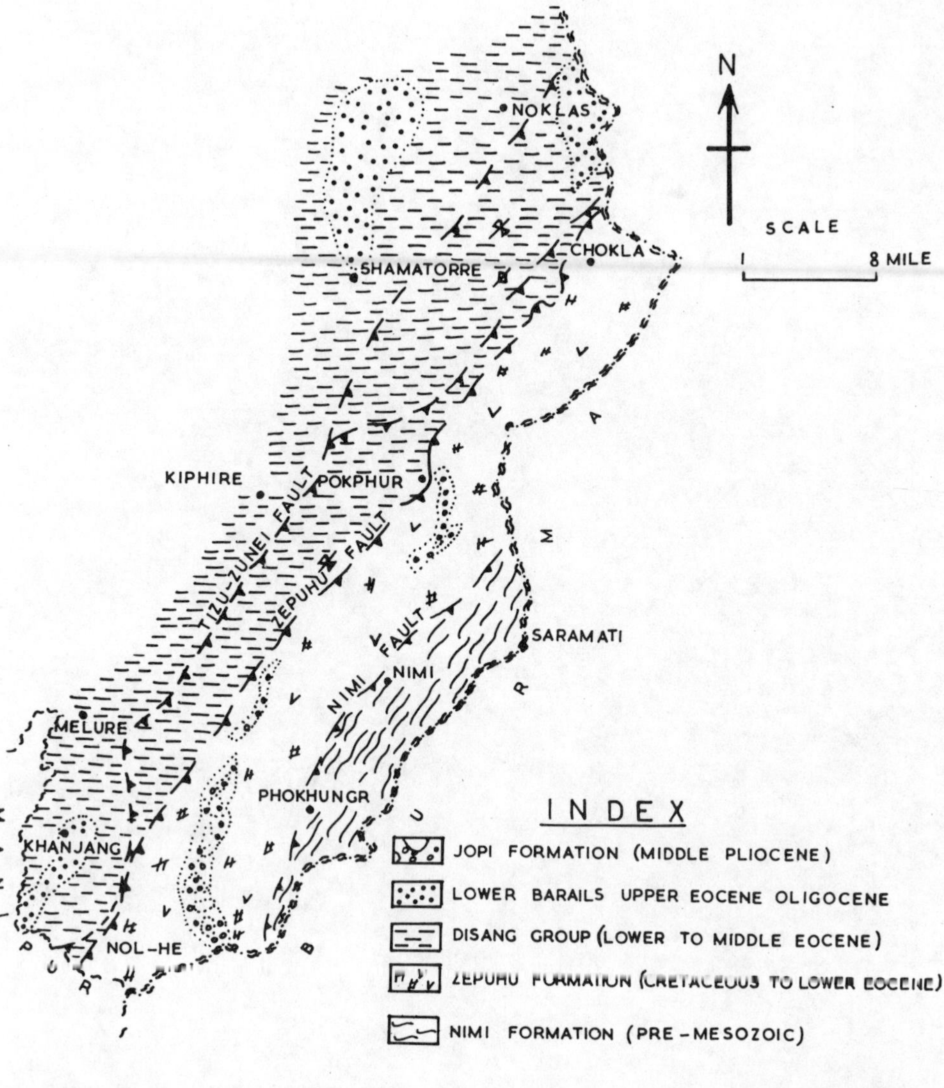


Fig. 1 Geological map of ophiolite belt of Nagaland

# CHARACTER OF THE SEDIMENTARY FACIES-BELTS AND THE SEDIMENTARY MODEL OF THE SUBSIDENCE BELT IN THE YARLUNG ZANGBO-XIANGQUAN RIVER

Wang Lian-cheng, Wang Dong-an
(Institute of Geology, Acadimia Sinica)

The subsidence belt of the Yarlung Zangbo-Xiangquan River is situated between the Gandise granite belt toward north, and the eastern Himalayan crystalline belt and the central and western parts of the southern Tethys Himalayan belt toward the south (Fig. 1). The subsidence belt extends roughly east-westwards about 1,600 km long and over 100 km wide.

The subsidence belt is characterized by its complex geological structure, its huge thickness of lithologic associations, and its well developed ophiolite. Therefore, it is of great interest among geologists over the world. This paper is based on the study on the characteristics of the main sedimentary facies-belts and with a tentative discusion on its geological history and evolution.

## TYPE AND CHARACTERISTICS OF SEDIMENTARY FACIES-BELTS

In this region the Mesozoic and Cenozoic strata are distributed widely and the Paleozoic strata are rare. Based on the sedimentary features, the sedimentary series of this subsidence belt can be subdivided into six sedimentary facies-belts. They are: flysch-like facies-belt, bathyal or abyssal sandstone and shale facies-belt, radiolarian siliceous rock facies-belt, olistostrome facies-belt, flysch facies-belt and molasse facies-belt. All of them are trending in an east-west direction and distributed regularly (Figs. 2,3).

### FLYSCH-LIKE FACIES-BELT

It is widely distributed in the eastern part of this subsidence belt, but pinching out gradually in the west. It consists mainly of sandy slate, muddy sandstone and slate. They belong mainly to terrigene sediments, volcanic materials are rarely found, lithology is rather simple and fossils are rare. It is over 5,000 m in thickness (Fig. 4).

Both sandy slates and muddy sandstones contain illite and clastic quartz and feldspar. The clays and organic materials are distributed as bands, the clastics are not well sorted and medium in roundness, but often they are elongated and deformed due to later metamorphic effect. Diagenetic and epigenetic coarse pyrite crystals that cut through beddings (Photo 1) are seen. The shales and sandstones are well bedded (Photo 2 ) and sole markings are found sometimes.

## BATHYAL OR ABYSSAL SANDSTONE AND SHALE FACIES-BELT

This facies-belt consists mainly of massive grayish black silty shales, siliceous shales and quartz sandstone with chert beds as well as lenticular intercalations of tuffaceous sandstones and limestones. In general, the nature of the composition of the rocks are rather complex. Volcanic materials are plenty, fossil ammonites, belemnoids are abundant. The thickness is over 2,000 m (Fig. 5).

The sorting and roundness of silicarenites in the clastic materials are of medium grade, the cements are made of siliceous and some ferruginous materials (Photo 3 ).

## RADIOLARIAN SILICEOUS ROCK FACIES-BELT

This facies-belt consists mainly of siliceous rocks and siliceous shales of various colours with intercalations of basic rocks (Photo 4, 5). Under scanning electron microscope the siliceous rocks are made of aggregates of minute quartz crystals. Their thickness is several hundred meters.

## OLISTOSTROME FACIES-BELT

It is situated on the south of the ultrabasic rocks and siliceous rock facies-belt. The matrix of these olistostromes consists of muddy graywackes and a lot of big or small blocks of various composition.

On the southern side of Zazala (Photo 6), many huge blocks bulge out of the surface. They are variegated siliceous rocks, basic rocks, sandstones and limestones. We believe that most of the blocks were derived from the radiolarian siliceous rock facies-belt and its adjacent layers in the north or some rocks were brought from older strata. These large blocks are sharply distinguished from their surrounding matrix (Photo 7), although some of the small blocks are not quite well shown up.

We consider that they were derived from a sedimentary origin, because they are also exposed in some stratigraphic horizons, situated in the south of ultrabasic rock zone, but they do not contain any constituents of ultrabasic rocks.

## FLYSCH FACIES-BELT

This facies-belt is distributed in the central part of the subsidence belt. It consists of thin interbeded graywackes (Photo 11) and shales with some lenses of sandy conglomerates and limestones. Occasionally, the beddings of the rock are partly twisted (Photo 8). Horizontal (Photo 9) and graded bedding are present. Some sole markings are recognized in sandstones(Photo 10). The whole thickness is over 1,500 m.

The composition of the flysch is mainly granitic clastics and intermediate or acidic volcanic rocks. We consider that they have been brought about from the southern margin of the Eurasian continent.

## MOLASSE FACIES-BELT

This facies-belt lying discordanty between the Gandise granite, ultrabasic rocks and siliceous rock belt, is mainly consisted of variegated conglomerates,sandstones and some shales. The whole thickness is more than 3,000 m. Shales of the upper part bear fossils of Oligocene and Miocene age . The main peak of Gandise Range, Mt. Kangrinboqe is composed of nearly flat lying conglomerates and sandstones of over 2,000 m in thickness (Photo 12 ). Photos 13, 14 are photographs of the variegated conglomerate.

## THE EVOLUTIONARY HISTORY OF THE SEDIMENTARY MODEL OF THE YARLUNG ZANGBO-XIANGQUAN RIVER SUBSIDENCE REGION

Since the Early Mesozoic, Tethys Himalayas began to split into a "south zone" and a "north zone"(1). Within the south zone limestones and arenaceous rocks of Mesozoic - Eocene were deposited upon Paleozoic basement. Whereas in the north zone, besides the thick Mesozoic - Eocene sediments, there are also many intrusive and eruptive rocks. The tectonic movements in the north zone were also more intensive and complex than those in the south zone. These features are similar to the model put forward by Dietz, etc.(4), that some miogeosynclines-eugeosynclines occur in pairs along continental margin. The south zone is characterized by miogeosyncline, but the north zone is similar to an eugeosyncline, especially after Jurassic period.

The transformation of ocean into land in the subsidence region underwent a complex history of evolution. The Tethyan ocean continually changed both in the width and depth and bottom topography, finally, the ocean entirely withdrew from there , and this region became land and rose as a part of the Qinghai-Xizang Plateau.

The development and evolution of this region, which proceeded under a certain background of tectonic movements was closely related to the following tectonic events, i.e. the Indian Plate continually drifted towards the north, thrusted under the Eurasian continent and finally, both the south and the north continent came combined together.

The evolutionary history of this region can be divided into 5 stages as follows and a sketch of the sedimentary model is shown in Fig. 6.

1. The flysch-like series in Triassic-Early Jurassic. There was the wide Tethyan ocean between the Indian and the Eurasian continents during this period. The sea basin was in a relative quiet state. There were neither strong crustal movement, nor large-scale magmatic activity, and large quantities of terrigenous materials were rapidly carried into the Tethys by turbidity current from time to time. As a result, a very thick flysch-like series was deposited in abyssal reductive environment.

2. The bathyal or abyssal sandstone and shale series formed in Middle and Late Jurassic-Early Cretaceous. The Indian Plate drifted further north. Strong compression force made the crust fold. The great difference in relief of submarine topography began to occur and volcanic activites became more intensive. At that time the sedimentary environment tended to be more complex and changeable. The environments of some parts of this region favoured the growth of planktons and benthos. Very thick series of

bathyal and abyssal grey-black shale and siliceous shale were formed. In addition to the series mentioned above, some shallow marine quartzose sandstones were also formed. The deposits contain a large amount of volcanic materials apart from stable terrigenous materials and fossils.

3. The radiolarian siliceous rock series in oceanic trenches and flysch formed in Late Cretaceous. During this transitional period from Early Cretaceous to Late Cretaceous, a severe crust movement occurred. In the southern margin of the Eurasian continent, a deep fault was formed and the oceanic crust subducted under the Eurasian continental crust. A trench which was steep in the north and sloping in the south was formed nearby the subducted zone. The basic lava flowed out and a large quantity of siliceous materials gathered. They formed the radiolarian siliceous rocks and basic volcanic rocks of hundred meters in thickness. Owing to the intrusion of acidic magma and extrusion of volcanic rocks in the southern margin of the Eurasian continent, the embryo of the Gandise Range was formed at that time. As the sea advancing north, the narrow trough trending nearly E - W, where the thick typical flysch was deposited, formed in the southern side of the Gandise Range.

4. The olistostrome formed in the middle and late stage of Late Cretaceous. As subduction continued, the radiolarian siliceous rock series and the nearby strata were folded, and uplifted. Folding and uplifting made these rock bodies break into various sizes. The gravity led to the occurrence of large scale submarine slides and avalanches, and a variety of rock blocks fell into the sea bottom and formed the olistostrome.

5. The molasse formed in the middle and late stages of Late Cretaceous. The Indian and Eurasian continents gradually approached to each other. The compressed ultrabasic rocks are outcropped along the fractures. Finally, the two continents came together and collided in Eocene, and the sea entirely retreated from this region. The successive compressing force made the whole subsidence region fold and uplift. Magmatic activties were still very intensive. The Gandise Range gradually rose, the molasse which mainly consists of detrital rocks was deposited on its southern side. In addition, the small scale molasse formed at both sides of the ultrabasic rock and siliceous rock belt and overlay unconformably upon them; the subsidence region entirely changed into a land and intense orogeny ended. Although, sometimes there were still uplifting, faulting, this region became more or less stablized since then.

## References


(1) The Xizang Scientific Expedition, Academia Sinica (ed.), Reports of the Scientific Expedition to Mt. Qolmolangma (1966-1968), Geology. Science Press (1974) (in Chinese).

(2) Gansser, A., The ophiolitic melange, a world - wide problem on Tethyan examples. Eclogae Geological Helvetiae, 67, 479-507 (1974).

(3) Hsü, K. J., Melanges and their distinction from olistostromes. Spec. Pub. Soc. Econ. Paleo. Miner., 19, 321-333 (1974).

(4) Dietz, R. S., Holden, J. C., Collapsing continental rises : Actualistic concept of geosynclines a Review. Spec. Pub. Soc. Econ. Paleo. Miner., 19, 14-25 (1974).

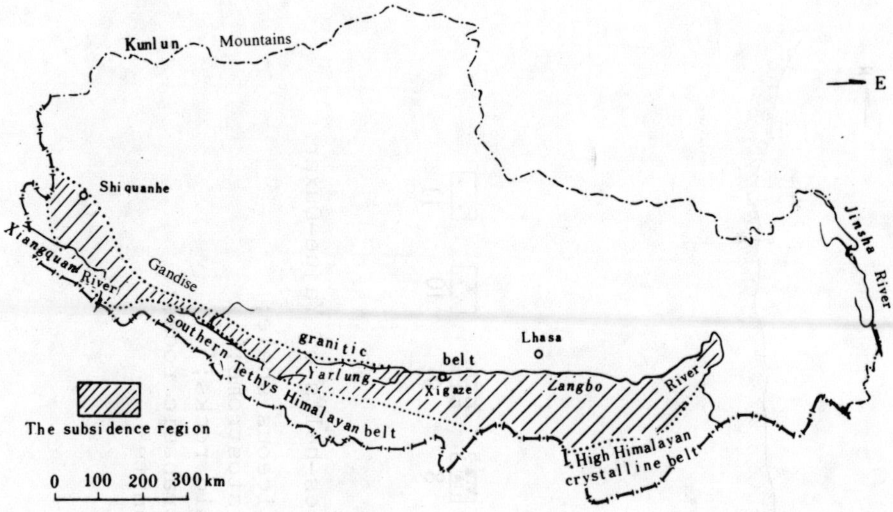


Fig. 1 Location map of the subsidence region of Yarlung Zangbo-Xiangquan River

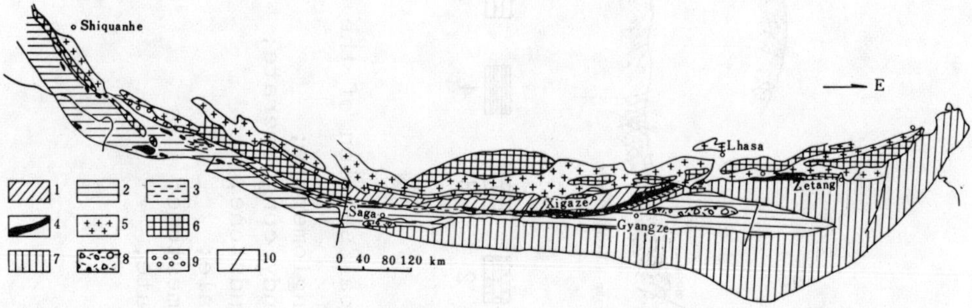


Fig. 2 Sketch of facies make up of the Yarlung Zangbo-Xiangquan River subsidence belt

1. Flysch facies-belt;
2. Bathyal or abyssal sandstone and shale facies-belt;
3. Radiolarian siliceous rock facies-belt;
4. Ultrabasic rocks;
5. Granite;
6. Pre-Triassic rocks;
7. Flysch-like facies-belt;
8. Olistostrome facies-belt;
9. Molasse facies-belt;
10. Fault

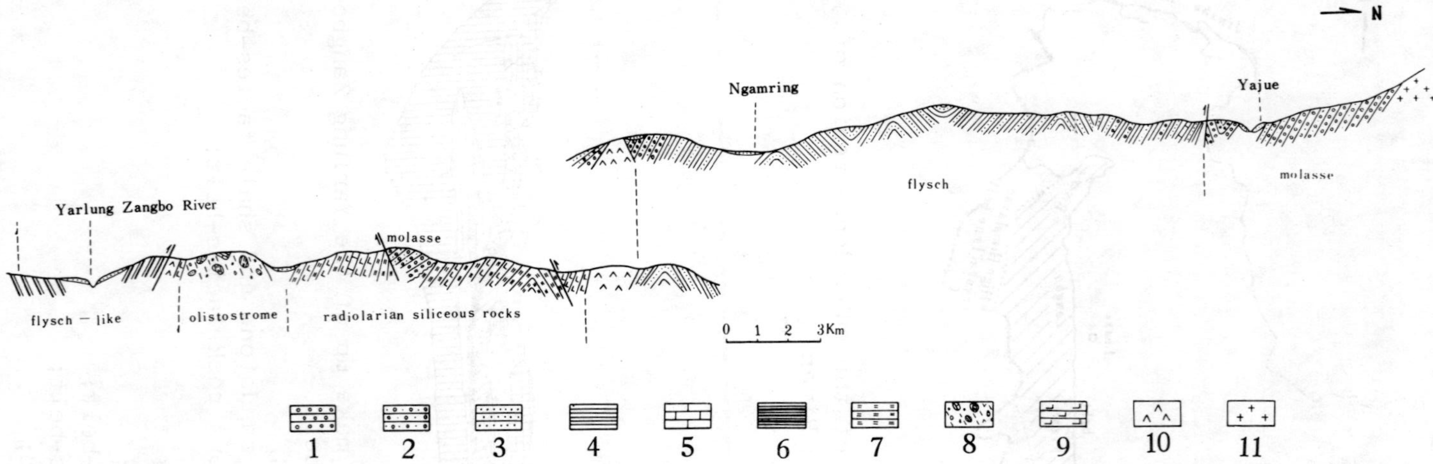


Fig 3 Geological section of the sedimentary facies-belts of Yajue-Quxar

1. Conglomerate;
2. Sandy conglomerate;
3. Sandstone;
4. Shale;
5. Limestone;
6. Slate;
7. Siliceous shale;
8. Olistostrome;
9. Basic rocks;
10. Ultrabasic rocks;
11. Granite

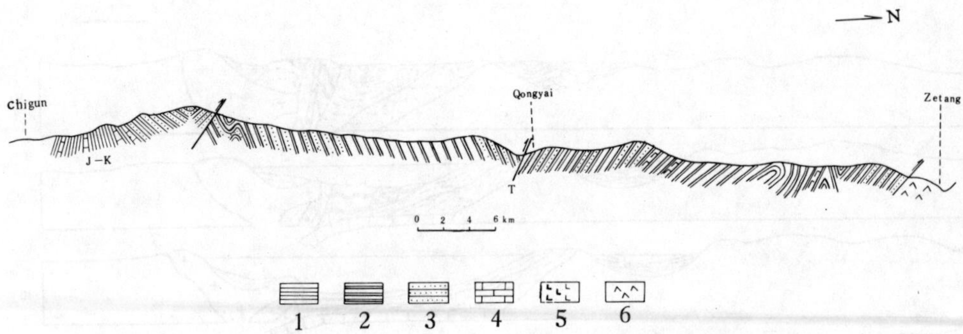


Fig. 4 Sketch of geological section of the flysch-like facies-belt from Zetang-Chigun

1. Shale;
2. Slate;
3. Sandstone;
4. Limestone;
5. Diabase;
6. Ultrabasic rocks

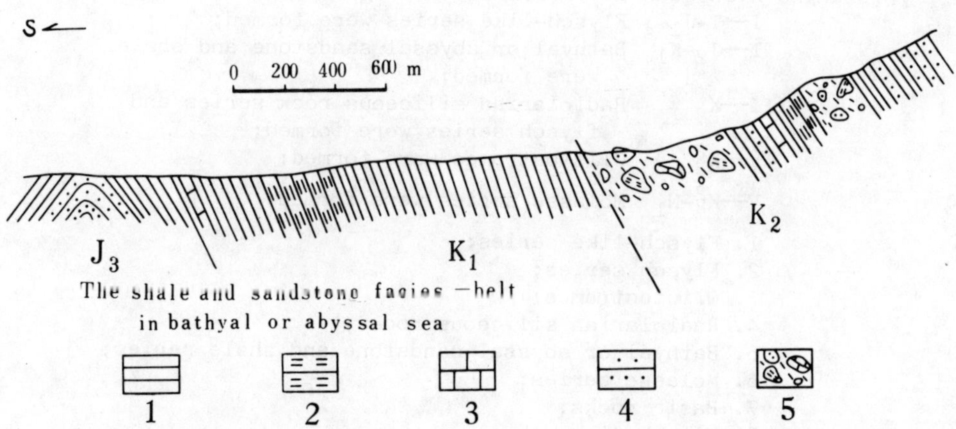


Fig. 5 Geological section of the northern Gyabrag Valley

1. Shale;
2. Siliceous shale;
3. Limestone;
4. Sandstone;
5. Olistostrome

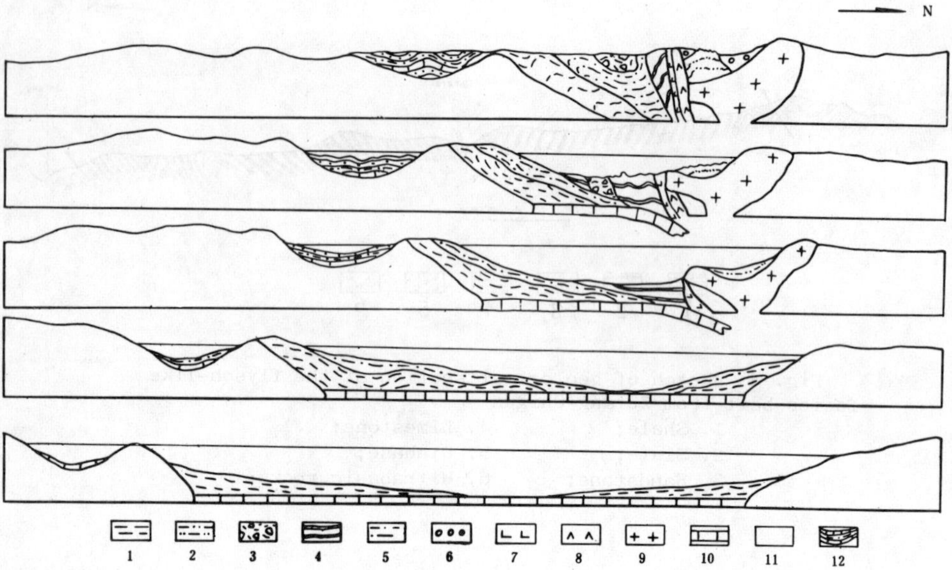


Fig.6 Sketch of the sedimentary model in the Yarlung Zangbo-Xiangquan River subsidence region

I—$T$-$J_1$ Flysch-like series were formed;

II—$J_2$-$K_1$ Bathyal or abyssal sandstone and shale were formed;

III—$K_2$ Radiolarian siliceous rock series and flysch series were formed;

IV—$K_2^2$ Olistostromes were formed;

V—$K_2^2$-$N_1$ Molasse series were formed;

1. Flysch-like series;
2. Flysch series;
3. Olistostromes;
4. Radiolarian siliceous rocks;
5. Bathyal or abyssal sandstone and shale series;
6. Molasse series;
7. Basic rocks;
8. Ultrabasic rocks;
9. Granite;
10. Oceanic crust;
11. Pre-Triassic basement;
12. Southern Tethys Himalayan zone

Photo 1. Sandy slate, Crossed nicols, X 37, Qonggyai

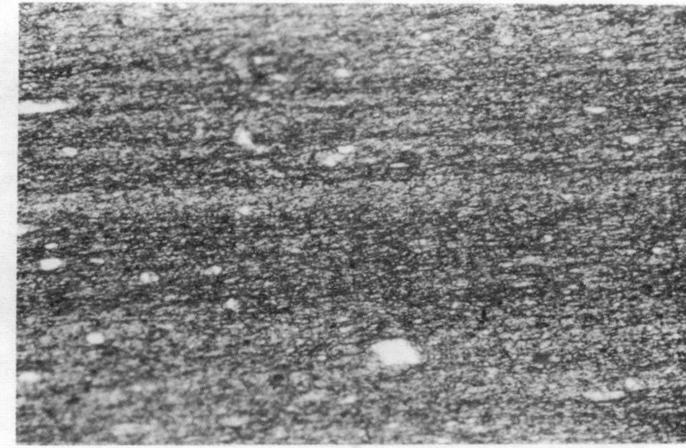

Photo 2. Shale of horizontal bedding, Lower nicol, X 66, Qonggyai

Photo 3. Quartz-sandstone, Crossed nicols, X 33 Weimei

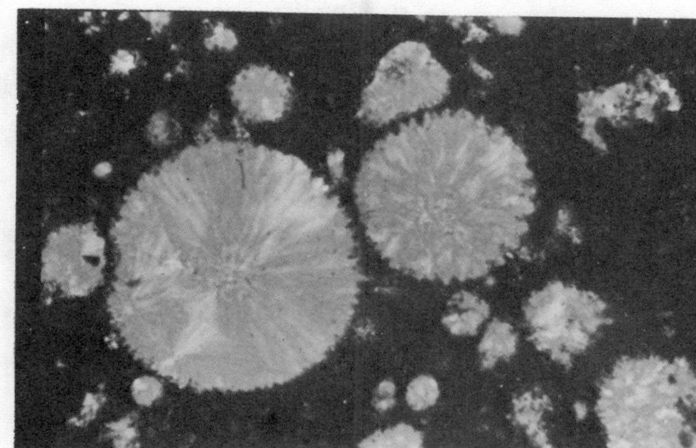

Photo 4. Radiolarian siliceous rock, Crossed nicols, X 270, Gyangze

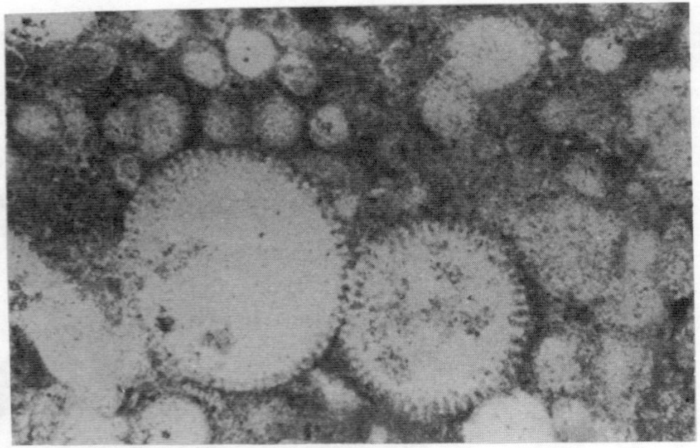

Photo 5. Radiolarian 'silliceous rock, Lower nicol, X 110, Xigaze

Photo 6. View of olistostrome,southern slope in Zazala

photo 7. The large block of siliceous rock in olistostrome, southern slope in Zazala

Photo 8. Folded flysch series, Sangsang

Photo 9. Interbedding of graywacke and shale in flysch, East Mountain in Xigaze

Photo 10. Sole mold of flysch, The north of Ngamring

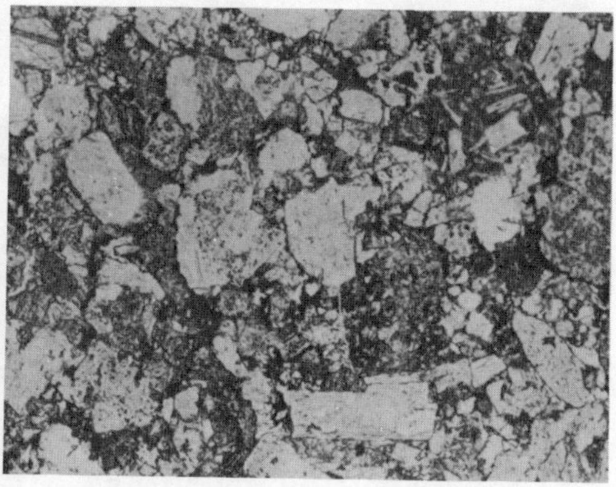

Photo 11. Graywacke, Lower nicol, X 37, Xabgeding

Photo 12. Mt. Kangrinboqe

Photo 13. Variegated conglomerate, the north of Ngamring

Photo 14. Variegated conglomerate, the north of Ngamring

# LITHOLOGICAL FEATURES AND GEOLOGICAL SIGNIFICANCE OF YARLUNG ZANGBO RIVER OPHIOLITE BELT AND TRENCH SEDIMENTS IN XIZANG PLATEAU

Cao Rong-long
(Institute of Geochemistry, Academia Sinica)

There is little doubt that the highest landmass on the earth, Xizang Plateau, has a geological setting of India-Eurasia continental collision. Various plate models have been presented, and opinions are different for many important problems, such as : Where is the site of the subduction zone, and in what direction is it thrust downwards? Which type the margin of Eurasia Continent is, is it an island arc before collision? Are the Yarlung Zangbo River ophiolites similar to current models of oceanic lithosphere, representing the slice of oceanic crust? And are they emplaced tectonically along a suture zone? These important scientific problems can be solved with more reliable new evidences now, since we found the trench sediments and discoverd their relation with ophiolite in 1979.

## CONGDU TRENCH SEDIMENTS

In Fig. 1 we have a geological map of this district. Yarlung Zangbo River runs from west to east, along which is the famous ophiolite belt. At Tangga, about 18 km northeast of Lhaze, a typical section of ophiolite has been found. Later, we have traced their relation with the country rock. There are only two different stratigraphic systems known before. One in the north is a flysch deposit named Xigaze Group of Late Cretaceous age. The other in the south is thick red radiolaria chert called Lhaze Group, also Late Cretaceous in age. But at Congdu, about 20 km southeast of Xigaze, we have found a thick sequence of trench turbidite associated with pillow lava and submarine volcanic conglomerate. They crop out to the south of Xigaze flysch and are in contact with ultramafic rocks. At the same place we have also found that the basal serpentinite overthrusts northwards on the trench turbidite.

A complete section of trench sediments at Congdu is shown in Fig. 2. From contact with serpentinite upwards it can be divided into four units. They are : lower turbidite, cherty volcanic conglomerate, upper turbidite and basaltic pillow lava. The thickness of the sequence is about 700 m, no bottom or top of the sequence can be found. It also shows that in the whole sequence no shallow water sediments such as limestone, shale, and quartz sandstone have been discovered.

The gravels of volcanic conglomerate are mostly basaltic, very different in size, from centimeters to more than 5 m. The cement of the gravels is radiolaria-bearing chert. It might be one type of melanges developing in trench.

The pillow lava is at the top of the trench sequence, overlying conformably the turbidite.

The trench turbidite is shown in Photo 1. It is well layered, with a suite of rhythmic deposits of graded bedding. Usually, for every rhythm, the coarse-grained greywacke is at the base, upwards it gives way to the finer material, and the chert and volcanic tuff rests at the top, showing a good rhythmic pattern. It was suggested by Bouma in 1962 that a sequence of turbidite would make up a fix characteristic succesion of five intervals or layers, known as Bouma cycle.

Layer A of Bouma cycle at the base is a coarse interval, and we have the coarse-graind greywacke. It is dark green, granularity is about 1.5 mm in diameter. The particles are composed mainly of rock fragments, such as : serpentinite, basalt, and chert. Small amount of minerals are pyroxene, olivine, feldspar, quartz, and magnetite. The serpentinite rock fragments usually contain chromite grains which have been identified by laser microspectral analyzer. Fig. 3 is triangular diagrams to display the composition of the rock. In upper diagram the selected components are rock fragments, pyroxene, and feldspar. The rock fragments are more than 80 %. The lower diagram displays the percentage of the three major rock fragments, chert about 52 %, serpentinite approximatly 30 %, and basalt 18 %. The average granularity is about 1.5 mm with a down order for chert -basalt-serpentinite according to their different specific gravity. It is believed that they are trench-source materials derived from scraping the wall of the underthrust oceanic lithosphere during subducting. Thus we call it "ophiolitic greywacke", because the association of rock fragments is similar to that of ophiolite.

Upwards it gives way to the gabbroid greywacke rich in pyroxene and plagioclase which belongs to the interval B of Bouma cycle, called lower parallel lamination. It is dark gray, well-bedded struture. The darker bands are rich in magnetite. Under microscope, pyroxene and plagioclase are principal minerals, feldspar more than 75%, pyroxene about 20%, magnetite is very common, quartz only 2 %. A small amount of olivine and amphibole have been identified by X-ray diffraction. The average granularity of constituent minerals is 0.2 mm, with a maxium of 0.5 mm. The minerals, in decreasing order of granularity, are quartz, feldspar, and pyroxene. The gabbroid greywacke is also of arc-trench source, because the rock fragments in small amount are same as the ophiolite. The large amount of ferromagnesian minerals, plagioclase and magnetite are derived from basic igneous rocks of granular nature, such as gabbro.

Above the gabbroid greywacke, there is fine grained greywacke, or a mixture of fine grained greywacke, chert and tuff, representing the interval C of Bouma cycle, known as current ripple lamination. There are various types of cross lamination. The rock is composed mainly of fine clastic minerals, siliceous material, and volcanic ash. They can mix in various ratio in this layer C. The grain size of fine fragmental materials is about 0.06 mm with a down order for radiolaria, feldspar, and pyroxene.

Layer D is called the upper parallel lamination, they are well interbedded chert, tuffaceous chert, cherty tuff, showing good parallelism. The bedded cherts are red or green, rich in radiolaria.

At the upmost part of turbidity cycle there is a volcanic tuff bed, corresponding to the top pelitic interval E. It is white in colour, like chalk in appearence, 30-50 cm in thickness, consisting mainly of volcanic ash and some radiolaria-bearing chert. It usally forms the top layer for each Bouma cycle.

Briefly, the major features of trench sediments are of deep-water origin, arc-trench derived materials and deposited by turbidity currents. In view of the fact that it has not been clearly documented before, and they are entirely different from Xigaze Group, so we consider it is necessary to give it a new name "Congdui Formation" according to the place where the section was found.

The discovery of trench sediments may possibly solve some debatable problems. It may indicate the site of the margin of Eurasian Continent; it may indicate the site of the subduction zone, downthrusting northwards under Eurasian Continent. Also, it marks the presence of a fossil island arc borderd by the trench. Besides, it was suggested by A. Gansser that along the Indus suture zone the deep-water marine sediments of a former basin already completely disappeared during Himalayan orogeny. But now we will confirm that they did not completely disappear along Yarlung Zangbo River subduction zone.

## XIGAZE GROUP — ARC-TRENCH GAP SEDIMENTS

For the purpose of further clarification, it is necessary to compare the Congdu trench sediments with the adjacent Xigaze flysch, as well as the red chert of Lhaze Group. Fig. 4 shows three stratigraphic columns. In the centre it is trench turbidite of Congdu Formation mentioned above. On the left is the Xigaze Group. It was divided into four formations. The lower part, named Qiuwu Formation, about 500 m. is basal conglomerate, overlying the granite. Its gravels mostly are granitoid. The upper part of this formation is coal system, composed of sandstone, shale, and coal beds. The second one is Qiabuling Formation, also about 500 m, red sandstone, conglomerate with yellow sandy shale. The third, called Shangzugang Formation, are thick beds of limestone, about 200 m. The uppermost named Ngamring Formation or Xigaze Formation, about 900 m, is flysch of interbedded sandstone and shale, with limestone bands. It must be emphasized that the sandstone and shale in flysch usually contain plant fossils. So, it is clear that they are partly submarine and partly subaerial deposits and they are quite different from Congdu Formation. It is believed that the Xigaze flysch might be " arc-trench gap" sediments. The right column is Lhaze Group, composed mainly of red radiolarite and siliceous shale, with limestone bands and siliceous conglomerate. It has no turbidite like Congdu Formation. Obviously they were deposited on abyssal plain of an ocean basin, and might be a part of oceanic crust from which ophiolite formed together with basalt and serpentinite.

Now we possibly can reconstruct the plate relation in southern Xizang before continental collision. Fig. 5 shows that passing through Congdu there is a trench, represented by Congdu Formation. The subduction zone of Tethys might be near the trench, it downthrusted northwards under The Eurasian Continent. To the north there might be an arc-trench gap, marked by Xigaze Group. To the south distribute the thick red chert of Lhaze Group, representing pelagic deposits of an ocean basin.

## YARLUNG ZANGBO RIVER OPHIOLITES

A typical section of ophiolite at Tangga in Lhaze district shows the sequential association of ophiolite, from north to south they are basal tectonized harzburgite, altered basalt, and red radiolarian chert. Contact of serpentinite with the overlying basalt, as well as the occurrence of chert shows that they are dipping southwards (Fig. 6). The thick radiolaria chert belongs to Lhaze Group.

It is clear that the ophiolites are lithologically similar to current models of oceanic crust and upper mantle, representing the slice of Tethyan ocean basin.

Photo 2 shows the relation of ophiolite with trench sediments at Congdu. The basal serpentinite overthrusts northwards on the trench turbidite. There, the basal ultramafic rock is dipping to the south, while the turbidites are dipping northwards, showing thrust drag at the contact. It is suggested that the ophiolites were emplaced tectonically along the suture zone where Indian Continent and Eurasia Continent were welded together, because the trench sediments mark the margin of Eurasian Continent.

## CONCLUSION

In summary, we would like to propose a four-stage model to explain the plate history for Himalayan geology (Fig. 7).

In the early stage of Tethyan plate convergence, a trench developed at the margin of Eurasian Continent over the subduction zone, and there are an island arc and an arc-trench gap associated with it. But at the same time, the coast of Indian Continent had only shelf sediments.

As the convergence continued, the slice of oceanic crust as wedge might thrust upwards.

In the late stage of convergence, the ophiolite characteristic of trinity might obduct onto the Eurasian Continental margin.

Finally, when collision occurred, the ophiolites, as the remnants of the intervening ocean basin, were emplaced tectonically along the suture zone, overthrusting northwards on the trench sediments of Congdu Formation.

## ACKNOWLEDGMENTS

The writer wishes to express his thanks to Dr. Chen Xian-pei for his suggestions and help in the work of the turbidites.

## References

(1) Condie, K. C., Plate tectonics and crustal evolution. Pergamon Press, Inc. pp. 175-199 New York (1976).

(2) Windley, B. F., The evolving continents. John Wiley & Sons pp. 247-257 London (1977).

(3) Jacobs, J. A., Russell, R. D., Wilson, J. T., Physics and geology. McGraw-Hill Kogakusha, Ltd, Tokyo Japan., Second Edition, (1974).

(4) Dewey, J. F., Bird, J. M., Mountain belts and the new global tectonics. Jour. of Geophysical Research, 75(14), 1625-1648 (1970).

(5) Mattson, P. H., Pessagno, Jr. E. A., Helsey, C. E., Outcropping layer

A and A"correlatives in the greater Antilles. in Memoir 132, The Geological Society of America, "Studies in earth and space sciences". 57-67 (1972).
(6) Gansser, A., The great suture zone between Himalaya and Tibet, A preliminary account, in "HIMALAYA", Sciences de la Terre, 181-192 (1977).

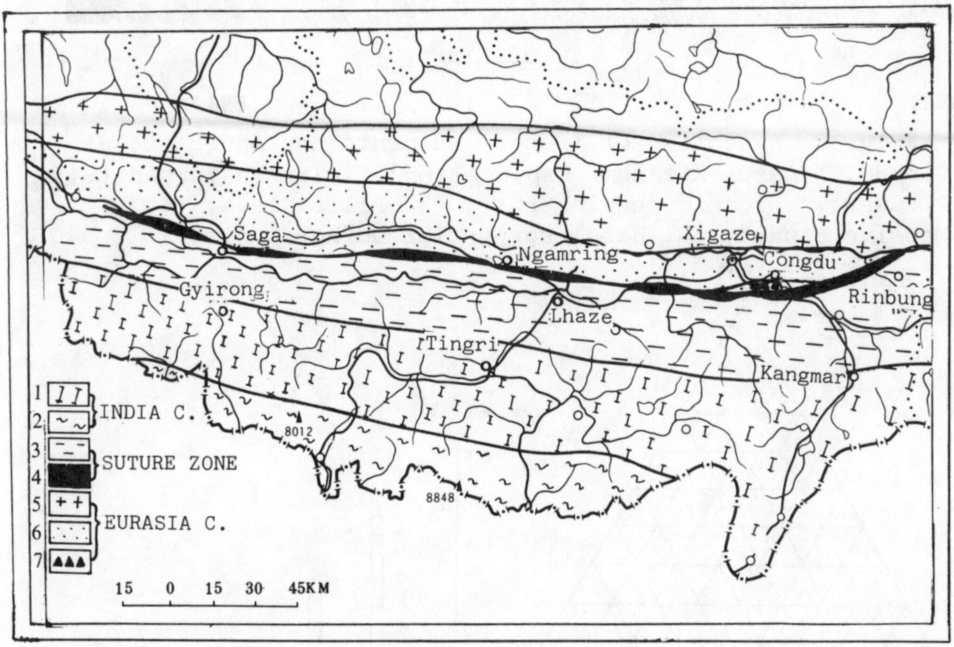


Fig. 1 Geological map of Yarlung Zangbo River ophiolite belt

Explanation: 1. Biostrata (O-E);
2. Gneiss (AnZ),
3. Slate, Phyllite (T-K);
4. Ophiolite ($K_2$);
5. Granitoids;
6. Xigaze Group flysch ($K_2$);
7. Congdu Formation trench sediments ($K_2$)

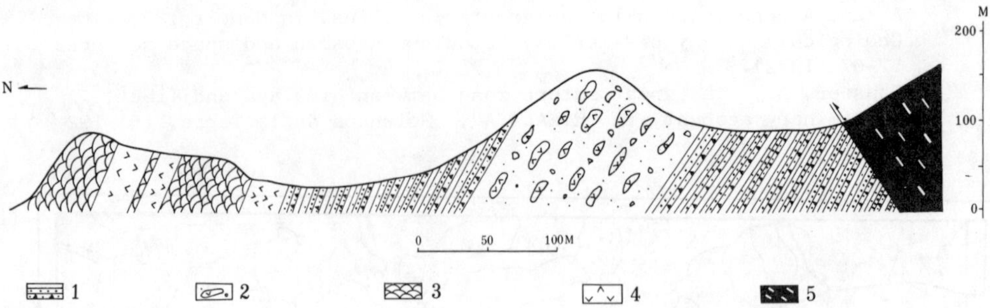


Fig. 2 A section of trench sediments and ophiolite at Congdu, Xigaze, Xizang. Explanation: 1. Trench turbidite; 2. Cherty volcanic conglomerate; 3. Pillow lava; 4. Massive basalt; 5. Basal serpentinized harzbugite

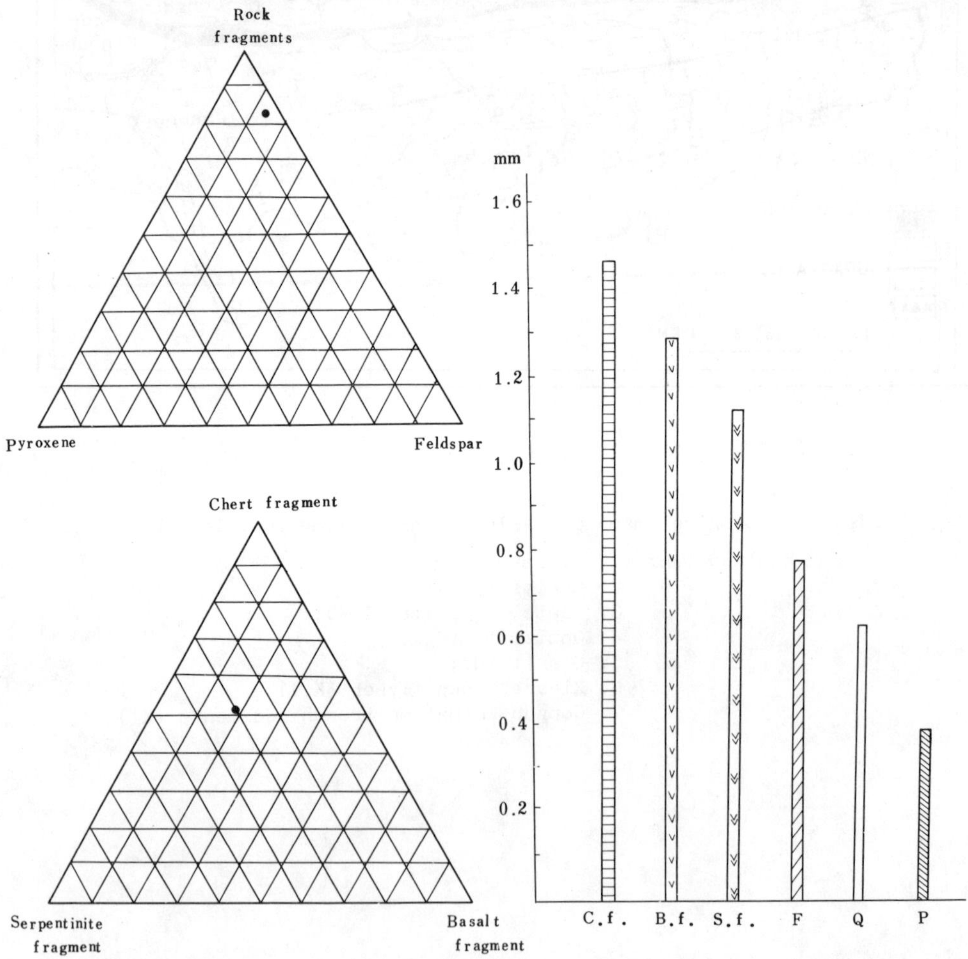


Fig. 3 The composition and granularity of the coarse-graind ophiolitic greywacke

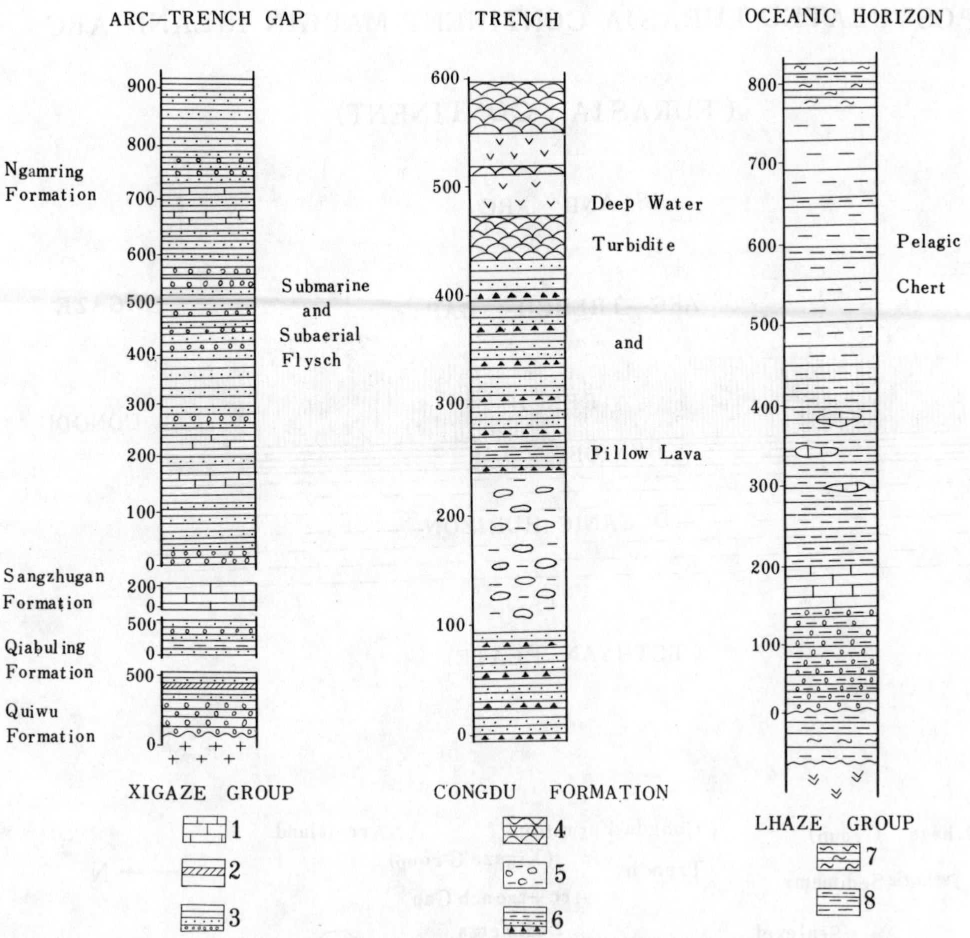


Fig. 4 Comparison of sedimentary types of Congdu formation with Xigaze group and Lhaze group

Explanation: 1. Basal conglomerate, sandstone and shale;
2. Coal system;
3. Limestone;
4. Interbedded greywacke, chert and tuff;
5. Cherty volcanic breccia;
6. Pillow and massive lava;
7. Radiolaria chert;
8. Phyllite

POSTULATED EURASIA CONTINENT MARGIN ISLAND ARC

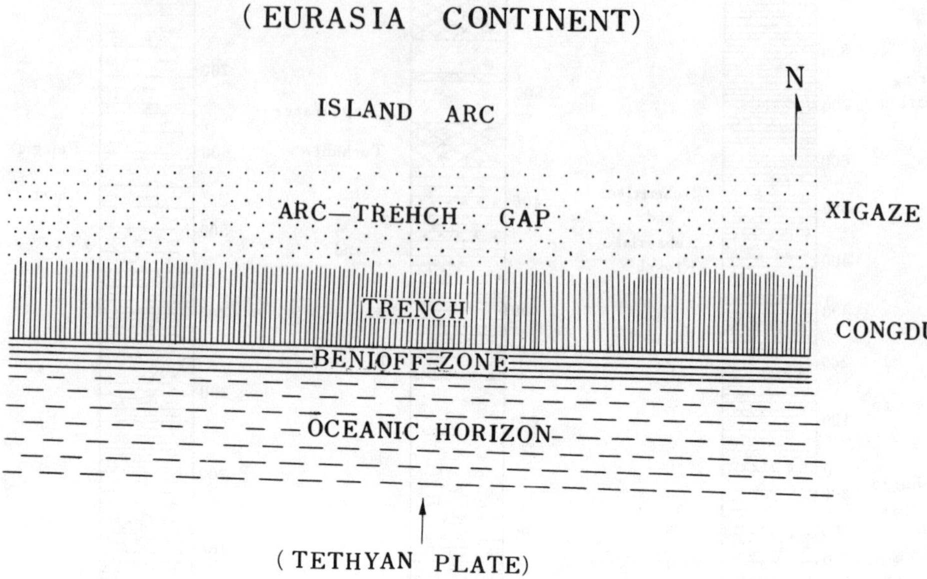


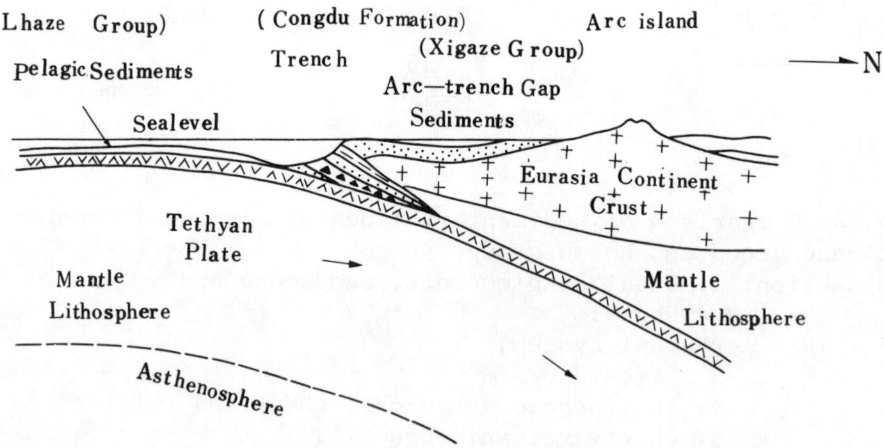


$K_2$ IN AGE: BEFORE CONTNENTS COLLISION

Fig. 5 Reconstruction of plate relation in southern Xizang before continental collision

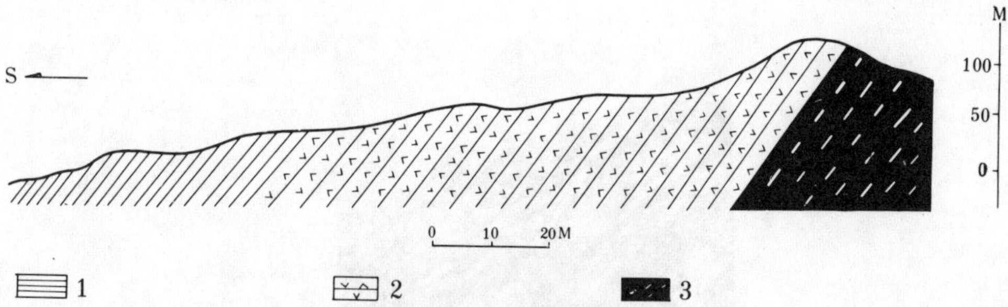


Fig. 6 A section of ophiolite at Tangga, Lhaze, Xizang
Explanation: 1. Red radiolaria chert; 2. Altered basalt; 3. Basal tectonized harzburgite

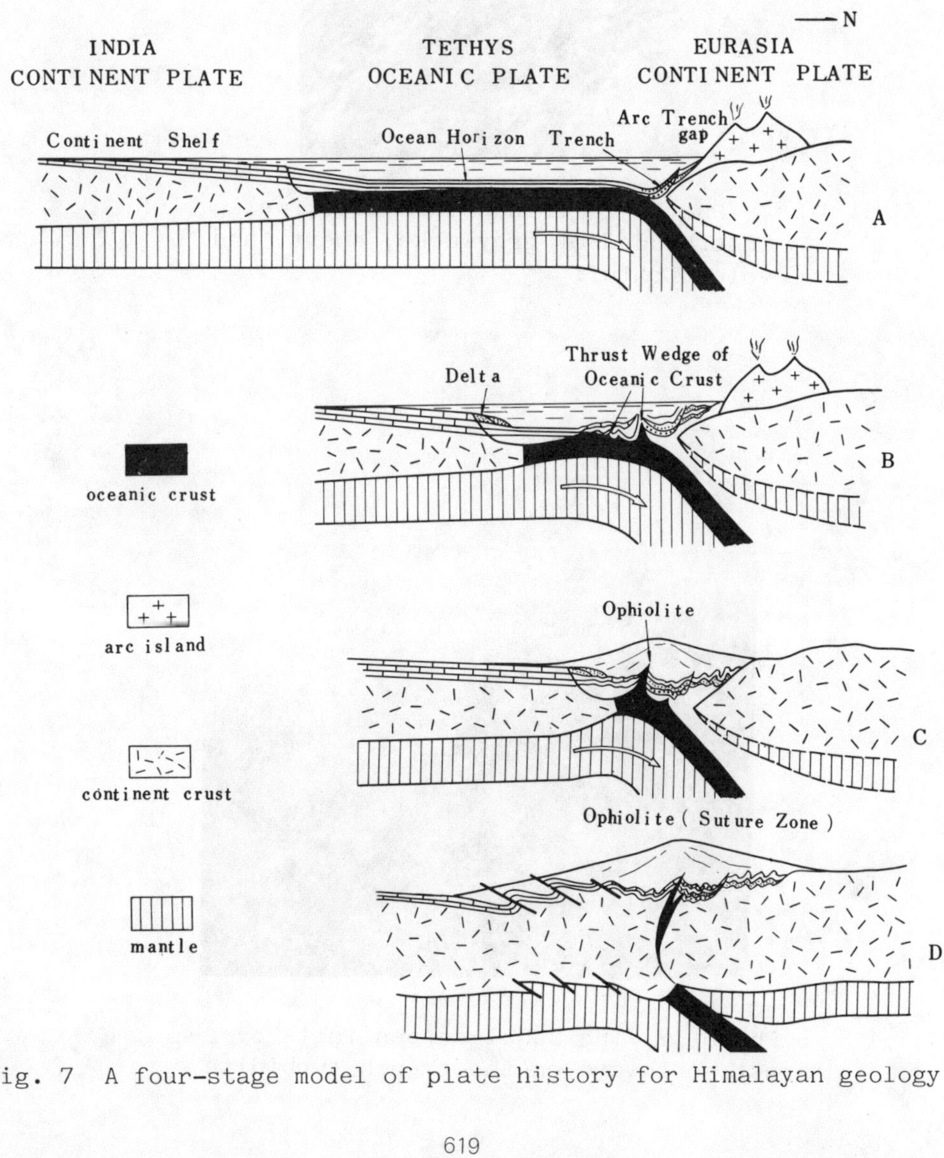


Fig. 7 A four-stage model of plate history for Himalayan geology

Photo 1 Trench turbidite, well interbedded of greywacke, chert, and volcanic tuff

Photo 2 The basal serpentinite overthrust northward on the trench turbidite

## OPHIOLITES AND OPHIOLITIC MELANGES

R.M. Shackleton
Dept. of Earth Sciences
Open University

The Zangbo ophiolites of Tibet clearly lie on a suture between two plates which collided. The tectonic situation of many ophiolites elsewhere is much less clear. I shall discuss some of the problems of ophiolite emplacement with particular reference to the Oman ophiolite and those in the late Precambrian of Egypt and Sudan.

Definition: The definition which was adopted by the Penrose conference on ophiolites (Anomymous 1972) is now generally accepted. Ophiolite complexes are characterised by the association of pillow basalts, dolerites (typically occurring as sheeted dyke complexes), gabbros and peridotites (dominantly tectonized harzburgites).

Source: Ophiolites have been interpreted as slices of crust and upper mantle from the ocean floor, or from back-arc basins, arc-trench gaps or even island arcs. Discrimination between these several proposed sources has mostly been based in geochemical criteria (Miyashiro 1973, Pearch 1975, Richard and Allègre 1980).

Back-arc basins and oceans seem to be intrinsically different (remnant basins such as the Bering Sea being included with the oceans). There is no evidence that a back-arc basin can develop into a true ocean. On the contrary oceans seem to develop from continental rifts which may be recognised by the characteristic alkali volcanics, rift evaporites and rift faulting whereas a back-arc basin develops <u>after</u> the volcanic arc behind which it lies. Existing back-arc basins are relatively small (less than 1500 km wide), non-linear and rapidly formed. It is doubtful whether any ophiolite can be conclusively shown to be derived from a back-arc basin although geochemical evidence often suggests such a source. The diagnostic <u>preceding</u> calc-alkaline vulcanicity has seldom been recognised. The short time interval between the formation of ophiolites and their obduction shows that the parts of the ocean floor to be obducted were newly formed and still close to a spreading ridge.

<u>Structure and emplacement</u>

The segments or slices of oceanic crust which are now found on the continents are supposed to have been incorporated in the continental fold belts by a process called obduction (Coleman 1971), that being the 'mechanism by which large slabs of dense oceanic crust and mantle and detached from the evolving oceanic lithosphere and tectonically incorporated into orogenic belts' (Dewey 1976). The implication is that the slabs were detached and forced upwards while the slab from which they were derived was subducted downwards. This is similar to the more explicitly stated hypothesis of flake tectonics (Oxburgh 1974) in which an ophiolite is thought to represent a flake split off from the descending slab and thrust upwards. It is doubtful whether any example of flake tectonics has been convincingly described. The field evidence does not always prove that ophiolite slabs are in fact detached. Especially in the case of long-continuous slabs along sutures, such as the ophiolites on the Zangbo suture in Tibet, or the ophiolite underlying the Great Valley sequence in California, it seems likely that they are still attached, probably to the overriding plate. If so the leading edge of that plate has been uplifted because light continental crust has been carried down, or packed beneath it, rather than being upthrust.

In more numerous cases, ophiolites do occur as detached slabs or lenses. If so the detachment may be the result of thrusting (i.e. it is tectonic) or it may be the result of gravity sliding within an olistrostrome. One criterion for distinguishing between these two is that thrusts cut up formation in the direction of thrusting. It is because the thrust at the base of the Oman ophiolite (Semail nappe) appears to cut down-section, and then up again through the sequence, on the northeast side (Glennie et al. 1974 ) that it has been interpreted as an immense gravity slide, as it must be if the basal thrust really does cut down-section and then up again as the map suggests. The problem is then that the Semail nappe is so thick, supposedly originally about 15km, that it could only have slid into an enormously deep trench. This seems so improbable that one is forced to doubt the evidence that the basal thrust really does cut down-section on the northeast side. It is possible that the dislocations which bring the sub-ophiolite sediments against the higher members of the ophiolite sequence are not parts of the basal thrust but later faults of a different character.

In many places ophiolites occur in ophiolitic melanges (Gansser 1974). The origin of these ophiolitic melanges is still by no means thoroughly understood: The term melange was originally used, by Greenly (Greenly 1919) for a chaotic breccia, of regional extent, in the Monian (late Precambrian) of Anglesey in North Wales. Greenly thought that this Anglesey melange was tectonic in origin and some authors, notably K.J. Hsü, have used the term melange with the implication of a tectonic origin. Most subsequent authors have preferred to use the term descriptively without genetic significance. The Monian melange, which Greenly thought to be tectonic, has since been interpreted as an olistostrome or series of olistostromes, produced by submarine sliding (Shackleton 1969, Wood 1974). The best-known olistromes, or breccias produced by submarine gravity sliding, are those in the Apennines

(Abbate et al. 197 0). The *argille scagliose* of the Apennines is an ophiolitic olistostrome which is shown, by mapping, drilling and field evidence to be intercalated as an extensive sheet (or series of sheets) within a stratigraphic sequence. Some characteristic features of these olistostromes are the angularity rather than lenticularity of the olistoliths, the presence of re-entrant angles, the evidence of soft-sediment deformation, the admixture of a variety of blocks from different sedimentary environments as well as various igneous rocks, the lack of penetrative deformation in the matrix and the stratigraphic rather than tectonic contacts with normal sediments above and below. Ophiolitic melanges arc essentially similar to other olistostromes except that they contain olistoliths of ophiolite.

Ophiolitic melanges are characteristically associated with subduction zones. They often occur underneath ophiolite slabs which are thought to be tectonically emplaced. They may contain exotic blocks, such as those beneath the Semail nappe in Oman and beneath the Zangbo ophiolite in Tibet, and these exotic blocks may be much older than the associated ophiolite and therefore presumably derived from a part of the ocean originally far distant from the ophiolite source - perhaps too far distant to have slid under the influence of gravity. For all of these reasons, there must be some doubt whether gravity sliding alone can explain the ophiolitic melanges. It may be that the fragmentation and transport of the blocks and slabs involved, at least in some cases, both gravity sliding and tectonic transport. Note that tectonic transport necessarily involves an overriding plate or thrust mass.

A criterion which may help in distinguishing obducted ophiolite slabs from slabs which have slid within an olistostrome is that obducted slabs are commonly underlain by a metamorphic aureole which indicates a very rapid downward decrease in temperature below the slab, during its emplacement. Such metamorphic aureoles are known under the Semail ophiolite (Oman), Vourinos (Greece), Neyriz (Iran), Klamath Mountains (Oregon), the Catalina schist terrain (California), the Bay of Islands ophiolite (Newfoundland), the New Caledonia and Papua New Guinea ophiolites, the Jungbwa nappe (Himalayas) and other ophiolite slabs.

## Late Precambrian (Pan-African) ophiolites in northeast Africa

Very many complete or partial ophiolite sequences have now been recognized in the late Precambrian of Saudi Arabia and northeast Africa (Leblanc 1976, Bakor et al. 1976, Shackleton 1977). In the Eastern Desert of Egypt most if not all of the ophiolites occur in a melange of regional extent (Shackleton et al. 1980) whereas in Saudi Arabia and N.E. Sudan the ophiolites occur in linear zones suggestive of sutures. The ophiolitic melange in Egypt is thought to be an olistostrome rather than a tectonic melange because there is no evidence of an overriding ophiolitic slab or nappe; the melange seems to be intercalated as a layer within normal (non -melange) sediments and volcanics (above) over a very large area; it is thin (a few Km) relative to its extent (at least 10 000 $Km^2$) and the low strength of the generally pelitic matrix makes it mechanically impossible for the mass to have been moved by a push from behind; it is associated with thinner non-ophiolitic melange units which are clearly small olistostromes; it contains olistoliths from different sedimentary environments, as well as very rare granitic and other igneous blocks. We conclude that this late Precambrian ophiolitic melange moved down-slope under the influence of gravity and that the ophiolite masses contained in it slid from an ophiolite which was being obducted somewhere to the southeast.

## Conclusion

The emplacement of ophiolites often involves gravity sliding as part of the process and ophiolitic melanges, typical of subduction complexes, probably originate as olistostromes before being tectonized when they are overridden by obducted ophiolite slabs.

## References


(1) Abbate, E., Bertollotti, V & Passerini, P. Olistostromes & Olistoliths. Sediment. Geol. 4, 3/4, 521-577 (1970).

(2) Anonymous, Penrose Field Conference (1972): Ophiolites. Geotimes 17/12, 24-25 (1972).

(3) Bakor, A.R., Gass, I.G., and Neary, C.R. Jabal al Wask, north west Saudi Arabia, an Eocambrian Back-arc ophiolite. Earth Planet Sci. Lett. 30, 1-9 (1976).

(4) Coleman, R.G. Plate tectonic emplacement of upper mantle peridotites along continental edges. J. Geophys. Res. 76, 1212-1222 (1971).

(5) Dewey, J.F. Ophiolite obduction. Tectonophysics. 31, 93-120 (1976).

(6) Gannser A. The Ophiolitic Melange a World-wide Problem on Tethyan Examples. Eclogae. Geol. Helv. 67,479-509 (1974).

(7) Glennie, K.W., Boeuf, M.G.A., Hughes-Clarke, M.W.H., Moody-Stuart, M., Pillar, M.F.H. & Reinhardt, B.M. The Geology of the Oman Mountains, Konink, Nederlands. Geol. Mijnbouwkundig Genoot. Verh. 31 1-423 (1974).

(8) Greenly, E. The Geology of Anglesey. Mem. Geol. Surv. G.B. 2 vols (1919).

(9) Leblanc, M. Proterozoic oceanic crust at Bou Azzer. Nature 261, 34-35 (1976).

(10) Miyashiro, A. The Troodos ophiolite Complex was probably formed in an island arc. Earth Planet. Sci. Lett. 19, 218-224 (1973).

(11) Oxburgh, E.R. The Plain Man's Guide to Plate Tectonics. Proc. Geol. Ass. 85, 299-358 (1974).

(12) Pearce, J.A. Basalt geochemistry used to investigate past tectonic environments on Cyprus. Tectonophysics 25, 48-67 (1975).

(13) Richard, P. & Allègre, C.J. Neodymium and Strontium isotope study of ophiolite and orogenic lherzolite petrogenesis. Earth Planet. Sci. Lett. 47, 65-74 (1980).

(14) Shackleton, R.M. The Pre-Cambrian of North Wales in The Pre-Cambrian and Lower Palaeozoic Rocks of Wales ed. A. Wood. Univ. of Wales press. Cardiff (1969).

(15) Shackleton, R.M. Possible late-Precambrian Ophiolites in Africa and Brazil. Ann. Rep. Res. Inst.Afr. Geol. 20, 3-7 Leeds Univ (1977).

(16) Shackleton R.M., Graham, R.H., Fitches, W.R. & Ries A.C. Late Precambrian ophiolitic melange in the Eastern Desert of Egypt. Nature, 285, 472-474 (1980).

(17) Wood, D.S. Ophiolites, melanges, blueschists and ignimbrites: early Caledonian subduction in Wales? In Dott, R.H. Jn. & Shaver, R.H. (eds). Modern and ancient geosynclinal sedimentation. Soc. Econ. Pal. Min. Spec. Pub. 19, 334-44.

# UPPER MANTLE STRUCTURE UNDER THE TIBETAN PLATEAU

L. Knopoff and F.-S. Chang
Institute of Geophysics and Planetary Physics
University of California, Los Angeles

Amid the outpouring of discussions of the genesis of the crust of the Tibetan Plateau as a consequence of the collision between the Indian and the Asiatic Plates, little attention has been paid to the dynamics of the mantle beneath the crust. One of the clues to the history of the Tibetan Plateau subsequent to the initial contact between the two plates is to be found in the mantle below the Plateau surface: is it partially molten, as in many tectonically active areas [1,2], or is it at temperatures well below the solidus as under most shields [2,3], or is it in some state intermediate to these extremes?

In the absence of direct observations of seismic waves under appropriate conditions, the most suitable method to study the upper mantle of a remote region is to investigate the dispersion of surface waves across it. The disadvantages of surface wave methods are well-known: one is unable to isolate the effects of lateral inhomogeneities on a horizontal scale of the order of the wavelengths and one is unable to resolve sharp discontinuities in vertical structure. Despite the evident defects, surface wave methods remain the only effective means at this time of investigating the mantle of the earth in remote areas. The depths to which such studies give useful information about the earth's interior depend on the longest wavelengths used; with the seismographs of the World Wide Standard Seismographic Network (WWSSN) the depth range is of the order of 200 km or so.

Because of the absence of long-period seismographs within Tibet, we have recourse to 'single-station' methods of dispersion analysis, in which we measure the phase or group delay properties of surface waves between an earthquake and the seismograph. Except in rare instances, the earthquake and the seismograph locations do not span homogeneous regions, so corrections for the regions outside the one under study must be applied. Most frequently, the regions outside the one under study also have unknown cross-sections, so these too must be investigated by similar methods, merely to obtain the corrections for the first region. The most advantangeous method of accomplishing this task is to consider the entire region within a network of stations and a network of earthquakes as having an unknown, laterally inhomogeneous cross-section and to solve simultaneously for the cross-section of each subregion to minimize the surface wave travel time residuals. This method has been used effectively at very long periods both on a global scale[4,5,6] and on a regional scale[7].

With regard to the choice between phase and group velocity methods, it can be remarked that phase velocities can be determined with greater accuracy, can be determined over a broader range of periods for the same earthquake, and give better structural resolution except near the group velocity extremum [8] than group velocities. If one enlarges the region of investigation to include regions outside the Tibetan plateau in the study, then one can employ the longer period signals available from larger earthquakes. The method of intersecting earthquake-station paths minimizes bias in the results due to effects of lateral heterogeneity on seismic profiles transversing the region in only one direction or from the use of only one earthquake.

We use WWSSN recordings of nine events listed in Table 1. Also listed in the Table are the seismograms used in the analysis and the period range of surface waves available on each. The Rayleigh phase wave travel times in these period ranges have been measured by Chang[9]. With respect to the least-squares models to be described below, the residual surface wave travel times are normally distributed with a standard deviation of

$$2/3 \max (7, 0.1T) \text{ sec}$$

with T the period of the surface waves. The phase travel times from the Xingtai earthquake to MSH were uniformly and significantly early with respect to the final models over the period range 38 to 69 sec and these data were not used; the reason for the early arrivals remains a puzzle. Many of the shorter period data, at 30 and 38 sec for example had large residuals, which we believe is testimony to the inadequacy of our crustal models. We have chosen to use our standard crustal models and have refrained from using these shorter period data in the inversions, except where indicated in Table 1.

The event-paths that sample the Tibetan plateau are starred in Table 1. To increase sampling of the Plateau, events 5-9 were subsequently added to the initial analysis. The narrow period range available from the Tashkent shock and the Lop Nor nuclear explosion means that these events probably do not contribute significantly to the sampling of the Tibetan Upper Mantle.

The regionalization we have chosen (Fig. 1) consists of 6 regions which we tentatively identify as I. Precambrian Shields (including the Baltic, Ukrainian, Siberian and Indian Shields), II. Seismically Stable Sedimentary Platforms (including the Siberian and Russian Platforms and the Arabian Peninsula), III. Folded Mountain Belts (including the Himalayan-Zagros-Turkish mountain zone as well as the Yakutsk-Magadan region of Eastern Siberia), IV. The Qinghai-Xizang Plateau, V. The Mongolian-Sinkiang seismically active region and VI. The quasi-stable region of Southeastern China. Most of these regions have been selected on the basis of similarities in their age, seismic stability or tectonic setting. Since we have ignored short-period surface wave data in the inversions, we are obliged to make some assumptions regarding crustal structure; we standardize the crustal structures largely from the available literature, using explosion seismology, P-delays, similarity to other regions where crustal structure is known and short-period surface wave data where available. We have found our solutions to be stable with regard to 'small' variations in the locations of the boundaries between the regions.

If we choose to parameterize each region by a 'lid' to a low-velocity channel (LVC), a LVC and perhaps a sub-channel region, we find that this leads to an underdetermined system of equations with far too many model parameters for the data set we have. The number of model parameters is 5 for each region, being the three S-wave velocities in each layer and the two layer thicknesses. We therefore have 30 unknowns in all. The number of independent degrees of freedom in the data is substantially lower than this and cannot be increased significantly by adding more earthquake-station pair paths with these seismographs; only additional redundancy is introduced by adding more event pairs. Thus the number of model parameters has to be reduced, which we do by making geophysically plausible assumptions or arguing that the assumptions we do make are adequate as long as the reader pursues a policy of 'caveat emptor'; these two methods may be the same in the long run.

For the purposes of this paper, we make the following assumptions regarding the model cross-sections in the interests of reducing the number of model parameters:

1) the bottom of any LVC is at the same depth across the entire Eurasian continent
2) the S-wave velocity in the lid to any LVC is the same everywhere across Eurasia
3) the S-wave velocity in any LVC is the same everywhere across Eurasia
4) the S-wave velocity below the LVC is assigned and fixed.

This set of assumptions reduces the number of model parameters to 9, these being the thickness of the lid in each of the 6 regions, and the common S-wave velocities in the lid and LVC and the depth to the bottom of the LVC. Below the lower boundary of the LVC we have put a 'standard' deep mantle structure that is similar to model 1066B. Rayleigh wave velocities are calculated by a program due to Schwab and Knopoff [10,11]. We assume that density fluctuations and P-wave velocity deviations from the starting model do not affect the inversion significantly, as long as these deviations are small. For the purposes of the present discussion, the most important result of applying a linear inverse procedure[12] to these data with a model thus parameterized is that regions I and II have very thick lids extending to such great depth that any LVC is not likely to be resolved well by the present data. This means that we can assume that the sedimentary platforms can be considered to have shield-like upper mantle structure and hence to be homomorphic with the shields of region I. We therefore reduce the number of geographic regions by one as the result of combining regions I and II into a single large region which we call region IA.

The above change in regionalization reduces the number of model parameters to 8. In most of the other variations in regionalization and parameterization that we have tried, the results of the inversions all are about the same. In one of these variations that we choose to mention here, the Himalayan part of region III has been reassigned to the Tibetan Plateau, which is region IV; no substantial differences in the results of the inversion have been obtained.

We report here only one of the results of these numerous inversions. In this variation we have used a regionalization as described above in which we have a large region IA spanning northern Eurasia and a region IV which is enlarged to include the Himalayas. The parameterization involves only 7 parameters since we assume, <u>in this version</u>, that no channel is present under region IA to the depth of the start of the subchannel region. This assumption is equivalent to asserting that no significant amount of partial melting exists under the rather older, cooler shields.

The results of the inversion are as follows:

Depth from surface to top of LVC under region III...100±21 km
Depth from surface to top of LVC under region IV...239±57 km
Depth from surface to top of LVC under region V... 69±15 km
Depth from surface to top of LVC under region VI... 63±12 km
S-wave velocity in lid to LVC across Eurasia......4.56±.03 km/s
S-wave velocity in LVC across Eurasia.............4.33±.03 km/s
Depth to top of subchannel from surface.......... 273± 5 km

The noteworthy ingredients of these results are:

1) The S-wave velocity in the lid is rather lower than the value for the other shields of the world. This is annoying because it may imply a different geochemical origin of the Asiatic shields than those of the other continents. This item is still under investigation.

2) The lid thickness for the Zagros-Turkic deformation belt is appropriate to the dehydration of amphibole as a universal mantle process for regions of the world with moderate continental heat flows.

3) The similarity of regions V and VI is surprising because of the apparently different seismic regimes of the two regions. Again this item is being investigated separately. The lid thicknesses for these regions are characteristic of regions of high heat flow and significant seismic activity.

4) The lid to the channel under the Tibetan Plateau is extremely thick. In no way can this interface have a depth as shallow as 100 km or less. However our resolution at depths of the order of 240 km is so poor that it is likely that no channel exists under the Tibetan Plateau at all; the lid may extend, shieldlike, all the way to the subchannel boundary at 270 km or thereabouts. Indeed the velocity contrast between the lid and the subchannel is so small in our models that in regions such as IA and IV, the latter being the Tibetan Plateau, the high velocity lid extends to the greatest depths resolved by our data, without any perceptible major velocity discontinuity.

Conclusion 4) implies that the upper mantle under the Tibetan Plateau is at temperatures well below the solidus to significant depths, and, as implied above, is a region that has all the appearances of an ancient, precambrian shield. The most plausible model to describe the historical antecedents of this interpretation of our data is that the present day Tibetan upper mantle is the extension of the shield upper mantle presently under India. We assume that underthrusting of the Indian Plate, including its shield mantle continues to this day. The cold mantle of the Indian shield has displaced northward and eastward the mantle material that was located under the Tibetan region prior to the collision of the two plates. In this case, it would appear that the northernmost edge of the present-day Tibetan uplift marks the farthest extent of the paleocene Indian Plate.

The absence of a significant Lg phase across Tibet[13] is possibly the result of frictional heating of the crust at the top of the underthrust Indian plate; the heat developed by the collision may still be trapped in the crust and may not yet have reached the surface or the mantle below. The crustal heat should cause an anomalously low velocity layer in the mid-crustal cross-section of the Tibetan region.

## References


(1) Knopoff, L., Schlue, J.W., Tectonophysics, 15, 157-163 (1972).
(2) Biswas, N.N., Knopoff, L., Geophys. J. Roy. Astron. Soc., 36, 515-540 (1973).
(3) Knopoff, L., Tectonophysics, 13, 497-519 (1972).
(4) Toksöz, M.N., Anderson, D.L., J. Geophys. Research, 73, 3801-3806 (1966).
(5) Dziewonski, A.M., Geophys. J. Roy. Astron. Soc., 22, 289-325 (1970).
(6) Kanamori, H., Phys. Earth and Planet. Interiors, 2, 259-275 (1970).
(7) Leeds, A.R., Knopoff, L., Kausel, E.G., Science, 186, 141-143 (1974).
(8) Knopoff, L., J. of Geophysics, 43, 509-510 (1977).
(9) Chang, F.-S., The Structure of the Upper Mantle of Eurasia and Tectonic Interpretation from a Study of Rayleigh Wave Dispersion, Ph.D. Thesis, University of California, Los Angeles (1979).
(10) Schwab, F., Knopoff, L., Bull. Seismol. Soc. Amer., 60, 321-344 (1970).
(11) Schwab, F.A., Knopoff, L., Fast Surface Wave and Free Mode Computations, in Methods of Computational Physics, vol. 11, B.A. Bolt, Editor, Academic Press (1972).
(12) Jackson, D.D., Geophys. J. Roy. Astron. Soc., 28, 97-109 (1972).
(13) Ruzaikin, A.I., Nersesov, I.L., Khalturin, V.I., Molnar, P., J. Geophys. Research, 82, 307-316 (1977).

Table 1.

| No. | Region | Location | Date | Time | $M_S$ | Station Period Range (sec.) | | | |
|---|---|---|---|---|---|---|---|---|---|
| 1. | Red Sea | 27.7°N 34.0°E | 31/ 3/69 | 07:15:54 | 6.8 | ANP | 50-167* | MAN | 69-167 |
| | | | | | | BAG | 38-250 | SHL | 100-167 |
| 2. | Xingtai | 37.3°N 114.9°E | 7/ 3/66 | 21:29:17 | 6.8 | ATU | 38-139 | POO | 50-139 |
| | | | | | | KEV | 38-100 | STU | 50-139 |
| | | | | | | LAH | 30- 69* | TAB | 30-119 |
| | | | | | | NDI | 30- 50* | | |
| 3. | Kamchatka | 51.8°N 156.8°E | 26/12/64 | 14:30:29 | 6.5 | CHG | 50-227 | NDI | 100-227* |
| | | | | | | HOW | 38-227 | QUE | 38-250 |
| | | | | | | IST | 38-208 | SHL | 38-227 |
| | | | | | | JER | 100-167 | | |
| 4. | Lena River | 78.2°N 126.6°E | 25/ 8/64 | 13:47:21 | 6.8 | ATU | 50-100 | KEV | 38-227 |
| | | | | | | HKC | 50-115 | NDI | 30-250 |
| | | | | | | HLW | 69-167 | NHA | 30-250 |
| | | | | | | HOW | 38-250* | SHI | 30-250 |
| 5. | Tashkent | 41.2°N 69.3°E | 25/ 4/66 | 23:22:53 | 5.0 ($m_b$) | NDI | 30- 38 | SHL | 30- 50* |
| 6. | Lop Nor | 40.9°N 89.4°E | 14/10/70 | 07:29:59 | 4.6 ($m_b$) | KBL | 30- 69 | NDI | 30- 69 |
| | | | | | | SHL | 30- 69* | | |
| 7. | Yunnan | 26.1°N 103.2°E | 13/ 2/66 | 10:44:41 | 5.9 | ANP | 30-100 | SEO | 30- 50 |
| | | | | | | BAG | 30- 50 | SHK | 30-139 |
| 8. | Hindu Kush | 36.4°N 71.1°E | 6/ 6/66 | 7:46:16 | 6.8 | ANP | 50-139* | HKC | 69-119* |
| 9. | Hindu Kush | 35.1°N 72.9°E | 28/12/74 | 12:11:44 | 6.2 | ANP | 38-119* | | |


## Acknowledgements

This research was supported by grant number EAR78-23841 of the National Science Foundation. Publication number 2104, Institute of Geophysics and Planetary Physics, University of California, Los Angeles.

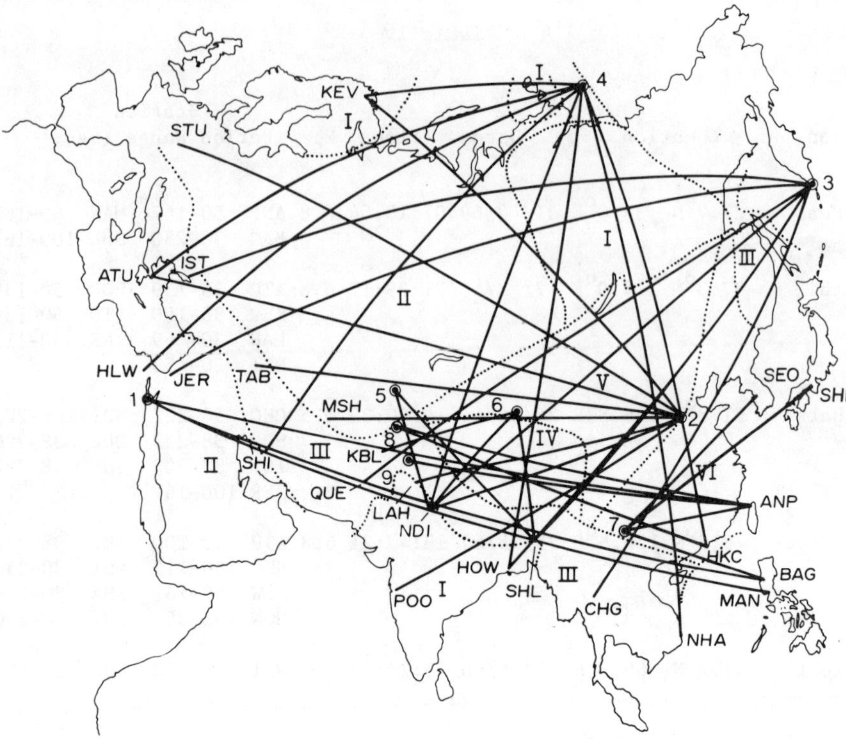


Figure 1

Map of Eurasia showing regionalization (Roman numerals) and earthquake (Arabic numerals) -- station (3-letter code) paths.

CHARACTERISTICS OF GEOPHYSICAL FIELDS AND PLATE TECTONICS OF THE QINGHAI-XIZANG PLATEAU AND ITS NEIGHBOURING REGIONS

Teng Ji-wen*
(Institute of Geophysics, Academia Sinica)

No geophysical work has ever been done in the past on the Qinghai-Xizang Plateau. In recent years, a considerable amount of geophysical investigations have been carried out on the Qinghai-Xizang Plateau, including geomagnetic and airborne magnetic surveys, paleomagnetic study, gravity, geothermics, seismicity, explosion seismology and magnetotelluric observations. These are of great values in the study of the structure and uplift of the plateau and their relation with plate tectonics.

## GEOMAGNETIC AND AIRBORNE MAGNETIC SURVEYS

The location and distribution of magnetic anomalies in the Himalayas are consistent with the trend of the arcuate mountain range. Anomalies in total intensity, vertical intensity and magnetic inclinations are all negative and their anomalous values are increasing from south to north gradually.

The magnetic total intensity in the Qinghai-Xizang Plateau region (Fig. 1) shows a series of bands in the east-west direction, i.e. the band of negative anomalies extending from Gandise to Nyainqentanglha to Hengduan Mts.(Meridional Range), the band of positive anomalies (several tens to 200 gammas) extending from Zhongba crossing the lower waterdivide in south Xizang, reaching the great bend of Yarlung Zangbo River, and the Himalayan negative anomalies with small amplitude and variations. Along both sides of Yarlung Zangbo River, for example, near Dagzhuka, the difference of positive and negative magnetic anomalies amounts to as much as 1,300 gammas.

We have found from the aeromagnetic data that there is a belt of high positive anomalies up to 150-450 gammas in the Yarlung Zangbo River area (Fig. 2) and the intensity reduces rapidly to the north. It is interpreted as the effect of ultra-basic rocks.

---

* Joint work by Wang Shao-zhou, Zhu Zhi-wen, Su De-yuan, Zhou Wen-hu, Yang Bing-ping, Chen Guang-ming, Xiang Yong-ren, Yao Zhen-xing, Zhang Li-min, Xu Zhen-wu, Xan Guo-hui, Xiong Shao-bo, Wei Si-yu and others.

Both the ground and airborne magnetic data reveal that the northern and southern parts of the Yarlung Zangbo River differ from each other in lithology, material composition, and deep structures.

## PALEOMAGNETIC STUDIES

The paleomagnetic studies from late Cretaceous to Paleocene formations show that, the latitudes of paleomagnetic poles for Biru, Lhunzhub, Lhasa and Zhamo range from $58^{\circ}$-$69^{\circ}$ N (Fig. 3a). Here the pole positions distribute uniformly in the group of pole positions for Eurasian Plate. The latitudes of the paleomagnetic poles for Tulong, Nyalam, Tingri, Gyangze and Yadong in the south of Yarlung Zangbo River, range from $23^{\circ}$-$31^{\circ}$N (Fig. 3b). Although the locations of paleomagnetic poles are different from those in the north of the Yarlung Zangbo River, yet they are similar to those in the Deccan Plateau of India and the Salt Range area of Pakistan. The difference of paleolatitudes for areas of the north and south of the Yarlung Zangbo River amounts to $50^{\circ}$. The average paleolatitude of northern Yarlung Zangbo River area is $27^{\circ}$N, similar to the present latitude, but in the southern area, the paleolatitude is $20^{\circ}$-$24^{\circ}$S in average. This indicates that the Yarlung Zangbo River is a boundary between different plates.

The relative positions of the Indian Plate (Fig. 4) in different geological periods suggest that about 38 my ago, Indian Plate moved rapidly(1) to the north at a rate of about 100 mm/yr, but the rate slowed down with time and the present rate is 50-60 mm/yr. The Indian Plate has moved northward about 5,000 km relative to Eurasian Plate

## GRAVITY MEASUREMENTS

The map of Bouguer anomaly for the Qinghai-Xizang Plateau reveals that surrounding the plateau there exists high gradient variation zones, but in the centre of the plateau the gravity field varies gently (smoothly). The northern Himalayan area is covered by negative anomalies. In the Qomolangma area, the Bouguer anomaly is -310 mgal and toward the center it reaches a minimum value of about -550 mgal (Fig. 5). Between the Yadong-Pagri area, the gradient amounts to 5.6 mgal/km.

The isostatic anomaly near Qomolangma amounts to +120 mgal (Fig. 6). To its south, near Kathmandu and Darjeeling Mt., the isostatic anomaly is close to zero; to the north of the Yarlung Zangbo River it is also close to zero. It is concluded, therefore, that the uplift of the Himalayan mountain region is due to the collision and compression of the two plates as well as the isostatic compensation of the crust-mantle materials. Under the large-scale tectonic effect the uplift of the Himalayan mountain system and the movement of mass at depth still continue.

Fig. 7 is the profile of the crustal thickness of Golmud-Amdo-Himalayas-Ganges plain according to the gravity data. The crust is getting thicker from the south to the north, it rises in the Kunlun mountains area; in the Qilian mountains area, the crustal thickness is from 50 to 52 km and in the north of the Qilian mountains there is an abrupt change in the crustal thickness. Inside the plateau it changes gently.

## GEOTHERMAL ACTIVITIES

Throughout the plateau there are striking geothermal activities(2) with more than 600 hydrothermal locations. Hydrothermal explosions, geysers and boiling springs are mainly located along the Yarlung Zangbo River and further south. To the east of the plateau, from Tengchong to western Yunnan Province not only there are wide spread hot springs, with the temperature of some exceeding the boiling point, but also recent volcanic activities. Here is one of the most active geothermal zones in China

From the high geothermal activities, the extremely thick crust, the existence of the low velocity layer and the low Bouguer anomaly of gravity, the low Q values, the small weak Lg phase of earthquakes and the reduction of the high frequency components of seismic waves in this area in comparison with those in its vicinity, we can conclude that melting or partial melting of the material within the crust and upper mantle must have taken place.

In Damxung area the Qp value of small earthquakes is 150, the Bouguer anomaly is as low as -550 - -570 mgal, the low stress drop ranges from 5 to 15 bar. These suggest that the temperature of the crust in Yangbajain area is relatively high. The temperature gradient in some shallow (tens of meters) wells in Yangbajain thermal field reaches $2.9^{\circ}$-$3.3^{\circ}$c/m (100 times higher than that of normal temperature gradient). It may be ascribed to the local hydrothermal activity or magmatic chambers.

## SEISMICITY AND THE FAULT PLANE SOLUTIONS

Earthquakes with magnitude more than 7 are located in the surrounding areas of the plateau. In the plateau itself shallow earthquakes (Fig. 8) with focal depths less than 70 km dominate. Earthquakes with medium focal depth are concentrated at the eastern and western extremes of the arcuate mountain system and are rare in the central part.

In Damxung, Lhasa and Yarlung Zangbo area the focal plane of shallow earthquakes inclines to the south (Fig. 9a) and the angle is large. The profile of focal depth distribution crossing Tianshan Mt., Tanggula Range to the Ganges plain (Fig. 9b) indicates, that at the northern and southern Yarlung Zangbo River the focal planes dip southward and northward respectively. On the northern rim of the Ganges plain the focal plane of the shallow earthquakes inclines northward which is in agreement with the results of Tetsuo Santo(9). Here the scattered intermediate earthquakes are shallower than those in the Yarlung Zangbo River area, but they are mainly concentrated in the Himalayan zone.

In the Hindu Kush, the intermediate earthquakes occur mostly at depths of about 200 km and the maximum depth is 300 km (Fig. 10). The focal plane in this area inclines to the north(3). It inclines to the south in the Pamirs area, and forms a "V"-shaped configuration.

The mountain arc in northwestern Burma is a zone of intermediate earthquakes with a maximum depth of 200 km (Fig. 11). The inclination of the plane of foci is about $40^{\circ}$-$45^{\circ}$ eastward.

In Kashmir, Kathmandu, and near Gonhati of India, the depths of the scattered intermediate earthquakes range from 70 to 90 km, while they are 82-180 km near Gyangze and Lhasa. This can not be simply explained by the southward or northward inclination of the focal planes. Owing to the compression between the Eurasian and Indian Plates, strong deformations and a series of faults were generated in the Himalayan area. The focal planes dip towards each other but whether they can penetrate into the mantle or not remains to be studied.

In Qinghai-Xizang Plateau, the axes of the principal compressive stress are NS and NE and the angle of the axis of the compressional stress relative to the horizontal plane is mostly less than 40°. This suggests the domination of the horizontal compression (Fig. 12). In the Himalayan seismic zone the result obtained from the fault plane solutions indicates thrust faulting*. In the Ganges plain near Delhi normal faults are found. Inside Xizang Plateau there are a series of parallel and bended left lateral strike slip faults. The direction of the compressional stress changes with the trend of arcuate structure, so that the distribution of stress is constrained by the direction of the arcuate structure.

## MAGNETOTELLURIC DEEP SOUNDING

Five stations along a profile of over 500 km are chosen for MT measurement in Nagqu, Nyetang, Gyangze, Pagri and Yadong. A five-layer model of the resistivity distribution is assumed and the total thickness of the upper four layers of Nagqu is 66 km, but at Pagri it is 57 km. At the depth of 10 km there is a low resistivity layer (Table 1).

## EXPLOSION SEISMOLOGICAL STUDY

The thickness of the earth crust in Qaidam Basin of Qinghai Province is 52 km(4). The velocity at the top of the mantle is 8.2 km/sec. The Damxung-Yadong profile on the Xizang Plateau is perpendicular to the strike of the tectonic system in this area. We made nine explosions in lakes** . The length of the profile is about 430 km.

Explosions in the lakes indicate that the crust of the plateau is extremely thick. To the north of the Yarlung Zangbo River the thickness is 70-73 km and to the south from Yamzho Lake to Yadong, 68-45 km (Fig. 13). The stratified crust includes a low-velocity layer in its lower part which is 10 km in thickness and 5.64 km/s in velocity, its depth at the north of the Yarlung Zangbo River is 42-46 km and at the south of the river is 45-29 km. The crust upwarps to the south. Along the Yarlung Zangbo River there is a steeply dipping rift zone with slightly southward inclination, the width of which is about 30-50 km.

As illustrated in Fig. 14, the crustal thickness is not uniform and not symmetrical in NS direction. The Himalayan area is a zone of abrupt changes in crustal thickness. After collision of the two plates, the shortening and thickening of the crust led to the formation of a series of folds and mountain systems (Fig. 15).

---

* Yan Jia-quan et al., Recent tectonics of the Qinghai-Xizang Plateau.

** Teng Ji-wen et al., Explosion seismological study for velocity distribution and structure of the crust and upper mantle from Damxung to Yadong of the Xizang Plateau.

## THE BOUNDARY REGION BETWEEN THE INDIAN PLATE AND THE EURASIAN PLATE--THE TRANSITIONAL BELT

The Qinghai-Xizang Plateau is one of the most important regions for intra-plate earthquakes, and is different from other seismic areas of the world in the character of seismic activities. Inside the continents, the junction of the two plates is not a narrow suture, but a broad zone, amounting to nearly 300 km in width (Fig. 16). Intense tectonic deformation, earthquakes, hydrothermal and magmatic activities and abrupt changes in the geophysical field are found in this zone. We may call this zone a transition belt.

Inside this transition belt the gravity field has a high gradient and the crust is not in isostatic equilibrium. On the map of total magnetic intensity there is a series of bands of positive and negative anomalies in the east-west direction on both sides of Yarlung Zangbo River.

To the north and south of the Yarlung Zangbo River, there is a zone of abrupt change in crustal thickness. In the granites the recrystallized biotite is found at the lower watershed in South Xizang. This is a belt of ophiolite accompanied with melange.

Historical records show that there were frequent occurrences of strong earthquakes near Yadong, Lhunze , Lhozhag, Zanda and Shannan. The focal planes of shallow earthquakes between Yarlung Zangbo River and the north of the Ganges plain are dipping in opposite directions toward each other. The depth of focal distribution formed a "V"-shaped region. The low velocity layer in the crust exists in this region and the Bouguer anomaly is negative. The transitional zone is characterized by the high hydrothermal activities and the depth of Curie isotherm is 18-30 km according to the airborne magnetic data. The Yarlung Zangbo River is a regional anomalous zone of high heat flow.

It can be suggested that the Himalayan zone is a transitional belt between the collided plates. The northern boundary of this zone is the Yarlung Zangbo River and the southern boundary is the northern rim of the Ganges plain.

The influence of collision and compression of two big plates extended southeast to Sichuan and Yunnan(5) (Fig. 17), and northeast to Shanxi, Gansu, Ningxia and Qinghai(6) Provinces of China (Fig. 18). The analysis of tectonic stress field throughout China (Fig. 19) shows that the effect of the stress field inside the plateau after the collision of the two plates is far-reaching.

Qinghai-Xizang Plateau has an extremely thick crust, but it is not a double-layered one. The south and the north of the Yarlung Zangbo River belong to the Indian and Eurasian Plates respectively. Accordingly, the focal planes of shallow earthquakes of the two plates in this region dip southward and northward in opposite directions respectively. As a result of the process of collision the crust was shortened and thickened and the earth's surface was therefore uplifted.

In Yarlung Zangbo River-Himalayan region, strong geothermal activities are the causes of low Q value and weak Lg waves. Here, melting and partial melting probably exist in the crustal media.

We are very grateful to Professor Fu Cheng-yi for his kind guidance and encouragement.

References

(1) Molnar P. et al., Cenozoic tectonics of Asia; Effects of a continental collision. Science, 189, 419-426 (1975).
(2) Tong Wei et al., The Himalayan Geothermal Belt. Acta Scientiaram Naturalium Universitatis Pekinensis, No. 1, (1978) (in Chinese).
(3) Huan Wen-lin et al., The distribution of earthquake foci and plate tectonics on the Qinghai-Xizang Plateau. Acta Geophysica Sinica, 23 (3), 269-280 (1980) (in Chinese).
(4) Teng Ji-wen et al., Deep reflected waves and the structure of the earth crust of the eastern part of Chaidam Basin. Acta Geophysica Sinica, 17 (2), 122-135 (1974) (in Chinese).
(5) Kan Rong-ju et al., Present tectonic stress field and its relation to the characteristics of recent tectonic activity in southwestern China. Acta Geophysica Sinica, 20 (2), 96-109 (1977) (in Chinese)
(6) Li Yu-long, Primary analyses on the new and present-day tectonic stress field and its power source of the Shansi-Gansu-Ningxia-Qinghai region Northwestern Seismological Journal, (4), 1-11 (1979).
(7) Tetsuo Santo, Seismicity around the Himalaya Science 41 (4), 231-244 (1971) (in Japanese).

Table 1

| Nagqu | | Pagri | |
|---|---|---|---|
| The total thickness of four layers (km) | The depth of low resistivity zone (km) | The total thickness of four layers (km) | The depth of the low resistivity zone (km) |
| 66 | 6 | 57 | 12 |

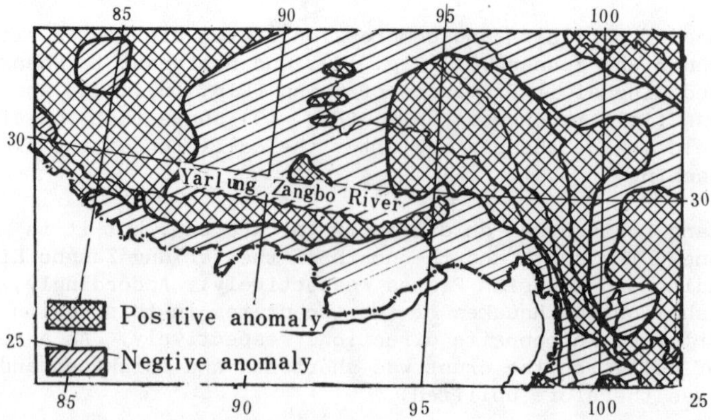


Fig. 1 Anomalies in total intensity of the Qinghai-Xizang Plateau.

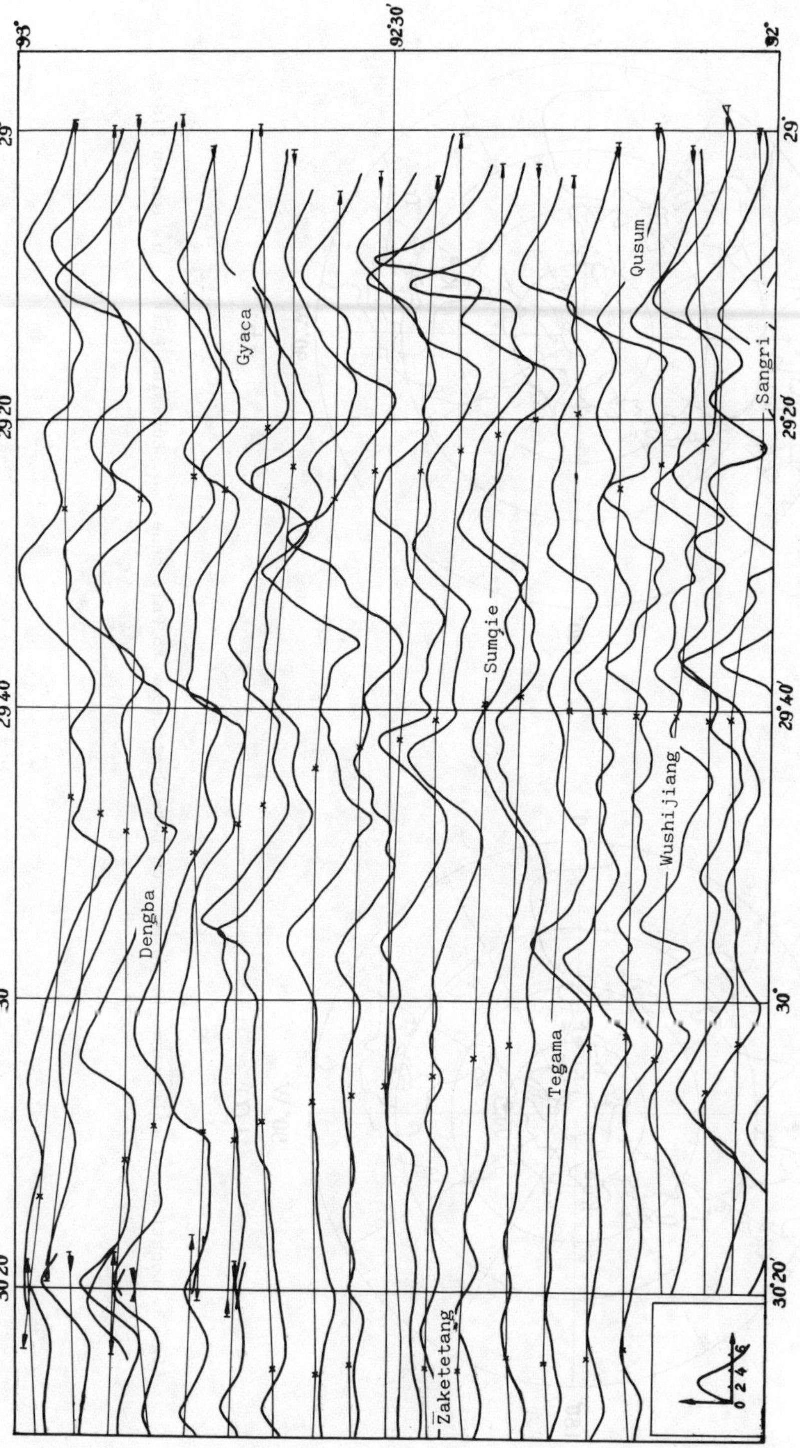


Fig. 2 Map of aeromagnetic anomaly in the south of Qinghai-Xizang Plateau.

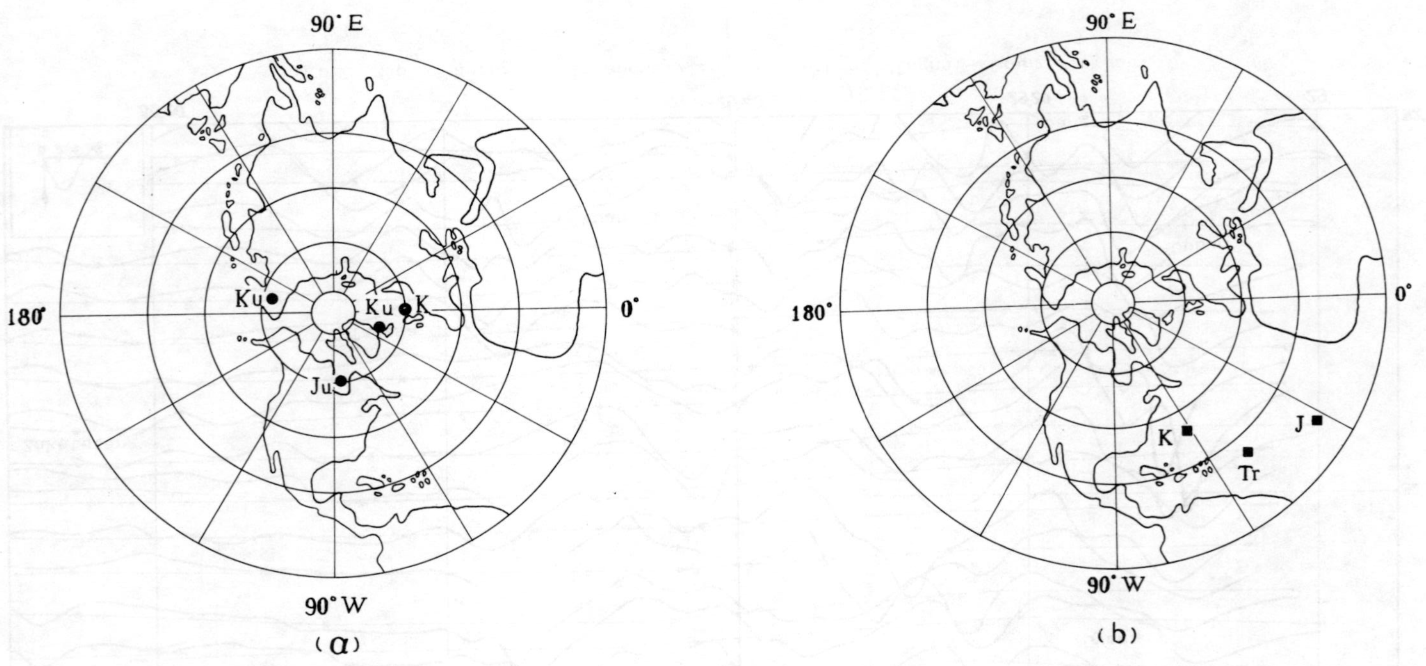


Fig. 3 Locations of paleomagnetic poles from Late Cretaceous to Paleocene. a. Eurasian Plate; b. Indian Plate.

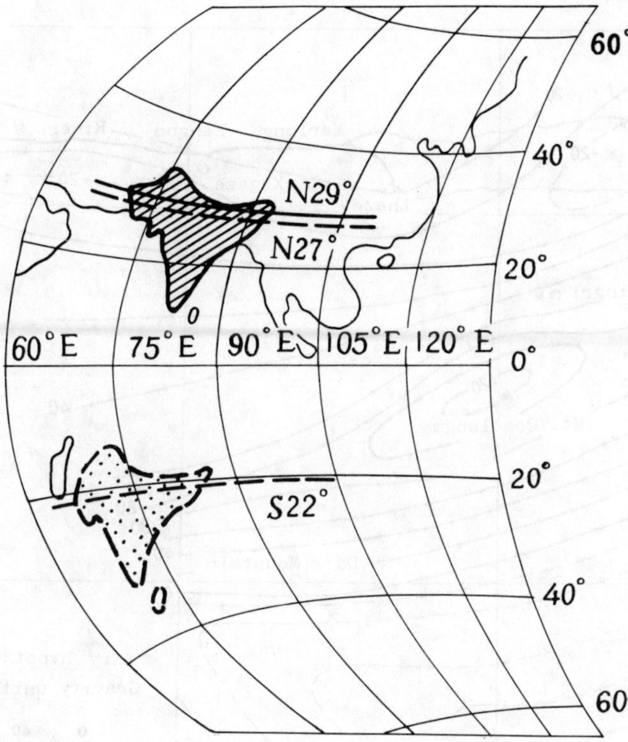


Fig. 4 Simplified map of the relative positions of the Indian Plate during its drift northeastward in different geological periods.

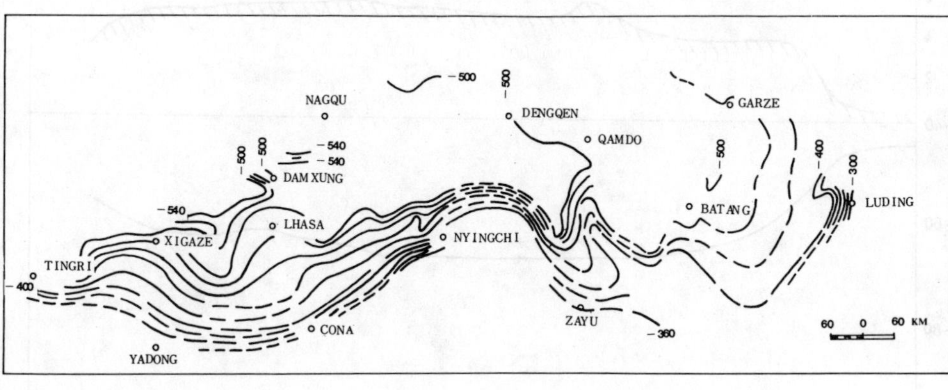


Fig. 5 The map of Bouguer anomalies in Qinghai-Xizang Plateau.

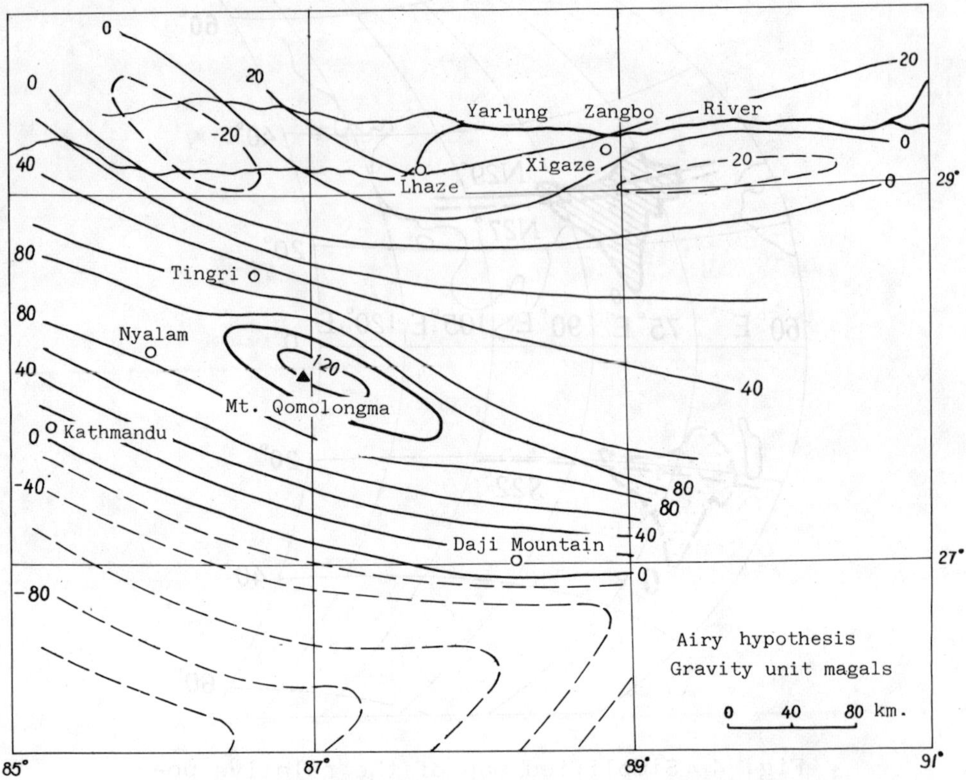


Fig. 6 Isostatic anomaly of the Himalayan region.

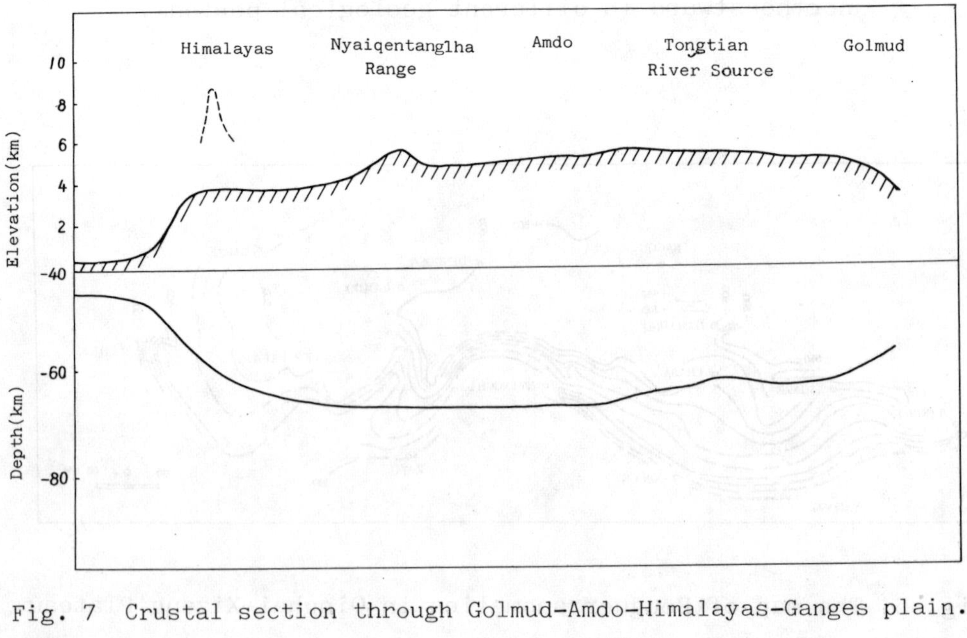


Fig. 7 Crustal section through Golmud-Amdo-Himalayas-Ganges plain.

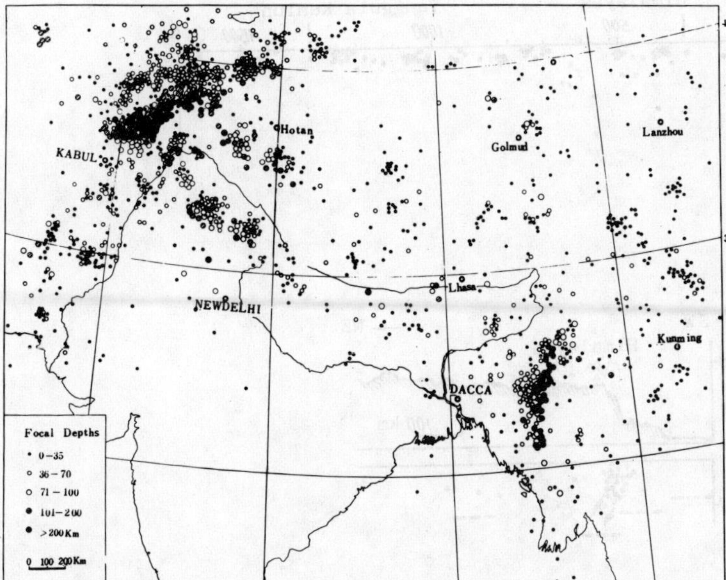


Fig. 8 Map of earthquakes distribution in Qinghai-Xizang Plateau and its neighbouring regions.

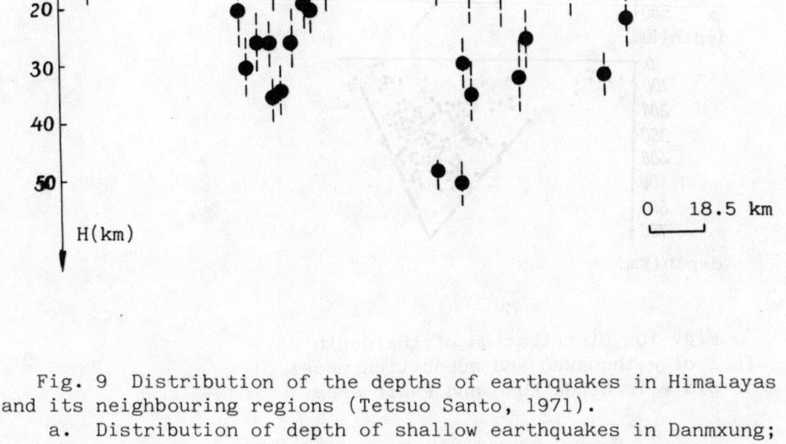


Fig. 9 Distribution of the depths of earthquakes in Himalayas and its neighbouring regions (Tetsuo Santo, 1971).

a. Distribution of depth of shallow earthquakes in Danmxung;

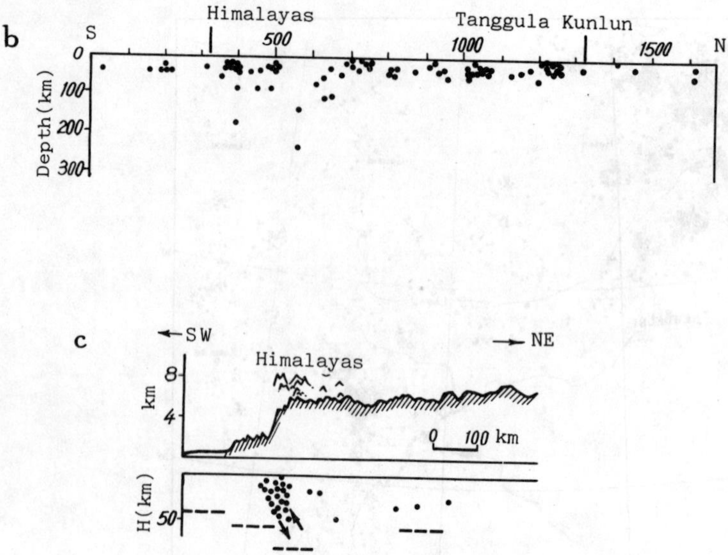


Fig. 9 Distribution of the depths of earthquakes in Himalayas and its neighbouring regions (Tetsuo Santo, 1971).
b. Section in Altun Mountain-Himalayas;
c. Southern foot of the Himalayas.

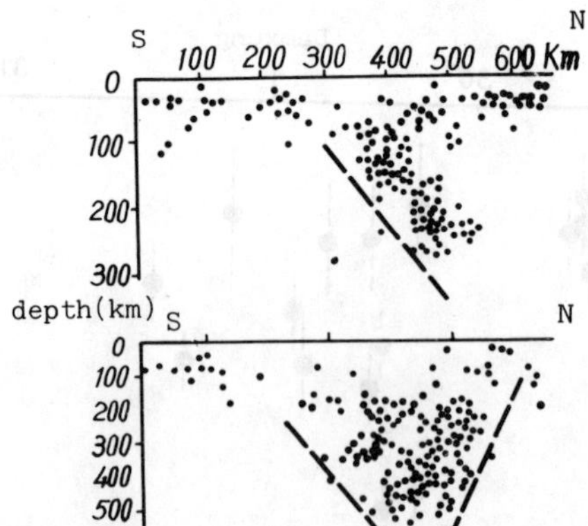


Fig. 10 Distribution of the depth of foci of eathquakes and subduction model of plate in Hindu Kush and Pamir area.

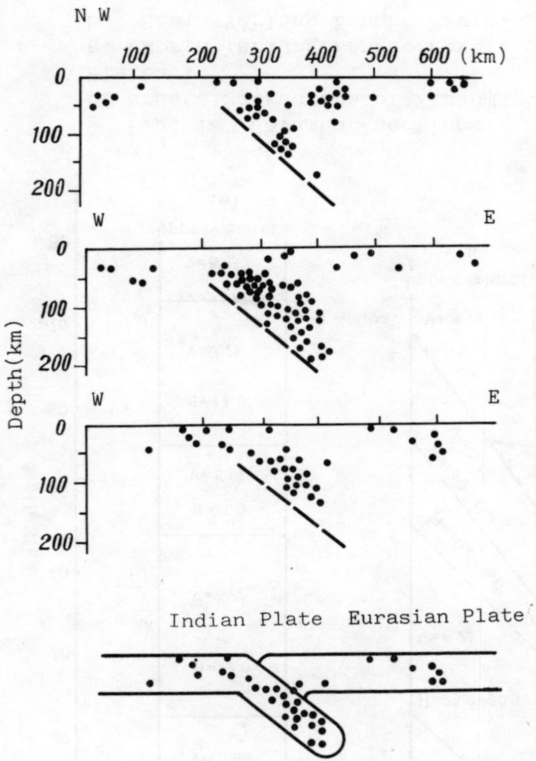


Fig. 11 Depths of foci of earthquakes and subduction zone in sections crossing the arcuate mountain system of India-Burma regions.

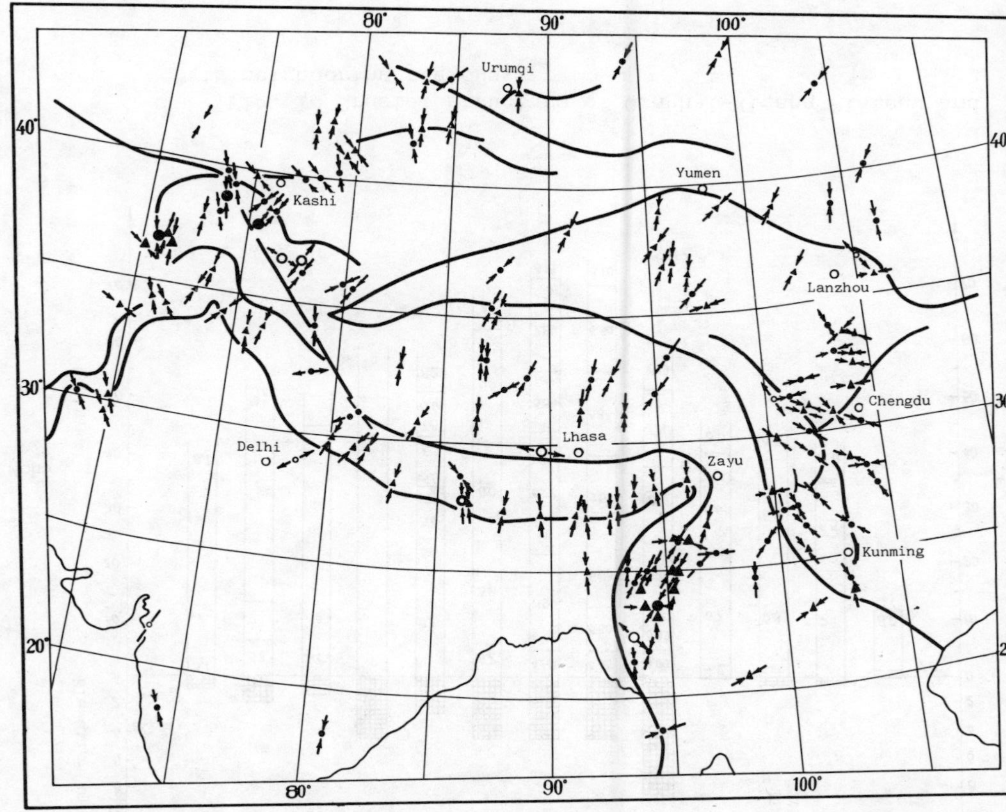


Fig. 12 Horizontal projections of principal compressive axes of shallow earthquakes in Qinghai-Xizang Plateau.

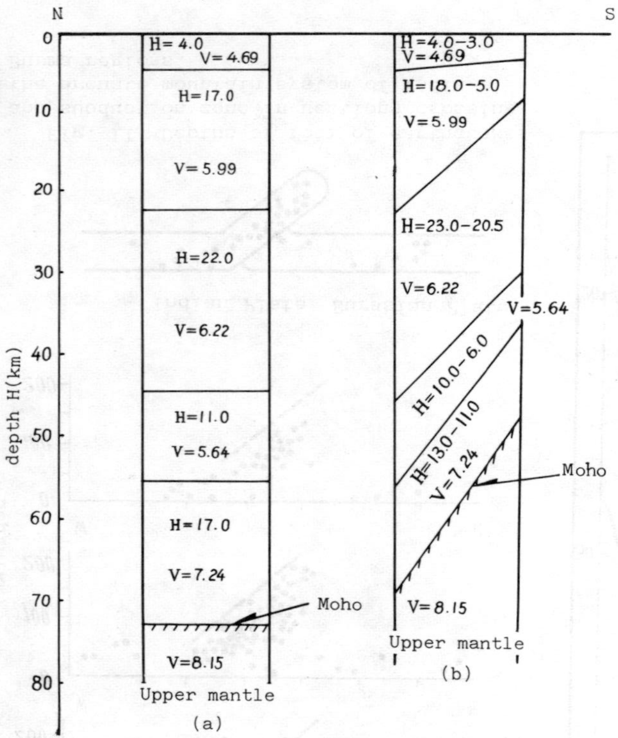


Fig. 13 Columnar sections of crustal structure of Xizang Plateau.
a. North Yarlung Zangbo River;
b. South Yarlung Zangbo River.

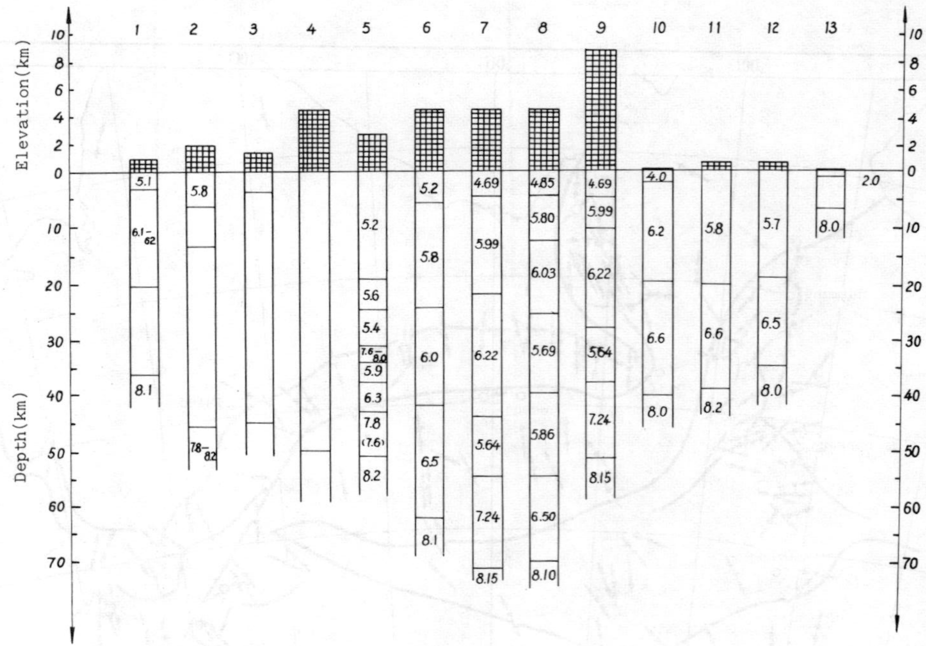


Fig. 14 Crustal structure of Qinghai-Xizang Plateau and its neighboring regions.

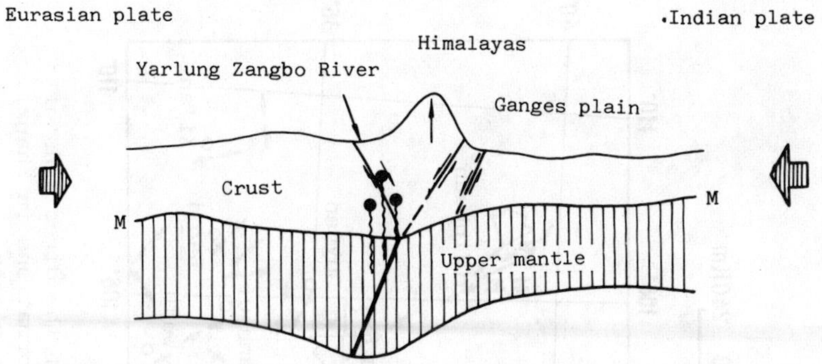


Fig. 15 Simplified map of the crustal model and formation of the extremely thick crust in consequence of the collision and compression of the Indian Plate and Eurasian Plate.

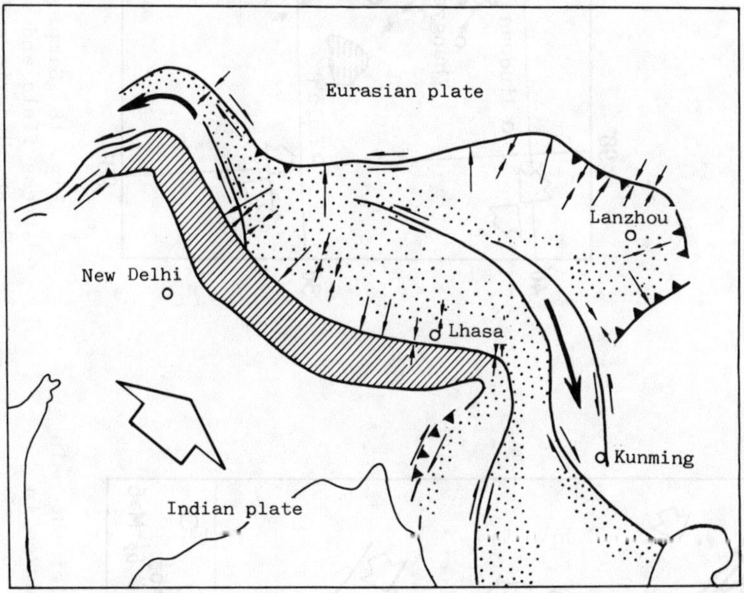


Fig. 16 Simplified map of the transition belt between Indian Plate and Eurasian Plate

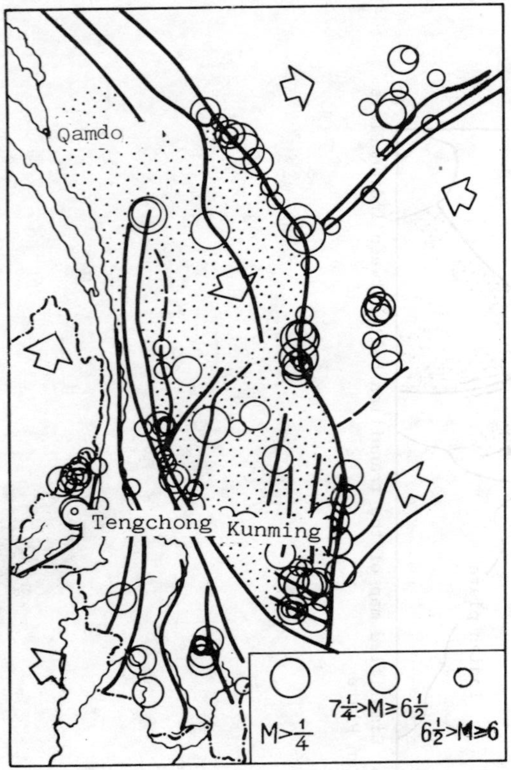


Fig. 17 Distribution map of the main rift belts and strong earthquakes in Sichuan and Yunnan regions.

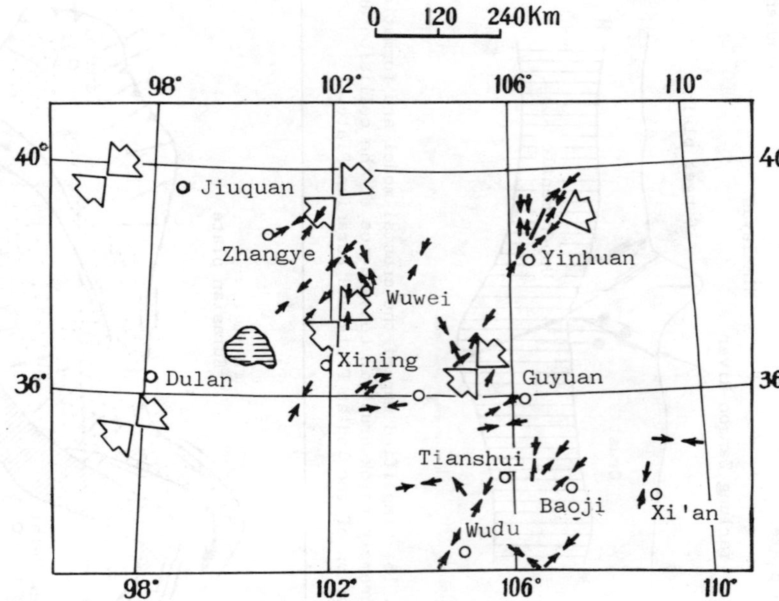


Fig. 18 Simplified map of the directions of stress field and seismic deformations in Shanxi, Gansu, Ningxia and Qinghai Provinces.

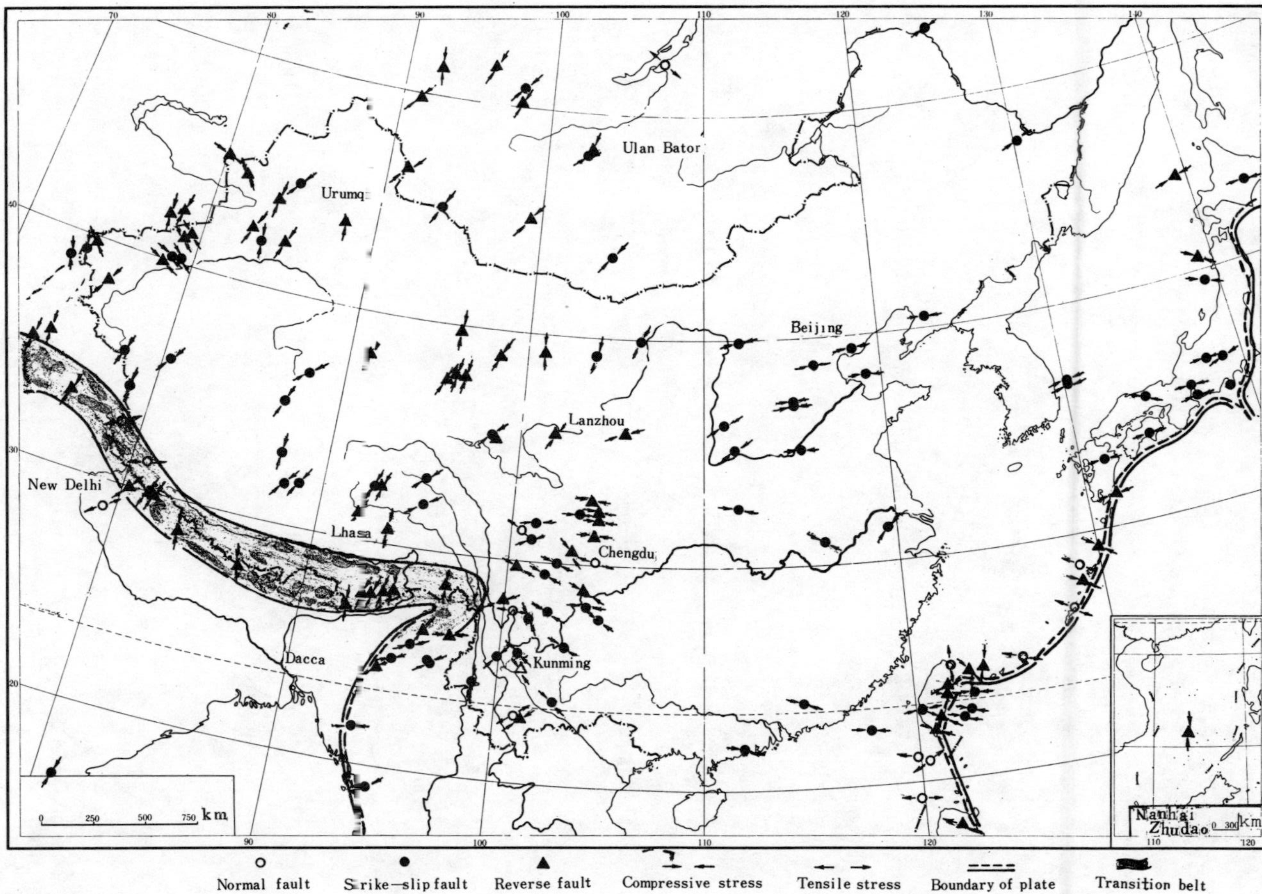


Fig. 19 Horizontal projections of principal compressive stress axes for shallow earthquakes in China.

# THE DISTRIBUTION OF EARTHQUAKE FOCI AND PLATE TECTONICS IN THE QINGHAI-XIZANG PLATEAU AND ITS VICINITY

Huan Wen-lin   Wang Su-yun   Shi Zhen-liang   Yan Jia-quan
(Institute of Geophysics, State Seismological Bureau)

In present paper, we attempt to discuss the features of plate tectonics by investigating the distribution of earthquake foci as well as some geological structure and geophysical results. The area studied is from 15° to 42° N and 60° to 108° E.

## THE DISTRIBUTION OF EARTHQUAKE FOCI

Based on the catalogue of earthquakes registered in Chinese seismological stations and data from ISC (1969-1976)(1,2), we selected those earthquakes with more accurately determined parameters to compile a map of distribution of earthquake foci in the Qinghai-Xizang Plateau and its vicinity (Fig. 1). From Fig. 1, we can see that in this region not only the shallow earthquakes are widely distributed, but also the intermediate shocks are regularly distributed. These intermediate shocks occurred not only in the Hindu Kush and the India-Burma Mountain Arc as indicated by the previous authors (3), but also in the Indus River and the Yarlung Zangbo River and the south of them as well as in the interior of the Asian continent such as the Pamir, the West Kunlun Mountains, the northern margin of the Qaidam Basin, and the southern margin of Tianshan Mountain. They often concentrate at one side of some deep seated faults, and form a series of inclined focal zones together with shallow earthquakes (Fig. 2). We have taken a series of profiles perpendicular to the seismic zones. The profiles were taken within 200 km on each side of the focal projection for the Yarlung Zangbo River and Qaidam Basin zone, within 100 km for the west Kunlun and Tianshan zone, and within 50 km for other regions. Following, a brief description of each region is given.

### Hindu Kush Zone

There are two profiles A-A' and B-B' across the Hindu Kush area. From profile A-A' we can see that the zone of foci is north-dipping at an angle of 45°. The greatest focal depth is about 300 km. Profile B-B' passes through both the Hindu Kush and the Pamir areas. From profile B-B' we can see there is a V-shaped plane of foci, it may be shown that the north-dipping Hindu Kush zone and the south-dipping Pamir zone are very close to and even contact with each other in depth (Fig. 3).

### India-Burma Mountain Arc Zone

Three profiles C-C', D-D' and E-E' pass through this area. From these profiles, we can see that the zone of foci is east-dipping at an angle of 45°. The greatest depth is about 180 km (Fig. 3).

## Indus Plate Convergence Zone

The profiles F-F', G-G' and H-H' pass through this area. From these profiles, it can be seen that the zone of foci dips southward beneath the Himalayas at an angle of $40^{o}$-$60^{o}$. The greatest depth are more than 200 km. Beneath the Ganges plain, there has occurred an intermediate earthquake about 280 km in depth. Whether the zone of foci continuously extends beneath the Ganges plain merits further attention (Fig. 4).

## Yarlung Zangbo River Plate Convergence Zone

Although only a few intermediate shocks were recorded in this area, from the profile I-I' we can see that a zone of foci which dips southward at an angle of 70° probably exists here. The deepest focus reaches 230 km (Fig. 5). Certainly more sufficient evidences should come from the further accumulation of seismic data.

## Pamir Area

From profiles J-J' and K-K', we can see that there is a zone of foci which dips southeastward at an angle of $60^{o}$ beneath the Pamir. The greatest depth is about 300 km (Fig. 6).

## West Kunlun Mountains Zone

From profiles L-L' and M-M' in Fig. 6, it can be seen that there is a zone of foci which is south-dipping at an angle of $45^{o}$ beneath the northern margin fault of the West Kunlun. The greatest depth is about 220 km.

## Northern Margin of the Qaidam Basin

In this area, intermediate shocks were few and distributed along the southern side of the northern margin fault of the Qaidam Basin. From profile N-N' in Fig. 6, it can be seen that the zone of foci is south-dipping at an angle of $60^{o}$. The greatest depth is about 110 km.

## Southern Margin of Tianshan Mts.

The profiles J-J'. K-K' and O-O' pass through this area. From these profiles, we can see there is also a zone of foci which is south-dipping at an angle of $60^{o}$ beneath the southern margin fault of Tianshan. The greatest depth is about 150-200 km (Fig. 6 and Fig. 7).

## Himalayas

Beneath the Himalayas, besides the intermediate earthquake zone which is south-dipping from the Indus-Yarlung Zangbo River plate convergence zone, there may be a north-dipping focal zone with lower dip angle and shallow depth beneath the main boundary faults of the Himalayas as may be seen from profiles a-a' and b-b'. The zone of foci dips northward at an angle of $30^{o}$ and the greatest depth is about 60 km on profile a-a' across Nepal; $35^{o}$ and 70 km on profile b-b' across the eastern end of the Himalayas (Fig. 8).

From above, in the plateau and its vicinity there are three inclined intermediate earthquake zones distributed successively from south to north. The first zone is located at the convergence zone of the Indian and Eurasian Plates, such as Hindu Kush, Indus-Yarlung Zangbo River and India-Burma Mountain Arc; the second zone is located at the Pamir, the West Kunlun and the northern margin of the Qaidam Basin; the third zone is located at the southern margin of Tianshan. The first zone is the greatest, and the scale of the other two has gradually decreased (Fig. 2). But there is only a shallow focal zone on the southern side of the Himalayas.

## PLATE TECTONICS

### The Three Lithospheric Subduction Zones

If we interpret the above mentioned zones of foci as the recent active lithospheric plate subduction zones, then there are three zones from the Himalayas to the Tianshan Mts. (Fig. 9).

The first zone lies in the plate convergence zone between the Indian and Eurasian plates. It covers the Hindu Kush, the Indus River, the Yarlung Zangbo River and the India-Burma Mountain Arc zones. The Hindu Kush subduction zone is north-dipping and the India-Burma Mountain Arc zone is east-dipping. This reveals that both sides of the Indian subcontinent underthrusts beneath the Eurasian continent, while the Indus and the Yarlung Zangbo River zones are south-dipping, showing that during collision most part of the north Indian subcontinent upthrusts over the Eurasian Plate, and the southern margin of the Asian continent is forced to subduct beneath the Himalayas on the northern margin of the Indian Plate. The second zone covers the Pamir, the West Kunlun subduction zones and the Qaidam deep fault zone. These zones are all south-dipping. The third zone lies in the southern margin of Tianshan and is south-dipping, too. It can be seen that the scale of the first zone is the greatest, while those of the other two have gradually decreased.

If we regard the depth of the subduction zone as the part disappeared into the mantle, the Asian continent along the above three zones has shortened at least 300-700 km since the India-Eurasia collision. The subducted depth is greater in the western part than that in the eastern part.

### The Strike Slip Fault

The plane distribution of the intermediate earthquake focal zones is discontinuous (Figs. 1,2). It indicates recent activity of many strike slip faults. These faults have cut the subduction zones into several segments (Fig. 10). For example, between the Indus and the Yarlung Zangbo River subduction zones, there is the Karakorum right lateral strike slip fault, relatively displacing the former about 250-300 km northward to the latter. The West Kunlun and the Qaidam zones were displaced about 900 km by the Altun left lateral strike slip fault(4).It may be an evidence of the powerful NE push of the Qinghai-Xizang Plateau (4,5).

The extensive tectonic framework which consists of the above three subduction zones and some strike slip faults can be interpreted as a result of the India-Eurasia collision.

## DISCUSSION

### On the Origin and Development of South-dipping Subduction Zones in the Asian Continent

A lot of data shows that the Indian subcontinent and the Asian continent were separated by the Tethyan ocean in Cretaceous period. Many authors consider that the Tethyan ocean was subducted northward beneath the Asian continent. But we think that the subduction zone of the Tethys was probably located at the northern margin of the Indian subcontinent and dipped southward. This is based on the south-dipping focal zone existed and south-dipping ophiolite suite along the plate convergence zone(6) , the northern Himalayan geosyncline with deep sea facies sediments located at the northern margin of the Indian subcontinent and other geological and geophysical evidences as well (Fig. 11).

According to these, we assume that the Asian continent is relatively stationary, but the Indian subcontinent continually drifted northward since Cretaceous, and the Tethys was forced to be subducted gradually southward beneath the northern margin of the Indian subcontinent. In the later period of the early Eocene the collision between the two continents took place. Afterwards the Indian subcontinent kept on upthrusting vigorously northwards, the marginal parts of the Asian continent was forced to continue underthrusting southwards following the Tethyan subduction zone. The effect of the northward strong compression of Indian Plate extended far into the interior of the whole plateau. The result was that some great, deep faults between the lithospheric blocks previously formed near the plate boundary became activated again. Because of strong northward up-thrusting of the southern sides of the faults, the continental crust beneath the lower sides of the faults was pushed into the mantle, causing different scales of the subduction zones.

### On the Uplifting of the Himalayas

There are many view-points about the origin of the Himalayan tectonics. We believe that the Indian Plate upthrusted northward over the Eurasian Plate and take the Himalayas as a southward reversing anticlinorium (7). The thrust faulting only exists in the crust of the southern slope of the Himalayas as demonstrated by the northward dipping shallow focal zones beneath the southern slope of the Himalayas and other geological and geophysical data. We propose a model in which the Himalayas were formed (Fig. 12) by the folding of the northern margin of the Indian subcontinent. Since the collision between the northward drifting Indian subcontinent and the Asian continent during the later Eocene and early Oligocene, the marginal parts of the Indian subcontinent have strongly folded, and at the early stage the huge Himalayan anticlinorium and the depressions at its southern slope were formed, owing to the stopping of the south-dipping subduction zones of the Asian continent. As the folding activity developed, the main central faults were created along the southern side of the Himalayan anticlinorium in Miocene. Because of the continuous northward drifting of the Indian continent, the Himalayan anticlinorium uplifted drastically and was reversed southward in Pleistocene, thus generating the great main boundary faults and the huge southward nappe tectonics along the southern margin.

We would like to express our sincere thanks to Professors Fu Cheng-yi, Gu Gong-xu and Ma Xing-yuan for their kind guidance and excouragement.

## References

(1) State Seismological Bureau, The Concise Catalogue of Earthquakes in China. Seismological Press (1977) (in Chinese).

(2) I.S.C., Regional Catalogue of Earthquakes (1969)-(1976).

(3) Tetsuo Santo, Seismicity around the Himalaya. Science, 41 (4), 231-244 (1971) (in Japanese).

(4) Huan Wen-lin, She Zhen-liang, Yan Jia-quan, Wang Su-yun. Characteristics of the recent tectonic deformations of China and its vicinity. Acta Seismologica Sinica, 1 (2), 109-120 (1979) (in Chinese).

(5) Molnar P., The Collision between India and Eurasia. Scientific American, 236 (4), 31-41 (1977).

(6) Lee Chun-yu, Tectonic evolutions of some mountain ranges in China, as tentatively interpreted on the concept of plate tectonics. Acta Geophysica Sinica, 18 (1), 52-73 (1975) (in Chinese).

(7) Yeh Hung et al., The analysis of the recent tectonic stress of the Himalaya Mountains Arc and its vicinities. Scientia Geologica Sinica, (1), 32-48 (1975) (in Chinese).

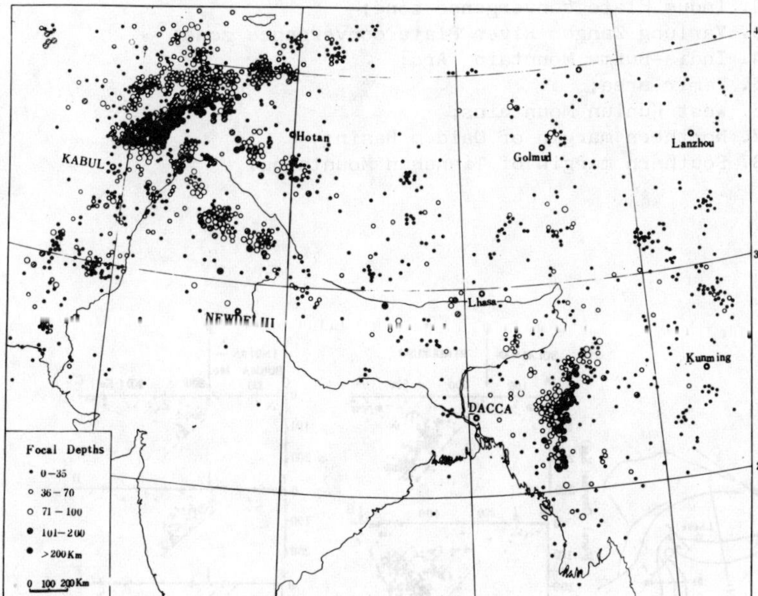


Fig. 1 The map of distribution of earthquake foci in the Qinghai-Xizang Plateau and its vicinity.

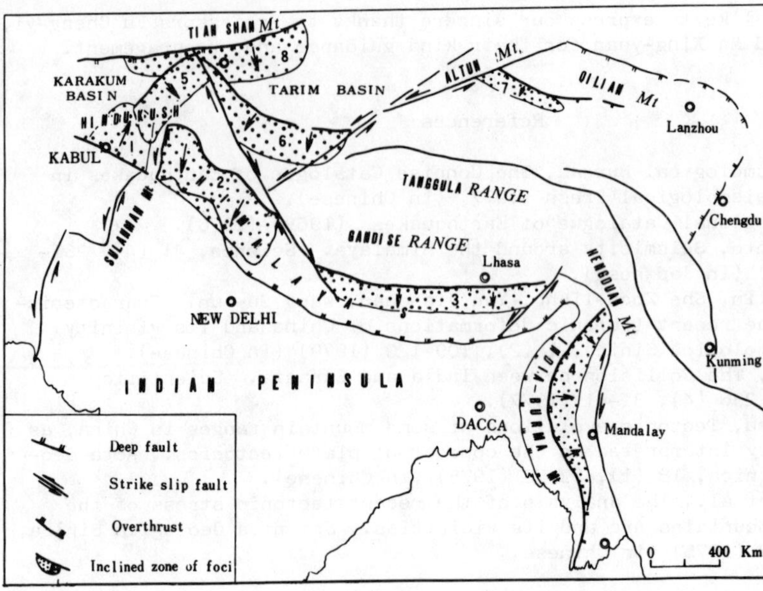


Fig. 2 The distribution of intermediate earthquake zones.

1. Hindu Kush area;
2. Indus Plate convergence zone;
3. Yarlung Zangbo River Plate convergence zone;
4. India-Burma Mountain Arc;
5. Pamir area;
6. West Kunlun Mountains;
7. Northern margin of Qaidam Basin;
8. Southern margin of Tianshan Mountains.

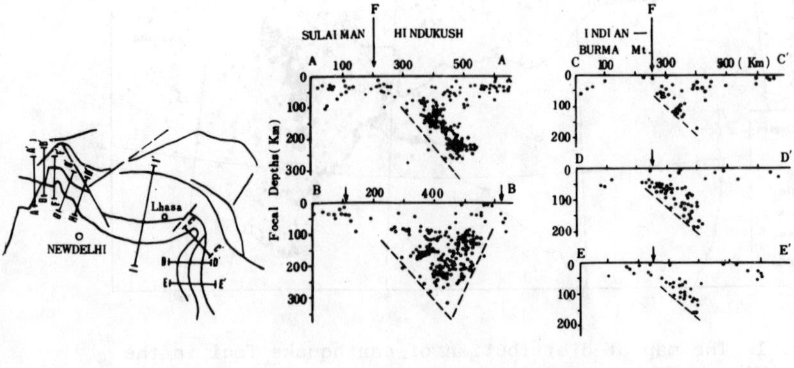


Fig. 3 The profiles of foci across the Hindu Kush and the India-Burma Mountain Arc.

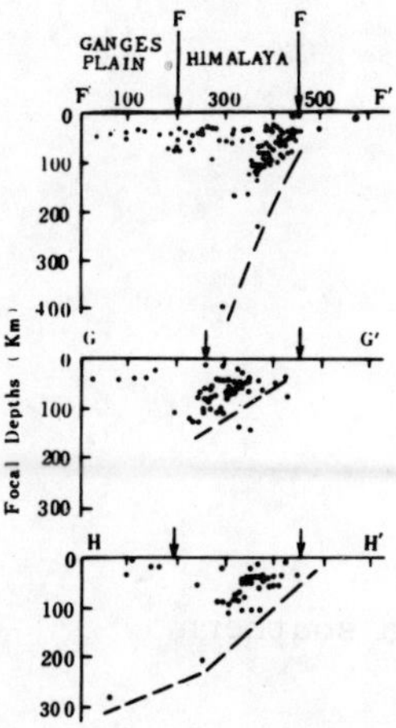


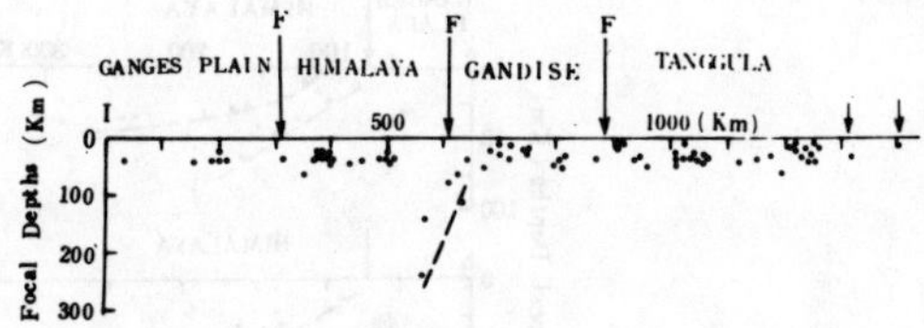


Fig. 5 The profile of foci across the Yarlung Zangbo River Plate convergence zone.

Fig. 4 The profile of foci across the Indus Plate convergence zone.

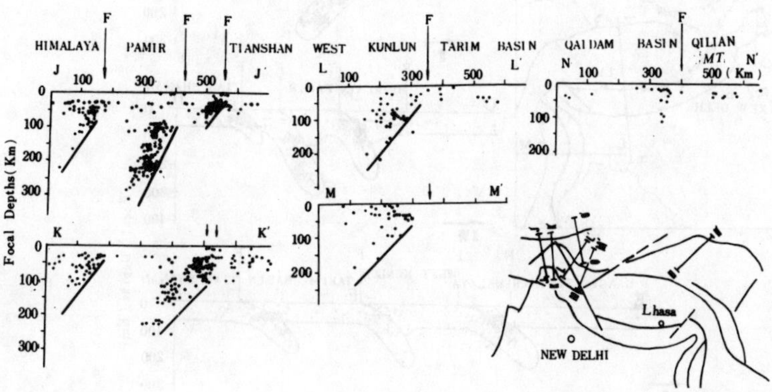


Fig. 6 The profile of foci in the northern Qinghai-Xizang Plateau.

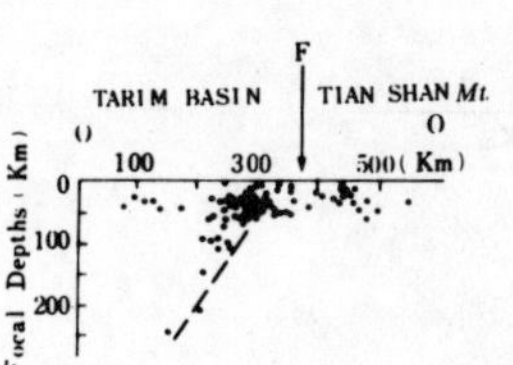


Fig. 7 The profile of foci across the southern margin of Tianshan Mountains.

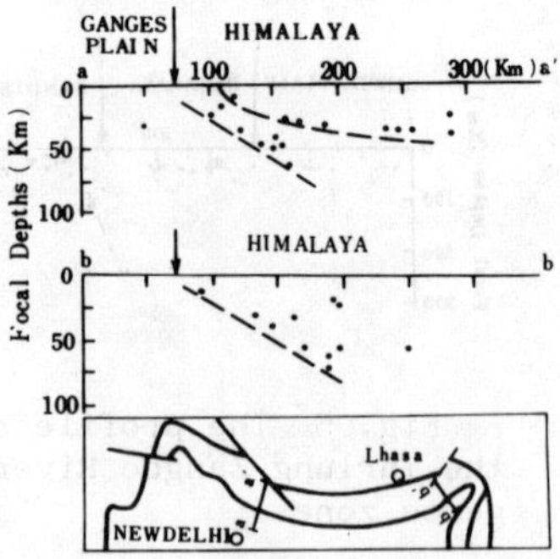


Fig. 8 The profile of foci across the southern Himalayas.

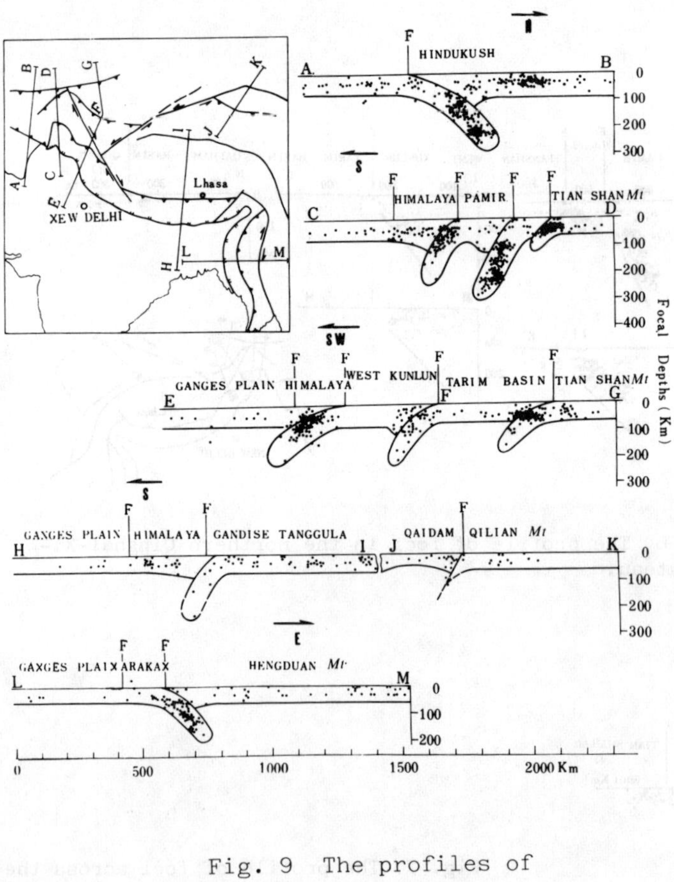


Fig. 9 The profiles of three subduction zones.

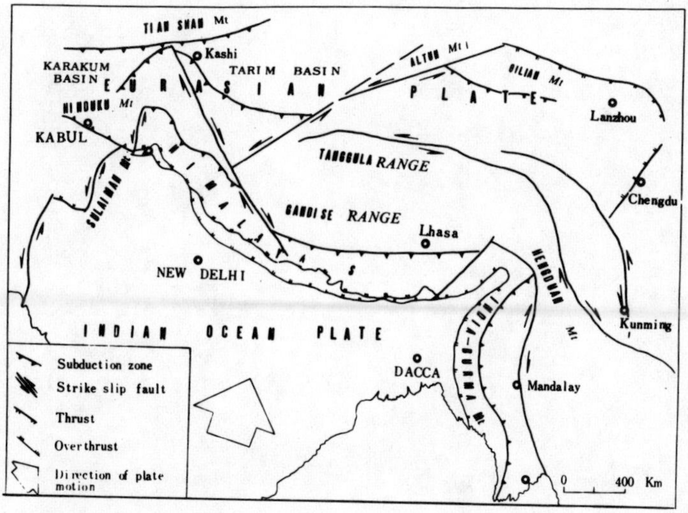


Fig. 10 The plate tectonics in the Qinghai-Xizang Plateau.

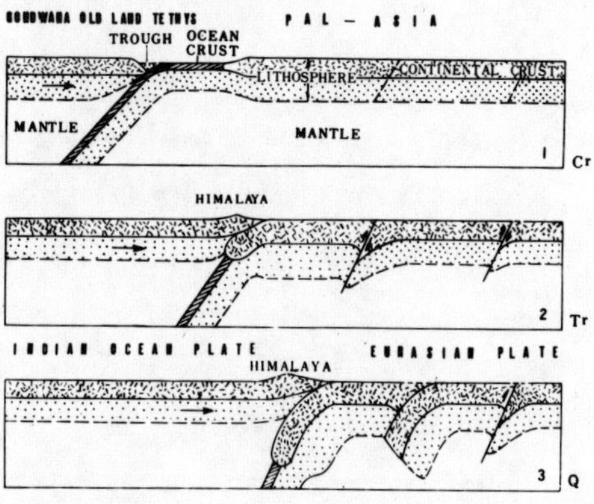


Fig. 11 The model on the origin of south-dipping subduction zone within Asian continent.

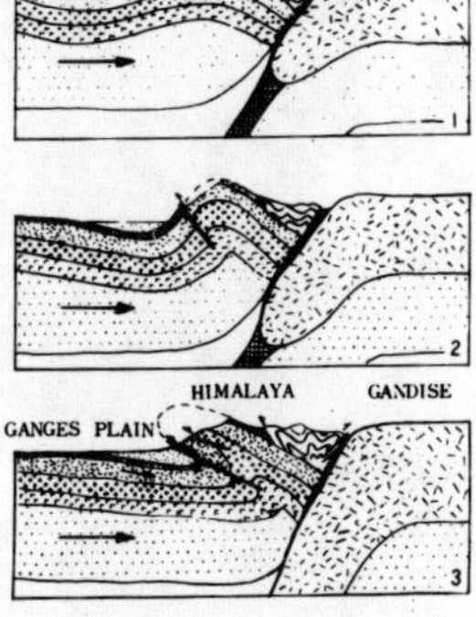


Fig. 12 About the uplifting of the Himalayas.

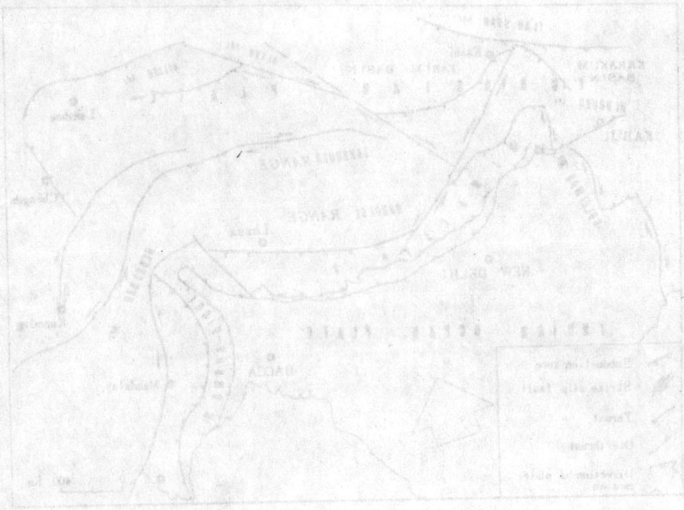

# GEOIDAL FEATURES AND CRUSTAL STRUCTURE IN CENTRAL ASIA

Antonio Marussi
(Istituto di Geodesia e Geofisica, University of Trieste)

1. The study of the motion of artificial geodetic satellites has permitted to outline in recent times a model of the actual outer gravitational field of the Earth. The features of the field are due to the attraction of matter within the Earth; although the density distribution of such matter is by no means uniquely determined by the knowledge of the outer gravitational field, the model still provides most valuable information on the constitution of the Earth's interior.

Space methods based on the use of geodetic satellites involve the exploration of the field in its totality with a great number of satellites orbiting in a wide range of distances and inclinations; the model of the gravitational field thus obtained acquires therefore a global character inasmuch as it describes with the same degree of reliability continental and oceanic, accessible and unaccessible areas, averaging out at the same time disturbances of purely local origin.

Since the model determined purely by space methods ignores the short wavelength anomalies of local origin, a more complete description of the field demands the integration of satellite data with surface gravity observations obtained by conventional methods. Here again the surface gravity data that are introduced represent values averaged over extensive areas, so much that only regional features will contribute to the integrated model.

The model of the gravitational field of the Earth is usually represented in two equivalent forms: elevations of the geoid with respect to the surface of a best fitting ellipsoid, or gravity anomalies computed with respect to the same ellipsoid.

Maps of the geoid obtained from satellite data only, and maps derived from the integration with surface data have been both produced in recent times. The latest published models are the models GEM 10B (satellite data only) and GEM 10B integrated with 1° x 1° surface gravity data published by NASA-Goddard Space Flight Center, 1980. Reference will be made in the following to the integrated GEM 10B model (42) (Plate I).

2. Density anomalies located in the crust produce short wavelength undulations of the geoid. If isostatic equilibrium is assumed to prevail over vast regions, topography and crustal anomalies account little for the long wavelength undulations; the cause of such undulations is therefore to be sought in density anomalies located deeper in the mantle or possibly in the core. The comparison of the global model obtained by satellite methods, with the regional and local features of the detailed gravity field obtained by surface methods, provides therefore the tool for separating in principle the effects of anomalous density distributions in the crust from those located in the mantle.

In the following an attempt is made to correlate the long wavelength disturbances in the gravity field for the Asian Continent with the surface morphology; and a working hypothesis is advanced for the functional dependence of the crustal tectonics from the density anomalies in the mantle (Plate II).

3. The geoidal map of Eurasia shows two marked rises of the geoid: a western one centered in the northern Atlantic Ocean, and an eastern one located in the Pacific area, culminating in Melanesia. The two rises are connected with an uninterrupted geoidal ridge extending along south-eastern Europe, Asia Minor, Iran, Afghanistan, Tibet, and up to the eastern coast of Asia.

The map further shows a very prominent downwarping of the geoid in the Indian Ocean south of Cape Comorin, from which two marked troughs extend northwards, both affecting the Asian Continent. The western trough extends along the Arabic Sea, and is flanked to the west by the eastern coast of the African Continent, the Arabian Peninsula and the Sulaiman Range, continuing in Central Asia along the zone of the Balkash and Raikal Lakes. The eastern trough penetrates in the continent along the Gulf of Bengal, and continues northwards in Yunnan and NE China.

4. If we attempt to establish a correlation between the geoidal features and the crustal features in the Asian Continent, it appears that in general the ridges of the geoid follow belts showing compressional traits, the seat of folding and thrusting processes; whereas the geoidal downwarpings are related to regions characterized by extensional traits, the seat of rifting processes.

In fact, the geoidal ridge originating in the Northern Atlantic, follows the whole Alpine-Himalayan orogenic belt, affecting the Western Mediterranean, the Alps, the Balkans, Anatolia and Caucasus, the mountain system of Iran, the Hindu Kush, Karakorum, Himalaya and Tibet, ending at the eastern coast of the Asian Continent.

On the other hand, the great downwarping of the geoid that has its apex in the Indian Ocean south of Cape Comorin affects the whole of the Indian Peninsula where extensional processes are widely witnessed by the Cambay graben, the Narbada-Son fault, the fractured zone of the Deccan traps outflow, and by a number of fractures having the trend transverse to the Himalayas that have been recognized in recent times. The western slopes of the Indian Ocean depression are furthermore related to the extensional regions of East Africa and Arabia dominated by vertical movements of the crust, large scale rifting and volcanism.

The Indian Ocean geoidal downwarping further shows two remarkable branches, a western and an eastern one, penetrating deeply into the Asian Continent; both branches appear again related to extensional features of the crust.

5. The western branch is associated with the Owen fracture in the Arabian Sea and affects the depression of the lower Indus Valley; the Valley might be interpreted as a graben interposed between the Indian and the Baluchi platforms. Extensional features are to be found on the western side of the Valley, as well as on the eastern. They are witnessed in the Seistan and Registan depressions, the Zhob-Katawaz trough, the intermontane basins of Kabul and Jalalabad, the tertiary-quaternary volcanics of Seistan, Baluchistan and Nawar, and the Chaman transcurrent faults system with the related magmatism.

The geoidal downwarping narrows at the crossing of the Alpine-Himalayan ridge and continues on the opposite side of the ridge with a NE trend along the extensional arc marked by the Tadjik depression, the Ferghana and the Issik Kul basins, the Balkash, the Zaisan and the Junggar (Tzungarian) depressions north of the Tien Shan, the Uvs Noor depression, the Baikal rift system and the upper Lena Valley north of the Altai.

It is noteworthy that in this view the Tadjik-Baikal arc appears to be connected with the extensional zone of East Africa and Arabia. A strain pattern having this trend had already been noted previously (22).

6. The Tadjik-Baikal arc has been subject to epiplatform activation characterized by upswelling of the crust, normal and transcurrent faulting, intense rifting, magmatism and volcanism. A low velocity, high attenuation layer has been recognized thanks to seismological studies in the upper mantle beneath the rift zone (12) possibly related to a layer of low-density anomalies and to high temperatures.

Epiplatform activation started with weak tectonic movements in the Eocene, continuing up to Early Pliocene; the activity greatly increased in the Middle Pliocene and continued intensively during Pleistocene on Holocene up to the present (34). Positive vertical movements are still present and seismic activity high. Fault plane solutions indicate tensional stresses at right angle to the arc (41).

The NE section of the arc is characterized by the spectacular Baikal Rift System displaying a number of deep grabens, faults, high heat flow, basic effusives, and volcanism.

The SW section, separated from the former by the Bolnai fault, is instead characterized by reliefs with linear structures; intense differential vertical movements originate here high mountain ridges and wide basins.

The contrasting tectonics of the two sections may be justified by the brittle nature of the crust in the former, and by the interaction of the extensional stresses of the epiplatform process with the compressional stresses still present in the pre-existing Altai and Tien Shan orogenic belts.

7. Such interaction may have its most spectacular expression in the Pamir Syntaxis, where the extensional arc crosses the Alpine-Himalayan compressional belt (Plate III).

Detailed geomorphological studies (7), (8), (9), (29), (30) have demonstrated that the whole of the Syntaxis in which Tien Shan, Kun Lun, Karakorum and Hindu Kush converge, is divided into blocks that have undergone separately intense upwards or downwards vertical movements of the order of several thousands of meters; the uplifts account for the exceptional mean elevation of the syntaxial zone.

The map of the vertical velocity gradients (7) shows a marked line crossing the Syntaxis at the 73° meridian, which separates a western group of blocks having higher uplift velocities from an eastern one. The differentiation between the two groups is reflected in the morphology, and is recognizable in the LANDSAT imageries (6). Remarkably, the separation line has the trend of the extensional arc, crossing the folded belt of the Pamirs at a high angle. The line shows furthermore a great concentration of crustal seismicity.

The same trend of the extensional arc becomes apparent in the Syntaxis by inspecting the map of isostatic gravity anomalies. As has been pointed out previously (38), (39) the isostatic gravity anomalies show two distinct regions separated by an ideal line continuing the trend of the Chaman fault, joining the city of Kabul with the lake Zorkul and the lake Karakul. The same line determines the eastern boundary of a high velocity block that has been evidenced by seismological studies (35), (60), (61). The block is located between the boundary of the Tadjik depression and the line just described, thus underlying the western Transalai and the western Pamirs; part of it is occupied by the nest of intermediate Hindu Kush earthquakes.

8. The eastern branch of the Indian Ocean geoidal downwarping follows the Ninetyeast fracture in the Gulf of Bengal and penetrates in the continent along the Irrawaddy valley. After crossing the eastern Himalayan syntaxial zone, the geoidal low follows the eastern slopes of the Qinghai-Xizang (Tibet) Plateau affecting the NS trending tectonic zone of deep faults extending between Kunming in Yunnan and Xichang in Sichuan, which displays a series of grabens with intense magmatism, seismicity and heat flow (62). The Yunnan-Sichuan tectonic zone appears to be further connected with the broad depression of the Sichuan Pendi flanked to the NW by the Longmen Shan fault which on its turn continues in the spectacular Shaanxi and Shanxi graben system. The graben contours the eastern side of the Ordos basin; the western side of the basin is flanked by the Yinchuan graben (49).

The whole of the described arc is marked by very high seismicity (32).

9. Research based on the study of LANDSAT satellite imageries has provided in recent times much additional knowledge on the morphology of Central Asia (6), (21), (49), (54), (66). Of special interest to us are the studies on the hitherto little known areas of the Qinghai-Xizang (Tibet) Plateau that have evidenced most recent extensional features (41), (43).

The analysis of satellite imageries has indeed shown a number of normal faults and grabens with SW-NE trend in NW Tibet (Lake Gozha Co area), with S-N trend in Central Tibet (between 82° and 88°E) and with SSW-NNE trend in Eastern Tibet (Nam Co area). Some of these faults extend southwards of the Yarlung Zangbo (Tsangpo) suture zone reaching the great Himalaya, as for example the Takkhola fault. Some of the grabens are filled with flat alluvial deposits or lakes. The faults are dated as Late Pliocene and Pleistocene.

A large number of fractures and faults having the same trend transverse to the Himalayan tectonics and parallel to the great faults discovered in the basement of the Ganga (Ganges) basin and on the Indian shield were already known from previous field studies (56); they may be possibly related to the system of fractures recently discovered in Tibet.

All evidence indicates that Tibet has undergone EW extension since Late Cenozoic and fault plane solutions confirm that tension with EW axis is still acting presently (41), (43).

It should be noted that the trend of the Nanga Parbat-Haramosh accident is in harmony with the trend of the fractures on the Plateau and across the Himalayas; the analogy may suggest a causal relationship between the extensional regime and the generation of the Nanga Parbat-Haramosh diapire.

Abundant evidence of young volcanism has been furthermore revealed by the analysis of satellite imageries especially between longitudes 80 to 93. In the south, volcanics cover more than 60% of an area adjoining the Indus-Zangbo suture zone, extending 1,000 kilometers in the EW and 400 kilometers in the NS directions. In the north, sporadic volcanic areas occur in the west, but a huge area comprised between longitudes 86 to 89 and latitude 34 to 37 shows abundant volcanic cover.

Some of the volcanics in the south may be as late as Cretaceous. Others show very young geomorphic features and are very unlikely to be older than Late Neogene.

Basalts seem to be uncommon; most of the volcanics are andesites, dacites and rhyolites. Their composition and extent are attributed to wide-spread partial melting of the lower part of the overthickened crust of the Plateau. Himalayan granites 9 to 12 MY old are part of the same magmatism (28).

10. As noted above, the undulations of the geoid are related to deep-seated density anomalies, upheavals corresponding to higher and downwarpings to lower than normal densities in the mantle.

The functional relationship that has been noted in the Asian Continent between higher mantle densities and prevailing compressional stresses (geosynclinal regime), and between lower mantle densities and prevailing tensional stresses in the crust (epiplatform regime) is instead conjectural.

There is general agreement that the presence of lighter material whether due to chemical, state, phase or thermal differentiation, would originate an uprising current in the mantle and consequently a dome-like uplift in the interface with the asthenosphere. Such uplift would on its turn generate divergent horizontal currents in the asthenosphere dragging the crust apart from the axis or the apex of the upheaval. Depending upon the viscosity of the asthenosphere, gravity sliding of the crust may alternatively be invoked to explain the extensional regime thus generated. The described mechanism requires the further functional relationship between extensional regime, positive vertical movements of the crust, and higher rate of heat flow, as is generally recognized in the regions subject to epiplatform activation.

Conversely, the presence of heavier material would originate a sinking current in the crust and consequently a depression in the mantle-asthenosphere interface that may be at the origin of a geosyncline. Such depression would originate convergent horizontal currents in the asthenosphere dragging the crust towards its axis, thus creating the compressional condition typical of the geosyncline regime. Here again gravity sliding might be involved in the mechanism.

It has been noted (50), (65) that epiplatform episodes are of relatively short duration and generally follow epigeosyncline episodes; and this appears to be the case in Central Asia. Occasionally they may affect the same region, in which case high mean elevations are to be expected as the result.

11. According to the views expressed so far, the Qinghai-Xizang (Tibet) Plateau appears comprised between the Himalayan compressional belt and the extensional districts of the Tadjik-Baikal and the Yunnan-Shanxi arcs. Whether the depression of the Tarim Pendi is to be included in the Tadjik-Baikal extensional system is a matter of further consideration. A positive answer would be suggested by the presence of the West Kun Lun-Altun Shan-North Qilian Shan transcurrent faults system that borders the northern slopes,

of the Plateau, and by the accompanying gentle low of the geoid that seems to connect along the 40° parallel the western with the eastern branch of the geoidal depression.

## References


(1) Academia Sinica, Inst. of Geophysics, Map of the Distribution of Strong Earthquakes in China. Beijing (1976).

(2) Afghan Geological and Mines Survey, Mineral Resources of Afghanistan (With Geological, Tectonic and Magmatic Maps). Kabul (1977).

(3) Anderson, Don L., The Deep Structure of Continents. Journ. Geoph. Res., 84 (B13), (1979).

(4) Artemjev, M. E., Artyushkov, E. V., Structure and Isostasy of the Baikal Rift and the Mechanism of Rifting. Journ. Geoph. Res., 76 (5), (1971).

(5) Banghar, A. R., Sykes, L. R., Focal Mechanisms of Earthquakes in the Indian Ocean and Adjacent Regions. Journ. Geoph. Res., 74 (2), (1969).

(6) Bartole, R., Structural Lineaments of the Central Asian Orogenic Syntaxis from LANDSAT Imageries. Rend. Acc. Naz. Lincei, S. VIII, 64 (5, 6), (1978).

(7) Beloussov, T. P., Tectonic Movements in the Pleistocene-holocene and Seismicity. Nauka, Moscow (1976), (in russian).

(8) Beloussov, V. V., Reisner, G. I., Rudich, E. M., Sholpo, V. M., Vertical Movements of the Earth's Crust on the Continents. Geophysical Surveys, 1, (1974).

(9) Beloussov, V. V., Mountain Formation in the History of the Earth's Tectonosphere. Geotettonica delle zone orogeniche del Kashmir Himalaya-Karakorum-Hindu Kush-Pamir. Acc. Naz. Lincei, Roma (1976).

(10) Billington, S., Isacks, B. L., Barazangi, M., Spatial Distribution and Focal Mechanism of Mantle Earthquakes in the Hindu Kush-Pamir Region: A Contorted Benioff Zone. Geology, 5, (1977).

(11) Buchroithner, M. F., Progress in the Geology of Afghanistan. Zbl. Geol. Paläont., 1 (5/6), (1979).

(12) Bugaevskij, G. N., The Baikal Rift Zone as an Example of the Relation between Surface Structure and Composition of the Mantle. Akademija Nauk, Institut Zemnoj Kory, Novosibirsk (1977), (in russian).

(13) Butovskaya, Ye. M., Volvovsky, I. S., Krasnopevtseva, G. N., Pak, V. A., Tal-Virsky, B. B., Seismic Parameters and Crustal Structure of the Pamirs-Himalayas Region Along the DSS Profile Frunze-Srinagar. Symposium "The Lithosphere-asthenosphere Interaction, its Role and Tectonic Processes", Leningrad (1978).

(14) Carey, S. W., The Expanding Earth. Elsevier Scient. Publ. Co., Amsterdam (1976).

(15) Chang Wen-yu, Chung Chia-yu, Yeh Hung, Cong Bo-lin, Preliminary Note on the Origin and Development of Rock-Fracture and its Bearing on Earthquakes. Acta Geol. Sinica, (1), (1975).

(16) Chinese Acad. Geol. Sc., Tectonic Map of the People's Republic of China. Peking (1975).

(17) Desio, A., Marussi, A., On the Geotectonics of the Granites in the Karakorum and Hindu Kush Ranges. Intern. Geol. Congress, Copenhagen (1960).

(18) Desio, A., Some Geotectonic Problem of the Kashmir Himalaya-Karakorum-Hindu Kush and Pamir Area. Geotettonica delle zone orogeniche del Kashmir Himalaya-Karakorum-Hindu Kush-Pamir, Acc. Naz. Lincei, Roma (1976).

(19) Ding Guoyu, Li Yongshan, Seismicity and the Recent Fracturing Pattern of the Earth Crust in China. Acta Geol. Sinica, 53 (1), (1979).

(20) Dzievonski, A. M., Hager, B. H., O'Connell, R. J., Large-Scale Heterogeneities in the Lower Mantle. Journ. Geoph. Res., 82 (2), (1977).
(21) Ebblin, C., Tectonic Lineaments in Karakorum, Pamir and Hindu Kush from ERTS Imageries. Rend. Acc. Naz. Lincei, S. VIII, 60 (3), (1976).
(22) Ebblin, C., Deformations in the Area of the Pamirs-Himalayan Syntaxis and Strain Patterns on the Earth's Surface. Boll. Geod. e Sc. Affini, 37 (2/3), (1978).
(23) Florensov, N. A., The Role of Riftogenesis in the Geological History of the Earth. Akademija Nauk S.S.S.R., Institut Zemnoj Kory, Novosibirsk (1977), (in russian).
(24) Hamrabaev, I. H., Pak, V. A., Yusuphodjaev, H. I., The Peculiarities of Geophysical Fields and Deep Structure of the Pamir and Tien Shan. Geotettonica delle zone orogeniche del Kashmir Himalaya-Karakorum-Hindu Kush-Pamir. Acc. Naz. Lincei, Roma (1976).
(25) Institute of Geomechanics, Chinese Academy of Geological Sciences, On Tectonic Systems. Scientia Sinica, 20 (2), (1977).
(26) Kaila, K. L., Krishna, V. G., Roy Chowdhury, K., Hari Narain, Structure of the Kashmir Himalaya from Deep Seismic Sounding. Journ. Geol. Soc. of India, 19 (1), 1-20 (1978).
(27) Kailasam, L. N., Plateau Uplift in Peninsular India. Conference on Plateau Uplifts, Flagstaff, Ed. Lunar and Planetary Inst., Houston, Texas (1978).
(28) Kidd, W. S. F., Widespread Late Neogene and Quaternary Calc-alkaline Volcanism on the Tibetan Plateau. EOS Trans. Amer. Geoph. Union, 56, (1975).
(29) Krestnikov, V. N., Nersesov, I. L., Tectonic Structure of the Pamirs and Tien Shan with Relation to the Mohorovicic Discontinuity. Soviet Geol., (11), (1962), (in russian). Transl. Intern. Geology Rev., 6 (8).
(30) Krestnikov, V. N., Stange, D. V., On the Zone of the Pamir-Himalaya Deep Faults. Fizika Zemli, (7), (1977), (in russian).
(31) Lang, J., Evolution Géodynamique de Bassins Intramontagneux Tertiaires d'Afghanistan Central (Bamyan, Yakwalang, et Vallée du Ghorband). Mém. H.S., Soc. Géol. de France, Paris, (8), (1977).
(32) Li, G. P., Li, Y. S., The Features of Mesozoic and Cenozoic Tectonic Development in Eastern China and their Relationship to Seismicity. Sci. Geol. Sinica, (3), (1973).
(33) Li Jijun, Wen Shixuan, Zhang Qingsong, Wang Fubao, Zheng Benxing, Li Bingyuan, A Discussion on the Period, Amplitude and Type of the Uplift of the Qinghai-Xizang Plateau. Scientia Sinica, 22 (11), (1979).
(34) Logatchev, N. A., Florensov, N. A., The Baikal System of Rift Valleys. Tectonophysics, 45, (1978).
(35) Lukk, A. A., Vinnik, L. P., Tectonic Interpretation of the Deep Structure of Pamir. Geotektonika, Moscow, (5), (1975), (in russian).
(36) Marussi, A., Le anomalie della gravità lungo la catena del Karakorum-Hindu Kush. Rend. Acc. Naz. Lincei, S. VIII, 35 (5), Roma (1963).
(37) Marussi, A., Geophysics of the Karakorum. Italian Expeditions to the Karakorum (K2) and Hindu Kush, II. Geophysics, I, E. J. Brill, Leiden (1964).
(38) Marussi, A., Gravity in the Karakorum. Geotettonica delle zone orogeniche del Kashmir Himalaya-Karakorum-Hindu Kush-Pamir, Acc. Naz. Lincei, Roma (1976).
(39) Marussi, A., Gravity, Crustal Tectonics and Mantle Structure in the Central Asian Syntaxis. Intern. Geodynamics Commission Symposium, Peshawar (1979).
(40) McGinnis, L. D., Gravity Fields and Tectonics in the Hindu Kush. Journ. Geoph. Res., 76 (8), (1971).
(41) Molnar, P., Tapponnier, P., Active Tectonics of Tibet. Journ. Geoph. Res., 83 (B11), (1978).

(42) NASA, Global Detailed Gravimetric Geoid Based upon a Combination of the G.S.F.C. GEM 10 Earth Model and 1°x1° Surface Gravity Data. NASA-Goddard Space Flight Center, Greenbelt, Md. 20771, USA (1977).
(43) Ni, J., York, J. E., Late Cenozoic Tectonics of the Tibetan Plateau. Journ. Geoph. Res., 83 (B11), (1978).
(44) Nowroozi, A. A., Seismicity of Afghanistan and the Hindu Kush Seismic Zones. Geotettonica delle zone orogeniche del Kashmir Himalaya-Karakorum-Hindu Kush-Pamir. Acc. Naz. Lincei, Roma (1976).
(45) Petrusevskij, B. A., The Indus-Pamir Zone; One of the Most Important Transversal Lineaments in Asia. Bjulleten Moskovskogo Obscestva Ispytatelej Prirody, (5), (1977), (in russian).
(46) Popov, V. I., Tal-Virskij, B. B., Popov, A., The Nalivkina Transasiatic Rift Belt. Uzbek Akademy of Science, Tashkent (1978), (in russian).
(47) Saxena, M. N., The Himalaya. A Modified Arc System and the Gravity Tectonics. Tectonic Geology of the Himalaya, P. S. Saklani Ed., Today and Tomorrow's Printers and Publishers, New Delhi (1978).
(48) Shatskii, N. S., Tectonic Map of the U.S.S.R. and Neighboring Regions. U.S.S.R. Acad. of Sc., Moscow (1956).
(49) Shih Chen-liang, Huan Wen-lin, Yan Jia-quan, Wang Su-yun, Wang Chihkang, On Recent Tectonics of China and its Vicinity. Terrestrial and Space Techniques in Earthquake Prediction Research. Fr. Vieweg & Sohn, Braunschweig-Wiesbaden (1978).
(50) Shlesinger, A. E., Relationship between Nongeosyncline and Epigeosyncline Uplifts and Mountain Building. Conference on Plateau Uplifts, Flagstaff, Ed. Lunar and Planetary Inst., Houston, Texas (1978).
(51) Sirokova, E. I., Detailed Study of Tensions and Fractures in the Foci of the Central Asian Earthquakes. Fizika Zemli, (11), Moscow (1974), (in russian).
(52) Stöcklin, J., Structural Correlation of the Alpine Ranges between Iran and Central Asia. Mém. H.S., Soc. Géol. de France, Paris, (8), (1977).
(53) Tapponnier, P., Francheteau, J., Necking of the Lithosphere and the Mechanics of Slowly Accreting Plate Boundaries. Journ. Geoph. Res., 83, (1978).
(54) Tapponnier, P., Molnar, P., Active Faulting and Cenozoic Tectonics of the Tien Shan, Mongolia and Baykal Regions. Journ. Geoph. Res., 84 (B7), (1979).
(55) Terman, M. J., Tectonic Map of China and Mongolia. U. S. Geol. Survey, Reston (1973).
(56) Valdiya, K. S., Himalaya Transverse Faults and Folds and their Parallelism with Subsurface Structures of North Indian Plains. Tectonophysics, 32, 353-386 (1976).
(57) Vesanen, E., Teisseyre, R., Gadomska, B., Tuominen, H., Mäki, M. L., Some Remarks on the Relevance of Seismicity, Seismic Chimneys and Fracture Systems to Geodynamics. Publ. Inst. Geoph. Polish Academy of Sc., Warszawa, 93, (1975).
(58) Vesanen, E., Teisseyre, R., Symmetry and Asymmetry in Geodynamics. Geophysics, Helsinki, 15:2, (1978).
(59) Vinnik, L. P., Lukk, A. A., Horizontal Inhomogeneities of the Upper Mantle in Central Asian Regions of Platform Activation. Izv. Fiziki Zemli, (7), (1975), (translated U.D.C. 550.34).
(60) Vinnik, L. P., Lukk, A. A., Nersesov, I. L., Nature of the Intermediate Seismic Zone in the Mantle of Pamir-Hindu Kush. Tectonophysics, 38, (1977).
(61) Vinnik, L. P., Lukk, A. A., Upper Mantle Structure and the Nature of the Pamirs-Hindukush Intermediate Seismic Zone. Symposium "The Lithosphere-asthenosphere Interaction, its Role and Tectonic Processes", Leningrad (1978).

(62) Wang Ji-yang, Geothermal Studies in China. Inst. of Geology, Academia Sinica, 1979.
(63) Wellman, H. W., Active Wrench Faults of Iran, Afghanistan and Pakistan. Geol. Rundschau, 55, 1966.
(64) Yahnshin, A. L., Tectonic Map of Eurasia. U.S.S.R. Acad. of Sc., Moscow, 1966.
(65) Yahnshin, A. L., The Role of Epiplatform Mountain Building in the Development of the Earth's Crust. Conference on Plateau Uplifts, Flagstaff, 1978. Ed. Lunar and Planetary Inst., Houston, Texas.
(66) York, J. E., Cardwell, R., Ni, J., Seismicity and Quaternary Faulting in China. Bull. Seism. Soc. America, 66, 6, 1976.
(67) Zhang Wenyou, Ye Hong, Zhong Jiayou, On "Fault Blocks" and "Plates". Scientia Sinica, 22, 12, 1979.
(68) Zorin, Yu. A., Florensov, N. A., On Geodynamics of Cenozoic Uplifts in Central Asia. Conference on Plateau Uplifts, Flagstaff, 1978. Ed. Lunar and Planetary Inst., Houston, Texas.

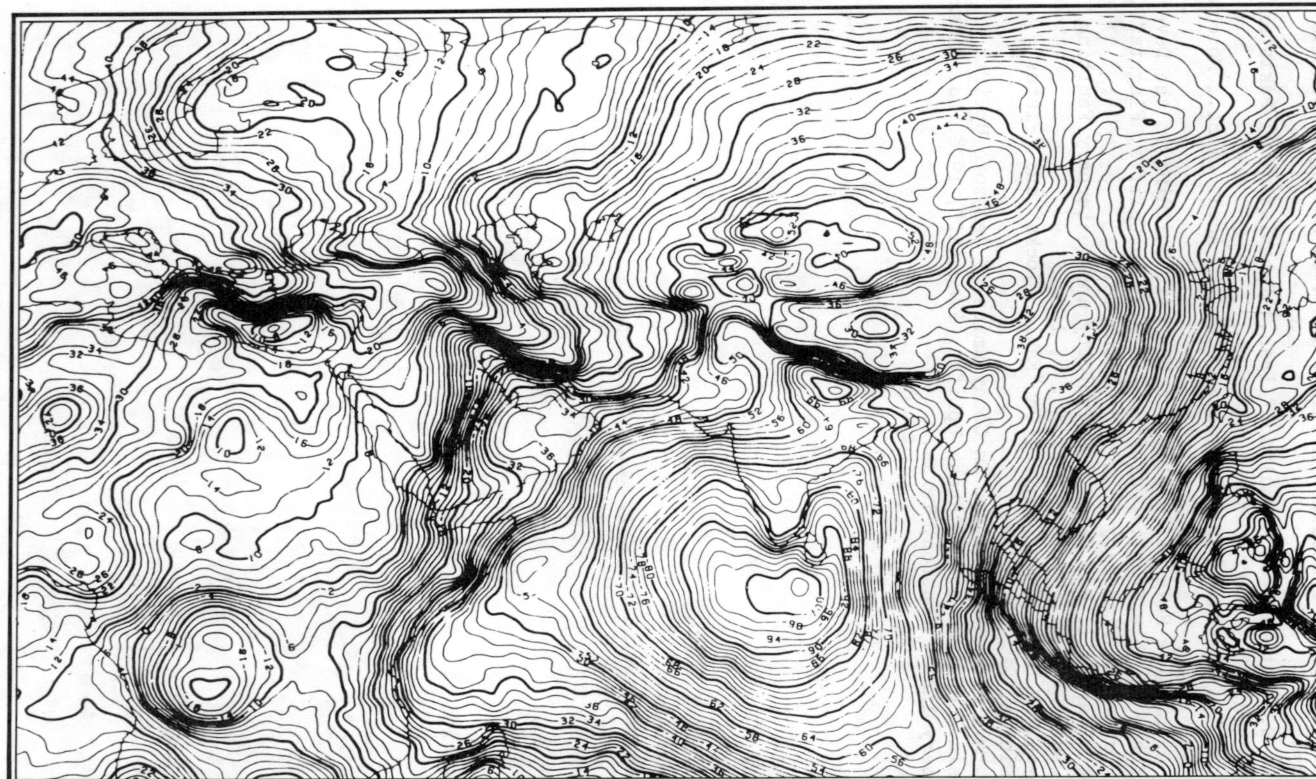

Plate I
Goddard Space Flight Center Global Detailed Gravimetric Geoid Based upon a Combination of the GEM 10B Earth Model and 1° x 1° Surface Gravity Data (1980).

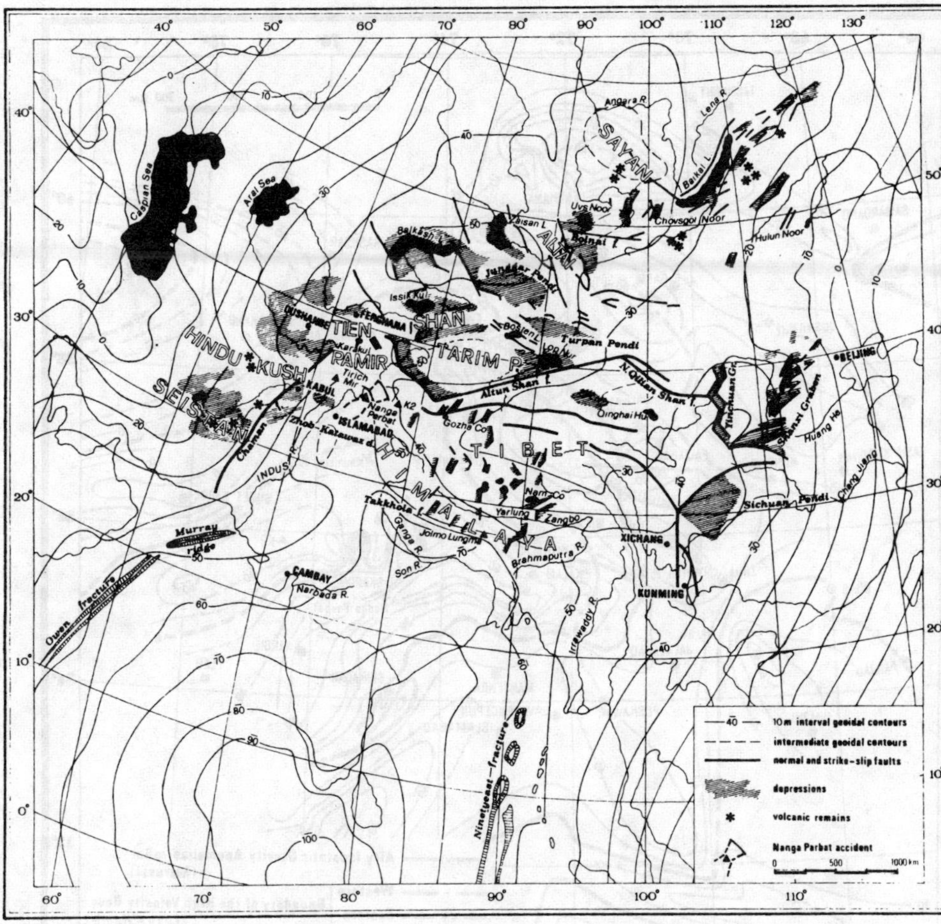


Plate 11
Gravimetric Geoid and Geomorphic Structures in Central Asia (for Tibet from Molnar and Tapponnier and for China from Shih Chen-liang et al. and Chang Wen-yu et al., simplified).

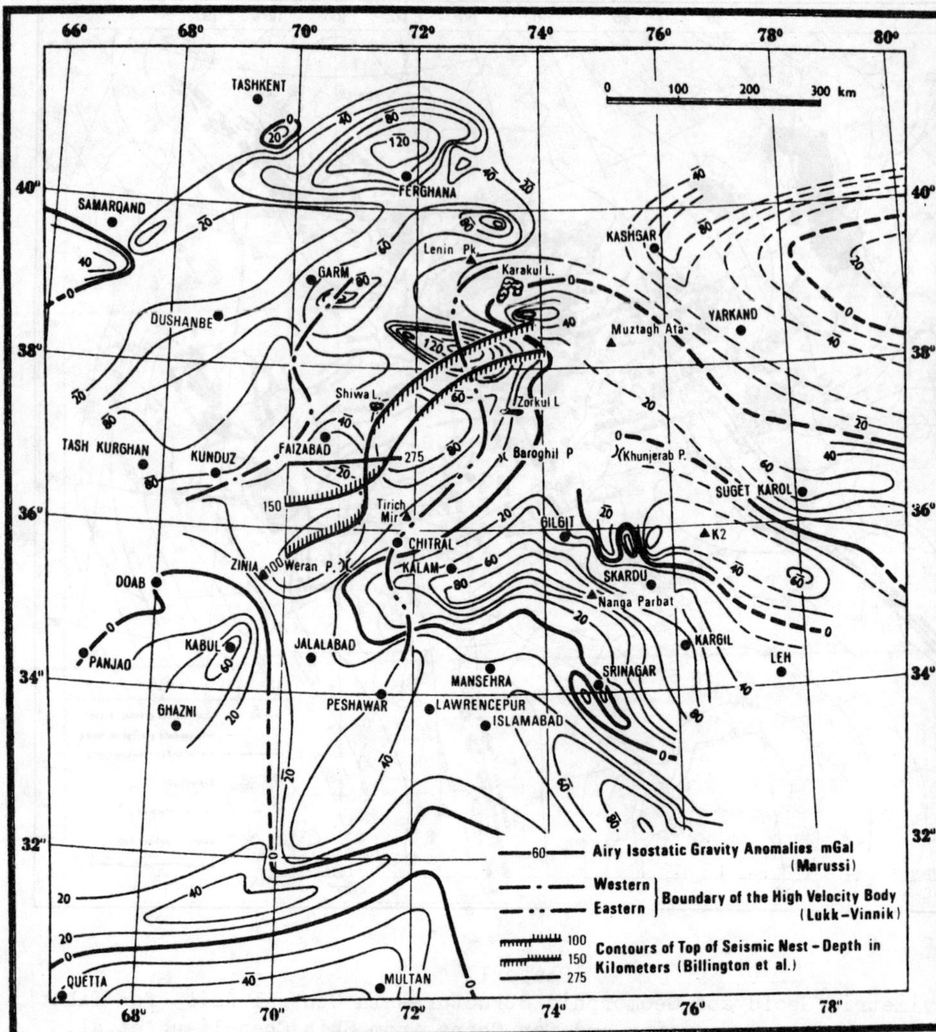


Plate III
Isostatic Gravity Field and Focal Zone of Intermediate Earthquakes at the Pamir Syntaxis.

# CHARACTERISTICS OF THE GRAVITY FIELD AND THE CRUSTAL STRUCTURE IN THE EASTERN AND CENTRAL REGIONS OF THE XIZANG PLATEAU

Zhou Wen-hu Yang Zhan-shou

Zhu Hong-bin Wu Li-gao

(Institude of Geophysics, Academia Sinica)

## ABSTRACT

Gravity data were obtained from the eastern and central regions of the Xizang Plateau in 1975-1977. In this paper maps of Bouguer gravity anomalies and isostatic anomalies for this region are given. A further study was made for correlation of Free air gravity anomalies and Bouguer gravity anomalies with altitudes of the observed points.

The results show that to the south and north of Yarlung Zangbo River the characteristics of the gravity field are different.

1. The maximum negative Bouguer gravity anomaly is less than 550 mgals.

2. Positive isostatic anomalies occur in the Himalayan region and the maximum value is + 120 mgals in the main ridge belt of Himalayas. The northern Yarlung Zangbo River almost reaches an isostatic ecuilibrium.

3. In the Himalayan region the correlation of Free air anomalies and Bouguer anomalies with the altitudes are different with that of the northern Yarlung Zangbo River.

4. The bottom boundary of crust dips from south to north in the Himalayan region. The bottom surface of the crust is flat and very deep in the back-lands of the Xizang Plateau north of Lhasa. In the vicinity of Zayu the bottom surface of the crust has a trough.

## INTRODUCTION

The Qinghai-Xizang Plateau has long been known as the roof of the world. The Himalayan region is one of the most active areas in modern tectonic movements. To study the characteristics of the gravity field in this area is of great significance to the research on the collision of the Indian Plate with the Eurasian. The gravity data used in this paper mainly comes from the work in the area shown as Figure 1 by the members of the Comprehensive Scientific Expedition during 1975-1977. The area is between $86^{\circ}$ and $114^{\circ}$E, and $27^{\circ}30'$ and $32^{\circ}$N. Besides, we also collected gravity data in Mt. Qomolongma region surveyed in 1966 and 1968, and the gravity data between Golmud and Lhasa surveyed by the Chinese State Geodetic Bureau in 1960. For the survey from 1975 to 1977, we used two gravity meters made in China. The longitudes, latitudes and the altitudes of gravity points were all taken from the 1:100,000 scale topographic map which was published in 1971. The terrain corrections for all gravity data were calculated on a computer using a 30 X 30 km square grid method. For isostatic corrections, within the Hayford's $O_2$ zone, the method of summing square columns was adopted, while outside of the $O_2$ zone, tables of 18-1 zone over the whole world by P.L. Kivioja and W.A. Heiskanen(1) were used, taking 30 km as the normal crustal thickness. In the valley region of the eastern part of the Qinghai-Xizang Plateau, the values of terrain corrections are about 50 mgals, while in the northern part of the Xizang Plateau the corrections are only a few mgals.

The normal gravity formula used is

$$r_{\circ} = 978049\ (1+0.0052844\sin^2\varphi - 0.0000059\sin^2 2\varphi)$$

The final errors are chiefly due to errors in the altitudes of the gravity points, when the 1:100,000 scale topographic maps were used. The station errors were estimated within 200 m to 300 m. thus, on the steep hillside, altitude errors may reached 40 mgals. Thus, in the final results of Free air anomalies, the maximum error at some points may reach about ± 12 mgals. The maximum error of Bouguer and isostatic anomalies at some points may reach about ±15 mgals. The absolute errors seem to be large. However, comparing with the gravity anomalies of the whole plateau, the relative errors in southern mountain region and northern Qinghai-Xizang Plateau may reach about 5% and 3%, respectively. Despite of these relative errors, the data may still reflect the basic features of the Qinghai-Xizang Plateau.

## CHARACTERISTICS OF THE GRAVITY FIELD AND THE CRUSTAL BOTTOM SURFACE

Figure 2 gives contours of equal Bouguer gravity anomalies with 50 mgals interval. The map has the following features:

1. The Bouguer gravity anomalies in the whole area are lower than -300 mgals, the minimum value being less than -550 mgals.

2. The contour of equal Bouguer gravity anomalies in the western part appears to trend east-west. But starting from Zetang the trend turns to the northeast, while in the vicinity of Zhamo it begins to bend toward the southeast. To the west of Zayu it becomes almost north-south.

There is a closed -500 mgals contour of Bouguer anomaly north of the Yarlung Zangbo River with very large areal extent. Within this contour the gravity anomaly value is lower than -500 mgals.

3. Greater gradients of Bouguer anomalies appear on the southern slope of the Himalayas, for instance, in the Jalpaiguri of India. Gravity values here are -170 mgals, dropping approximately with gradient -0.6 mgal/km toward the north. From Yadong, Bouguer gravity anomaly values begin to drop rapidly, from -250 mgals to -320 mgals in Pagri. Its gradient is -5.6 mgals/km. The areas of the big bend of the Yarlung Zangbo River, for instance, Nyingchi and Tangmai, and other regions from Qionglai to Ludin are zones of larger gradients of gravity anomalies. The gradients are -4 mgals/km and -4 mgals/km, respectively.

Figure 3 is a map of isostatic anomalies with contours appearing on the southern side of the Yarlung Zangbo River, extending from Zayu on the east and Nyalam on the west. The largest positive anomaly values are found on the main ridge of the Himalayas at Mt. Qomolongma. Its value is about +120(2) mgals, decreasing towards the south and north, positive isostatic anomaly areas are situated right at the Himalayas. To the north of the Yarlung Zangbo River, the values of isostatic anomalies are near zero. Only in local areas their values are negative, but small.

Figure 4a and Figure 4b show the relationship between the Free air and Bouguer anomalies and the altitudes of the gravity points for the vast areas north of Yarlung Zangbu River, south of Amdo and west of the Qionglai Mt. From east to west the observation points are located on three straight lines. Geotectonically they coincide with the Songpan-Garze, Sanjiang and Lhasa fold belts. The slopes of straight lines show that in this region, the Free air anomalies vary with the topography. They were chiefly influence by the rise of the topographic surface. From Figure 4a we can see that relationship between Free air anomalies and altitudes is expressed by 0.112 mgal/m.

Figures 5a and 5b are relationships of the Free air and Bouguer anomalies and the altitudes of gravity stations on the northern slope of the Himalayas.

From Figure 5b it can be seen that from the main ridge of the Himalayas down to the Yarlung Zangbo River, though the topographic surface descends yet the Bouguer anomalies increase very little. It possibly shows that the Moho discontinuity deepens in this region.

Figures 6a and 6b show the relations between the Free air and Bouguer anomalies and the altitudes of the gravity points on the southern slope of the Himalayas. These relations may be expressed by the following formulas:

$$\Delta g_{FA} = -298 + 0.099H(m) \qquad \Delta g_{FB} = -276 - 0.027H(m)$$

Here $\Delta g_{FA}$ is Free air anomaly and $\Delta g_{FB}$ is Bouguer anomaly.

These results are similar to those obtained on the southern slope of the Himalayas over gravity points higher than 700 m by M.N. Qureshey(3) in 1971 and by Masarn Kono in 1974 in Nepal. They indicate that when the topography rises, the Moho discontinuity possibly deepens.

Table 1 is the comparision of the data of Figures 4a to 6b. From the intercepts and the slopes listed in the table, we can see that tectonically the Himalayan area and north of Yarlung Zangbo River are not alike.

Figure 7 illustrates the difference in relationship between $1^{\circ}x1^{\circ}$ mean Bouguer anomaly values and the altitudes of the Himalayan region and the region north of Lhasa. The relations may be expressed by

$$\Delta g_h = 440 - 0.198H(m) \quad \text{and} \quad \Delta g_L = -169 - 0.079H(m).$$

Where $\Delta g_h$ is the anomaly for the Himalayan region and $\Delta g_L$ is that for the region north of Lhasa.

The relation shows that the crust of Himalayan region inclines at a large angle.

Figure 8 is the profiles of the topography, depths of Moho discontinuity and the Bouguer anomalies from Yadong to Golmud.

An approximate outline of the crustal bottom boundary in this region was obtained with least square fitting of the Bouguer gravity anomaly.

## CONCLUSION

Through analysis of the gravity fields data we note the following significant points:

1. The characters of the gravity field are not the same between the Himalayan region and northern Xizang, where the least value of Bouguer anomaly is less than -550 mgals. In the Himalayan region the gravity field is a transition zone. The Bouguer anomalies decrease from -300 mgals south of the zone to -500 mgals in the Yarlung Zangbo River region.

2. In the Himalayan region, the maximum positive isostatic anomaly appears in the main mountain ridge belt, indicating insufficiency of the compensating "mountain root". However, the gravity field is isostatically compensated over a very large region in northern Xizang.

3. From Figure 8 and Figure 2, the shape of the Moho discontinuity as reflected by Bouguer anomaly is that under the Himalayas, it is downwarping, and north of the Yarlung Zangbo River, it is flat and very deep. But, in the vicinity of Zayu, a trough occurs at the bottom surface of the crust. It might be related to the northeastward pushing by the Indian Plate at the bend of the Yarlung Zangbo River and to the appearance of strike-slip faults on the surface.

## References


(1) Kivioja P.L., Heiskanen W.A., Topographic-isostatic reduction maps for the world for the Hayford jou  18-1, Airy-Heiskanen System. T 30 km. Helsinki Publ. Isostat. Inst. Assoc. Geod., No.35 (1961).

(2) Tang Bo-xiong et al., Isostatic gravity anomalies in the area of central portion of the Himalayas. In this Proceedings of Symposium on Qinghai-Xizang (Tibet) Plateau, Academia Sinica (1981).

(3) Qureshy M.N., Geophysical investigation in the Himalayas. Himalayan Geology, 1, 165-177 (1971).

(4) Masarn Kono, Gravity anomalies in East Nepal and their implications to the crustal structure of the Himalayas. Geophys. J. Roy. Astr. Soc., 39, 283-300 (1974).

(5) Teng Ji-wen et al., Explosion seismic study for velocity distribution and structure of the crust and upper mantle from Damxung to Yadong of Xizang Plateau. In this Proceedings of Symposium on Qinghai-Xizang (Tibet) Plateau, Academia Sinica (1981).

Table 1 The comparison of factors of Figs. 4a to 6b

| No | Type of anomaly | Unit of tectonic | Intercept mgal/m | Slope mgal/km | Relative factor | Mean anomaly | Number of measurements |
|---|---|---|---|---|---|---|---|
| 1 | FREE-AIR | Lhasa fold belt | -526 | 114 | 0.576 | -34 | 106 |
| 2 | | Sanjiang fold belt | -449 | 108 | 0.955 | -48 | 144 |
| 3 | | Songpan Ganze fold belt | -398 | 108 | 0.977 | -25 | 26 |
| 4 | | The west of Chengdu plain | 34 | -326 | 0.829 | -145 | 16 |
| 5 | | (north slope) The region of Himalayas | -564 | 140 | 0.816 | 2 | 98 |
| 6 | | (south slope) The region of Himalayas | -298 | 99 | | | |
| 1 | BOUGUER ANOMALY | Lhasa fold belt | -523 | -3.3 | 0.142 | -523 | 106 |
| 2 | | Sanjiang fold belt | -439 | -6.00 | 0.159 | -461 | 144 |
| 3 | | Songpan Ganze fold belt | -444 | 9.00 | 0.301 | -410 | 26 |
| 4 | | The west of Chengdu plain | 79 | -509 | 0.905 | -210 | 16 |
| 5 | | (north slope) The region of Himalayas | -616 | 41.5 | 0.345 | -463 | 98 |
| 6 | | (south slope) The region of Himalayas | -276 | -27 | | | |

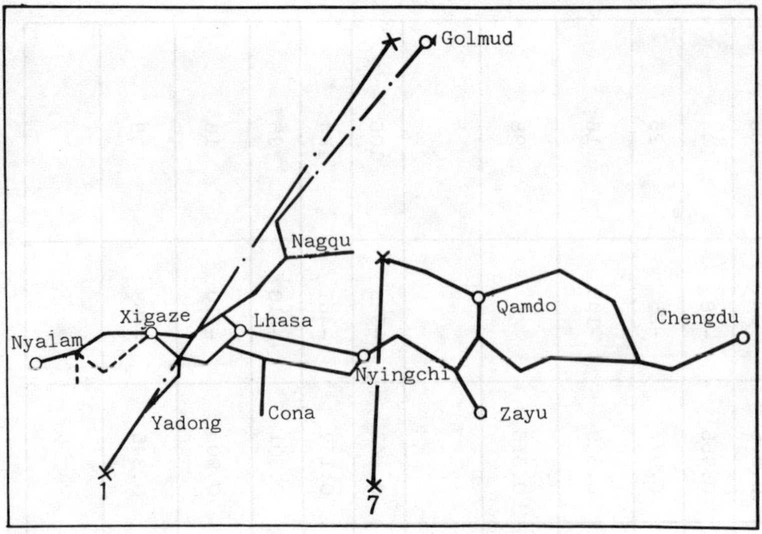


Fig. 1 The profiles of a gravity surveys in the Xizang Plateau

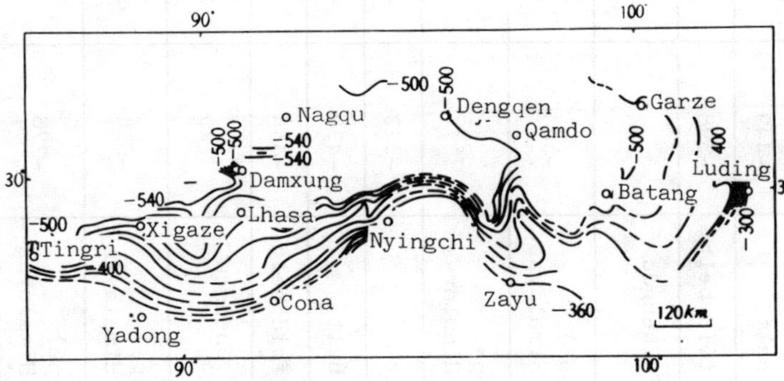


Fig. 2 The map of isostatic anomalies

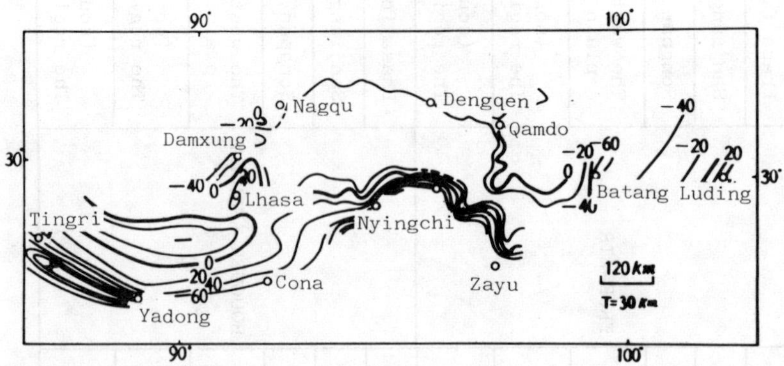


Fig. 3 The map of isostatic anomalies

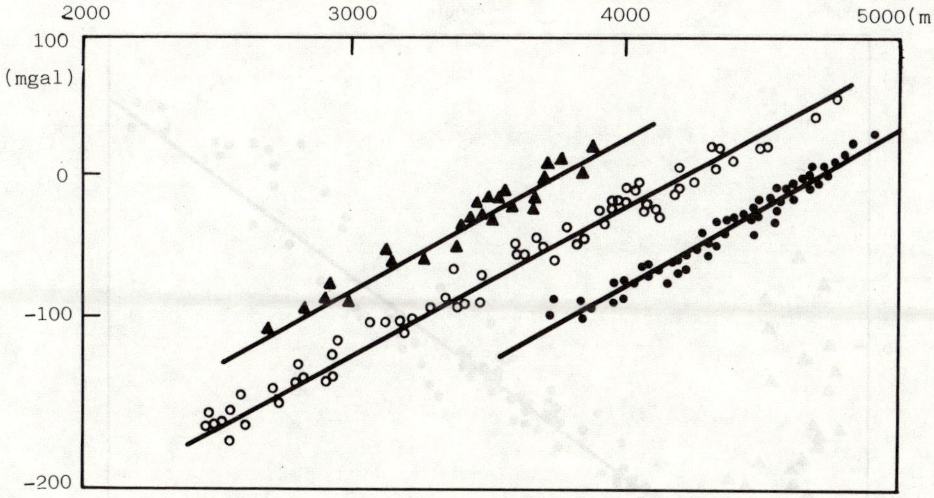


Fig. 4a The relationship between the Free-air anomalies and the observed altitudes in the north of Yarlung Zangbo River.

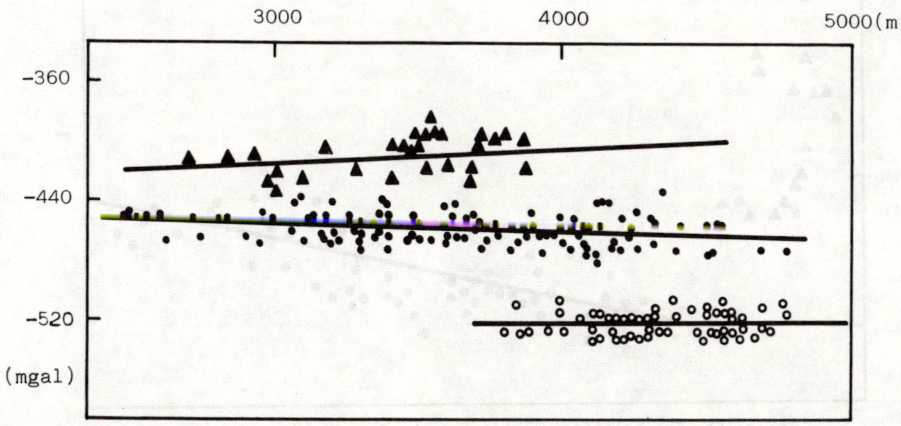


Fig. 4b The relationship between the Bouguer anomalies and the observed altitudes in the north of Yarlung Zangbo River.

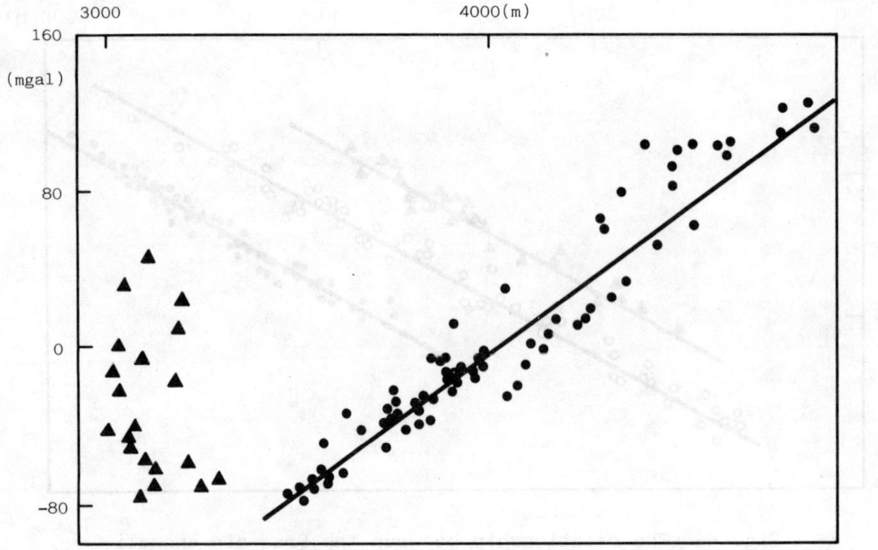


Fig. 5a The relationship between the Free-air anomalies and the observed altitudes in the northern slope of the Himalayas.

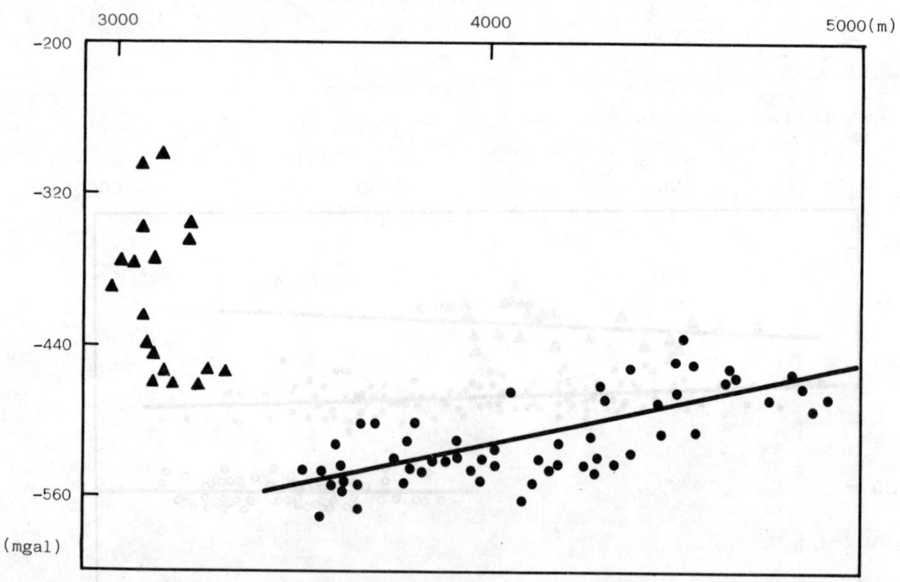


Fig. 5b The relationship between the Bouguer anomalies and the observed altitudes in the northern slope of the Himalayas.

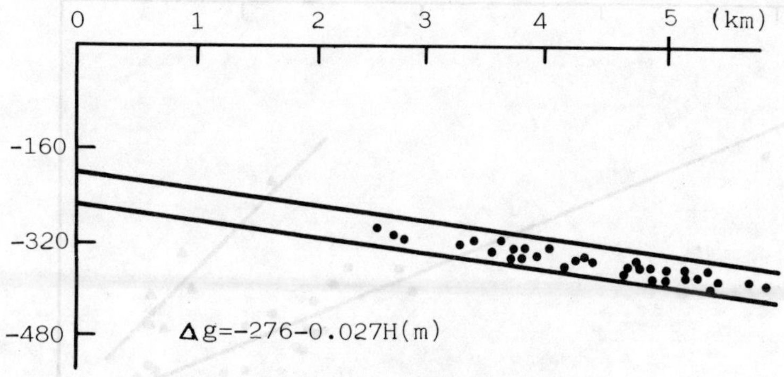


Fig. 6a The relationship between the Free-air anomalies and observed altitudes in the southern slope of the Himalayas.

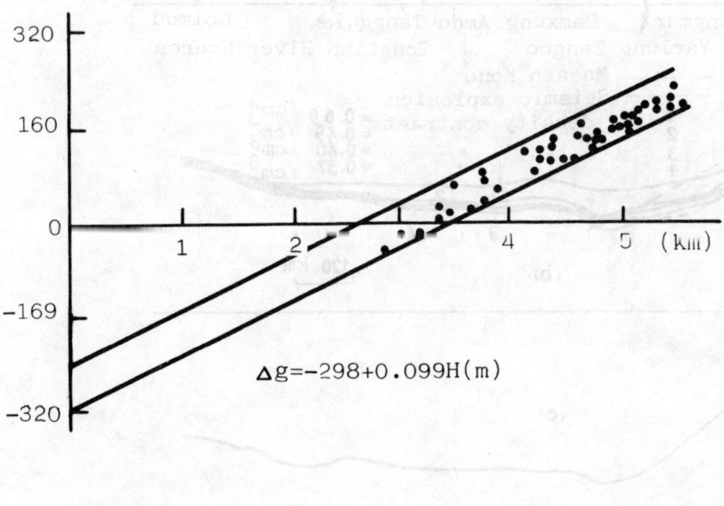


Fig. 6b The relationship between the Bouguer anomalies and the observed altitudes in the southern slope of the Himalayas.

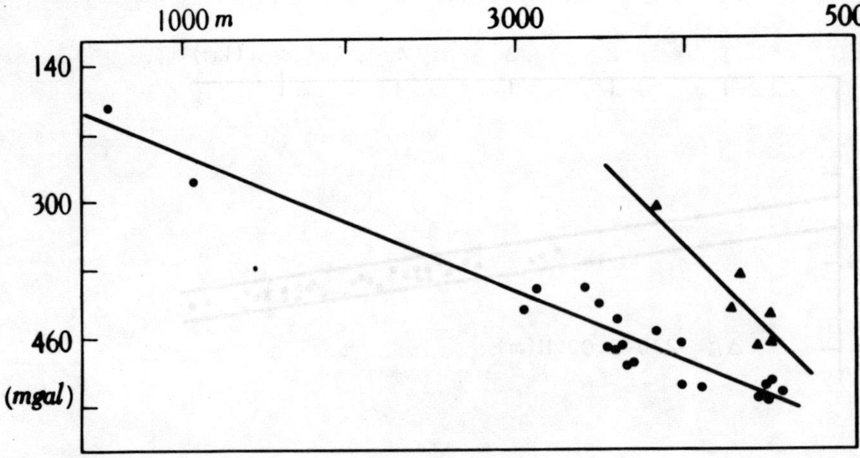


Fig. 7 The relationship between the mean Bouguer anomalies of 1°X 1° and mean observation altitudes of 1°X 1°.

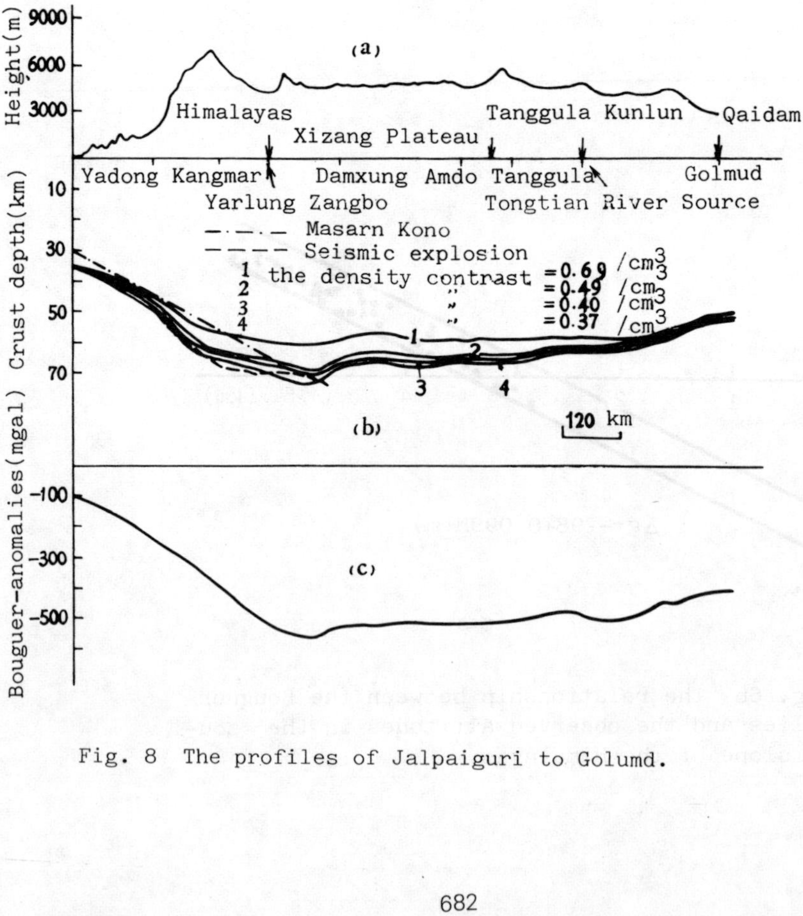


Fig. 8 The profiles of Jalpaiguri to Golumd.

# ISOSTATIC GRAVITY ANOMALIES IN THE CENTRAL PORTION OF THE HIMALAYAS

Tang Bo-xiong
(Seismological Bureau of Yunnan Province)

Liu Yuan-long   Zhang Li-min   Zhou Wen-hu   Wang Qian-shen
(Institute of Geophysics, Academia Sinica)

## ABSTRACT

In this article, the isostatic gravity anomalies in the central portion of the Himalayas is calculated on the basis of Airy-Heiskanen's isostatic hypothesis. The map of the isostatic gravity anomalies of the region is given under the condition of normal crustal thickness T=30 km. It can be seen clearly from this map that the Himalayan region appears to have a positive isostatic anomaly. In the vicinity of the Mt. Qomolangma, the isostatic anomaly is about 120 mgal. Hence we can infer that the crust in this region has not reached the stage of isostatic equilibrium. These positive isostatic anomalies are considered to be caused by the lack of mountain root in the Himalayas.

The Bouguer gravity anomaly in Mt. Qomolangma region as postulated by some scientists in the past, does not correspond to the rule that the higher the elevation the smaller the Bouguer gravity anomalies(1). By topographic and isostatic reduction to the gravity observation data in this region, the result shows that the isostatic gravity anomaly in Mt. Qomolangma region is larger than that of the value within the Qinghai-Xizang Plateau, by about 90 mgal or so. This shows that the Himalayas are not compensated isostatically.

## THE METHOD USED TO CALCULATE ISOSTATIC GRAVITY ANOMALY

In this paper, the Airy Heiskanen isostatic hypothesis is mainly adopted to process the reduction. As for the reduction the parameters were chosen as follows: the normal crustal thickness is 30 km, corresponding to that at zero elevation; the density used for terrain correction is $2.67 g/cm^3$; the density contrast between the crust and mantle is $0.6 g/cm^3$. In addition, the Vening-Meinez regional compensation and the Hayford's isostatic hypothesis were applied to five gravity stations in the reduction.

The Vening-Meinez's regional reduction table(2), the Airy-Heiskanen's local isostatic compensation reduction table (3-6), the Hayford's reduction table(7) as well as the topographic reduction table compiled by ourselves were adopted in our isostatic compensation reduction. The topographic maps used here were in three scales of 1:100,000; 1:200,000; and 1:1,000,000.

## THE RESULTS OF CALCULATIONS

In this region, 47 gravity stations were chosen to go through the isostatic compensation reduction. For 5 stations, we made not only the Airy-Heiskanen's isostatic reduction for various normal crustal thicknesses T=30, 40, and 60 km, but also the Vening-Minez's regional isostatic reduction, in which the normal crustal thicknesses T=30 km and T=40 km, and the radii of isostatic compensation region R=58.1, 116.2 and 232.4 km were adopted. We also made the Hayford's isostatic reduction in which the depth of compensation was taken as 113.7 km. All these results are shown in Table 1 and Fig. 1.

The accuracies for the field gravity measurements are as follows: for the elevation, it is about 10 m, for the co-ordinates 500-1,000 m, and for gravity observations $\pm0.2$ mgal. The maximum value of topographic reduction reaches 53 mgal while those of 7 stations are 30 mgal.

In Fig. 2 and Table 1, three points should be noted, first when different T values were adopted for isostatic reduction, the resulting values of isostatic gravity anomalies are not the same. The maximum value of isostatic gravity anomaly is reached when the crustal thickness is taken as 30 km, the minimum when it is taken as 60 km. The maximum difference may reach 50 mgal. Second, the effect of different T values on the isostatic gravity anomalies in the Himalayan region and within Qinghai-Xizang Plateau is different. The difference of values in the Himalayan region may reach 50 mgal, while it is smaller, only 13 mgal, within the Qinghai-Xizang Plateau. The cause of such a phenomenon is that, the mean elevation variation of the reduction zones at the station in the Himalayan region (station no. 219) are greater than those of the stations within the Qinghai-Xizang Plateau (station no. 105). Fig. 3 and Fig. 4 clearly show that if the elevation of every zone is 5,000 m, the differential reduction value is 14 mgal when the normal crust thickness varies from 30 km to 60 km, and is only 1 mgal, when T varies from 30 km to 40 km. If the elevation of every zone is 1,000 m, the differential reduction value then is 7 mgal when normal crust thickness varies from 30 km to 60 km, and 1 mgal when it varies from 30 km to 40 km. It also indicates that, when the difference of elevation between the zones is not large, the effects due to variation of T are comparatively small. Now the mean elevation of every reduction zone in stations of the Himalayan region varies greatly, so the effects of depth variation are large. Third, no matter what type of reduction we adopt, the isostatic anomalies are all positive in the Himalayan region, its maximum value reaches 120 mgal; in the inner region of Qinghai-Xizang Plateau generally it fluctuates around zero (negative in the area of Yarlung Zangbo River at station no. 213, due to the effects of shallow geological structure). This suggests that the Qinghai-Xizang Plateau is in isostatic equilibrium, but the Himalayan region is not.

## DISCUSSION

### Isostasy of the Qinghai-Xizang Plateau

From Fig. 1 we can see that in the Himalayas, there is a positive isostatic anomaly, whose maximum value is about 120 mgal. It is about 90 mgal higher than those of the inner region of Qinghai-Xizang Plateau and the Ganges River Plain in India. The Himalayas about 120 km wide, and several hundreds km long are neither in complete compensation nor in isostatic equilibrium. On the other hand, the Qinghai-Xizang Plateau as a whole seems to be isostatically in equilibrium.

## The Origin of the Positive Isostatic Gravity Anomalies in the Himalayas

Fig. 5 is the crustal model along a profile for Ganges River-Yadong-Tanggula Range and its gravity anomalies. The cross "x" represents the results of deep seismic sounding(8). The Moho discontinuity curve is calculated by the method of "compressed mass plane"(9)from gravity data. This points out that the crustal boundary derived by gravity data roughly corresponds to reality. It can also be seen in the figure that the crustal boundary does not down-warp underneath Himalayas, showing that there is no mountain root, or only small "mountain root" under the Himalayas. From the data of deep seismic sounding, no high density layer (higher velocity layer) within the crust is found. As for the origin of the positive isostatic anomalous values in the Himalayan region, it is not mainly caused by heterogeneous densities such as the thickening of the basalt layer in the crust, but caused by lack of mountain root.

Fig. 6 shows the topography, crustal boundary and gravity anomalies along the profile from Darjeeling to Rinbung. The curve is the crustal boundary deduced from Bouguer gravity anomalies by using the method of "compressed mass plane". The dotted line is deduced from the topographic elevations by assuming complete compensation. As shown in the figure, there are discrepancies between the two in the Himalayan region, but in other regions they are in agreement with each other. The gravity anomalies represented by the curve are derived from the so-called "mountain root" -- that is the difference of the crustal boundaries from these two kinds of data. Notation "." indicates the isostatic anomalies deduced by the observed data which correspond closely to the above anomalies. This also shows that the origin is caused by the lack of "mountain root", instead of by the heterogenous materials within the crust.

In the light of the principle of isostasy, those areas with positive isostatic anomalies should subside. In fact, they are rising continuously in the Himalayas(10). It shows that there is a strong lifting force which is greater than that of the isostatic compensation force in the Himalayas, and this force makes the Himalayas going upward, instead of downward. this force is commonly ascribed to the collision of the Indian Plate against the Eurasian Plate. Fig. 5 is a possible model for the Himalayan region. MBF is the main boundary fault. MCT is the main central thrust fault. ISZ is the suture zone of the two plates. When the Indian Plate moved northward and collided with the Eurasian Plate, the Himalayas were thrusted up toward the south along the fault planes. This is followed by the continuous uplift of the Himalayas up to the present; topographically it is steep on the south but gentle on the north.

## References


(1) Masaru Kono, Gravity anomalies in East Nepal and their implications to the crustal structure of the Himalayas, Geophy. Jour. R. Astr., 39, 283-299 (1974).

(2) Vening-Meinez F.A. Tables for the regional and local isostatic reduction (Airy system) (1941).

(3) Klaki P., Kivioja L., Heiskanen W.A. Topographic isostatic reduction maps for the world for Hayford zones 18-1, Airy-Heiskanen system, T-30 km, Helsinki, Publ. Isostat. Inst. Int. Assoc. Geod. 35 (1961).

(4) Heiskanen W., Nuoto U. Topographic-isostatic world maps of the effect of the Hayford zones 10, 9, 8 and 7-1, Helsinki. Publ. Isostat.

Inst. Int. Assoc. Geod., 3 (1938).

(5) Niskanen E., Kivioja E. Topographic-isostatic world maps of the effect of the Hayford zones 10-1 for the Airy-Heiskanen and Pratt-Hayford systems, Helsinki, Publ. Inst. Int. Assoc. Geod., 27 (1951).

(6) Heiskanea W. New isostatic tables for the reduction of the gravity values calculated on the basis of Airy's hypothesis, Helsinki. Publ. Inst. Int. Assoc. Geod., 2 (1938).

(7) Hayford J.F., Bowie W. The effect of topography and isostatic compensation upon the intensity of gravity (1912).

(8) Teng Ji-wen et al., Explosion seismic study for velocity distribution and structure of the crust and upper mantle from Damxung to Yadong of Xizang Plateau, in this Proceedings of Symposium on Qinghai-Xizang (Tibet) Plateau, Academia Sinica,(1981).

(9) Liu Yuan-lung et al., Inversion of gravity data by use of a method of "compressed mass plane" to estimate crustal structure. Acta Geophysica Sinica, 20 (1), 59-69 (1977) (in Chinese).

(10) Chang Cheng-fa et al., Tectonic features of the Mount Qomolangma region in southern Tibet, China. Scientia Geological Sinica, (1), 1-12 (1973)(in Chinese).

Table 1 Hayford, Airy-Heiskanen, Vening-Meinez Isostatic gravity Anomalies.

| Station | Free-air anomalies (mgal) | Bouguer anomalies (mgal) | Isostatic anomalies(mgal) | | | | | | | | | |
|---|---|---|---|---|---|---|---|---|---|---|---|---|
| | | | T = 30 KM | | | | T = 40 KM | | | | T = 60.0 KM | D = 113.7 KM |
| | | | R = 0 | 58.1 | 116.2 | 232.4 | R = 0 | 58.1 | 116.2 | 232.4 | R = 0 | |
| 219 | 397 | −304 | 113 | 99 | 85 | 49 | 79 | 81 | 74 | 54 | 62 | 91 |
| 14 | 83 | −392 | 75 | 75 | 63 | 33 | 57 | 54 | 46 | 22 | 38 | 84 |
| 213 | 43 | −523 | −44 | −29 | −41 | −60 | −56 | −51 | −56 | −68 | −71 | 30 |
| 65 | 27 | −494 | 8 | 13 | 11 | 3 | −3 | −3 | −2 | −14 | −5 | 31 |
| 106 | 7 | −505 | −4 | 2 | 1 | 4 | −9 | 9 | −1 | 0 | −12 | 35 |

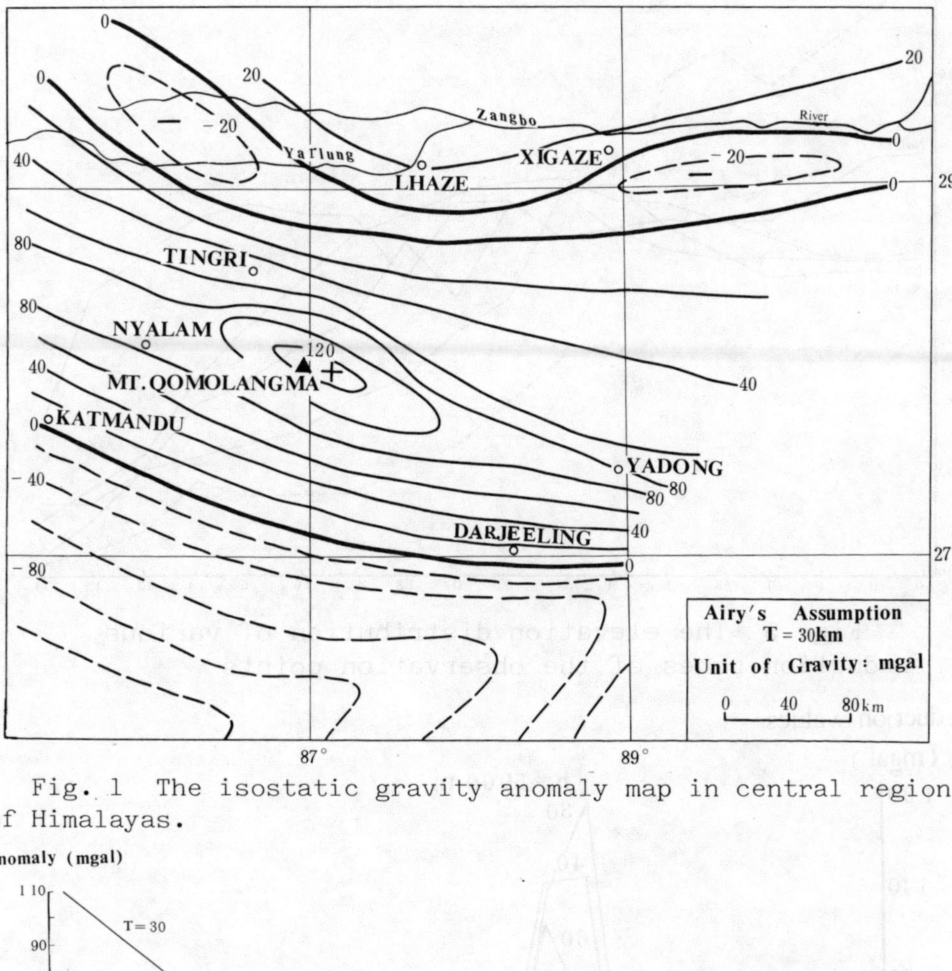


Fig. 1 The isostatic gravity anomaly map in central region of Himalayas.

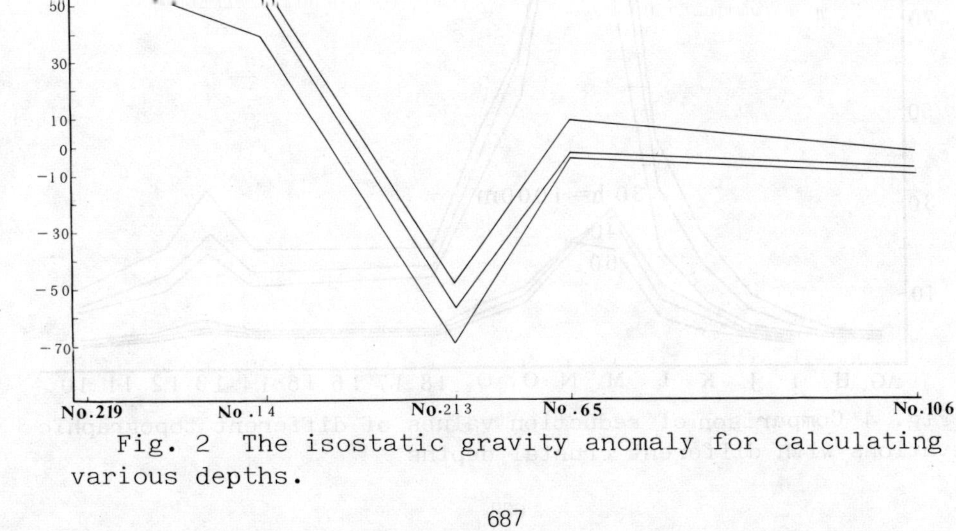


Fig. 2 The isostatic gravity anomaly for calculating various depths.

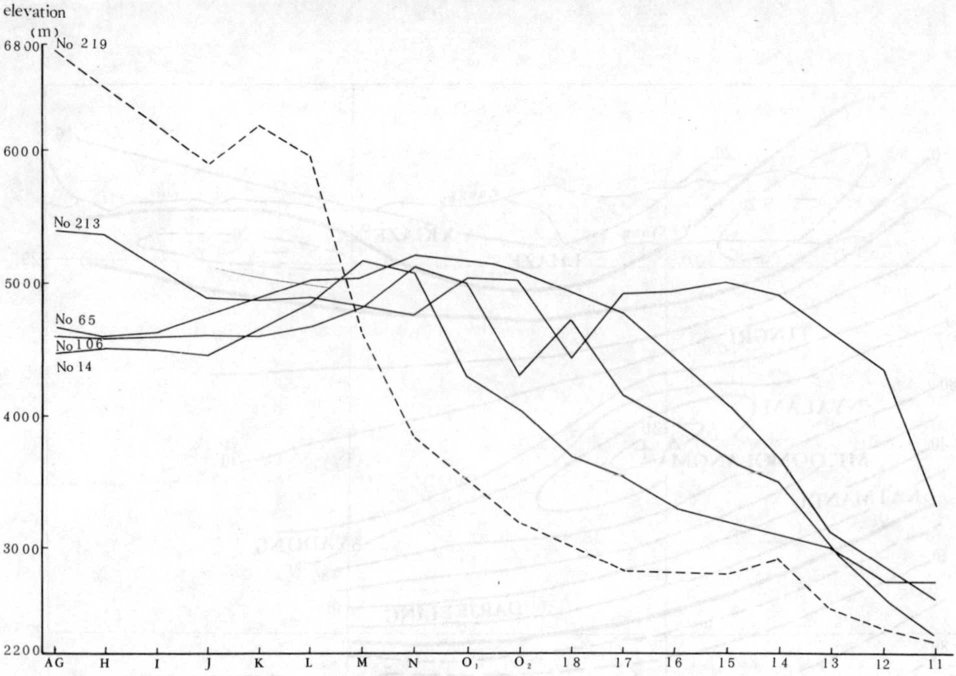


Fig. 3 The elevation distribution of various reduction zones of the observation points

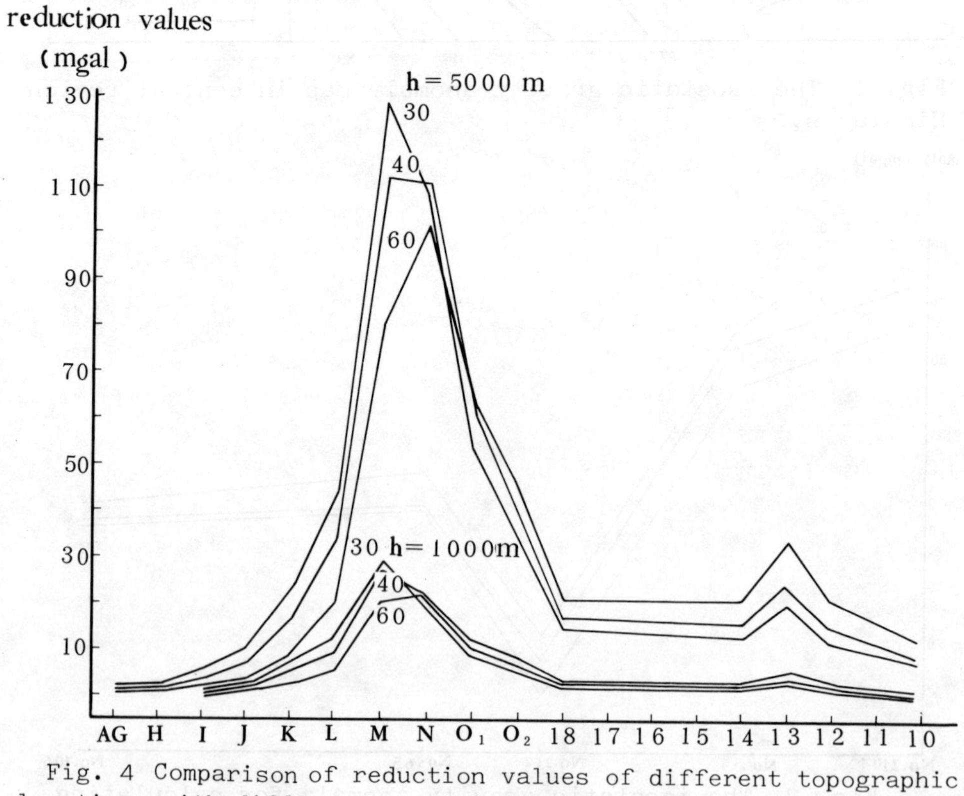


Fig. 4 Comparison of reduction values of different topographic elevations with different crustal depths

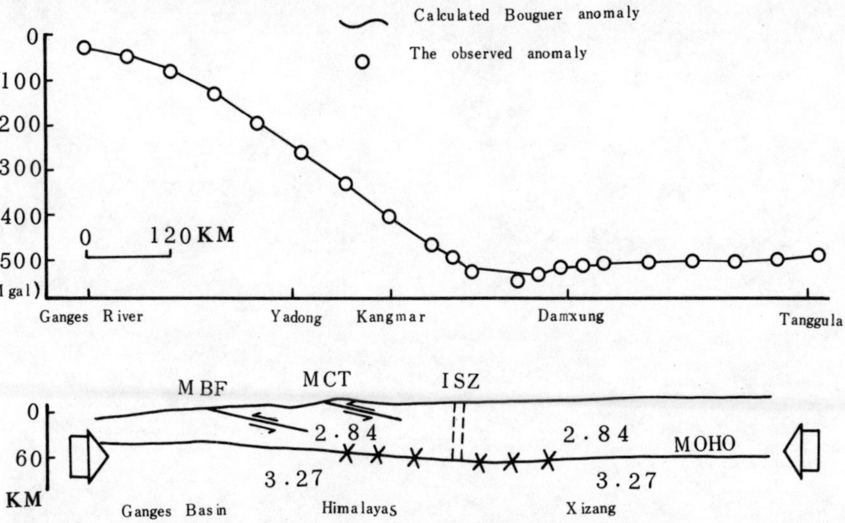


Fig. 5 The crustal model and its gravity anomalies along the profile of Ganges River–Yadong –Tanggula Range
the Moho discontinuity calculated from gravity data, "X" the results of deep seismic sounding (after Teng Ji-wen et al.)

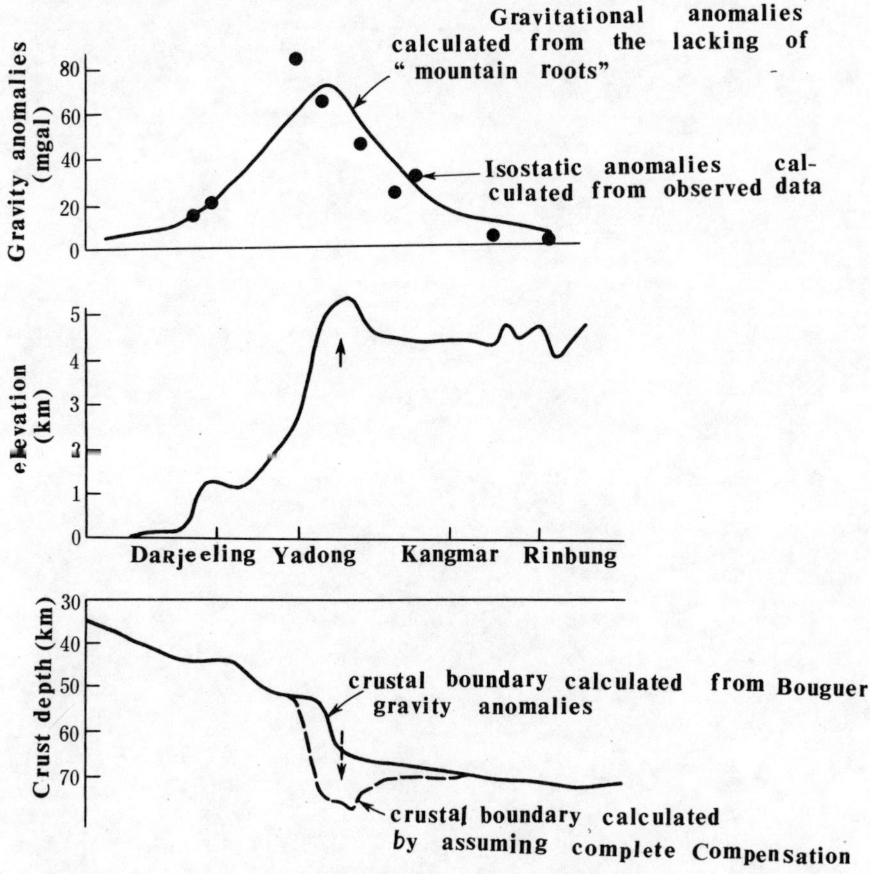


Fig. 6 The topography, crustal boundary and gravity anomalies through Darjeeling to Rinbung

# EXPLOSION SEISMOLOGICAL STUDY FOR VELOCITY DISTRIBUTION AND STRUCTURE OF THE CRUST AND UPPER MANTLE FROM DAMXUNG TO YADONG OF THE XIZANG PLATEAU


Teng Ji-wen, Xiong Shao-po, Sun Ke-zhong, Yin Zhou-xun, Yao Hung, Chen Li-fang, Mu Teng, Lai Ming-hui, Wu Ming-chu, Su De-yuan, Wang Shao-zhou, Huang Wen-jian, Ou Ren-sheng, Hao We-cheng, Shao An-min

(Institute of Geophysics, Academia Sinica)

Gao En-yuan

(Changchun Geological Institute)

Wang Meng-lin

(Institute of Computing and Techniques, Ministry of Geology)

Lin Zhong-yang, Qu Ke-xin

(Institute of Geophysics, State Seismological Bureau)


The study of the crustal structure of this region was made by explosions under water in lakes on the Xizang Plateau (Yamzho , Puma, Nam Lakes).

## OBSERVATION SYSTEM

Locations of shotpoints and stations in the Damxung-Yadong profile are shown in Fig. 1. The profile is perpendicular to the strike of the tectonic trend (Fig. 2). Explosions were made under water in lakes and the maximum explosion depth was from 22 to 47 meters. There were nine explosions with four kinds of charges 3, 5, 9 and 15 tons. The length of the profile was about 430 km. The interval between receiver points was about 10 km, with a few reaching 30 or 50 km.

Analogue tape recording system was used, with dynamic range about 50 db.

## WAVE CHARACTERISTICS

The seismogram obtained for the section of the profile from Yamzho Lake to Yadong (Fig. 3) manifests the complexity of the wave characters from various deep interfaces in the Xizang Plateau. The main wave phases can be divided into six groups which may be denoted by $t_1$, $t_2$, $t_3^o(t_3)$, $t_4^o$, $t_5^o(t_5)$ and $t_6^o(t_6)$ (Figs. 3,4).

The results of correlation(1) processing for the profile (N 790) are shown in Fig. 5. The effect of correlation processing on correlation coefficient section (a), correlation weighted section (b) and normalized correlation coefficient section (c) is evident. From these diagrams, we can see that the random disturbance is greatly suppressed. Using the least square method, we processed these data of wave arrivals and obtained a velocity of the top of the upper mantle, 8.05 km/sec. and the crustal thickness, 69.7 km which is nearly equal to that obtained from reflected waves.

The squared record section of amplitude (Fig. 6) accentuates the difference between large and small amplitudes and makes the phases clearer.

The section diagram obtained after velocity scanning shows that the corresponding stacked curves can be obtained with the step distance 0.05 km/sec., when the scanning velocity is 7.5-8.5 km/sec. is given. When the scanning velocity is equal to the apparent velocity of the effective wave, signals become the strongest, because of stacking of the same phase, for example head wave $t_6$, (Fig. 7) from the top of the upper mantle.

The frequency spectrum of the reflections of the profiles N790 and N812 is given in Fig. 8. The dominant frequencies are given in the following tables (Table 1 and 2). It is seen that the average dominant frequency of each wave phase on N812 is lower than that on N790. It may be caused by the combined effects of the seismic waves crossing the Yarlung Zangbo River rift and the Yangbajain geothermal region.

The kinematic characters indicate, that the velocity of head wave $V_e$, does not change with distance (Fig. 9). The time difference of head waves from neighboring shot points along profiles N790 and N926 also does not change with distance (Fig. 10), but the apparent velocity of the reflected wave decreases rapidly with distance (Fig. 11). The time difference of the reflected waves along the profiles N812 and N910, increases with distance (Fig. 12). Fig. 13 shows the combined travel-time curve of the profiles of N790, N910 and N926.

According to the kinematic and dynamic characters of the waves mentioned above, it can be assumed that $t_1$, $t_2$ are head waves from sedimentary rock layers and $t^o_3 (t_3)$, $t^o_4$, $t^o_5 (t_5)$, $t^o_6 (t_6)$ are reflections and head waves from the interior of the crust and the top of the upper mantle respectively.

## VELOCITY DISTRIBUTION IN THE CRUST AND UPPER MANTLE

Table 3 shows the average velocities of various waves reflected in the crust, the average velocity of the crust being 6.23 km/sec.

Table 4 shows the layer velocities derived from inversion by damping least square method, according to a multilayered homogeneous model (dip angle not equal to zero). The larger velocities of the crust are 4.69, 5.99, 6.22, 5.64, 7.24 km/sec. respectively. The velocity at the top of the upper mantle is 8.15 km/sec.

## THE CRUSTAL SECTION AND MODEL OF THE CRUST

The crustal structure of Xizang Plateau is a multilayered medium with alternate high and low velocities (Fig. 14). From the north of Yarlung Zangbo River to Nam Lake, the thickness of the sedimentary layer $R_2$ ranges from 4-5 km, the depth of interface $R_3$ is 20-22 km, the depth of $R_4$ is 42-46 km, the depth of $R_5$ is 54-56 km, and the depth of Moho is 70-73 km. The inclinations of all the interfaces are very small ($1^o$-$3^o$). In the region to the south of the Yarlung Zangbo River, all the interfaces of the crust become shallow and upwarping towards the south: the depth of $R_2$ is 3-4 km, the depth of $R_3$ is 22--8 km, the depth of $R_4$ is 45-29 km, the depth of $R_5$ is 55-35 km, the depth of Moho is 68-45 km. The maximum thickness of the crust in Xizang is 73 km, the thickness of the crust in the Himalayas and its eastern extension is 45 km.

The depth of the low velocity layer is 42-46 km to the north of the Yarlung Zangbo River and 45-29 km to the south. The average thickness of the low velocity layer is 10 km with velocity of 5.64 km/sec.

The characters of time-distance curves of various wave phases at different distances, apparent velocity $v_e$ , variations of waveforms and amplitudes etc. all show that along the Yarlung Zangbo River there is a deep rift zone, the width of which is about 30-50 km, with a steep angle slightly dipping towards the south.

It has been ascertained above that a low-velocity layer exists in the lower crust of the Xizang Plateau. The calculation of $V$-$t_o$ (Fig. 15) velocity spectrum of reflection along profile N790 by computer gives the same result.

The average crustal models of the Xizang Plateau from north of Yarlung Zangbo River to Damxung and Himalayas to the south of Yarlung Zangbo River are given in Fig. 16.

The extremely thick crust was caused by the results of the collision of the Indian Plate against the Eurasian Plate, and during the process of continuous pressing the horizontal shortening took place on a large scale in the crust.

## CONCLUSION

The following three main results are obtained:

1. According to the kinematics and dynamics of the seismic waves, we have obtained the crustal structure and the velocity distribution of Xizang Plateau. The crust here is extremely thick, and is composed of alternate layers with high and low velocities. From the north of Yarlung Zangbo River to Damxung region, the interfaces of various layers in the crust change gently, and the thickness of the crust is 70-73 km. In the southern region of Yarlung Zangbo River, the thickness becomes smaller. In Doqen lake - Tüna zone (Himalayan arcuate mountain system) it is about 45 km.

2. Along the Yarlung Zangbo River, there is a great fault zone extending down to the top of the upper mantle with a steep angle slightly dipping towards the south. The difference of geophysical fields between the two sides of Yarlung Zangbo River is evident. Here is the northern boundary of the transitional belt between the two colliding plates.

3. There is a low velocity layer in the lower crust. To the north of Yarlung Zangbo River, the depth of this layer is 42-45 km; to the south, it is 45-29 km. The thickness of the low velocity layer is about 10 km, with the layer velocity 5.64 km/sec.

## References

(1) Merkel R.H., Alexander S.S., Use of correlation analysis to interpret continental margin ECOOE refraction data. J.G.R., 74 (10) (1969).

(2) Hinj E., Kaminski W., Stain A., Results of a seismic-refraction profile from the Hoher meissner to the North German Plain. Explosion seismology in Central Europe 257-267 (1977).

(3) Sattegges J., A method of computing true interval velocity from expanding spread data in the case of arbitrary long spread and arbitrarily dipping plane interfaces, Geophysical Prospecting, 13 (2) 306-318 (1965).

(4) Gangi A.F., Sung T.Y., Traveltime curves for reflections in dipping layers. Geophysics, 41 (3) (1976).

Table 1

| N790 | Frequency (Hz) / Observation region (km) | $t^o_3$ | $t^o_4$ | $t^o_5$ | $t^o_6$ |
|---|---|---|---|---|---|
| Yamzho Lake-Yadong | 130 (inside) | 4.60 | 4.0 | 4.4 | 4.4 |
| | 130 (outside) | 4.60 | 4.7 | 4.7 | 4.5 |

Table 2

| N812 | Frequency (Hz) / Observation region (km) | $t^o_4$ | $t^o_5$ | $t^o_6$ |
|---|---|---|---|---|
| Yamzho | Lake - Damxung | 3.6 | 3.4 | 3.6 |

Table 3

| Average velocity (km/sec.) \ Profile | N790 | N812 | N910 | N914 | N926 | Average velocity (km/sec) |
|---|---|---|---|---|---|---|
| Earth's surface $-t_2$ | 4.84 | 4.69 | 4.60 | 4.62 | 4.69 | 4.69 ± 0.049 |
| $t^o_3$ | 5.86 | 5.80 | 5.80 | 5.84 | 5.83 | 5.82 ± 0.024 |
| $t^o_4$ | 6.00 | 5.98 | 5.97 | 5.99 | 6.03 | 5.99 ± 0.03 |
| $t^o_5$ | 5.99 | 5.78 | 5.45 | 5.59 | 5.95 | 5.87 ± 0.156 |
| $t^o_6$ | 6.23 | 6.29 | 6.13 | 6.29 | 6.18 | 6.23 ± 0.05 |

Table 4

| Profile \ $V_e$ (km/sec.) | Earth's surface $-t_2$ | $t_2-t^o_3$ | $t^o_3-t^o_4$ | $t^o_4-t^o_5$ | $t^o_5-t^o_6$ |
|---|---|---|---|---|---|
| N790 | 4.84 | 6.07 | 6.17 | 5.31 | |
| N812 | 4.69 | 6.04 | 6.17 | 5.50 | 7.24 |
| N910 | 4.60 | 5.94 | 6.28 | 5.91 | 7.44 |
| N914 | 4.62 | 5.94 | 6.28 | 5.89 | 7.44 |
| N926 | 4.69 | 5.95 | 6.19 | 5.61 | 6.84 |
| Average layer velocity (km/sec.) | 4.69 ± 0.062 | 5.99 ± 0.054 | 6.22 ± 0.05 | 5.64 ± 0.204 | 7.24 ± 0.25 |

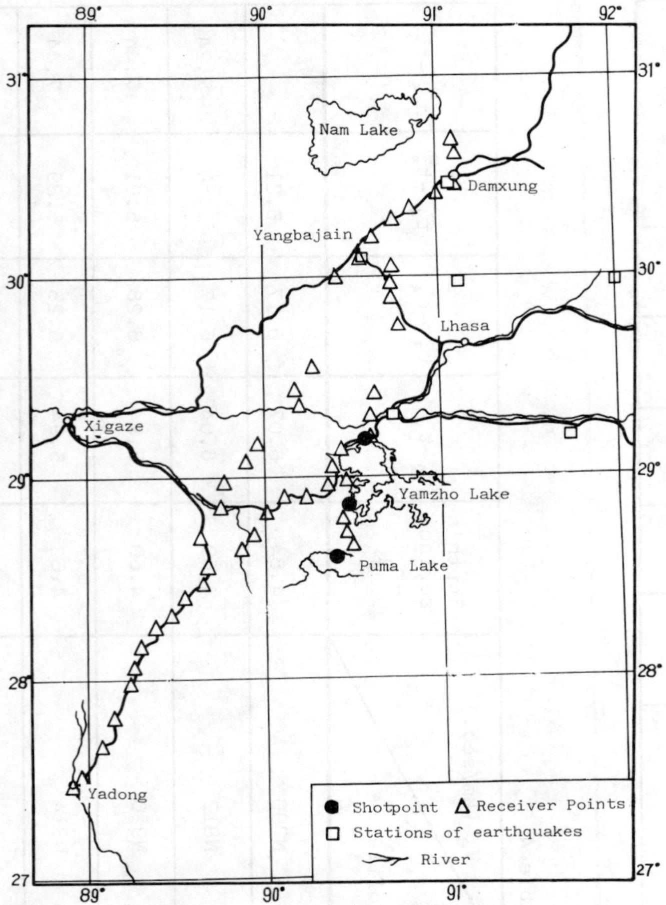


Fig. 1 Locations of shotpoints and stations in Damxung-Yarlung Zangbo River-Yadong profile.

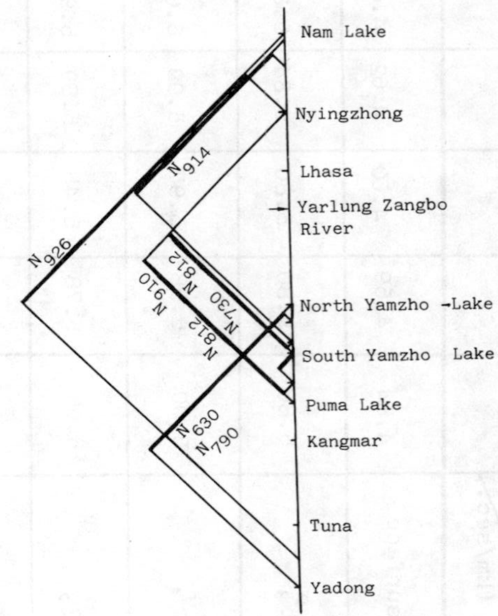


Fig. 2 Lay - out of the observation system of Damxung Yadong profile.

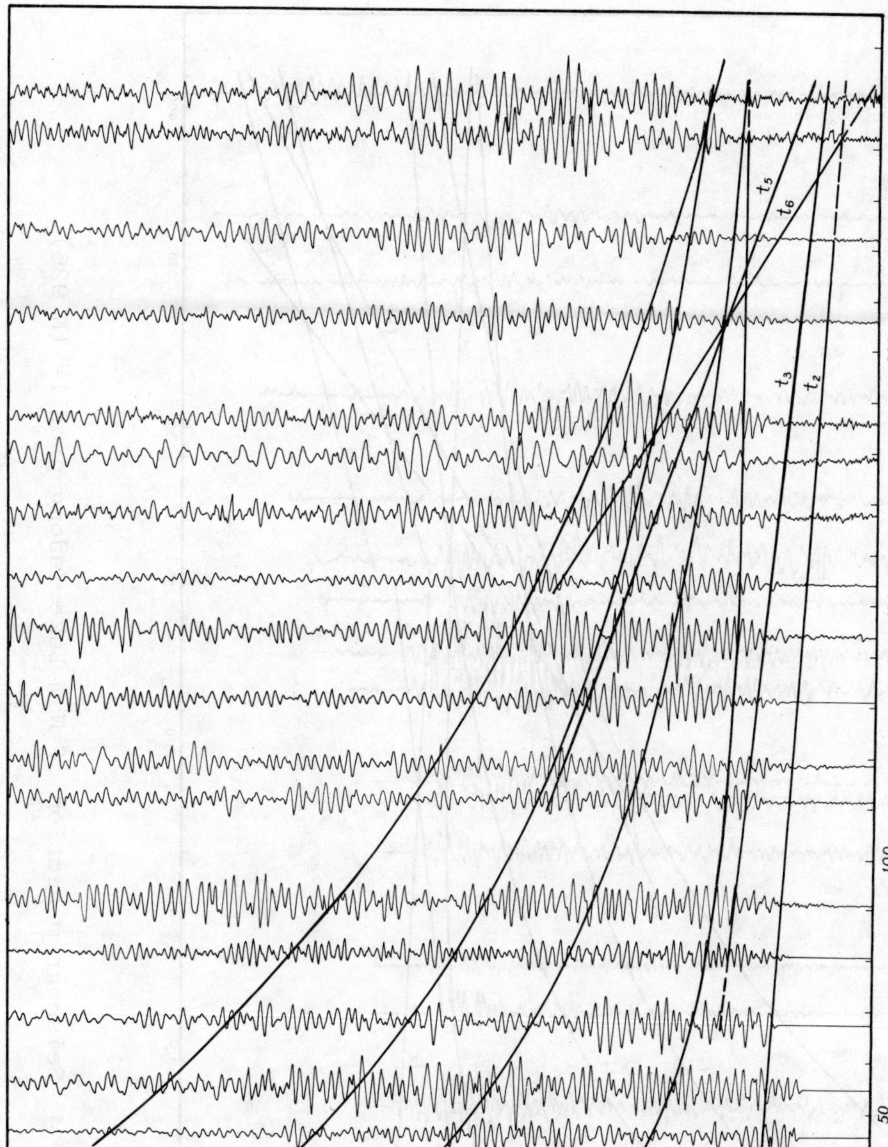


Fig. 3 Seismogram section of the profile in Yamzho Lake – Yadong (N 790).

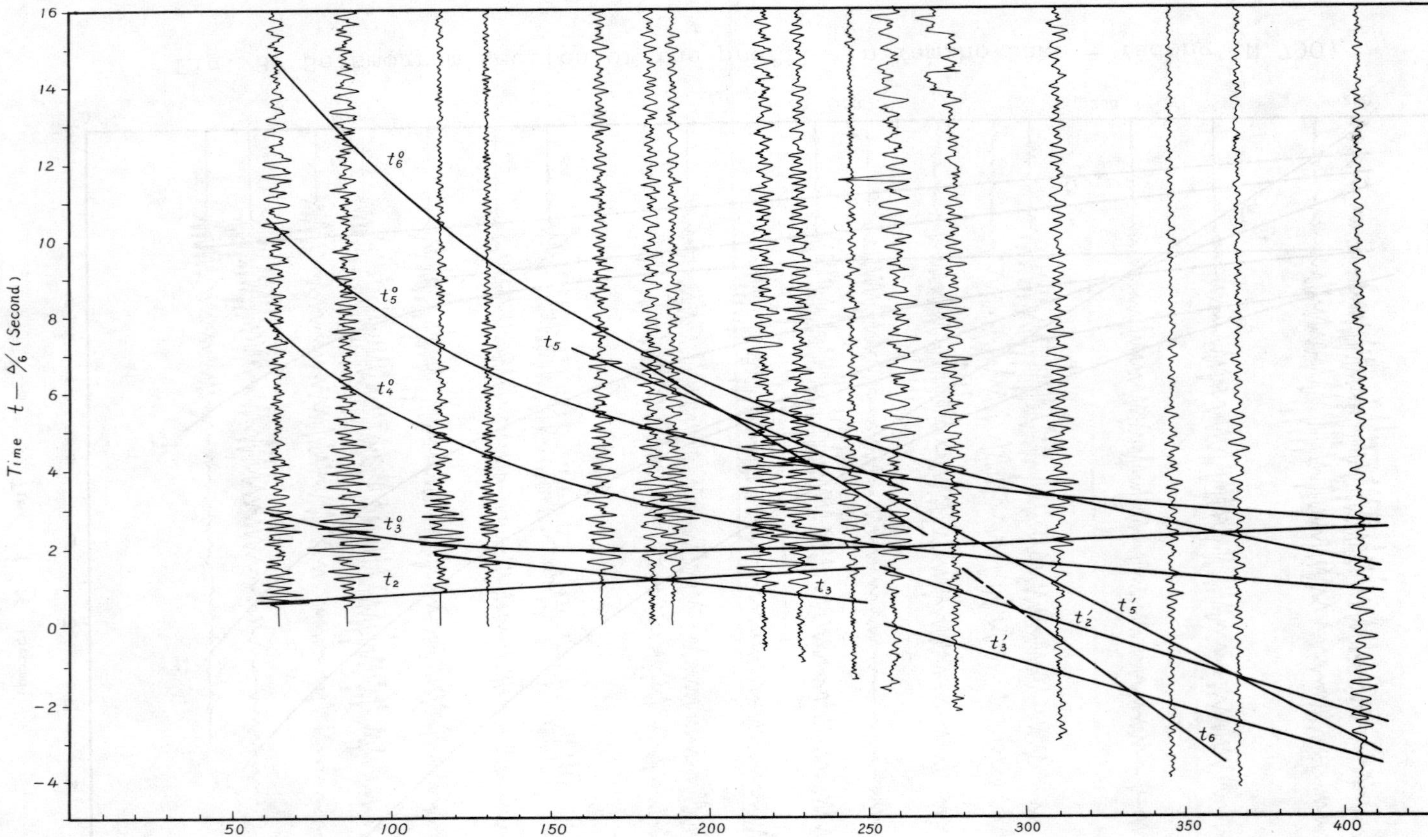


Fig. 4 Seismogram section of Nam Lake-Yadong profile (N 926).

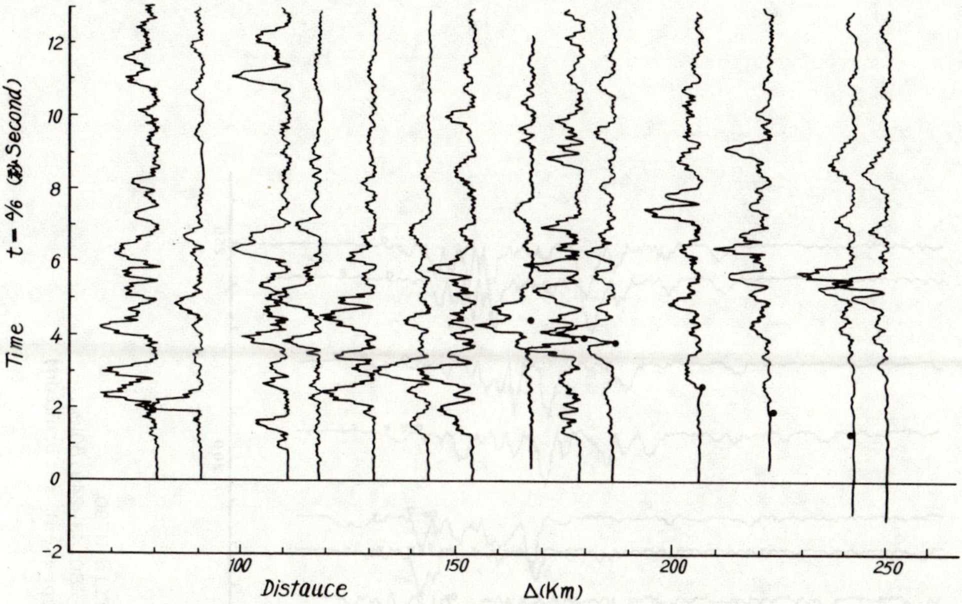


Fig. 5 The results of correlation processing for the proiile N790

(a) Correlation coefficent section of N790 Δ=80.43-249.80km (mark "•" represents the arrival position of head wave from Moho)

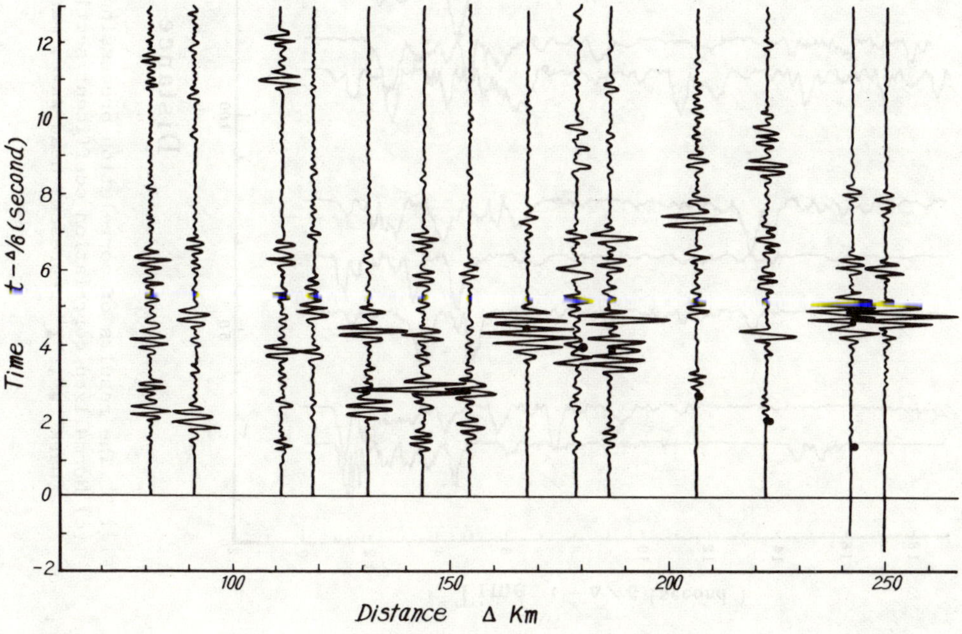


Fig. 5 The results of correlation prcessing for the profile N790.

(b) Correlation weighted section of N790 Δ=80.43-249.80km (mark "•" represents the arrival of head wave from Moho)

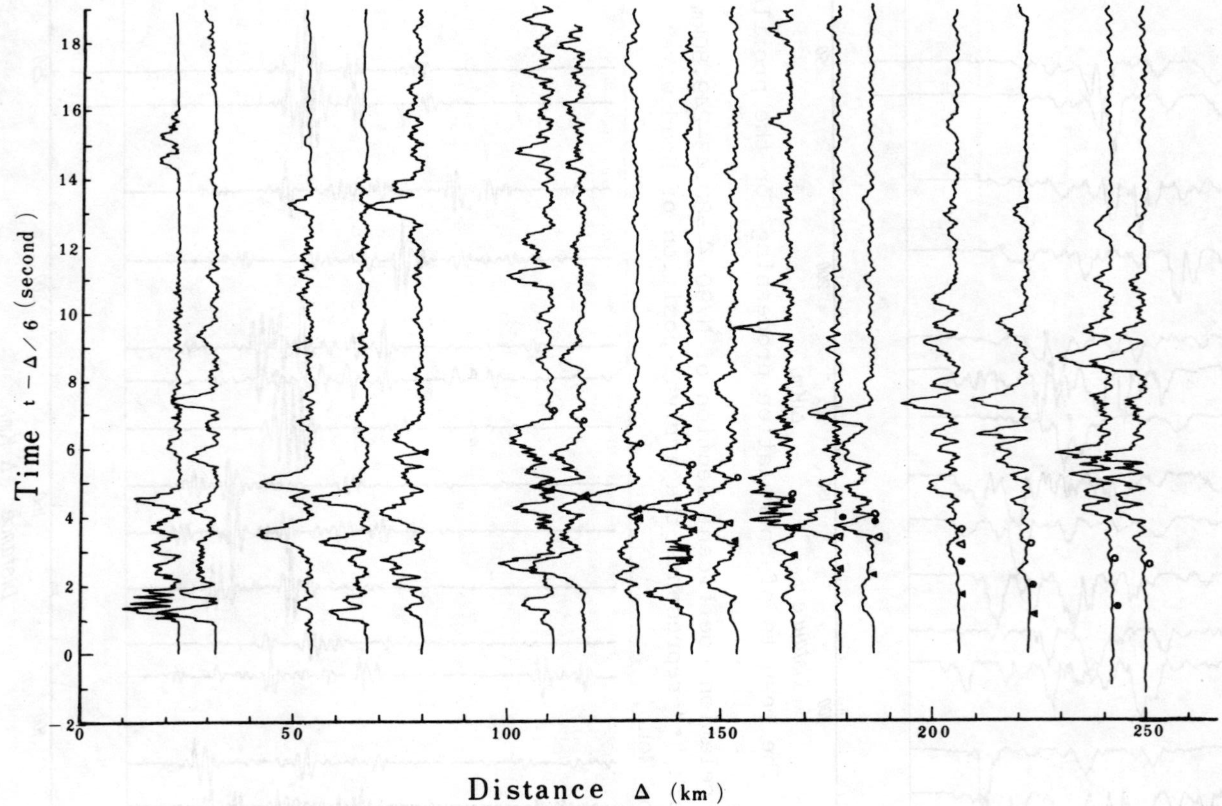


Fig. 5 The results of correlation processing for the profile N790
(c) Normalized correlation coefficent section of N790 Δ =80.43-249.80km.
(mark "●", "▲" represent head wave, mark "o", "Δ" represent reflection)

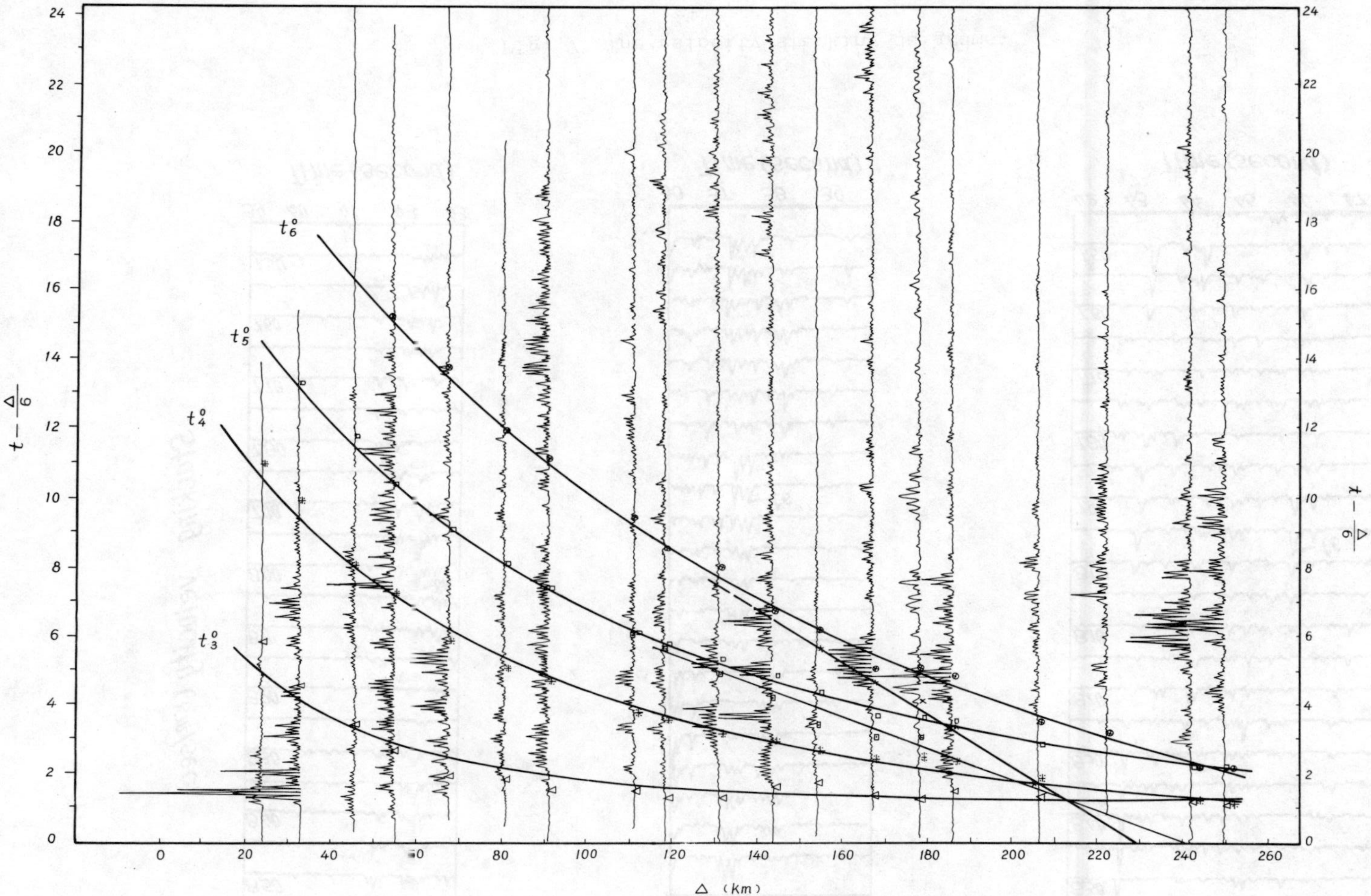


Fig. 6 Squared record section of amplitude in profile (N790).

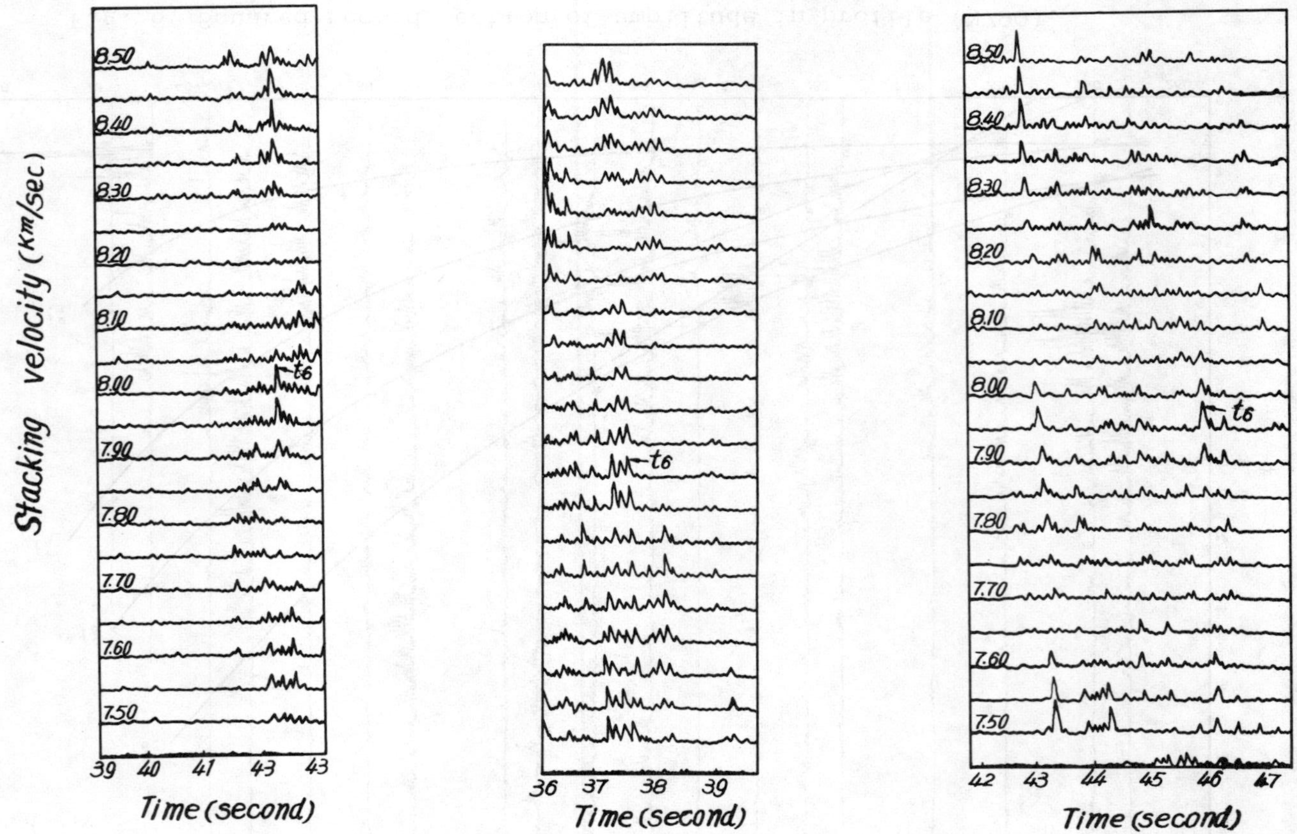


Fig. 7 The velocity stacking daigrams.

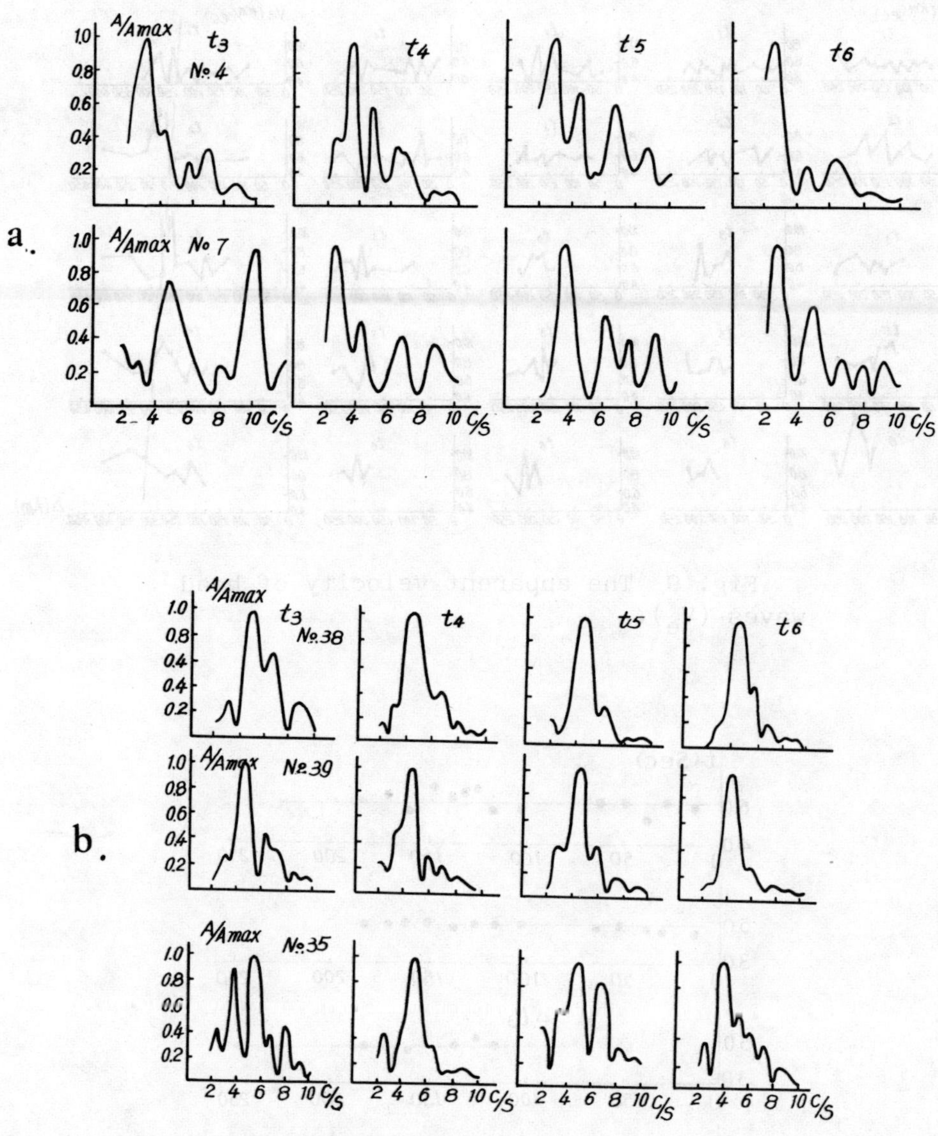


Fig. 8 Frequency spectrum of the reflections of profiles N790 (a) and N812 (b)

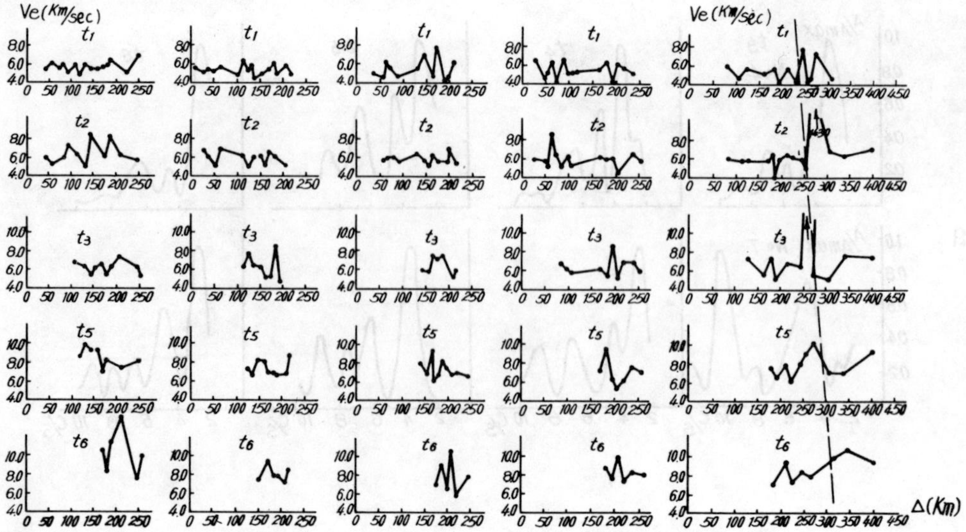

Fig. 9 The apparent velocity of head waves ($V_e$).

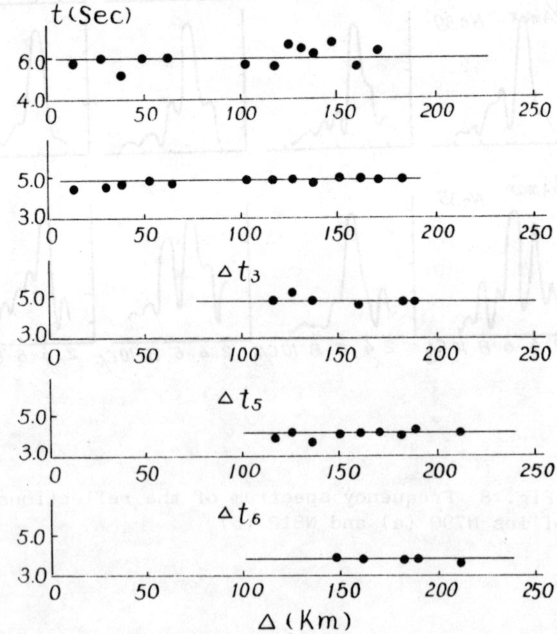

Fig. 10 The time difference of timedistance curves of head wave

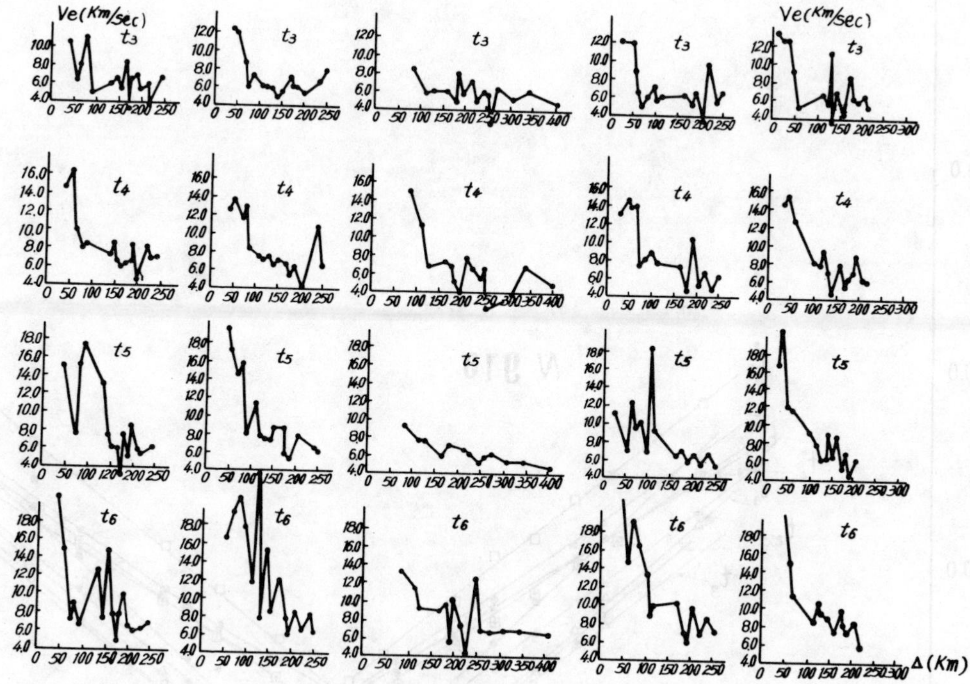

Fig. 11 The apparent velocity of reflected waves $V_e$

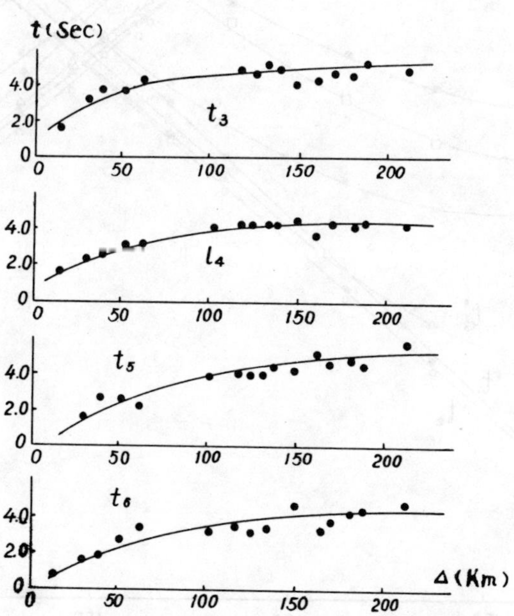

Fig. 12 The time difference for the timedistance curves of reflected waves.

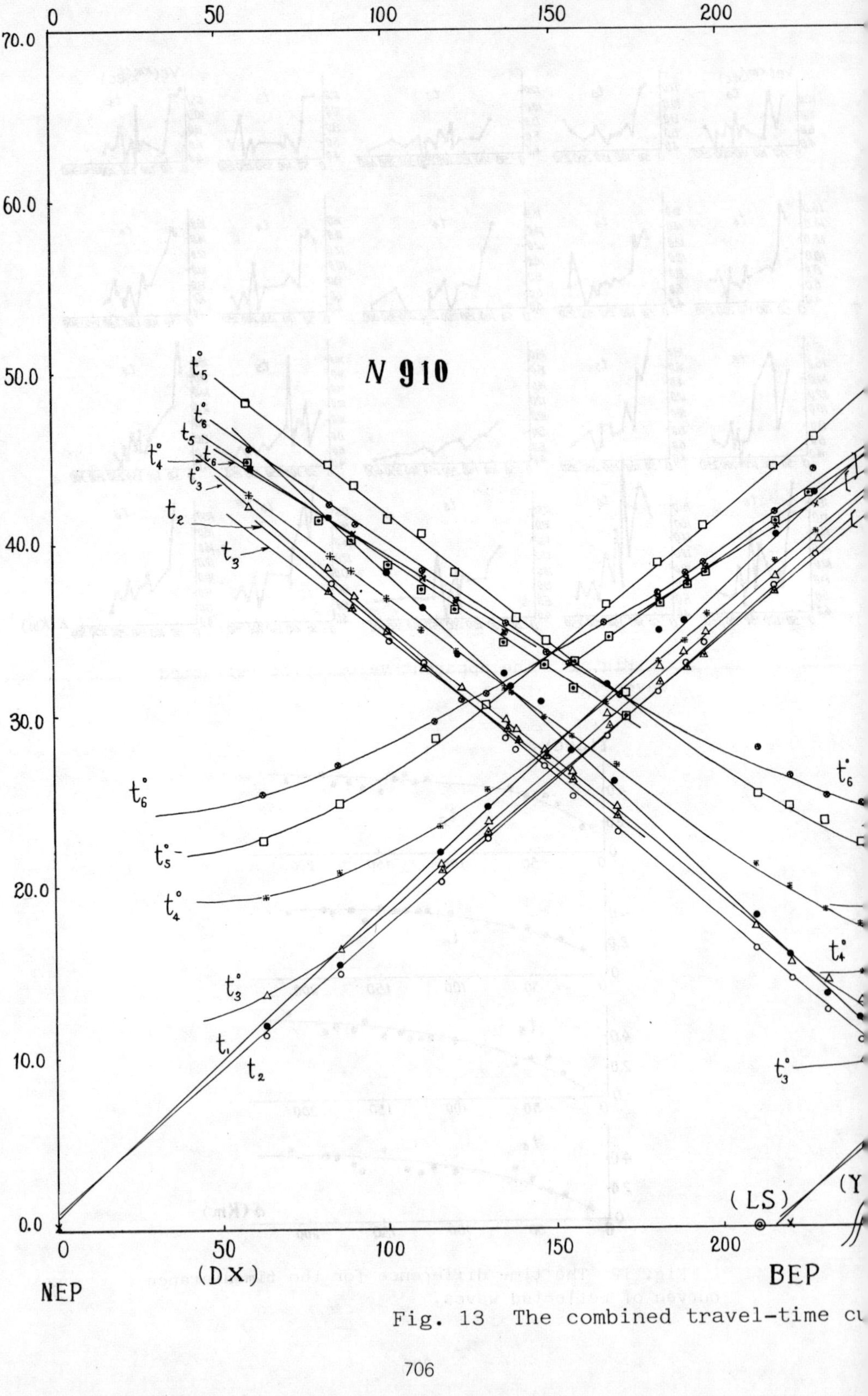


Fig. 13 The combined travel-time cu

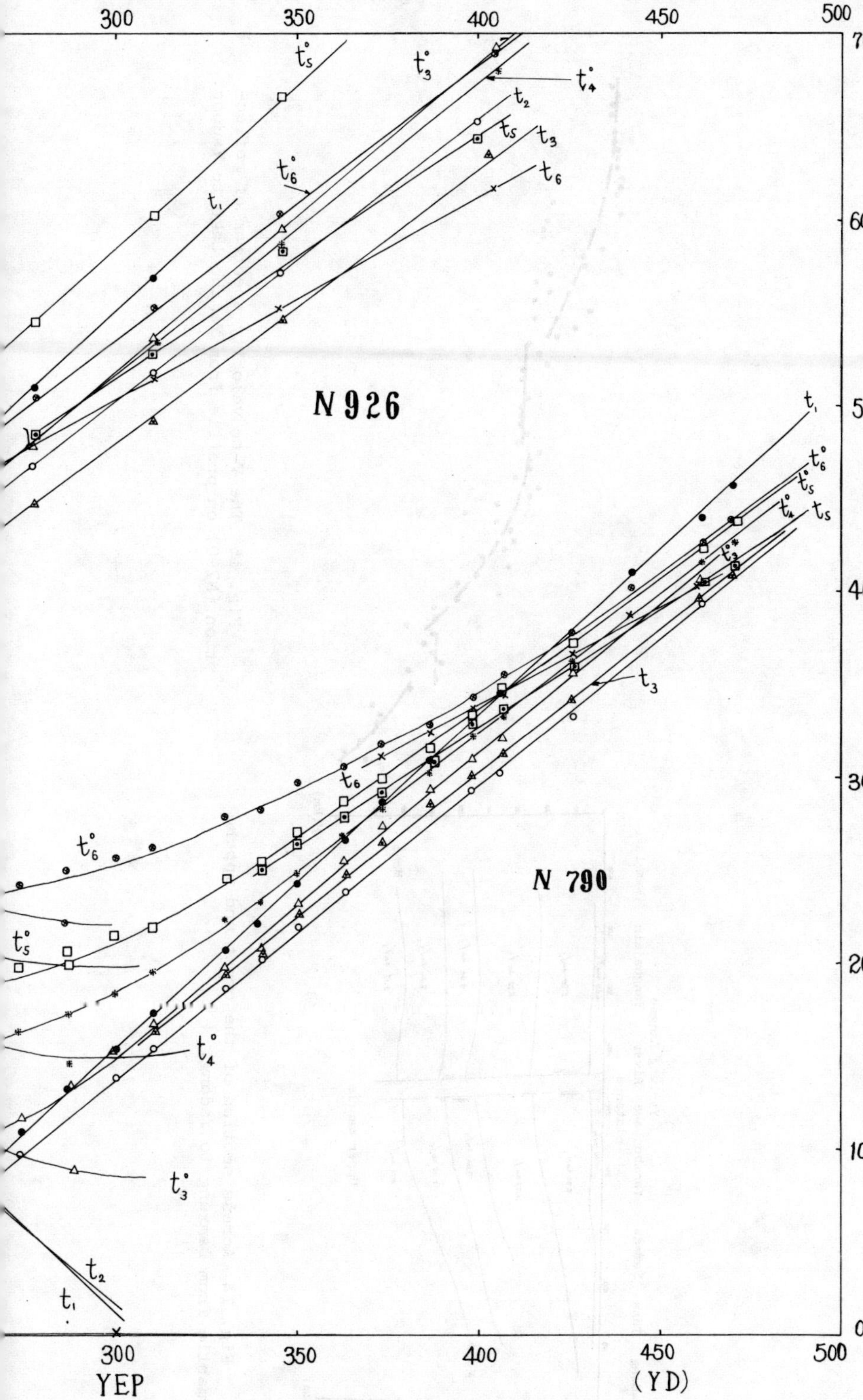


the profiles N790, N910 and N926.

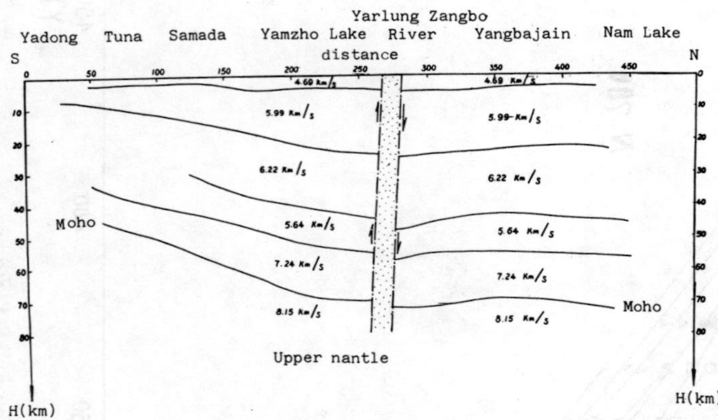


Fig. 14 Across section of the crust and upper mantle from Damxung to Yadong.

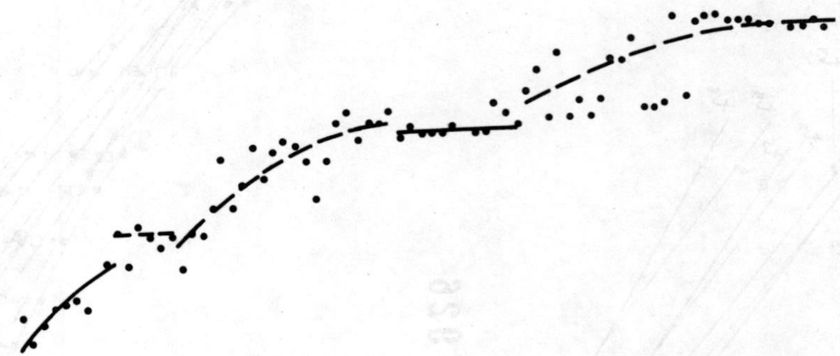

Fig. 15 The V-to velocity spectrum of reflection (N790) on profile from Yamzho Lake to Yadong.

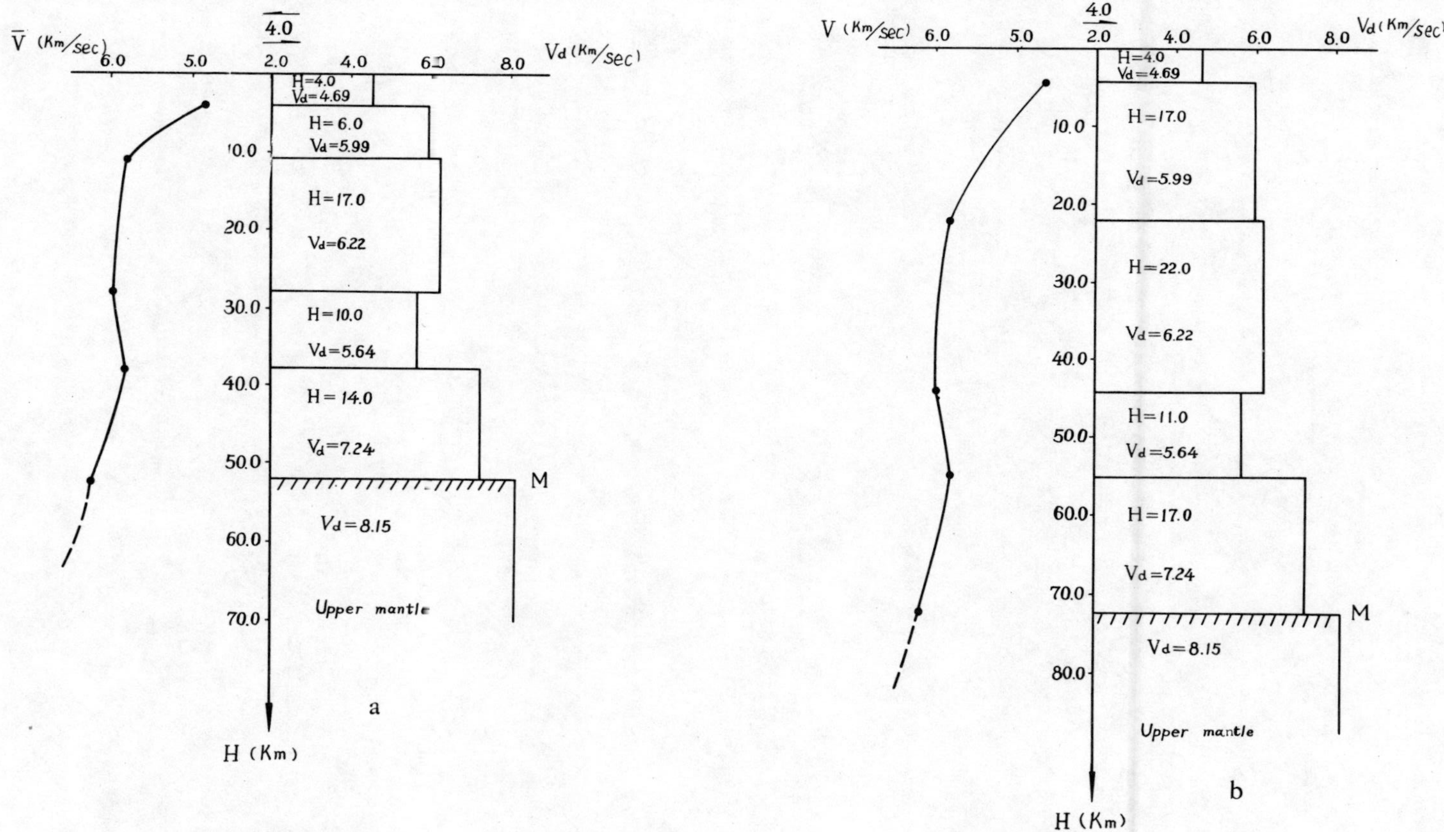


Fig. 16 Seismic crustal models and velocity distribution in Xizang Plateau
a. Crustal model to the north of Yarlung Zangbo River;
b. Crustal model of Himalayas to the south of Yarlung Zangbo River.

# CRUSTAL AND UPPER MANTLE STRUCTURE OF QINGHAI-XIZANG PLATEAU AND ITS SURROUNDING AREA

Ta-liang Teng
(Department of Geological Sciences,
University of Southern California,
Los Angeles, California, USA 90007)

## ABSTRACT

Multiple-filter technique and generalized linear inversion are applied to group velocity data derived from properly rotated three-component digital surface waves from Seismological Research Observatory stations. Repeated group velocity measurements over the same paths were made possible by using large aftershocks of several great (M > 7) earthquakes that recently occurred in China. Specific attention is paid to the crustal and upper mantle structure of the Qinghai-Xizang plateau and its comparison with those of neighboring tectonic provinces. The results demonstrate the strong lateral heterogeneity in the crust and upper mantle of the Chinese mainland. A particularly thick crust and an upper mantle of relatively low velocity are found underlying the Qinghai-Xizang plateau that corroborates the continental collision between India and Eurasia.

## INTRODUCTION

One of the most outstanding landscapes of grandiose scale on the earth, the Himalayas and the Qinghai-Xizang plateau, has long held the interest of earth scientists. Many recognized that in this vast region, having the highest as well as the youngest mountain ranges in the world, lies an important key to a major tectonic process that took place in the recent past. Besides the surface expressions, the plate collision between India and Eurasia must have produced fundamental modifications in the crust and upper mantle underlying this giant plateau. There is great interest to delineate these modifications. A detailed delineation will undoubtedly provide useful constraints on models proposed for the collision process. Among geophysical tools, surface wave dispersion analysis is one of the most effective means for the crustal and upper-mantle delineations. Knowledge of the crustal thickness and velocity structure under the Qinghai-Xizang plateau from other surface wave dispersion studies is limited. Early works by Stonely[1], Tandon and Chandhury[2], Sara[3], Zeng *et al.*[4], Gupta and Narain[5], and Negi and Singh[6] have given some gross descriptions of the earth's crust underneath this plateau. Most of their results generally represent the average crustal structure over long surface wave paths through which the plateau only accounts for a short segment. More detailed works were carried out by Bird and Toksöz[7], Tung[8], and Chun and Yoshii[9]. Their results were among the first to give information on the plateau's layering crust.

In order to have a surface wave study that more directly reflects the crustal and upper mantle structure of a tectonic province, it is important to obtain as close to pure-path dispersion data as possible. Moreover, repeated dispersion measurements over the same surface wave path are desirable to improve the data reliability. Our work begins by tectonically dividing China into four major provinces here referred to as subplates (Fig. 1). Each subplate has a similar tectonic past, therefore, it can be considered to a first approximation as a single tectonic unit over which the lateral change of crustal and upper mantle structure is relatively small. Among the four subdivisions in figure 1, the Qinghai-Xizang plateau forms a subplate by itself, and this paper shall concentrate on the derivation of its crustal and upper mantle structure. To obtain pure-path dispersion data, we take advantage of (1) several large ($M > 7$) earthquakes that have occurred near the boundaries of these subplates and (2) two high-quality Seismological Research Observatory (SRO) stations located near boundaries of these subplates. Therefore, a reasonable pure-path configuration is closely approximated. We note that tectonically the western extension of the Qinghai-Xizang subplate should also include northern Pakistan and Afghanistan, for the entire orogeny belt was formed during the collision process. Besides the two paths crossing the Qinghai-Xizang plateau, we also studied three other paths that sample the surrounding three subplates so that a comparative analysis can be made to obtain information on lateral variations of crustal and upper mantle structure in China.

## DATA AND METHOD

Table 1 gives all earthquake sources used. They are large aftershocks of the three damaging earthquakes that occurred in China in 1976. A number of sources are used for each path in order to obtain an estimate on the uncertainty of the observed dispersion data. All events in this study were recorded at two SRO stations located in Mashad, Iran (MAIO) and Taipei, Taiwan (TATO). SRO stations operate at high gain, thus lowering the detection threshold for seismic events and making small-magnitude events ($M \sim 5$) useful for long- and short-period surface wave dispersion analysis. Using smaller events would also avoid the difficulty in dispersion analysis often caused by large multiple and propagating sources. Since the SRO data are digitally recorded, pure Love and Raleigh waves are conveniently obtained by rotating the data coordinates. An example of a seismic event, before and after rotation, is given in figures 2 and 3. Surface wave data are windowed and subjected to a multiple-filtering analysis. A typical contoured result of the relative amplitude of surface wave energy arrivals over the path Yunnan-Mashad for both vertical and radial Rayleigh waves is shown in figures 4 and 5. The dots in these figures represent the average group velocities over the filter band. Both fundamental mode and the first higher mode energies are clearly present. Dispersion data extracted from these contour plots for the two paths through the Qinghai-Xizang subplate are given in figures 6-8. Along the path Sichuan-Mashad, five events each with two (vertical and radial) Rayleigh components are used to obtain the observed dispersion data given in figure 9. Thus in figure 9, every observed dispersion data point is the average of ten independent measurements. The standard errors as a result of these repeated measurements range from 3-5%. The averaged dispersion data for the path Yunnan-Mashad are given in figure 10. For the purposes of comparison, three other paths have also been studied. Their averaged observed dispersion data are given in figures 11-14. A visual examination of these observed dispersion data immediately reveals that those paths traversing through the Qinghai-Xizang plateau give substantially lower group velocity values. This anomalously low group velocity extends to a period as long as

80 s, suggesting that the anomalous structure of the Qinghai-Xizang plateau extends down to the upper mantle.

A generalized linear inversion method first introduced by Backus and Gilbert[10] is applied to the observed dispersion data. Detailed descriptions of this method have been presented by Jackson[11], Wiggins[12], and Crosson[13]. The basic logic of the inversion theory and some experience in its numerical applications can be found in Pines et al.[14]. Initial models used for the inversion were based on the results of Tung[8]. The data were first inverted for both shear velocity and density. During the inversion process, large instabilities of the density parameter were found in the crust. This was corrected by changing the starting models to ones that more closely approximated the dispersion data, and at the same time compressional velocity was also included as an inversion parameter. As discussed by Wiggins[12], Der et al.[15], and Jackson and Burkhard[16], the partial derivatives for shear velocity are larger than those for compressional velocity or density. Therefore, the resolution for shear velocity is considerably better than for the other two parameters.

## RESULTS AND DISCUSSION

Upon application of a generalized linear inversion to the observed dispersion data (shown by open squares in figures 9-14), we have obtained the best-fit models corresponding to each path with theoretical group velocities for each of these models given as open circles in the respective figures. For the two paths crossing the Qinghai-Xizang plateau, the first higher mode data are available and they were also entered into the inversion process. Love waves were not used in the inversion due to the absence of amplitude maxima in the dispersion curves between 30 s and 60 s. Earlier studies by Thatcher and Brune[17] and James[18] have discussed the problem of interference of Love waves crossing inhomogeneous media. Tung[8] attributed this problem on the Qinghai-Xizang plateau to Love waves following a non-least time path. A possible structural implication is that there may be structures along the path corresponding to wavelengths of Love waves, causing part of the wave energy to travel along a non-least time path with slower velocities. Resulting models are labeled SM1 for the Sichuan-Mashad path, and YM1 for the Yunnan-Mashad path. Numerical values of the two models are given in Tables 2 and 3. In the inversion process, the compressional velocity and the density values were practically unchanged as they are not very sensitive parameters for the group velocity. Initial and final values of the shear velocity are tabulated to show the changes as required by the data. To compare the crustal and upper mantle structure of the Qinghai-Xizang plateau with its surrounding regions, we have a summary plot given in figure 15 where the shear velocity structures of the models SM1 and YM1 are shown together with three other models resulting from the dispersion data along the three other paths: TM1 for Tangshan-Mashad, TT1 for Tangshan-Taipei, and SC1 for Sichuan-Taipei. The comparison in figure 15 shows a distinctly thick crust for the Qinghai-Xizang plateau, with lower crust extending to a depth of about 70 km. From the width of the resolving kernels it is only possible to resolve layers of 10-15 km in the upper crust, 15-20 km in the middle crust, and 25-30 km in the lower crust. Resolution in the mantle is poor, a low-velocity layer in the upper mantle can only be inferred. Additional data for periods much longer than 120 s and good Love waves as well as higher mode data would significantly increase the resolving power of the upper mantle structure. The shear wave velocity structure for both paths crossing the Tibetan plateau is similar. A 70 km thick crust may be broken into four layers. The first layer has shear velocities that range between 2.87 and 3.06 km/s, suggesting a layer approximately 15 km thick of partly sedimentary origin. Tung[8] and Mu et al.[19] discussed the existence of thick sedimentary layers in the Mount Jolmo Langma region of the Himalayas. Both Birch[20] and Simmons[21] reported shear velocities for sedimentary rocks at 10- to 15-km depth of approximately 3.00-3.08 km/s. A second layer has shear wave velocities ranging from 3.20 to 3.40 km/s and is approximately 15 km thick, suggesting a layer of granitic or equivalent composition. A third layer has shear velocities between 3.40 and 3.70 km/s is approximately 15-20 km thick, and also seems to consist of granite. For comparison, Birch and Bancroft[22], Birch[20], and Simmons[21] reported shear velocities for granite under 3-5 kbars of pressure of 3.45-3.70 km/s. Simmons reported that granite with a shear wave velocity of 3.79 km/s at pressures of 10 kbars. The fourth layer with shear wave velocities between 3.85 and 3.95 km/s corresponds to granite at pressures of approximately 20 kbars, which would correspond to depths of 60-70 km.

The dispersion curves and shear wave velocity profile for the Tibetan plateau are in agreement with those values presented by Tung[8] and Bird and Toksöz[7]. Chun and Yoshiii[9], using Rayleigh and Love waves for many different paths across the Qinghai-Xizang pleateau, reported a crustal thickness of approximately 70 km and observed a velocity reversal in the middle crust. Owing to a 15- to 20-km resolution in the middle crust it is doubtful a velocity reversal could be observed for data over the two paths presented.

Because of the small amount of information derived from inverting Rayleigh waves in the upper mantle, the above structural conclusions are only tentative. The upper mantle shear velocities are lower than either the Gutenburg earth model or the Canadian Shield. The velocity at the top of the mantle is approximately 4.4 km/s, with indications of a low-velocity layer at about 90-km depth. Since the low-velocity layer can only be inferred, the existence of extremely low velocities of 4.1 km/s is tentative. Low shear velocities in the upper mantle suggests a high degree of partial melting and a low resistance to deformation, Anderson et al.[23]

## References


(1) Stonely, R., Rayleigh waves in a medium with two surface layers: Mon. Nat. Roy. Astron. Soc. Geophys. Supp. 7, pp. 71-75 (1955).

(2) Tandon, A. N., and Chandhury, H. M., Seismic waves from high yield atmospheric explosions: Ind. J. Met. Geophys. 14, pp. 283-301 (1963)

(3) Sara, B. P., $M_2$ or first shear mode continental Rayleigh waves from Russian nuclear explosion of 30 October 1961: Ind. J. Mt. Geophys. 16, pp. 270-280 (1965).

(4) Zeng, J. S. and Sung, Z. A., Phase velocity of Rayleigh waves in China: Acta Geophysica Sinica 12, no. 2, pp. 148-165 (1963).

(5) Gupta, H. K. and Narain, H., Crustal structure in the Himalayan and Tibetan plateau region from surface wave dispersion: Bull. Seism. Soc. Am., v. 57, no. 2, pp. 235-248 (1967).

(6) Negi, J. G. and Singh, V. P., Love wave dispersion analysis for crustal structure of laterally inhomogeneous Himalayas: Bull. Seism. Soc. Am., v. 63, no. 3, pp. 1163-1172 (1973).

(7) Bird, P. and Toksöz, M. N., Strong attenuation of Rayleigh waves in Tibet: Nature, V266, pp. 163-165 (1977).
(8) Tung, J. P., The surface wave study of crustal and upper mantle structure of Mainland China: Ph.D. dissertation, University of Southern California, 248 pp. (1974).
(9) Chun, K. Y. and Yoshii, Crustal structure of the Tibetan pleateau: A surface wave study by a moving window analysis: Bull. Seism. Soc. Am., v. 67, pp. 735-750 (1977).
(10) Backus, G. and Gilbert, F., Uniqueness in the inversion of inaccurate gross Earth data: Phil. Trans. Roy. Soc. (London), Ser. A266, pp. 123-192 (1970).
(11) Jackson, D. D., Interpretation of inaccurate, insufficient, and inconsistent data: Geophys. J. Roy. Astron. Soc. 28, pp. 97-109 (1972).
(12) Wiggins, R. A., The general linear inverse problems: Implications of surface waves and free oscillations for Earth structure: Rev. Geophys. Space Phys. 10, pp. 251-285 (1972).
(13) Crossen, R. S., Crustal structure modeling of earthquake data 1, simultaneous least squares estimation of hypocenter: J. Geophys. Res. 81, pp. 3030-3046 (1976).
(14) Pines, I., Teng, T. L., Rosenthal, R., and Alexander, S. S., A surface wave dispersion study of the crustal and upper mantle structure of China, J. Geophys. Res., vol. 85, pp. 3829-3844 (1980).
(15) Der, Z., Masse, R. and Landisman, M., Effects of observational errors on the resolution of surface waves at intermediate distances: J. Geophys. Res. 75, pp. 3399-3409 (1970).
(16) Jackson, D. D. and Burkhard, N. R., Density and surface wave inversions: Geophys. Res. Let. v. 3, no. 11, pp. 637-638 (1976).
(17) Thatcher, W. and Bruen, J. N., Higher mode interference and observed anomalous apparent Love wave velocities: J. Geophys. Res. 74, pp. 6603-6611 (1969).
(18) James, D. E., Anomalous Love wave phase velocities: J. Geophys. Res. 76, pp. 2077-2083 (1971).
(19) Mu, E. Z., Yin, J. X., Wen, S. X., Wang, Y. G. and Zhang, B. G., Stratigraphy of the Mount Jolmo Lunga region in Southern Tibet: Chinese Scientia Geologica Sinica, no. 1, pp. 13-25 (1973).
(20) Birch, F., Some geophysical applications of high pressure research in solids under pressure: edited by W. Paul and D. M. Warschauer; McGraw-Hill Book Company, New York (1963).
(21) Simmons, G., Velocity of shear waves in rocks to 10 kilobars: J. Geophys. Res. 69, no. 6, p. 1123 (1964).
(22) Birch, F., and Bancroft, D., The effect of pressure on the rigidity of rocks: J. Geol., 46, pp. 59-83 113-141 (1938).
(23) Anderson, D. L., Sammis, C. and Jordan, T., Composition of the mantle and core, the nature of the Solid Earth: Edited by E. L. Robertson, McGraw Hill, p. 677 (1972).

TABLE 1. Pertinent Earthquakes Information Used in this Study

| Location | Origin Time* | Station | M | Wave Type | Component | Path | Distance, km |
|---|---|---|---|---|---|---|---|
| 32.8°N, 104.2°E Sichuan | Aug. 16, 1976 1406:45.9 | MAIO | 6.1 | Rayleigh | vertical radial | Sichuan-Mashad | 4080 |
| 32.9°N, 104.2°E, Sichuan | Aug. 19, 1976 1249:47.7 | MAIO | 5.4 | Rayleigh | vertical radial | Sichuan-Mashad | 4078 |
| 32.8°N, 104.2°E, Sichuan | Aug. 21, 1976 2149:54.2 | MAIO | 6.1 | Rayleigh | vertical radial | Sichuan-Mashad | 4095 |
| 32.5°N, 104.2°E, Sichuan | Aug. 23, 1976 0330:7.6 | MAIO | 6.2 | Rayleigh | vertical radial | Sichuan-Mashad | 4088 |
| 32.5°N, 104.2°E, Sichuan | Sept. 1, 1976 0106:51.8 | MAIO | 5.1 | Rayleigh | vertical radial | Sichuan-Mashad | 4090 |
| 24.3°N, 98.6°E, Yunnan | May 31, 1976 0508:28.5 | MAIO | 5.5 | Rayleigh | vertical radial | Yunnan-Mashad | 3956 |
| 24.2°N, 98.7°E, Yunnan | July 3, 1976 1633:23.1 | MAIO | 5.3 | Rayleigh | vertical radial | Yunnan-Mashad | 3967 |
| 39.8°N, 118.6°E, NE China | July 28, 1976 1535:55.3 | TATO | 5.4 | Rayleigh Love | vertical radial transverse | Tangshan-Taipei | 1661 |
| 39.8°N, 118.9°E, NE China | July 30, 1976 2123:13.8 | TATO | 5.4 | Rayleigh Love | vertical radial transverse | Tangshan-Taipei | 1670 |
| 39.6°N, 117.8°E, NE China | Aug. 1, 1976 2053:53.6 | TATO | 4.6 | Rayleigh Love | vertical radial transverse | Tangshan-Taipei | 1656 |
| 39.9°N, 118.8°E, NE China | Sept. 6, 1976 1702:01.5 | TATO | 4.8 | Rayleigh | vertical | Tangshan-Taipei | 1672 |
| 39.9°N, 118.0°E, NE China | July 29, 1976 0101:04.1 | MAIO | 5.1 | Rayleigh Love | vertical transverse | Tangshan-Mashad | 5119 |
| 39.8°N, 117.8°E, NE China | July 30, 1976 2123:13.8 | MAIO | 5.4 | Love | transverse | Tangshan-Mashad | 5072 |
| 39.6°N, 117.9°E, NE China | Aug. 2, 1976 0916:00.5 | MAIO | 4.4 | Rayleigh | vertical | Tangshan-Mashad | 5053 |
| 39.7°N, 118.5°E, NE China | Aug. 8, 1976 0109:12.4 | MAIO | 4.9 | Rayleigh Love | vertical transverse | Tangshan-Mashad | 5100 |

TABLE 1. Pertinent Earthquakes Information Used in this Study (Continued)

| Location | Origin Time* | Station | M | Wave Type | Component | Path | Distance km |
|---|---|---|---|---|---|---|---|
| 40.2°N, 118.9°E, NE China | Aug. 8, 1976 2241:34.3 | MAIO | 5.1 | Rayleigh | vertical | Tangshan-Mashad | 5118 |
| 39.6°N, 118.5°E, NE China | Aug. 14, 1976 1602:44.5 | MAIO | 4.7 | Rayleigh<br>Love | vertical<br>transverse | Tangshan-Mashad | 5103 |
| 32.9°N, 104.2°E, Sichuan | Aug. 19, 1976 1249:47.7 | TATO | 5.4 | Rayleigh | vertical | Sichuan-Taipei | 1902 |
| 33.0°N, 104.2°E, Sichuan | Aug. 22, 1976 1319:53.0 | TATO | 4.7 | Rayleigh | vertical | Sichuan-Taipei | 1884 |
| 32.5°N, 104.2°E, Sichuan | Sept. 1, 1976 0106:51.8 | TATO | 5.1 | Rayleigh<br>Love | vertical<br>transverse | Sichuan-Taipei | 1886 |
| 28.0°N, 100.0°E, Sichuan | Sept. 3, 1976 0957:28.5 | TATO | 5.2 | Love | transverse | Sichuan-Taipei | 1854 |

TABLE 2. Model Fit for Path Szechwan-Mashad

| Layer | H, km | $\alpha$, km/s | $\beta$,* km/s | p, g/cm | $\beta$,† km/s |
|---|---|---|---|---|---|
| 1 | 15 | 5.55 | 2.80 | 2.70 | 3.06 |
| 2 | 15 | 5.60 | 3.20 | 2.90 | 3.28 |
| 3 | 25 | 6.40 | 3.30 | 3.00 | 3.67 |
| 4 | 20 | 6.60 | 3.80 | 3.10 | 3.98 |
| 5 | 25 | 7.50 | 4.45 | 3.30 | 4.39 |
| 6 | 25 | 7.20 | 4.20 | 3.35 | 4.31 |
| 7 | 25 | 7.20 | 4.10 | 3.36 | 4.18 |
| 8 | 25 | 7.20 | 4.20 | 3.37 | 4.21 |

*Initial model.
†Final model.

TABLE 3. Model Fit for Path Yunnan-Mashad

| Layer | h, km | α, km/s | β,* km/s | p, g/cm | β,† km/s |
|---|---|---|---|---|---|
| 1 | 10 | 5.50 | 2.80 | 2.70 | 2.91 |
| 2 | 10 | 5.60 | 3.10 | 2.90 | 3.29 |
| 3 | 10 | 5.65 | 3.20 | 2.92 | 3.40 |
| 4 | 10 | 6.25 | 3.30 | 3.00 | 3.56 |
| 5 | 15 | 6.45 | 3.40 | 3.05 | 3.54 |
| 6 | 15 | 6.60 | 3.80 | 3.10 | 3.90 |
| 7 | 20 | 7.45 | 4.45 | 3.30 | 4.44 |
| 8 | 20 | 7.20 | 4.20 | 3.34 | 4.19 |
| 9 | 25 | 7.20 | 4.15 | 3.35 | 4.10 |
| 10 | 25 | 7.20 | 4.15 | 3.36 | 4.10 |
| 11 | 25 | 7.20 | 4.20 | 3.37 | 4.15 |

*Initial model.
†Final model.

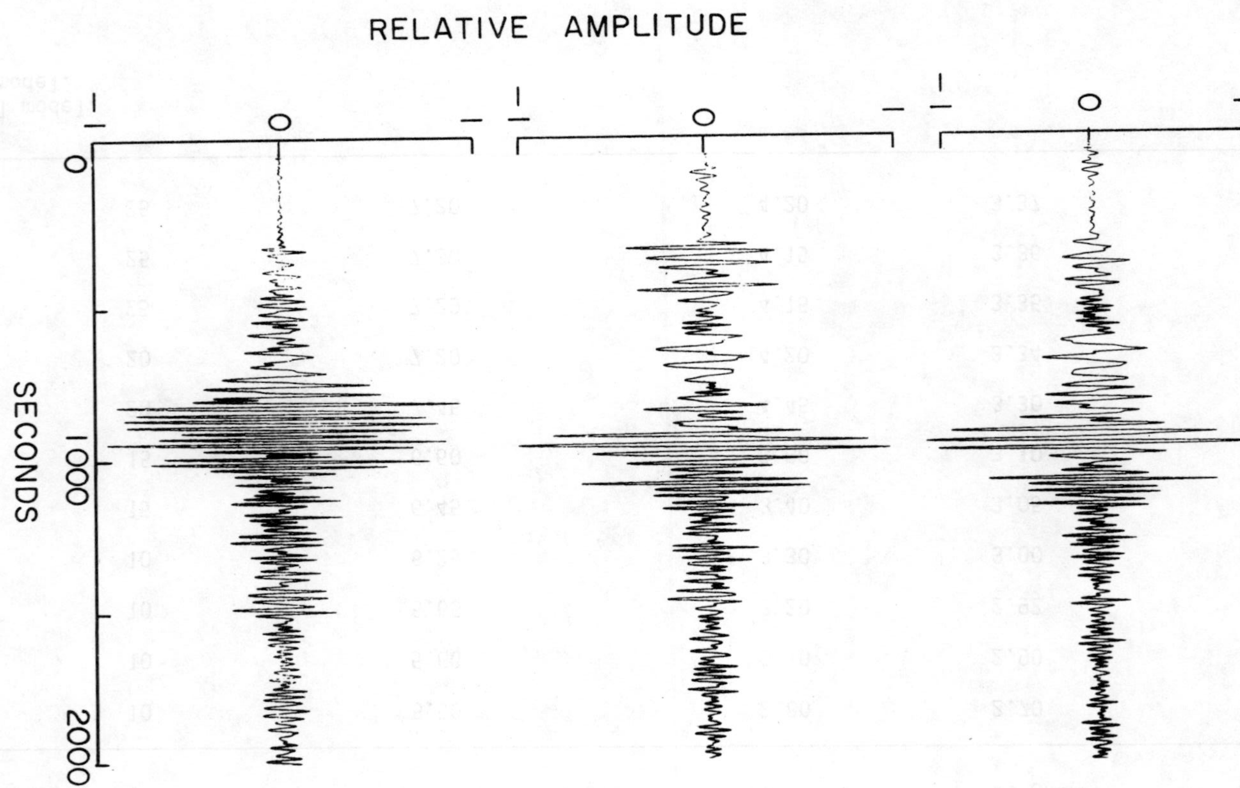


Fig. 1 Surface wave paths (triangles are earthquakes, circles are SRO stations).

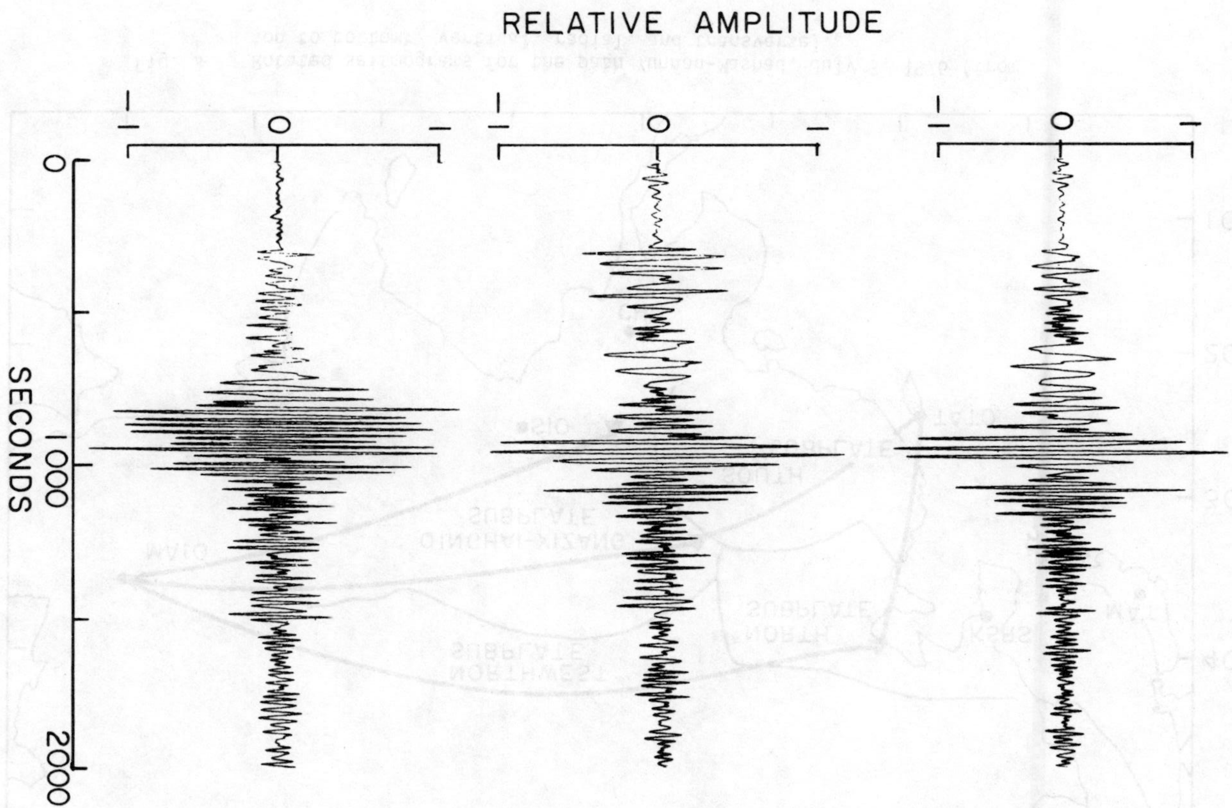


Fig. 2 Unrotated seismograms for the path Yunnan-Mashad, July 3, 1976 (from top to bottom: vertical, NS, and EW).

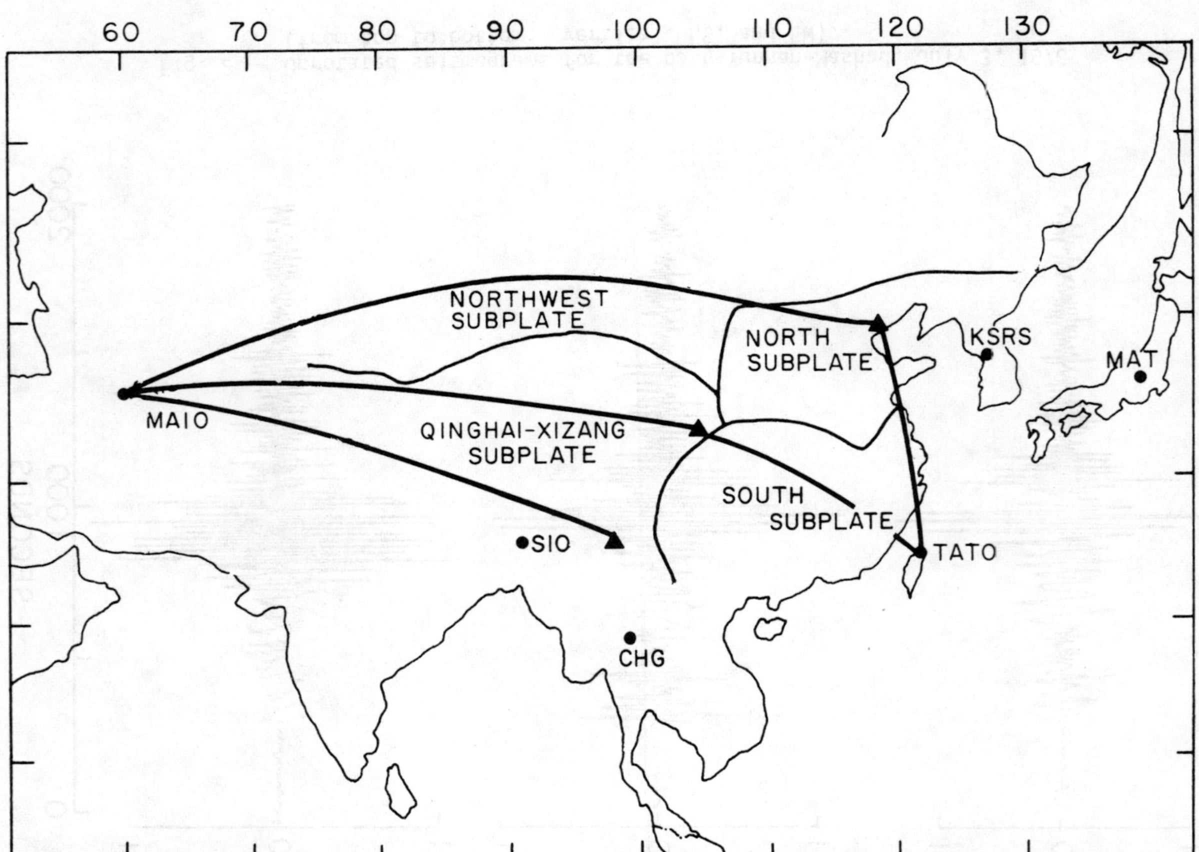


Fig. 3 Rotated seismograms for the path Yunnan-Mashad, July 3, 1976 (from top to bottom: vertical, radial, and transverse).

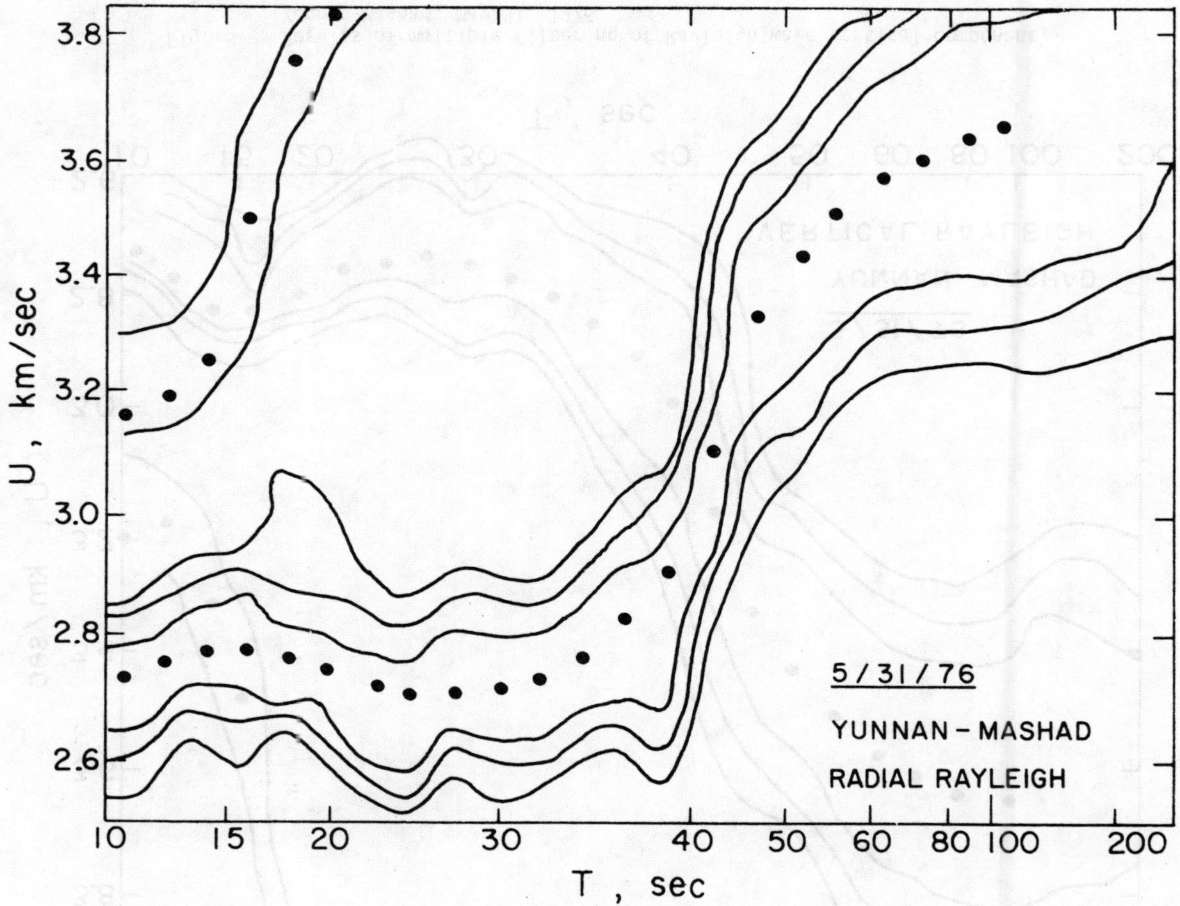


Fig. 4 Results of multiple filtering of Rayleigh wave radial component, Yunnan-Mashad, May 31, 1976.

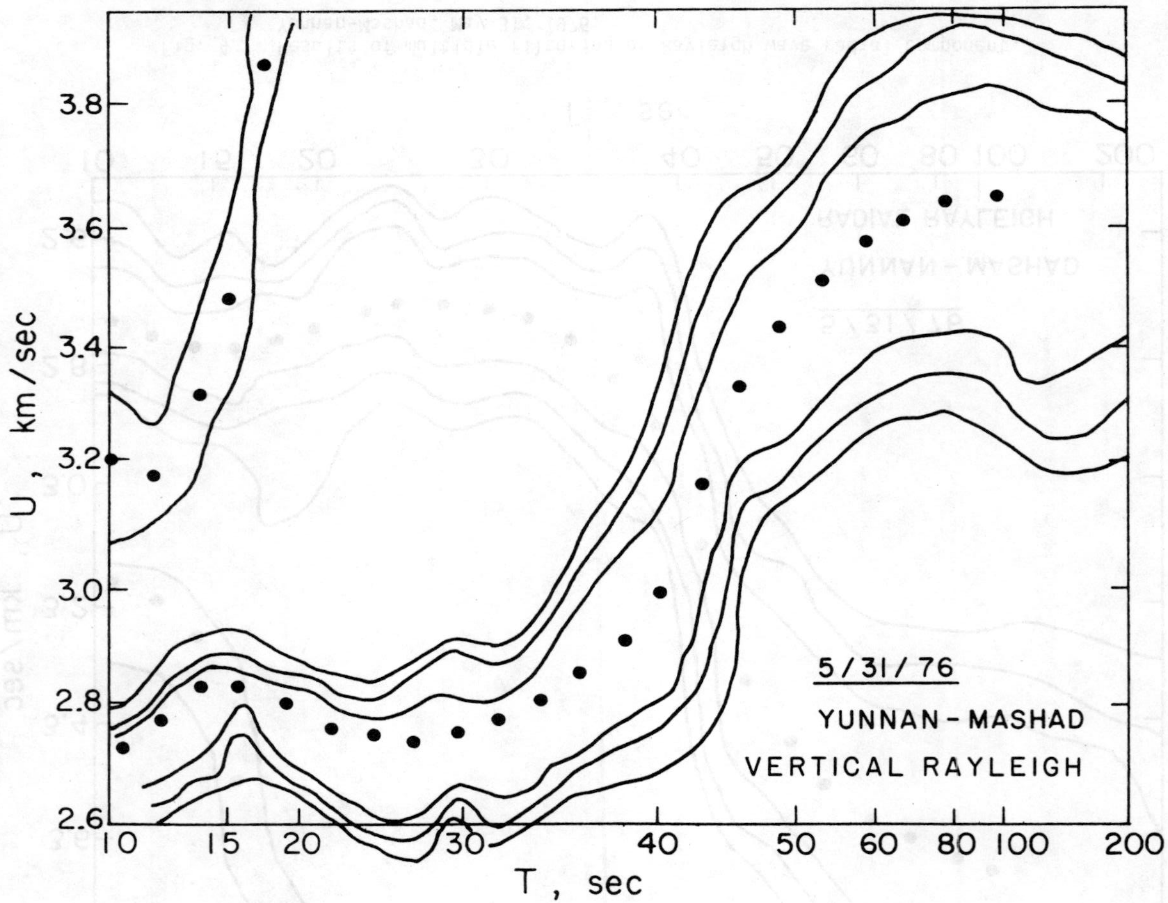


Fig. 5 Results of multiple filtering of Rayleigh wave vertical component, Yunnan-Mashad, May 31, 1976.

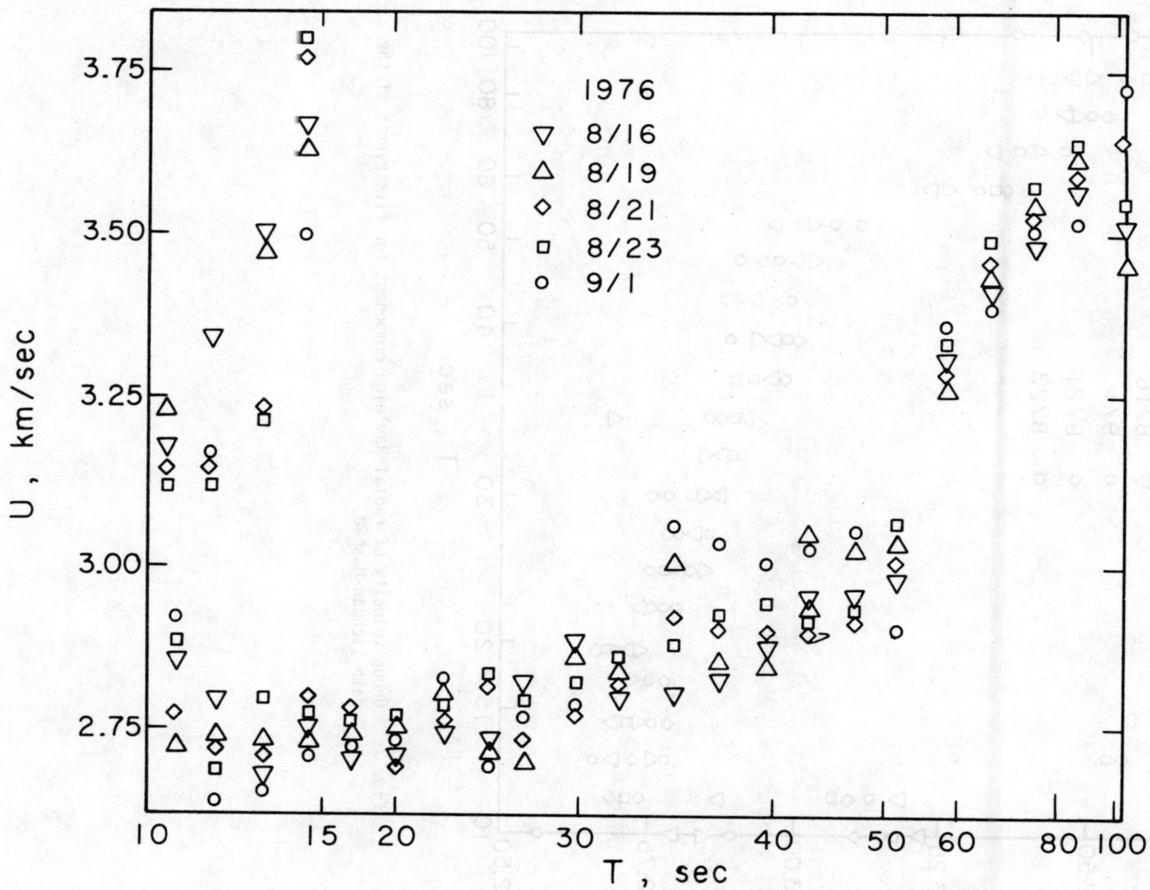


Fig. 6 Group velocity of radial Rayleigh component for five events of the path Sichuan-Mashad.

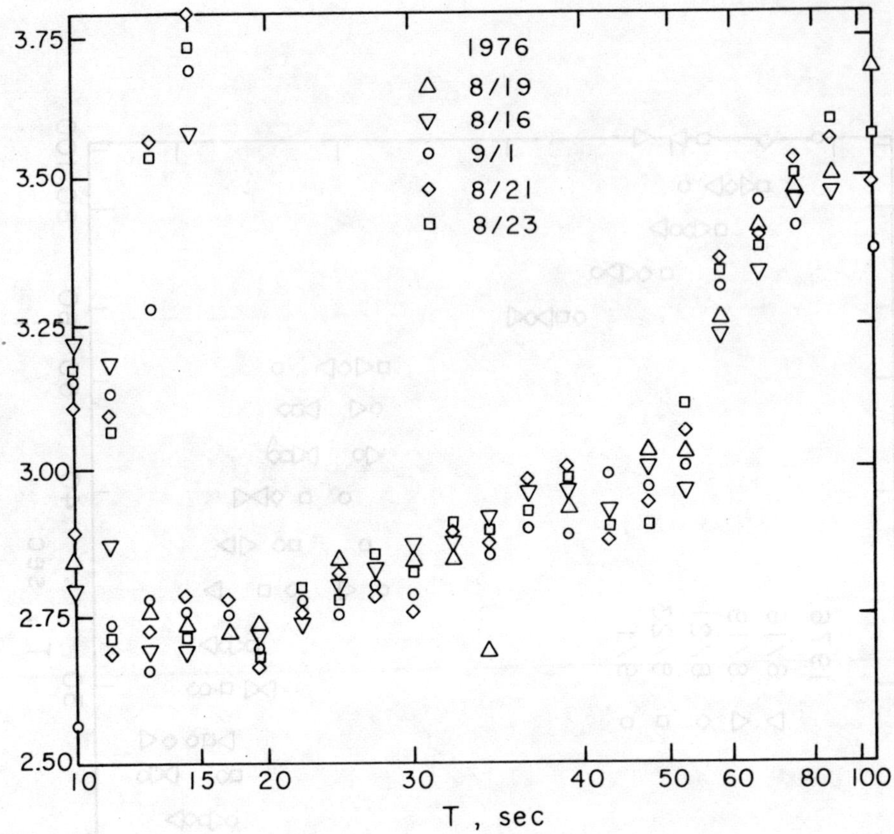


Fig. 7 Group velocity of radial Rayleigh component for five events of the path Sichuan-Mashad.

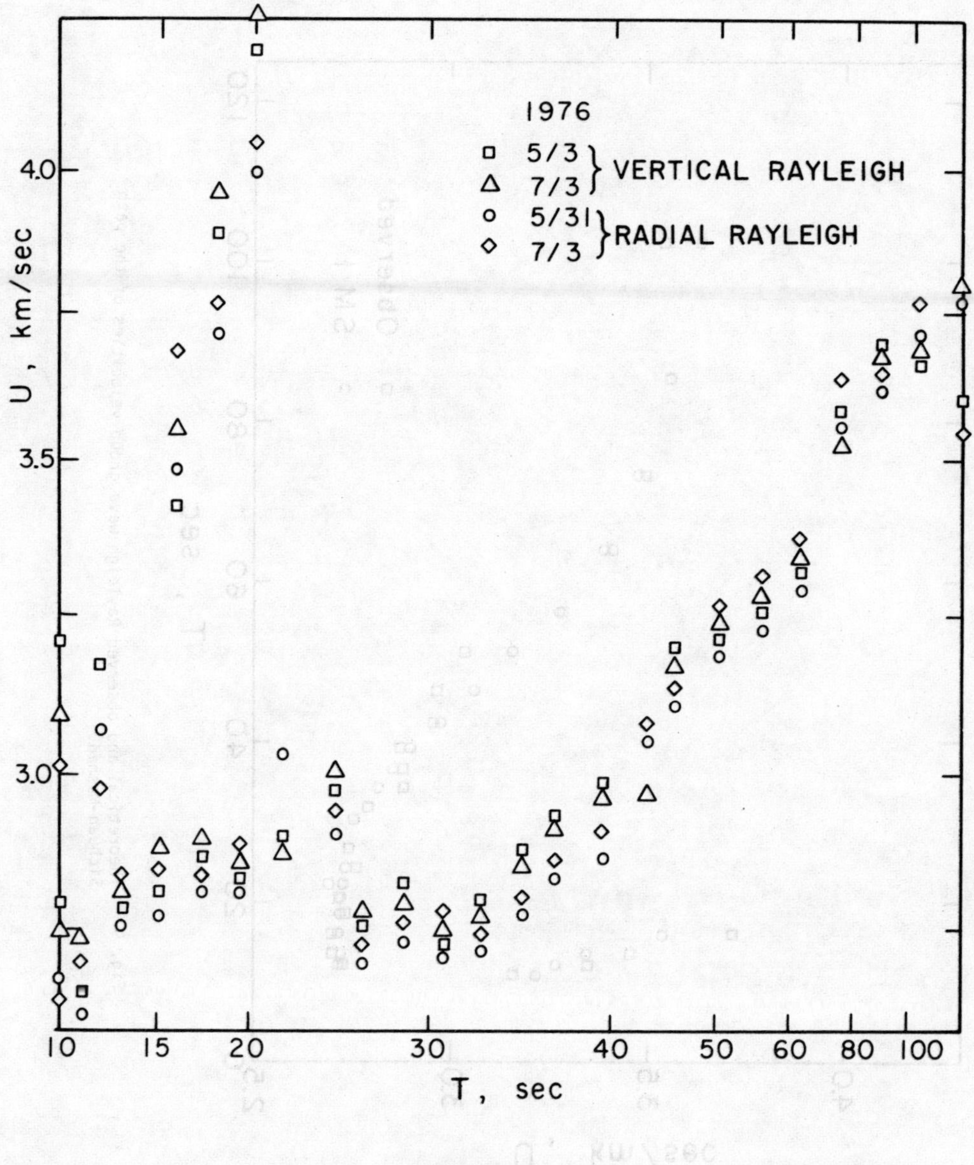


Fig. 8 Rayleigh wave group velocity for three events of the path Yunnan-Mashad.

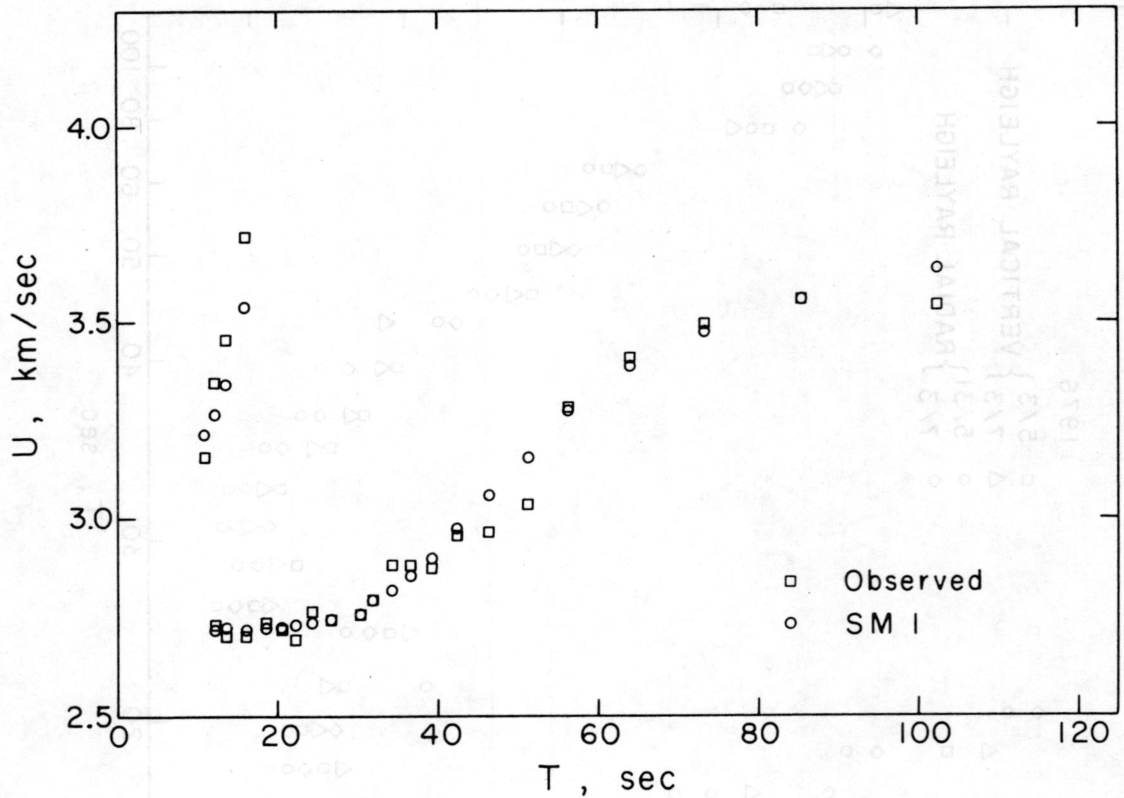


Fig. 9 Theoretical and observed Rayleigh wave group velocities of the path Sichuan-Mashad.

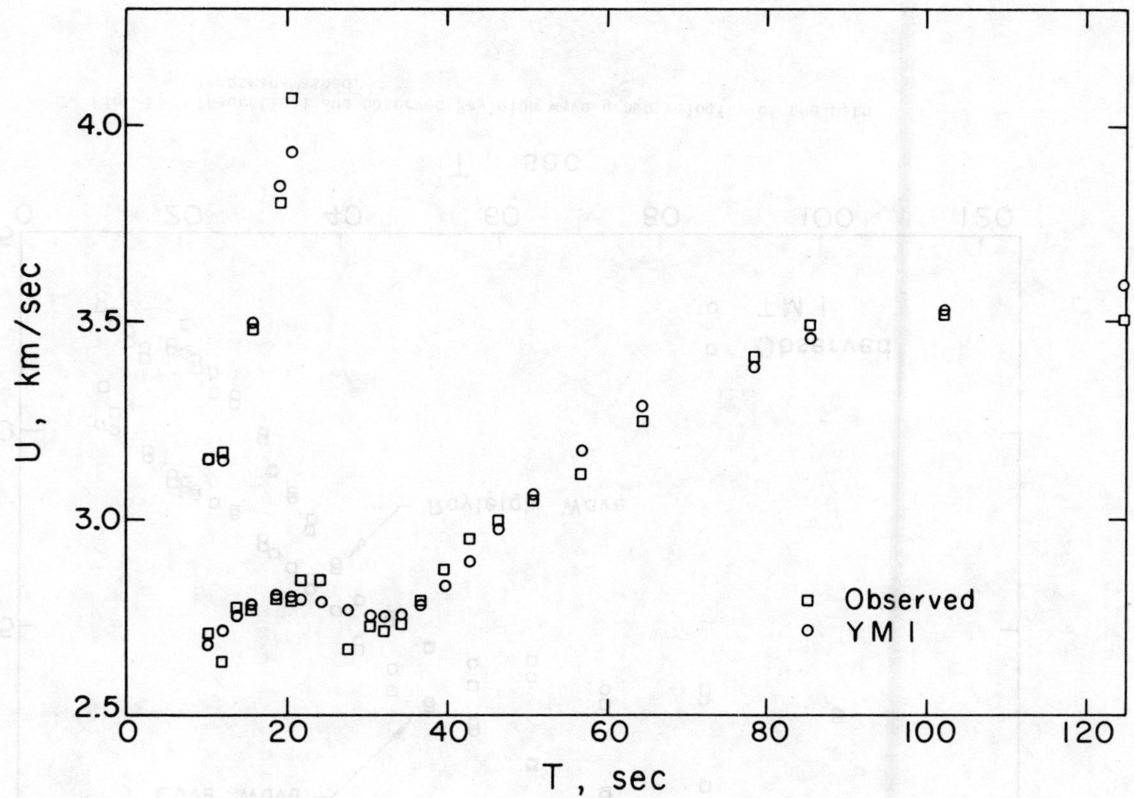


Fig. 10 Theoretical and observed Rayleigh wave group velocities of the path Yunnan-Mashad.

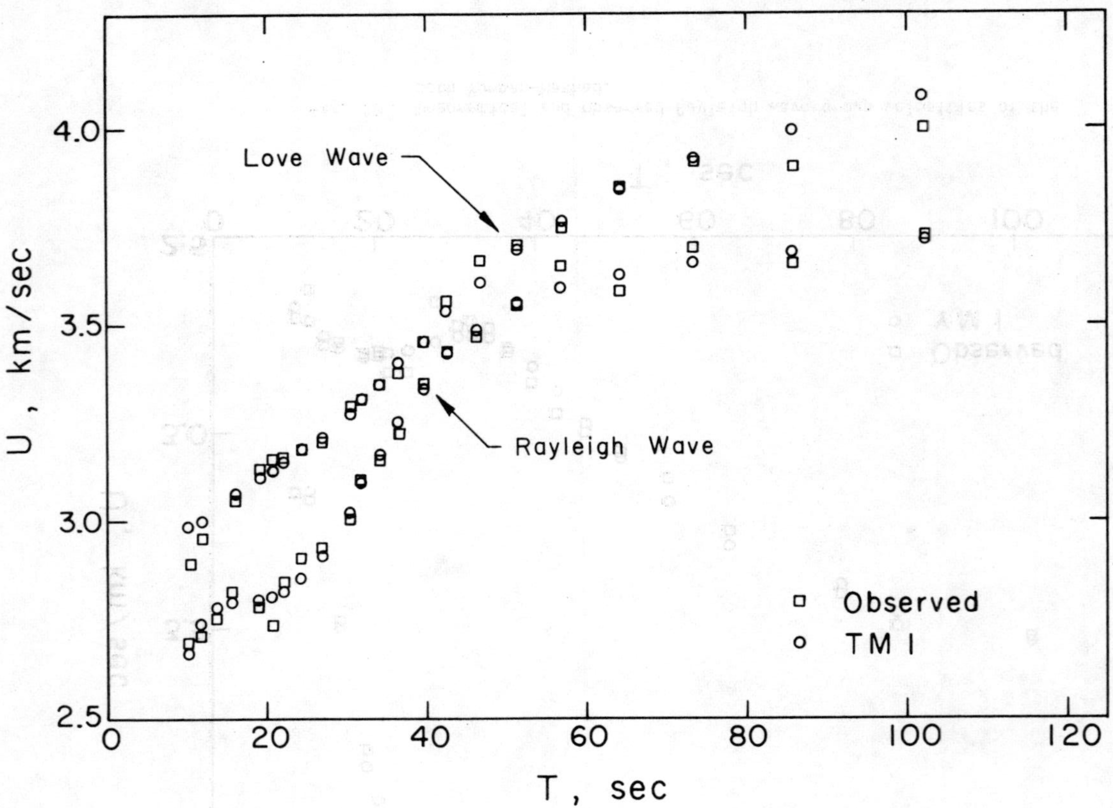


Fig. 11 Theoretical and observed Rayleigh wave group velocity of the path Tangshan-Mashad.

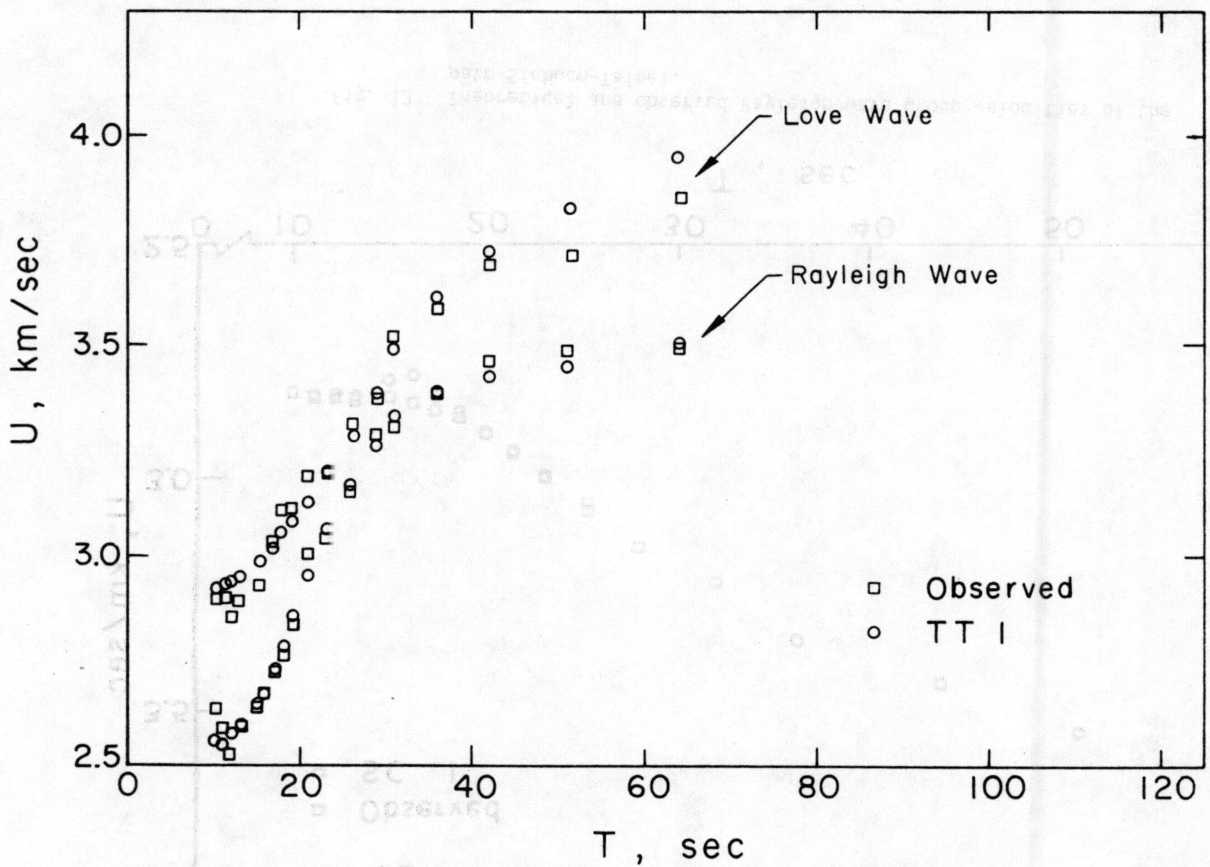


Fig. 12 Théoretical and observed Rayleigh wave group velocities of the path Tangshan-Taipei.

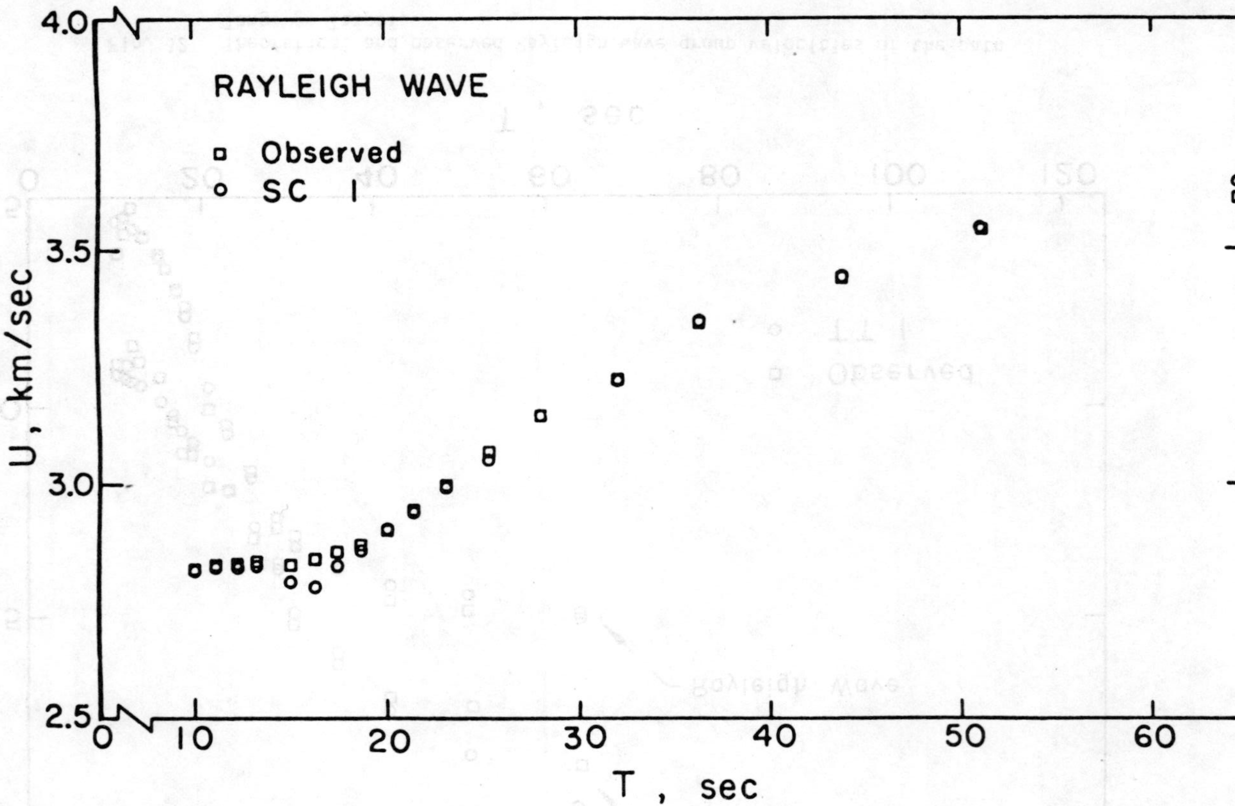


Fig. 13 Theoretical and observed Rayleigh wave group velocities of the path Sichuan-Taipei.

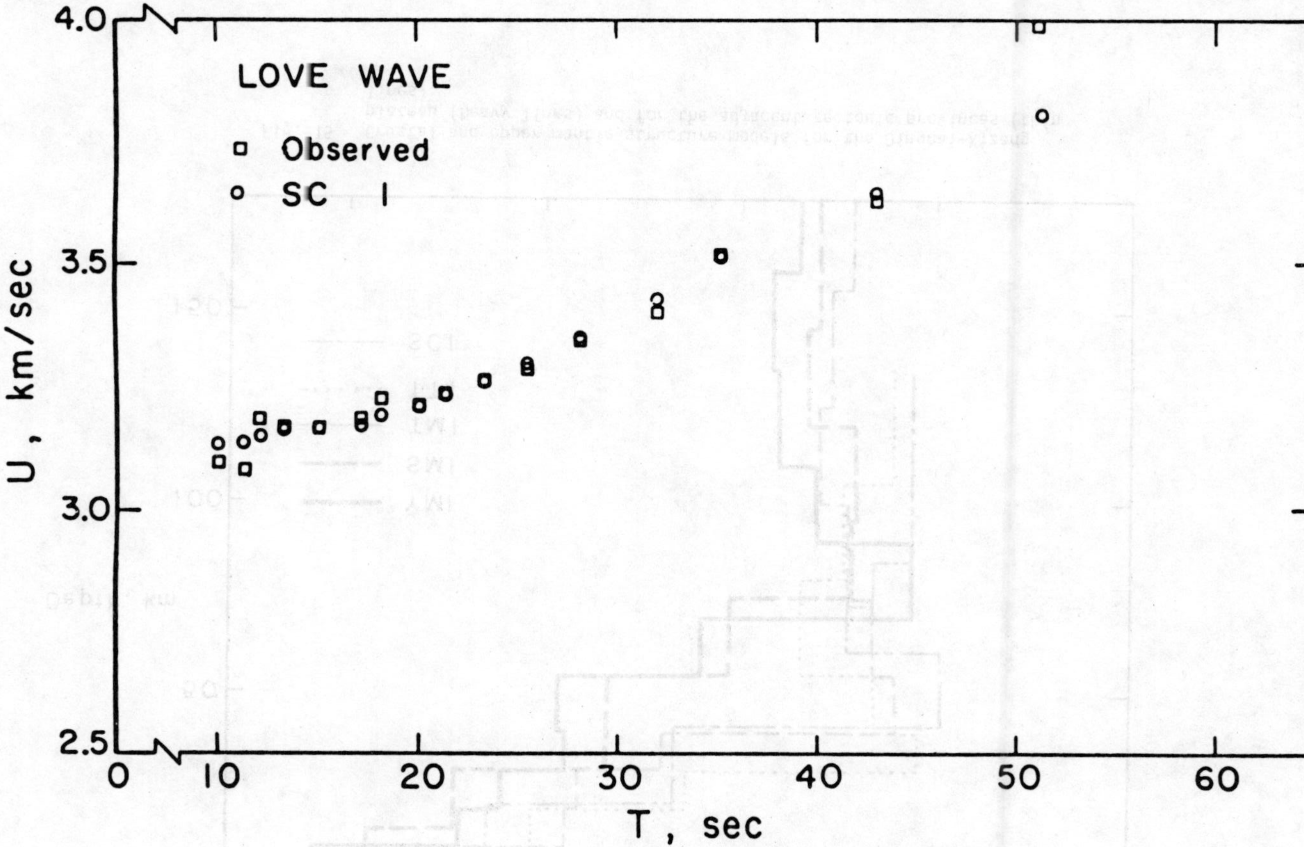


Fig. 14 Theoretical and observed Love wave group velocities of the path Sichuan-Taipei.

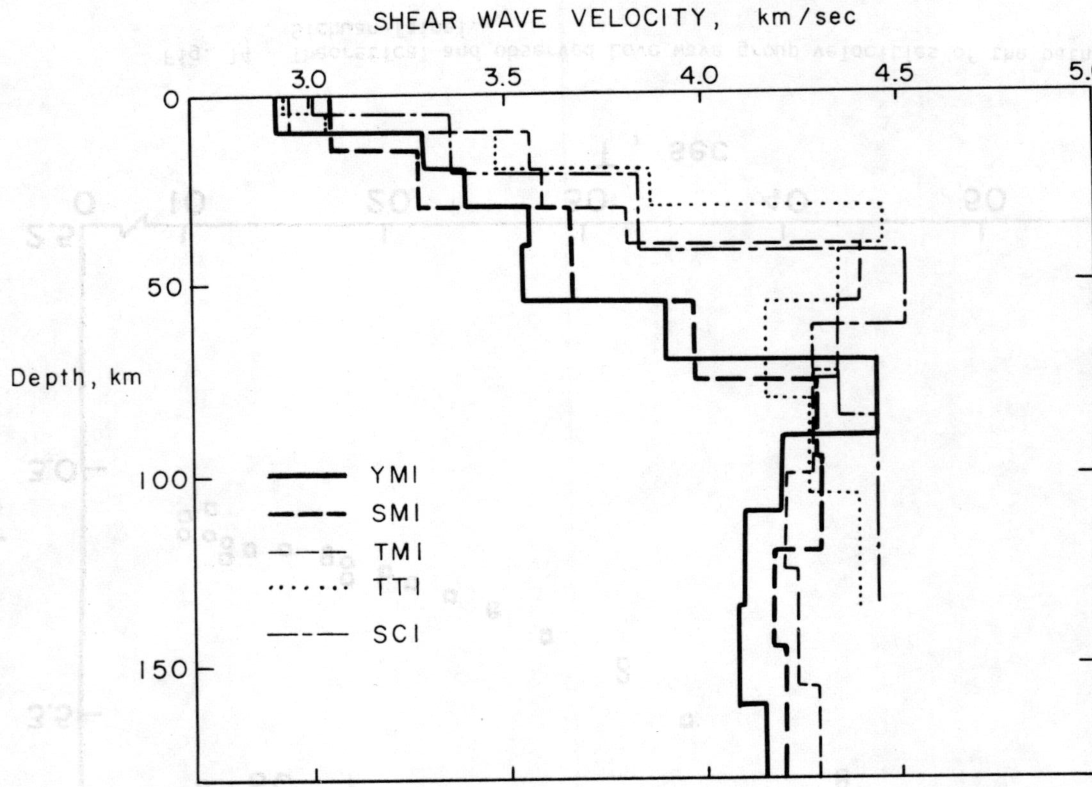


Fig. 15 Crustal and upper mantle structure models for the Qinghai-Xizang plateau (heavy lines) and for the adjacent tectonic provinces (thin lines).

ON THE GROUP VELOCITY OF RAYLEIGH WAVES AND THE CRUSTAL STRUCTURE OF QINGHSI-XIZANG PLATEAU

Yao Zhen-xing*     Li Bai-ji**     Liang Shang-hong*

Zhu Pei-ding*     Zhang Li-min*

## INTRODUCTION

Up to now, we know very little about the crust-mantle structure of Qinghai-Xizang Plateau. Most of the published results only provided the estimated value that the crust thickness is about 70 km (1,2,3). Among the recent results, K.Y. Chun and T. Yoskii reported the TP-4 model characterized by a low velocity zone between 30 to 40 km(4). On the other hand, the model reported by I. Pines and T.L. Teng has no low velocity zone in the crust(5). However, considerable parts of the path of the surface waves they used are outside Qinghai-Xizang Plateau. In our paper, shear-wave velocity profile of Qinghai-Xizang Plateau was derived from the group velocity dispersion, by use of the surface waves passing just within the plateau region.

## SURFACE WAVE ANALYSIS

Data. Qinghai-Xizang Plateau is located in the south of the northern Altun and Qilian mountains, and in the west of Longmen and Hengduan (Meridional Range) Mountains. Its southwest boundary is Himalayas. The data used are Rayleigh wave recordings from 20 events. Magnitude range of the earthquakes is from 5.2 to 6.6. The largest hypocentral depth is 53 km. The recording stations are Lhasa, Urumqi, Gaotai and Kunming. The recording instrument is of the Kirnos type. Distances from epicenter are from 726 km to 2,680 km. As shown in Figure 1, most of the seismic surface-wave paths are within this region. We attempted to divide the surface-wave paths into two groups, one directing from the north to south, and the other from west to east. But the results obtained from these two groups agree well.

Method and theory. The group velocity dispersions of the Rayleigh waves were obtained using the single stations, by the multiple filtering process(6,7).

We inverted Rayleigh wave group velocity by means of the generalized linear inversion(8,9), assuming that the medium is made up of elastic, isotropic and homogeneous parallel layers. The observed group velocities W which are related to unknown model parameter x, and period T may be written as

---

* Institute of Geophysics, Academia Sinica.

** Seismological Institute of Yunnan Province.

$$W=f(x,T) \tag{1}$$

The theoretical group velocities were calculated using the "fast" form of Knopoff's method(10).

Let $\Delta W$ be the difference between the theoretical and observed data.

$$\Delta W=A\Delta x \tag{2}$$

where A is a matrix of partial derivatives of group velocities with respect to unknown parameters, calculated by the method discribed by Rodi, et al(11).

According to Lanczos decomposition(12), the matrix A may be written as

$$A=USV^T \tag{3}$$

$$S=\begin{pmatrix} s_1 & & & & 0 \\ & s_2 & & & \\ & & \cdot & & \\ & & & \cdot & \\ 0 & & & & s_n \end{pmatrix}$$

A suitable inverse is given by

$$A^+=VS^+U^T \tag{4}$$

$$S^+=\begin{pmatrix} s_1^+ & & & & \\ & s_2^+ & & & \\ & & \cdot & & \\ & & & \cdot & \\ & & & & s_n^+ \end{pmatrix} \qquad s_i^+=\begin{cases} 1/s_i & s_i \neq 0 \\ 0 & s_i=0 \end{cases}$$

A generalized solution is represented as

$$\Delta \hat{x}=A^+ \Delta W \tag{5}$$

when A is ill conditioned, the solution depends on P largest nonzero eigenvalues, which are given by equation

$$\Delta x_i = \sum_{j=1}^{p} v_{ij}/s_j \qquad i=1,2, \ldots n \tag{6}$$

$$\mathrm{Var}\,(\Delta \hat{X}_i)=\sigma^2 \sum_{j=1}^{p} (v_{ij}/s_j)^2 \tag{7}$$

Equation (7) is one estimate of variance of the model.

Form equation (2) and (5). We can get equation

$$\Delta \hat{x}=A^+A\,\Delta x=R\,\Delta x \tag{8}$$

It indicates that solution $\Delta\hat{x}$ is a weighted sum for all possible solutions. If R is an identity matrix, then $\Delta\hat{x}$ is a unique solution. The degree of an identity matrix to which R approximates is a measure of the resolution for the model. Rows of R are known as resolving kernels.

## RESULTS AND DISCUSSIONS

1. Rayleigh wave group velocity dispersion curves. Figure 2 shows the dispersion curve obtained, which is drawn from the mean values from the multiple filtering using 20 recordings of vertical component. In addition to the fundamental mode of Rayleigh waves, there are also given some observed data of the first higher mode. The observed data of Rayleigh wave group velocity are shown in Table 1.

2. Initial model. We can not find all the parameters of the medium, owing to limited informations and errors of observed data involved. The calculation of partial derivatives shows the shear-wave velocities have the largest effect on the surface-wave dispersion. So we only inverted shear-wave velocity parameters. The initial values can be estimated from the corresponding-wave velocities, obtained from explosion seismic surveys in Xizang Plateau(13). The transformation is based on the assumed ratio of 3 of the compressional to shear-wave velocities. The chosen densities were held fixed in inversion. The crust is divided into 15 layers, each of a thickness from 4 to 5 km. The periods of the observed data were up to 56 seconds only. So we have to assume that the depths and velocities on M-discontinuity and interface below it are known.

3. Result of inversion. Resulting shear wave velocity model for Qinghai-Xizang Plateau is shown in Figure 3. The dashed line in the figure denotes the initial model. Solid line was derived by inversion. From this figure, the crust seems to be divided into six layers, with depths of 4, 13, 27, 40 and 55 km. Among these, the interface at 55 km depth is hard to be clearly recognized from our data.

The curves in Figure 4 indicate the resolving kernel for each layer. The greater the depth, the lower and wider the peak shape for the resolving kernel becomes. This width represents the depth range over which solution $\hat{x}$ is the weighted mean value over the depth of actual solution x.

## CONCLUSION

The group velocity dispersion curves of Rayleigh waves through most of the Qinghai-Xizang Plateau were obtained, using the multiple-filtering process. The observed period for the fundamental mode ranges from 5.0 to 56.0 seconds. For the first higher mode, the periods are from 10 to 16 seconds. By means of these data, we obtained a five-layer crustal model with a low velocity layer at a depth between 27 to 40 km. The crustal thickness is 70 km. Resolution in lower crust is poor owing to the limitation of the bandwidth of the recording instrument.

## References


(1) Tseng Jung-sheng, Sung Zi-an, Phase velocities of Rayleigh waves in China, Acta Geophysica Sinica, 12, 148-165 (1963) (in Chinese).

(2) Song Zhong-he, Tan Cheng-ye, On the crustal thickness in China evaluted by the group velocity of Rayleigh and Love waves, Acta Geophysica Sinica, 14, 33-44 (1965) (in Chinese).

(3) Liu Yuan-lung, Wang Chien-shen, Wu Chuan-zhen, Zhou Wen-hu, A preliminary study of the crustal structure and the geological significance of the central portion of the Himalayan Mountain range, Acta Geophysica Sinica, 20, 143-149, (1977) (in Chinese).

(4) Chun K.Y., Yoshii T., Crustal structure of the Tibetan Plateau: A surface-wave study by a moving window analysis. Bull. Seism. Soc. Am., 67, 735-750, (1977).

(5) Pines I., Teng T.L., Crustal and upper mantle velocity and Q structures of Mainland China, Tech. Report, Geophysical Laboratory, Univ. Southern Calif., (1977).

(6) Dziewonski A., Bloch S., Landisman M., A technique for the analysis of transient seismic signals, Bull. Seism. Soc. Am., 59, 427-444 (1969).

(7) Li Bai-ji, Shi Jie-shan, Song Zi-an, Zeng Rong-sheng, Digital processing for seismic surface wave dispersion, 20, 283-298 (1977). (in Chinese).

(8) Wiggins R.A., The general liner inverse problems: Implications of surface waves and free oscillations for earth structure, Rev. Geophys. Space Phys., 10, 251-285, (1972).

(9) Backus G., Gilbert F., Uniqueness in the inversion of inaccurate gross earth data, Phil. Trans. Roy. Soc (London), Ser. A266, 123-192, (1970).

(10) Schwal F.A., Knopoff L., Methods in computational physics, 11, 87-180 Academic Press (1972).

(11) Rodi W.L., Glover P., Li T.M.C., Alexander S.S., A fast accurate method for computing group-velocity partial derivatives for Rayleigh and Love modes, Bull. Seism. Soc. Am., 65, 1105-1114,(1975).

(12) Langzos C., Linear Differential operators, Chap. 3, D. Van Nostrand Co., London, (1961).

(13) Teng Ji-wen et al., The explosion seismic study for velocity distribution and structure of the crust and upper mantle from Damxung to Yadong of Xizang Plateau, Acta Geophysica Sinica, (in press).

Table 1 The observed data of Rayleigh-wave group velocity (km/sec.)

| Period (sec.) | Fundamental mode | | First higher mode | |
|---|---|---|---|---|
| | mean vel. | S.D. | mean vel. | S.D. |
| 56.10 | 3.09 | 0.11 | | |
| 50.0 | 2.95 | 0.15 | | |
| 44.6 | 2.89 | 0.12 | | |
| 39.7 | 2.85 | 0.08 | | |
| 35.4 | 2.83 | 0.08 | | |
| 31.6 | 2.81 | 0.09 | | |
| 28.1 | 2.80 | 0.11 | | |
| 25.1 | 2.81 | 0.09 | | |
| 22.3 | 2.84 | 0.08 | | |
| 19.9 | 2.85 | 0.10 | | |
| 17.7 | 2.88 | 0.11 | | |
| 15.8 | 2.90 | 0.12 | 3.60 | 0.12 |
| 14.1 | 2.92 | 0.13 | 3.45 | 0.08 |
| 12.6 | 2.91 | 0.11 | 3.38 | 0.13 |
| 11.2 | 2.87 | 0.09 | 3.29 | 0.08 |
| 10.0 | 2.87 | 0.08 | 3.22 | 0.05 |
| 8.9 | 2.85 | 0.08 | | |
| 7.9 | 2.81 | 0.09 | | |
| 7.1 | 2.78 | 0.10 | | |
| 6.3 | 2.75 | 0.11 | | |
| 5.6 | 2.71 | 0.10 | | |
| 5.0 | 2.69 | 0.09 | | |

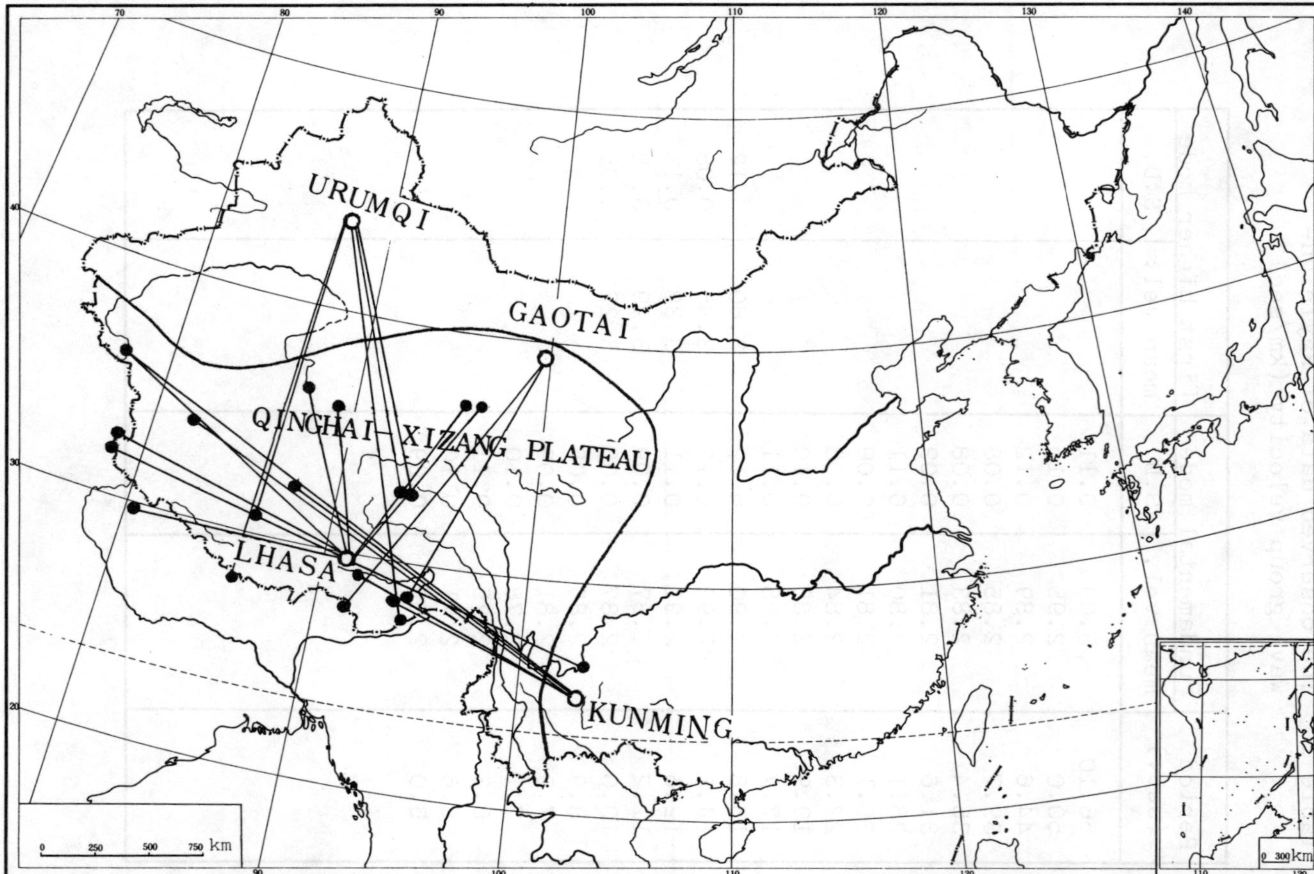


Fig. 1 Recording stations, epicenters, and the paths for Rayleigh waves

o recording station ● epicenter

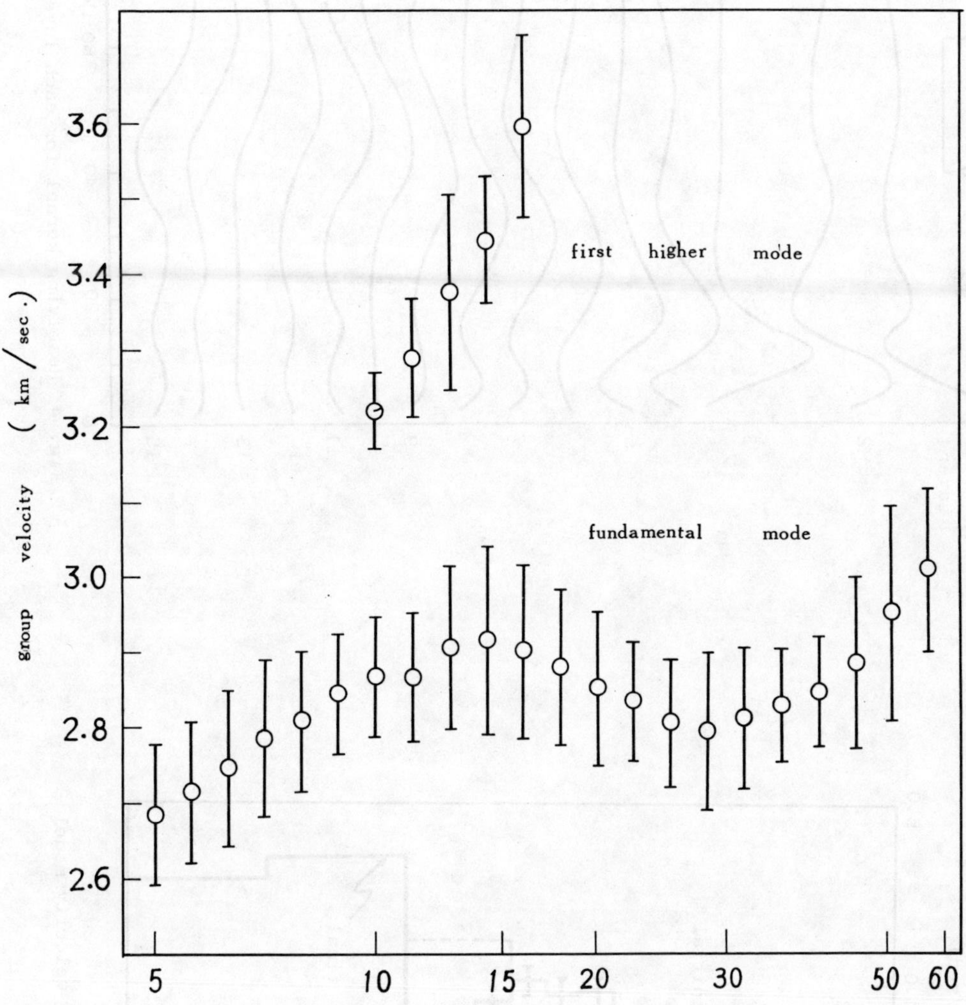


Fig. 2 The observed group velocity dispersion curves of Rayleigh waves. Error bars represent the standard deviations of observed velocities

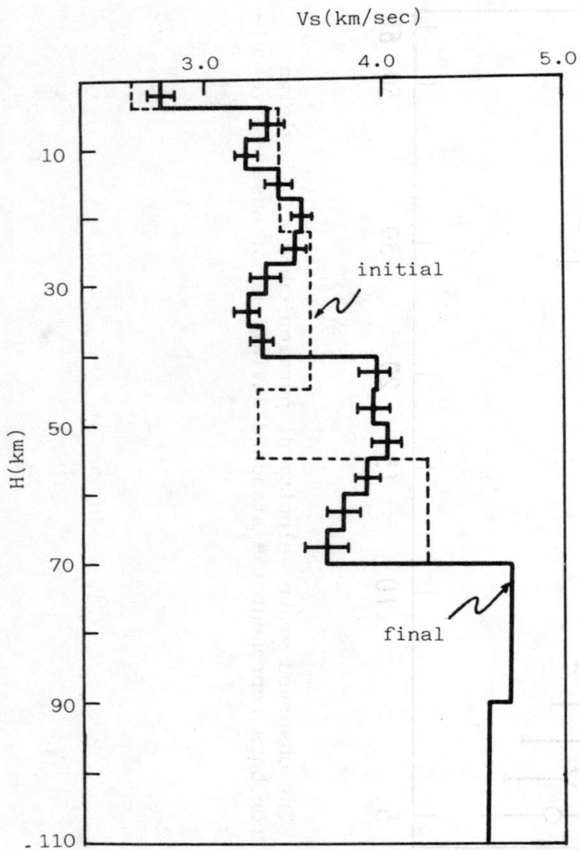


Fig. 3 The shear - wave velocity model

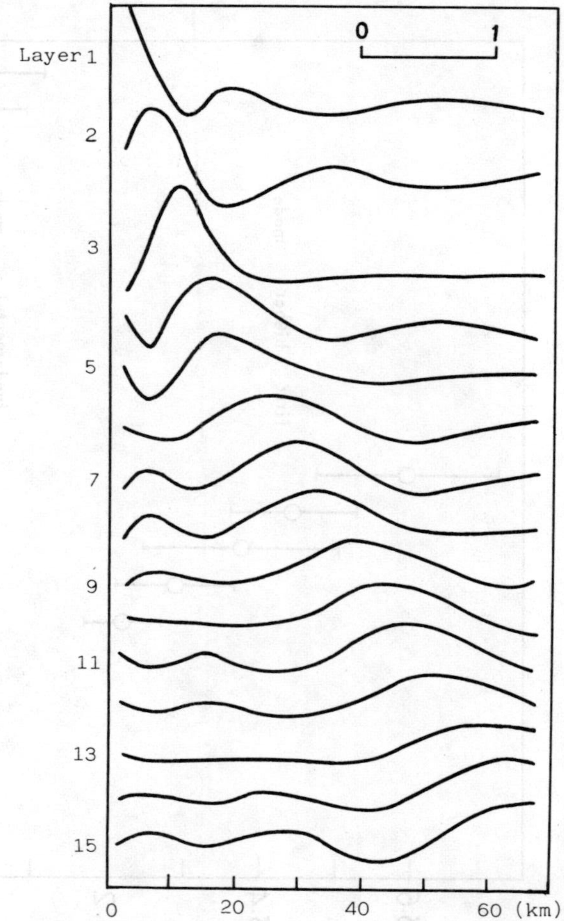


Fig. 4 Resolving kernel for each layer

# CONTINENTAL COLLISIONS ANALAGOUS TO THAT FORMING THE QINGHAI-XIZANG PLATEAU

Kevin Burke, J.F. Dewey, W.S.F. Kidd and A.M.C. Şengör
Department of Geological Sciences
State University of New York at Albany
Albany, New York 12222 U.S.A.

## INTRODUCTION

Papers presented at this symposium have provided, for the first time, a coherent picture of the present state and geological development of the Qinghai-Xizang plateau as a response to the Cenozoic collision of India with Asia along the Indus-Yarlung Zangbo suture zone.

The importance of the collisional process in tectonic evolution has been appreciated since Wilson (1) pointed out that recognition of the plate structure of the lithosphere indicated that, through much of earth history, oceans have been opening in some places and closing in others.

The process of ocean closing is complex and diverse and the variety of phenomena associated with collision are well illustrated in the Alps and the Mediterranean. A distinctive set of collisional phenomena developing after suturing involves continental thickening, partial melting and both intrusive and extrusive magmatic activity (2). In the Qinghai-Xizang plateau we interpret the extent of these processes to be very generally indicated by: (I) the area of very high ground (> 5 km) and (II) the envelope of areas of geothermal activity in the Himalaya and on the plateau (Tong Wei and Zhang Mingtao, this volume).

The exact mechanism of continental thickening and partial melting is as yet poorly understood and future studies of the Qinghai-Xizang plateau are most likely to throw light on the processes that operate. It is clear that thickening the continent in response to convergence will bring its lower part into a realm where isotherms are normally hot enough to induce partial melting but the thickened continent will be cold and conduction alone could only warm it very slowly. Some combination of conduction and convective heat transfer seems essential and the role of the mantle is as yet uncertain. Some workers (e.g. Toksöz, this volume) have suggested that the occurrence of an unusually hot uppermost mantle may influence partial melting of the continent after collision.

## ANALOGUES OF TIBET

There are numerous areas on earth where igneous rocks whose formation has involved partial continental melting are distributed over a wide area on one side of a suture zone that marks a former continental collision. In some of these areas the time of suturing can be shown to have preceded by a few million years the episode of widespread igneous activity. We have interpreted such areas as analogues of the present Tibetan plateau (2 , 3) Table 1 lists a selection of these areas in order of increasing age. No effort has been made to be exhaustive in the list. Although some of the areas are better known than others, the main aim of the compilation is to show that Tibet-style continental reactivation phenomena are and have been (at least since the Proterozoic) a relatively common element in earth history.

Figure 1 is a map of the world illustrating the distribution of the sutures and Tibet-style terrains listed in the table. These areas constitute a substantial proportion of all continental crust, and we suggest that Tibetan style reactivation represents a significant stage in continental evolution involving the separation of high-level rocks enriched in water and large ion lithophile elements from anhydrous granulitic lower crust.

## References


(1) Wilson, J.T., Static or mobile earth: the current scientific revolution, in Gondwanaland Revisited: New evidence for continental drift. Proc. American Philos Soc., 112, 309-320 (1968)

(2) Dewey, J.F., Burke, K.C.A., Tibetan, Variscan and Precambrian basement reactivation: Products of continental collision, Jour. Geology 81, 783-792 (1973)

(3) Burke, K., Dewey, J., The lost oceans of Africa, Jour. Min. Geol. Nigeria Min. Geol. Met. Soc. 6, 75 (1971)

(4) Şengör, A.M.C., Kidd, W.S.F., Post-collisional tectonics of the Turkish-Iranian Plateau and a comparison with Tibet, Tectonophysics 55, 361-376 (1979)

(5) Trümpy, R., et al., Geology of Switzerland, A Guidebook, Wepf and Co. Publishers, Basel (1980)

(6) Bögel, H., Schmidt, K., Kleine Geologie der Ostalpen, Ott Verlag, Thun (1976)

(7) Fujita, K., Pre-Cenozoic tectonic evolution of northeast Siberia, Jour. Geology 86, 159-172 (1978)

(8) Şengör, A.M.C., Mid-Mesozoic closure of Permo-Triassic Tethys and its implications, Nature 279, 590-593 (1979)

(9) Şengör, A.M.C., Yilmaz, Y., Ketin, I., Remnants of a pre-late Jurassic ocean in northern Turkey: fragments of Permo-Triassic Palaeo-Tethys? Geol. Soc. America Bull. 91 (Part I), 599-609 (1980)

(10) Ivanov, S.N., Per filiev, A.S., Poutchkov, V.N., Les traits principaux de la structure géologique de l'Oural, in La Chaîne Varisque d'Europe moyenne et occidentale, CNRS Paris, 571-582.

(11) Hatcher, R.D., Jr., Tectonics of the western Piedmont and Blue Ridge southern Appalachians: Review and speculation, Am. Jour. Sci. 278, 276-304 (1978)

(12) Wickham, J., Roeder, D., Briggs, G., Plate tectonics models for the Ouachita foldbelt, Geology 4, 173-176 (1976)

(13) King, P.B., Marathon revisited, in Symposium on the geology of the Ouachita Mountains 1, 41-69 (Arkansas Geological Commission (1977)

(14) Burke, K., Dewey, J.F., Orogeny in Africa, in African Geology, Ibadan Univ. Press, Ibadan, 583-608 (1972)

(15) Şengör, A.M.C., Butler, J.C., The Llano Uplift, central Texas: a Proterozoic example of continental collision? Geol. Soc. America Abst. Progs. 9, 72-73 (1977)

(16) Hoffman, P., Dewey, J.F., Burke, K., Aulacogens and their genetic relation to geosynclines with a Proterozoic example from Great Slave Lake, Canada, in Soc. Econ. Paleontologists and Mineralogists Spec. Pub. 19, 38-55 (1974)

TABLE 1

CONTINENTAL COLLISIONS WITH TIBETAN-STYLE REACTIVATION

| | COLLISIONAL EVENT | SUTURE ZONE | AGE | KEY REFERENCES |
|---|---|---|---|---|
| 1 | Arabia-Eurasia | Bitlis-Zagros | Began in Mid-Miocene, continues | (4) Şengör and Kidd (1979) |
| 2 | Alps | Arosa-Matrei zones and equivalents | Late Middle Eocene | (5) Trümpy et al. (1980)<br>(6) Bögel and Schmidt (1976) |
| 3 | India-Eurasia | Indus-Yarlung Zangbo | Middle to Late Eocene | (2) Dewey and Burke (1973)<br>Numerous papers in this volume |
| 4 | Verkmoyansk | Chersky | End Jurassic | (7) Fujita (1978) |
| 5 | Cimmerian Continent-Laurasia | Various Palaeo-Tethyan sutures | Late Triassic to Late Jurassic | (8) Şengör (1979)<br>(9) Şengör et al. (1980)<br>Şengör (this volume) |
| 6 | Hercynian | Lizard-Harz Ortegal-Massif Central | Upper Carb. Mid-Carb. | (2) Dewey and Burke (1973) |
| 7 | Eurlaurentia-Asia | Urals | Early Permian | (10) Ivanov et al. (1977) |
| 8 | Eurlaurentia-Gondwanaland | S. Appalachian Ouachita-Marathon | Late Carb. | (11) Hatcher (1978)<br>(12) Wickham et al. (1976)<br>(13) King (1977) |
| 9 | Pan African | Numerous sutures | Latest Pre-cambrian Locally into Lowest Paleozoic | (14) Burke and Dewey (1972) |
| 10 | Grenville | Llano (Texas) | $1x10^9$ years | (2) Dewey and Burke (1973)<br>(15) Şengör and Butler (1977) |
| 11 | Coronation | Wopmay | $1.8x10^4$ years | (16) Hoffman et al., (1974) |

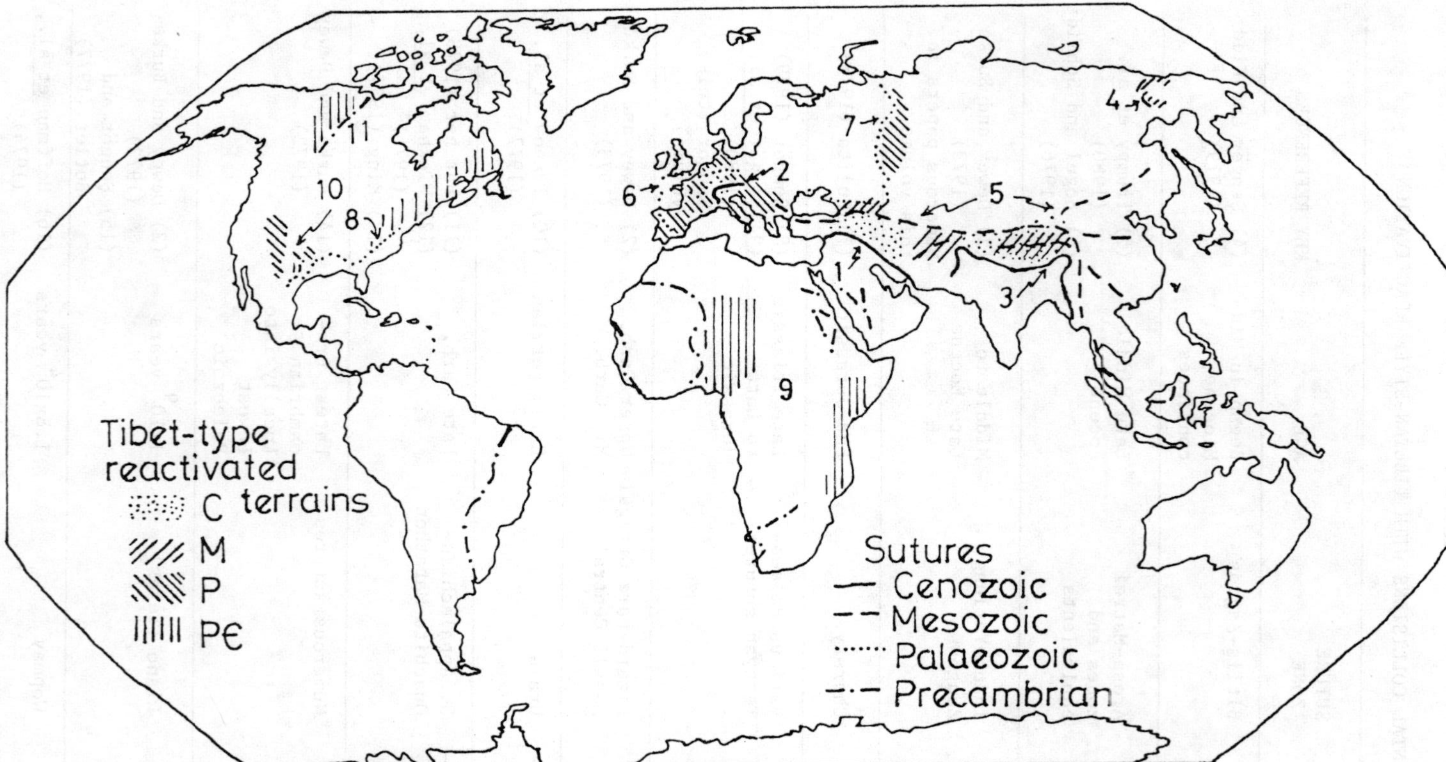


Figure 1. Location of selected Tibetan-type reactivated terrains and suture zones with ages ranging from Cenozoic to Proterozoic. Numbers correspond to those in Table 1.

# ON THE STRUCTURAL FEATURES IN THE CENTRAL PART OF XIZANG AND OBDUCTION OF THE INDIAN PLATE

Fei Ding    Cheng Shi-xing    Hao Chun-rong
Yin Shu-jie    Ren Rui    Dong Yun
Ma Jun-ru    Chang Xue-ju    Wang Men-ying
(Aero Geophysical Exploration Division, Ministry of Geology)

The aeromagnetic surveys were conducted in the central part of Xizang from 1969 to 1971. The surveyed area lies between $29^{o}-32^{o}30'$ N latitudes and $88^{o}-93^{o}$E longitudes, about 200,000 km in area. Nuclear precession magnetometer was used which is accurate to ± 5 gamma. The direction of the surveys is SN with an interval of about 5 km. Flight altitude was about 2,000 m, and the maximum height over the valley of Yarlung Zangbo River was about 3,000-4,000 m.

The Yarlung Zangbo rift, which may be a suture zone between Indian and Eurasian Plates, passes through the southern part of the surveyed area, and the fault of Bangong Lake-Denqqen-Nujiang River (north Xizang fault) which also may be a suture line passes through the northern part of the surveyed area.

## MAGNETIC FIELD AND THE BASEMENT

On the map of magnetic field, there are many local (positive) anomalies and linear (positive) anomalies which are caused by igneous rocks. The background magnetic field of the surveyed area is basically smooth and negative, usually of -25 gamma - -50 gamma. But in the south-eastern part the value is a little greater, amounting to -50 gamma - -100 gamma. The magnetic background is caused by the magnetic basement rocks (Pre-Mesozoic group) which are composed of comparatively soft supergene para-rocks.

Aero-magnetic surveys were also conducted in the central and southern part of Qinghai and Qaidam Basin. If the surveyed areas are viewed as a whole, the direction of its extension is approximately perpendicular to the strike of the regional structure and passes through the central and northern part of the Qinghai-Xizang Plateau. The results show that the background magnetic fields of the different areas are identical and mainly negative. Their magnetic basements are therefore composed of the same kind of rocks, which may be regarded as soft ones.

On the other hand, the background magnetic field of Tarim Basin, which is situated to the north of Qinghai-Xizang Plateau, is smooth and mainly positive. It differs from the background fields of the above mentioned area. The basement rocks are mainly composed of hypometamorphic rocks, and the basement is therefore rigid.

In the Indian Shield, the ancient hypometamorphic rocks outcrop extensively. This is the dominant tectonic unit on the south of Yarlung Zangbo rift. It is regarded as rigid.

After the collision of the two continents, the southern edge of the soft Eurasian Plate was pushed and compressed northward by the rigid Indian Plate, resulting in shortening and thickening of the crust (the width of the effected area exceeds over 1,000 km). Under the effect of isostatic compensation, the area where the crust was thickened, was uplifted and became a plateau. But in the Indian Plate, the width of the effected area is less than 150 km, and the increased thickness in the southern edge is only half of that of the Indian Shield.

## SEVERAL MAJOR FAULTS

North Xizang fault is not easily discerned in the aeromagnetic map (Fig. 1), which only shows the discontinuous anomaly zone of post-Jurassic ultrabasic rocks. Especially in the northwest of the surveyed area, not even the isolate anomalies can be seen. Rock anomalies distributed along the fault are less than 200 gamma in intensity, and less than 5 km in width. It indicates that there are no magnetic intrusions of large scales. As mentioned above, the magnetic background is the effects of the magnetic basement. The background anomalies on both sides of the fault are consistent and joined together on the NW part of the surveyed area, which indicates the magnetic basement on both sides of the fault is consistent, too. It can hardly be imagined that the basement of different plates which were once far apart could be so consistent. So this fault can not be a suture.

Yarlung Zangbo rift anomaly zone is very clear on the magnetic map. It is a spectacular feature in the world because of its high intensity (200-450 gamma), excellent continuity and great extension. Within the surveyed area, the length of the anomaly zone is nearly 500 km, and the rift certainly extends far out of the surveyed area. This is really a very great fault. It can also be seen on the map that the linear anomalies are truncated, which indicates the existence of faults. Yarlung Zangbo rift, a suture as attested by the existence of ophiolites, was strike-slipped by Jongnag-Rinbung fault and Zetang fault (Figs. 2, 3).

The Yarlung Zangbo magnetic anomalies are about 15-20 km in width. Though some basic-ultrabasic rocks are distributed along the fault, they are not responsible for these anomalies, because they are small and discontinuous. We are inclined to think that these anomalies are caused by deep-seated intermediate-acid rocks of a large scale.

There is another anomaly-zone yet, and roughly parallels to the Yarlung Zangbo anomaly-zone, but is separated about 30 km from it. Its intensity is smaller, but shows good continuity and considerable length. We shall call it North Yarlung Zangbo fault.

In the central part of the surveyed area, the Xainza-Nam Lake fault and Baingoin-Sangba fault are also characterized by their linear anomalies. They are the southern boundary of the area of basic-ultrabasic rocks of North Xizang. These two faults are greater in scale and significance than the North Xizang fault.

## INCLINATION OF THE YARLUNG ZANGBO RIFT

This is the most important question relating to whether the Indian Plate is obduction or subduction. Most of the geologists conjectured it is inclined to the north. But we think otherwise. The reasons are as follows:

## Aeromagnetic Information

The Yarlung Zangbo anomaly zone lies nearly 30° N latitude, with normal geomagnetic dip of 45° approximately. The relation between the maximum and minimum values of an anomaly depends on the dip angle of the magnetized body. The magnetized body is inclined to the south when the absolute value of the minimum of negatative anomaly is greater than that of the maximum of anomaly. In this altitude, line anomaly No. 148 as illustrated in Fig. 4 gives a larger minimum than the maximum. So the magnetized body -- intermediate-acid rocks along the rift -- is inclined southward.

## Geological Information

Based on a large number of geological observations, it was found that the fault plane, in general, is inclined southward with 40°-70° dip angle and at a few places, with 20°-30° dip angle. In a nappe outlier in eastern Zetang, the dip angle of a reverse fault plane is only 10° approximately. In some places, the Yarlung Zangbo rift is consisted of several thrusts, which are inclined to the south, forming an en echelon structure. The basic-ultrabasic rocks which lie adjacent to the rift and connects with it genetically are inclined to the south.

## Information from Deep Seismic Sounding

A fractured zone inclined to the south about 30-50 km wide adjacent to the Yarlung Zangbo River is found on a Yadong-Damxung deep seismic sounding profile. This seems to correspond to the Yarlung Zangbo rift and the north Yarlung Zangbo fault. On the approximately north- south profile, from Tokotokur in U.S.S.R., crossing Pamir, Karakorum, Indus, High Himalaya to Srinagar (this profile is only 30 km away from the nearest Chinese border), there are deep-seated faults on both sides of the Indus flysch zone. They are the west extension of the Yarlung Zangbo rift and north Yarlung Zangbo fault respectively. Both fault planes are inclined to the south. The depth of sounding exceeds 80 km.

## Focal Mechanism Solutions in the Himalayan Region

B.K. Rastogi once has studied seven focal mechanism solutions of west Himalayan region. Three of them are located on the north of Indus-Yarlung Zangbo River, and two of the three which are located northwest near Hindu Kush, are of intermediate depths (about 100 km). From fault plane solutions, they are both thrust faults which could extend downwards to 100 km deep along the inclined fault zone. The three fault plane solutions mentioned above indicate underthrusting towards SSW and SW.

## Focal Profile

There is a focal zone of intermediate earthquakes along the Indus-Yarlung Zangbo River. The focal Zone is inclined to the south on the near NS profile in the central part of Xizang, with $65^{o}$-$70^{o}$ dip angle, extending 230 km deep approximately.

As mentioned above, various geological and geophysical data indicate that the fault plane of Yarlung Zangbo rift is inclined to the south.

## OBDUCTION OF INDIAN PLATE

In the paired geosyncline area,the eugeosyncline is located near the ocean side, some may even cross the ocean crust, while miogeosyncline is only located near the continent or island side. There seems to be a causal relation between the movement of the plate near the convergent boundary and the development of eugeosyncline (the eugeosynclinal type sediments, which are rather thick, compile in submarine trenches. Sometimes, uplift occurs between the eugeosyncline and miogeosyncline, which is usually called a "ridge". There is an uplift in Tethys Himalayan geosyncline area which is somewhat similar to the Cona-Saga anticlinal zone. The nucleus of the anticline is granite adjacent to Kangmar. On the southern side, the Upper Triassic series is 1,056 m thick and the Jurassic-Paleogene series is 5,500 m thick. They are miogeosynclinal type of sediments. On the northern side, there are three suites of sediments of eugeosynclinal type, Upper Triassic (nearly 10,000 m thick), Upper Jurassic-Lower Cretaceous (3,450-5,400 m thick), and Upper Cretaceous (greater than 9,000 m thick). These are intermediate -- basic volcanic rocks of bathyal facies and abyssal facies (trench) -- silicolites containing a great amount of radiolaria and accumulation of flysch.

We recognized that the old Benioff zone is located at the northern edge of Indian Plate which is separated from Gondwana land, because the Tethys Himalayan geosyncline was laid on the foundation of the Gondwana basement.

The Indian Plate has drifted northward over 5,000 km since Triassic. The inclination of the fault plane of the Yarlung Zangbo rift indicates the inclination of the Benioff zone. This movement indicates that the Indian Plate is upthrusting (Fig. 5). During the Upper Triassic to Eocene, the Indian Plate upthrusted to the north over the Tethys oceanic crust and ultimately made the Tethys between the continents closed. After the collision of two continents, the obduction movement ended, which afterwards was replaced by the compression of the rigid Indian block towards the southern edge of the Eurasian continent.

## FORMATION OF QINGHAI-XIZANG PLATEAU

The cause of the formation of Qinghai-Xizang Plateau is of multiple origin:

1. Qinghai-Xizang Plateau began to uplift from the northern part and then extended to the south. This can be seen from the fact that the age of the latest marine sediments in the northern part of Yarlung Zangbo River trends to be younger toward the south (from north to south, Permian, Triassic, Jurassic, Cretaceous in proper order). From the inclination of plane of thrust and asymmetric fold in the central and northern part of the plateau, it can also be seen that the early principal compressive stress was to the south.

2. After collision, the Indian Plate was pushed northward and compressed.This is probably the most important factor for the formation of the plateau.

3. Xizang region is characterized by strong geothermal activities. There may be partial melting of the crust which cause the expansion of rocks and thickening of the crust.

4. Thickness of the crust in High Himalayan mountains is approximately 45 km. In the vicinity of Mt. Qomolangma, the isostatic anomaly is about +120 mgal (calculated under the condition of normal crustal thickness T=30 km). Hence, we can not explain the uplift by using of isostatic compensation. The uplift must be caused by thickening of the crust which is due to collision of the two plates.

Our point of view is that Himalayas are located on the upper wall of the Benioff zone, uplifted and folded after the collision (upthrusting to the south), constituting the southern edge of Qinghai-Xizang Plateau.

Qinghai-Xizang Plateau today, is just a geological-geomorphic unit placed between the rigid Indian shield and Tarim-Alxa block.

We wishes to acknowledge our gratitude to Professor Fu Cheng-yi for his kind advice and encouragement.

## References

(1) Fei Ding, Cheng Shi-xing, Hao Chun-rong. The structural characteristics of central Xizang and the problems of obduction of the Indian Plate. Abstracts of Papers (The Symposium on the Geological Sciences of the Qinghai-Xizang Plateau. Sponsored by the Ministry of Geology, 1979) Geological Institute of Qinghai-Xizang Plateau, Chinese Academy of Geological Sciences, 141 (1980).

(2) Xu Bao-wen, Ba Den-zhu, Zhang Yi-zhi. A preliminary discussion on the giant structural belt of the Yarlung Zangbo River, Abstracts of Papers (The Symposium on the Geological Sciences of the Qinghai-Xizang Plateau, Sponsored by the Ministry of Geology, 1979). Geological Institute of Qinghai-Xizang Plateau, Chinese Academy of Geological Sciences, 52 (1980).

(3) Teng Ji-wen et al. Explosion seismic study for velocity distribution and structure of the crust and upper mantle from Damxung to Yadong of Xizang Plateau. In this Proceedings of Symposium on Qinghai-Xizang (Tibet) Plateau, Academia Sinica (1981).

(4) Beloussov V.V. The structure of lithosphere along deep seismic sounding profile through Tianshan-Pamir-Karakorum-Himalaya. Soviet Geology, (1979).

(5) Rastogi B.K. Earthquake mechanism and plate tectonics in the Himalaya region. Tectonophysics, 21 (1/2), 47 (1974).

(6) Huan Wen-lin et al. The distribution of earthquake foci and plate tectonics on the Qinghai-Xizang Plateau. In this Proceedings of Symposium on Qinghai-Xizang (Tibet) Plateau, Academia Sinica (1981).

(7) Tang Bo-xiong et al. Isostatic gravity anomalies in the area of central portion of the Himalayan mountain range. In this Proceedings of Symposium on Qinghai-Xizang (Tibet) Plateau, Academia Sinica, (1981).

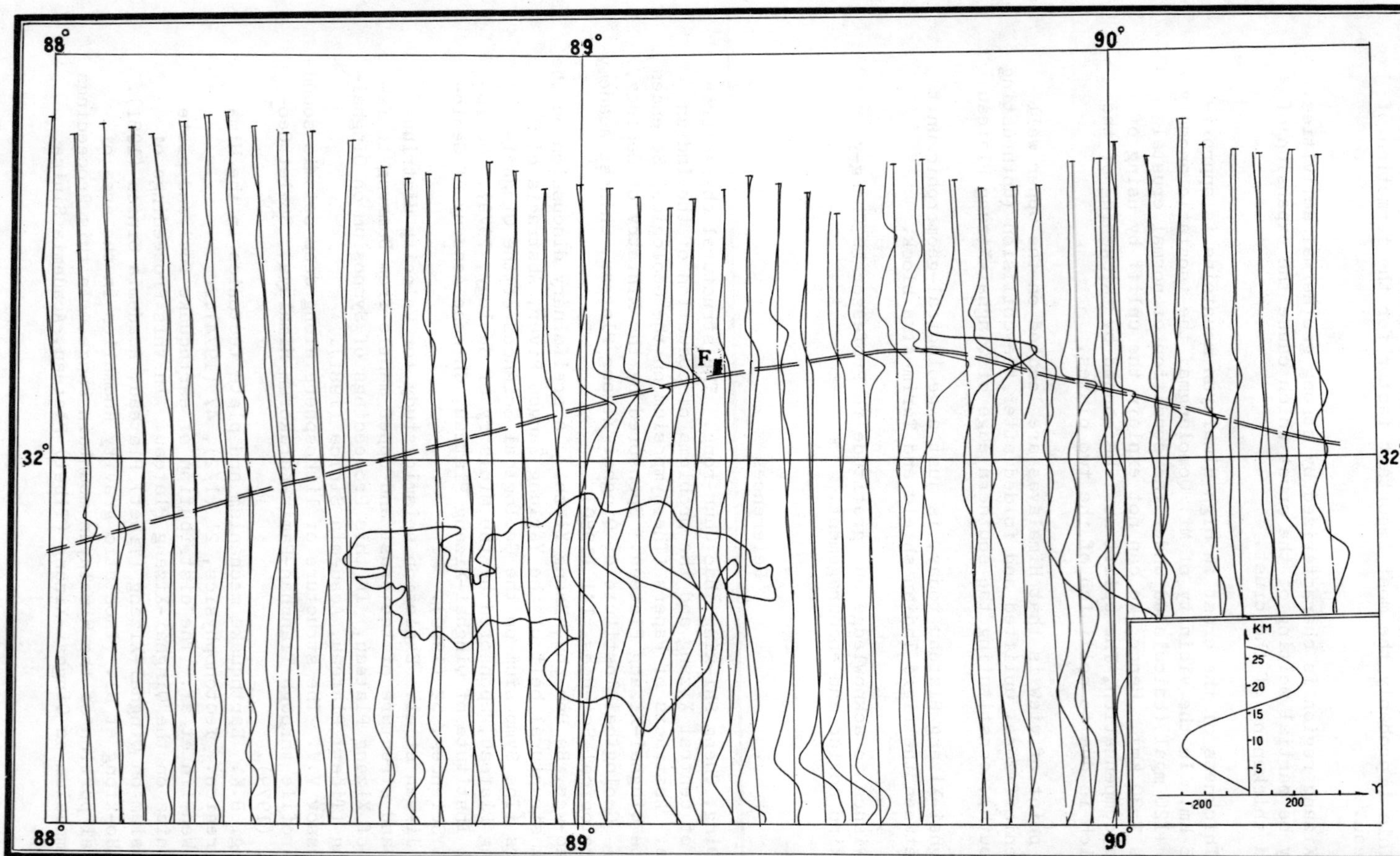

Fig. 1 Aeromagnetic profiles in the northwestern part of the surveyed area.
$F_1$ – North Xizang fault.
The figure shows that the north Xizang fault is not clear.

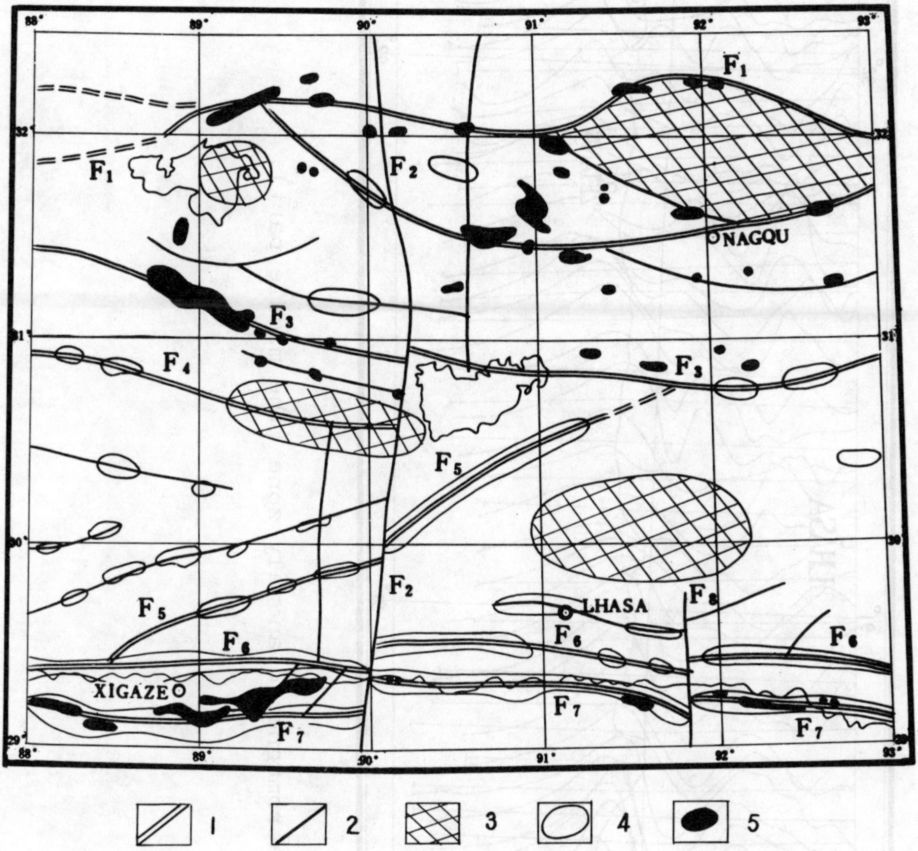


Fig. 2 Geological feature map of the central part of Xizang.

1. deep-seated fault;
2. fault;
3. rigid block (composed of acid-intermediate rocks);
4. acid-intermediate rock;
5. ultrabasic rock.

$F_1$. North Xizang fault;

$F_2$. Jongnag-Rinbung fault;

$F_3$. Baingoin-Sangba fault;

$F_4$. Xianza-Nam Lake fault;

$F_5$. Damxung-Namling fault;

$F_6$. North Yarlung Zangbo fault;

$F_7$. Yarlung Zangbo rift;

$F_8$. Zetang fault

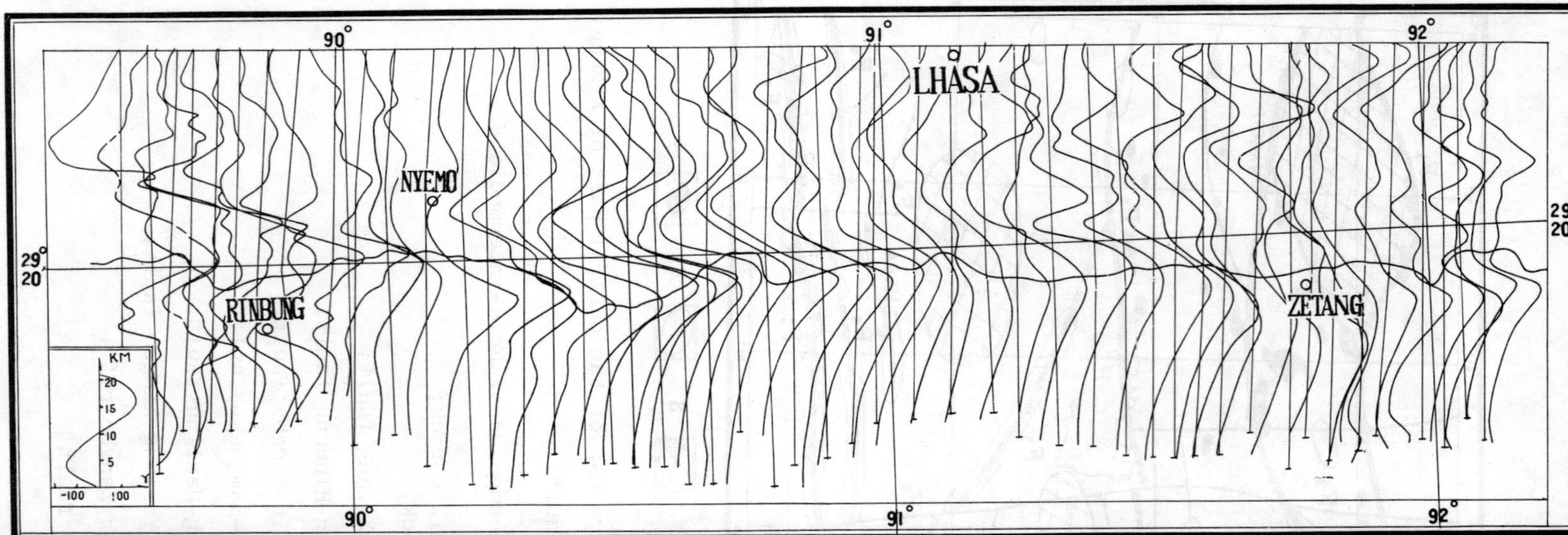


Fig. 3 Yarlung Zangbo aeromagnetic anomaly zone (the middle part).

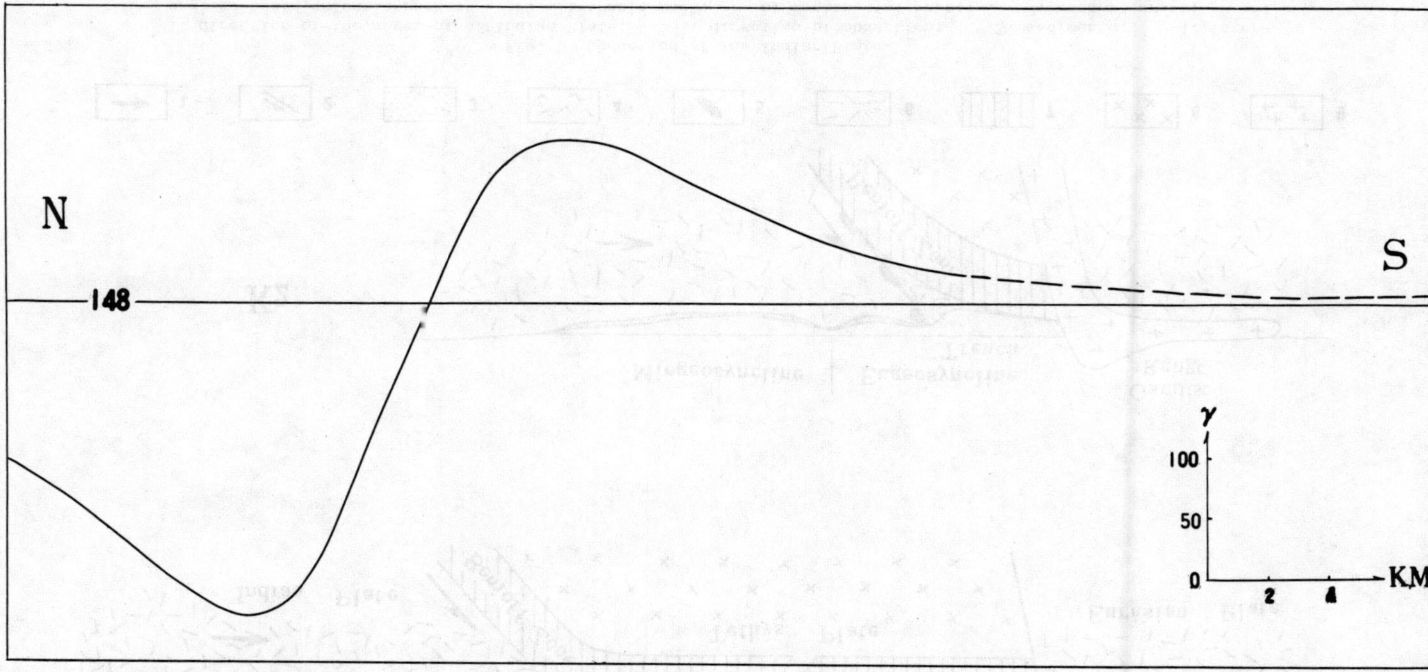


Fig. 4 Aeromagnetic anomaly profile (line 148).

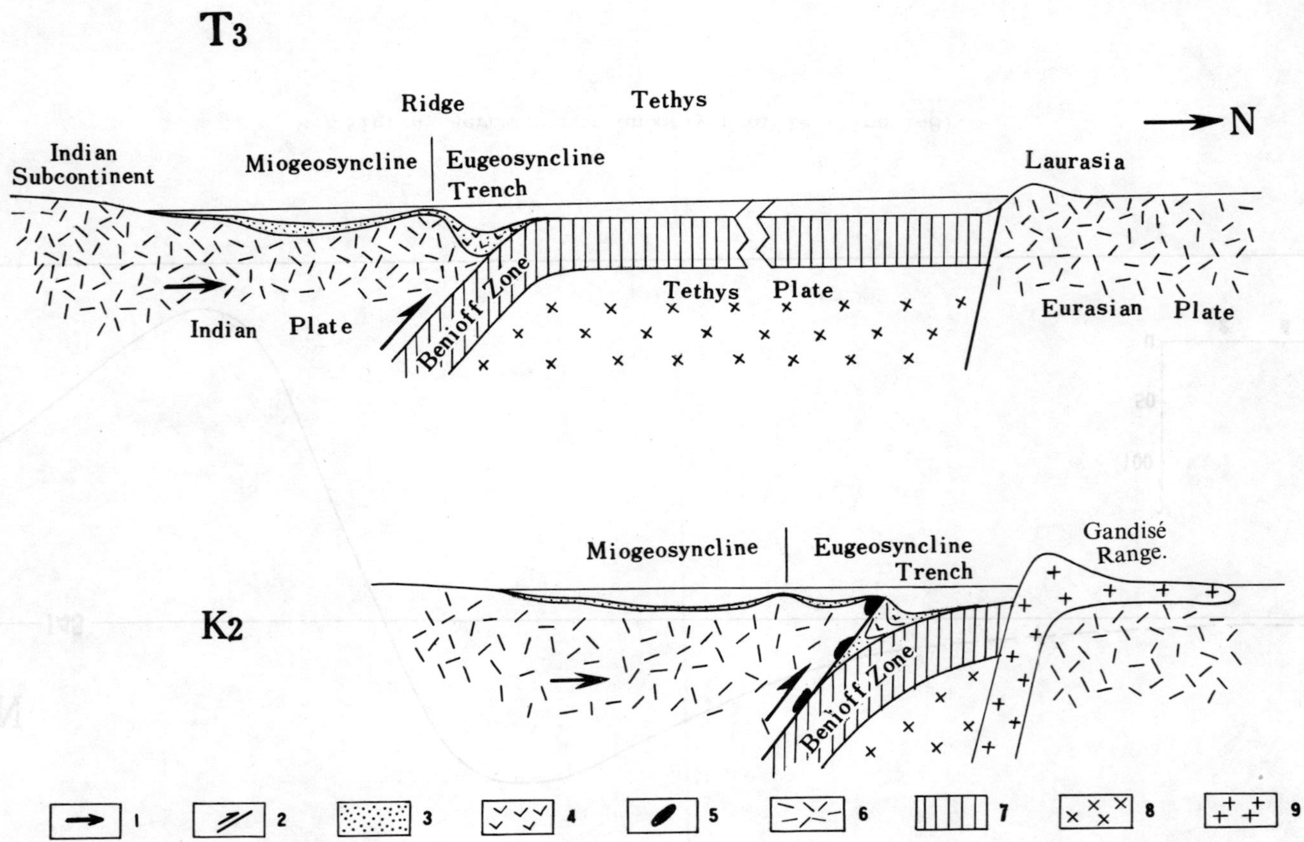


Fig. 5 Obduction of the Indian Plate.
1. direction of the movement of Indian Plate; 2. direction of obduction; 3. sediments; 4. intermediate-basic volcanic rocks; 5. ultrabasic rock; 6. continental crust; 7. oceanic crust; 8. upper mantle; 9. acid-intermediate rocks: The upper diagram shows the obduction in $T_3$. The lower shows the

# CONSTRAINTS OF THE AMOUNT OF NORTH-SOUTH SHORTENING IN TIBET DURING THE CENOZOIC

Peter Molnar and Wang-Ping Chen

Department of Earth and Planetary Sciences, Massachusetts Institute of Technology, Cambridge, Massachusetts 02139, U.S.A.

Paul Tapponnier

Institute de Physique du Globe, Universite de Paris VII, Paris, France

## Abstract

We have made reconstructions of the relative positions of India and Eurasia at various times in the late Cretaceous and Cenozoic. These reconstructions are made by combining the known histories of sea floor spreading in the Indian Ocean between India and Africa and in the Atlantic Ocean between Africa and North America and between North America and Eurasia. Between about 70 my and 40 or 50 my ago, India moved rapidly north towards Eurasia at more than 100 mm/yr. Since about 50 my ago this rate has been only about 50 mm/yr. The slowing down accords well with the inferences that collision between India and Eurasia occurred approximately 40 to 50 my ago and that it is difficult to subduct continental crust. Since 40 to 50 my ago, there has been 2000 to 3000 km of convergence between India and Eurasia, that presumably must have caused crustal shortening in Asia. It is likely that most of it occurred in Tibet and further north of Asia. Paleomagnetic data from late Cretaceous-early Tertiary rocks in Southern Tibet accord well with this inference. When compared with the late Cretaceous-early Tertiary paleomagnetic pole for Eurasia, the pole for southern Tibet indicates 2000 km of convergence between them. This convergence must occur north of the Indus Zang-Po suture and therefore not by underthrusting of India between Tibet.

There are two basic ideas for how the Tibetan plateau formed. In one, India has underthrust a large distance beneath Eurasia causing a doubling of the crust (Argand, 1924; Powell and Conaghan, 1973; and others). In the other, there has been a more uniform north-south shortening of the crust by folding and thrust faulting (Dewey and Burke, 1973). In both cases there would have to have been considerable northward motion of India with respect to Eurasia. In the former case, however, there would be little northward motion of southern Tibet with respect to Eurasia since the collision with India, whereas in the latter, there would be considerable such northward motion but little motion between India and southern Tibet since the collision. We have used the known histories of the Atlantic and Indian Oceans to calculate the positions of India with respect to Eurasia at different times in the past (Molnar and Tapponnier, 1975). We have used the results of paleomagnetic studies in southern Tibet (Zhu Xiangyuan, et al., 1977) to place constraint on the motion of southern Tibet with respect to Eurasia (Molnar and Chen, 1978).

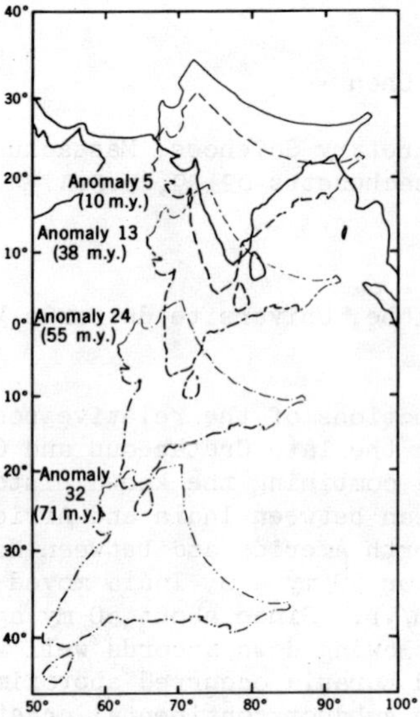


Figure 1. Position of India with respect to Eurasia at different times in the past, corresponding to magnetic anomalies in the oceans. Eurasia is arbitrarily assumed fixed. The northern boundary of India is drawn at the Himalayan foothills because we cannot know how much shortening has occurred in the Himalaya. At the time of anomaly 5, India has been rotated 6.5° about a pole at 23.0°N, 33.9°E; at anomaly 13, 27.1° about a pole at 18.8°N, 35.6°E; at anomaly 24, 42.2° about a pole at 14.8°N, 22.5°E; and at anomaly 32, 69.4° about a pole at 6.3°N, 23.7°E.

Figure 1 shows calculated positions of India with respect to Eurasia. These positions were calculated assuming rigid plate motion, using the evolution of the North Atlantic of Pitman and Talwani (1972), of the Central Atlantic of Francheteau (1973), and of the Indian Ocean of Fisher et al. (1971), McKenzie and Sclater (1971), and Molnar and Francheteau (1975). The times correspond with well defined magnetic anomalies: anomaly 5 (10 my), anomaly 13 (38 my), anomaly 24 (55 my), and anomaly 32 (71 my), using Heirtzler et al.'s (1968) scale for the younger periods, and a modification similar to that of Sclater et al. (1973) for the older periods. The uncertainties in these reconstructions are inherently smaller than those made using paleomagnetism for two reasons. First, the positions of the magnetic lineations can be determined much more precisely than can paleomagnetic poles. Second, because we are concerned with relative motion, there is no need to assume a paleomagnetic pole for Eurasia.

The relative motion of India with respect to Eurasia can be shown by plotting the distances of the northeast and northwest corners of India at different times from their present positions (Figure 2). Notice that in the later Cretaceous and early Tertiary, India moved rapidly (>100 mm/yr) towards Eurasia. During the last 40 my, however, the velocity has been close to

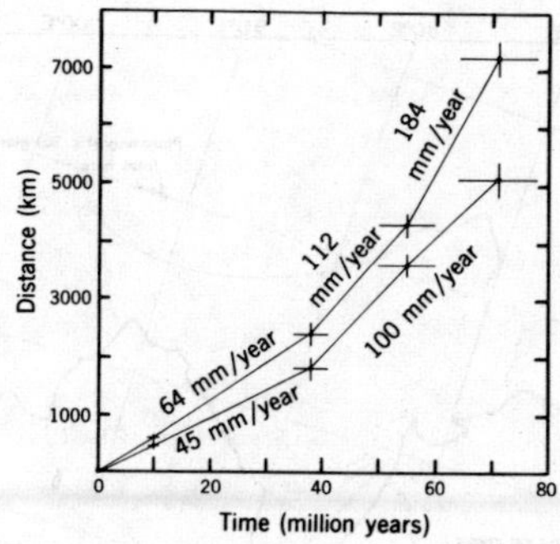


Figure 2. Distance from present position of northeast and northwest tips of India (Fig. 1) as a function of time. Uncertainties in the abscissa are from Minster et al., (1974) for Anomaly 5 and from our estimates of the uncertainties in the reconstructions for other anomalies. Uncertainties in the ordinate are assumed to be 10 percent.

50 mm/yr. The collision between Eurasia and India is poorly dated, but is thought by many to have occurred 40 to 50 my ago. The change in rate at that time is consistent with the idea that continental lithosphere, with its thick layer of light crust, is too buoyant to be subducted deep into the aesthenosphere. When the oceanic lithosphere between India and Eurasia was eliminated during the collision, the resistance to India's motion towards Eurasia presumably increased and slowed the rate of convergence.

Note that if the collision occurred 40 to 50 my ago, then since that time, India has moved 2000-3000 km northwards toward Eurasia (Figure 2). We presume that this northwards motion has been absorbed partly by a deformation of the northern edge of India to form the Himalaya but largely by a deformation of Eurasia (Molnar and Tapponnier, 1975; Tapponnier and Molnar, 1977).

Zhu Xiangyuan et al. (1977) obtained a paleomagnetic pole from samples of a late Cretaceous-early Tertiary red bed sequence in southern Tibet. The paleomagnetic pole is distinctly different from those of the Deccan Traps in India and shows that this formation was not part of India when it was magnetized (Zhu Xiangyuan et al., 1977). The paleomagnetic pole (67.5°N, 283.1°E) is also distinctly different from Irving's (1977) late Cretaceous-early Tertiary pole for Eurasia (75.0°N, 172.5°E). The difference indicates that southern Tibet has moved northwards 23° with respect to Eurasia since the late Cretaceous (Molnar and Chen, 1978). We presume that this movement has occurred since the collision with India.

Figure 3 shows the location of the samples and the positions of them relative to Eurasia. The plot is an oblique Mercator projection about Irving's (1977) pole for Eurasia 65 my ago. Also shown are the positions of India 38 and 55 my ago. Because we cannot know where the northern margin of India lay, we show the boundary of the Himalaya foothills. Note, however, that at the time of the collision India lay between the positions 38 and 55 my ago, but that the northern margin was probably 300 to 600 km north of the Himalayan foothills. Thus, the reconstructions suggest that there has been little relative motion between India and southern Tibet since the collision,

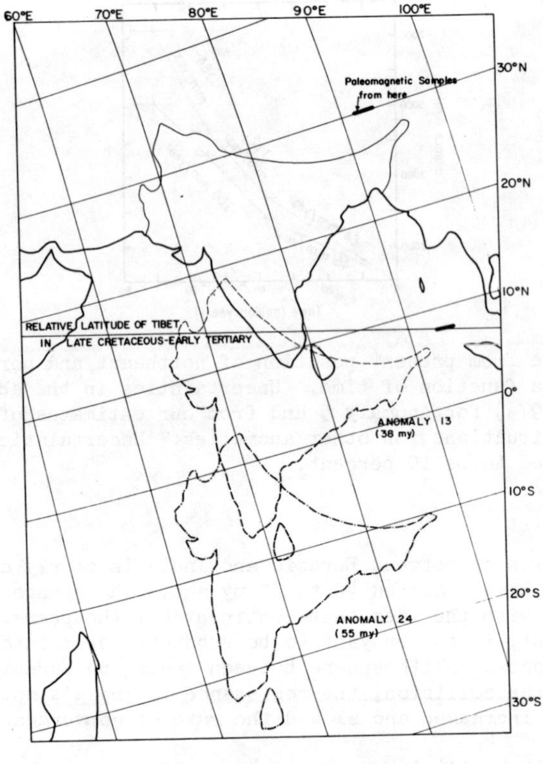


Figure 3. Oblique Mercator projection of southern Asia with pole at 75°N, 172.5°E, Irving's (1977) pole for Eurasia at 65 Myr ago. The possible positions of Tibet relative to Eurasia assuming that the rocks were magnetized 65 Myr ago are shown, and the dark bar at the present site of Tibetan samples is rotated to its paleolatitude at 65 Myr ago to show rotation of site about an axis through it. India is reconstructed (Fig. 1) to its positions at the times of anomaly 13 (38 Myr) and 24 (55 Myr). Note that the northern margin of India probably lay 300 to 600 km north of the present front. Therefore, the collision would have occurred before 38 Myr ago.

but that both have moved a large distance with respect to Eurasia. Thus, they imply that India has not underthrust Tibet a long distance and instead suggest that Tibet has been shortened in a north-south direction. Part of this shortening probably manifested itself by crustal thickening, but part also probably occurred by lateral movement of crustal material out of the way of the converging continental masses of India and Eurasia. Finally, one should note that the uncertainties in both the reconstructions of India and in the paleomagnetic data do not permit a categorical denial of large scale underthrusting of India. Further paleomagnetic work would be especially useful to resolve this question.

Acknowledgements. This work is largely the result of continued support by the National Science Foundation of the U.S.A. Molnar also thanks the John Simon Guggenheim Foundation for support during the preparation of this report.

## References


Argand, E., 1924: La tectonique de l'Asie, 13th International Geological Congress, Brussels, 1922, Comptes rendu, I, Pt. 5, 171-372.

Dewey, J.F., and K.C.A. Burke, 1973: Tibetan, Variscan and Pre-Cambrian basement reactivation: Products of a continental collision, J. Geol., 81, 683-692.

Fisher, R.L., J.G. Sclater, and D.P. McKenzie, 1971: Evolution of the central Indian ridge, Western Indian Ocean, Geol. Soc. Amer. Bull., 82, 553-562.

Francheteau, J., 1973: Plate tectonic model of the opening of the Atlantic Ocean south of the Azores, in: Implications of Continental Drift to the Earth Sciences, Runcorn, S.K., and D.H. Tarling, eds., 197-202, Academic Press, London, New York.

Heirtzler, J.R., G.O. Dickson, E.M. Herron, W.C. Pitman, III, and X. Le Pichon, 1968: Marine magnetic anomalies, geomagnetic field reversals, and motions of the ocean floor and continents, J. Geophys. Res., 73, 2119-2136.

Irving, E., 1977: Drift of the major continental blocks since the Devonian, Nature, 270, 304-309.

McKenzie, D.P. and J.G. Sclater, 1971: The evolution of the Indian Ocean since the Late Cretaceous, Geophys. J. Roy. Astron. Soc., 24, 437-528.

Minster, J.B., T.H. Jordan, P. Molnar, and E. Haines, 1974: Numerical modelling of instantaneous plate tectonics, Geophys. J. Roy. Astr. Soc., 36, 541-576.

Molnar, P. and W.P. Chen, 1978: Evidence for large Cainozoic crustal shortening of Asia, Nature, 273, 218-220.

Molnar, P. and J. Francheteau, 1975: The relative motion of "hot-spots" in the Atlantic and Indian Oceans during the Cenozoic, Geophys. J. Roy. Astr. Soc., 43, 763-774.

Molnar, P. and P. Tapponnier, 1975: Tectonics of Asia: Consequences and implications of a continental collision, Science, 189, 419-426.

Pitman, III, W.C., and M. Talwani, 1972: Sea-floor spreading in the North Atlantic, Geol. Soc. Amer. Bull., 83, 619-646.

Powell, C.M., and P.J. Conaghan, 1973: Plate tectonics and the Himalayas, Earth Planet. Sci. Lett., 20, 1-12.

Sclater, J.G., R.D. Jarrard, B. McGowran, and S. Garther, Jr., 1974: Comparison of the magnetic and biostratigraphic time scales since the late Cretaceous, in: Initial Reports of the Deep Sea Drilling Project, XXII, 381-386, U.S. Government Printing Office, Washington, D.C..

Tapponnier, P. and P. Molnar, 1977: Active faulting and Cenozoic tectonics of China, J. Geophys. Res., 82, 2905-2930.

Zhu Xiangyuan, Chung Liu, Sujuan Ye, and Jinlu Liu, 1977: Remanence of red beds from Linzhou, Xizang and the northward movement of the Indian plate, Scientia Geol. Sinica, 1, 44-51.

# CONSTRAINTS ON THE SEISMIC WAVE VELOCITY STRUCTURE BENEATH THE TIBETAN PLATEAU

Wang-Ping Chen and Peter Molnar

Department of Earth and Planetary Sciences, Massachusetts Institute of Technology, Cambridge, Massachusetts 02139, U.S.A.

Abstract

We combine observations of group and phase velocity dispersion of Rayleigh waves, of the wave form of a long period $P_L$ phase, of $P_n$ and $S_n$ velocities from unreversed refraction profiles using earthquakes, and of teleseismic S-P travel time residuals to place bounds on the seismic wave velocity structure of the crust and upper mantle under Tibet. From long period surface wave dispersion alone, the Tibetan crustal thickness can be from 55 km to 85 km, with corresponding uppermost mantle shear wave velocities of about 4.4 km/s and 4.9 km/s, respectively. The $P_n$ and $S_n$ velocities were determined to be $8.12 \pm 0.06$ km/s and $4.8 \pm 0.1$ km/s respectively using travel time data recorded at Lhasa from earthquakes in and on the margins of Tibet. With these estimates of the uppermost mantle velocity and with the surface wave dispersion, the crustal thickness is most likely to be between 65-80 km with an average shear wave velocity in the crust less than 3.5 km/s. A synthesis of one $P_L$ wave form does not provide an additional constraint to the velocity structure but is compatible with the range of models given above. Measurements of both teleseismic S wave and P wave arrival times for nine earthquakes within Tibet show unusually large intervals between P and S compared with the Jeffreys-Bullen Tables. We infer that most of the large interval is due to delays in the S wave travel times. These delays are not a consequence of errors in the Tables, because S wave and P wave arrival times from eight earthquakes in the nearby Himalaya agree quite well with the Tables. A minimum delay of the S waves of two to three seconds exists even if the focal depth of Tibetan events were only 5 km, while Himalayan events were 33 km. Thus the $P_n$ and $S_n$ velocities apparently do not reflect high velocities in the mantle to a great depth beneath Tibet.

A knowledge of the crustal and upper mantle structure of Tibet is fundamental for understanding the evolution of the plateau and its thermal structure. Because of its remoteness, however, field investigations are limited and have been restricted to those by Chinese scientists. Accordingly we have used seismic data, largely from stations and agencies outside of China to investigate the seismic wave velocity structure beneath the plateau. The results reported here are only a brief summary of a study reported in detail elsewhere (Chen and Molnar, 1980).

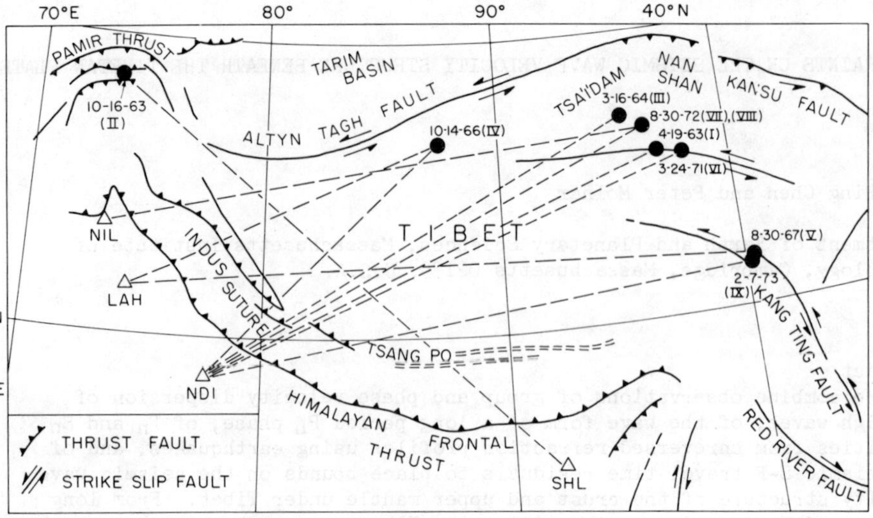


Figure 1. Map of Tibet showing earthquake (dots) - station (triangles) combinations used for surface wave dispersion and the $P_L$ phase observation. Events I and II generated large long-period Rayleigh waves. Note the difference in the paths to stations from these two events. Event I was used for the $P_L$ phase observation also. Major strike-slip and thrust faults (see legend) were also plotted following Molnar and Tapponnier (1975).

We have used four types of seismic waves to examine the structure - surface waves, $P_L$ (a leaky mode), refracted body waves - $P_n$ and $S_n$, and S waves recorded at teleseismic distances. Data from each places different constraints on the structure. We have attempted to use all such data available to us as of early 1979. Although each does not give a unique structure, when taken together certain general aspects of the structure seem inescapable - the crust is thick (~70 km), there is a high velocity zone beneath it, and further below there is a relatively thick low velocity zone. Although we have used as extensive a data set as we could, the reader will note that there is a great deal that can be done to improve the measured velocities and to examine lateral heterogeneity that could underlie the plateau.

<u>Surface Wave Dispersion</u>. Using stations in India and Pakistan and earthquakes within or on the margins of Tibet we have measured dispersion curves for both Love and Rayleigh waves (Figure 1). We have confined the study to such short paths because we do not think that we can reliably regionalize the structure using longer paths that traverse other structures. Although we have attempted to use all such possible data for Tibet since the installation of the World Wide Standardized Seismograph Network (WWSSN), the data both scatter and are inadequate to define the dispersion curves for the entire period range from 15 to 90 seconds. Thus the data are not of the same quality as for some other regions, but we think that they include essentially all reliable extant data.

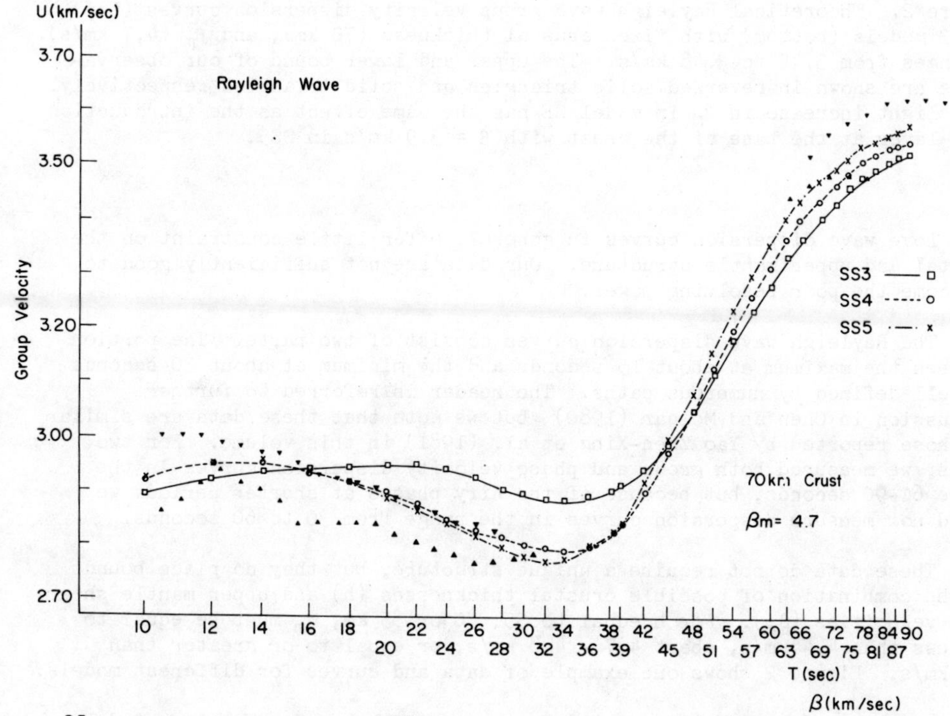
U(km/sec)
Rayleigh Wave
Group Velocity
3.70
3.50
3.20
3.00
2.70
SS3
SS4
SS5
70km Crust
βm= 4.7
10 12 14 16 18 20 22 24 26 28 30 32 34 36 38 39 42 45 48 51 54 57 60 63 66 69 72 75 78 81 84 87 90
T (sec)

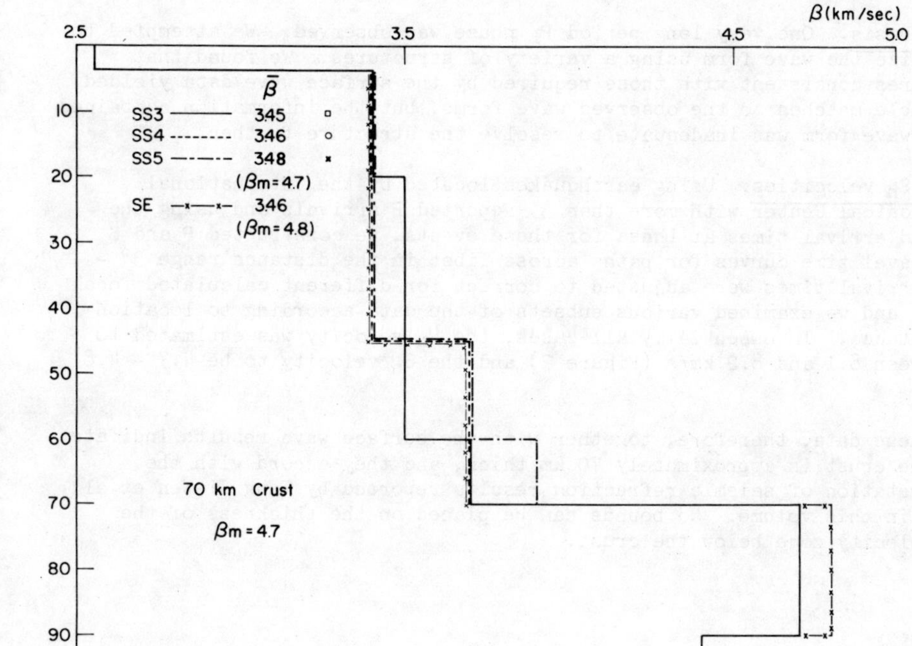
β(km/sec)
2.5
3.5
4.5
5.0
β̄
SS3 3.45
SS4 3.46
SS5 3.48
(βm=4.7)
SE 3.46
(βm=4.8)
70 km Crust
βm = 4.7
10
20
30
40
50
60
70
80
90
100
Depth (km)

Figure 2. Theoretical Rayleigh wave group velocity dispersion curves (top) for 3 models (bottom) with fixed crustal thickness (70 km), and $\beta_m$ (4.7 km/s). $\bar{\beta}$ ranges from 3.45 to 3.48 km/s. The upper and lower bound of our observations are shown in reversed solid triangles and solid triangles respectively. The slight increase in $\beta_m$ in model SE has the same effect as the introduction of a layer at the base of the crust with $\beta$ = 3.9 km/s in SS5.

Love wave dispersion curves in general, offer little constraint on the crustal and upper mantle structure. Our data are not sufficiently good to overcome the poor resolving power.

The Rayleigh wave dispersion curves consist of two parts. The portion between the maximum at about 15 seconds and the minimum at about 30 seconds is well defined by numerous paths. The reader is referred to further discussion in Chen and Molnar (1980), but we note that these data are similar to those reported by Yao Zhen-Xing et al. (1981) in this volume. For two paths, we measured both group and phase velocity dispersion curves in the range 60-90 seconds, but because of the Airy phases at shorter periods we could not measure dispersion curves in the range from 30 to 60 seconds.

These data do not require a unique structure, but they do place bounds on the combination of possible crustal thicknesses (h) and upper mantle shear wave velocities ($\beta_m$). For h equal to 55, 70 or 85 km, $\beta_m$ must be equal to or less than 4.4 km/s, about 4.7 - 4.8 km/s, or equal to or greater than 4.9 km/s. Figure 2 shows one example of data and curves for different models.

$P_L$ Synthesis. One very long period $P_L$ phase was observed. We attempted to synthesize the wave form using a variety of structures. We found that structures consistent with those required by the surface wave data yielded acceptable matches to the observed wave forms, but the information contained in the wave form was inadequate to resolve the structure further.

$P_n$ and $S_n$ velocities. Using earthquakes located by the International Seismological Center with more than 35 reported P arrivals and using the reported arrival times at Lhasa for these events, we constructed P and S wave travel time curves for paths across Tibet in the distance range 3° - 17°. Arrival times were adjusted to correct for different calculated focal depths, and we examined various subsets of the data according to location and distance. In essentially all cases, the $P_n$ velocity was estimated to be between 8.1 and 8.2 km/s (Figure 3) and the $S_n$ velocity to be 4.7 - 4.8 km/s.

These data, therefore, together with the surface wave results indicate that the crust is approximately 70 km thick, and they accord with the interpretation of seismic refraction results reported by Teng Ji-Wen et al. (1981) in this volume. No bounds can be placed on the thickness of the high velocity zone below the crust.

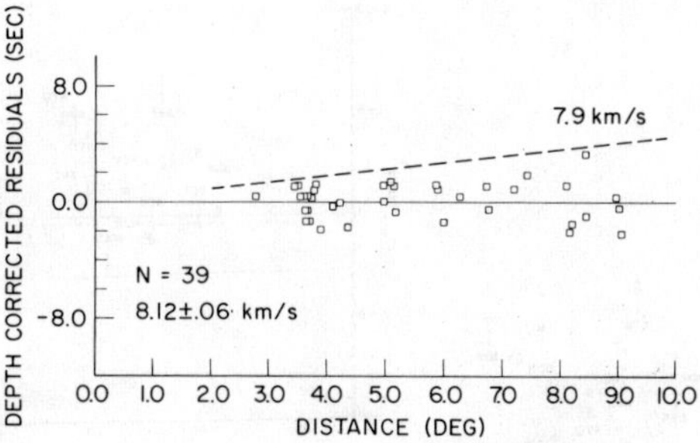


Figure 3. Plots of $P_n$ travel time residuals normalized for a focal depth of 33 km vs. distance to LHA for earthquakes north of the Indus-Tsang Po suture. The distance range is about 3°-10°. The $P_n$ velocity (8.12±0.06 km/s) and travel time residuals were obtained by a least-square fitting of the 39 observations assuming a Moho depth of 70 km. The dashed line shows the slope of a curve to which the data would be parallel if the $P_n$ velocity were 7.9 km/s.

S Wave Travel Time Residuals. We measured P and S wave arrival times for 9 events in Tibet and 8 events in the Himalayas (Figure 4). Because the events are located using P waves, the P wave residuals average near zero. The S waves, however, are usually late. Those from Tibet are especially late - an average of 2 - 7 seconds later than the Jeffreys-Bullen travel times for each event. For Himalaya events, the average delay is from 0 to 3 seconds. Thus the S waves from Tibetan events are later than those from Himalayan events, and the delay cannot be ascribed only to errors in the J-B curves.

If the S wave delay beneath Tibet is due to relatively low velocity in the depth range 100-240 km, then the velocity there is 4-5% lower than a shield structure. Compared with that of Dziewonski (1971), the average velocity in this depth range would be about 4.3 km/s.

Summary and Geological Implications. Taken together these data indicate that the crust is thick (~70 km) and is underlain by a relatively high velocity zone, which in turn is underlain by a relatively low velocity zone. Presumably the high elevation of the plateau results from isostatic compensation of the thickened crust. The thick crust must have formed since the Cretaceous, because there are marine limestones covering much of it. Yet, it seems very difficult to uplift the entire plateau in a short time (since the Pliocene), because that would involve the transfer of an enormous mass of crustal material since that time. For a crust of twice the normal thickness, it would call for at least 1000 km of northward motion of India in 5 my (20 cm/yr).

We think that the relatively high velocities can be reconciled with a temperature at the Moho of about 800°C (see Chen and Molnar, 1980). This temperature is high enough to melt lower crustal materials if they are slightly wet, but not mantle materials. Possibly some of the volcanics on Tibet are due to melting of the crust.

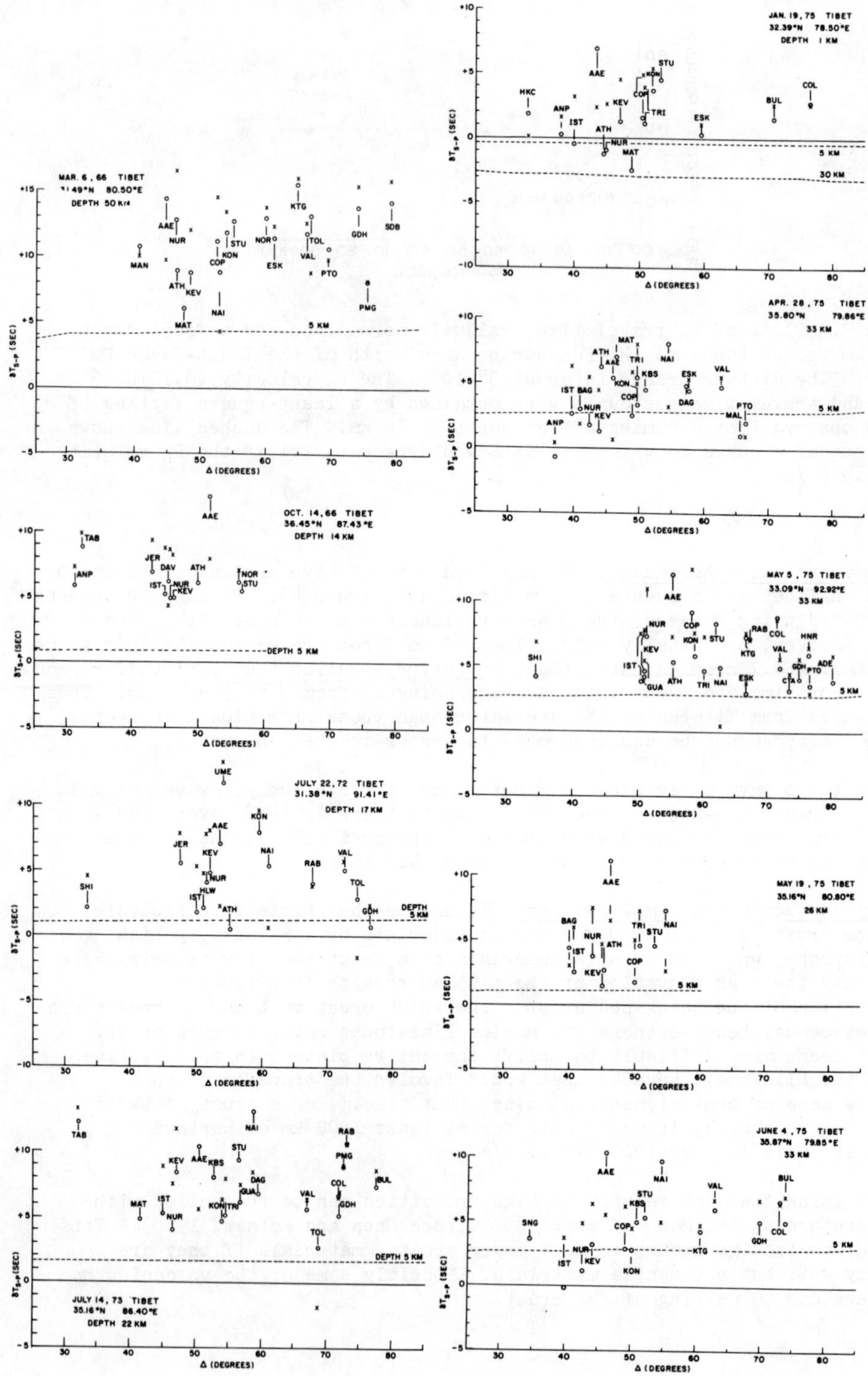

MAR. 6, 66 TIBET
31.49°N 80.50°E
DEPTH 50 KM
OCT. 14, 66 TIBET
36.45°N 87.43°E
DEPTH 14 KM
JULY 22, 72 TIBET
31.38°N 91.41°E
DEPTH 17 KM
JULY 14, 73 TIBET
35.16°N 86.40°E
DEPTH 22 KM
JAN. 19, 75 TIBET
32.39°N 78.50°E
DEPTH 1 KM
APR. 28, 75 TIBET
35.80°N 79.86°E
33 KM
MAY 5, 75 TIBET
33.09°N 92.92°E
33 KM
MAY 19, 75 TIBET
35.16°N 80.80°E
26 KM
JUNE 4, 75 TIBET
35.87°N 79.85°E
33 KM
δT_{S-P} (SEC)
Δ (DEGREES)

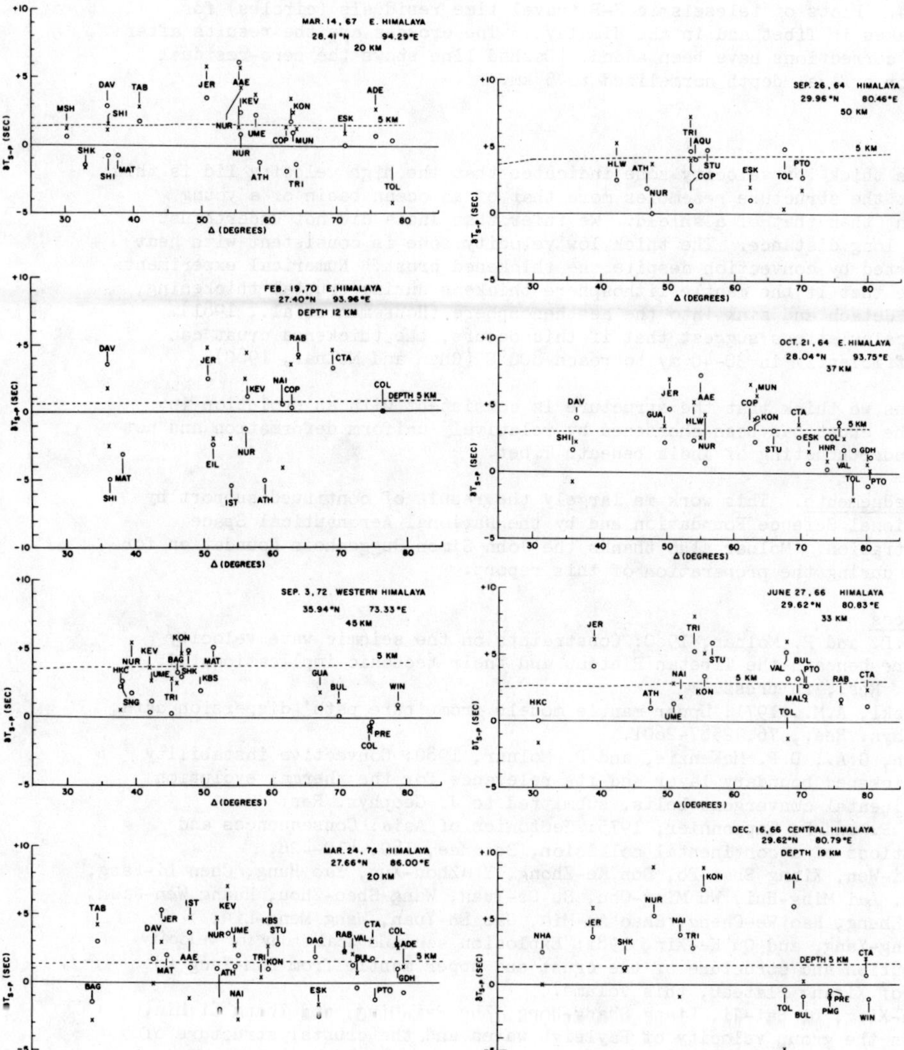

MAR. 14, 67 E. HIMALAYA
28.41°N 94.29°E
20 KM
SEP. 26, 64 HIMALAYA
29.96°N 80.46°E
50 KM
FEB. 19, 70 E. HIMALAYA
27.40°N 93.96°E
DEPTH 12 KM
OCT. 21, 64 E. HIMALAYA
28.04°N 93.75°E
37 KM
SEP. 3, 72 WESTERN HIMALAYA
35.94°N 73.33°E
45 KM
JUNE 27, 66 HIMALAYA
29.62°N 80.83°E
33 KM
MAR. 24, 74 HIMALAYA
27.66°N 86.00°E
20 KM
DEC. 16, 66 CENTRAL HIMALAYA
29.62°N 80.79°E
DEPTH 19 KM
δT S-P (SEC)
Δ (DEGREES)

Figure 4. Plots of teleseismic S-P travel time residuals (circles) for earthquakes in Tibet and in the Himalaya. The crosses are the results after station corrections have been added. Dashed line shows the zero-residual level for a focal depth normalized to 5 km.

The thick low velocity zone indicates that the high velocity lid is thin and that the structure resembles more that of an ocean basin or a young continent than that of a shield. We infer that India did not underthrust Tibet a long distance. The thick low velocity zone is consistent with heat transported by convection despite the thickened crust. Numerical experiments indicate that if the mantle lithosphere thickens during crustal thickening, it will detach and sink into the aesthenosphere (Houseman et al., 1981). Simple calculations suggest that if this occurs, the thickened crust can warm sufficiently in 30-40 my to reach 800°C (Chen and Molnar, 1980).

Thus we think that the structure is consistent with an evolution in which the crust has been shortened by relatively uniform deformation and not by an underthrusting of India beneath Tibet.

Acknowledgements. This work is largely the result of continued support by the National Science Foundation and by the National Aeronautical Space Administration. Molnar also thanks the John Simon Guggenheim Foundation for support during the preparation of this report.

References

Chen, W.P. and P. Molnar, 1980: Constraints on the seismic wave velocity structure beneath the Tibetan Plateau and their tectonic implications, J. Geophys. Res., in press.

Dziewonski, A.M., 1971: Upper mantle models from 'pure path' dispersion data, J. Geophys. Res., 76, 2587-2601.

Houseman, G.A., D.P. McKenzie, and P. Molnar, 1980: Convective instability of a thickened boundary layer and its relevance for the thermal evolution of continental convergent belts, submitted to J. Geophys. Res.

Molnar, P. and P. Tapponnier, 1975: Tectonics of Asia: Consequences and implications of a continental collision, Science, 189, 419-426.

Teng, Ji-Wen, Xiong Shao-Po, Sun Ke-Zhong, Yin Zhou-Xun, Yao Hung, Chen Li-Fang, Mu Teng, Lai Ming-Hui, Wu Ming-Chu, Su De-Yuan, Wang Shao-Zhou, Huang Wen-Jian, Ou Ren-Sheng, Hao We-Cheng, Shao An-Min, Gao En-Yuan, Wang Meng-Lin, Liu Zhong-Yang, and Qu Ke-Xin, 1981: Explosion seismological study for velocity distribution and structure of the crust and upper mantle from Dam-Xung to Yadong of Xizang Plateau, this volume.

Yao Zen-Xing, Li Bai-Ji, Liang Shang-Hong, Zhu Pei-Ding, and Zhang Li-Min, 1981: On the group velocity of Rayleigh waves and the crustal structure of Qinghai-Xizang Plateau, this volume.

# A MAGNETOTELLURIC STUDY ON THE EARTH'S CRUST AND UPPER MANTLE AT NAGQU AND PAGRI, XIZANG PLATEAU

Chen Guang-ming Ren Ai-hua Li Zong-shun
Li Hong-zheng Yang Zhong-wu Huang Bo-min
(Institute of Geophysics, Academia Sinica)

Shi Ying-jun
(Chengdu Geological College)

## INTRODUCTION

Xizang Plateau has been the focus of attention of scientists over the world. In 1977, the magnetotelluric (MT) measurements were made for the first time in Xizang Plateau by use of home-made equipments. The MT method is one of the principal means of studying the earth's electrical structure.

## THE FIELDWORK

The block diagram of our MT measurement system is shown in Fig. 1. It consists of five parts: (1) a three-component fluxgate magnetometer with a resolution of 0.2 gamma and temperature coefficient of less than 1 gamma /°C; its fundamental is the same as described by Trigg (1970); (2) a telluric device to measure the two horizontal components of the geoelectric field (Ex, Ey); the earthed electrodes of four lead sheets with spacing of several hundred meters forming a cross-shaped array; (3) an XWX-204 Type electronic potentiometer with ink recording; (4) a time device providing accurate timing information; (5) a power supply with zinc-air battery.

The MT measurements were carried out in Xizang Plateau during June-September, 1977. The locations of our observation sites are shown in Fig. 2. Mountains, rivers and other geological features run nearly from east to west. The five sites chosen are, from north to south: Nagqu, Nyetang, Gyangze, Pagri and Yadong, stretching over 500 km crossing the geological structure. We shall report here only the results obtained in Nagqu and Pagri. The others will be published elsewhere.

## DATA ANALYSIS

### General Consideration

From records obtained at each sites, several data sections of over two hours duration were selected visually for digitizing. The main criterion for selecting data sections is the presence of more abundant spectrum components for the four horizontal components of the electromagnetic field. Hz was not used due to its small variation and low resolution of the three-component fluxgate magnetometer. The selected data sections were digitized by hand or by machine. The interval of digitization is to ensure a sample point per 4 seconds, so that periodic signals of 10 seconds could be obtained. After the preprocessing of editing, digitizing, interpolation, prewhitening, Hanning window and Fourier transformation, the required parameters were obtained.

## Estimate of the Magnetotelluric Tensor Impedance

For the analytical model used in the analysis of the MT data, the conductive earth is assumed to occupy the half space $z \geqslant 0$, as shown in Fig. 3. The positive sense of the orthogonal coordinates X and Y is northward and eastward respectively. Electromagnetic excitation is assumed to be provided by plane waves propagation vertically downward.

Following the study carried out by other workers (Cantwell and Madden, 1960; Bostick and Smith, 1962), the frequency domain relations of surface electromagnetic field are as follows:

$$\{E\} = [Z]\{H\} \tag{1}$$

and

$$\{H_z\} = \{Y_z\}\{E\} , \tag{2}$$

where

$$\{E\} = \{E_x \; E_y\}^T \tag{3}$$

and

$$\{H\} = \{H_x \; H_y\}^T \tag{4}$$

are the column vectors representing the tangential electric and magnetic field and

$$[Z] = \begin{bmatrix} Z_{xx} & Z_{xy} \\ Z_{yx} & Z_{yy} \end{bmatrix} \tag{5}$$

is the tensor impedance of rank two. It is a function of position and frequency.

$$\{Y_z\} = \{Y_{zx} \; Y_{zy}\} \tag{6}$$

is a tensor admittance function. Both $[Z]$ and $\{Y_z\}$ were estimated from the measured field data.

If the coordinate axes are rotated clockwise by the angle $\theta$, as in Fig. 3, then in the primed coordinate system the tensor impedance and admittance are

$$[Z'] = [R][Z][R]^T \tag{7}$$

and

$$\{Y_z'\} = [R]\{Y_z\}[R]^T , \tag{8}$$

where

$$[R] = \begin{bmatrix} \cos\theta & \sin\theta \\ -\sin\theta & \cos\theta \end{bmatrix} \tag{9}$$

is the Cartesian rotation matrix.

Sims (1969) has recognized the loci of the $[Z'(\theta)]$ and $\{Y_z'(\theta)\}$ elements in the complex impedance plane as a function of rotation angle $\theta$ are generally elliptical. The direction of the elliptical major axis, i.e. the principal direction of the tensor impedance, is obtained from

$$\left| Z'_{xy} + Z'_{yx} \right|_{max} \longrightarrow \theta_0 \tag{10}$$

Then from equation (7) we obtain the tensor impedance in the principal direction.

## Determination of the Apparent Resistivity and Ellipticity Index

After computing the tensor impedance along the principal direction, we can obtain the principal apparent resistivities and ellipticity index

$$\rho_{xy} = 0.2T \left| Z'_{xy} \right|^2 , \quad (11)$$

$$\rho_{xy} = 0.2T \left| Z'_{yx} \right|^2 \quad (12)$$

and

$$\beta(\theta_0) = \frac{(Z'_{xx} - Z'_{yy})}{(Z'_{xy} + Z'_{yx})} , \quad (13)$$

where $\rho_{xy}$ and $\rho_{yx}$ represent the apparent resistivities, T, the period, and $\beta(\theta_0)$, the ellipticity index which is a three-dimensionality indicator. $\beta = 0$ for a two-dimensional earth, whereas $\beta = 1$ for a three-dimensional earth. If there is a pseudo strike direction in the structure of the observational sites, $\rho_{11}$ and $\rho_{\perp}$ are used to represent respectively the apparent resistivity parallel and perpendicular to the pseudo strike direction. The dip-axis estimates are defined by the x' axis for $\theta = \theta_{z_0}$, obtained from

$$\left| Y'_{zy}(\theta) \right|_{max} \longrightarrow \theta_{z_0} \quad (14)$$

The strike direction is perpendicular to dip-axis. Because of the maximum admittance in Y' direction, this direction is regarded as strike one, tallying with the actual situation. The principal direction of the tensor impedance defined by $\theta_0$ is parallel or perpendicular to the dip-axis, depending upon the conductive construction. In quadrauts I and IV, $\Delta\theta = |\theta_0 - \theta_{z_0}|$ is defined, thus

$$\rho_{11} = \rho_{yx}(\theta_0) \quad \text{if} \quad \Delta\theta \leq 45°$$

and

$$\rho_{11} = \rho_{xy}(\theta_0) \quad \text{if} \quad \Delta\theta > 45° ,$$

$\rho_{\perp}$ proving the contrary.

## Inversion

Each conductivity profile is obtained by inverting an impedance-period relation into an estimate of conductivity-depth relation. This inverted conductivity estimate is taken to be representive of the true conductivity profile beneath the recording site.

In inverting MT data by computers, we use the method of Maquardt. The basic principle of this method is to satisfy the condition (Wu, 1968):

$$W = \sum_i (\rho_{ci} - \rho_{ai})^2 = \text{minimum}, \quad i = 1, 2, \ldots K, \quad (15)$$

where $\rho_{ai}$ is the observed apparent resistivity with period Ti and $\rho_{ci}$ is the model-computed apparent resistivity with the same period, which is a function of the layer-resistivities, layer-thicknesses and period.

The minimum of W occurs when the equations

$$\frac{\partial W}{\partial RH_j} = 0 \qquad \left[ j = 1, 2, \ldots (2n-1) \right] \tag{16}$$

are satisfied, where $RH_j$ are 2n-1 parameters for an n-layer model.

After linearization of equations (16), and using damping factor, we obtain a set of linear equations

$$( [P]^T [P] + DI ) \{\Delta RH\} = [P]^T \{R\} \quad , \tag{17}$$

where $R_i = \rho_{ci} - \rho_{ai}$ , $P_{ij} = \partial\rho_{ci} / \partial RH_j$, D is damping factor and I is a unit matrix.

We solved equations (17) for the increment $\Delta RH_j$ for each layer parameter, and repeated the iteration until the condition (15) attained the necessary precision.

In the program we adopted two kinds of damping factors (contraction D/U and amplification DU) so as to make the direction and speed of convergence to meet our demands.

## RESULTS

Figure 4 shows the variations of the principal directions of the impedance with the periods at Nagqu and Pagri. The abscissae represent the periods, and the angles away from the north direction represent the principal directions. From the figure we can see, that the principal direction roughly tends north-south, and $\theta_o < 45^o$. The geological structure of the studied region is nearly in west-east direction, therefore $\theta_{zo}$ can be taken as zero, and $\Delta\theta < 45^o$. As a result, $\rho'_{11} = \rho_{yx}(\theta_o)$.

Figure 5 is the variations of $\rho_{11}$ with period at two sites. The ordinates give $\rho_{11}$ in $\Omega$-m and the abscissas, period in second. The ranges of the period are from 10 seconds to 2000 seconds. Small points show the computed results from our measurements, and the curves show the results of model computing.

Figure 6 give the relations between ellipticity index and period at two sites. From the figure we can see, that the ellipticity indices are less than 0.6. Thus we can approximately regard the medium as a two-dimensional one. Along the direction of the strike, $\rho_{11}$ is homogeneous. So we can make a one-dimensional inversion to $\rho_{11}$ by above-mentioned inversion method. As a result we have obtained two five-layer models representing electricity features under Nagqu and Pagri respectively. Their resistivity sections are shown in Fig. 6, the ordinates represent layer resistivities in $\Omega$-m, the abscissas denote the depths in km. The parameters for two models are shown in Tables 1 and 2.

From our study we have come to the following conclusions:

1. The total thickness of the upper four-layer at Nagqu is 66 km but at Pagri 57 km. They approximately agree with other geophysical results if they are taken to be the thickness of the earth's crust;

2. The trends of the conductivity distributions at the two sites are good accord. They all belong to low-high-low type;

3. There is a low resistivity zone in the earth's crust at both sites, the depth of which is about 10 km.

The results mentioned above are similar to those electrical characteristics obtained from other regions. It reveals the presence of a low resistivity zone in the earth's crust and that the value of resistivity and the depth of the layer of low resistivity zone show no great difference. For example, Laird and others (1970) reported a zone of low resistivity of 20 ohm-meters, which is at nine kilometers in depth and is forty-seven kilometers thick. It is situated in central Texas of United State.

There is a variety of explanations in respect to low resistivity zone within the earth's crust. But it is usually considered that the formation of such a zone is connected with the dehydration of rocks or the partial melting of rocks. This question will be left for future study.

## DISCUSSION

1. From Figure 5 it can be seen that our data points have large dispersion, which may influence the uniqueness of the solution to some extent. This may be caused by the stability and temperature drift of our instruments.

2. For the data processing we adopt the two-dimensional tensor impedance method, which is based on two-dimensional medium. But as seen from the diagram of ellipticity variation with period, certain approximation can be derived by doing so. When the ellipticity is close to zero, the earth's medium approximates to two-dimensional, but when the ellipticity is close to unity, the latter approximates to three-dimensional.

3. Due to the low resolution and sensitivity of the instruments, we have the data only within the period of 10-2,000 seconds, lacking the data of the short period below 10 seconds and the long period above 2,000 seconds. This influences the precision of explanations to some extent, the data of short period influencing upper layers and the data of long period influencing lower layers.

4. Hz component is an important parameter for judging the strike of the structure. This component is not used because of the low resolution of the instruments. To such an region with obvious structure as Xizang, Hz component may be ignored, whereas it may be important to regions with complex structures.

## ACKNOWLEDGEMENTS

Several teachers and students of the Chengdu Geological College and Changchun Geological College and other members of our instrument group all took part in the fieldwork. Colleagues of Institute of Geology, State Seismological Bureau and Wuhan Geological College provided the original programs of computer. They were consulted in compiling the data processing and inversion program.

Also, our work enjoys warm encouragement and support from Professors Gu Gung-xu, Chen Zong-ji and Chu Gan-kun. Careful check and instruction were given to the paper. We hereby express our thanks.

## References


(1) Bostick F.X. Jr., Smith H.W., Investigation of large-scale inhomogeneities in the earth by the magnetotelluric method. Proc. IRE, 40, 2339-2346 (1962).

(2) Cantwell T., Madden T.R., Preliminary report on crustal magnetotelluric measurements. J. Geophys. RES., 65, 4202-4205 (1960).

(3) Laird C.E., Bostick F., One dimensional magnetotelluric inversion technigues. Technical Report No. 101, Electronics research centre, University of Texas, Austin. Texas, AD 719392 (1970).

(4) Sims W.E., Methods of magnetotelluric analysis, PhD dissertation, The University of Texas (1969).

(5) Trigg D.F., Serson P.H., Camfield P.A., Solid state electrical recording magnetometer. Publ. Earth Phys. Branch, 41 (5), Ottawa (1970)

(6) Wu F.T., The inverse problem of magnetotelluric sounding. Geophysics, 33, 972-979 (1968).


Table 1 Best Fit Model
Parameters for Nagqu

| Layer | Resistivity ($\Omega$ m) | Thickness (km) | Interface Depth (km) |
|---|---|---|---|
| 1 | $4.85 \times 10^{0}$ | 1.4 | 1.4 |
| 2 | $7.74 \times 10^{2}$ | 4.9 | 6.3 |
| 3 | $2.17 \times 10^{0}$ | 20.2 | 26.5 |
| 4 | $2.29 \times 10^{2}$ | 39.5 | 66.0 |
| 5 | $1.44\ 10^{0}$ | | |

Table 2 Best Fit Model
Parameters for Pagri

| Layer | Resistivity ($\Omega$ m) | Thickness (km) | Interface Depth (km) |
|---|---|---|---|
| 1 | $1.93 \times 10^{1}$ | 7.1 | 7.1 |
| 2 | $9.85 \times 10^{2}$ | 5.2 | 12.3 |
| 3 | $1.54 \times 10^{0}$ | 34.5 | 46.8 |
| 4 | $2.03 \times 10^{2}$ | 10.1 | 56.9 |
| 5 | $1.10 \times 10^{-1}$ | | |

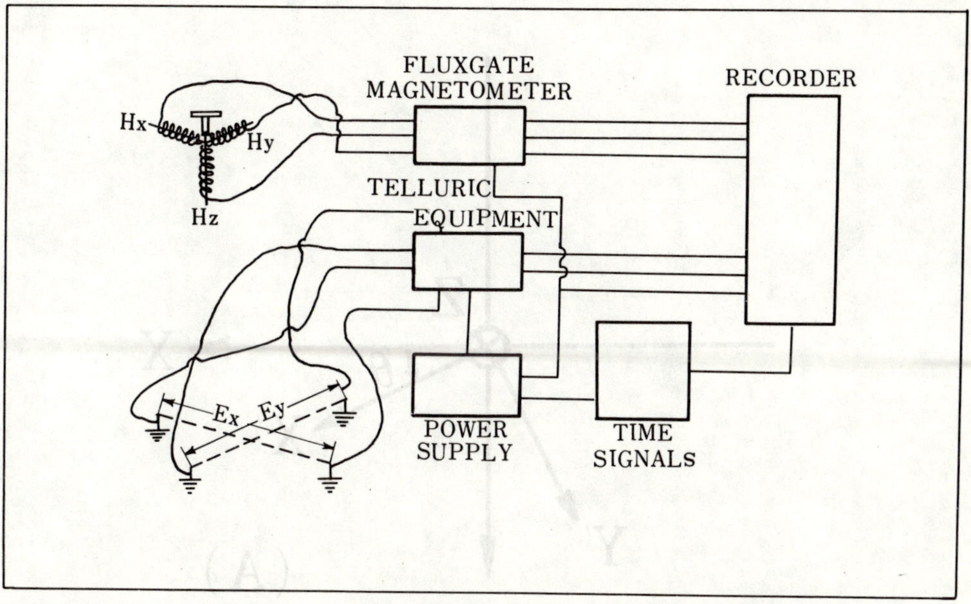


Fig. 1. Block diagram of the magnetotelluric measurement

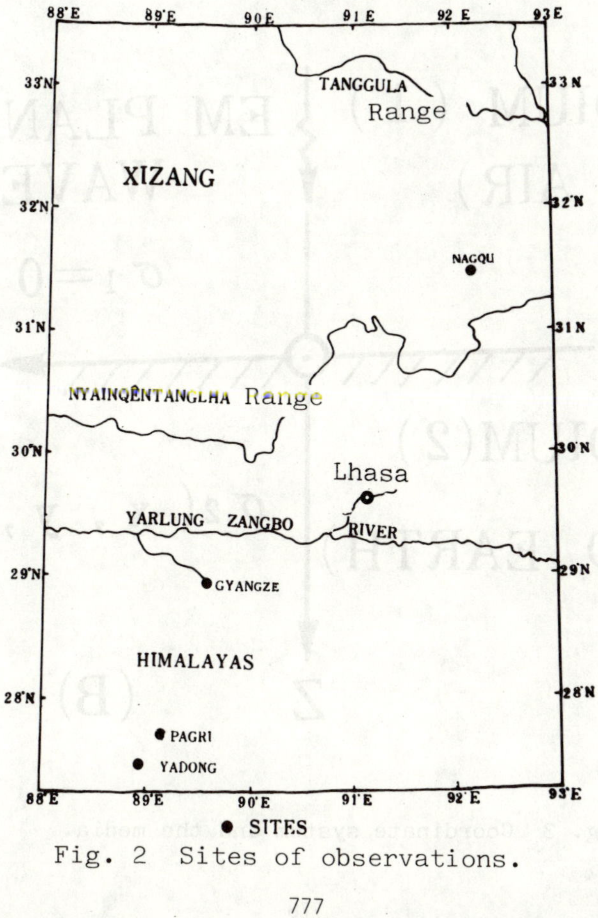


Fig. 2 Sites of observations.

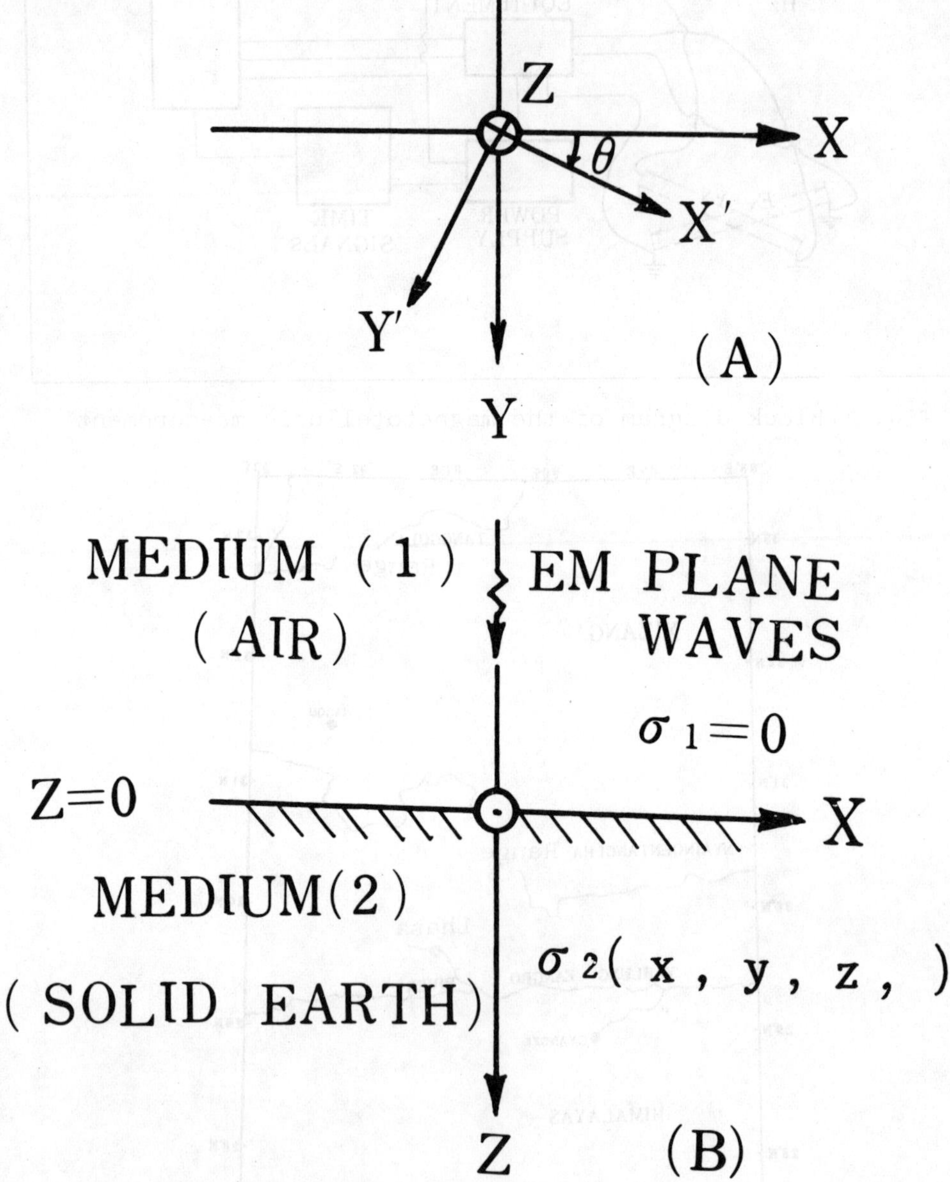


Fig. 3 Coordinate system and the media.

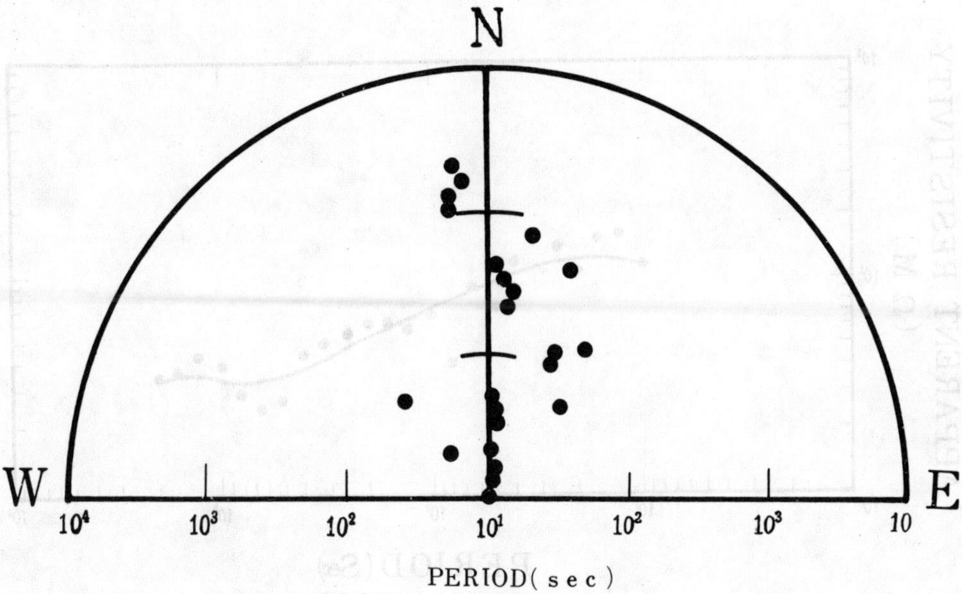


(a) NAGQU

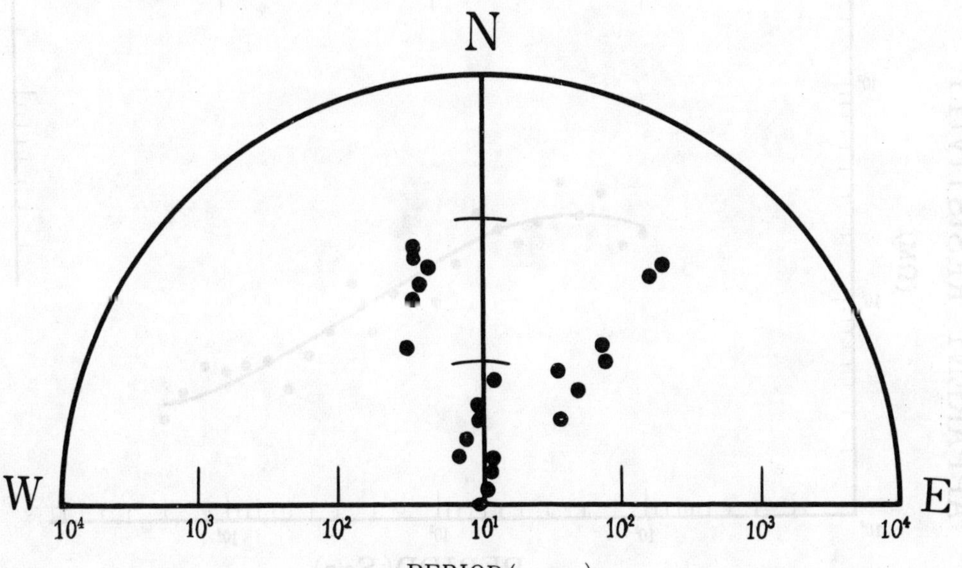


(b) PAGRI

Fig. 4 The variation of the azimuth of the major resistivity axis

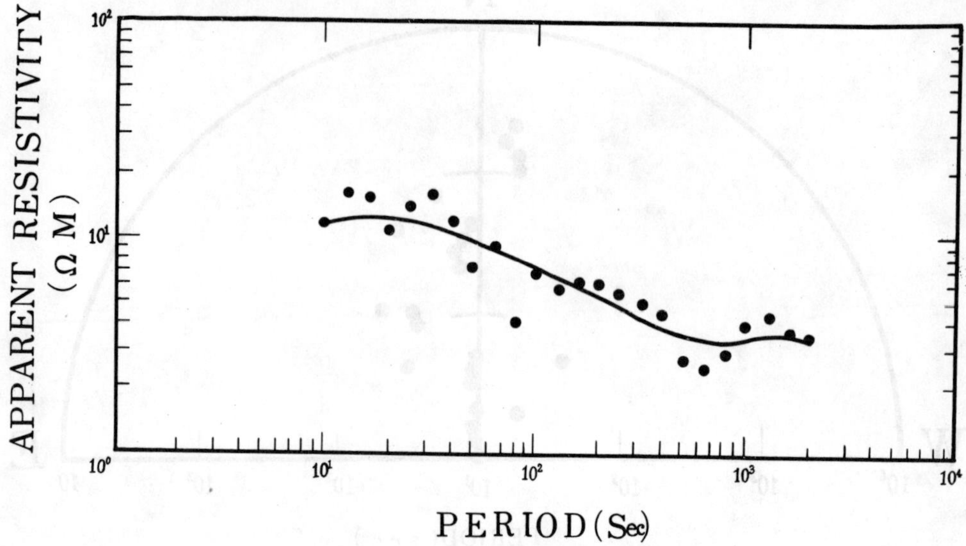


(a) NAGQU

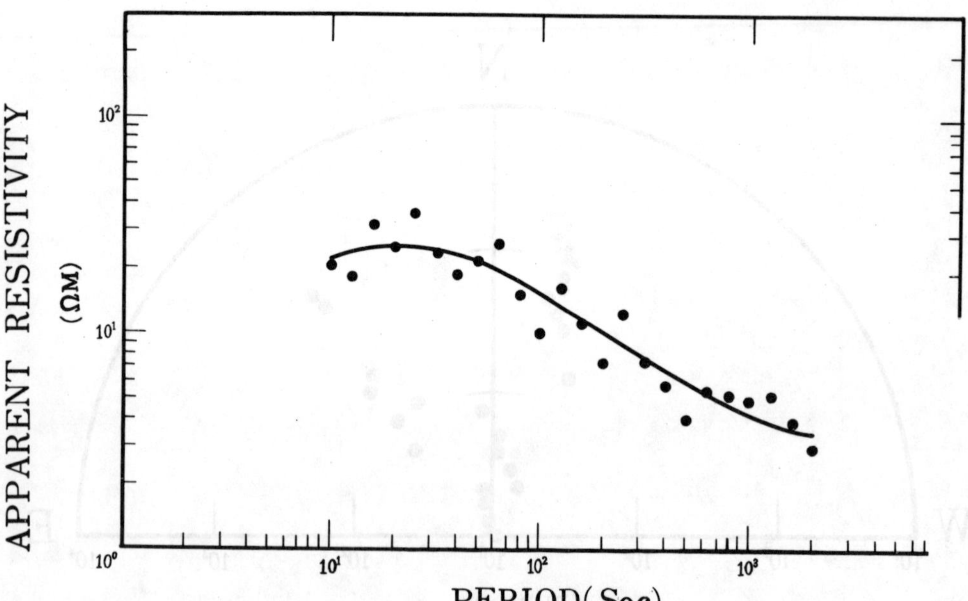


(b) PAGRI

Fig. 5 Data and model apparent resistivity.

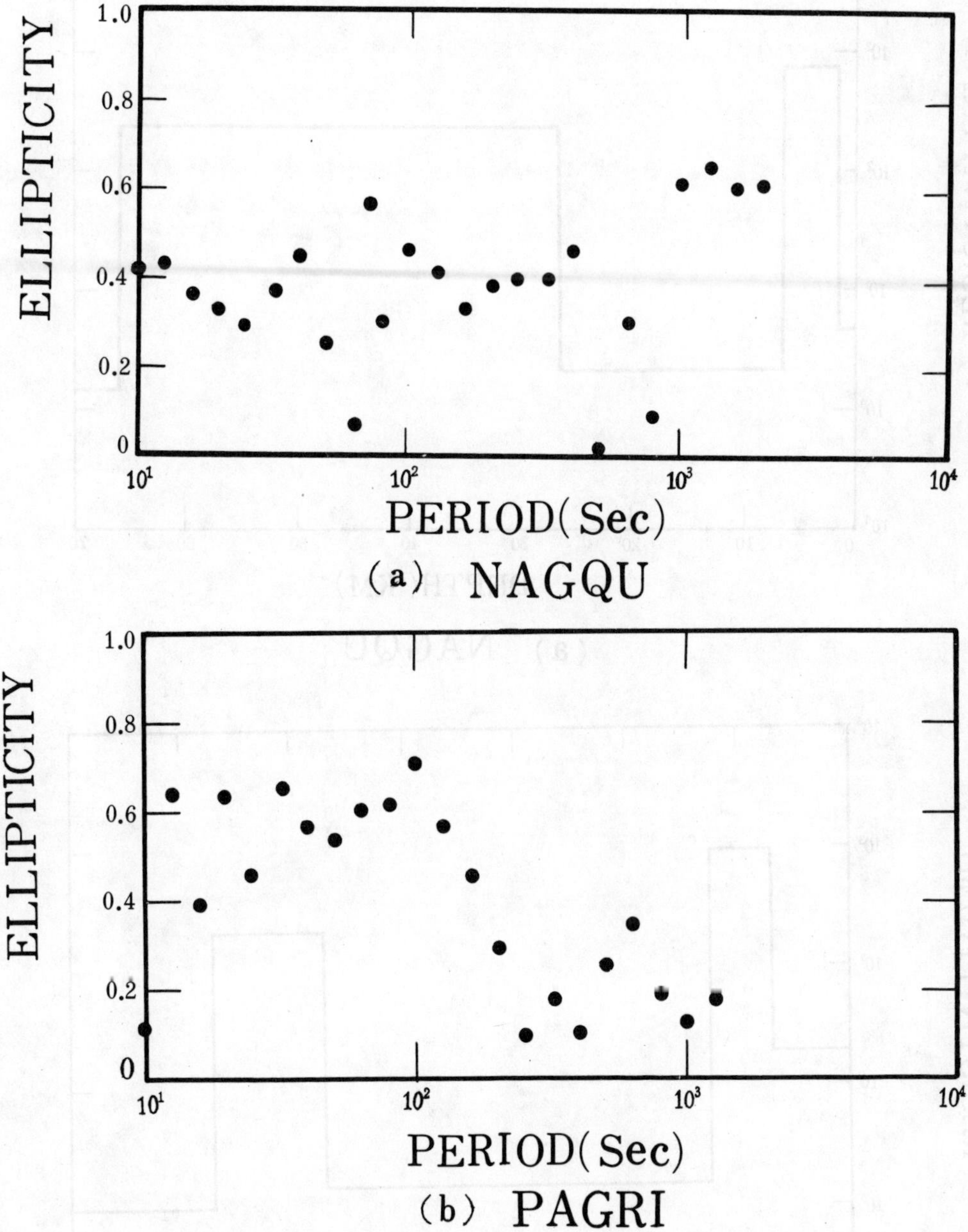


Fig. 6 The variation of the ellipticity index with period.

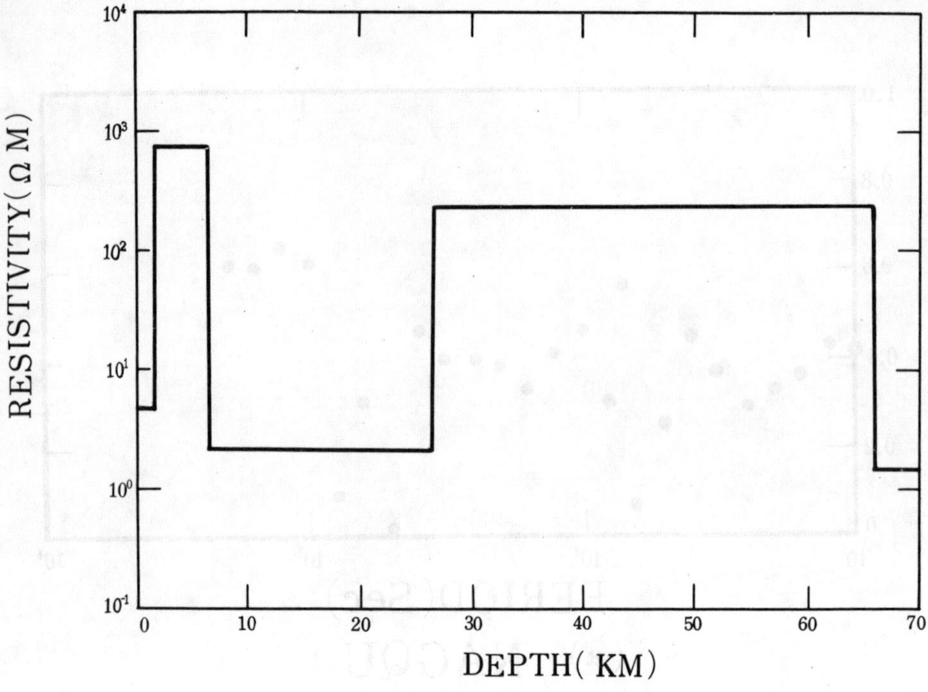


(a) NAGQU

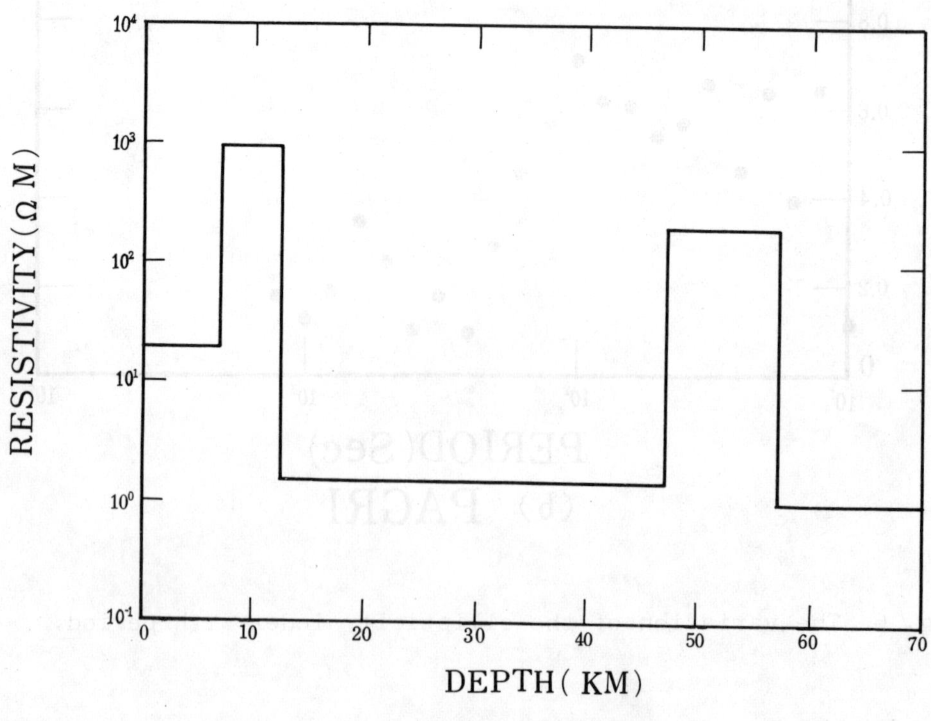


(b) PAGRI

Fig. 7 Best fit model resistivity profile.

# Q-VALUES FOR SURFACE WAVES IN THE EASTERN REGION OF THE XIZANG PLETEAU

Feng Rui He Zheng-qin
(Institute of Geophysics, State Seismological Bureau)

The quality factor Q is an important geophysical parameter for studying the earth's inelastic nature. Studies of Q variations are widely applied to describe the physical and chemical state of the media in the earth.

## METHOD AND DATA

For surface waves, the spectral amplitude at the epicentral distance can be expressed by the following formula:

$$A(\omega, \theta, \Delta) = C_o I(\omega)\ S(\omega)\ R(\theta)\ G(\Delta)\ e^{-\gamma\Delta}$$

where, $\theta$ is the azimuth of the seismic ray radiated from the hypocenter, $S(\omega)$ the hypocentral spectrum, $R(\theta)$ the azimuthal distribution of seismic wave radiation, $I(\omega)$ the frequency response of the instruments, $C_o$ the constant of spectral amplitude, $G(\Delta)$ the geometrical scattering factor of surface waves. If two stations are used, the formula can be simplified as follows:

$$Q(\omega) = \frac{\pi}{UT}(\Delta_2 - \Delta_1) \Big/ \ln\left(\frac{A_1\sqrt{\sin\Delta_1}}{A_2\sqrt{\sin\Delta_2}}\right)$$

$A_1$ or $A_2$ is the spectral amplitude of each station corrected by the instrumental response, and U is group velocity.

For this paper, seismic recordings were available from 5 Chinese Standard Stations at Chengdu, Lanzhou, Kunming, Gaotai and Lhasa (Fig. 1) The attenuation of Rayleigh waves between each of 4 stations and Lhasa was determined. The Q-values thus obtained are strictly a measure of the physical parameters in the Xizang Plateau. The Kirnos-seismometers are used. The earthquakes selected here are nearly on the great circle path between 2 stations (Table 1).

## RESULTS

The spectral amplitudes from 4 paths are shown in Fig. 2. The period range is from 7 to 51 seconds. 108 Q-values obtained for Rayleigh waves are shown in Fig. 3. The standard deviations of the average Q-values are generally small except for periods shorter than 20 seconds.

The following characteristics merit further attention:

1. The Xizang Plateau is a region with very low Q-values in the crust. The average value for 9 seconds is 376, and for 51 seconds it is 31.

These values are lower than those of normal continental regions which are generally in the range of a few hundreds (Fig. 4).

We can campare them with the Q-values in the North American Continent obtained by Tsai and Aki (1969). The Q-values are about 400-800. Bird and Toksöz (1977) estimated the upper limit of Q for Rayleigh waves in the Xizang Plateau by taking the difference at 20 seconds to be zero and using Xianggang (Hongkong) as the reference station. They obtained Q about 50 at 50 seconds, and up to 100 or more for the shorter period. Singh and Gupta (1979) determined Q for an earthquake in the middle of the Xizang Plateau by using WWSSN data. The Q-values obtained by them are still low. Although less than half of the paths of most of the rays are actually located within the Xizang Plateau.

2. The $Q_R$ versus frequency curve reveals that the maximum Q-value appeares at 20-30 seconds. This $Q_R$ decreased at shorter or longer periods, is related to the softness of medium in the upper crust and the low velocity layer in the upper mantle. However, $Q_R$ at these periods in the Xizang Plateau are still lower than those in the normal crust; Besides, there is a low velocity layer in the crust of the Xizang Plateau. From all these, it is reasonable to speculate that in the deep crust of the plateau there probably are partially molten materials which absorb the energy of elastic waves, thus causing the strong attenuation of seismic waves at such a great depth.

3. At the shorter period the deviation of Q-values is very significant. Q-values are from 200 to 600 which are normal values in the crust. Perhaps the structures of the upper crust along these 4 paths are very different; there are strong lateral reflections and refractions for the seismic waves.

In 1965 D.L. Anderson suggested the MM8-model which fits the Q distribution in the upper mantle; the theoretical Q-values are shown by the dotted line in Fig. 3. We can compare it with the Q-model by Tsai (1969) for fitting the Q-values in North American Continent. Evidently, the theoretical Q-values of the MM8 are much closer to those observed by us in the Xizang Plateau. But the following modification should be made for the MM8-model in order to make the theoretical values better fit the observations. That is: Q-value in the first layer (i.e. crust) should be smaller (about 100-400) and the thickness of the low-Q layer in the upper mantle should be thicker. The modified model perhaps will be better suited to the Q-model in the Xizang Plateau (Fig. 5).

## DISCUSSION

The characteristics of low Q-value and low velocity in the Xizang Plateau are caused by the peculiarity of deep structures in this region. The Xizang Plateau is not only thicker in the crust, but also significantly different in the vertical and lateral structures. The low velocity layer existing in the crust may be contributed to the high heat flux in the Xizang Plateau, which has caused the partially molten layers. Furthermore, the low density of rocks, soft medium and the intense upheaval movements of the plateau are important factors too.

## References

(1) Feng Rui et al., Crustal structure in China from surface waves. Acta Seismologica Sinica (in press) (in Chinese).
(2) Anderson D.L. et al., Attenuation of seismic energy in the upper mantle. J.G.R., 70 (6), 1441-1448 (1965).
(3) Bird P., Toksöz M.N., Strong attenuation of Rayleigh waves in Tibet Nature, 266, 161-163 (1977).
(4) Sigh D.D., Gupta H.K., Source mechanism and surface-wave attenuation studies for Tibet earthquake of July 14, 1973. Bull. Seism. Soc. Amer., 69 (3), 737-750 (1979).
(5) Tsai Y.B., Aki K., Simultaneous determination of the seismic moment and attenuation of seismic surface waves. Bull. Seism. Soc. Amer., 59 (1), 275-287 (1969).


Table 1 Earthquake parameters

| No. | Date | O(G.M.T.) | Lat. | Lon. | h | M |
|---|---|---|---|---|---|---|
| 1 | 1959.10.24 | $23^h40^m37^s$ | $41.6_N$ | $70.0_E$ | 0 | |
| 2 | 1960.10.13 | 14 52 34.7 | 54.8 | 161.2 | | |
| 3 | 1961. 5.16 | 21 45 25 | 30.4 | 132.0 | 14 | 5.7 |
| 4 | 1961.11.15 | 07 17 06 | 42.8 | 145.5 | 0 | 6.5 |
| 5 | 1973 6.15 | 11 20 51.5 | 53.5 | 161.5 | 33 | 6.4 |

Table 2 The average Q-values for Rayleigh waves

| T(sec) | f(Hz) | $Q_R$ | RMS | T(sec) | f(Hz) | $Q_R$ | RMS |
|---|---|---|---|---|---|---|---|
| 7.9 | 0.126 | 179 | 52 | 20.5 | 0.049 | 95 | 79 |
| 8.8 | 0.114 | 376 | 305 | 22.3 | 0.045 | 150 | 91 |
| 9.8 | 0.102 | 180 | 105 | 24.4 | 0.041 | 137 | 127 |
| 10.9 | 0.092 | 173 | 143 | 26.9 | 0.037 | 64 | 26 |
| 11.9 | 0.084 | 70 | 33 | 28.4 | 0.035 | 123 | 29 |
| 12.8 | 0.078 | 120 | 70 | 30.1 | 0.033 | 91 | 49 |
| 13.8 | 0.072 | 145 | 64 | 32.0 | 0.031 | 102 | 57 |
| 14.6 | 0.068 | 146 | 100 | 34.1 | 0.029 | 68 | 20 |
| 15.5 | 0.065 | 136 | 61 | 36.6 | 0.027 | 77 | 101 |
| 16.5 | 0.061 | 99 | 40 | 39.4 | 0.025 | 39 | 16 |
| 17.7 | 0.056 | 69 | 42 | 42.7 | 0.023 | 36 | 26 |
| 18.3 | 0.055 | 95 | 47 | 46.5 | 0.022 | 32 | 19 |
| 19.7 | 0.051 | 182 | 135 | 51.2 | 0.019 | 31 | 17 |

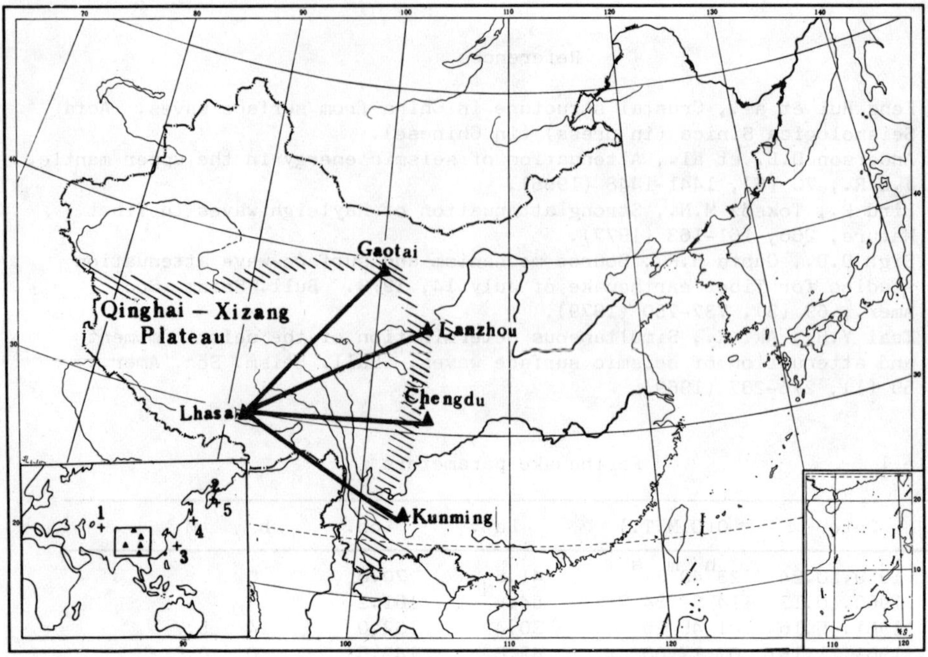


Fig. 1 The distribution of the stations and earthquakes.
▲ Station;
+ Epicenter.

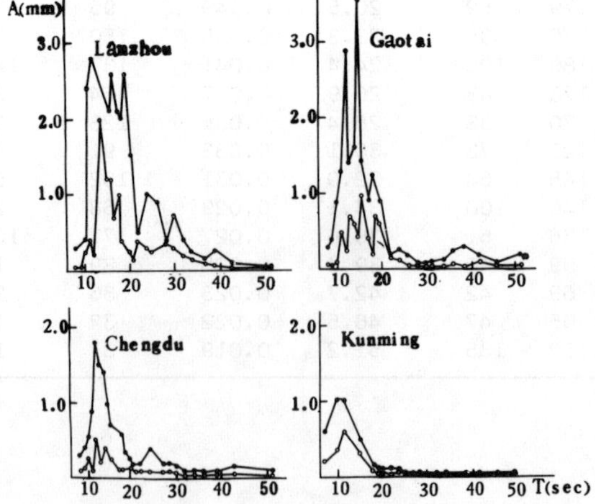


Fig. 2 Spectral amplitudes of every station (The circles are values of Lhasa station, dots are others).

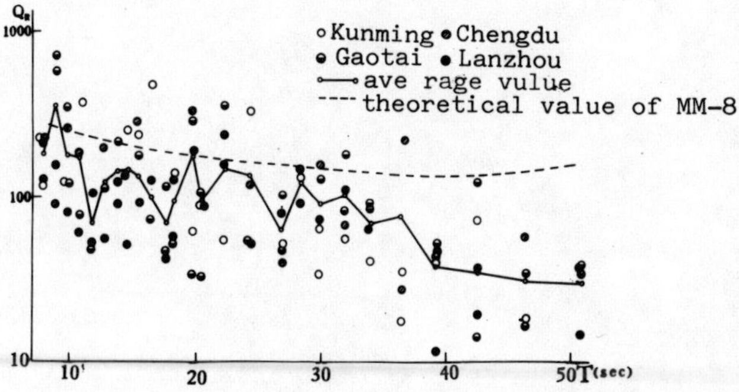


Fig. 3 Observational Q-values for Rayleigh waves in the eastern region of the Xizang Plateau.

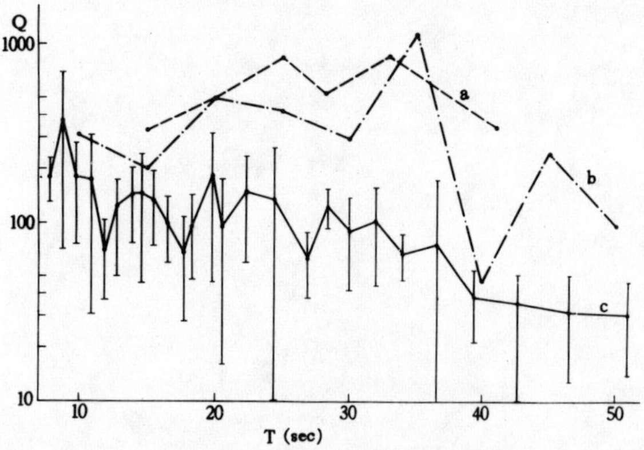

Fig. 4 Q-values for Rayleigh waves observed in different regions (a. North America(5); b. Results from a Xizang earthquake (4); c. The Xizang Plateau).

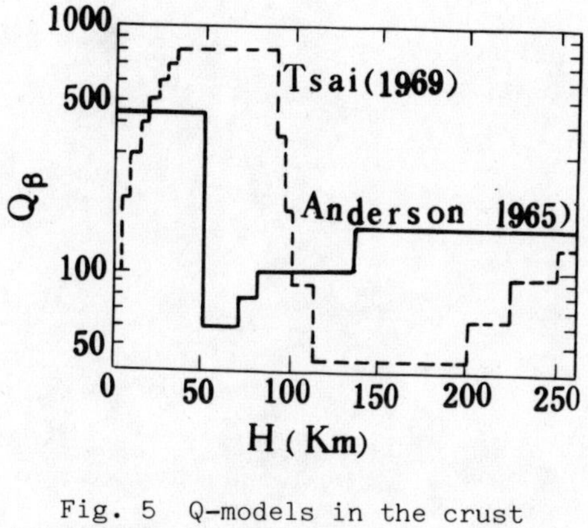


Fig. 5 Q-models in the crust and upper mantle.

# RECENT TECTONICS IN THE QINGHAI-XIZANG PLATEAU

Yan Jia-quan, Shi Zhen-liang, Wang Su-yun, Huan Wen-lin
(Institute of Geophysics, State Seismological Bureau)

## ABSTRACT

Based on the data of earthquake fault plane solutions, seismogeology, velocity structure of seismic wave in the crustal and upper mantle etc., we have investigated recent tectonic problems which are related to the origin of the plateau. In this paper, the data of fault plane solutions are given. We used data of the large earthquakes during 1950-1977, of which 135 events left good records. At the same time, we have also made use of the results obtained by both Chinese and foreign authors(3-7). All results are plotted in Fig. 1.

## THE CHARACTERISTICS OF TECTONIC DEFORMATION

According to the fault plane solution, seismogeological data and satellite imagery, we have compiled the map of recent tectonic deformation in this area (Fig. 2). The deformation characteristics can be classified into four categories:

1. The crustal shortening is represented by thrust faults and folded strata. The distribution of earthquakes with thrust faulting on the plateau is wide-spread and the Mesozoic and Cenozoic strata are widely distributed too, thus showing the extensive existence of crustal shortening over the whole plateau. This phenomenon is mostly obvious in the Himalayan folded zone. Tertiary and early Pleistocene strata are strongly folded and are accompanied with large-scale magma intrusion. The fault plane solutions of 15 earthquakes in the southern slope of Himalayas are all thrust or overthrust faultings, and they are consistent with the gigantic nappe structure in this area. The Qilian, Liupan and Longmen fold belts in the northeastern margin of the plateau are all thrust faulting, too. In the Qilian fold belt, it can be seen that Paleozoic metamorphic rocks overthrusted upon the Cretaceous or even upon the Quaternary "Jiu-quan" conglomerates in many places, and the younger strata have been strongly folded. The crustal shortening inside the plateau is not stronger than that along the edge of it. The perpendicular differential motion accompanied by the crustal shortening also has similar features, that is, the motion on the edge of the plateau is the strongest, but considerably stable inside the plateau.

2. Horizontal displacements represented by strike-slip faulting, mostly occur along the deep faults between the blocks in the plateau. The most important strike slip faults are as follows: Karakorum fault zone, Altun fault zone, Xianshui River-Anning River-Eastern Yunnan fault zone etc. Especially, the Altun fault zone with left lateral strike-slip faulting and the Longmen fault with right lateral strike-slip faulting, have demonstrated, that crustal material of the plateau is pushed forward in northeast direction. In the inner parts of the plateau, there also

exist a series of left lateral strike-slip fault zones which are nearly parallel and arcuated bending. It should be noted, that Jinsha River-Honghe River fault zone is right lateral strike-slip faulting, but in the east of the fault zone up to Anning River-Eastern Yunnan fault zone, most of the earthquakes are left lateral strike-slip faulting in nearly S-N orientation. A few of them are characterized by thrust faults with NE-SW direction. All of these show that while the crustal materials of this area are pushed south-eastward there exists crustal shortening.

3. Phenomena accompanying crustal tension. In the areas of shallow earthquakes by normal faulting there are many hot springs. At both sides of Yarlung Zangbo River there are also many famous hot springs, and depressed lakes as well. These have shown that at the same time of the crustal shortening in the plateau, the crustal tension also occurred.

4. Deformation in the upper mantle. The depths of intermediate earthquakes in the plateau are less than 300 km. They are mostly concentrated in the areas of Hindu Kush mountain and Naga mountain. There are 7 intermediate shocks that occur between $75^{\circ}$-$92^{\circ}$E (east longitude), of which three are with the normal faulting, three with strike-slip faulting, and only one is thrusting. The diversity of fault plane solutions of intermediate earthquakes shows that deformation process in the upper mantle at the plate contact zone is very complicated. Especially, the intermediate earthquake on September 14 1976 near Lhasa have been well determined for the two nodal planes of the fault plane solution, it is normal faulting with strike-slip faulting component. This shows tension also occurs in the upper mantle.

## THE CHARACTERISTICS OF TECTONIC STRESS

When the stress state in the interior of the plate is investigated, the direction of principal stress axis is very important. The statistical result obtained from equivalent stresses of the earthquake fault plane solution can be regarded as representative of regional tectonic stress field. Fig. 3 shows the horizontal projections of the principal compressive stress-axes of the fault plane solutions. The principal compressive stress axes are plotted for the earthquakes with strike-slip faulting and thrust faulting, but the principal tensional stress axes are plotted only for normal faulting.

In the region west of Zayü-Lanzhou line, that is from the Himalayas to Qilian Mt. the orientations of principal stress axes almost range from SN to NE. There are fault plane solutions of 49 shallow shocks characterized by strike-slip faulting or thrust faulting, the average location of the emergent points of the principal compressive stress axes is orientated NE $22^{\circ}$ with the dip angle of $7^{\circ}$; but the emergent points of principal tensional stress axes are around the equator with the pole at the above-mentioned average location. Therefore, in the wide region from the Himalayas to Qilian Mt. there exist rather consistently NNE horizontal compressive stress and tensional stress which are perpendicular to each other.

In the region east of the Zayü-Lanzhou line, the directions of the compressive stress axes varied regularly; i.e. from Liupan Mt. to Honghe River fault zone the compressive stress axes orientate gradually from NE to SE. But in the west region of the Honghe River fault zone the directions of the compressive stress axes are NE and are consistent with those of the northern region of the Himalayas.

## A BRIEF DESCRIPTION ON THE CRUST AND UPPER MANTLE

The regional feature of the recent tectonic deformation is not only related to regional stress field, but also to the structures of the crust and upper mantle. We have obtained P-wave velocity of 8.0-8.1 km/sec at the top of the upper mantle near Lhasa from natural earthquake data. An intermediate earthquake with Ms=5.3 and h=107 km occurred near Lhasa on Sept. 14 1976, the P and S-wave velocities near the focus are 8.1 km/sec and 4.58 km/sec respectively. We have also sorted out some results obtained by other Chinese and foreign authors from seismic body waves (8-13) , to plot the NE profile crossing Lhasa (Fig. 4). These data have shown that the crust and upper mantle of the plateau have the following features:

The crustal thickness is more than 50 km, the greatest thickness about 70 km; the seismic wave velocity in the upper part of the crust is slower than those in Indian Peninsula and Alxa block, and there also exists a low velocity layer in the lower part of the crust; the seismic wave velocity at the top of the mantle of the plateau is slower than that in the Indian Peninsula, but faster than that in island arc volcanic regions. Meanwhile, the intermediate earthquakes occur there, too, thus showing the material of the top upper mantle belongs to the lithosphere.

## DISCUSSION ON THE DYNAMIC MODEL

On the problem of tectonic dynamic model of the plateau, some authors have presented many interesting view-points. We think that the recent tectonics of the plateau is not only related to the collision of the Indian Plate, but also to the horizontal heterogeneities in the lithosphere. The process of tectonic deformation in the plateau can be described as the "forging die process" used in machinery building. We regard the Indian subcontinent as an active "convex die", while the passive "concave die" consists of the Tarim, Alxa, Ordos and Sichuan massives etc.; the lithosphere of the plateau serves as a "work piece" which is slightly heated (The slight heating indicates the plasticity of its surface and margin is higher than that of its inner part). Due to the collision of the Indian Plate, first, along the contact zone at which the Himalayas are located, strong tectonic deformations have occurred, secondly, the lithosphere of the plateau has shortened in the direction of the compressive stress axis, so that the relief heightened and the crust thickened, the strata folded and thrust faulting earthquakes occurred. At the same time, arcuated bends and lateral displacements of the crustal material of the plateau took place, and the pushing velocity eastward was faster in the south than in the north, forming a series of left lateral strike-slip faults and forcing the crustal material of the east region of Jinsha River -- Honghe River fault to be pushed southeastward from the breach of the "forging die" (Fig. 5).

Under the strong horizontal compressive stress, the crustal material in the plateau also has dilated perpendicularly to the compressive stress axis. Because of the dilatation, a large number of tensional cracks were formed in the lithosphere. This provides the channels and sites for the hydrothermal minerals to rise and concentrate, thus resulting in magma intrusions, low seismic wave velocity and high thermal flux in the crust and also causing seismic activity of normal faulting and lots of depressed lakes and thermal springs.

## References


(1) Yan Jia-quan, Shi Zhen-liang, Wang Su-yun, Huan Wen-lin, Some features of the recent tectonic stress field of China and environs. Acta Seismologica Sinica, 1 (1), 9-24 (1979) (in Chinese).

(2) Huan Wen-lin, Shi Zhen-liang, Wang Su-yun, Yan Jia-quan, Characteristics of the recent tectonic deformations of China and its vicinity. Acta Seismologica Sinica, 1 (2), 109-120 (1979) (in Chinese).

(3) Shirokova E.I., Detail study of stresses and fractures in the foci of the Middle Asia earthquakes, Fizika Zemli (11), 22-36 (1974) (in Russian).

(4) Chandra U., Seismicity earthquake mechanisms and tectonics along the Himalayan Mountain Range and vicinity. Phys.Earth Planet Inter. 16 (2), 109-131 (1978).

(5) Ichikawa M. et al., Focal mechanisms of earthquakes occurring in and around the Himalayan and Burmese Mountain Belts. Papers in Meteorology Geophysics, 23 (3), 149-162 (1972).

(6) Sykes L.R., Sbar M.L., Intraplate earthquakes, lithospheric stresses and the driving mechanism of plate tectonics. Nature, 245 (5404), 298-302 (1973).

(7) Yeh Hung et al., The Analysis of the recent tectonics stress of the Himalaya Mountain Arc and its vicinities. Scientia Geologica Sinica, (1) 32-48 (1975) (in Chinese).

(8) Teng Chi-wen et al., Deep reflected waves and the structure of the earth crust of the eastern part of Chaidam basin. Acta Geophysica Sinica, 17 (2), 122-135 (1974) (in Chinese).

(9) Zhang Cheng et al., The crustal thickness of Gansu Province and neighbour regions. North-western Seismological J., 1 (2), 22-26 (1979) (in Chinese).

(10) Tandon A.N., A study of the crustal structure beneath the Himalayas from body waves. Pure Appl. Geophysics, 111 (10), 2207-2215 (1973).

(11) Verma G.S., Structure of the foot-hills of the Himalayas. Pure Appl. Geophysics, 112 (1), 18-26 (1974)

(12) Kaila K.L., Crustal structure in the Himalayan foot-hills area of North India, from P-wave data of shallow earthquakes. Bull. Seism. Soci. Am., 58, 597-612 (1968).

(13) Teng Ji-wen et al., Explosion seismic study for velocity distribution and structure of the crust and upper mantle from Damxung to Yadong of Xizang Plateau, in this Proceedings of Symposium on Qinghai-Xizang (Tibet) Plateau. Academia Sinica (1981).

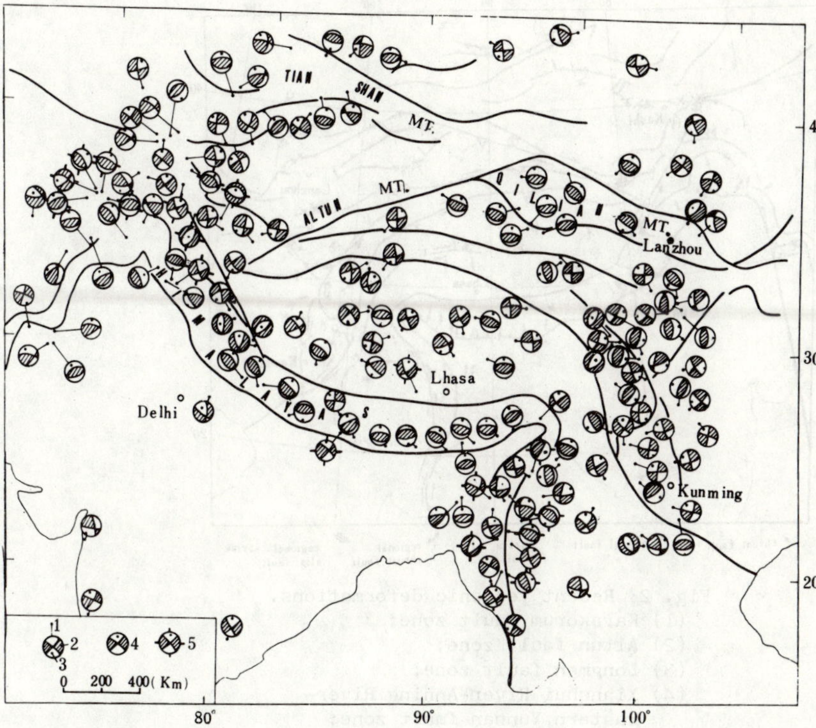


Fig. 1 Distribution of earthquake fault plane solutions.

1. Epicenter of earthquake;
2. First motion upward;
3. The emergent point of P-axis;
4. Shallow earthquake;
5. Intermediate earthquake.

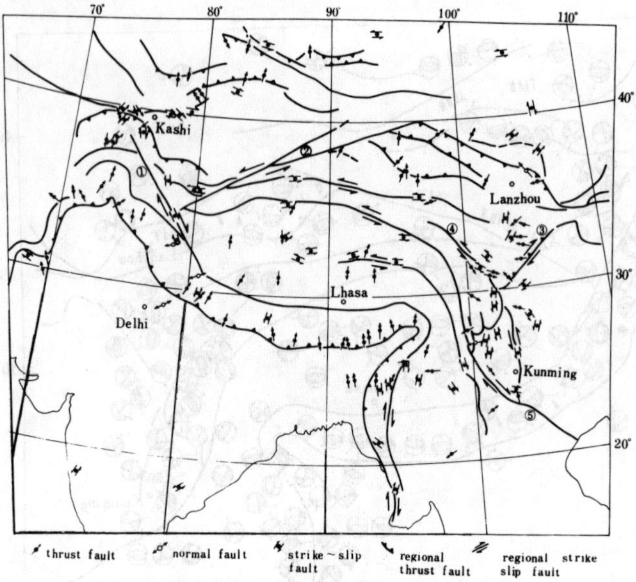


Fig. 2 Recent tectonic deformations.
(1) Karakorum fault zone;
(2) Altun fault zone;
(3) Longmen fault zone;
(4) Xianshui River-Anning River-Eastern Yunnan fault zone;
(5) Jinsha River-Honghe River fault zone.

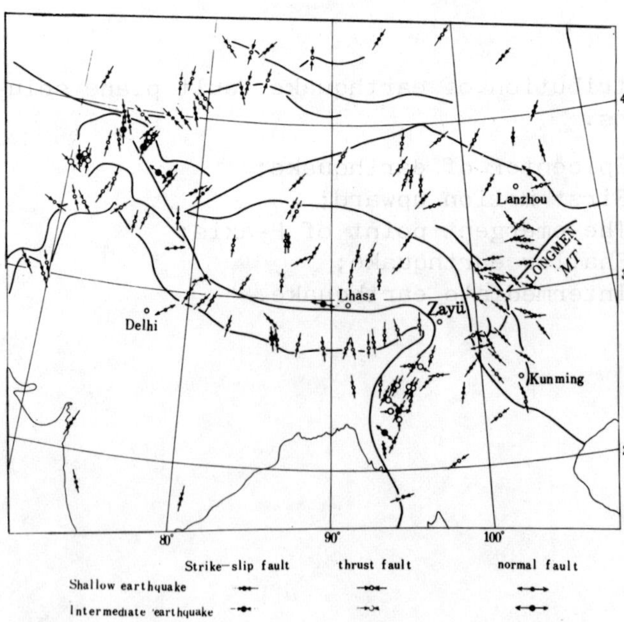


Fig. 3 Horizontal projections of principal compressive stress axes.

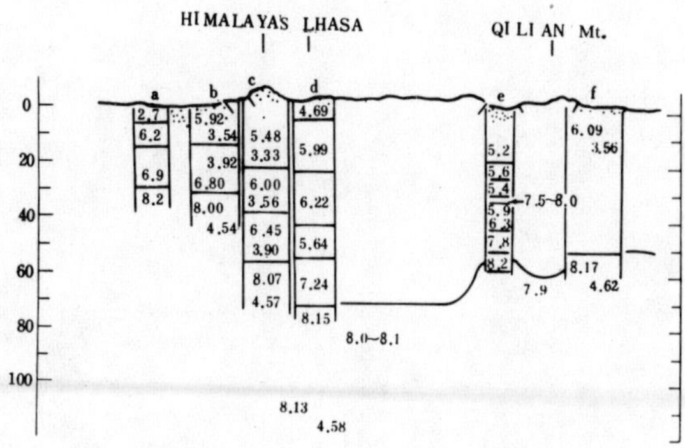


Fig. 4 Profile of the crustal structure references of the alphabets indicated in the figure are a(12); b(11); c(10); d(13); e(8); f(9) respectively.

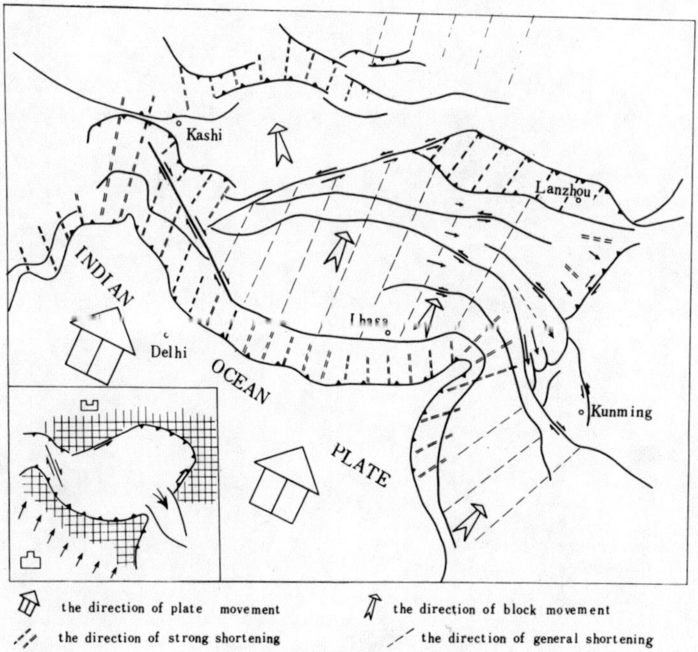


Fig. 5 Dynamic model.

Fig. 4 Profile of the crustal structure. The references of the alphabets indicated in the figure are a(12); b(11); c(10); d(5)(4); e(6); f(9) respectively.

the direction of plate movement — the direction of block movement

the direction of crustal thickening — the direction of crustal shortening

the vector of lateral movement

Fig. 5 Dynamic model.

# SOME CHARACTERISTICS OF SMALL EARTHQUAKES IN THE ZHAMO AND DAMXUNG DISTRICTS IN XIZANG PLATEAU

Yang Bing-ping, Shu Pei-yi, Zhang Zhi-he
(Institute of Geophysics, Academia Sinica)

## ABSTRACT

Regional seismological networks were set up in Zhamo and Damxung districts in 1976 and 1977. On the basis of the data thus obtained, the spacio-temporal parameters of the small earthquakes which occurred there were obtained by use of Wadati method. A disc-shaped dislocation model was adopted for the focus of a small earthquake. After considering the effects of absorption in the transmitting medium and the frequency characteristics of seismographs, focal parameters of the earthquakes in the time domain were determined.

The results are as follows:

1. Small earthquakes are characterized by intense local concentration and shallow focal depths. The focal planes of the small earthquakes dip towards the south in Damxung region.
2. The mean stress drop of the earthquakes ranges from a few bars to tens of bars.
3. The $Q_p$ values for longitudinal waves in that region are relatively low, with an average value of about 150.
4. At the depth of about 20 km in the upper crust, the average velocity of longitudinal waves is relatively low, with an average $\overline{V}p$ value of 5.65 km/sec.

## INTRODUCTION

It is well known that Xizang Plateau is both strongly active in recent geological tectonics and also in seismic activity. Studying the crustal structure and geophysical field in the plateau is of great importance in understanding the origin of the uplift of the plateau and the continental plate tectonics. In recent years and especially from 1973-1977, large-scale multidiciplinary scientific investigations have been carried out. Local seismic network were set up in Zhamo and Damxung areas of Xizang Plateau, from 1976-1977. This paper gives the preliminary analysis of the data of small earthquakes obtained in both areas.

## THE TIME-SPACE PARAMETERS OF SMALL EARTHQUAKES

The networks were equipped with short-period seismographs with amplification of about one hundred thousand times. The earth medium can be considered as homogeneous within a not very large area which is under the control of the local network. We have located 98 small earthquakes with magnitudes $M_L$ ranging from 1.5 to 4.5 by use of Wadati method. Also, the depths of 38 shocks were determined.

The distributions of small earthquake epicenters are shown in Figures 1 and 2.

From Fig. 2, it is seen that the epicenters of small earthquakes in Damxung area, were concentrated on the west of Yangbajain and on the south of Zhigung. Yangbajain is located at the crossing of a NE-SW striking fault and a nearly EW striking fault, and Zhigung is at the east end of the latter.

Figs. 3 and 4 are projections of the hypercenters of small earthquakes on north-south vertical cross-sections in Zhamo and Damxung respectively. In Fig. 3, the section for Zhamo goes just along E 95° longitude with a length of about 160 km, while in Fig. 4, the section for Damxung goes along E 91° longitude with a length of about 200 km.

As shown in Figs. 3 and 4, the focal depths of small earthquakes are not so deep, most of them are within 30 km. In the northern side of the Yarlung Zangbo River, the focal plane dips steeply from north to south in Damxung area.

## THE FOCAL PARAMETERS OF SMALL EARTHQUAKES

After taking into account effects of absorption in the transmitting medium and the seismograph's frequency response(1-2), we have analyzed the observed half-periods of the first motions of the P waves and calculated the source parameters (namely, source radius, seismic moments, average dislocations and stress drops) and the Q-values of the medium by using Brune's formula, and a disc— shaped dislocation model being assumed.

### The Relations between the Seismic Moments 2a and Magnitude $M_L$

Many people have studied great shallow earthquakes, not only from the microscopic seismic data, but also from the macroscopic seismic data, while for the small ones, no macroscopic seismic data seem to be available. The source dimensions are obtained from the analysis of recording data. Fig. 5 gives the relations between the source dimensions and magnitudes of the small earthquakes.

It can be seen from Fig. 5 that a logarithmic linear relation exists between the source dimensions and magnitudes of the small earthquakes, but the data points are rather scattered. This shows the relation between the two can not be fully described by a simple formula. Here a rough linear relation is assumed and can be expressed as

$$\lg(2a)=0.5M_L-1.5$$

where a is radius of the disc-shaped shear dislocation source in km. The result obtained here is in agreement with the statistical results of great earthquakes by previous investigators.

### The Relations between the Average Dislocations $\bar{D}$ and the Magnitudes

The relations between the two is shown in Fig. 6, which can be expressed as

$$\lg\bar{D}=0.72\ M_L-2.1$$

where the unit of $\bar{D}$ is in centimeter.

## The Relations between the Seismic Moments $M_0$ and the Magnitudes

As shown in Fig. 7, there is a linear relation between the logarithms of seismic moments and magnitudes. In Fig. 8, there are additional data obtained by other authors besides the data analyzed in this article(2-7). The statistical relation is given as follows

$$l_g m_0 = 1.2M + 17.5$$

where the unit of $m_0$ is in dyne•centimeter.

## The Distribution of Stress Drop of Small Earthquakes.

Figs. 9 and 10 show the stress drop distributions of small earthquakes. In Fig. 9 it is seen that the stress drop of small earthquakes are 10-30 bars in Zhamo area, and in Fig. 10, the stress drop is 5-15 bars in Damxung area.

## THE VELOCITY OF P WAVE OF SMALL EARTHQUAKES

According to the time-distance curves of small earthquakes observed from Zhamo and Damxung areas, the average velocity of seismic P waves in the upper part of the crust in both areas has been determined. The $\overline{V}_p$ velocity is relatively low, about 5.65 km/sec as shown in Fig. 11.

The group velocity value of short period Rayleigh waves in Xizang Plateau is much lower than that in other regions, as determined by some investigators using the seismic surface waves data. And also the velocity of seismic shear waves passing through the plateau is rather low too. Furthermore, the seismic phase $L_g$ in the seismograms is quite weak and characterized by the lack of high frequency components. Besides the average velocity of 5.65 km/sec of P waves in the upper part of the crust as obtained from the small earthquake data, the quality factor of the regional transmitting medium has been calculated for the P waves. The average $Q_p$ is about 150.

## CONCLUSION

From the data in both areas, we see that the small earthquakes are of shallow depths. In Damxung, there is a local concentration of shocks and the focal depth increases from north to south on the northern side of the Yarlung Zangbo River. Low values of $\overline{V}_p$ and medium values of $Q_p$ in both areas seem to indicate that the crust there is not quite rigid and probably somewhat shattered causing considerable energy dissipation.

## References

(1) Chen Pei-shan et al., The stress field of the Beijing-Tientsin-Tangshan-Zhangjiakou area before and after the Tangshan earthquake of July 28, 1976. Acta Geophysica Sinica, 21 (1), 57-65 (1978) (in Chinese).

(2) Chen Yun-tai et al., The determination of source parameters for small earthquakes in Qiaojia and Shimian and the estimation of potential earthquake danger. Acta Geophysica Sinica, 19 (3), 206-233 (1976) (in Chinese).

(3) Chen Pei-shan et al., A study of the earthquake faulting process and earthquake prediction in the light of fracture mechanics. Acta Geophysica Sinica, 20 (3), 185-202 (1977) (in Chinese).
(4) Zheng Zhi-zhen et al., The focal parameters of microearthquakes in the Zhangjiakou-Huairou region. Acta Geophysica Sinica, 19 (3), 234-238 (1976) (in Chinese).
(5) Aki K., Earthquake mechanism, Tectonophysics, 13 (1-4), 423-446 (1972).
(6) Eaton J.P. et al., Aftershocks of the 1966 Parkfield-Cholame, California, earthquake: A detailed study, Bull. Seism. Soc. Amer., 60 (4), 1151-1197 (1970).
(7) Aki K., Analysis of the seismic coda of local earthquakes as scattered wave. J.G.R., 74 (2), 615-631 (1969).

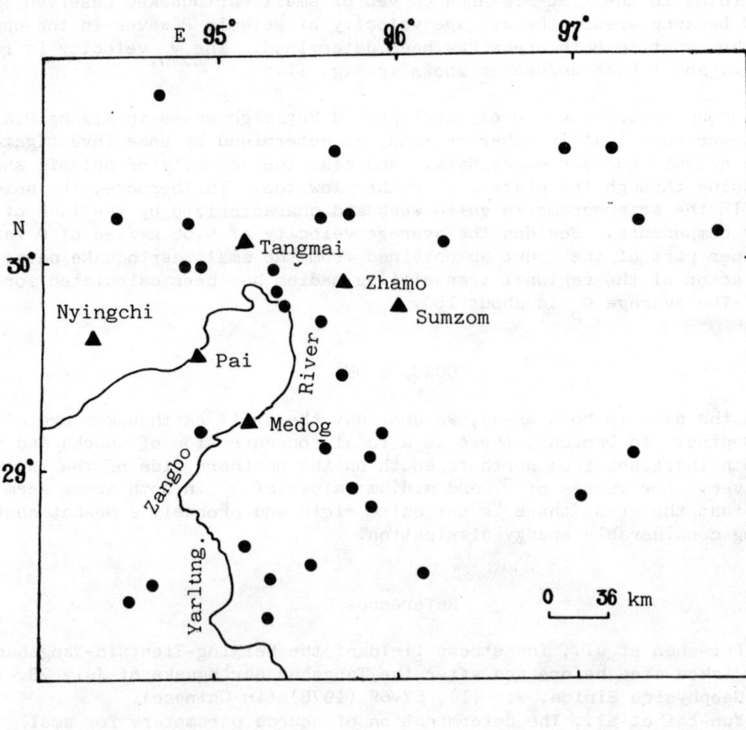


Fig. 1 The distribution of epicenters of small earthquakes in Zhamo area.
▲ The seismic station.

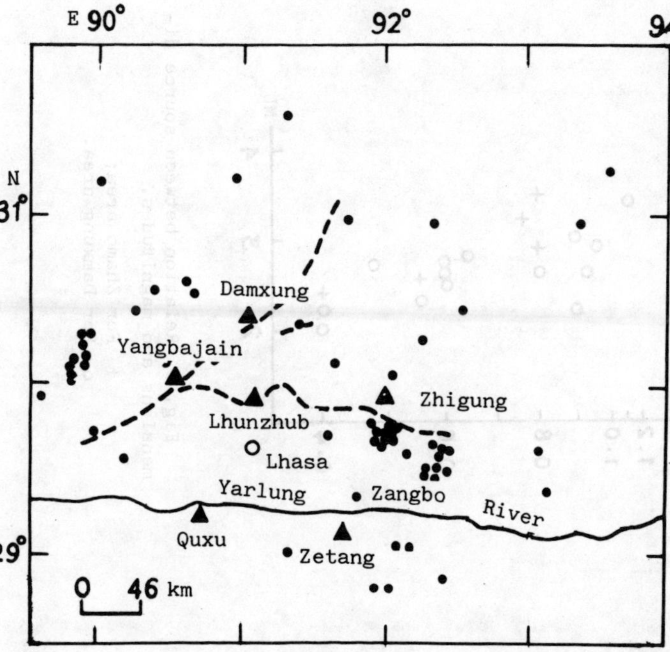


Fig. 2 The distribution of epicenters of small earthquakes in Damxung area.
---- The fault.

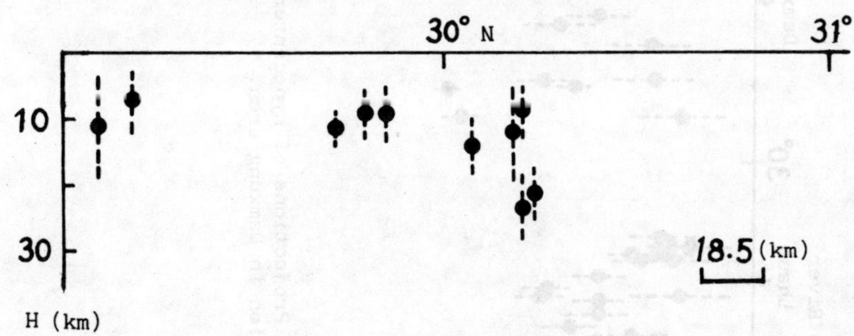


Fig. 3 Projections of hypocenters on a NS vertical cross-section in Zhamo area. The dashed lines indicate the precision of the focal depths.

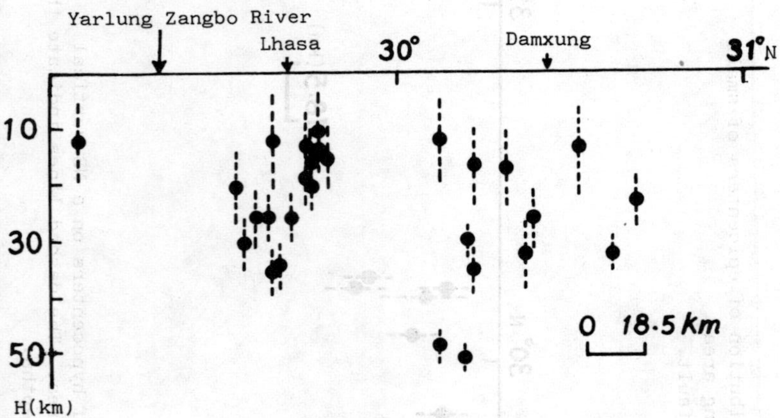


Fig. 4 Projections of hypocenters on a NS vertical cross-section in Damxung area.

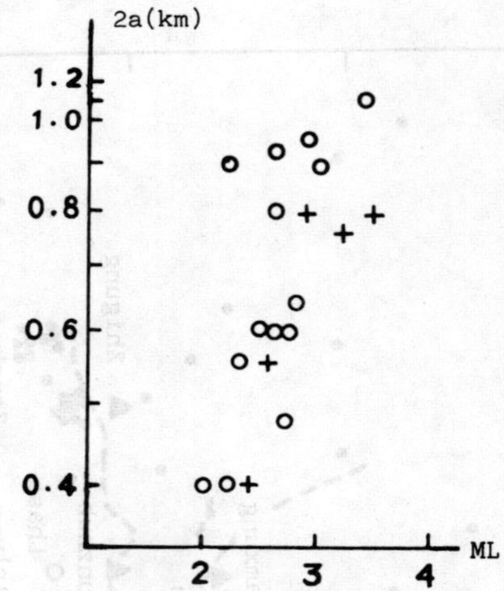


Fig. 5 Relation between source dimensions and magnitudes.

+ For Zhamo area;
o For Damxung area.

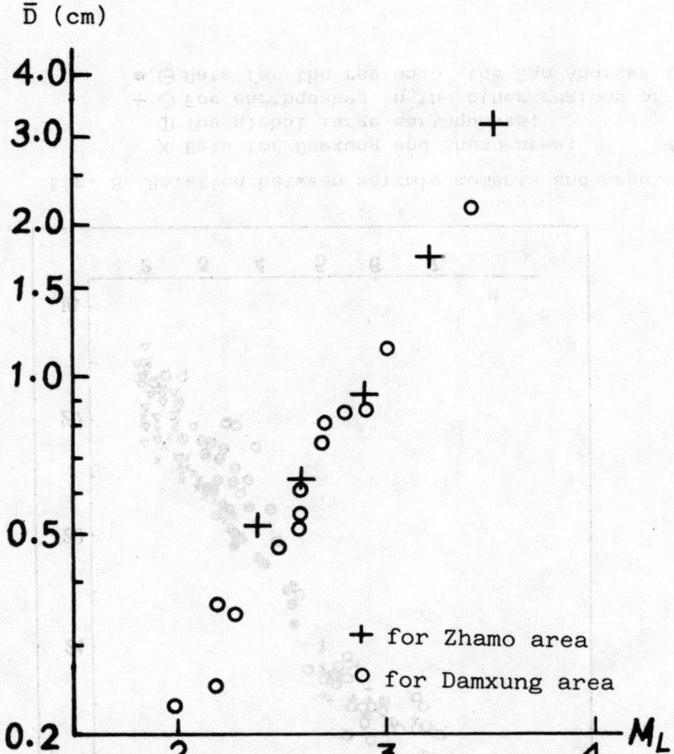


Fig. 6 The relation between the average dislocations and the magnitudes.

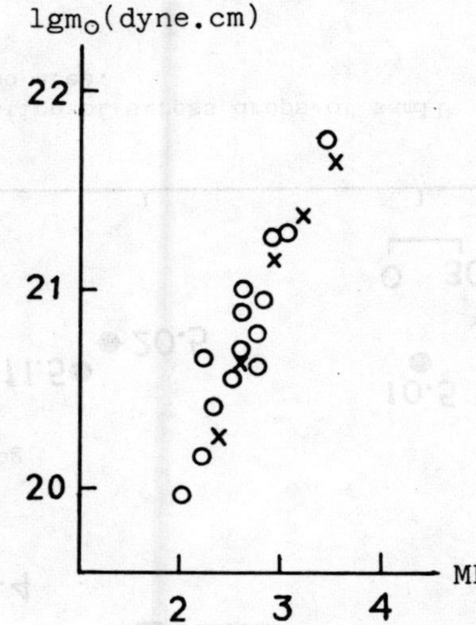


Fig. 7 Relation between seismic moments and magnitudes.

× For Zhamo area;
o For Damxung area.

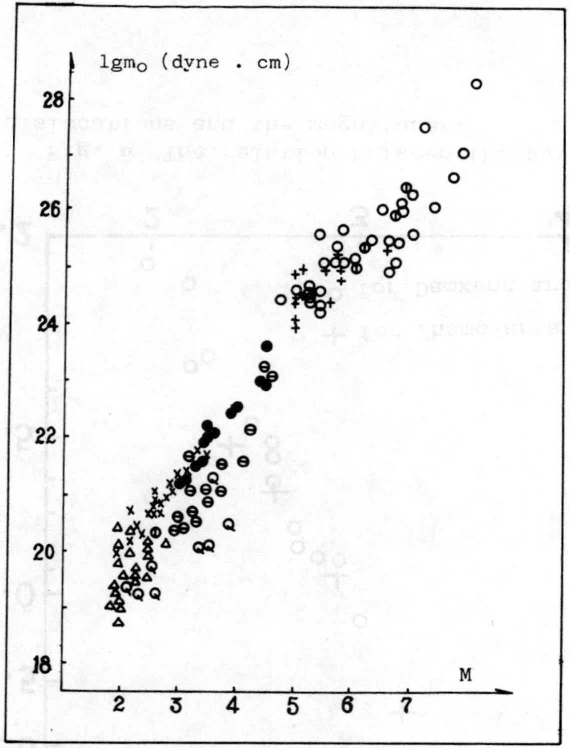


Fig. 8 Relation between seismic moments and magnitudes.

× Data for Damxung and Zhamo area; ○ Data for Shimian and Qiaojia area;
⦶ The global large earthquakes; Data for Zhangjiakou-Huairou area;
+ ○ For earthquakes in the other regions of China;
● ⊖ Data for the region of the San Andreas fault.

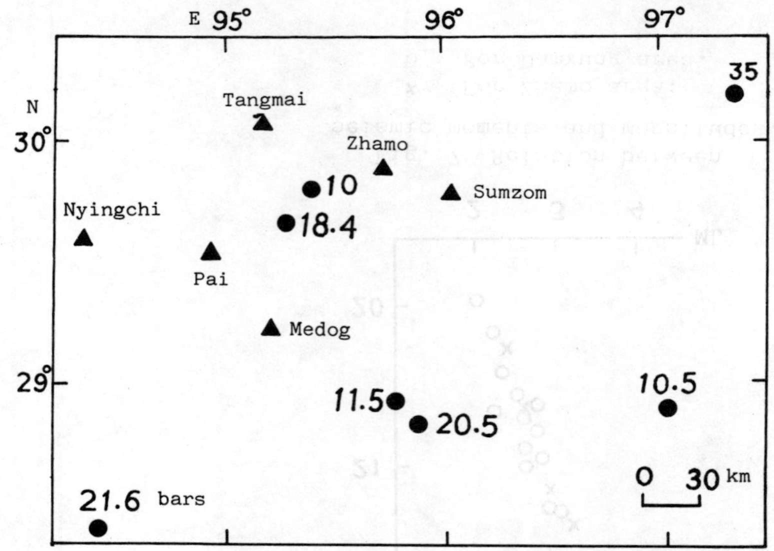


Fig. 9 The distribution of stress drops of samll earthquakes in the Zhamo area.

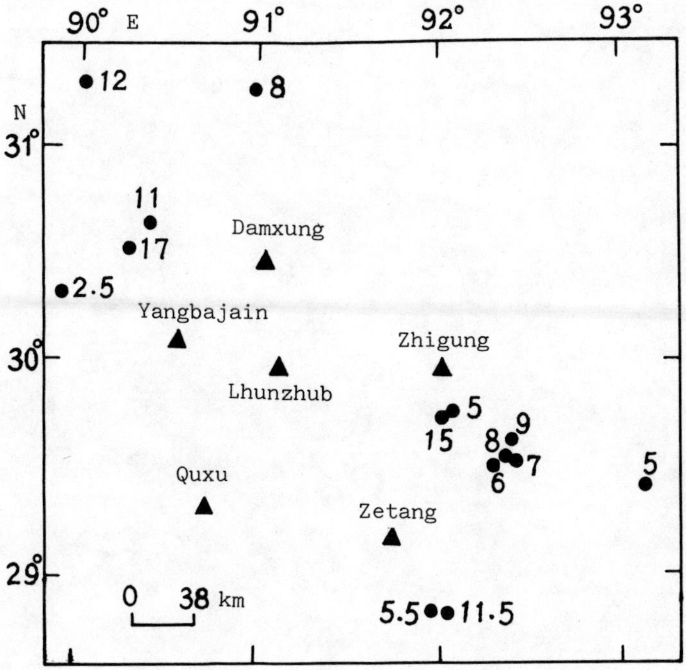


Fig. 10 The distribution of stress drops of small earthquakes in the Damxung area.

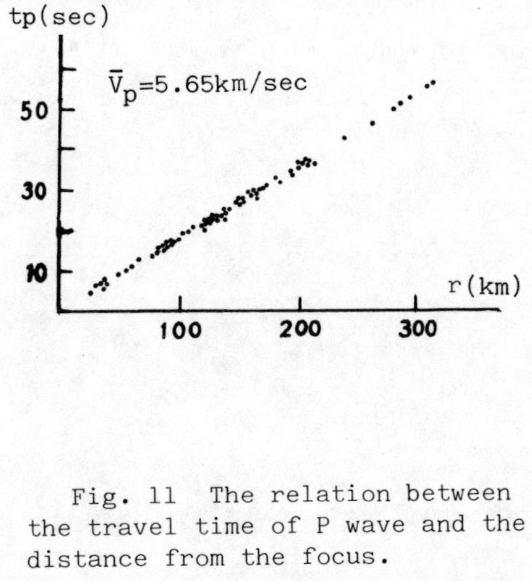


Fig. 11 The relation between the travel time of P wave and the distance from the focus.

# DISCUSSION ON FORECASTING GREAT EARTHQUAKES (M≥8) IN THE QINGHAI-XIZANG PLATEAU AND ITS VICINITY

Qin Bao-yan Kuo Zeng-jian Liu Guang-yuan
(Lanzhou Institute of Seismology, State Seismological Bureau)

Recent tectonic movement in the Qinghai-Xizang Plateau has been very intensive. Since the beginning of this century six great earthquakes (M≥8) in the Qinghai-Xizang region have occurred. In this paper we attempt to discuss some seismic characteristics and related phenomena before the occurrence of a great earthquake with a view of finding some clues or possible precursors of a forthcoming major event.

## SPATIAL DISTRIBUTION OF MAJOR EARTHQUAKES (M≥8) IN THE QINGHAI-XIZANG REGION

According to the catalog of great earthquakes, five epicenters among the above-mentioned six great earthquakes were located along the plateau's boundary. These events are the Kashmir earthquake (M=8) on April 4, 1905, the Haiyuan earthquake (M=8.5) on December 16, 1920, the Gulang earthquake (M=8) on May 23, 1927, the India-Nepal border earthquake (M=8.3) on January 15, 1934 and the Zayu earthquake (M=8.5) on August 15, 1950. Among them only one event occurred in the plateau's interior, namely, the Damxung earthquake (M=8) on November 18, 1951. The spatial distribution of these earthquakes is shown in Figure 1. A probable cause of the stronger seismic activities along the boundary than those in the plateau's interior, is probably the superposition of strong differential vertical movement on the horizontal movement along the plateau's boundary.

## SEISMIC GAPS BEFORE GREAT EARTHQUAKES (M≥8)

Of the six great earthquakes mentioned above, four of them occurred in Chinese territory and the other two were beyond the borders of our country. Except for the 1905 Kashmir earthquake (M=8) which occurred at the beginning of this century and lacked historical accounts, the other five great earthquakes all appeared to have their own seismic gaps before their occurrence. They are shown in Figures 2-6. These earthquakes (M=5-7) around the gaps are taken from Chinese data(1-2),while the others were referred to foreign data(3-4). Since there are abundant historical accounts of the Haiyuan and Gulang regions, it is possible to obtain the durations of the seismic gaps before Haiyuan and Gulang earthquakes to be 282 years and 262 years respectively. According to the durations and areas of the seismic gaps, we can roughly estimate the magnitude of the forthcoming great earthquakes.

## OCCURRENCE OF EARTHQUAKES ($6 \le M < 7$) IN THE BORDER REGION OF QINGHAI, XINJIANG AND XIZANG 1 OR 2 YEARS BEFORE A GREAT EARTHQUAKE

There were modern and Quaternary volcanic activities in this border region. For example, there were eruptions of Amapaxia, Qiemo and Bamaogiongzong volcanoes and Amapaxia volcano erupted in 1951(5). According to historical accounts, these volcanoes erupted more than once. In this volcanic region and its vicinity some smaller earthquakes ($6 \le M < 7$) have occurred since the beginning of the century. Comparing with the great earthquakes in the Qinghai-Xizang region, the earthquakes in this region were usually of M6-7 which occurred 1 or 2 years before the occurrence of the great earthquake. For instance in this region there were three earthquakes (M=6¼) which occurred 2-7 months before the Haiyuan earthquake (Fig. 7); two earthquakes (M=6, 6¾) in this border region occurred 9-11 months before the Gulang earthquake (Fig. 8). And there were two earthquakes (M=6¼) in this border region, occurred 2 years and 3 months before the Zayu earthquake and the earthquake at the Indian-Nepal border respectively (Figs. 9, 10).

The relation between the above mentioned seismicity in volcanic regions and the great earthquakes along the plateau's boundary is possibly due to the fact that they were both controlled to a certain extent by the unified movements of the upper mantle beneath the plateau. Due to the weakness of the rock strength in the volcanic region, the occurrence of smaller earthquakes in this region preceded the major ones along the plateau's boundary.

It must be pointed out that in the whole of the Qinghai-Xizang Plateau there is no M6 shot in the time interval from the occurrences of M6-7 earthquakes in the volcanic region (the region surrounded by the dotted line in Figs. 7-10) to occurrences of great earthquakes (except the Zayu earthquake).

In the plateau, the occurrence of smaller earthquakes ($6 \le M < 7$) had no obvious correlation with the occurrence of great earthquakes. We noticed that some M 6 shots in the volcanic region were not followed by great earthquakes ($M \ge 8$), instead, they were followed by elevated seismic activity consisting of several earthquakes with magnitudes greater than 7.

## 1 TO 3 YEARS BEFORE THE GREAT EARTHQUAKES ($M \ge 8$) THERE USUALLY OCCURRED M6 EARTHQUAKES ALONG AN EXTENSION OF THE SEISMOGENETIC STRUCTURAL TREND

Along an extension of the seismogenetic structural trend 1 to 3 years before great earthquakes, there usually occurred M 6 earthquakes along the plateau's boundary. For example, there was an earthquake (M=6) in 1917, 3 years before the Haiyuan earthquake, which was located in the WNW extension of the seismogenetic structural trend of the Haiyuan great earthquake, namely, the northwestern part of Qilian Mountain, the border region with Altun Mountain. Also another earthquake (M=6) occurred in Qilian Mountains near Har Lake along the WNW extension of the seismogenetic structural trend 2 months before the Gulang earthquake of 1927. Before the Indian-Nepali border earthquake an M~6 earthquake occurred, whose location agreed with the extension of the seismogenetic structural trend of the earthquake in 1934. The Zayu great earthquake occurred in the southern arc of Qinghai-Xizang Plateau. About half a year before the great earthquake, there was an earthquake (M=6) in its northeast. All these M 6 earthquakes occurred before the great earthquakes located along the plateau's boundary as shown in Fig. 11.

## THE APPARENT CORRELATION BETWEEN THE GREAT EARTHQUAKES IN THE QINGHAI-XIZANG PLATEAU AND THE EARTHQUAKE ACTIVITY ALONG THE WESTERN BOUNDARY OF THE PACIFIC PLATE

According to the point of view of some seismologists(6,7),there is a correlation among global earthquakes (M≥8). We found that the great earthquakes (M≥7.7) in the Qinghai-Xizang Plateau respond to the activity of great earthquakes (M≥8) along the western boundary of the Pacific Plate, and in most cases, the latter were preceded by the former as shown in Figs. 12, 13. For comparison, we give some other great earthquakes (M≥8) which occurred in other regions of China and Mongolia in Figs. 12, 13. From the viewpoint of prediction, the correlation between great earthquakes of two regions should merit some attention. However, in this paper we do not attempt to make any hypothesis as to the cause of this correlation because of the scantiness of data.

## CORRELATION OF SOME PHENOMENA WITH THE GREAT EARTHQUAKES IN THE QINGHAI-XIZANG PLATEAU

It is interesting to note that the frequency of global volcanic activity(8) and polar wobble amplitude approached peak values(9) during 1947-1952. In the same period, the activity of great earthquakes in the Qinghai-Xizang Plateau also reached the climax. It might not be incidental that a volcanic eruption, which is very rare, occurred in the northern part of the plateau in 1951. Fig. 14 gives some statistical facts.

Finally, we shall offer some speculations for predicting the future great earthquakes in the Qinghai-Xizang Plateau.

1. The possibility of occurrence of future great earthquakes is larger along the plateau's boundary than in its interior.

2. More attention should be paid to the study of the gaps surrounded by earthquakes of magnitude 5½ - 7, in which great earthquakes (M=8) will probably take place.

3. Serious studies should be made on the seismic activity in the northern part of the plateau, namely, in the volcanic region with a view to predict great earthquakes in the near future.

The above criteria should be applied collectively so that we may achieve a better result for the prediction of great earthquakes (M≥8) in the Qinghai-Xizang Plateau, while volcanic activities and polar wobbles are only additional referential factors for prediction work in this region.

## References


(1) State Seismological Bureau. The Concise Catalogue of Earthquakes in China. Seismological Press, 11-23 (1977) (in Chinese).

(2) Great Earthquake Chronology of China , Science Press, 440-562 (1956) (in Chinese).

(3) Gutenberg,B., Richter,C.F., Seismicity of the Earth and Associated Phenomeno, Princeton University Press, 65-69 (1954).
(4) Medvegef, S.F., Engineering Seismology, State Press, (Moscow), 39-42 (1962) (in Russian).
(5) Shinihtsun,F.M., New information about volcanic activity in Asia, 89-90 (U.S.S.R.), Nature (9), (1954).
(6) Benioff, H., Global strain accumulation and release as revealed by great earthquakes, Bull. Geol. Soc. Am. 62, 331-338 (1951).
(7) Mogi K., Migration of seismic activity, Bull. of the Earthquake Research Institute, University of Tokyo, 46, part 1, 53-72 (1968).
(8) Vlodawiz,V.E., Volcanic activity on the earth in historical time, Volcanology Conference, Science Press (Moscow), 1, 7-17 (1966).
(9) Weng Wen-po, Variation of latitude and polar motion, Acta Geophysica Sinica, 7 (1), 1-11 (1958).

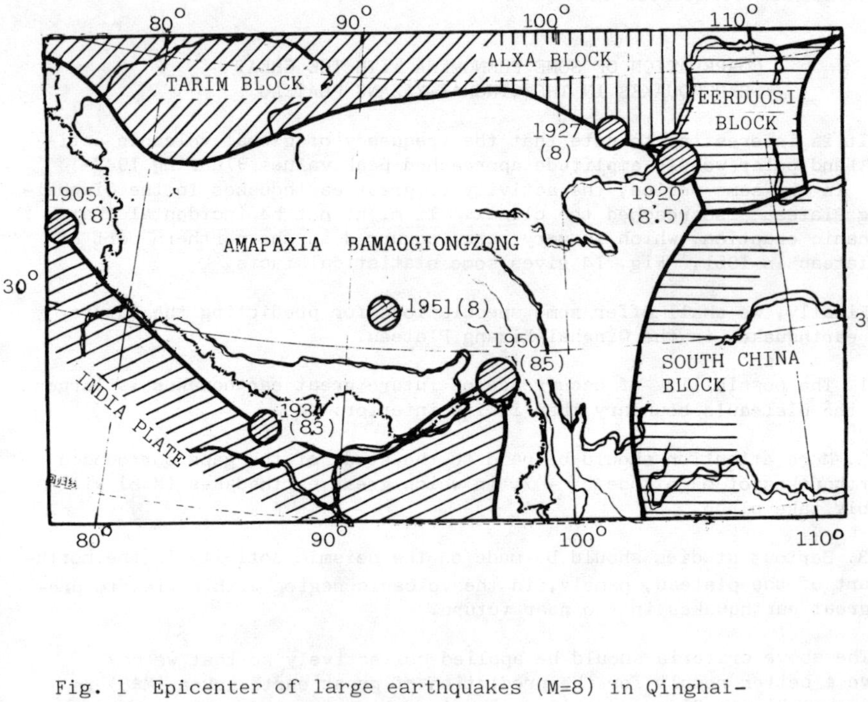


Fig. 1 Epicenter of large earthquakes (M=8) in Qinghai-Xizang Plateau

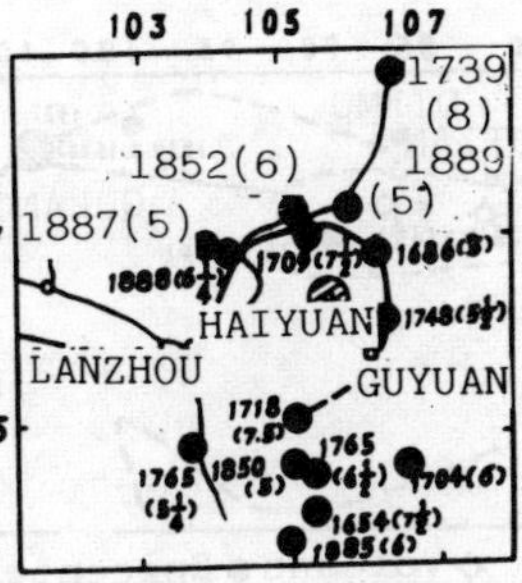


Fig. 2 The seismic gap before the Haiyuan earthquake of 1920

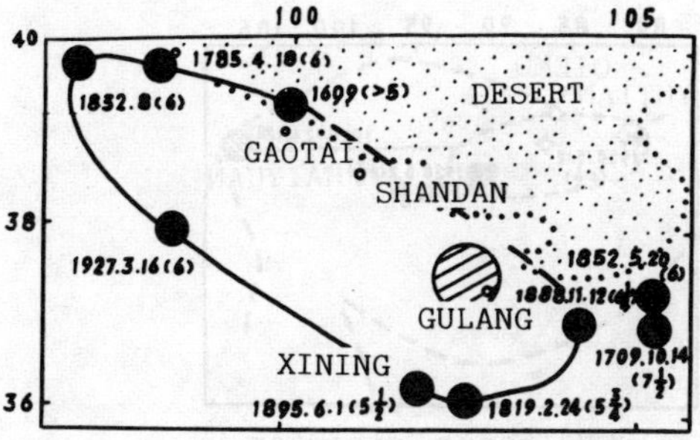


Fig. 3 The seismic gap before the Gulang earthquake of 1927

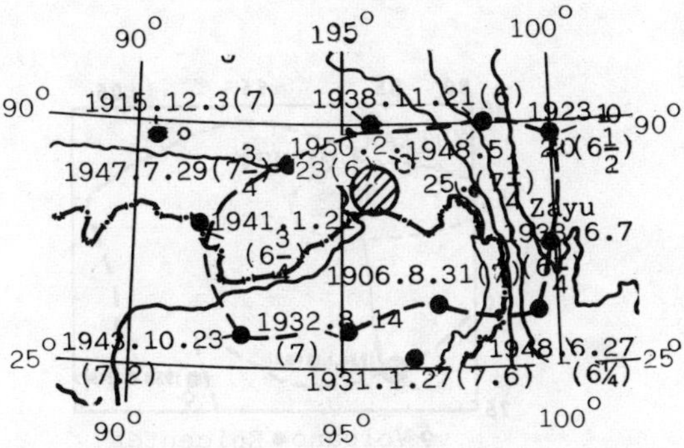


Fig. 4 The seismic gap before the Zayu earthquake of 1950

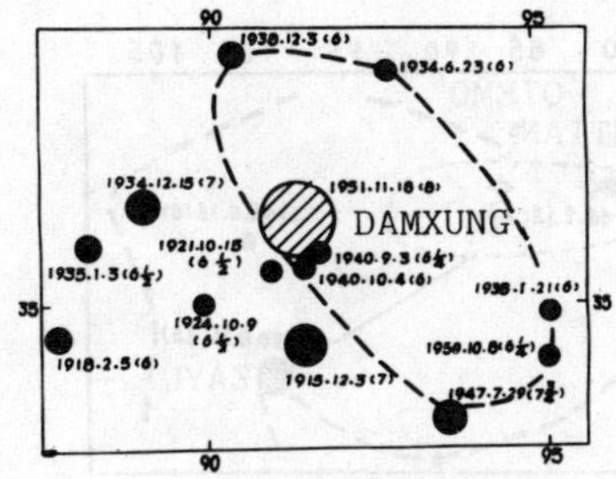


Fig. 5 The seismic gap before the Damxung earthquake of 1951

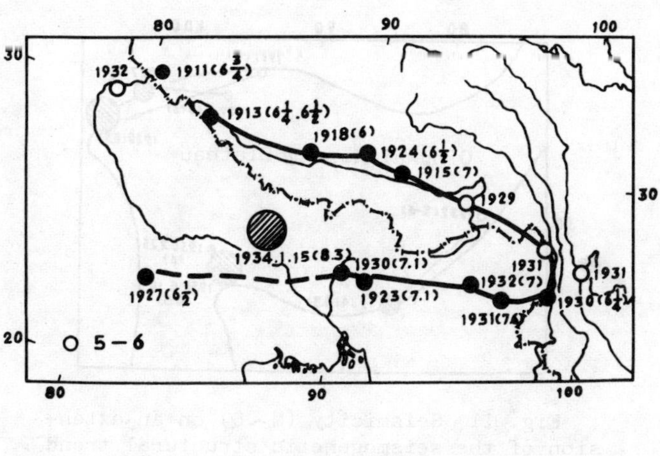


Fig. 6 The seismic gap before India-Nepal earthquake (M=8.3) of 1934

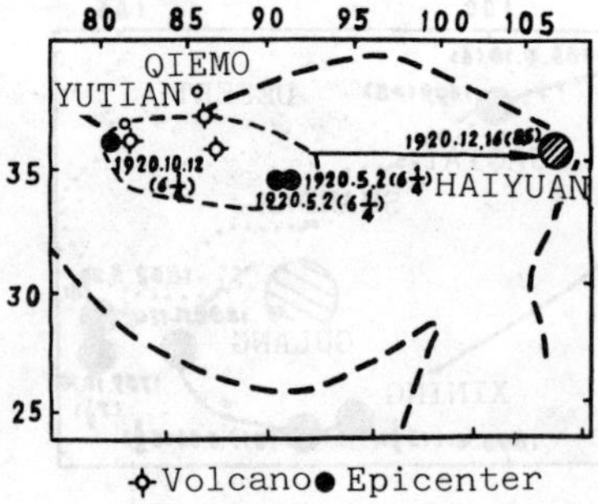


Fig. 7 The seismic activity (6≤M<7) of the volcanic region and its vicinity before the Haiyuan earthquake of 1920

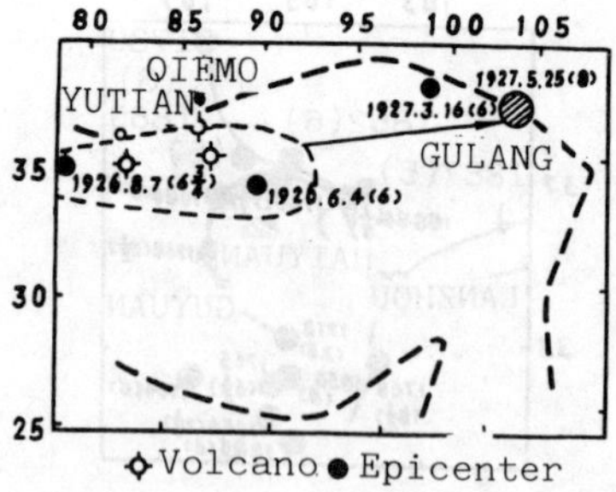


Fig. 8 The seismic activity (6≤M<7) of the volcanic region and its vicinity before the Gulang earthquake of 1927

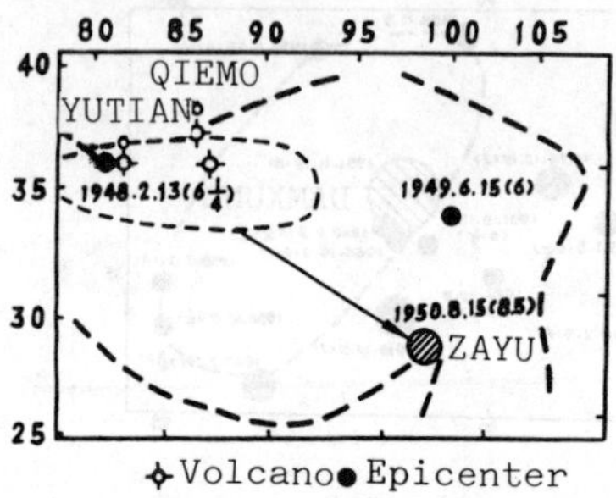


Fig. 9 The seismic activity (6≤M<7) of the volcanic region and its vicinity before the Zayu earthquake of 1950

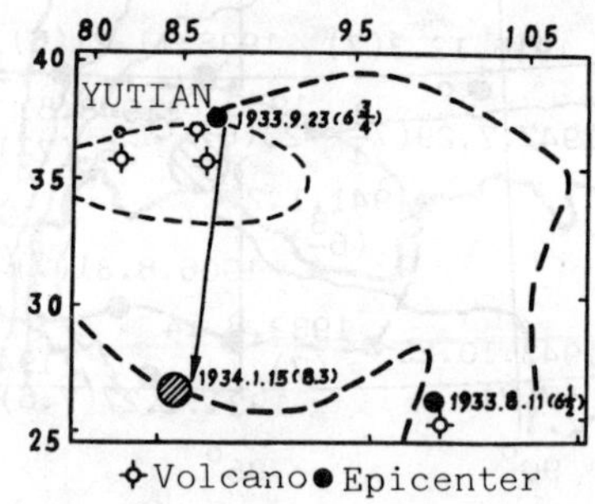


Fig. 10 The seismic activity (6≤M<7) of the volcanic region and its vicinity before the India-Nepal border earthquake of 1934

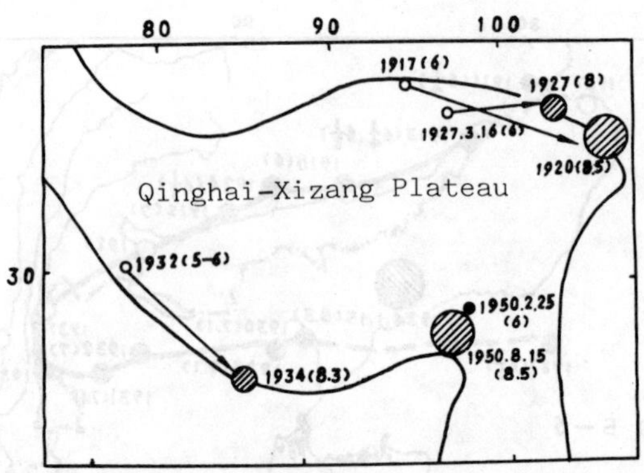


Fig. 11 Seismicity (M∼6) on an extension of the seismogenetic structural trend in epicenter before great earthquakes along the plateau's boundary

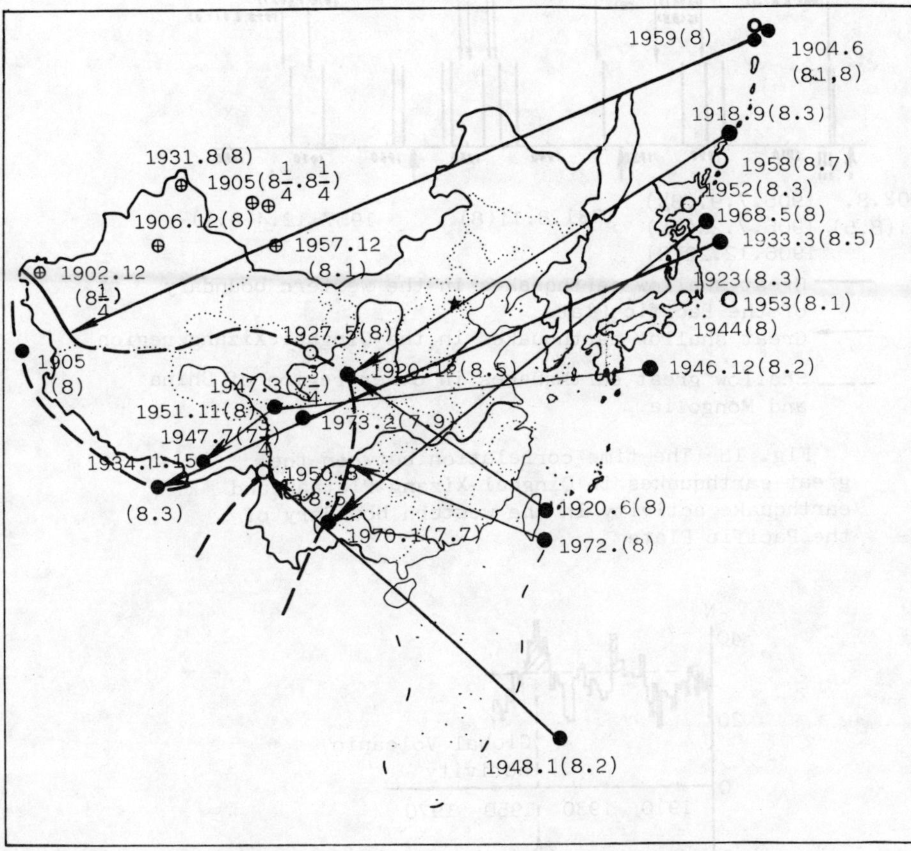


● Well correlation
○ Not well correlation or not obvious correlation
⊕ The great earthquakes in other region of China and Mongolia

Fig. 12 Apparent correlation between the great earthquakes in Qinghai-Xizang Plateau and the earthquake activity in the western boundary of the Pacific Plate

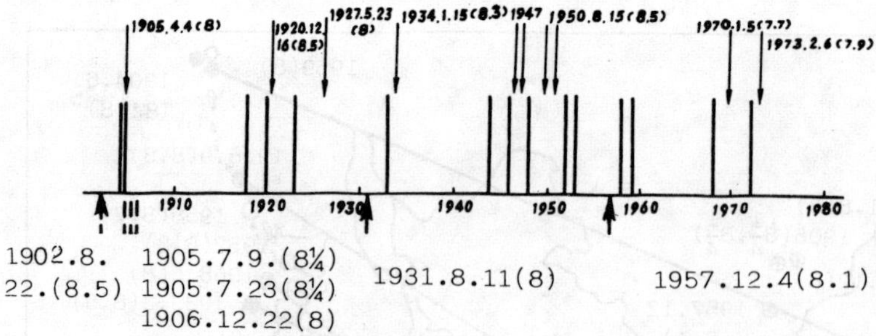


——— Great shallow earthquakes in the western boundary of the Pacific Plate

——→ Great shallow earthquakes in the Qinghai Xizang region

- - - Shallow great earthquakes in other region of China and Mongolia

Fig. 13 The time correlation between the great earthquakes in Qinghai-Xizang Plateau and earthquake activity in the western boundary of the Pacific Plate

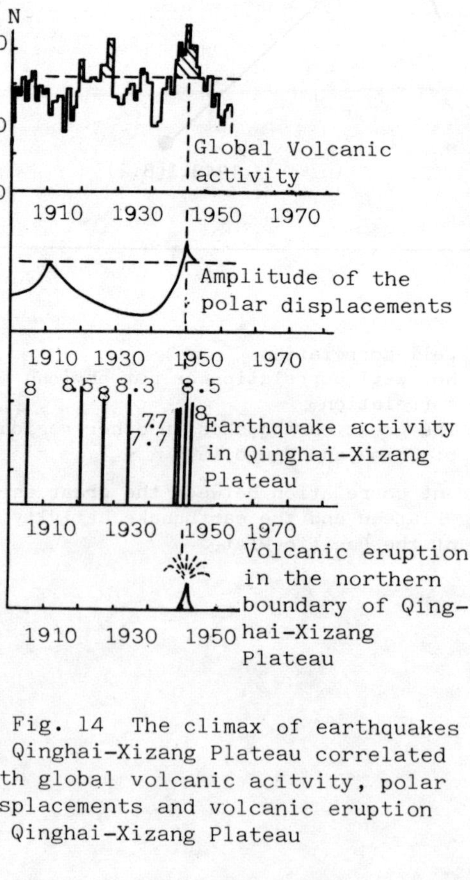


Fig. 14 The climax of earthquakes in Qinghai-Xizang Plateau correlated with global volcanic acitvity, polar displacements and volcanic eruption in Qinghai-Xizang Plateau

# FINITE ELEMENT ANALYSIS FOR RECENT QINGHAI-XIZANG PLATEAU UPLIFTING AND ITS RELATION TO SEISMICITIES

Luo Huan-yan* Xu Yu-jian Song Hui-zhen
Sun Jun-xiu Guo Cai-hua Zhang Ru-ze
Gu Zhi-juan
(Institute of Geology, State Seismological Bureau)

## ABSTRACT

The current deformation and stress state in a simplified tectonic profile crossing northeasterly the Xizang Plateau, about 2,000 km long and 80 km deep, have been analyzed using finite element method. Numerical results show that the Indian Plate motion is the driving force to disturb the gravity equilibrium leading to a lateral density contrast, but the horizontal compression resulting from the isostatic adjustment is the mechanism for plateau uplift and earthquake development, and that stresses often concentrate at the main fault ends near the low velocity zone.

Electrical conductivity measurements of 2-mica gneiss under temperature and pressure conditions equivalent to about 40 km deep, together with the optical determination and the X-ray diffraction analysis, indicate that the mechanical weakening due to the dehydration, accompanied with lattice breaking of muscovite and the thermal cracking of quartz may be the mechanisms of forming the low velocity-low resistivity zone in the plateau crust, but the shear movement existing along the low velocity zone seems to be the source of raising its temperature.

## BACKGROUND AND GUIDING THOUGHTS

Most geologists and geophysicists have paid attention to the driving mechanism of the recent uplift of Himalayas for a long time. Based on a large amount of research work of earth scientists at home and abroad in the plateau region during past decades, we have recognized that recent uplift of Himalayas is interconnected with the evolution of plateau. More detailed discussion can be found in papers(1,2,8,10,11,16,17,18)

In order to perform numerical analysis and to conduct experimental studies, we propose some concepts for supplementation:

1. Himalayas are a huge and complex system of arc structure, its southeastern side is the "Assam Wedge" consisting of large shear belts of Hengduan Mt. (Meridional Mts.), Naga-Arakan Mountains and Ninety East Ridge of Bengal Bay, its northwestern side is the "Pamir Node" consisting of large shear belts of Sulaiman-Quetta and Owen of Arabian Sea. Arc itself is also a complex flake structure which is nearly horizontal.

2. Topographical contrast of both the southern and the northern boundaries of the plateau is remarkable. The southern Deccan plateau and the northern Tarim-Qaidam basins are like "firm rock walls".

---

* Huan-yen Loo

3. In the folded and faulted ranges surrounding the plateau, such as Kunlun, Altun, Qilian and Western Qinling, new evidences of horizontal slip can be found frequently. Further field work on geomorphology, geology, geochronology and seismology is then needed earnestly.

4. The surface tectonic morphology of plateau reflects the deep-seated tectonic pattern.

5. The exposed ultrabasic rocks and other magmatic rocks containing high temperature minerals are aligned along the regional lineaments.

These tectonic features represent only the episodic formation, deformation, magmatism and metamorphism of the plateau. If combining them with the geophysical data, it should be able to study the causes of plateau uplift.

In the past decade, the plate tectonic theory has been a unifying concept in many branches of earth sciences and has been a frame work for understanding the evolution of oceanic crust, but in the continent, there still many problems remain unsolved, for instance, the structural complexities resulting from the superimposing of tectonic events throughout the long geological time span, the large deformation and the intensive fractures during the Meso-Cenozoic periods, are quite different from the rigid body hypothesis of the plate theory.

Recently, Chun (1977) and Teng (1980) have justified by the results of surface wave study and seismic explosive sounding respectively that within the plateau crust, there is a wide existence of low velocity zone varying in depth from 30 km in the south to 50 km in the north and changing in thickness from 6 to 10 km, and that most large earthquakes were located above the weak zone or at its ends. It may be a controlling factor for the development and migration of large earthquakes.

The ancient Indian subcontinent began to subduct beneath the Asian Plate with an estimated convergence rate of more than $10^{cm}/yr$. (Chun, 1977). The recent mode of the plate motion, either shortening or subducting, after continental collision has been under much debate in recent years. However, both gravity studies(1)and deep seismic soundings(13)have revealed a maximum elevation difference of more than 40 km along the Moho discontinuity in the transition zone between the Ganges plain and the Himalayas, which initiates a lateral density contrast and may be the driving mechanism for local stress condition responsible for plateau lifting and earthquake developing.

In this paper, we attempt to correlate the plate motions and the local isostatic adjustment with the neotectonic stress of plateau region, to deal with the causes of recent plateau uplift and the geological condition causing large earthquakes, and to conduct experimental studies for the formation of the low velocity zone.

## ASSUMPTIONS AND BOUNDARY CONDITIONS FOR NUMERICAL ANALYSIS

Starting from the isostatic adjustment, assume the crust to be elastic, the density and mechanical parameters varying with depth in order to take account of the temperature effect, but heating processed from shear-strain, phase change and radioactivity being neglected for the time being.

The location of a profile, about 2,000 km long and 80 km deep, crossing the plateau northeasterly from Ganges Plain to Qaidam Basin, is labelled A-A as shown in Fig. 1. Based on the available geological and geophysical data(1,13,16), the simplified geological setting is described in Fig. 2 and supposed to be representative. We assume the faults and the low velocity zone to be plastic and they are treated with reduced elastic constants.

The boundary conditions for numerical analysis are prescribed as follows: 1. Ground surface be free; 2. Horizontal zero displacement constrained at the northern end of the profile; 3. Vertical displacement along the bottom of the same profile be zero; 4. The southern end be fixed for the case of body force acting alone but be relaxed for the other case of body force and Indian Plate motion acting simultaneously.

## NUMERICAL RESULTS

Case 1 -- gravitational force acting alone:

Due to the lateral density contrast resulting from the difference ($>$40 km) of Moho discontinuity level as mentioned before, the current deformation and stress state determined by a two dimensional finite element method are shown in Fig. 2. The surface displacements are similar to the plateau topography with complementary subsidences adjacent to the uplifts on either side; the lower crust just above the mantle uplift is under tension but the uppermost crust of the same location is under compression; the transitional zone is highly compressed and its uppermost material moves upward and outward leading to over-thrusts along Himalayas. The stresses often concentrate at the main fault ends near the low velocity along which shear strain exists, but the stress concentration would not occur, if the fault intersects the weak zone or the fault end is quite far away from it. The crust structural pattern is closely relevant to stress distribution and associated with the development and the migration of many large earthquakes of M$\geqslant$6 within the continental crust, because of the plastic deformation and energy absorption or consumption of both the fault and the low velocity zone.

Case 2 -- Indian Plate motion and body force acting simultaneously:

The effects of superimposing the average rate of Indian Plate motion on the southern end of the numerical model is only subjected to stress amplification which falls off rapidly with horizontal distance from the end; particularly, the additional pushing force coming from Indian Plate does not make too much difference in displacement beyond the collision zone. Evidently, the plate motions take only the responsibility of causing the crust of the Ganges Plain buckling and the mantle upwelling so as to disturb the gravity equilibrium and thus to initiate the lateral density contrast, but these motions are not associated directly with the local stress system relevant to plateau uplifts and seismicities.

## EXPERIMENTAL STUDIES ON THE FORMATION OF LOW VELOCITY ZONE

From the numerical results, it is realized that the low velocity zone is significant in stress adjusting and transferring, it is then worth to dig deeply into its formation mechanism. As pointed out by Condie (1976) that highly conducting zones correlate well with low velocity zones, we have tested several different rocks such as basalt, granite, hornblende and gneiss, under temperature and pressure equivalent to about 40 km deep in a solid-medium apparatus, but the 2-mica gneiss, collected from Himalayas, is more probable constituent of the low velocity zone. As shown in Fig. 3, the slope of the electrical conductivity verse temperature curve changes around 650°C and then the resistivity drops rapidly to a minimum value of several ohm -m, it seems to be microstructure and also temperature-dependent, because the optical study of thin sections of rocks and the X-ray diffraction analysis of powdered crystals illustrate the evidences of the crack of quartz and the dehydration reaction of muscovite, accompanied with lattice break and phase transition. The comparison of these phenomena with the wave velocity variation (Kato, 1978) of serpentine under high temperature and high pressure conditions suggest that mechanical weakening due to lattice breaking and thermal cracking of individual minerals may be the most possible mechanisms for developing the low seismic velocity -- low resistivity zone, but the heating shear-strain determined numerically along the weak zone, may be the source to raise the temperature and to cause thermal activation of minerals.

## SUMMARY

1. Indian Plate motion is the driving force to disturb the gravity equilibrium leading to a lateral density contrast which is the mechanism for both the recent plateau uplifting after collision and the earthquake development during the localized isostatical adjustment.

2. Earthquakes generated within the continental crust are governed by the geological patterns, such as the extent of fault systems, the distribution of low velocity zone and the level of Moho discontinuity. The migrations of earthquake are controlled by these weak portions as well.

3. To study the formation and evolution of the Xizang Plateau from the viewpoint of crustal shortening and thickening, a 3-dimensional visco-plastic numerical model coupling with thermal energy equation(14,15), must be established so as to reflect more clearly the deep-seated effects and the large areal extents.

## References


(1) Chang Chun-fa, Cheng Hsi-lan, Some tectonic features of Mt. Jolmo Lungma area, southern Tibet, China. Scientia Sinica, 6 (2), 257-265 (1973).

(2) Chang Chun-fa, Zheng Xi-lan, Pan Yu-sheng, The geological history, tectonic zonation and origin of uplifting of the Himalayas, 1-17 (1977).

(3) Chun K.Y., Crustal structure of the Tibetan Plateau: A surface-wave study by a moving window analysis, B.S.S.A., 67 (3), 735-750 (1977).

(4) Condie K.G., Plate tectonics and crustal evolution,Pergamon Press (1976).

(5) Gansser A., The geology of Himalayas, New York, Wiley Interscience (1964).

(6) Gansser A., The Alps and the Himalayas, International Geo. Cong., 22nd, Report, V.11 (1964).

(7) Gansser A., The Indian Ocean and the Himalayas: a Geological Interpretation, Ecologae Geol. Helv., 59 (2) (1966).

(8) Gill W.D., The Stratigraphy of the Siwalik Series in the Northern Potwar Punjab, Pakistan, Q.J.G.S., 107 (428), 393 (1952).

(9) Kato S., Vp Measurement of serpentine rocks at 5 km and 600°C, Lecture preprints, Japan Seismol. Soc., No. 2 (1978) (in Japanese).

(10) Loo Huan-yen, Discussion on driving forces and mechanisms of earthquakes in China from the viewpoint of earth's crust dynamics, Seismology and Geology, 1, 1-10 (1979).
(11) Sahni S., The Himalayan uplift since the advant of the man: its cultistorical significance, Current Science, 5 (2), 57-62 (1936).
(12) Santo T., Classification of earthquakes from the view of plate tectonics -- a Supplement, Bull. Intern. Seism. and Earthquake Eng., 11, 21-31 (1973).
(13) Teng Ji-wen, et al., Explosive seismic studies on velocity distribution and crust structure along Damxung-Yadong profile of Xizang Plateau, In this Proceedings of the Symposium on Qinghai-Xizang (Tibet) Plateau, Academia Sinica (1981).
(14) Toksoz M.N., P. Bird, Modelling of temperature in continental converence zones, Tectonophysics, 41, 181-193 (1977).
(15) Toksoz M.N., A.T. Hsui, Numerical studies of back-arc convection and the formation of marginal basins, Tectonophysics 50, 177-196 (1978).
(16) Yan Jia-quan et al., Recent tectonics in Qinghai-Xizang Plateau, In this Proceedings of the Symposium on Qinghai-Xizang (Tibet) Plateau Academia Sinica, (1981).
(17) Yang Li-hua, Liu Tung-sheng, On the neotectonics movements in the Mount Jolmo Lungma region, Scientia Geologica Sinica, No. 3, 210-220 (1974) (in Chinese).
(18) Zhao Xi-tao, On the recent uplift of the Himalayas, Scientia Geologica Sinica, No. 2, 243-252 (1975) (in Chinese).

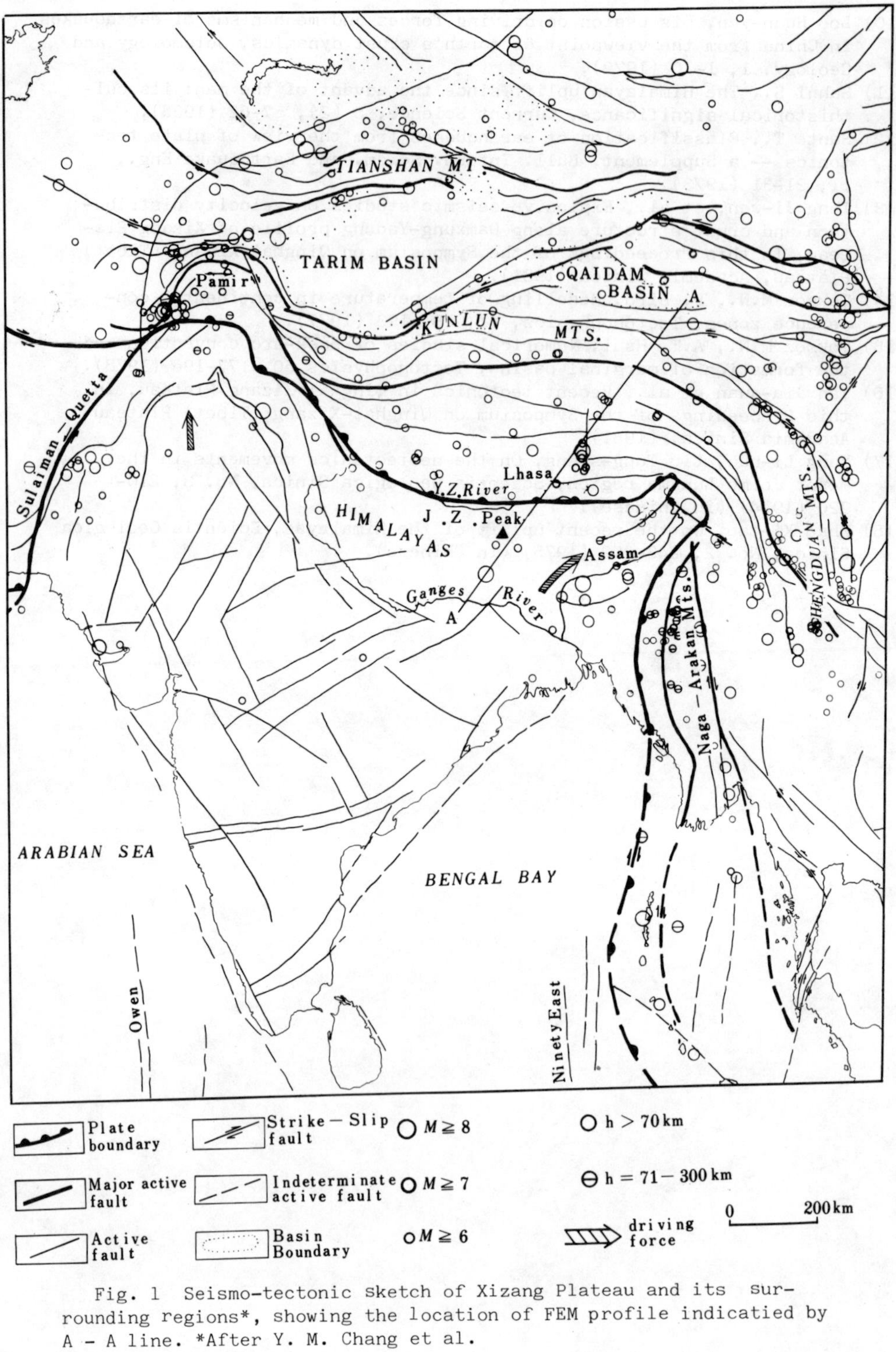


Fig. 1 Seismo-tectonic sketch of Xizang Plateau and its surrounding regions*, showing the location of FEM profile indicatied by A — A line. *After Y. M. Chang et al.

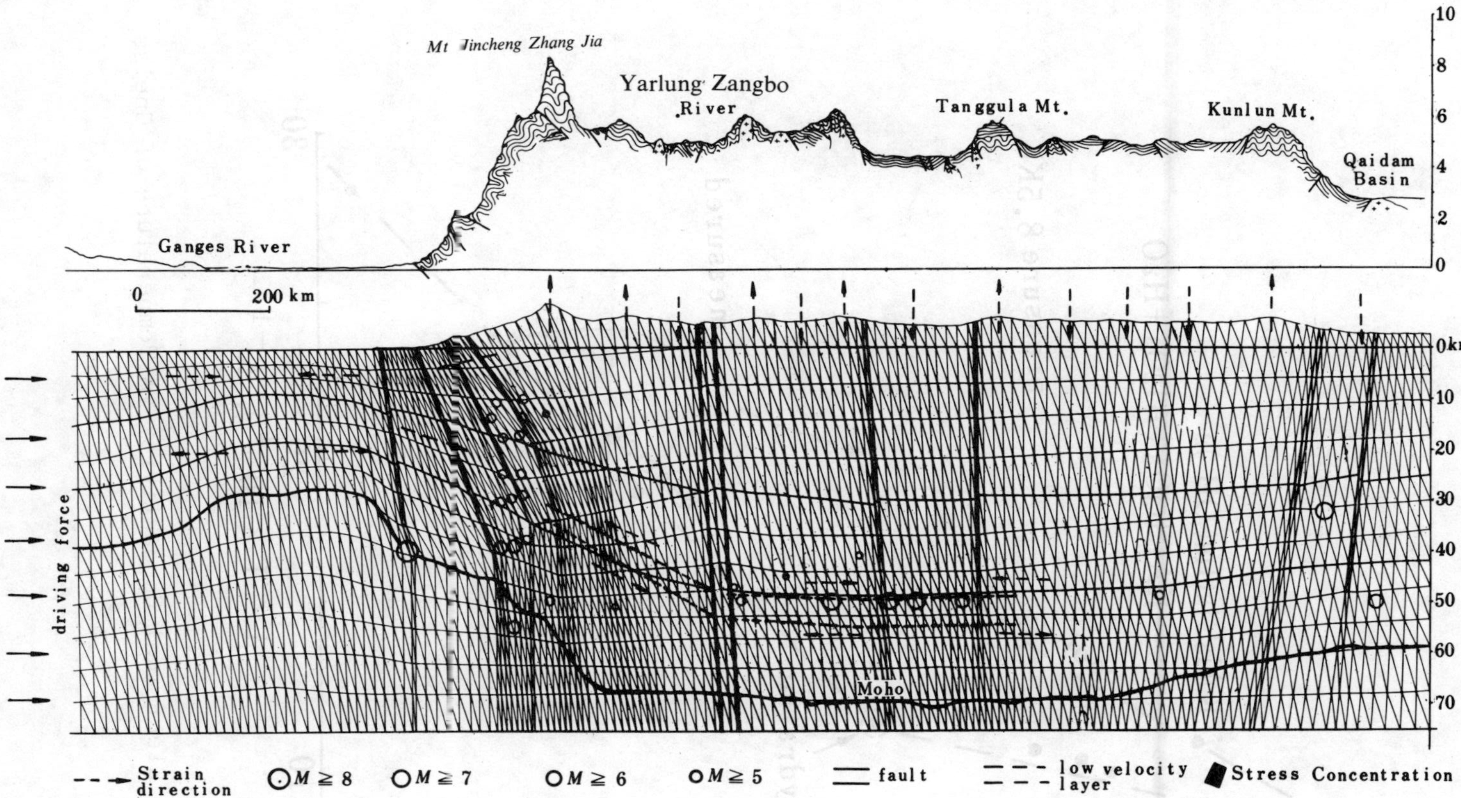


Fig. 2 Finite element meshes of a simplified tectonic profile of Xizang Plateau

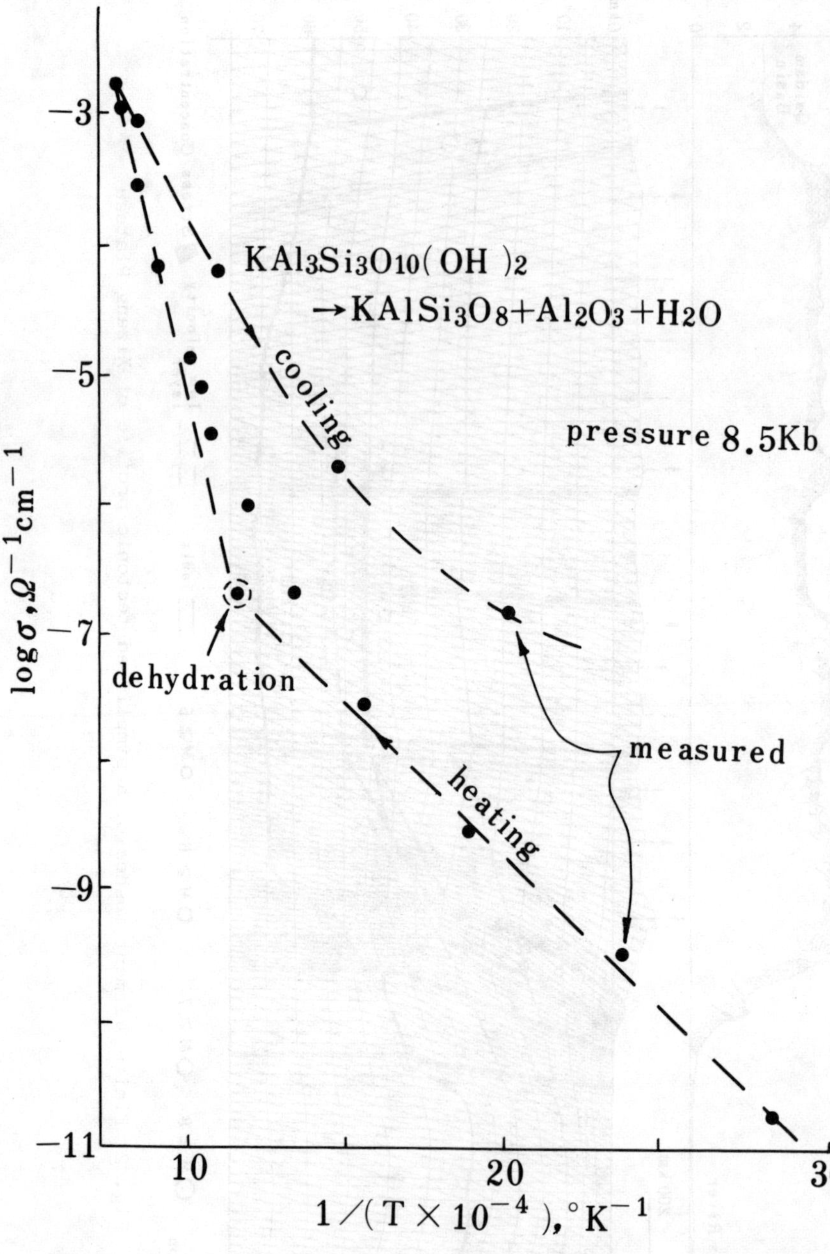


Fig. 3 Electrical Conductivity vs Temperature of Gneiss

# THE AVERAGE STRESS FIELDS OF ZHAMO AND DAMXUNG REGIONS IN THE XIZANG PLATEAU

Zhang Li-min Su Pei-yi Yang Bing-ping Zhang Zhi-he
(Institute of Geophysics, Academia Sinica)

The Himalayas in the Xizang Plateau is considered to be originated from the collision between the Indian Plate and the Eurasian Plate. The forces acting on these two plates are still active over a very large region, so that the average stress fields in the plateau are closely related to these plate forces.

Our work is to find the average stress field based on the data of small earthquakes observed in the regional seismological network and to analyse the relation between the stress fields and the plate motion.

In 1976 and 1977 the geophysical group of the Comprehensive Scientific Expedition to the Qinghai-Xizang Plateau, Academia Sinica, established a temporary seismological network including six stations in Zhamo and Damxung regions, and obtained the first set of regional seismological data in that region.

The sites were so chosen that the region must be tectonically active and the number of small earthquakes recorded in a few months would be adequate for our research. In addition, logistic conditions must also be considered.

Zhamo region (Fig. 1) is situated on the sharp river bend where the Yarlung Zangbo River changes its course from east-west to north-south and the mountain ranges also change their direction at the same place. There is a zone of deep earthquakes in the eastern part of Burma where the distribution of earthquake foci is approximately north-south. In that region, we attempted to record such earthquakes having clear first motions which can be used for correct determinations of epicenters and focal depths. Thirty-one such earthquakes have been recorded.

Damxung region (Fig. 2) is situated near the zone of collision between the two plates. There are many small earthquakes and the hydrothermal activities are rather intense. In this region, we have recorded sixty-four earthquakes which can be used for the determinations of epicenters and focal depths, and all of these earthquakes have clear first motions.

In our analysis, we used the method of combined fault-plane solution for small earthquakes.

Fig. 3 shows the projection of the upper hemisphere of the combined fault-plane solution obtained from the data on 177 first motion patterns of 64 earthquakes in Damxung region. The axis of the main compressive stress is nearly north-south. The angle of elevation is comparatively small, being 23 degrees.

Many large earthquakes have occurred in Damxung region. Fig. 4 is the mechanism solution of Damxung earthquake with magnitude 7.5 in 1952. The axis of the main compressive stress is nearly north-south, too. The angle of elevation is comparatively small, being 11 degrees. It is similar to the result of the combined fault-plane solution for small earthquakes.

Fig. 5 is the mechanism solution of Maizhokunggar earthquake with magnitude 5.2 in 1968. By the records from our network, the direction of the axis of the main compressive stress also appears nearly north-south. The angle of elevation is relatively small, being 13 degrees. It is similar to the result of combined fault-plane solution for small earthquakes.

The above mentioned facts show that in Damxung region the axes of the main compressive stress are nearly north-south. The angles of elevation are comparatively small. It means that the interaction between the plates is dominantly horizontal compression.

Fig. 6 is the projection of the upper hemisphere of the combined fault-plane solution which is based on 67 first motion patterns of 31 earthquakes in Zhamo region. The points are widely scattered in a haphazard manner. It is very difficult to make a unique fault-plane solution. This shows that the mechanisms of small earthquakes are not consistent, owing to the particular tectonics of this region.

Zhamo region is situated on the sharp river bend where the Yarlung Zangbo River changes its course. The interaction between the Indian Plate and Eurasian Plate is changed from north-south collision to east-west underthrusting. The tectonic movement is very complicated. Therefore, it is inappropriate to treat all earthquakes as under the action of a unified stress field. These earthquakes may be divided into two groups, A and B. Group A belongs to the case of collision in north-south direction. Group B belongs to the case of underthrusting in east-west direction. The combined fault-plane solution has been made for each group.

Fig. 7 is the combined fault-plane solution for group A. The axis of the main compressive stress is nearly north-south. The angle of elevation is 14 degrees. It is similar to the result of Damxung region.

Fig. 8 is the combined fault-plane solution for group B. The axis of the main compressive stress is nearly east-west. The angle of elevation is large, being 51 degrees, in accord with the circumstance of underthrusting.

In Zhamo region, there occurred the famous Zayu earthquake of magnitude 8.5 in 1950. However, it was found later by further detailed investigations that the meizo-seismic area was at Medog instead of at Zayu.

Fig. 9 is the mechanism solution of Zayu earthquake. The axis of the main compressive stress is nearly north-south. The angle of elevation is 13 degrees, approximately the same as the result of group A.

In Fig. 10, solid line shows the horizontal projection of the upper hemisphere of the main compressive stress axis. The dashed line shows the projection of the lower hemisphere. In Zhamo region, the directions of the main compressive stress of large earthquakes are scattered, but there is a tendency of changing from north-south to east-west.

In conclusion, we may summarize our principal results:

1. The axes of the main compressive stress of the average stress fields in Damxung region and part A of Zhamo region are roughly in the north-south direction. The angles of elevation of these axes are small, being 23 degrees and 14 degrees respectively. This indicates that the interaction between the plates is dominantly horizontal compression.

2. In Zhamo region, the direction of axis of the main compressive stress is changing from north-south to east-west. The angle of elevation is larger, being 51 degrees. This seems to correspond to the underthrusting condition. Such a result is basically in accord with the model of mutual collision of the two plates in north-south direction and underthrusting in both the eastern and the western ends.

3. The Zayu earthquake, Damxung earthquake and Maizhokunggar earthquake all occurred just within or nearby the network which we established recently. Their mechanism solutions are in good agreement with the combined fault-plane solutions. This implies that all earthquakes whether large or small are all controlled by the interaction between the Indian Plate and Eurasian Plate.

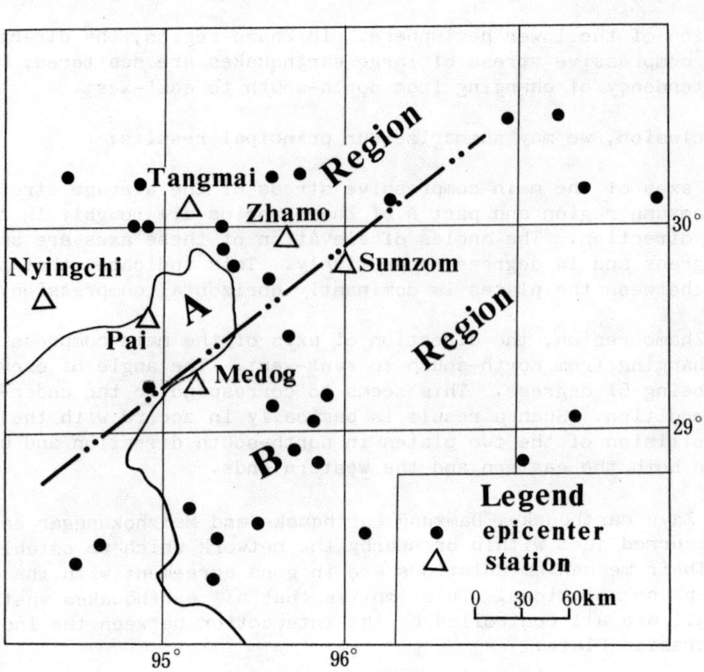


Fig. 1 Distribution of epicenters and stations in Zhamo region

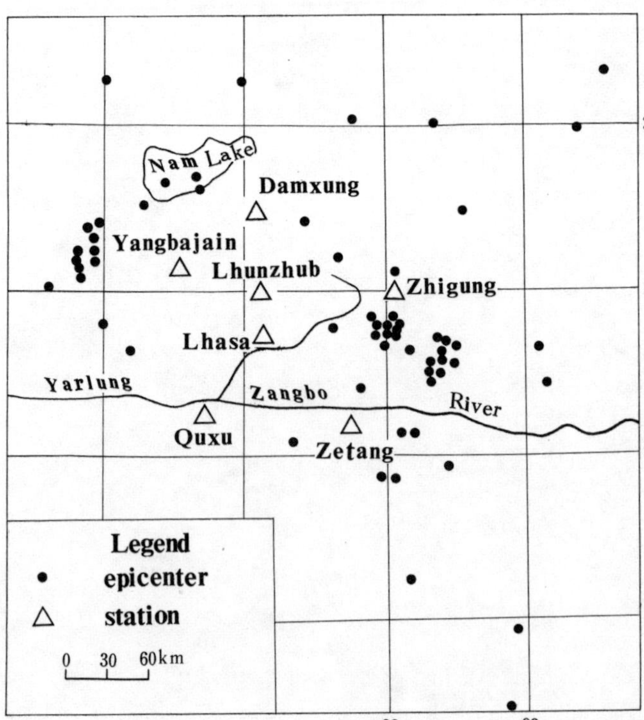


Fig. 2 Distribution of epicenters and stations in Damxung region

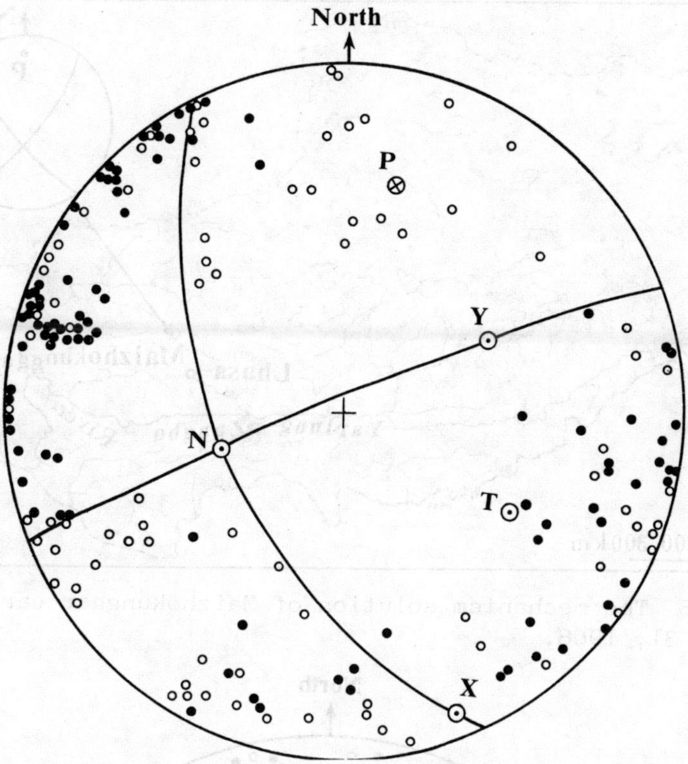


Fig. 3 The projection of the upper himisphere of the combined fault-plane solution in Damxung region

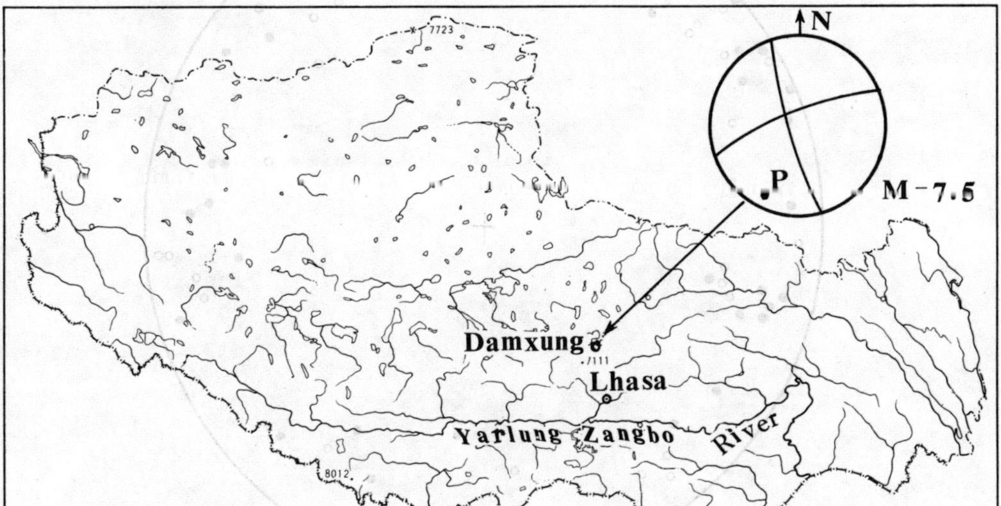


Fig. 4 The mechanism solution of Damxung earthouake on August, 18, 1952.

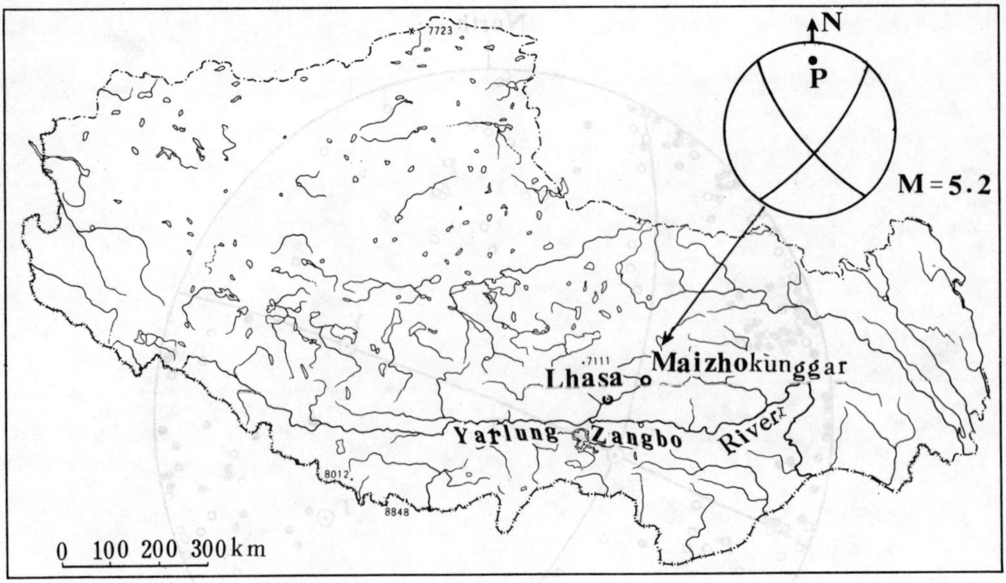


Fig. 5 The mechanism solution of Maizhokunggar earthquake on January, 31, 1968.

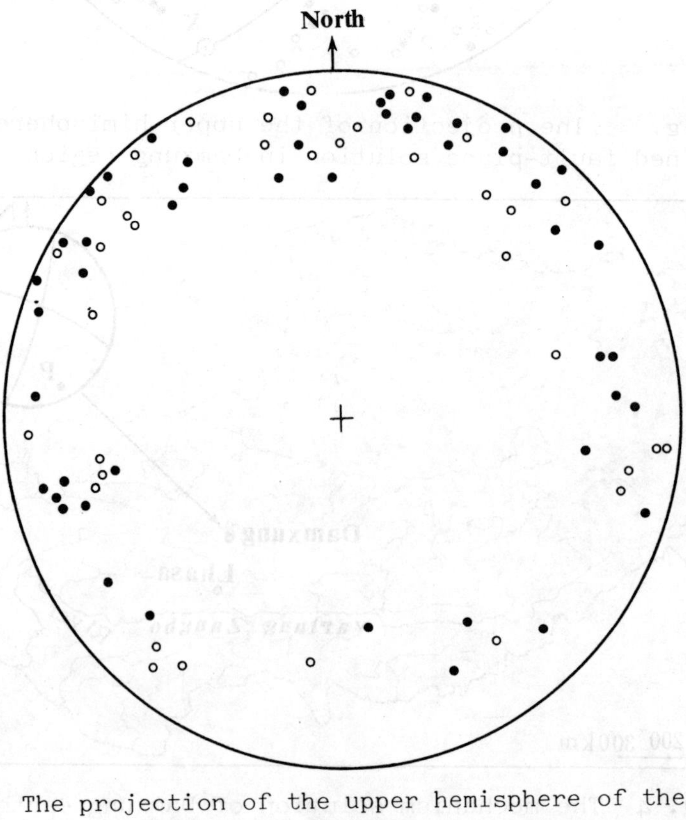


Fig. 6 The projection of the upper hemisphere of the combined fault-plane solution in Zhamo region.

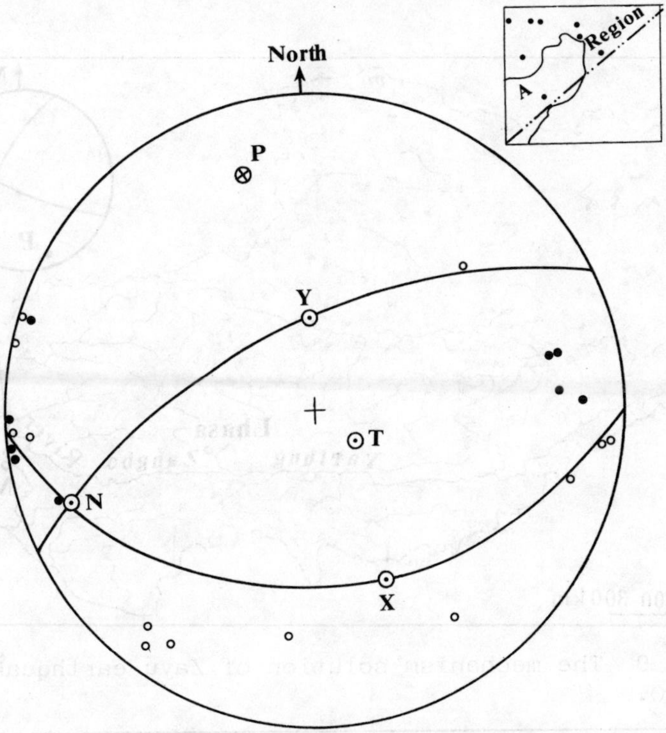


Fig. 7  The projection of the upper heimisphere of the combined fault-plane solution for group A.

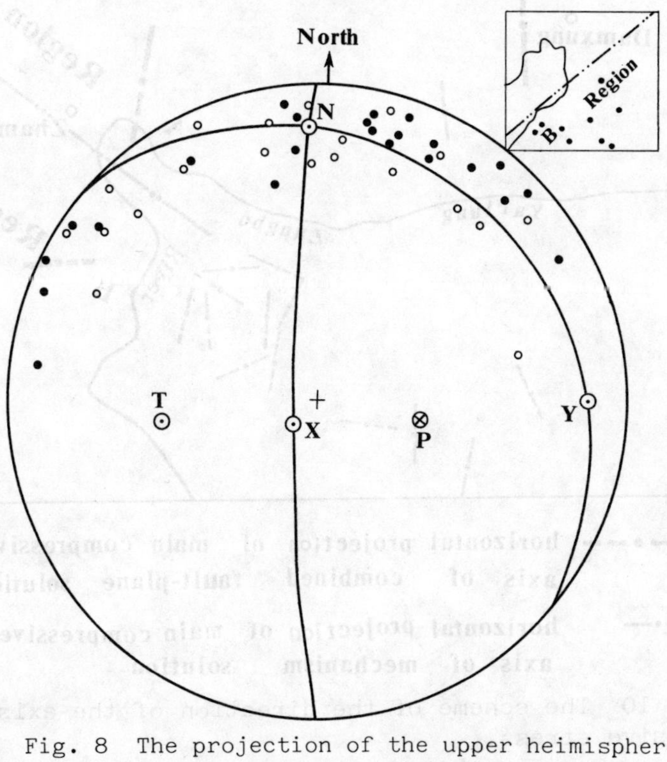


Fig. 8  The projection of the upper heimisphere of the combined fault-plane solution for group B.

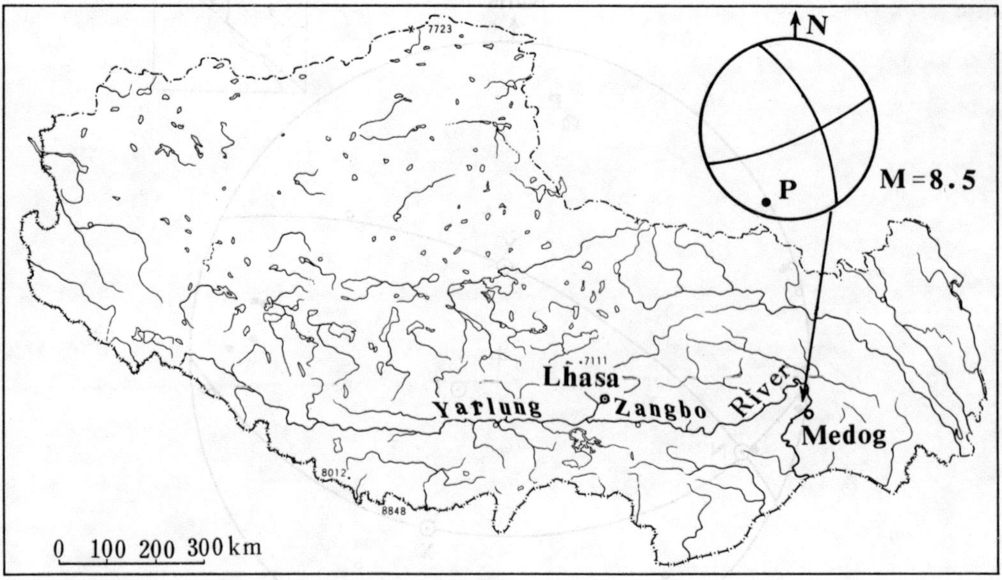


Fig. 9 The mechanism solution of Zayu earthquake on August, 15, 1950.

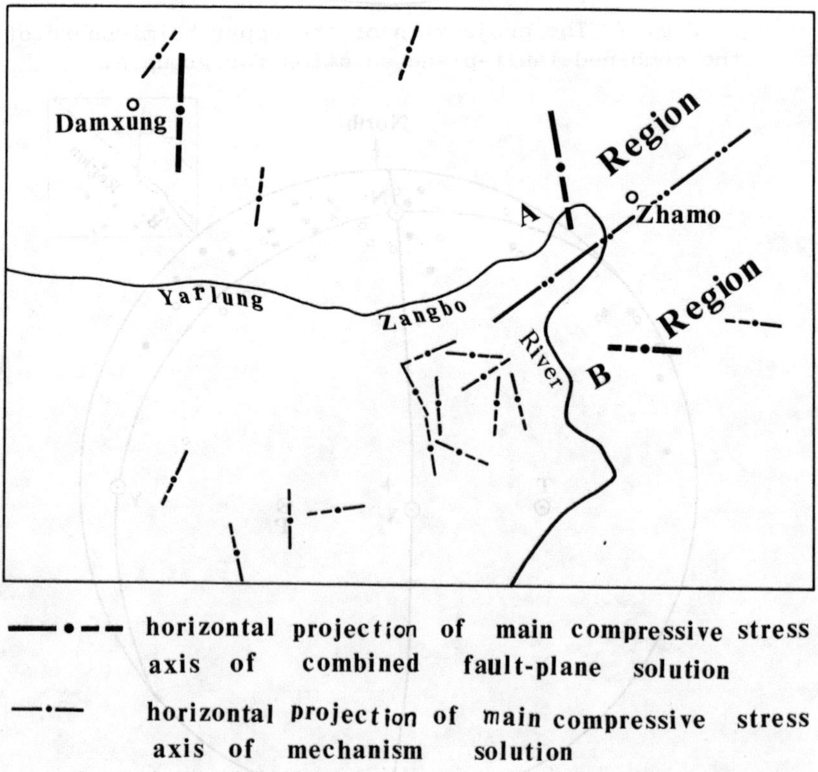


Fig. 10 The scheme of the direction of the axis of the main compressive stress.

# A PRELIMINARY STUDY OF THE P-VELOCITIES OF THE YARLUNG ZANGBO OPHIOLITES AND TENTATIVE RECONSTRUCTION OF THE STRUCTURE OF THE TETHYS OCEANIC CRUST

Wang Chi-yuen, Mei Hou-jun, and Chen Yong
(Department of Geology and Geophysics, University of California, Berkeley)
(Institute of Geochemistry, Academia Sinica, Guiyang)
(Institute of Geophysics, State Seismological Bureau, Beijing)

## INTRODUCTION

Between the world's greatest mountain range (the High Himalayas) and the largest plateau (the Tibetan) lies the Yarlung Zangbo ophiolite belt (Figure 1). It extends intermittently for a total length of 1400 km and is by far the most extensive exposure of a single ophiolite suite. It is generally considered a fragment of the oceanic lithosphere of the disappeared Tethys Ocean, abducted to the earth's surface, and now exposed at an elevation of 4 to 6 km above sea level.

Studies of many other ophiolites around the world (e.g., Poster, 1973; Petersen et al., 1974; Salisbury and Christensen, 1978) showed that rocks of the ophiolite suite display velocities similar to those commonly observed in oceanic refraction studies. Salisbury and Christensen (1978) showed that the velocity structure reconstructed along a traverse across the Bay of Island ophiolite complex in Newfoundland is "indistinguishable from that of normal oceanic crust." Thus an exclusive opportunity to reconstruct the vanished Tethys oceanic lithosphere lies in the Yarlung Zangbo ophiolite belt. In the present study, we report determinations at high pressures of the velocities of P-waves in a suite of rocks collected from the Yarlung Zangbo ophiolite belt. We then attempt to reconstruct the seismic velocity structure of the Tethys oceanic crust by combining field evidence for the said ophiolite belt with the high-pressure laboratory data.

## ROCKS AND OCCURRENCES

In the Yarlung Zangbo ophiolite zone, radiolarian chert, pillow basalt spilite flows, and ultramafic rocks occur in great abundance. The base of the ophiolite belt is in fault contact with a thick sequence of turbidite of the Cretaceous age known as the Xigaze group. Breccias and boulders of the ultramafics and volcanics of the ophiolites, on the other hand, are found in a molasse deposit of the Oligocene-Miocene time. Most radiolarian fossils in the red chert had lost their detailed features as a result of secondary recrystallization. At present, nothing more can be said about the age of abduction than that it occurred sometime between the late Cretaceous and the early Tertiary, even though suggestions have been made that the abduction process may have taken place concurrently with the onset of collision between the Indian subcontinent and the Eurasian continent, some 40 million years ago.

The geological cross section at North Mountain near the old town of Saga, Saga County, may be used to illustrate the sequence of occurrence of the various rock types in this ophiolite belt (Figure 2). A body of ultramafic rocks, mainly harzburgite, extending for 22 km in the EW direction and varying in width from 0.5 to 3 km, is in fault contact on its northern boundary with the Xigaze group. Small veins of the ultramafic body were seen to cut the Xigaze formation. Blocks of sandstone from the Xigaze formation were found as inclusions in the ultramafics. South of the ultramafic body is a gabbroic body up to 600 m in thickness. Diabasic dikes cut into a thick sequence of spilite flows and chert occurring further southward. South of the ophiolite complex, Triassic shale and phyllite are exposed.

According to the cross sections near Saga and other localities, the uppermost layer of the ophiolite complex appears to be a thick sequence of spilite flows, intercalated with limestone lens, thin-bedded sandstone, and spilite breccia. The spilite is composed of fine crystals ($\leq$ 0.1 mm) of epidote, chlorite , actinolite, and sodium plagioclase. There is about 35 % epidote, 25 % chlorite, 25% plagioclase, and 10% actinolite.

Below the spilite flows often occurs a sequence of thin-bedded red chert. Near Bainang, the sequence is about 800 m in thickness. The chert is purplish in color and is rich in radiolarian fossils. The beds are nearly vertical or dip steeply southward, and they strike nearly EW. Another sequence of spilite which occurs, along with pillows, below the radiolarian chert is somewhat coarser (0.1 - 0.4 mm) than the flows which occur above the chert formation. It consists mostly of augite and sodium plagioclase, with a small amount of magnetite. It occurs either as pillow lavas or as flows with an ophitic texture, and there are fine variations in its constitution; it is dense spilite near North Mountain in Saga, basalt with amygdules and andesitic aggregate between Que Wuza and Kula mountain pass, olivine basalt near Zetang, andesitic basalt near Ngamring, and pyroxene basalt near Qusum.

Numerous diabasic and gabbroic dikes cut across the various formations of the ophiolite complex. The rocks are medium grained (0.2 - 1.0 mm) and are composed of common augite (28%), hornblende (25%), and plagioclase (45%). Gabbros also occur in thick layers above the ultramafic rocks; typical localities for such occurrences are at the North Mountain near Saga and at the South Moutain near Ngamring. The gabbros are fine-grained with distinct flow texture. Near Ngamring the exposed gabbroic body is about 1.7 km in thickness and about 6 km in length.

The ultramafic rocks occur in numerous lenticular bodies of various sizes. There are nine large bodies, each extending in the EW direction for several tens of km and varying in thickness from 1 to 3 km. These form a series of peaks, the tallest reaching 6 km above sea level. The ultramafic bodies are tectonized and are in fault contact with the other rock units; the contacts are generally vertical and run in the EW direction. The rocks are mostly harzburgite, with minor amounts of dunite and all are serpentinized to various degrees. The sample collected from Lannga Co Lake in Burang County was a harzburgite. It consists predominantly of crystals of olivine and orthopyroxene, with minor amounts of clino-pyroxene,chrome spinel, and serpentine. The composition of the olivine crystals is $Fo_{92}Fa_8$, with grain sizes between 0.40 and 3.10 mm. The crystals have a composition of $En_{89}Fs_{11}$ (i.e., bronzite) and grain sizes between 0.30 and 3.05 mm. Ultramafic cumulates occur only locally and in small quantity near Ngamring. For this reason, they are not included in our laboratory measurements nor in the reconstruction for the general structure of the Tethys oceanic crust.

## VELOCITY DETERMINATION

For the laboratory determinations of velocity, rock samples were collected from the Yarlung Zangbo ophiolite complex at seven different structural attitudes across the geological sections. These include one block of the radiolarian chert, one block of spilite from the upper level and another from the lower level, one block of gabbro, one block of diabase, and one block of harzburgite. The locations where these rock samples were collected are shown in Figure 1. The selection of sites was determined by freshness of samples, absence of pervasive fractures, and simplicity in the local structures.

Cylindrical cores in three mutually perpendicular directions were 1.9 cm in diameter and about 4 cm in length. The ends of the cylinders were ground flat and perpendicular to the axis. The drilling and grinding processes were all carried out with water as the cooling agent. The fabricated cylinders were then left in room condition for a month or so to dry. After that, each sample was jacketed in copper foil to seal the sample from the fluid medium during the high pressure experiment.

The high pressure system was a conventional one, with a pressure vessel connected to an air pump for rapid increase of pressure and a hand pump for fine adjustment. Since we are only interested in the upper 10 km or so of the Tethys oceanic plate, the pressure range of this study was limited to 3 kb. Determinations of pressure was done by using a Heise gauge, calibrated with a dead weight gauge, at a precision of 0.1%.

The determination of velocity through the rock samples was done by the use of a conventional ultrasonic method described by Birch (1960). The piezoelectric transducers used in the present study were made of barium titanate ceramic with a natural frequency of 1 Mhz. Accuracy in the velocity determinations was about 1%. The experimental determinations of velocity, together with the density of each rock, are listed in Table 1.

As noted previously, the harzburgite sample used in this study was slightly serpentinized. The serpentinization, common in many ophiolites, is thought to take place during or after abduction (e.g., Christensen, 1972; Wenner and Taylor, 1973; Margaritz and Taylor, 1974). The effect of serpentinization is to produce a drastic lowering of velocity in the ultramafic rocks (Birch, 1961; Christensen, 1966) and thus complicates the reconstruction of the upper mantle velocity using laboratory determination of velocities of these rocks. For this reason, the structural reconstruction will be limited to the crustal section only.

It should also be noted that there is significant anisotropy in the velocities through the red chert and the harzburgite samples. For the red chert, the results show that it is transversally isotropic, with the slowest direction perpendicular to bedding; the difference between the slowest and the fastest velocities reaches about 0.3 km/sec. The harzburgite sample shows an even greater anisotropy in the velocities, with the greatest difference at about 0.4 km/sec. On the other hand, the gabbro and diabase are remarkably isotropic in their elastic properties. The two spilite samples appear quite similar as far as their velocities are concerned.

## RECONSTRUCTION OF SEISMIC VELOCITY STRUCTURE

Ideally the velocity structure of the Tethys oceanic crust could be easily reconstructed if a complete geological cross section of the Yarlung Zangbo ophiolite belt were given and if the velocities of elastic waves through the various rocks occurring in this cross section were determined at the appropriate pressure and temperature. However, the Yarlung Zangbo ophiolites have been severely tectonized, and the "normal" sequence of rock units has been disrupted. In order to piece together a complete cross section of the rock units composing the Tethys oceanic plate, we examined the geological information from twelve cross sections across the Yarlung Zangbo ophiolite belt. We then assembled the less disrupted units in these cross-sections to form an "average" cross section for the Yarlung Zangbo ophiolite belt as shown in Figure 3, which may be taken as a model to represent the crust of the disappeared Tethys Ocean. Much more work would be necessary to improve this model, but for the present we shall assume it to be acceptable and will reconstruct its seismic velocity structure.

For the range of depth of concern here, the seismic velocity is largely a function of rock types and pressure (e.g., Salisbury and Christensen, 1978). The effect of temperature is of minor importance and may be justifiably neglected in this preliminary study. Using the constructed geological column shown in Figure 4 and the densities of rocks given in Table 1, the lithospheric pressure at any depth in the Tethys oceanic crust may be computed by

$$P(d) = 0.5 \text{ kb} + \int_0^d g\,\rho(z)dz$$

where d is depth, g is gravity, and ρ is density. A constant pressure of 0.5 kb is added to account for the overburden pressure of the water column in the Tethys Ocean. In this computation it is assumed that the state of stress in the Tethys oceanic crust is not far from being hydrostatic; the uncertainty arising from this assumption is not significant.

With pressure so computed, the seismic velocity at any particular depth in the Tethys oceanic crust can then be interpolated from the experimental velocity data listed in Table 1. The reconstructed velocity structure is shown in Figure 4. The most distinct feature of this structure is the presence of a low velocity layer with velocity of 5.5 km/sec with a thickness of 0.8 km, sandwiched between a fairly uniform layer of higher velocity of 6.3 km/sec. Near the bottom of this layer, another layer (1.8) km of higher velocity (6.5 km/sec) is present. An unconsolidated sedimentary layer of a few hundred meters thick originally on top of the present crustal section was probably missing. Thus the total crustal thickness is probably about 5 km, below which velocity rises to that for the oceanic upper mantle.

Also shown in Figure 4 is the range of velocities for the present oceanic crusts, as revealed by seismological means (Houtz, 1980; Houtz, et al., 1980). We see that the velocity structure for the Tethys oceanic crust is in general agreement with those for the present oceanic crusts. The absence of fine structures in the seismological velocity profiles may be due to a lack of resolution in the seismological methods. On the other hand, the possibility that the crustal model given in Figure 3 needs to be significantly improved cannot be ruled out. Nevertheless, we may conclude that the reconstructed seismic velocity structure of the disappeared Tethys oceanic crust does not show significant differences from those for the present oceanic crusts. Improvement and refinement in the analysis must await further field and laboratory studies on the rocks and structures of the Yarlung Zangbo ophiolite belt.

## ACKNOWLEDGMENTS

The velocity determinations were made while one of us (Chen Yong) was visiting the University of California at Berkeley. The first author thanks the University of California, Berkeley, for a grant which supported his trip to Tibet in the summer of 1980, and part of the laboratory research. We thank Xie Hong-sen and Ta-liang Teng for hand delivery of the rock samples used in this study. Len Vigus was most helpful in maintaining the high-pressure system.

## References

(1) Birch, F., The velocity of compressional waves in rocks to 10 kilobars, 1, Jour. Geophys. Res., 65, 1083-1102 (1960).

(2) Birch, F., The velocity of compressional waves in rocks to 10 kilobars, 2, Jour. Geophys. Res., 66, 2199-2224 (1961).

(3) Christensen, N.I., Elasticity of ultrabasic rocks, Jour. Geophys. Res., 71, 5921-5931 (1966).

(4) Christensen, N.I., The abundance of serpentinites in the oceanic crust, Jour. Geol., 80, 709-719 (1972).

(5) Houtz, R.E., Crustal structure of the North Atlantic on the basis of large airgun-sonobuoy data, Geol. Soc. Amer. Bull., 91, 406-413 (1980).

(6) Houtz, R.E., Windisch, C., Muranchi, S., Changes in the crust and upper mantle near the Japan-Bonin Trench, Jour. Geophys. Res., 85, 267-274 (1980).

(7) Magaritz, M., Taylor, H., Oxygen and hydrogen isotope studies of serpentinization in the Troodos ophiolite complex, Earth Planet. Sci. Lett., 23, 8-14 (1974).

(8) Peterson, J.J., Fox, P.J., Schreiber, E., Newfoundland ophiolites and the geology of the oceanic layer, Nature 247, 194-196 (1974).

(9) Poster, C.K., Ultrasonic velocities in rocks from the Troodos massif, Cyprus, Nature, 243, 2-3 (1973).

(10) Salisbury, M.H., Christensen, N.I., The seismic velocity structure of a traverse through the Bay of Island ophiolite complex, Newfoundland, an exposure of oceanic crust and upper mantle, Jour. Geophys. Res., 83, 805-817 (1978).

(11) Wenner, D.B., Taylor, H.P., Jr., Oxygen and hydrogen isotope studies of the serpentinization of ultramafic rocks in oceanic environments and continental ophiolite complexes, Amer. Jour. Sci., 273, 207-239 (1973).

Table 1. Velocity of compressional waves as function of pressure

| Rock(sample) No. | Density | Velocity (km/sec) | | | | | | | | | | |
|---|---|---|---|---|---|---|---|---|---|---|---|---|
| | | p=10 | 50 | 100 | 200 | 500 | 800 | 1000 | 1500 | 2000 | 2500 | 3000 |
| Harzburgite(Za74) | | | | | | | | | | | | |
| 1 | 3.25 | 7.40 | 7.43 | 7.46 | 7.48 | 7.53 | 7.54 | 7.56 | 7.56 | 7.56 | 7.58 | 7.58 |
| 2 | 3.25 | 7.06 | 7.11 | 7.15 | 7.16 | 7.20 | 7.29 | 7.32 | 7.32 | 7.34 | 7.34 | 7.36 |
| 3 | 3.26 | 7.25 | 7.41 | 7.36 | 7.42 | 7.41 | 7.45 | 7.45 | 7.50 | 7.50 | 7.52 | 7.54 |
| mean | 3.25 | 7.27 | 7.32 | 7.33 | 7.35 | 7.38 | 7.42 | 7.44 | 7.47 | 7.48 | 7.48 | 7.49 |
| Diabase(Za95) | | | | | | | | | | | | |
| 1 | 2.97 | 6.48 | 6.57 | 6.57 | 6.63 | 6.64 | 6.65 | 6.65 | 6.67 | 6.67 | 6.68 | 6.70 |
| 2 | 2.97 | 6.32 | 6.53 | 6.55 | 6.59 | 6.63 | 6.66 | 6.66 | 6.67 | 6.67 | 6.68 | 6.68 |
| 3 | 2.97 | 6.30 | 6.38 | 6.48 | 6.54 | 6.60 | 6.61 | 6.62 | 6.62 | 6.62 | 6.64 | 6.64 |
| mean | 2.97 | 6.37 | 6.48 | 6.53 | 6.59 | 6.62 | 6.64 | 6.64 | 6.65 | 6.66 | 6.67 | 6.67 |
| Diabasic gabbro(Za192) | | | | | | | | | | | | |
| 1 | 2.90 | 6.02 | 6.24 | 6.28 | 6.30 | 6.37 | 6.40 | 6.41 | 6.43 | 6.44 | 6.45 | 6.45 |
| 2 | 2.90 | 6.25 | 6.28 | 6.31 | 6.33 | 6.38 | 6.41 | 6.43 | 6.44 | 6.44 | 6.46 | 6.47 |
| 3 | 2.90 | 6.36 | 6.40 | 6.40 | 6.43 | 6.47 | 6.48 | 6.49 | 6.50 | 6.52 | 6.52 | 6.52 |
| mean | 2.90 | 6.21 | 6.31 | 6.33 | 6.35 | 6.41 | 6.43 | 6.44 | 6.46 | 6.47 | 6.48 | 6.48 |
| Basalt(Zc217) | | | | | | | | | | | | |
| 1 | 2.90 | 6.01 | 6.06 | 6.22 | 6.26 | 6.28 | 6.30 | 6.30 | 6.31 | 6.32 | 6.33 | 6.33 |
| 2 | 2.89 | ---- | ---- | ---- | ---- | ---- | ---- | ---- | ---- | ---- | ---- | ---- |
| 3 | (2.84) | 5.85 | 5.97 | 6.09 | 6.15 | 6.16 | 6.21 | 6.23 | 6.27 | 6.28 | 6.32 | 6.32 |
| mean | 2.89 | 5.93 | 6.01 | 6.15 | 6.20 | 6.22 | 6.25 | 6.26 | 6.29 | 6.30 | 6.32 | 6.33 |
| Spilite(Zc130) | | | | | | | | | | | | |
| 1 | 2.87 | ---- | ---- | 6.26 | 6.28 | 6.31 | 6.32 | 6.34 | 6.40 | 6.43 | 6.43 | 6.43 |
| 2 | 2.87 | 5.79 | 5.98 | 6.04 | 6.08 | 6.12 | 6.13 | 6.17 | 6.22 | 6.24 | 6.25 | 6.29 |
| 3 | 2.87 | 6.16 | 6.22 | 6.27 | 6.32 | 6.35 | 6.37 | 6.39 | 6.40 | 6.41 | 6.43 | 6.43 |
| mean | 2.87 | 5.97 | 6.10 | 6.19 | 6.23 | 6.26 | 6.27 | 6.30 | 6.34 | 6.36 | 6.37 | 6.38 |
| red chert(Zc126) | | | | | | | | | | | | |
| 1 | 2.61 | 4.76 | 5.01 | 5.18 | 5.27 | 5.50 | 5.55 | 5.61 | 5.68 | 5.70 | 5.72 | 5.73 |
| 2 | 2.64 | 4.68 | 4.79 | 4.94 | 5.11 | 5.39 | 5.65 | 5.65 | 5.72 | 5.77 | 5.78 | 5.79 |
| 3 | 2.63 | 4.70 | 4.95 | 5.03 | 5.12 | 5.31 | 5.33 | 5.34 | 5.41 | 5.46 | 5.48 | 5.50 |
| mean | 2.63 | 4.71 | 4.92 | 5.05 | 5.17 | 5.40 | 5.51 | 5.53 | 5.60 | 5.64 | 5.66 | 5.67 |

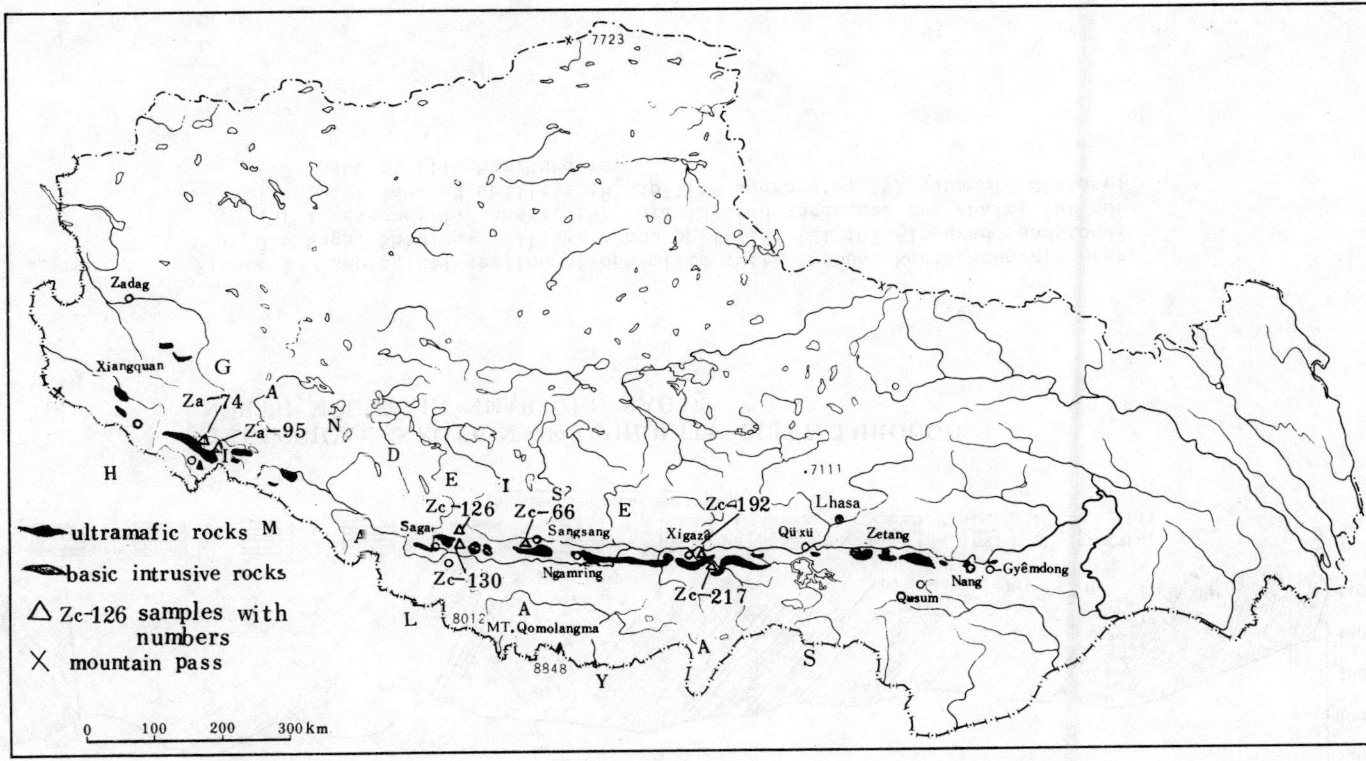


Figure 1. Sketch map showing the occurrence of the Yarlung Zangbo ophiolite belt and the locations from where the samples used in the present study were collected.

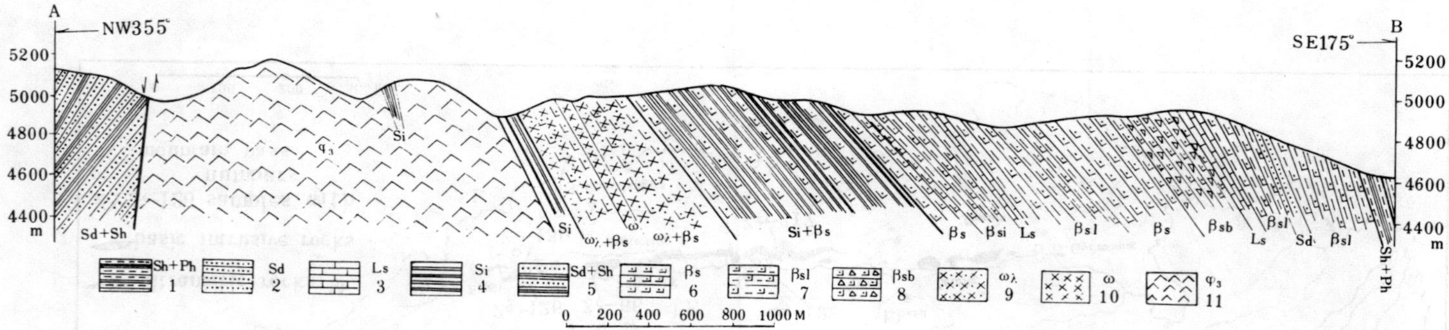


GEOLOGICAL SECTION OF OPHIOLITE SUITE THROUGH NORTH MOUNTAIN NEAR OLD SAGA

Figure 2. Geological section of ophiolite suite through North Mountain near Old Saga. Shadings: (1) shale and phyllite; (2) thin-layered sandstone; (3) limestone; (4) chert; (5) interlayered sandstone and shale; (6) spilite; (7) layered spilite; (8) spilite aggregate; (9) gabbroic diabase; (10) gabbro; (11) harzburgite.

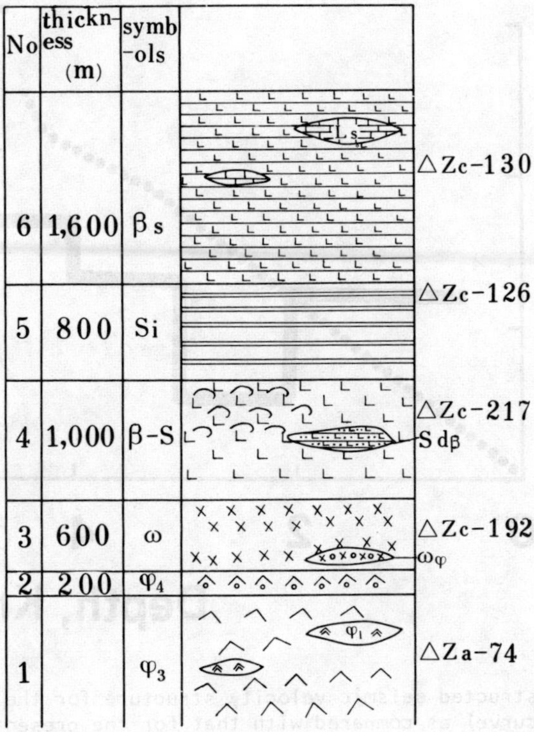


Figure 3. Stratigraphic column of rock units for the Yarlung Zangbo ophiolite zone as reconstructed from twelve geological cross-sections.

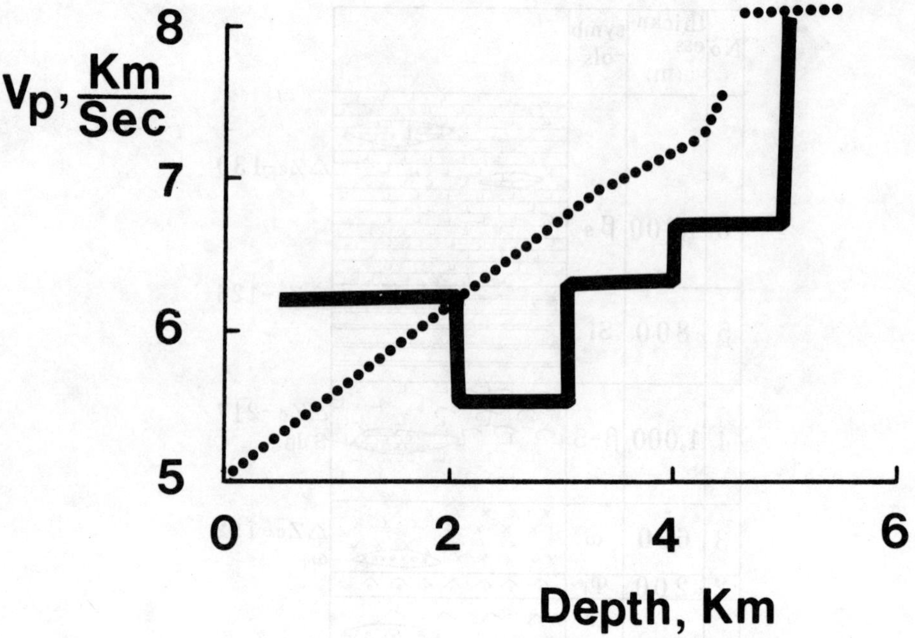


Figure 4. A reconstructed seismic velocity structure for the Tethyan oceanic crust (solid curve) as compared with that for the present ocean (dotted curve, modified from Houtz's Atlantic crust of age of 55 million years). The bottom of the ocean is taken to be depth zero. It is assumed that a top layer of sediments 500 m thick is absent.

# CHARACTERISTICS OF GEOTHERMAL ACTIVITIES IN XIZANG PLATEAU AND THEIR CONTROLLING INFLUENCE ON PLATEAU'S TECTONIC MODEL

Tong Wei
(Department of Geology, Beijing University)

Zhang Ming-tao
(Integrated Survey Commission of Natural Resources, Academia Sinica)

There are three kinds of surface manifestations of the endogeneous heat activities, i.e. volcanic eruption, hydrothermal manifestation and the conductive heatflow to the earth's surface. Xizang itself is lacking in active volcanism, but the hydrothermal activities are widespread over the whole plateau.

Hydrothermal activities in Xizang was first reported by T.G. Montgomerie, who visited the boiling spouter now named Bibilong during his visit in 1871-1872, and he greatly admired the spouter's spectacular scenery(1). S. Hedin twice traveled through this plateau during 1906-1908 and reported several hot springs(2). S. Pranavananda also observed four hot springs located nearby the Mapam Lake in western Xizang(3). When participating in the integrated scientific expedition to the Qinghai-Xizang Plateau during 1973-1976 we observed in situ the hydrothermal manifestations and their geologic settings over the plateau, measured their natural heatflows, collected 279 springwater, some sinter and efflorescences samples and analyzed the H-, O-and S-isotopic ratios. According to the field inspection and the laboratory studies we confirm the existence of a powerful geothermal belt in southern Xizang which we call Himalayan Geothermal Belt (HGB).

The HGB is not only a potential energy resource to the Xizang Region, but also plays a significant role in Xizang Plateau's tectonic modelling studies.

## THE CHARACTERS OF THE GEOTHERMAL ACTIVITIES IN XIZANG

Volcanism is the strongest geothermal surface manifestation, but no active volcano has been found in Xizang up to now. All the Cenozoic volcanism can be only observed in the north of Xizang, where it occurs in two zones, one of which lies at the northern foot of the Gandise Range, and mainly consists of dacite, andesite, rhyolite and some basalts, which, according to petrochemistry and the plate theory, belong to the maturated island arc or continental edge volcanics. Most of these are Eocene in age and give no evidence for its continuing action. Therefore, it might not serve as a heat source to the active hydrothermal activity. Another zone is located in the territory between the Gandise Range and the Kunlun Mts. The youngest volcano of China is precisely located in the rugged Kunlun, the latest eruption of which took place on May 27, 1951. This is the volcano Ka-er-da-xi (Uygur language, which means the fire stone). The volcanics of the northern zone are alkaline or calc-alkaline. It is found that all the northern volcanic fields are substantially younger than the southern ones.

During the field expedition we paid special attention to the hydrothermal activities occurring in Xizang because they are exceedingly widespread over the whole plateau. Types of the surface manifestation are quite diverse and many of them are quite vigorous. Many boiling or hot-springs contain some special chemical constituents and are of particular hydrochemical types. In addition, the hydrothermal activities in southern and northern Xizang are quite different.

To our knowledge, there are more than 600 hydrothermal areas in this plateau, of which 354 were investigated in situ during our field observations(4) and about 70 percent of these areas are concentrated in southern Xizang, between the south of the Gandise and Nyainqentanghla Ranges and the north of Himalayan Arc, in other words, just in the so-called HGB, which stretches along both sides of Yarlung Zangbo River.

The most vigorous hydrothermal manifestations consist of 11 hydrothermal explosion areas, 3 high-temperature geysers and 28 boiling springs. The largest swarm of explosion craters is located in Qupu geothermal field, where a violent explosion occurred on November 12, 1975. One witness of this event told us: "Suddenly an appalling roar from Qupu startled our flocks which ran off in all directions. Turning round, I saw a darkish grey mushroom cloud of soil and dust rising to several hundreds of meters. It resembled a nuclear explosion which I had seen in a film. The mushroom gathered in the sky until it was slowly dispersed by the wind." Another typical case occurred in Yangbajain geothermal field, where the latest explosion took place on December 4, 1977. This event was closely related to the drilling activity of test-hole Y-1 and witnessed by many geothermal experts. It is believed that the large hot lake with an area about 7,350 $m^2$ and a maximum depth of 16.1 m might be originated from more than one hydrothermal explosion. In southern Xizang the hydrothermal explosion activities are not only very young but also occur very frequently. Excellent examples are observed in Kurme, where up to 20 small explosions have occurred in one year and in general, one can expect 4 or 5 events per year now. We also notice that there are 7 hydrothermo-explosion areas in the south of Yarlung Zangbo River, and 4 are located in the north of this river. But all the geysers only occur in the north of Yarlung Zangbo. The natural geysers in China can be found in Xizang only, of which Tagejia is the largest one, but since 1975 four drillholes, at least, have changed into geysering eruption in Yangbajain geothermal field.

All the high-temperature hydrothermal areas are located in HGB (Fig. 1). The temperatures of the hydrothermal springs outside HGB are far below the local boiling points.

One of the interesting problems is whether some abyssal (or juvenile) constituents are present in the hydrothermal fluids of Xizang. It is very difficult in answering such a question, but nearly all the Xizang's boiling springs discharge sodium-chloride or bicarbonate-sodium-chloride waters with some $CO_2$ and $H_2S$. In addition, the hot waters are rich in boron, lithium, rubidium, cesium, mercury, antimony, fluorine and arsenic. In general, the $HBC_2$-content in springwaters is about 10-30 ppm, but in more than 58 hydrothermal areas, this exceeds 100 ppm, and in 5 areas the $HBO_2$ content exceeds 1,000 ppm. The maximum content of Cs is more than 50 ppm(4). The hydrothermal alteration and mineralization with cinnabar, stibnite and pyrite are often observed in some high-temperature hydrothermal areas. It is worth noting that the salt lakes are widespread over the whole northern Xizang Plateau and they are rich in boron and lithium, but the springwaters discharged from the north are very poor in these elements.

## CONTRIBUTIONS OF AFOREMENTIONED CHARACTERS TO THE XIZANG PLATEAU'S TECTONIC MODELLING STUDIES

Thermal springs are nearly omnipresent upon the earth's surface, but according to plate theory, the hydrothermal areas manifesting on the earth surface may be broadly classified into intraplate hydrothermoactivities and the plate-marginal ones. The intensity of the intraplate hydrothermal activity is relatively low. The plate-marginal hydrothermal areas (Fig. 2) seem to show the following characteristics:

1. Distributing zonaly and extending to thousand kilometers with width only tens or hundreds of kilometers.

2. Often associating with active or recent volcanism and the magma pockets may serve as the direct heat sources for those hydrothermal activities;

3. The hydrothermal manifestations are widespread, while the intraplate hydrothermal areas are relatively sparse;

4. The intensity of the plate-marginal activity is comparatively higher than those of the intraplate ones and the high-temperature manifestations such as hydrothermal explosion, geysers and boiling springs (including fumaroles and the sulfur flowers) are only present in the plate-marginal areas. Consequently, there occur considerably high heat flows and high geothermal gradient;

5. In general, the springwaters discharged from plate-marginal zones often contain some special constituents and are of particular hydrochemical types;

6. The plate-marginal geothermal belts are often also seismic zones.

In spite of lacking in active volcanism, we do believe that the active hydrothermal areas in southern Xizang are belonging to the plate-marginal geothermal belt. In other words, there is a boundary zone between two different plates. This is an important fact which must be considered when any tectonic model about the Xizang Plateau is to be constructed.

In general, the geothermal activity in a convergent zone is originated from the relative motion between the downgoing plate and the upthrusting plate. Such a motion leads to partial melting inside the crust and some convection of the upper-mantle materials. The plate-marginal geothermal belts are invariably located in the upthrusting side, and the subducting side is a cold one. This is the case of the HGB. As stated above, the high-temperature hydrothermal areas in Xizang are located in both sides of Yarlung Zangbo River and most of the hydrothermal explosions occur in the south of this river. We suggest that the active suture zone which is intimately associated with the origin of HGB might only be the Indo-Gangetic

Deep Trough. The Yarlung Zangbo River has nothing to do with the HGB but only a fossil suture zone at present. This is the second controlling factor which must be considered when constructing any tectonic model for the Xizang Plateau.

In the light of these basic facts we propose a conceptual model for the evolution of the hydrothermal activities occurring in Xizang. We suggest that when the Pangea was disintegrated in Jurassic, the Indian Plate moved northeastward (case A in Fig. 3). The Tethys oceanic crust progressively subducted along the Gandise Arc beneath the Eurasian Plate. When such a motion intensively progressed, the volcanism and hydrothermal activity behind the Gandise Arc began, which is precisely the hydrothermal activities witnessed by us in northern Xizang in nowadays (case B in Fig. 3). From Eocene to Oligocene, the Tethys crust wholly disappeared and the Indian and Eurasian continents came into direct collision along the Yarlung Zangbo zone, and the downgoing of the Tethyan oceanic crust stopped. Consequently, the hydrothermal activities behind the Gandise Arc became weaker and weaker (case C in Fig. 3). At the same time the Indian Plate continued moving to the north and under their great thrust, the Indian continental crust dipped beneath the Xizang's crust along the Indo-Gangetic Deep Trough. Such a process could result in partial melting inside the Xizang's crust and the convection of the upper-mantle materials. At last the HGB appeared and the hydrothermal activities behind the Gandise Arc declined (case D in Fig. 3). Some areas might have retrograded to salt lakes under such environment as in northern Xizang Plateau. To many of them only huge travertine deposits and a lot of low-temperature warm springs with small water flows remained. It is evident, from the great scale of the fossil travertine, that the historical hydrothermal activities might have been stronger than those of the present. But the evolution model of the hydrothermal activities in northern Xizang might be complicated by the important phenomena such as the occurrence of the active volcano named Ka-er-da-xi in Kunlun. We believe that the plateau's crust upthrusted upon the Tarim Massif (case D in Fig. 3), which might cause the Ka-er-da-xi event and produce the superposed effect upon the hydrothermal activities in northern Xizang.

In contrast with those mentioned above, salt lakes and the large-scale travertine deposits are conspicuously lacking in southern Xizang, but the hydrothermal explosions there occur rather frequently. We tend to believe that the HGB is a typical developing geothermal belt.

## ACKNOWLEDGEMENTS

We greatly appreciate the long term cooperation with colleagues in our group in Xizang's geothermal investigation. We gratefully acknowledge Professors Fu Cheng-yi, Wang Ren and Chen Zong-ji for their invaluable encouragements and reading of the manuscript.

References

(1) Montgomerie, T.G., Narrative of an Exploration of the Namcho, or Tengri Nur Lake, in Great Tibet, made by a Native Explorer, during 1871-2: Royal Geog. Soc. (London) Jour., 45, 315-330 (1875)
(2) Hedin, S.A., The Southern Tibet, Stockholm, Lithographic Institute of the General Staff of the Swedish Army, 3, (1917).
(3) Pranavananda, S., The Source of the Brahmaputra, Indus, Sutlej and Karnali with Notes on Manasaravar and Rakas Tal, Geog. Jour., 93, 126-136 (1937).
(4) Tong Wei, Zhang Mingtao, et al., Geothermals beneath Xizang (Tibetan) Plateau, The series of the Scientific Expedition to Qinghai-Xizang Plateau, Science Press (in press) (in Chinese).
(5) Zhang Ming-tao, Guo Guo-uing, Hydrothermal Explosions in Xizang (Tibet) Plateau, in this Proceedings of Symposium on Qinghai-Xizang (Tibet) Plateau. (1981)


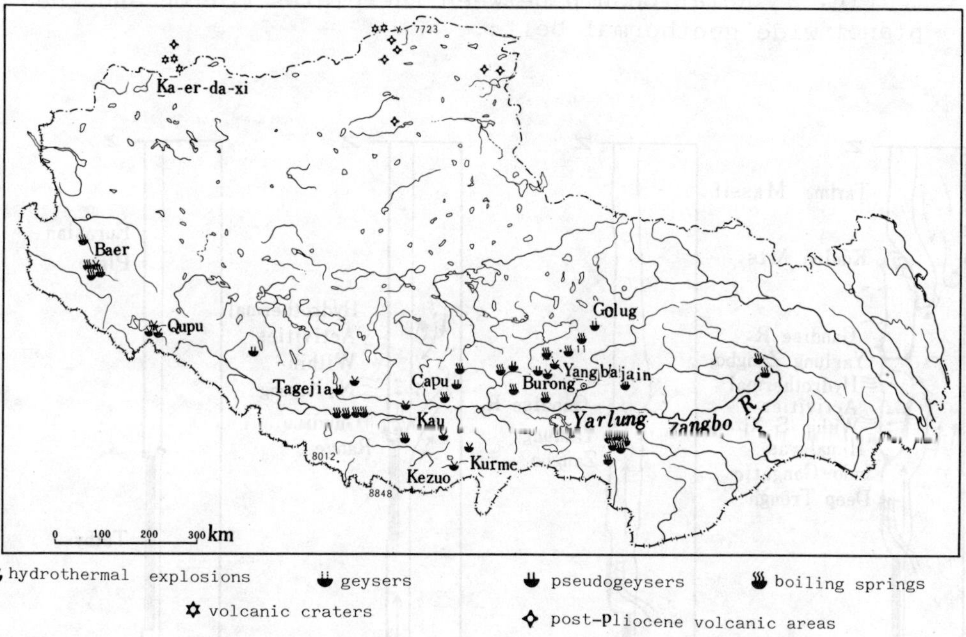


Fig. 1 Geographical distribution of the high temperature hydrothermal areas in Xizang.

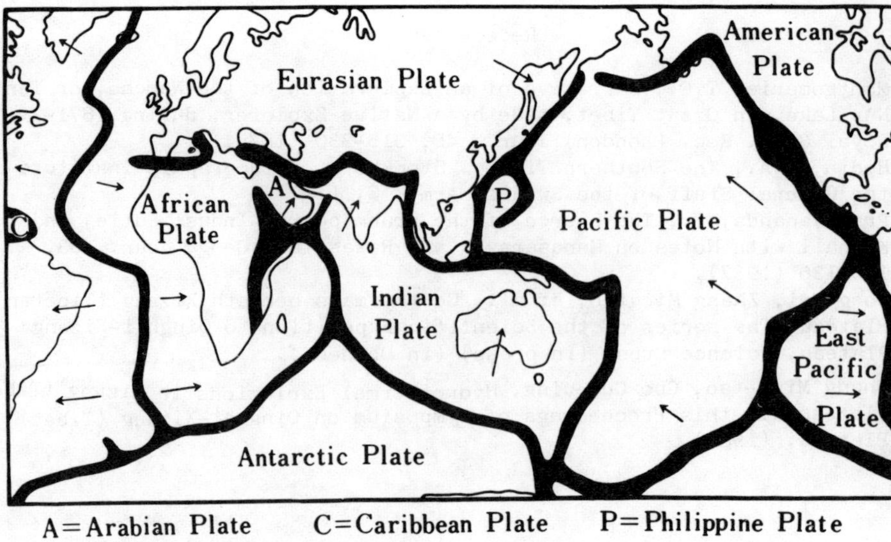


Fig. 2 Relationship between the plates motion and the planet' wide geothermal belts.

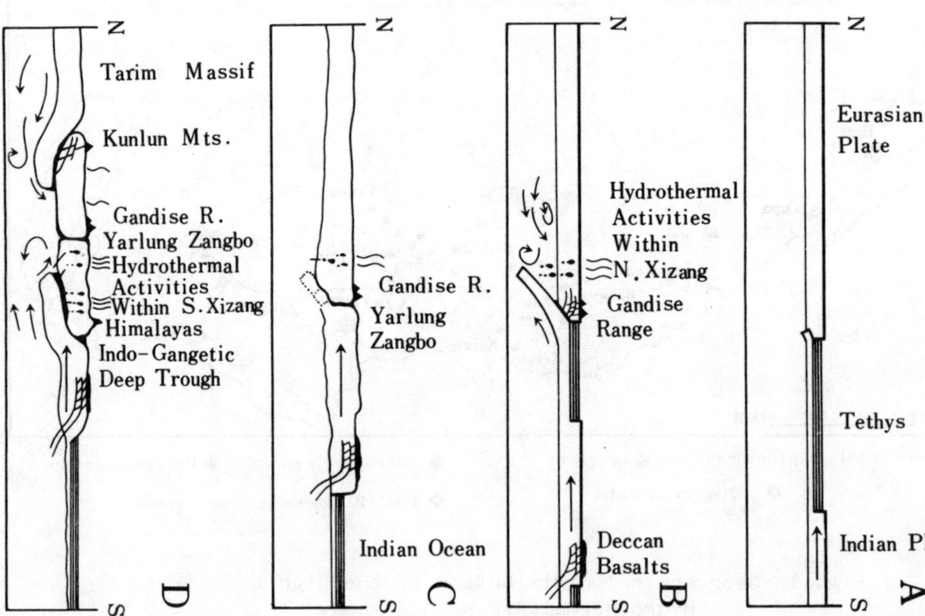


Fig. 3 Conceptual model of evolution of the hydrothermal activities within Xizang

# CRUSTAL EVOLUTION AND THERMAL STATE OF TIBET

M. Nafi Toksöz, W. Roger Buck
Department of Earth and Planetary Sciences
Massachusetts Institute of Technology
Cambridge, Massachusetts 02139

Albert T. Hsui
Department of Geology
University of Illinois
Urbana, Illinois 61801

## ABSTRACT

The crustal structure and seismic attenuation under Tibet have been studied by seismic body and surface waves. The crust is about 70 km thick. Average velocities in the crust are lower than average continental values. The attenuation in the lower crust (below about 50 km) is very high. The Rayleigh waves in the period range of 30-50 seconds are highly absorbed. Shear wave Q values may be as low as $Q_s = 10$ in the lowermost part of the crust. The above properties may be explained by the thickening of the crust through internal deformations and elevated temperatures. We have examined the thermal consequences of such an evolution. The initial conditions are those determined by the subduction processes that have preceded the Himalayan episode of continental collision. Internal heating due to redistribution of crustal radioactivity contributes significantly to the thermal regime. The effects of shear strain heating during crustal thickening are less significant. In a period of about 40 m.y., temperatures in the lower crust will exceed 700°C. This could product partial melting and result in excessive attenuation of shear waves.

## INTRODUCTION

Tibet is the largest continental plateau on the earth. It covers approximately 700,000 km$^2$ with an average elevation of 5 km. Its formation and present state are in part tied to the collision of India with Eurasia. In this paper we propose a model for the evolution of the Tibetan Plateau based on geological and geophysical data and we test the thermal consequences of such an evolution using finite difference calculations.

Our model implies that before the continental collision Tibetan lithosphere was heated from underneath by induced mantle convection. After collision crustal thickening occurred by internal deformation; the heat flux from the mantle and radioactive heat sources in the crust raised the temperatures in the lower crust and caused partial melting. This model is consistent with the geophysical data. In the paper we first review these data and then describe the results of the thermal calculations.

## GEOPHYSICAL DATA AND STRUCTURE OF THE TIBETAN CRUST

The data we consider in constructing our models come from gravity, magnetics and seismic data.

The best coverage of gravity data for Tibet comes from satellite data. Because of its large area and great elevation, Tibet stands out as a major feature on satellite gravity maps. In Fig. 1 the free air gravity map of Tibet and vicinity based on the GEM-8 model is shown[1]. The Tibetan anomaly can be modeled by an isostatically compensated or partially compensated density distribution. Satellite data alone cannot reveal the degree of isostatic compensation since the anomalies in the mantle contribute strongly to the measured field. Limited land-based data[2,3] favor compensated models. The great elevation of Tibet, if explained as the result of the isostatic compensation of the 70 km thick crust, requires a very low average crustal density $\rho_c \gtrsim 2.8\ g/cm^3$. An alternate explanation[4] is that only part is due to isostatic compensation and part is due to the compression of the area by the continued convergence of India and Asia.

Magnetic field measurements from satellites show a very large negative anomaly over Tibet (Fig. 1). Magnetic data only see crustal anomalies[5]. Inversion of the data in terms of crustal magnetization (Mayhew, 1980, personal communication) shows the entire plateau region to be a fairly uniform area of slightly lower than normal magnetization. The far western region of the plateau is a region of particularly low magnetization. A possible cause for the generally low level of total magnetization of the crust is that the depth to the Curie isotherm, taken to be about 500°C[5], is less than for normal crust. Recent calc-alkaline volcanics, which have been found over much of Tibet[6,7,8], suggest elevated temperatures and possible partial melting in the crust. Abundant hot springs are found throughout southern Tibet and some active hydrothermal areas are found in northern Tibet, but many areas there now appear to be cooler than in the past[9].

The most specific information about the crustal properties under Tibet comes from seismic data. Although there are relatively few earthquakes and stations in Tibet, seismic waves propagating across the Tibetan Plateau provide information about the average properties. Rayleigh waves from many paths crossing Tibet have been analyzed by a number of investigators to obtain phase and group velocity curves[10,11,12,13,14]. These data can be complemented by some body wave data[11], refraction results[15], and attenuation measurements [10,16].

In this study we inverted the phase and group velocity data, together with the refraction data, to obtain crustal structure and velocities. Some examples of Rayleigh waves crossing Tibet are shown in Fig. 2. The source is an explosion at Lop Nor. Note that as the fraction of the Tibetan segment increases in the total path, both dispersion and attenuation characteristics change. For example, Rayleigh waves recorded at Kabul (KBL) have well-defined long-period (T ≥ 40S) waves but these are strongly attenuated across Tibetan paths leading to New Delhi (NDI) and Shillong (SHL). This suggests that strong seismic attenuation is taking place in the lower crust under Tibet. The attenuation as a function of period for the two paths from Lop Nor which cross Tibet is shown in Fig. 3.

The Rayleigh wave phase and group velocity data from all available paths are averaged and inverted to obtain a crustal velocity model (Tubman, 1980, personal communication). The data and theoretical fit are shown in Fig. 4. In the inversion, the compressional wave velocities were specified from the refraction data for central Tibet[15]. Shear wave velocities, densities and crustal thickness were iterated to obtain the best fit.

The velocity profiles of the final model are shown in Fig. 5. Important features of the model are: (a) the crust is 70 km thick and this is verified independently by both the refraction and surface wave data; (b) the average P and S wave velocities in the crust are low; (c) there may be a "lower" velocity layer below a depth of 45 km, although surface waves alone can resolve neither the thickness nor the actual velocity of this layer; and (d) velocities below the Moho, although at 70 km depth, correspond to typical continental upper mantle velocities ($V_p$ = 8.15 km/sec, $V_s$ = 4.6 km/sec). This model suggests that the crust under Tibet is thickened not by stacking two layers of a continental crust but by a complex process that might involve plastic deformation and vertical mixing.

The attenuation data complement the velocities for determining the crustal properties. Fitting the Rayleigh wave attenuation data of Bird and Toksöz[10], part of which is shown in Fig. 3, requires an increased attenuation in the crust below about 45 km depth, although the thickness of this zone and the attenuation ($Q_\beta^{-1}$) value cannot be determined independently. The lowered velocities along with the increased attenuation suggest partial melting in the lower crust below the 45 km depth. Feng and He[16] have also come to this conclusion for the eastern part of the plateau. Both the velocities and attenuation and the implied partial melting will serve as constraints for the thermal calculations in the next section.

## EVOLUTION OF TIBET

How Tibet formed and how it evolved to its present state is an important problem. We will study this theoretically using finite difference calculations. Our models will start from the time of subduction of the oceanic plate on the leading edge of the Indian plate. Then it will go through the collision of India with Eurasia. After this, consequences of compression and crustal thickening in Tibet will be modeled in detail.

The possible sequence of events in the collision of the Indian and Eurasian plates is shown schematically in Fig. 9[4,17,18,19].

The different stages of this process are:

1. The oceanic slab subducts under a continental lithosphere which may be "proto Tibet".

2. Subduction induces convection in the asthenosphere and heats the lithosphere from underneath.

3. The collision of India and Asia occurs and the Indian suture develops. Once the collision occurs, continental crust of India cannot be subducted because of its low density. The continued motion of India induces shortening of Tibet with the heated lower crust and lithosphere deforming plastically.

4. As Tibet rises the stresses required to lift it further become greater. After a critical elevation is reached, thrusting occurs on the colder Indian side of the plate. At this stage the Main Central Thrust (MCT) develops.

5. Subduction under the MCT continues until buoyancy forces become large. Then the Main Boundary Thrust forms.

6. During these times the convergence zone is still moving northward; Tibetan crust is thickening and becoming hotter at depth. The top half of the crust, however, remains elastic and transmits stress to Asia.

The regional thermal model based on this sequence of events was carried out using finite difference calculations with both convection and conduction included. The regional temperature profiles are shown in Fig. 7. The model shows that broken oceanic slab is still cool. Temperatures under the Himalayas are normal or below normal except in local regions of high strain heating. Lithospheric temperatures are higher than normal under Tibet. The geologic history is certainly much more complex and there are indications of different dips of subduction in different parts of the plateau[20]. A southward dip of subduction would somewhat alter the mantle calculations, but the crustal calculations include a range of possible mantle temperatures wide enough to encompass this possibility. The detailed calculation for Tibet focuses on the crust.

The thickening of Tibet is modeled as being due to strain as the crust is thickened by internal deformation. The thermal history of the crust is modeled using a finite difference solution of the equation of heat conduction which includes the option of having material moving up and down due to internal deformation. To do this the crustal grid point spacing is increased while thickening is going on. The strain heating is turned off at the end of thickening.

Strain heating for an area being thickened is difficult to estimate and almost certainly involves both shear strain heating along faults and heating due to ductile flow. We treat only the case of heating along faults. This overestimates the heating if ductile flow occurs in the lower crust. The amount of heat produced in thickening is calculated for thrust faults dipping at 45°. The faults are taken to be distributed uniformly through the crust. Each grid point in the model gets the same amount of strain heat during the period of thickening. Two cases are considered in the modeling: $\tau = 1$ kb and $\tau = 2$ kb ($\tau$ = shear stress). In both cases the average heat production rate is less than that due to the radioactive heat sources.

The amount and distribution of radioactive elements in the crust is another important input to the models. Again two cases are tested. One has two layers in the crust of equal thickness, but with three times the abundance of heat producing elements in the upper crust, which is considered to be the normal case. The other case is for homogeneous distribution of radioactive elements. This homogenization could occur during the thickening by the mixing of the layers. In both cases the same average crustal heat production rate is assumed. It is taken to be $7.1 \times 10^{-6}$ ergs/cm$^3$.

In the modeling both a normal mantle temperature gradient and higher mantle temperatures are considered. The scheme includes 30 km of mantle lithosphere which is pushed down and heated by the crust. No heat sources are put in the mantle material, but at the base of the mantle layer a fixed temperature boundary condition is imposed to simulate the effect of convection. The boundary condition at the surface is that it is maintained at 0°C. The initial condition for all cases is the steady state temperature solution for the particular crustal arrangement tested.

Four examples of these models are shown in Fig. 8 which gives the depth to the isotherms shown against time during and after the thickening of the crust. Partial melting is taken to occur when lower crustal temperatures are above 700°C. We see that with uniform crustal radioactive abundances (1 layer crust), (Fig. 8a,b,d), low shear stress and normal mantle heat flux, we do get partial melting in the lower crust. We then see that concentrating the radioactive heat sources in the upper half of the crust (Fig. 8c), even with doubled strain heating, leads to significantly lower temperatures. Next we see that the time span of the thickening has little effect on the final temperature profile (as illustrated in Fig. 8a). Of course, if most of the thickening took place in the last part of the 40 million year period since the collision, the crustal temperatures would be lower. Finally, the effect of a hotter mantle, giving slightly less than twice the mantle heat flux, greatly increases the lower crustal temperatures.

## CONCLUSIONS

Available refraction data and the Rayleigh wave phase and group velocities independently and jointly require about a 70 km thick crust under Tibet. Average compressional and shear velocities in the crust are low and there is relatively little change of velocities with depth. The above properties suggest that the crust is thickened by some mechanism that may have involved vertical mixing. The high attenuation in the lower crust implies high temperatures and possibly partial melting below about 45 km depth.

The thermal consequences of the crustal thickening due to internal deformations, calculated using a finite difference scheme, indicate that redistribution of crustal radioactivity contributes most significantly to the heating of the thickened crust. Heat flux from the mantle and shear strain heating during crustal thickening are other factors that affect the temperatures, although the effect of shear strain heating is less important. In about 40 m.y., temperatures in the lower crust may rise by about 200°C over typical lower crustal temperatures. This could produce partial melting in the lower crust, and can explain the low velocities and high attenuation of seismic waves.

In the middle and upper part of the crust, the temperatures are about the same or less than would be seen in normal crust. To explain the regions of high heat flow which are implied by hot springs, recent volcanics and magnetic data, upward material and heat transport is required. The partial melting of the lower crust may allow hot material to move as diapirs into the upper regions of the crust, thus heating these areas. This may also contribute to the general manner in which radioactive elements are concentrated upward in the crust.

## References


(1) Khan, M.A., Comparative evaluation of recent global representations of the earth's gravity field. Geophys. J. R. Astr. Soc., 16, 535-553 (1976).

(2) Ambolt, N., Reports from the scientific expedition to the northwestern provinces of China under the leadership of Dr. Sven Hedin. Publ. 30 (II), Geodesy 2, Tryckeri, Attiebolaget Thule Stockholm (1948).

(3) Zhou Wen-hu, Yang Zhan-shou, Zhu Hong-bin, Wu Li-gao, Characteristics of the gravity field and the crustal structure in the eastern and middle regions of the Xizang Plateau. Proc. Symposium on Qinghai-Xizang Plateau, 79-80 (1980).

(4) Molnar, P., Tapponier, P., Tectonics of Asia: consequences and implications of a continental collision. Science, 189, 419-426 (1975).

(5) Wasilewski, P.J., Thomas, H.H., Mayhew, M.A., The Moho as a magnetic boundary. Geophys. Res. Lett., 6, 541-544 (1979).

(6) Burke, K.C.A., Dewey, J.F., Kidd, W.S.F., The Tibetan plateau, its significance for tectonics and petrology. Geol. Soc. Amer. Abstr. Programs, 6, 1027-1028 (1974).

(7) Kidd, W.S.F., Widespread late Neogene and Quaternary alkaline volcanism on

the Tibetan plateau (abstr.). EOS, Trans. Am. Geophys. Un., 56, 453 (1975).
(8) Norin, E., Exploration in western Tibet. Report from the Scientific Expedition to the Northwestern Provinces of China under the Leadership of Dr. Sven Hedin, Publ. 29, (III), Geology 7, Tryckeri, Attiebolaget Thule, Stockholm (1946).
(9) Tong Wei, Zhang Mingtao, Characteristics of geothermal activities in Xizang Plateau and their controlling influence on Plateau's tectonic model. Proc. Symposium on Qinghai-Xizang Plateau, 90 (1980).
(10) Bird, P., Toksöz, M.N., Strong attenuation of Rayleigh waves in Tibet. Nature, 266, 161-163 (1977).
(11) Chen, W.P., Seismic studies of central Asia: some characteristics of earthquake mechanisms and seismic wave velocity structure beneath the Tibetan Plateau. Ph.D. Thesis, Massachusetts Institute of Technology, Cambridge, Massachusetts (1979).
(12) Chen, W.P., Molnar, P., Constraints on the seismic wave velocity structure beneath the Tibetan Plateau. Proc. Symposium on Qinghai-Xizang Plateau, 84 (1980).
(13) Knopoff, L., Upper mantle under Tibet. Proc. Symposium on Qinghai-Xizang Plateau, 75 (1980).
(14) Teng Ta-liang, Crustal and upper mantle structure of Qinghai-Xizang Plateau. Proc. Symposium on Qinghai-Xizang Plateau, 82-83 (1980).
(15) Teng Ji-wen, Xiong Shao-po, Sun Ke-zhong, Yin Zhou-xun, Yao Hung, Chen Li-fang, Mu Teng, Lai Ming-hui, Wu Ming-chu, Su De-yuan, Wang Shao-zhou, Huang Wen-jian, Ou Ren-sheng, Hao We-cheng, Shao An-min, Gao En-yuan, Wang Meng-Lin, Lin Zhong-yang, Qu Ke-xin, Explosion seismic study for velocity distribution and structure of the crust and upper mantle from Damxung to Yadong of Xizang Plateau. Proc. Symposium on Qinghai-Xizang Plateau, 81-82 (1980).
(16) Feng Rui and He Zheng-gin, Q-value for surface waves in the eastern region of Xizang Plateau. Acta Geophysica Sinica, 23, 291-297 (1980).
(17) Bird, P., Thermal and mechanical models of continental convergence zones: Zagros and Himalayas. Ph.D. Thesis, Massachusetts Institute of Technology, Cambridge, Massachusetts (1976).
(18) Toksöz, M.N., Bird, P., Formation and evolution of marginal basins and continental plateaus. In Island Arcs, Deep Sea Trenches and Back-Arc Basins, M. Talwani and W.C. Pitman III, eds., Maurice Ewing Series, 1, 379-393, American Geophysical Union, Washington, D.C. (1977).
(19) Toksöz, M.N., Bird, P., Modelling of temperatures in convergence zones. Tectonophysics, 41, 181-193 (1977).
(20) Huan Wen-lin, Wang Su-yun, Shi Zhen-liang, Yan Jia-quan, The distribution of earthquake foci and plate motion on the Qinghai-Xizang Plateau. Acta Geophysica Sinica, 23, 269-280 (1980).
(21) Regan, R.D., Cain, J.C., Davis, W.M., A global magnetic anomaly map. J. Geophys. Res., 80, 794-802 (1975).

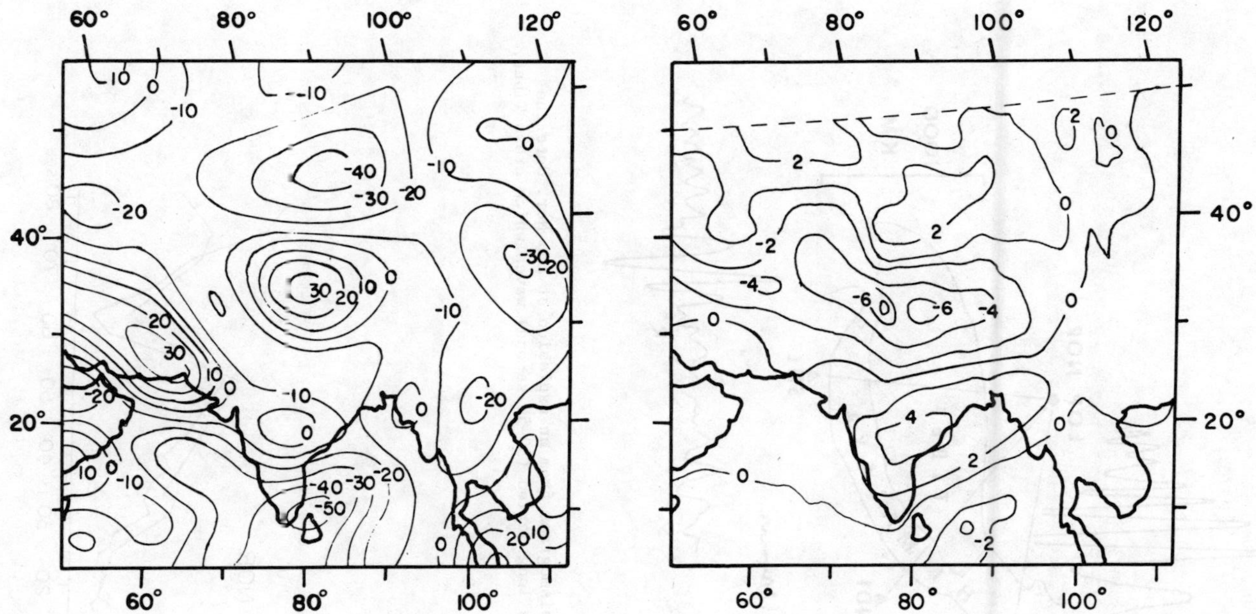

Figure 1. Comparison of gravity and magnetic anomalies over Tibet as measured by satellites. The heavy lines outline the landmasses and the latitudes and longitudes are also shown. On the left the free air gravity anomalies in milligals taken from the NASA Goddard Earth Model-8 are shown. The magnetic anomalies in gammas, for 500 km altitude(21), are shown at the right. Magnetic data above 50° north latitude are not reliable.

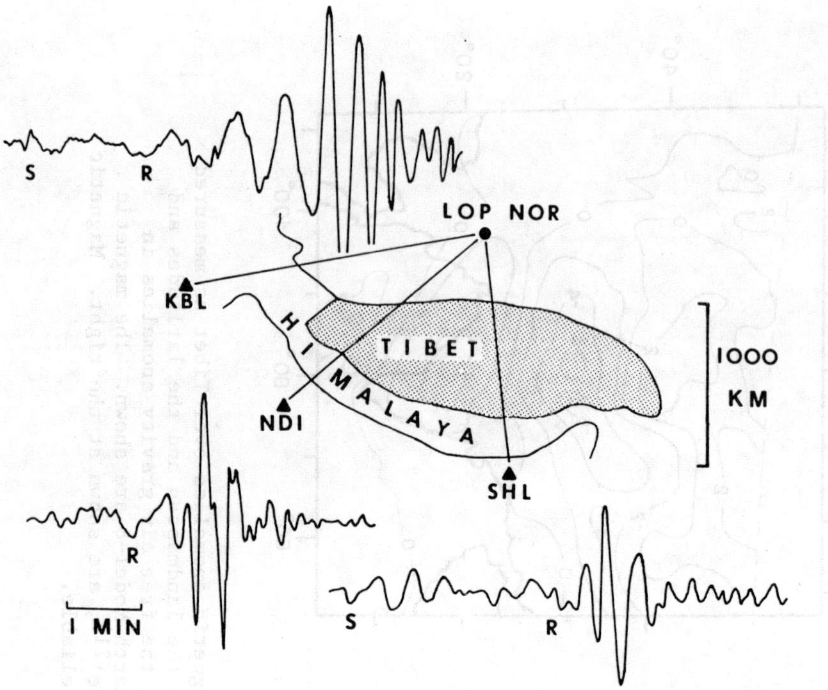


Figure 2. Rayleigh wave seismograms from an explosion at Lop Nor. Note the greater attenuation of the long period (30-50 second) waves which cross Tibet.

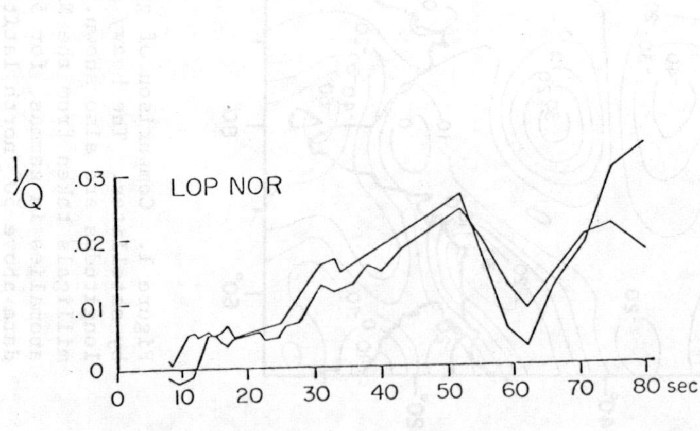


Figure 3. Observed attenuation (1/Q) of Rayleigh waves crossing Tibet for the Lop Nor explosion.

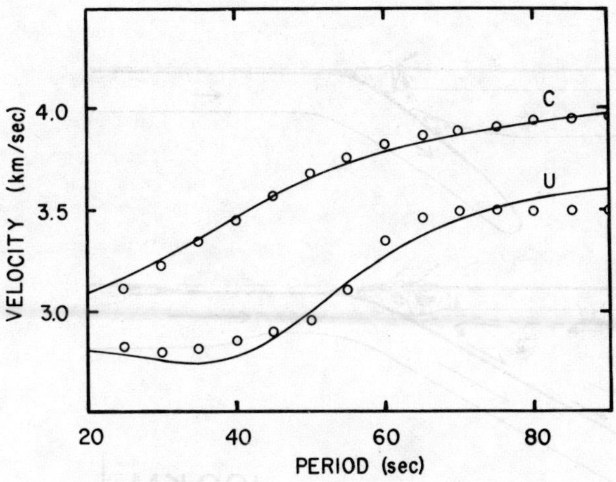


Figure 4. Rayleigh wave phase velocity (c) and group velocity (u) for the average of all paths crossing Tibet.

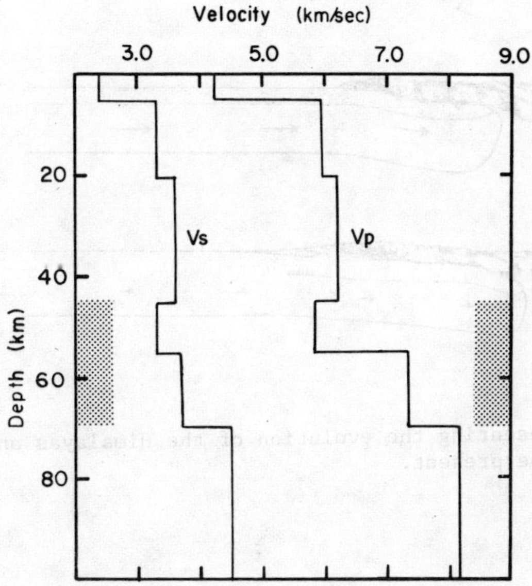


Figure 5. Crustal S-wave velocity model resulting from the inversion of Rayleigh wave data combined with the refraction P-wave profile (15) shown at right. Dotted area indicates the region of high attenuation.

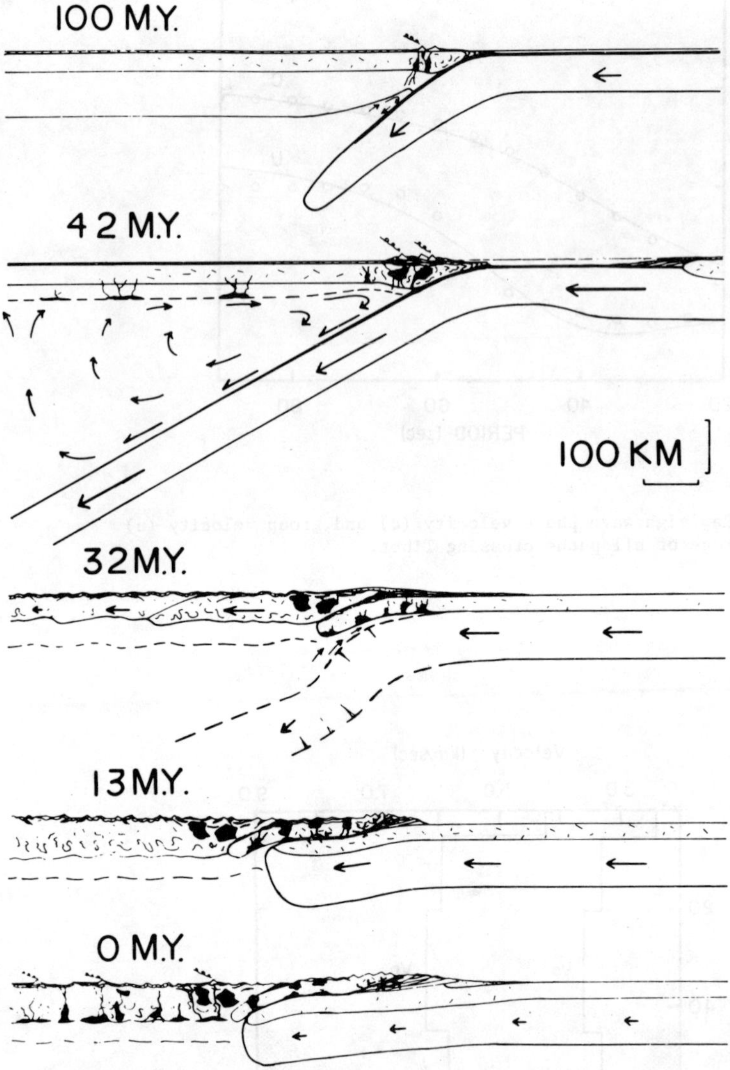


Figure 6. Schematic representing the evolution of the Himalayas and Tibet from the Cretaceous to the present.

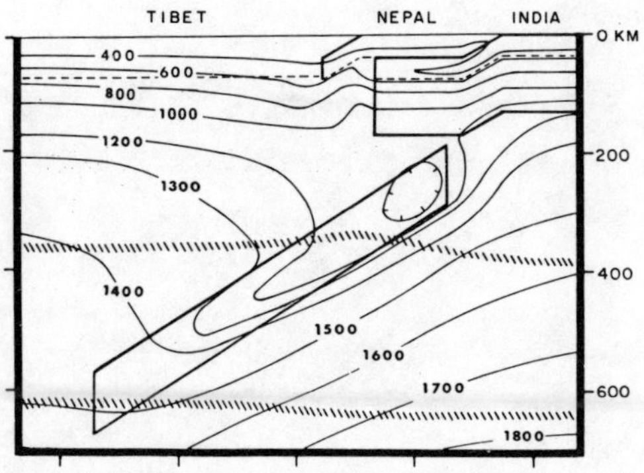


Figure 7. Present day temperatures in the Himalayan-Tibetan region that would result from the sequence of events shown in Fig. 6. The Moho is shown as a dashed line and upper mantle phase boundaries are shown as diagonal slashes. The broken slab is bounded by solid lines and is sinking. Note the general depression of temperatures under the Himalayas as a result of subduction of cold material. Temperatures are above normal under Tibet. From Toksöz and Bird[19].

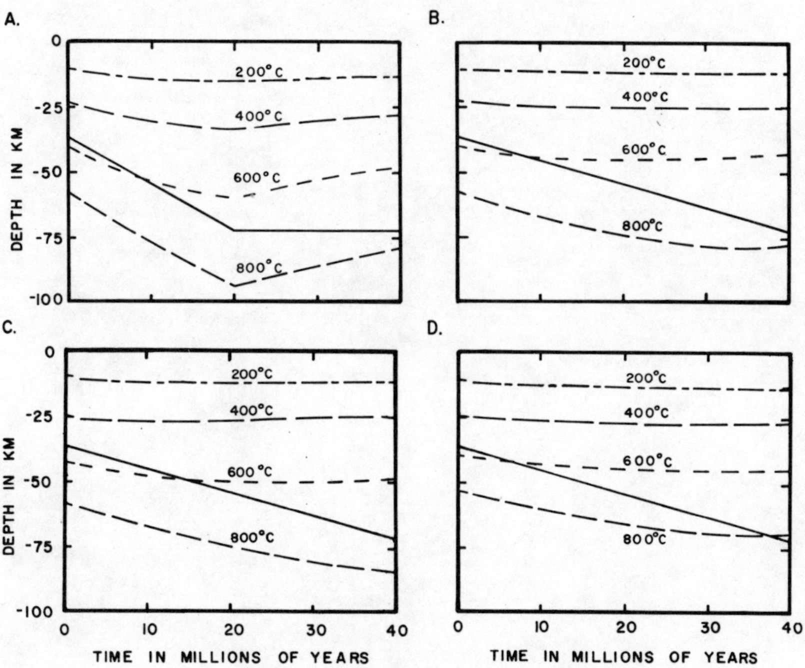


Figure 8. Results of the crustal thermal calculations. Dashed lines are depth to isotherms against time since the collision of India and Asia. Solid line is the crust-mantle boundary. All cases have the doubling in crustal thickness occurring in 40 million years except case A which has it taking 20 million years. Case A is for a 1 layer crust and 1 kb shear stress. Case B is for a 1 layer crust and 2 kb shear stress. Case C shows that a 2 layer crust even with 2 kb shear stress leads to much lower temperatures. Case D has one third more mantle heat flux and thus higher temperatures in the lower crust.

# PRELIMINARY RESEARCHES ON HYDROTHERMO-EXPLOSION ACTIVITIES AND THEIR EXPLOSION MECHANISM IN XIZANG

Zhang Ming-tao
(Integrated Survey Commission of Natural Resources, Academia Sinica)

Guo Guo-ying
(Department of Geology, Beijing University)

Hydrothermo-explosion is one of the most violent hydrothermo-activities. This spectacular phenomenon has been reported only in a few countries in the past such as United States, New Zealand, Japan and Italy, but eleven hydrothermal areas with active or recent hydrothermo-explosions in Xizang were discovered during our field inspection in 1975-1976 (Fig. 1).

It is believed that the main factor inducing hydrothermo-explosion might be the sudden flashing of high-temperature water steaming within the shallow reservoir, which consequently break out the over-lying layers and on the ground level a round-like or irregular-shaped crater often forms. Most of the mud and gravels thrown out of crater may be filled back to the bottom of the crater and cover the heat reservoir, or forms a round constructional body around the rim of the crater. Generally there remains one or more tunnels connecting with the reservoir, while the hot water gushes out continuously and often forms a hot pool or lake at the ground surface (Photo 1).

The most violent hydrothermo-explosions occurred in recent years took place in Qupu hydrothermal field. It is just located nearby the suture zone between Indian subcontinent and Eurasian continent. There the number of recognizable explosion craters are over 30. An enormous explosion occurred on Nov. 12, 1975. Several eye witnesses proved that: "The explosion made a great noise, and a huge black air column went up high into the sky, reaching a height of about 800-900 m. until a mass of cloud was wafted away by wind." The diameter of this explosion crater is about 25 m. (Photo 2), but the biggest one occurred in Qupu is about 100 m. where the mud and gravels accumulated around the crater formed a round rim about 10-18 m high. The bottom of the crater is filled with water and become a hot lake. A hot river flows from southeast across this lake. The forming date of the crater is unknown, but it looks very young. (Photo 3).

A hot lake is located in Yangbajain of Damxung county. Its area is about 7,350 $m^2$, and the maximum depth is about 16.1 m. The concentrated isobaths of the lake bottom lie north to east and north to west (Fig. 2). It is evident that they are in accordance with the trend of the regional structure line. Nearby the hot lake, the shallow heat reservoir may be buried about only 30 m. in depth (Fig. 3). In a drill hole of 60 m. of depth, the measured temperature is 165°C. According to the geometric figure of the lake, the geologic conditions, and topographic characteristics, we suggest that the hot lake was also formed by the past hydrothermo-explosion. The cover of the heat reservoir at the bottom of the lake was thrown to the surface along the fault during the explosion.

On Dec. 4, 1977, a latest explosion event occurred in drillhole Y-1 of Yangbajain geothermal field and formed a big crater about 10-12 m in diameter. Energy released from this event was calculated using the method of D.E. White and it is equivalent to about 3 tons TNT. The initial temperature of the explosion is thought to be about 170°C.

In general, the characteristics of the hydrothermo-explosion occurred in Xizang are as follows:

1. At least 8 hydrothermo-explosion areas have been found since 1950, and most of the explosions are of the intermediate or small scales, but their frequencies are very high. To our knowledge up to the present, the Kurme hydrothermo-explosion occurs often 4-5 times per year. Sometimes there are as many as 20 events per year. The population in Xizang is sparse. It is thought that there are still many explosive events which remain to be discovered.

2. Most hydrothermo-explosion areas are located in the fault valleys with altitude about 4,500 m. Around which there are high snowy peaks. During the early Quaternary period when the plateau was continuously uplifted, more glacial period occurred with the accumulation of thick and porous glacial products, which in late Quaternary was covered with dense alluvial-lacustrine deposits, the thickness of which is about tens of meters. In the hydrothermo-explosion areas, the sinters are not very developed up to now. Thus, it seems that the hydrothermal explosions in Xizang are rather young events and the buried depths of these reservoirs are relatively shallow.

3. The hydrothermo-explosions over the whole world invariably have occurred around the recent or active volcanic areas. But active volcano is not present in south Xizang. Therefore, the authors suggest that the heat source of Xizang hydrothermo-explosions might be the magmatic bodies which have intruded into the shallow crust.

4. Some hydrothermo-explosion craters act as a quiet hot pool and deposited a large sinter terrace. Such a pool might evolve into underground reservoir of the geysers. Fig. 4 shows the evolution model (Fig. 4): First, the hydrothermo-explosion crater was formed. Second, the crater was evolved into a hot pool and the sinter deposited. Third, the mouth of this pool gradually contracted into the shape of a jar because around the water surface, the temperature was low in comparison with other parts and the sinter growth was faster. Fourth, the underground water reservoir was formed and a geyser might arise.

The underground reservoir is a necessary condition for the formation of the spring-like geyser. Thus, there is certain transitional relationship between hydrothermo-explosion and geyser in Xizang. The distribution of the hydrothermo-explosions is along both sides of the continental sūture zone, the Yarlung Zangbo River, but the geysers are only in the northern side. That means the history of the hydrothermal activities in the northern side is comparatively older.

The genetic mechanism of hydrothermo-explosion in Xizang can be summarized as follows:

1. Due to the intensive activities of the heat source, in the reservoir bottom the hot water mass forms. It pass through the porous glacial deposits and rise to the critical position of flash. Such a process will easily induce the explosion.

2. The upper part of the shallow reservoir may be reduced in its permeability, or even more subjected to self-sealing due to mineral-deposits or hydrothermal alteration. In such cases the upflowing heat can not be easily conducted to the surface and must be stored in shallow reservoir. Thus the hydrothermal-explosion might occur easily.

3. The underground channel may be destroyed or blocked up with the collapsed debris due to the violently turbulent water flow, which facilitate the accumulation of heat. This is also a factor of explosion, but the scale of which is often small.

4. Two small hydrothermal explosions occurred in Yangbajain geothermal field during 1975-1977, one of these two craters is just located on the drill site of Y-1, and another is located in the neighborhood of Y-2. The authors suggest that these events were closely related to uncontrolled blowout of these drillholes which might form underground cavities just beneath the drillhole and weaken the overlying caprocks.

5. According to the observation in Qupu, the bottoms of some big craters are often recovered with mud and gravels thrown out during explosions. The margin between the new and old caprocks is the weak zone where even small explosions can easily occur.

6. The earthquake may also be the inducing factor of explosions in Xizang.

## References


(1) Muffler L.J.P., White D.E., Truesoell A.H., Hydrothermal explosion craters in Yellowstone National Park, Geological Society of America Bulletin, 82, (1-3), 723-740 (1971).

(2) Tong Wei, Zhang Ming-tao et al., Geothermals beneath Xizang (Tibetan) Plateau Science Press (1980) (in press, in Chinese).

Fig. 1 Geographical distribution of the hydrothermo-explosion areas in Xizang

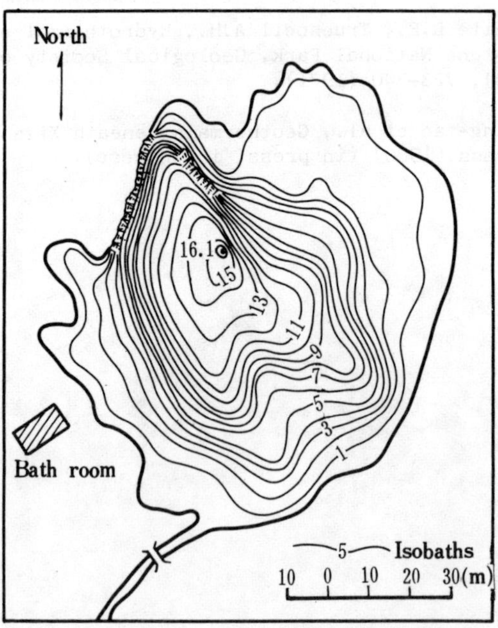


Fig. 2 The isobaths map of the hot lake bottom in Yangbajain

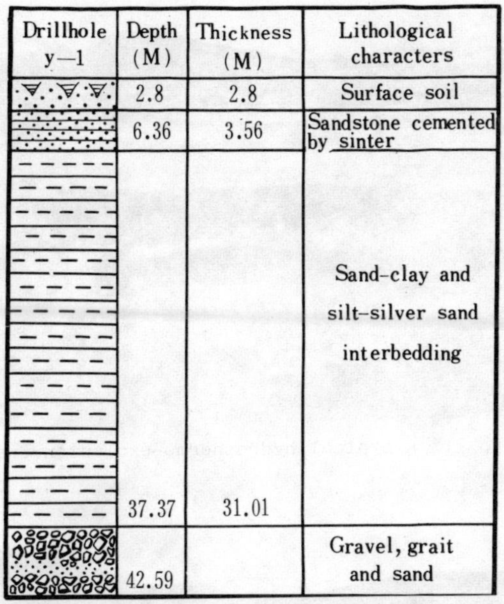


Fig. 3 The section map of drillhole Y-1 in Yangbajain

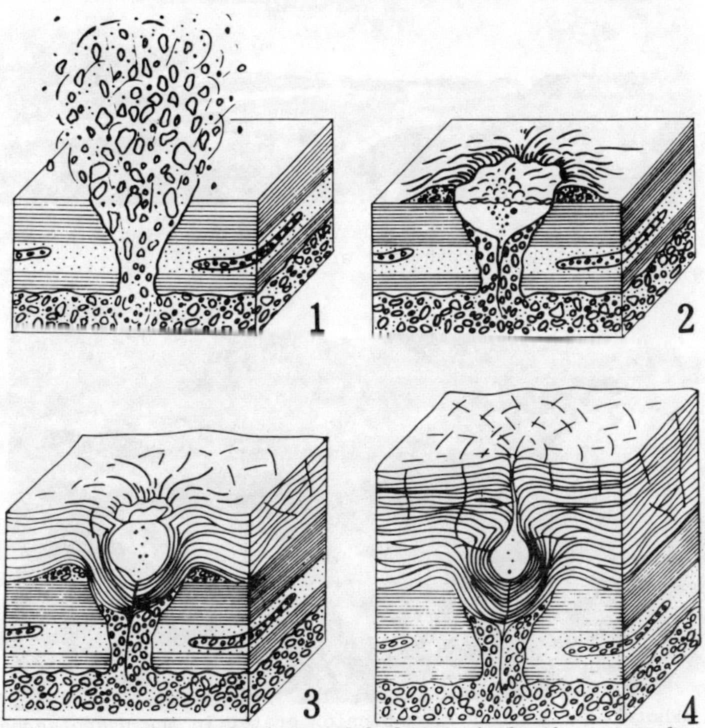


Fig. 4 The evolution model from a hydrothermo-explosion crater to a underground water reservoir.

Photo 1 A typical hydrothermo-explosion crater

Photo 2 Here an enormous explosion occurred on Nov. 12, 1975 in the hydrothermal area of Qupu

Photo 3 The biggest explosion crater in the hydrothermal area of Qupu

# DISTRIBUTION OF GEOTHERMAL ACTIVITY AND CHARACTERISTICS OF GEOPHYSICAL FIELDS ON THE XIZANG PLATEAU

Wei Si-yu Teng Ji-wen

Yang Bing-ping Hu Zhong-yi

(Institute of Geophysics, Academia Sinica)

## ABSTRACT

The Xizang Plateau is very intense in geothermal activities and the crustal structure is very complex and earthquakes are frequent.

Based on geothermal data and various other geophysical field measurements in the region the authors have made certain studies and analyses. A unified interpretation of the nature of the deeper part of the crust is given. It indicates that the cause of geothermal distribution and its activity on the Xizang Plateau is the high temperature in the crustal media and the existence of melting or partial melting matter in the crust. The geothermal activity is affected by the deep crustal structure and material migration as well as the geothermal energy provided by the heat evolved during the collision process of the Indian Plate and Eurasian Plate and magmatic intrusion.

## INTRODUCTION

Xizang Plateau has been formed in recent times. In elevation, it is the highest zone of hydrothermal activity in the world. Its activity is no less intense than that of regions of recent and present volcanic activities. These surface phenomena reflect the peculiar state of the materials in the deep crust and the structural difference of its media. Thus on the basis of the geothermal point of view and the nature of various geophysical fields it is significant to investigate the deep structural background of geothermal activity and the plate tectonics of the Xizang Plateau.

## THE CHARACTERISTICS OF HYDROTHERMAL DISTRIBUTION ON THE XIZANG PLATEAU

Through several years of field survey on hydrothermal regions of the plateau by Academia Sinica, it has been found(1) that there exists intense and extensively distributed surface hydrothermal activity. They are mainly found north of the Himalayas, south of the Nyainqentanglha Range and Gandise Range as well as the Yarlung Zangbo River valley. Over 600 hydrothermal localities have been found altogether (Fig. 1). There are no less than 20 kinds of hydrothermal display including all types of hot springs of the world, such as hydrothermal explosions, geysers, boiling springs and so on. The temperature of the hot springs varies from 35°C to more than 100°C. Up to the present 11 hydrothermal explosions, 3 geysers and 42 boiling springs have been discovered. But no boiling spring is found in the Indian subcontinent to the south, Gandise Range and Nyainqentanglha Range to the north, even those with temperatures higher than 55°C are rare and the strength of hydrothermal activity is very different on both sides of the Yarlung Zangbo River, behind the Himalayan Arc. From the surface distribution of the hot springs they do not distribute in active volcanic regions which seems to differ from those of other countries such as Italy, New Zealand and U.S.A. So far as we know these hydrothermal phenomena are closely related to contemporary tectonic movements and magmatic activities. Obviously, this great geothermal activity belt which exists within the transitional zone of collision and compression of the Indian and the Eurasian Plates is also closely related to the global geothermal zones (Fig. 2)(1).

## HYDROTHERMAL DISTRIBUTION AND CHARACTERISTICS OF GEOPHYSICAL FIELDS ON THE XIZANG PLATEAU

### Hydrothermal Distribution and Earthquake Activity

All the earthquakes on the plateau are shallow with depth less than 70 km. Statistical analysis of historical and present earthquake data shows that there are about 70 earthquakes with magnitude greater than 6. The epicenters are mainly restricted to the Himalayan zone. Intermediate depth earthquakes are mostly located in Hindu Kush and Burma at the east and west ends of the Himalayan Arc, (Fig. 3)(2) indicating concentration of stress at these two points. From the distribution of epicenters along the arcuated Himalayan mountain system it can be seen that they are consistent with the hydrothermal activity zone (Fig. 4). Such consistency suggests that earthquake and hydrothermal activities are related to each other. They are the results of crustal movement and migration of materials at depths.

### The Characteristics of Propagation of Seismic Waves on the Xizang Plateau

By analyses of the seismograms recorded at (Figs. 6, 7) stations in China, it is found that the characteristics of propagation of seismic waves penetrating through the Xizang Plateau are different from those unpenetrating ones, such as the case through eastern China.

The distribution of earthquakes and stations are shown in Fig. 5. From the seismograms in Fig. 6, we can see the fifth one is recorded at the Guangzhou station in Guangdong Province, from the Nanhua earthquake of Yunnan Province. The sixth and seventh seismograms are respectively recorded at the Lanzhou station in Gansu Province and the She Shan station near Shanghai from the Jinghong Dong earthquake, Yunnan Province. It can be seen that the Lg wave recorded in Eastern China is characterized by its high energy and high frequency content.

Fig. 7 shows seismograms of seismic waves penetrating through the Xizang Plateau. The first one is a record of the Nanhua earthquake of Yunnan, the second one is from the Jinghong Dong earthquake of Yunnan, both were recorded at the Lhasa station in Xizang. The third one is from an earthquake occurred on the Sino-Nepalese border. The fourth one is from the Jinghong Dong earthquake of Yunnan. These two seismograms were recorded at the Urumqi station, Xinjiang Province. From Fig. 7, we can see that the Lg waves penetrating through the plateau have the characteristics of low energy and containing no high frequency component.

In order to make more studies for the propagation of Lg waves, we have collected some results of Lg waves studies for Xizang and its neighbouring regions by foreign authors(3). The seismograms shown in Fig. 8 are from 1 to 9, Lg waves not penetrating through the plateau regions. Evidently the energy of the Lg wave is high and the high frequency component presents clearly. For seismograms 10 to 14 the seismic waves pass around the margins of the plateau, Lg wave energy becomes comparatively low. For seismograms 15 to 19 the Lg waves penetrate through the entire plateau and their energies are very low or even disappear, and with no high frequency components. In addition, the Lg waves on the plateau and its vicinity are characterized by different types (Fig. 9) for different regions. The Lg wave outside the plateau recorded by Soviet stations is very clear and its energy high. But the Lg waves from the Xizang Plateau, the Assam region of India and the western end of the arcuated Himalayas are very low in energy and even disappear. The above-mentioned characteristics of Lg wave propagation, show that the property of the crustal media under the Xizang Plateau are different from other regions. Lg waves are strongly absorbed by the crustal media under the plateau. This might be caused by the presence of melted or partial melted materials in the crust of the plateau.

## The Absorption of Media in the Crust and Upper Mantle of the Xizang Plateau

Basing upon the $\bar{Q}p$ value obtained by the P wave data of small earthquakes from regional seismic networks it can be seen(4,5) that the average $\bar{Q}p$ value is 560 in the Shimian region of Sichuan, east of the plateau. The $\bar{Q}p$ value is 620, in Qiaojia of Sichuan. Prior to the great earthquake occurred in the Haicheng region of Liaoning, the $\bar{Q}p$ value is 550. But the average $\bar{Q}p$ value of P wave derived from small earthquakes recorded by the networks in Zhamo and Damxung regions is 150. This further shows that the seismic body wave is also highly absorbed in the Xizang Plateau.

## The Relation between the Hydrothermal Activity Distribution and the Gravity and Magnetic Anomalies

The Qinghai-Xizang Plateau is characterized by the lowest gravity Bouguer anomalies. The periphery of the plateau is a belt of high gravity gradient change. The Himalayan geothermal belt has an obvious negative Bouguer anomaly value(6). It is -550 milligal at the Yangbajain geothermal field. The existence of the hydrothermal region shows perhaps that the density of material in the crust of the Xizang Plateau is very small. Moreover, the Himalayan zone also shows an intense aero-magnetic anomaly of 150 to 450 gamma. The depth of burial of the Curie isothermal surface is about 18-30 km(7). Owing to the natural flux of heat flow being 116,000 kcal/sec at Yangbajain region the heat flow strength will be 575 microcal. $cm^{-2}$. $sec^{-1}$. It shows that there exists movement of high temperature materials, melting and migrating in the crust. Combining with the crustal model derived from explosion seismology on the plateau it can be seen that there exists a low velocity layer with thickness about 10 km at the depth of 40 km(8). All of these are thought to be related to aero-magnetic anomaly, geothermal activity and gravity anomalies. It indicates the existence of hot materials in the crust. According to the low Bouguer anomalies at Yangbajain and the geothermal gradient which is 2.9-3.3°C/km measured in several drilling wells of tens of meters in depth it is suggestd that the heat flow value in the region is 100 times higher than the normal value in various parts of the world. This evidently is caused by the heat flow activity in local regions. In the Yangbajain region magma chamber might exist.

## DISCUSSION

Based on the above-mentioned relationship between geothermal distribution and geophysical fields as well as geology it can be shown that high temperature and the existence of molten materials and magma chamber and so forth beneath the Xizang Plateau cause the strong hydrothermal activity, seismic activity, low Q value, low energy of S and Lg waves, absence of high frequency components in seismic waves, strong aeromagnetic anomaly, variation of depth of burial of the Curie isothermal surface, strong negative Bouguer gravity anomalies and existence of low velocity layer of 10 km in thickness. All these characteristics of the geophysical fields and also geochemical features (such as the presence of B, As, Li, Rb, Cs etc.) show that they are due to the melting or partial melting of materials in the crust of the plateau. Obviously the crust and upper mantle structure in this region has a certain peculiarity and the materials at depth is in migration.

These characteristics of the geophysical fields can also be observed in the Himalayan geothermal activity zone and its neighbouring regions. They are resulted from the collision and compression of the Indian and the Eurasian Plates. The collision and compression caused the concentration of radioactive substances and abrupt changes of thermal flow and temperature fields. During the process of compression, the substance in the crust and the upper mantle was forced to up-well. Thus it constitutes strong hydrothermal activities, frequent occurrence of small earthquakes, and some intermediate depth earthquakes such as the Lhasa and Gyangze earthquakes which occurred in this belt of depths at 82 and 40 km. Consequently, it also results in the formation of a crust of tremendous thickness and continuous uplift of the Himalayas. According to the regional characteristics of peculiar geophysical field and surface hydrothermal display it shows that hydrothermal explosions, geysers, boiling springs, and a great many hot springs, have the same features of the boundaries of the tectonic plates in the world. The Himalayan granite has higher Sr87/Sr86 ratio which is a consequence of the northward movement of the Indian Plate, underthrusting beneath the Eurasian Plate.

## References

(1) Tong Wei et al., Characteristics of geothermal activities in Xizang Plateau and their controlling influence on Plateau's tectonic model. In this Proceedings of Symposium on Qinghai-Xizang (Tibet) Plateau, Academia Sinica (1981).

(2) Huan Wen-lin et al., The distribution of earthquake foci and plate tectonics in the Qinghai-Xizang Plateau and its vicinity. In this Proceedings of Symposium on Qinghai-Xizang (Tibet) Plateau, Academia Sinica (1981).
(3) Ruzaikin A.I. et al., Propagation of Lg and lateral variations in crustal structure in Asia. J.G.R., 82 (2), 307-316 (1977).
(4) Chen Yun-tai et al., The determination of source parameters for small earthquakes in Qiaojia and Shimian and the estimation of potential earthquake danger. Acta Geophysica Sinica, 19 (3), 206-233 (1976) (in Chinese).
(5) Zhu Chuan-chen et al., Source parameters for small earthquakes and the quality factor of the medium before and after the Haicheng earthquake. Acta Geophysica Sinica, 20 (3) 222-231 (1977).
(6) Zhou Wen-hu et al., Characteristics of the gravity field and the crustal structure in the eastern and middle regions of the Xizang Plateau. In this Proceedings of Symposium on Qinghai-Xizang (Tibet) Plateau Academia Sinica, (1981).
(7) Pan Zuo-shu et al., The characteristics of aeromagnetic anomalies over southern Xizang and the structure of the earth's crust. In this Proceedings of Symposium on Qinghai-Xizang (Tibet) Plateau, Academia Sinica (1981).
(8) Teng Ji-wen et al., Explosion seismic study for velocity distribution and structure of the crust and upper mantle from Damxung to Yadong of Xizang Plateau. In this Proceedings of Symposium on Qinghai-Xizang (Tibet) Plateau, Academia Sinica (1981).

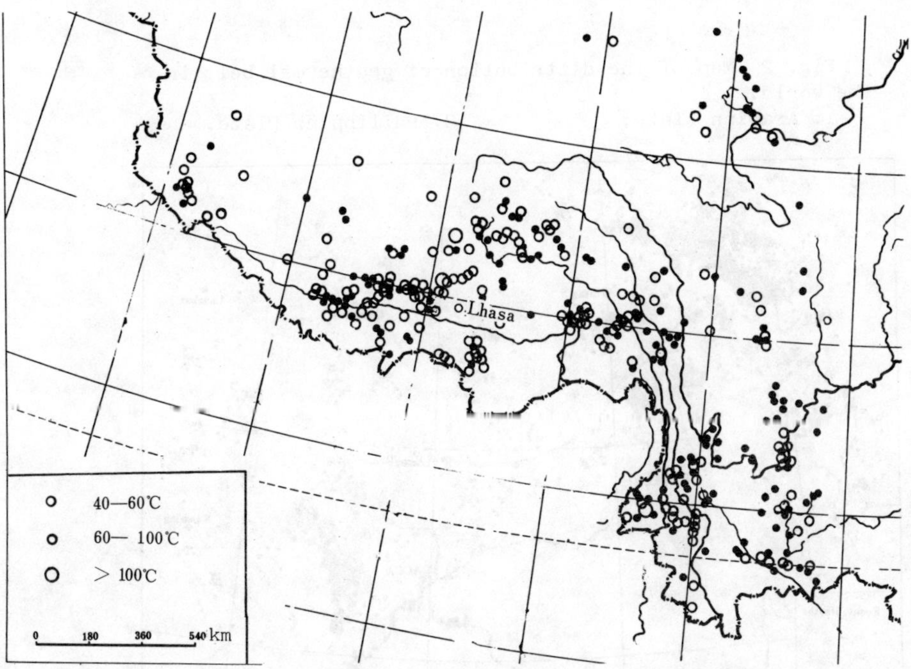


Fig. 1 Distribution of hydrothermal activities in Xizang Plateau.

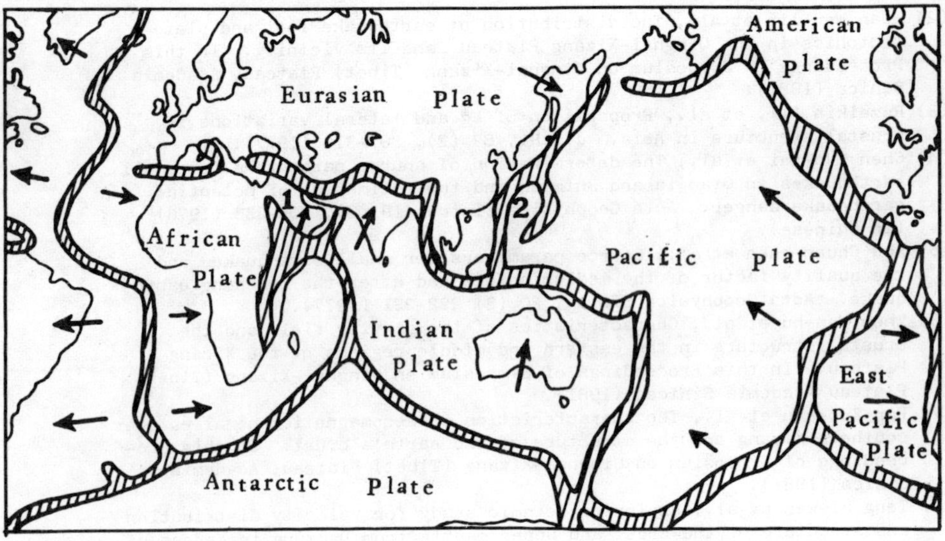


1. Arabian plate 2. Philippine plate

Fig. 2 Map of the distribution of geothermal belt in the world.
1. Arabian Plate; 2. Philippine Plate.

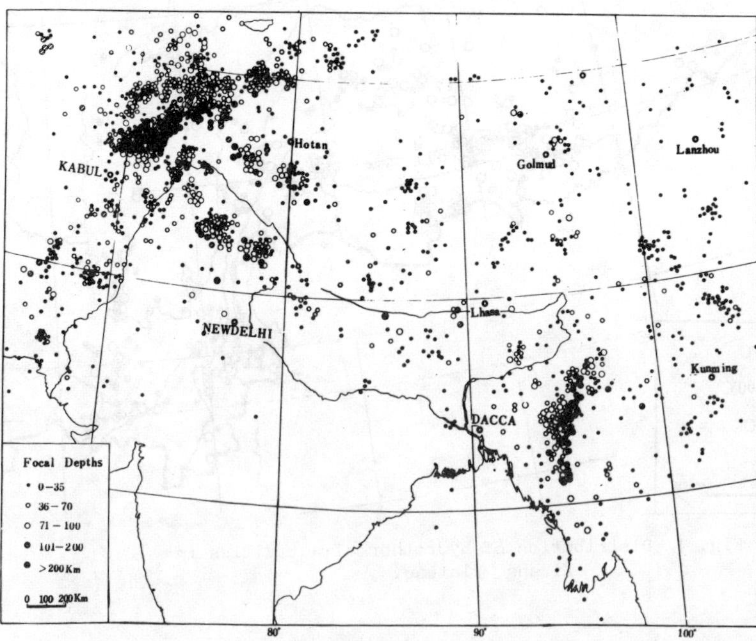


Fig. 3 Map of earthquakes distribution in the Qinghai-Xizang Plateau and its neighbouring regions.

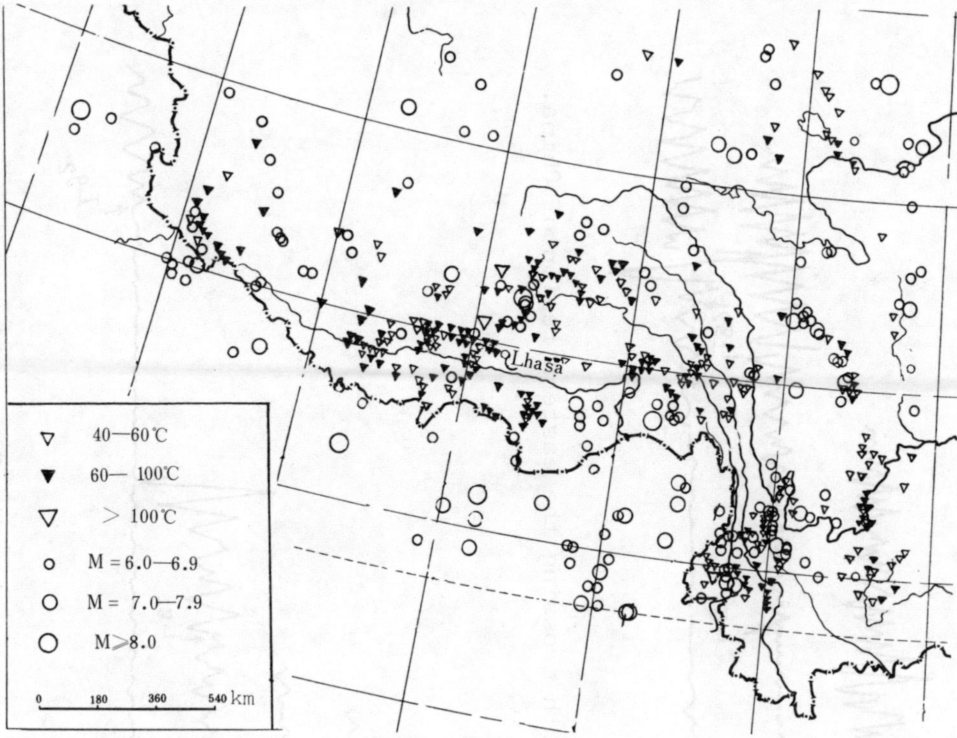


Fig. 4 Map of the distribution of hydrothermal regions and earthquakes in Xizang Plateau.

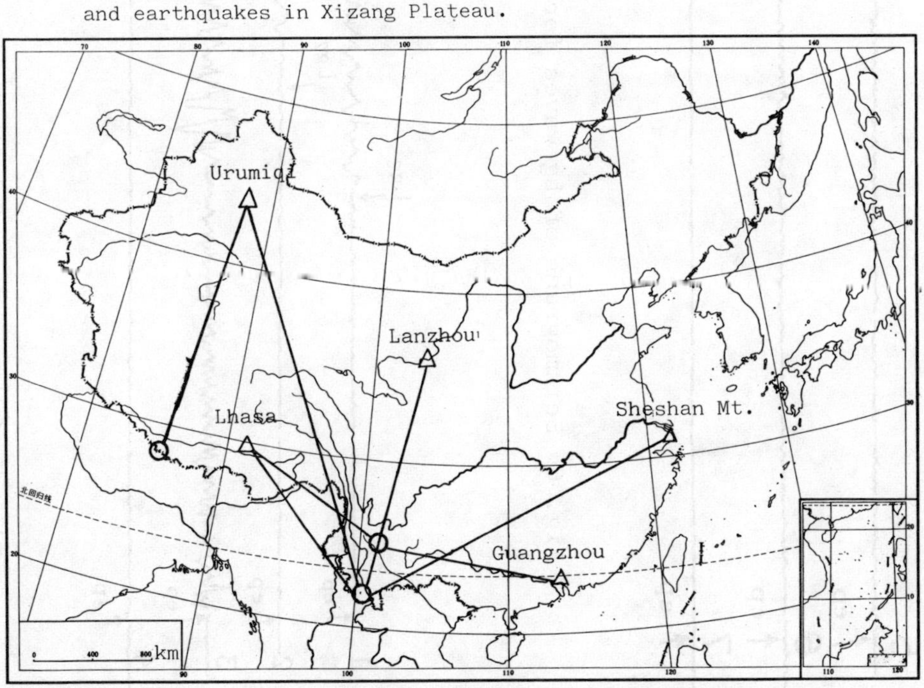


Fig. 5 Map of distribution of earthquakes (Nanhua of Yunnan, Jinghong Dong of Yunnan, Sino-Nepalese border) and stations.

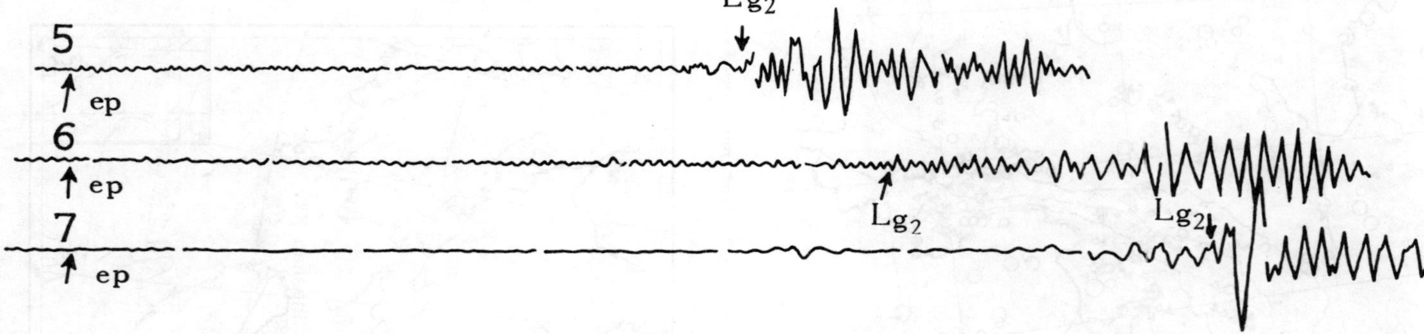


Fig. 6 Seismograms of Lg waves for a path crossing the eastern region of China.

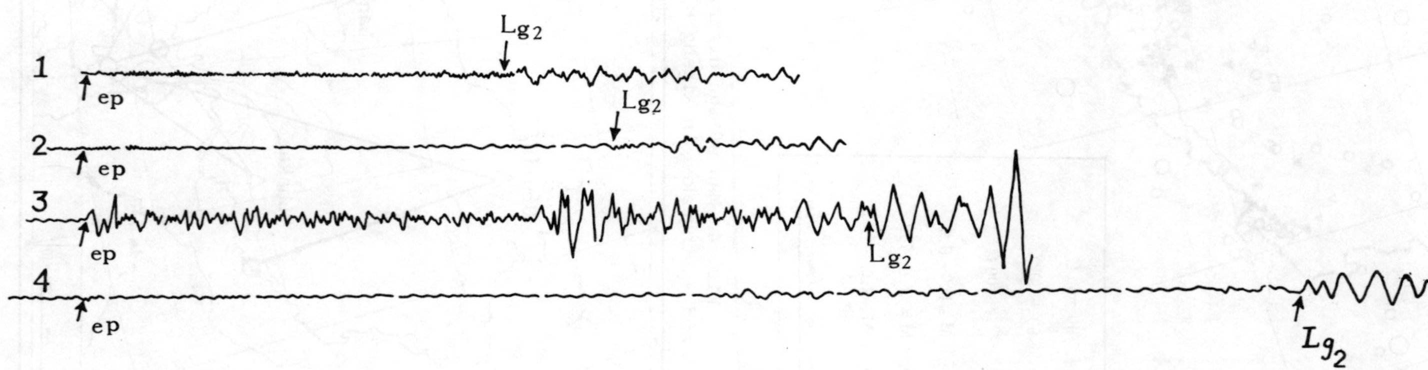


Fig. 7 Seismograms of Lg waves for a path crossing the Xizang Plateau.

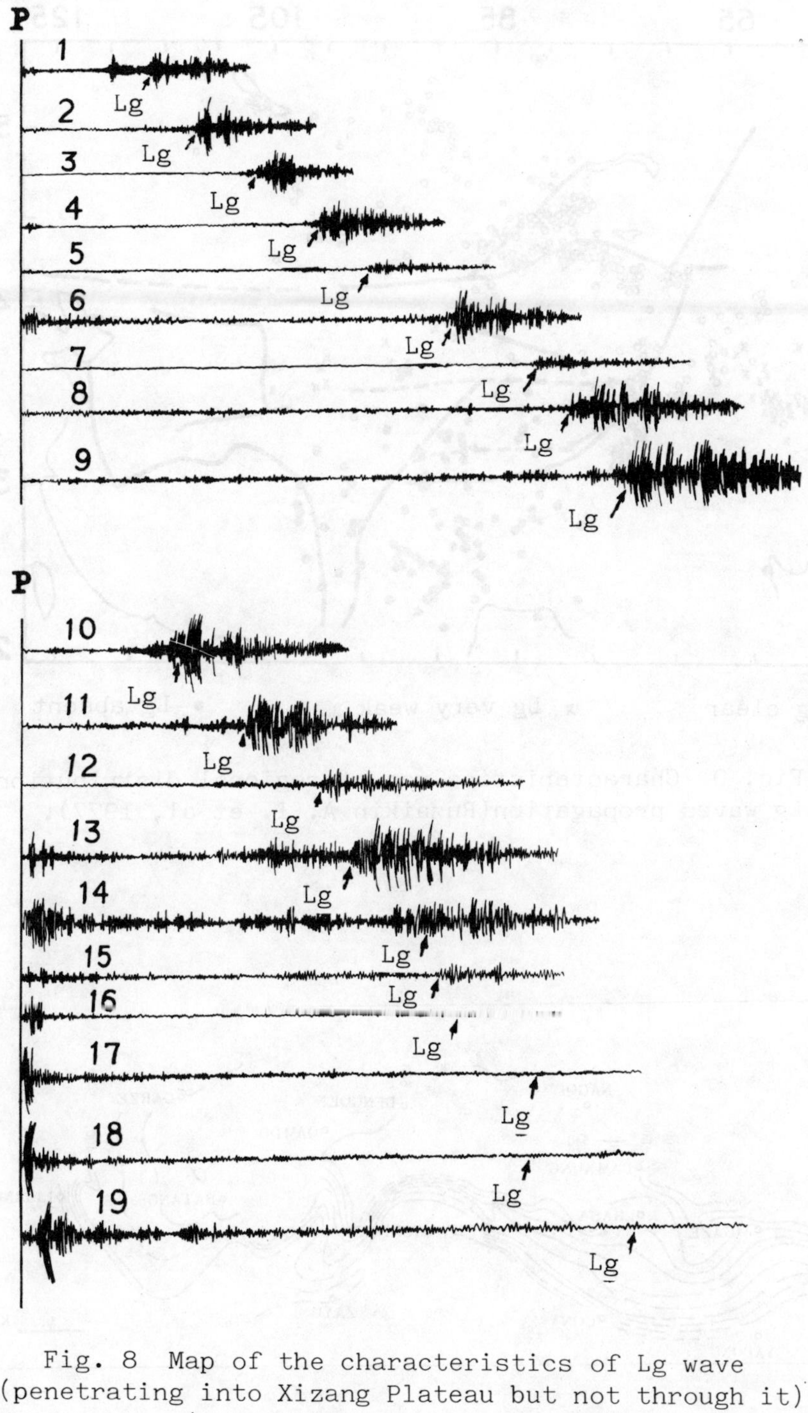


Fig. 8 Map of the characteristics of Lg wave (penetrating into Xizang Plateau but not through it). (Ruzaikin A. I. et al, 1977).

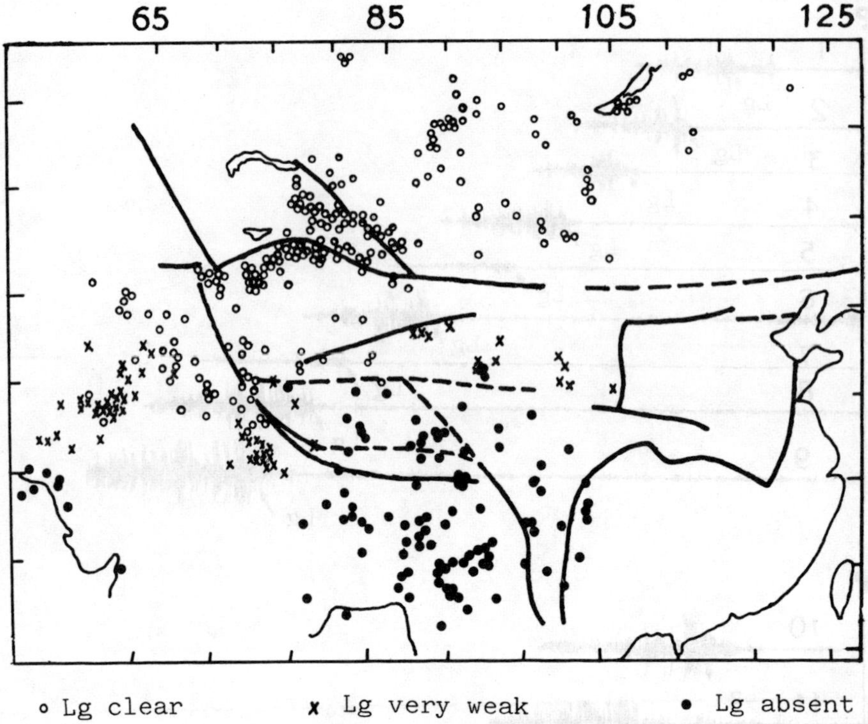


Fig. 9 Characteristics map of regional distribution of Lg waves propagation(Ruzaikin A. I. et al, 1977).

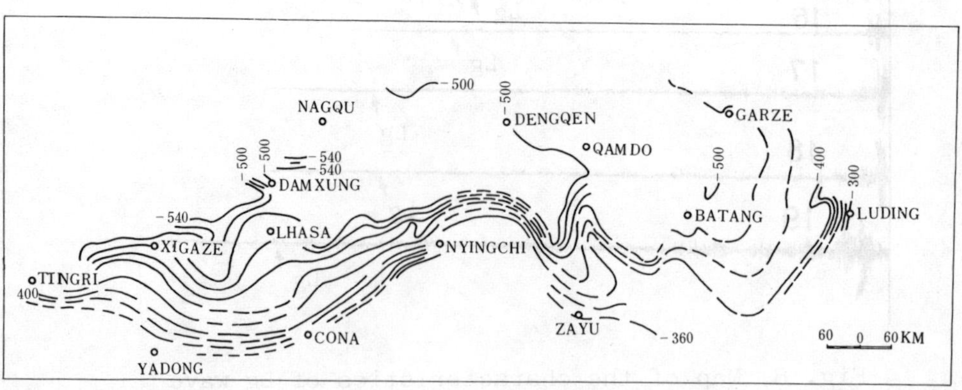


Fig. 10 The map of Bouguer anomalies in Qinghai-Xizang Plateau.

# ON THE THERMAL HISTORY OF COMPRESSING PLATES WITH SPECIAL REFERENCE TO THE XIZANG PLATEAU

Zhu Bing-quan
(Institute of Geochemistry, Academia Sinica)

The Xizang Plateau is situated on the convergent margin between the Indian and Eurasian plates. The inter-collision and compression of the two continents caused the increase of crustal thickness, thus leading to the redistribution of radioactive energy under the earth's surface. The energy redistribution in return gave rise to the change of mode of plate existence. The Xizang Plateau is characterized by its extremely thick crust (70-73 km)(1), comparatively thin slab ( the attenuation of Lg waves) and high heat flow, precisely distinct from normal continental and oceanic plates. The purpose of this paper is to establish a model for the temperature profile in the Xizang Plateau with the temporal variation, and to give a reasonable explanation to the time sequence and mechanism of magmatic evolution, the uplifting rate of the plateau and its elevation, and plate thickness, based upon the thermal conductive model in conjuction with the data on isotopes and trace elements.

## THE TEMPERATURE PROFILE OF COMPRESSING CONTINENTAL PLATES

The knowledge of heat generation in various geospheres under normal continental plates is important to establish the initial state of continental plates before compression.

Studies on heat flow in the last decade(2,3,4) have proved that there exists a linear correlation between the earth's surface heat flow and heat generation for global normal continental plates, including North American, Indian-Australian and Eurasian plates (Fig. 1). It is indicated from the slope and intercept of the line in Fig. 1 that radioactive elements under the normal continents mostly concentrate heterogeneously in the upper crust within 8 km, while the heat flow contribution beneath the lower crust seems equivalent within the various regions. Therefore, we can divide the geosphere structure into the upper crust, the lower crust, the upper mantle lithosphere and the asthenosphere. Listed in Table 1 are the contents of radioactive elements, and the heat flow contribution in the individual geospheres calculated on the basis of various types of rocks. If adopting 1.32 HFU as the mean value of continental heat flow, the calculated value of mantle heat flow should be equal to 0.61 HFU, in concurrence with the estimation by other workers(5,6). In terms of the radioactive energy distribution, the solution of thermal conductive equation showing the temperature within the plates as a function of depth and time, has been made as follows:

$$T'(x',t') = \sum_{n=1}^{\infty} \frac{2\cdot(-1)^{n+1}\cdot a^2}{(n\pi)^3\ KTm} \Big[ H_1(1-\cos n\pi\varepsilon_1) + H_2(\cos n\pi\varepsilon_1 - \cos n\pi\varepsilon_2) + H_3(\cos n\pi\varepsilon_2 - (-1)^n)\Big] \cdot (1-e^{-(n\pi)^2 t'}) \cdot \mathrm{Sin}\ n\pi x' + 1 - x' \qquad (1)$$

where T', t' and x' are dimensionless temperature, time and depth; $\varepsilon_1$ and $\varepsilon_2$ are the ratios of the thickness of the upper crust and the crust to that of plate (a) ; $H_1$, $H_2$ and $H_3$ represent the heat generations in the upper crust, the lower crust and the upper mantle, respectively; Tm is the temperature of the asthenosphere (1,200° C) ; K is the thermal conductive coefficient (0.006 cal/cm °C. sec). The boundary and initial conditions are given by $T'(o,t') = Tm$, $T'(1,t') = 0$, $T'(x,o) = 1 - x'$ . In the case of plate compression, a was a variable, while the mantle heat flow maintained constant. a, $\varepsilon_1$ and $\varepsilon_2$ were all variables if the collision took place in the lithosphere.

## THE THERMAL HISTORY OF PLATE AND THE TIME SEQUENCE OF THE EVOLUTION OF MAGMATIC ROCKS

In order to establish the temporal relationship between the thermal history of plate and the time sequence of magmatic evolution, we must, first of all, ascertain the time when the intercollision between the Indian and Eurasian plates took place, and the relative moving rate.

Fig. 2 shows the relationship between the relative distance of the two plates from each other, and the time, based upon the K-Ar ages and the palaeomagnetic data(7,8). From this figure, it is indicated that the intercollision between the two plates seems to have taken place at about 42-45 m.y. ago, and the relative moving rate after the collision is estimated to be 5 cm/yr. The histogram for the isotopic age frequency also indicates the dramatic increase of magmatism and metamorphism during the postcollision.

In view of this time sequence, the crust of the Qinghai-Xizang Plateau has at least shortened by 2,000 km and thickened in two fold, subsequently setting a compressing interval from the Himalayas to the Kunlun-Altun. Thus, the temperature-depth curves of different ages in the thermal equilibrium state (equilibrium time greater than 20 m.y.) can be established for the plateau (Fig. 3). It is suggested that remolten granites could not be generated in the crusts of normal continental plates (less than 600°C) before the plate compression, instead, the remelting could occur within the range from the lower crust to the upper crust only when the crust thickened up to 45 km due to the subduction or the collision. The data on isotopes and trace elements in the granitoid rocks of different ages in this region also illustrate that the magmatism developed gradually from the upper mantle towards the upper crust.

The relatonship between the Eu anomalies of REE in various rock types and their ages (Fig.4) shows that the plutonic rocks emplaced at the early stage have a $\delta$ Eu value of 0.95(9), characteristic of mantle derivation, while the granites of the later stage have a $\delta$ Eu of 0.36, indicating their crustal origin. Fig. 5 shows the relationship between $TiO_2/P_2O_5$ of rocks and their ages. For the rocks formed during the pre-collision, their $TiO_2/P_2O_5$ ratios fall within the range of mantle partial melting (3-10), and for those formed during the post-collision, the ratios are within the limits of crust partial melts. In order to confirm the conclusion that the tonalite and granodiorite are of mantle origin, a plot of $TiO_2$ against $P_2O_5$ in these rocks according to a partial melting expression is given in Fig. 6. Their good linear correlation and zero intercept indicate that the rocks are resulted from partial melting within the mantle with $TiO_2/P_2O_5$ to be 10-15. The variation of $Al_2O_3/TiO_2$ with time (Fig. 7) similarly indicates that the ratio is close to that of the mantle (20.4)(10) before the collision, and deviates greatly from the mantle value after the collision (40-372). It can be seen from the tracing diagram of Pb isotopes (Fig. 8) that the late granites are similar to pre-Cambrian base rocks in their Pb isotopic composition, indicating their base remelting origin. However, the early rocks show an isotopic composition close to that of the mantle. The strontium isotopic composition of rocks also supports this conclusion. The early granodiorites have a initial $^{87}Sr/^{86}Sr$ ratio of 0.7036, being typical of the mantle, while the initial strontium ratio of the later granites is as high as up to 0.7478, indicating their crust derivation.

## THE COOLING AGES OF MINERALS AND THE UPLIFTING RATE OF THE PLATEAU

A considerable number of biotite and K-feldspar cooling ages have been determined on samples from the Xizang Plateau. Although they are unable to be representative of the formation ages of the rocks, the dating of individual minerals by different methods reveals the thermal history of the rock system. Thus, they do have close relationship with the variation of temperature profile within the plate and the uplifting rate of the plateau. The Ar closure temperature for biotite and K-feldspar is estimated to be $220 \pm 40^{o}C$ and about $160^{o}C$, respectively(11). The depth ( Zc (t) ) at which the Ar closure temperature reached at different time can be calculated in the light of the temperature-depth curves mentioned above. Therefore, based on the cooling age ($t_{Ar}$), the mean uplifting rate ($\bar{V}$) of the rock system can be worked out by a simple expression as follows:

$$\bar{V} = Z_B (t) / t_{Ar} \quad \ldots\ldots\ldots\ldots\ldots\ldots \quad (2)$$

The calculated average uplifting rate for the individual parts of the Xizang Plateau is shown in Fig. 9. It can be seen that the Himalayan Region in the neighbourhood of Mt. Xixabangma, Mt. Qomolangma and northwestern Himalayas of India, has a higher uplifting rate ranging from 0.6 to 0.85 mm/yr. At the bend of the Yarlung Zangbo River, the uplifting rate seems the highest up to 1.1 mm/yr. However, in the vast interior areas of the plateau, the uplifting rate is within the range of 0.36 - 0.48 mm/yr.

## THE PLATE THICKNESS OF THE QINGHAI-XIZANG PLATEAU AND ITS UPLIFTING ALTITUDE

From equation (1), we may obtain an expression of mantle heat flow ($F(0, \infty)$) in the equilibrium state, and subsequently deduce a universal equation expressing the relationship between plate thickness and crustal thickness:

$$F(o, \infty) = -K \frac{2T'(x'.\infty)}{2x} . Tm \Big|_{x'=1} \quad (3)$$

$$a = \frac{2KTm - H_1 b^2 - H_2 (c^2 - b^2)}{2F(o, \infty)} \quad (4)$$

where b and c represent the thickness of the upper crust and the crust respectively. In the pre-formation of the crust, i.e. in the existence of normal oceanic plate, $F(0, \infty) = F(1, \infty)$, equation(4) becomes $a = KTm/F(1, \infty)$ which deduced by S. T. Grough(12). As the mean heat of flow of normal oceanic plate is equal to that of normal continental plate (1.1-1.32 HFU), the oceanic plate should be 54.5-65.5 km in thickness. After the formation of the crust by the gradual fractionation of the mantle, $F(1, \infty)$ became a constant, $F(0, \infty) = F(a, \infty) - H_1 b - H_2 (c-b)$. Therefore, the plate thickness increased with thickening crust from 54.5 km to 109 km, corresponding to the thickness of normal continental plate with a crust thickness of 33 km, as shown on Curve 1 in Fig. 10. However, in the situation of compressing plate, the mantle heat flow maintained constant during the convergent process. The plate thickness reduced with increasing crustal thickness as shown on Curve 2 in Fig. 10. As for the interior of the Qinghai-Xizang Plateau, its crustal thickness is estimated to be 70-73 km, and then, the plate thickness would be about 75-80 km. Due to crustal thickening and plate thinning, the plateau gradually uplifted under the action of isostatic equilibrium. According to the calculated plate thickness of the Qinghai-Xizang Plateau, the elevation has been estimated to be 4.5-5.7 km in terms of isostatic equilibrium equations, well consistent with the observed elevation only with an exception of Mt. Qomolangma region which is out of isostatic equilibrium state. In accordance with equation (4), if the parameters adopted are precise, then the crustal thickness would by no means exceed over 74 km, otherwise the crustal materials would have melted and were brought back to the asthenosphere.

## CONCLUSION

1. The thermal history model indicates that the source regions of magmatic evolution could move from the upper mantle towards the upper crust, in good agreement with the geochemical and isotopic data.

2. The uplifting rate of the individual parts of the plateau is within the range of 0.36 - 1.10 mm/yr during the post-collision.

3. Deduced from the thermal conductive model, an expression showing the plate thickness as a function of crustal thickness has been obtained. In the light of this expression, the plate thickness of the Xizang Plateau is approximately 76-80 km. The uplifting of the plateau is mainly attributable to plate thinning.

## References

(1) Teng Ji-wen, et al., Explosion seismological study for velocity distribution and structure of the crust and upper mantle from Damxung to Yadong of Xizang Plataeu. In this Proceedings of Symposium on Qinghai-

Xizang Plateau, Academia Sinica (1981).
(2) Lachenbruch, A. H., Crustal temperature and heat production: Implications of the linear heat-flow relation. Jour. Geophys. Res., 75(17), 3291-3300 (1970).
(3) Rao R. U. M., Radioactive heat generation and heat flow in the Indian shield, Earth Planet. Sci. Lett., 30(1), 57-64 (1976).
(4) Geothermal Group, Institute of Geology, Academia Sinica, Collected Works of Geothermal Research, Science Press, 12-13 (1978).
(5) Pollack H. N., Chapman D. S., Mantle heat flow, Earth Planet. Sci. Lett., 30(1), 57-64 (1977).
(6) Ringwood A. E., Composition and petrology of the earth's mantle. McGraw-Hill, New York, 213-249 (1975).
(7) Duncan R. A., $^{40}Ar$-$^{39}Ar$ geochronology of basalts from ocean basin--some successes with samples from a seismic ridge. Short Papers of Fourth International Conference, Geochronology, Cosmochronology, Isotope Geology, 100-102 (1978).
(8) Molnar P., Tapponnier P., Cenozoic tectonics of Asia: effects of a continental collision, Science, 189(4201), 419-426 (1975).
(9) Zhao Zhenhua, et al., Geochemistry of REE in intermediate-acidic intrusive rocks in Southern Xizang, In this Proceedings of Symposium on Qinghai-Xizang Plateau,Academia Sinica (1981).
(10) Sun S. S., Hesbitt R. W., Chemical heterogeneity of the Archaean mantle composition of the earth and mantle evolution, Earth Planet. Sci. Lett., 35(3), 429-448 (1977).
(11) Harrison T. M. et al., Thermal models and cooling histories from fission-track, K-Ar, Rb-Sr and U-Pb mineral dates, northern coast plutonic complex. Short Papers of Fourth International Conference, Geochronology, Cosmochronology, Isotope Geology, 167-170 (1978).
(12) Crough S. T., Numerical and approximate solutions for lithospheric thickening and thinning, Earth Planet. Sci. Lett. 31(3), 397-402 (1976).
(13) Armstrong R. L., Lead and Strontium Isotopic Evolution. Rev. Geophys. 6(1), 175-199 (1968).
(14) Bickle M. J., Heat loss from the earth:a constraint on Archaean tectonic from the relation between geothermal gradients and the rate of plate production. Earth Planet. Sci. Lett. 40(3), 301-315 (1978).
(15) Heier K. S. et al., Concentration of radioactive elements in deep crustal material. Geochim. Cosmochim. Acta. 29(1), 53-61 (1965).
(16) Fisher D. E., Terrestrial uranium, heat flow and cosmochemistry. Short Papers of the Fourth International Conference, Geochronology, Cosmochronology, Isotope Geology, 109 (1978).
(17) Walcott R. I., Flexural Rigidity, thickness and viscosity of lithosphere, Jour. Geophys. Res. 75(20), 3941-3954 (1970).
(18) Smith P. J., Oceanic lithosphere thickness with age. Nature, 253(5487), 91 (1975).

Table 1 The contents of radioactive elements, heat generation and heat flow contribution in various geospheres

| Geospheres | Rock type | U (ppm) | Th (ppm) | K (%) | Heat generation (HGU) | Normal continental heat flow (HFU) | Oceanic heat flow (HFU) |
|---|---|---|---|---|---|---|---|
| Upper crust rich in U, Th and K | Global granites (average) (13) | 4.75 | 18.5 | 3.79 | 7.0 | | |
| | Granitoids of Xizang Plateau (average) | 3.1 | 19.0 | 2.95 | 6.0 | 0.48–0.56 | ~ 0.08 |
| | Pre-Cambrian rocks of Indian Shield (10) | | | | 6.25–7.0 | | |
| | Archaean Shield (14) | | | | 4 – 6.7 | | |
| Lower crust | Granulites (15) | 0.39–0.88 | 0.93–5.1 | 2–3 | 1 – 1.8 | | |
| | Basalts (13) | 0.2 | 0.4 | 0.5 | 0.3 | 0.15 | ~ 0.02 |
| Upper mantle | Abyssal tholeiites(16) | 0.078 | 0.2 | 0.106 | 0.13 | < 0.09 | ~ 0.07 |
| (Lithosphere) | Peridotites(13) | 0.001 | 0.0018 | 0.004 | ~ 0 | | |
| Asthenosphere | Pyrolite, lherzolite (6) | 0.016–0.026 | 0.047–0.1 | 0.012–0.024 | 0.036–0.067 | 0.65 | ~ 1.2 |

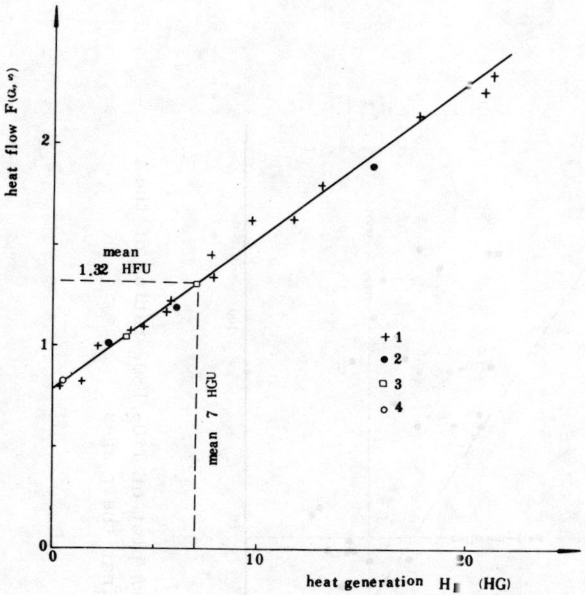


Fig. 1 A plot of the earth's surface heat flow against the heat generation ir the surface rocks for global normal continental plates
1. North America; 3. India
2. Australia; 4. Chira

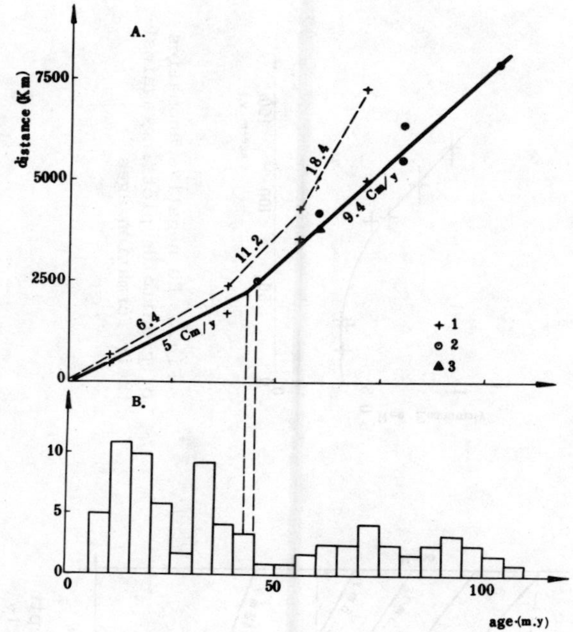


Fig. 2 A. A plot of the relative distance between the Indian and Eurasian plates v.s. the time on the basis of the K-Ar ages and paleomagnetic data; B. A histogram of isotopic age frequency for the Xizang Plateau
1. Paleomagnetic anomaly
2. K-Ar ages of the upper mantle
3. K-Ar ages and paleolatitude

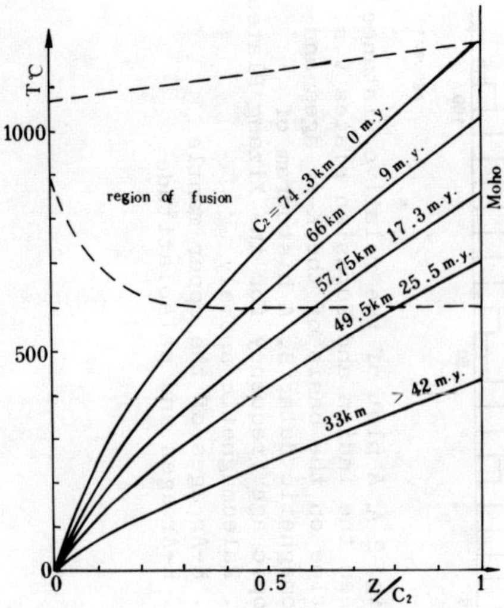


Fig. 3 Temperature-depth curves for the differently thick crusts in the Xizang Plateau

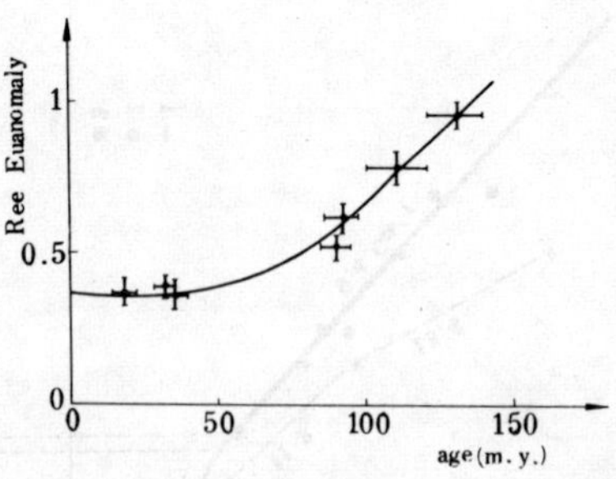


Fig. 4 Eu negative anomalies of granitoids plotting against their formation ages

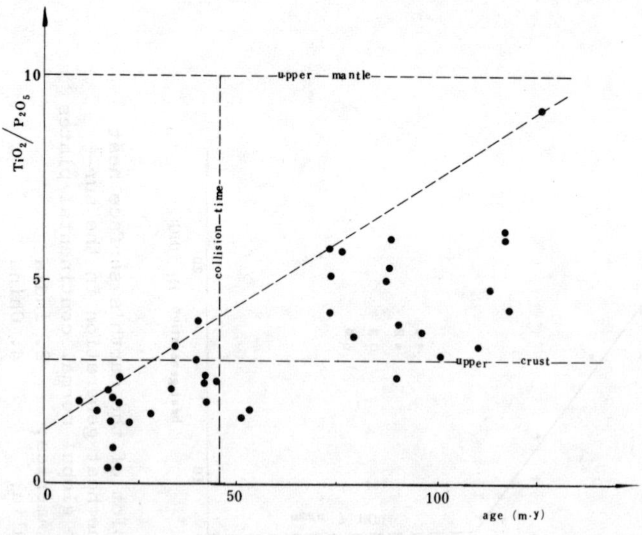


Fig. 5 A plot of $TiO_2/P_2O_5$ ratios of the rocks against their ages

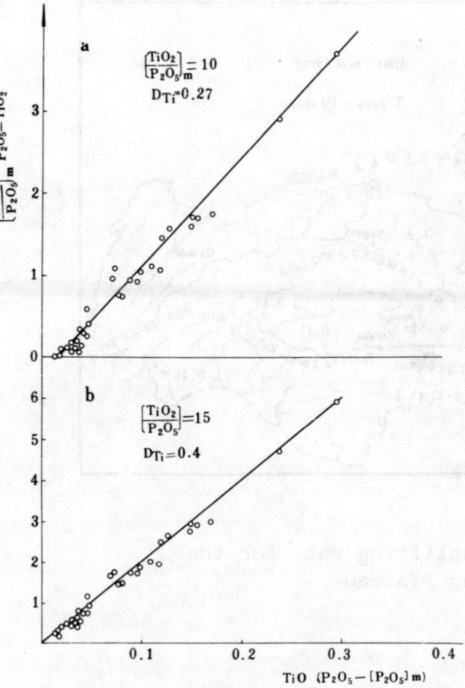


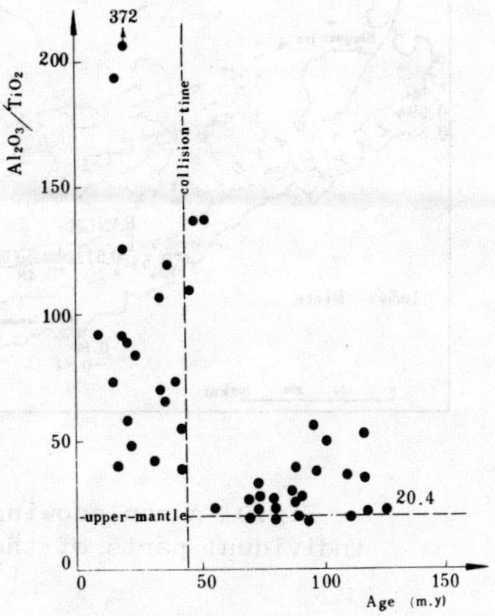


Fig.6 The correlation diagram between $TiO_2$ and $P_2O_5$ for granodiorites in Xizang, based upon a partial melting expression

Fig.7 The $Al_2O_3/TiO_2$-isotopic age plot for granitoid rocks

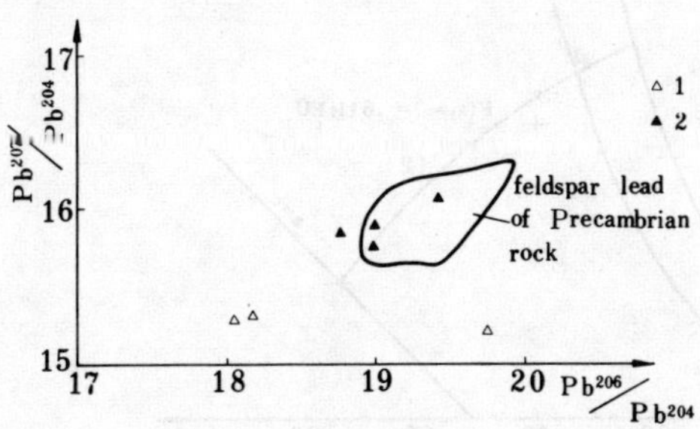


Fig.8 The $Pb^{207}/Pb^{204}$—$Pb^{206}/Pb^{204}$ tracing diagram for feldspars in granitoids and associated galena from Xizang
(1) Pre-collision; (2) Post-collislon

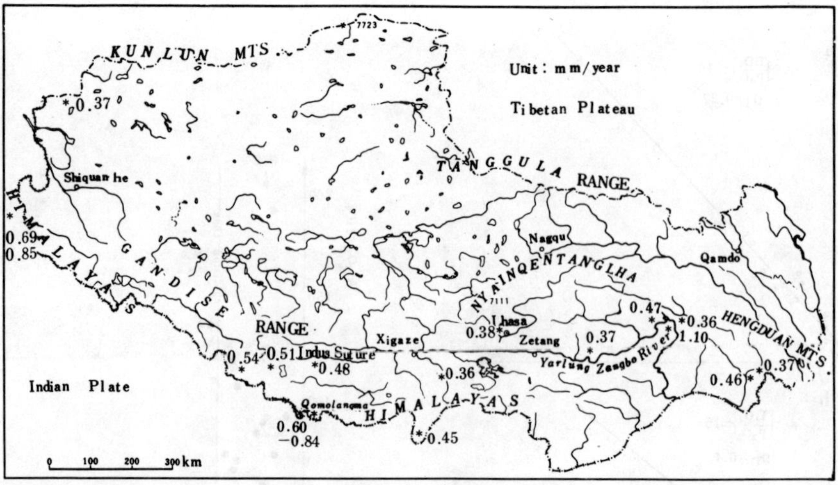


Fig.9 A map showing the uplifting rate for the individual parts of the Xizang Plateau

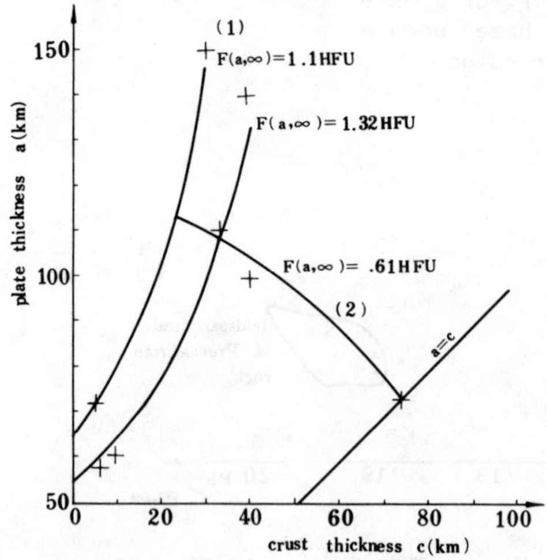


Fig.10 A theoretical diagram showing the relationship between plate thickness and crustal thickness. The crosses represent the observed values by different methods(17, 18)

# ISOTOPIC VARIATIONS IN MANTLE AND CRUSTAL HELIUM: APPLICATIONS TO THE TIBET - YUNNAN GEOTHERMAL SYSTEMS

H. Craig

Isotope Laboratory
Scripps Institution of Oceanography
University of California, San Diego
La Jolla, California, USA 92093

Helium has two stable isotopes, mass 3 and mass 4, which vary in their ratio by more than a factor of 3000 in terrestrial materials. In atmospheric helium, which is used as a reference standard for isotopic studies, the 3/4 ratio is $1.40 \times 10^{-6}$. Since both isotopes of helium are continually escaping from the atmosphere to space, this atmospheric ratio is fixed by the ratio of input fluxes from the crust and mantle to the atmosphere, divided by the ratio of the first-order rate constants for escape from the atmosphere:

$$R_A = (^3He/^4He)_{ATM} = (\phi_3/\phi_4)/(k_3/k_4).$$

Here $\phi_i$ is the total flux into the atmosphere for each isotope, and $k_i$ is the escape rate constant. At steady-state atmospheric composition the input and escape fluxes are equal:

$$\phi_3 = k_3N_3$$

$$\phi_4 = k_4N_4$$

where $N_3$ and $N_4$ are the atmospheric inventories of $^3He$ and $^4He$.

Helium 4 is generated continually in the crust and mantle by radioactive decay of U and Th, which also generates the heat flux from the earth. Assuming that the $^4He$ and heat flux to the earth's surface are in proportion to their production rates, $\phi_4$ is of the order of $10^6$ atoms/cm$^2$sec. Most of this input is from the continental crust, the exact fraction depending on the distribution of U and Th in continental crust, oceanic crust, and the mantle. However, some of the $^4He$ flux from the mantle is original helium which has been stored there since the formation of the earth; this is known because the flux of $^3He$ is almost entirely from this primordial undegassed reservoir.

Helium 3 has several sources in the atmosphere, crust, and mantle. A small fraction of the total flux, about 0.5 atoms/cm$^2$sec, is produced directly in the atmosphere by the cosmic-ray bombardment which also produces tritium, the mass 3 hydrogen isotope (this isotope subsequently decays to $^3He$). An even smaller flux originates in the crust, primarily in the continental crust, by reaction of neutrons with lithium:

$$^{6}Li(n,\alpha) \rightarrow {}^{3}H(\beta^{-}) \rightarrow {}^{3}He$$

this flux is of the order of 0.1 atoms/cm$^2$sec. The mean $^3He/^4He$ in the so-called "crustal helium" or radiogenic helium, is ~ $10^{-7}$ or less, i.e. approximately 0.1 x the atmospheric 3/4 ratio or less. The relationship between atmospheric and crustal helium isotope ratios was determined by L. T. Aldrich and A. O. Nier in 1946, 12 years after the discovery of natural $^3He$ by L. W. Alvarez and R. Cornog in 1938.

The third and major source of $^3He$ in the earth is the primordial $^3He$ inventory in the mantle, a component of the still undegassed initial volatile content of the earth after accretion. This "primordial" $^3He$ is recognized by the high $^3He/^4He$ ratio in helium originating from the mantle, from 10 to 20 times greater than the ratio in atmospheric helium, and more than 100 times greater than the ratio in crustal helium. The discovery of this primordial helium component was announced simultaneously by US and USSR scientists in 1969, after it had been observed in helium injected into the deep Pacific Ocean from ridge-crest spreading centers by W. B. Clarke, M. A. Beg, and H. Craig, and in volcanic gases from Kamchatka by B. A. Mamyrin, I. N. Tolstikhin, G. S. Anufriev, and I. L. Kamenskii. The flux of this primordial $^3He$ to the earth's surface, as determined from the distribution in helium dissolved in the world oceans, is about 4 atoms/cm$^2$sec, which is an approximate agreement with the estimated escape rate from the atmosphere. The flux is an order of magnitude greater than the production of $^3He$ by cosmic rays and is the dominant source of terristrial $^3He$.

Since the discovery of primordial $^3He$, helium isotope ratios have been extensively mapped in the ocean, in geothermal and volcanic areas, and in oceanic and continental volcanic rocks and xenoliths, primarily by US, USSR, and Japanese scientists (N. Takaoka and colleagues). Accordingly, we now recognize, in addition to crustal and atmospheric helium, three rather well-defined mantle helium components in characteristic petrologic and tectonic environments.

1. Mid-ocean ridge basalt (MORB) helium, with a 3/4 ratio 8 to 9 times the atmospheric He ratio is found ubiquitously at oceanic spreading centers in tholeiitic basalts. Although the interiors of the basalt "pillows" are quickly stripped of their original gases by penetration of seawater after extrusion of the basalt flows, the primordial helium is retained in the gases trapped in the quenched glass rims of these pillows. We have also observed this MORB helium in the hydrothermal vents along mid-ocean ridges, where it emerges in submarine hot springs. In 1979 we were able to sample these fluids directly with a submersible, obtaining helium emerging from hydrothermal vents at 350°C, at a depth of 2500 m on the East Pacific Rise at 21°N. The 3/4 ratio of helium emitted in oceanic hydrothermal fluids is ~ 8.6 times the atmospheric ratio in the Red Sea geothermal brines, and about 8.0 times atmospheric in the hydrothermal fluids emitted at the Galapagos Rift and the 21°N site on the East Pacific Rise. Basalts themselves contain very uniform $^3He/^4He$ ratios in samples from spreading centers in the western Pacific (Lau Basin and Mariana Trough), in the eastern Pacific (East Pacific Rise, Galapagos Rift), the

Mid-Atlantic Ridge, and Indian Ocean. The uniformity of the MORB ratio despite a large range in absolute He content in these basalts (more than a factor of 50) demonstrates the existence of a well-mixed upper mantle source in which the $^3He/^4He$ and $^3He/(U + Th)$ ratios are very uniform throughout this reservoir. We believe this reservoir to be the primary source for the $^3He$ flux to the atmosphere, by means of intrusion of tholeiitic basalt at oceanic spreading centers and subsequent extraction of helium and other volatiles by hydrothermal circulation of seawater through the basalts. Recently we have shown that large fluxes of $H_2$, $CH_4$, and $CO_2$ accompany this helium flux from the oceanic ridge-crests; the $CO_2$ emitted in these regions has a carbon isotope ratio very similar to the ratio in diamonds and carbonatites.

2. "Hot-spot" helium, associated with regions proposed as the surface expression of deep-mantle plumes, probably from the undifferentiated lower mantle, is even higher in $^3He/^4He$ ratio, ranging from 15 to 20 times the atmospheric ratio. We have observed this characteristic helium type at Kilauea Volcano, in the central portion of the Ethiopian Rift Valley, and at Yellowstone Park in the continental U.S. The Russian workers have observed such helium in hot springs in Iceland, and we have made similar observations there. This helium is clearly a more primitive component, less degassed and less "depleted" in $^3He$, and we associate it with the widely-postulated "undepleted" lower mantle reservoir. Its occurrence at the oceanic "hot spots" (Kilauea and Iceland) is not surprising, but the fact that it is found in rhyolitic continental regions such as Yellowstone Park in the U.S., essentially undiluted with crustal helium, is quite unexpected, and indicates that the thick continental crust in such areas is quite transparent to the flux of mantle volatiles. (This implies that such helium might be observed in the Tibetan geothermal areas, despite the very thick crust in this area).

3. "Subduction zone" or convergent-margin helium characteristically has $^3He/^4He$ ratios from 6 to 8 times the atmospheric ratio. We have mapped this type of helium in thermal fluids in a number of convergent plate margin terrains: Lassen Volcano in the U.S. Cascades, Hakone Volcano in Japan, Pagan and Agrigan Volcanoes in the Mariana Arc, New Zealand, Alaska, and other areas. USSR scientists have observed similar helium in Kamchatka and the Kuriles. The range of 6 - 8 times the atmospheric ratio, distinctly less than in MORB helium, is probably characteristic of subduction zones, at least in the Circum-Pacific plate margins where all studies to date have been made. The importance of the $^3He$ component in these areas is very great: it identifies the regions in which a significant mantle volatile flux is occurring (so that other volatiles such as $CO_2$, $CH_4$, etc. can be studied in this context), as opposed to areas where purely "crustal" fluids are circulating, and it provides a valuable input to the problem of magma sources in subduction areas, i.e. the relative contributions of descending oceanic crust, the mantle wedge above the downgoing slab, and sediments embedded on the slab itself. Considerable variability in helium isotope ratios is encountered in such areas, including values as low as 1.4 times atmospheric in Central American andesitic areas, and these variations must be associated with differing proportions of mantle and crustal helium, as well as with possible mantle heterogeneity.

We have also measured helium in samples from the Tatun Volcano Group area in northern Taiwan along the Chingshan fault where Yu-Chia Chung of our laboratory sampled hot springs and wells associated with this andesitic volcanic region. This area also has relatively low $^3He/^4He$ ratios, with a maximum value of about 2.7 times atmospheric in the hot springs. This is another example of the rather large variations found in the convergent plate-margin terrains and indicates the need for very detailed studies to understand the meaning of this heterogeneity.

Two very important areas for future study are clearly the Himalayan suture zone in Tibet along the Himalayan fold system, with its thick continental crust, and its extension along the West Yunnan fold system in southwest China, especially in the region of the Ten-chung geothermal area associated with relatively young volcanoes east of the leading edge of the Indian Plate. The study of such areas is fundamental for comparison of the extent of mantle volatile penetration to the surface in continental arcs (continent-continent collisions), transitional (continent-oceanic) arcs, and purely intra-oceanic arcs such as the Mariana arc. In our laboratory, my colleagues J. E. Lupton, W. Rison, J. A. Welhan, R. Poreda, and I are currently studying the helium and rare gas isotope ratios, and associated isotopic ratios and chemistry of other volatiles ($CO_2$, $H_2O$, $CH_4$, $H_2$, $N_2$), in an attempt to understand the very diverse sources and occurrences of these components, both in geothermal and volcanic fluids and gases, and in the rocks themselves. These studies are not only of interest for purely geological problems, they have practical applications for exploration for geothermal resources (i.e. deep vs. shallow reservoirs), and for earthquake prediction with possible fluid-phase precursors, i.e. isotopic and chemical variations associated with pre-seismic tectonic effects.

## References

The following papers are recent references on variations in helium isotope ratios in terrestrial rocks and volcanic gases.

(1) Craig, H., Lupton, J. E., Helium-3 and mantle volatiles in the ocean and the oceanic crust, in The Sea, vol. 7: The Oceanic Lithosphere, Wiley - Interscience, (in press).

(2) Craig, H., Lupton, J. E., Horibe, Y., A mantle helium component in Circum-Pacific volcanic gases, in Terrestrial Rare Gases, Advances in Earth and Planetary Sciences vol. 3, 3-16 (1978).

(3) Tolstikhin, I. N., A review: Some recent advances in isotope geochemistry of light rare gases, ibid. 33-62.

(4) Craig, H., Lupton, J. E., Primordial neon, helium, and hydrogen in oceanic basalts, Earth Planet. Sci. Lett. 31, 369-385 (1976).

(5) Craig, H., Lupton, J. E., Welhan, J. A., Poreda, R., Helium isotope ratios in Yellowstone and Lassen Park volcanic gases, Geophys. Res. Lett. 5, 897-900 (1978).

(6) Welhan, J. A., Craig, H., Methane and hydrogen in East Pacific Rise hydrothermal fluids, Geophys. Res. Lett. 6, 829-831 (1979).

GEOCHEMICAL CHARACTERISTICS OF HOT FLUIDS IN XIZANG

Shi Hui-xin
(Analysis and Prediction Center, State Seismological Bureau)

Two main regions of strong hydrothermal activity in China--Xizang and Taiwan belong to active zones of different types at the borders of plates. Xizang is at the junction of Eurasian Plate and Indian Plate, whereas Taiwan is in a chain of Ryukyu-Taiwan-Philippines islands arc system. Modern tectonic movements in these two regions are very much different.

For the purpose of comparison we have studied the data in regions of the western border of the Pacific Ocean. In these regions oceanic crust underthrusts the continent and causes increase of heat flow in the Benioff zone on the side of continental crust. Melted magma, geothermal liquids, and volatile matter migrate upwards. Various fluids gush out of the deep crust, forming magnificent active volcanos, fountains, geysers, boiling springs, and hot springs. These hot fluids are not only of high temperature, but also contain characteristic geochemical ingredients. Among those listed below, water and gases are discharged in great quantities in active zones of the western border of the Pacific plate. They are rather rare in stable inland (Table 1).

1. Strongly acidic water. Its temperature may reach 340°C or more. This kind of water is distributed in areas with modern volcanic activity, such as Kamchatka Pen., Japan, Tatum volcanic region in Taiwan of China, and New Zealand.

2. Pure chloride-sodium water with low mineralization. Milligram-equivalent percentages of chlorine ion and sodium ion are more than 90%. Boron, lithium and arsenic are enriched. This kind of water is neutral and alkaline and is distributed in central zones of discharge of hot fluids.

3. $CO_2$ fluids. They are mainly from warm and cold springs.

4. Superhot brine. Its origin is influenced by the high heat flow and it is being replenished by water of marine origin.

Based on the characteristics of composition of isotopes of some inertial gases, it is inferred that there may be fluids coming from the mantle. For example, ratio of helium isotopes and ratio of argon isotopes in some hot fluids are close to that in the mantle.

Xizang is in the zone of collision of two continental plates. It is generally thought that there was subduction of the Andes type in the early period in this area. But subduction of oceanic crust has ended and now a collision of two continental plates still continues. Yarlung Zangbo River is in a suture zone. Thermal activities of the two periods were superposed.

Hydrothermal activity develops on both sides of the suture zone, extending as far as the Himalayas (Fig. 1), whereas in regions of western border of Pacific Ocean, geothermal anomaly and hydrothermal activity develop only on the trench side, which is toward the continent.

Hot water in Xizang area is mainly of the chloride type, but the chlorine ion is usually less than 60%*. Other anions are bicarbonate, carbonate and sometimes also sulphate ions. The water is neutral or alkaline and is enriched with boron, fluorine, lithium, arsenic and silicon.

Another hot water is of the bicarbonate type. Milligram-equivalent percentage of bicarbonate ion is more than that of chlorine ion. It is neutral or weakly alkaline and is also enriched with fluorine, boron and silicon.

In some springs, carbonate ions may be the main anion, forming strongly alkaline soda water (pH near 10) owing to the escape of abundant $CO_2$ when pressure decreases during eruption. In these springs, water is also enriched with fluorine, boron, lithium, ammonium, arsenic and silicon.

Sulphate water occurs in some individual regions with gypsum layer. This kind of water is alkaline and is formed probably as a result of leaching of gypsum. In these springs, water is also enriched with fluorine, boron, ammonium and silicon.

To sum up, in hot water on both sides of Yarlung Zangbo River sodium ion or the total content of sodium and potassium ions are more than 90%, chlorine ion is usually less than 60%. Bicarbonate ion may reach 1 g/l or more. Type of water may be Cl-$HCO_3$, $HCO_3$-Cl or $HCO_3$ which are neutral and alkaline, and may also be of the type $CO_3$ which is strongly alkaline. Mineralization of water is low. Temperature is near or higher than boiling point. Other cations with rather high contents may be potassium, lithium, rubidium, cesium, strontium, barium, ammonium and mercury. Among anions, boron, arcenic, fluorine and silicon are enriched. Boron exists as metaboric acid.

Comparing the geochemical characteristics of deep-formed fluids in various regions along borders of plates, we see that they are quite different from each other (Table 2). Superhot strongly acidic water, brine, pure chloride-sodium water, which are well developed in regions of island arcs - trenches, are undeveloped in the suture zone. Chloride-bicarbonate and bicarbonate-chloride water and $CO_2$ fluids are extremely well developed in the suture zone. As seen on Fig. 2, distribution of zones of carbon dioxide discharges are quite similar to the distribution of major zones of seismicity. Xizang, Mediterranean--Asia Minor and California constitute three regions of concentrated discharges of $CO_2$ fluids. In Xizang mostly, hot and warm fluids occur, which is different from the other two regions, where cold $CO_2$ fluids are dominant.

---

* Data of chemical composition of water are obtained by Ministry of Geology and the Comprehensive Scientific Expedition to the Qinghai-Xizang Plateau, Academia Sinica.

## DISCUSSION

1. Geochemical characteristics of deep-formed fluids in regions with different dynamic histories along borders of plates are quite different. So investigation of the geochemistry of fluids may be the approach for geodynamic research.

2. In the suture zone, deep-formed fluids are enriched with $CO_2$. It is related probably with high stress in the crust.

3. In vast area on both sides of the suture zone, discharged hot waters are all neutral and alkaline, of chloride-bicarbonate or bicarbonate-chloride type, with similar compositions of micro-elements. It indicates that water is probably not related with individual intrusives, but is related with deep seated heat sources of relatively uniform composition. Geophysical investigation shows that there may be a partially melting layer in the crust. Formation of the above-mentioned characteristics of chemical composition of deep-formed fluids is related probably with the boiling and degasification of this melting layer under reducing conditions. This does not, however, rule out the possibility that water may be related with individual intrusives in some individual regions.

4. Investigation of fluid-geochemistry is very useful for the study of the dynamics of plates and earthquake activity in this area. Investigations of stable isotopes, such as helium, argon, carbon, hydrogen and oxygen are especially fruitful.

## References


(1) Barnes I. et al., Global distribution of carbon dioxide discharges and major zones of seismicity. U.S. Geological Survey (1978).

(2) Shi Hui-xin, A note on carbon-dioxide discharging zones and their seismicities in China. Seismology and Geology, 1 (2), 86-93 (1979).

(3) An Ke-shi et al., Geochemical characteristics of the geothermal field at Yangbajain, Tibet. Hydrogeology and Engineering Geology, (1), 14-18 (1980).

(4) Zhang Ming-tao, A preliminary study on the geothermal field at Gobdü. Geographical Knowledge, (5), (1977) (in Chinese).

(5) Zhang Ming-tao, Beautiful geyser at Dagejia. Geographical Knowledge, (6), (1977) (in Chinese).

(6) He Shi-chun, Hot spring in Tanggula. Geographical Knowledge, (8), (1977) (in Chinese).

(7) State Geological General Bureau, Institute of Hydrogeology and Engineering Geology, Zones of distribution of hot groundwater in China. Treatise on geology for international exchanges, (5), 34-45 (1979).

Table 1 Some kinds of characteristic hot water in regions of island arcs --trenches

| Strongly acidic water | Superhot brine | Pure Cl-Na water with mineralization < 4 | $CO_2$ fluids |
|---|---|---|---|
| $Cl^-$ type<br>pH 0.2--4 | $M27.25\ \frac{Cl98}{(Na+K)61Ca38}$<br>(U.S.S.R.) | $M2.7\ \frac{Cl95}{Na89}$<br>(U.S.S.R.) | Gas with $CO_2$ >90%; water with dissolved $CO_2$ > 1000mg/l or $HCO_3^-$ >1000 mg/l, pH < 8.3 |
| $SO_4^{2-}$ type<br>pH 1--4.5 | $M76.75\ \frac{Cl99}{(Na+K)81Ca17}$<br>(Japan) | $M3.0\ \frac{Cl87}{Na88}$<br>(Japan) | |
| Main gaseous components:<br>$H_2S$, $CO_2$ | $M20.45\ \frac{Cl99}{(Na+K)54Ca44}$<br>(Japan) | $M3.7\ \frac{Cl97}{Na88}$<br>(New Zealand) | |

Table 2 Comparison of hot water in various regions at the border of plates

| | Strongly acidic water | Superhot brine | Superhot pure Cl-Na water | Superhot $Cl-HCO_3$, $HCO_3-Cl$ water | $CO_2$ fluids |
|---|---|---|---|---|---|
| Regions of island arcs -- trenches | Well developed | | | | |
| Regions of sutures | Undeveloped | | | Well developed | Extremely well developed |

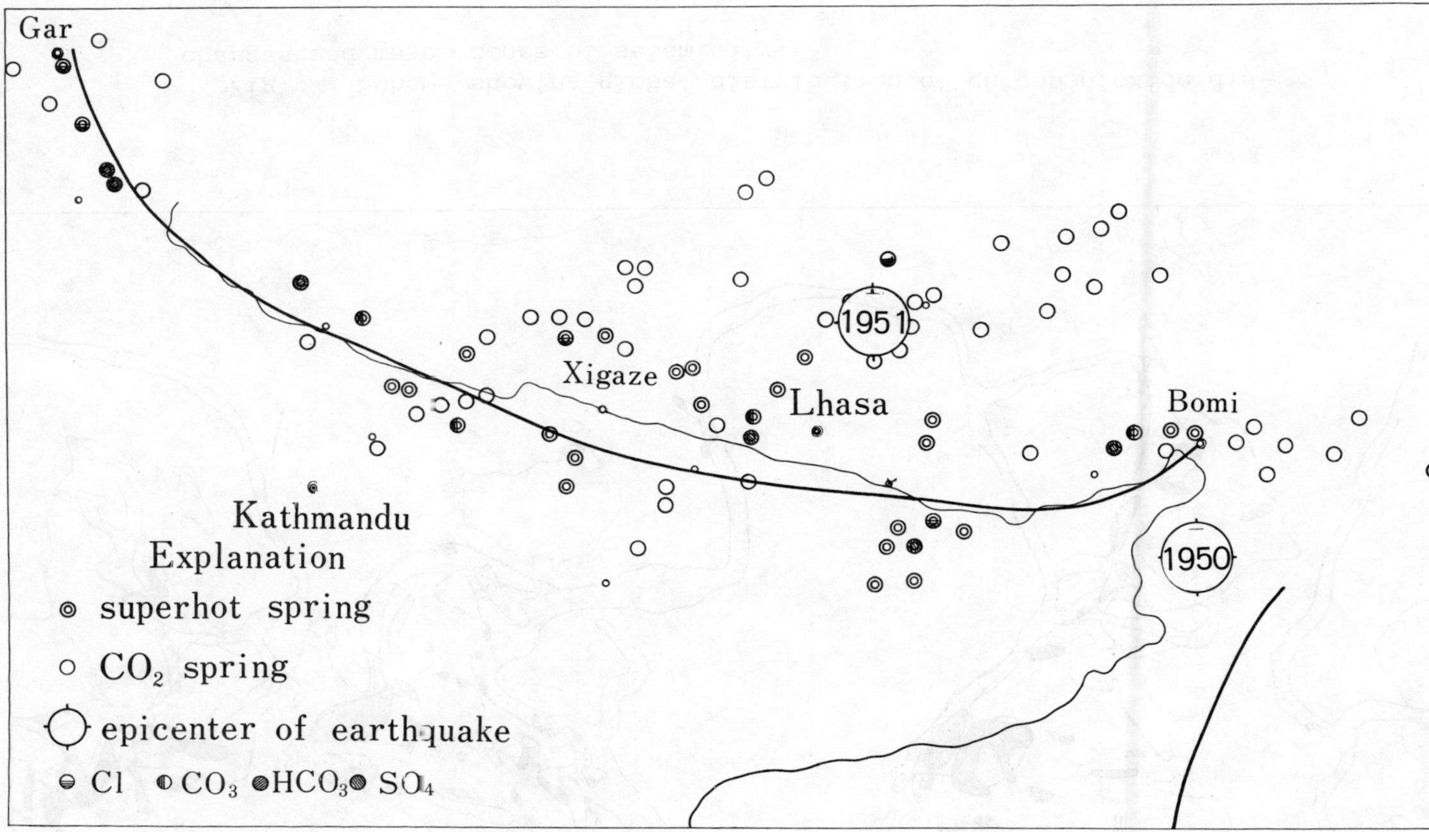


Fig. 1 Map showing superhot and carbon dioxide springs in Xizang.

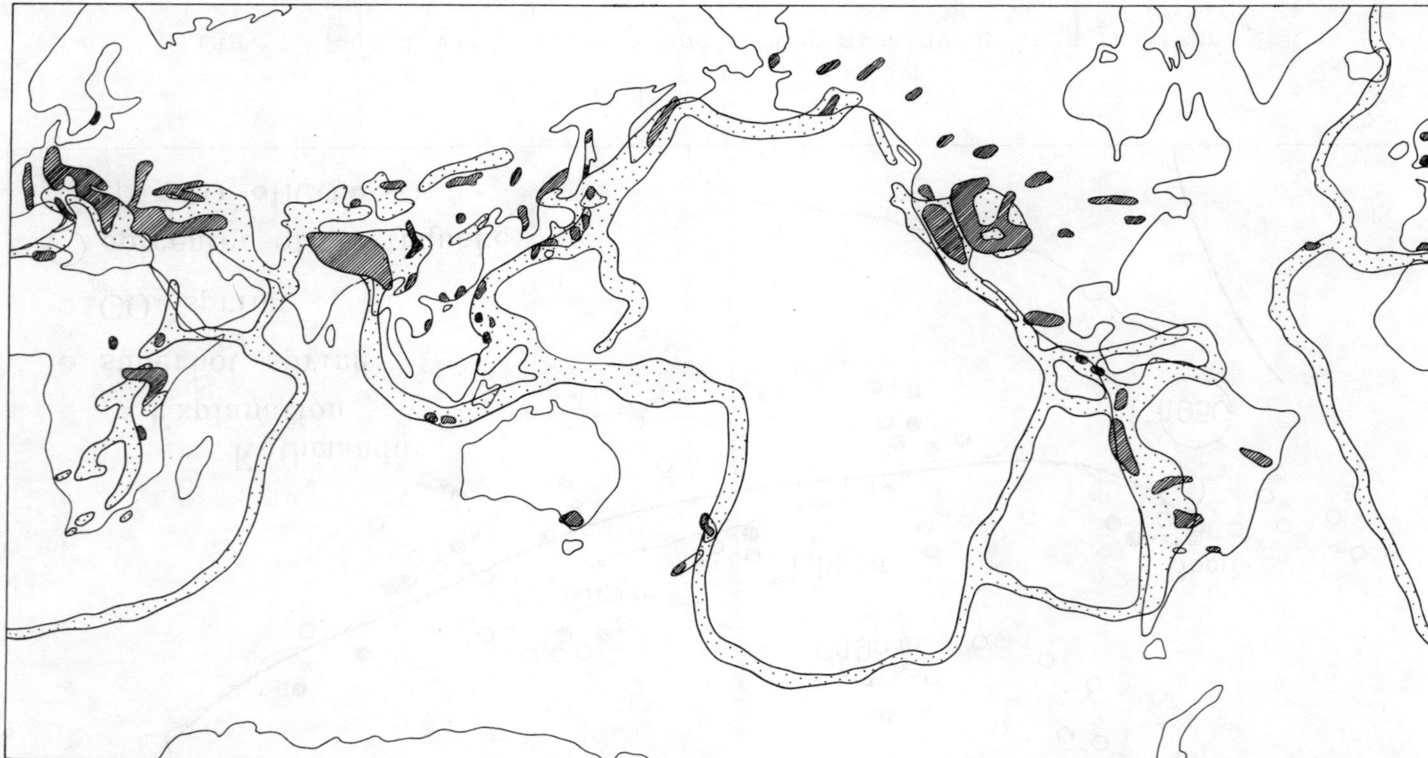

Fig. 2 Scheme showing global distribution of carbon dioxide discharges and major zones of seismicity.

## PRELIMINARY STUDIES ON HYDROTHERMAL GEOCHEMISTRY OF XIZANG

Zhang Zhi-fei    Zhu Mei-xiang    Liu Shi-bin
(Department of Geology, Beijing University)

During geothermal investigations from 1973 to 1976, we collected more than 400 water samples, most of which are from hot springs, 24 gas samples emitted from thermal springs, and a lot of sinter or travertine. Based on the examination of these samples, some characters of hydrothermal geochemistry in Xizang were revealed.

The pH values of hot waters in most geothermal areas are neutral or somewhat alkaline and similar to those in most geothermal areas in the world. Some geothermal areas in the world are associated closely with active volcanoes and give acid or strong acid hot water. In Xizang, however, hot water with pH value less than 4 has not been found. Hot waters with degree of mineralization less than 2 grams per liter make up 76.2% of the total. The highest degree of mineralization of 30.7 grams per liter occurs in saline hot springs with a temperature of 41.4°C from Yanjing, Markam County, Qamdo. The real hot brine has not been found yet in Xizang.

Xizang's hot waters can be divided into five hydrochemical types, depending on the percentage of equivalent concentration of major ions within hot water, and simultaneously taking into consideration of the significance of boron content in hot water (Table 1). The sodium chloride type water in Xizang may be further subdivided into two varieties whose genetic processes may be different. The variety poor in boron may be generated from filtering salt strata by meteoric water. The variety rich in boron is similar to that of some recent volcanic type high-temperature geothermal field. Their genesis may be closely related to the active magmatism. Such variety has drawn great attention from investigators of many countries and given rise to some special terminology, such as "geyser water", "non geyser Na-Cl type water"(1), "high-temperature hot water system"(2), "volcano water system"(3) and "fluid within reductive zone related to process of volcano directly"(3), etc. The bicarbonate-sodium chloride type water is very similar to boron-abundant sodium chloride type water in composition, except the percentage of equivalent concentration for $HCO_3$ and $CO_3$ is dominant in the anions and geographically accompanies with the latter. It seems that the generation of both types are the same, so some authors put them both into sodium chloride type. These two types make up 41% of the hydrothermal areas in Xizang. The distribution of these areas is shown in Fig. 1.

Xizang's hot waters are rich in B, As, Li, Rb and Cs. The contents of these elements in many samples are several to tens times their average abundance in the crust except rubidium. See Table 2.

Distribution of these elements in various chemical types of water is very different. In general, their concentrations in boron-abundant sodium chloride type and bicarbonate-sodium chloride type are higher than those in other types by about one order. See Fig. 2

The concentrations of these elements in hot water from Xizang are also higher than those in some major geothermal areas abroad. Informations related to B and Li are given in Fig. 3. Each point represents one area. Most values from other countries are of the sodium chloride type. In Xizang, the point represents the average value of the boron-abundant sodium chloride type and bicarbonate-sodium chloride type. It is seen from Fig. 3, that Xizang occupies a position of rather higher value and its close neighbours is Puga of Kashmir, which is located in the same geological environment as Xizang. Although the B- and Li- concentrations in hot water discharged from some areas are higher than those in Xizang, but that two elements contents of each gram total dissolved solids (TDS) are generally lower than those in Xizang, because their TDS are even higher. In addition, As, Rb, Cs are similar to the condition of B and Li mentioned above.

In order to analyze the feature of geographical distribution of concentrations of these elements in hot water, we have made a trend analysis for the boron concentration. Since the contents of B, As, Li are correlated with that of Cl, the trend of geographical distribution of the concentrations for boron in hot water will give useful indication of the others.

The result of trend analysis is shown in Fig. 4. In this map the contour interval is not constant. Seven grades are designed and their unit is ppm. The conspicuous feature shown in Fig. 4 is that the high value belt of $HBO_2$ concentration runs along both sides of Yarlung Zangbo River and Shiquan River, on the northern flank of the Himalayas.

This feature of geographical distribution not only appears in $HBO_2$ contents of hot water, but also in other respects of hydrothermal geochemistry. Sinter-occurring hydrothermal areas are less and spread along Shiquan River-Yarlung Zangbo River. Sinter can generally indicate a higher temperature of subsurface. Subsurface temperatures of 193 hydrothermal areas were estimated by Na-K-Ca geochemical geothermometer and 49 of which are over 150°C, making up 25.4% of the total amount. The distribution trend of these high temperature hydrothermal systems is also concentrated on both sides of Shiquan and Yarlung Zangbo Rivers.

To sum up, all the phenomena, such as distribution of boron-abundant sodium chloride type waters and bicarbonate-sodium chloride type waters, distribution of sinter, and high-temperature hydrothermal systems or spread of high value belt of $HBO_2$ content, along Shiquan-Yarlung Zangbo Rivers, effectively delineate the Himalayan Geothermal Belt and determinate its direction(22) of extension.

What kinds of geological process are under way so as to enrich B, As, Li, Rb and Cs in hot waters in Xizang in such a high level and in such a regularity of geographical distribution? L.S. Balashov has pointed out that it is favourable for enrichment of these elements in hot water in Quaternary volcanoe areas and strong earthquake areas(23). In Iceland(21) and Tengchung district of western Yunnan there were Quaternary volcanos and strong earthquakes, but the B, As, Li, Rb and Cs concentrations are considerably low*. Contrarily, these elements in hot water from southern Xizang are enriched to a very high degree but Quaternary volcano has not been found throughout the southern part of Xizang up to the present.

* unpublished data.

A.J. Ellis and W.A.J. Mahon emphasized that enrichment of boron in hot water results from the contact of hot water with sedimentary rocks.(15). The average abundance of boron in sedimentary rocks is greatly higher than that in the crust and some boron-abundant sodium chloride water occurs in some sedimentary rock areas in southern Xizang, but not all of the type do so. In Qamdo Prefecture in eastern part of Xizang there are many hot springs from the sedimentary rock areas, but the B concentration in them is rather low.

We would like to infer that formation of both types is related to metamorphism and remelt of sedimentary rocks subducted to the deeper part of the crust. The abundances of C, Cl, S, B, As, Li, Rb, Cs in sedimentary rocks are higher than those in crystalline rocks by several times up to more than one hundred times(24). All these elements are soluble and volatilisable. When the sedimentary rocks subduct to the deeper part of the crust and the conditions of temperature and pressure are changed greatly, these rocks could be subjected to metamorphism or even remelted and turned into crystalline rocks. During such processes, these elements could be dissolved or volatilized by the super- critical steam from magma and then carried into deep ground water.

The geological background of the Himalayan Geothermal Belt are metamorphism and remelting. This belt runs along the Yarlung Zangbo River which is the position of the subduction of Indian Plate under the Eurasian Plate before Oligocene. After collision of the two continents the subduction zone have turned into a suture zone, but the pressing of Indian continent to Eurasia has been going on. This results in a series of thrust faults, which are newly subduction zone with latitudinal strike and dipping to the north. These thrusts become younger and younger from north to south.

Due to the subduction of these faults, the sedimentary rocks with a thickness of ten kilometers(25)has been buried in the depth of the crust. Since the thicknesses of both sedimentary rocks and the crust in southern Xizang are very large, the processes of metamorphism and remelting will be easy to enrich those elements in hot water somewhat higher than those in other regions. This is our explanation on the occurrence of the boron-abundant sodium chloride type water and bicarbonate-sodium chloride type water, both of which are rich in B, As, Li, Rb and Cs, within Himalayan Geothermal Belt. This explanation have yet to be proved by the studies on petrochemistry as well as the high-temperature and high-pressure tests.

## ACKNOWLEDGEMENTS

The work of chemical analyses for water samples was shared by colleagues of Integrated Survey Commission of Natural Resources, Academia Sinica; Qinghai Institute of Saline Lake, Academia Sinica; Institute of Hydrogeology and Engineering-geology, the Chinese Academy of Geological Sciences; Department of Geography, and Geochemical Section, Department of Geology, Beijing University. The programming of trend analysis was made by Xu Zhen-bang and Lin Wan-zhi of Department of Geology, Beijing University. The authors wish to thank them all.

## References


(1) White D.E. et al., Chemical composition of subsurface waters, Geol. Survey Prof. Paper 440-F (1963).

(2) White D.E., Geochemistry applied to discovery, evaluation and exploitation of geothermal energy resources, Geothermics, Spec. Issue 2, 1, 58-80 (1970).

(3) Truesdell A.H., Summary of section III, geochemical techniques in exploration, Proc. U.N. Symp. Develop. Use geothermal resources, 2nd, San Francisco, Calif. 1, liii-lxxix (1975).

(4) Li Tong, Chemical element abundances in the earth and it's Major shells. Geochimica, 3, 167-173 (1976).

(5) Robertson E.C. et al., Hydrothermal activity in southwestern Montana, Proc. U.N. Symp. Develop. Use geothermal resources, 2nd, San Francisco, Calif. 1, 553-562 (1975).

(6) Mariner R.H., et al., Geochemistry of thermal waters in Long Valley, Mono County, California, 81 (5), 792-800 (1975).

(7) White D.E., Thermal and mineral waters of nonmeteoric origin, California Coast Ranges, Geol. Soc. Amer. Bull., 84, 547-560 (1973).

(8) Helgeson H.C., Geology and thermodynamic characteristics of the salton sea geothermal system, Amer. Jour. Sci. 366, 129-166 (1968).

(9) Reed M.J., Geology and hydrothermal metamorphism in the Cerro Prieto geothermal field, Mexico, Proc. U.N. Symp. Develop. Use geothermal resources, 2nd, San Francisco, Calif. 1, 539-549 (1975).

(10) Banwell C.J., et al., Geothermal exploration in Mexico, Geothermics, Spec. Issue 2, 1, 27-40 (1970).

(11) Romagnoli P. et al., Hydrogeological characteristics of the geothermal field of Ahuachapán, El Salvador, Proc. U.N. Symp. Develop. Use Geothermal Resources, 2nd, San Francisco, Calif. 1, 563-571 (1975).

(12) D'Archimbaud J.D., et al., Geothermal exploration progress at Bouillante in Guadeloupe, Proc. U.N. Symp. Develop. Use geothermal resources, 2nd, San Francisco, Calif. 1, 105-108 (1975).

(13) Arango E.E., et al., Preliminary study on the Ruiz Geothermal Project (Colombia), Geothermics, Spec. Issue 2, 1, 43-56 (1970).

(14) Cusicanqui H., et al., The geochemistry of the El Tatio geothermal field, Northern Chile, Proc. U.N. Symp. Develop. Use geothermal resources, 2nd, San Francisco. Calif. 1, 703-711 (1975).

(15) Ellis A.J. et al., Chemistry and Geothermal Systems, Academic Press, New York (1977).

(16) Shanker R., et al., Geothermal exploration of the Puga and Chumathang geothermal fields, Ladakh, India, Proc. U.N. Symp. Develop. Use geothermal resources, 2nd, San Francisco, Calif. 1, 245-258 (1975).

(17) Kurtman F., et al., Geothermal energy possibilities, their exploration and evaluation in Turkey. Proc. U.N. Symp. Develop. Use geothermal resources, 2nd, San Francisco, Calif. 1, 447-458 (1975).

(18) Dominco E., et al., Characteristics of Greek geothermal waters. Proc. U.N. Symp. Develop. Use geothermal resources, 2nd, San Francisco, Calif. 1, 109-122 (1975).

(19) Baldi P. et al., Geothermal research in western Campania (southern Italy): chemical and isotopic studies of thermal fluids in the Campi Flegrei, Proc. U.N. Symp. Develop. Use geothermal resources, 2nd, San Francisco, Calif., 1, 687-697 (1975).

(20) Boldizar T., et al., Hydrogeology of the Pannonian geothermal basin, Proc. U.N. Symp. Develop. Use geothermal resources, 2nd, San Francisco, Calif., 1, 297-304 (1975).

(21) Tomasson J. et al., A hydrological model for the flow of thermal water in southwestern Iceland with special reference to the Reykir and Reykjavik thermal areas, Proc. U.N. Symp. Develop. Use geothermal

resources, 2nd. San Francisco, Calif. 1, 643-648 (1975).
(22) Tong Wei et al., The Himalayan geothermal belt, Acta Scientiarium Naturalium Universitatis Pekinensis, 1, 76-88 (1978).
(23) Balashov L.S., Rare elements in thermal ground water, Proc. U.N. Symp. Develop. Use geothermal resources, 2nd, San Francisco, Calif. 3, 2187-2195 (1975).
(24) Li Tong et al., The abundance of chemical elements in the earth's crust and its major tectonic units, Acta Geologica Sinica, 45,1, 82-92 (1965).
(25) Mu En-zhi, et al., The stratigraphy of the Mt. Qomolangma district in the southern part of Xizang, China, Scientia Sinica, 1, 59-71 (1973).

Table 1 Hydrochemical types of hot waters in Xizang

| Type | Number |
|---|---|
| 1. Sodium chloride type | |
| (1) poor in boron variety | 9 |
| (2) rich in boron variety | 32- |
| 2. Bicarbonate - sodium chloride type | 82- |
| 3. Bicarbonate type | 28 |
| 4. Sulphate type | 33 |
| 5. Bicarbonate - sulphate type | 93 |

Table 2

| | | B ($HBO_2$) | As | Li | Rb | Cs |
|---|---|---|---|---|---|---|
| Average abudance in the crust(4) | ppm | 7.6 (30.9) | 2.2 | 21 | 78 | 1.4 |
| Number of hydrothermal areas where the concentration is over the average abundance in the crust | | 110 | 20 | 7 | - | 50 |
| Maximum concentration | ppm | (1965) | 125.6 | 50 | 11.04 | 58 |
| Times of the maximum concentration to the average abundance in the crust | | 64 | 57 | 24 | - | 41 |

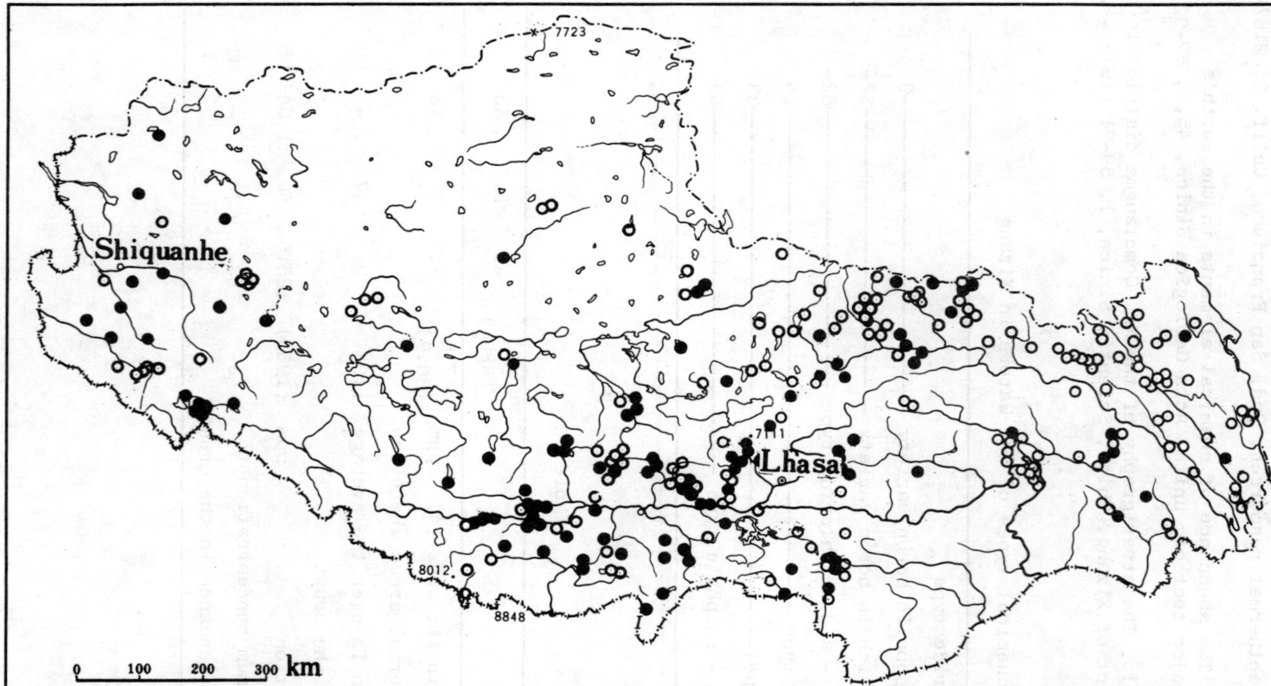


Fig. 1 Map for distribution of boron-abundant sodium chloride type and bicarbonate-sodium chloride type waters (shown by solid cycle) and other types (shown by open cycle) from Xizang.

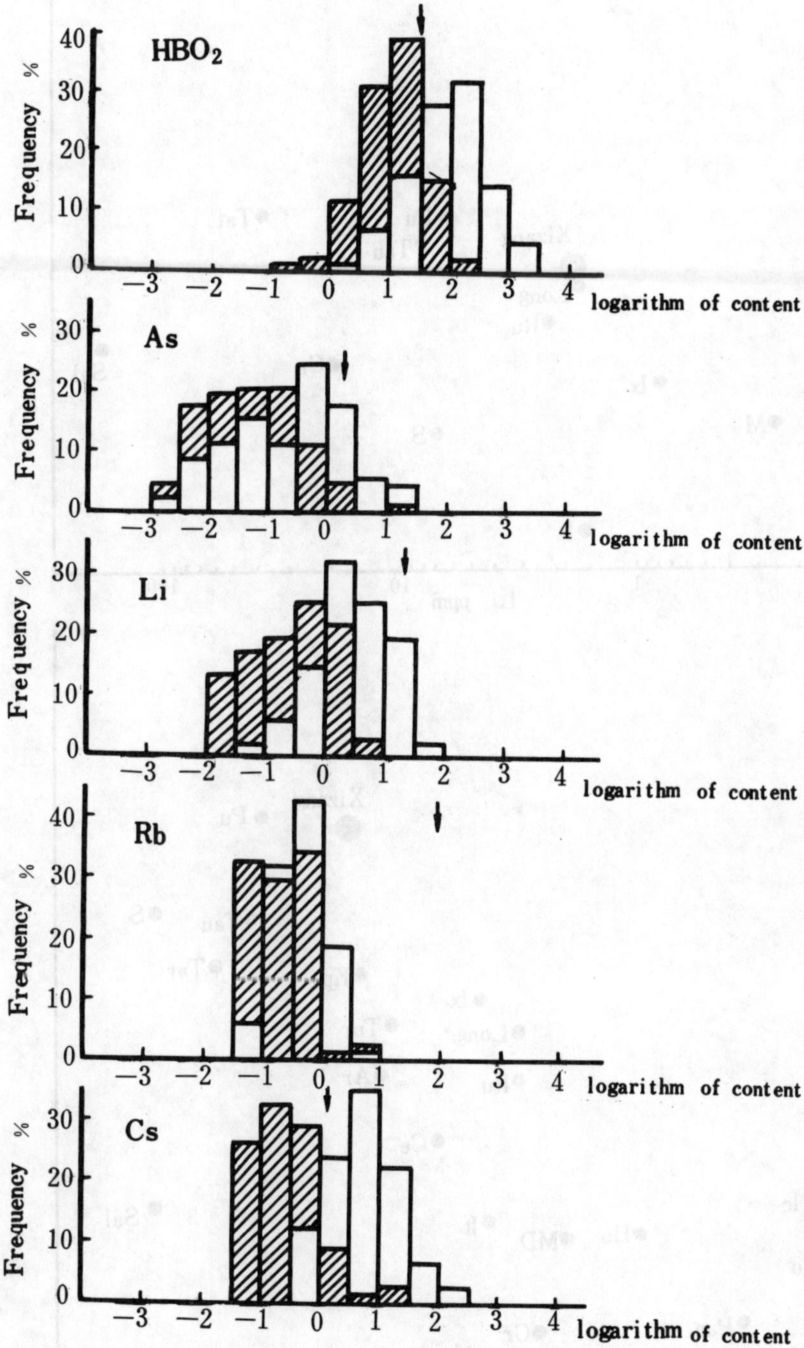


Fig. 2 Difference of concentration for HBO$_2$, As, Li, Rb, Cs, respectively, in various hydrochemical types of hot water. The open column represents boron abundant sodium chloride type and bicarbonate-sodium chloride type, and the column with shadow represents other types. The position indicated by short arrow represents the average abundance in the crust.

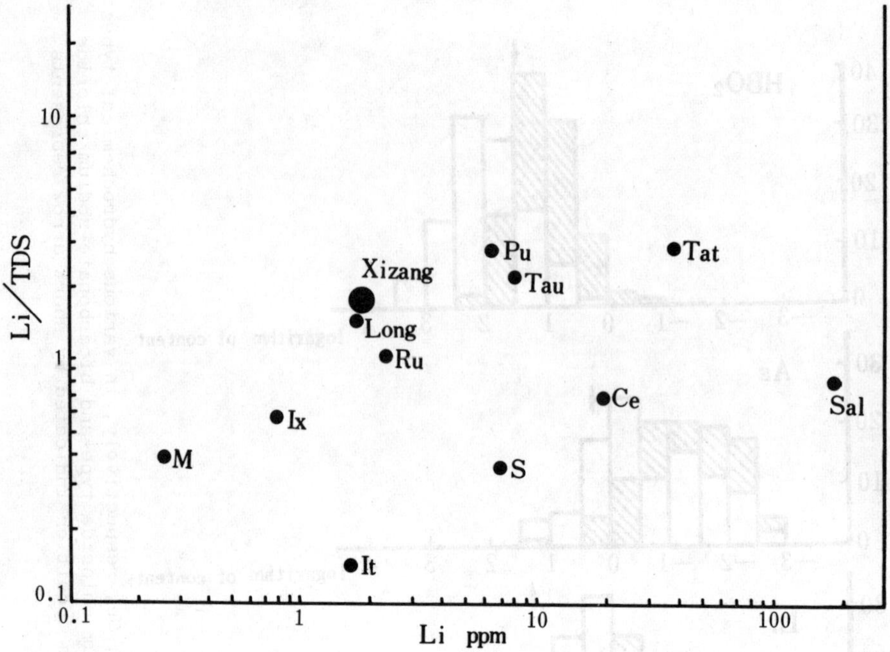

Li/TDS
Li ppm
Xizang
Long
Ru
Pu
Tau
Tat
Ix
M
S
Ce
Sal
It
0.1
1
10
100


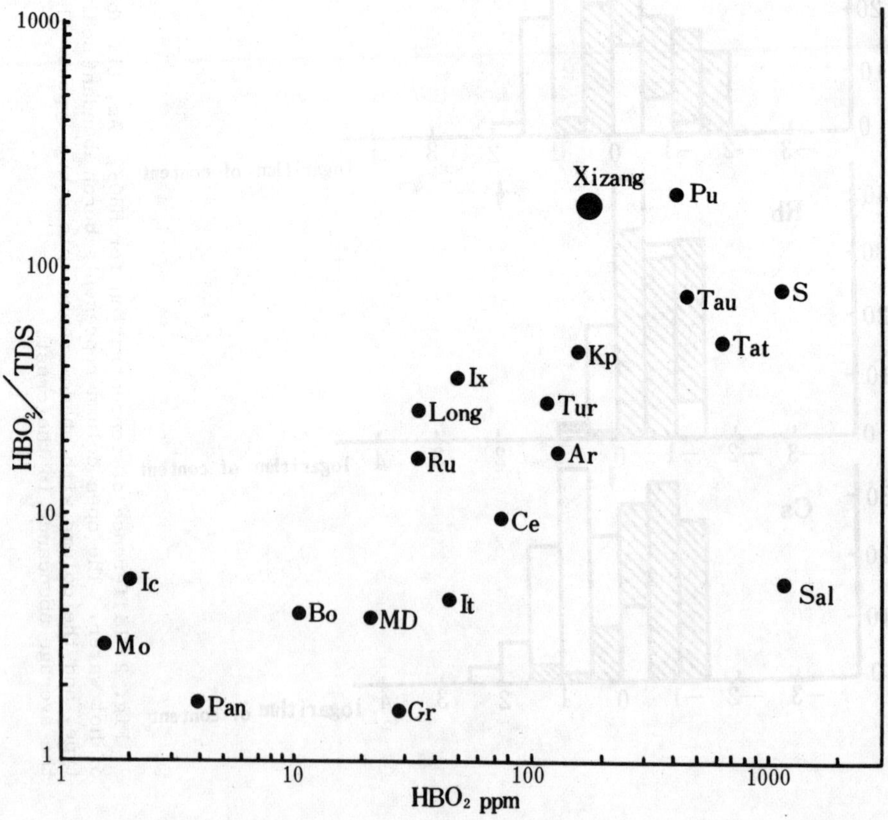

HBO₂/TDS
HBO₂ ppm
Xizang
Pu
Tau
S
Tat
Kp
Ix
Long
Tur
Ru
Ar
Ce
Ic
Sal
It
Bo
MD
Mo
Pan
Gr
1
10
100
1000

Fig. 3 Comparative diagrams for B, Li contents in hot water, respectively, from some major hydrothermal areas throughout the world. The signs of each point and its reference are shown as follows; Mo, Southwestern Montana, USA(5); Long, Long Valley, California, USA(6); S, Coast Range, California, USA(7); Sal, Salton Sea, California, USA(8); Ce, Cerro Prieto, Mexico(9); Ix, Ixtlan, Mexico(10); Ar, Ahuachapan, El Salvador(11); Bo, Bouillante, Guadeloupe(12); Ru, Ruiz, Colombia(13); Tat, El Tatio, Chile(14); Tau, Taupo Volcanos Belt, New Zealand(15); Pu, Puga and Chumathang, Kashmir(16); Tur, Kizildere and western part of Turkey(17); Gr, Greek(18); MD, Makhachkala, Dagestan, USSR(15); Kp, Pauzhetsk, Kamchatka, USSR(15); It, western Campania, southern Italy (19); Pan, Pannonian Basin, Hungary(20); Ic, Southwestern Iceland (21).

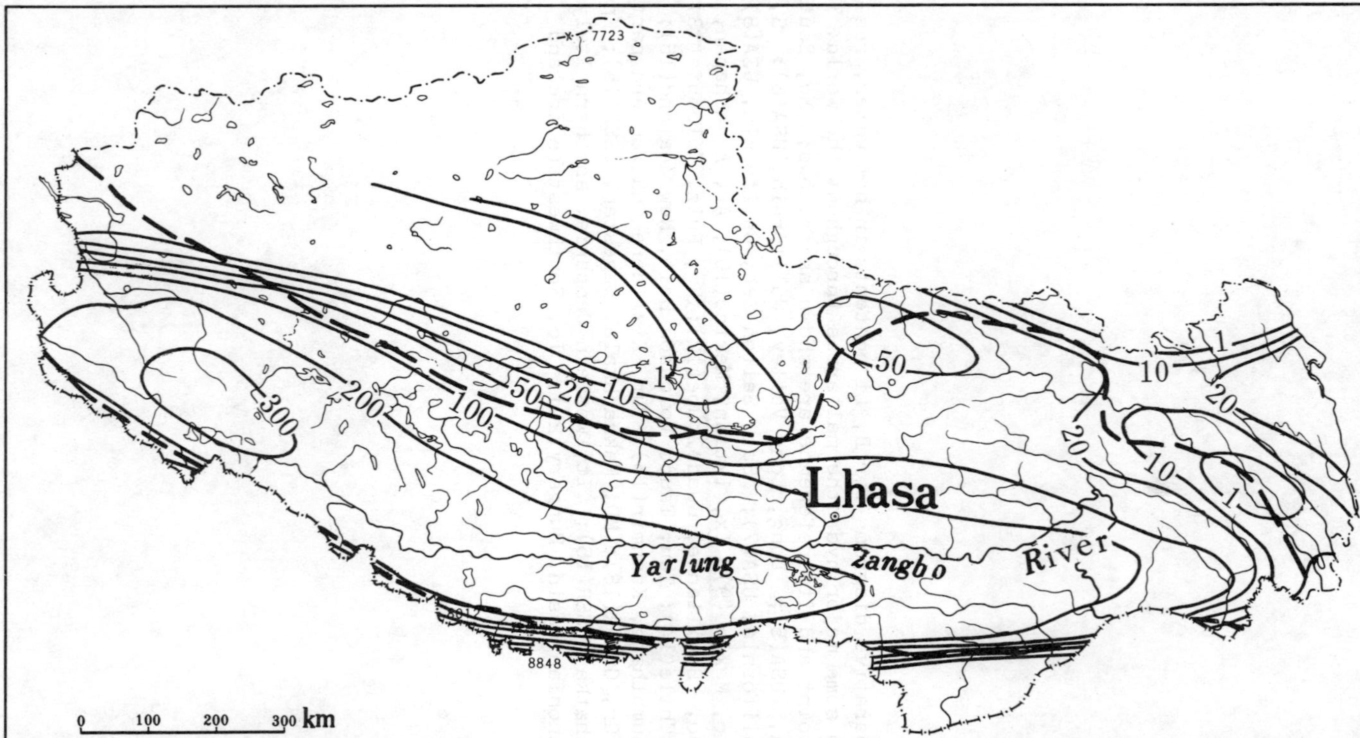


Fig. 4 Sketch isogram of trend analysis for $HBO_2$ in hot water from Xizang.

## ESTIMATION OF EXPLOITATION POTENTIALITY OF THE YANGBAJAIN GEOTHERMAL FIELD

Guo Guo-ying    You Mao-zheng    Liu Shi-bin
(Department of Geology, Beijing University)

Yangbajain geothermal field is one of the 40 high temperature hydrothermal areas in the Himalayan geothermal belt. It is located in the middle part of the belt and is the largest field found so far in our country.

The field is situated at the centre of the Cenozoic Yangbajain graben basin. The elevation is about 4,300 m. Nyainqentanglha Range and Tang Mountain tower over the north-western and southeastern side of the basin respectively, the peaks of the mountains are perennially covered with snow and glaciers. The basin has developed along the Nyainqentanglha fault and Danguolin-Maruguo fault for a long time. In this basin, Pleistocene moraine, glacio-fluvial sediments and fluvial, lacustrine sediments are accumulated, forming the shallow layered reservoir and cap rock of the geothermal field. In the shallower part of the crust below the basin or its adjacent area, relatively recent magmatism might have taken place to form a strong thermal source of the geothermal field. Zangbu River flows through the Yangbajain geothermal field from the southwest to the northeast, which is an ideal cooling source within the field.

Yangbajain geothermal field is a high-temperature hot-water field. According to the geochemical geothermometer, it is estimated that the reservoir temperature is about 200°C to 220°C, the highest may reach 275°C. The highest temperature actually measured in the drillhole is 172°C

In 1975, for the first time we measured and calculated the natural heat flow using comparative method in the Yangbajain geothermal field and thus estimated the exploitation potentiality of the field. The in situ test was continued in 1976. Natural heat flow is found to be about 11.61 x $10^4$ kcal/sec. The estimation of the exploitation potentiality is about 15.5 x $10^4$ kw.

Yangbajain geothermal field is about 90 km from Lhasa, in a flat terrain and has good transport facilities. In order to solve the shortage of energy resource in Lhasa, geological exploration has been carried out and a pilot geothermal power plant has operated for several years. This year, two turbo-generating sets of 3,000 kw will be constructed and the power will be transmitted to Lhasa.

### THE NATURAL HEAT FLOW OF YANGBAJAIN GEOTHERMAL FIELD

There are many different manifestations in the field. Hydrothermal activities are young and intensive. For convenience, the water-steam outcrops within the field was divided into 17 manifestation areas (Fig. 1).

Warm and hot springs spread all over the field. There are many boiling springs with temperatures of about $84^oC$ to $86^oC$. The boiling spring within the second manifestation area is the most typical. The boiling springs at the third and fourth manifestation areas are very concentrated. The fumaroles and the steaming grounds are the outcrops of natural steam. The hot ditch of sulfur deposits is typical of steaming grounds. The temperature at the depth of half a metre reaches $86^oC$. The steaming grounds are mainly distributed over the second, third, fourth, fifth, sixth and seventh manifestation areas. The shallow temperature gradient at the steaming grounds reaches as high as $3^oC$ per metre.

There is a hot lake in the field with area of about 7,350 $m^2$, its maximum depth is 16.1 m. The temperature of the water changes with seasons and lies between $45^oC$ to $57^oC$. The flow rate changes between 33.31 l/sec and 53 l/sec. The temperature of the lake water from top to bottom is rather uniform. This hot lake might be a product of previous hydro-thermo-explosions.

There are no natural geysers at Yangbajain, but after exploration drilling, artificial geysering holes arise. There appeared a geysering boiling water pond at the southern side of drillhole No. 2. Its temperature and flow rate change periodically. The Yangbajain drillholes No. 6 and No. 7 are also geysering drillholes. From aerial photographs the area of vegetation anomaly is about 5.8 $km^2$ . From geophysical data, the contour of minimum resistivity of ten ohmmetres gives an area of about 7.2 $km^2$ . It also coincides with the area of anomalous snow melting. The whole area of the thermal field would be much larger and may reach 15 $km^2$ .

There are four major types of manifestations: 1. hot water discharged by springs, evaporation, heat conversion losses from water surfaces of ponds, lakes and marshes; 2. water vapour losses from fumaroles and steaming grounds; 3. discharge of seeping through the Zangbu River bed; 4. conduction through the earth surface. Although the last one may not be an important source of heat loss, it may not be negligible.

Direct measurements and testings were performed on heat losses through each of these ways and the relative curves of heat output versus water temperature were constructed.

The heat loss from the springs was computed by the following equation:

$$Q_{hs} = C_w \gamma V_{hs} (\theta_{hs} - \theta_o),$$

where $\theta_o$ is the ambient temperature which is $10^oC$, $\theta_{hs}$ is the measured temperature of hot spring, $V_{hs}$ is the flow rate of hot water, $C_w$ is the specific heat of the water, and $\gamma$ is the density of the water. Evaporation and heat conversion losses from the water surface of ponds, lakes and marshes were determined by measuring the evaporated quantity at different water temperatures from which relative curve of heat loss at different water temperatures was established (Fig. 2).

Heat carried away by vapour from steaming grounds was measured by calorimeter. The soil temperatures at the depths of 2.5, 7, 10 and 15 cm were observed by means of a thermometer. From these observations, a family of temperature versus depth curves at different heat flow rates were prepared ( Fig. 3).

At the surface temperature anomaly area, thermocouple was inserted to 0.8 m depth to measure conductive heat flow. The rate of heat flow ranges from about 0.3 cal/$m^2$ sec. to 0.6 cal/$m^2$ sec., with the exception of only a few individual points where it is larger than 1 cal/$m^2$ sec. The value of 0.5 cal/$m^2$sec. was adopted in the computation.

Heat loss of seepages through the bed of the Zangbu River were computed from the differences of the temperatures and flow rates at the upper and lower section of the Zangbu River, directly measured on Sept. 27, 1976 (Table 1).

With the help of direct measurements and testing data, natural heat outputs of each of the seventeen manifestation areas, Zangbu River and surface temperature anomaly areas were computed. The results are summarized in Table 2.

The computed result shows that the natural heat flow of Yangbajain geothermal field is about 11.6 x $10^4$ kcal/sec., in which heat flow carried by steam is about 4.2 x $10^4$ kcal/sec., which is about 36.6 per cent of the total heat flow.

The seepages to Zangbu River contribute a considerable portion of the total heat flow, because Zangbu River flows through the strong manifestation areas at the southeastern part of the field. Heat flow carried away by the river is 1.2 x $10^4$ kcal/sec., amounting to about 10% of the total heat flow.

## ESTIMATION OF THE EXPLOITATION POTENTIALITY

In order to estimate the exploitation potentiality of the field, we must discuss the problem of the heat source. A definite answer is not yet possible at present, but we might make some very significant inference from the researches of geothermo-geological data. The highest temperature logged at the shallow reservoir is 172°C in the holes. The heat source obviously can not be the terrestrial heat flow.

The representative type of water at Yangbajain geothermal field is of NaCl type. The ratio between sodium and chloride is larger than two. Thus, the possibility of the origin of chloride ion from the leaching or connate water is eliminated. The important volatile component boron of magmatic differentiation and typical component lithium, rubidium of pegmatite are quite rich in the hot fluids. The arsenic, antimony contents are anomalous, too. It is worthy to note that $H_2S$ gas emerges from rock fissures at the hot ditch and natural sulfur deposits continually. The analyses of sulfur isotope show that the sulfur might be originated directly from the deep crust or even the upper mantle. All these show that there is a possible magmatic pocket at depth under the Yangbajain basin. It may be the strong heat source of the Yangbajain geothermal field.

Experiences obtained in other parts of the world show that geothermal fields usually have magmatic pockets and the exploited heat flow is several times larger than the natural heat flow. In the light of the experiences of the Wairakei geothermal field of New Zealand, we adopted four times of the natural heat flow and eight per cent thermal effectiveness to be the exploitation potentiality of Yangbajain geothermal field and obtained an estimate of $15.5 \times 10^4$ kw.

## ACKNOWLEDGEMENT

We especially appreciate the long-time cooperation with the colleagues of our group for Xizang's geothermal investigation. We are grateful to Professors Wang Ren, Fu Cheng-yi and Chen Zong-ji for their invaluable advice and revision of the manuscript.

## References

(1) Thompson G.E.K., Banwell C.J., Dawson G.B., Dickinson D.J., Prospecting of hydrothermal areas by surface thermal surveys, UN Conference on New Sources of Energy, Rome, Proceedings, G/54 (1961).

(2) Lloyd E.F., Geology and Hot Springs of Orakeikorak, New Zealand Geological Survey Bulletin 85, Department of Scientific and Industrial Research, 100-103 (1972).

(3) Tong Wei, Zhang Ming-tao et al., Geothermals beneath Xizang (Tibetan) Plateau, Science Press, (in press, in Chinese).

Table 1 Hydrometric data of Zangbu River

| Station | Average temperature °C | Average flow rate $m^3$/sec. |
|---|---|---|
| Lower section | 9.3 | 10.5 |
| Upper section | 8.44 | 10.12 |

September 27, 1976

Table 2 Natural heat flow of Yangbajain geothermal field

| Types | Heat Flow kcal/sec. | % | Mass flow kg/sec. |
|---|---|---|---|
| Steaming ground | 42,400 | 36.6 | 68.2 |
| Spring and evaporations | 55,000 | 47.3 | 751.0 |
| Conduction | 6.500 | 5.6 | |
| Seepages of Zangbu River | 12,200 | 10.5 | 159.0 |
| Total | 116,100 | 100.0 | 978.2 |

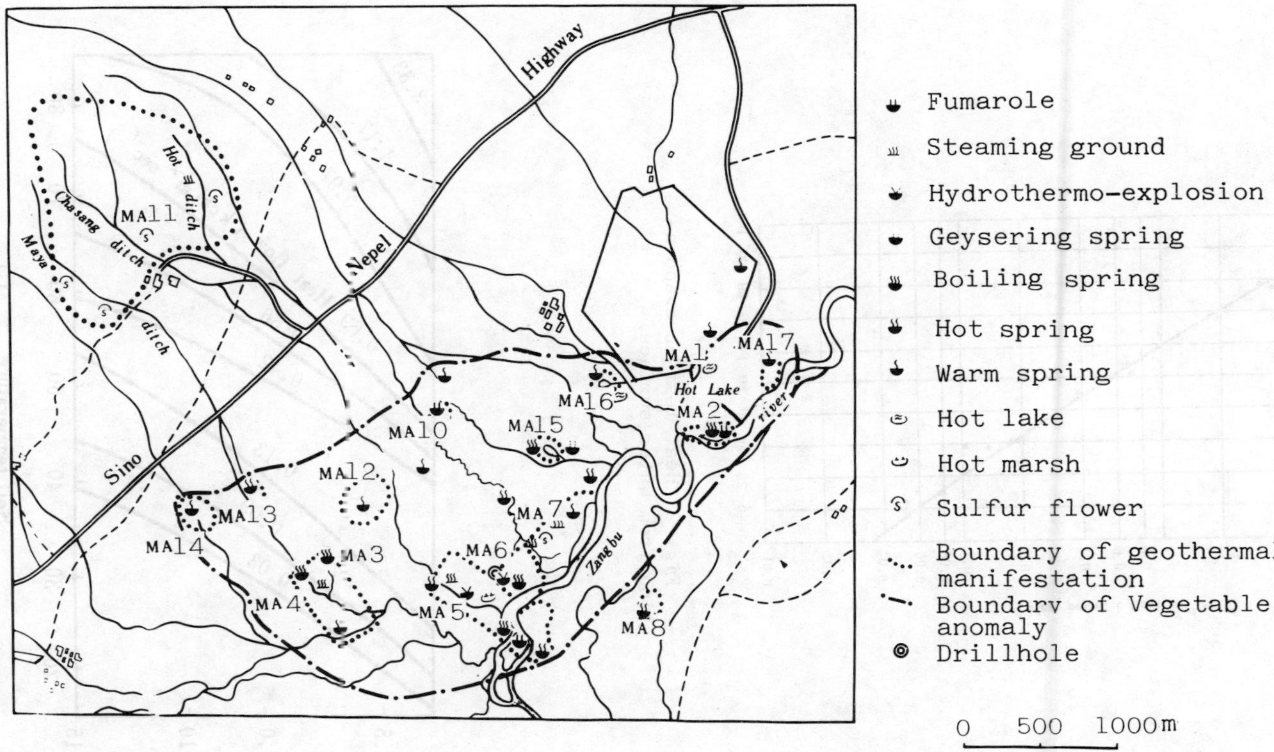


Fig. 1 Distribution of geothermal manifestation at Yangbajain geothermal field.

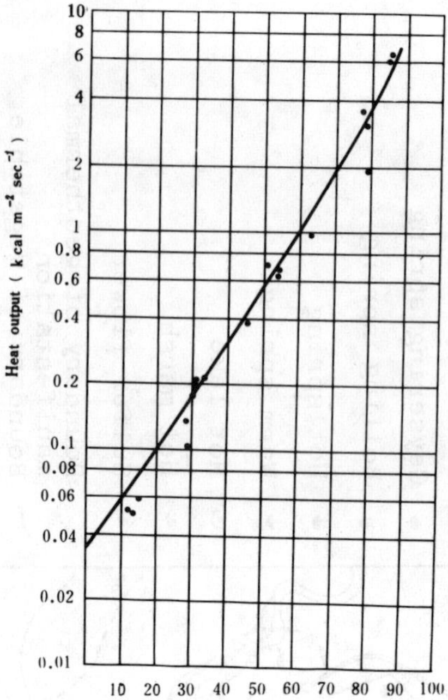


Fig. 2 Heat lost as evaporating from water surface.

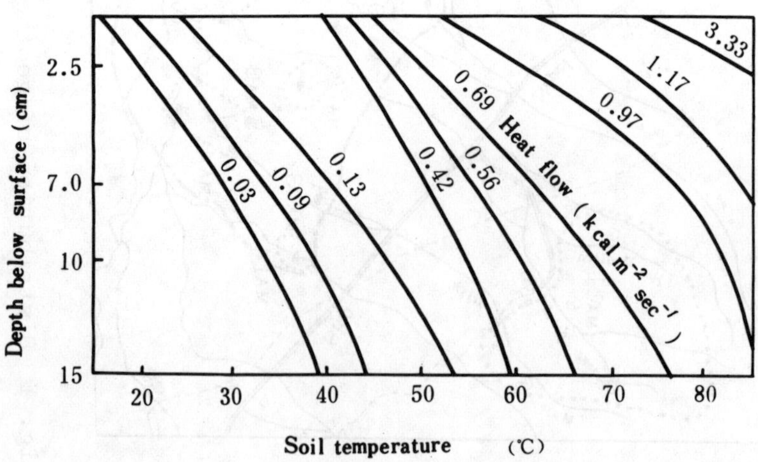


Fig. 3 Heat flow through Yangbajain soils as related to temperature and depth.

# EFFLORESCENCE IN THE GEOTHERMAL AREAS OF XIZANG AND ITS GEOLOGICAL SIGNIFICANCE

Zhu Mei-xiang    Tong Wei    You Mao-zheng
(Department of Geology, Beijing University)

In the geothermal areas of Xizang some white substance in the form of powder is often found around the fumaroles, on the surface around the vents of hot springs and on the cliffs near the springs as well as on the surface of sinters and travertines. They are similar to the saline soils in salt marshes, whose difference from the latter lies in the fact that they are deposits of spring water. They generally occur in forms of efflorescence, crust, flaggy or ring. And occasionally stalactites or saline bubbles are found in the specific places. Most of these deposits of spring water are soluble salts which we call efflorescence.

Efflorescences are widespread in the hydrothermal areas of Xizang, but distributed sporadically. As their grains are so fine, it is hard to be observed under the microscope, and X-ray diffractions are mainly used in their examinations. They are also studied by microscopic observation, immersion method and scanning electron micro-analysis. Based on our observation, we have, for the first time, gained clear idea of the composition of efflorescence and the relations among their various components and the strength of hydrothermal activities. We have further discovered the interrelation between the salt lakes in Xizang and the hydrothermal activities. Following we will discuss three problems: the composition of efflorescence; their generative features and the relationship between the hydrothermal activities and the salt lakes of Xizang.

## COMPOSITION OF EFFLORESCENCE

According to the X-ray powder analyses of over 50 samples of efflorescence taken from the hydrothermal areas of Xizang the salt minerals in this area are primarily determined as follows: simple sulfates (thenardite, epsomite and gypsum); carbonates (trona, thermonatrite and calcite); borates(9)(borax, tincalconite, kernite, hungtsaoite(10) and inderite); complex sulfates (potash alum, soda alum, ammonia alum, alunite and alunogen); fluorite, quartz, etc. The first three kinds of the above minerals are often found in salt lakes of Northern Xizang. Two or three out of such minerals as trona, thenardite and halite are often associated with each other. In a few areas they are associated with borax. This group of composition widely occurs in the Nagqu area. In the Qamdo area it is widespread too. Epsomite and gypsum occur also in Qamdo area.

The association of halite, hungtsaoite and kernite occur in Yangbajain basin and its vicinity. Each of such minerals as calcite, quartz, fluorite, tincalconite, inderite and thermonatrite is found only in one place. Complex sulfates are originated only from high-temperature hydrothermal areas, and so far they have never been discovered in salt lakes.

Similar to the deposited minerals of the salt lakes, the mineral association of efflorescences, except complex sulfates, fluorite and quartz, is closely related to the chemical types of spring water: (1) Almost all the efflorescences deposited from boron-abundant water contain boron minerals. But the occurrence of either boron or other borate minerals in different areas depends on the strength of hydrothermal activities. Generally speaking, borax only occurs in low-temperature hydrothermal areas, while other borate minerals often occur in high-temperature hydrothermalareas. For example, the efflorescences found in the hydrothermal areas of Nagqu, which are rich with warm spring and hot spring, contain only slight amount of borax. Kernite and hungtsaoite occur in the high-temperature hydrothermal areas of Yangbajain and inderite occurs in Labulang of Southern Xizang. Of course, the deposition of different borate minerals also depends on the alkalinity of solution. In this respect we have not made a thorough study yet; (2) Most of the cation in spring water of hydrothermal areas in Xizang are sodium. Most of the cations of efflorescence minerals are sodium too. As to the occurrence of sulfate, carbonate or haloids, it depends on the amount of cation contained in spring water. This is only a qualitative deduction. In the process of the spring water's passing through earth layer by infiltration and rising up to the surface, there must be an ion exchange between the spring water and the earth layer. As a result, the composition of the deposition is not merely dependent on the components of the spring water.

Complex sulfates with aluminum have nothing to do with the chemical types of spring water. They invariably occur in those areas where there are very strong hydrothermal activities. For example, in the steam ground and fumaroles in Yangbajain and in the hydrothermal explosion areas of Kau and Qupu. There is no or little aluminum in the spring water of Xizang, and the content of aluminum is far from being adequate to deposit the complex sulfates. Aluminum is derived from the soft beds of sandy soil. Ammonia and sulfur in these minerals are derived from the geothermal fluid. During the upflowing of the fluid to the surface, $H_2S$ is oxidized into $SO_4$. Such is in agreement with the view-point of A. Bencini(3) and W.A.J. Mahon(4) in respect of the source of $H_2S$ and $NH_4$ in high temperature hot spring.

## FEATURES OF GENERATION OF EFFLORESCENCE

Although the efflorescences are deposits of spring water, they generally are not directly deposited round the vents of spring. The mineralization of hot spring water in hydrothermal areas is mostly very low (generally lower than 200 mg/l). Such concentration cannot make the minerals deposit. Efflorescences will be formed on the surface only when spring water passes through soft beds by capillarity, rises to the surface and continuously evaporates and concentrates in an arid climate(5). Thus the efflorescences have the following characteristics: their crystals are minute, they are distributed by piecemeal and occur usually in forms of efflorescence and crust. Most of the minerals have been found with perfect crystals.

The grain size of the efflorescence mineral is commonly 0.002--0.004 mm. It is difficult to observe its formal features under the binocular microscope, but most of the minerals have been found with perfect crystals under the scanning electron microscope (see Photos 1-4). It is especially worth while to mention alunogen, which is a common mineral found near the spring of some volcanic areas in other countries. For instance, the acid springs near the solfatara holes of the Hekla in

Iceland(6) are widely distributed, and often associated with gypsum. But this is rarely seen in our country. We find that alunogen often occurs in the high-temperature hydrothermal areas of Xizang. And it occurs mainly in the forms of efflorescence on the cliffs, of thin flags on the ground, of stalactites along the margin of cliffs, and of prismatic aggregates in the soil bed 10 cm below the surface. The crystals of alunogen in the first and the fourth forms have been found to be exactly of the same shape under the scanning electron microscope (Photos 3,4). and they differ only in grain size.

## RELATIONSHIP BETWEEN HYDROTHERMAL ACTIVITIES AND SALT LAKES OF XIZANG

We find that the efflorescences are somewhat similar in composition to that of the salt lakes. First, they both contain the same minerals, such as thenardite, epsomite and trona. Second, rich boron and lithium are contained in both the salt lake water and the spring water. This is a commonly distinguished feature in their chemical composition. Borate minerals are deposited from both kinds of water too. Owing to their different environments, they have different deposited minerals. For example, the components of the minerals in salt lakes are complex; the types of borate minerals are diverse; their crystals are bulky; their generations are well developed, and their sedimentary thickness is large . In contrast, the components of the efflorescences in hydrothermal areas are simple; the crystal grain is minute; the minerals of the same effloscences are usually the products of the same generation; their sedimentary thickness is very small and they are distributed by piecemeal. Except the aforementioned similarities, the salt lakes are closely related to the hydrothermal activities in space, that is, salt lakes and saline lakes of Northern Xizang often go together with hydrothermal activities. Warm spring, hot spring and stretches of fossil travertins are usually found nearby some of the salt lakes. For example, in Zige Tang Lake, the field extends east-west for 10 km; there are 10 areas of manifestation in the geothermal field. Quite a number of spring vents are directly located at the bottom of Zige Tang Lake. Another example is Chagcam Caka which is a famous lake area on the Heihe-Ngari Road. The Qinghai Institute of Saline Lakes of the Academia Sinica made a detailed investigation there in 1977. It extends east-west for 30 km. The debris of tall and big travertins occur along the lake shore intermitently. Hot spring gushes out from the lake bottom. The temperature of the mixture of the spring water with the lake water may be over 40°C.

The aforementioned facts show that in the history of the development of the salt lakes in the inland drainage of Northern Xizang, the alimentation of the thermal water at depth has played an important role for the accumulation of saline materials in the lake. Throughout the world, there are some lakes rich in boron such as those in western United States and in Kashmir. It is generally inferred that boron in these lakes is partially originated from hot spring. The case of the Dead Sea directly shows the alimentary relation of the hot springs to the salt lake(7) . In comparison with the facts mentioned above, the evidence of the relation between the salt lakes and the hydrothermal activities in Northern Xizang is even more conclusive. The alkali pool of the Chongsar geothermal area is a typical example. It extends east-west for 5-6 km and may be divided into four manifesting areas (Fig. 1). The highest temperature at the spring vent is 78.5°C. The alkali pool is located in the north-east of this geothermal field, occurring in an elliptic form with its diameter being 60 m or so, which covers an area of about 2,300 m to a depth of only 0.3-0.5 m. No saline formation has been observed around the lake. The lake-water table has apparent ebb and flow, but it has nothing to do with atmospheric precipitation. The degree of mineralization is 95.33 g/l. Except the atmospheric water, there is not any running water recharging into the alkali pool spring area. And a lot of springs on the lake bottom exist. Thus the source of the lake water can only be subsurface hot water. The hot water gushing from the subsurface does not flow away but is only consumed by evaporation and saline components accumulate little by little in the lake basin. The alkali pool has come into being as it is today.

The salt lakes in Northern Xizang are growing and they are rich in boron, but there is little boron in the spring water; there is plenty of boron in the thermal water of Southern Xizang, but no salt lakes exist there (except the Targyai Lake) because Southern Xizang belongs to the exterior drainage. This distinquished feature of Southern Xizang which belongs to Himalayan Geothermal Belt can be found as far as the Mediterranean. For instance, Turkey and Tuscany of Italy are famous for their abundant boron. The geothermal steam of Tuscany contain 100-500 mg/l boron. The borate minerals are directly deposited at the vents of the fumaroles or in the coagulative water and boiling pool by the fumaroles(8). No salt lakes exist there, which is similar to the case in the hydrothermal areas of Southern Xizang.

## SUMMARY

At present there are strong hydrothermal activities in Southern Xizang, spring waters of which are rich in boron, but no salt lakes exist there. In Northern Xizang the hydrothermal activities are weak at the present. There is little boron in the spring water, but the salt lake in Northern Xizang are growing and exceedingly rich in boron. However, the extent and scale of hydrothermal activity in the history of Northern Xizang were larger than those of the present time. The salt lakes in this area with much boron in them may be regarded as important remains left over by the hydrothermal activity in geological time. The geothermal fluid has been the important recharge source of some components in salt lakes. Under specified conditions, some constituents contained in hot water may be the only source of the materials in salt lakes, such as the alkali pool is.

## ACKNOWLEDGEMENTS

The authors are greatly indebted to Professor Peng Zhi-zhong, Mrs. Qu Yi-hua and Mr. Zheng Mian-ping for the identification of the minerals included in this paper, and also to Mr. Zhang Peng-xi of the Qinghai Institute of Saline Lakes, Academia Sinica for his collaboration.

## References

(1) Kostov I., Mineralogy, Tr. ed. by Peter G.E., James P., Edinburgh, Oliver Boyd (1968).

(2) Winchell A.N., Winchell H., Elements of Optical Minerlogy, N.Y. John Wiley & Sons, Inc. 4th ed part 2 (1951).

(3) Bencini A., Dushi V., Martini M., Geochemistry of thermal springs of Tuscany (Italy), Chemical Geology 19 (3), 229-252 (1977).

(4) Mahon W.A.J., Chemistry in the exploration and exploitation of hydrothermal systems, Geothermics Spec. Issue 2, 2, pt. 2, 1310-1322 (1970).
(5) Lippmann F., Sedimentary Carbonate Minerals, Springer-Verlay, Berlin Heidelberg New York (1973).
(6) Barth T.F.W., Volcanic geology, Hot Springs and Geysers of Iceland, Carnegie Institution of Washington (1950)
(7) Bentor Y.K., Some geochemical aspects of the Dead Sea and the question of its age, Geochim. et Cosmochim. Acta 25, 239-260 (1961).
(8) Barbier, E., Fanell M.,Non-electrical uses of geothermal energy, Prog Energy and Combust. Sci. 3, Issue 2, 73-103 (1977).
(9) Xie Xian-de, Zheng Mian-ping, Liu Lai-bao, Borate Minerals, Science Press (1965).
(10) Qu Yi-hua, Xie Xian-de, Qian Zi-qiang, Liu Lai-bao, Hungtsaoite, Acta Geologica Sinica, 44 (3), 351-356 (1964).

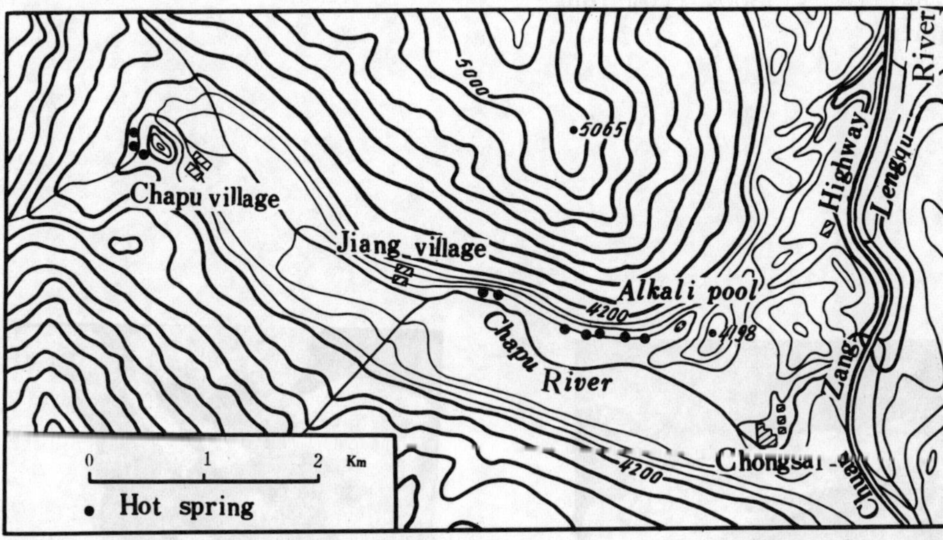


Fig. 1 Spacial relationship between the Chongsar geothermal area and the alkali pool.

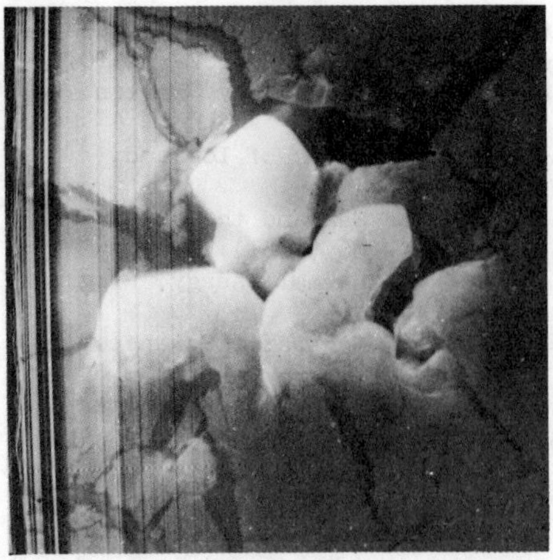

Photo 1 Hungtsaoite occurring as pseudohexagonal schistose and cubic crystal of halite, x1,300, from Yangbajain.

Photo 2 Pseudocubic crystals of alunites, x1,800, from Yangbajain.

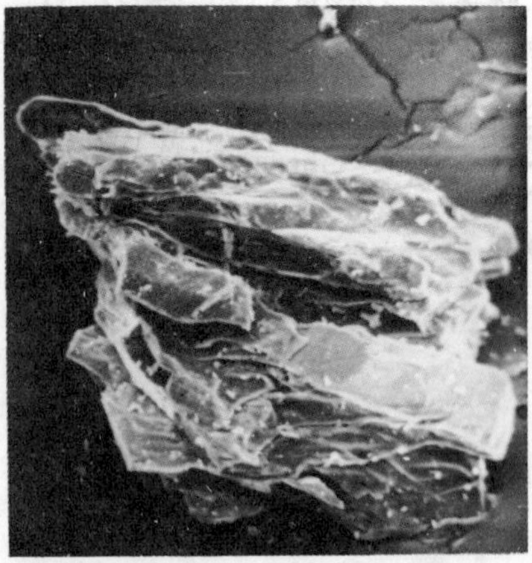

Photo 3 Alunogen in parallel groups of plate crystals, x73, sampled beneath the soil bed with depth 10 cm, Yangbajain.

Photo 4 Alunogen, x1460, from the cliffs in Yangbajain.

# HYDROGEN AND OXYGEN ISOTOPE STUDIES ON GEOTHERMAL WATERS IN XIZANG

Zheng Shu-hui   Zhang Zhi-fei   Hou Fa-gao
Shen Min-zi
(Department of Geology, Beijing University)

Ni Bao-ling
(Department of Chemistry, Beijing University)

## INTRODUCTION

A large number of deuterium and oxygen-18 analyses were made on water samples all over the world(1,2). Many people have applied this technique to the study of stable isotope variation in the hydrologic cycle(3). In particular, such measurements have been performed on thermal water to settle the problems of the origin of water and also to trace the recharge areas(4-8) . The first deuterium and oxygen-18 measurements on water from Xizang have been reported in the region of north Mt. Qomolangma at altitudes from 4.5 to 7 km(9). This paper summarizes the results obtained from the thermal water in Xizang and gives more detailed results in Yangbajain geothermal field and Gobdü geothermal field.

## EXPERIMENTAL TECHNIQUES

A $CO_2$-$H_2O$ equilibration method was used, with MAT CH-5 mass spectrometer for the $O^{18}$ analyses. The reproducibility of the $O^{18}$ analyses was determined by double preparation and double analyses of samples. The deviation is ±0.15‰. The analytical method for the hydrogen isotope has been applied in the laboratory using the uranium conversion for preparation of $H_2$ and MM602D mass spectrometer for the isotope analysis. The reproducibility of the $\delta$D value determined by double preparation and analysis of samples is ±1.5‰.

## $\delta$D AND $\delta^{18}$O VALUES OF THERMAL WATERS IN XIZANG

Thermal water samples from 34 geothermal areas in Xizang were collected. Fig. 1 shows the sampling points for thermal waters and the $\delta$D values in parenthesis. According to the $\delta$D value, four districts can be roughly divided. The first district is located in the south of the main ridge of Himalayas and in the southeastern part of Xizang. $\delta$D(‰) is approximately -100. The other districts are located within the Xizang Plateau. The division here coincides roughly with the climatic zones in Xizang. In general, all of the waters in geothermal areas is collected from surface and the $\delta$D value of thermal water is often similar to that of meteoric water. So we may expect that $\delta$D value should decrease from south to north because of the effects of latitude and altitude, but in fact it changes in the reverse direction, $\delta$D value increasing from south to north. The possible reason to explain the phenomenon is the influence of arid climate. In Xizang Plateau, the higher the latitude and altitude, the stronger the aridity. The ratio of annual evaporation to annual precipitation is greater than 15.0 in northern Xizang and heavy isotope concentrated in the surface water by strong evaporation.

We have determined $\delta$D value in waters of some inland lakes in Xizang (see Tables 1,2), the $\delta$D(‰) value lying between -40 - -70. From available data it indicates that most of the recharging water to the thermal area in Xizang may not come from the lake water. In general due to the evaporation effect, the $\delta$D value in thermal water is higher than that in the recharging water.

Comparing the $\delta$D results obtained from the thermal waters in different thermal areas along the Yarlung Zangbo River, we notice a characteristic distribution of $\delta$D value. On both sides of the river, $\delta$D of the thermal waters in different thermal areas varies symmetrically. The lowest values of $\delta$D (‰) (-160) are obtained in the middle part of the district and increase gradually to the east or west. The origin of this phenomenon has not yet been explained.

Fig. 2 shows the $\delta$D and $\delta^{18}$O plot of thermal waters in different hydrothermal areas in Xizang. The points of surface water collected are near the meteoric line, while the points of thermal water are irregular and scattered.

## $\delta$D and $\delta^{18}$O Studies in Yangbajain Thermal Field

Yangbajain thermal field is situated 90 km northwest from Lhasa. This is a powerful hydrothermal area. Fig. 3 shows the sampling points in Yangbajain geothermal field, waters being collected both from drill-holes and hot springs. In Fig. 4 isotopic data are shown for all waters analysed in this area. The thermal waters are moderately enriched in the heavy isotope along a line whose slope is about 3. The isotopic compositions of recharging ground water are obtained by extrapolation of the mixing line. The $\delta$D(‰) value of recharging water is about -150, and the $\delta^{18}$O(‰) value is about -20. It seems that the recharging water is not the surface water from the vicinity of Yangbajain, but might be a ice-snow water with lower values of $\delta$D and $\delta^{18}$O coming from distant places.

The origin of water in the geothermal area is of interest. As we see in Fig. 4, the mixing line may indicate an effect due to high temperature evaporation of about 70-90$^{\circ}$C. In Fig. 7 the linear relation between $\delta^{18}$O and [Cl] of waters of hot springs suggests that the waters are predominantly meteoric in origin but contain a small amount of NaCl rich deep fluid of a high heavy isotope ratio differing from the meteoric water.

## $\delta$D AND $\delta^{18}$O STUDIES IN GOBDU GEOTHERMAL FIELD

The isotopic and chemical investigations of thermal waters in Gobdü geothermal field are similar to those of the Yangbajain. Fig. 5 shows the sampling points and the location of Gobdü geothermal field. The plot of $\delta$D and $\delta^{18}$O in Fig. 6 also shows the high temperature evaporation effect. In Fig. 7 it is interesting to compare the behavior of thermal waters in Nuo hydrothermal area with others in Gobdü. It shows a different relation between $\delta^{18}O$ and [Cl] of thermal waters in Nuo area. It is obvious that thermal waters in Nuo area belong to a different hydrothermal system.

## ACKNOWLEDGEMENTS

We would like to thank Qiu Chun-yi and Zhang Zhong-lu of Institute of Geology, Seismological Bureau for mass-spectrometric determination of deuterium and thank Fan Yun-qi of Nanjing Institute of Geography for kindly supplying us the lake water samples.

## References

(1) Craig H., Isotopic variations in meteoric waters, Science, 133, 1702-1703 (1961).
(2) Dansgaard W., Stable isotopic in precipitation. Tellus. 16, 436-463 (1963).
(3) Friedman I. et al., The variation of the deuterium content of natural waters in the hydrologic cycle. Rev. Geophys, 2, 177-244 (1964).
(4) Arnason B., Hydrothermal systems in Iceland traced by deuterium, Geothermics, 5, 125-151 (1977).
(5) Sakai H. et al., Stable isotopic studies of Japanese geothermal systems. Geothermics, 5, 97-124 (1977).
(6) Giggenbach W., Isotopic composition of waters of the Broadlands geothermal field. New Zealand J. Sci., 14, 959-970 (1971).
(7) Craig H., Isotopic composition and origin of the Red Sea and Salton Sea geothermal brines. Science, 154, 1544-1548 (1966).
(8) Panichi C., Gonfiantini R., Environmental isotopes in geothermal studies. Geothermies, 6, 143-161 (1978).
(9) Zhang Qing-lian, Analysis for minor isotopes deuterium and oxygen-18 in natural waters. Kexue Tongbao, 3, 208-212 (1980).

Table 1 Isotopic data for thermal waters in Xizang

| No. | Locality | $\delta D$ (SMOW) | $\delta^{18}O$ (SMOW) |
|---|---|---|---|
| 5 | Quzhen rakong, Rutog | -80 | -12.7 |
| 11 | Namru, Gar | -128 | -16.2 |
| 15 | Sogdoi II, Gar | -117.5 | -12.2 |
| 24 | Xungba, Gegya | -120 | -14.3 |
| 28 | Anbu, Burang | -131 | -17.0 |
| 29 | Yarmenzha Burang | -131 | -15.9 |
| 30 | Danguoqisa, Burang | -129 | -16.1 |
| 37 | Kunggyü, Burang | -140 | -16.1 |
| 38 | Marme, Gerze | -134 | -17.9 |
| 43 | Xagangjam, Gerze | -128 | -16.9 |
| 44 | Tagejia, Ngamring | -149 | -18.6 |
| 46 | Lungmare, Xainza | -114 | -14.9 |
| 50 | Gamlai, Saga | -148 | -15.8 |
| 52 | Lablang, Ngamring | -145 | -15.0 |
| 62 | Quxam, Zham | -96 | -13.3 |
| 83 | Nyixia, Tingri | -164 | -21.6 |
| 89 | Amu Lake, Xainza | -101 | -15.1 |
| 102 | Chaga, Xaitongmoin | -162 | ----- |

Table 1 ( Continued )

| No. | Locality | δD (SMOW) | $\delta^{18}O$ (SMOW) |
|---|---|---|---|
| 124 | Sertang, Namling | -153 | -20.5 |
| 135 | Gyinkar, Gyangze | -145 | -19.4 |
| 145 | Latogka, Damxung | -141 | -14.8 |
| 147 | Yangbajain, Damxung | -146 | -18.2 |
| 151 | Burong, Damxung | -159 | -19.0 |
| 160 | Dagmo, Lhozhag | -117 | -17.0 |
| 175 | Golug, Nagqu | -133 | -15.7 |
| 184 | Nuo, Comai | -133 | -15.2 |
| 207 | Nagqu, Nagqu | -135 | -15.6 |
| 222 | Tuduo, Comai | -147 | -18.8 |
| 237 | Caka, Biru | -122 | -16.6 |
| 255 | Ribug, Sog | -128 | -16.3 |
| 256 | Erlian, Bomi | -96 | -14.4 |
| 303 | Jang, Baxoi | -137 | ----- |
| 309 | Puzang II, Zayü | -98 | -13.8 |
| 319 | Zhixinong, Qamdo | -133 | -13.3 |

Table 2 Isotopic data for surface waters and lake waters in Xizang

| No. | Locality | δD (SMOW) | $\delta^{18}O$ (SMOW) |
|---|---|---|---|
| | Meteoric water of Yangbajain (Aug. 1977) | -111 | -15.7 |
| 124 | Gariqiong, up stream of Zhangbu River | -113 | -16.0 |
| 125 | Streamlet, north of the Hot Lake | -120 | -17.0 |
| 126 | Streamlet, east of Yebague | -114 | -16.1 |
| 127 | Streamlet, east of the Luganqu | -116 | -16.1 |
| 128 | Luganqu streamlet | -114 | -16.2 |
| L-1 | Nam Lake (surface) | -66 | ----- |
| L-2 | Yamzhog Lake (surface) | -72 | ----- |
| L-3 | Serling Lake | -49 | ----- |
| L-4 | Zige Tang Lake | -60 | ----- |
| L-5 | Yargedong Lake | -56 | ----- |
| L-6 | Azhiqikekur, southern Xinjiang | -44 | -7.1 |
| L-7 | Near south of the Nam Lake | -79 | -9.0 |

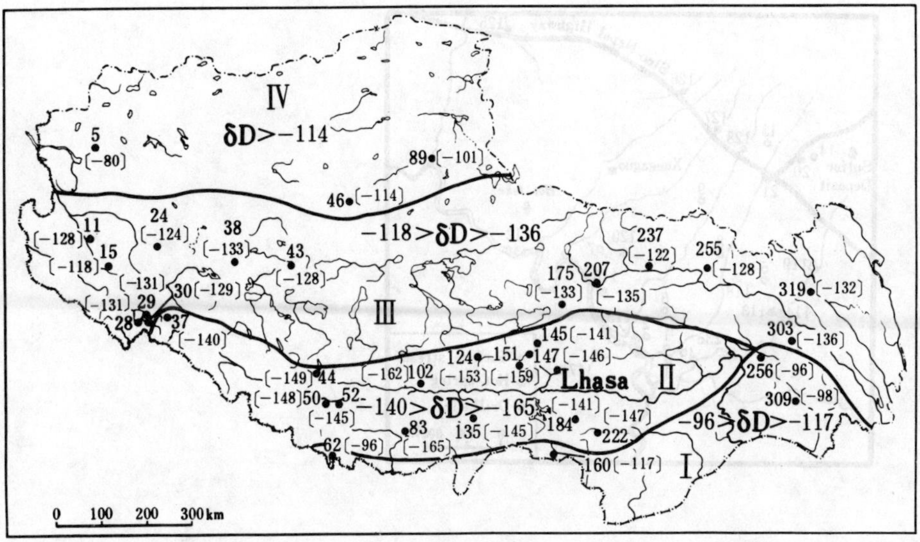


Fig. 1 Sampling points for thermal water in the hydrothermal areas in Xizang. The $\delta$ D(‰) values given in parenthesis are mean value of the respective resources. Places see Table 1, 2.

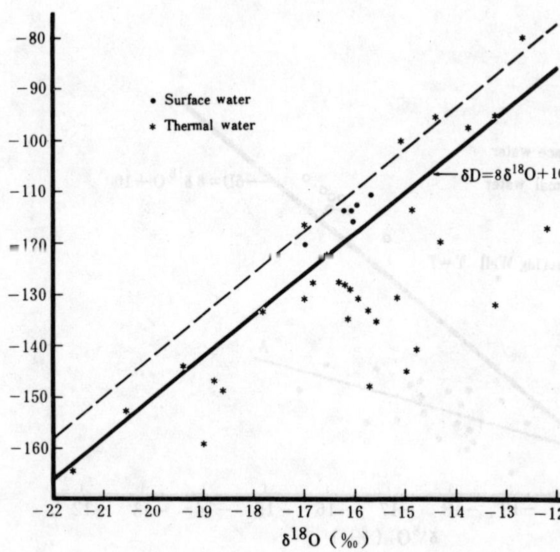


Fig. 2 $\delta$ D vs $\delta^{18}$O plot of thermal water in Xizang.

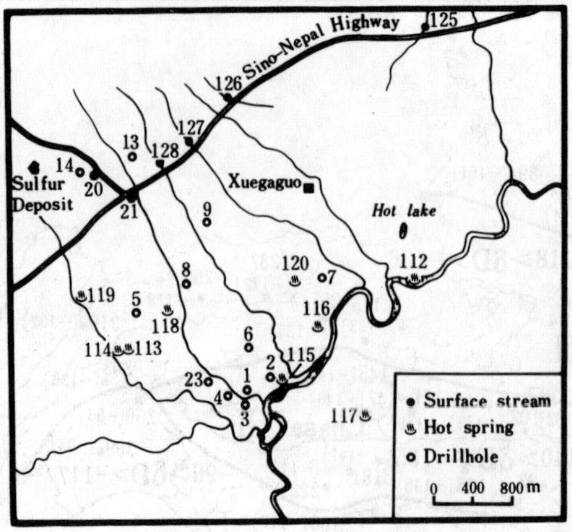


Fig. 3 Sampling points for thermal water and surface water in Yangbajain geothermal field.

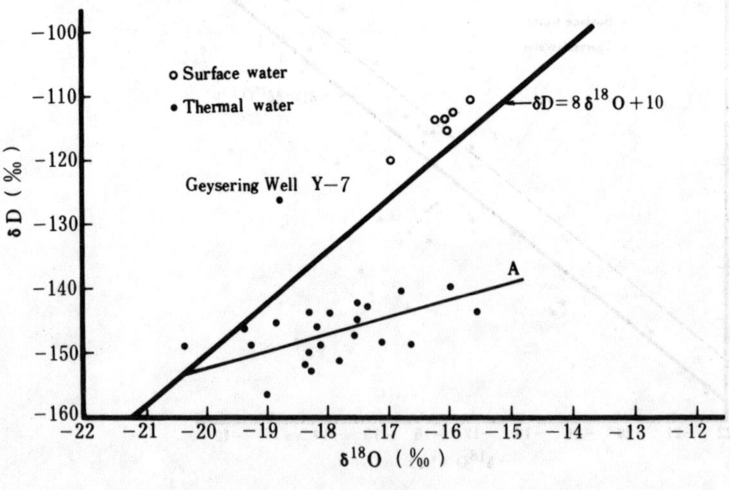


Fig. 4 $\delta$ D vs $\delta^{18}$O plot of thermal waters in Yangbajain geothermal field.

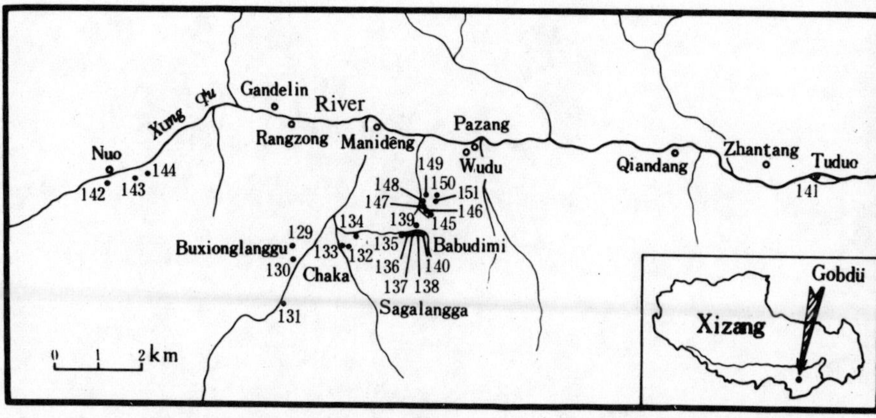


Fig. 5 Sampling points for thermal water in Gobdü geothermal field.

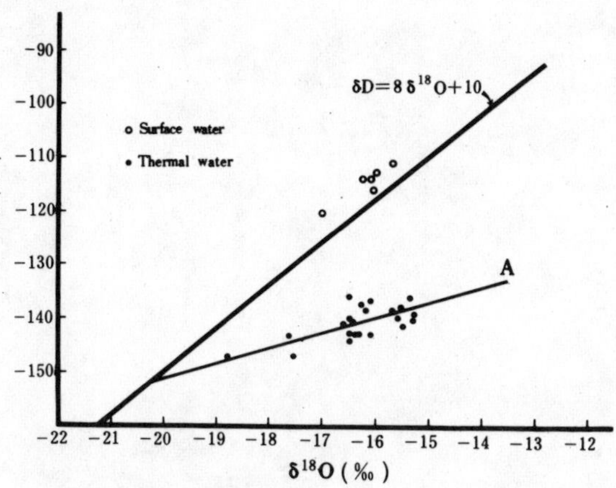


Fig. 6 $\delta$ D vs $\delta^{18}$O plot of thermal waters in Gobdü geothermal field.

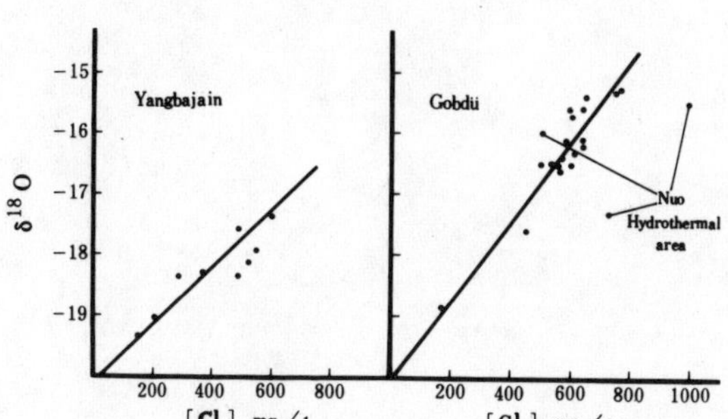


Fig. 7 $\delta^{18}$O vs (Cl) plot of thermal waters in Yangbajain and Gobdü geothermal fields.

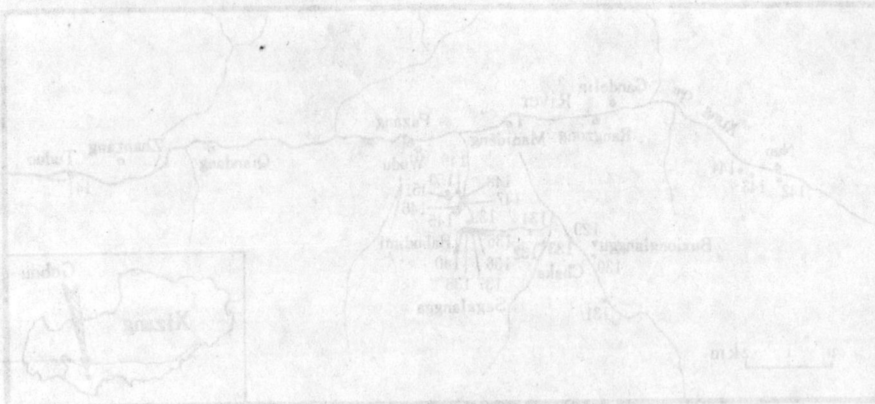




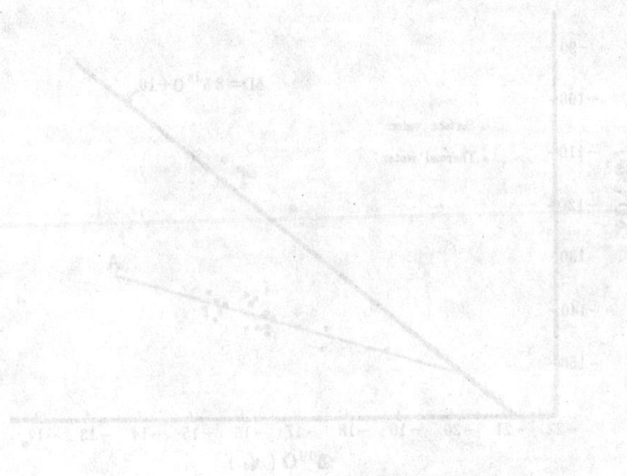

# THE SETTING OF GEOTHERMAL ACTIVITIES IN XIZANG AND DISCUSSION ON THE HEAT SOURCE PROBLEM

Liao Zhi-jie
(Department of Geology, Beijing University)

Xizang is one of the best known geothermal regions in China, where there are about 600 geothermal areas, more than 300 of which have been reconnoitred by us during 1973-1976(1).

We have confirmed that an active geothermal belt, the Himalayan geothermal belt, is extending about 2,000 km along the Yarlung Zangbo suture. The belt runs from the northern slopes of the Himalayas to the northern slopes of the Gandise and Nyainqentanglha Ranges. In the narrow region, we have discovered some most extensive manifestations including 11 hydrothermal explosion areas, 3 geysers, many boiling springs and quite a number of hot springs and warm springs (Fig. 1). In this area, the magmatic activity is very strong and shows a zonal distribution.

The late Yanshan-early Himalayan granite is distributed along the Gandise Range. Obviously, it is an igneous intrusion of an ancient island arc.

The Himalayan granite can be found throughout this belt, and its age is about 40-10 m.y. old. The rock is getting younger and younger from north to south. Many youngest tourmaline granites intruded into the High Himalayas have a high isotopic ratio of Sr (0.7478)(2). This probably means that they were generated by remelting of the crust.

Since Cenozoic, the metamorphism, migmatization and granitization of crystalline rocks of the Himalayas have taken place. Moreover, there are two volcanic zones in Xizang since Cenozoic: one is distributed approximately on the northern slope of Gandise Range, and the other is distributed in Kunlun Mts. The former is called the "Gandise Volcanic Zone" or "South Zone" for short and the latter is called the "Kunglun Volcanic Zone" or "North Zone".

The area in the "South Zone" that is covered by volcanic rocks is about 90,000 km$^2$ and there are no obvious cones. The rocks present include rhyolite, dacite, andesite, basalt, ignimbrite and their pyroclastics. The age is about 50 m.y. old. Most of the rocks belong to the calc-alkali series.

Volcanos in the "North Zone" often show cones, and do not form a typical normal series. Latite is the dominant petrological type, but at Bamoqionzong, peralkaline is dominant. The age of these volcanos may be Neocene - Holocene.

The calc-alkali series of volcanic rocks of the "South Zone" were produced by Indian Plate underthrusting, and the "North Zone" were produced by Xizang Plateau upthrusting to the Tarim massif. The volcanic zones have different ages and are large in scale.

The Himalayan geothermal belt embraces the "Gandise Volcanic Zone" and "Xizang Tethyan Zone". The hydrothermal activities are very strong on the southern side of the suture which is different from Island Arc of Western Pacific Ocean.

There are no presently active volcanos in South Xizang, but the Gandise volcanic rocks which extruded 40 m.y. ago are too old to become the heat source of geothermal activities while the younger northern zone is too far away from the geothermal belt to be the heat source. Even so, the setting of geothermal activities can still be explained by using plate theories.

It is generally acknowledged that the Himalayas are results of continental collision. The main problem here is whether or not the Indian continent was thrusting under Asia along the Main Boundary Thrust (MBT), because the light continental crust can not sink into the mantle. To answer this question we should examine the geophysical data.

From the Ganges Plain to the Xizang Plateau, the Bouguer gravity anomalies decrease from -50 mgal. to -550 mgal. This means that the light materials have large thickness below the surface of Xizang Plateau. The isostatic anomaly is about -120 mgal. at Ganges Basin, but changes to +120 mgal. at Mt.Qomolangma. Inside the Xizang Plateau, the isostatic anomaly fluctuates around zero.

Intermediate-deep earthquakes mainly occur at both ends of the Himalayan arc, i.e. at Kashmir and Burma. In the Himalayas and Xizang Plateau, intermediate earthquakes are very rare, but the shallow earthquakes are distributed along the Himalayas and to the north.

Focal planes of earthquakes dip differently in different regions. In Nepal and Kumaon Himalayas, the hypocenters are distributed near the Main Central Thrust with a smooth dip angle and the depths of foci are about 20-50 km. The shallow earthquakes might be caused by slip of the MBT. But in the Punjab region, there are a lot of intermediate and shallow earthquakes with a focal plane dipping about $35^{o}$.

There are some intermediate earthquakes with depths of less than 200 km along and north of the Himalayas. South of the range, these mantle earthquakes have near north-south principle stress axes because of the northward movement of the Indian continent. The mantle earthquakes in the Xizang Tethyan Zone have focal depths of 140-180 km. The fact that there are mantle earthquakes inside the Xizang Plateau means that the Indian Plate subducts under the Xizang Plateau, but the fact that the mantle earthquakes are very few implies that the temperature is high beneath the Xizang crust. An earthquake with a depth 80 km near Lhasa (29.8N/89.6E) has a very low Q (25) and many shallow earthquakes also have low Q (150) that are associated with high temperatures beneath Xizang(3,4).

According to Ruzaikin et al. (1977)(5), the Lg phase is very weak for paths crossing the Xizang Plateau between the area south of $35^{o}$N latitude and north of the Himalayas.

A seismic refraction profile was made by the Institute of Geophysics, Academica Sinica from Yadong to Nam Lake and it was found that there is a thick layer with low velocity (5.64 km/s) at a depth of about 40 km, probably because of the high temperature state in the crust(6).

In the light of these data, it is necessary to review the geological history during continental collision which can be divided into following stages:

1. Owing to the breakup of Pangaea, the Indian Plate drifted northward, Tethyan Sea subducted gradually under the Eurasian Plate. Andesitic magmas associated with island arc volcanos extruded on or intruded in Gandise region until late Eocene, so that the crust of Xizang came into heat.

2. By the end of the Eocene, the Indian continent crust came into collision with Euroasian continent to produce the Yarlung Zangbo suture. When continental crust moved into subduction zone, the crust of subducting plate was detached and deformed and pushed up to a mountain range. The temperature beneath Xizang further rose and many little granitic bodies intruded into Eocene volcanic rocks in Gandise region.

3. As plate convergence continued, the crust was itself underthrusted by more continental crust to create thick crust and many large thrusts. Thus, the Main Central Thrust formed and was operating strongly during Miocene. The bimica granites intruded into Mt. Lhagukanri are the products generated by remelting of continental crust due to subduction of southern crust along Main Central Thrust.

4. Since Pliocene, Main Boundary Thrust formed to replace Main Central Thrust. The thickness of crust was uninterruptedly growing increasing the buoyancy to support the high topography of Xizang Plateau. The slip along the Main Boundary Thrust caused the remelting of the crust which made the tourmaline granites intrude into the High Himalayas and brought about further heating of the Xizang crust. The Main Boundary Thrust is a newly subduction zone for convergence of continents. It can be indicated by the gravity profile and earthquake distribution, many aspects of which look like an island arc of the Western Pacific Ocean. Of course, its subduction rate is probable very slow now. High temperatures beneath Xizang are indicated by the scarcity of intermediate- and deep-focus earthquakes, the presence of a low velocity layer, very weak Lg phase and low value of Q.

It is entirely possible that many buried igneous intrusions are still intruding into the shallow crust in Himalayan Geothermal Belt, as the subduction continues: Therefore, the background of geothermal activities is that as the subducted zone migrating southward after collision the Xizang crust has been uninterruptedly growing and heated.

High temperature in the Xizang crust is an essential condition in forming the Himalayan Geothermal Belt, but it is not sufficient to form the geothermal field within this belt, because temperature in the shallow crust are too low to supply the local reservoirs with enough heat. It is very interesting to note that there are no presently active volcanos in Southern Xizang but it is difficult to imagine that there are no igneous intrusions in the shallow crust.

If a buried igneous intrusion is present in the shallow crust, it could occur in late middle Pleistocene. The temperature of the cooling igneous intrusion is unknown. In general, the cooling of an igneous intrusion is relatively rapid in the geological history and depends on its size and permeability. The temperatures in and around a cooling igneous intrusion of various shapes have been calculated. Jaeger (1965)[7] pointed out that if a plane sheet of two dimensions and infinite lateral extent is intruded at a temperature which exceeds that of the country rock by $T_o$, the excess temperature $T_m$ on its mid-plane at time t after intrusion is approximately:

$$T_m = dT_o(kt)^{-1/2}$$

If the effect of latent heat is considered, $T_o$ may be replaced by $(T_o+L/c)$, where L is the latent heat of solidification and c is the specific heat of the magma.

For the deeply buried spherical intrusion of radius d, the temperature $T_m$, at the centre after time t is approximately:

$$T_m = \frac{1}{6}\pi d^3 T_o(\pi kt)^{-3/2}$$

Based on this equation, the relative curve of temperature variation of igneous intrusions with various sizes and ages can be computed (Fig. 2).

If we suppose that the deeply buried igneous intrusion is 500,000 years old (middle Pleistocene) and has a radius of 7 km, then the excess temperature of the centre of intrusion will now be $T_o/2$ or $(T_o+L/c)/2$. These intrusions could occur during late middle Pleistocene, when violent folding and faulting took place, but are not yet exposed.

## References


(1) Tong Wei et al., Geothermals beneath Xizang (Tibet), Science Press, 170 (1980) (in Chinese).

(2) Tu Guang-chi et al., Characteristics and evolution of granitoids of South Xizang (Tibet). In this Proceedings of Symposium on Qinghai-Xizang (Tibet). Academia Sinica (1981).

(3) Zhang Li-min et al., The Q-value of the medium of Tibetan Plateau around Lhasa Region. Acta Geophysica Sinica, 22 (3), 303-310 (1979).

(4) Yang Bing-ping et al., Some characteristics of small earthquakes in the Zhamo and Daxung districts of Xizang Plateau. In this Proceedings of Symposium on Qinghai-Xizang (Tibet) Plateau, Academia Sinica, (1981).

(5) Ruzaikin, A.I. et al., Propagation of Lg and lateral variation in crustal structure in Asia. J. Geoph. Res. 82, 307-311 (1977).

(6) Teng Ji-wen et al., Explosion seismic study for velocity distribution and structure of the crust and upper mantle from Damxung to Yadong of Xizang Plateau. In this Proceedings of Symposium on Qinghai-Xizang (Tibet) Plateau, Academia Sinica (1981).

(7) Jaeger, J.C., Application of the theory of heat conduction to geothermal measurements, in Terrestrial Heat Flows, Geoph. Monog. Ser., No. 8, 7-23 (1964).

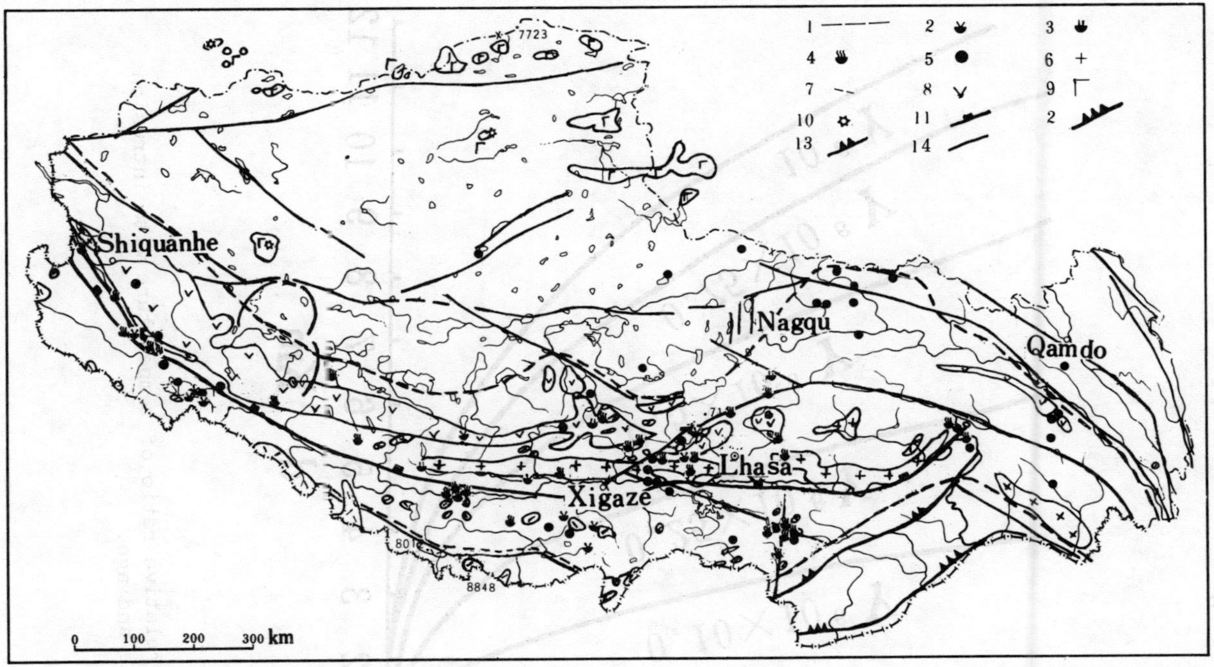


Fig. 1 The geothermal geology sketch map of Xizang.

1. Boundary of Himalayan geothermal belt;
2. Hydrothermal explosion;
3. High temperature geyser;
4. Boiling spring;
5. Hot spring ( > 70 C);
6. Late Yanshan and early Himalayan granite;
7. Himalayan granite;
8. Eocene-Oligocene volcanic rocks;
9. Pliocene-Holocene volcanos;
10. Volcanic cones;
11. Suture zone;
12. MBT (Main Boundary Thrust);
13. MCT (Main Central Thrust);
14. Fault.

$$Tm = \frac{1}{6}\pi d^3 \, To \, (\pi Kt)^{-\frac{2}{3}}$$

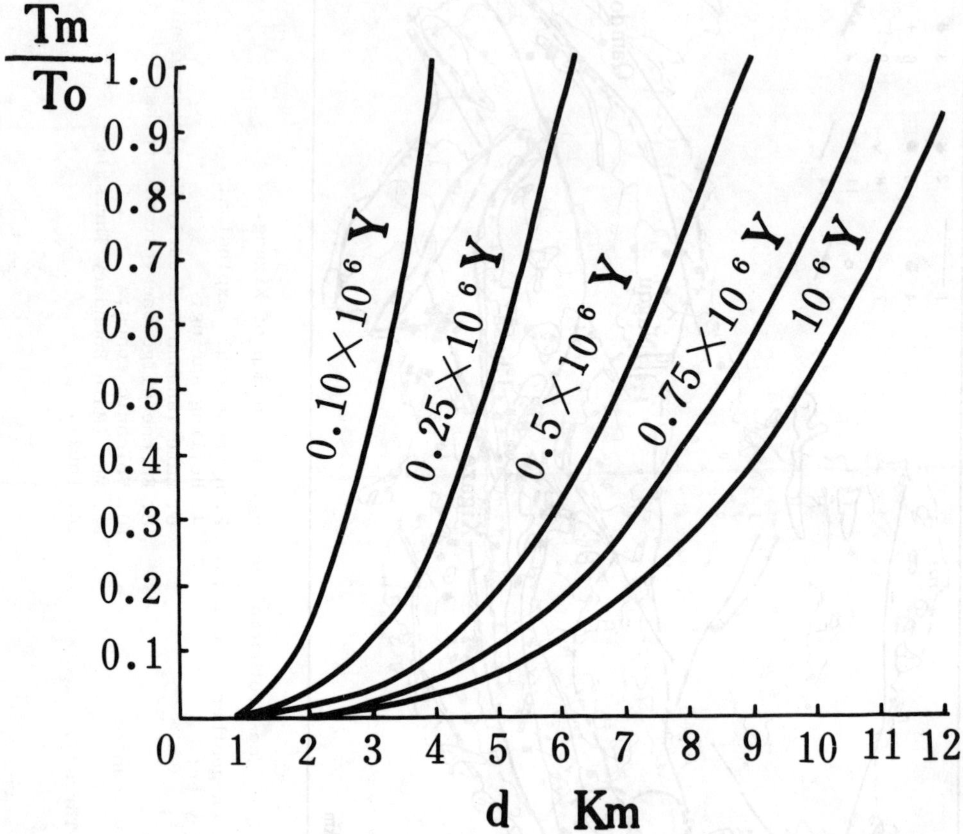


Fig. 2 Ideal relative ratio of temperature with intrusion of different size and age.

# PALAEOMAGNETIC OBSERVATION IN XIZANG AND CONTINENTAL DRIFT

Zhu Zhi-wen    Zhu Xiang-yuan    Zhang Yi-ming
(Institute of Geophysics, Academia Sinica)

Since 1960's the rapid development of plate tectonics theory has made a revolutionary advance in earth sciences. The main point of the theory is that the earth's crust is composed of several plates and interactions between these plates are the main cause of tectonic movements(1). At present many geophysicists use this theory to explain crustal movements and the problems connected with suture. This paper deals with some palaeomagnetic results of rock samples collected from Xizang and their relation to the position of suture line.

## SAMPLING

We collected oriented rock samples located on both sides of the Yarlung Zangbo River in 1975, 1976 and 1979. The geological formations studied are as follows (see Fig. 1): Pliocene sediments in Pagri; Palaeocene and Late Cretaceous red sandstone in Lhünzhub; Cretaceous red sandstone in Biru; Late Cretaceous granodiorite in Zhamo; Cretaceous limestone in Tingri; Late Jurassic limestone in Lhasa; Late Jurassic slate in Gyangze; Triassic sandstone in Tulong; Pre-Cambrian quartz sandstone in Nyalam; Early Cambrian mica-gneiss in Yadong.

## MEASUREMENTS AND DATA

All the rock samples were demagnetized by A.C. field and the remanent magnetism measured by LAM-24 astatic magnetometer. We use Fisher(2) statistics to deal with these data. The results are given in Table 1.

## DISCUSSION

A great number of determinations of palaeomagnetic poles from different parts of the world show that there are two situations for the distribution of pole positions of the same age: the poles of the same plate are closed to each other; but those of different plates are rather far apart (see Fig. 2). On the basis of this fact, palaeomagnetic poles of Xizang are compared with those of Eurasia and India.

We can see in Figure 3 that the pole positions for Zhamo, Lhünzhub and Biru all fall into the pole group for Eurasia, which shows that in Cretaceous and Palaeocene they were located on Eurasia. On the contrary, the pole position for Tingri is grouped with that of India, which indicates that Tingri was on the Indian Plate.

Although the pole positions for Jurassic are rather scattered, we can still see that poles for Eurasian and Indian plates are grouped (Fig. 4) and separated far from each other. From this figure, we can see that the Lhasa limestone has always been on the Eurasian Plate and Gyangze slate was once on the Indian Plate.

In the same way, we argue that Tulong sandstone was on the Indian Plate in Triassic (see Fig. 5).

Comparisons are also made on Cambrian strata. Although their pole positions are largely scattered (Fig. 6), we have noticed that poles for Eurasia lie from latitudes $0^{\circ}$-$25^{\circ}$ while for the Indian Plate, they are in latitudes higher than $30^{\circ}$. The pole position for Yadong mica gneiss, Malani rhyolite of Vindhyan system and Kaimur quartzite in India are close to each other. Although the Nyalam quartz sandstone and sandstone of Rawa series and Bhander series are different in longitudes, their latitudes are nearly the same. This shows Nyalam and Yadong were both on the Indian Plate.

Continents have been drifting and carried with them the palaeomagnetic poles. According to axial geocentric dipole hypothesis the average pole position is located at the geographical pole. As the age of strata gets older, their pole position gradually move away from the north geographic pole.

Palaeomagnetic poles for Zhamo, Lhünzhub, Biru from Cretaceous to Paleocene were not far from the north geographic pole (Fig. 7). This implies that Eurasian Plate has not drifted far since Cretaceous. On the other hand, poles for Tingri of Cretaceous, Gyangze of late-Jurassic and Tulong of Triassic were far from the north geographic pole and near the equator. This suggests that these strata have moved a large distance, which coincides with the long-distance drift of the Indian Plate.

On the basis of the above analysis, we conclude that the strata of Zhamo, Lhünzhub, Biru and Lhasa belong to Eurasia and the strata of Tingri, Gyangze, Tulong, Yadong and Nyalam belong to the Indian Plate, and we suggest that the suture line between the two plates should be located in the area of the upper Indus River and the Yarlung Zangbo River, see Figure 8.

Himalayas is one of the youngest mountain range in the world. This mountain range and the related Qinghai-Xizang Plateau have suffered severe tectonic movements. The beds were folded, rotated and inclined. These tectonic movements make palaeomagnetic sampling difficult. However, these movements can not change the angles between remanent magnetization and the bedding planes of formations. Once the planes are determined, we can use of formula $\text{ctg}\ P=\frac{1}{2}\text{tg}J$ to calculate the colatitude P of the sampling locality (where J is the inclination of primary remanent magnetization).

From Figure 9, we can see that the mean palaeolatitude of Zhamo, Lhunzhub, and Biru is $27^{\circ}$N which is near to the present colatitude. This means that the southernmost part of Eurasia has no obvious displacement since Cretaceous. However, Tingri, Gyangze, Tulong have palaeolatitudes of $21^{\circ}$S, $23^{\circ}$S and $24^{\circ}$S respectively. The variation of the latitudes was slow from Triassic to Cretaceous, but became rapid since Cretaceous.

Our work shows that strata from the northern and southern sides of the Yarlung Zangbo River belong originally to different plates. From the variation of the palaeolatitude, the average speed of northward drifting of the Indian Plate is estimated to be less than 5.5cm/yr.

## References

(1) Fu Cheng-yi, Continental drift sea-floor spreading and plate tectonics. Science Press, 33-35 (1972) (in Chinese)

(2) Tarling D.H., The principles and applications of palaeomagnetism Chapman and Hall, London 65-74 (1971).

(3) Molnar P. et al., Cenozoic tectonics of Asia: Effects of a continental collision. Science, 189, 419-426 (1975).


Table 1 Some relevant data for the samples

| No. | area | locality (degrees) | | geological period | number |
|---|---|---|---|---|---|
| | | latitude | longitude | | |
| 1 | Pagri | 27.75 | 89.10 | Tp | 42 |
| 2 | Zhamo | 29.24 | 95.86 | Ku | 18 |
| 3 | Lhünzhub | 29.90 | 91.25 | Ku-Tpa | 42 |
| 4 | Biru | 31.50 | 93.70 | K | 41 |
| 5 | Tingri | 28.60 | 87.20 | K | 13 |
| 6 | Lhasa | 29.65 | 90.05 | Ju | 21 |
| 7 | Gyangze | 28.93 | 89.75 | Ju | 11 |
| 8 | Tulong | 28.60 | 86.10 | Tr | 26 |
| 9 | Nyalam | 28.20 | 89.95 | Pc | 13 |
| 10 | Yadong | 27.53 | 89.00 | Є1 | 31 |

Table 1 ( Continued )

| mean direction (degrees) | | palaeo-latitude (degrees) | pole position (degrees) | | $\alpha_{95}$ (degrees) | $\delta_m$ (degrees) | $\delta_p$ (degrees) | K |
|---|---|---|---|---|---|---|---|---|
| declination | inclination | | latitude | longitude | | | | |
| 7.8 | 42 | 65.6 | 82 | 152 | 7.3 | 8.9 | 6 | 10 |
| 30.4 | 52 | 57.7 | 63.8 | 171.4 | 5 | 6.9 | 4.7 | 48 |
| 338 | 40.3 | 67 | 69 | 347 | 5 | 6.5 | 3.8 | 20 |
| 325 | 44.5 | 64 | 59 | 2.4 | 4 | 5 | 3 | 30 |
| 331 | -37 | 111 | 33 | 299 | 16.5 | 19 | 11 | 13 |
| 175 | 2 | 89 | 59 | 280.4 | 16.6 | 16.6 | 8.3 | 5 |
| 292.3 | -40 | 113 | 7 | 329 | 11 | 13 | 8 | 22 |
| 318 | -41.8 | 114 | 23.7 | 308 | 6.3 | 7.7 | 4.7 | 22 |
| 248.6 | -50.6 | 121 | 31.5 | 154 | 6 | 8 | 5.4 | 41 |
| 2.4 | 37 | 69 | 82.7 | 250.3 | 8.6 | 8.6 | 5 | 13 |

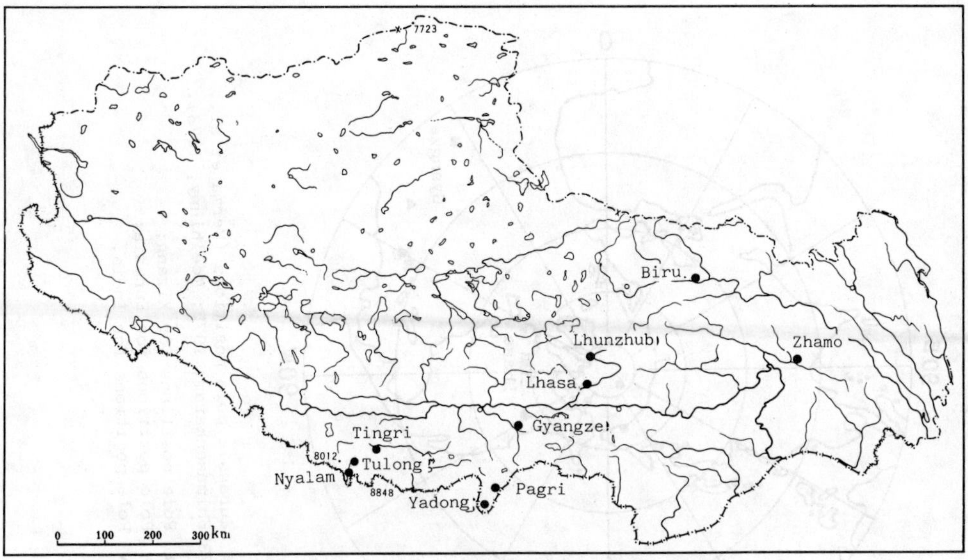


Fig. 1 Sites for the collected samples.

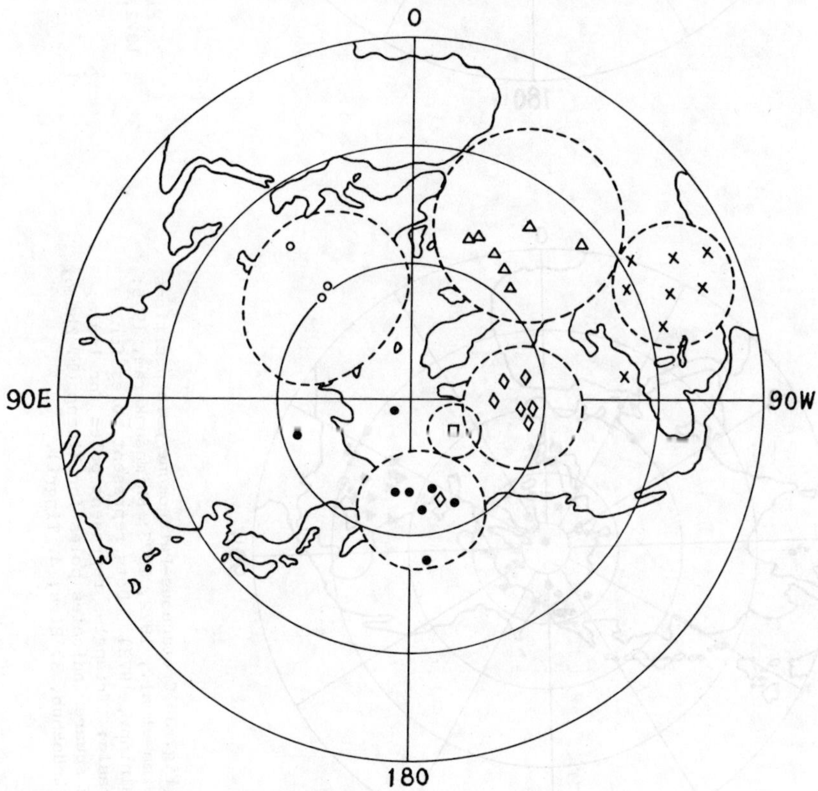


Fig. 2 Mesozoic palaeomagnetic pole positions for six continents (Gromme et al., 1967).

● North America, □ South America, ◊ Africa,
o Antarctica, × India, △ Australia ,

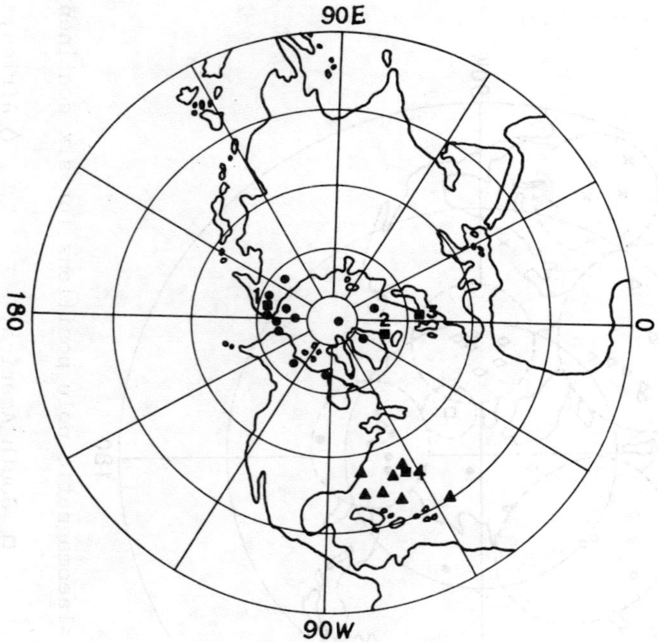


Fig. 3 Cretaceous-Palaeocene pole positions (Verma et al., 1972, 1973; Bhimasankaram, 1975; Mcelhinny, 1973). Dots represent poles for Eurasia; Triangles represent poles for India; The square indicates poles for Xizang (1. Zhamo, 2. Lhünzhub, 3. Biru, 4. Tingri).

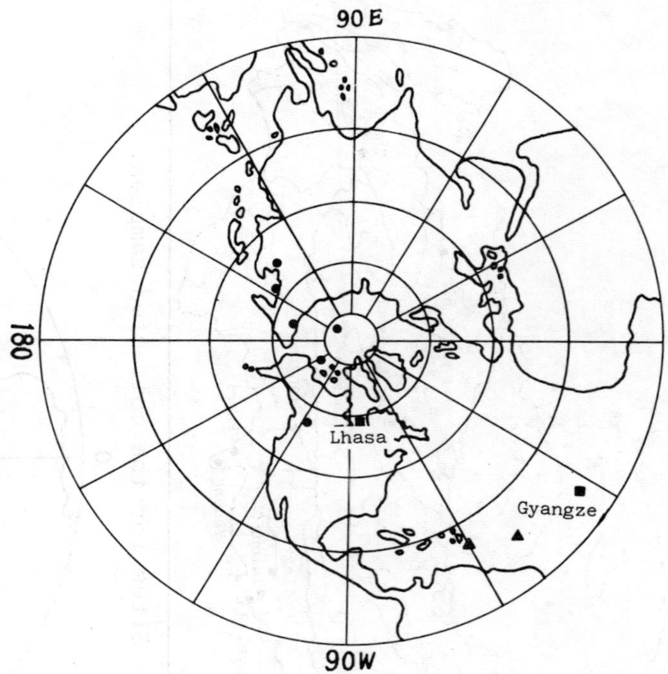


Fig. 4 Jurassic pole positions (Verma et al., 1972, 1973 Bhimasankaram, 1975; Mcelhinny, 1973).

■ Pole positions for Xizang;
● Pole positions for Eurasia;
▲ Pole positions for India.

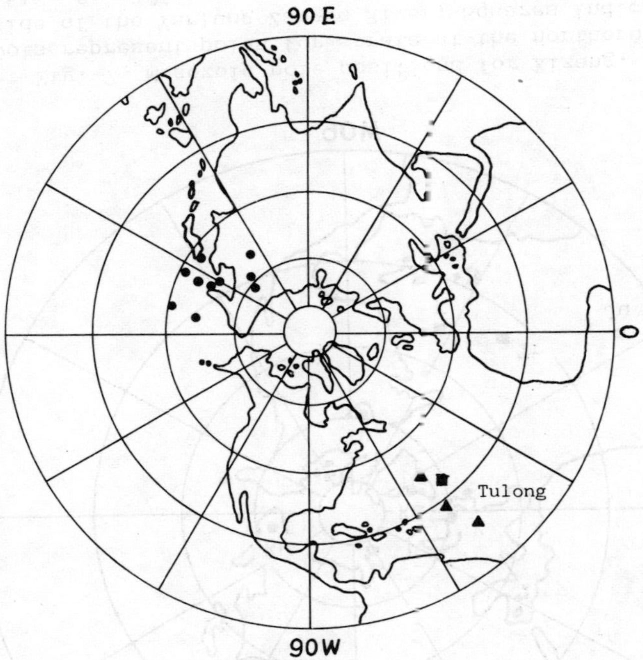


Fig. 5  Triassic pole positions (Verma et al., 1972, 1973; Bhimasankaram, 1975; Mcelhinny, 1973).

■ Pole position for Xizang;
● Pole position for Eurasia;
▲ Pole position for India.

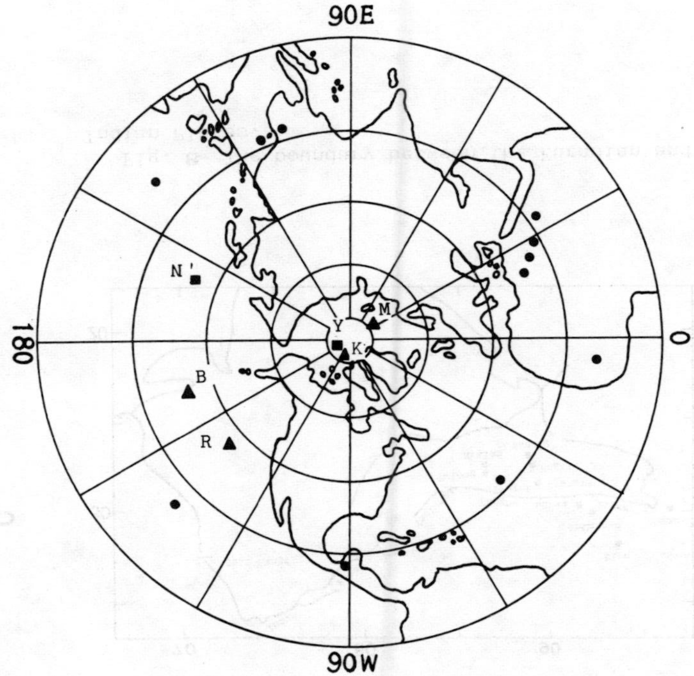


Fig. 6  Pole positions for Late Precambrian to Early Cambrian (Verma et al., 1972, 1973; Mcelhinny, 1973).

N–Nyalam, B–Bhander, R–Rawa,
Y–Yadong, K–Kaimur, M–Malani.

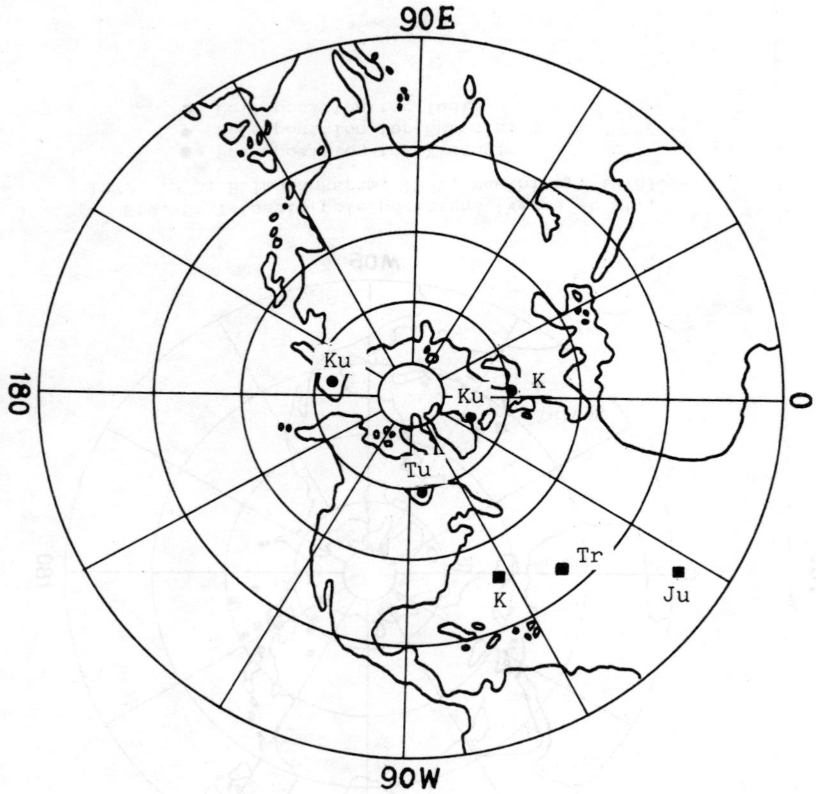


Fig. 7 Mesozoic pole positions for Xizang. Dots represent poles for strata at the northern side of the Yarlung Zangbo River; Squares indicate poles for strata at the southern side of the Yarlung Zangbo River.

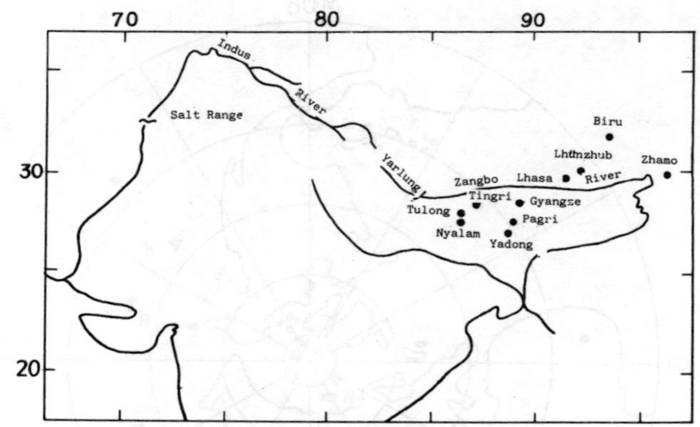


Fig. 8 The boundary between the Eurasian and Indian Plates.

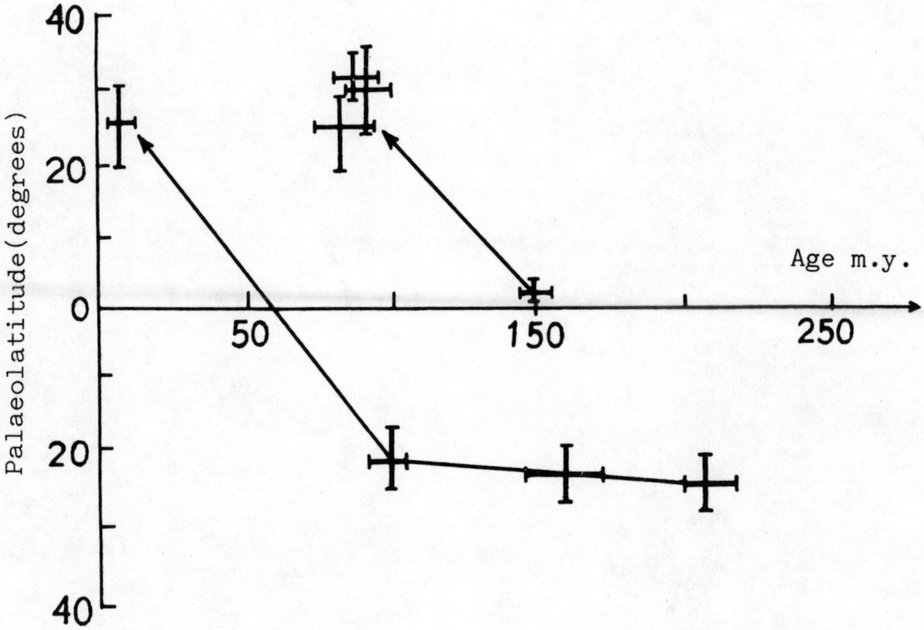


Fig. 9 The variation of the palaeolatitudes with geological periods.

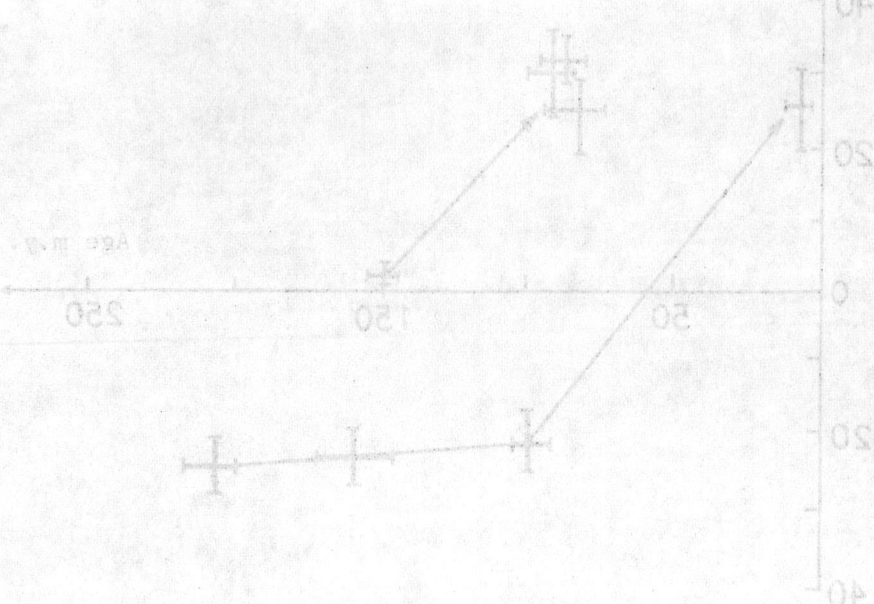




# THE CHARACTERISTICS OF AEROMAGNETIC ANOMALIES AND THE STRUCTURE OF EARTH'S CRUST OVER SOUTHERN QINGHAI-XIZANG PLATEAU

Pan Zuo-shu   Shen Jiang-sheng   Wang Xiao-pei
(Department of Geophysics, Changchun Geological Institute)

## INTRODUCTION

The investigated region covers the Gandise-Nyainqentanglha folding series, the Himalayan folding series and the interface of the two series -- the collision zone of Yarlung Zangbo River(1).

The magnetic anomalies in this district are mainly caused by igneous intrusions. The magnetic inclination is about 44° and the strikes of most anomalies are approximately E-W. Therefore the anomalies of the total field anomaly $\Delta$ T are complicated.

## METHODS OF PROCESSING AND INTERPRETATION

The interesting anomalies are those caused by gigantic geological structures at great depths. All the processing methods used, therefore, belong to low-pass filtering. They are reduction to pole, upward continuation, pseudo-gravimetric anomalies and matched filtering(2).

Two interpretation methods by computer are adopted. One is calculation of the top depths of the magnetic sources. The other is determination of the parameters of magnetic bodies by Marquardt optimization method.

## THE ANALYSIS OF THE DEEP STRUCTURE

The anomalies can be divided into four parts (Fig. 1): the east, the west, the south and the Yarlung Zangbo River. The anomalies of the western part are slow positive anomalies and its average value is about 150 gammas. The interpreted magnetization of the basement is about 0.00024 e.m.u.; it approximates to the magnetism of most granitic intrusions of the district. The characteristics of the anomalies are very simple, and show that this is a relatively stable area with granitic basement.

The eastern part (Fig. 2) shows complicated anomalies with negative background. This is a region where the earth's crust is quite brittle and the magmatic activitics are very intense.

The fourth part is a zone of stronger anomalies striking E-W, which coincides with the collision zone of Yarlung Zangbo River. The aeromagnetic anomalies show that the whole collision zone is a gigantic zone of rocks of ultrabasic type. In the western section of that zone, it is divided into a south and a north branch and the north one is larger than the south.

The south of Yarlung Zangbo River is the southern part. It consists of only gentle anomalies. On both sides of Yarlung Zangbo River, the characteristics of the anomalies are quite different, implying different geological structures.

The problems related with the structures of the crust are analysed as follows:

## The Inclination of the Collision Zone at the Yarlung Zangbo River

The inclination of the collision zone can be expressed as the inclination of the zone of rocks of ultrabasic type. In the western section of the anomaly zone of Yarlung Zangbo River, the two flanks of the anomaly curve are symmetrical (the slops of the tangent of the inflection points are the same) (Fig. 3). Therefore, it is thought that the zone of the ultrabasic rock type in the western section is approximately vertical.

The characteristics of the eastern section of that anomaly zone can be expressed by the 78th profile reduced to pole (Fig. 4). The zone of rocks of ultrabasic type is somewhat inclined toward north. The result of curve fitting shows that the inclination of the eastern section is more than 60°. Therefore, the collision zone should be described by the vertical model.

## The Top Depths of Magnetic Sources and the Depths of Curie Point Isotherm

The depths of the top and bottom surface of magnetic bodies determined by Marquardt optimization method are listed in Table 1. Fig. 5 is the map of the top depths of the gigantic intrusions computed from the pseudo-gravimetric anomaly. In the whole district, the top depths vary from 0 to 16 km. They are shallow in the west and at Yarlung Zangbo River and deep in the east and on both sides of Yarlung Zangbo River. The deepest part is located in the north of the central part.

The results of curve fitting and interpretation show that in the western section of the rocks of ultrabasic type, the top depth of the magnetic body is 5 km and that of the bottom depth is 33 km (from the surface) (Fig. 3). In the east section, the top depth is 8 km and bottom depth 18 km (Fig. 4). The bottom depths of the gigantic magnetic bodies should correspond to the depth of Curie point. Therefore, the difference in the height between the two Curie point isotherms of the eastern and the western sections may reach 10 km. The result of curve fitting of line 83 (Fig. 6) shows that the bottom depth of the magnetic body is also about 20 km. So, the Curie point isotherm of igneous intrusions is deep in the west and shallow in the east.

There is a positive anomaly at the northern end of 30-50 lines. There is a belt of negative anomaly in the south of that positive anomaly and the negative one compresses the positive one to both sides (Fig. 7). It shows that the geological body causing the negative anomaly is larger than that causing the positive anomaly. If in the crust, the rock is of layered structure, then the geological reason causing negative anomaly can be interpreted like this: the rising of the magma at that place made the temperature of rocks reach Curie point and lose its magnetism, so the magnetic field of that place is lower than that of the neighbouring places. Therefore, this is a rising region of the Curie point isotherm. In Fig. 2, it can be found that there is a anomalous zone striking NE-SW to the west of Damxung and it extends toward the southwest to Namling. This is a fracture zone that becomes the channel of the heat source of magma in the rising region of Curie point isotherm. So some heat anomalies distribute along the fracture zone. The results of interpretation are shown in Fig. 8.

## The Relation between the Two Plates

Gandise complex intrusions distribute in a wide area to the north of Yarlung Zangbo River. According to the characteristics of the anomalies (Fig. 2), it belongs to intermediate and acid intrusions. Comparing Fig. 2 with Fig. 1, it is seen that as far as the single intrusive bodies are concerned they are small and scattered. And the farther away from the Yarlung Zangbo valley, the more scattered and acid the igneous intrusions are.

The Bouguer gravity anomaly (Fig. 9) shows* that the position of steep dipping zone of the Moho-surface is located under the Himalayas instead of the Yarlung Zangbo River. The existence of the steep dipping zone of the Moho-surface shows that the crust and the mantle have not reached isostasy so that the Himalayan region is still in activity.

Combining the gravimetric, magnetic and geological data, our general ideas are as follows: The region including Yarlung Zangbo River and Himalayas is a consuming zone formed by the collision of the two plates. The collision zone at Yarlung Zangbo River is only one of the structures in that zone instead of the whole. There are various modes of consumption in that zone, such as rising of Himalayas, deepening and steepening of Moho-surface, forming of the fracture zone, compression zone and collision zone, etc. (Fig. 10).

## ACKNOWLEDGEMENTS

The authors wish to thank the Aerogeophysical Brigade and the Institute of Computing Technique of Ministry of Geology for their support. The authors also wish to thank the teachers of Changchun Geological Institute -- Xu Zhong-xin and Wang Shu-yu for their helps.

## References

(1) Group of K-Ar Geochronology, Institute of Geology, Academia Sinica, K-Ar dating and division of the Himalayan movement in southern Xizang, Scientia Geologica Sinica (1) (1979) (in Chinese).
(2) Hou Zong-chu. The transformation of potential field for the

---

* Provided by Institute of Geophysics, Ministry of Geology.

interpretation system of the magnetic and gravity anomalies, Geophysical and Geochemical Prospecting No. 3 (1979) (in Chinese).

Table 1 The result of Marquardt optimization method

| Number of lines | Number of points | The top depths (km) | The bottom depths (km) | $J \times 10^{-6}$ e.m.u. |
|---|---|---|---|---|
| 0-34 | 0-35 | 6.5 | 30.0 | 1,550 |
| 0-34 | 36-92 | 6.0 | 24.0 | 240 |
| 44-67 | 15-34 | 11.6 | 22.0 | 3,440 |
| 67-90 | 50-80 | 13.0 | 21.0 | 2,880 |
| 67-90 | 15-45 | 11.0 | 21.0 | 2,880 |
| 93-116 | 45-92 | 15.0 | 18.0 | 2,390 |
| 25-57 | 55-92 | 8.0 | 22.0 | 2,400 |

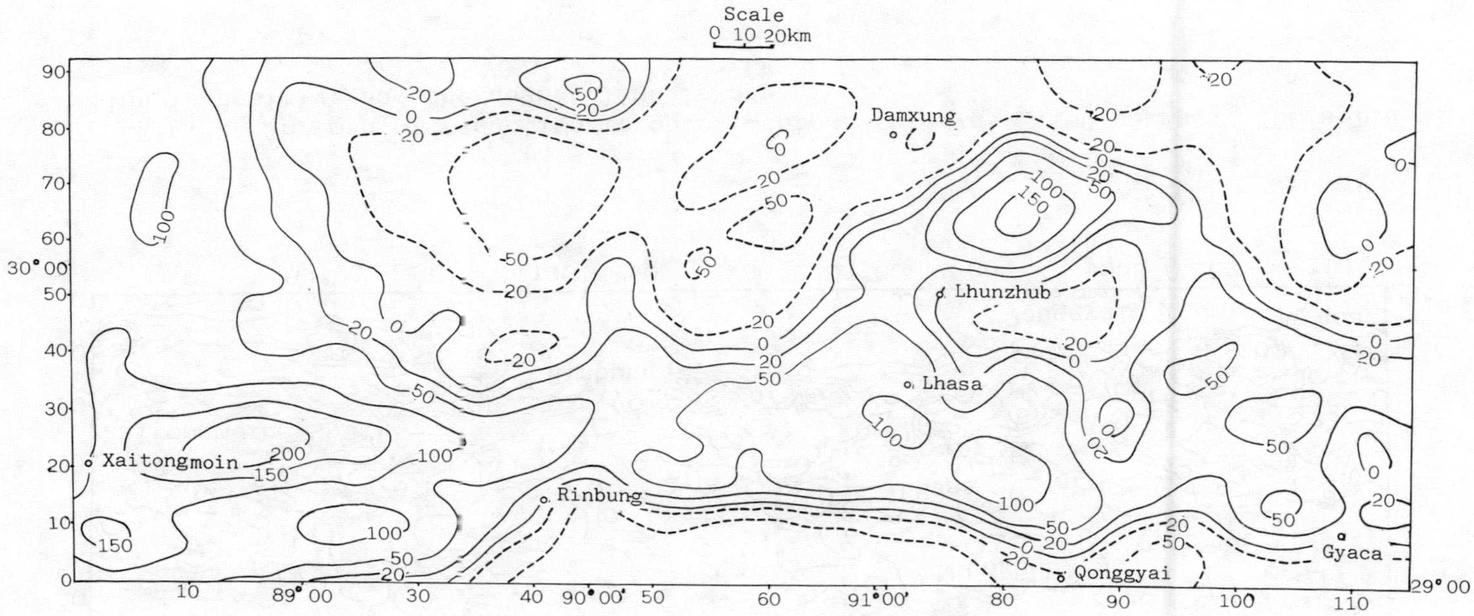


Fig. 1 Upward continuation at 10 km of the field reduced to pole (Contour interval 0, ± 20, ± 50, ± 100, ± 200 gammas. The solid line is positive and the dashed line is negative).

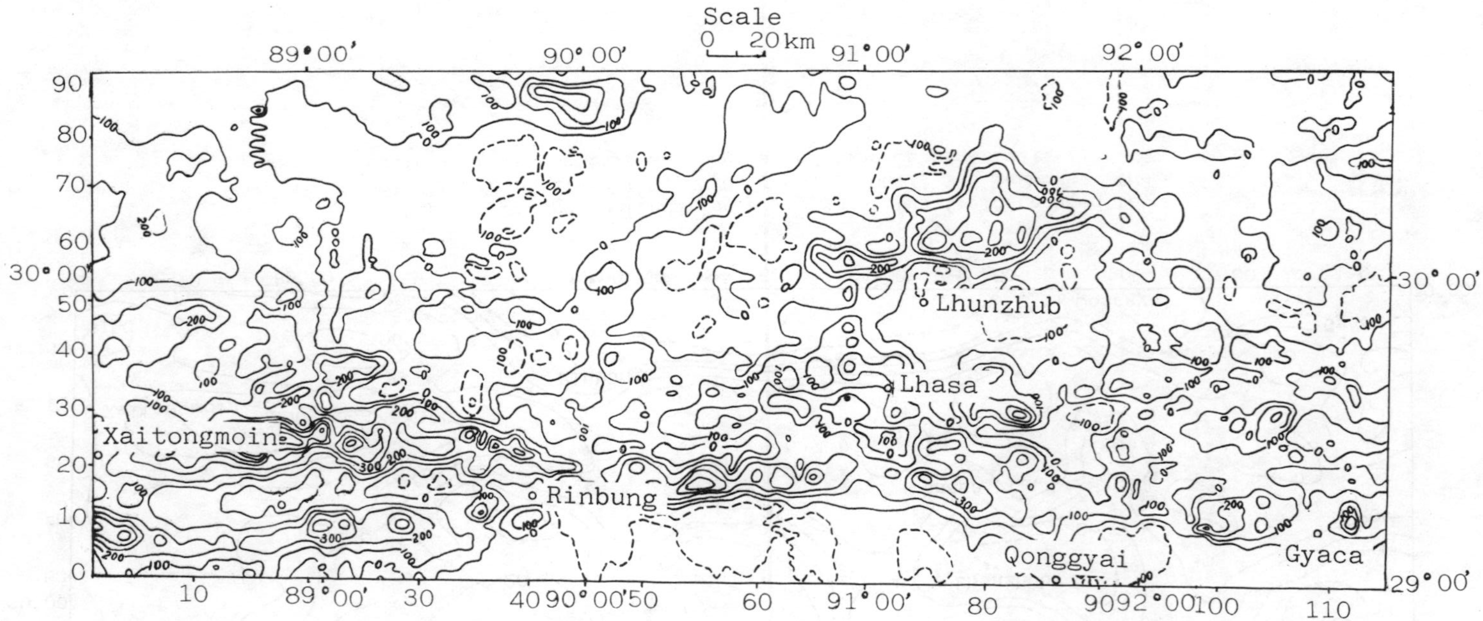


Fig. 2 The map of reduction to pole (Contour interval $\pm$ 100 gammas. The solid line is positive and the dashed line is negative).

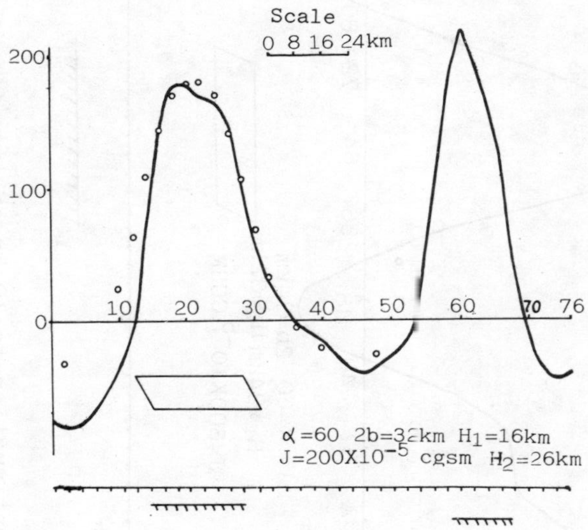


Fig. 3 Upward continuation at 20 km of line 16 reduced to pole and interpretation cross section.

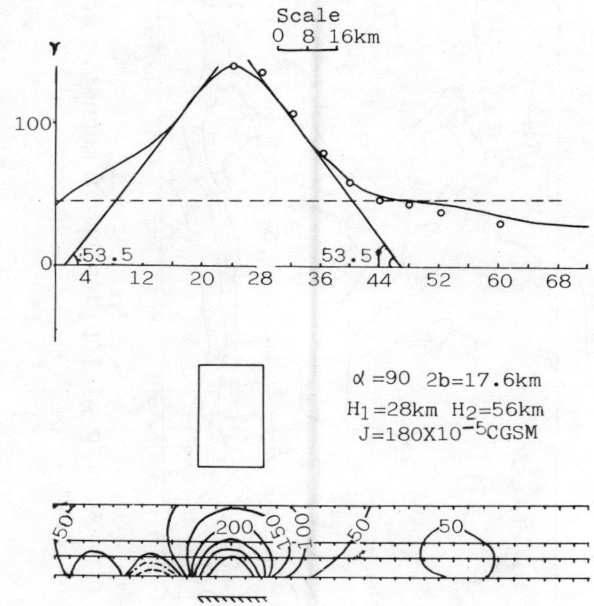


Fig. 4 Upward continuation 5 km of line 78 reduced to pole and interpretation cross section.

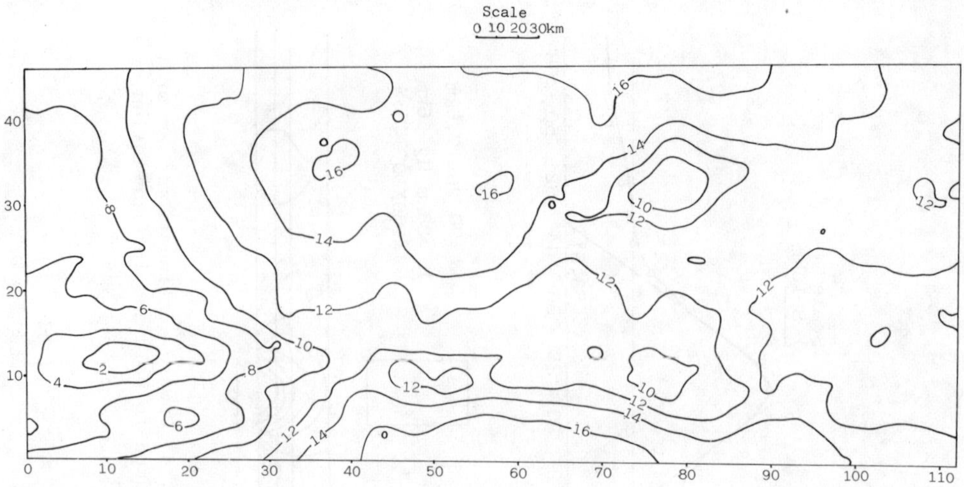


Fig. 5 The contour map of top depths of magnetic sources (Contour interval 2 km).

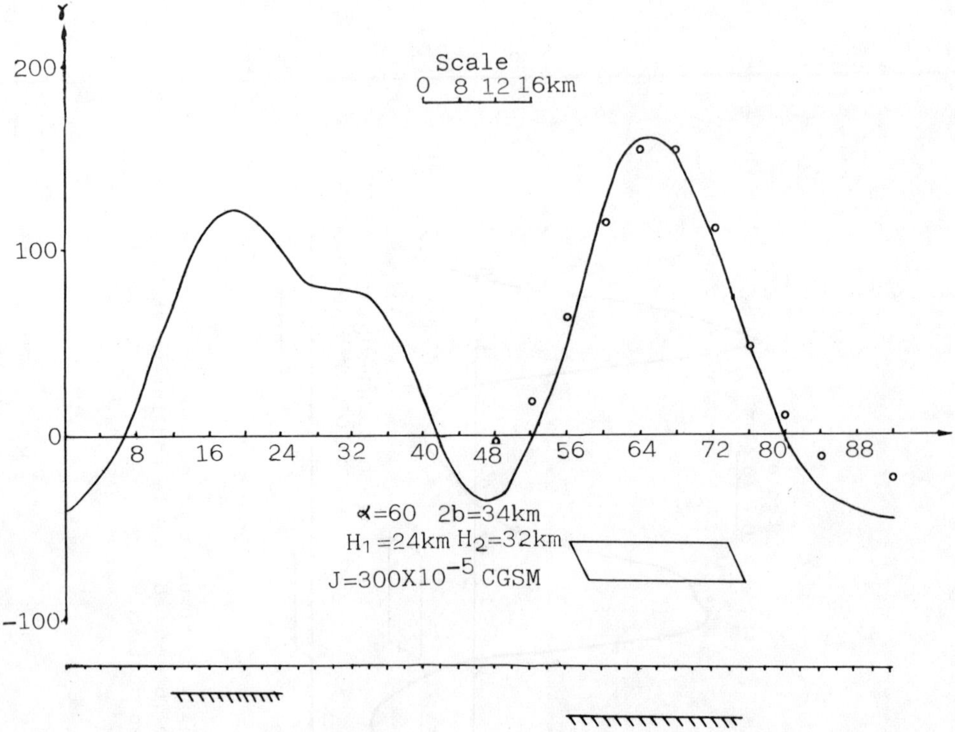


Fig. 6 Upward continuation to 10 km of line 83 reduced to pole and interpretation cross section.

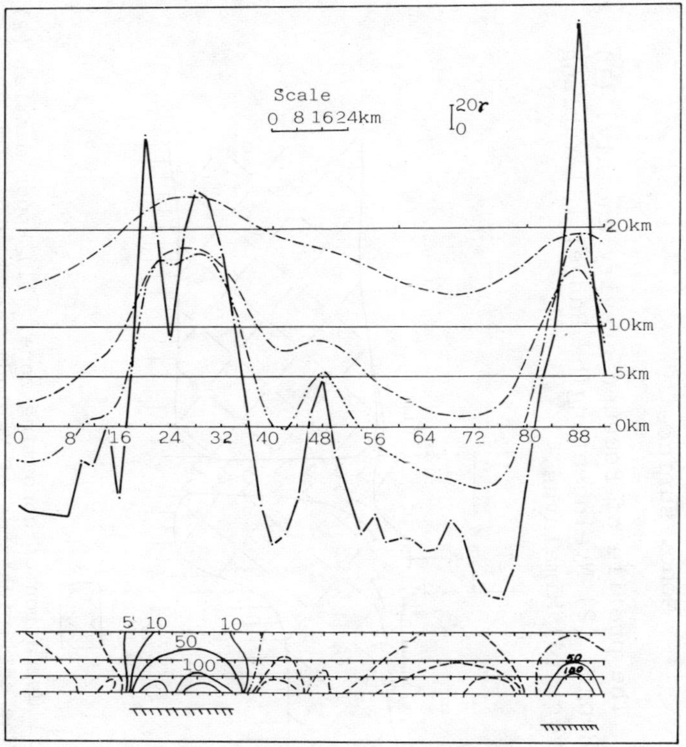


Fig. 7 Upward continuation to different heights (0, 5, 10, 20 km) of line 43 and interpretation cross section.

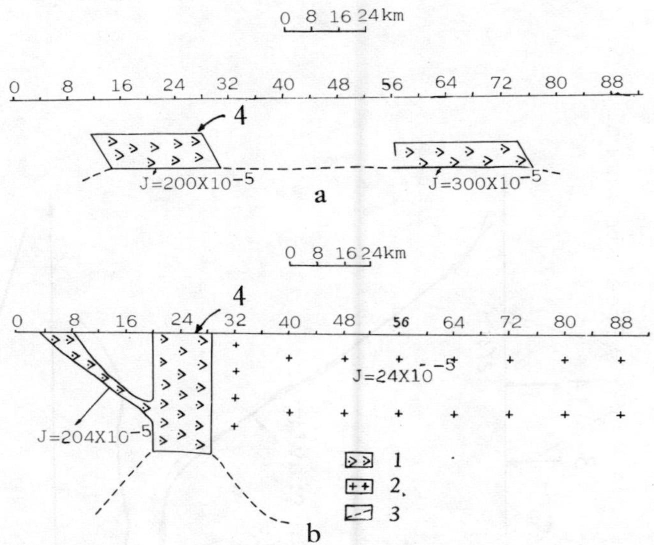


Fig. 8 The illustration of Curie point isotherm of eastern and western parts.

(a) The results interpreted from the lines 45-96;
(b) That from lines 1-16.
(1) The rocks of ultrabasic type;
(2) The rocks of granitic type;
(3) Curie point isotherm;
(4) Collision zone.

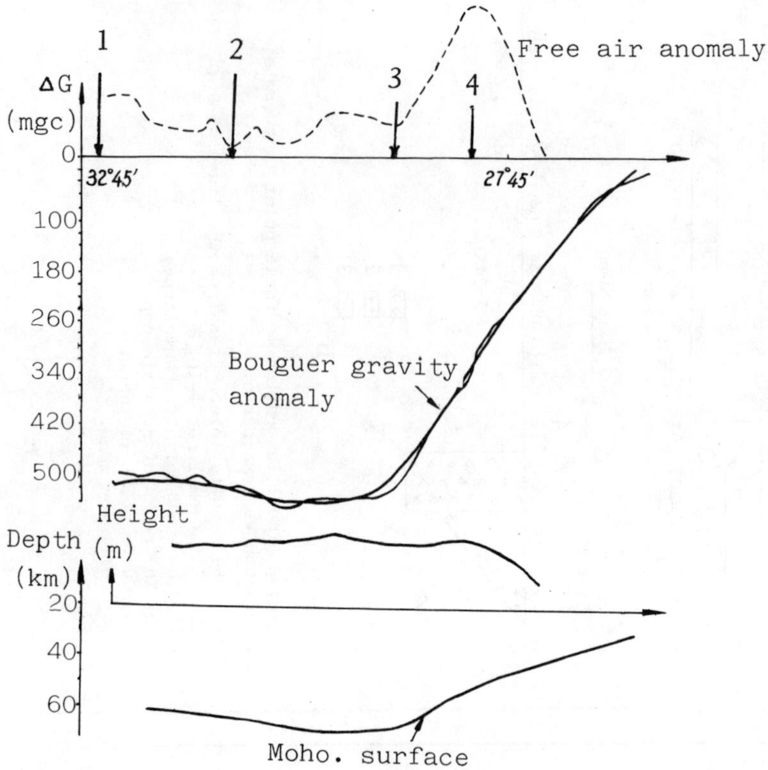


Fig. 9 The profile of Bouguer gravity anomaly. (1) Tanggula Range; (2) Nyainqentanglha Range; (3) Yarlung Zangbo River; (4) Himalayas.

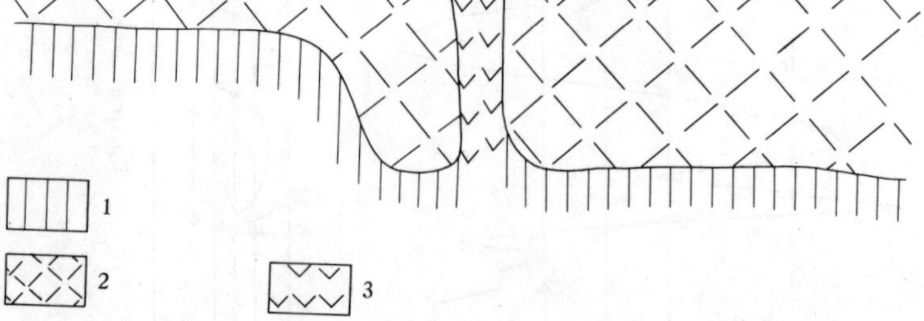


Fig. 10 The illustration of consuming zone. (1) Upper mantle; (2) The crust; (3) The collision zone of Yarlung Zangbo River (The rocks of ultrabasic type); (4) Consuming zone; (5) Himalayas.

# THE INDIA-ASIA COLLISION: A SUMMARY OF PALAEOMAGNETIC CONSTRAINTS

Chris T. Klootwijk
Research School of Earth Sciences
Australian National University **

## ABSTRACT

Comparison of new palaeomagnetic data from extrapeninsular Indo-Pakistan with data from the peninsula and from south central Asia indicates: 1. Collision of continental Indo-Pakistan and south central Asia or off-shore island arcs began during late Palaeocene - early Eocene times at equatorial to low northern latitudes (less than 10°N). Both continental masses became jammed at about the Eocene - Oligocene time boundary. Counterclockwise rotational underthrusting of Indo-Pakistan along the Main Central Thrust beneath its former leading edge may have proceeded over an angle of 10 to 15 degrees. 2. Postcollisional northward movment of Indo-Pakistan resulted in more than 2500 km indentation of south central Asia, which may have been taken up in part by lateral displacement of lithospheric blocks within southern Asia in the sense of Molnar and Tapponnier's model.

## INTRODUCTION

Palaeomagnetic constraints will be discussed on:

- Timing and location of the India-Asia collision.
- The former northern extent of Indo-Pakistan.
- Postcollisional deformation of south central Asia and northern Indo-Pakistan.

This discussion is based on results (Table 1) from: (A) Recent palaeomagnetic studies in extrapeninsular Indo-Pakistan; from northeastern Baluchistan, the Salt Range, Kashmir and the Krol Belt, the Nepalese Himalaya, Chitral and Ladakh (1-9). (B) The outer Pamir and Transalai and southern Tibet (5, 10-12). (C) Peninsular Indo-Pakistan; from the lower Tertiary Deccan Traps, from the Lower Cretaceous Rajmahal Traps and from upper Palaeozoic to Mesozoic Gondwana sediments (13,14). (D) Late Cretaceous and Tertiary palaeolatitude observations from DSDP cores on the Indian plate; from the Ninetyeast Ridge and from the Arabian Sea (13,15,16).

## THE INDIA-ASIA COLLISION, TIMING AND PALAEOLOCATION

Late Palaeozoic to early Tertiary palaeomagnetic data from peninsular and extrapeninsular Indo-Pakistan are available in reasonable quantity and quality, but Tertiary data from the peninsula are lacking.

**Now at: Departement des Sciences de la Terre, Université Paris 7.

This is unfortunate, because such data may constrain Indo-Pakistan's Tertiary northward drift and the timing and location of the India-Asia collision. Estimates of the timing of this collision, based on interpretation of seafloor spreading data and the geological record, range from late Palaeocene to late Eocene - early Oligocene:
(A) Molnar and Tapponnier analyzed the India-Asia movement from seafloor spreading data of the Indian and the Atlantic Ocean, and interpreted a distinct slowdown in convergence rate, from more than 10cm/yr to about 5cm/yr at about 40 my ago, to represent collision (17). (B) Powell and Conaghan concluded to an earlier pre-Middle Eocene collision on basis of the geological record, subsequently refined to collision at about 55 my ago from analysis of the absolute northward movement of the Indian plate (18-21). This analysis was based on relative motion data from the Indian Ocean, India versus Antarctica versus Australia, and on Australian palaeomagnetic data for palaeo-latitudinal positioning of the Indian plate. A decrease in rate of northward movement, from 15 to 20 cm/yr to about 5 cm/yr at about 55 my ago was interpreted to represent collision. This decrease may have been overestimated, however, because of an inconsistency in the Australian palaeomagnetic data used (15,16).

Palaeomagnetic data from northern Indo-Pakistan and south central Asia, and direct palaeomagnetic control on Indo-Pakistan's Tertiary northward drift may further constrain these estimates. The Tertiary part of the Indian apparent polar wander part (APWP) is based only on a set of pole positions (figs. 1, 2) determined from palaeolatitude observations on DSDP cores from the Indian plate in combination with seafloor spreading data, and on early Tertiary palaeomagnetic results from northeastern Baluchistan (fig. 1A) (21,22). Direct palaeomagnetic constraints on the palaeolocation of the collision zone come from the Tibetan Sedimentary Series (TSS) of the Nepalese Himalaya south of the Indus-Tsangpo suture zone (ITS), and from a region in Ladakh immediately north of the ITS which is representative for a former island arc off-shore south central Asia. The TSS data have been interpreted as secondary components acquired upon collision, which finds support in a K-Ar overprint (50-60 my) observed throughout central and eastern Nepal (23). The Ladakh data, likewise, have been interpreted as secondary components acquired upon collision shortly before closure of the K-Ar system at about 50 my ago. These palaeomagnetic constraints on the Tertiary northwards movement of Indo-Pakistan and on the palaeolocation of the collision zone are compared in fig. 2A. Pole positions for secondary "collision" components from Ladakh and from the TSS, not corrected for local rotations around vertical axes, form a swath which intersects the Indo-Pakistan APWP in the 50-60 my interval. This clearly supports Powell and Conaghan's model of initial collision at about the Palaeocene-Eocene time boundary.

Palaeomagnetic control on Indo-Pakistan's Tertiary northward drift and on timing and location of its collision with Asia is summarized in fig. 3A. This figure indicates fast northward motion during late Mesozoic and early Tertiary till about 55 my, which is in good agreement with seafloor spreading evidence. By that time Indo-Pakistan had moved into equatorial latitudes and initial collision with south central Asia or off-shore island arcs (at least in the west) was established. Indo-Pakistan's northward movement continued at a slower rate between about 55 my till about 40 my, when the ITS became immobilized and both continental blocks jammed. More than 1500 km of convergence may have occurred during this period. Subsequent northward movement of Indo-Pakistan till present is taken up by indentation of southern Asia, by crustal shortening within the Himalaya and by intracontinental underthrusting along the Main Central Thrust (MCT).

## FORMER NORTHERN EXTENT OF INDO-PAKISTAN (POST-NEOTETHYS)

Currently prevalent models for the Tertiary evolution of the Tibetan Plateau and its notable twice average crustal thickness; (a) Dewey and Burke's basement reactivation model (24), (b) Molnar and Tapponnier's lateral displacement model (17,25) and (c) Powell and Conaghan's (18,19,21,26) large scale underthrusting model, differ widely in their estimates for the former northern extent of Indo-Pakistan (post-Neotethys). These not necessarily mutually exclusive models, can be constrained by palaeomagnetic data from peninsular and extrapeninsular Indo-Pakistan, and from Eurasia (5,13,27). Model (c) in particular can be constrained on basis of palaeomagnetic results from the TSS of Nepal.

Seafloor spreading data show that postcollisional convergence of Indo-Pakistan with Asia was characterized by a counterclockwise rotational movement of Indo-Pakistan with respect to Asia (17,20,21,28). This may have led to eastwards more pronounced underthrusting along the MCT (19,29). An expected eastwards increase in differential movement between Indo-Pakistan and its former leading edge can be tested palaeomagnetically, because of the favourable circumstances that this former leading edge is exposed and can be studied, whereas differential "rotational" movement can be determined more precisely than "translational" movement.

Palaeomagnetic results thought to be representative for this leading edge were obtained from the TSS of the Thakkhola region in the central Himalaya (fig. 4). Pole positions for primary magnetic components (early Aptian to Permian) and a secondary component acquired upon early Tertiary collision are compared in fig. 1 with the Indo-Pakistan APWP. The TSS pole path is similar in shape to the Indo-Pakistan APWP, both showing a conspicuous Triassic-Jurassic loop, probably reflecting opening of the Neotethys with reversal of Gondwanaland's movement towards south, and extensive Jurassic-Cretaceous and Tertiary segments. The TSS pole path, however, is displaced to the west of the Indo-Pakistan APWP. The displacement corresponds to a difference in declination of about 15 degrees for the Permian to Early Cretaceous data and about 10 degrees for the early Tertiary data. This 10 to 15 degrees difference is interpreted as the approximate lower and upper limits for the magnitude of a differential rotational movement between the TSS of the Thakkhola region and Indo-Pakistan (7). Interpretation of this differential rotation leads to a as yet speculative model, which can be tested by palaeomagnetic studies from other regions of Indo-Pakistan's former leading edge. The TSS pole path is based on data from three structural units which are well continuable along strike into regions outside the Thakkhola graben. Rotational underthrusting along the MCT of Indo-Pakistan beneath at least part of its former leading edge seems therefore a more plausible interpretation than local rotations within the TSS of the Thakkhola region. Palaeomagnetic results from Kashmir indicate more complex movements in the western part of this former leading edge (5). The surface expression of the MCT and the structural trends of the TSS, however, are well continuable along strike throughout the central Himalaya. It is, therefore, tempting to speculate that this 10 to 15 degrees mismatch may be found throughout at least the central Himalayan segment of Indo-Pakistan's former leading edge. The pivot point for such a rotational underthrusting has to be sought for in NW. Indo-Pakistan and may be taken in first approximation at the western end of the Indus-Kohistan seismic zone (fig. f) (29,30). This zone forms the subsurface extension of the Main Boundary Thrust - Main Central Thrust zone beyond the Western Himalayan Syntaxis. Under these assumptions the 10 to 15 degrees mismatch corresponds to an eastwards increasing magnitude of underthrust over 200 to 350 km at the longitude of the sampled locality in central Nepal (fig. 4). Surface evidence along the MCT, gravity data, and faultplane solutions from the Himalayan region show consensus for an about 15 degrees northward dip of the MCT and of Indo-Pakistan's underthrusted frontal slab (31-35). If of uniform dip, the frontal part of this underthrusted slab reaches depths from 60 km north of Mount Kailas to 120 km north of Lhasa at the "15 degrees" line in fig. 4. A relation between this underthrusted slab and the calcalkaline volcanism in southern Tibet (fig. 4) has been suggested previously assuming a Neogene age for the latter (6,36). New radiometric data point, however, to an early Tertiary or older age for these volcanics (37,38). If so, their coincidence with the surface projection of the subducted continental slab (fig. 4) must be purely fortuitous.

The above model for intracontinental underthrusting along the MCT leads to a minimum estimate of Indo-Pakistan's northern extent (post-Neotethys; fig. 5, lines 4 and 5) which is based also on generally accepted values for crustal shortening within the Himalaya (17,31,35, 39,40). This minimum estimate is in good agreement with seafloor spreading evidence from the eastern Indian Ocean (41,42). These data indicate that before breakup of eastern Gondwanaland in the Early Cretaceous a continental mass most probably northeastern Indo-Pakistan extended at least as far north as the Cape Range fracture zone (CRF) bordering the Exmouth Plateau to the south. That is up to the eastwards continuation of line 5 in fig. 5.

## POSTCOLLISIONAL DEFORMATION

### A) Rotational deformations

Palaeomagnetic constraints on postcollisional rotations within south central Asia and northern Indo-Pakistan are summarized in fig. 6. The apical angles indicate the magnitude and sense of discrepancies between observed declinations in south central Asia and northern Indo-Pakistan, and declinations expected in these regions assuming no postcollisional deformation. The western limb of the Hindu Kush -

Pamir - Karakorum syntaxial bend shows counterclockwise rotations over up to 75 degrees since the Palaeogene, whereas post-Palaeogene clockwise rotations over up to about 3Ø degrees in the Transalai suggest a pattern of clockwise rotations in its eastern limb. These opposing rotations clearly indicate the postcollisional origin of this syntaxis. Within northern Indo-Pakistan there is evidence for contrasting supracrustal rotations on opposite limbs of the Western Himalayan Syntaxis; an about 4Ø degrees clockwise rotation of at least part of the Krol Belt and of the Kashmir Basin (figs. 2B, 5), and less conclusive evidence for a counterclockwise rotation over maybe more than 2Ø degrees of the Salt Range/Potwar Plateau unit (5). In northeastern Baluchistan the Loralai Range has rotated clockwise over fifty degrees (figs 2B, 5), whereas the northern part of the Kirthar Range does not appear to have rotated. Preliminary palaeomagnetic results from the southern Sulaiman Range are as yet confusing in their interpretation, and do not decisively discriminate between inwards movement through counterclockwise rotational bending or through en-echelon shear (1). In accordance with the regional tectonics the palaeomagnetic data from the Kirthar Range and the Loralai Range can be interpreted as an inward movement of the Sulaiman lobe over a shallow decollement, guided by NW-SE directed basement trends or possibly by regional occurrence of Infra-Cambrian evaporites (43,44).

It is open to further study whether to observed 5Ø degrees rotation of the Loralai Range may be representative for a wider region of the Sulaiman lobe. It is a tempting but oversimplified speculation that correction for such an uniform rotation relocates the Zhob Valley ultramafics near the Chaman fault zone, which can be interpreted as the original zone of obduction and relocates the Zhob-Kakar flysch at the Makran side of the Chaman fault zone (21). A further correction for late Tertiary northward movement, while positioned on the northward moving Indo-Pakistan subcontinent, relocates the Zhob-Kakar flysch much nearer to the lithologically similar north Makran flysch. Such a reconstructed position may identify the Zhob-Kakar flysch as part of the arc-trench gap sedimentary sequence, deposited on the oceanic frontal part of the Afro-Arabian plate subducting since the Late Cretaceous underneath the Makran - Ras Koh - Chagai region (43,45-47).

B) Crustal shortening

The palaeolatitudinal position and the reconstructed northern margin of Indo-Pakistan (fig. 5) at the time of initial collision with Asia are shown in fig. 3B. Indo-Pakistan is relocated according to the "collision" component observed in the TSS. The contemporaneous palaeoposition of the juxtaposed southern boundary of south central Asia is not constrained, other than by early Tertiary ("collision") palaeomagnetic data from the "Ladakh" island arc (full line), and to some extent by Upper Cretaceous to lower Tertiary red bed data from southern Tibet (dashed line). The stippled area indicates the magnitude of postcollisional crustal shortening within northern Indo-Pakistan. The vertically striped area indicates the magnitude of postcollisional convergence within south central Asia. This is based on a mismatch of 2500 to 3000 km between the palaeomagnetically observed palaeolatitude of collision, represented by Indo-Pakistan's reconstructed northern boundary along the ITS, and the expected about 30 degrees N palaeolatitude of the ITS assuming no postcollisional deformation within south central Asia (27). Palaeomagnetic results from locations within central Asia's southern rim, i.e. Ladakh, the eastern Hindu Kush and southern Tibet, likewise show an about 3000 km discrepancy between observed and expected palaeolatitudes (fig. 7).

The above about 3000 km mismatch may be accounted for by: a) convergence within the collision zone, b) crustal shortening through crustal thickening, and c) lateral displacement of lithospheric blocks within southern Asia out of the way of indenting Indo-Pakistan (17). The magnitude of each of these effects is not properly established as yet. There are, however, some palaeomagnetic constraints on the magnitude of the first two effects in particular in the western part of the collision zone, which allow for an indirect estimate of the third effect. Former existence of an island arc has been suggested in the western part of the collision zone, i.e. Ladakh and possibly also in Kohistan (48,49). The extreme case that the total India-Asia convergence in this western region, between initial contact (at about 55 my ago) and the distinct slow down in convergence at about 40 my ago was taken up fully by convergence within the collision zone- closure of marginal basins, deformation of non-matching opposing continental lithosphere- would account for a maximal magnitude of about 1500 km for the first effect. Estimates for crustal thickening within the Pamir account for 200 to 300 km of crustal shortening (17). The balance of more than 1000 km is in good agreement with palaeomagnetic evidence for about 1000 km crustal convergence since the Palaeogene north of the Pamir - Transalai zone (fig. 7), and can be interpreted in support of Molnar and Tapponnier's lateral displacement model.

## References


(1) Klootwijk, C.T., Nazirullah, R., De Jong, K.A. and Ahmed, H., A palaeomagnetic reconnaissance of northeastern Baluchistan, Pakistan, J. Geophys. Res., in press.

(2) McElhinny, M.W., Palaeomagnetism of the Cambrian Purple Sandstones from the Salt Range, West Pakistan, Earth Planet. Sci. Lett., 8, 149-156 (1970).

(3) Wensink, H., The palaeomagnetism of the Salt Pseudomorph beds of Middle Cambrian age from the Salt Range, West Pakistan, Earth Planet. Sci. Lett., 16, 189-194 (1972).

(4) Wensink, H., The palaeomagnetism of the Speckled Sandstones of Early Permian age from the Salt Range, Pakistan, Tectonophysics, 26, 281-292 (1975).

(5) Klootwijk, C.T., A summary of palaeomagnetic data from extrapeninsular Indo-Pakistan and south central Asia: Implications for collision tectonics, in: Structural geology of the Himalaya, Saklani, P.S., ed., Today and Tomorrow Publishers, Delhi, 307-360 (1979).

(6) Bingham, D.K. and Klootwijk, C.T., Palaeomagnetic constraints on Greater India's underthrusting of the Tibetan Plateau, Nature, 284, 336-338 (1980).
(7) Klootwijk, C.T. and Bingham, D.K., The northern extent of India 3: Palaeomagnetic data from the Tibetan Sedimentary Series, Thakkhola region, Nepal Himalaya, Earth Planet. Sci. Lett., in press.
(8) Klootwijk, C.T. and Conaghan, P.J., The extent of Greater India 1: preliminary palaeomagnetic data from the Upper Devonian of the Eastern Hindu Kush, Chitral (Pakistan), Earth Planet. Sci. Lett., 42, 167-182 (1979).
(9) Klootwijk, C.T., Sharma, M.L., Gergan, J., Tirkey, B., Shah, S.K., and Agarwal, V.K., The extent of Greater India 2: Palaeomagnetic data from the Ladakh Intrusives at Kargil, northwestern Himalaya, Earth Planet. Sci. Lett., 44, 47-64 (1979).
(10) Burtman, V.S. and Gurariy, G.Z., Character of folded areas in the Pamirs and Tien-Shan from the geophysical data, Geotectonics, 1, 90-92 (1973).
(11) Bazhenov, M.L., Burtman, V.S. and Gurariy, G.Z., Palaeomagnetic results for the Palaeogene of the outer Pamirs and Transalai, Dokl. Akad. Nauk. SSSR., 242, 1137-1139 (in Russian), (1978).
(12) Zhu Xiangyuan, Liu Chun, Ye Sujuan and Lin Jinlu, Remanence of red beds from Linzhou,Xizang, and the northward movement of the Indian plate, Sci. Geol. Sinica, 1, 44-51 (1977).
(13) Klootwijk, C.T., A review of palaeomagnetic data from the Indian fragment of Gondwanaland, in: Geodynamics of Pakistan, Farah, A. and De Jong, K.A., eds., Geol. Surv. Pakistan, Quetta, 41-80 (1979).
(14) Klootwijk, C.T. and Radhakrishnamurty, C., Phanerozoic palaeomagnetism of the Indian plate and the India-Asia collision, in: Int. Comm. Geodynamics, Final report WG 10, McElhinny, M.W. and Valencio, D., eds., in press.
(15) Peirce, J.W., The northward motion of India since the Late Cretaceous, R. Astron. Soc. Geophys. J., 52, 277-311 (1978).
(16) Klootwijk, C.T. and Peirce, J.W., India's and Australia's pole path since late Mesozoic and the India-Asia collision, Nature, 282, 605 607 (1979).
(17) Molnar, P. and Tapponnier, P., Cenozoic tectonics of Asia: effects of a continental collision, Science, 189, 419-426 (1975).
(18) Powell, C.McA. and Conaghan, P.J., Plate tectonics and the Himalayas, Earth Planet. Sci. Lett., 20, 1-12 (1973).
(19) Powell, C.McA. and Conaghan, P.J., Tectonic models of the Tibetan Plateau, Geology, 4, 727-731 (1975).
(20) Johnson, B.D., Powell, C.McA. and Veevers, J.J., Spreading history of the eastern Indian Ocean and Greater India's northward flight from Antarctica and Australia, Geol. Soc. America Bull., 87, 1560-1566 (1976).
(21) Powell, C.McA., A speculative tectonic history of Pakistan and surroundings: some constraints from the Indian Ocean, in: Geodynamics of Pakistan, Farah, A. and De Jong, K.A., eds., Geol. Surv. Pakistan, Quetta, 5-24 (1979).

(22) Powell, C. McA., Johnson, B.D. and Veevers, J.J., A revised fit of East and West Gondwanaland. Tectonophysics, 63, 13-29 (1980).
(23) Bordet, P., Colchen, M., Krummenacher, D., Le Fort, P., Mouterde, R. and Remy, M., Rechêrches géologiques dans l'Himalaya du Nepal, région de la Thakkhola, C.N.R.S., Paris, 279 pp, (1971).
(24) Dewey, J.F. and Burke, K.C.A., Tibetan, Variscan and Precambrian basement reactivation: products of continental collision, J. Geol., 81, 683-692 (1973).
(25) Molnar, P. and Tapponnier, P., Active tectonics of Tibet, J. Geophys. Res., 83, 5361-5375 (1978).
(26) Veevers, J.J., Powell, C.McA. and Johnson, B.D., Greater India's place in Gondwanaland and in Asia, Earth Planet. Sci. Lett., 27, 383-387 (1975).
(27) Irving, E., Drift of the major continental blocks since the Devonian, Nature, 27Ø, 3Ø4-3Ø9 (1977).
(28) Sclater, J.G. and Fischer, R.L., Evolution of the east central Indian Ocean with emphasis on the tectonic setting of the Ninety East Ridge, Geol. Soc. America Bull., 85, 683-7Ø2 (1974).
(29) Colchen, M. and Le Fort, P., Some remarks and questions concerning the geology of the Himalaya, in: Colloq. Int. C.N.R.S., Ecologie et Géologie de l'Himalaya 2, 131-137 (1977).
(30) Armbruster, J.G., Seeber, L. and Jacob, K.H., The northwestern termination of the Himalayan mountain front: Active tectonics from microearthquakes, J. Geophys. Res., 85, 269-282 (1978).
(31) Le Fort, P., Himalayas: The collided range. Present knowledge of the continental arc, Am. J. Sc., 275-A, 1-44 (1975).
(32) Le Fort, P., pers. comm., (1980).
(33) Kono, M., Gravity anomalies in east Nepal and their implication to the crustal structure of the Himalayas, R. Astron. Soc. Geophys. J., 39, 283-3ØØ (1974).
(34) Warsi, W.E.K. and Molnar, P., Gravity anomalies and plate tectonics in the Himalaya, in: Colloq. Int. C.N.R.S., Ecologie et Géologie de l'Himalaya 2, 463-478 (1977).
(35) Molnar, P., Wang-ping Chen, Fitch, T.J., Tapponnier, P., Warsi, W.E.K. and Wu, F.T., Structure and tectonics of the Himalaya: A brief summary of relevant geophysical observations, in: Colloq. Int. C.N.R.S., Ecologie et Géologie de l'Himalaya 2, 269-294 (1977).
(36) Gansser, A., The great suture between the Himalaya and Tibet: A preliminary account, Colloq. Int. C.N.R.S., Ecologie et Géologie de l'Himalaya 2, 181-191 (1977).
(37) Gansser, A., pers. comm., (1980).
(38) Zhang Yuquan, pers. comm., (1980).
(39) Gansser, A., Himalaya, in: Mesozoic-Cenozoic orogenic belts: data for orogenic studies, Spencer, A.M., ed., Geol. Soc. London, Spec. publ., 4, 267-278 (1974).
(40) Mattauer, M., Sur le mechanisme de formation de la schistosite dans l'Himalaya, Earth Planet. Sci. Lett., 28, 144-154 (1975).
(41) Larson, R.L., Mutter, J.C., Diebold, J.B., Carpenter, G.B. and Symonds, P., Cuvier Basin: A product of ocean crust formation by Early Cretaceous rifting off western Australia, Earth Planet. Sci. Lett., 45, 1Ø5-114 (1979).

(42) Johnson, B.D., Powell, C.McA. and Veevers, J.J., Early spreading history of the Indian Ocean between India and Australia, Earth Planet. Sci. Lett., 47, 131-143 (1980).
(43) Hunting Survey, Reconnaissance geology of part of western Pakistan, Government of Canada, Toronto, 550 pp, (1960).
(44) Sarwar, G. and De Jong, K.A., Arcs, oroclines and syntaxes: The curvatures of mountain belts in Pakistan, in: Geodynamics of Pakistan, Farah, A. and De Jong, K.A., eds., Geol. Surv. Pakistan, Quetta, 341-350 (1979).
(45) Farhoudi, G. and Karig, D.E., Makran of Iran and Pakistan as an active arc system, Geology, 5, 664-668 (1977).
(46) Jacob, K.H. and Quittmeyer, R.C., The Makran region of Pakistan and Iran: trench-arc system with active plate subduction, in: Geodynamics of Pakistan, Farah, A. and De Jong, K.A., eds., Geol. Surv. Pakistan, Quetta, 305-317 (1979).
(47) Arthurton, R.S., Sarwar Alam, G., Anisuddin Ahmed, S. and Saeed Iqbal, Geological history of the Alamry-Mashki Chah area, Chagai district, Baluchistan, in: Geodynamics of Pakistan, Farah, H. and De Jong, K., eds., Geol. Surv. Pakistan, Quetta, 325-331 (1979).
(48) Frank, W., Gansser, A. and Trommsdorff, V., Geological observations in the Ladakh area (Himalayas); A preliminary report, Schweiz, Mineral. Petrogr. Mitt., 57, 89-113 (1977).
(49) Khan Tahirkheli, R.A., Mattauer, M., Proust, F. and Tapponnier, P., The India-Asia suture zone in northern Pakistan: Synthesis and interpretation of recent data at plate scale, in: Geodynamics of Pakistan, Farah, A. and De Jong, K.A., eds., Geol. Surv. Pakistan, Quetta, 125-130 (1979).
(50) Gansser, A., The geology of the Himalaya, Interscience Publishers, London, 289 pp, (1964).
(51) Gansser, A., The significance of the Himalayan suture zone, Tectonophysics, 62, 37-52 (1980).
(52) Smith, A.G. and Hallam, A., The fit of the southern continents, Nature, 225, 139-144 (1970).
(53) Griffiths, J.T., Revised continental fit of Australia and Antarctica, Nature, 249, 336-338 (1974).
(54) Veevers, J.J. and Cotterill, D., Western margin of Australia: evolution of a rifted arc system, Geol. Soc. America Bull., 89, 337-355 (1978).

TABLE 1 SUMMARY OF PALAEOMAGNETIC DATA (see figs. 1, 2, 3, 6, 7)[1)]

| Rock Unit | Sampling Area (o) | Age | DIRECTION Decl (o) | Incl (o) | k | α95 (o) | N | SOUTH POLE POSITION Lat (o) | Long (o) | dp (o) | dm (o) | E95 (o) | Pole |
|---|---|---|---|---|---|---|---|---|---|---|---|---|---|
| A) Primary magnetization data Indo-Pakistan (figs. 1a, 2)[2)] | | | | | | | | | | | | | |
| DSDP cores | Indian Ocean | 5-22.5my(20) | | | | | 2 | 78.5S | 114.5E | | | | 1 |
| DSDP cores | Indian Ocean | 25-35my(30) | | | | | 4 | 72.5S | 136E | | | 3.5 | 2 |
| DSDP cores | Indian Ocean | 35-45my(40) | | | | | 3 | 67.5S | 132E | | | 12 | 3 |
| DSDP cores | Indian Ocean | 45-55my(50) | | | | | 4 | 57.5S | 121.5E | | | 9 | 4 |
| Sanjawi Lst. | 30.2N 68.2E | T (pa-e) | 22 | +15 | 13.5 | 14.5 | (67) | 59.5S | 22E | 7.5 | 15 | | 5' |
| As above 3) | | | 332 | +15 | 13.5 | 14.5 | (67) | 55.5S | 123.5E | 7.5 | 15 | | 5 |
| Brewery Lst | 30.1N 66.6E | Tpa | 331.5 | +8.5 | 8.5 | 22 | (7) | 52.5S | 118.5E | 11 | 22 | | 6 |
| DSDP cores | Indian Ocean | 55-65my(60) | | | | | 5 | 40.5S | 120E | | | 10.5 | 7 |
| Deccan Traps upper normal epoch | | 60-65my | | | | | 32 | 39.5S | 101.5E | | | 5.5 | 8 |
| Deccan Traps lower normal epoch | | 60-65my | | | | | 47 | 32.5S | 97.5E | | | 3 | 9 |
| Tirupati Sst. | 16.8N 81.2E | Kl | 332 | -52.5 | 52.3 | 11 | (43) | 33.5S | 109.5E | 10.5 | 15 | | 10 |
| Tirupati Sst. | 16.8N 81.2E | Kl | 330 | -58 | 443 | 4 | 4(65) | 28S | 107.5E | 4.5 | 6 | | 11 |
| DSDP cores | Indian Ocean | 65-75my(70) | | | | | 2 | 29S | 111.5E | | | | 12 |
| Satyavedu beds | 13.5N 79.9E | Kl | 321 | -58 | 80 | 4 | (13) | 26.5S | 113E | 4.5 | 6 | | 13 |
| Sylhet Traps | 25.0N 91.0E | Kl?? | 332 | -59 | | 7 | (25) | 20S | 113.5E | 8 | 10.5 | | 14 |
| Goru Fm.-Parh Lst. | 30.2N 68.2E | K(l-u) | 359.5 | -40.5 | 32.5 | 3 | (76) | 36.5S | 69.5E | 2 | 3.5 | | 15' |
| As above 3) | | | 309.5 | -40.5 | 32.5 | 3 | (76) | 17.5S | 117E | 2 | 3.5 | | 15 |
| Goru Fm.-Parh Lst. | 30.1N 66.6E | K(l-u) | 307 | -43.5 | 16.5 | 4 | (86) | 14.5S | 115E | 3 | 5 | | 16 |
| Rajmahal Traps | 24.5N 87.6E | 100-105my | 316 | -65 | 85 | 2.5 | 48(294) | 7.5S | 116.5E | 3 | 3.5 | | 17 |
| Loralai Lst. | 30.2N 68.2E | Jm | 338 | -43 | 9 | 8 | (41) | 31S | 91.5E | 6 | 10 | | 18' |
| As above 3) | | | 228 | -43 | 9 | 8 | (41) | 2S | 128E | 6 | 10 | | 18 |
| Chiltan Lst | 30.1N 66.6E | J(m-u) | 104 | 38 | 17 | 10.5 | (30) | 0.5S | 131.5E | 7.5 | 12.5 | | 19 |
| Australia 4) | | Jm | | | | | 3 | 21.5S | 130.5E | | | 9 | 20 |
| Australia 4) | | Jl | | | | | 2 | 17S | 130.5E | | | | 21 |
| Antarctica 4) | | Jm | | | | | 5 | 17S | 124.5E | | | 14.5 | 22 |
| Parsora beds | 23.5N 81.5E | Tru | 320 | -39 | 200 | 5.5 | 3(48) | 30S | 125E | 4 | 7 | | 23 |
| Pachmarhi beds | 22.8N 78.8E | Tru | 117 | +49 | 33 | 4.5 | 3(31) | 10S | 130.5E | 4.5 | 6 | | 24 |
| Kamthi beds | 20.0N 79.0E | Pu-Trl | 101.5 | 58.5 | | 6.5 | 22(105) | 4N | 128.5E | 7 | 9.5 | | 25 |
| Mangli beds | 20.5N 79.0E | Pu-Trl | 100 | 63 | 43.5 | 4.5 | 2(23) | 7N | 124.5E | 5.5 | 7 | | 26 |
| Panchet beds | 23.9N 85.2E | Pu-Trl | 110.5 | 69 | 49 | 6 | (13) | 7N | 120.5E | 8.5 | 10 | | 27 |
| Kamthi beds | 20.0N 79.0E | Pu-Trl | 79.5 | 61.5 | 63 | 2.5 | 6(57) | 21N | 129.5E | 3 | 3.5 | | 28 |
| Talchir beds | 21.5N 78.9E | P-C | 66 | 59 | 109.5 | 3 | (23) | 31.5N | 134.5E | 3.5 | 4.5 | | 29 |
| B) Secondary magnetization data Kashmir and Krol Belt (fig.2B) | | | | | | | | | | | | | |
| NW.Kashmir 5) | 34.4N 74.2E | T(l-u) | 334 | 26.5 | | | 2 | 59S | 130E | | | | 30 |
| NW.Kashmir 5) | 34.4N 74.2E | T(l-u) | 16 | 42 | 49 | 18 | 3 | 73S | 26E | 13.5 | 22 | | 31 |
| C.&E.Kashmir (R.polarity) 6) | | T(l-u) | 192.5 | -38 | 74 | 7 | 7 | 73S | 30E | 5 | 8.5 | | 32 |
| As above (N.polarity) 6) | | Tl | 21 | 30.5 | 49.5 | 17.5 | 3 | 64S | 22E | 11 | 20 | | 33 |
| As above (N.polarity) 6) | | Tl | 207 | -27 | 84 | 9 | 4 | 59S | 17.5E | 5.5 | 10 | | 34 |
| Krol Belt 7) | 30.2N 78.4E | T(l-u) | 13.5 | 35 | 97.5 | 8 | 5 | 73.5S | 26.5E | 9 | 5 | | 35 |
| Krol Belt 8) | 30.2N 78.4E | Tl | 23.5 | 23 | 67.5 | 9.5 | 5 | 62S | 23.5E | 10 | 5.5 | | 36 |
| C) Secondary magnetization data Ladakh (fig. 2A)[9)] | | | | | | | | | | | | | |
| Dras Flyschoids | 34.4N 76.9E | Tl | 24.5 | 0 | 24.5 | 6.5 | (21) | 48.5S | 38E | 3.5 | 6.5 | | 37 |
| Dras Flyschoids | 34.4N 76.9E | Tl | 14.5 | 1.5 | 38 | 5 | (23) | 53.5S | 53E | 2.5 | 5 | | 38 |
| Dras Flyschoids | 34.4N 76.9E | Tl | 357.5 | 2.5 | 27.5 | 5.5 | (27) | 57S | 81.5E | 2.5 | 5.5 | | 39 |
| Dras Flyschoids | 34.4N 76.9E | Tl | 350 | 2.5 | 37.5 | 4.5 | (31) | 55.5S | 94.5E | 2 | 4.5 | | 40 |
| Indus Molasse | 34.4N 76.9E | Tl | 354 | 2.5 | 15.5 | 6.5 | (34) | 56.5S | 88E | 3 | 6.5 | | 41 |
| Indus Molasse | 34.4N 76.9E | Tl | 325 | 11 | 6 | 19.5 | (12) | 47S | 133E | 10 | 19.5 | | 42 |
| Indus Molasse | 34.4N 76.9E | Tl | 312.5 | 9 | 7 | 15.5 | (15) | 37S | 143.5E | 8 | 15.5 | | 43 |
| Ladakh Intrusives | 34.6N 76.1E | ~50my | 3 | 14.5 | 53.5 | 4 | 5(27) | 62.5S | 69.5E | 2 | 4 | | 44 |
| Ladakh Intrusives | 34.6N 76.1E | ~50my | 355 | 19.5 | 13.5 | 8.5 | 5(23) | 65S | 88E | 4.5 | 9 | | 45 |

TABLE 1 SUMMARY OF PALAEOMAGNETIC DATA (ctd)

| Rock Unit | Sampling Area | Age | DIRECTION Decl | Incl | k | α95 | N | SOUTH POLE POSITION Lat | Long | dp | dm | E95 | Pole |
|---|---|---|---|---|---|---|---|---|---|---|---|---|---|
| | (o) | | (o) | (o) | | (o) | | (o) | (o) | (o) | (o) | (o) | |
| 10) | | | | | | | | | | | | | |
| D) Primary and Secondary Magnetization data Tibetan Series, Nepal Himalaya (fig. 1B) | | | | | | | | | | | | | |
| Kagbeni Sst. 10) | 28.7N 83.7E | Tl | 2 | -12 | 16 | 8 | (22) | 55S | 80E | 4 | 8 | | 46 |
| Thinigaon Lst. 10) | 28.7N 83.7E | Tl | 6.5 | -9.5 | 39 | 6 | (17) | 56S | 72E | 3 | 6 | | 47 |
| Jomosom Lst. 10) | 28.7N 83.7E | Tl | 346 | -4 | 35 | 2.5 | (77) | 56.5S | 110E | 1 | 2.5 | | 48 |
| Kagbeni Sst. 10) | 28.7N 83.7E | Tl | 347.5 | -5 | 16.5 | 8.5 | (18) | 56.5S | 106.5E | 4.5 | 8.5 | | 49 |
| Dzong Sst. 10) | 28.7N 83.7E | Tl | 348 | -6.5 | 23.5 | 3.5 | (67) | 56S | 106E | 2 | 3.5 | | 50 |
| Jomosom Qzt. 10) | 28.7N 83.7E | Tl | 346 | -16 | 32 | 4.5 | (36) | 50.5S | 106E | 2.5 | 4.5 | | 51 |
| Dzong Sst. | 28.7N 83.7E | L.Aptian | 326 | -61.5 | 12 | 5 | (78) | 12S | 109E | 6 | 7.5 | | 52 |
| Kagbeni Sst. | 28.7N 83.7E | Wealden | 333.5 | -56.5 | 15.5 | 6.3 | (34) | 19.5S | 106E | 6.5 | 9.5 | | 53 |
| Kagbeni Sst. | 28.7N 83.7E | Wealden | 322.5 | -55.5 | 14 | 8 | (24) | 16S | 114.5E | 8.5 | 11.5 | | 54 |
| Lumachelle Fm. | 28.7N 83.7E | Dogger | 308 | -61 | 12 | 8.5 | (27) | 4.5S | 119.5E | 9.5 | 12.5 | | 55 |
| Jomosom Lst. | 28.7N 83.7E | Lias | 307 | -59.5 | 10.5 | 10 | (22) | 5S | 121.5E | 11.5 | 15 | | 57 |
| Jomosom Lst. | 28.7N 83.7E | Rhaetian | 32.3 | -48 | 11 | 8 | (32) | 22S | 118.5E | 7 | 10.5 | | 58 |
| Thinigaon Lst. | 28.7N 83.7E | Norian | 324.5 | -43 | 10 | 10 | (22) | 26.5S | 119.5E | 8 | 13 | | 59 |
| Thinigaon Lst. | 28.7N 83.7E | Ladin.-Carnian | 329 | -47 | 21 | 5.5 | (35) | 26S | 114E | 4.5 | 7 | | 60 |
| Thini Chu-Ice Lake Fm. | 28.7N 83.7E | M?Permian | 291.5 | -63.5 | 10 | 7.6 | (41) | 13.5S | 104E | 2 | 2.5 | | 61 |
| | 11) | 11) | | | | | | | | | | | |
| E) Primary magnetization data Pamir, Transalai and Hindu Kush (figs.6,7) | | | | | | | | | | | | | |
| S.Darvaz | 37.9N 70.1E | To | 305 | 30 | | | | 37S | 151.5E | | | | 62 |
| Darvaz | 37.9N 70.1E | K(l-u) | 311 | 47 | | | | 48.5S | 161.5E | | | | 63 |
| Chil'Dara | 38.8N 70.6E | T(e-m) | 346 | 40 | | | | 70S | 111.5E | | | | 64 |
| Khipsiun | 38.9N 70.8E | To | 329 | 44 | | | | 62S | 145.5E | | | | 65 |
| Kyzylart | 39.5N 73.2E | T(e-m) | 43 | 43 | | | | 51.5S | 8N | | | | 66 |
| Transalai | 39.5N 73.2E | Ku | 93 | 56 | | | | 20.5S | 48W | | | | 67 |
| Khatyn-Kanysh | 39.4N 73.3E | T(e-o) | 35 | 38 | | | | 55S | 4E | | | | 68 |
| Lhasa | 30.0N 91.3E | Ku-Tl | 355.5 | 15.5 | | | | 67.5S | 103E | | | | 69 |
| Chistral | 36.2N 72.5E | Du 12) | 318 | -6.5 | 14 | 16.5 | (7) | 34.5S | 126E | 8.5 | 16.5 | | 70 |
| | 13) | | | | | | | | | | | | |
| F) Primary magnetization data Salt Range (fig.6) | | | | | | | | | | | | | |
| Nilawan-Sardahi Gorge | 32.6N 72.5E | Pl | 85.5 | 42.5 | 113 | 6.5 | 6(61) | 16.5N | 143.5E | 5 | 8 | | 71 |
| Nurpur-Kushab Road | 32.7N 72.7E | Pl | 108.5 | 53.5 | 278 | 7.5 | 3(25) | 4.5N | 124.5E | 7 | 10.5 | | 72 |
| Salt Pseudomorph | 32.2N 73.0E | Cm | 221.5 | 39.5 | 52 | 9.5 | 6(43) | 22S | 32E | 7 | 11.5 | | 73 |
| Purple Sandstones | 32.7N 73.0E | Cl | 218 | 31.5 | 15 | 12 | 1(10) | 28.5S | 31E | 10.5 | 17 | | 74 |

1) For references, see summaries (1,5,7,13,14).
2) Si=sites, Sa=samples.
3) Corrected for clockwise rotation over 50 degrees of the Loralai Range with respect to the Indian Shield.
4) Mean pole transferred to the Indian plate according to seafloor spreading data (1).
5) Mean secondary component NW.Kashmir, of post-folding origin.
6) Mean secondary component C.&E.Kashmir, of post-folding origin.
7) Mean secondary component Krol Belt, of post-folding origin.
8) Mean secondary component Krol Belt, of pre-folding origin.
9) Probably acquired upon collision at about 50 my ago.
10) Probably acquired upon collision at about 50 to 60 my.
11) Directions and locations read from original diagrams and maps (5, 10, 11).
12) Possibly a secondary (collision) magnetization of early Tertiary origin (14).
13) Not corrected for a suggested counterclockwise rotation over maybe more than 20 degrees (5).

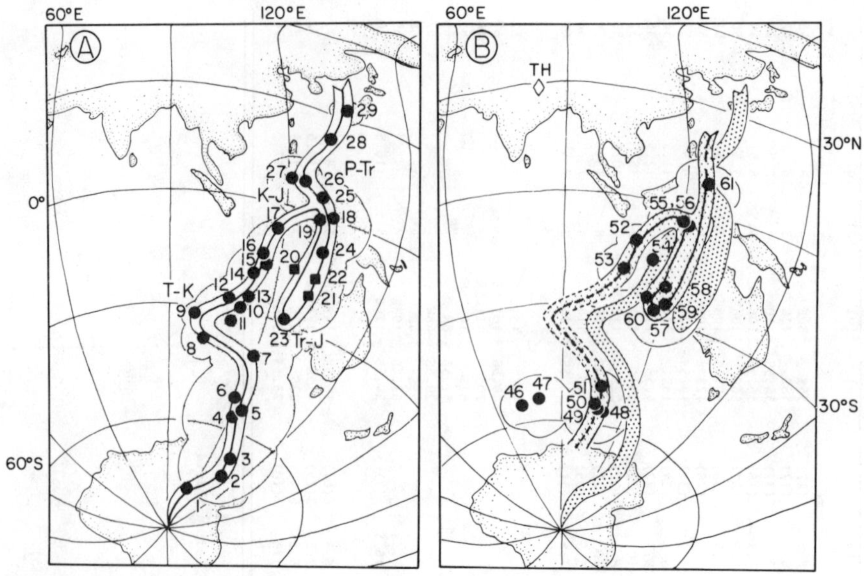

Fig. 1 Comparison of the Indo-Pakistan APWP (A) with the pole path for the TSS (B), Th= sampling locality. Pole numbers according to Table 1. The envelope of the ellipses of confidence (95% level) is given.

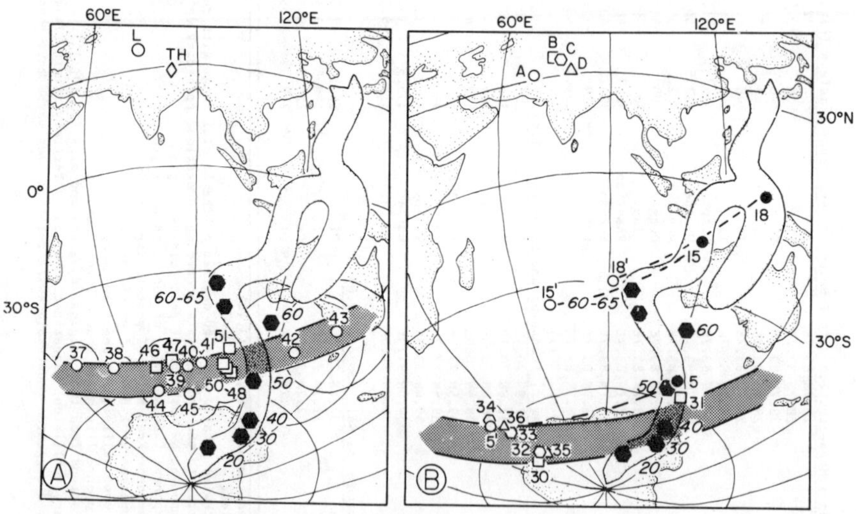

Fig. 2 A) Comparison of pole positions for the "collision" components observed in Ladakh (circles) and in the TSS (squares) with the Indo-Pakistan APWP. Pole numbers according to Table 1. Numbers in italics indicate stratigraphic or radiometric mean ages (my) for Tertiary pole positions on the Indo-Pakistan APWP.

B) Comparison of (secondary) pole positions for the Kashmir Basin (hexagons) and for the Krol Belt (triangles), and (primary) pole positions for the Loralai Range (Baluchistan, circles) with the Indo-Pakistan APWP. Pole numbers and ages as in A).

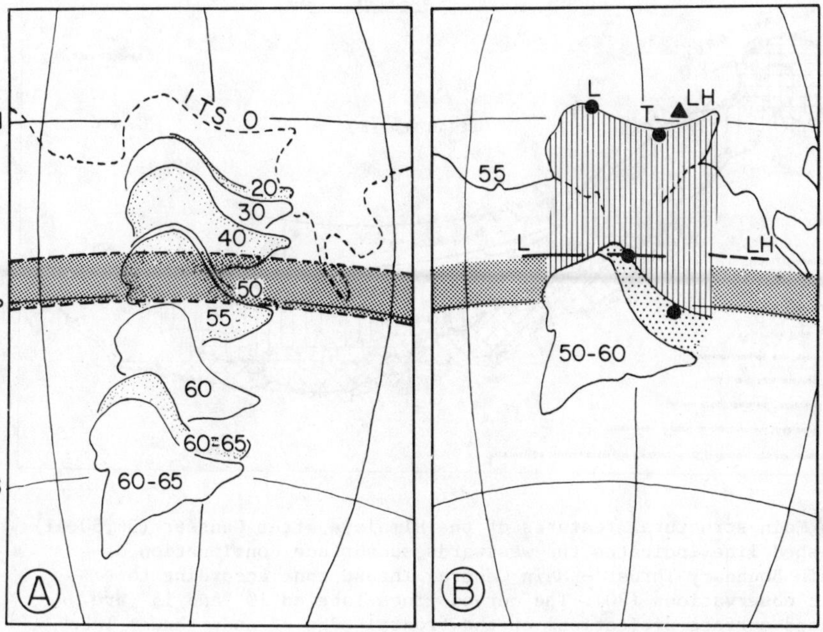


Fig. 3 A) Successive palaeopositions of Indo-Pakistan according to Tertiary pole positions (Table 1, numbers indicate mean ages in my). The heavy shaded zone indicates the palaeolatitude of initial collision (Fig. 2A). Southern Asia and the ITS are shown for comparison in present-day shape and position.

B) Comparison of Indo-Pakistan's and Asia's palaeolatitudinal position upon initial collision (about 55 m.y. ago). India is positioned according to the "collision" component from the TSS (dot T), which is in good agreement with the contemporaneous palaeoposition of the Ladakh Intrusives (heavy line, Table 1 pole 44), which are now situated directly north of the ITS (dot L). The dashed line indicates the palaeoposition for Upper Cretaceous-lower Tertiary red beds in southern Tibet, north of Lhasa (triangle Th, Table 1 pole 69). Asia's southern boundary and the ITS are shown under assumption of no postcollisional deformation within south central Asia. The heavy shaded zone indicates the palaeolatitude of initial collision. The magnitude of crustal convergence within southern Asia and of crustal shortening within northern Indo-Pakistan are indicated by the vertically striped and stippled zones respectively.

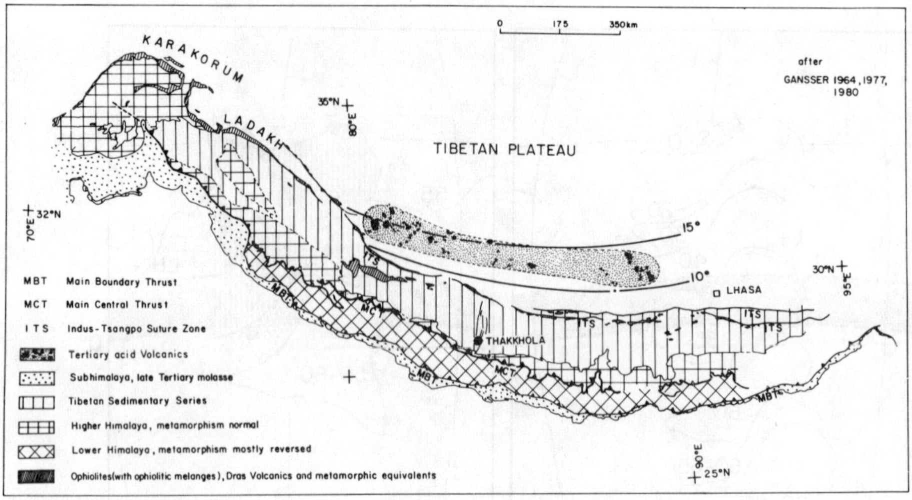


Fig. 4 Main structural features of the Himalaya after Gansser (36,50,51) The dashed line indicates the westwards subsurface continuation of the Main Boundary Thrust - Main Central Thrust zone according to seismic observations (30). The curved lines labeled 1∅$^{o}$ and 15$^{o}$ are estimated surface projections of the frontal edge of underthrust Indo-Pakistan.

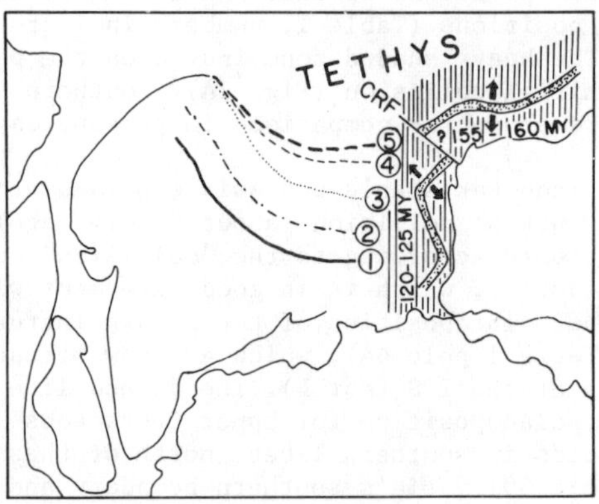


Fig. 5 The northern extent of Indo-Pakistan within the updated Smith and Hallam reconstruction (52,53). 1= present Himalayan front, 2= the ITS, 3= a reflection of the ITS separated from 2 by a 3∅∅ km wide belt, representing a) 2∅∅ km crustal shortening within the Himalayan region, and b) 1∅∅ km continental subduction along the ITS and northwards translational underthrusting along the MCT. The zone between line 3 and line 4 or line 5 represents the contribution to northern Indo-Pakistan corresponding with rotational underthrusting along the MCT over 1∅ or 15 degrees respectively. Rift zones (stippled) and assumed continental lithosphere (vertical stripes), with ages of commencement of spreading.

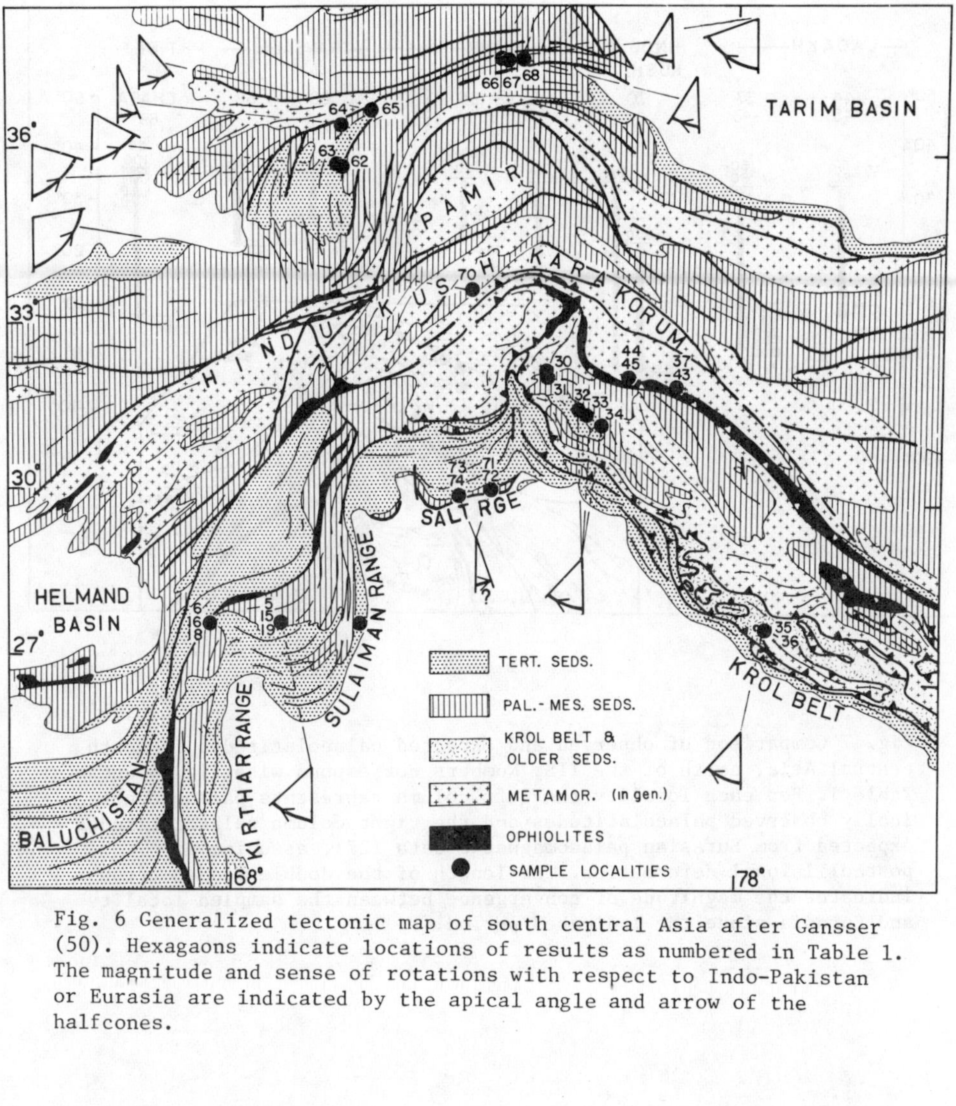


Fig. 6 Generalized tectonic map of south central Asia after Gansser (50). Hexagaons indicate locations of results as numbered in Table 1. The magnitude and sense of rotations with respect to Indo-Pakistan or Eurasia are indicated by the apical angle and arrow of the halfcones.

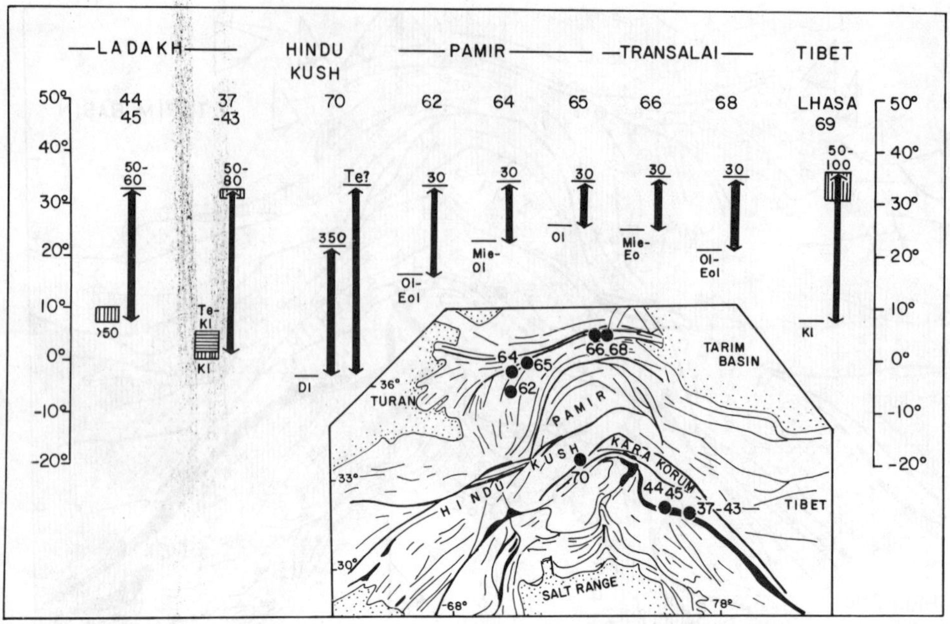


Fig. 7 Comparison of observed and expected palaeolatitudes in south central Asia, north of the ITS. Numbers correspond with fig. 6 and Table 1. For each locality the left column represents palaeomagnetically observed palaeolatitudes and the right column palaeolatitudes expected from Eurasian palaeomagnetic data (27), assuming no postcollisional deformation. The length of the double-headed arrow indicates the magnitude of convergence between the sampled locality and Eurasia since the indicated period.

## THERMO-RHEOLOGICAL PROCESSES IN THE HIMALAYAN OROGENESIS AND THE UPLIFT OF THE XIZANG PLATEAU (A PRELIMINARY ANALYSIS)

Chen Zong-ji* Li Hong-ji
(Institute of Geophysics, Academia Sinica)

Plate collisions are being studied all over the world; they have many similarities but also many specific pecularities. For instance, the impingence of the Pacific Plate on the Eurasian Plate has resulted in the subduction and the most energetic seismic belt along the Pacific, whereas the Indian Eurasian collision has resulted in the highest mountain range (the Himalayas), the thickest plate (the Xizang Plate) forming the highest roof of the world. The writers are of opinion that the different manifestations of the collision can be attributed to regional factors. So in this paper the anomalous case of the Himalayas and the exceptional thickness of the Xizang crust (about 72 km) will be attributed to thermo-rheological processes generated by a mantle of extra ordinary high heat anomaly which is believed to be active since the formation of the crust: this finding is based on an analysis of avaliable geophysical, geological and geochemical data. This evaluation is supported by finite elements computations and the corresponding estimation of viscosities not higher than $5.10^{23}$ poises. A further specific feature is the large depression in the gravimetric geoid with its apex south of Cape Comorin with a minimum of -110 mgal(5). These are believed to be two crucial factors, which make the result of the collision so completely different from those of other collisions. The authors only try to analyse the thermo-rheological processes underlying the orogenesis and tectonogenesis and since most of the published geological, geophysical and geochemical data are only known for Southern Xizang, the analysis is restricted to this important part of the Xizang Plateau.

### MANIFESTATIONS OF EXTRA ORDINARY HEAT ANOMALY

In the following, the authors will describe some hydrothermal manifestations and some $Q_R$ values, magnetic field data, geological data, which are valuable for the analysis of the thermo-rheological processes active in the past and at present.

#### Current Hydrothermal Manifestations

The hydrothermal explosion sites, geysers and hot wells are distributed in a long belt stretching on both sides along the Yarlung Zangbo zone in the south bordered by the MCT; in addition, there is a branch extending to the Northeast (Fig. 1). These hydrothermal manifestations give one the impression that there may exist an enormous mass of hot rocks under this belt, situated not far from the surface. This gigantic heat reservoir may be supplied by magma intrusions from the upper mantle. As it will be further discussed in our computational analysis, upward heat transfer is

---

* Tan Tjong-kie

presumably supported by convection currents in the upper crustal layers; the flow of the groundwater will be made possible by the increase in permeability of the rocks due to cracking and void formation.

### Overall Low $Q_R$ Values

In order to obtain the $Q_R$ values for Rayleigh waves propagating through the Xizang Plateau, four seismic stations, one in Lhasa and the other three just located beyond the eastern border of the Xizang Plateau have been engaged(2). The results are shown in Fig. 2; it shows us that the overall $Q_R$ values for all the periods (Curve C) is much lower than the corresponding values of the North American continent (Curve A) and the other continents (Curve B). Furthermore, the $Q_R$ values decrease with the period much quicker than those of the North American and the other continents. These valuable data will be evaluated as follows: the dissipation of seismic waves may be caused by seismic energy dissipation due to frictional effects, faults, cracks, fissures and viscous energy dissipation; this viscous dissipation increases with the temperature. As cracks and fissures are only present at the upper crustal layers but will disappear with the depth due to higher hydrostatic pressures, the overall lower $Q_R$ values is an indication that the overall temperatures in the Xizang crust and upper mantle may be higher than those in the North American and the other continents. Further from the fact that the $Q_R$ decrease much quicker with the depth than those of the other continents, it can be inferred, that the temperature increases more rapidly with the depth. As the coefficients of heat conduction in all the crustal continents are known to vary slightly, it can further be deduced that there may exist an extra ordinary heat anomaly in the mantle, whereas heat sources will pressumably be present within crust in the form of magma intrusions, frictional heat sources, regions of strong radio-active decay, etc.

### Negative Magnetic Anomalies

Negative anomalies have been measured in a long zone stretching along the northern part of the Himalayas(12); no Chinese measurements have been made outside the Chinese territory. These negative anomalies (Fig. 3) may be ascribed to the presence of thermally demagnetised or partially demagnetised rocks due to high temperatures near the surface. This is a possibility which merits further investigations, as it is apparently compatible with the presence of recent tourmaline granitic intrusions and presumed high heat in these regions.

## VOLCANICS, MAGMA INTRUSIONS, METAMORPHOSIS

The above three points are indications of geothermal activity at present and give one an impression of the possible existence of an enormous heat reservoir not far away from the surface. This allocated reservoir is presumably fed from heat sources deeper in the crust and upper mantle. In order to analyse it, the authors will now study the structural map of the Himalayas and Xizang(3) (Fig. 4). The following specific features will be noticed:

1. The transhimalayan granites north of the Yarlung Zangbo zone (late Cretaceous to Eocene age) A;

2. A chain of Neocene to recent acid volcanics concentrated in a chain extended from 80° E to 88° E south of the transhimalayan lineament B;

3. Dras volcanics at Dras and Kargil running in an arch to the north intruded by granites C;

4. Granites and gneiss uplift concentrated in a belt south of the Yarlung Zangbo zone D;

5. Further south the Miocene tourmaline granites E.

The belts and chains A.B.D.E. at a large scale are nearly running parallel to the Himalayas and show one the sites of magma activities in the past. As they are parallel to the Himalayas they must be related to its orogenesis and mutually correlated with the acting stress field. The most recent ones are the chains of granite-gneiss metamorphosis D and further to the south the chains of Miocene tourmaline granite intrusions E. Both are just situated between the MCT and the Yarlung Zangbo zone, i.e. the geothermal activities at present are continuations of those processes of anomalous heat in the past.

## ABNORMAL CRUSTAL THICKENING

Recent gravity and explosion seismology investigations have shown that the thickness of the Xizang crust is circa 72 km(7,8).This is 35 km thicker than the crust of the neighbouring provinces. So we are inclined to believe that this may be attributed to local factors such as higher stress intensity in this region, large creep deformations due to the abnormal low rigidity and viscosity of the Xizang crust, intense folding, and the possible accretion of materials from currents upwelling in the mantle. Under low hydrostatic pressures up to a few kb, rocks behave as brittle materials and creep proceeds by microfracturing. Under sufficiently high shearing stresses and increased temperatures due to uprising granitic magmas, horizontal cracks were gradually formed in the direction of the maximum principal stresses. They coalesced gradually and form continuous horizonal planes of separation; ultimately the originally monolythic layers were disintegrated into a packet of much thinner layers. Further increase of stress and creep could lead first to buckling and then to folding. This folding increased with the time as well in amplitudes as in intensity (Fig. 5). Thereby, the process was facilitated by lubricating effects due to flysch and other shistose materials especially at high temperatures.

The middle and lower parts of the crust are ductile and deformed by viscous flow along planes of easy glide. Low viscosities lead to large deformations. The present maximal value of the viscosity is estimated in the order of $5.10^{23}$ poises, whereas it decreases exponentially with the temperature. At the start of the thickening process after the collision the viscosities might be still lower.

## THERMO-RHEOLOGICAL ANALYSIS

It is helpful to devise a physical model which can explain the thermal and deformational features described above and can give an estimation of the order of magnitude of the viscosities in the crust. For this purpose we make use of the gravity crustal-model shown in Fig. 6(7) and will base our computations on the assumption of linear visco elasticity; as only stationary states are studied we can fulfill with a temperature dependent Newton viscosity. The crustal thickness is circa 72 km and the slope of the Moho is 8 to 10 degrees. The Main Central Thrust (MCT) and the Main Boundary Thrusts (MBT) are taken nearly parallelly to this slope. This model is considered to reflect the present profile of the Himalayas and the Xizang Plateau. The stationary temperature distribution is computed starting from initial boundary conditions. The temperature at the lower boundary of the crust varies from 800°C in the south to 1,200°C in the deepest part. Temperatures of 1,200°C are assessed in a very narrow zone along the MCT and MBT in their lower portions; whereas the temperatures in these sliding zones are reduced to 800°C and lower in the middle and higher portions. These zones may be considered to be heated by several causes such as magma intrusion (flowing magma has temperatures from 800° to 1,200°C), the decay of highly concentrated radio active materials and the friction along the sliding parts of the crust. The computed isotherms and heat flow distribution are shown in Fig. 7; it can be noticed that the maximal heat flow of 3.6 HFU precisely appears in the belt of high hydrothermal activities. Note further that precisely in these regions, temperatures of 650°C to 800°C occur already at depths varying from 15 to 20 km, hence it can be inferred that at these depths granites may be in a state of melting; here molten granitic magmas can be readily injected in fissures under the existing hydrostatic pressures, which can be estimated at 4 to 5.3 kbars.

In order to study the stress field and the field of the rates of displacements in the crust we have made some finite elements computations. Thereby temperature dependent viscosities are assumed according to the well known formula: $\eta = \eta_0 \exp(-Q/RT)$ the viscosity varying with the depth is shown in Fig. 8; note that the estimated curve for Xizang lies much lower than that for the normalcrust(11).The magnitude of the viscosity $\eta_0$ of course must be choosen that the computed rates of displacement must approach the observed values as closely as possible i.e. 5 cm/yr and 3 cm/yr for the horizontal motion of the Indian Plate and the Xizang Plateau respectively and 0.45 cm/yr for the vertical uprise of the top of the Himalayas. The computed displacement rates for a deformation rate of 5 cm/yr of the Indian Plate is shown in Fig. 9.

The observed data can quite well be matched by the computed values when it will be taken $\eta_0 \sim 5.10^{23}$ poises, thus a factor 50-100 lower than the known values for a normal crust(11). Consequently the crust of the Xizang Plateau must obviously be at a temperature much higher than the normal crust (Fig. 8). In the above paragraph we have studied four different phenomena from which a high heat anomaly for the crust and the upper mantle can be deduced; this computational experiment provides a further support for this hypothesis. From these computations the following additional conclusions may be drawn:

1. The trajectories of normal stress and shearing stress are bent upwards (Fig. 10 and 11) that the stresses now concentrate in more upper layers. This result may give an explanation for the phenomenon that the larger majority of earthquakes in the Xizang Plateau has occurred since historical times at low focal depths of 10 to 35 km(4);

2. The temperatures of 600° to 800°C is attained under the hydrothermal belt extending along the Yarlung Zangbo zone and north of the MCT at depths only varying from 10 to 15 km. As granites have melting temperatures from 650°C to 850°C locally can be considered to be in the molten

state under an estimated hydrostatic presure of 2.7 to 4 kb.In this heterogeneous region and in the higher situated portions, the shearing stresses amount to 1 to 2 kb and presumably fractures and cracks must have been formed So granitic magma can be injected into these fissures and due to the low viscosities it will gradually move upwards bringing the upper layers to higher temperatures. Hence, under the favourable influence of high temperature and pressure granites can be metamorphised into gneisses. This may be an explanation for the formation of abundant gneissic masses south of the Yarlung Zangbo zone up to the MCT.

3. Due to the cracking of the rock layers near the surface the overall permeability of the rocks in these hydrothermal belts will be so large that crustal water convection can occur resulting in a quicker and more efficient dissipation of heat energy in the form of powerful hydrothermal explosions, geysers and hot wells as described earlier.

## References


(1) Chinese Academy of Geologic Sciences, Geologic Map of Asia (1:5000000) The Cartographic Publishing House (1975).

(2) Feng Rui et al., Q-value for surface waves in the region of Xizang Plateau. In this Proceedings of Symposium on Qinghai-Xizang (Tibet) Plateau, Academia Sinica (1981).

(3) Gansser A., The great suture zone between Himalaya and Tibet. A preliminary account. Colloque Int. C.N.R.S. No. 268 Ecologie et Geologie de L'Himalaya, 181 (1977).

(4) Huan Wen-lin et al., The distribution of earthquake foci and plate tectonics of the Qinghai-Xizang Plateau. In this Proceedings of Symposium on Qinghai-Xizang (Tibet) Plateau, Academia Sinica (1981).

(5) Marussi A., Geoidal features and crustal structure in Central Asia. In this Proceedings of Symposium on Qinghai-Xizang (Tibet) Plateau, Academia Sinica (1981).

(6) Molnar P., Chen W.P., Fitch T.J., Taponnier P., Warsi W.E.K., Wu F.T., Structure and tectonics of the Himalayas; a brief summary of relevant geophysical obsverations. Coll. Int. C.B.R.S. No. 268, Ecol. et Geol. de Himalaya: 269 (1977).

(7) Tang Bo-xiong, Liu Yuan-long et al., Isostatic gravity anomalies in the central portion of the Himalayas. In this Proceedings of Symposium on Qinghai Xizang (Tibet) Plateau, Academia Sinica (1981).

(8) Teng Ji-wen et al., Explosion seismic study for velocity distribution and structure of the crust and upper mantle from Damxung to Yadong of Xizang Plateau. In this Proceedings of Symposium on Qinghai-Xizang (Tibet) Plateau, Academia Sinica, (1981).

(9) Toksöz, M.N., Bird P., Tectonophysics of the continuing Himalayan orogeny, Coll. Int. C.N.R.S. No. 268, Ecol. et Geol. de Himalaya , 443 (1977).

(10) Tong Wei et al., The Himalayan Geothermal belt, Acta Scientarium Naturalium Universitas Pekinensis, (1) (1978).

(11) Weertman J., The creep strength of the earth's mantle, Rev. Geoph. and Space Phys. 8, 1266 (1970).

(12) Xu Zhen-wu et al., The characteristics of the geomagnetic anomalies for the Qinghai-Xizang Plateau (unpublished report) (1980).

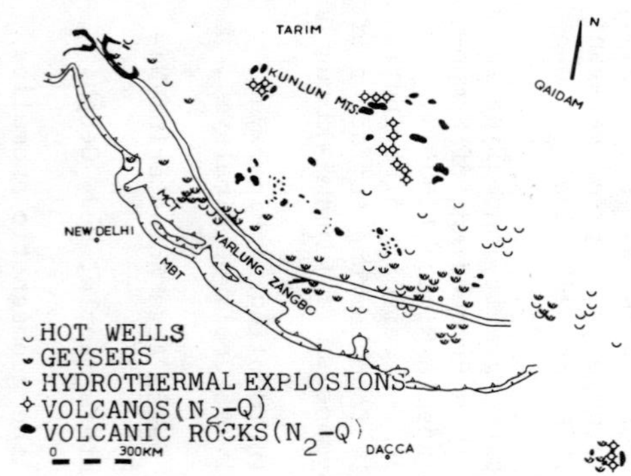


Fig. 1 Distribution of hot wells, geysers, hydrothermal explosion sites and volcanics (Tong et al., 1978).

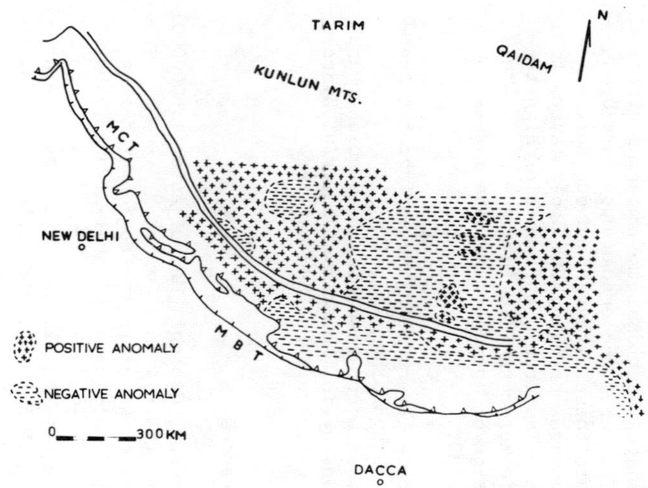


Fig. 3 Total geomagnetic intensity (Partly reproduced from the figure by Xu et al., 1980).

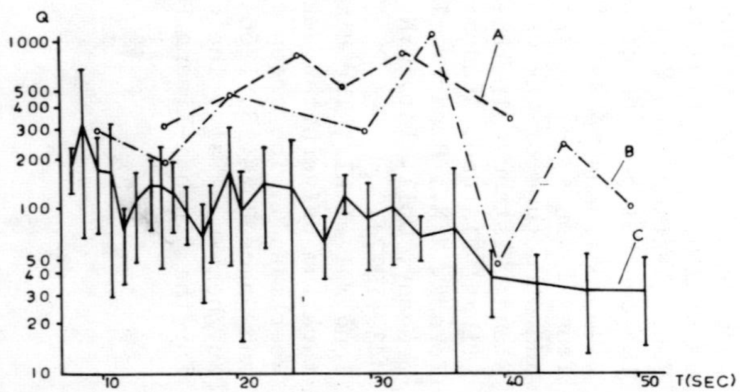


Fig. 2 $Q_R$-values for Rayleigh waves observed in different regions.

A. North America;
B. Results from a Xizang earthquake, monitored over all the continents;
C. Xizang Plateau (Feng Rui et al, 198).

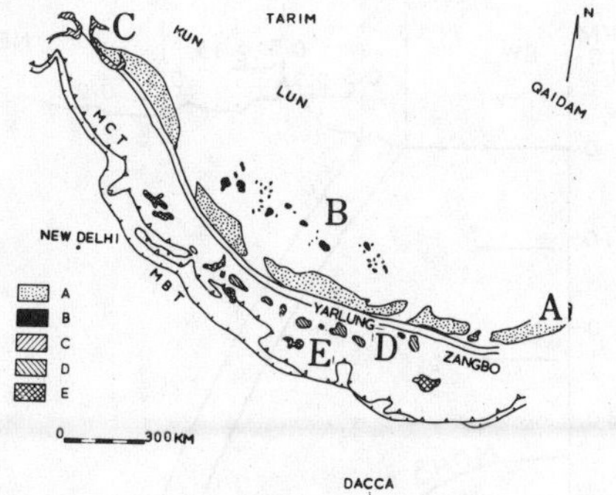


Fig. 4 Distribution of granites, gneisses and volcanics.

A. Transhimalayan granites, Ladakh granite and Karakorum;
B. Neogene to recent acid volcanics;
C. Dras volcanics and metamorphic equivalents;
D. Granite and gneiss uplifts in Tethys Himalaya;
E. Neogene tourmaline granites (Collected after data by Chin, A. C. Geol. Sciences, 1976 and Gansser, 1977).

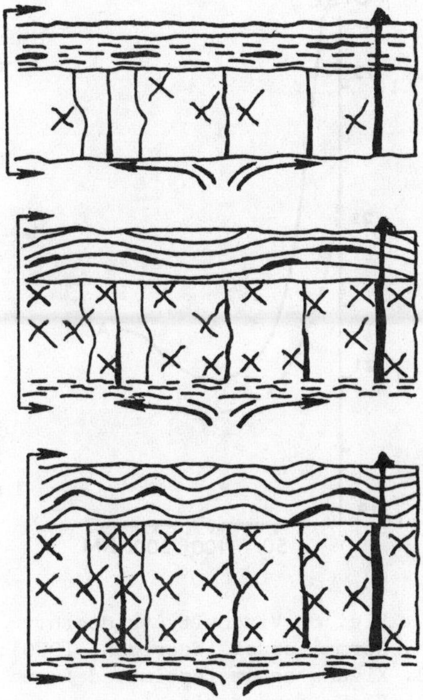

Fig. 5 Cartoon for crustal thickening by intense folding of the upper layers, creep deformation of the middle and lower layers and possible accretion from upwelling currents.

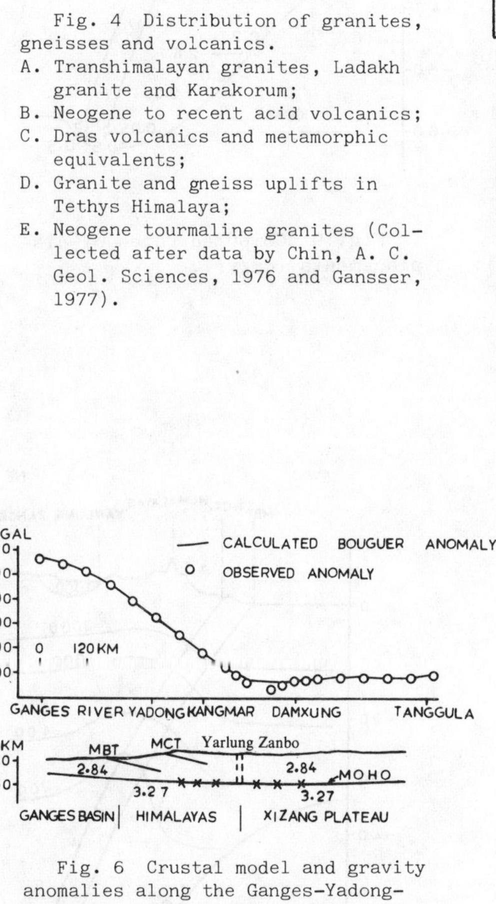


Fig. 6 Crustal model and gravity anomalies along the Ganges-Yadong-Tanggula profile (Tang and Liu et al., 1980).

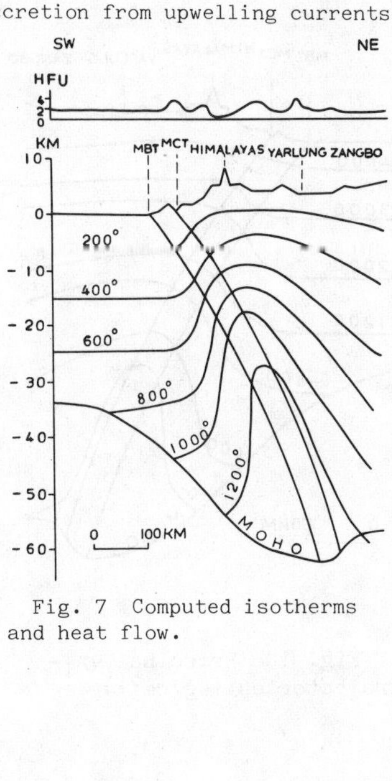


Fig. 7 Computed isotherms and heat flow.

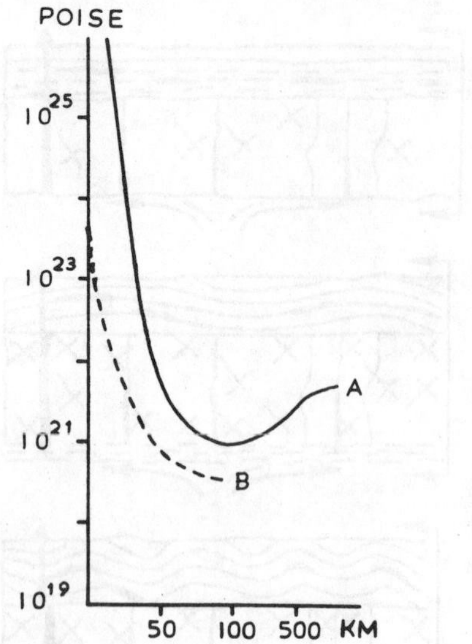


Fig. 8 Viscosity vs depth.
A. Normal crust (Weertman, 1970);
B. Xizang crust (estimated from F. E. computations).

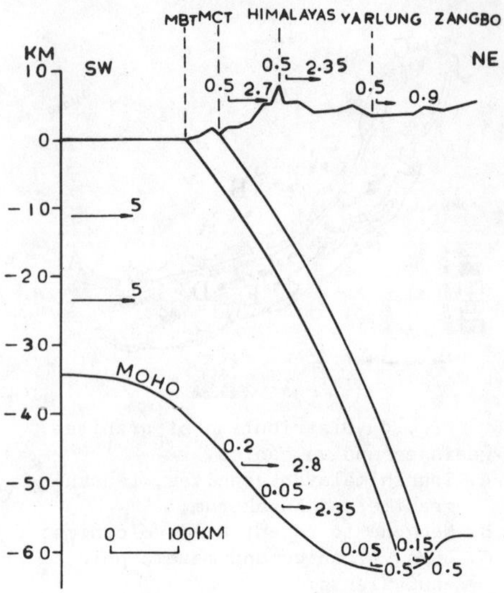


Fig. 9 Computed rates of displacements (cm/yr).

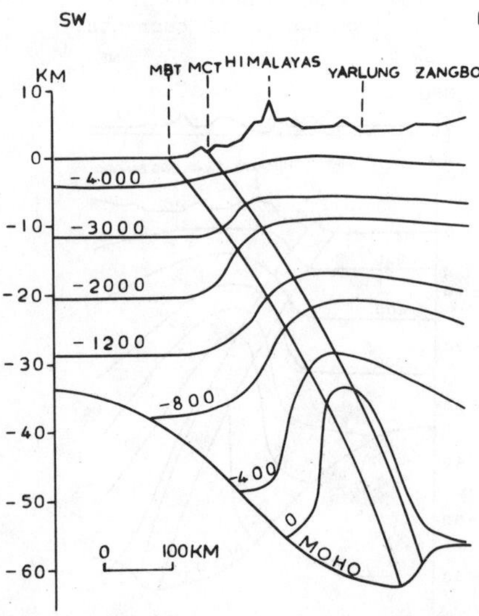


Fig. 10 Principal $\sigma_1$-trajectories (kg/cm$^2$).

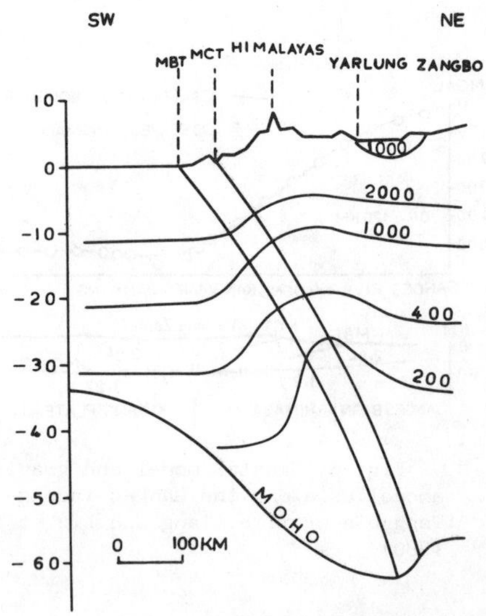


Fig. 11 Maximum shearing stress $\tau_{max}$ (kg/cm$^2$).